Fernando L. Alvarado

Stability of Large Electric Power Systems

OTHER IEEE PRESS BOOKS

Clearing the Air: The Impact of the Clean Air Act on Technology
Edited by John C. Redmond, John C. Cook, and A. A. J. Hoffman
Active Inductorless Filters
Edited by Sanjit K. Mitra
A Practical Guide to Minicomputer Applications
Edited by Fred F. Coury
Power Semiconductor Applications, Volume I: General Considerations
Edited by John D. Harnden, Jr. and Forest B. Golden
Power Semiconductor Applications, Volume II: Equipment and Systems
Edited by John D. Harnden, Jr. and Forest B. Golden
Semiconductor Memories
Edited by David A. Hodges
Minicomputers: Hardware, Software, and Applications
Edited by James D. Schoeffler and Ronald H. Temple
Digital Signal Processing
Edited by Lawrence R. Rabiner and Charles M. Rader
Laser Theory
Edited by Frank S. Barnes
Integrated Optics
Edited by Dietrich Marcuse
Literature in Digital Signal Processing: Terminology and Permuted Title Index
Edited by Howard D. Helms and Lawrence R. Rabiner
Laser Devices and Applications
Edited by Ivan P. Kaminow and A. E. Siegman
Computer-Aided Filter Design
Edited by George Szentirmai
Key Papers in the Development of Information Theory
Edited by David Slepian
Technology and Social Institutions
Edited by Kan Chen
Key Papers in the Development of Coding Theory
Edited by E. R. Berlekamp

Stability of Large Electric Power Systems

Edited by
Richard T. Byerly

Advisory Engineer
Power Systems Planning Department
Westinghouse Electric Corporation

Edward W. Kimbark

Head, System Analysis Unit
Branch of System Engineering
Bonneville Power Administration

A volume in the IEEE PRESS Selected Reprint Series,
prepared under the sponsorship of the
IEEE Power Engineering Society.

IEEE
PRESS

The Institute of Electrical and Electronics Engineers, Inc. New York

International Standard Book Numbers: Clothbound: 0-87942-037-5
Paperbound: 0-87942-038-3

Library of Congress Catalog Card Number 74-81759

PRINTED IN THE UNITED STATES OF AMERICA

Contents

Introduction 1

Part I: Mathematical Models and Apparatus Description for Systems Analyses 9

Digital Simulation of Synchronous Machine Transients, *D. W. Olive* (*IEEE Transactions on Power Apparatus and Systems*, August 1968) 13

Dynamic Models of Turbine Generators Derived From Solid Rotor Equivalent Circuits, *R. P. Schulz, W. D. Jones, and D. N. Ewart* (*IEEE Transactions on Power Apparatus and Systems*, May/June 1973) 20

Effects of Synchronous Machine Modeling in Large Scale System Studies, *P. L. Dandeno, R. L. Hauth, and R. P. Schulz* (*IEEE Transactions on Power Apparatus and Systems*, March/April 1973) 28

Generator Models Established by Frequency Response Tests on a 555 MVA Machine, *G. Manchur, D. C. Lee, M. E. Coultes, J. D. A. Griffin, and W. Watson* (*IEEE Transactions on Power Apparatus and Systems,* September/October 1972) 37

Experimental Determination of Exact Equivalent Circuit Parameters of Synchronous Machines, *Y. Yu and H. A. M. Moussa* (*IEEE Transactions on Power Apparatus and Systems*, November/December 1971) 45

Computer Representation of Excitation Systems, *IEEE Committee Report* (*IEEE Transactions on Power Apparatus and Systems*, June 1968) 50

Excitation System Dynamic Characteristics, *IEEE Committee Report* (*IEEE Transactions on Power Apparatus and Systems*, January/February 1973) 55

Experience with High-Speed Rectifier Excitation Systems, *K. R. McClymont, G. Manchur, R. J. Ross, and R. J. Wilson* (*IEEE Transactions on Power Apparatus and Systems*, June 1968) 67

A Rotating Thyristor Excitation System for Hydroelectric Generators, *E. C. Hartung, E. H. Lenfest, and G. R. Meloy* (*IEEE Transactions on Power Apparatus and Systems*, September/October 1972) 74

Althyrex Excitation System with Power System Stabilizer, *M. L. Crenshaw, W. J. Miller, R. P. Schulz, and M. Temoshok* (*IEEE Paper 70 CP 563-PWR*) 82

Design of a Power System Stabilizer Sensing Frequency Deviation, *F. W. Keay and W. H. South* (*IEEE Transactions on Power Apparatus and Systems*, March/April 1971) 90

A High Initial Response Brushless Excitation System, *T. L. Dillman, J. W. Skooglund, F. W. Keay, W. H. South, and C. Raczkowski* (*IEEE Transactions on Power Apparatus and Systems*, September/October 1971) 97

A Simplified Analysis of the No-Load Stability of Mechanical-Hydraulic Speed Control Systems for Steam Turbines, *M. A. Eggenberger* (*ASME Paper 60-WA-34*) 103

Detailed Hydrogovernor Representation for System Stability Studies, *D. G. Ramey and J. W. Skooglund* (*IEEE Transactions on Power Apparatus and Systems*, January 1970) 121

Dynamic Models for Steam and Hydro Turbines in Power System Studies, *IEEE Committee Report* (*IEEE Transactions on Power Apparatus and Systems*, November/December 1973) 128

Dynamic Modeling of Loads in the Stability Studies, *M. H. Kent, W. R. Schmus, F. A. McCrackin, and L. M. Wheeler* (*IEEE Transactions on Power Apparatus and Systems*, May 1969) 139

Steady-State Voltage Power Characteristics for Power System Loads, *R. B. Adler and C. C. Mosher* (*IEEE Paper 70 CP 706-PWR*) 147

Behavior of Loads During Voltage Dips Encountered in Stability Studies—Field and Laboratory Tests, *F. Iliceto, A. Ceyhan, and G. Ruckstuhl* (*IEEE Transactions on Power Apparatus and Systems*, November/December 1972) 154

System Load Dynamics—Simulation Effects and Determination of Load Constants, *IEEE Committee Report* (*IEEE Transactions on Power Apparatus and Systems*, March/April 1973) 164

Electromechanical Equivalents for Use in Power System Stability Studies, *J. M. Undrill, J. A. Casazza, E. M. Gulachenski, and L. K. Kirchmayer* (*IEEE Transactions on Power Apparatus and Systems*, September/October 1971) 174

Part II: Dynamic Stability 187

Concepts of Synchronous Machine Stability as Affected by Excitation Control, *F. P. deMello and C. Concordia* (*IEEE Transactions on Power Apparatus and Systems*, April 1969) 189

Block Diagram and Torque-Angle Loop Analyses of Synchronous Machines, *I. Nagy* (*IEEE Transactions on Power Apparatus and Systems*, July/August 1971) 203

Dynamic System Stability: Part I—Investigation of the Effect of Different Loading and Excitation Systems, *M. K. El-Sherbiny and D. M. Mehta* (*IEEE Transactions on Power Apparatus and Systems*, September/October 1973) 212

Large System Simulation

New Techniques for the Cálculation of Dynamic Stability, *D. W. Olive* (*IEEE Transactions on Power Apparatus and Systems*, July 1966) 221

Dynamic Simulation of Interconnected Systems, *D. G. Ramey and R. T. Byerly* (*Power Industry Computer Applications Conference Proceedings*, 1967) 232

Equipment and System Modeling for Large-Scale Stability Studies, *C. C. Young* (*IEEE Transactions on Power Apparatus and Systems*, January/February 1972) 242

Fast Transient Stability Solutions, *H. W. Dommel and N. Sato* (*IEEE Transactions on Power Apparatus and Systems*, July/August 1972) 253

Case Studies

Dynamic Stability of the Peace River Transmission System, *H. M. Ellis, J. E. Hardy, A. L. Blythe, and J. W. Skooglund* (*IEEE Transactions on Power Apparatus and Systems*, June 1966) 261

Influence of Excitation and Speed Control Parameters in Stabilizing Inter-System Oscillations, *O. W. Hanson, C. J. Goodwin, and P. L. Dandeno* (*IEEE Transactions on Power Apparatus and Systems*, May 1968) 276

Four Corners Project Stability Studies, *R. G. Farmer, R. H. Hartley, M. H. Kent, and L. M. Wheeler* (*IEEE Paper 68 CP 708-PWR*) 284

Damping of Power Oscillations in Salient-Pole Machines with Static Exciters, *R. T. Byerly, F. W. Keay, and J. W. Skooglund* (*IEEE Transactions on Power Apparatus and Systems*, July/August 1970) 295

Power System Stabilizers: Installation, Testing, and Experience

Field Tests of Dynamic Stability Using a Stabilizing Signal and Computer Program Verification, *R. M. Shier and A. L. Blythe* (*IEEE Transactions on Power Apparatus and Systems*, February 1968) 308

Excitation Control to Improve Powerline Stability, *F. R. Schleif, H. D. Hunkins, G. E. Martin, and E. E. Hattan* (*IEEE Transactions on Power Apparatus and Systems*, June 1968) 316

Effect of High-Speed Rectifier Excitation Systems on Generator Stability Limits, *P. L. Dandeno, A. N. Karas, K. R. McClymont, and W. Watson* (*IEEE Transactions on Power Apparatus and Systems*, January 1968) 325

Improved Stability With Low Time Constant Rotating Exciter, *H. R. Perry, J. F. Luini, and J. C. Coulter* (*IEEE Transactions on Power Apparatus and Systems*, September/October 1971) 337

Power System Stabilizer: Field Testing and Digital Simulation, *A. D. Gerhart, T. Hillesland, Jr., J. F. Luini, and M. L. Rockfield, Jr.* (*IEEE Transactions on Power Apparatus and Systems*, September/October 1971) 343

Experience with Supplementary Damping Signals for Generator Static Excitation Systems, *W. Watson and G. Manchur* (*IEEE Transactions on Power Apparatus and Systems*, January/February 1973) 349

Static Exciter Stabilizing Signals on Large Generators—Mechanical Problems, *W. Watson and M. E. Coultes* (*IEEE Transactions on Power Apparatus and Systems*, January/February 1973) 354

Part III: Transient Stability 363

Series Capacitors

The Use of Series Capacitors to Obtain Maximum EHV Transmission Capability, *G. D. Breuer, H. M. Rustebakke, R. A. Gibley, and H. O. Simmons, Jr.* (*IEEE Transactions on Power Apparatus and Systems*, November 1964) 365

Improvement of System Stability by Switched Series Capacitors, *E. W. Kimbark* (*IEEE Transactions on Power Apparatus and Systems*, February 1966) 378

Series Capacitor Switching to Quench Electromechanical Transients in Power Systems, *R. H. Webster, A. Mane, and O. J. M. Smith* (*IEEE Transactions on Power Apparatus and Systems*, March/April 1971) 387

The Prediction and Control of Self-Excited Oscillations Due to Series Capacitors in Power Systems, *L. A. Kilgore, L. C. Elliott, and E. R. Taylor* (*IEEE Transactions on Power Apparatus and Systems*, May/June 1971) 393

Self-Excited Torsional Frequency Oscillations With Series Capacitors, *C. E. J. Bowler, D. N. Ewart, and C. Concordia* (*IEEE Transactions on Power Apparatus and Systems*, September/October 1973) 400

Single-Pole Switching

Suppression of Ground-Fault Arcs on Single-Pole-Switched EHV Lines by Shunt Reactors, *E. W. Kimbark* (*IEEE Transactions on Power Apparatus and Systems*, March 1964) 408

A Method for Reducing Dead Time for Single-Phase Reclosing in EHV Transmission, *H. A. Peterson and N. V. Dravid* (*IEEE Transactions on Power Apparatus and Systems*, April 1969) 414

Digital Analysis of Single-Pole Switching on EHV Lines, *R. B. Shipley, H. J. Holley, and D. W. Coleman* (*IEEE Transactions on Power Apparatus and Systems*, August 1968) 421

Single-Pole Switching on TVA's Paradise-Davidson 500-KV Line-Design Concepts and Staged Fault Test Results, *L. Edwards, J. W. Chadwick, Jr., H. A. Riesch, and L. E. Smith* (*IEEE Transactions on Power Apparatus and Systems*, November/December 1971) 430

Fast Load Control

Improved Reliability of Bulk Power Supply by Fast Load Control, *R. H. Park* (*American Power Conference Proceedings*, 1968) 445

Improvement of Power System Stability by Changes in the Network, *E. W. Kimbark* (*IEEE Transactions on Power Apparatus and Systems*, May 1969) 459

Fast Valving as an Aid to Power System Transient Stability and Prompt Resynchronization and Rapid Reload After Full Load Rejection, *E. W. Cushing, Jr., G. E. Drechsler, W. P. Killgoar, H. G. Marshall, and H. R. Stewart* (*IEEE Transactions on Power Apparatus and Systems*, November/December 1971) 468

Modern Stability Aids for Calvert Cliffs Units, *W. A. Morgan, H. B. Peck, D. R. Holland, F. A. Cullen, and J. B. Ruzek* (*IEEE Transactions on Power Apparatus and Systems*, February 1971) 479

Part IV: DC Lines 489

Calculation of Dynamic Behavior of Generators Connected to a DC Link, *J. J. Cladé and H. Persoz* (*IEEE Transactions on Power Apparatus and Systems*, July 1968) 491

Digital Representation of a DC Transmission System and its Controls, *J. F. Clifford and A. H. Schmidt, Jr.* (*IEEE Transactions on Power Apparatus and Systems*, January 1970) 503

Power Systems Stability Considerations with Dynamically Responsive DC Transmission Lines, *J. J. Dougherty and T. Hillesland, Jr.* (*IEEE Transactions on Power Apparatus and Systems*, January 1970) 511

Part V: Load Shedding 523

Application of Underfrequency Relays for Automatic Load Shedding, *H. E. Lokay and V. Burtnyk* (*IEEE Transactions on Power Apparatus and Systems,* March 1968) 525

Frequency Actuated Load Shedding and Restoration—Part I Philosophy, *R. M. Maliszewski, R. D. Dunlop, and G. L. Wilson* (*IEEE Transactions on Power Apparatus and Systems*, July/August 1971) 533

Frequency Actuated Load Shedding and Restoration—Part II: Implementation, *S. H. Horowitz, A. Politis, and A. F. Gabrielle* (*IEEE Transactions on Power Apparatus and Systems*, July/August 1971) 541

Part VI: Effects of Trends of Future Generator Designs 551

Effects of Trends in Large System Steam Turbine Driven Generator Parameters on Power System Stability, *C. Concordia and P. G. Brown* (*IEEE Transactions on Power Apparatus and Systems*, September/October 1971) 552

Effects of Future Turbine-Generator Characteristics on Transient Stability, *H. E. Lokay and P. O. Thoits* (*IEEE Transactions on Power Apparatus and Systems*, November/December 1971) 560

Bibliographies 565

Author Index 567

Subject Index 569

Editors' Biographies 573

Stability of Large Electric Power Systems

Introduction

Stability considerations have been recognized as an essential part of power system planning for a very long time. Fundamental papers on the subject, a group of which are listed in Table I, were published in the TRANSACTIONS OF THE AIEE nearly 50 years ago. Within a few years, stability studies of new generation and transmission facilities had become routine. Over most of the history of power system growth in the United States and Canada, acceptable stability performance has been achieved without great difficulty. Approximately within the last decade, however, changes have occurred which require power system engineers to devote much more of their effort to this problem.

Depending on individual experience and viewpoint, there are no doubt various explanations for the increased concern with stability. The underlying cause, however, would appear to be the very extensive interconnection of power systems with greater dependence on firm power flow over ties. This magnifies the undesirable consequences of instability and complicates the analytical processes through which acceptable system behavior is assured.

By the late 1960's, most of the generators in the United States and Canada, with the exception of certain isolated systems, were connected to one of two large transmission networks, one in the east and one in the west.[1] The primary motivation for interconnection is to obtain the economic benefits of new large-scale generation and transmission facilities.

Interconnection can also improve reliability through mutual support in emergencies. However, improved reliability is realized only if the bulk transmission system remains intact through such emergencies. If ties between systems are not strong enough to maintain synchronism in the face of disturbances, the interconnected system breaks up and the ability to provide support to areas which become isolated vanishes.

The possible consequences of instability in an interconnected system were dramatized by the Northeast Power Failure [1]-[3] of 1965. Changes in design and operation since that event virtually assure that such widespread failure will not happen again, even when system separation does occur. Even so, integrity of a bulk transmission system is of primary importance.

TABLE I

SOME EARLY SYNCHRONOUS STABILITY PAPERS FROM THE TRANSACTIONS OF THE AIEE

Vol. XLIII, 1924.

"Superpower Transmission—Economics and Limitations of the Transmission System of Extraordinary Length," Percy H. Thomas, pp. 1-15.

"Some Theoretical Considerations of Power Transmission," C. L. Fortescue and C. F. Wagner, pp. 16-23.

"Power Limitations of Transmission Systems," R. D. Evans and H. K. Sels, pp. 26-38.

"Experimental Analysis of Stability and Power Limitations," R. D. Evans and R. C. Bergvall, pp. 39-58.

"The Limitations of Output of a Power System Involving Long Transmission Lines," E. B. Shand, pp. 59-70.

"Discussions," pp. 71-103.

Vol. XLIV, 1925.

"Power System Transients," V. Bush and R. D. Booth, pp. 80-103.

"Stored Mechanical Energy in Transmission Systems," J. P. Jollyman, pp. 935-937.

"Fundamental Considerations of Power Limits of Transmission Systems," R. E. Doherty and H. H. Dewey, pp. 972-984.

"Transmission Stability: Analytical Discussions of Some Factors Entering into the Problem," C. L. Fortescue, pp. 984-1003.

Vol. XLV, 1926.

"An Investigation of Transmission-System Power Limits," C. A. Nickle and F. L. Lawton, pp. 1-21.

"Steady-State Stability in Transmission Systems," Edith Clarke, pp. 22-41.

"Practical Aspects of System Stability," Ray Wilkins, pp. 41-50.

"Studies of Transmission Stability," R. D. Evans and C. F. Wagner, pp. 51-80.

Discussions, pp. 80-94.

Vol. XLVI, 1927.

"Synchronous Machines—III. Torque-Angle Characteristics Under Transient Conditions," R. E. Doherty and C. A. Nickle, pp. 1-17.

Stability behavior of a large power system is difficult to predict. Size is in itself a complication. In the Northeast Power Failure, for example, the systems most directly affected included hundreds of generators, with hundreds more in neighboring systems less affected by the instability. Such large systems are subjected to a wide range of operating conditions, depending on load level and equipment availability, and may

[1] Ties between east and west were first closed in 1967. These have very limited capacity in terms of the systems they interconnect and do not provide firm power. They are subject to frequent tripping.

undergo a limitless variety of disturbances involving faults, loss of generation, loss of transmission facilities, and loss of load. The inherently nonlinear nature of ac power transmission limits the ability to extrapolate from one situation to another with confidence. Thus, the minimum effort required to determine the stability characteristics of any large system is substantial, and choosing between design alternatives or establishing operating guides becomes a time-consuming and costly undertaking.

The difficulties of interconnected system stability are compounded by the fact that certain systems exhibit tendencies toward dynamic instability. Although power system oscillations are at best only lightly damped, the assumption of positive damping has in the past been used to simplify studies and has been validated by operating experience. In the eastern United States, there is still no general evidence of dynamic instability. However, in the west, in parts of the midwest, and in parts of Canada, dynamic instability is a problem. Power oscillations may be initiated by some disturbance or occur spontaneously, and may on occasion increase in magnitude to the point of causing system separation. The common feature of systems displaying such behavior is long-distance transmission. Load centers tend to be widely separated and partially supplied by remote generation with large angular displacements between remote generators and those near load centers. Ties between load areas are also long, and interchange of power requires relatively large angle differences, frequently resulting in still larger angle separations between some remote generator groups and load center machines with which they must synchronize. The overall result is reduced capacity for synchronizing power flow between machine groups and low natural oscillation frequencies that are poorly damped.

It is the intent of this book to present, by means of reprints, the stability aspects of the recent growth of large power systems in the United States and Canada. The book is intended to provide a reference for practicing engineers. It should also be useful to students and teachers in the identification of an industry problem, and may help achieve perspective through organization and editorial comment.

Selection of Papers

It is interesting to note some statistics regarding stability papers. An IEEE bibliography [4] published in 1965 lists just 35 AIEE and IEEE Transactions papers over the 30-year period from 1935 through 1964. Then, abruptly, the number jumps from 2 in 1964 to 18 in 1965, if one includes PICA[2] conference papers, and has remained at a high level to the present time. Even allowing for the general increase in the number of published papers in all fields, there is no doubt that power system engineers have been much more occupied with stability considerations in the past decade than they were previously.

The reader should be aware of the fact that the papers in this book are selected from a much larger set. The set includes all of the papers in the IEEE Transactions on Power Apparatus and Systems for the years 1966 to the present. It includes the AIEE and IEEE Transactions papers in the IEEE report "Power System Stability Bibliography, 1935–1965." The set also includes all of the papers in the PICA conference records, and a very large number of conference papers from the general meetings of the Power Engineering Society of recent years. This entire set was examined in the selection process, and the papers chosen are each intended to document in some way the impact of the stability problem on the power systems of the United States and Canada. To provide depth and diversity of opinion, most topics are represented by several papers. Some papers from sources outside the set described above are in the book to fill specific needs, but there has been no exhaustive search of other published literature beyond that already mentioned.

The limitations of size do not permit reprinting all of the applicable papers from the sources searched. There are many excellent papers omitted for this reason alone, and the Editors can only offer sincere regrets to their authors. It should also be observed that there is a large body of IEEE literature from recent years outside the scope of this book. This generally consists of research papers which, however excellent, have not had a noticeable influence on the design of apparatus, the design or operation of systems, or the analytical methods employed by the industry. Representative bibliographies are included for two categories of papers, namely, those dealing with optimal generator control and those dealing with the so-called direct methods of determining stability.

Background Reading

The reprints in this book generally require the reader to be familiar with power system stability. There are a number of English language textbooks devoted entirely or in part to the subject, some of which are listed in Table II. The following is a brief presentation of the elementary concepts of synchronous stability.

Synchronous Stability—Concepts and Definitions

Power system generators connected through a transmission network must run in synchronism, that is, at the same average speed. Automatic load and frequency control systems and individual machine speed-governing systems tend to keep generator speeds, and consequently speed differences, within narrow bounds, but it is the effect of variational power flow through the network which forces speed differences to be zero on the average. If any generator runs faster than another, the angular position of its rotor relative to that of the slower machine will continue to advance as long as the speed difference exists, and its generated voltage will likewise advance in phase relative to the voltage of the slower machine. The resultant phase difference, within limits, shifts a load from the slow machine to the fast one, tending to reduce the speed difference.

The shift in load between generators is a nonlinear function of the difference in rotor angles, and above a certain angle difference, nominally 90°, the incremental load shift due to

[2]Power Industry Computer Applications.

TABLE II

TEXTBOOKS IN ENGLISH ON POWER SYSTEM STABILITY

Power System Stability, Vol. I: Steady-State Stability, S. B. Crary. New York: Wiley, and London: Chapman and Hall, 1945.
Power System Stability, Vol. II: Transient Stability, S. B. Crary. New York: Wiley, and London: Chapman and Hall, 1947.
Power System Stability, Vol. I: Elements of Stability Calculations, E. W. Kimbark. New York: Wiley, and London: Chapman and Hall, 1948.
Power System Stability, Vol. II: Power Circuit Breakers and Protective Relays, E. W. Kimbark. New York: Wiley, and London: Chapman and Hall, 1950.
Power System Stability, Vol. III: Synchronous Machines, E. W. Kimbark. New York: Wiley, and London: Chapman and Hall, 1956.
Transient Phenomena in Electric Power Systems, V. A. Venikov. New York: Pergamon, 1964.
Electrical Transmission and Distribution Reference Book. East Pittsburgh, Pa.: Westinghouse Electric Corp., 1964, ch. 13.
Computer Methods in Power System Analysis, Glenn L. Stagg and Ahmed H. El-Abiad. New York: McGraw-Hill, 1968, ch. 10.
Electric Energy Systems Theory, O. I. Elgard. New York: McGraw-Hill, 1971.

incremental angle change reverses, and the forces which tended to reduce speed differences become forces tending to increase speed differences. This, in essence, is the loss-of-synchronism phenomena. Because levels of kinetic energy change, recovery of synchronous operation is uncommon. Moreover, as machines lose synchronism, currents and voltages vary over wide ranges, and protective apparatus should operate, and normally will, to trip affected generators and lines. Loss of synchronism can occur between one machine and the rest of the system or between groups of machines, with synchronism perhaps maintained within those groups.

Power system stability is primarily concerned with variations in speeds, rotor positions, and generator loads. It focuses attention on the transmission network, since it is the network, more than power plant or system controls, which provides for the power shifts between generators required to maintain synchronism. Generating plant controls do have some influence, however. Automatic generator excitation systems modify power flow through the network by their influence on generator magnetic flux levels, and turbine speed-governing systems affect the power delivered by prime movers to generators.

Power system engineers have found it useful to identify three types of unstable behavior, designated steady-state instability, dynamic instability, and transient instability. There are no universally accepted precise definitions of this terminology, and there is no intent to imply definitions in what follows. Instead, the purpose is simply to describe generally several ways in which instability occurs.

Steady-state instability is a possible but improbable event on large power systems. In the simplest hypothetical two-machine system, loss of synchronism will occur if an attempt is made to operate with an angle separation between machines greater than 90°. For real multimachine systems, large angle differences also tend toward steady-state instability, but it is not possible at this time to formulate steady-state stability limits for large systems in terms of angle differences. Between widely separated machines on a large system, angle differences may greatly exceed 90° with no threat of steady-state instability. On the other hand, depending on the location and the characteristics of system loads, steady-state instability may occur with angle differences less than 90°. In those systems where this kind of instability is a genuine hazard, operators are frequently able to recognize limiting conditions in terms of some gradual change, such as a sagging bus voltage, and alter operation in time to avoid instability. If limiting conditions are exceeded, rates of change increase enormously, and loss of synchronism with system breakup can occur within seconds.

Dynamic instability is more probable than steady-state instability, or at least is more common in existing North American power systems. Small speed deviations occur continuously in normal operation with corresponding variations in angle differences and generator loads. If the variations resulting from any initial change diminish with time, the system is said to be dynamically stable. Conversely, if these variations, in the form of oscillations, increase with time, the system is dynamically unstable. Due to the effects of nonlinearities, such oscillations may be limited at some magnitude, or they may increase to the point of loss of synchronism and system breakup unless some intervention occurs.

Transient instability may occur on any system subjected to a major disturbance. There is no limit to the kinds of disturbances which can occur, but a fault on a heavily loaded line which requires opening the line to clear the fault is usually of greatest concern. The tripping of a loaded generator or the abrupt dropping of a large load may also cause instability. Usually, disturbances alter the system at least temporarily so that the subsequent steady-state operation will be different from that prior to the disturbance. There is then the necessity for that altered system to be stable in its new steady state. There is also the possibility that the altered system will be dynamically unstable, and that oscillations subsequent to the disturbance will be sustained, or eventually increase in magnitude to the point of causing system breakup. Most often, however, instability following a disturbance is due to the relatively large transient deviations in speeds, rotor angle differences, and power shifts initiated by the disturbance. Loss of synchronism can occur quite rapidly, within 1 s for a generator close to a transmission line fault. In the Northeast Power Failure of 1965, transient instability was initiated by tripping a 230-kV transmission circuit, and within about 7 s, the Northeast had become isolated from the larger eastern network, and had subdivided into four major isolated subsystems.

ELEMENTARY ANALYTICAL VIEWPOINTS

A basic understanding of synchronous stability can be acquired using simple analytical methods. The reader should be advised that these methods cannot be readily extended to large systems and are presented only to provide insight.

Generator and Infinite Bus

Consider a remote generator connected radially to a major substation of a very large system. Assume that the voltage at the substation is so strongly influenced by large nearby generation that it is independent of events at the remote generator. A single line diagram of this system is shown in Fig. 1. A sta-

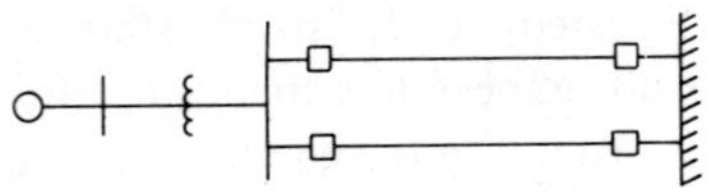

Fig. 1. Single line diagram for machine and infinite bus system.

bility model for this generator consisting of three equations will be described.

Acceleration Equation

The combined inertias of the generator and its prime mover will be accelerated by any unbalance in the applied torques. Neglecting certain losses, one may write

$$J \frac{d\omega}{dt} = T_M - T_E. \tag{1}$$

Symbols are interpreted as follows:

T_M : prime mover torque, megajoules/radian

T_E : electrical torque, megajoules/radian

ω : speed, radians/second

J : moment of inertia for generator and prime mover, megajoules · second^2/radian^2.

It is customary to normalize (1) using the inertia constant H. The latter is defined to be the kinetic energy at rated speed ω_R divided by the generator mva base S_B:

$$H = \frac{\frac{1}{2} J \omega_R^2}{S_B}. \tag{2}$$

Substituting for J in (1),

$$\left(\frac{1}{\omega_R}\right)\left(\frac{d\omega}{dt}\right) = \left(\frac{1}{2H}\right)\left(\frac{\omega_R}{\omega}\right)\left(\frac{\omega T_M - \omega T_E}{S_B}\right). \tag{3}$$

Each of the products of speed and torque is power in megawatts, so that division by S_B yields power in per unit on that base. The symbols P_M and P_E will be used for these per unit variables. Speed divided by rated speed is per unit speed and will be designated $\overline{\omega}$. The ratio of rated speed to speed is always quite close to unity in practical cases, and that ratio on the right-hand side of (3) can be set to unity as a simplification. The acceleration equation then becomes

$$\frac{d\overline{\omega}}{dt} = \frac{P_M - P_E}{2H}. \tag{4}$$

Power Flow Equation

It is convenient here to assume that P_M is equal to the initial prime mover power throughout the time interval of analysis. This ignores speed-governing effects. Acceleration is then entirely due to variations in P_E. An equation for P_E can be written beginning with the positive-sequence diagram of Fig. 2, in which all resistances are assumed to be negligible. The remote generator is represented by a complex voltage E' behind an internal generator reactance X_D', the transformer by a reactance X_T, the transmission lines by a reactance X_L, and the infinite bus by a complex voltage E. All of these are per unit quantities.

Let the total reactance of the circuit be designated X:

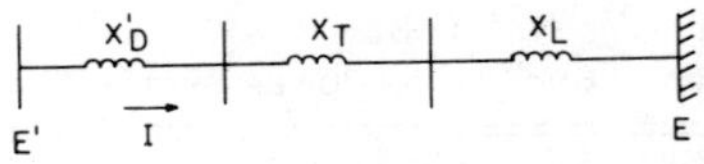

Fig. 2. Positive sequence circuit diagram for machine and infinite bus system.

$$X = X_D' + X_T + X_L. \tag{5}$$

Write the complex voltages in polar form using the following notation:

$$E' = V' e^{j\theta'} \tag{6}$$

$$E = V e^{j\theta}. \tag{7}$$

Define the voltage angle difference

$$\delta = \theta' - \theta. \tag{8}$$

With a little effort, it can be shown that

$$P_E = \frac{VV' \sin\delta}{X}. \tag{9}$$

Angle Difference Equation

Ignoring variations in generator rotor flux, the speed induced voltage V' may reasonably be assumed to have constant magnitude.

Since the induced voltage is caused by flux which is fixed in relationship to the generator rotor, the relative phase of the voltage will be advanced for rotor speeds higher than synchronous, and retarded for speeds less than synchronous. Mathematically,

$$\frac{d\theta'}{dt} = N(\omega - \omega_R) \tag{10}$$

where N is the number of pole pairs on the generator rotor. Let the rated electrical frequency in radians/second be

$$\omega_0 = N\omega_R. \tag{11}$$

Then (10) may be written

$$\frac{d\theta'}{dt} = \omega_0(\overline{\omega} - 1). \tag{12}$$

Finally, since the angle θ of the infinite bus voltage is assumed to be constant,

$$\frac{d\delta}{dt} = \omega_0(\overline{\omega} - 1). \tag{13}$$

Summary of Basic Equations

A stability model for the remote generator consists of the following three equations:

$$\frac{d\overline{\omega}}{dt} = \frac{P_M - P_E}{2H} \tag{4}$$

$$P_E = \frac{VV' \sin\delta}{X} \tag{9}$$

$$\frac{d\delta}{dt} = \omega_0(\overline{\omega} - 1). \tag{13}$$

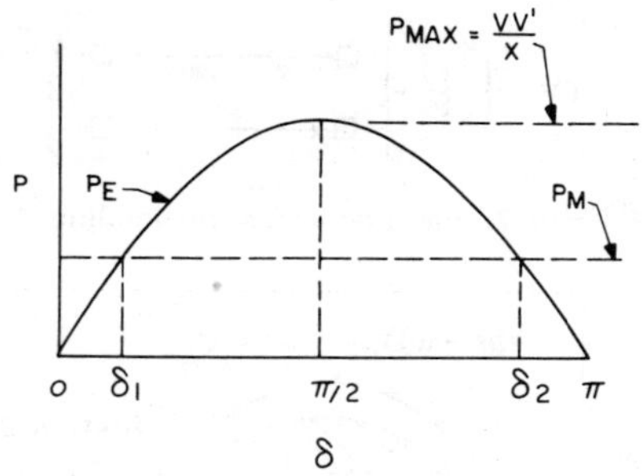

Fig. 3. Power-angle diagram.

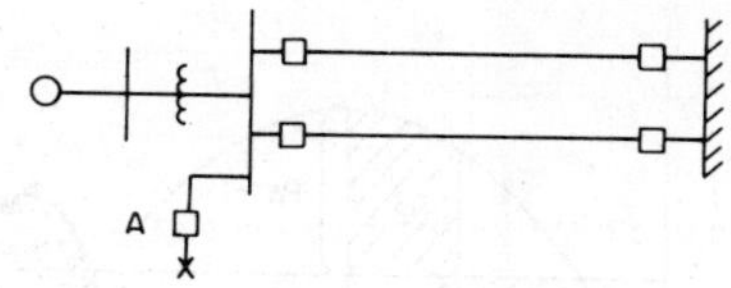

Fig. 4. Fault at transformer high-side bus.

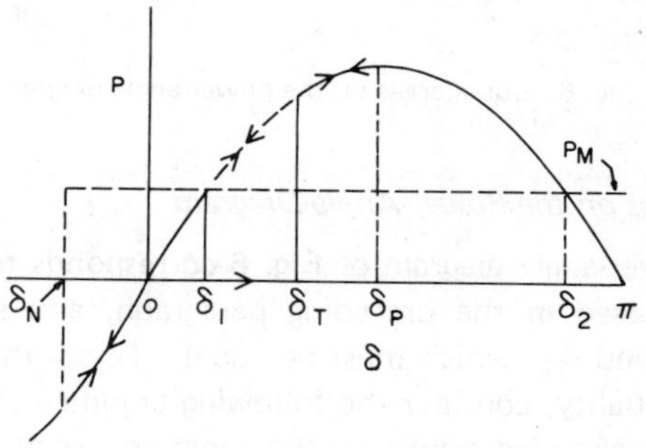

Fig. 5. Electrical power and angle variations during and subsequent to fault of Fig. 4.

As an alternative, a single equation can be obtained by differentiating (13) and using that result and (9) in (4):

$$\frac{d^2\delta}{dt^2} = \frac{\omega_0}{2H}\left[P_M - \frac{VV'\sin\delta}{X}\right]. \tag{14}$$

Power-Angle Diagram and Steady-State Stability

A graphical plot of (9) is called a power-angle diagram. The diagram can be plotted over any range of angles, but it is usually sufficient to show it over the interval $0 \leq \delta \leq \pi$. Such a plot is shown in Fig. 3. This diagram serves a number of purposes, the first of which is to define stable and unstable equilibrium angles. A horizontal line is drawn on the diagram at a power level equal to P_M, the prime mover power. This horizontal line also represents the electrical power at equilibrium, i.e., no acceleration. The remote generator voltage angle may lead the infinite bus voltage angle by either of the two values δ_1 or δ_2 at which P_M intersects P_E. The angle δ_1 is a stable equilibrium angle because small variations in δ about δ_1 produce torques on the generator which tend to restore the angle to δ_1. For example, if δ slightly exceeds δ_1, the electrical power P_E given by the curve will exceed the prime mover power P_M, and the generator will tend to slow down, reducing δ. A similar argument for restoration of equilibrium may be made for values of δ slightly less than δ_1. On the other hand, for δ slightly greater than δ_2, P_E is now less than P_M, and the generator will tend to speed up. This further increases δ, so that δ_2 is not a stable equilibrium. If δ is less than δ_2, the generator power exceeds the prime mover power so that the generator tends to slow down, and tends toward δ_1, the stable equilibrium.

Stable equilibrium requires that the slope of the power-angle curve be positive at the operating angle. The power P_{max} represents the stable limit of power transfer between the remote generator and the infinite bus. A practical operating limit is considerably less than P_{max} if stability is to be maintained under transient conditions.

Faults and Transient Stability

Attention will now be directed toward the behavior of this stability model during and subsequent to a three-phase fault at the remote generator transformer high-side bus. A one-line diagram showing the fault location is presented in Fig. 4. Let the instant at which the fault occurs be the reference for time, that is, $t = 0$, with the fault cleared by tripping the breaker A when $t = T$. This leaves the transmission system intact subsequent to fault clearing, and to that extent simplifies the analysis.

During the fault, the electrical power is zero, and (4) and (13) are easily integrated to obtain the speed and angle at the fault clearing time T:

$$\overline{\omega} = 1 + \frac{P_M T}{2H} \tag{15}$$

$$\delta_T = \delta_1 + \frac{\omega_0 P_M T^2}{4H}. \tag{16}$$

Fig. 5 shows the effect of the fault in terms of the power-angle diagram. Initially, P_E is equal to P_M with an angle separation of δ_1. During the fault, the electrical power is zero while the angle δ increases from δ_1 to δ_T. Simultaneously, the generator speed increases from unity to $\overline{\omega}_T$ with a corresponding increase in kinetic energy. When the fault is cleared, the electrical power is greater than the prime mover power as indicated by the curve at the angle δ_T, so that the generator experiences a decelerating torque and its speed begins to return to normal. At some peak angle δ_P, the generator speed will recover its normal value, but the net torque on the generator is in the direction of deceleration so that the machine will continue to slow down, with the speed now dropping below normal. This results in a diminishing angle. When the angle reaches δ_1, the net torque on the generator will again be zero, but the speed will be less than normal, so the angle will continue to diminish. Below δ_1, the torque on the machine reverses, and the resulting acceleration begins to increase speed toward its normal value. At some minimum angle δ_N, normal speed again occurs, but the net torque now causes the machine to speed up and the angle to increase. According to the stability model established for this generator, this oscillating motion will continue indefinitely since no provision for damping has been included. Assuming dynamic stability, such oscillations will damp out eventually, and operation at the equilibrium angle will be restored. For the model behavior described, no loss of synchronism occurs, and the system is stable.

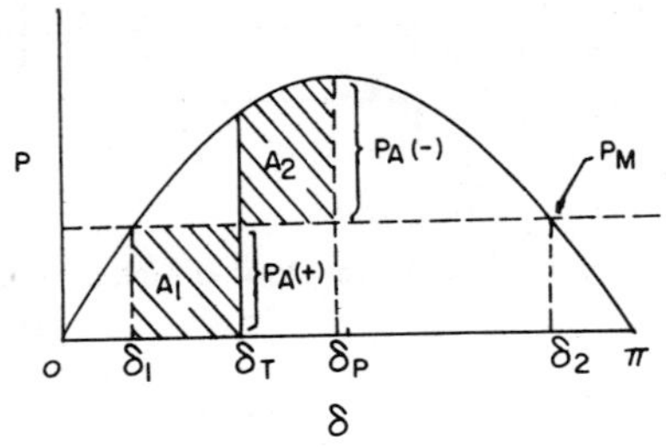

Fig. 6. Equal areas on the power-angle diagram.

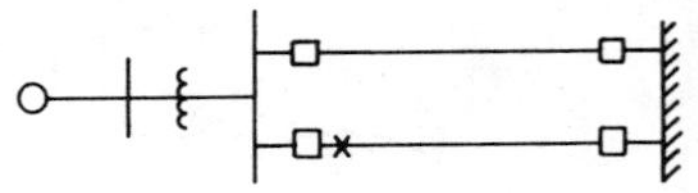

Fig. 7. Fault on a transmission line.

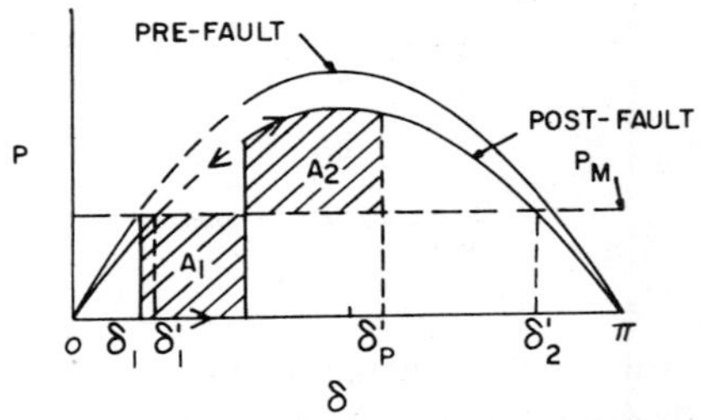

Fig. 8. Power-angle diagrams for pre-fault and post-fault conditions.

Equal Areas on the Power-Angle Diagram

The power-angle diagram of Fig. 6 corresponds to the transient discussed in the preceding paragraph, and shows two areas, A_1 and A_2, which must be equal. To see the necessity for this equality, consider the following argument. Let P_A be the net accelerating power on the generator, so that (4) may be written

$$2H\frac{d\overline{\omega}}{dt} = P_M - P_E = P_A. \tag{17}$$

Equations (17) and (13) imply a differential relationship between $\overline{\omega}$ and δ which, eliminating the independent variable time, can be written

$$2\omega_0 H(\overline{\omega} - 1)\, d\overline{\omega} = P_A d\delta. \tag{18}$$

Let the right-hand side of (18) be integrated over the interval $\delta_1 \leq \delta \leq \delta_P$. The left-hand side vanishes for this interval since the upper and lower limits of integration are equal. That is to say, the initial per unit speed at the angle δ_1 is unity, and the per unit speed at the peak angle swing δ_P is also unity. Hence

$$\int_{\delta_1}^{\delta_P} P_A d\delta = 0. \tag{19}$$

This integral may be separated into two parts,

$$\int_{\delta_1}^{\delta_P} P_A d\delta = \int_{\delta_1}^{\delta_T} P_M d\delta + \int_{\delta_T}^{\delta_P} (P_M - P_{\max}\sin\delta)\, d\delta \tag{20}$$

so that

$$\int_{\delta_1}^{\delta_T} P_M d\delta = \int_{\delta_T}^{\delta_P} (P_{\max}\sin\delta - P_M)\, d\delta. \tag{21}$$

The left-hand side of (21) is the area A_1 of Fig. 6, and the right-hand side is A_2.

Critical Clearing Time and the Equal Area Criterion

From (16) it is clear that a fault of longer duration than that assumed would have resulted in a larger value for the angle at clearing time, δ_T. In Fig. 6, A_1 would have been larger, and since A_2 must equal A_1, δ_P would have increased. Note, however, that the speed of the remote generator must be reduced to its normal value before the angle δ reaches δ_2 or synchronism with the infinite bus will be lost. To see this, assume that some speed greater than unity exists with $\delta = \delta_2$. This speed will shortly cause δ to be greater than δ_2, where the net accelerating power on the generator will be positive, further increasing the speed.

The angle δ_2 is frequently referred to as the critical angle, and the fault clearing time which makes A_1 equal to A_2 in Fig. 9 is called the critical clearing time. For this relatively simple stability model, the critical clearing time can be calculated using (21) and (16). Critical clearing times for faults in real systems are much more difficult to obtain, involving a trial and error process with digital computer simulation. This can be quite expensive, so that critical clearing times may not often be determined with precision. Even so, the notion that a critical clearing time exists is important.

Factors Which Influence Transient Stability

The diagram of Fig. 7 depicts a more realistic fault location than that of Fig. 4 since it occurs on an operating transmission circuit. Clearing the fault requires removal of the line from service, and assuming equal reactance in each line, increases the post-fault reactance to

$$X_{PF} = X'_D + X_T + 2X_L. \tag{22}$$

This reduces the power transmission capability and increases the risk of instability. Fig. 8 shows two power-angle diagrams corresponding to the pre-fault and post-fault transmission capacities. Note that there will be new post-fault operating and critical angles, δ'_1 and δ'_2, and more importantly, that the peak angle δ'_P moves closer to the critical angle δ'_2 than in the case previously discussed. This constitutes a reduction in the margin of stability, and leads to the consideration of other factors which tend to reduce the margin of stability. Symbols are to be interpreted in terms of Fig. 6.

1) Increasing the generator initial load P_M increases δ_1 and A_1, thus increasing δ_P. The angle δ_2 is also reduced.

2) Increasing the fault duration increases A_1, hence δ_P.

3) Increasing the pre-fault impedance increases δ_1, hence δ_P. The critical angle δ_2 is also reduced.

4) Reducing the generator and prime mover inertia increases δ_T, hence A_1 and δ_P.

It is important to note that the effects of these factors is not linear, and that for small stability margins, the reduction in margin becomes extremely sensitive to unfavorable changes in

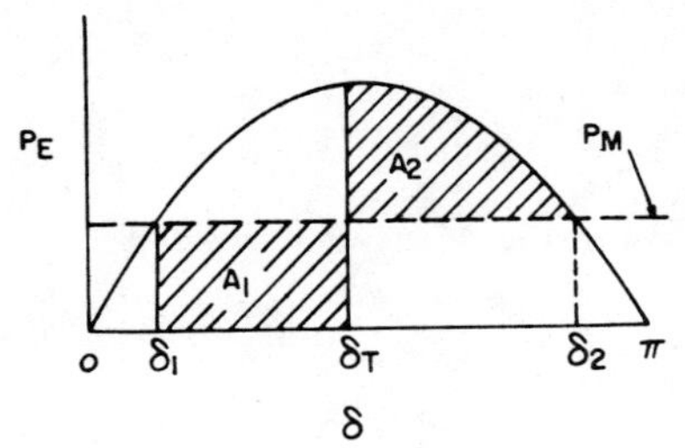

Fig. 9. Equal areas for critical clearing time.

these important factors. The implication is that real systems must operate with sufficient margin to allow for variations in operating conditions and uncertainties in models and data if secure operations are to be achieved. Unfortunately, there are neither definitions nor standards for stability margins applicable to real systems.

References

[1] "Prevention of power failures, Vol. I: Report of the commission," Federal Power Commission, 1967.

[2] "Prevention of power failures, Vol. II: Advisory committee report. Reliability of electric bulk power supply," Federal Power Commission, 1967.

[3] "Prevention of power failures, Vol. III: Studies of the task groups on the northeast power interruption," Federal Power Commission, 1967.

[4] "Power system stability bibliography, 1935–1965," IEEE Committee Rep., 1965.

Part I

Mathematical Models and Apparatus Description for Systems Analyses

Stability analyses of large electric power systems depend almost entirely on digital-computer simulation of system dynamic behavior. Simulation implies the existence of mathematical models for a variety of apparatus, data files which contain model parameters for specific power systems, and computer programs. Much of the stability activity of the past decade has been devoted to these various aspects of simulation. This section contains papers which pertain primarily to mathematical models used in stability studies. Models for generators, generation excitation systems, turbine speed-governing systems, loads, and system equivalents are discussed.

The central computational requirement for stability studies is the repeated calculation of the varying electrical power at each generator. This calculation utilizes models for the transmission network, for loads, and for generators. The transmission network is represented by its positive-sequence equivalent circuit. Bulk loads at transmission system buses are usually represented by various functions of bus voltage and frequency. Until the 1960's, it was common practice to represent generators as constant voltage magnitude sources connected to the network through constant impedances. This generator representation has come to be known as the classical representation.

It has long been recognized that the classical model has certain limitations. It is not applicable to dynamic studies where interest centers on the damping of oscillations. To a large extent, damping is determined by variations in generator flux levels which are excluded by the assumption of constant internal voltage magnitudes. Its use may also lead to significant errors in transient stability studies, especially those involving severe three-phase faults with substantial generator demagnetization during the fault. The representation of changing generator flux levels, of course, requires the representation of automatic excitation systems as well.

Turbine speed-governing systems play an important role in the behavior of isolated systems which have either excess load or generation in significant amounts. They also influence damping in dynamic stability studies, and have some effect on transient stability behavior. Automatic generation control and turbine load control systems are not represented on the hypothesis that they respond too slowly to significantly influence stability characteristics. Steam generator controls and steam system dynamics upstream from the governor-controlled valves are excluded from stability studies for the same reason.

The functional block diagram of Fig. I-1 indicates the power apparatus normally represented in stability studies. This specific arrangement of mathematical subsystems is typical, but there are, of course, alternatives. It is particularly important to call attention to the interface between the generator and the network. This interface consists of the two coordinate transformation blocks. For network calculations, the generator internal voltage, terminal voltage, and current are represented as positive-sequence phasors. On the generator side of the transformations, these quantities are represented by the projections of their positive-sequence phasors on the direct and quadrature magnetic axes of the machine. Because of saliency, the generator output impedance used with the network varies as the rotor angle moves relative to the positive-sequence coordinate axes.

Authors of papers dealing with mathematical models of synchronous machines for system studies generally assume that the reader is familiar with the basic machine theory covered in numerous textbooks. Current stability papers on machines deal largely with equivalent circuits, with methods for obtaining parameters, with the derivation of differential equations for generator internal voltages, and with methods

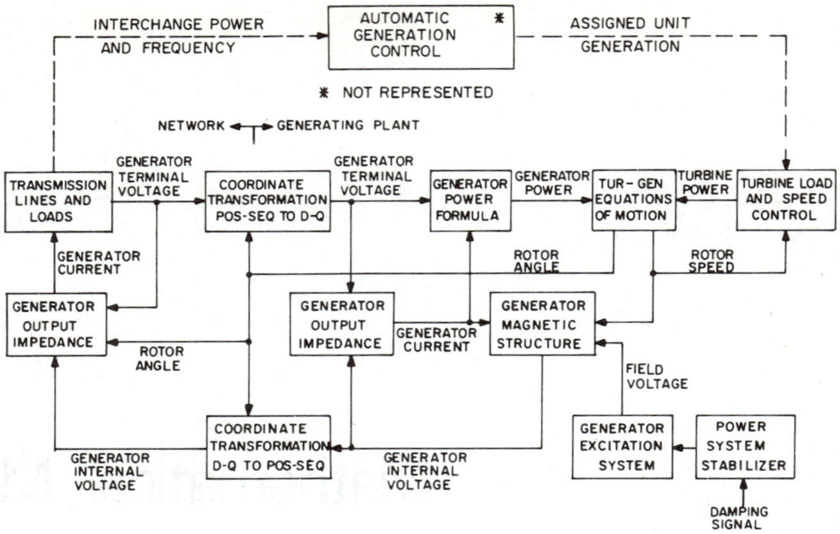

Fig. I-1

for handling saliency in the positive-sequence network calculations.

The paper "Computer Representation of Excitation Systems" is the most general reference available for excitation system models. It includes mathematical models suitable for a variety of excitation systems, definitions for symbols, and typical parameters for commonly encountered equipment. Considerable emphasis in recent years has been placed on improved stability performance by means of fast-acting excitation systems, and a number of papers deal with descriptions and models for relatively new apparatus. For systems with poor damping, it is frequently possible to improve performance by means of devices called power system stabilizers. During transients, these devices act through the generator excitation systems and alter the normal voltage regulating function. Several papers in this section describe power system stabilizers and their mathematical models.

The paper "Dynamic Models for Steam and Hydro Turbine in Power System Studies" is a general reference for speed-governing system models. It also includes mathematical models, definitions, and typical model parameters. Two papers which describe steam and hydro speed-governing systems in more detail are included.

The basic mathematical structures used in modeling generating plant apparatus are derived from physical laws. Testing of apparatus serves primarily to verify assumptions and to obtain model parameters. Mathematical models for bulk system loads cannot, in general, be obtained in this way because of the complex way in which loads may interact when connected to a common supply system. Load models are strongly dependent on system tests, and three papers which describe tests and test results are included. A fourth paper presents laboratory test results for individual air conditioning, fluorescent lighting, and induction motor loads.

Perhaps the most perplexing problem in the stability analysis of large electric power systems lies in determining how much of the system must be represented in detail in order to have reliable study results. The costs of collecting and verifying data for larger systems and the costs of simulation motivate detailed representation for the smallest possible system consistent with study objectives. Selecting the part of the system to be represented in detail and finding an equivalent for the remainder is presently a matter for engineering judgment. The paper "Electromechanical Equivalents for Use in Power System Stability Studies" describes the results of recent research efforts in this area.

Space limitations have excluded many excellent papers dealing with models for stability studies, some of which are listed in the following bibliography.

Supplementary Bibliography

Synchronous Machines

[1] IEEE Committee Report, "Bibliography of rotating electric machinery for 1948–1961," *IEEE Trans. Power App. Syst.*, vol. 83, pp. 589–606, June 1964.

[2] ——, "Bibliography of rotating electric machinery for 1962–1965, *IEEE Trans. Power App. Syst.*, vol. PAS-87, pp. 679–689, Mar. 1968.

[3] IEEE Bibliography and Publicity Subcommittee of the IEEE Rotating Machinery Committee, "Bibliography of rotating electrical machinery for 1966–1968," *IEEE Trans. Power App. Syst.*, vol. PAS-89, pp. 1293–1307, July/Aug. 1970.

[4] IEEE Committee Report, "Recommended phasor diagram for synchronous machines," *IEEE Trans. Power App. Syst.*, vol. PAS-88, pp. 1593–1610, Nov. 1969.

[5] I. M. Canay, "Causes of discrepancies on calculation of rotor quantities and exact equivalent diagrams of the synchronous machine," *IEEE Trans. Power App. Syst.*, vol. PAS-88, pp. 1114–1120, July 1969.

[6] W. B. Jackson and R. L. Winchester, "Direct- and quadrature-axis equivalent circuits for solid-rotor turbine generators," *IEEE Trans. Power App. Syst.*, vol. PAS-88, pp. 1121–1136, July 1969.

[7] E. F. Fuchs and E. A. Erdelyi, "Determination of waterwheel alternator transient reactances from flux plots," *IEEE Trans. Power App. Syst.*, vol. PAS-91, pp. 1795–1802, Sept./Oct. 1972.

[8] ——, "Determination of waterwheel alternator steady-state reactances from flux plots," *IEEE Trans. Power App. Syst.*, vol. PAS-91, pp. 2510–2527, Nov./Dec. 1972.

[9] ——, "Nonlinear theory of turboalternators—Part I: Magnetic fields at no-load and balanced loads," *IEEE Trans. Power App. Syst.*, vol. PAS-92, pp. 583–592, Mar./Apr. 1973.

[10] —, "Nonlinear theory of turboalternators—Part II: Load dependent synchronous reactances," *IEEE Trans. Power App. Syst.*, vol. PAS-92, pp. 592–600, Mar./Apr. 1973.

[11] J. F. Szablya and J. M. Bressane, "Transfer function of A.C. machines," *IEEE Trans. Power App. Syst.*, vol. PAS-91, pp. 177–186, May 1972.

Excitation Systems and Related Devices

[12] H. C. Barnes, J. A. Oliver, A. S. Rubenstein, "Alternator-rectifier for cardinal plant 724-MVA generator, "*IEEE Trans. Power App. Syst.*, vol. PAS-87, pp. 1189–1198, Apr. 1968.

[13] G. L. Wilson and W. S. Morgan, "Stability study and modification of the cardinal plant alternator-rectifier exciter," IEEE paper 69CP149-PWR.

[14] A. P. Mane, R. H. Webster, and O. J. M. Smith, "Rotor velocity measurement from power balance analog," *IEEE Trans. Power App. Syst.*, vol. PAS-90, pp. 434-441, Mar./Apr. 1971.

[15] T. H. Einstein, "Generalized representation of generator excitation power requirements," *IEEE Trans. Power App. Syst.*, vol. PAS-91, pp. 1840–1847, Sept./Oct. 1972.

[16] W. F. Wright, R. Hawley, and J. L. Dineley, "Brushless thyristor excitation systems," *IEEE Trans. Power App. Syst.*, vol. PAS-91, pp. 1848–1854, Sept./Oct. 1972.

[17] IEEE Committee Report, "Excitation system dynamic characteristics," *IEEE Trans. Power App. Syst.*, vol. PAS-92, pp. 64–75, Jan./Feb. 1973.

Speed-Governing Systems

[18] L. M. Hovey, "Optimum adjustment of hydro governors on Manitoba hydro system," *AIEE Trans. Power App. Syst.*, vol. 81, pp. 581–587, Dec. 1962.

[19] M. Birnbaum and E. G. Noyes, "Electro hydraulic control for improved availability and operation of large steam turbines," presented at the ASME-IEEE Nat. Power Conf., Albany, N.Y., Sept. 19–23, 1965.

[20] M. Leum, "The development and field experience of a transistor electric governor for hydro turbines," *IEEE Trans. Power App. Syst.*, vol. PAS-85, pp. 393–402, Apr. 1966.

[21] F. R. Schleif and A. B. Wilbor, "The coordination of hydraulic turbine governors for power system operation," *IEEE Trans. Power App. Syst.*, vol. PAS-85, pp. 750–756, July 1966.

[22] J. M. Undrill and J. W. Woodward, "Nonlinear hydro governing model and improved calculation for determining temporary droop," *IEEE Trans. Power App. Syst.*, vol. PAS-86, pp. 443–452, Apr. 1967.

[23] L. O. Long, "Joint control of electric governors for hydraulic turbines," *IEEE Trans. Power App. Syst.*, vol. PAS-87, pp. 1754–1762, Sept. 1968.

[24] C. Concordia, "Effect of prime-mover speed control characteristics on electric power system performance," *IEEE Trans. Power App. Syst.*, vol. PAS-88, pp. 752–756, May 1969.

Loads

[25] D. S. Brereton, D. G. Lewis, and C. C. Young, "Representation of induction-motor loads during power-system stability studies," *AIEE Trans. Power App. Syst.*, vol. 76, pp. 451–461, Aug. 1957.

Equivalents

[26] H. E. Brown, R. B. Shipley, D. Coleman, and R. E. Nied, Jr., "A study of stability equivalents," *IEEE Trans. Power App. Syst.*, vol. PAS-88, pp. 200–207, Mar. 1969.

[27] A. Chang and M. M. Adibi, "Power system dynamic equivalents," *IEEE Trans. Power App. Syst.*, vol. PAS-89, pp. 1737–1744, Nov./Dec. 1970.

[28] J. M. Undrill and A. E. Turner, "Construction of power system electronmechanical equivalents by modal analysis," *IEEE Trans. Power App. Syst.*, vol. PAS-90, pp. 2049–2059, Sept./Oct. 1971.

Special Relays and Instruments

[29] F. R. Schleif, L. W. Loyd, Jr., R. W. World, and W. B. Gish, "A swing relay for the east-west intertie," *IEEE Trans. Power App. Syst.*, vol. PAS-88, pp. 821–825, June 1969.

[30] IEEE Working Group Report, "Application of instrumentation for the evaluation of power system dynamic performance," *IEEE Trans. Power App. Syst.*, vol. PAS-91, pp. 1379–1382, July/Aug. 1972.

[31] C. L. Clemans and W. R. Roemish, "A Power rate relay for system applications," *IEEE Trans. Power App. Syst.*, vol. PAS-92, pp. 122–127, Jan./Feb. 1973.

Digital Simulation of Synchronous Machine Transients

DAVID W. OLIVE, SENIOR MEMBER, IEEE

Abstract—**When synchronous machine transients in an extensive system are to be accurately calculated, many problems arise for which the information required to resolve them is not available. The resulting accuracy will depend upon the assumptions used not only in the development of the machine model but also in the way in which the incomplete model is solved. The primary problems involve the manner in which the machine terminal relations are related to those of the system and in the handling of the various torques produced in the transient operation. The state of the art of machine analysis is now such that synchronous machine transients are fairly well understood and calculations may be reasonably comprehensive. The system representation is, however, not capable of accepting the information available from the machine with the result that a satisfactory marriage is difficult. Some of the inadequacies of large scale stability analysis as it now exists and some of the precautions and planning that should be observed before embarking upon extensive studies or in drawing conclusions from such studies are covered here, in a philosophic and general manner.**

Introduction

As THE SIZE and complexity of electric power systems increase because of the press of economics and population, the desire to predict system behavior more accurately has also increased. Fortunately, the digital computer has given the engineer the ability to predict in situations where complexity would have been too great before. The stability problem and its calculation will be the specific example used here. Although this determination has stimulated varying degrees of interest over the years, it is currently enjoying new popularity, not only because systems are more massive but also because there is now a possibility of obtaining accurate answers.

In the first flush of digital computation those parts of the system that were well understood tended to become more detailed while other parts remained as they had been represented in an earlier era. Also, analyses previously geared to a simple model were not always applied correctly for the transient behavior of the more complicated one. It is therefore time to reinspect the problem and to look at what can and cannot be done; to see if we are perhaps attempting ten-place accuracy answers with two-place accuracy data and to see what, if anything, needs to be done to improve the accuracy of transient stability determinations.

This paper will not attempt to rewrite any digital program or to draw conclusions regarding the worth of any study. Because systems differ so widely, such conclusions would be not only highly suspect but also highly arbitrary, drawn from no data whatsoever. It is merely intended to discuss aspects of machine–system interaction with a view towards ways in which accuracy might be improved.

Paper 31 C 69, recommended and approved by the Power System Engineering Committee of the IEEE Power Group for presentation at the PICA Conference, Pittsburgh, Pa., May 15–17, 1967. Manuscript submitted February 6, 1967; made available for printing February 19, 1968.

The author is with the University of Nebraska, Lincoln, Neb.

Synchronous Machine Equations

The equations of a synchronous machine may be derived straightforwardly from a linear two-pole model. The stator is considered to be three identical, symmetrically placed, lumped windings called a, b, and c. The rotor windings are four unequal lumped windings, each placed with its axis 90° around the periphery. The two windings in the one axis (direct) will be called f and x and the two in the other (quadrature) axis g and y. Winding f represents the field winding while windings x, g, and y are ficticious windings inserted to account for, respectively, subtransient behavior in the direct axis and transient and subtransient behavior in the quadrature axis. Although the resulting two windings in the quadrature axis will both have iron paths and can, with some justification, be replaced with one winding; this development will at least start with the symmetrical machine.

The assumptions made in the development of the model are:

1) all inductances are independent of current (saturation is neglected);

2) all self- and mutual inductances may be represented as constants plus a simple sine variation of the rotor angle θ_r or $2\theta_r$;

3) all distributed windings may be adequately represented as lumped windings;

4) the effects of currents flowing in the iron parts of the rotor or in damper windings may be represented by three lumped rotor coils as previously described.

Since extensive documentation of the following procedure exists, only the steps involved will be given here. The Kirchoff's law equations describing this model contain time-varying parameters but they may be eliminated by transforming the stator quantities from the a,b,c representation to the d, q, o representation. Rotor quantities are invariant in the transformation.

$$C = \frac{2}{3}\begin{array}{c|ccc} & a & b & c \\ d & \cos\omega_r t & \cos\omega_r t - 120^\circ & \cos\omega_r t + 120^\circ \\ q & -\sin\omega_r t & -\sin\omega_r t - 120^\circ & -\sin\omega_r t + 120^\circ \\ o & \frac{1}{2} & \frac{1}{2} & \frac{1}{2}. \end{array} \quad (1)$$

Note that $\theta_r = \omega_r t$ is also used in the transformation. If it is not, the transformation fails to remove the time-varying coefficients in the differential equations and the object of the transformation is defeated.

Under normal synchronous machine operating conditions, the transformation procedure is designed to convert the three-phase stationary stator winding carrying alternating currents into two quadrature windings carrying direct currents and rotating at synchronous speed. Under transient conditions when the rotor is not rotating at a 60-Hz synchronous speed, the frequency of currents supplied to the stator are not 60 Hz and switching discontinuities occur. All winding currents then cease to be purely

Reprinted from *IEEE Trans. Power App. Syst.*, Aug. 1968.

direct. Regardless, the resulting linear differential equations are readily solvable provided only that ω_r may be considered constant over the solution range. The above-mentioned behavior modifies the driving voltages of the machine but not the performance equations.

The transformation simplifies the solution of the machine equations. However, it will be shown, after the derivation is complete, that this transformation can become a pitfall in the process of relating machine performance to the rest of the system.

The equations become

$$v_d = -R_a i_d + \frac{d\lambda_d}{dt} - \omega_r \lambda_q \tag{2}$$

$$v_q = -R_a i_q + \frac{d\lambda_q}{dt} + \omega_r \lambda_d \tag{3}$$

$$v_o = -R_a i_o + \frac{d\lambda_o}{dt} \tag{4}$$

$$v_f = R_f i_f + \frac{d\lambda_f}{dt} \tag{5}$$

$$v_g = R_g i_g + \frac{d\lambda_g}{dt} \tag{6}$$

$$v_x = R_x i_x + \frac{d\lambda_x}{dt} \tag{7}$$

$$v_y = R_y i_y + \frac{d\lambda_y}{dt} \tag{8}$$

$$\lambda_d = -L_d i_d + L_{afm} i_f + L_{axm} i_x \tag{9}$$

$$\lambda_q = -L_q i_q + L_{aym} i_y + L_{agm} i_g \tag{10}$$

$$\lambda_o = L_o i_o \tag{11}$$

$$\lambda_f = -\frac{3}{2} L_{afm} i_d + L_{ff} i_f + L_{fx} i_x \tag{12}$$

$$\lambda_g = -\frac{3}{2} L_{agm} i_q + L_{gg} i_g + L_{gy} i_y \tag{13}$$

$$\lambda_x = -\frac{3}{2} L_{axm} i_d + L_{xx} i_x + L_{fx} i_f \tag{14}$$

$$\lambda_y = -\frac{3}{2} L_{aym} i_q + L_{yy} i_y + L_{gy} i_g \tag{15}$$

where L_d and L_q are the direct and quadrature axis synchronous reactance.

Two other equations are necessary: first the mechanical motion equation

$$T_i = \frac{2GH}{\omega_o^2} \frac{d^2\delta}{dt^2} + K\left(\omega_o + \frac{d\delta}{dt}\right) + T_e \tag{16}$$

and second, the developed electromagnetic torque T_e

$$T_e = \frac{3}{2}(\lambda_d i_q - \lambda_q i_d). \tag{17}$$

These equations are sufficient to solve for machine behavior when used in conjunction with a terminal system where v_d, v_q, i_d, and i_q may be correctly related. Note that all equations given are not necessarily in per unit.

Equations (2)–(16) are given in terms of the flux linkages, currents, and the various inductances of the model windings (either real or ficticious). Some of the parameters may be measured physically but, since many consist of interactions between iron circuits, it might be expected that the values would be difficult to obtain and highly variable due to the nonlinearities. These equations differ in nomenclature but actually are the equations originally derived by Park.

It has become common practice to discuss synchronous machine behavior in terms of a set of parameters other than the Park variables. These are the direct and quadrature axes voltages and currents and the steady state, transient and subtransient inductances, and associated time constants. The reason for this was presumably to standardize nomenclature, to attempt to eliminate some of the measurement problems by allowing parameter determination from a group of reasonably simple tests, and to relate the behavior to a set of more familiar variables. This result has not been entirely achieved. First, the behavior of all machines has been forced into the mold of this model; second, if the resulting equations are to be simple, further assumptions concerning the relationships between certain machine variables must be made; and third, the ability to obtain accurate values from the simple tests is doubtful. In particular, measurements of quadrature axis quantities are subject to error.

Nevertheless, since this is by far the most common basis for machine analysis, because more people are familiar with these variables and because standards exist for their definition, this set of equations will be given.

The further assumptions which must be made if the equations are to be put into this form are:

1) The following relationships must exist between the rotor and stator coil inductances in each axis.

$$\frac{L_{axm} L_{fx}}{L_{xx} L_{afm}} = \frac{L_{aym} L_{gy}}{L_{yy} L_{agm}} = 1.$$

Another way of stating this is that any transient change in stator currents will be reflected initially in the damper windings and will not affect the field. This assumption appears to be valid.

2) The time constants of the two rotor coils in each axis should differ by at least a factor of ten. This is true in the direct axis since the equivalent damper time constant is normally about (1/100)th of the field winding time constant. It is not a very good assumption for the windings in the quadrature axis and therefore often only one quadrature winding is used. The resulting quadrature axis behavior would then have only the subtransient (or transient) period rather than the two periods associated with the direct axis. Because of the difficulties in measuring quadrature axis values this will be assumed hereafter. The y coil will be assumed open ($i_y = 0$) and the resulting quantities will be called subtransient quantities.

The following definitions will apply.

$$e_{q1} = \omega_o L_{afm} i_f, \quad e_{q2} = \omega_o L_{axm} i_x, \quad e_d = -\omega_o L_{agm} i_g$$

$$e_q'' = \frac{\omega_o L_{axm}}{L_{xx}} \lambda_x, \quad e_q' = \frac{\omega_o L_{afm}}{L_{ff}} \lambda_f, \quad e_d'' = \frac{\omega_o L_{agm}}{L_{gg}} \lambda_g$$

$$T_{do}'' = \frac{X_d' - X_d''}{X_d - X_d''} \frac{L_{xx}}{R_x}, \quad T_{do}' = \frac{L_{ff}}{R_f}, \quad T_{qo}'' = \frac{L_{gg}}{R_g}.$$

Then

$$v_d = -R_a i_d + (X_q i_q + e_d)\frac{\omega_r}{\omega_o} + \frac{d\lambda_d}{dt} \quad (18)$$

$$v_q = -R_a i_q - (X_d i_d - e_{q1} - e_{q2})\frac{\omega_r}{\omega_o} + \frac{d\lambda_d}{dt} \quad (19)$$

where

$$\lambda_d = -L_d i_d + (e_{q1} + e_{q2})/\omega_o$$

$$\lambda_q = -L_q i_q - e_d/\omega_o$$

and

$$e_q' = -(X_d - X_d')i_d + e_{q1} + \frac{X_d - X_d'}{X_d - X_d''}e_{q2} \quad (20)$$

$$e_q'' = -(X_d - X_d'')i_d + e_{q1} + e_{q2} \quad (21)$$

$$e_d'' = (X_q - X_q'')i_q + e_d \quad (22)$$

$$\frac{de_d''}{dt} = -e_d/T_{qo}'' \quad (23)$$

$$\frac{de_q''}{dt} = -\frac{e_{q2}}{T_{do}''}\frac{X_d' - X_q''}{X_d - X_d''} \quad (24)$$

$$\frac{de_q'}{dt} = (V_f - e_{q1})/T_{do}' \quad (25)$$

$$T_e = e_q''i_q + e_d''i_d - i_d i_q(X_d'' - X_q'') \quad (26)$$

$$T_a = T_i - T_e = \frac{2H}{\omega_o}\frac{d^2\delta}{dt^2} + K'\frac{d\delta}{dt}. \quad (27)$$

Equations (18)–(27) are in per unit based upon the machine rating. One per unit field voltage gives one per unit open-circuit voltage on the air gap line, and one per unit torque is based on rated power at synchronous speed.

The constant K' in (27) is not related to the constant K in (16). K' is called the "system damping factor" and is inserted for completeness. No good way is known to us for the calculation of this factor and it is usually chosen from a knowledge of system behavior. The most important part of K' relates the amount of load picked up by this machine when the frequency changes and so it would appear that K' should be calculated continuously throughout any study. It might be approximated on the basis of an assumed torque angle diagram for this machine represented as a machine–infinite bus system.

Either of the two sets of equations may be used in the solution of the problem. Since they are basically the same, there should be no discernable difference in the answers.

External System

At this point it is necessary to postulate the representation of the system external to the synchronous machines. Theoretically, a set of differential equations (a very large set) could be written, many of which would be nonlinear. The equations would give the behavior of all loads, lines, and transformers to any and all changes in voltage or frequency. While this is possible it is clearly impractical, and for purposes of this discussion the external system will be assumed to be in the form generally associated with transient stability studies.

It will be assumed that the system is a collection of constant impedances and possibly a collection of sources which represent loads that do not fit the fixed-impedance requirement. This implies that the system is a steady-state system and has been evaluated at a constant frequency and a fixed voltage level. If it is to be of use, however, the assumption must be made that, when frequency and voltage variations from the base are small, this system also represents transient behavior accurately. The attempts to improve the transient representation have centered primarily on the loads themselves. At the present time sources simulating loads have usually been voltage controlled for, although frequency undoubtedly plays an important part, this variation is less well known. Needless to say, frequencies appreciably different from the base could give ridiculous results.

A digital solution requires a certain method of calculation. At the present time one of two general types is used: either a matrix method or the nodal–iterative method. The comments made here can be extended to either, although the extension to nodal–iterative techniques is much more difficult. It will be assumed here that a matrix formulation (either Y or Z) is used. The reason for this choice is very important, as will be seen. It is very simple to convert the entire system with the exception of the machine in question into a Thevenin equivalent system. It then becomes possible to investigate a succession of one machine–infinite bus systems. This facilitates convergence when iteration is required and approximations when these must be made.

This external system representation is then designed to approximate the fundamental frequency behavior of the system with a minimum of complexity in data requirements and to present it in a form suitable for maximum usefulness in computations.

Total System

A procedure must be available to utilize the system data given as real–imaginary matrix quantities in conjunction with the machine data given in terms of d–q quantities. It may be shown easily that, in general, the machine cannot be represented as a voltage behind a fixed impedance. Even in certain special cases (e.g., when $X_d'' = X_q''$) when it appears that it might be possible, the representation would be correct only for steady-state operation and might yield incorrect behavior during transient conditions.

Therefore, machine and system data must be kept separate and voltages (or currents) simulating the machine in the system must be calculated on the basis of the predicted flux linkage voltage behavior of the machine. This discussion will then distinguish between the flux linkage voltages of the machine and the matrix voltages used in the determination of Thevenin's equivalent systems or in the calculation of the behavior of the external system.

These comments do not apply to machines represented as a constant voltage behind transient direct axis impedance. The performance of these machines has already been limited by the assumption that the machine does in fact behave as a voltage behind a fixed impedance.

This section will consider the behavior of a synchronous generator-impedance-equivalent generator system subject to a disturbance. The problem resolves into two facets:

1) the calculation of the correct accelerating torque or power and from this the determination of the new pole angle at a later time

2) the calculation of the true changes in the flux linkage voltages over this time interval.

The first facet has been cussed and discussed so extensively that at least two philosophies have developed.

1) The positive-sequence output to the system is calculated from the positive-sequence terminal voltage and current (or matrix relations). This is then the fundamental torque produced by the machine. To this is added approximations for the many extraneous torques obtained in transient operation. The result is subtracted from input torque to obtain accelerating torque. "Approximations" is a key word. Torques due to induction motor/generator action between the damper windings and the positive-sequence field, or braking due to negative-sequence fields or direct currents in the armature, or torques due to nonfundamental frequency currents in the machine cannot be calculated exactly because of a lack of information. They may be approximated on the basis of an induction motor equivalent circuit and the use of a one machine–infinite bus system equivalent. This procedure has been related in the literature.

2) Since λ_d and λ_q are the true flux linkages in the machine and i_d and i_q are the exact currents, the equation for the developed electromagnetic torque, (26), must be the total torque. Some of this will appear as the output to the system but some will be converted into the extraneous torques and will be lost in nonfundamental frequency behavior. Thus the one calculation gives the desired accelerating torque.

Lest one be led to believe that there should be no real choice in the philosophies, the second facet must be examined—that is, how best to calculate the changes in linkage voltages and in the currents.

The inverse of the d–q transform states

$$i_a = i_d \cos \omega_r t - i_q \sin \omega_r t$$

$$i_b = i_d \cos \omega_r t - 120° - i_q \sin \omega_r t - 120°$$

$$i_c = i_d \cos \omega_r t + 120° - i_q \sin \omega_r t + 120°.$$

If i_d and i_q are direct currents then the conversion into phasor notation states that

$$I_a = I_d + jI_q.$$

Therefore, i_d and i_q play the role of components of the armature current and the input into the fundamental frequency system is then I_a, as just given. However, if i_d and i_q are not pure direct currents then the fundamental frequency input to the system cannot be considered $i_d + ji_q$ but will be, in general, some smaller number. To illustrate this, consider the following simple example.

A machine is considered to be connected through an inductance L_a to an infinite bus. All resistances will be neglected. The flux linkages at time 0^- are then

$$\lambda_d = L_{afm} i_f + L_{axm} i_x - (L_d + L_a) i_d$$

$$\lambda_q = L_{agm} i_g + (L_q + L_a) i_q$$

$$\lambda_f = -3L_{afm} i_d/2 + L_{ff} i_f + L_{fx} i_x$$

$$\lambda_x = -3L_{axm} i_d/2 + L_{fx} i_f + L_{xx} i_x$$

$$\lambda_g = -3L_{agm} i_q/2 + L_{gg} i_g.$$

It will be assumed that some disturbance occurs which changes either the infinite bus voltage or the inductance between machines or both. Also the angular velocity of the three-phase bus voltages corresponds to the velocity of the machine and v_d and v_q are the projections of the bus voltage on the machine axes at and after time zero. Then

$$v_d = \frac{d\lambda_d}{dt} - \lambda_q \omega_r, \quad v_q = \frac{d\lambda_q}{dt} + \lambda_d \omega_r$$
$$v_f = \frac{d\lambda_f}{dt}, \quad \frac{d\lambda_x}{dt} = \frac{d\lambda_g}{dt} = 0 \tag{28}$$

or λ_f, λ_g, λ_g are constant and λ_d and λ_q must be

$$\lambda_d = v_q/\omega_r + (\lambda_{do} - v_q/\omega_r) \cos \omega_r t + (\lambda_{qo} + v_d/\omega_r) \sin \omega_r t$$

$$\lambda_q = -v_d/\omega_r + (\lambda_{qo} + v_d/\omega_r) \cos \omega_r t - (\lambda_{do} - v_q/\omega_r) \sin \omega_r t$$

where λ_{do} and λ_{qo} are the 0^- values of the direct and quadrature axis stator flux linkages. The currents may then be obtained:

$$i_d = C_1[e_q'' - v_q - (\omega_r \lambda_{do} - v_q) \cos \omega_r t - (\omega_r \lambda_{qo} + v_d) \sin \omega_r t]$$

$$i_q = C_2[v_d - e_d'' - (\omega_r \lambda_{qo} + v_d) \cos \omega_r t + (\omega_r \lambda_{do} - v_q) \sin \omega_r t].$$

The armature current will then be the well-known form consisting of dc, fundamental, and second harmonics.

$$i_a = K_1 + K_2 \sin \omega_r t + \beta + K_3 \sin 2\omega_r t + \gamma.$$

Note that the fundamental frequency current into the system is not given by $i_d + ji_q$ but by the dc part of that expression. The developed torque will be of the form

$$T = C_3 + C_4 + C_5 \sin 2\omega_r t + \zeta.$$

C_3 is due to the constant components of λ and i and is the fundamental torque, C_4 is due to the variable components of λ and i and is so assigned to dc braking and to the second harmonics in the armature current. The other term is a relatively high-frequency variable torque.

Now assume that the rate of change of stator flux linkages is neglected. Then

$$\lambda_d = v_q/\omega_r, \qquad \lambda_q = -v_d/\omega_r,$$

$$i_d = C_1(e_q'' - v_q), \quad i_q = C_2(v_d - e_d'')$$

and T equals C_3.

Therefore, neglecting stator flux linkage changes causes the currents to remain direct and not only is $i_a = i_d + ji_q$ but the value obtained is the correct fundamental part of the true current obtained by a more rigorous analysis. However, i_d and i_q calculated here are not the true machine currents, since the variable component is missing. Also the fundamental torque calculation is the correct magnitude even though information on the extraneous torques is lost.

Needless to say, all sorts of time constants must be inserted if the true behavior is to be obtained. These equations are intended to furnish insight into the problem with a minimum of the complexity which would mask the results in a more precise calculation.

This is a fairly restricted model since it has been assumed that the bus and machine are in synchronism. Assume now that the bus voltages are at a frequency corresponding to ω_o rad/s so that the v_d and v_q used in (28) become

$$v_d \cos (\omega_o - \omega_r)t - v_q \sin (\omega_o - \omega_r)t \quad \text{for } v_d$$

$$v_d \sin (\omega_o - \omega_r)t + v_q \cos (\omega_o - \omega_r)t \quad \text{for } v_q.$$

Now, v_d and v_o are the instantaneous projections of the 0^+ voltage v on the machine d, q axes. In the same fashion

$$\lambda_d = \lambda_{do} \cos \omega_r t + \lambda_{qo} \sin \omega_r t + v(\sin \omega_r t - \alpha + \sin (\omega_o - \omega_r)t + \alpha)/\omega_o$$

$$\lambda_q = -\lambda_{do} \sin \omega_r t + \lambda_{qo} \cos \omega_r t + v(\cos \omega_r t + \alpha - \cos (\omega_o - \omega_r)t + \alpha)/\omega_o$$

$$i_d = C_1 [e_q'' - \omega_o\lambda_{do} \cos \omega_r t - \omega_o\lambda_{qo} \sin \omega_r t - v(\sin \omega_r t - \alpha + \sin (\omega_o - \omega_r)t + \alpha)]$$

$$i_q = C_2[-e_d'' + \omega_o\lambda_{do} \sin \omega_r t - \omega_o\lambda_{qo} \cos \omega_r t - v(\cos \omega_r t + \alpha - \cos (\omega_o - \omega_r)t + \alpha)].$$

The armature current is of the form

$$i_a = K_4 + K_5 \sin \omega_r t + \eta + K_6 \sin \omega_o t + \sigma + K_7 \sin (2\omega_r - \omega_o)t + \rho + K_8 \sin 2\omega_r t + \nu.$$

The choice of what part of $i_d + ji_q$ should be used is somewhat open to question due to the presence of three sinusoids of almost the same frequency.

The developed torque becomes

$$T = C_6 + C_7 \sin (\omega_o - \omega_r)t + \gamma + C_8 \sin \omega_r t + \zeta + C_9 \sin \omega_o t + \beta + C_{10} \sin 2\omega_r t + \varphi.$$

C_7 contains the primary component of torque but it is now slowly time-dependent as would be expected of the output to the system. C_6 contains torque components attributable to direct currents, second harmonics, and induction motor/generator action in the system.

If the stator flux linkage changes are neglected,

$$\lambda_d = (v/\omega_r) \sin (\omega_o - \omega_r)t + \alpha$$

$$\lambda_q = (-v/\omega_r) \cos (\omega_o - \omega_r)t + \alpha$$

$$i_d = C_1[e_q'' - v \sin (\omega_o - \omega_r)t + \alpha]$$

$$i_q = C_2[-e_d'' + v \cos (\omega_o - \omega_r)t + \alpha].$$

Inspection of the two sets of equations will show that the same type of results are obtained here as in the previous example. The almost synchronous current behavior is the same whether rate of change of stator flux is neglected or not and the fundamental torque gives the same result. Again, extraneous torques will not appear and the total current is not correct.

Up to now balanced system operation has been assumed and despite this the assumptions and their complexity are becoming burdensome. The mind boggles at what would happen with unbalanced conditions. The state of the art is such that the behavior of the machine could be found for a fully described external system. Relating this behavior to a system representation of the type assumed here appears staggering. The assumptions necessary would multiply like rabbits.

Summary and Conclusions

It is perhaps time to stop and dwell upon the implications. This paper has attempted to take a closer look at the assumptions that go into a transient stability study of a large system. While it is theoretical in nature it does point out some pitfalls and possible improvements. However, and this is of supreme importance, when a large number of assumptions are made it does not follow that the errors attributed to each assumption must be cumulative. The final test must always be a comparison to a physical system. The extensive stability programs available have demonstrated that the methods used do yield satisfactory results for a wide variety of studies. The conclusions must then temper the theory with practicality.

1) It may be stated flatly that synchronous machine transient behavior may be calculated as precisely as desired *provided* the system external to the machine is adequately described. This will require much more than steady-state data and easily may require more information than is necessary for the machines. The present procedures for stability calculation are somewhat analogous to hooking a racehorse representation of the machine to a plow representing the system. The capabilities and data available for machine calculation are not being fully utilized.

2) The treatment of saturation has been neglected. This factor is known to be very important in machine operation and must be included if precise calculations are desired. Various ways have been advanced for doing this and while no serious comparison of the methods has been attempted they all yield reasonable results.

The presence of voltage regulating or speed regulating equipment will modify the equations by changing the field voltage and/or input prime mover torque, but their presence or absence is not pertinent to this discussion.

3) The two philosophies on torque calculation may now be examined more completely.

If the terminal output plus miscellaneous torque procedure is used, then rate of change of stator flux linkages *must* be neglected. The resulting calculations for voltages and currents ignore any behavior except the fundamental or near fundamental and if all torques are approximated with reasonable care the accelerating torque is correct. Since the machine behavior equations are derived on the correct basis we are forced to predict the time behavior of this fundamental on the basis of changes in only the fundamental component or to find the correct values of i_d and i_q for use in the equations. The former is inaccurate and may not agree with a more rigorous calculation while the latter is approximate and difficult. Since only positive-sequence behavior is of interest, the use of $i_a = i_d + ji_q$ is justified.

If internal torque linkages are used and rate of change of stator flux linkages are not neglected then the single torque calculation contains all machine torques. Now problems exist in the calculations for voltages, currents, and the way in which the voltages change. In addition $i_a \neq i_d + ji_q$ and difficulties are encountered in relating the machine to the system.

There will be virtually no difference between the two methods except during severe transients. During the transient period neither one appears particularly attractive and it must be a personal choice. Probably no real calculation advantage would exist for either method.

The first method does appear to have more freedom, however. Since all miscellaneous torques are calculated, they can be excluded if so desired, or others may be included. The torque which might be excluded is that caused by dc damping. If the fault that created the disturbance were a progressive one which became a three-phase fault in a very short time, the difference in positive-sequence behavior in comparison to a simultaneous three-phase fault would be difficult to detect. However, offset currents may not exist in the progressive case. Likewise, offset currents may fail to exist in unbalanced faults and yet negative sequence effects may be desired.

The equations derived in the examples can form the basis for any necessary corrections in the actual calculations. It seems reasonable to assume that data from more precise calculations will divide in the same way as they do in these simple examples.

This assumption makes it possible to calculate the miscellaneous torques or corrections for use in the basic machine equation.

When the system is completely represented such that all behavior to any frequency is taken into account then the second method given will be far superior to the first.

4) This analysis points out a possible answer to the question of which machines should be represented in detail for a study. It can be seen that when i_d and i_q are direct currents then the output to the system is always correctly given as $i_d + ji_q$ and that no torques other than the fundamental component exist. Therefore, if any switching or faults create little or no dc offset in the generator output currents that generator will not need subtransient circuit representation. If a correct system damping factor is used it might even be represented as constant voltage behind transient direct axis reactance. (Positive-sequence damping action should be simulated in some way.)

5) In the dynamic period, an analysis of the positive-sequence behavior is adequate. The subtransient circuits will not be of extreme importance except as they influence the induction motor/generator damping (positive-sequence damping).

6) If subtransient effects are to be calculated as discussed here, study cost must increase. Data requirements are larger, study preparation is more difficult, smaller calculation intervals are required, so running time goes up. The added cost must be charged primarily to this early period since this degree of refinement is not required in the dynamic region. It is perhaps time to reassess the role of input data accuracy upon stability studies.

7) We would vigorously defend the ability of a digital program to accurately compare two alternate proposals such as two different generators at a point in the system. This involves calculations using what might be called an inadequate representation but it is consistently inadequate. More soul searching should be done before guaranteeing that the answer obtained by the study would be reproducable by test. The assumptions necessary preclude this degree of certainty.

8) It may be pointed out, with justification, that much of this discussion has been picayune. Reasonable studies involve small deviations from synchronism and no good reason has been given to seriously question the answers that have been obtained in thousands of studies. However, reliable system operation has recently received great impetus and as control devices get better it is entirely within the realm of possibility that rotor control may allow the pulling into step of a generator running out of synchronism, or that subsynchronous operation of portions of a system might be allowed. If this situation does develop and extensive systems are involved, the points raised here will be important and their resolution will be crucial to the accurate calculation of system behavior.

Nomenclature

The symbols used correspond insofar as possible to standard definitions. Other symbols used are

- v_j applied voltage to circuit j
- λ_j flux linkages of circuit j
- R_j resistance of circuit j
- T_i prime mover input torque
- T_e machine electromagnetic torque
- T_a accelerating torque
- ω_o base angular velocity
- ω_r angular velocity of rotor, $\omega_o t = \omega_r t + \delta$
- δ angle of machine direct axis
- K' system damping coefficient
- L_{ij} inductance between rotor circuits i and j
- L_{ajm} inductance between stator circuit a and rotor circuit j (maximum value)
- C_j, K_j arbitrary constants, used for illustrative purposes, that may be calculated from data given. (All Greek letters used, with the exception of δ, fall in this same category.)

References

[1] R. H. Park, "Two-reaction theory of synchronous machines, generalized method of analysis, pt. I," *AIEE Trans.*, vol. 48, pp. 716–730, July 1929; pt. II, *AIEE Trans.*, vol. 52, pp. 352–355, June 1933.

[2] E. W. Kimbark, *Power System Stability*, vol. 3. New York: Wiley, 1956.

[3] W. A. Lewis, *The Principles of Synchronous Machines*. Chicago, Ill.: Illinois Institute of Technology Press, 1949.

[4] C. Concordia, *Synchronous Machines*. New York: Wiley, 1951.

[5] H. E. Lokay and R. L. Bolger, "Effect of turbine-generator representation in system stability studies," *IEEE Trans. Power Apparatus and Systems*, vol. PAS-84, pp. 933–942, October 1965.

Discussion

W. Janischewskyj and **K. Prabhashankar** (University of Toronto, Toronto, Ont., Canada): The discussers would like to complement the author on his lucid inspection of the complex topic of desirable models in the study of power system stability with the aid of digital computers. In this connection we would like to comment on the author's observation on the improvement of computational speed when changes of flux linkages in Park's model of the synchronous machine are neglected. In our studies of this particular point [6], a 50-fold increase in the step size (from 0.001 to 0.05 second) was found possible as a result of elimination of $p\psi_d$ and $p\psi_q$ terms. There exists an undesirable aspect of this elimination, and this fact is stressed by Dr. Olive as well as by other authors [7]. By this procedure a portion of transient process in the machine is omitted, and the associated information is lost. Fortunately, when mechanical transients of power systems are studied, the procedure may be made to lead to some advantages.

First, the resultant computed torque lacks a damping component, which, under certain conditions, may be significant [8]. Important effects of the omitted damping torque, however, may be accounted for separately by computation of the copper loss in rotor circuits of the machine.

Second, transients in computed stator currents are absent. While such an omission would be unacceptable when determining the transient behavior of electrical variables in a power system, there exists no such restriction when mechanical oscillations of machines are of interest. It is customary in such studies to represent the network in its steady state, the brevity of electrical transients in the network being used as the argument for this procedure. Such treatment requires existence of purely sinusoidal voltages and currents at all nodes of the network including those at which synchronous machines are connected. In this respect, the omission of $p\psi_d$ and $p\psi_q$ actually helps to achieve the desired end.

References

[6] K. Prabhashankar and W. Janischewskyj, "Digital simulation of multimachine power systems for stability studies," *IEEE Trans. Power Apparatus and Systems*, vol. PAS-87, pp. 73–80, January 1968.

[7] C. C. Young and R. M. Webler, "A new digital computer program for predicting dynamic performance of electric power systems," presented at the 5th PICA Conf., Pittsburgh, Pa., 1967.

[8] K. Prabhashankar and W. Janischewskyj, "Digital simulation and analysis of power system dynamic performance: I—Effects of synchronous machine amortisseur circuits and control mechanisms," presented at the IEEE Winter Power Meeting, New York, N. Y., January 28–February 2, 1968.

Manuscript received February 25, 1968.

David W. Olive: The calculation of machine transients has two facets. First, the construction of an accurate model; second, the precise calculation of the behavior of the model. The model selected determines programming complexity, data collection problems, the accuracy possible and, to some extent, the calculation interval size. The method of calculation is crucial in the determinations of interval size and precision. This paper has been primarily concerned with the second facet.

Manuscript received May 13, 1968.

I appreciate the comments of Prof. Janischewskyj and Dr. Prabhashankar and I would compliment them for the work they have done which so aptly illustrates comments in this paper.

The exclusion of $p\psi_d$ and $p\psi_q$ terms in the equations does have a dramatic effect on permissible calculation interval since the associated high-frequency and dc behavior are thus ignored. It has an adverse effect in that accurate reinsertion of torques may be necessary. We have the problem of models within models.

I must concur, however, that at this stage in stability analysis the procedure of neglecting $p\psi_d$ and $p\psi_q$ and correcting for torques appears to yield sufficiently accurate results and is very effective in the reduction of study costs.

DYNAMIC MODELS OF TURBINE GENERATORS DERIVED FROM SOLID ROTOR EQUIVALENT CIRCUITS

Richard P. Schulz William D. Jones Donald N. Ewart
General Electric Co., Schenectady, New York

Abstract—A method is described for determining the structure and coefficients of dynamic models of turbogenerators from detailed equivalent circuits of the d- and q-axes of a solid rotor generator. These models are intended for analyses of power system dynamics in cases where it is desirable to have a very accurate representation of the synchronous machine. The models have been used to study power system hunting or dynamic stability, first swing or transient stability, load rejection overvoltages, resynchronization of cross-compound turbine-generator sets, field current and voltage during out-of-step operation, voltage dip and other phenomena associated with machine operation on power systems. The paper reviews some of the experience obtained in system studies and in comparisons between test results and computer simulations.

INTRODUCTION

The equations and computer representations normally used to study synchronous machine behavior are based on a classical machine consisting of discrete physical windings on the stator and rotor.(1) The direct and quadrature-axis equivalent circuits which represent the classical machine are shown on Figure 1. The elements represent armature leakage inductance, L_ℓ, stator-rotor mutual inductances, L_{ad} and L_{aq}, field resistance and leakage inductance, R_{fd} and L_{fd} and amortisseur resistance and leakage inductance, R_{1d}, L_{1d}, R_{1q} and L_{1q}.

It has been recognized that the solid iron rotor used in modern steam-turbine generators provides multiple paths for circulating eddy currents which act as distributed amortisseurs under dynamic conditions. Further, these equivalent amortisseurs have mutual coupling with the field winding in the direct axis which is not common with the armature. Studies[2] have shown that the classical model does not provide enough flexibility for use in system studies in those special cases when the behavior of the synchronous machine and its controls must be represented very accurately. Thus, a new model is needed for accurate simulation of solid rotor turbine-generators during transients introduced by the power system and abnormal operation.

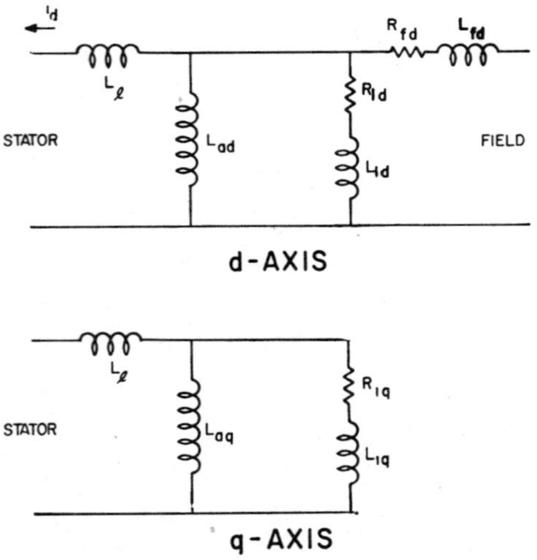

Fig. 1. Equivalent Circuits of Classical Machine Model

Paper T 72 515-5, recommended and approved by the Rotating Machinery Committee of the IEEE Power Engineering Society for presentation at the IEEE PES Summer Meeting, San Francisco, Calif., July 9-14, 1972. Manuscript submitted February 15, 1972; made available for printing May 4, 1972.

A recent paper(3) described a set of direct and quadrature-axis equivalent circuits for turbine-generators with solid iron rotors. This will be referred to as the Jackson-Winchester model. This set of equivalent circuits is very complex, providing a capability to make calculations including the effects of the important current carrying paths in the rotor. The complexity is apparent in Figure 2 (Figure 18 of Ref. 3) which illustrates the direct-axis only. A similar circuit was presented for the quadrature axis.

The Jackson-Winchester model was not intended for use in time-domain simulations, Indeed, assuming a given saturation level, some of the inductances are functions of the square-root of slip, and would be very difficult to represent using conventional integration algorithms. However, it is possible to determine the magnitude and phase (or real and imaginary components) of the impedances of the elements of an equivalent circuit of the form shown on Figure 3. Here sL_ℓ represents the impedance of the armature leakage inductance as a function of slip or the Laplace operator, s. Likewise sL_{ad} and sL_{aq} represent the impedance of the armature-rotor mutual inductance; $Z_{fkd}(s)$ represents a mutual impedance between the field and rotor body; $Z_{fd}(s)$ represents the field impedance and $Z_{kd}(s)$ and $Z_{kq}(s)$ represent the impedances of the iron and any non-excited windings on the rotor.

The remainder of this paper describes the new equivalent circuit based on these complex impedances, how the coefficients are determined, how the circuit is transformed into a model suitable for use in computer simulations and concludes with examples of how the simulation model has been compared with tests.

EQUIVALENT CIRCUIT

To produce a time domain simulation model, it is convenient to start from a physical model containing discrete elements. The development of such a model in this case began from an analysis for many machines of the frequency response of the impedance elements shown on Figure 3, in particular, those which are indicated as being functions of s. Figure 4 illustrates how the magnitude of these elements varies with slip frequency for a given generator. Note that the data from which these curves were produced came from the Jackson-Winchester model.

Since the objective is to develop a discrete element equivalent circuit it is helpful to review briefly how discrete-element branches would behave on plots such as Figure 4. A branch consisting of simply a resistor would have an impedance whose magnitude is independent of slip and thus would plot as a straight horizontal line on Figure 4. Likewise, a simple inductor would plot as a straight line with positive slope of 45° on Figure 4. A series resistance-inductance would have the characteristics of the resistor at low frequencies and the inductor at high frequencies; thus it would plot like the curve labeled $Z_{fd}(s)$. Parallel combinations of resistance and inductance have similarly identifiable characteristic shapes when viewed on the axes of Figure 4.

Based on observations of plots $Z_{fkd}(s)$, $Z_{fd}(s)$, $Z_{kd}(s)$ and $Z_{kq}(s)$ for dozens of machines, it became apparent that a discrete-element model for the solid-iron-rotor generator as shown on Figure 5 would be adequate for most studies which require an accurate model of the machine. The model shown on Figure 5 will be referred to as the Solid Iron Rotor Model (SIRM).

The selection of numerical values for the discrete elements of SIRM aside from L_ℓ, L_{ad}, and L_{aq} is done by a curve-fitting procedure.

Reprinted from *IEEE Trans. Power App. Syst.*, May/June 1973.

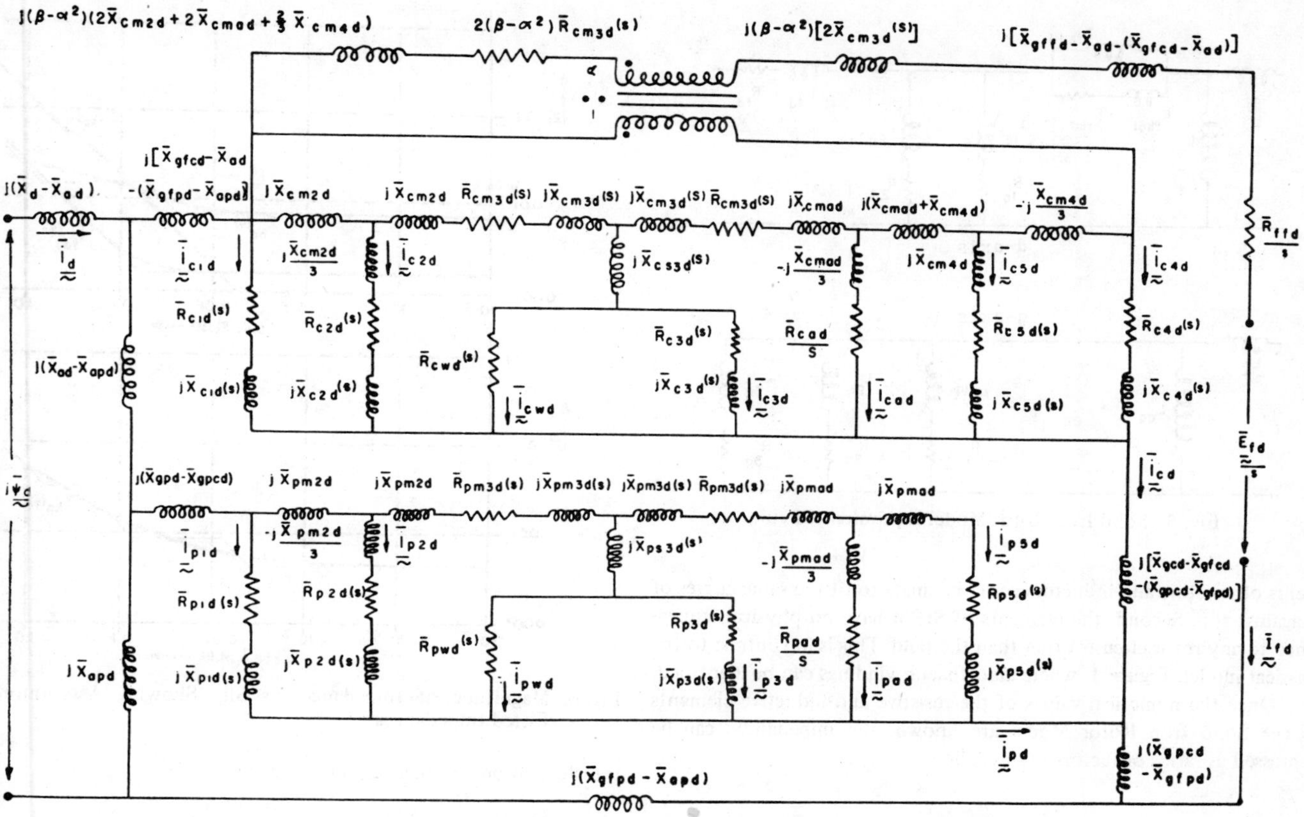

Fig. 2. Direct Axis Equivalent Circuit (Jackson-Winchester Model)

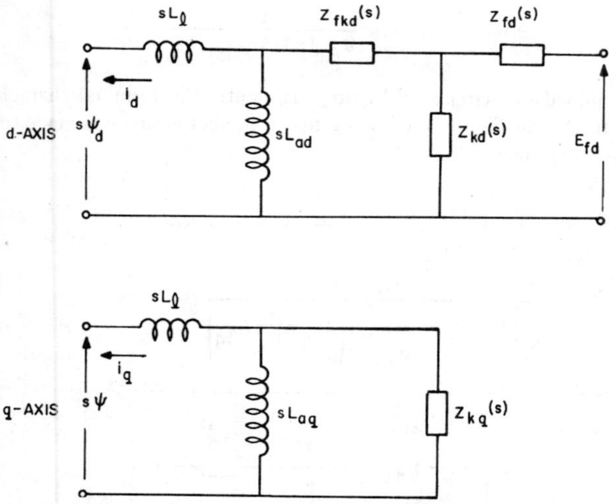

Fig. 3. Structure of Equivalent Circuits of Solid Iron Rotor Model

The elements L_{fkdo}, L_{fkdl} and R_{fkd} are obtained from $Z_{fkd}(s)$. The elements R_{fd} and L_{fd} are obtained from $Z_{fd}(s)$. The elements L_{1d}, R_{1d}, L_{2d} and R_{2d} are obtained from $Z_{kd}(s)$ and the elements L_{1q}, R_{1q}, L_{2q}, R_{2q}, L_{3q} and R_{3q} are obtained from $Z_{kq}(s)$. As an example of this identification process, consider the shape of the impedance $Z_{fd}(s)$ in Figure 4. At low ranges of slip frequency it is horizontal. At the left edge of the figure, where slip is 0.00001 pu the phase angle of $Z_{fd}(s)$ is extremely small and lagging, of the order of $-0.05°$. This region is identified as one where resistance effects dominate; the resistance has a value equal to $|Z_{fd}|$. At higher values of slip, i.e. for slip greater than about 0.2 p.u. (12 Hz), the magnitude of Z_{fd} is increasing proportionally to slip, and its phase is approaching $-90°$. This region is identified as being dominated by an inductance whose value is equal to the intercept of the 1 pu slip axis and a straight line fitted to Z_{fd} with 45° slope in this region of slip. The fitting straight lines are indicated in Figure 6 for $Z_{fd}(s)$ and the other impedances.

Similarly, very close approximations of the other elements can be obtained by such curve-fitting procedures.

Two points are of interest. One, the form of SIRM shown on Figure 5 is only one of several which might be chosen. Other arrange-

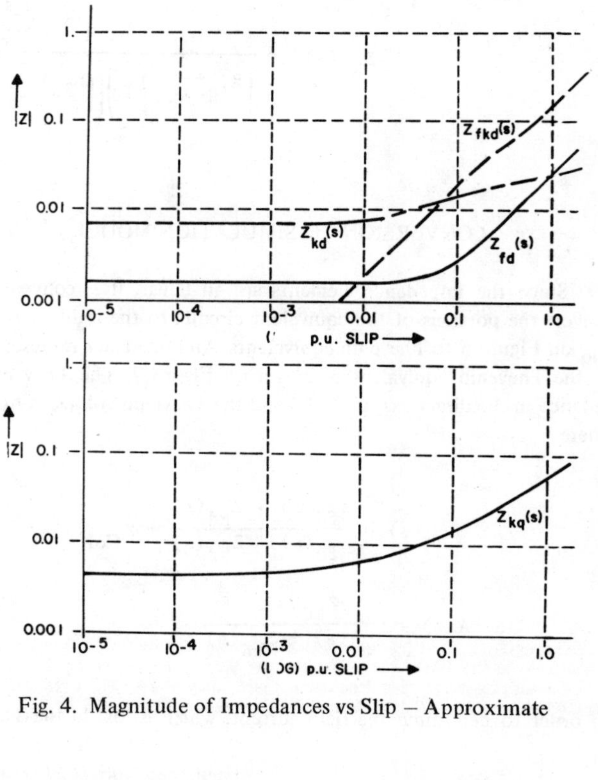

Fig. 4. Magnitude of Impedances vs Slip – Approximate

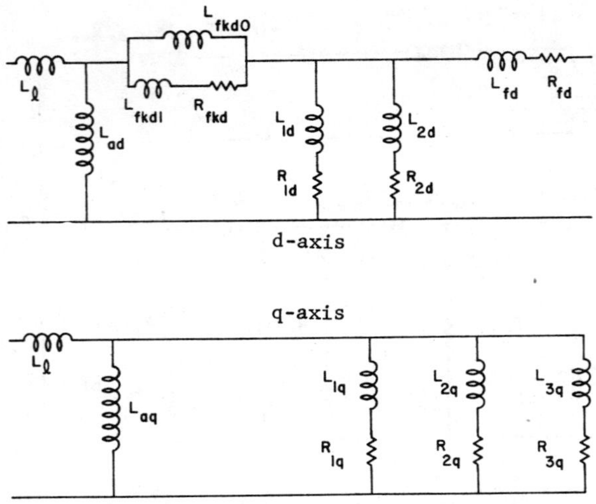

Fig. 5. Solid Iron Rotor Model Equivalent Circuit

ments of resistors and inductors could be made to fit the same curves of Z against slip. Second, the elements of SIRM have no physical equivalence to any rotor circuits other than the field. This is in contrast to the classical model, Figure 1, where amortisseur windings can be identified.

Once the numerical values of the resistive and inductive elements of the Solid Iron Rotor Model are known, the impedances can be expressed as ratios of factors in s as follows:

$$Z_{fd}(s) = R_{fd} + \frac{s}{\omega_o} L_{fd}$$

$$Z_{fkd}(s) = \frac{\left(\frac{s}{\omega_o} L_{fkdo}\right)\left(R_{fkd} + \frac{s}{\omega_o} L_{fkd1}\right)}{R_{fkd} + \frac{s}{\omega_o}\left(L_{fkdo} + L_{fkd1}\right)}$$

$$Z_{kd}(s) = \frac{\left(R_{1d} + \frac{s}{\omega_o} L_{1d}\right)\left(R_{2d} + \frac{s}{\omega_o} L_{2d}\right)}{R_{1d} + R_{2d} + \frac{s}{\omega_o}\left(L_{1d} + L_{2d}\right)}$$

$$Z_{kq}(s) = \frac{\left(R_{1q}+\frac{s}{\omega_o} L_{1q}\right)\left(R_{2q}+\frac{s}{\omega_o} L_{2q}\right)\left(R_{3q}+\frac{s}{\omega_o} L_{3q}\right)}{\left(R_{1q}+\frac{s}{\omega_o} L_{1q}\right)\left[\left(R_{2q}+\frac{s}{\omega_o} L_{2q}\right)+\left(R_{3q}+\frac{s}{\omega_o} L_{3q}\right)\right]+\left(R_{2q}+\frac{s}{\omega_o} L_{2q}\right)\left(R_{3q}+\frac{s}{\omega_o} L_{3q}\right)}$$

CONVERSION TO SIMULATION MODEL

Since the impedance elements are all linear, it is convenient to convert the portions of the equivalent circuits to the right of L_{ad} and L_{aq} on Figure 5 to Thevenin equivalents. An impedance representation of the Thevenin equivalent is shown on Figure 7. The Thevenin impedance in the direct axis is $Z_{eq}(s)$ and the Thevenin voltage is $A(s)E_{fd}$, where

$$Z_{eq}(s) = \frac{Z_{fd}(s) \cdot Z_{kd}(s)}{Z_{fd}(s) + Z_{kd}(s)} + Z_{fkd}(s)$$

$$A(s) = \frac{Z_{kd}(s)}{Z_{fd}(s) + Z_{kd}(s)}$$

In order to determine the field current, which is lost in the Thevenin

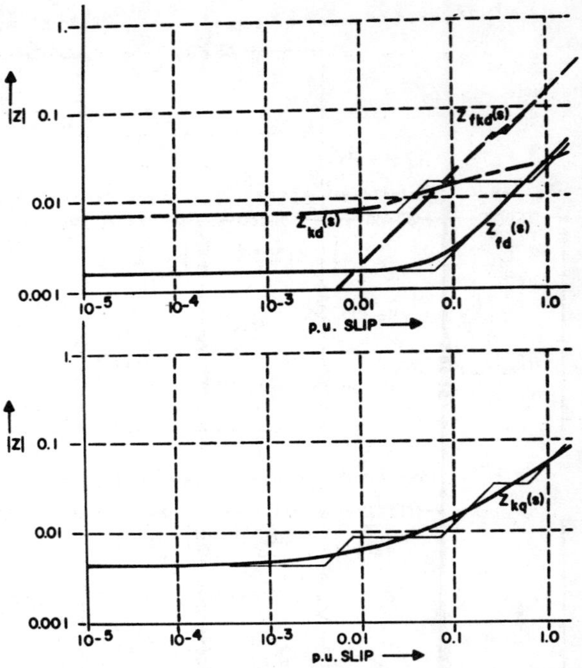

Fig. 6. Magnitude of Impedances vs Slip Showing Assymptotes of Fitted Impedances

transformation, another relation is needed:

$$I_{fd} = A(s) * i_{eq} + B(s) * E_{fd}$$

where

$$B(s) = \frac{1}{Z_{fd}(s) + Z_{kd}(s)}$$

The impedance circuit on Figure 7 is identical in terminal characteristics to that in Figure 3 assuming that a perfect match was made to the impedance elements.

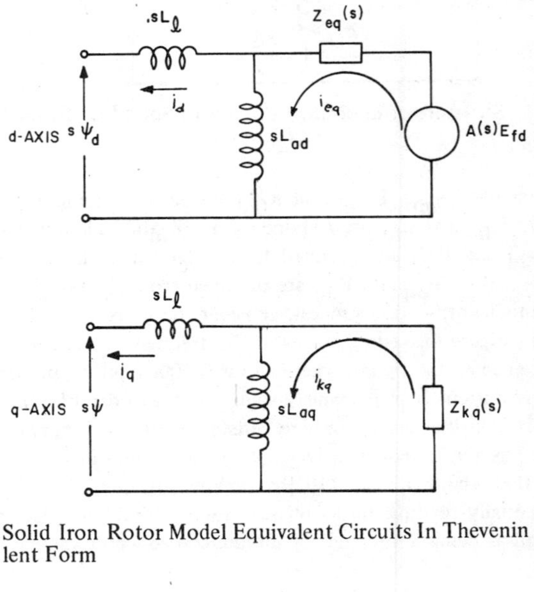

Fig. 7. Solid Iron Rotor Model Equivalent Circuits In Thevenin Equivalent Form

If $Z_{eq}(s)$ and $Z_{kq}(s)$ can be broken into the following form:

$$Z_{eq}(s) = R_{eqo} + \frac{s}{\omega_o} L_{eq}(s)$$

$$Z_{kq}(s) = R_{kqo} + \frac{s}{\omega_o} L_{kq}(s)$$

then the following flux linkage and voltage equations can be written which describe the machine:

d-axis

$$\psi_{eq} = \psi_{ad} + i_{eq} L_{eq}(s)$$

$$\psi_{ad} = (i_{eq} - i_d) L_{ad}$$

$$\psi_d = \psi_{ad} - i_d L_\ell$$

$$\frac{s}{\omega_o} \psi_{eq} = A(s)\, e_{fd} - i_{eq} R_{eqo}$$

$$e_{fd} = \frac{R_{fd}}{L_{ad}} E_{fd}$$

q-axis

$$\psi_{kq} = \psi_{aq} + i_{kq} L_{kq}(s)$$

$$\psi_{aq} = (i_{kq} - i_q) L_{aq}$$

$$\psi_q = \psi_{aq} - i_q L_\ell$$

$$\frac{s}{\omega_o} \psi_{kq} = -i_{kq} R_{kqo}$$

These equations can be described in block diagram form as shown[4] in Figure 8. Note that saturation has been accounted for by modifying the values of the mutual inductances L_{ad} and L_{aq} by a function of the net air gap flux linkages[5].

The stator voltage equations can be written in one of two forms as follows:

Exact:

$$e_d = \frac{s}{\omega_o} \psi_d - \frac{\omega}{\omega_o} \psi_q - r_a i_d$$

$$e_q = \frac{s}{\omega_o} \psi_q + \frac{\omega}{\omega_o} \psi_d - r_a i_q$$

or Approximate:

$$e_d \doteq -\psi_q - r_a i_d$$

$$e_q \doteq \psi_d - r_a i_q$$

It has been shown empirically(6) that the approximate form is adequate for studies of synchronous stability.

Returning to Figure 5 or Figure 7, it is possible to combine the impedance elements in the direct and quadrature axes respectively into expressions for $L_d(s)$ and $L_q(s)$, the direct and quadrature axis inductances as a function of slip. Similarly, G(s) can be defined so that

$$\psi_d = G(s)\, e_{fd} - L_d(s)\, i_d$$

$$\psi_q = -L_q(s)\, i_q$$

where

$$L_d(s) = L_\ell + \frac{\omega_o L_{ad} \cdot Z_{eq}(s)}{s L_{ad} + Z_{eq}(s)\omega_o}$$

$$L_q(s) = L_\ell + \frac{\omega_o L_{aq} \cdot Z_{kq}(s)}{s L_{aq} + Z_{kq}(s)\,\omega_o}$$

EXPERIENCE WITH THE SOLID IRON ROTOR MODEL IN STUDIES AND TESTS

The Solid Iron Rotor Model was first implemented in an analog computer simulation. The first published comparison between test and simulation results was reported in 1962(7). Those tests determined the decay of field current and stator flux and voltage following field breaker trip from rated terminal voltage open circuit operation of a solid iron rotor steam turbine generator. The computed values of stator voltage and field current were shown to closely match the test values.

On two occasions dynamic stability tests were made with computer stability analyses made before and after the tests.(8,9) The tests were conducted with the tested generating unit connected to its load through deliberately weakened transmission systems; the test disturbances were step changes in the automatic voltage regulator set point. In the first set of tests,(8) two test machines were involved in the form of a cross-compound unit; in the second set of tests(9) a single machine with its tandem-compound turbine was tested. In both studies a close agreement was obtained between test and simulation results using the solid iron rotor model.

It is of interest that following the second set of tests, simulations were made of the generator using the classical model shown on Figure 1. It was not possible to obtain as good an agreement for stator and field voltages as was obtained using the Solid Iron Rotor Model, although the rotor angle damping matched test results well. Thus, it is generally not necessary to use the more exact model when making system stability studies, but it is desirable when making detailed generator or excitation system design calculations.

A set of analytical studies and tests were made of the electrical resynchronizing of a large cross-compound turbine generator set;(10) the analytical portion of the study was carried out with the Solid Iron Rotor Model. The tests and studies were made of the procedure for returning a cross-compound set to operation following disconnection from the system and loss of excitation. As the two machines coast down their electrical speeds diverge. The procedure investigated the pulsating torques, currents and voltages following application of excitation to one of the units while their electrical speeds differed up to 0.4 per unit. It was concluded that the Solid Iron Rotor Model was capable of accurately representing the generators during the resynchronizing operation. Quantities compared included stator currents, induced field currents, torques and time to resynchronize.

More recently frequency response tests were made of a large turbine generator to determine generator models directly by test methods.(11) These tests were made by sinusoidally modulating the generator field voltage over a range from 0.005 Hz to 10 Hz. Measurements were made of the resulting magnitude and phase relations between the generator field voltage, field current, stator voltage and stator current and resolved into direct and quadrature axis components of the stator quantities. These tests were made at no load and near full load on the unit. The values of $L_d(s)$ and G(s) calculated from the test measurements were found to closely agree with values of the same quantities calculated by the Jackson-Winchester equivalent circuit.(12)

We are seeking additional opportunities for further verification. As an example, the effect of saturation of the iron under various loading conditions should be studied. In studying phenomena where the saturation effects are believed to be significant, a model might be required in which individual elements of the equivalent circuit of Figure 5 are retained and modified to reflect the changes in permeability. Also, experimental verification of the quadrature axis operational inductance would be of interest relative to present definitions and test methods for determining transient values of quadrature axis reactances and time constants.

CONCLUSIONS

A new direct-and quadrature axis model for simulating solid-iron rotor generators has been described. Techniques for obtaining the coef-

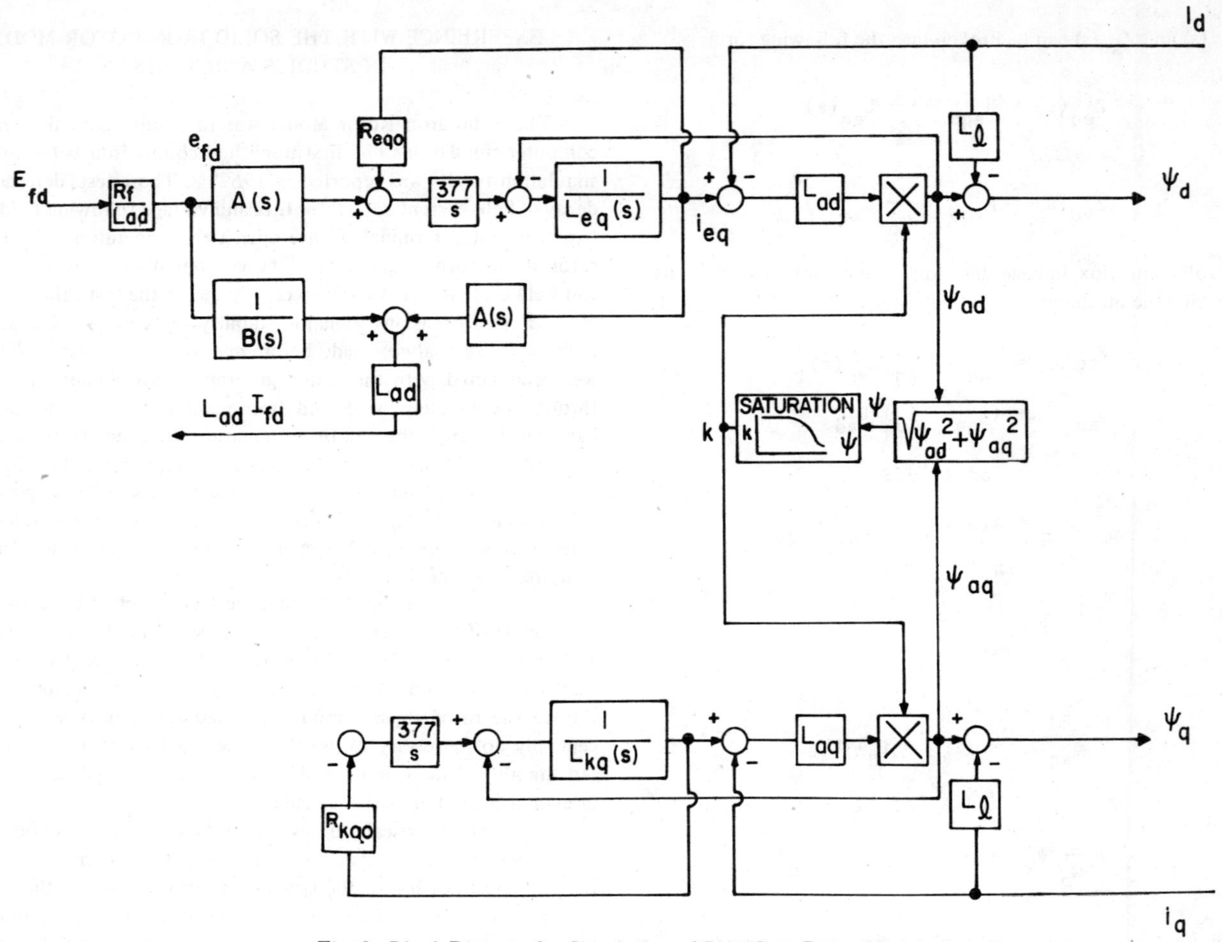

Fig. 8. Block Diagram for Simulation of Solid Iron Rotor Model

ficients of the model in terms of a very detailed model of the machine impedance is presented. Reference is made to previous studies which have used this model in which good comparison has been obtained between tests and simulations.

NOMENCLATURE

All quantities are in per unit unless noted.

ψ_d, ψ_q	armature flux linkages—direct and quadrature axis components
i_d, i_q	armature current, direct and quadrature axis components
e_d, e_q	armature voltage, direct and quadrature axis components
$L_d(s)$, $L_q(s)$	generator inductances in operational form
G(s)	operational transfer function between E_{fd} and ψ_d
r_a	armature resistance
s	Laplace operator (sec^{-1})
δ	angle between quadrature axis and infinite bus, radians
ω_o	base frequency, radians per second.
L_ℓ, L_{ad}, L_f, L_{kd}, L_{aq}, L_{kq}, r_f, r_{kq}, r_{kd}	elements of synchronous machine equivalent circuit
E_{fd}	generator field voltage on equal L_{ad} base
e_{fd}	generator field voltage on MVA base
$Z_{fd}(s)$, $Z_{fkd}(s)$, $Z_{kd}(s)$, $Z_{kq}(s)$	operational impedances of the solid rotor machine detailed equivalent circuit
A(s), B(s)	transfer functions in the time domain model
$Z_{eq}(s)$, $L_{eq}(s)$, $L_{kq}(s)$	impedances and inductances obtained by algebraic reduction of $Z_{fkd}(s)$, $Z_{kd}(s)$, $Z_{fd}(s)$ and $Z_{kq}(s)$
R_{eqo}, R_{kqo}	Resistive components of $Z_{eq}(s)$ and $Z_{kq}(s)$, respectively at extreme low slip

REFERENCES

[1] C. Concordia, "Synchronous Machines," Wiley, New York, 1951.

[2] P. L. Dandeno, R. L. Hauth, and R. P. Schulz, "Effects of Synchronous Machine Modeling in Large Scale System Studies", paper submitted for 1972 IEEE Summer Power Meeting.

[3] W. B. Jackson and R. L. Winchester, "Direct and Quadrature-Axis Equivalent Circuits For Solid-Rotor Turbine Generators," IEEE Transactions on Power Apparatus & System, vol. PAS-88, No. 7, July, 1969, pp. 1121–1136.

[4] D. N. Ewart, R. P. Schulz, "FACE Multi-Machine Power System Simulator Program; PICA Conference May 18–21, 1969. Conference Record No. 69 C1-PWR, pp. 133–153.

[5] F. P. deMello, G. W. Walsh, "Reclosing Transients In Induction Motors With Terminal Capacitors; AIEE Transactions, vol. 79, Pt. III, Feb. 1961, pp. 1206–1231.

[6] D. N. Ewart, F. P. deMello, "A Digital Computer Program For The Automatic Determination of Dynamic Stability Limits," IEEE Transactions on Power Apparatus Systems, vol. PAS-86, No. 7, pp. 867-875, July, 1967.

[7] J. B. McClure, M. Temoshok, "Performance of Large Steam Turbine-Generators On A Modern Power System" CIGRE Paper No. 108, 1962.

[8] D. N. Ewart, G. L. Landgren, M. Temoshok, and W. W. Walkley, "Stability Studies and Tests on 532 MW Cross-Compound Turbine-Generator Set," IEEE Transactions on Power Apparatus & Systems, vol. PAS-84, No. 4, April, 1965, pp. 338–543, disc. June 1965, pp. 526–8.

[9] R. P. Schulz, M. Temoshok, and R. G. Farmer, "Dynamic Sta-

bility Tests at Four Corners," Proceedings 1969 American Power Conference, vol. 31, pp. 872–882.

[10] R. K. Lay, E. H. Lenfest, M. Temoshok, and R. L. Winchester, "Electrical Resynchronizing of a Large-Cross-Compound Turbine-Generator Set: Part I – Analytical Study" IEEE Transactions of Power Apparatus & Systems, vol. PAS-88, No. 7, July 1969, pp. 1137–1146.

[11] G. Manchur, D. C. Lee, M. Coultes, J. D. Griffin, W. Watson, "Generator Models Established by Frequency Response Tests on a 555 MVA Machine," IEEE Paper T 72 136-5 presented at the 1972 Winter Power Meeting New York; to be published in the IEEE Transactions on Power Apparatus and Systems.

[12] R. P. Schulz, C. E. J. Bowler, W. D. Jones, discussion of ref. above; to be published in the IEEE Transactions on Power Apparatus and Systems.

Discussion

Swarn S. Kalsi, (General Electric Company, Philadelphia, Pa. 19124): The classical machine model of Fig. (1) and the new solid iron rotor model of Fig. (5) have the following differences:

i) The new solid iron rotor model has two more circuits in each axis than the classical machine model.

ii) The elements of the new solid iron rotor model are obtained from plots of secondary complex impedances shown in Fig. 3. Once these elements are evaluated, they are considered fixed and linear.

The coefficients of differential equations representing machine behavior during a transient, are dependent on the complex impedances $Z_{eq}(S)$ and $Z_{kq}(S)$. During the integration process by a step-by-step method, it is necessary to recalculate the coefficients at each step. This consumes a large part of computing time.

Since elements of the solid iron rotor model (Fig. 5) are fixed and linear, by making appropriate approximations, it may be possible to rearrange the operational impedance functions $L_d(S)$ and $L_q(S)$ as follows:

Direct Axis Operational Impedance:

$$\frac{1}{X_d(p)} = \frac{1}{X_d} + \left(\frac{1}{X'_d} - \frac{1}{X_d}\right)\frac{pT'_d}{1+pT'_d} + \left(\frac{1}{X''_d} - \frac{1}{X'_d}\right)\frac{pT''_d}{1+pT''_d} + \left(\frac{1}{X'''_d} - \frac{1}{X''_d}\right)\frac{pT'''_d}{1+pT'''_d}$$

Quadrature Axis Operational Impedance:

$$\frac{1}{X_q(p)} = \frac{1}{X_q} + \left(\frac{1}{X'_q} - \frac{1}{X_q}\right)\frac{pT'_q}{1+pT'_q} + \left(\frac{1}{X''_q} - \frac{1}{X'_q}\right)\frac{pT''_q}{1+pT''_q} + \left(\frac{1}{X'''_q} - \frac{1}{X''_q}\right)\frac{pT'''_q}{1+pT'''_q}$$

Variables on the right hand-side of above equations can be obtained in terms of elements of Fig. (5), or alternatively determined directly from the admittance plot by a curve fitting process similar to that of Fig. (6).

The above impedance functions $X_d(p)$ and $X_q(p)$ represent a synchronous machine rotor by three fictitious windings on each axis. Differential equations for presenting such a model, including the effect of $p\psi_d$ and $p\psi_q$ terms, may be written by extending the induction machine model described in Refs. (1) and (2). A possible arrangement of equations could be as follows:

$$\begin{bmatrix} pi_d \\ pi_q \\ p\psi_f \\ p\psi_{kd1} \\ p\psi_{kd2} \\ p\psi_{kq1} \\ p\psi_{kq2} \\ p\psi_{kq3} \end{bmatrix} = \begin{bmatrix} A11 & \omega A12 & A13 & A14 & \omega A15 & \omega A16 & \omega A17 & A18 & A19v_d & A110 \\ \omega A21 & A22 & \omega A23 & \omega A24 & \omega A25 & A26 & A27 & A28 & A29v_q & 0 \\ A31 & A32 & A33 & A34 & A35 & A36 & A37 & A38 & 0 & 1 \\ A41 & A42 & A43 & A44 & A45 & A46 & A47 & A48 & 0 & 0 \\ A51 & A52 & A53 & A54 & A55 & A56 & A57 & A58 & 0 & 0 \\ A61 & A62 & A63 & A64 & A65 & A66 & A67 & A68 & 0 & 0 \\ A71 & A72 & A73 & A74 & A75 & A76 & A77 & A78 & 0 & 0 \\ A81 & A82 & A83 & A84 & A85 & A86 & A87 & A88 & 0 & 0 \end{bmatrix} \begin{bmatrix} i_d \\ i_q \\ \psi_f \\ \psi_{kd1} \\ \psi_{kd2} \\ \psi_{kq1} \\ \psi_{kq2} \\ \psi_{kq3} \\ 1 \\ v_t \end{bmatrix}$$

The advantage of the above arrangement is that none of the coefficients A11, A12, etc. change unless system impedances are altered.

The merits of including or excluding the effect of $p\psi_d$ and $p\psi_q$ terms in the calculations depend upon the type of fault studied. In general, most of the disturbances initiated in the excitation system are slowed by the large reactance of the field winding, and therefore, $p\psi_d$ and $p\psi_q$ terms could be neglected. On the other hand, if the fault is on the primary side, like a 3-phase short circuit, it may not be possible to disregard the effect of these terms.

REFERENCES

[1] S. S. Kalsi, D. D. Stephen and B. Adkins, "Calculation of System Fault Currents Due to Induction Motors", Proc. IEE, Vol. 118, No. 1, January 1971, pp. 201–215.

[2] S. S. Kalsi and B. Adkins, "Transient Stability of Power Systems Containing Both Synchronous and Induction Machines", Proc. IEE, Vol. 118, No. 10, October 1971, pp. 1467-1474.

Manuscript received July 26, 1972.

P. L. Dandeno and **W. Watson** (Ontario Hydro, Toronto, Ont., Canada): It has been the author's objective to develop generator representations which are simple enough for use in multi-machine transient stability studies and which, at the same time, provide acceptable accuracy. Their simplified model is derived from the detailed analytical method presented in the Jackson-Winchester paper (3).

Investigations at Ontario Hydro have had the objective of developing similar equivalent circuits, but based on field measurements on the generator. Initial measurements for a machine operating on and off the line were reported in ref. 11 of the subject paper. The results are presented in the form of plots of operational impedance in the direct and quadrature axes in figure 1. The Jackson-Winchester model can be utilized to derive similar operational impedance plots and we should be interested in a comparison of the analytical and field tests results.

More recent work at Ontario Hydro has included further frequency response tests, with the turbine-generator (555.5 MVA, 0.9 p.f.) connected to the system; in addition low voltage impedance measurements have been made with the same generator at standstill, and disconnected from the unit step-up transformer. The results of the former tests are still under analysis. The results of the standstill impedance measurements are also shown in figure 1. These values were calculated from voltage and current measurements at the stator terminals over a wide range of frequencies. The d.c. resistance of the stator winding was subtracted from the measured impedance as shown in the following example:

For a frequency of 0.1 hertz, the impedance measurement of two stator windings connected in series was:

$$Z_{stator} = 0.00525 \angle 14.5^\circ \text{ ohms}$$

Thus the impedance per winding at 0.1 Hz was, in Cartesian form

$$Z = 0.00254 + j0.00068 \text{ ohms}$$

To obtain the operational expression $L_d(s)$ we note that, at 0.1 Hz (0.001667 p.u. Hz)

$$Z_d(s) = R_a + j(0.001667)L_d(s),$$

Manuscript received August 2, 1972.

where R_a is the stator resistance. Thus $R_a + j(0.001667)L_d(s) = 0.00254 + j0.00068$

Noting that R_a = 0.002 ohms by d.c. measurement, and assuming that the frequency correction at 0.1 Hz can be neglected,

$$L_d(s) = \frac{0.00068 - j0.00054}{0.001667} = \frac{0.666}{1 + j0.794} \text{ ohms}$$

or

$$L_d(s) = 0.522 \angle -38.4° \text{ ohms.}$$

For the generator in question base ohms were $\frac{Kv^2}{MVA}$ or

$$\frac{24^2}{555.5} = 1.038 \text{ ohms,}$$

giving the impedance of

$$L_d(s) \text{ at } 0.1 \text{ Hz} = 0.505 \angle -38.4° \text{ p.u.}$$

This particular expression relates the p.u. changes in the direct axis currents to impressed p.u. changes in direct axis voltages at the particular frequency of 0.1 Hz. Similar expressions may be obtained at other frequencies to give the curves shown in Figure 1, both in the direct and quadrature axis.

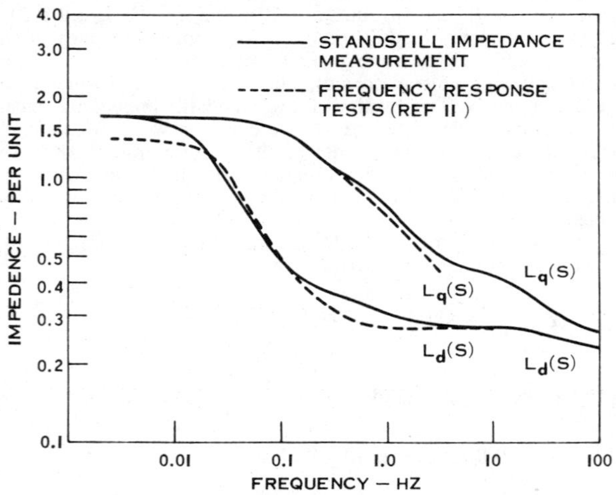

Fig. 1. Comparison of Standstill Terminal Impedance Measurements and On-Line Frequency Response Tests

The maximum current passed through the stator winding was 20 amps peak. It is noteworthy that the low frequency values of $L_d(s)$ and $L_q(s)$ thus obtained are approximately 15% lower than the unsaturated values of synchronous impedance for this machine. We presently speculate that this discrepancy may result from the low level of magnetic saturation used in the tests and from hysteresis effects. Would the authors care to comment on the effect of repeating the tests with some direct current present in the rotor?

It appears that the measured frequency response curve can be closely approximated by three time constants. Thus, a representative operational expression for the 555.5 MVA generator, obtained under other loaded conditions (not shown in figure 1) might be approximated to

$$L_d(s) = 1.50 \frac{(1 + 2.1s)(1 + .38s)(1 + .05s)}{(1 + 6.0s)(1 + .60s)(1 + .06s)}$$

This would yield direct axis equivalent circuit with three rotor circuits, including a circuit representative of the field winding. Knowledge of the leakage reactance L_ℓ and L_{ad}, the reactance of armature reaction, are necessary to calculate these equivalent rotor circuits. These quantities are usually available; in order to obtain the equivalent circuit the authors have shown in figure 5 of their paper, the $L_{fkd}(s)$ and $R_{fkd}(s)$ values must be calculated. We believe that these operational mutual reactances between the field and the rotor iron would be extremely difficult for the utility power system analyst to obtain, and would be interested in the authors' comments on this matter.

Richard P. Schulz, William D. Jones, and **Donald N. Ewart:** We thank the discussors for bringing additional considerations into the discussion of the Solid Iron Rotor Model of Turbine generators.

Dr. Kalsi points out that the modeling of solid rotor turbine generators is quite similar to the modeling of induction motors, particularly of rotor circuits, and ties in some of the British literature on analytic techniques for induction machinery. His first points are that the Solid Iron Rotor Model has two more circuits in each axis than the classical machine model and that the elements of the Solid Iron Rotor Model are determined from secondary complex impedances Fig. 3; these summarize the development of the SIRM. His comments on the coefficients of the differential equations for transient analysis of the machine introduce the matter of time domain simulation with this model. The model shown in block diagram of fig. 8 of the paper indicates operational expressions such as A(s) and $1/L_{eq}(s)$ that represent constant coefficient differential equations. These equations are implemented by integration after this manner.

For the set

$$y = A(s)x = \frac{N(s)}{D(s)}x$$

where

$$\frac{y}{x} = \frac{n_0 + n_1 s + n_2 s^2}{d_0 + d_1 s + d_2 s^2}$$

$$\left(\frac{d_0}{s^2} + \frac{d_1}{s} + d_2\right)y = \left(\frac{n_0}{s^2} + \frac{n_1}{s} + n_2\right)x$$

$$y = \frac{1}{d_2}\left[\frac{n_0 x - d_0 y}{s^2} + \frac{n_1 x - d_1 y}{s} + n_2 x\right]$$

Thus the operational expressions of fig. 8 may be directly implemented in a linear manner with constant coefficients.

We agree with the comments that Dr. Kalsi presents regarding the inclusions of the effect of $P\psi_d$ and $P\psi_q$ terms. A longer discussion of these effects can be found in references 13 and 14.

Mr. Dandeno and Mr. Watson have been working to obtain an advanced machine model, with particular emphasis on obtaining the data or parameters for such a model directly from tests which might be made in either the factory or the generating station. We would like to continue to support this development activity which is related to the goals of a working group of the IEEE Power Systems Engineering Committee and to the Rotating Machinery Committee, "Determination of Solid Rotor Synchronous Machine Stability Study Constants."

We feel that development of methods for determining these parameters from test must be viewed in light of the accuracy and precision obtainable in the measurements. In the example cited by the discussors, the dc measurement of the armature resistance has only one significant figure, yet it is subtracted from the nearly equal resistance determined at 0.1 hertz. If one assumes the dc resistance to be 10% higher i.e., that r_a equals 0.0022 ohms, then the value of $L_d(s)$ at 0.1 hertz will be 14% lower in magnitude and 12° less in phase shift. Consequently small errors in measurement may readily be magnified by the subtractions inherent in determination of the elements of the Solid Iron Rotor Model by test methods. This result, that determination of the parameters from tests requires subtraction of nearly equal numbers, presents a difficult challenge in determination of the solid iron rotor model parameters by test methods. The effort to obtain test calibration of analytically determined machine values should be continued and encouraged, with necessary attention to overcome these measurement problems. The development of simple, accurate test methods would be a valuable contribution to the industry.

We feel that if Mr. Dandeno and Mr. Watson were to repeat the tests they have described with the dc field winding excitation in addition to the stator ac excitation, the conditions would approach more closely normal machine operating conditions. We have begun some investigation of the results they reported, but have no firm results.

Manuscript received October 12, 1972.

REFERENCES

[13] R.P. Schulz, "Synchronous Machine Modeling," IEEE paper C 72 616-1, a talk prepared for the Symposium "Adequacy and Philosophy of Modeling: System Dynamic Performance" San Francisco, July 9–14, 1972

[14] J.L. Dineley, and A.J. Morris "Synchronous Generator Transient Control: Part I – Theory and Evaluation of Alternative Mathematical Models" IEEE PICA 1971 Conference, pp. 182–190

EFFECTS OF SYNCHRONOUS MACHINE MODELING IN LARGE SCALE SYSTEM STUDIES

Paul L. Dandeno
Ontario Hydro, Toronto, Ontario
and Chairman, Northeast Power Coordinating
Council Task Force on System Studies

Ronald L. Hauth Richard P. Schulz
General Electric Co.
Schenectady, New York

Abstract—A study was made of the accuracy of utility power system stability simulations, as affected by the complexity of the synchronous generator models and the data used with the models. The investigation was performed by simulating the operation of a representative multimachine power system with many combinations of disturbance, initial load, system stiffness, model detail and model data. The findings from this study provide guidelines for the selection of the simplest computational machine models for use in large scale stability analyses, with the goal of minimizing the cost of computer usage while assuring sufficient accuracy of the results. Recommendations are made for selecting machine data more suitable for use in stability studies than the standard machine data.

INTRODUCTION

An important step in power system planning is the examination of dynamic and transient stability characteristics of alternative system designs. This examination generally involves the time simulation of the behavior of many generators and their controls using a digital computer stability program. The computation cost of this process is a function of the complexity with which the power system elements are modeled. This paper discusses how the accuracy obtained in large scale system stability studies is related to the complexity of models used for synchronous machines in the system and the data used with those models.

The generator data presently used in stability studies are based on ANSI Standards Section C42.10. The transient and subtransient direct and quadrature axis reactances and time constants in the ANSI standards are based on assumed terminal short circuit conditions. The standards implicitly assume that there are two rotor circuits in each axis of the machine; the field, one d-axis amortisseur, and two q-axis amortisseurs. Recently a very complex two-axis set of equivalent circuits of a solid iron turbine generator was presented [1]. The complexity of this model is judged too great to allow its practical use in multimachine simulation studies of large power systems. A model of the solid iron rotor generator has been derived [2] from the complex Jackson and Winchester model. The derived model, which is more complex than the models designed to use the ANSI data, has 7 rotor circuits: three q-axis amortisseurs, two d-axis amortisseurs, the field and a mutual impedance between the field and d-axis amortisseurs. This derived model provides dynamic responses equivalent to the Jackson-Winchester equivalent circuit models for time simulations by accounting for the effects of rotor body currents implied in the equations relating terminal (stator, field) quantities.

This paper is based upon a study of the effects of machine modeling [3] sponsored by a Working Group of the Northeast Power Coordinating Council (NPCC). The NPCC Working Group's interest in this area was prompted by a prior examination of these problems by the IEEE Computer and Analytical Methods Subcommittee and its CAPS Working Group. The NPCC study used the derived 7-rotor circuit model as a bench mark to determine the accuracy of several machine models, less complex than the bench mark. The less complex models used two sets of generator data. One set was the ANSI standard data and the second set was derived from the frequency characteristics of the Jackson and Winchester model. The study included 79 simulation tests of transient stability and dynamic stability in multimachine and single machine configurations. The paper also discusses the influence of other factors on simulation accuracy, including the equivalent system stiffness, the magnitude of the disturbance, the initial power factor of the generator, and whether the excitation system has high initial response.

Paper T 72 514-8, recommended and approved by the Power System Engineering Committee of the IEEE Power Engineering Society for presentation at the IEEE PES Summer Meeting, San Francisco, Calif., July 9-14, 1972. Manuscript submitted February 15, 1972; made available for printing May 4, 1972.

SCOPE OF THE STUDY

The influence of generator modeling complexity on the accuracy of stability study results varies with many factors. The dynamic behavior of a real generator varies in a nonlinear way with the electrical load on the generator. Therefore, a model chosen to represent the generator must be accurate over a wide range of operating conditions, e.g., real power output and excitation. A real generator's dynamic performance also varies with the transmission system to which it is connected and the electrical proximity between it and others of comparable size in the system. If it is closely coupled to other generators of near equal size, its observed dynamic behavior when subjected to a test disturbance will be different from its behavior when electrically remote from other generators. These considerations, plus others such as the severity of disturbance (fault, loss of line or generation) and the type of excitation system (conventional-rotating machine or high initial response) assumed with the generator, influence the accuracy requirements of a given generator model.

Transmission Systems Modeled

Two transmission system models were used; a detailed 39-bus, 46-line, 10-machine model and a single machine connected to an infinite bus through a series impedance line. The second model was derived from the detailed model. The first was the model of a 345 KV transmission system typical of the New England area shown in Figure 1a. It is the same system used by several investigators on the ERC-RP-90 research project. [4] All system loads were assumed to be constant impedance loads for this study. One of the ten machines was an equivalent power source representing parts of the U.S.-Canada interconnected system. That equivalent machine was called USCAN and its rotor axis was chosen as the reference axis for the angles of the nine other machines.

The test unit, which was modeled in nine ways and to which disturbances were applied, was assumed to occupy either of two sites in the system of Figure 1a: a "remote generation" site or a "closely coupled" site. The site at Bus 329 was electrically remote from the nine remaining generators, i.e., the impedance "looking into" bus 329 was about 0.5 p.u. on the test generator's 800 MVA base. The "closely coupled" site was selected as bus 322 since there the test unit was electrically closely coupled with the nontest unit at bus 323 with the impedance "looking into" bus 322 about 0.3 p.u. on the test generator's MVA base. The intent of studying these two sites was to show the effects of differing system stiffness on the accuracy obtained with the simpler models. The combination of low natural system damping and the multifrequency behavior of the 10-machine system simulations, however, made it difficult to measure the small differences in damping between the results obtained using the different machine models. Therefore, a second system configuration was devised which did not possess a multifrequency behavior.

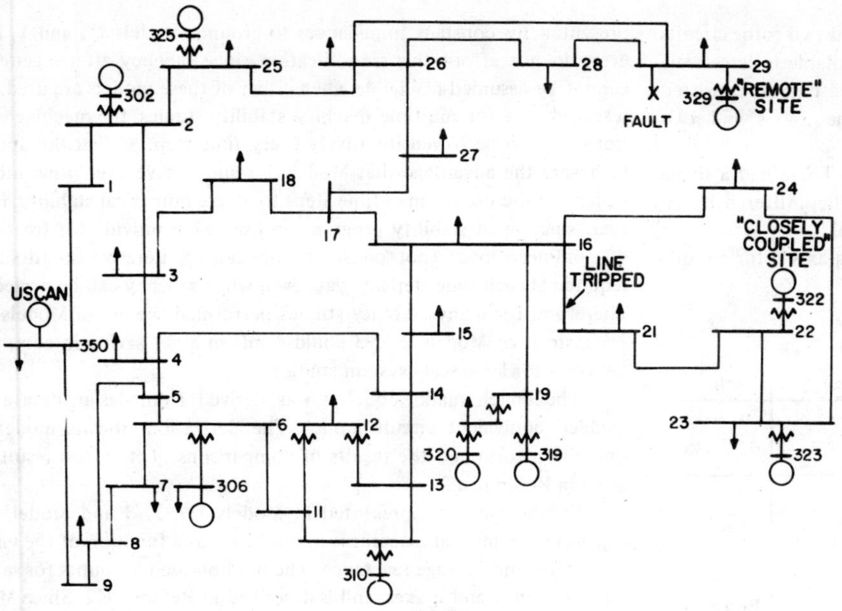

39-Bus, 46 line, 10 Machine System
Figure 1a.

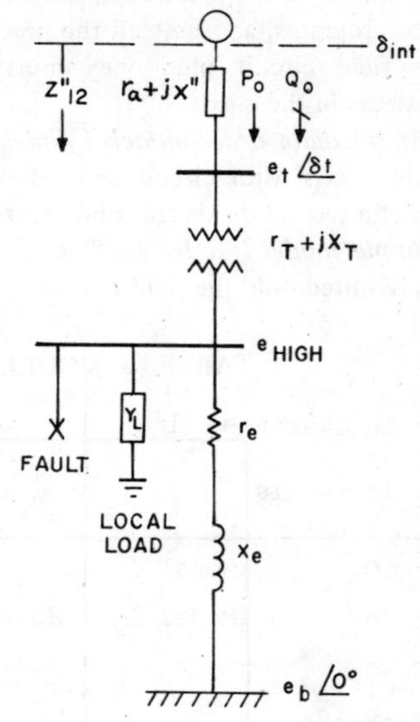

One-machine-infinite-bus system
Figure 1b.

The second system configuration studied was the one-machine-infinite bus system model shown in Figure 1b. Three versions of the one-machine system were studied: Nominal line, short line and long line as characterized by their electrical stiffness. In order to relate this system model to the 39 bus system, the nominal line version was made electrically equivalent to the 39-bus system when looking into that system from bus 329. The transfer impedance (Z''_{12} as defined in Figure 1b) across the single machine system was varied over approximately a nine to one range to determine the effect of natural frequency or system stiffness on the accuracy of the simpler models. This change was made by varying X_e and R_e from a range of 0 (short line) to .014 + j.0556 (nominal line) to .161 + j.332 (long line) on a 100 MVA base. The load Y_L was adjusted to match power factor of the 39 bus system and to allow stable power transfer over the long line.

The one-machine-infinite bus system was also used for studying the influence of modeling complexity and data differences on computed critical clearing times.

Types of Analyses – Disturbances

The study considered numerous multimachine simulations with various combinations of generator models, generator data sets, initial loadings, excitation systems and system disturbances made using the 10 machine and one machine transmission systems discussed above. Most of these simulations were performed by imposing a stub fault applied on the high voltage side of the test generator's station transformer. This stub fault was cleared approximately one-half cycle before the critical clearing time and the system response was computed for a three-second period after the application of the fault. The swing curves obtained using different generator models were then compared and the differences due to modeling complexity and/or data differences were noted.

Small-disturbance time simulations were performed to observe whether inclusion of more or less complex models of the solid rotor eddy current (iron) effects made a significant difference in those cases. The small-disturbance studies were performed using the 39 bus system wherein the line between buses 16 and 21 was tripped out, without a fault, to perturb the system. Swing curves were computed using different models and the results were compared.

The effect of modeling complexity and data differences on computed values of critical clearing times was also investigated using the simple one-machine-infinite bus system in Figure 1b. The system configuration assumed was the "nominal line" version of the one machine system, which was the electrical equivalent of the "remote" site of the 10-machine system, and generator power factor was shifted by modifying the initial voltages at the machine terminals and infinite bus.

The time domain simulations described above were performed using the FACE multimachine program described in Reference 5 and the multimachine simulation program in Reference 6.

To confirm the damping performance observed in the time domain simulations, several dynamic stability studies were performed using the program described in Reference 7. That program determined the power limits of dynamic stability, defined in the P, Q plane; i.e. real power versus reactive voltampere outputs of test generator. The power system model used in those cases was the nominal line version of the system in Figure 1b. The Nyquist criterion was used to detect the onset of instability assuming infinitesimal perturbations of the test machine's internal angle.

Prime Mover and Excitation System Models

No prime mover dynamics were considered in any of the simulation studies performed. All the generators, test and nontest units, were assumed to have a constant input torque.

Two excitation systems were considered in the time-domain simulations performed. The test unit was supplied with either a conventional rotating exciter system or a high initial response thyristor-type exciter. The excitation systems on the nontest units of the 10-machine system were the same as those of the ERC study. [4] All of the one-machine infinite bus simulation runs were made with the conventional exciter on the test unit since early 10-machine runs showed the high initial response excitation system partly masked the differences between the results obtained with different machine models. Power System Stabilizers were not modeled with either system.

Generator Models

The machine models studied were from the most complex to the simplest:

The bench mark (Model 4) [2] This model included the field circuit and six rotor body circuits, three in each axis. The performance with this model was used as a reference to asses the accuracy of the other models.

High order model (Model 3) This model included two rotor circuits in each axis; it is the most complex model now available in large scale stability programs, and uses all the presently defined data. It consisted of the field circuit plus one amortisseur in the d-axis and two amortisseurs in the q-axis.

Intermediate order models (Models 1½ and 2) These had a single quadrature axis rotor circuit and a field circuit. They differed by inclusion of a second d-axis rotor body circuit in Model 2.

Simple model (Model 1) This model had no q-axis rotor circuits and represented only the field circuit.

TABLE I – MODELS STUDIED

Order in d-axis →		1	2	3	4
Order in q-axis ↓		f	f, d_1	f, d_1, d_2	f, d_1, d_2, fkd
0	0	Model 1			
1	q_1	Model 1½	Model 2		
2	q_1, q_2		Model 3		
3	q_1, q_2, q_3				Model 4

f = field circuit; d_1, d_2 = d-axis amortisseurs;
q_1, q_2, q_3 = q-axis amortisseurs; fkd = mutual between field and d-axis amortisseurs

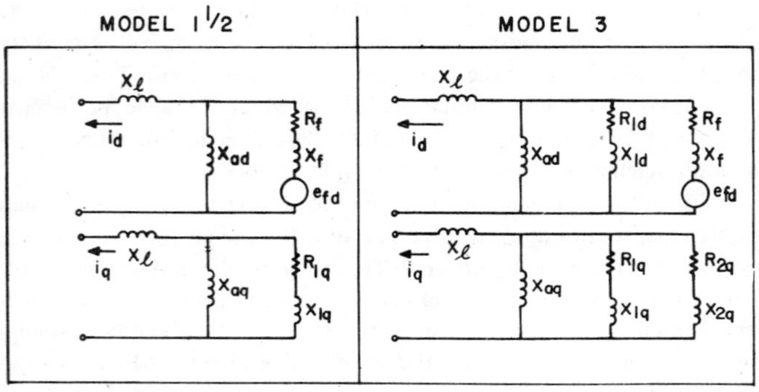

Data required
$x_\ell, x_d, x'_d, T'_{do}, x_q, x'_q, T'_{qo}$

Data required
$x_\ell, x_d, x'_d, x''_d, T'_{do}, T''_{do}, x_q, x'_q, x''_q, T'_{qo}, T''_{qo}$

Figure 2

The relative differences in order of complexity between the models are illustrated in Table I. As indicated in Figure 2, implementation of Model 1½ requires two principal time constants T'_{do} and T'_{qo}. In contrast, Model 2 requires the use of T''_{do} in addition to T'_{do} and T'_{qo} due to its added complexity in the d-axis. One goal of this study was to find a model which gave reasonably accurate results without incurring excessive computing costs. Potential computational cost savings were anticipated with use of Model 1½ since it could be implemented without the need for very small integration time steps which are often required when small time constants such as T''_{do} and T''_{qo} are modeled. Using Model 1½, the computation may be three to five times faster if not limited by models of other equipment. Model 1 also promised to be inexpensive to use for the same reasons.

The data required for implementing Model 3 is indicated in Figure 2. Notice that subtransient time constants and reactances are needed with that model. Since the numerical values of X''_q and X''_d are nearly equal for round solid rotor synchronous machines subtransient saliency is often neglected by setting X''_q equal to X''_d for use in stability studies. This assumption affords a computational simplification in the stability study of multiple machine power systems if system loads can be represented by constant impedances to ground. Models 1½ and 1, however, do not afford this simplification since saliency effects generally cannot be assumed negligible when either of these models are used. The result is that for multiple machine stability studies, the machine equations have to be solved iteratively every time step. At first this appears to negate the advantage that Models 1½ and 1 have over other models which require use of small time steps to insure numerical stability. However, since most stability programs in use today provide for treatment of nonlinear loads (not constant impedance), iterative solutions are required at each time step anyway, even when saliency can be neglected. Therefore, for many stability studies performed the use of Models 1 or 1½ instead of Models 2 or 3 could result in a net saving in computer-use costs for large scale system studies.

The bench mark Model 4 was derived from design data and a detailed equivalent circuit model. The derivation of this model, its simulation form and the results of comparisons of it to test results are given in Reference 2.

Saturation was represented in Models 1½, 2, 3 and Model 4 by adjusting the mutual reactances in both axes as a function of the virtual voltage behind leakage reactance. The method used accounts for saturation in both d and q axes, and is described in Reference 2. Since Model 1 lacked q-axis amortisseurs, saturation effects for it were assumed as a function of E'_q, proportional to field flux linkages. The different treatments of saturation caused the initial steady-state rotor angle of the Model 1 simulation runs to differ slightly from the steady state angle for similar conditions with other models.

Generator Data

Two representative generators were chosen for which comprehensive solid rotor equivalent circuit data was available. One was a two-pole 3600 rpm generator, typical of fossil fuel thermal units. The second was a four-pole 1800 rpm generator typical of nuclear powered units. Both units had an 800 MVA rating; the data for the 3600 rpm unit data is given in Table II.

TABLE II
GENERATOR DATA FOR 3600 RPM UNIT

	Direct Axis					Quadrature Axis				
MODEL	X_d	X'_d	X''_d	T'_{do}	T''_{do}	X_q	X'_q	X''_q	T'_{qo}	T''_{qo}
3	1.75	.285	.24	4.0	.029	1.68	.47	.24	.540	.053
1½	1.75	.285	---	4.0	----	1.68	.47	---	.540	----
1	1.75	.285	---	4.0	----	1.68	---	---	----	----
3A	1.75	.285	.24	5.2	.011	1.68	.47	.24	1.96	.053
2A	1.75	.285	.24	5.2	.011	1.68	.24	---	.540	----
1½A	1.75	.275	---	5.2	----	1.68	.47	---	1.96	----
1A	1.75	.240	---	5.2	----	1.68	---	---	----	----
1B	1.75	.275	---	5.2	----	1.68	---	---	----	----

Two sets of numerical data were used for each test unit. The first set was the standard data as defined by ANSI standard C42.10 and normally supplied by generator manufacturers. This set includes four time constants and seven reactance values: T'_{do}, T'_{qo}, T''_{do}, T''_{qo}, and x_d, x'_d, x''_d, x_q, x'_q, x''_q, x_ℓ. The second or modified set of data was obtained by adjusting one or more of the eleven standard constants to more closely agree with the Model 4 data.

Model numbers shown with the letter A or B appended to them indicate the use of modified values of reactances and time constants; otherwise standard data was used. For example, Model 3A was identical in form (Figure 2) to Model 3. However, whereas standard data values, T'_{do}, T''_{do}, T'_{qo}, x'_d, etc. were used with Model 3, some of these data were modified numerically for use with Model 3A. The data were modified to force the dynamic terminal characteristics $L_d(j\omega)$ and $L_q(j\omega)$, of the equivalent circuits for the simpler models to approximate the dynamic terminal characteristics of the bench mark Model 4. The reasons for considering modified data are considered next.

The terminal characteristics of a synchronous machine can be described in terms of the d-axis and q-axis impedances, as viewed looking into terminals of the two circuits.

$$Z_d(s) = R_d(s) + s\,L_d(s) \quad . \quad . \quad . \quad . \quad . \quad . \quad . \qquad (1)$$

$$Z_q(s) = R_q(s) + s\,L_q(s) \quad . \quad . \quad . \quad . \quad . \quad . \quad . \qquad (2)$$

These equations are operational expressions of the complex impedances and s is the familiar Laplace operator. If s is set equal to $j\omega$, the relationships describe the variation of impedances Z_d and Z_q with slip frequency $j\omega$ for sinusoidal slip frequency currents flowing in the equivalent circuits. The inductive terms $L_d(s)$ and $L_q(s)$ relate the flux linkages ψ_d and ψ_q to the currents i_d and i_q, respectively. These relationships account for the effects of field and rotor body currents on the terminal (stator) characteristics of the generator.

The operational expressions for $L_d(s)$ and $L_q(s)$ are of importance in describing why the modified data was required. The operational expression for $L_d(s)$ can be generally expressed as a ratio of polynomials in s. The behavior of the solid iron rotor generator was represented by a solid iron rotor model in the FACE program, Ref. 5. The d-axis inductance of the model may be represented by a ratio of factored polynomials

$$L_d(s) = X_d \frac{(1+As)(1+Bs)(1+Cs)(1+Ds)}{(1+T_1s)(1+T_2s)(1+T_3s)(1+T_4s)} \quad . \quad . \qquad (3)$$

The order of numerator and denominator polynomials determines the d-axis order of Model 4. Model 3 has second order inductances in both axes so an approximate second order expression for its d-axis operational inductance, $L_d(s)$ would be:

$$L_d(s)_{approx} = X_d \frac{(1+Es)(1+Fs)}{(1+T_5s)(1+T_6s)} \quad . \quad . \quad . \quad . \quad . \qquad (4)$$

It is clear that $L_d(s)$ in (4) can never equal $L_d(s)$ in (3) unless, by chance, two of the numerator coefficients equal two of the denominator coefficients. However, E, F, T_5 and T_6 in (4) can be chosen so that a "best fit" of magnitude and phase of $L_d(j\omega)$ is achieved over a frequency range of interest.

Model 3, using standard data, results in an approximate expression for $L_d(s)$ as:

$$L_d(s)_{MOD3} = X_d \frac{(1+Gs)(1+Hs)}{(1+T'_{do}s)(1+T''_{do}s)} \quad . \quad . \quad . \quad . \quad . \qquad (5)$$

where $G \cong \frac{x'_d}{x_d} T'_{do}$ and $H \cong \frac{x''_d}{x'_d} T''_{do}$.

In general the "best fit" coefficients do not all equal the corresponding standard (Model 3) coefficients. Thus Model 3 (standard) data is generally not a "best fit" to Model 4 data. The "best fit" data in Equation 4 was determined from a study of the Model 4 data. That "best fit" data (E, F, T_5 and T_6) was called the "modified" d-axis data for use with a second order d-axis circuit. By following a similar "best fit" approach for arriving at the coefficients of $L_q(s)$ using a second order q-axis circuit, Model 3A was defined. Similar reasoning was used to establish modified data for the other models.

Generally, the reactances of the modified data closely match the reactances of the standard data. The differences between some of the time constants and their "best fit" counterparts were significant. For example, it was found that the principal d-axis time constant value derived from the solid rotor equivalent circuit data was nearly thirty percent higher than T'_{do} for both sample generators. The principal q-axis time constant, so derived, was more than double the value of T'_{qo} for both sample generators. The results obtained in the stability studies were found to be quite sensitive to these differences in data.

RESULTS

The results presented in this paper are representative of the total body of results of the study. [3] For brevity figures are given for the 10-machine large disturbance multiswing stability studies and the critical clearing time studies. Brief descriptions of results are given for the three other types of analyses described above.

Large Disturbance Multiswing Studies–Ten Machine System

Qualitative comparisons of the 10-machine system results are presented here with the aid of Figures 3 and 4. Figure 3 shows a comparison between results obtained with Models 1, 1½, and 3 (using standard data) and the bench-mark results obtained using Model 4. These results were obtained with the 3600 rpm test unit located at the "remote" site and initially operating at 1 p.u. power (800 MW). The conventional excitation system was assumed in use on the test unit. The voltage profile on the system was tilted to obtain the indicated range of initial reactive output and consequent range of excitation. Power output was maintained at 800 MW from the test unit. The fault clearing time, shown for each of the operating power factor conditions, was chosen as approximately 0.5 cycles less than the critical clearing time obtained using Model 4 for each power factor condition considered.

The excitation level on the test generator increases, moving from left to right in Figure 3. The accuracy of results obtained using Models 1½ and 3, relative to the Model 4 results, increase as excitation level increases. Model 1 also becomes more accurate but its ranking in Figure 3c became unclear, because of the different initial angle.

Figure 3 also illustrates that increased model complexity did not always result in increased accuracy of results when standard data was used.

Figure 3 illustrates the difference between Model 4 results and those obtained using the standard machine data in the simpler models. Figure 4 shows results from the same simulation studies except modified values of time constants and reactances, derived from the solid rotor equivalent circuit, were used in place of the standard quantities.

In general, the results obtained using the modified data were in closer agreement to the bench mark Model 4 results than the results obtained using standard machine data. Model 3A consistently gave results in close agreement with Model 4. Model 1½A was also a good approximation to Model 4 but was more sensitive to changes in excitation level than was Model 3A. Model 1B was also quite faithful to the Model 4 results except its lack of amortisseur circuit representation causes it to be less reliable in predicting damping performance. The evidence for that point is most clear in Figure 4c. No apparent improvement was found in damping performance of Model 1B over that for Model 1. Similar results were observed when the 1800 rpm unit was studied with Models 3, 3A and the bench mark Model 4.

Using a high initial response exciter with the test unit had the effect of masking differences between the results obtained using the different machine models. This may be illustrated by comparing Figures 5 and 6. In those figures the swing curves (angles) obtained using Models 1A, 2A, 3A and bench mark Model 4 are shown. The swing curves for the various models are more closely grouped when a high initial response exciter was used (Figure 5) than they were with a conventional exciter (Figure 6). The masking effect is most evident in those figures for Model 2A. The large difference between Model 2A results and Model 4 results can be explained in terms of the value of the operational inductances at the swing frequency.

Figures 5 and 6 also show how the computed terminal voltage is affected by differences in model complexity.

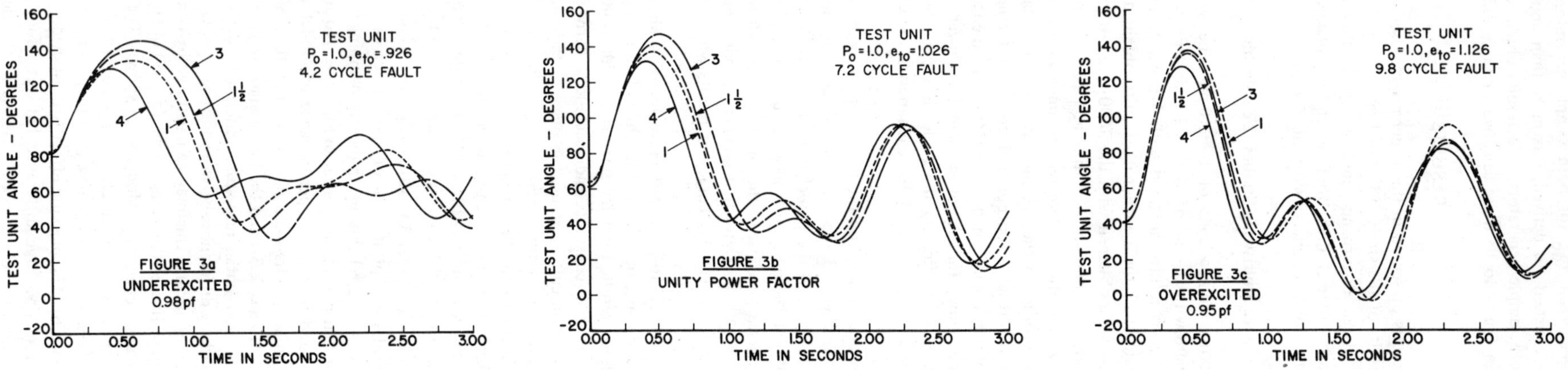

Fig. 3. Swing curves from 10-machine, multiswing simulations – Using standard data for test unit

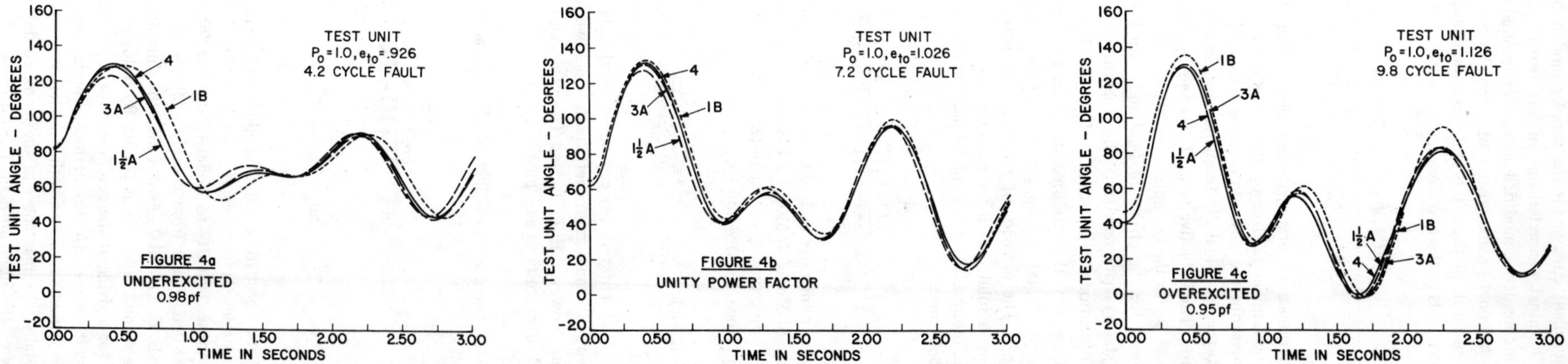

Fig. 4. Swing curves from 10-machine, multiswing simulations – Using modified data for test unit

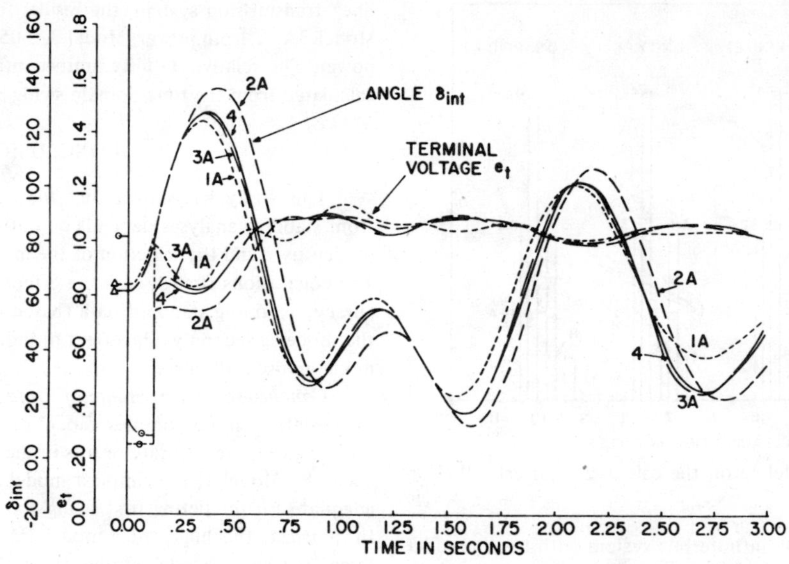

Fig. 5. Using a high initial response exciter on test unit. Test unit in "remote site", initial p.f. = unity

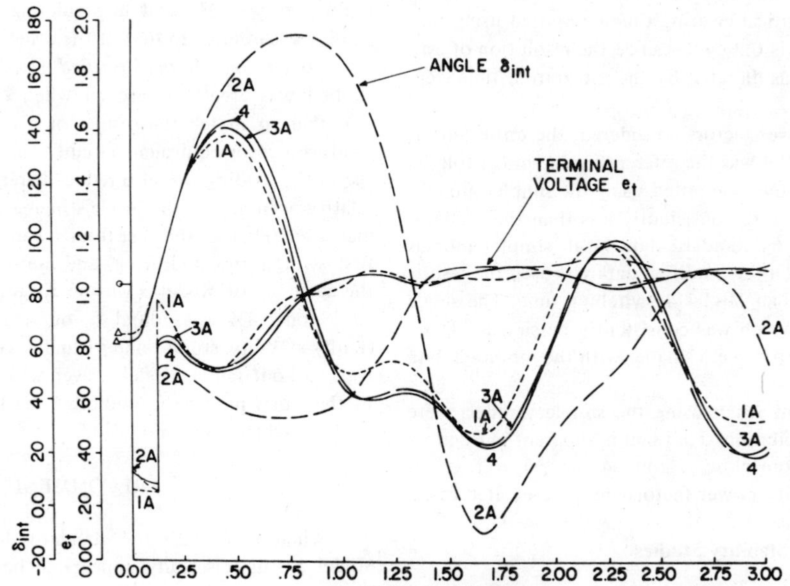

Fig. 6. Using a conventional exciter on the test unit. Test unit in "remote site", initial p.f. = unity

Large Disturbance Multiswing Studies–One Machine System

To better quantify the effects of machine model on damping of rotor angle swings, several fault cases were run with the one machine to infinite bus system. This system has only one mode of rotor angle oscillations and show the damping differences more clearly than the multimodal swings of the multimachine system.

Models 1B, 1½A, and 4 were studied together with three system stiffness levels and three test unit initial power factor levels, as discussed in the Scope. The "short line" and "long line" cases were run at unity power factor, and the nominal reactance line was used at 0.98 p.f. underexcited, unity p.f., and 0.95 p.f. overexcited machine conditions. The "short line" represented a system reactance of about twelve percent on the machine's MVA base, which corresponds to a practical lower limit on system reactance. The "long line" case constituted the "weakest" system considered with a tie line reactance of about 2.7 p.u. on the machine's MVA base. Two significant observations were made from these results. First, the differences in damping performance were small, as in the 10-machine simulations. Model 3A was not even considered since the differences between Model 1½A and the bench-mark model were so small. The second point is that these differences changed significantly when the system stiffness was varied. In all but one case the results using Model 1½A exhibited less damping than Model 4. That one exception was in the case of the "long line" where Model 1½A exhibited more damping than Model 4. This apparent reversal in the damping error exhibited by Model 1½A can be explained in terms of the variation of swing frequency from case to case, and the models' operational inductances.

Critical Clearing Time Studies

The effects of model complexity and data on computed values of critical clearing time for fault conditions were studied by applying a three phase stub fault on the high side bus of the test generator's trans-

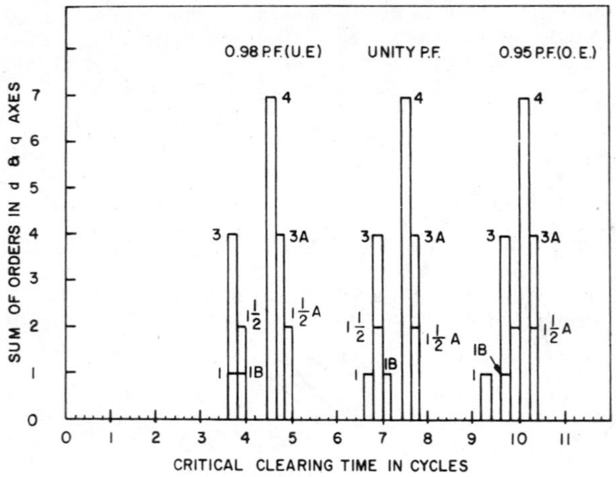

Fig. 7. Effect of machine modeling on the calculation of critical clearing times

former using the single machine infinite bus system with the "nominal line" configuration. The results are illustrated in Figure 7. The ordinate in that figure is a measure of the complexity of the models and corresponds to the sum of d-axis and q-axis orders. For each model considered, a vertical bar was drawn with its horizontal distance from the origin corresponding to the critical clearing time computed using that model. The width of each bar is 0.2 cycles since the resolution of estimating critical clearing time was dictated by the integration time step, which was 0.2 cycles.

For each of the three power factors considered, the critical clearing time computed using Model 4 was the reference value in the following discussion. The switching time computed using the simpler models, when standard data was used, were consistently less than the Model 4 results. That is, the results using standard data in all simpler models were pessimistic. Generally, when the modified data was used the results were optimistic, i.e., greater than Model 4 switching time. The single exception was for Model 1B, which was consistently pessimistic. For a given model complexity, the errors were smaller with the modified data than with the standard data.

The relative errors evident when using the simpler models were smaller for higher levels of excitation. That fact is apparent in Figure 7 where the vertical bars are more closely grouped for the overexcited cases than they are for the unity power factor and underexcited cases.

Small Disturbance, Multiswing Stability Studies

Multiswing time simulations were performed on the 39 bus, 10 machine system where the disturbance was caused by tripping out the line between buses 16 and 21 identified in Figure 1a. In those cases the test unit was assumed to occupy the "closely coupled" site. Only Models 3, 3A, 1½A and the bench mark Model 4 were considered in these studies. The purpose for the small disturbance runs was to provide simulations of 5 seconds in length for analyzing the difference in damping supplied by the various models.

In these tests, the results with Models 3, 3A and 1½A were very accurate and not significantly different one from another. The results were only slightly dependent on the choice of exciter (conventional or high initial response) on the test unit.

Damping Studies – Frequency Domain

A quantitative measure of the damping performance, for small disturbance analyses, was also sought using the dynamic stability computer program described in Reference 7 to find the dynamic stability limits of the one machine infinite bus system and to evaluate the damping obtained with several models at one realistic operating condition. The absolute stability limit and a family of relative stability limits was found for each of Models 3, 3A, and 1½A. Using the "nominal line" transmission system, the results for the models at unity p.f. were: Model 3A, 2.1 p.u. power; Model 3, 2.05 p.u.; and Model 1½A, 2.14 p.u. power. The relative stability limits confirmed the relative damping ratio calculated from the time domain swing curves, $\zeta \cong 0.04$.

CONCLUSIONS

This study has shown that the accuracy of the results obtained from stability analyses depends on both the complexity of the machine models used and the selection of the machine data used in these models. The conclusions first discuss the effect of modeling complexity on accuracy, assuming standard data (based on ANSI standards) is used with the models. Secondly, the effect of modifying the machine data for the model in use is discussed.

Conclusions With Standard Data: When using the standard machine data, a more complex model of a synchronous machine did not always give more accurate results in the stability studies performed. For example, Model 1, the simplest model studied, was shown to be quite adequate for predicting first swing stability of a system in some cases. In contrast, the high order model (Model 3) was less accurate in the same first swing study. Model 1½ with intermediate complexity gave reasonable accuracy, better than Model 3 and comparable to Model 1.

The study showed that the relative accuracy of the models was significantly affected by the initial power factor of the machine. Also, the accuracy of Model 1 in predicting damping in multswing stability studies was inferior to that of the other models.

Conclusions With Modified Data: The accuracy of all models studied was greatly improved when the standard machine data was modified to match the results of a frequency response analysis of a solid iron rotor equivalent circuit. The dependence of the accuracy on the initial loading was also reduced greatly. With the modified data, the relative accuracy of the models increased when model complexity was increased. That is, the best results were obtained with Model 3 for both first swing (critical clearing) and multiple swing stability analyses; and the accuracy of Model 3 was nearly unaffected by initial power factor.

Model 1½ was found to be very accurate for large disturbance (fault) stability studies and virtually exact for small disturbances (line trip, without fault). Model 1, even with the modified data, was inferior to the other models in predicting the long-term damping performance of the machine being modeled.

RECOMMENDATIONS

When used with the standard data, the machine model called Model 1½ in this study appears to be sufficiently accurate for use in multiswing transient stability studies. Because it requires use of only the larger time constants T'_{qo} and T'_{do}, Model 1½ may allow the use of larger computational time steps than those needed with models requiring the use of T''_{do} and T''_{qo}. Therefore, Model 1½ is recommended for use in large scale stability studies where significant computer-usage cost savings may be realizable without serious sacrifices in accuracy.

The use of machine data modified to suit the machine model chosen results in a significant improvement in accuracy of all models considered in the study. The modified data was derived from frequency response characteristics of the detailed reference model which are not generally available. This suggests that further work is required to define new data for use in stability studies and to devise general methods for computing these data for a specific generator. This effort must be undertaken on an industry-wide basis resulting in a revision of the standard definitions for data used in power system stability studies.

When used with the appropriate modified data, Model 3 was found to be an excellent model for the stability analyses performed in the study. Model 3 is therefore recommended for use in the more critical stability studies performed by the industry after the new stability study data has been defined and made available. These more critical stability studies might include situations with one or more of these conditions: a machine operating underexcited or in some other weakly coupled situation or, a fault occurring very close to machine under study or, the machine being modeled is considered to be a significant element of the system.

ACKNOWLEDGEMENTS

Many people from several organizations contributed to the work from which this paper is drawn. The authors gratefully acknowledge the particular contributions of: the Northeast Power Coordinating Council and its Task Force on System Studies, which was chaired by Halbert E. Pierce, Jr. during the time the work was started and accomplished; to the NPCC Review Committee with which the authors worked: Ernest G. Neudorf, Edward M. Gulachenski and George C. Loehr; and to engineers of the General Electric Company for their advice. Dr. Prabha Kundur of Ontario Hydro also assisted the review committee by providing results for several of the tests.

REFERENCES

[1] W. B. Jackson, and R. L. Winchester, "Direct and Quadrature-Axis Equivalent Circuits For Solid-Rotor Turbine Generators," IEEE Transactions on Power Apparatus & System, vol. PAS-88, No. 7, July, 1969, pp. 1121-1136.

[2] R. P. Schulz, W. D. Jones and D. N. Ewart "Dynamic Models of Turbine Generators Derived from Solid Rotor Equivalent Circuits," paper submitted for the 1972 IEEE Summer Power Meeting.

[3] R. L. Hauth and R. P. Schulz, "Effects of Synchronous Machine Modeling in Large Scale System Studies," final report for Northeast Power Coordinating Council (NPCC) Task Force on System Studies, System Dynamic Simulation Techniques Working Group, Dec. 1, 1971.

[4] J. M. Undrill, and A. E. Turner, "Construction of Power System Electromechanical Equivalents by Modal Analysis," IEEE Transactions on Power Apparatus and Systems, vol. PAS-90, no. 5, Sept./Oct. 1971, pp. 2049-2059.

[5] D. N. Ewart, R. P. Schulz, "FACE Multimachine Power System Simulator Program." PICA Conference May 18-21, 1969. Conference Record No. 69 C1-PWR, pp. 133-153.

[6] Prabhashankar Kundur, "Digital Simulation and Analysis of Power System Dynamic Performance," Ph.D Thesis, University of Toronto; May, 1967.

[7] D. N. Ewart, F. P. deMello, "A Digital Computer Program For The Automatic Determination of Dynamic Stability Limits," IEEE Transactions on Power Apparatus Systems, vol. PAS-86, No. 7, pp. 867-875, July, 1967.

Discussion

A. M. El-Serafi and **M. A. Badr** (University of Saskatchewan, Sask., Canada): The Authors are to be commended for their interesting study.

The use of the conventional equivalent circuit representation, as shown in Fig. 2 of the paper, does not accurately represent the synchronous machine as far as rotor quantities are concerned. The measured variations of the rotor currents following a three phase sudden short circuit of a turbo-alternator were found to differ greatly from those calculated using this equivalent circuit [1]. This is due to the fact that such an equivalent circuit is based on an approximate application of the per-unit system, in which the couplings between the rotor circuits on one axis are assumed equal. A more accurate representation of the synchronous machine has been found by modifying this conventional equivalent circuit [1].

Do the Authors think that taking this point into consideration will appreciably affect the accuracy of the machine models used for their studies?

REFERENCE

[1] I. M. Canay, "Causes of discrepancies on calculation of rotor quantities and exact equivalent diagrams of the synchronous machine", IEEE Transactions, PAS-88, No. 7, pp. 1114-1120, July 1969.

Manuscript received July 31, 1972.

Swarn S. Kalsi (General Electric Company, Philadelphia, Pa. 19124): The authors should be commended for presenting a paper examining the suitability of different models of a synchronous machine for multimachine studies. Models 1, 1½, and 2 are well known and Model 2 is widely accepted [Ref. 1, 2] as an accurate model of a synchronous machine. However, due to the complex phenomena of the quadrature axis, a solid rotor synchronous machine may be adequately represented by the Model 3. Before attempting to complicate the machine models any further, I feel that Models 1 to 3 should first be comprehensively tested against test results of a single machine system. Theoretically, to achieve exact duplication of conditions inside a machine, an infinite number of circuits must be used in each axis of the equivalent machine. Addition of more circuits increases the complexity of the model and the Model 4 is an example of this type. Moreover, the most crucial aspect of machine modeling is to devise methods for evaluating parameters which are good for calculating the machine performance for all operating conditions.

The discusser developed a model [Ref. 3] which is equivalent to paper's Model 3. Few methods for calculating the machine parameters were suggested, but the two important test methods are:

Method A: Short Circuit Test at Machine Terminals – equivalent to standard parameters as defined in ANSI Standard C42-10.

Method B: Static Impedance Test – based on an operational function similar to Eqn. (3) of the paper.

Transient and sub-transient reactances calculated by above methods agreed, but transient and sub-transient short-circuit time constants did not. Ratio of sub-transient short-circuit time constant measured by Method-A to that measured by Method-B, was about 3. It must be noted that the Jackson-Winchester Model (Ref. 1 of the paper) is equivalent to Method-B and parameters determined by this method exhibit more damping than those determined by Method-A. Depending upon the type of disturbance studied, two sets of parameters give different results. Using parameters determined by the terminal short-circuit method in the Model-2, good agreement was observed between test and calculated swing curves [Ref. 2]. Referring to Fig. (3) of the paper, I presume that swing curve 4 was calculated using parameters obtained from the Jackson-Winchester Model and the other three curves are calculated with standard parameters. Since a sudden short circuit generates a fast transient, the higher damping effect incorporated in parameters calculated by the Jackson-Winchester Model, severely damped swing curve 4. I feel that swing curve 3 (calculated with the Model-3 using standard parameters) should be more accurate. This difference of opinion can be resolved by comparing calculated result against test result, preferably on a single machine system because in a multi-machine system, the swing curve of each machine is affected by energy transferred to and from other machines.

While making calculations for curves in Fig. 3 to 6, I wonder if the following voltage components were considered.

i) Stator Transients – $p\psi_d$ and $p\psi_q$ terms in the primary voltage equations.

ii) Voltage drop in system reactance (X/wo pI) due to rate of change of current.

Omission of these voltages caused a considerable error in the calculations of a three-machine system [Ref. 1]. On the other hand, if the stator transients and sub-transient saliency are neglected in the Model-2, it is possible to write machine and system equations together in terms of phasors [Refs. 1, 2]. This arrangement saves considerable computing time, since an integrating time step as large as 0.5 cycle can be used without causing numerical instability, and the results are still within 1% of those obtained with integrating time step of 0.1 cycle.

REFERENCES

[1] S. S. Kalsi and B. Adkins, "Transient Stability of Power Systems containing both Synchronous and Induction Machines", Proc. IEE, Vol. 118, No. 10, October 1971, pp. 1467-1474.

[2] R. G. Harley and B. Adkins, "Calculation of the Angular Back Swing Following a Short Circuit of a loaded Alternator", Proc. IEE, Vol. 117, No. 2, February 1970, pp. 377-386.

[3] S. S. Kalsi, D. D. Stephen, and B. Adkins, "Calculation of System Fault Current Due to Induction Motors", Proc. IEE, Vol. 118, No. 1, January 1971, pp. 201-215.

Manuscript received July 26, 1972.

G. Shackshaft (Central Electricity Generating Board, London, England): The authors present a fascinating account of studies to assess the accuracy of several machine models. Having been concerned with stability

Manuscript received August 7, 1972

analysis at the Central Electricity Generating Board in England for a number of years, I am particularly interested to see that the authors' recommendations with regard to Model 3 line up exactly with the multi-machine program which we are using at present [1], and which we find to be satisfactory.

Until recently we used Model 2, but as a result of comparisons with system tests carried out over many years and particularly those carried out at Northfleet Power Station [2] recently, we have now abandoned this model. Its main deficiency is in its inadequate representation of the quadrature axis, which is particularly important at leading power factor operation.

The authors will be aware that we have laid great stress on the importance of validating our models by carrying out large scale system fault-throwing tests; very complex models, like the Jackson and Winchester model, have never been used in the development of our analytical methods.

At present, we also have the problem which the authors discuss with reference to Model 3, i.e. that of obtaining adequate data; however, I do not expect that this will be difficult to overcome. Data is clearly crucial in such advanced models and I feel that I must express my unease about the use of the Jackson and Winchester model to modify manufacturers data. In my view, one may be deluding oneself on the absolute accuracy of the models by matching operational impedances with the Jackson and Winchester model, whilst ignoring important practical factors such as the saturation of machine leakage reactances under high current conditions. The following table illustrates this effect as measured on a 120 MW turbo-generator at Northfleet Power Station [2].

Type of Test	VAr Rejection Test	Short Circuit at 50% Volts	Short Circuit at 100% Volts
Approx. stator current change (p.u.)	0.5	2.0	5.0
x'_d (p.u.)	0.478	0.27	0.245
x''_d (p.u.)	0.288	0.19	0.17
T'_d (seconds)	1.7	0.97	0.63
T''_d (seconds)	0.083	0.046	0.026

I conclude that saturation of leakage flux paths is an important consideration in stability studies such as those described in the paper, in which the current variations are very large, and should be obliged if the authors could tell me whether or not the accuracy of the Jackson and Winchester model has been tested under such current conditions and if so, with what results.

REFERENCES

[1] P. Humphreys and M. R. Payne, "Development of Digital Program for the Simulation of Power-Angle Disturbances by Comparison with System Tests" paper to Fourth Power Systems Computation Conference at Grenoble, 1972.

[2] G. Shackshaft and R. Neilson "Results of Stability Tests on an Underexcited 120 MW Generator" Proceedings I.E.E., February 1972, Vol. 119, pp. 175-188.

Paul L. Dandeno, Ronald L. Hauth, and Richard P. Schulz: We welcome the discussors' comments and are pleased to continue the discussion with them.

Manuscript received September 18, 1972.

There seems to be some misunderstanding about our intent in including model 4 within the study reported in this paper. Model 4, which is exactly the Solid Iron Rotor Model reported in reference 2, was used as a standard of comparison for the lower order models. We are not proposing that its structure be implemented in large scale stability programs, but rather we are advocating that models 1½ and 3 be used in large scale stability programs, depending principally upon the data available. The data used for the lower order models was either that available directly from the manufacturer as calculated according to the present standards, or that which was determined through the use of the full Jackson-Winchester model.

Dr. El-Serafi and Mr. Badr note that neither models 1½ nor 3 accurately represent the synchronous machine as far as rotor circuit currents are concerned. We concur; our intent was to evaluate the adequacy of lower order models for representing terminal characteristics for those machines as elements of a large scale power system. In the course of the study, the variations of field current were examined for the several models for several of the disturbances; these comparisons were not made for all the disturbances since the fundamental intent of the study was the representation of the generator as it affected the overall system dynamics. Those comparisons of field current which were made, showed significant differences between field current waveforms obtained with the various models. The discussors point out that Dr. Canay presented a modified conventional equivalent circuit which allowed the accurate determination of rotor circuit currents. The form of the Solid Rotor Iron Model used as model 4 is a similar equivalent circuit, and provides a realistic determination of rotor circuit currents.

In reply to Dr. Kalsi, we note that the simulation programs used (Reference 5, 6) did not include the $P\psi_d$, $P\psi_q$ terms in the stator voltage equations nor did the programs include the rate of change of energy storage in the transmission system elements' equations. These inclusions would have significantly increased the cost of the simulations, particularly when studying the 39 bus, 46 line, 10 machine system shown in Fig. 1A. There have been recent recommendations [Ref. 11] for the inclusion of the $p\psi$ terms for machines and transmission networks in power system simulation. A recent paper [12] shows the effect of ignoring these terms to be very small, as Dr. Kalsi notes.

We are pleased that Dr. Shackshaft's experience in his choice of model 3 over model 2 confirms that which we recommend. Dr. Shackshaft has been instrumental in the pursuit by CEGB of large scale stability testing. We have no direct experience similar to those tests reported by Dr. Shackshaft; i.e., 50% var rejection, and short circuits from 50% and 100% initial voltage, on large generators. The extent and nature of the various test confirmations we have of Solid Iron Rotor Model are indicated in the latter part of a companion paper, reference 2, which was given IEEE no. T 72 515-5. We are interested in the implications of saturation on machine leakage reactances that he has reported. The method used for representing saturation effects in the Solid Iron Rotor Model used for Model 4 is shown in Fig. 8 of reference 2. It can be appreciated that this means of including saturation does have an effect upon transient and subtransient reactances and time constants, although they may not be of the order that Dr. Shackshaft reported.

The discussors and we concur in the need to obtain appropriate data for power system studies. A working group of the P.E.S. Committees on Power System Engineering and Rotating Machinery has been recently formed called the "Working Group on Determination of Solid Rotor Synchronous Machine Stability Study Constants". Its goal is to review the determination of reactances and time constants (or inductances and resistances) and saturation data of generators and, ultimately, to prepare standards for determining the data.

REFERENCES

[11] R. P. Schulz, "Synchronous Machine Modeling", IEEE paper C 72 616-1, a talk prepared for the Symposium "Adequacy and Philosophy of Modeling: System Dynamic Performance" San Francisco, July 9-14, 1972.

[12] J. L. Dineley, and A. J. Morris "Synchronous Generator Transient Control: Part I – Theory and Evaluation of Alternative Mathematical Models" IEEE PICA 1971 Conference, pp. 182-190.

GENERATOR MODELS ESTABLISHED BY FREQUENCY RESPONSE TESTS ON A 555 MVA MACHINE

Gerald Manchur, David C. Lee, M. E. Coultes, J. D. A. Griffin, Wilfred Watson
The Hydro-Electric Power Commission of Ontario
Toronto, Canada

Abstract–An accurate model of a turbogenerator is required for excitation and transient stability studies. Frequency response tests have been made on a 555 MVA unit to determine direct and quadrature axis transfer functions under operating conditions. Digital and analog computations were performed on line in the course of system tests. The performance of the model derived from the test results compares satisfactorily with the behaviour observed in service.

INTRODUCTION

The development of accurate model representations for turbogenerators has been the subject of study by many investigators[1-14].

Analytic and experimental methods have been applied and saturation effects have sometimes been included. A basic difficulty in the application of the detailed equivalent circuits that have been developed, is that the required generator data is usually either unavailable to the user or unknown. Experimental results, on the other hand, generally apply only to the particular unit tested, which is often quite small and cannot be extrapolated with any certainty to units of different size or construction.

The need for an accurate model representation of a solid rotor turbo alternator arose as a consequence of the problems experienced in commissioning of a static exciter stabilizing signal, utilizing shaft speed as an input, on a 555 MVA thermal unit. Control loop oscillation occurred at a torsional mechanical frequency of the turbogenerator set whenever the stabilizing loop was closed and parameters adjusted to yield the required damping of generator power oscillations. (This problem and its solution are discussed in another paper[17].) To investigate this mode of instability, simulation was attempted using a linearized generator model based on Park's equations[16], with one amortisseur circuit represented in the direct and quadrature axes. It was found, however, that this model did not satisfactorily simulate the observed behaviour at the torsional frequencies, thus making it necessary to improve the representation before the investigation could be continued. It was decided that the data should be obtained from tests on the actual generator, under conditions closely approximating those in service. It was also evident that such data would be extremely useful in other areas of investigation such as synchronous machine stability and application of optimal control.

Paper T 72 136-5, recommended and approved by the Rotating Machinery Committee of the IEEE Power Engineering Society for presentation at the IEEE Winter Meeting, New York, N.Y., January 30-February 4, 1972. Manuscript submitted September 16, 1971; made available for printing December 13, 1971.

APPLICATION OF FREQUENCY RESPONSE TESTS

Frequency response measurements made by introducing small sinusoidal perturbations about a given operating condition are ideally suited for tests under realistic conditions on a power system. A static exciter with its low time constant and low control power requirement provides a convenient method of introducing the perturbations. At the higher perturbing frequencies, the input signal may be increased up to the ceiling capabilities of the exciter whilst still only resulting in small oscillations of field current and other quantities in the machine.

Measurement of relative magnitudes and phase of the variations in machine quantities permits derivation of the transfer functions with a high degree of accuracy and the frequency range utilized in the tests defines the minimum area of validity of the results. Saturation and other non-linear effects are accounted for by operating the machine at an appropriate loading during the tests.

BASIC MODEL REPRESENTATION

$$E_{fd} = I_{fd} R_{fd} + \frac{p}{\omega_o} \psi_{fd} \quad (1)$$

$$\psi_{fd} = Z_1(p) I_{fd} + Z_2(p) I_d \quad (2)$$

$$\psi_d = Z_3(p) I_{fd} + Z_4(p) I_d \quad (3)$$

$$\psi_q = Z_5(p) I_q \quad (4)$$

$$E_q = \frac{p}{\omega_o} \psi_q + \frac{\omega}{\omega_o} \psi_d - R_a I_q \quad (5)$$

$$E_d = \frac{p}{\omega_o} \psi_d - \frac{\omega}{\omega_o} \psi_q - R_a I_d \quad (6)$$

Where $Z_1(p)$ to $Z_5(p)$ are operational impedance expressions which include the effects of the rotor iron circuits.

In developing a block diagram from these equations, the assumption is made that the mutual coupling between the direct and quadrature axis circuits due to rotor iron saturation is negligible. Writing equations 1 to 6 in their small perturbation forms and eliminating the field flux linkages from the expressions results in the block diagram shown in Figure 1, where the transfer functions $G_1(p)$ to $G_5(p)$ to be determined by tests are functions of the operational impedances Z_1 to Z_5. Base field current and voltage were chosen to conform to the "common Xad base"[2].

It may be noted that elimination of field current results in a model in terms of the machine operational impedances $X_d(p)$ and $X_q(p)$. The retention of field current, however, provides for better resolution of test data and gives greater insight into the coupling effects between the field circuits and the stator and rotor iron circuits. It also provides essential information for studies which may require field current as a control signal such as in optimal control applications.

THREE TESTS REQUIRED TO OBTAIN TRANSFER FUNCTIONS

For the representation of Figure 1, three tests are required to determine the transfer function blocks $G_1(p)$ to $G_5(p)$. These are, letting $p = j\omega_s$:

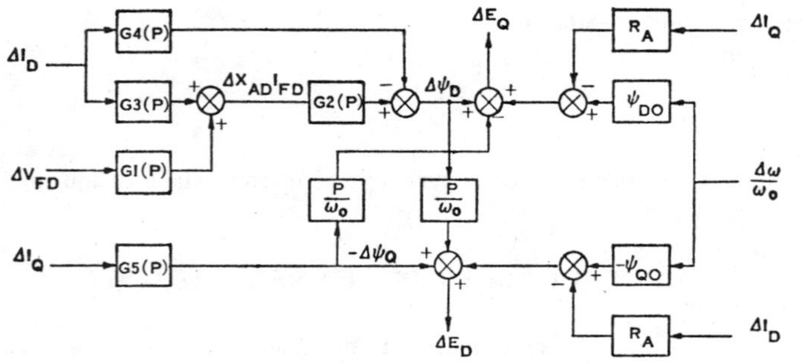

Fig. 1. Generalized Small Signal Model of Synchronous Machine

Test 1: Generator off line at rated speed and voltage.

Measure $\Delta E_q, \Delta\omega, \Delta I_{fd}, \Delta V_{fd}$

$$\Delta\psi d = \Delta E_q - E_{qo}\frac{\Delta\omega}{\omega_o} \tag{7}$$

$$G_1(j\omega_s) = \frac{\Delta X_{ad}\, I_{fd}}{\Delta V_{fd}} \tag{8}$$

$$G_2(j\omega_s) = \frac{\Delta\psi_d}{\Delta X_{ad}\, I_{fd}} \tag{9}$$

Test 2: Generator on line at no load, rated speed and voltage.

Measure $\Delta E_q, \Delta E_d, \Delta\omega, \Delta I_d, \Delta I_{fd}, \Delta V_{fd}$,

$$\Delta\psi_d = \frac{\Delta E_q + j\frac{\omega_s}{\omega_o}\Delta E_d - E_{qo}\frac{\Delta\omega}{\omega_o} + j\frac{\omega_s}{\omega_o}R_a\Delta I_d + R_a\Delta I_q}{1-\left(\frac{\omega_s}{\omega_o}\right)^2} \tag{10}$$

$$G_3(j\omega_s) = \frac{\Delta X_{ad}\, I_{fd} - G_1(j\omega_s)\,\Delta V_{fd}}{\Delta I_d} \tag{11}$$

$$G_4(j\omega_s) = \frac{G_2(j\omega_s)\,\Delta X_{ad}\, I_{fd} - \Delta\psi d}{\Delta I_d} \tag{12}$$

Test 3: Generator on line at full load

Measure $\Delta E_d, \Delta E_q, \Delta\omega, \Delta I_d, \Delta I_q$,

$$\Delta\psi_q = \frac{-\Delta E_d + j\frac{\omega_s}{\omega_o}\Delta E_q + \left(E_{do} + j\frac{\omega_s}{\omega_o}E_{qo}\right)\frac{\Delta\omega}{\omega_o} + j\frac{\omega_s}{\omega_o}R_a\Delta I_q - R_a\Delta I_d}{1-\left(\frac{\omega_s}{\omega_o}\right)^2} \tag{13}$$

$$G_5(j\omega_s) = \frac{-\Delta\psi_q}{\Delta I_q} \tag{14}$$

It may be noted that transfer function blocks $G_3(j\omega_s)$ and $G_4(j\omega_s)$ are not determined independently by these test methods, as they involve $G_1(j\omega_s)$ and $G_2(j\omega_s)$ in their calculation. For this reason it is necessary to perform tests 1 and 2 under similar generator operating conditions so that magnetic conditions in the rotor and stator remain the same for the two tests.

INSTRUMENTATION TECHNIQUES

The instrumentation setup is shown pictorially in Figure 2 and in block diagram form in Figure 3. Much of the signal conditioning was performed in an analog computer and additional operational amplifier circuits. A diagram of these signal processing circuits is shown in Figure 4.

A gear wheel on the end of the generator shaft, magnetic pickups and a frequency transducer were used to obtain a signal proportional to changes in machine speed ($\Delta\omega$). Power and reactive power were transduced by conventional Hall effect transducer circuits and terminal voltage was derived from a three phase rectifier bridge circuit. Field voltage and current were obtained from existing transducers.

The exciting signal ($E_s \sin\omega_s t$) was generated by the analog computer and fed as an input to the voltage regulator.

In order to derive direct and quadrature axis voltages and currents, a set of three-phase voltages with fixed phase relationship to generator rotor position was provided from an auxiliary generator mounted on the shaft. This set of three phase voltages was used to generate two more sets of three phase voltages, one in phase with the quadrature axis of the generator and one in phase with the direct axis. In addition a reference voltage magnitude signal was produced. These signals were then combined with terminal voltages and currents to produce direct and quadrature axis voltages and currents. Steady state components of these signals were biased off to allow the amplification of the change components. The steady state components were measured, to establish initial conditions and for scaling, while the change components were output to recording equipment and through an analog-to-digital converter to a small digital computer.

The digital computer* with its associated analog to digital converter** was set up to take 200 samples of each of 7 analog machine quantities and the reference signals $E_s \sin\omega_s t$ and $E_s \cos\omega_s t$ (see Figure 3). The time interval between each set of 9 samples was determined by an external oscillator connected to the computer's interrupt bus, and was adjusted in each case such that the 200 sets took an integral

Fig. 2. Test Instrumentation on Site

* Digital Equipment Corporation PDP8 (4096 12-bit words; 1.5 microsecond cycle time).

**35 microsecond conversion time, 12-bit precision; 10 channel analog multiplexer at input.

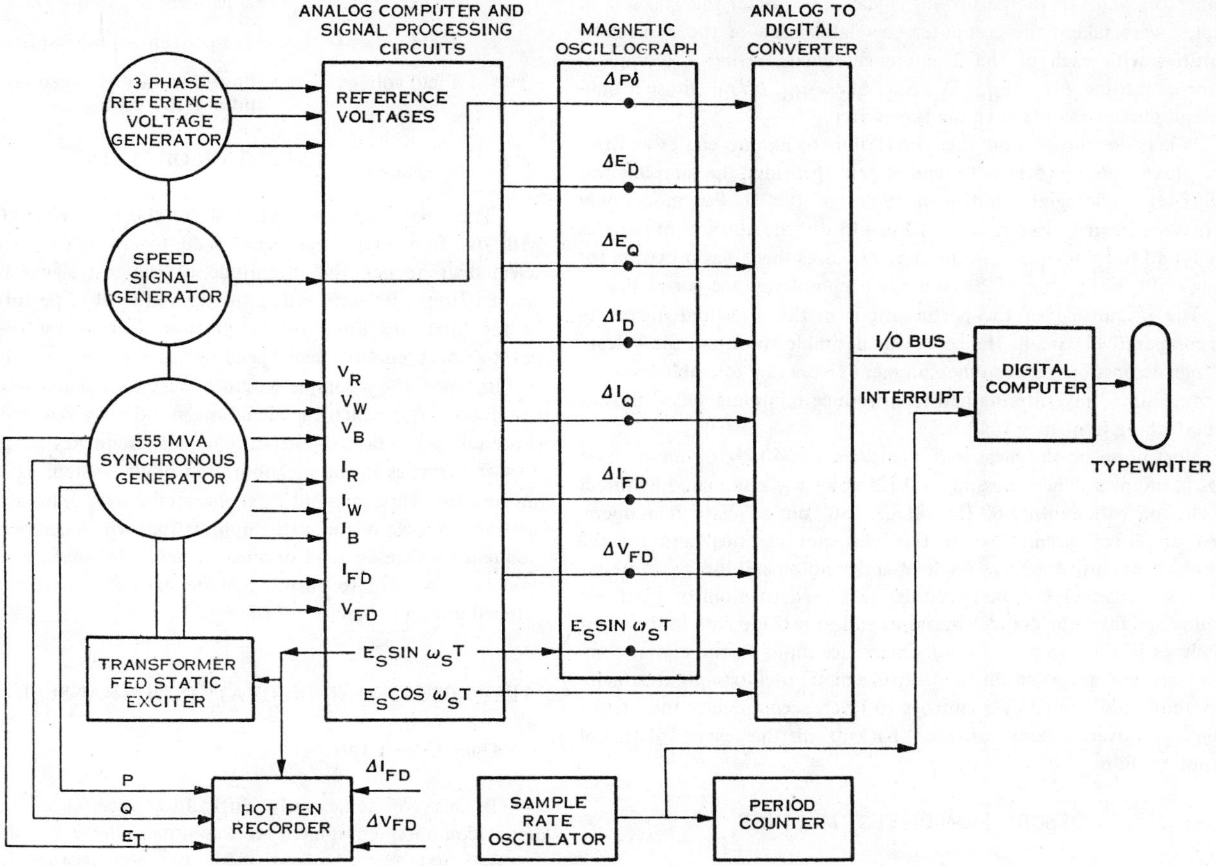

Fig. 3. Flow Diagram of Analog and Digital Recording Systems

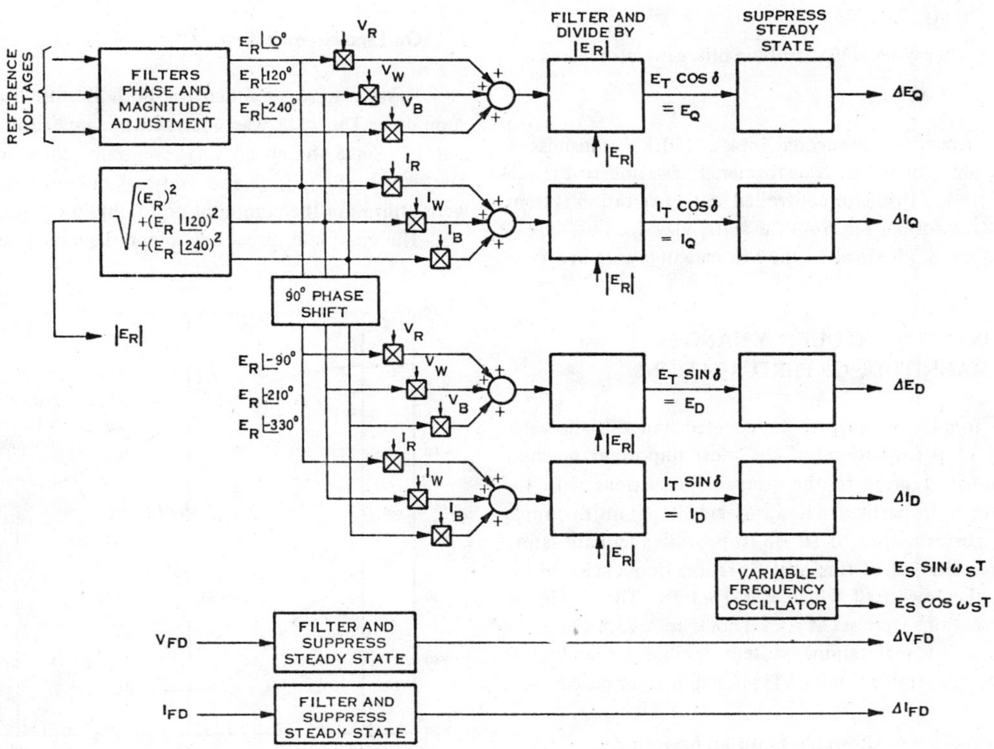

Fig. 4. Signal Generation and Processing Circuits

number of cycles of the perturbing signal "ω_s". After the 200 sets of samples were taken, the computer correlated each of the 7 machine quantities with each of the 2 reference signals, giving 7 (complex) phasor quantities, $\Delta\omega$, ΔE_d, ΔE_q, ΔI_d, ΔI_q, ΔI_{fd}, ΔV_{fd}, whose magnitude and phase was output to the typewriter.

It may be shown that this correlation technique causes additive noise, distortion and offsets to tend to zero (provided the sampling was made over an integral number of cycles of "ω_s"). For moderately severe cases (Signal/noise ratio = −10 to −15 db) the 200 sets of samples were found to be adequate; for more severe cases there was provision for taking additional groups of 200 sets and accumulating the correlations.

The magnitude of the perturbations in the measured quantities was comparatively small. Hence, it was desirable to obtain the "cleanest" signals possible to keep measurement errors to a tolerable level. At the same time, any filtering had to introduce minimum phase shift at the perturbing frequency (ω_s).

Most noise on the measured quantities was 60 Hz or one or more of its harmonics. Where possible, 60 Hz noise was eliminated by direct cancellation with another 60 Hz voltage, 180° out of phase in an operational amplifier circuit. When this was not practical and for the harmonics, extensive use was made of active notch and low pass filters.

A six channel hot pen recorder was used to monitor pertinent machine variables for control purposes and to provide data in the event of failures in the computer setup. Quantities input to the digital computer were also recorded on an electromagnetic oscillograph.This facilitated amplitude scaling of quantities to levels acceptable to the analog-to-digital converter and provided backup in the event of digital computer failure.

DESCRIPTION OF TEST MACHINE

The test machine was a 3600 rpm tendem compound solid rotor turbo alternator with water cooled stator, hydrogen cooled rotor and having the following ratings and characteristics.

MVA = 555.55

kV = 24

pF = 0.90

H = 3.5 kW-sec/kVA (turbine plus generator)

Xdu = 2.0 pu

The machine output was connected into a 230 kV transmission network through a unit connected transformer. Excitation to the machine was supplied from a thyristor controlled static excitation system, with the excitation transformer fed from the station service bus.

Two additional units, identical to the test machine were in service during the tests.

LIMITS ON FREQUENCY RANGE AND MAGNITUDE OF PERTURBATIONS

Limits on the frequency range to be covered and the allowable levels of amplitude of perturbations of the most important machine variables were established prior to the tests in discussions with the generator manufacturer. In particular it was decided to limit the upper frequency range of perturbations to 10 Hz to provide adequate safety margin for possible excitation of torsional vibration frequencies of the turbogenerator set, the lowest of which was 16.3 Hz. The maximum amplitudes of perturbations were set at a level consistent with obtaining measurable signals whilst maintaining system oscillations and shaft torques well below acceptable limits. Typical limits imposed were as follows:

Quantity	Allowable Limit on Magnitude of Perturbation
Terminal voltage	±2 per cent of rated
Power	±5 per cent of rated
Field current	±2.5 per cent of full load current
Field voltage	allowed to vary between positive and negative ceiling

EXECUTION OF TESTS

The frequency range selected for the tests was 0.005 to 10 Hz, with five frequency steps per decade, logarithmically spaced. At each selected frequency the magnitude of the test signal was slowly increased from zero until either the desired level of perturbations or one of the limits discussed in the previous section was reached. At this point a short length of high speed record was obtained on the recording oscillographs and a sample length of test data channeled to the digital computer, after which the signal magnitude was reduced to zero. This procedure was repeated until the complete frequency range was covered. Test frequencies in the neighborhood of the system natural frequency and the first torsional and its subharmonic were approached with extra caution, because of the high amplifications which can occur at resonant frequencies. One channel of sensitive generator speed measurement was used as a monitor to ensure that no excessive shaft torques were developed during the course of the tests.

TEST RESULTS AND DERIVATION OF TRANSFER FUNCTIONS

1. Open Circuit Test

Because of some early difficulties experienced in aligning the digital computer equipment, it was not available for this test and all necessary data was obtained on the hot pen recorder. Since there are no changes in machine speed or angle during this test:

$$\Delta\psi_d = \Delta E_q = \Delta E_t$$

Calculation of $G_1(j\omega_s)$ and $G_2(j\omega_s)$ follows directly from equations 8 and 9. Test data and the corresponding derived transfer functions are shown in Figures 5 and 6.

2. On Line No-load Test

Results were obtained on both the oscillograph and the digital computer. The data was analyzed by application of equations 10, 11 and 12. Since the speed was essentially constant, $\Delta\omega$ was negligible. The values of $G_1(j\omega_s)$ and $G_2(j\omega_s)$ obtained from the preceding test were utilized in the computation to obtain $G_3(j\omega_s)$ and $G_4(j\omega_s)$.

The complete direct axis block diagram thus obtained is shown in

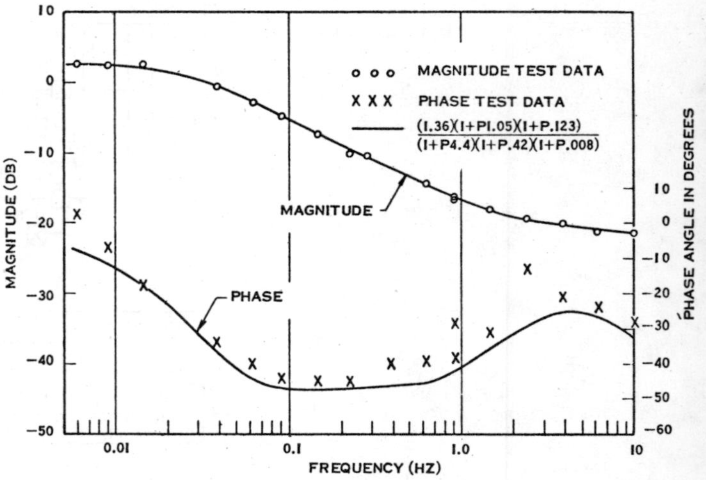

Fig. 5. Bode Diagram for Transfer Function $G_1(j\omega_s)$

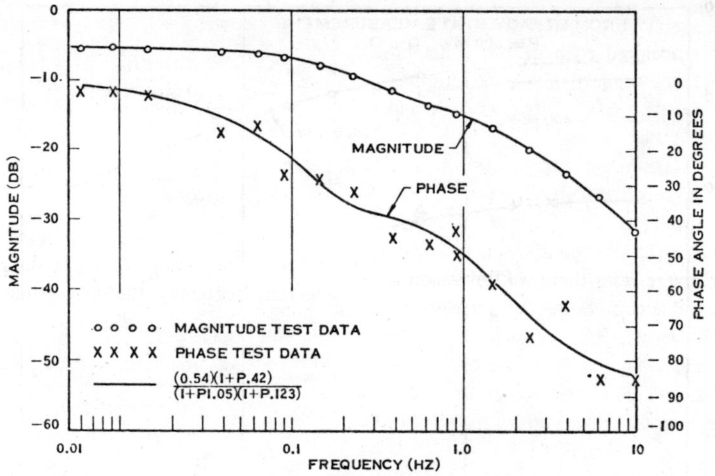

Fig. 6. Bode Diagram for Transfer Function $G_2(j\omega_s)$

Figure 7. Also shown in this figure is a reduced model with ΔI_{fd} eliminated and the quadrature axis voltage ΔE_q is seen to be a function of field voltage, and direct axis current ΔI_d, multiplied by direct axis operational impedance $X_d(p)$.

3. Full Load Test

All the results on this test were obtained from the on-line digital computer. As the magnitude of the perturbations fell off at the higher frequencies, the sinusoidal variations on the magnetic oscillograph were almost entirely obscured by noise. However, the correlation techniques employed on the digital data produced satisfactory and consistent results at frequencies up to 10 Hz in the direct axis and up to about 2.5 Hz in the quadrature axis.

Due to failure of a pulverizer motor during this test the first part of the test run (high frequencies) was made at a load of 500 mW and the remainder at 400 mW.

A plot of the transfer function $G_5(j\omega_s) = [X_q(p)]$ is shown in Figure 8. The actual transfer function has not been evaluated owing to the lack of data at the higher frequencies.

Torque/Exciter Control Input Relationship

The hot pen recordings on the full load test were analyzed to provide an alternative model involving a direct relationship between electrical torque (or power) output and exciter control signal input. This was the most useful relationship for the study of problems of application of exciter stabilizing signals[17]. This relationship is shown in the lowest block of Figure 9 which is a reduced model of the generator for full-load operation. The remainder of the blocks in this diagram were derived from the known inertia of the machine and external system impedances.

DISCUSSION OF RESULTS

Results from the tests indicated that determination of direct axis transfer functions of a large synchronous machine by frequency response methods of the type described in this paper is feasible. With due attention paid to instrumentation and recording techniques, good signal to noise ratios were maintained in the measured variables over the entire

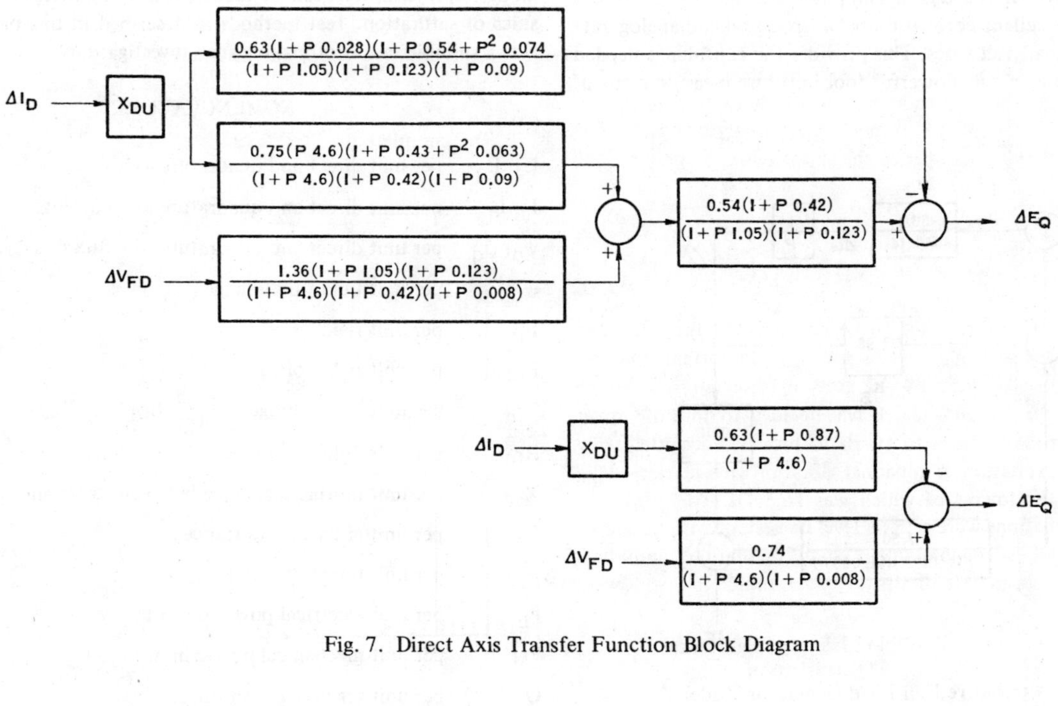

Fig. 7. Direct Axis Transfer Function Block Diagram

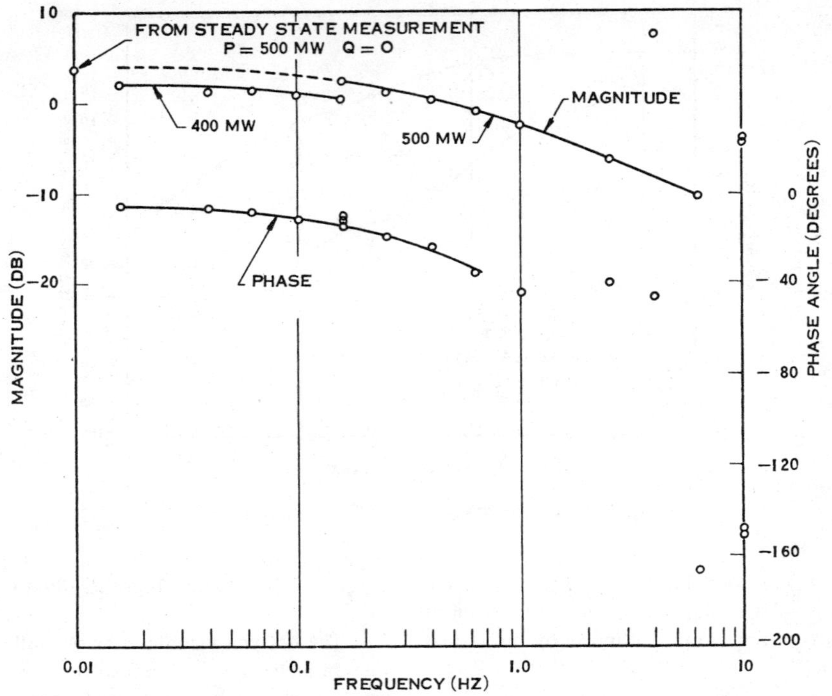

Fig. 8. Quadrature Axis Transfer Function $G_5(j\omega_s)$

test frequency range. Data at higher test frequencies could readily have been obtained.

Quadrature axis data, on the other hand, was difficult to obtain by these test methods. Analysis revealed that there was a rapid drop off in the magnitude of the quadrature axis variables with frequency, due to smaller angle changes at the higher frequencies. Since the field perturbations are all in the direct axis, quadrature axis quantities can only change through changes in angle. An alternative frequency response test method has since been formulated to obtain quadrature axis (and direct axis) data which involves perturbations on the generator terminal axis, rather than on the direct axis. This may be accomplished by using a second machine to provide the perturbations and it is planned to verify this with further tests in the near future.

The performance of the digital computer during these tests was very encouraging. Excellent correlation between digital and analog data was obtained in virtually all cases. This provides the confidence needed to make increased use of this powerful tool in future measurements of this and similar nature.

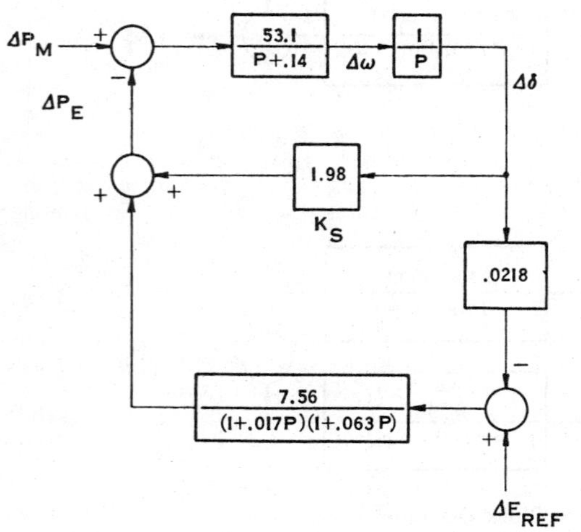

Fig. 9. Alternative Full Load Generator Model

CONCLUSIONS

Frequency response test methods provide an excellent means of deriving small signal generator models under normal service conditions. Although considerable effort was expended in development of instrumentation techniques, in performing the tests and in analysing the results, the excellent correlations which have been obtained between the derived models and field tests provide some justification for this effort. Both the DQ and alternate models have been used in various studies to date, in addition to their use in designing supplementary speed stabilizing systems for these units.

Saturation was found to be significant in both the D and Q axis under normal service conditions. Further improvements in generator model representations can be made possible by investigating the mechanics of saturation. Test methods as described in this paper could provide a suitable means of making such investigations.

NOMENCLATURE

E_d, E_q	per unit direct and quadrature axis voltages
I_d, I_q	per unit direct and quadrature axis currents
ψ_d, ψ_q	per unit direct and quadrature axis flux linkages
ψ_{fd}	per unit field flux linkage
I_{fd}	per unit field current
E_{fd}	per unit field voltage
V_{fd}	per unit field voltage $= \frac{X_{ad}}{R_{fd}} E_{fd}$
R_{fd}	per unit field resistance
X_{ad}	per unit mutual reactance between stator and field
R_a	per unit armature resistance
E_t	per unit terminal voltage
P_E	per unit electrical power output
P_M	per unit mechanical power input
Q	per unit reactive power output

ω_o rated generator frequency in rad/sec

ω instantaneous generator frequency in rad/sec

ω_s perturbation frequency in rad/sec

f_s perturbation frequency in Hz

δ instantaneous rotor angle relative to vector rotating at synchronous speed, in rad

Δ small perturbation about a mean value

p differential operator $\frac{d}{dt}$

Subscript o signifies steady state value of variable at operating point.

REFERENCES

[1] A. W. Rankin, "Per Unit Impedances of Synchronous Machines", AIEE Trans Vol 64, pp 569-573, August 1945.

[2] A. W. Rankin, "Per Unit Impedances of Synchronous Machines", Ibid, pp 839-841.

[3] A. W. Rankin, "The Direct and Quadrature Axis Equivalent Circuits of Synchronous Machines", Ibid, pp 861-868.

[4] C. Concordia, A. J. Wood, "An Analysis of Solid Rotor Machines", Part I, AIEE Trans Vol 78, pp 1657-1665, February 1960.

[5] C. Concordia, "Synchronous Machine With Solid Cylindrical Rotor – Part II", Ibid, pp 1651-1657.

[6] W. B. Jackson, R. L. Winchester, "Direct and Quadrature – Axis Equivalent Circuits for Solid-Rotor Turbine Generators", IEEE Trans Vol PAS-88, pp 1121-1136, July 1969.

[7] I. M. Canay, "Causes of Discrepancies on Calculation of Rotor Quantities and Exact Equivalent Diagrams of the Synchronous Machine", Ibid, pp 1114-1120.

[8] G. R. Slemon, "Analytic Models for Saturated Synchronous Machines", Ibid, Vol PAS-90, pp 409-417, March 1971.

[9] B. Adkins, P. Bharali, "Operational Impedances of Turbo-Generators With Solid Rotors", IEEE Proc Vol 110, pp 2185-2199, 1963.

[10] S.K. Sen, B. Adkins, "Application of Frequency Response Method to Electrical Machines", Ibid, Vol 103C, pp 378-391, 1956.

[11] J. D. Edwards, "New Method of Measuring 2-Axis Quantities", Ibid, Vol 115, pp 1493, 1968.

[12] H. Kaminosono, K. Uyeda, "New Measurement of Synchronous Machine Quantities", IEEE Trans Vol PAS-87, pp 1908-1918, November 1968.

[13] Y. Y. Kazovskii, G. G. Rogozin, "Test Determination of the Response Characteristics of Large Turbo-Generators", Electric Technology, USSR, Vol 1-4, 1963.

[14] M. P. Kostenko, et al, "Large Modern Highly Utilized Turbine and Waterwheel Generators, Their Cooling Systems, Characteristics and Parameters", CIGRE, Vol 11, Paper 129, 1964.

[15] Test Procedure for Synchronous Machines, IEEE Publication 115, 1965.

[16] C. Concordia, Synchronous Machines, New York, Wiley, 1951.

[17] W. Watson, M. E. Coultes, "Mechanical Problems Caused by Static Exciter Stabilizing Signals on Large Generators". To be presented at 1972 IEEE Winter Power Meeting.

Discussion

A. M. El-Serafi and **H. S. Khalil** (University of Saskatchewan, Saskatoon, Sask., Canada): We would like to compliment the authors on this paper which deals with discovering practical transfer functions of a large synchronous machine under actual operating conditions. In our discussion we would like to emphasize that the authors have also used Park's transformation in their approach. However, the representation of the damper winding and the effect of the solid rotor in their measured transfer functions is more accurate than that obtained by using two equivalent d- and q-axis circuits. It has been also mentioned in this paper that a model using the two equivalent d- and q-axis circuits did not satisfactorily simulate the behavior of the machine at torsional frequencies. In our opinion, the measured transfer functions in this paper are still suffering from the same disadvantage since the range of the frequency used in the tests was below the torsional one. Moreover, the range of the frequency used in deriving the transfer functions is small. We believe that this range is not enough to give accurate informations about the transfer functions under investigation, especially the quadrature-axis one G5. This is due to the fact that the effect of the damper winding is usually more pronounced behind the range of the frequency used. We think that the expressions of the transfer functions may be greatly simplified if the terms of the speed changes in these functions will be neglected. This is usually an acceptable assumption. As a last point, we would like to emphasize that the transfer functions derived for the machine under investigation are valid only for small disturbances around the operating point, and consequently it is not suitable to be used in the cases of large disturbances, e.g., in transient stability studies.

Manuscript received February 22, 1972.

Richard P. Schulz, Colin E. J. Bowler, and **William D. Jones** (General Electric Company, Schenectady, N.Y. 12345): We are pleased that the authors have taken the initiative to plan, perform and report these tests. We feel that tests made under load by frequency response techniques like these will provide new useful information to check, complement and extend synchronous machine analysis. These tests have provided us with another verification of the calculations of impedances and transfer functions made with our solid rotor equivalent circuit technique. They have also provided a check of calculating effects of saturation. We would like to encourage other utilities to undertake similar tests and would welcome the opportunity to work with them.

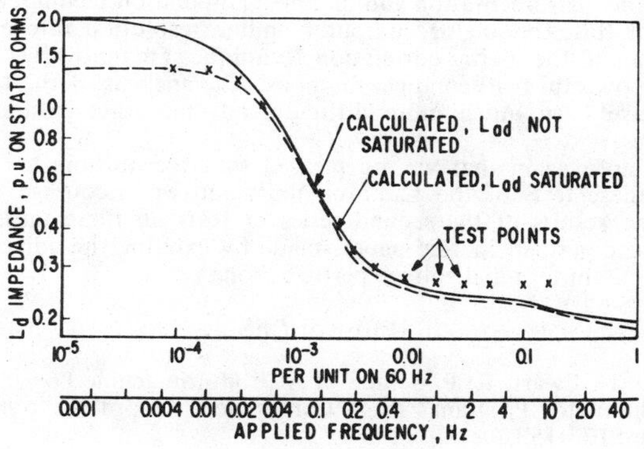

We worked with the authors on pretest calculations and witnessed the execution of the tests. As our part of the cooperative venture we provided the Hydro Electric Power Commission of Ontario calculated values of the operational axis impedances $Z_d(s)$, $Z_q(s)$, and of the transfer function between field voltage and d-axis flux, G(s). The impedance calculations and calculations of the expected heating induced in the rotor from the harmonic currents were made using the solid rotor equivalent circuit program cited as the author's reference 6. After the test results were reduced by the authors, there were discrepancies between the calculated and measured values of the operational impedances. On reviewing the test conditions, we concluded that since the initial calculations had been made using unsaturated mutual reactances, L_{ad} and L_{aq}, the principal difference was due to saturation. We recalculated the impedance using modified values of the mutual inductances L_{ad} and L_{aq}; the modifications were made by multiplying the inductances by the incremental permeability or slope in the open circuit saturation curve corresponding to the total air gap flux linkages, $\sqrt{\psi_{ad}^2 + \psi_{aq}^2}$.

This method of accounting for the effects of saturation is the same that we use for round rotor machine simulations. (A1) The operation impedances calculated with the saturated values of mutual inductance match the test values very well. This correlation is shown in fig. 1 for the magnitude of $Z_d(s)$; similar good correlation was found for the phase and magnitudes of $Z_d(s)$, and G(s). The solid line is the value of $Z_d(s)$ calculated using the unsaturated value of L_{ad}; the dashed line is calculated using the saturated value; and the points are the test values. The abscissa of the plot is the same as the authors', per unit slip frequency. The ordinate is stator impedance expressed on a logarithmic scale in per unit ohms on the rated MVA, kV stator impedance base; we prefer this to the equivalent scale expressing per unit ohms in decibels. We are heartened by the test confirmation of the calculated transfer relations and would like to have further tests to obtain a broad verification.

The authors state that generator data is unavailable for their equivalent circuits. For the most popular generator model having four rotor circuits (field d-axis amortisseur and two q-axis amortisseurs) data is generally available from the manufacturer and is usually given to the

Manuscript received February 17, 1972.

utility for stability studies. We have found this modeling to be satisfactory for studying effects such as torsional stability.

We are somewhat puzzled by the reduction of the test data to operational expressions of the transfer functions. In particular, we note that the numerator terms of the G3 and G4 terms of the d-axis block diagram in fig. 7 are oscillatory; we would expect that these terms would have only real poles and zeros. Also, it would seem that the pair of expressions G1 and G2 should have between them one fewer numerator terms than denominator terms so that G(s) would have a numerator order one less than its denominator order. These expectations came from the reduction of the equations of the machine to transfer functions, as in references A2 and A3.

The reduction of the more complete d-axis block diagram of fig. 7 to the simpler diagram does not seem to us to be straight forward. Would the authors please describe their method and assumptions in the reduction?

We are also puzzled by the intent of the authors to use test data taken from very low frequencies up to 10 hertz in the study of stabilizer effects on torsional vibration resonances, the lowest of which occurs at 16 hz. This frequency region, 10-20 hertz, includes the impedance magnitude and phase changes associated with the transient-to-subtransient reactance range. We agree with the authors that the rotor iron effects in this range may modify the effect of the excitation system on machine dynamics; both we and they include these effects in eigenvalue analyses of these problems. Would the authors explain how the test data are used in their stability analyses?

The authors should be particularly commended for their development of the instrumentation and on-line manipulation techniques. Both aspects of this, the analog computer on-line resolution into d, q axis quantities and the digital correlation techniques are innovative applications of powerful mathematical methods. The analysis of the test data would have been much more difficult and inaccurate without these techniques.

We note again that we are pleased that the authors have documented these tests of the Lambton units and we encourage them to report the results of the second series of tests on these units which they have described in this paper, made by exciting the adjacent machine to obtain terminal voltage perturbations.

REFERENCES

[A1] D. N. Ewart, R. P. Schulz "FACE Multimachine Power System Simulator Program" PICA Conference, Record No. 69C1-PWR pp. 133-153.

[A2] C. Concordia "Synchronous Machines" Wiley & Sons, 1951, pp. 54-57.

[A3] B. Adkins "The General Theory of Electrical Machines", London: Chapman and Hall, 1964, pp. 112-124.

Gerald Manchur, David C. Lee, M. E. Coultes, J. D. A. Griffin, and **Wilfred Watson**: We thank the discussers for their interest in the paper and will attempt to answer some of the questions they have posed.

Manuscript received April 25, 1972.

Both discussions bring up the question of the validity of extrapolating the results of the frequency response tests for studies at frequencies which were beyond the range of the maximum test frequency, i.e., the torsional frequencies. The validity of this approach is justified by the following reasoning. The upper frequency of prime interest was the first torsional which occurred at 16 Hz or at 1.6 times the maximum test frequency. Inspection of Figure 1 of the discussion by Messrs. Schulz et al shows the calculated value of Xd(s) changes very little over this narrow frequency range of 10 to 16 Hz. This might be expected from physical considerations as well, since this region reflects the impedance characteristics of the solid iron rotor and it would be difficult to visualize this impedance exhibiting any marked discontinuities over a narrow range of frequency.

A further justification is that the extrapolated data for both the DQ model and the alternative model, when used to study the excitation of torsional oscillations in the turboalternator set, produced excellent correlations with field test results. This provided confidence that the data was at least valid up to the first torsional mode. Since higher torsional modes were not observed in field measurements, the adequacy of the models cannot be substantiated for the higher torsionals.

Messrs. Schulz et al note that generator data is generally available from the manufacturer in considerable detail for use in stability studies. When applied to turboalternators, it is important to consider how this data is derived. Measured or calculated data derived from frequency response curves under normal operating conditions, when transformed into appropriate equivalent circuits, can be expected to provide accurate results. Inaccuracies may arise when obtaining stability data from test code definitions or from considering unsaturated machine conditions.

Messrs. Schulz et al raise some points as to the form and reduction of the detailed D-axis model. It is quite possible that the numerator of G3(p) should not have an oscillatory term. The amount by which it is oscillatory, however, is very small and is no doubt due to small errors in the data used in the determination of this block. It should be remembered that G3(p) and G4(p) require G1(p) and G2(p) for their determination, which were obtained from a different test under slightly different initial operating conditions. The reduced model of the D-axis is not a reduction of the detailed model but is derived independently from the measurements. Algebraic reduction of the detailed model shows, however, that the two are essentially equivalent. The presence of a numerator term in G(p), if smaller than its companion denominator term, would be outside the range of detection at the highest test frequency of 10 Hz.

We concur with Messrs. El-Serafi and Khalil that the upper range of test frequency of 10 Hz is not sufficient to completely describe rotor iron effects in a turboalternator. As indicated in the paper, this choice was dictated by the desire to stay clear of the torsional resonant frequencies. It was also indicated that this method of testing presents some limitations in obtaining quadrature axis data at the higher frequencies. As this effort was to a large extent an exploratory one, these deficiencies are as gratifying as they are disappointing since they suggest the necessary direction in which improvements should be made in the test methods and procedures for obtaining synchronous machine data.

The extension of such small signal data to large signal disturbances cannot readily be made until more information is known about the mechanics of rotor iron saturation and how these change with the machine operating conditions. It is hoped that field data such as this will supplement theoretical considerations to bring such knowledge to the forefront in shorter time.

EXPERIMENTAL DETERMINATION OF EXACT EQUIVALENT CIRCUIT PARAMETERS OF SYNCHRONOUS MACHINES

Yao-nan Yu
Senior Member, IEEE

H.A.M. Moussa
Student Member, IEEE

The University of British Columbia

Abstract—For stability studies of large power systems, accurate representation of the synchronous machine is required. As pointed out by Canay[1], the conventional equivalent circuits for synchronous machines do not give accurate computed field voltage and current values. He suggested several circuits and showed good agreement between his test and calculated results. His circuit parameters were calculated from design.

Questions arise: How to determine accurate circuit parameters from simple field tests and how to choose the equivalent circuits. The circuits are not unique because of different base volts, base amperes and circuit elements.

This paper reports a systematic procedure to determine these circuit parameters from simple field tests. In addition to IEEE test code, an extra test will be needed. To exhibit the meaning of the exact equivalent circuit parameters it is also necessary to re-derive the circuits.

d-AXIS EXACT EQUIVALENT CIRCUITS

Applying Park's transformation, the d-axis voltage equations of a synchronous machine in MKS unit can be written in the form

$$\begin{bmatrix} V_d - U_d \\ V_F \\ 0 \end{bmatrix} = \left[\begin{bmatrix} R_a & & \\ & R_F & \\ & & R_D \end{bmatrix} + p \begin{bmatrix} X_d & X_{aF} & X_{aD} \\ \frac{3}{2}X_{Fa} & X_F & X_{FD} \\ \frac{3}{2}X_{Da} & X_{DF} & X_D \end{bmatrix} \right] \begin{bmatrix} I_d \\ I_F \\ I_D \end{bmatrix} \tag{1}$$

The X-matrix is not symmetric[2]. Here all X's are reactances of single-phase excitation except X_d which is of three-phase excitation. The numerical coefficient 3/2 and hence the asymmetry of the matrix originally come from the a,b and c three-phase excitation on the stator side and the F or D single-phase excitation on the rotor side.

The matrix form itself suggests that the per unit reactances must, and voltages and currents may be defined as follows

$$\begin{aligned}
x_{dF} &= X_{aF}\frac{I_{FB}}{V_n} & x_{dD} &= X_{aD}\frac{I_{DB}}{V_n} & x_{FD} &= X_{FD}\frac{I_{DB}}{V_{FB}} \\
x_{Fd} &= (\tfrac{3}{2}X_{Fa})\frac{I_n}{V_{FB}} & x_{Dd} &= (\tfrac{3}{2}X_{Da})\frac{I_n}{V_{DB}} & x_{DF} &= X_{DF}\frac{I_{FB}}{V_{DB}} \\
x_d &= X_d\frac{I_n}{V_n} & x_F &= X_F\frac{I_{FB}}{V_{FB}} & x_D &= X_D\frac{I_{DB}}{V_{DB}} \\
r_a &= R_a\frac{I_n}{V_n} & r_F &= R_F\frac{I_{FB}}{V_{FB}} & r_D &= R_D\frac{I_{DB}}{V_{DB}} \\
i_d &= I_d/I_n & i_F &= I_F/I_{FB} & i_D &= I_D/I_{DB} \\
v_d &= V_d/V_n & v_F &= V_F/V_{FB} & &
\end{aligned} \tag{2}$$

The matrix of x's is not necessarily reciprocal. To make it reciprocal[3], the following constraints must be imposed,

$$\tfrac{3}{2}V_n\, I_n = V_{FB}\, I_{FB} = V_{DB}\, I_{DB} \tag{3}$$

Paper 71 TP 63-PWR, recommended and approved by the Rotating Machinery Committee of the IEEE Power Engineering Society for presentation at the IEEE Winter Power Meeting, New York, N. Y., January 31–February 5, 1971. Manuscript submitted September 8, 1970; made available for printing November 16, 1970.

resulting in

$$\begin{aligned}
& x_{dF} = x_{Fd} = \frac{X_{aF}}{Z_n}\left(\frac{I_{FB}}{I_n}\right), \quad x_{dD} = x_{Dd} = \frac{X_{aD}}{Z_n}\left(\frac{I_{DB}}{I_n}\right) \\
& x_{FD} = x_{DF} = \frac{2}{3}\frac{X_{FD}}{Z_n}\left(\frac{I_{FB}}{I_n}\right)\left(\frac{I_{DB}}{I_n}\right) \\
& x_d = \frac{X_d}{Z_n}, \quad x_F = \frac{2}{3}\frac{X_F}{Z_n}\left(\frac{I_{FB}}{I_n}\right)^2, \quad x_D = \frac{2}{3}\frac{X_D}{Z_n}\left(\frac{I_{DB}}{I_n}\right)^2 \\
& r_a = \frac{R_a}{Z_n}, \quad r_F = \frac{2}{3}\frac{R_F}{Z_n}\left(\frac{I_{FB}}{I_n}\right)^2, \quad r_D = \frac{2}{3}\frac{R_D}{Z_n}\left(\frac{I_{DB}}{I_n}\right)^2 \\
& Z_n = V_n/I_n
\end{aligned} \tag{4}$$

The d-axis voltage equations can be written now in per unit as

$$\begin{bmatrix} v_d - u_d \\ v_F \\ 0 \end{bmatrix} = \begin{bmatrix} r_a + px_d & px_{dF} & px_{dD} \\ px_{Fd} & r_F + px_F & px_{FD} \\ px_{Dd} & px_{DF} & r_D + px_D \end{bmatrix} \begin{bmatrix} i_d \\ i_F \\ i_D \end{bmatrix} \tag{5}$$

One of the general d-axis equivalent circuit corresponding to (5) is as Fig. 1. The general circuit reduces to Canay's Fig. 5 if one set

$$x_{Fd} = x_{Dd} \tag{6}$$

On the other hand if one set

$$x_{Fd} = x_{Dd} = x_{FD} \tag{7}$$

Fig. 1 reduces to Canay's Fig. 6(b), which is the simplest and closely resembles the conventional equivalent circuit, Fig. 2.

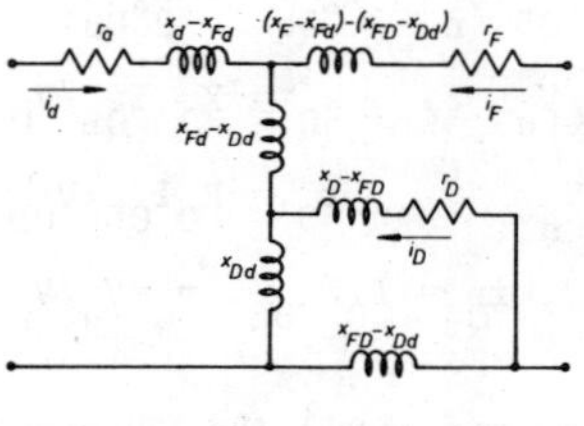

Fig. 1. General d-axis circuits.

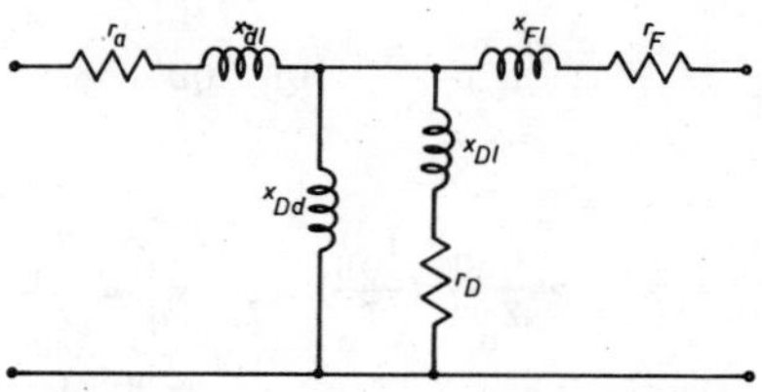

Fig. 2. Simplified d-axis circuit.

Note that $x_{d\ell}$, $x_{F\ell}$ and $x_{D\ell}$ are no more leakage reactances. They are defined as

$$x_{d\ell} = x_d - x_{Fd}, \quad x_{F\ell} = x_F - x_{Fd}, \quad x_{D\ell} = x_D - x_{FD} \tag{8}$$

Reprinted from *IEEE Trans. Power App. Syst.*, Nov./Dec. 1971.

The following information, although not needed in determination of parameters from field tests, are useful in design. The current ratios in (4) determined by (7) are as follows

$$\frac{I_{FB}}{I_n} = \frac{3}{2}\frac{X_{aD}}{X_{FD}} , \quad \frac{I_{DB}}{I_n} = \frac{3}{2}\frac{X_{aF}}{X_{FD}} \tag{9}$$

Then the circuit parameters of Fig. 2 can be expressed by winding parameters when (9) is substituted into (4) and (8),

$$\begin{aligned}
x_{d\ell} &= \frac{X_d}{Z_n} - \frac{3}{2}\frac{X_{aF}}{Z_n}\frac{X_{aD}}{X_{FD}} \\
x_{F\ell} &= \frac{3}{2}\frac{X_{aD}}{X_{FD}}\left(\frac{X_F}{Z_n}\frac{X_{aD}}{X_{FD}} - \frac{X_{aF}}{Z_n}\right) \\
x_{D\ell} &= \frac{3}{2}\frac{X_{aF}}{X_{FD}}\left(\frac{X_D}{Z_n}\frac{X_{aF}}{X_{FD}} - \frac{X_{aD}}{Z_n}\right) \\
x_{FD} &= x_{Dd} = x_{Fd} = \frac{3}{2}\frac{X_{aF}}{Z_n}\frac{X_{aD}}{X_{FD}} \\
r_a &= \frac{R_a}{Z_n} , \quad r_F = \frac{3}{2}\frac{R_F}{Z_n}\left(\frac{X_{aD}}{X_{FD}}\right)^2 , \quad r_D = \frac{3}{2}\frac{R_D}{Z_n}\left(\frac{X_{aF}}{X_{FD}}\right)^2
\end{aligned} \tag{10}$$

q-AXIS EXACT EQUIVALENT CIRCUITS

The q-axis voltage equations for synchronous machine in MKS unit are as follows

$$\begin{bmatrix} V_q - U_q \\ 0 \end{bmatrix} = \left[\begin{bmatrix} R_a & \\ & R_Q \end{bmatrix} + p \begin{bmatrix} X_q & X_{aQ} \\ \frac{3}{2}X_{Qa} & X_Q \end{bmatrix} \right] \begin{bmatrix} I_q \\ I_Q \end{bmatrix} \tag{11}$$

The X-matrix is again not symmetric[2]. While X_q is a reactance of three-phase excitation, X_{aQ}, X_{Qa} and X_Q are of single-phase excitation. The matrix form suggests the following definitions of per unit reactances, voltages and currents

$$\begin{aligned}
x_{qQ} &= X_{aQ} I_{QB}/V_n , \quad x_{Qq} = (\tfrac{3}{2}X_{Qa}) I_n/V_{QB} \\
x_q &= X_q I_n/V_n , \quad x_Q = X_Q I_{QB}/V_{QB} \\
r_a &= R_a I_n/V_n , \quad r_Q = R_Q I_{QB}/V_{QB} \\
i_q &= I_q/I_n , \quad i_Q = I_Q/I_{QB}, \quad v_q = V_q/V_n, \quad u_q = U_q/V_n
\end{aligned} \tag{12}$$

To make the x-matrix reciprocal[3], the following constraint must be imposed

$$\frac{3}{2} V_n I_n = V_{QB} I_{QB} \tag{13}$$

resulting in

$$\begin{aligned}
x_{qQ} &= x_{Qq} = \frac{X_{aQ}}{Z_n}\left(\frac{I_{QB}}{I_n}\right) , \quad x_q = \frac{X_q}{Z_n} \\
x_Q &= \frac{2}{3}\frac{X_Q}{Z_n}\left(\frac{I_{QB}}{I_n}\right)^2 , \quad r_Q = \frac{2}{3}\frac{R_Q}{Z_n}\left(\frac{I_{QB}}{I_n}\right)^2
\end{aligned} \tag{14}$$

The per unit q-axis voltage equation now can be written as

$$\begin{bmatrix} v_q - u_q \\ 0 \end{bmatrix} = \begin{bmatrix} r_a + px_q & px_{qQ} \\ px_{Qq} & r_Q + px_Q \end{bmatrix} \begin{bmatrix} i_q \\ i_Q \end{bmatrix} \tag{15}$$

The general q-axis equivalent circuit corresponding to (15) is as Fig. 3.

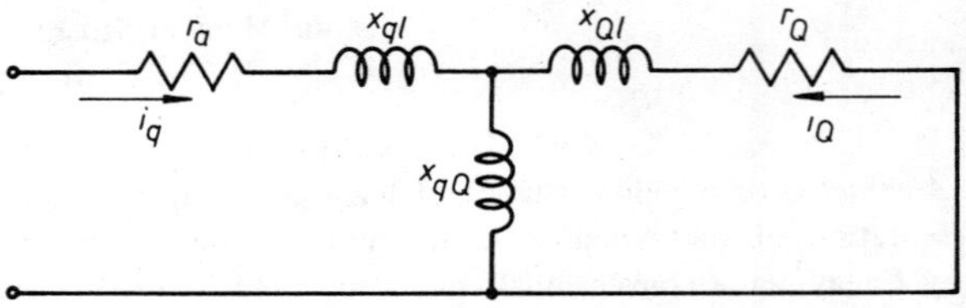

Fig. 3. General q-axis Equivalent Circuit.

where

$$x_{q\ell} = x_q - x_{qQ} , \quad x_{Q\ell} = x_Q - x_{Qq} \tag{16}$$

Although x_{qQ} of Fig. 3 exactly represents the mutual reactance and $x_{q\ell}$ and $x_{Q\ell}$ the leakage reactances, mathematically however the branch reactance x_{qQ} can be set equal to x_q or x_Q resulting in two simplified equivalent circuits. Figs. 4 and 5 respectively.

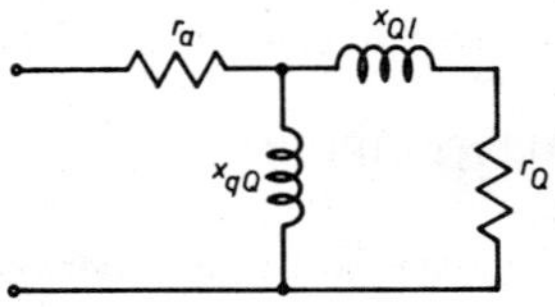

Fig. 4. q-Axis Circuit, ($x_{qQ} = x_q$).

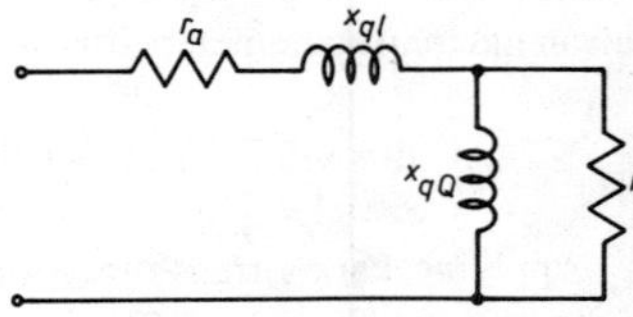

Fig. 5. q-Axis Circuit ($x_{qQ} = x_Q$)

The parameters of these two circuits can be easily determined from field tests. They can also be expressed in terms of winding parameters: Fig. 4, $x_{qQ} = x_q$

From (14) one has

$$\frac{I_{QB}}{I_n} = \frac{X_q}{X_{aQ}}$$

Hence

$$\begin{aligned}
x_{Q\ell} &= x_Q - x_{Qq} = \left(\frac{2}{3}\frac{X_Q}{X_{aQ}}\frac{X_q}{X_{aQ}} - 1\right)x_q \\
r_a &= \frac{R_a}{Z_n} , \quad r_Q = \frac{2}{3}\frac{R_Q}{Z_n}\left(\frac{X_q}{X_{aQ}}\right)^2
\end{aligned} \tag{17}$$

Fig. 5, $x_{qQ} = x_Q$

From (14) one has

$$\frac{I_{QB}}{I_n} = \frac{3}{2}\frac{X_{aQ}}{X_Q}$$

Hence

$$\begin{aligned}
x_{q\ell} &= x_q - x_{qQ} = \left(1 - \frac{3}{2}\frac{X_{aQ}}{X_q}\frac{X_{aQ}}{X_Q}\right)x_q \\
r_a &= \frac{R_a}{Z_n} , \quad r_Q = \frac{3}{2}\frac{R_Q}{Z_n}\left(\frac{X_{aQ}}{X_Q}\right)^2
\end{aligned} \tag{18}$$

CIRCUIT PARAMETERS IN TERMS OF CONVENTIONAL PARAMETERS

From IEEE test code[4] eight conventional d-axis parameters, i.e., r_a, x_d, x_d', x_d'', T_d', T_d'', T_{do}' and T_{do}'' can be determined. There are two identities.

$$\frac{T_d'}{T_{do}'} = \frac{x_d'}{x_d} , \quad \frac{T_d''}{T_{do}''} = \frac{x_d''}{x_d'} \tag{19}$$

To determine the seven d-circuit parameters of Fig. 2, with IEEE test code, an extra test is necessary. It is suggested to measure a newly defined parameter x''_{do} or a damper time constant T_D.

There are five independent equations for the two sets of parameters[5],

$$x_d = x_{d\ell} + x_{Dd} \quad , \quad x'_d = x_{d\ell} + \frac{x_{Dd}x_{F\ell}}{x_{Dd}+x_{F\ell}}$$

$$x''_d = x_{d\ell} + \frac{x_{Dd}x_{F\ell}\,x_{D\ell}}{x_{Dd}x_{F\ell} + x_{F\ell}x_{D\ell} + x_{D\ell}x_{Dd}} \tag{20}$$

$$\omega_o r_F T'_{do} = x_{Dd} + x_{F\ell}$$

$$\omega_o r_D T''_{do} = x_{D\ell} + \frac{x_{Dd}\,x_{F\ell}}{x_{Dd} + x_{F\ell}}$$

besides r_a. If x''_{do} is separately determined, then we have another equation

$$x''_{do} = x_{d\ell} + \frac{x_{Dd}x_{D\ell}}{x_{Dd} + x_{D\ell}} \tag{21}$$

The solutions of the circuit parameters are

$$x_{d\ell} = x - \sqrt{x^2 - \frac{x'_d x''_d (x_d - x''_{do}) - x_d x''_{do}(x'_d - x''_d)}{(x_d - x''_{do}) - (x'_d - x''_d)}}$$

where

$$x = \frac{x_d x''_d - x'_d x''_{do}}{(x_d - x''_{do}) - (x'_d - x''_d)} \tag{22}$$

and

$$x_{Dd} = x_d - x_{d\ell}, \quad x_{F\ell} = \frac{x'_d - x_{d\ell}}{x_d - x'_d}\, x_{Dd}$$

$$x_{D\ell} = \frac{(x'_d - x_{d\ell})(x''_d - x_{d\ell})}{x'_d - x''_d}$$

$$r_F = \frac{1}{\omega_o T'_{do}}\,\frac{x_{Dd}^2}{x_d - x'_d} \tag{23}$$

$$r_D = \frac{1}{\omega_o T''_{do}}\,\frac{(x'_d - x_{d\ell})^2}{x'_d - x''_d}$$

Next if T_D is separately determined, we have another equation

$$\omega_o T_D = x_D / r_D \tag{24}$$

instead of (21). The solutions are (23) and

$$x_{d\ell} = x'_d - \sqrt{(x_d - x'_d)(x'_d - x''_d)\,\frac{T''_{do}}{T_D - T''_{do}}} \tag{22a}$$

The current ratio I_{FB}/I_n of (4) can now be determined, but not I_{DB}/I_n since there is no way to measure R_D because of the short circuit. The voltage ratio V_{FB}/V_n can then be determined from (3).

The q-circuit parameters can be easily determined. For Fig. 4 we have

$$x''_q = \frac{x_{qQ}x_{Q\ell}}{x_{qQ} + x_{Q\ell}} \quad , \quad \omega_o T''_{qo} r_Q = x_{Q\ell} + x_{qQ} \tag{25}$$

The solutions are

$$x_{qQ} = x_q \quad , \quad x_{Q\ell} = \frac{x_q x''_q}{x_q - x''_q} \quad , \quad r_Q = \frac{1}{\omega_o T''_{qo}}\,\frac{x_q^2}{x_q - x''_q} \tag{26}$$

For Fig. 5 we have

$$x''_q = x_{q\ell} \quad , \quad \omega_o T''_{qo} r_Q = x_{qQ} \quad , \quad x_q = x_{qQ} + x_{q\ell} \tag{27}$$

The solutions are

$$x_{q\ell} = x''_q \quad , \quad x_{qQ} = x_Q = x_q - x''_q \quad , \quad r_Q = \frac{1}{\omega_o T''_{qo}}(x_q - x''_q) \tag{28}$$

EXTRA TESTS TO DETERMINE T_D AND x''_{do}

Two test methods are suggested to determine T_D and one to determine x''_{do}. All methods were tested in laboratory.

Determination of T_D from a Varying Slip Test

The rotor is driven at various speeds. Positive sequence voltages are applied to the armature winding with the field open. From phase voltage-current ratio equivalent reactances $x_d(s)$ and $x_q(s)$ are approximately determined. Replacing r_D by r_D/s in Fig. 2, the imaginary part of the circuit impedance is a function of slip s as follows

$$x_d(s) = x_d - \frac{x_{dD}^2 x_D}{r_D^2}\,\frac{1}{(\omega_o T_D)^2 + (\frac{1}{s})^2} \tag{29}$$

or

$$\frac{1}{x_d - x_d(s)} = \frac{r_D^2}{x_{dD}^2 x_D}\left(\frac{1}{s}\right)^2 + \frac{x_D}{x_{dD}^2} \tag{30}$$

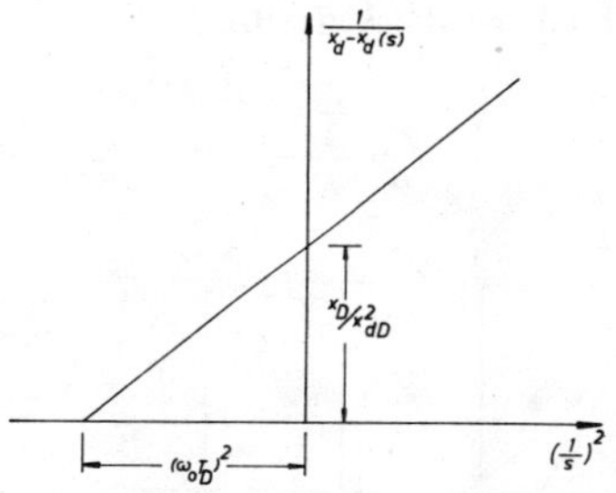

Fig. 6. Determination of T_D from Slip Test.

which can be plotted as Fig. 6 for the determination of T_D. For the calculation an accurate value of x_d from open and short circuit tests must be used.

Determination of T_D From Decaying Current Test

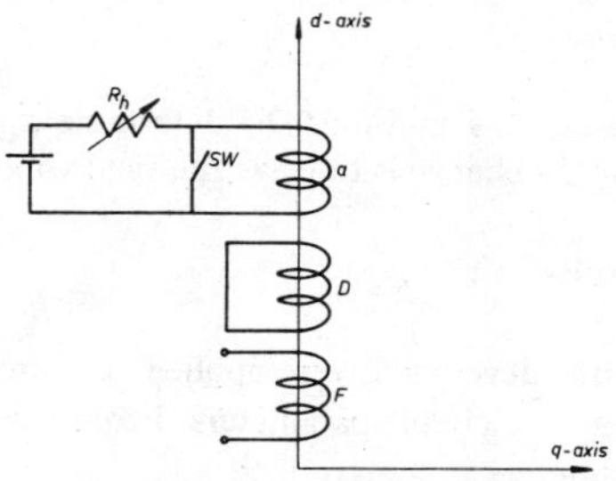

Fig. 7. Connection for the Decaying Current Test.

Kaminosono and Uyeda's indicial response method[6] is modified to determine T_D. Since a clean step voltage is hard to obtain, a decaying current is used instead. Apply a constant current to one phase winding in the d-axis position and then suddenly short circuit the armature terminals with the switch Sw in Fig. 7. The rheostat R_h protects the power supply.

The voltage equations for Fig. 7 in Laplace transform are

$$\begin{bmatrix} 0 \\ 0 \end{bmatrix} = \begin{bmatrix} r_a + sL_d & sL_{dD} \\ sL_{dD} & r_D + sL_D \end{bmatrix} \begin{bmatrix} I_a(s) \\ I_D(s) \end{bmatrix} - \begin{bmatrix} L_d & L_{dD} \\ L_{dD} & L_D \end{bmatrix} \begin{bmatrix} i_{ao} \\ o \end{bmatrix} \tag{31}$$

where i_{ao} is the initial current of armature winding. The solution of $I_a(s)$ can be written in a convenient form

$$I_a(s) = \frac{s + \frac{1}{T'_D}}{(s + \frac{1}{T_1})(s + \frac{1}{T_2})} i_{ao} \tag{32}$$

where

$$T'_D = T_D(1-K^2_{Dd}) \quad , \quad T_1T_2 = T_dT'_D \quad , \quad T_1+T_2 = T_d+T_D \tag{32a}$$

and

$$T_D = L_D/r_D \quad , \quad T_d = L_d/r_a \quad , \quad K^2_{Dd} = L^2_{Dd}/L_dL_D \tag{32b}$$

$I_a(s)$ of (32) can be resolved into two components

$$I_a(s) = \frac{i_{10}}{s + \frac{1}{T_1}} + \frac{i_{20}}{s + \frac{1}{T_2}} \tag{33}$$

and it can be shown that the initial component current ratio

$$\frac{i_{10}}{i_{20}} = \frac{T_2}{T_1} \frac{T_1-T'_D}{T'_D-T_2} \tag{34}$$

From T_D', T_1+T_2 of (32a) and (34), the following solutions are obtained

$$T_d=(i_{10}T_1+i_{20}T_2)/i_{ao} \quad , \quad T'_D=T_1T_2/T_d, \quad T_D=(i_{10}T_2+i_{20}T_1)/i_{ao} \tag{35}$$

T_1, T_2, i_{10} and i_{20} are determined from a semilog plot as Fig. 8. The T_d value from (35) should be checked with

$$T_d = \frac{\frac{2}{3}x_d}{\omega_o r_a} \tag{36}$$

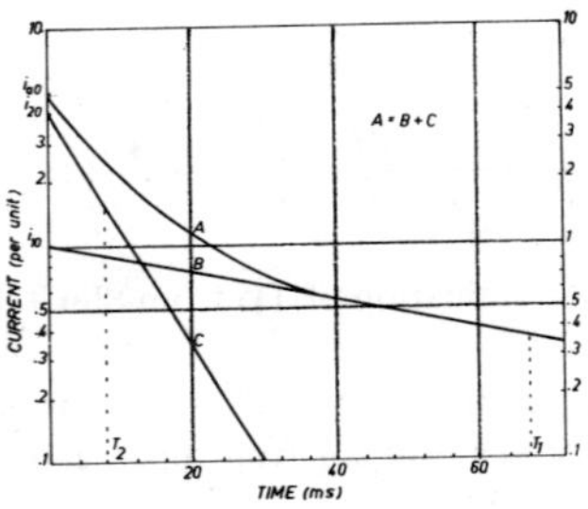

Fig. 8. Resolving Decaying Current into Two Components.

Determination of x''_{do}

Dalton and Cameron's method[7] to determine x''_d is adapted to determine x''_{do}. The only change is to leave the field winding open.

Laboratory Test Results

The methods thus developed were applied to a small synchronous machine to determine the circuit parameters. From IEEE test code the following d-axis parameters are determined.

$$r_a = 0.72\Omega, \; x_d = 16.2\Omega, \; x'_d = 2.74\Omega, \; x''_d = 2.42\Omega$$

$$T'_{do} = 0.27s, \; T''_{do} = 0.027s$$

The per unit values can be obtained when the base ohm Z_n is chosen. From extra tests the following are determined

A $T_D = 0.049$ s (varying slip test)

B $T_D = 0.055$ s (decaying current test)

C $x''_{do} = 8.18\ \Omega$ (adapted Dalton and Cameron)

The computed results of d-circuit parameters in ohms are as follows

	x_{Dd}	$x_{d\ell}$	$x_{F\ell}$	$x_{D\ell}$	r_F	r_D
A	15.8	0.40	2.75	14.6	0.182	1.66
B	15.5	0.68	2.38	10.9	0.176	1.28
C	15.9	0.33	2.84	15.5	0.184	1.76

The discrepancy in results of B is attributed to the difficulty of resolving the decaying current into components. The field resistance R_F is 70Ω and the current and voltage ratios are

$$I_{FB}/I_n = 0.0625 \quad , \quad V_{FB}/V_n = 24$$

For the q-axis

$$x_q = 9.71\Omega \quad , \quad x''_q = 7.2\Omega$$

are determined by conventional methods and

$$T''_{qo} = 0.0165 \text{ s}$$

by a decaying current method similar to Fig. 7. The computed results of q-axis parameters are

Fig. 4 $x_{qQ} = 9.71\ \Omega$, $x_{Q\ell} = 27.8\ \Omega$, $r_Q = 6.05\ \Omega$

Fig. 5 $x_{q\ell} = 7.2\ \Omega$, $x_{qQ} = 2.51\ \Omega$, $r_Q = 0.407\ \Omega$

CONCLUSION

An extra test with the IEEE test code is needed to determine the d-circuit synchronous machine parameters. Three methods were suggested, a varying slip test or a decaying current test to determine the D-damper time constant T_D or an adapted Dalton and Cameron's method to determine the open field subtransient reactance x''_{do}. No extra test is needed to determine the q-circuit parameters. Laboratory tests show close agreement of results for the first and the third methods but not for the second method.

A step by step derivation of the exact equivalent circuits for synchronous machines from voltage equations in MKS units is given, in order to clarify some basic concepts of winding and circuit parameters which were often buried in per unit computation and coordinate transformation.

So far the methods have not been applied to large synchronous machines. Neither the paper did include the case of solid rotor machines. The next step of investigation is to incorporate the accurate circuit parameters in multi-machine power system dynamics studies.

NOMENCLATURE

ω_o	synchronous angular velocity; 377 rad/s
p	differential operator; $(d/dt)/\omega_o$
suffices a,d,q	armature a-phase, d-axis, and q-axis windings
suffices F,D,Q	rotor field, d-axis damper and q-axis damper windings
V's, v's	applied voltages in V, and per unit
U's, u's	rotational voltages in V, and per unit
I's, i's	currents in A, and per unit
V_n, I_n	base armature volt, ampere
V_{FB}, V_{DB}, V_{QB}	base field, D-winding and Q-winding volts
I_{FB}, I_{DB}, I_{QB}	base field, D-winding and Q-winding amperes
R's, r's	winding resistances in Ω, and per unit
X's, x's	self and mutual reactances in Ω, and per unit
L's	self and mutual inductances, H
Z_n	armature base ohm, Ω
x_d, x'_d, x''_d	d-axis synchronous, transient and subtransient reactances
x''_{do}	newly defined open field d-axis subtransient reactance

x_q, x_q''	q-axis synchronous, subt. reactances
T_d', T_d''	short circuit d-axis transient and subtransient time constants, s
T_{do}', T_{do}''	open circuit d-axis transient and subtransient time constants, s
T_D	d-axis damper winding time constant, s
T_{qo}''	open circuit q-axis subtransient time constant, s

ACKNOWLEDGEMENT

The authors wish to acknowledge the financial support of National Research Council of Canada through Grant A3626, and Dr. H. Chinn and Mis Bev Harasymchuk's help in preparing this paper.

REFERENCES

[1] I. M. Canay, "Causes of Discrepancies on Calculation of Rotor Quantities and Exact Equivalent Diagrams of the Synchronous Machine." IEEE Transactions, PAS-88 No. 7, July 1969, pp. 1114-1120.

[2] C. Concordia, "Synchronous Machines" (book), John Wiley, 1951, eqts. (21) and (24), pp. 14-15.

[3] A. W. Rankin, "Per-unit Impedances of Synchronous Machines," I. AIEE T. Vol. 64 pp. 569-573, 1945 II. AIEE T. Vol. 64 pp. 839-841, 1945.

[4] "Test Procedure for Synchronous Machines," IEEE Publ. 115, 1965.

[5] B. Adkins, "The General Theory of Electrical Machines," (book), Chapman and Hall 1964, pp. 122-123.

[6] H. Kaminosono and K. Uyeda, "New Measurement of Synchronous Machine Quantities," IEEE Transactions, PAS-87, No. 11, Nov. 1968, pp. 1908-1918.

[7] F. K. Dalton and A. W. Cameron, "Simplified Measurement of Subtransient and Negative Sequence Reactances in salient-pole Synchronous Machines," AIEE Trans., PAS-71, pp. 752-757, Oct. 1952.

Discussion

I. M. Canay (Brown, Boveri and Cie., Baden, Switzerland): I would like to express a sincere welcome to this paper by Prof. Yao-nan Yu and Mr. H.A.M. Moussa, in which once again reference is made to the inapplicability of the conventional equivalent circuit of the direct axis. The traditional equivalent circuit diagram is indeed similar to the circuit diagrams Fig. 2 or Fig. 6b of [1] provided that, x_{dl} is equal to the armature leakage reactance $x_{a\sigma}$. As the authors also emphasize, x_{dl} in these diagrams is not identical with the leakage reactance of the armature. To avoid any probable confusion, I would suggest that this be denoted as characteristic reactance x_c, consequently, $x_c \equiv x_{dl}$.

Prof. Yu and Mr. Moussa go a step further and illustrate how the circuit parameters of equivalent diagrams could be computed from measurements. This is very important, when one, starting with the measured values of x_d, x_d', x_d'', T_d' and T_d'' investigates, the field current and voltage or stability of the machine, taking a very fast voltage regulator into consideration. Such analyses could often procure an incorrect result, because with the supposition $x_c = x_{a\sigma}$, the rotor quantities being influenced by the voltage regulator are already falsified.

But the equivalent diagram in Fig. 2 (or Fig. 6b in [1]) is only valid for an idealised synchronous machine, thus only a close approximation for machines with totally laminated poles and a damper winding. When the field winding of a synchronous machine with solid rotor core remains open, the rotor resistance is replaced by an eddy current resistance r_{fE} of the pole sides (Fig. D1), as explained in [1]. In practice this influence, as confirmed by our measurements, occurs to a certain degree in all machines. There are even machines in which, in consequence of this effect, the values of x_{do}'' and x_d'' differ only insignificantly.

The hereby proposed method is based on x_{do}'', i.e., on a measurement during which the field winding remains open. Consequently, this procedure is only valid for an ideal machine, i.e., without the r_{fE}-effect and only approximately for machines with totally laminated poles. Furthermore, I believe that the degrees of saturation during the measurement of x_d'' (saturated) and x_{do}'' (unsaturated) are different and have a great effect on the accuracy of measurement.

On grounds of the stated reasons, we prefer in Brown Boveri & Co. to ascertain the circuit parameters x_c, x_{Dc}, x_{fc}, r_f and r_D out of only one sudden short-circuit test, whereby the field winding always remains short-circuited. x_d, x_d', x_d'', T_d' and T_d'' are ascertained as usual from ASA. In addition to this, the ratio "a" is computed from the behavior of field current decay with respect to time as follows (Fig. D2):

Manuscript received January 8, 1971.

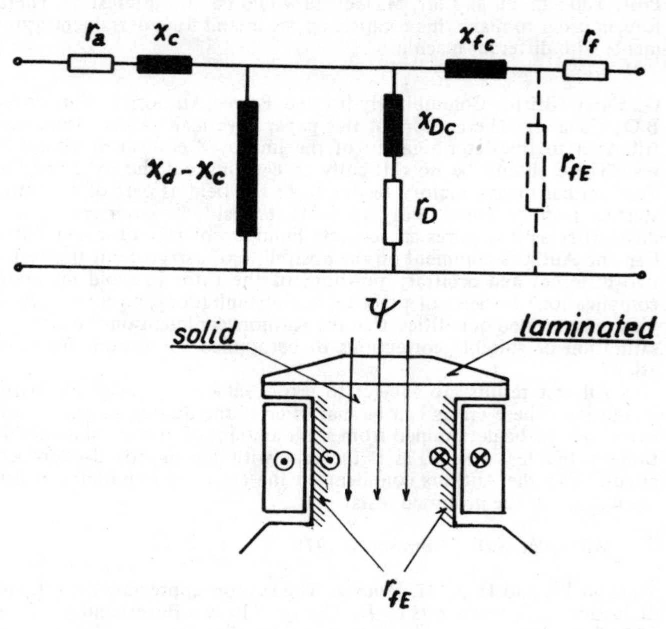

Fig. D1. General d-axis Diagram

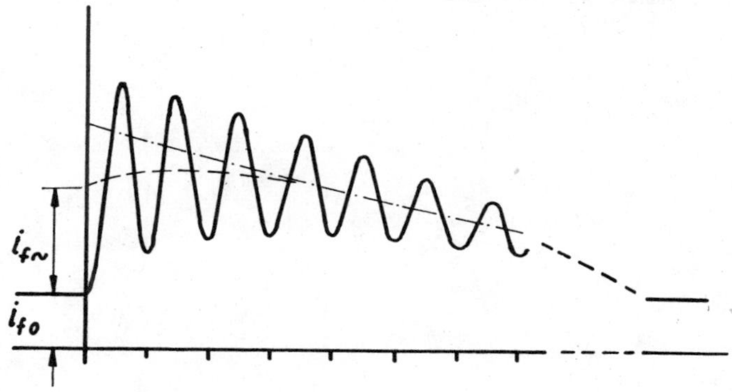

Fig. D2. Decay of the field-current during short-circuit

$$a = \frac{i_{f\sim}(t=+0)}{i_{fo}(t=-0)} \tag{1}$$

Consequently,

$$x_c = x_d'' \frac{x_d - x_d' - a x_d'}{x_d - x_d' - a x_d''} \tag{2}$$

For other circuit parameters,

$$x_{Dc} = \frac{(x_d''-x_c)(x_d'-x_c)}{x_d'-x_d''} \tag{3}$$

$$x_{fc} = \frac{(x_d'-x_c)(x_d-x_c)}{x_d-x_d'} \tag{4}$$

$$r_D = \frac{x_d-x_c+x_{Dc}}{\omega T_D} \quad \text{with} \tag{5}$$

$$T_D = T_d'' \frac{x_d'}{x_d''} \frac{(x_d-x_c)(x_d'-x_d'')+(x_d'-x_c)(x_d''-x_c)}{(x_d'-x_c)^2}$$

$$r_f = \frac{x_d-x_c+x_{fc}}{\omega T_f} \quad \text{with} \quad T_f = T_d' \frac{x_d}{x_d'} - T_D \tag{6}$$

A comparison between the above mentioned method and the one from Prof. Yao-nan Yu and Mr. M. Moussa would be very interesting. Therefore, in order to make this comparison, we intend to undertake measurements with different machines.

G. Furst (British Columbia Hydro and Power Authority, Vancouver, B.C., Canada): The authors of this paper have made a significant contribution to the determination of the improved equivalent circuit by test. There should be no difficulty in carrying out the "Varying Slip Test" either during factory testing or in the field as part of the initial startup testing. The "Decaying Current Test", however, may cause difficulties as it requires an accurate lining up of the rotor and stator. Can the Authors comment on the possibility of carrying out this test at two different and arbitrary positions of the rotor to avoid the above complication? Because of practical testing limitations, both tests would yield unsaturated quantities. Can the Authors comment on the effect of saturation or suitable corrections to be applied to account for saturation?

All test results are subject to errors inherent to available instrumentation. These errors can be magnified if the desired physical quantities have to be determined from a fair amount of numerical manipulation of the test results, as is the case with the improved equivalent circuit. Can the Authors comment on the accuracy required in instrumentation for the proposed tests?

Manuscript received February 16, 1971.

Yao-nan Yu and **H. A. M. Moussa:** The authors appreciate the valuable discussions and comments by Dr. Canay of Brown Boveri and Mr. Furst of B. C. Hydro.

Manuscript received March 19, 1971.

We agree that iron circuit has a dominant effect on solid rotor machines, and that more accurate representation with additional rotor circuits is necessary to cover a wide-range frequency effect. In such a case more tests would be required, and methods remain to be developed to determine all the parameters. However, for laminated rotor machines, the iron effect is relatively small and is eventually included in the "equivalent" damper circuits.

We noted that Dr. Canay used the standard sudden short-circuit test method to determine all the parameters, but did not include the iron circuit effect r_{fE}. We also noted that equations (3) through (6) of the discussion are based on Fig. 2 which is exactly the same as Fig. 6b of his paper [1].

Mr. Furst raised a practical problem in conducting the decaying current test. Since the mutual inductance of a-phase winding with the d-axis damper winding is a function of cos α and that with a q-axis damper winding it is a function of sin α, where α is the angle between the a axis and d axis, as long as α is kept small, say $\pm 5°$, the effect of α is not noticable. However, α cannot be too large; otherwise it will have noticable coupling effect of "a" and the q-axis damper windings introducing another time constant in the test results.

What Mr. Furst's stated on accuracy and instrumentation is true. But we did not have any difficulty in our tests. Saturation could be included in analysis [8]. What we were trying to measure were the unsaturated values. Also we did not include a particular circuit to represent the iron effect.

[8] M. Riaz, "Analogue Computer Representation of Synchronous Generators in Voltage-Regulation Studies", AIEE Transactions on Power Apparatus and Systems, pp. 1178-84, December 1956.

Computer Representation of Excitation Systems

IEEE COMMITTEE REPORT

***Abstract*—The availability of large digital computers has permitted more extensive computation of power system stability, a subject of increasingly greater importance. This paper suggests a common nomenclature and control system representation of the various excitation systems now available. It can be used to define input data requirements for computer programs, and can provide a consistent format in which manufacturers can respond to requests for excitation system data to be used for system studies.**

Paper 31 TP 67-424, recommended and approved by the Power Generation Committee of the IEEE Power Group for presentation at the IEEE Summer Power Meeting, Portland, Ore., July 9–14, 1967. Manuscript submitted April 7, 1967; made available for printing May 1, 1967.

Members of the Working Group of the Excitation Systems Subcommittee of the IEEE Power Generation Committee are: R. R. Bast, *Chairman;* A. Hauspurg, F. W. Keay, R. L. Krahn, P. R. H. Landrieu, K. R. McClymont, A. S. Rubenstein, J. W. Skooglund, G. I. Stillman, M. Temoshok, and H. S. Wilson.

Introduction

Power system stability has been the subject of intensive study in the United States and Canada since the 1920s and 1930s when the first large hydroelectric installations were being developed. Long lines and relatively slow circuit breakers and relay operation made transient stability a serious problem. Analytical studies generally considered no-control system response and generators were represented as constant voltage behind transient reactance. For many years, this procedure was followed and machine swings were calculated step-by-step with an ac network calculator.

The availability of large digital computers in the late 1950s permitted more rapid and economical computations, and many of the ac network calculator functions were transferred to computer programs. Included were system load flow and transient stability. The first digital programs merely performed the same functions as the ac network calculator (constant voltage behind transient

Reprinted from *IEEE Trans. Power App. Syst.*, June 1968.

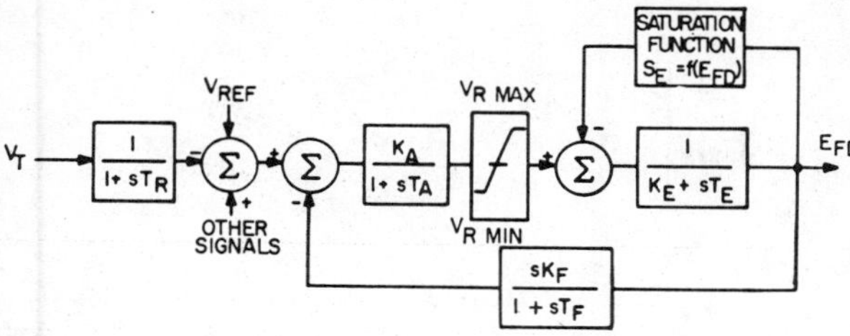

Fig. 1. Type 1 excitation system representation, continuously acting regulator and exciter.

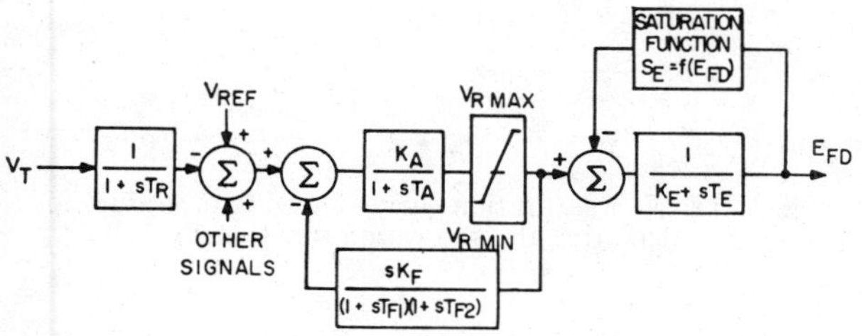

Fig. 2. Type 2 excitation system representation, rotating rectifier system.

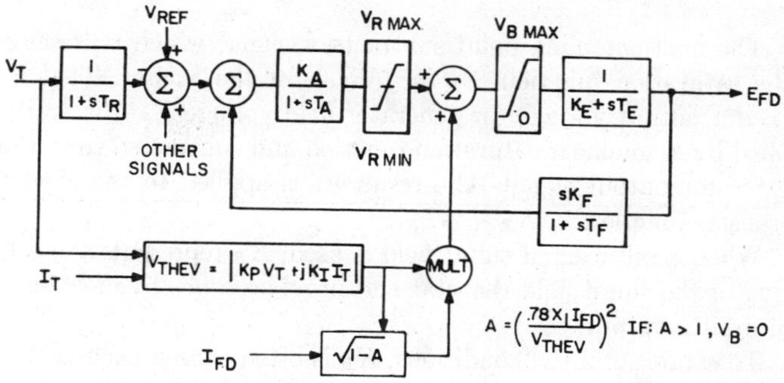

Fig. 3. Type 3 excitation system representation, static with terminal potential and current supplies.

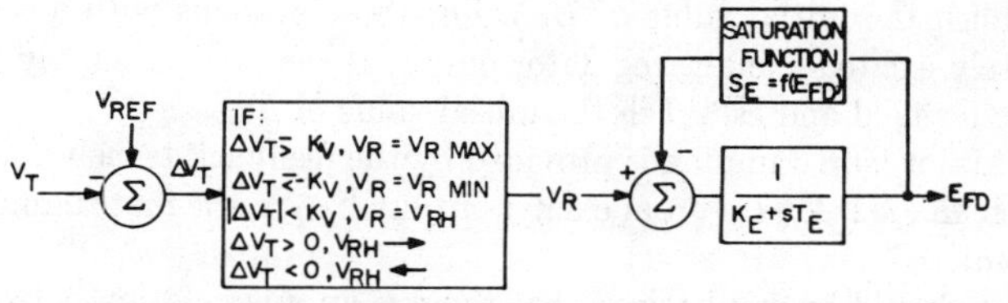

Fig. 4. Type 4 excitation system representation, noncontinuously acting regulator. *Note:* V_{RH} limited between $V_{R\,\min}$ and $V_{R\,\max}$. Time constant of rheostat travel T_{RH}.

reactance), but did permit representation of more machines and large power networks.

Additional development of digital programs in the 1960s then included the representation of machine control systems such as the speed governor and excitation systems. It is no longer necessary to assume a constant voltage behind a reactance and more realistic and accurate machine representations are now available.

A number of individuals and organizations worked concurrently, but independently, to develop these programs. This resulted in excitation system representations that are generally similar but with some significant differences. Fortunately, the differences are primarily in the form of input data, rather than control system representation.

The purpose of this paper is to suggest a common nomenclature and control system representation of the various excitation systems now available. This can then be used to define input data requirements for computer programs and provide a consistent format whereby manufacturers can respond to users' requests for excitation system data. It should be strongly emphasized that the present discussion deals with the representation of excitation systems solely for system studies.

Basic Excitation System Types

After considering the different types of excitation systems now in service in the United States and Canada and contemplated for the immediate future, the Working Group has defined four excitation system types to be used in computer representations (Figs. 1–4). These should be adequate to represent all modern systems. Some older systems must be approximated with one of the four types or it must be assumed that the machine has fixed excitation. The symbols used in the four excitation system representations are listed in the Appendix.

Per Unit System

In the development of the excitation system block diagrams, it has been necessary to establish a per unit voltage base. For the following discussion, one per unit generator voltage is defined as rated voltage. One per unit exciter output voltage is that voltage required to produce rated generator voltage on the generator air gap line.

Type 1—Continuously Acting Regulator and Exciter

The excitation system designated type 1 is shown in Fig. 1. The type 1 excitation system is representative of the majority of modern systems now in service and presently being supplied. This includes most continuously acting systems with rotating exciters such as:

Allis Chalmers	Regulex regulator
General Electric	Amplidyne regulator
	Alterrex
	Alterrex-thyristor
Westinghouse	Mag-A-Stat regulator
	Brushless (1967 on)
	Rototrol
	Silverstat regulator
	TRA regulator.

Fig. 1 shows the significant transfer functions which should be included for satisfactory representation in computer studies. Many other system types may be represented if excitation system ceiling voltage is assumed to be independent of generator terminal conditions.

The transfer functions and nomenclature of type 1 will be described in detail, referring to Fig. 1. V_T is the generator terminal voltage applied to the regulator input. The first transfer function is a simple time constant T_R representing regulator input filtering. For most systems, T_R is very small and may be considered to be zero.

The first summing point compares the regulator reference with the output of the input filter to determine the voltage error input to the regulator amplifier. Most computer programs do not require an input of V_{REF}, but rather internally calculate the proper value by assuming V_T at $t = 0$ is at the proper value.

The second summing point combines voltage error input with the excitation major damping loop signal.

The main regulator transfer function is represented as a gain K_A and a time constant T_A. Following this, the maximum and minimum limits of the regulator are imposed so that large input error signals cannot produce a regulator output which exceeds practical limits.

The next summing point subtracts a signal which represents the saturation function, $S_E = f(E_{FD})$, of the exciter. That is, exciter output voltage (or generator field voltage E_{FD}) is multiplied by a nonlinear saturation function and subtracted from the regulator output signal. The resultant is applied to the exciter transfer function $1/(K_E + sT_E)$.

When a self-excited shunt field is used, K_E represents the setting of the shunt field rheostat and provides a positive feedback of exciter output.

To establish initial conditions, K_E is often chosen such that it is equal in magnitude to the saturation function at the initial value of E_{FD}. At this value, the shunt field exactly compensates for exciter saturation and no regulator output is required to establish the initial value of E_{FD}. For those systems with a separately-excited exciter, regulator output is required to supply the exciter field and establish the initial value of E_{FD}.

Major loop damping is provided by the feedback transfer function $sK_F/(1 + sT_F)$ from exciter output E_{FD} to the first summing point.

It should be emphasized that there is an interrelation between exciter ceiling $E_{FD\ max}$, regulator ceiling $V_{R\ max}$, exciter saturation S_E, and K_E. The following expression must be satisfied under steady-state conditions:

$$V_R - (K_E + S_E)E_{FD} = 0, \quad E_{FD\ min} \leq E_{FD} \leq E_{FD\ max}. \quad (1)$$

(The sign of K_E is negative for a self-excited shunt field.)

At ceiling, or $E_{FD} = E_{FD\ max}$,

$$V_{R\ max} - (K_E + S_{E\ max})E_{FD\ max} = 0. \quad (2)$$

K_E is always specified, either as input data or program logic, to permit automatic calculation. In addition, of the remaining three constants $V_{R\ max}$, $S_{E\ max}$, and $E_{FD\ max}$, the specification of any two establishes the third. Since different programs may use different input data, care should be taken to make sure that (2) is satisfied.

For some systems employing noncommutating-type exciters, the minimum value of E_{FD} is zero and cannot be negative.

Saturation Function

The exciter saturation function S_E is defined by the Working Group as a multiplier of exciter output E_{FD} to represent the increase in exciter excitation requirements because of saturation. Fig. 5 illustrates the calculation of a particular value of S_E. At a given exciter output voltage, the quantities A and B are defined as the exciter excitation to produce the output voltage on the constant–resistance–load saturation curve and air gap line, respectively.

$$S_E \equiv \frac{A - B}{B}.$$

The exciter constant–resistance–load saturation curve has been used in this definition of S_E.

Different computer programs have represented exciter saturation in different ways but, in general, the saturation function can be defined by two points. To be consistent, the procedure suggested is to establish two voltages at which to specify S_E and use them as data for computer input. The form of the saturation function is not defined here, but rather considered to be a part of the particular computer program used.

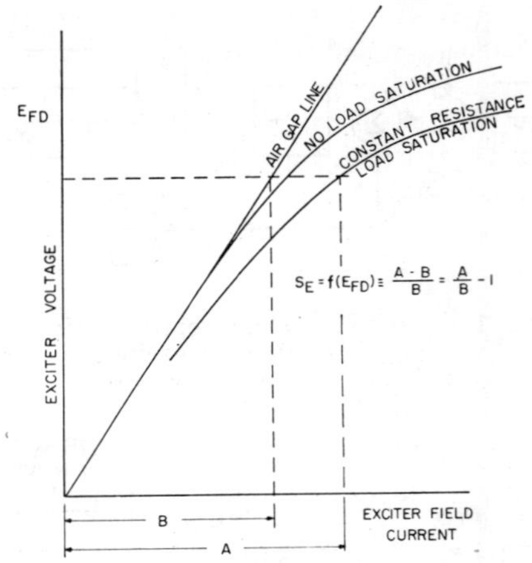

Fig. 5. Exciter saturation curves showing procedure for calculating the saturation function S_E.

It is suggested that S_E be specified at the following voltages:

Voltage	Saturation
$E_{FD\ max}$	$S_{E\ max}$
$0.75\ E_{FD\ max}$	$S_{E\ 0.75\ max}$

Type 1S—Controlled-Rectifier Systems with Terminal Potential Supply Only

A special case of type 1 is a system employing an excitation source from terminal voltage with controlled rectifiers only. A system of this type responds quickly, but with ceiling voltage proportional to generator terminal voltage. Referring to Fig. 1, $V_{R\ max}$ is not assumed constant but rather is computed as a proportion of terminal voltage.

$$V_{R\ max} = K_P V_T.$$

In general, the constants for this type of system are such that $K_E = 1$, $T_E = 0$, and $S_E = 0$.

Type 2 Excitation System—Rotating Rectifier System

The type 2 system, Fig. 2, applies for units with the major damping loop input from the regulator output. An example is the Westinghouse brushless system put into service up to and including 1966. The transfer function includes one additional time constant to compensate for the exciter which is not included within the damping loop. Other characteristics of type 2 systems are similar to type 1.

Type 3 Excitation System—Static with Terminal Potential and Current Supplies

Some static systems cannot be represented by type 1 or 1S because generator terminal current is used with potential as the excitation source. An example of such a system is the General Electric SCPT. The type 3 system has been developed to represent this particular static system. The manufacturer should be contacted to establish the applicability of the type 3 representation to other excitation systems.

Fig. 3 gives the transfer functions making up a type 3 system. The regulator transfer functions are similar to type 1, up to and including the regulator output limiter ($V_{R\ max}$–$V_{R\ min}$). The following summing point combines the regulator output with the signal representing the self-excitation from the generator terminals. K_P is the coefficient of the shunt excitation supply proportional to terminal voltage. Similarly, K_I is the coefficient of the supply obtained from terminal current transformers. The multiplier (MULT) accounts for the variation of self-excitation with change in the angular relation of field current I_{FD} and self-excitation voltage V_{THEV}.

The $V_{B\ max}$ limiter sets the excitation system output to zero when $A > 1$, that is, when the field current exceeds the excitation output current. Excess generator field current bypasses the excitation supply by flowing through the output rectifier.

The transfer function $1/(K_E + sT_E)$ represents the constants of the excitation transformer. Damping is provided by the feedback transfer function $sK_F/(1 + sT_F)$.

Type 4 Excitation System—Noncontinuously Acting

The systems discussed previously are representative of the modern high-gain, fast-acting excitation sources. The type 4 system is used to represent other systems, in particular those that were used immediately before the development of the continuously acting excitation systems. Examples are:

General Electric	GFA4 regulator
Westinghouse	BJ30 regulator.

These systems respond at basically two different speeds, depending upon the magnitude of voltage error. For small errors, adjustment is made with a motor-operated rheostat. Larger errors cause resistors to be quickly shorted or inserted and a strong forcing signal applied to the exciter.

Fig. 4 illustrates this action. The exciter representation is similar to the previously described systems except that no major damping loop is represented.

Depending upon the magnitude of voltage error ΔV_T, different regulator modes are specified. If voltage error is larger than the fast raise/lower contact setting K_V (typically five percent), $V_{R\ max}$ or $V_{R\ min}$ is applied to the exciter, depending upon the sign of the voltage error. For a voltage error less than K_V, the exciter input equals the rheostat setting V_{RH}. The rheostat setting is adjusted up or down, depending upon the sign of the error. The time constant representing the slow adjustment of exciter field voltage is T_{RH}.

Other Systems

Several excitation systems in use on older units and some smaller-sized turbine generators currently being installed have not been discussed. They include some systems no longer in manufacture. Because their use represents only a small fraction of the generating capacity in service or being installed, no unique excitation system type has been designated for their representation. They can be represented by type 1 or type 4 systems with the proper choice of constants.

Other Regulator Input Signals

In the last several years increased emphasis on the study of, and system design to improve, dynamic stability has resulted in the use of other regulator input signals in addition to terminal voltage. These signals are chosen to provide positive damping of power system oscillations to improve generator stability and damp tie line oscillations.

Some of these signals are: accelerating power, speed, frequency and rate-of-change of terminal voltage. When used, they are added, at the voltage reference summing point, to the terminal voltage error and are indicated in Figs. 1–3 as *other signals.* Usually the stabilizing signal is inserted through a transfer function providing gain adjustment and a lead-lag compensation for phase shifting.

Since the use of stabilizing signals is still in the development stage, the Working Group does not recommend a specific representation at this time.

TABLE I

TYPICAL CONSTANTS OF EXCITATION SYSTEMS IN OPERATION TODAY ON 3600-R/MIN STEAM TURBINE GENERATORS

(Excitation system voltage response ratio—0.5 p.u.)

Symbol	Self-Excited Exciters, Commutator or Silicon Diode, with Amplidyne Voltage Regulators	Self-Excited Commutator Exciter with Mag-A-Stat Voltage Regulator	Rotating Rectifier Exciter with Static Voltage Regulator
T_R	0.0 to 0.06	0.0	0.0
K_A	25* to 50*	400	400
T_A	0.06 to 0.20	0.05	0.02
$V_{R\ max}$	1.0	3.5	7.3
$V_{R\ min}$	−1.0	−3.5	−7.3
K_F	0.01 to 0.08	0.04	0.03
T_F	0.35 to 1.0	1.0	1.0
K_E	−0.05	−0.17	1.0
T_E	0.5	0.95	0.8
$S_{E\ max}$	0.267	0.95	0.86
$S_{E\ 0.75\ max}$	0.074	0.22	0.50

* For generators with open-circuit field time constants greater than four seconds.

TYPICAL EXCITATION SYSTEM CONSTANTS

Table I lists typical excitation system constants supplied to the Working Group by manufacturers, current as of January 1967.

To obtain data for a specific generator, the manufacturer should be contacted. It should be emphasized that the data of Table I are generalized and do not necessarily represent any specific generator.

It is not intended that the representations reflect the response of individual components comprising the excitation system or that the data of Table I be used for confirmation of excitation system performance. When used as part of a system study, the representations provide the proper response of the generator excitation system.

Most generators operating in the United States have been supplied by three manufacturers. The manufacturers have furnished similar excitation systems but with sufficient differences in design and operation to warrant individual numerical representations. Commutator exciters have been furnished with rotating-amplifier voltage regulators, with magnetic-amplifier voltage regulators, and with combinations of magnetic-amplifier and rotating-amplifier voltage regulators. Recent exciters utilizing silicon diodes, in place of the commutator, have deviated more in design concepts. One design retains a self-excited exciter with silicon diodes and a rotating amplifier regulator; another design is a rotating-rectifier exciter with pilot-exciter supply to a controlled-rectifier voltage regulator. There are slight differences in operation design and in details that affect methods of representation. Table I provides a comparison of the numerical variations that can be expected for similar equipments, due to the different philosophies of representation.

Differences may be observed in the variety of time constants, gains, and feedback. The large differences are due to methods of representation and of the per unit systems used. Each of the typical systems illustrated has been provided for similar applications and designed to existing industry standards.

The purpose of Table I is to indicate that even though the block diagram representations of the various excitation systems and the identification of the various portions of the diagrams are the same, there is a marked difference in the numbers that represent the actual systems. Each is correct if consistency is maintained in the per unit system and the constants of the overall regulator–exciter system.

Appendix

Nomenclature

E_{FD}	exciter output voltage (applied to generator field)
I_{FD}	generator field current
I_T	generator terminal current
K_A	regulator gain
K_E	exciter constant related to self-excited field
K_F	regulator stabilizing circuit gain
K_I	current circuit gain of type 3 system
K_P	potential circuit gain of type 1S or type 3 system
K_V	fast raise/lower contact setting, type 4 system
S_E	exciter saturation function
T_A	regulator amplifier time constant
T_E	exciter time constant
T_F	regulator stabilizing circuit time constant
T_{F1}, T_{F2}	regulator stabilizing circuit time constants (rotating rectifier system)
T_R	regulator input filter time constant
T_{RH}	rheostat time constant, type 4 system
V_R	regulator output voltage
$V_{R\,max}$	maximum value of V_R
$V_{R\,min}$	minimum value of V_R
V_{REF}	regulator reference voltage setting
V_{RH}	field rheostat setting
V_T	generator terminal voltage
V_{THEV}	voltage obtained by vector sum of potential and current signals, type 3 system
ΔV_T	generator terminal voltage error.

EXCITATION SYSTEM DYNAMIC CHARACTERISTICS
IEEE COMMITTEE REPORT

ABSTRACT

This tutorial paper describes various analytical methods for specifying and analyzing excitation system response. Its purpose is to relate the complex-frequency parameters of transfer function blocks to the excitation system time response. Included are discussions of transfer function blocks, Bode plots, S-plane plots and real-time response.

INTRODUCTION

In 1968, an IEEE Committee Report[1] was published in which several "standardized" representations of excitation systems were recommended for use in computer programs. The representations given there are in block diagram format, each major block representing in quasi-physical fashion the transfer function of a major component or group of components of the excitation system.

The transfer functions used in these representations are in the complex-frequency domain because the mathematical description of the physical relations are much easier to handle when written in terms of complex frequency, while the engineer is primarily interested in the phenomena as they occur in the time domain. Indeed, the computer programs, in which the excitation system representations are used, are written so that their outputs are in the real-time domain. It may not be clear how the complex-frequency parameters of the transfer-function blocks are related to the time-response characteristics of an excitation system, and, conversely, of how desired time-response characteristics of an excitation system are related to the transfer functions, and, thereby, the design of an excitation system.

It is the purpose of this paper to help bridge any gap in understanding by correlating the parameters of the transfer functions used in the block-diagram representations with the frequency response of the transfer functions, and then to correlate this frequency response with the real-time response of an excitation system. The concepts used in making the correlations which follow herein were developed in the decade of the 1950's, but they may not be familiar to many people in the electric power industry. Hence, it is felt that it is appropriate at this time to make some of these concepts available in tutorial form. Technical papers are appearing in Power Apparatus and Systems which make use of some of these concepts and of techniques which have been developed from them.

The original draft of this report was written by F. W. Keay and G. I. Stillman, members of the Working Group of the Excitation Systems Subcommittee of the IEEE Power Generation Committee. Other Working Group members are: J. W. Skooglund, Chairman, M. L. Crenshaw, R. L. Krahn, J. F. Luini, G. Manchur, F. R. Schleif, N. W. Simons and H. S. Wilson.

MATHEMATICAL MODELS OF EXCITATION SYSTEMS

Block diagrams depicting the transfer functions of an element or group of elements of an excitation system constitute a mathematical model of the physical system. Models provide a means of evaluating excitation system performance. Normally, these mathematical models do not bear a one-to-one correspondence to physical components of the excitation system because such models would contain too much detail for general use. If the model exhibits the same gain and phase characteristics as the actual equipment, it is adequate. Gain and phase characteristics are complex frequency parameters to be discussed later.

Intermediate and simplified models are frequently derived from more complicated models using algebraic model reduction techniques. One of the important considerations arising out of any type of model reduction technique is that direct correspondence between the mathematical model parameters and the physical system is by and large lost. Thus, for example, a particular value of feedback gain in the major loop damping circuit in the mathematical model will not necessarily correspond to the measured value obtained from the physical system. This should be kept in mind when trying to compare field measurements of excitation-system component characteristics with values calculated from mathematical models, particularly when the latter have been reduced; some values calculated for reduced model parameters can differ widely from field measurements.

In brief, the mathematical model can be one having elaborate detail, which retains a one-to-one correspondence with the physical system, or it may be a reduced model which portrays the excitation system in its simplest form. The particular model used will be governed by the type of study for which it is intended.

FREQUENCY RESPONSE OF A CIRCUIT REPRESENTED BY A BLOCK DIAGRAM

Transfer Functions

A block diagram of a simple feedback control system with parameters that might correspond to a static excitation system is shown in Fig. 1. This block diagram is already in a form which is convenient for deriving an expression for the output voltage $V_o(s)$ as a function of the input voltage $V_i(s)$ (i.e., a mathematical model) and this will be done in a following section.

First, however, it is interesting to examine Fig. 1 more closely. It contains a circle labelled "summer," two blocks – each containing an equation, and several connecting lines whose arrows show the direction of signal flow. The "summer" combines incoming signals as shown by the signs and directs the resultant output signal to the upper block. The equation in the block is the transfer function of this block.

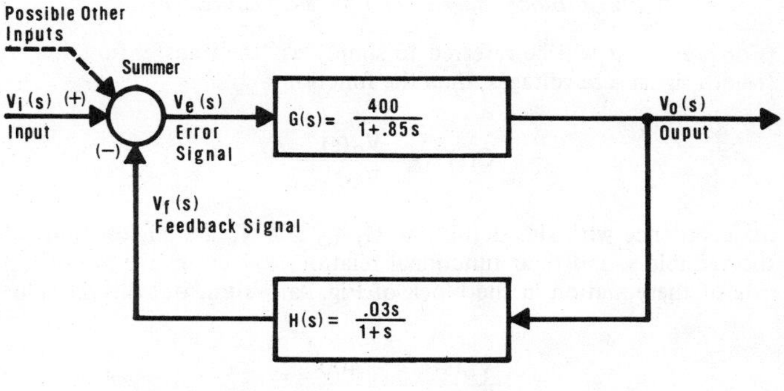

Fig. 1 Block diagram of simple feedback control system.

Fig. 2a shows the upper block with its input and output signals.

The input and output signals can be electrical, such as voltage and current, or electrical analogs of such physical quantities as force and acceleration, or they can be mixed – provided that units are consistent. In this paper, all the examples chosen happen to have input and output signals which are voltages, a situation which is often, but not always, encountered when modelling excitation systems.

The output transfer function is defined[2] as the ratio of the output to the input signal in standard automatic control terminology.

Paper T 72 590-8, recommended and approved by the Power Generation Committee of the IEEE Power Engineering Society for presentation at the IEEE PES Summer Meeting, San Francisco, Calif., July 9-14, 1972. Manuscript submitted February 17, 1972; made available for printing May 16, 1972.

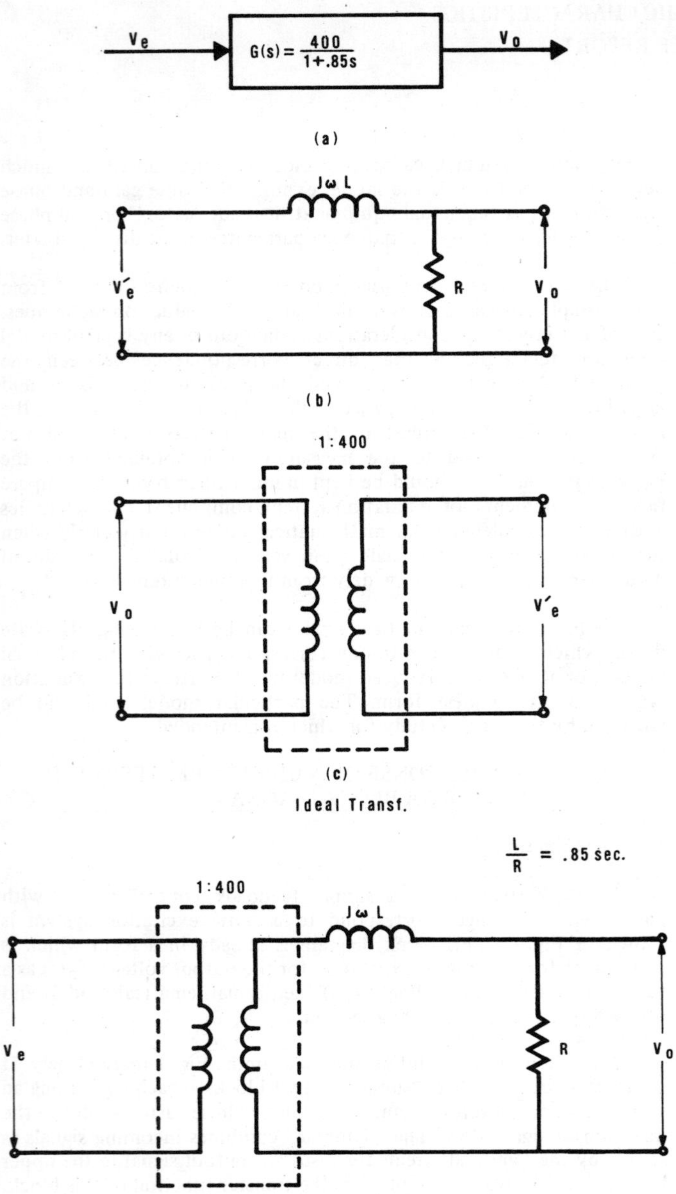

Fig. 2 Block diagram and physical equivalents

(For brevity, it will be referred to simply as "the transfer function"). If both signals are voltages, then the function

$$G(s) = \frac{V_o(s)}{V_e(s)}$$

in accordance with this definition. G, V_o and V_e are all functions of the variable s, and their functional relation is given by the right-hand side of the equation in the block of Fig. 2a. Writing out this relation:

$$\frac{V_o(s)}{V_e(s)} = \frac{400}{1 + .85s} \tag{1}$$

This is an abstract mathematical expression for a voltage ratio, and it would be satisfying to know what combination of physical elements could produce it. Consider the simple circuit of Fig. 2b which has two elements, R and jwl. By regarding this circuit as a potentiometer, the ratio of output to input voltage is:

$$\frac{V_o}{V'_e} = \frac{R}{R + jwl} = \frac{1}{1 + jw\frac{L}{R}} \tag{2}$$

For steady-state circuit analysis, s=jw. Discussion and proof of this fact may be found in any standard text, such as Chapter 5 of Reference 3 or Chapter 7 of "Analysis of Linear Systems" by David K. Cheng.

If the ratio L/R is chosen to be 0.85, then Eq. (2) becomes:

$$\frac{V_o}{V'_e} = \frac{1}{1 + .85s} \tag{3}$$

which is the same as Eq. (1) except for the "400" in the numerator. Next, consider the ideal transformer of Fig. 2c, with input voltage V_e, output voltage V'_e, and a transformation ratio of 1:400, so that

$$\frac{V'_e}{V_e} = 400$$

If the transformer is connected to the circuit of Fig. 2c, as is done in Fig. 2d, then the voltage output-to-input ratio is:

$$\frac{V_o}{V_e} = \frac{V'_e}{V_e} \times \frac{V_o}{V'_e} = 400 \times \frac{1}{1 + .85s} \tag{4}$$

which is identical with Eq. (1). The circuit of Fig. 2d, then is apparently capable of yielding the transfer function of the upper block of Fig. 1.

There are other circuits which can yield the same transfer function. Consider the circuit of Fig. 3a. The ratio of output to input voltage is:

$$\frac{V_o}{V'_e} = \frac{\frac{1}{jwC}}{R_1 + \frac{1}{jwC}} = \frac{1}{1 + jwCR_1} \tag{5}$$

If R_1 and C are so chosen that their product is 0.85, and remembering that s = jw during steady state analysis, Eq. (5) becomes:

$$\frac{V_o}{V'_e} = \frac{1}{1 + .85s} \tag{6}$$

which is identical with Eq. (3). Using the ideal transformer of Fig. 2c to supply the circuit of Fig. 3a, the complete circuit is shown on Fig. 3b. The circuits of Figs. 2d and 3b have identical transfer functions if

$$\frac{L}{R} = R_1 C$$

L/R and R_1C are the time constants of the circuits shown in Figs. 2b and 3a respectively.

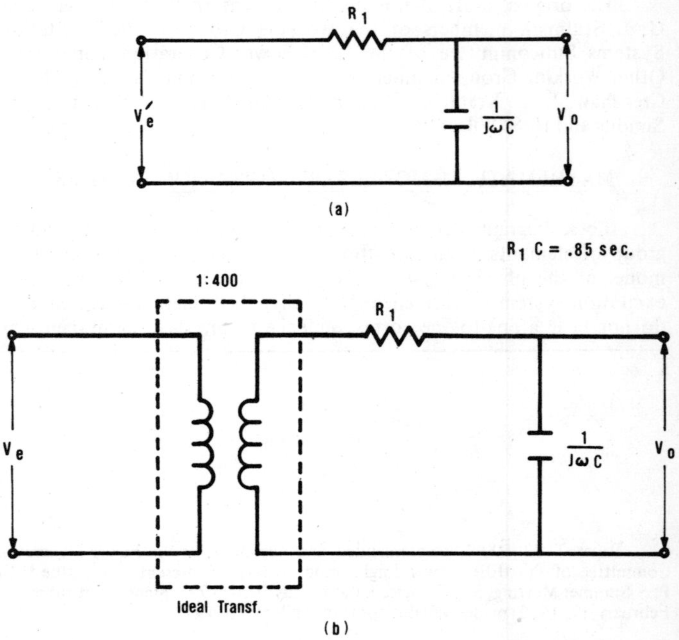

Fig. 3 Equivalent circuits of equation 3 and equation 4.

Referring to the lower block of Fig. 1, the transfer function H(s) is somewhat more complicated than G(s) because of the "s" in the numerator. What combination of physical elements can produce this transfer function? Consider the circuit of Fig. 4a. The output-input voltage ratio is:

$$\frac{V_f}{V'_o} = \frac{jwL}{R + jwL} = \frac{jw\frac{L}{R}}{1 + jw\frac{L}{R}} \tag{7}$$

Choosing L and R so that their ratio is unity and letting jw=s, Eq. (7) becomes:

$$\frac{V_f}{V'_o} = \frac{s}{1 + s} \tag{8}$$

Connecting the ideal transformer of Fig. 4b to the circuit of Fig. 4a yields the circuit of Fig. 4c whose output-input voltage ratio is:

$$\frac{V_f}{V_o} = \frac{0.03s}{1 + s} \tag{9}$$

which is exactly what is desired. The circuit of Fig. 4d will also yield this same transfer function. It should be noted that the reason for using V_o as the input voltage to the circuits of Figs. 4c and 4d and V_f as the output voltage is because the arrows of Fig. 1 specify this to be the case for the lower block.

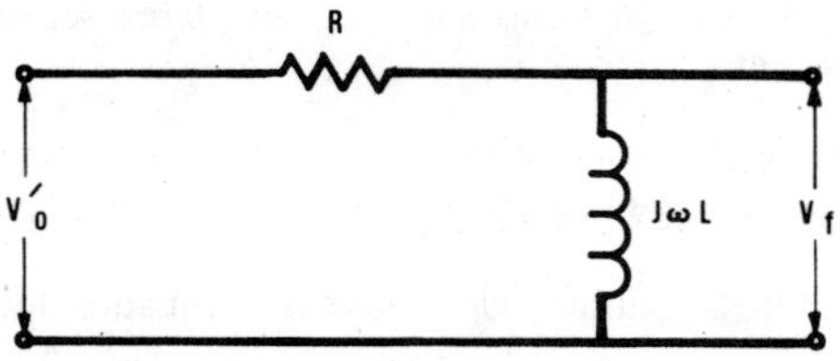

(a)

1 : 03

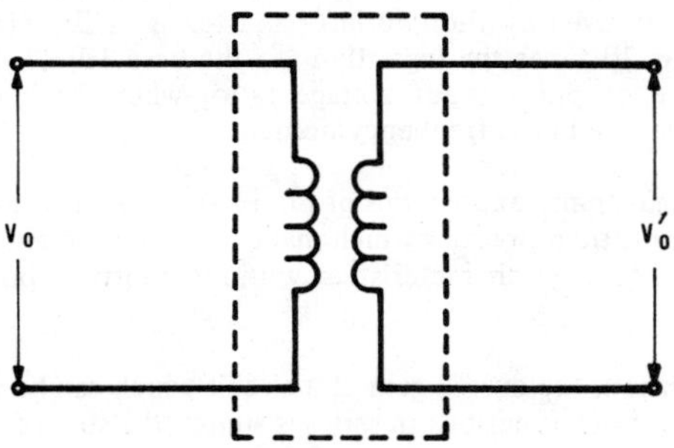

Ideal Transf.

(b)

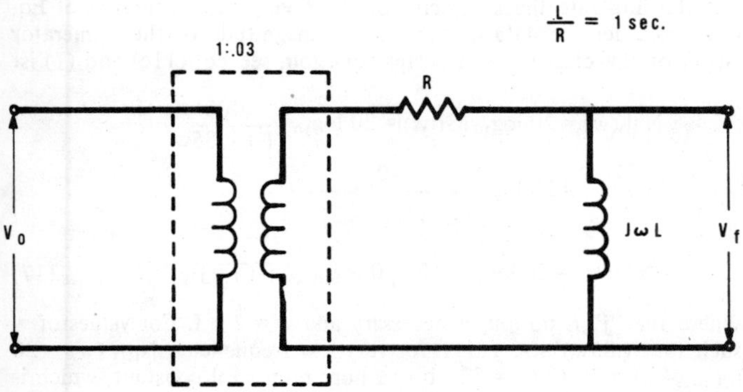

Ideal Transf

(c)

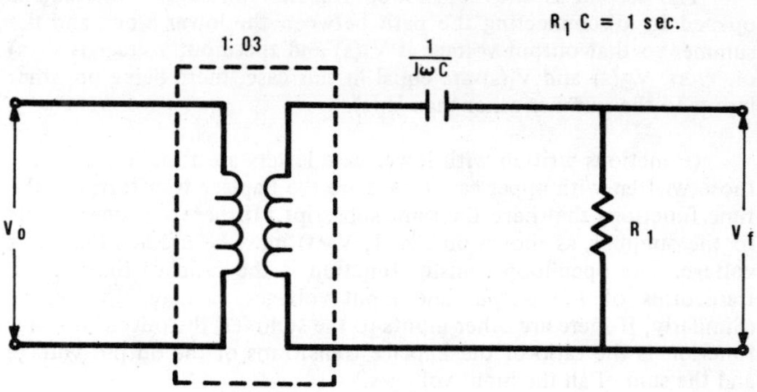

Ideal Transf

(d)

Fig. 4 Equivalent circuits of equation 9.

Laplace Transforms

Before proceeding to manipulate the transfer functions of the upper and lower blocks of Fig. 1 to obtain the overall transfer function for the feedback system shown there, the variable s and Laplace transforms will be discussed.

Despite the fact that both output and input voltages of either block are functions of time, their ratios G(s) and H(s) are expressed as functions of s rather than of t. For steady-state circuit analysis, s = jw and when the steady-state voltage ratios are written in the form of Eqs. (2), (5) or (7), they are generally familiar forms. In general, for steady-state or transient analysis, s is a complex frequency, usually written $s = \sigma + jw$; for steady-state analysis, $\sigma = o$, i.e., the real part of the frequency, is not significant for the solution.

Functions which are stated in terms of s are known as Laplace transforms of the associated time-functions. In Eq. (1), for example, $v_o(s)$ and $v_e(s)$ are Laplace transforms of time functions $v_o(t)$ and $v_e(t)$. The definition of "output transfer function" in the paragraph preceding Eq. (1) is not completely correct; the output transfer function is correctly defined as the ratio of the Laplace transform of the output signal to the Laplace transform of the input signal[2]

The Laplace transform is a mathematical transformation which converts time functions into complex-frequency functions in a manner analogous to the transformation of real numbers into logarithms. Just as logarithms permit the use of the simpler operations of addition and subtraction to replace multiplication and division, Laplace transforms permit the use of algebraic operations to replace those of integration and differentiation. Furthermore, transformation to the complex-frequency domain often clarifies the relationships between driving forces (voltages, in the case of excitation systems) and the response of circuits to these driving functions in a way which is not possible if the forces and responses are expressed only as functions of time. When mathematical operations are completed in the complex-frequency domain, the inverse transforms of the resulting expressions convert them to the time domain so that, for example, voltage responses can be characterized in terms of familiar time-domain concepts like rise time, overshoot, settling time, etc. Once again, the process of taking the inverse transform of a complex-frequency domain function in order to obtain its time-domain expression is analogous to taking the anti-logarithm of a "logarithm-domain" number in order to obtain its "real number domain" equivalent.

The processes of obtaining Laplace transforms of time functions and inverse transforms of complex-frequency functions both involve integrals, having infinite limits, of complex-variable functions. These processes are dealt with in many textbooks, and the transforms of commonly encountered time functions are usually tabulated there.

Transfer Functions of a Feedback System

The simple excitation system of Fig. 1 is a feedback control system. It has two transfer functions of major interest. The first is the "output" or "closed-loop transfer function" which is the ratio of the Laplace transform of the output voltage to the Laplace transform of the input voltage of the circuit as shown in Fig. 1, namely $V_o(s)/V_i(s)$.

The second is the "open-loop transfer function." The loop is opened by disconnecting the path between the lower block and the summer so that output voltage is $V_f(s)$ and the input voltage is $V_e(s)$ or $V_i(s)$, $V_e(s)$ and $V_i(s)$ are equal in this case, there being no other inputs to the summer other than $V_i(s)$.

(Functions written with lower case letters are functions of time; those written with upper case letters are the Laplace transforms of the time functions that have the same subscript.) If there are other inputs to the summer, as shown on Fig. 1, $V_e(t)$ must be used as the input voltage. The open-loop transfer function is the ratio of the Laplace transforms of the output and input voltages, namely $V_f(s)/V_e(s)$. (Similarly, if there are other inputs to the summer, the output transfer function is the ratio of the Laplaçe transforms of the output voltage and the sum of all the input voltages.)

The open-loop transfer function for Fig. 1 is:

$$G(s)H(s) = \frac{400}{1+.85s} \times \frac{.03s}{1+s} = \frac{12s}{(1+.85s)(1+s)} \qquad (10)$$

The closed-loop transfer function is more difficult to obtain. With the loop closed, the input to the upper block is $V_e(s)=V_i(s)-V_f(s)$, and the output from the upper block, and from the feedback systems, is $V_o(s)$, or:

$$V_o(s) = G(s)\left[V_i(s) - V_f(s)\right] \qquad (11a)$$

The input to the lower block is $V_o(s)$ and its output is $V_f(s)$, or:

$$V_f(s) = H(s)\,V_o(s) \qquad (11b)$$

Substituting the value of $V_f(s)$ from Eq. (11b) into Eq. (11a), and going through the necessary algebraic manipulations, a solution for $V_o(s)$ is obtained in terms of $V_i(s)$.

$$\frac{V_o(s)}{V_i(s)} = \frac{G(s)}{1+G(s)H(s)} \qquad (11c)$$

This is the desired closed-loop transfer function. Two points should be made here: (a) the open-loop transfer function is part of the denominator of the closed-loop transfer function, and (b) almost all feed-back systems can have their closed-loop transfer functions manipulated into the form of Eq. (11c), so that it is a good general form to study.

The Bode Diagram

The transfer functions of the previous section are expressions in the complex-frequency domain, and they determine the transient response of the feedback system in the time domain to, say, a step change in voltage. In other words, the transient response and the frequency response of a circuit are intimately related.

The transient response of a circuit or system is described, usually with a graph, showing how the magnitude of the output signal of the circuit or system varies with time for any given input signal, usually a step function or an impulse function. The frequency response is described by how the output signal varies as the frequency of a sinusoidal input signal is varied.

A number of methods are available for describing the frequency response of a circuit. One of these is the Bode diagram, named after Hendrick Bode, a Bell System research engineer who developed the technique. The Bode diagram will be used in this paper because in addition to providing a clear picture of how gain and phase angle between input and output signals of a linear system vary with frequency, it is also a convenient method for using measured frequency response data obtained from field or factory tests to construct representative transfer functions.

The general form of the equations for plotting the Bode diagram of transfer functions will be developed next and then applied to the closed-loop and open-loop transfer functions of Fig. 1 in the following two sections. Eq. (11c) shows that the general form of a transfer function of a feedback system consists of a numerator which is a function of s and a denominator which is a function of s. Calling the first function N(s) and the latter function D(s):

$$\frac{V_o(s)}{V_i(s)} = T(s) = \frac{N(s)}{D(s)} \qquad (12)$$

Where T(s) simply stands for "transfer function." For a steady-state sinusoidal input voltage, the complex frequency $s = \sigma + jw$, becomes the familiar simple frequency $s = jw$. The transfer function can then be written:

$$T(jw) = \frac{N(jw)}{D(jw)} \qquad (13)$$

Since T(jw), N(jw) and D(jw) are complex quantities, they can be written in terms of their magnitudes (gains) and phase angles. Eq. (13) becomes:

$$T(jw) = \frac{|N(jw)|\angle\phi_N(jw)}{|D(jw)|\angle\phi_D(jw)} \qquad (14)$$

Taking the natural logarithms of both sides of Eq. (14), it is possible to obtain separate expressions for the magnitudes and phase angles. Thus,

$$\ln|T(jw)|+j\phi_T(jw) = \ln|N(jw)|+j\phi_N(jw) - \ln|D(jw)|-j\phi_D(jw) \qquad (15)$$

Which, on equating real terms and imaginary terms separately, yield the two equations:

$$\ln|T(jw)| = \ln|N(jw)| - \ln|D(jw)| \qquad (15a)$$

$$T(jw) = \phi_N(jw) - \phi_D(jw) \qquad (15b)$$

Since $\ln = 2.3 \log_{10}$, multiplying both sides of equation 15a by 20/2.3 yields:

$$20\log_{10}|T(jw)| = T_{db}(jw) = 20\log_{10}|N(jw)| - 20\log_{10}|D(jw)| \qquad (16)$$

where each gain on the right-hand side is in decibels and $T_{db}(jw)$ stands for the transfer function gain in decibels. The definition of decibel gain is given by the two left-hand terms of Eq. (16): the gain in decibels is 20 times the logarithm (to the base 10) of the absolute value of the output-to-input voltage ratio, where both output and input voltages are in the frequency domain.

The logarithmic expressions of Eq. (16) possess certain especially simple asymptotic properties which make it possible to construct their complete frequency characteristics without plotting points in the usual sense.

In addition, the overall gain of a circuit which can be represented by several transfer functions in series is simply the sum of the gains of the individual transfer functions when the gains are in logarithmic form. Decibels are used rather than the "raw" logarithmic form because of the simplification of notation.

Bode Diagrams for Closed-Loop Transfer Functions

To illustrate the simplicity of the asymptotic properties of Eq. (16), consider the data of Fig. 1. The magnitude of the numerator N(jw) of the closed-loop transfer function, see Eq. (11c) and (1) is:

$$N_{db}(w) = 20\log_{10}|G(jw)| = 20\log_{10}\frac{400}{1+j.85w} = 20\log_{10}\frac{400}{\sqrt{1+(.85w)^2}}$$

$$N_{db}(w) = 20\log_{10}400 - 20\log_{10}\sqrt{1+(.85w)^2} \qquad (17)$$

Where the "j" is no longer necessary and $w = 2\pi f$. For values of w such that $(.85w)^2 \ll 1$, i.e., for very-low frequencies, $N_{db}(w) = 20 \log_{10} 400 = 20\,(2.6) = 52$ db – a pure numerical constant, which is called the d.c. or static gain of N_{db}.

For values of w such that $(.85w)^2 >> 1$, Eq. (17) becomes:

$$N_{db}(w) = 20 \log_{10} 400 - 20 \log_{10} (.85w) \tag{18}$$

So that the high frequency asymptote is a linear function of $\log_{10}(.85w)$. When $w = 1/.85$ Eq. (18) yields $N_{db}(w) = 20 \log_{10} 400$ which is the same value as the static gain, so the low frequency and high frequency asymptotes intersect at this point. Such an intersection is called a "break point." When $w = 10 \times 1/.85$, Eq. (18) yields $N_{db}(w) = 20 \log_{10} 400 - 20$, so that the slope of the high frequency asymptote is -20 db/decade, where a decade is any interval on the frequency scale covering a 10:1 frequency ratio.

With the low and high frequency asymptotes and their intersection all known, they can be plotted as shown on the upper curve of Fig. 5. Thus, a good approximation of the frequency response curve of the gain of N(jw) has been obtained without plotting points in the usual sense. One such point can be easily obtained, however, which permits the actual response curve to be sketched in more accurately.

When $w = \frac{1}{.85}$, Eq. (17) yields

$$N_{db}(w) = 20 \log_{10} 400 - 20 \log_{10} \sqrt{2} = 20 \log_{10} 400 - 3 \tag{19}$$

Thus where the two asymptotes intersect, $w = \frac{1}{.85}$, the actual curve is 3 db lower than the intersection of the asymptotes. The actual curve is also sketched in dotted on the upper curve of Fig. 5.

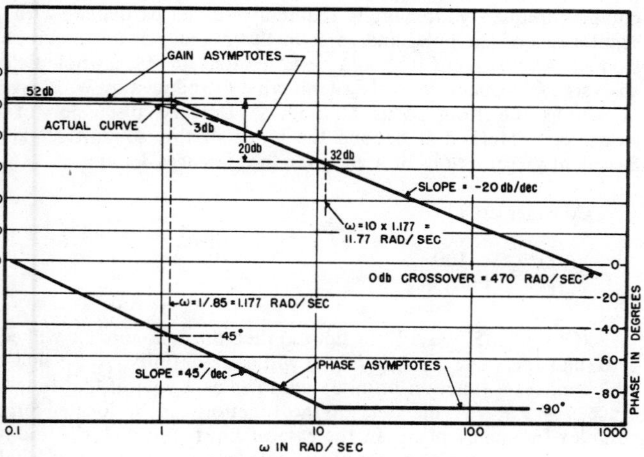

Fig. 5 Bode Plot of $N(jw) = \frac{400}{1 + .85jw}$

The frequency characteristics of the phase angle can also be drawn. Since,

$$N(jw) = \frac{400}{1 + j.85w} = \frac{400\angle 0°}{\sqrt{1 + (.85w)^2} \angle \tan^{-1}(.85w)}$$

the phase angle of N(jw) is

$$\phi_N(w) = -\tan^{-1} .85w \tag{20}$$

At the limit, when $w = 0, \phi_N(w) = 0°$; and when $w = \infty, \phi_N(w) = -90°$. When $w = 1/.85, \phi_N(w) = -45°$ and one approximation for the asymptote is to take the slope as -45°/decade. The phase angle asymptotes are shown on the lower curve of Fig. 5.

Next, consider the frequency characteristic of the denominator D(jw) of the closed-loop transfer function. From Eq. (11c), $D(jw) = 1 + G(jw)H(jw)$ so that:

$$20 \log_{10} D(jw) = D_{db}(jw) = 20 \log_{10} [(1 + G(jw)H(jw)]$$

Substituting the expressions from Fig. 1 for G(jw) and H(jw):

$$D_{db} = 20 \log_{10} \left| 1 + \frac{400}{1 + j.85w} \times \frac{.03jw}{1 + jw} \right| \tag{21}$$

The expression within the absolute magnitude lines becomes

$$\frac{.85(jw)^2 + 13.85(jw) + 1}{(1+j.85w)(1+jw)} = .85 \frac{(jw)^2 + 16.3(jw) + 1.177}{(1+j.85w)(1+jw)}$$

The roots of the numerator are found to be (jw) = -16.2216, -0.725249 so that the expression becomes

$$\frac{.85\,(jw + 16.22)(jw + .073)}{(1 + j.85w)(1 + jw)} \tag{22}$$

Taking the absolute magnitude of each of the bracketed terms of Eq. (22), permits Eq. (21) to be written as:

$$D_{db}(w) = 20 \log_{10} \frac{\sqrt{1 + (\frac{w}{16.22})^2} \quad \sqrt{1 + (\frac{w}{.073})^2}}{\sqrt{1 + (.85w)^2} \quad \sqrt{1 + w^2}} \tag{23}$$

The logarithmic form of Eq. (22) permits separation of terms:

$$D_{db}(jw) = 20 \log_{10} \sqrt{1 + (\frac{w}{16.22})^2} + 20 \log_{10} \sqrt{1 + (\frac{w}{.073})^2} - 20 \log_{10} \sqrt{1 + (.85w)^2} - 20 \log_{10} \sqrt{1 + w^2} \tag{24}$$

Table 1, below, gives the pertinent information required for a Bode plot of each term of Eq. (24).

TABLE I

Term of Eq. (24)	Value of Low-Freq. Asymptote (db)	Slope of Low-Freq. Asymptote	Breakpoint Freq. (rad/s.)	Slope of High-Freq. Asymptote (db/decade)
1	0	0	16.22	+20
2	0	0	.073	+20
3	0	0	1/.85 = 1.177	-20
4	0	0	1	-20

As an example of how the values in the Table were arrived at, consider the first term of Eq. (24):

(a) At low frequencies, i.e., when $\frac{w}{16.22} << 1$ the asymptote is 0 db.

(b) The intersection of the low and high frequency asymptotes, i.e., the breakpoint, occurs at $\frac{w}{16.22} = 1$ or $w = 16.22$

(c) At high frequencies, i.e., when $\frac{w}{16.22} >> 1$, the expression for the asymptote is $20 \log_{10} \frac{w}{16.22}$. When w is a decade away from the breakpoint, i.e., when $w = 10 \times (16.22)$, the value on the asymptote is $20 \log_{10} 10$ or 20 db, so the slope of the high frequency asymptote is +20 db/decade.

Fig. 6 shows the Bode plot of $D_{db}(w)$ and is obtained by adding together the plots of the individual terms for which the information is given in Table I.

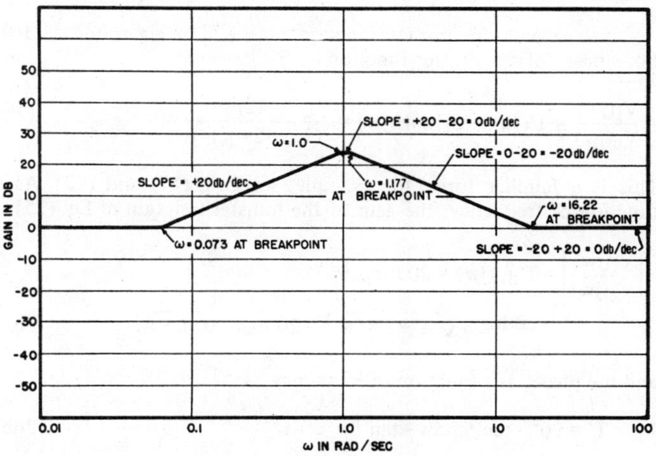

Fig. 6 Bode plot of Equation 24 (see Table I)

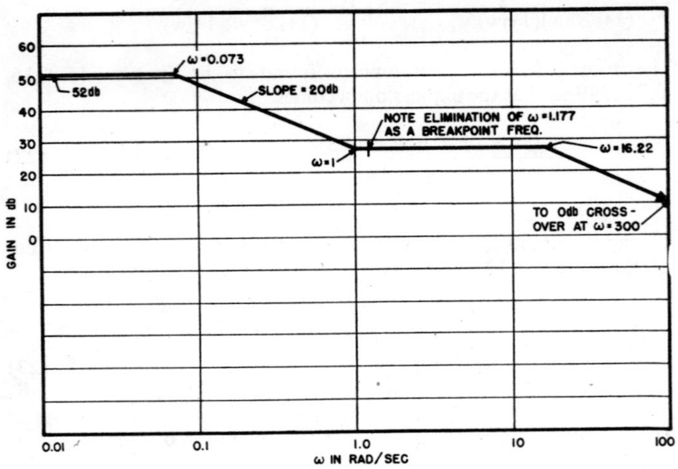

Fig. 7 Bode plot of Equation 16 (obtained by subtracting plot of Fig. 6 from plot of Fig. 5)

Finally, Fig. 7 shows the Bode plot of the output transfer function $T_{db}(w)$ and is obtained, in accordance with Eq. (16), by subtracting the Bode plot of $D_{db}(w)$ in Fig. 6 from the Bode plot of $N_{db}(w)$ of Fig. 5.

If the process of constructing the Bode plot of the output transfer function is reversed, i.e., if the frequency response of a suitable transfer function were known, then the parameters of the transfer function could be derived from the frequency response parameters – the breakpoints and the slopes. Therefore, if the values of the parameters of a block-diagram model of a piece of equipment are not known, then a convenient procedure would be to obtain, by field measurements, the response of the magnitude and phase angle of the output voltage as the frequency of the sinusoidal input voltage to the piece of equipment is varied. The ratio of output-to-input voltage as a function of frequency is, in effect, a field-test Bode plot. The asymptotes could then be approximated and the values of the d.c. gain and the breakpoint frequencies could be obtained, thereby permitting determination of the parameters in the block diagram.

The assumptions involved here are (a) that some idea of the basic blocks in the diagram is available, (b) that the system is linear, i.e., that none of the signals are so large as to cause any of the elements to saturate or become non-linear. If the signal is large enough to result in non-linear operation of the system, then the type of system model that we have discussed until now can be modified to take account of such non-linearities. The modifications consist of additional blocks in the block diagram called "limiters," or of additional feedback loops. Limiters and feedback loops to take account of saturation or other non-linearities are shown in the standardized excitation systems of Reference 1.

Bode Diagram of the Open-Loop Transfer Function

The open-loop transfer function of Fig. 1 is given in Eq. (10). For sinusoidal signals, this function is:

$$\frac{V_f(jw)}{V_e(jw)} = T'(jw) = G(jw)H(jw) = \frac{12jw}{(1+j.85w)(1+jw)} \tag{25}$$

This is a familiar form; for example, see Eqs. (17) and (22). As a function of frequency, the gain of the transfer function of Eq. (25) is

$$\left|\frac{V_f(jw)}{V_e(jw)}\right| = T'_{db}(jw) = 20\log_{10} 12w$$

$$- 20\log_{10}\sqrt{1+(.85w)^2} - 20\log_{10}\sqrt{1+w^2} \tag{26a}$$

and the phase, as a function of frequency, is:

$$T' = 90° - \tan^{-1}.85w - \tan^{-1}w \tag{26b}$$

The open-loop gain and phase frequency responses given by Eqs. (26a) and (26b) are plotted in Fig. 8.

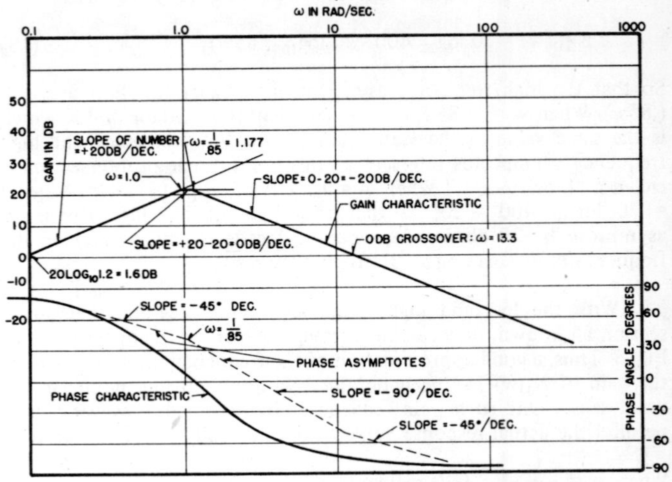

Fig. 8 Open-loop gain and phase frequency response.

COMPLEX FREQUENCY DOMAIN

The time-response of a circuit can be obtained by taking the inverse Laplace transform of the output signal expressed in the complex frequency domain but this is often an impractical procedure for a complicated circuit. To be able to take advantage of simpler methods that have been developed, more familiarity with the complex-frequency domain is required and, in particular, with the significance of the poles and zeros in this domain of the circuit being studied. At this point it should be noted that there is a whole field of analysis of feedback control systems (first introduced by W. R. Evans) known as the Root-Locus Method.[8,9] Detailed discussion of the Root-Locus Method is beyond the scope of this committee report. Excellent presentations of it are given in references 8 and 9.

Consider Eq. (1).

$$\frac{V_o(t)}{V_e(t)} = \frac{400}{1+0.85s} \tag{1}$$

If s = -1/.85 = -1.177 in Eq. (1) then the denominator will equal zero, and the value of $V_o(s)/V_e(s)$ will equal infinity. Any value of s which makes a function infinite, is known as a pole of that function. Since $s = \sigma + jw$, the pole of the function can be located on the complex-frequency plane. In the case of Eq. (1), $\sigma = -1.177$ and $jw = 0$, and the corresponding plot of the pole is shown as the cross labelled A on Fig. 9(a).

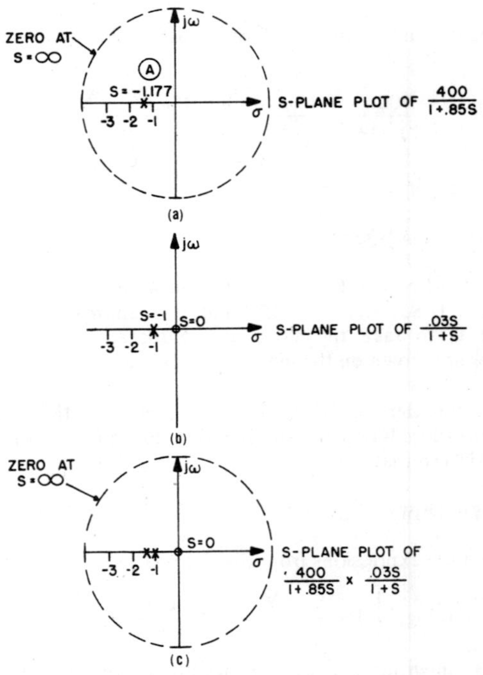

Fig. 9 S-Plane plots

If $s = \infty$ in Eq. (1), then the denominator will be infinite, and the value of $V_o(s)/V_e(s)$ will equal zero. Any value of s which makes a function zero is known as a zero of that function. In the case of Eq. (1), either σ can equal infinity, or jw can equal infinity. All such values of $\sigma + jw$ lie on a circle at infinity as shown on Fig. 9(a).

Next, consider Eq. (9).

$$\frac{V_f}{V_o}(s) = \frac{0.03s}{1+s} \tag{9}$$

Using the same reasoning as in the previous two paragraphs, this function has one zero and one pole. The zero is located at $\sigma = 0, jw = 0$ and is shown as a circle at the origin. The pole is located at $\sigma = -1, jw = 0$. These points are shown on Fig. 9(b).

Finally, consider Eq. (10).

$$G(s)H(s) = \frac{12s}{(1+.85s)(1+s)} \tag{10}$$

This equation has two zeros, at $s = 0$, $s = \infty$; and two poles, at $s = -1.177$ and at $s = -1$. The two zeros and two poles are shown on Fig. 9(c), and it should not be surprising on reviewing the relation between Eqs. (1), (9) and (10) that the poles and zeros of Fig. 9(c) are the combined poles and zeros shown on Figs. 9(a) and 9(b).

So far, the poles and zeros, except for those at infinity, that have been discussed all lie on the σ - axis. Poles and zeros can lie anywhere on the s-plane. Consider for example a function which has a denominator $s^2 + 2s + 5$. This can be factored into $(s + 1 + j2)(s + 1 - j2)$ so that the function will have poles at $s = -1 + j2$ and $s = -1 - j2$ as well as zeros all along the circle at infinity. This pair of poles is plotted on Fig. 10.

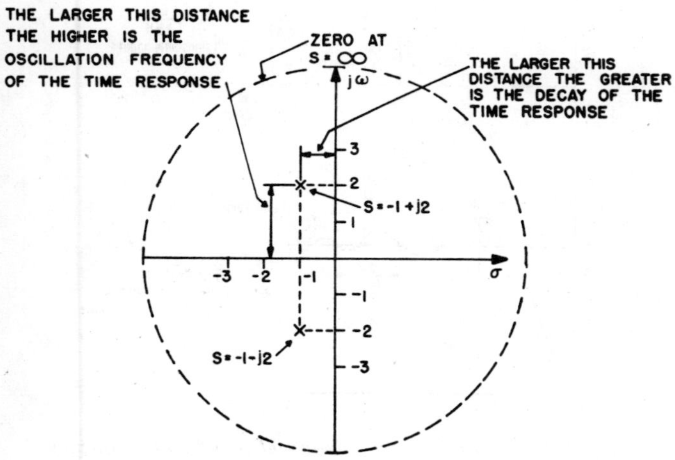

Fig. 10 S-Plane plot of $\frac{K}{S^2 + 2S + 5}$

Laplace transform theory tells us that the perpendicular distances of a pole from the σ -axis and the jw-axis have important physical implications. (See Fig. 10.) If the distance of the pole from the σ-axis is zero, i.e., if the pole is on the σ-axis, the time-response of the function is non-oscillatory; the greater the distance of the pole from the σ-axis, the higher will be the frequency of oscillation of the time-response function. If the distance of the pole from the jw-axis is zero, i.e., if the pole is on the jw-axis, the time-response of the function does not decay. The greater the distance of the pole from the jw-axis in the left half of the plane, the more rapid will be the decay of the time-response of the function; the greater the distance of the pole from the jw-axis in right half of the plane, the more rapid will be the growth of the time-response of the function. This latter information tells us that all poles must lie to the left of the jw-axis or on the jw-axis if the function is to be stable.

Laplace transform theory gives much more interesting and useful information about the s-plane, which could not possibly be covered even briefly in this paper. It is worthwhile, however, mentioning two theorems which are relevant to the next section.

1. Final Value Theorem

If the Laplace transform of a function is known, then by taking the limit of this function as s approaches the origin of the s-plane, the time-response of the function as t approaches ∞, i.e., its steady-state behavior, can be found. Formally, this is expressed as follows:

$$\lim_{s \to 0} s\,F(s) = \lim_{t \to \infty} f(t) \tag{27}$$

2. Initial Value Theorem

If the Laplace transform of a function is known, then by taking the limit of the transform as s approaches the circle at infinity, the time-response of the function as t approaches zero, i.e., its initial value, can be determined. Formally, this is expressed as:

$$\lim_{s \to \infty} s\,F(s) = \lim_{t \to 0} f(t) = f(0^+) \tag{28}$$

DETERMINING TIME RESPONSE FROM FREQUENCY RESPONSE

By transient response is meant the description in real time of how a circuit responds to a disturbance. Fig. 11 shows a typical time (or transient) response of the output of a system to a step increase in input to the system.[2] Aside from the initial and final steady-state values, four other quantities are of interest: (1) dead time or delay time; (2) rise time; (3) maximum overshoot or peak value of transient deviation, and (4) settling time.

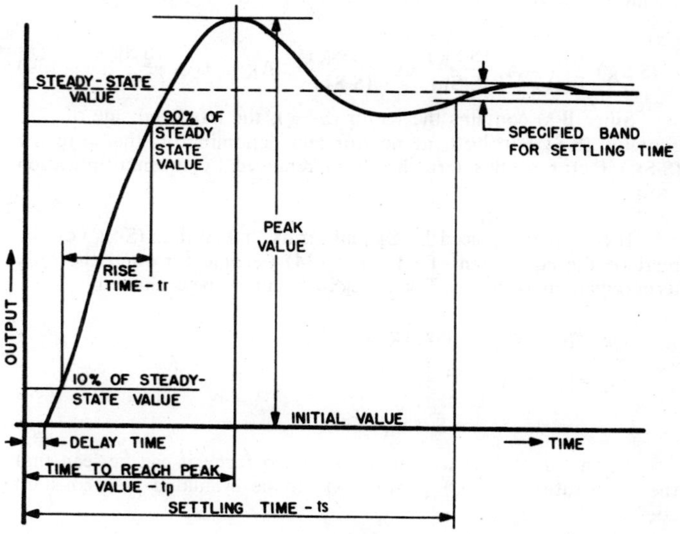

Fig. 11 Typical time response of system to a step increase in input.

Fig. 12 shows (a) an open-loop gain frequency response, (b) a closed-loop gain frequency response, and (c) a transient time response to a step-function input to the feedback system shown there. The quantities are self-explanatory from the figure, but t_p – the time to peak overshoot – contains in it both the delay time and the rise time and some small additional time.

The time response of the output of a feedback control system may be obtained by taking the inverse Laplace transform of the function found by multiplying the Laplace transform of the input function by the closed-loop transfer function of the feedback control system, i.e.

$$V_o(t) = \mathcal{L}^{-1}\left[V_i(s)\,\frac{G(s)}{1+G(s)H(s)}\right] \tag{29}$$

A general method of evaluating the inverse transform of a function is to express the function as a sum of partial fractions with constant coefficients.[3] The inverse transform of each of the partial fractions may then be obtained from tables of Laplace transforms. The complete inverse transform of the function is the sum of the inverse transforms of all of the partial fractions. The procedure described in the following is essentially the same as presented in chapter 4 of reference 3.

Equation (29) may be written as (30)

$$V_o(t) = \mathcal{L}^{-1}\ F(s)$$

In general F(s) is a rational algebraic function that can be written in the form:

$$F(s) = \frac{A(s)}{B(s)} \quad \frac{a_m S^m + a_{m-1} S^{m-1} \ldots + a_1 S + a_o}{S^n \quad + b_{n-1} S^{n-1} \ldots + b_i S + b_o} \tag{31}$$

To express F(s) as a sum of partial fractions, it is first necessary to determine the factors of the denominator. Unfortunately, as stated in the preceding section, this is a complicated and laborious procedure. A method of approximating these factors will be described later; however, for purposes of illustrating the inverse transformation procedure, it is sufficient to indicate them in literal form. Eq. (31) may then be written as:

$$F(s) = \frac{A(s)}{B(s)} = \frac{A(s)}{(S-S_1)(S-S_2)\ldots(S-S_K)\ldots(S-S_n)} \tag{32}$$

where $S_1, S_2 \ldots S_K \ldots S_n$ are factors of the denominator and may be real or complex quantities.

The fraction A(s)/B(s) can be expressed as a sum of partial fractions:

$$\frac{A(s)}{B(s)} = \frac{A_1}{(S-S_1)} + \frac{A_2}{(S-S_2)} + \ldots \frac{A_K}{(S-S_K)} + \ldots \frac{A_n}{(S-S_n)} \tag{33}$$

To evaluate the "typical" coefficient, A_K, both sides of Eq. (33) are multiplied by the denominator of the A_K term giving:

$$(S-S_K)\frac{A(s)}{B(s)} = A_1\frac{(S-S_K)}{(S-S_1)} + A_2\frac{(S-S_K)}{(S-S_2)} + \ldots A_K + \ldots + A_n\frac{(S-S_K)}{(S-S_n)} \tag{34}$$

Since B(s) contains the factor $(S-S_K)$, the left hand side of (34) contains $(S-S_K)$ in both numerator and denominator; therefore the $(S-S_K)$ factor of this term has been removed by the multiplication process.

If S is now replaced by S_K, all terms that contain $(S-S_K)$ on the right of the equal sign of equation (34) become zero and the only term remaining is the coefficient A_K, which is to be evaluated.

The equation defining A_K is:

$$A_K = \left[(S-S_K)\frac{A(s)}{B(s)}\right]_{S = S_K} \tag{35}$$

When Eq. (35) is written in expanded form, it can be seen that the substitution $(S=S_K)$ permits A_K to be evaluated as a constant coefficient:

$$A_K = \frac{A(S_K)}{(S_K-S_1)(S_K-S_2)\ldots(S-S_{K-1})(S-S_{K+1}) + \ldots (S_K-S_n)} \tag{36}$$

Equation (36) permits the evaluation of all the coefficients A_1 to A_n as K is set equal to all values from 1 to n, i.e. the literal coefficients arbitrarily assigned in Eq. (33) have been explicitly defined and can be readily evaluated.

The operation of the inverse transformation of F(s) is now simplified since the transformation of each term in the expansion of F(s) can be readily obtained from tables of transform pairs. A typical form is:

$$\mathcal{L}^{-1}\left[\frac{A_K}{(S-S_K)}\right] = A_K \epsilon^{S_K t}, 0 \leqslant t \tag{37}$$

The complete inverse transform of F(s) is then the sum of the inverse transform of each term:

$$V_o(t) = \mathcal{L}^{1}\ (F(s) = A_1\epsilon^{S_1 t} + A_2\epsilon^{S_2 t} + \ldots A_K\epsilon^{S_K t} + \ldots A_N\epsilon^{S_N t} \tag{38}$$

Frequently, the transient response of a feedback control system for a unit step input $(V_1(s) = \frac{1}{s})$ is of interest. In this case the evaluation of L^{-1} (F(s) leads to a time response of the form:

$$V_o(t) = A_o + A_1\epsilon^{S_1 t} + A_2\epsilon^{S_2 t} + \ldots A_K\epsilon^{S_K t} + \ldots A_N\epsilon^{S_N t}$$

It has been emphasized in the foregoing that the process of directly evaluating the inverse transform is a complicated one due to the difficulty in evaluating the factors of the denominator, i.e. the poles of the function. Fortunately, the work of Biernson and Chen in the mid-1950's (references 4-7) provides a means of approximating the transient response of a linear feed-back control system directly from the open-loop frequency response without the necessity of evaluating the complete inverse transform of the closed loop transfer function.

The approximation technique is based on estimating the values of the poles of the closed-loop transfer function from the system open-loop transfer function and estimating only a few of the coefficients of the approximate partial fraction expansion of the closed-loop transfer function. The approximate poles, along with the approximate coefficients, are used to estimate the principal characteristics (rise time, overshoot, and settling time) of the time response.

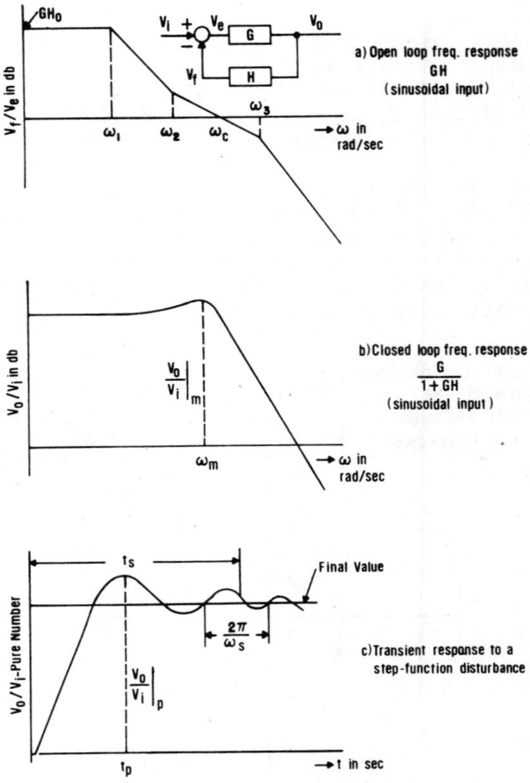

Fig. 12 Comparison of open-loop and closed-loop frequency response and transient response.

As previously shown, the closed-loop transfer function of a feedback control system, such as shown in Figs. 1, 12a, and 13 is:

$$\frac{V_o}{V_i}(s) = \frac{G(s)}{1 + G(s)H(s)} \tag{40}$$

For such a closed-loop system, having the characteristic equation 1 + G(s) H(s) = 0, the root locus is a plot in the s plane of all values of s which make G(s) H(s) a negative real number; those points which make the loop transfer function G(s) H(s) = -1 are roots.[2]

If G(s) is written as KG_1 (s) the equation satisfied along the root loci is:

$$KG_1(s)H(s) = -1 \tag{41}$$

As K approaches zero, G_1(s) H(s) must approach infinity or s must approach the poles of G(s) H(s). Thus, since the loci are plots of the closed-loop pole positions when the open-loop gain varies from zero to infinity, the loci start at the open-loop poles.[9]

Equation 41 indicates that a large value of K requires G_1(s) H(s) approaching zero or a value of s approaching a zero of G_1(s) H(s).

Thus, the loci terminate on the zeros of $G_1(s)\,H(s)$ as K approaches infinity.[9]

Figure 13 is a plot of the root-locus of the excitation system of Fig. 1 having the open-loop transfer function:

$$\frac{V_O}{V_e}(s) = \frac{.03KS}{(1+S)(1+.85S)(1+.02S)} \tag{42}$$

where K is nominally 400 in Fig. 1. The root locations for various values of gain (K) are indicated on Fig. 13 showing how the poles of the closed loop transfer function migrate as the open-loop gain changes. From the foregoing, it is seen that the closed-loop poles migrate in the s-plane from the open-loop poles to the open-loop zeros along the root loci as the open loop gain increases from zero to infinity.

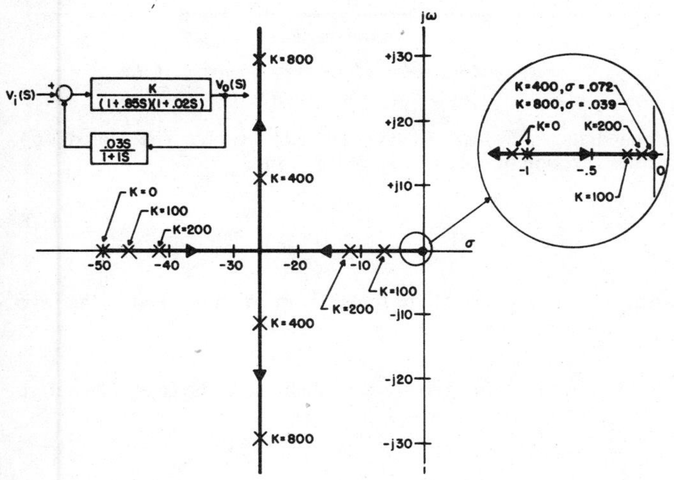

Fig. 13- Root loci for $\frac{V_O}{V_i}(s) = \frac{K(1+S)}{.017S^3 + .887S^2 + (1.87 + .03K)S + 1}$

The gain information offered by the asymptotes of the open-loop Bode plot enables one to determine quickly which closed loop poles are near the beginning, which are near the end, and which are near the middle of migration along the root loci.

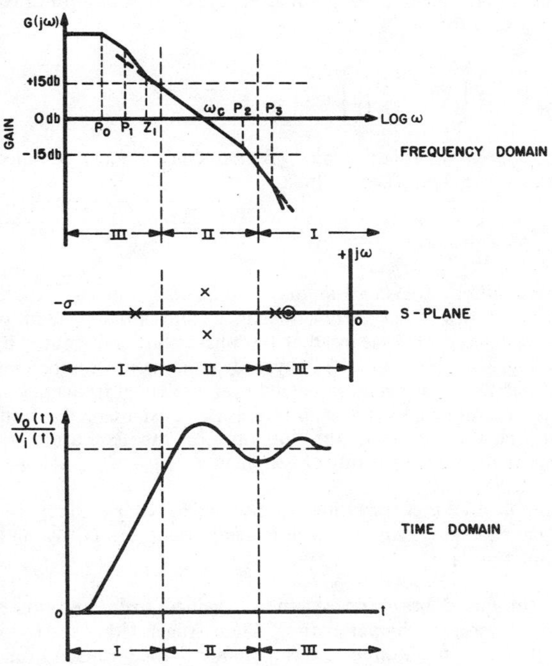

Fig. 14 Comparison of frequency domain, S-Plane, and time domain.

Using the Final Value and Initial Value theorems, Chen suggested that the closed-loop poles be divided into three groups corresponding to three ranges of the frequency response and time response characteristics, respectively, as shown in Fig. 14[6]. The Bode plot in Fig. 14 represents the open-loop frequency response of the feedback control system shown in Fig. 15. It should be noted that the ±15 db band selected to divide the frequency range into the three regions corresponding to the pole migration in the s-plane is arbitrary. It was suggested by Chen based on engineering experience in application of the Initial and Final Value theorems to many typical control systems.

In Figure 14, Range I corresponds to the locus of closed loop poles near the beginning of migration, i.e., near the open-loop poles. As shown in Fig. 14 Range I also corresponds to a sub-transient in the time domain and to the frequency response below the -15 db band on the Bode plot. In this range of the frequency response, $G(jw)H(jw) \ll 1$: therefore:

$$\frac{V_T}{V_R}(s) = \frac{G(s)}{1+G(s)H(s)} \approx G(s) \tag{43}$$

and the closed-loop pole ≈ -p3, the open-loop pole corresponding to the break-point on the Bode plot slightly above the 15 db line. In terms of the root-loci, this simply means that the closed loop pole which starts from -p3 has not moved far from s = -p3 for the gain in range I of the Bode plot. This closed-loop pole near -p3 corresponds to a transient term that decays rapidly and the overall system response is not affected significantly if the closed loop pole is assumed not to have moved from -p3 at all. Poles other than -p3 may be neglected in this range.

At the other extreme, Range III of Fig. 14 corresponds to closed loop poles near the end of migration, i.e., near the open-loop zeros (approaching steady-state). In the corresponding range of the frequency response, above the +15 db line on the Bode plot, $G(jw)H(jw) \gg 1$ and the following approximation may be written:

$$\frac{V_e}{V_R}(s) = \frac{1}{1+G(s)H(s)} \approx \frac{1}{G(s)H(s)} \tag{44}$$

From this approximation, it is evident that the closed loop poles of $\frac{V_T}{V_R}(s)$, which are the same as the poles of $\frac{V_e}{V_R}(s)$, are approximately the open loop zeros of G(s)H(s).

As shown on Fig. 14, one approximate closed loop pole is near the open loop zero, $s = -z_1$ in range III. The open-loop zero corresponds to the break point on the Bode plot above the 15 db line. In terms of pole migration along the root loci, this means that one closed loop pole has moved from $-p_1$ very close to $-z_1$ but has not quite reached $-z_1$. Since $s = -z_1$ is also a closed loop zero, there is a "dipole" (formed by a zero and a pole close to each other) near $s = -z_1$. The dipole has a significant effect on the settling time of the system since, as shown in Fig. 14, it corresponds to a small but long transient term producing a "tail" in the system response.

Chen has shown that in the case of a dipole close to the origin of the s plane the closed-loop pole is approximately located at:

$$S = -(1 + K')\,Z_1 \tag{45}$$

Where K′ is the gain obtained from the open-loop frequency asymptote plot at the break frequency corresponding to Z_1.

Finally, Range II of Fig. 14 corresponds to the locus of closed loop poles in the middle of migration. Range II also corresponds to the main transient including the rise time and overshoot. With respect to the Bode plot, it corresponds to the frequency response in the region of the cross-over frequency and within the ±15 db band.

The overall open-loop transfer function in terms of its poles and zeros may be written as:

$$G(s) = \frac{K''(S+Z_1)}{(S+P_O)(S+P_1)(S+P_2)(S+P_3)} \tag{46}$$

where $K'' = \frac{KP_OP_1P_2P_3}{Z_1}$ is the gain which satisfies Eg. (41) along the root loci.

In the mid-range of the frequency response it may be considered that $P_o < P_1 < Z_1 << |s| << P_3$ and hence the following approximation may be written:

$$G(s) \approx \frac{K''/P_3}{S(s+P_2)} \tag{47}$$

The approximate closed-loop transfer function is then:

$$\frac{G(s)}{1+G(s)} \approx \frac{K''/P_3}{S(S+P_2)+K''/P_3} \tag{48}$$

From this approximation, the approximate complex closed loop poles are obtained as the roots of the equation:

$$S^2 + P_2 S + \frac{K''}{P_3} = 0 \tag{49}$$

Approximation of the closed-loop transfer function within the ±15 db does not always result in a quadratic for the characteristic equation; however, for most stable systems, the approximation will usually not be higher than third order. A number of common variations of open-loop frequency response characteristics near the cross-over frequency along with their corresponding transfer functions are tabulated in reference 6.

The techniques for approximating the transient response from the frequency response can best be illustrated by applying them to an excitation control system. Fig. 15 represents a typical excitation control system off-line, (generator at no load) assuming that the synchronous machine may be represented by a single time constant. In the following, the approximate transient response is compared to the exact transient response calculated using a digital computer program.

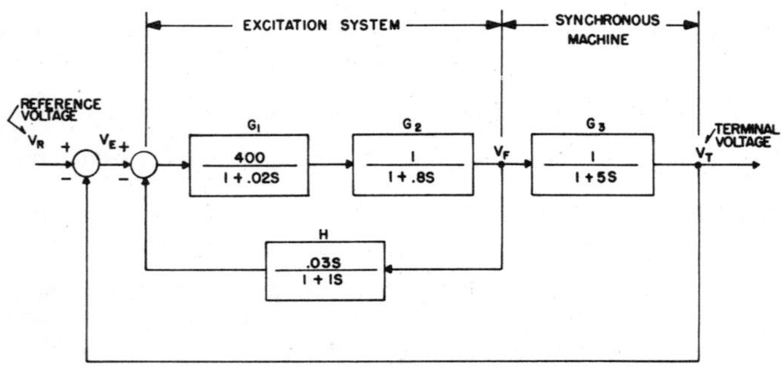

Fig. 15 Typical excitation control system at no load.

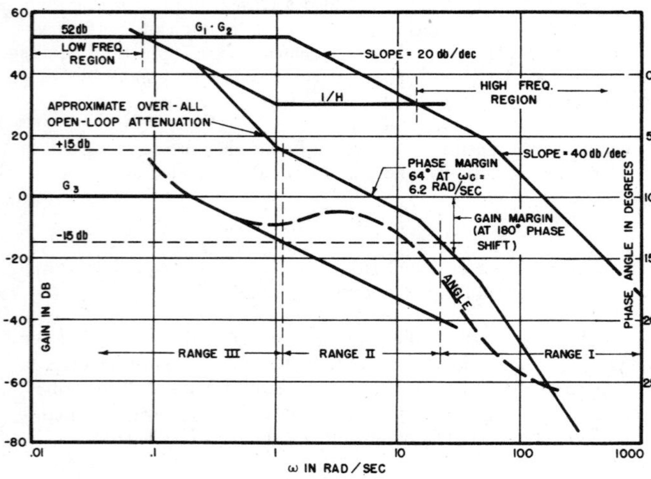

Fig. 16 Bode plot of the excitation control system of Fig. 15

Figure 16 is a Bode plot of the open-loop frequency response of the excitation control system of Fig. 15 and Fig. 17 is the corresponding closed-loop frequency response. In Fig. 16 a graphical technique has been used to obtain the equivalent of the sub-loop comprised of the excitation system including its stabilizer. The graphical technique for obtaining the equivalent of the sub-loop is based on approximating its closed loop transfer function at low,

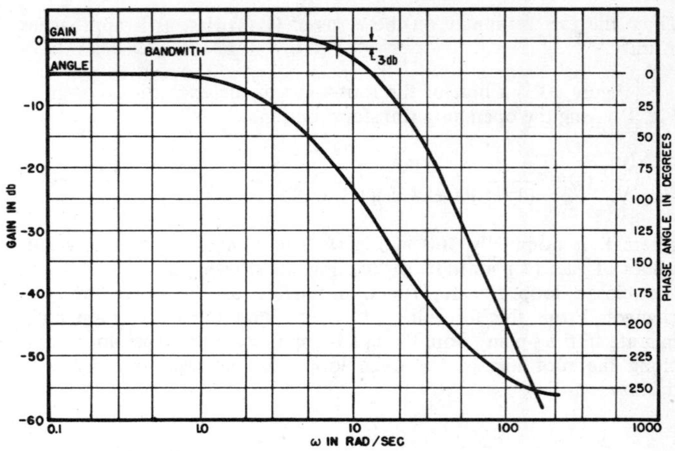

Fig. 17 Closed-loop frequency response of the excitation control system of Fig. 15

intermediate, and high frequencies, respectively. The closed-loop transfer function of the sub-loop of Fig. 15 is:

$$\frac{V_f}{V_e} = \frac{G_1 G_2}{1 + G_1 G_2 H} \tag{50}$$

where $\frac{V_f}{V_e}$, G_1 G_2, and H are functions of jw. (For brevity, the jw is omitted.)

At low frequencies, H is very small and $G_1 G_2 H << 1$, therefore

$$\frac{V_f}{V_e} \approx G_1 G_2 \tag{51}$$

At high frequencies of $G_1 G_2 H$ approaches zero as w approaches infinity, therefore the denominator of equation (50) approaches one and again:

$$\frac{V_f}{V_e} \approx G_1 G_2 \tag{52}$$

For an intermediate range of frequencies, it is helpful to rewrite equation (50) as:

$$\frac{V_f}{V_e} = \frac{1}{H}\left[\frac{G_1 G_2 H}{1 + G_1 G_2 H}\right] \tag{53}$$

In this intermediate range of frequencies, most systems are designed so that $G_1 G_2 H >> 1$, then:

$$\frac{V_f}{V_e} \approx \frac{1}{H} \tag{54}$$

And the resultant transfer function is dependent almost entirely on the characteristics of the feedback element and is independent of the forward elements. This means that the phase shift contributed by the time delays inherent in the physical equipment comprising the forward elements has been minimized over a selected frequency range, which is determined by the gain and time constant of the feedback element. This is, in fact, the function of the excitation system stabilizer in an excitation control system.

Graphically, the transition from low, to intermediate, to high frequency regions occurs at the intersections of the $G_1 G_2$ and 1/H plots. (See Fig. 16.)

In summary, based on experience gained from studying many transient response characteristics, Chen suggested the following empirical rules of thumb to supplement the transfer function approximations in estimating the transient response:

Range I:

$$\text{Delay time, } T_d \approx \frac{1}{\Sigma P_\iota} - \frac{1}{\Sigma Z_\iota} \tag{55}$$

In Eq. (55) the delay time attributed to a pair of complex poles is equal to $1/2\ W_n$, where W_n is the undamped natural frequency, regardless of its damping factor. This rule was established by Chen based on pictorial studies of the transient response of second-order oscillatory systems.[7] W_n is equal to the square root of the product of the imaginary part of a complex pole and twice its real part.

Range II:

$$\text{Rise time,}\quad T_r \approx 3/2\ W_c \tag{56}$$

Where W_c is the crossover frequency, i.e. the frequency where the gain is unity.(0 db)

The time to maximum overshoot and the first approximation to the maximum overshoot is obtained from the equation

$$f_d\,(t_p) = 2|A_d|\epsilon^{-\sigma at}\,\text{COS}(jw_at + \angle A_d) \tag{57}$$

In this equation $\sigma_a \pm jw_a$ are the approximate, dominant, complex closed-loop poles in this range. A_d is the coefficient (residue) of the term in the approximate partial fraction expansion of the closed-loop transfer function. A_d is evaluated by the method described earlier. Since the complex roots are conjugate complex functions, the sum of their imaginary parts is zero and the sum of the real parts is twice the real part of either. Thus, it is necessary to determine the magnitude and phase angle of only one term. In this approximation technique all other terms are neglected. The approximation neglecting all other terms is good only if the other poles are far from the dominant pair and/or those poles which are not far from the dominant pair are approximately cancelled by zeros. Otherwise, the effect of the other poles and zeros on A_d must be taken into account in evaluating A_d.

Range III:

The approximate settling time and a second approximation to overshoot is obtained from:

$$F_{p_1}\,(t) = A_t\epsilon^{-p_1t} \tag{58}$$

where p_1 is the approximate closed loop pole, approaching the open-loop zero in this range.

A_t is the coefficient (residue) of the term in the partial fraction expansion associated with the approximate pole of $-P_1$. Truxal[9] has shown that if the closed-loop pole and the closed-loop zero of the dipole are much closer to the origin than any other closed-loop poles and zeros, which is normally the case, the magnitude of the residue, A_t is given by:

$$A_t \approx 1/K' \tag{59}$$

where K′ is simply the gain at the break frequency on the open-loop Bode plot.

To estimate the settling time (t_s), the magnitude of f_t of interest, typically from one to five percent, may be inserted in equation (58) and the value of t_s solved for by:

$$t_s = \frac{\ln |f_t|/A_t}{P_1} \tag{60}$$

From the overall open loop straight line asymptotic plot of Fig. 16, the approximate transfer function obtained from the breakpoints is:

$$G(s) \approx \frac{400\,(1+s)}{(1+\frac{s}{.08})(1+\frac{s}{.2})(1+\frac{s}{16})(1+\frac{s}{50})}$$

Applying the transfer function approximation and the rules of thumb described in the foregoing yield the results shown in Table 2. Detail calculations of the approximation are given in the appendix.

TABLE II

Parameter	Approximate	Exact
Closed-Loop Poles	-1.16 -8 ±j 6.2 -50	-1.17 -12.1 ±j 3.35 -27
Delay Time		0
Rise Time	.24 sec.	.27 sec.
Maximum Overshoot	0.11	0.11
Settling time to within 1%	2.48 sec.	2.20 sec.

While the preceding approximation technique is useful in providing a means of estimating the transient response of a control system, its major value is in the insight it gives as to how the transient response and frequency response are related. When more accurate transient response data is required, particularly in the presence of one or more non-linearities in the control system, computer simulation of the system should be used.

Frequency Response Performance Criteria

In the foregoing, it was indicated that rise time, overshoot, and settling time are the principal performance criteria with respect to the transient or time response of a feedback control system.

The principal performance criteria with respect to the frequency response of a feedback control system include the bandwidth, phase margin and gain margin. These quantities are indicated on Figs. 16, and 17.

Bandwidth is the interval separating two frequencies between which both the gain and the phase difference (of sinusoidal output referred to sinusoidal input) remain within specified limits.[2] For excitation control systems and many of their components the lower frequency is zero; the higher frequency is usually defined as the frequency at which the closed-loop frequency response falls to $1/\sqrt{2}$ of its low frequency value (equivalent to 3 db below).

Bandwidth is related to the cross-over frequency of the open-loop frequency response of the system and to the system transient response (rise time) as shown earlier. Bandwidth is also indicative of the highest input frequency that can be handled by the system, i.e. it approximately describes the filtering characteristics of the system.

Phase and gain margin are both measures of the relative stability of a control system. A feedback control system will be unstable if the phase shift is greater than 180° at the frequency at which the gain is unity, i.e., at the crossover frequency on the Bode plot of the system open loop characteristic. Phase margin is equal to 180° minus the absolute value of the loop phase angle at a frequency where the loop gain is unity.[2] A phase margin of 40° or more is generally considered good design practice. Similarly gain margin is defined as the reciprocal of the gain at the frequency at which the phase angle reaches minus 180°.[2] A gain margin of 6 db or more is generally considered good design practice.[10]

SUMMARY

This paper has shown that mathematical models of excitation systems with the dynamic characteristics expressed in terms of frequency response provide a useful means for evaluating several aspects of excitation control system performance.

The accuracy of regulation and the stability of the excitation control system itself is immediately evident. Also, some insight into the approximate transient response characteristics may be readily attained from the frequency response characteristics.

Of prime importance is the evaluation of excitation system performance in an excitation control system connected to a power system. This performance as related to the frequency response characteristic of excitation systems will be described in a future committee report paper.

REFERENCES

1. IEEE Committee Report, "Computer Representation of Excitation Systems, "Power Apparatus and Systems, vol. PAS-87, pp 1460-1464, June 1968.

2. "Terminology for Automatic Control," ANSI C85.1-1963.

3. Harold Chestnut and Robert W. Mayer, Servo-mechanism and Regulating System Design, vol. I, (book), John Wiley & Sons, 1951.

4. G. A. Bierson, "Quick Methods for Evaluating the Closed-Loop Poles of Feedback Control Systems," AIEE Transactions, vol. 72, pt. II, pp 53-70, May, 1953.

5. G. A. Bierson, "Quick Methods for Evaluating the Closed-Loop Poles of Feedback Control Systems," AIEE Transaction, vol. 75, pt. II, pp 253-73, November, 1956.

6. K. Chen, "A Quick Method for Estimating Closed-Loop Poles of Control Systems," AIEE Transactions, Applications and Industry, vol. 76, pt. II, pp 80-87, May, 1957.

7. K. Chen, "Quasi-Linearization Techniques for Transient Study of Nonlinear Feedback-Control Systems," AIEE Transactions, vol. 74, pt. II, pp 354-365, 1955.

8. W. R. Evans, Control System Dynamics, McGraw Hill Book Co., Inc., New York, New York, 1954.

9. J. G. Truxal, Automatic Feedback Control System Synthesis, McGraw-Hill Book Co., Inc., New York, New York, 1955.

10. J. G. Truxal (Ed.) Control Engineers Handbook, McGraw-Hill Book Co., Inc., New York, New York, 1958.

APPENDIX

The transient response to a step change in input of the excitation control system of Fig. 15 is approximated as follows:

1. From the break frequencies of the open-loop Bode plot of Fig. 16:

$$G(s) \approx \frac{400\,(1+s)}{(1+\frac{s}{.08})(1+\frac{s}{.2})(1+\frac{s}{16})(1+\frac{s}{50})}$$

2. Within the ±15 db band of Fig. 16, the approximate closed-loop transfer function, from Eq. 48 is

$$\frac{G(s)}{1+G(s)} \approx \frac{\dfrac{400(.08)(.2)(16)(50)}{(1)(50)}}{s(s+16) + \dfrac{400(.08)(.2)(16)(50)}{(1)(50)}}$$

$$\frac{G(s)}{1+G(s)} \approx \frac{102.4}{S^2 + 16S + 102.4}$$

The approximate closed-loop poles are obtained from Eq. 49, i.e. setting the denominator of the above expression equal to zero and solving for the roots as

$$S = -8 \pm j6.2$$

The transient associated with these complex poles is given by Eq. 57 as

$$f(t) = 2|A_d|\epsilon^{-8t}\,\mathrm{COS}(6.2t + \angle A_d)$$

Considering that the other closed-loop poles and zeros have no effect on A_d, A_d is determined from the partial fraction expansion of the closed-loop transfer function as

$$A_d = \left[\frac{1}{S}(S+8+j6.2)\right]_{S=-8+j6.2}$$

where: $\frac{1}{S}$ is the Laplace transform of the step change in input

$$A_d = -\frac{8+j6.2}{j12.4} = -.815\,\angle{-52.2°}$$

$$f(t) = -1.63\,\epsilon^{-8t}\,\mathrm{COS}(6.2t - 52.2°)$$

The time, t_p, to maximum overshoot is obtained by setting the derivative of f(t) = 0 and solving for t_p.

$$f^1(t) = -1.63\left[-8\,\epsilon^{-8t}\,\mathrm{COS}\,(6.2t\text{-}52.2°)\text{-}6.2\,\epsilon^{-8t}\,\mathrm{SIN}(6.2t\text{-}52.2°)\right]$$

$$\tan(6.2t\text{-}52.2°) = -\frac{8}{6.2} = -1.29$$

$$6.2t - 52.2° = \tan^{-1} -1.29 = 180°\text{-}52.2°$$

$$t = 180° \times \frac{1\ \text{rad}}{57.3°} \times \frac{1}{6.2} = 0.51\ \text{sec.}$$

The first approximation to the overshoot is obtained by evaluating $f(t_p)$ at t_p

$$f(t_p) = 1.63\,\epsilon^{-8(.51)}\,\mathrm{COS}(6.2 \times .51 - 52.2°)$$

$$f(t_p) = .0205$$

3. Above the +15 db band of Fig. 16:

From Eq. (45) the approximate closed-loop pole is

$$P_1 \approx -(1 + \frac{1}{K'})\,z_1$$

where K^1 = 6.3 is the gain at the break-frequency, w = 1, on Fig. 16.

$$P_1 \approx -(1 + \frac{1}{6.3})(1) = -1.16$$

The transient associated with this pole is given by Eq. (58) as

$$f(t) = A_t\,\epsilon^{-1.16t}$$

$$A_t \approx \frac{1}{K'} = \frac{1}{6.3} = .159 \qquad \text{from Eq. [59]}$$

A second approximation to the peak overshoot is now obtained from

$$f(t_p) = .159\,\epsilon^{-1.16(.51)} = .0878$$

The total overshoot is then

$$.0878 + .0205 \approx 0.11$$

The settling time, t_s, for the output to come within 1 percent of steady-state value is obtained as follows

$$f(t) = .159\,\epsilon^{-1.16t_s} = .01 \text{ from Eq. (58)}$$

$$t_s = \frac{\ln \frac{.01}{.159}}{1.16} = 2.48 \text{ sec. from Eq. (60)}$$

4. Below the -15 db band of Fig. 16: the approximate closed-loop pole at P_3 = -50 gives rise to a small, quickly decaying, transient which is neglected.

5. Rise time is approximated from Eq. 56 as

$$t_r \approx \frac{3}{2} w_c \approx 1.5\,(\frac{1}{6.2}) = .24 \text{ sec.}$$

6. Delay time is approximated from Eq. 55 as

$$T_d \approx \frac{1}{2}\left[\frac{1}{\Sigma P_\ell} - \frac{1}{\Sigma Z_\ell}\right]$$

where the contribution of the complex póles is $\frac{1}{2w_n}$ and

$$w_n = \sqrt{(6.2)(2 \times 8)} = 9.96$$

$$T_d \approx \frac{1}{2}(\frac{1}{50} + \frac{1}{9.96} + \frac{1}{1.16} - 1) \approx 0$$

Experience with High-Speed Rectifier Excitation Systems

KENNETH R. McCLYMONT, SENIOR MEMBER, IEEE, GERALD MANCHUR, MEMBER, IEEE, R. JOHN ROSS, MEMBER, IEEE, AND RONALD J. WILSON, MEMBER, IEEE

***Abstract*—Excitation systems which utilize controlled rectifiers supplied directly from the generator terminals for the main exciter are in service at four remote generating stations. Due to the nature of the power system, these excitation systems were designed to withstand high ac overvoltages and to provide simultaneous control of a number of parallel generators. Experience in testing and operating these excitation systems is described.**

Introduction

WITHIN the last six years, Ontario Hydro has installed 550 MW of hydro generation located within 50 miles of James Bay on the Moose River system in Northern Ontario. This generation is incorporated into the main system over a single-circuit 500-kV transmission line. Seven of the ten generators on this system have been provided with exciters utilizing controlled rectifiers. This type of excitation system can provide a decided improvement in steady-state and transient stability limits because of its ability to change generator field voltage nearly instantaneously.

Paper 31 TP 67-19, recommended and approved by the Power Generation Committee of the IEEE Power Group for presentation at the IEEE Winter Power Meeting, New York, N. Y., January 29–February 3, 1967. Manuscript submitted November 1, 1966; made available for printing August 16, 1967.

The authors are with the Hydro-Electric Power Commission of Ontario, Toronto, Ont., Canada.

The seven generators are located in four separate generating stations and the excitation equipment has been provided by three manufacturers. One type of exciter employs single-anode mercury-arc rectifiers (ignitrons), two use multi-anode mercury-arc rectifiers, and the fourth uses silicon-controlled rectifiers (SCRs). Three excitation systems were placed in service in 1963, two in 1965, and two in 1966.

The paper proceeds as follows:

1) A brief description of the Moose River generation is given to provide a background for the special problems associated with these excitation systems.

2) The main features of the excitation systems using mercury-arc rectifiers and the major differences between these and a type using SCRs are described.

3) Special requirements for withstanding overvoltages and line energization are outlined.

4) Excitation system response and the telephone influence factor as obtained from test results are presented.

5) Maintenance requirements and forced outage experience are outlined.

Moose River Generation

Ontario Hydro's generation on the Moose River system is located about 450 miles north of the system load center at Toronto and about 50 miles south of James Bay. This generation

Reprinted from *IEEE Trans. Power App. Syst.*, June 1968.

TABLE I
SUMMARY OF CHARACTERISTICS OF GENERATORS AND RECTIFIER EXCITATION SYSTEMS

Station	Generator Rating MVA	Generator Rating PF	Number of Generators	Date Placed in Service	Number and Type of Valve	Exciter: Valve Cooling	Exciter: Rated Current, amperes	Exciter: Ceiling Voltage, volts	Exciter: Ceiling Voltage, per unit*
Otter Rapids	46	95	3	2 in 1961 and 1 in October 1963	conventional rotating exciters				
	46	95	1	October 1963	six single-anode mercury-arc	water	1000	528	3.94
Little Long	64	95	2	November and December 1963	two 6-anode mercury-arc per generator	air	1250	486	4.27
Harmon	68	95	2	May and July 1965	two 6-anode mercury-arc per generator	air	1250	865	7.93
Kipling	66	95	2	June 1966	132 SCRs per generator	air	1500	578	4.45

* Per unit base voltage is that required to produce rated generator terminal voltage on the air gap line at a field temperature of 120°C.

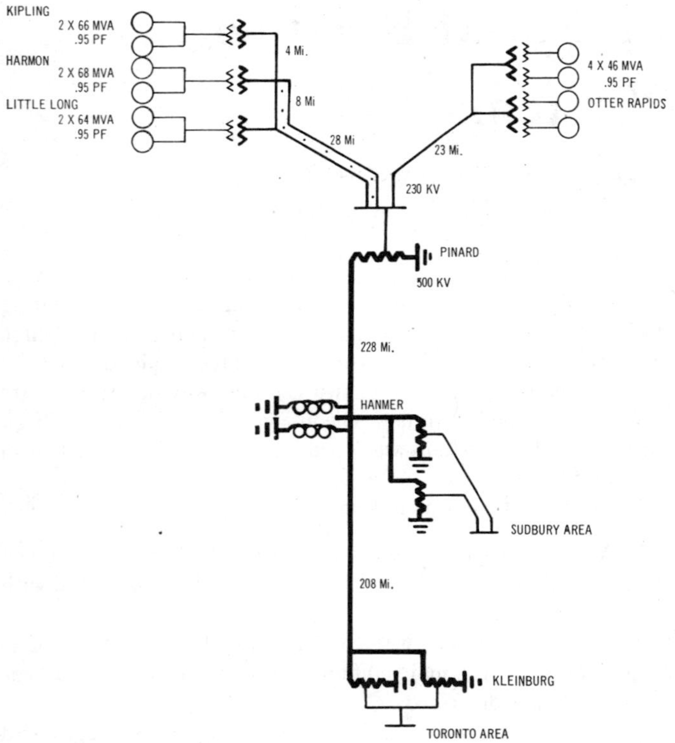

Fig. 1. Single-line diagram of 500-kV transmission.

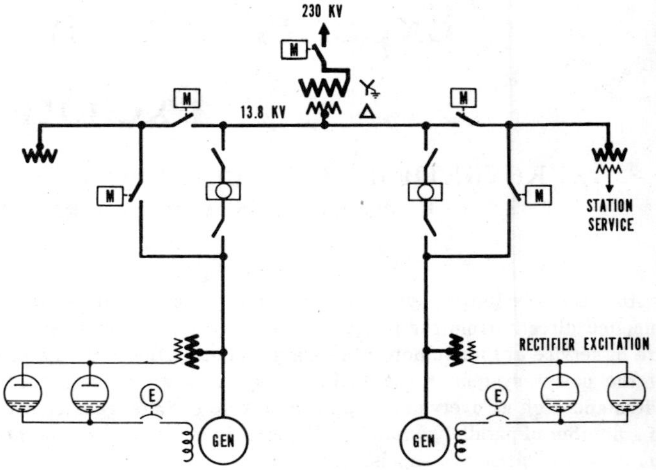

Fig. 2. Typical generating station arrangement.

and the 500-kV transmission system for incorporating it are shown in Fig. 1. Part of the 500-kV facilities were placed in service in 1966 and the remainder is scheduled for completion early in 1967. Prior to operation at 500 kV, the transmission was operated at 230 kV for incorporation of the generation.

The generation consists of four hydraulic stations having a total capacity of approximately 550 MW. This generation is designed for peaking operation and has a capacity factor in December of approximately 40 percent. Three of the four stations have two generators while one station, Otter Rapids GS, has four. Two of the generators at Otter Rapids GS were placed in service in 1961 and have conventional direct-driven rotating exciters with a rotating-amplifier voltage regulator. The remaining generation was placed in service in the period from the fall of 1963 to the summer of 1966. Seven of the eight generators, totalling 442 MVA, placed in service in this period were provided with rectifier excitation systems. Generator and exciter ratings are summarized in Table I.

The generating stations are connected on radial 230-kV transmission lines to a gathering station, Pinard TS, where the voltage is stepped up from 230 to 500 kV. The generation is carried on a single 500-kV transmission line to load centers at Sudbury and Toronto. The generating stations are unattended and are remotely controlled from Pinard TS.

DESCRIPTION OF EXCITATION SYSTEMS

Equipment Arrangement

In the usual arrangement, the rectifier transformers are connected directly to the 13.8-kV generator bus without breakers or fuses, as shown in Fig. 2. The rectifier transformers are the

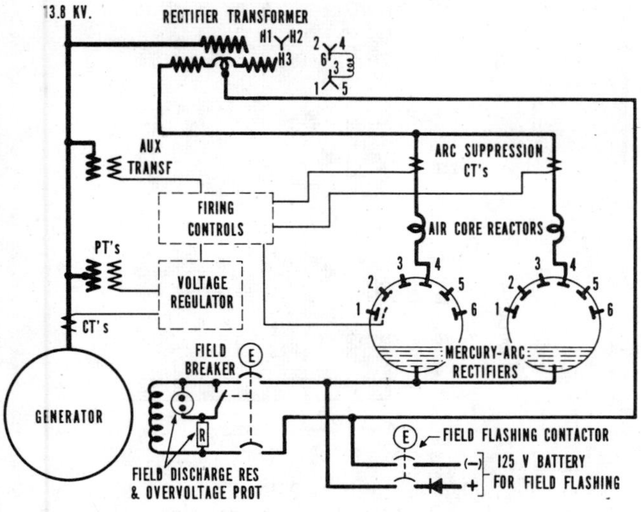

Fig. 3. Mercury-arc rectifier excitation system.

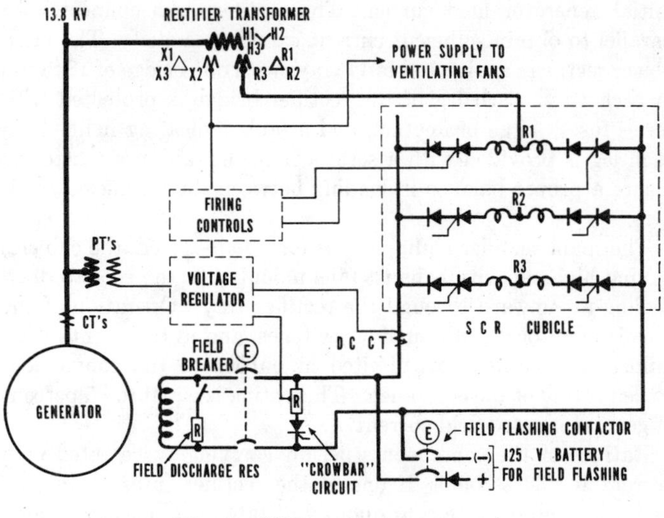

Fig. 4. SCR excitation system.

multiphase indoor type, immersed in a nonflammable synthetic liquid with forced air-cooling. They are connected to the generator terminals through an isolated phase bus. The power supply to the grid control circuits and other auxiliary equipment is taken from an auxiliary transformer supplied from the same bus or from an auxiliary winding of the rectifier transformer to insure a reliabile supply having the correct phase relationship.

Mercury-Arc Rectifiers

A typical schematic diagram for the multi-anode mercury-arc rectifiers is shown in Fig. 3. This arrangement does not provide for negative field current but does permit a negative field voltage of a magnitude approaching the positive ceiling. As the field voltage can be changed almost instantaneously both increases and decreases in field current can be extremely rapid. A nonlinear resistor is provided across the generator field to protect against excessive negative field voltages which could be induced in the rotor on the occurrence of pole slipping.

If an arc-back occurs, one of the valves conducts in the reverse direction and short-circuits two phases of the rectifier transformer. With the multi-anode rectifiers, arc-backs are suppressed by phasing back the grid control so the tubes can not fire for several cycles. At the end of this fixed period, normal firing control is resumed and usually the rectifiers will resume normal operation. However, the arc-back may be repeated. If two or three arc-backs occur in a period of five to ten seconds then the excitation is shut down. Manufacturers estimate that there will be about one arc-back per 6-anode tube per year of operation.

The single-anode rectifier circuit is similar to that of Fig. 3 except that anode breakers are used in each valve and the field breaker is not required. In addition to interrupting generator field current, the anode breakers are used to suppress arc-backs. In the event of an arc-back, the anode breaker on the tube having the reverse current is opened and then reclosed.

As the only outside source of supply for most of the generating stations is over a single 230-kV line, it was considered essential that the generators be able to start up without any supply from outside the station. This is accomplished by supplying the initial generator excitation from the 125-volt station battery which is also used for other control and alarm functions. On startup, contactors are closed to connect the station battery across the generator field through a diode, as shown in Fig. 3. When the generator voltage reaches 50 to 75 percent of rated value, depending on the particular excitation system, the main rectifiers take over. The diode prevents reverse current flowing into the battery. This startup procedure requires only about five seconds so that the drain on the battery is only a fraction of an ampere-hour.

Each multi-anode valve is provided with a fan for forced air-cooling, and cathode and anode heaters for temperature control. This equipment is controlled by temperature sensitive relays. The single-anode rectifiers have anode heaters only and are cooled by a water system provided with heaters.

The firing control and the control elements of the voltage regulator are static elements. Transistorized amplifiers having negligible time delays are employed in the control circuits. Much of the control switching is performed by static switching devices.

If one of the 6-anode valves must be removed from service, the remaining valve has sufficient capacity to supply rated generator field current and exciter ceiling voltage remains essentially unchanged. With one valve out of service, a limit circuit prevents field current from exceeding the capacity of the remaining valve. If one of the single-anode valves is removed from service, the remaining valves have sufficient capacity to supply rated generator field current; however, the exciter ceiling voltage is reduced by approximately 15 percent. The mercury-arc rectifiers have considerable inherent overload capacity and one of the multi-anode systems, for example, is capable of carrying twice rated current for 60 seconds.

Principal among the protective devices which can cause the generator to shut down are:

rectifier transformer protective relays
valve overload relays
multiple backfire
rotor overcurrent or overvoltage
excess cathode temperature
anode heater failure
grid control and voltage regulator power supply failure.

SCR

A schematic diagram of the SCR excitation system is shown in Fig. 4. In common with the mercury-arc rectifier excitation systems, this system does not provide negative field current and startup is accomplished using the station battery to provide the initial generator field current. The rectifiers are connected in parallel to obtain sufficient current carrying capacity. The main power rectifiers are housed in 11 trays with one bridge of 12 SCRs in each tray. Each leg of the rectifier bridge is protected with series fuses, surge protection, and a fault indicating light. Also, each leg is provided with a series-connected, air-core reactor to insure a proper balance in loading between the paralleled rectifiers.

The main rectifier equipment is forced air-cooled, the cubicles having high-capacity exhaust fans mounted in the roof to draw cooling air upward through the rectifier trays. Operation of the fans is monitored with air-flow switches, and in the event of fan failure the rectifiers are limited in output to the convection-cooled rating of the equipment. This rating is equal to 85 percent of generator rated field current.

Ratings of the components permit carrying twice rated field current for 30 seconds. If one of the rectifier fuses blows, the excitation can operate continuously at rated field current. Blowing of additional fuses will result in automatic shutdown.

A so-called *crowbar* circuit is provided to protect the rectifiers and generator against the effect of an excessive negative voltage which can occur if the generator falls out of step. If a negative voltage exceeding a preset value occurs, the rectifiers are turned off and an SCR, connected across the field so as to permit the flow of negative field current, is turned on. If the current flowing in this SCR exceeds 100 amperes, a protective device initiates tripping of generator and field breakers.

Since the SCRs do not arc-back, no arc-back protection is required. The following protections which operate to shut down the unit are provided:

rectifier dc short circuit
rectifier ac overcurrent
rectifier transformer fault pressure relay
excessive current in the crowbar circuit.

AC Overvoltage Withstand Requirements

Rejection of generator load by opening of the Hanmer 500-kV line will result in the generators overspeeding. While the relaying is arranged so that both ends of the 500-kV lines are opened at the same time, malfunctioning of the relays or breakers could result in failure of the Pinard breakers to open and leave the generator connected to the 500-kV line. With the rise in speed, the generators connected to the line become self-excited and extremely high generator voltages may occur.[1] While this occurrence is unlikely, it was considered that the excitation system should be conservatively designed to avoid failure under this condition since such a failure could lead to a lengthy generator shutdown. Therefore, specifications called for the excitation system to withstand for three seconds without damage, a generator terminal voltage having a fundamental frequency voltage up to 1.7 times rated with harmonic components giving peak voltages up to twice rated.

Line Energization and Group Control

To minimize switching surges on energizing the 500-kV line from the Moose River generation it was decided to make provision for reducing the generator voltage to about 60 percent of its rated value prior to energizing the line. To facilitate line energization and also voltage changes during normal operation a group control was installed at Pinard TS which permits simultaneous adjustment of the voltage setting on any number of preselected generators.

The units selected for a voltage change can be placed in either

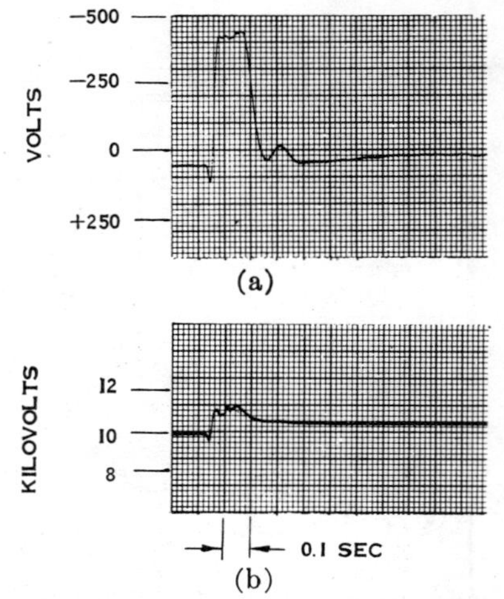

Fig. 5. Oscillograph of energizing a 228-mile 230-kV line. (a) Generator field voltage. (b) Generator terminal voltage.

the *continuous raise or lower* mode on a group-control console or in a *sudden raise or lower* mode. When placed in the appropriate mode, operation of a single control switch simultaneously energizes the motors which drive the voltage-adjusting rheostats on all the selected generators and thereby controls the voltage setting on these generators. The *continuous raise or lower* mode permits control of generator terminal voltage over the range from 72 to 110 percent of rating. The *sudden raise or lower* mode permits an instantaneous change of the generator voltage by about 15 percent. For energizing the transmission line, voltage on the generator is normally reduced to 72 percent by the continuous lower control, then dropped to 58 percent by the sudden lower control. The line is then energized at 58 percent voltage. To accommodate this line-energizing operation, the excitation systems were required to be capable of operation for five minutes at this reduced voltage.

Fig. 5 shows an oscillogram of the performance of the static excitation systems on line energization. In this operation the 500-kV line was being operated at 230 kV and the line charging requirement is approximately 100 MVA. Two 64-MVA generating units were used to energize the line.

Excitation System Response

The outstanding characteristic of a controlled rectifier excitation system using solid-state control equipment is its high speed of response. Performance tests carried out on these systems have shown that the response time to ceiling for a sudden change in voltage supply to the regulator input is less than 20 milliseconds on all except one type where the response time was approximately 65 ms.

Exciter ceiling voltage is approximately proportional to the average of the three phase-to-phase generator terminal voltages. Firing control and voltage regulator control circuits are designed to function correctly to obtain maximum available ceiling voltage for voltages as low as 25 to 50 percent of normal voltage. Therefore, these control circuits function correctly for virtually all faults on the 230- or 500-kV systems.

Other signals in addition to generator voltage are required to obtain the maximum improvements in dynamic and transient stability limits. To achieve these desired improvements, equipment which produced a signal proportional to change in generator speed was designed and installed on the rectifier excitation systems.[2]

TABLE II
SUMMARY OF FORCED OUTAGES—RECTIFIER EXCITATION SYSTEMS
(*Otter Rapids GS—One Unit*)

Type of Fault	November 1963–September 30, 1964 Forced Outages	Hours	October 1, 1964–September 30, 1965 Forced Outages	Hours	October 1, 1965–September 30, 1966 Forced Outages	Hours
1) Cooling equipment	2	4	0	0	0	0
2) Rectifier	1	1	3	36	0	0
3) Control circuit	6	274	1	17	0	0
4) Main transformer	0	0	0	0	0	0
	9	279	4	53	0	0
Total to September 30, 1966					13	332
Average outage per unit per year					4.5	115

Note: Rectifier faults include multiple backfires and failure to fire. Control circuit faults include breaks in the printed circuit, incorrect wiring, rectifier trips where the cause was unknown, and faults in conventional relays and contactors.

TABLE III
SUMMARY OF FORCED OUTAGES—RECTIFIER EXCITATION SYSTEMS
(*Little Long GS—Two Units*)

Type of Fault	November 1963–October 1964 Forced Outages	Hours	November 1964–October 1965 Forced Outages	Hours	November 1965–September 1966 Forced Outages	Hours
1) Cooling equipment	1	1	2	4	6	17
2) Rectifier	4	9	2	11	0	0
3) Control circuit	8	34	8	73	2	36
4) Main transformer	0	0	0	0	0	0
	13	44	12	88	8	53
Total to September 30, 1966					33	185
Average outage per unit per year					5.5	31

Note: Rectifier faults include multiple backfires and failure to fire. Control circuit faults include breaks in the printed circuit, incorrect wiring, rectifier trips where the cause was unknown, and faults in conventional relays and contactors

TELEPHONE INFLUENCE FACTOR—RADIO INTERFERENCE

American Standard C50 requires that the balanced telephone influence factor (TIF) for large synchronous generators shall not exceed 50. As rectifier circuits produce harmonic voltages in the ac supply system, there was some concern that the rectifier excitation systems might result in an excessive TIF. A review of the TIF measurements on recent hydraulic generators with conventional rotating exciters shows that the TIF normally falls in the range from 10 to 15. TIF measurements on the Moose River generation produced the following results:

	Generator on Open Circuit: Rectifier Exciter Fed from System	Generator on Open Circuit: Rectifier Exciter Fed from Generator
Little Long GS	12.4	27.4
Harmon GS	21.2	32.2

It is evident that there is a significant increase in TIF values with the exciter supplied from the generator terminals, but they are still well within ASA C50 requirements.

The possibility of radio interference on radio equipment located at the generating stations was also a source of concern. Some radio interference had been observed on radio equipment located within a 100-foot radius of the excitation equipment. The frequency modulated, 49-MHz equipment used as a backup voice carrier had severe interference. Also, it is suspected that interference from the exciters may have caused erroneous remote tripping over some 150-MHz communications channels. Tests are planned to investigate this possibility. The excitation systems do not interfere with the microwave communications equipment used for supervisory control, telemetering, relay protection, and voice communications.

OPERATING EXPERIENCE

A summary of the forced outages due to faults in excitation equipment at the four stations is listed in Tables II, III, and IV. Regarding the number of hours of down-time shown in the tables, it should be realized that in some instances the down-time was prolonged because the necessity for returning the unit to service was not as urgent as other work in progress. Also, the location of maintenance staff headquarters some distance from the plants tended to extend the down-time. The three excitation systems at Otter Rapids GS which employ rotating exciters with rotating amplifier voltage regulators had two forced outages for a 22.7-hour total outage time over the period from November, 1963, to September 30, 1966.

The record of performance at the two stations which have been in service for approximately three years indicates that performance has improved after the first two years of operation. Efforts of the maintenance staff to deal with troubles were initially hampered by lack of adequate instructions at a sufficiently early date.

TABLE IV

SUMMARY OF FORCED OUTAGES—RECTIFIER EXCITATION SYSTEMS
(Harmon GS and Kipling GS—Two Units Each)

	Harmon GS		Kipling GS	
	September 1965–September 30, 1966		August 1966–September 30, 1966	
Type of Fault	Forced Outages	Hours	Forced Outages	Hours
1) Cooling equipment			1	6
2) Rectifier	2	3		
3) Control circuit	11	55	5	223
4) Main transformer	3	121		
Total	16	179	6	229
Average outage per system per year	7.4	83		

Note: Rectifier faults include multiple backfires and failure to fire. Control circuit faults include breaks in the printed circuit, incorrect wiring, rectifier trips where the cause was unknown, and faults in conventional relays and contactors.

The faults encountered, mainly in auxiliary components, have been corrected by such action as providing relays or contactors with heavier contacts and supplying improved types of diodes, transistors, etc. Only one failure of a main rectifier has been experienced, that of a single-anode mercury-arc rectifier. One intermittent break in a printed-circuit board resulted in considerable trouble, causing six interruptions totalling 68 hours before the fault was identified.

A rectifier excitation system includes equipment required to handle the heavy field currents and high forcing voltages needed for a large alternator and also the precise and sensitive devices that are used to control these quantities. Hence, a wide range of maintenance skills and equipment is required.

Corrective maintenance of the control equipment is made easier and quicker by the use of plug-in components. However, this requires the stocking of a complete spare set of such components. Some trouble has been experienced with deteriorating contact conductance of some plug-in units. More attention to the method of supplying cooling air and to dust prevention measures may be necessary to overcome this difficulty.

If a mercury-arc rectifier cell is allowed to remain deenergized for a long period it will deteriorate. Hence it has not been usual to stock a spare rectifier cell at the site. Instead, manufacturers are relied upon to maintain spare cells at their factory or local parts depot.

There was concern that the use of static excitation systems was adversely affecting the generator slip rings and brushes. While some of the stations with static excitation systems have experienced unsatisfactory brush and slip ring performance, others have experienced no difficulties. No conclusive evidence has been found which links brush and slip ring performance with the type of excitation system used. It may be noted in this regard that in most Ontario Hydro installations the polarity of the slip rings is reversed at intervals of from three to twelve months in order that wear, surface film formation, and similar effects of currents direction may be equal on each ring and each set of brushes.

A large content of clay sediment in the rivers on this watershed makes it necessary to overhaul the pump supplying cooling water to the rectifier excitation equipment at Otter Rapids GS at intervals of two years. Some care is required to keep the conductance of the cooling water at a low level since there is a considerable voltage difference between points in the water circuit and excessive electrical current would heat the water.

CONCLUSION

Rectifier excitation systems utilizing controlled mercury-arc valves or SCR as part of the main exciter and supplied directly from the generator terminals in general have a response time to ceiling of about 20 milliseconds and provide excellent dynamic performance. The control circuitry for these systems tends to be considerably more complex than that of a conventional rotating excitation system. Despite the use of the latest control circuit hardware, forced outage rates have been high and some outages have been quite lengthy. It is felt however, that much has been accomplished in the effort to eliminate equipment faults and that remaining weaknesses can be corrected. In view of the steadily improving operating experience, it is considered that all four excitation systems will prove equal to or superior to rotating types in reliability and economy of maintenance.

REFERENCES

[1] P. L. Dandeno and K. R. McClymont, "Extra-high voltage system overvoltages following load rejection of hydraulic generation," *IEEE Trans. (Power Apparatus and Systems)*, vol. 82, pp. 49–57, April 1963.

[2] P. L. Dandeno, A. Karas, K. R. McClymont, and W. Watson, "Effect of high-speed rectifier excitation systems on generator stability limits," *IEEE Trans. Power Apparatus and Systems*, vol. PAS-87, pp. 190–201, January 1968.

Discussion

G. I. Stillman (American Electric Power Service Corporation, New York, N. Y.): In the section entitled *AC Overvoltage Withstand Requirements*, the authors mention the possibility of self-excitation of the generators when load is rejected by the opening of the 500-kV breakers at Hanmer, thereby leaving the generators in the north connected to the 228-mile undercompensated 500-kV line from Pinard to Hanmer. The excitation for all units is supplied by controlled rectifiers which do not, as the authors point out, permit the flow of negative field current. (This is true for all nonbridged excitation systems which utilize rectifiers, controlled or not.) When load is rejected as in the foregoing and the line charging current causes boosting ampere-turns in the direct and quadrature axes of the machine, the terminal voltage rises from this cause as well as from overspeeding of the unit. The voltage regulator detects the rise in terminal voltage and begins its attempt to reduce it by reducing field current, i.e., by reducing direct-axis excitation. When the current in the field is reduced to zero, the regulator is pegged since it cannot supply negative current. The regulator continues to hold excitation in the direct axis to zero so that the remaining factors which can cause overvoltages at the generator terminals are a) self-excitation in the quadrature axis, b) overspeeding of the unit, and c) increased charging current resulting from the increased frequency produced by the overspeeding generator.

Since the regulator under discussion has no *bucking* capability and since the overspeed for both hydro and thermal units is limited to reasonable finite values by the governor controls, it follows that the generator terminal voltages should also be limited to reasonable finite values. Have the authors performed either field or computer tests which show that the above factors lead to terminal voltages on their hydro units of the order of 2.0 per crest including harmonics?

A comparison between hydro units and thermal units shows that:

a) Hydro units can overspeed up to 130 percent of normal speed on load rejection while thermal units overspeed to about 110 percent of normal speed so that, due to this factor, overvoltages on hydro units are potentially greater than on thermal units.

Manuscript received February 13, 1967.

b) The quadrature-axis reluctance of hydro units is much greater than that of round-rotor thermal units, so that the self-exciting mmf in the quadrature axis of a hydro unit is correspondingly far less effective in producing overvoltages than the same mmf in the quadrature axis of a thermal unit.

With this in mind, have the authors performed field or computer tests for load rejection on any thermal units on their system? Do they have any comparative data concerning terminal overvoltages for hydro and thermal units?

The authors state that the excitation system is designed to function correctly to obtain maximum ceiling voltage for voltages as low as 25 to 30 percent of normal voltage. Should a 3-phase fault occur on the 230-kV systems in the north, voltages would presumably fall to values far below 25 to 30 percent. How do the generators obtain their excitation under such conditions in order to remain in synchronism?

As the authors point out under *Telephone Influence Factor*, rectifier circuits produce harmonic voltages in the ac supply system. Aside from interference with communications media, these harmonics produce rotating fields in the machine air gap. These fields rotate at their own synchronous speeds, and will induce voltages in the amortisseur windings on the rotor. Although these voltages are small, the currents they produce may be large because of the short-circuited nature of the amortisseur. This can result in rotor heating which can sometimes limit machine capability. Have the authors noticed such effects in any of their units?

The authors state that arc-back protection is not required since SCRs do not arc back. It is my impression that when excessive reverse voltage exists across an SCR, avalanche breakdown is a distinct possibility. Are the rectifier circuits used by the authors so designed that excessive reverse overvoltages cannot occur?

R. W. Lye and **F. E. Lucas** (Canadian General Electric Company, Ltd., Peterborough, Ont., Canada): The authors and their organization deserve considerable credit for their courage in committing the majority of units in their Moose River System to excitation systems of high performance using equipment which had limited application experience in this type of service. The relative remoteness of these sections poses further concern when considering new types of equipment.

The performance benefits predicted by computer studies and demonstrated in field tests are well documented in the companion paper.[2]

Much of the basic equipment provided by the three manufacturers has been proved in application for duties such as industrial drives but significant innovations have been made in tailoring this equipment to a different application and environment. This is especially so in the area of control performance.

The opportunity offered to the participating manufacturers will ensure not only the improved operating experience which the authors predict for these particular installations but will also give greater assurance of satisfactory performance for rectifier excitation equipments. Steady improvement in reducing outage times after the initial startup period is indicated.

The simplicity of the configuration of unit rectifier transformers and the use of battery field flashing is brought out in the paper. While the outages related to mercury-arc rectifiers do not show up as a significant problem in total outage times, the complexity of storage and maintenance of this type of unit clearly lends support to the use of thyristor equipment.

Manuscript received February 20, 1967.

K. R. McClymont, G. Manchur, R. J. Ross, and **R. J. Wilson:** Mr. Stillman's comments and questions show a keen awareness of the application of problems associated with this type of excitation system. He has raised a number of important questions.

When load is rejected by opening of the 500-kV breakers at Hanmer, leaving the generators in the north connected to the 228-mile line from Pinard to Hanmer, our studies have shown that with conventional excitation systems where the field current can reverse, the generator voltage at maximum overspeed can reach about 2.0 per unit crest including the effect of harmonics.[1] With rectifier excitation systems which do not permit negative field current, the maximum terminal voltage for these conditions is only about 1.5 per unit crest. Studies were not carried out to investigate the effect of systems employing both conventional and rectifier excitation systems. In the interests of conservatism it was assumed that the terminal voltages on a system employing both types of excitation system would approach the higher value. While the probability of reaching such high overvoltages is quite small, it has to be weighed against the effect of failure of a number of excitation systems if subjected to the overvoltage and the cost increase to provide for this overvoltage.

We do not have any comparative data concerning overvoltages for hydro and thermal units.

Typical generator and transformer reactances on the generator base are $X'du = 0.24$ per unit and $X_T = 0.1$ per unit. At full load the voltage behind transient reactance is approximately 1.1 per unit. Using transient reactance the generator terminal voltage will drop to $1.1 \times 0.1/(0.24 + 0.1) = 0.32$ per unit for a 3-phase fault at the transformer 230-kV terminals. However, the performance of the excitation system for a 3-phase fault at the terminals of the generator transformer is not particularly significant since these generators are connected radially to the terminal station at Pinard and are disconnected from the system for a fault at this location. Therefore, the closest critical 3-phase fault is at Pinard TS. For a 3-phase fault at this location minimum generator voltage is approximately 50 percent.

The harmonic currents required to supply the rectifier excitation systems are quite small and do not cause significant heating of the amortisseur bars. Normal exciter power requirements are only about 0.5 percent of generator rating and the harmonic currents supplying the exciter are only a fraction of one percent of rated generator currents. As Mr. Stillman points out, the rectifiers can increase the harmonic content of the generator voltage and this in turn will cause harmonic currents to circulate between the generator and the system. However, an analysis of the harmonic content of the generator waveform with and without rectifier excitation has shown that in the frequency range below 1000 hertz the increase in harmonic content due to the rectifier excitation is less than a tenth of one percent of rated voltage. The impedance of the system and generator will be sufficient to limit harmonic currents to very small values. While we have not actually measured the harmonic currents, there is no evidence of undue rotor heating from harmonic currents and we would not expect it. The generators are capable of carrying 5 to 10 percent negative sequence current continuously and there is no evidence that the rotor heating from harmonic currents will affect machine capability.

While the SCR circuits are designed to avoid reverse voltages which could cause breakdown, it is not possible to design mercury-arc circuits to eliminate arc-backs.

We appreciate the comments of Mr. Lye and Mr. Lucas who designed the excitation system using SCRs.

Futher investigation of the effect of radio interference due to the excitation systems indicates that none of the communications equipment is adversely affected and that the interference noted in the paper was attributable to other causes.

Manuscript received April 27, 1967.

A ROTATING THYRISTOR EXCITATION SYSTEM FOR HYDROELECTRIC GENERATORS

Edward C. Hartung
E. Harry Lenfest
General Electric Company
Schenectady, N.Y.

Glenn R. Meloy
US Corps of Engineers
Portland, Oregon

Abstract – A contemporary approach to the high response excitation system for hydroelectric generators is described. This approach eliminates the conventional mechanical commutator, the main generator field collector rings, and many field power connections and switching devices external to the generating unit. Excitation power generation, and commutation are accomplished by using an AC exciter and thyristor rectifier bridge directly connected to the generator shaft. These elements, in combination with a sophisticated static regulator depending only on electrically derived inputs, provide improved positive damping of electric power system oscillations by proper control of excitation. This system was first applied to three 142,105 KVA, 90 RPM generators now in commercial operation.

INTRODUCTION

Excitation systems in service on hydroelectric generators have been extremely reliable. Generator outages due to failure of elements of the excitation system are rare. Reliability is due to:

1. Conservative rating of elements, exciters, etc.
2. Low generator speed, minimizing maintenance of brushes.
3. Short direct connection of exciter output to the field collector rings, minimizing exposure of the excitation power connections and eliminating mechanical switching devices.
4. Voltage control element employing magnetic amplifiers.
5. Isolation of excitation system from voltage variations and transients in the AC power systems.
6. Transfer to manual operation on failure of automatic control, avoiding loss of excitation for such reason.

The use of solid state power components for rectification, amplificiation, and control now affords opportunity to further improve reliability and contribution to power system performance by increased speed and more sophisticated control. The design objective for the system described is to retain the proven functional features of the present direct-connected excitation systems, improving these as possible, and to better provide for supplementary stabilizing signals to improve power system performance through excitation control. Recently, improvements have been made on a piecemeal basis;-a solid state regulator to replace the amplidyne or magnetic amplifier control, a supplementary stabilizing control for addition as a separate package, and a solid-state exciter to replace the rotating exciter. Presented here is a systematic approach.

SYSTEM OPERATION

A block diagram of the excitation system is shown in Figure 1. The section enclosed by dashed lines is the rotating portion of the system. The remainder is divided into 2 independent regulating systems.

Paper 71 TP 588-PWR, recommended and approved by the Power Generation Committee of the IEEE Power Engineering Society for presentation at the IEEE Summer Meeting and International Symposium on High Power Testing, Portland, Ore., July 18-23, 1971. Manuscript submitted February 17, 1971; made available for printing May 12, 1971.

The exciter regulator maintains the exciter ac terminal voltage at a nearly constant level corresponding to ceiling voltage. The exciter output is then used to power the entire excitation system effectively isolating its operation from outside disturbances. Voltage build up can be accomplished using only the residual exciter flux. For convenience and to shorten the build-up time, flashing from the station battery supply is also incorporated

The generator regulator operates to maintain the generator terminal voltage at a nearly constant level as determined by the voltage adjustor setting and modified by the underexcited limit, line compensation (or droop depending on the desired connection) and a stabilizing signal derived from generator power.

The exciter has a rotating armature and stationery field winding. Typical exciter frequencies range from 16 to 30 Hz with a ceiling voltage of 1.5 p.u., based on generator rated field voltage.

The generator field rectifier is a double-way thyristor bridge circuit capable of inverting to reduce the generator field current. Since the voltage supplying the thyristor bridge is the ceiling voltage, the speed of response is a function of that ceiling (1).

The power supply and rectifier control included as part of the rotating element produce the necessary phase-controlled triggering signals in response to a dc control level supplied by the voltage regulator.

Generator field current measurement is obtained from current transformers located in the ac side of the generator field rectifier. The current level sensed is equivalent to the dc generator field current except for filtering losses which are typically less than 1% of the rated field current.

Not shown in Figure 1 is an overvoltage protection control which is combined with a field discharge resistor to protect the system in the event of out-of-step operation. The control and resistor provide full protection independent of the excitation system operation.

Self-excitation power as well as control and measurement signals are transferred across the air gap by collector rings and brushes sized for that application.

The exciter ac voltage is averaged, compared against the exciter voltage reference, and applied to the rectifier control and rectifier in such a manner as to maintain the exciter terminal voltage at the desired level. Lag-lead compensation is used to stabilize the voltage control. A current limit is provided in the exciter field control to protect the system.

Standard potential and current transformers are used for sensing the generator terminal conditions. The potential signal is averaged and compared directly with the operators voltage adjustment (voltage reference) in the automatic mode of operation. The potential and current signals are combined in a set of transducers to produce the Watt and Var signals for the complimentary functions. These complimentary signals are then added in proper phase and polarity to modify the basic voltage regulation.

In the manual mode of operation, the generator field current is compared with a separate manual voltage adjustment (reference) and applied to the rotating thyristor bridge through separate regulating elements.

Reprinted from *IEEE Trans. Power App. Syst.*, Sept./Oct. 1972.

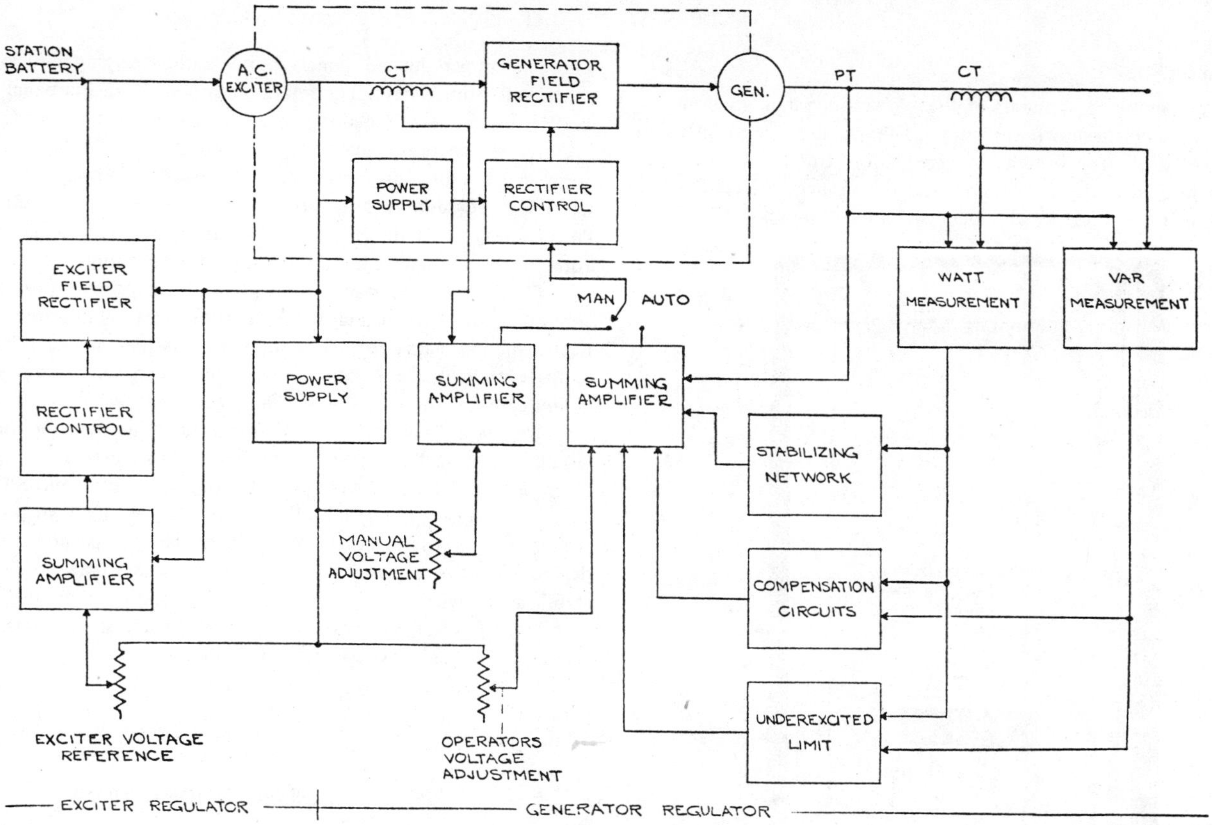

Fig. 1. System Block Diagram

The generator regulator has lag-lead compensation to provide stable, responsive operation. In the automatic mode, the power derived stabilization signal is added above a minimum load (typically 25%) to assist in damping power system oscillations (2).

SYSTEM HARDWARE

Rotating Elements

The ac exciter and rotating thyristor assembly are similar in dimension to a standard dc exciter. In the application illustrated here, one requirement was that the ac exciter be interchangeable with the dc exciters supplied for other machines of the same rating. The design standards and practices employed in producing this ac exciter are similar to those used for dc exciters for temperature rise, rating and general construction.[3,4]

The exciter rotor or armature provides attachment and support for the rotating thyristor assembly which overhangs the rotor diameter. The stator poles have an amortisseur winding to reduce the exciter commutating reactance and associated losses.

Figure 2 shows one exciter as assembled for factory testing. This exciter has a rating of 500KW, 375V at 90 RPM.

Normally a vertical machine, it is run horizontally for testing with temporary bearings and a power collector added in the position normally occupied by a Kaplan oil head when erected on its hydro-generator.

The open panels on the exciter housing permit a view of the rotating thyristor assembly. Also seen internally are the exciter collector rings providing transfer of the control and measurement signals and self-excitation power.

The rotating thyristor assembly is constructed on a solid flat insulating disc in a fan-configuration to allow forced-air cooling of the semiconductors. A top shroud piece, removable in sections, completes the low-profile assembly. The center of the thyristor mounting disc is open to allow cooling air to be drawn through the exciter by a fan mounted below the stator. In Figure 2, this fan is enclosed by an air baffle used in test.

Each thyristor is independently fused and arranged in parallel groups. The fuses are sized to open following the failure of a thyristor thereby allowing the system to continue in operation. The thyristor assembly is rated to allow continuous operation at ceiling with sufficient margin to permit the loss of 25% of the installed thyristors. No attempt has been made to provide an indication of fuse or thyristor failure. The maintenance procedure suggests semi-annual inspection and testing of the fuses.

Fig. 2. A 375V, 500KW, 90 RPM Rotating Thyristor Exciter

The power bridge circuit consists of 4 parallel thyristors for this rating. Each thyristor has a PRV rating 2.5 times the rms exciter voltage. The PRV value is coordinated with the maximum exciter output and overvoltage protection on the generator field bus. The system is self-contained and isolated from ac voltage variations or transients.

Excitation Cubicle

The remaining components of the excitation system are housed in the excitation cubicle shown in Figure 3. The overall cubicle dimensions are 36"x36"x90". The double, louvered doors front and back allow protected ventilation by convection. Cable entrances are provided on top, bottom and one side for flexible installation.

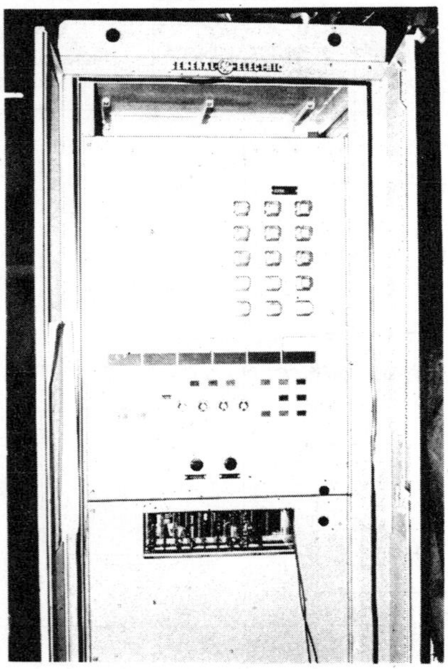

Fig. 3. Rotating Thryistor System Excitation Cubicle

The front doors, open in this view, expose 2 hinged panels which contain the system metering control and regulator circuits. The exciter field rectifier and other power components are mounted on a side panel in the cubicle. The overall cubicle size is determined by the current rating of the exciter field rectifier and internal cubicle temperature rise. This particular design has a maximum current rating of 150 amperes.

The upper hinged panel supports the relaying and operating controls for the system. The relays are mounted on the panel face for ease of inspection and maintenance. The manual and automatic motor-operated rheostats project through the front panel to allow hand operation. A slip-clutch arrangement in the rheostats prevents damage due to continuous motor operation and permits easy hand adjustment. Position sensing of the rheostats is accomplished electronically eliminating the need for mechanical switches in the assembly.

The lower hinged panel houses a printed circuit card assembly containing all of the regulator functions on replaceable printed circuits. Each function-regulated power supplies, amplifier, underexcited limit, rectifier control, stabilization, etc. – has been isolated on a plug-in circuit card for ease of calibration and maintenance.

The basic building block of this system has been the integrated-circuit operational amplifier. Forty-four of these circuits are used in the excitation cubicle to provide all of the regulating and control functions except for start-up and remote control relaying and high power control. The incorporation of these integrated circuits has provided a great degree of design capability and flexibility. The voltage references are stabilized to within .1% for all operating conditions by an integrated amplifier. Position switches on the voltage adjusting rheostats have also been replaced by these integrated circuits.

PERFORMANCE CONSIDERATIONS

In developing the rotating thyristor excitation system, it was felt that it should have system performance characteristics equal to or better than the widely used hydroelectric excitation system, the amplidyne-voltage regulator-dc exciter system. Accordingly two performance areas were defined for study using the FACE Multimachine simulator(5). These two areas were voltage recovery capability and dynamic stability performance.

The over-all transfer function block diagram for the rotating thyristor excitation system as used in the simulations is shown in Figure 4.

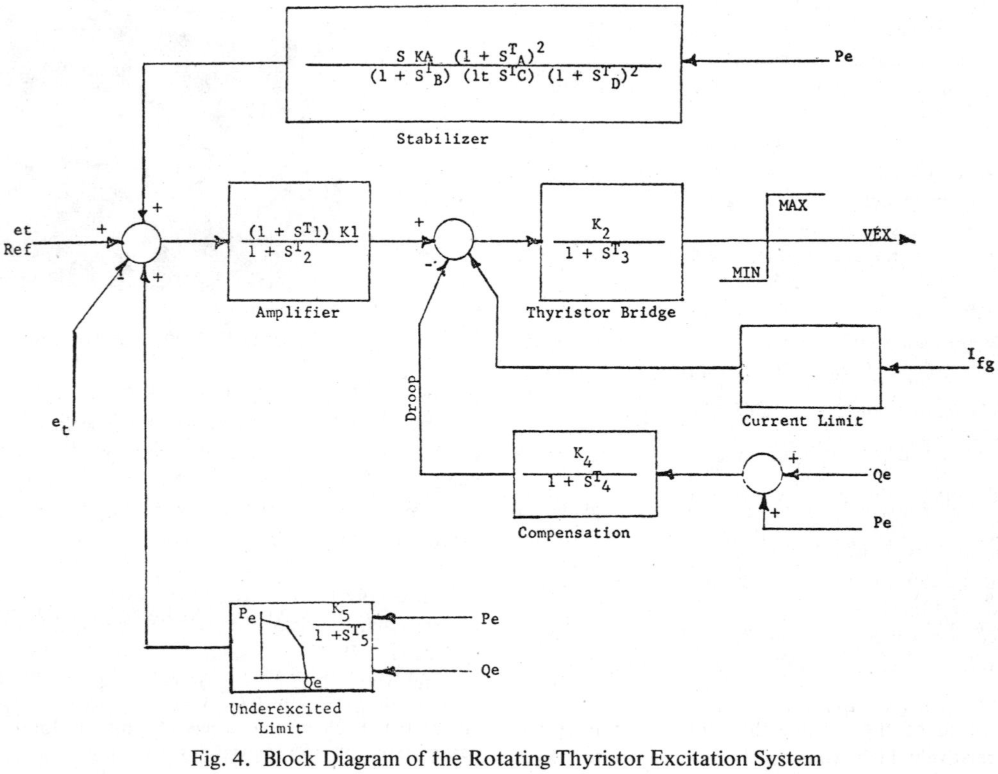

Fig. 4. Block Diagram of the Rotating Thyristor Excitation System

Voltage Recovery Capability

This criterion measures the ability of the generator with its excitation system to maintain terminal voltage when subjected to large changes in loading such as transmission system changes or load regulator system is to maintain the terminal voltage of the generator at the desired value.

The application of a large reactive load to an isolated machine provides a severe test of the ability of an exciter-regulator to maintain terminal voltage. This criterion also gives an indication of the transient or first swing stability performance of the excitation system. Figure 5 shows the relative performance of the two excitation systems following the application of a .48 p.u. reactive load.

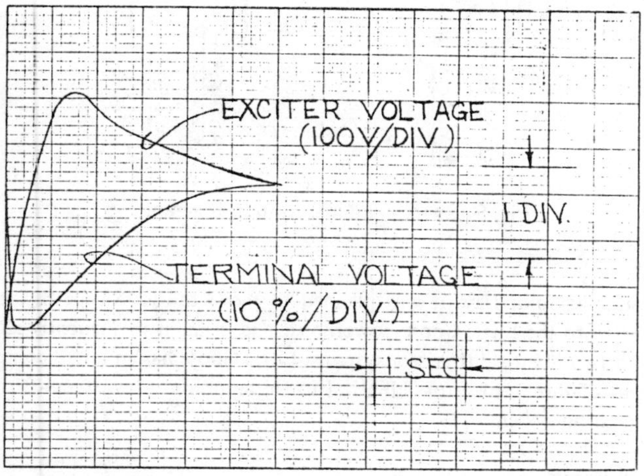

Fig. 5a. Amplidyne System Voltage Recovery to Reactive Load Pick-up

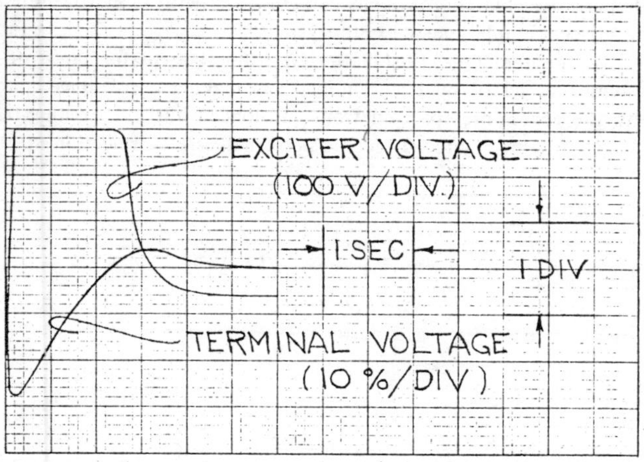

Fig. 5b. Thyristor System Voltage Recovery to Reactive Load Pick-up

Dynamic Stability Performance

The ability of a system to remain stable when subjected to the continuous small disturbances caused by the application and removal of loads, remote line-switching, etc., is evidence of dynamic stability.

Dynamic stability is referred to as the ability of the system to settle to a new operating condition following the withstanding of the first swing of a major disturbance. The role of the voltage regulator in maintaining dynamic stability is twofold, namely:

1. To hold the machine terminal voltage near the desired level during changing load conditions, and
2. To vary excitation voltage in proper time phase during system swings so as to provide positive rather than negative damping to these swings.

The high response speed of the rotating thyristor excitation system allows the use of relatively high gains in the voltage control loop. Dynamic stability of the machine is effected by a special stabilizing signal from appropriate power system quantities which provides positive damping of power system swings beyond that which is possible with more conventional excitation systems (2).

A signal from electrical power was chosen over other available options for the following reasons:

1. Good power transducers are readily available.
2. All quantities needed to measure electrical power are present in the excitation cubical for other reasons.

Figure 6 shows the relative dynamic performance of the amplidyne and thyristor excitation systems following a 3% change in terminal voltage set point. The effectiveness of the new excitation system in changing the terminal voltage in a fast well-damped manner is apparent from this figure.

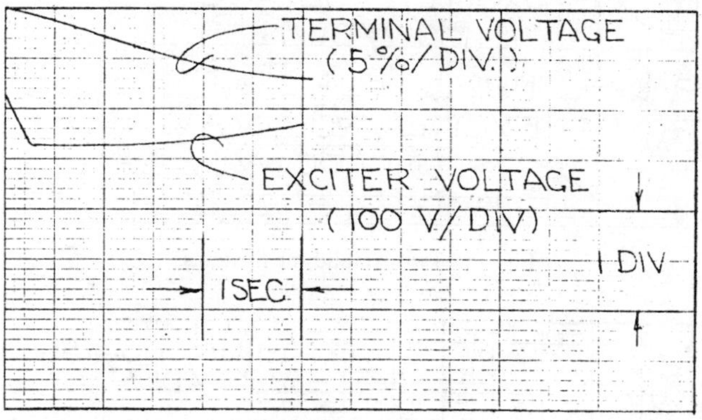

Fig. 6a. Amplidyne System Response to a Step of Reference Voltage

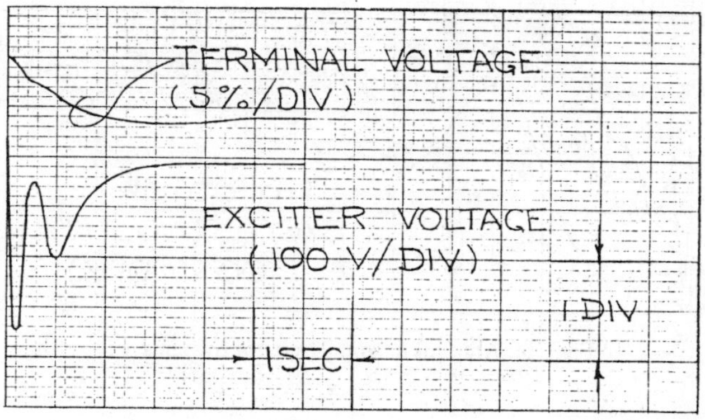

Fig. 6b. Thyristor System Response to a Step of Reference Voltage

The rotating thyristor system may be used on a machine in a power station having machines equipped with conventional excitation systems. An extreme case would seem to be that of two machines bussed together at their terminals with one machine having a conventional excitation system and the other machine having a rotating thyristor excitation system. Shown on Figure 7 is a comparison of the dynamic performance of this combination following a 3% step in voltage with the performance of an identical system except both machines have conventional excitation systems. It is apparent that equivalent performance is obtained either with the mixed excitation systems or two conventional excitation systems.

INITIAL INSTALLATION AND TESTING

Initial installation of the prototype of this new type excitation system was made on Unit 7 at the Corps of Engineers' John Day Powerhouse (6). Each of the initial 16 units at John Day is rated 142,105 KVA, 0.95 PF, 90 RPM, and 13.8 KV. The units are connected in pairs through the low-side windings of three-winding step-up transformers to the 500 KV transmission system. A one-line diagram of one transformer bank and its associated four units is shown in Figure 8. A partial elementary switchyard and system diagram is shown in Figure 9. The system diagram is shown as set up for system tests to be described later.

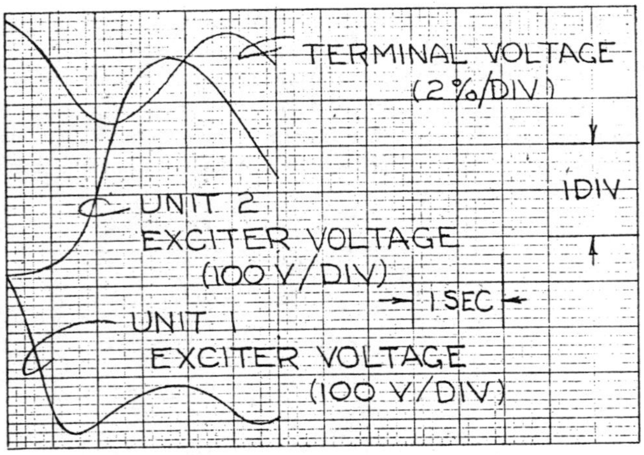

Fig. 7a. Two Amplidyne Systems

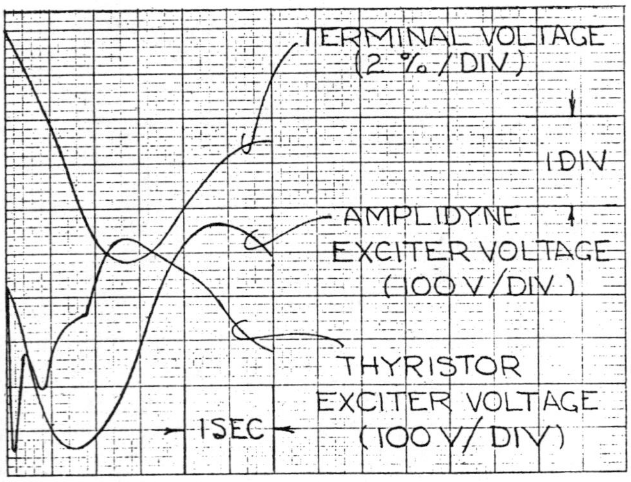

Fig. 7b. Amplidyne and Thyristor Systems

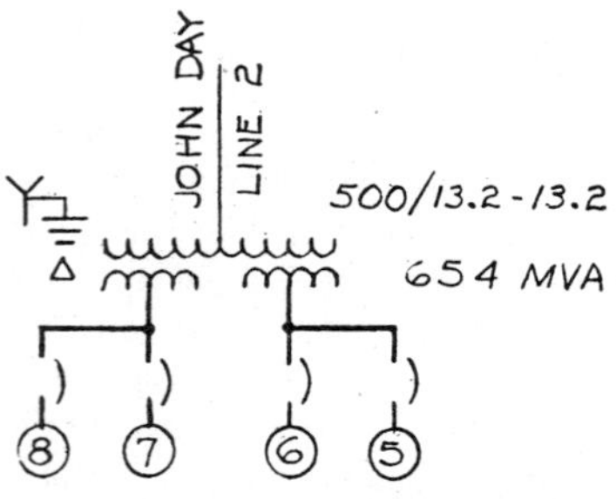

Fig. 8. Unit One-Line Diagram

It was desired to obtain confirmation of the design performance of the new system at the earliest convenient opportunity. The state of generator installation at John Day provided such an opportunity to evaluate practical excitation system operation and performance over a relatively long period of time. The unit connections at John Day provided a means of comparison of the new thyristor system against a standard NA143A amplidyne dc exciter system with a 2.0 p.u. response on a similar adjacent generating unit. Accordingly, the initial installation of Unit 7 at John Day incorporated the prototype of the new excitation system in place of the dc exciter. Subject to satisfactory test

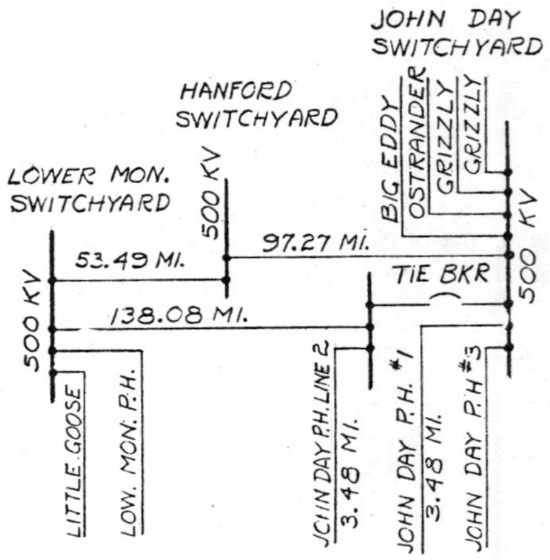

Fig. 9. Switchyard and System Diagram

results at John Day, it was planned to move the prototype thyristor system to the Little Goose powerhouse and install it permanently along with two identical thyristor systems. Little Goose has an initial complement of three generators identical to those at John Day except for the excitation system (6).

Test Results

In addition to normal operating behavior, tests were conducted to compare the excitation system performances with design criteria and the adjacent amplidyne-dc exciter system.

In the temporary installation at John Day, performance of the systems was tested by rejecting load and making step voltage changes with the generator separated from and loaded into the existing transmission system. Figure 10 records performance after rejection of 115%

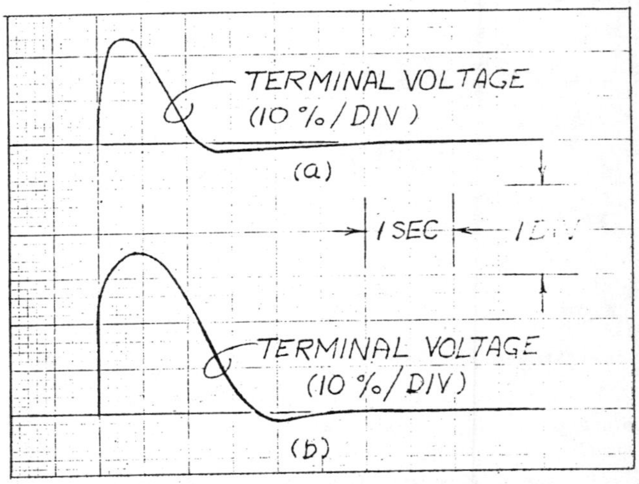

Fig. 10. Response to Generator Load Rejection
(a) Thryistor System
(b) Amplidyne System

load. With the amplidyne-dc exciter system, the voltage rise is 18%. The new thyristor excitation system limits the voltage rise to 12%. The 12% voltage rise is determined by the flux in the generator at the instant of load rejection and is, therefore, the lowest voltage rise that can be achieved with any excitation system. Figure 10 is not theoretically exact due to the limitations of test equipment recording the event. Figure 11 shows performance after a step change in terminal voltage reference with the generator open-circuited. The thyristor system stabilizes at the new voltage level in less than one-half the time required by the amplidyne system. The relative speed and forcing capability of the 2 excitation systems is apparent from the recorded field voltage traces. Figure 12 shows performance after a step change in terminal voltage reference with the generator on-line. MVARs have been plotted in this figure instead of terminal voltage to illustrate the relative performance. Again the thyristor system speed of response is apparent.

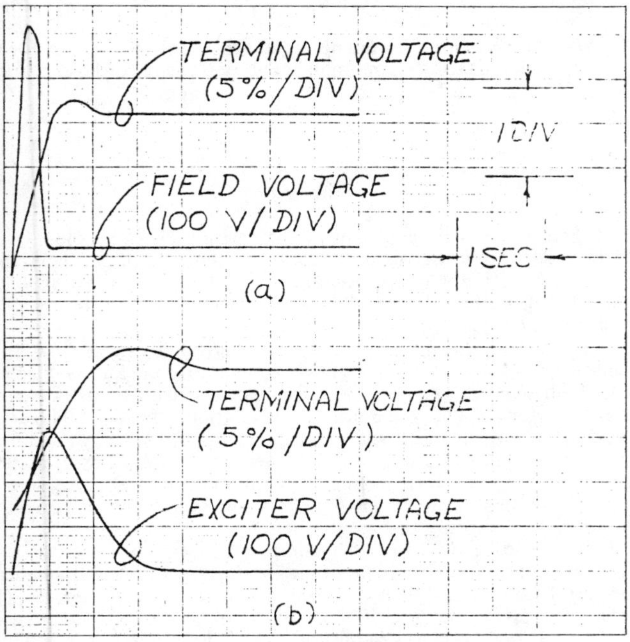

Fig. 11. Step Change Response Open-Circuited
(a) Thyristor System
(b) Amplidyne System

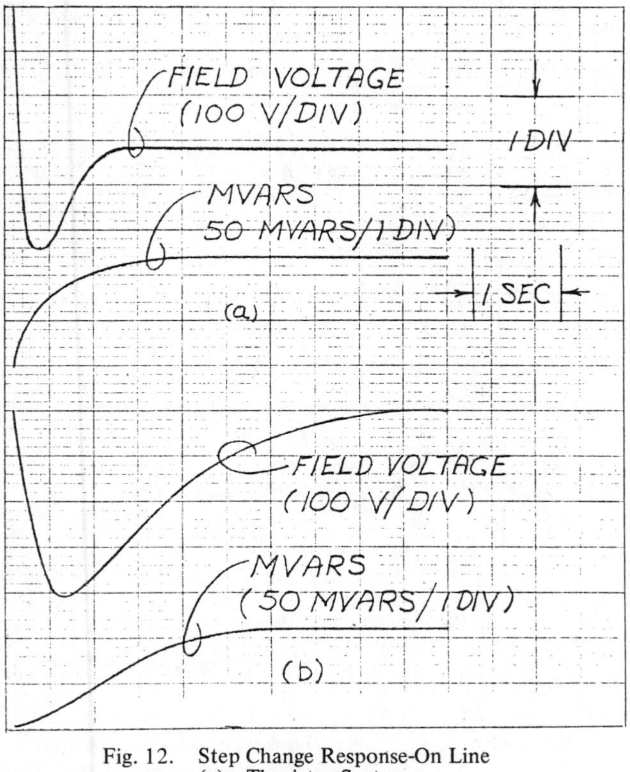

Fig. 12. Step Change Response-On Line
(a) Thyristor System
(b) Amplidyne System

System Testing By Line Switching

It was considered desirable to obtain additional performance data on the new excitation system by creating a known power system disturbance and monitoring the response of the excitation system and generator. To this end the John Day switchyard was set up in the manner indicated in Figure 9. Opening and closing of the switchyard tie breaker would create a change in power system impedance as seen from the terminals of Units 7 and 8. Under the loading conditions obtainable, the disturbance created by this type of line switching was not sufficient to obtain any conclusive results as to the merits of either system.

Evaluation of Prototype Test Results

The prototype performance, as compared against design objectives and compared with the amplidyne system, was encouraging. As with any brand new system, certain "bugs" were encountered. Although the performance of the system compared quite favorably with the theoretical results and was superior to the NA143A, its field-forcing capability presented a problem in mixed system operation. For system disturbances, the thyristor system had a tendency to do more than its share of correction initially and then back off as the rest of the generating units caught up. The power system stabilizer, which derives its input from changes in machine power, proved sensitive to normal scheduled load changes on the unit.

FINAL COMMERCIAL INSTALLATION

Performance of the prototype system matched design predictions sufficiently well to continue plans for permanent installation at Little Goose in conjunction with some equipment changes derived desirable from the operating experience gained at John Day.

General Equipment Changes

General changes were made in the overall equipment control design to:

1. Simplify operation by eliminating some redundant controls in the excitation cubicle and adding manual voltage adjustor indicator lamps.
2. Improve overall operation by removing ground loops, revising some circuit designs and changing component valves.
3. Add underexcited limit capability.

Power Stabilization Changes

Changes were incorporated to eliminate the interaction between the power stabilization and governor control which resulted in constant VAR excursions in response to governor action and large VAR excursions during planned load changes. The circuit changes reduced the low frequency response so that slow changes (less than 1 radian/sec) in power provided relatively small output from the stabilizer circuits.

Additional Testing

After completion of the above changes, additional tests were performed at the Little Goose project. Some of the tests made at John Day were repeated and verified that the redesign of some of the circuitry had improved operation. The interaction with governor action and planned load changes had been eliminated. The elimination of some ground loop contributed to a much better damped response to changes, both in reference signals and system conditions. The units exhibited loss of a tendency to "hog" all of the corrections to system changes.

Frequency response of the generator and excitation system was measured in accord with reference 7. The frequency response of the unit is shown in Figure 13 where the sinusoidal forcing signal is fed directly into the stabilizing network and the response is the terminal voltage of the generator at rated load. For comparison with frequency measurements of generators having speed-derived stabilizers, appropriate account must be taken of the 90° phase difference between power and speed and the difference in stabilizer gain required to achieve a similar damping effect. It is apparent that good compensation is ob-

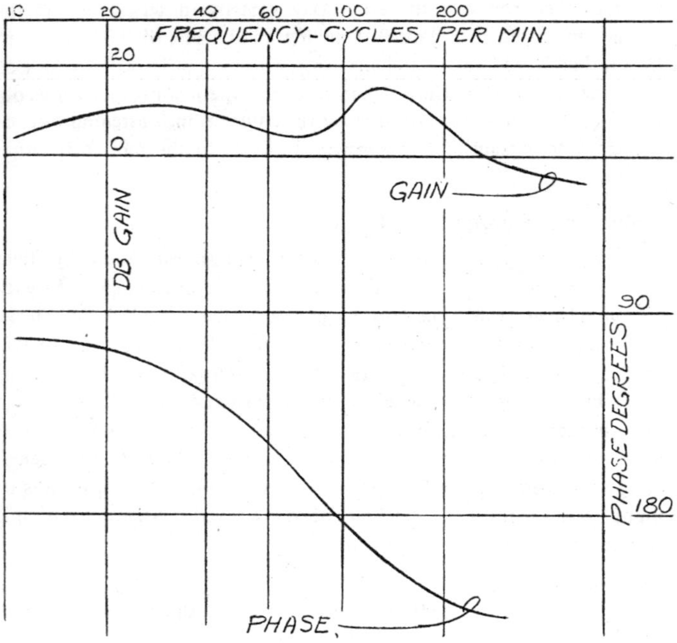

Fig. 13. Thyristor System Frequency Response

tained beyond the 20 CPM frequency currently of interest on the western systems. The reduced gain at lower frequencies is due in part to the attenuation included in the stabilizer to eliminate reaction to planned load changes. The dip in the gain characteristic at approximately 100 CPM is due to the loaded effect of the generator amortisseur.

CONCLUSIONS

The results of field testing and operating experience gained from the prototype installation and the first commercial installation of the new excitation system indicate that the design concepts have been realized. The problem of brush and collector ring maintenance has been virtually eliminated. System isolation long a bulwark of shaft connected exciters, has been retained and, in fact, improved upon by the use of exciter power for system auxiliaries and controls. High power control contactors, field rheostats, and the field breaker have been eliminated by a "state-of-the-art" design approach. Latest concepts in improvement of supplementary excitation control (power system stabilization) have been incorporated. The result has been a modern excitation system, incorporating all of the necessary auxiliary features and employing the latest technique for fast, stable, precise control of generator voltage.

REFERENCES

[1] IEEE Committee Report, "Proposed Excitation System Definition for Synchronous Machines", IEEE Paper 69TP154-PWR.

[2] F. P. DeMello and C. Concordia, "Concepts of Synchronous Machine Stability as affected by Excitation Control", IEEE Trans. Power Apparatus and Systems, Vol. PAS-88, PP. 316-329, April 1969.

[3] Electrical Standards Board, "Rotating Exciters for Synchronous Machines", ASA, C50.5, 1955.

[4] E. P. Smith, J. W. Gough, B. L. Goss, "Recent Developments of DC Motors and Generators for Metal Rolling Mill Main Drives". Iron and Steel Engineer, March 1968.

[5] D. N. Ewart and R. P. Schulz, "Face Multi-Machine Power System Simulator Program", Power Industry Computer Applications Conference May 18-21, 1969.

[6] Glenn R. Meloy, "The John Day Powerhouse", IEEE Trans. Power Apparatus and Systems, Vol. PAS 87, PP. 854-859, March 1968.

[7] Western Systems Coordinating Council, "Test Procedure for Coordination of Excitation Supplementary Control for Power System Damping".

Discussion

F. R. Schleif (U. S. Department of the Interior, Denver, Colo. 80225): These new types of thyristor excitation systems are of particular interest because of the superior potential they offer for improving voltage control and, with a supplementary stabilizer, improving transient and dynamic stability.

Since the dynamic stabilizer system deriving its signal from a watt transducer is somewhat unusual, we sought to compare its characteristics with those of the speed and frequency signal systems. A two-axis representation of the generator, its excitation system, stabilizer and line to an infinite bus were set up on an analog computer. Although performance of the generator and voltage regulator system closely matched that shown in the paper, results from the dynamic stabilizer were disappointing when using a reset time constant short enough to avoid var responses to changes in turbine torque.

It should be observed that, compared to a speed deviation signal the power deviation signal amplitude increases with rising frequency at the rate of about 20 dB per decade, and it leads by 90 degrees. (The polarity reversal reflecting the fact that a minus deviation of power to the system load produces a positive speed deviation of the generator should not be interpreted as the kind of lag which would reflect a diminishing gain characteristic.) If the power deviation signal is separated from the steady component with a short time constant (which amounts to differentiation), it has an additional rise of amplitude at 20 dB per decade and an additional 90 degrees of phase lead, making 180 degrees lead in all and amplitude rise of 40 dB per decade. The 180 degrees of phase shift can be accommodated by reversing the signal leads and that is correctly shown for this case by the (+) sign for feedback of the power derived signal on the authors' block diagram, Figure 4. However, because of the steeply rising gain we were unable to accomplish damping in the range needed without depending upon a limit circuit. Thus any damping was under a restricted set of conditions. If the disturbance were increased the damping effort could not increase correspondingly.

Much better results were obtainable with considerably different parameters to produce the transfer function

$$\frac{30 \cdot 3S}{(1 + 3S)} \; \frac{(1 + S)}{(1 + 10S)} \; \frac{(1 + 0.15S)}{(1 + 0.05S)} .$$

Here the reset time constant of 3 seconds has been made long enough to produce little attenuation or phase shift down to 1.9 radians per second, the natural swing frequency of the Pacific intertie. It is also important

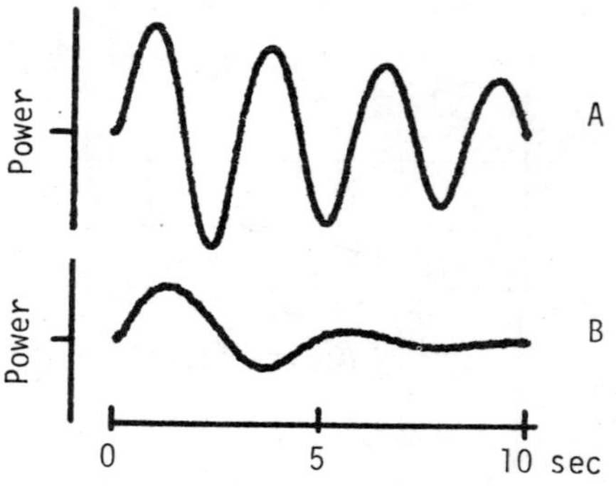

Fig. 1. Comparison showing damping obtainable
A. Reference system without dynamic stabilizer
B. With power signal driven stabilizer using the transfer function

$$\frac{30 \cdot 3S}{(1 + 3S)} \; \frac{(1 + S)}{(1 + 10S)} \; \frac{(1 + 0.15S)}{(1 + 0.05S)} .$$

Manuscript received August 2, 1971.

that the second term $(1 + S)/(1 + 10S)$ compensate the rising gain and leading phase inherent in the power deviation signal. Its time constants must be such as to produce a lag in the useful range rather than the lead sometimes needed with a speed signal on slower exciter systems. Accordingly, this signal is inserted with a sign opposite to that shown in Figure 4 of the paper. Results were quite good as shown in Figure 1 of this discussion, and response was proportional to the disturbance over a wide range.

However, it is inherent in the power signal that to produce useful damping at low swing frequencies the stabilizer will also produce var excursions from changes of turbine torque unless it is compensated by some means which distinguishes the difference, such as by a gate position signal converted to turbine torque by the turbine torque transfer function $(1 - \alpha T_W S)/(1 + 0.5\ \alpha T_W S)$ where α is the relative gate position and T_W is the water starting time.

The authors are to be commended upon an excitation system with excellent potential. The analyses in this discussion are offered toward exploiting that potential for best advantage to the power system.

Tore Petersson (ASEA, Västerås, Sweden): The authors have published an interesting paper on a sophisticated excitation system.

Concerning large hydro-electric generators, I doubt whether the real technical advantages of a rotating thyristor excitation system can justify the high cost compared with other systems. Within our company we are of the opinion that a static thyristor excitation system, uncompounded and supplied from the generator terminals, is usually the best solution for large low-speed generators, from the integrated economical and technical viewpoints.

The following has been experienced from several years use of the latter system:

1) The cost is usually lower than for a rotating d.c. exciter system, at low generator speeds. (The cost for a low-speed a.c. exciter and rotating thyristors will probably be much higher than for a d.c. exciter.)

2) With short power system fault-clearing times, which are usual today, the excitation performance is fully satisfactory for all conditions, if the excitation system is properly designed.

3) The system has been adopted more or less as a standard for large generators in some countries, e.g. Canada and Sweden.

4) The good performance and reliability have been proven by several years of operation with large generators up to 225 MVA.

Have the authors made any integrated economical and technical comparison of the two mentioned thyristor excitation systems?

Manuscript received August 5, 1971.

Edward C. Hartung, E. Harry Lenfest, and **Glenn R. Meloy**: We thank the discussers for their response to our paper. We offer the following in reply:

Messrs. F. R. Schleif and L. E. Eilts – We agree with the discussers comments in general. The transfer function originally planned was essentially the same as that suggested in this discussion. The VAR changes experienced with load changes were more than desired. The stabilizer transfer function was modified to reduce the interaction between the excitation system and the governor.

The turbines in the installation were the Kaplan type so that the turbine torque is a function of blade position as well as gate position. To use a suitable turbine torque signal for compensation would add significant complexity.

We anticipate that further work in this area will yield a stabilizing function which does exploit the full potential of this excitation system.

Mr. Tore Petersson – Our experience has shown that the cost of a completely static excitation system of reasonably comparable reliability and performance is higher than the cost of a shaft-driven excitation system. This of course reflects the standards to which the system is designed. Some of the factors involved are:

A. Ceiling capability at reduced system voltage.

B. Connection to the isolated phase bus.

C. Provision for disconnecting the excitation system for faults at input voltage. For larger generators connected through step-up transformers to EHV systems (500 kv in the case of Little Goose) this fault duty could require in excess of 2500 MVA interrupting capacity. Failure to provide adequately for fault interruption results in a hazard.

D. Connection between the power components of the excitation system and the main generator which generally include mechanically operated switching devices.

E. Space requirements in the powerhouse particularly in underground stations.

As stated in the introduction to our paper, excitation systems in use on hydroelectric generators have been extremely reliable. The purpose of developing the rotating thyristor excitation system has been to improve upon the performance of the system without sacrificing the characteristics which have been responsible for that reliability record. Briefly these are:

A. Isolation of the excitation system power components from voltage variations and transients in the AC power system.

B. Short direct connection of the exciter output to the main generator field with elimination of mechanical switching devices in this circuit.

An additional contribution of the rotating thyristor system is the elimination of power slip rings and brushes.

Manuscript received September 15, 1971.

ALTHYREX EXCITATION SYSTEM WITH POWER SYSTEM STABILIZER

M. L. Crenshaw
Member IEEE
General Electric Co.
Schenectady, N. Y.

W. J. Miller
Member IEEE
General Electric Co.
Waynesboro, Va.

R. P. Schulz
Member IEEE
General Electric Co.
Schenectady, N. Y.

M. Temoshok
Senior Member IEEE
General Electric Co.
Schenectady, N. Y.

ABSTRACT

The problem of providing electric power at the lowest cost has led utilities to use remote generation to take advantage of cost differences between energy transportation by high-voltage electrical transmission and other methods. The resulting systems usually have high transmission reactances with correspondingly reduced stability margins. In order to meet high reliability requirements it is necessary that stability be improved in every way possible.

A particular example of this situation occurs at the Four Corners Station of the Arizona Public Service Co. Unit No. 4, a 750 MW cross-compound steam turbine generator, has been placed in operation. A duplicate unit, No. 5, is scheduled for operation early in 1970. Excitation for each of the four generators of these cross-compound sets is supplied by an Althyrex excitation system, which consists of a shaft-driven alternator and stationary thyristors. A supplementary control circuit (power system stabilizer) is added to each excitation system to provide positive damping of rotor swings. A tachometer driven from the shaft of each turbine-generator rotor is utilized as a speed signal source.

This paper describes the excitation system and power system stabilizers. Test results and calculations are included to indicate expected performance of the excitation system and power stabilizer in operation.

INTRODUCTION

The special transient and steady state stability problems at the Four Corners Station influenced the selection of an Althyrex excitation system.[1,2] To fully capitalize on the high initial voltage response capability of this excitation system, these studies and more general studies[3,4] indicated that power system stabilizers would provide significant benefits in power system damping. This power system stabilizer has been identified in previous technical literature as "supplementary control" and "excitation stabilizer".[5] Computer studies indicated that shaft speed is an effective input for the power system stabilizer.

These damping benefit conclusions had been confirmed by tests performed with a conventional dc commutator excitation system on Unit No. 2 at the Four Corners Station.[6] Unit No. 2 is a 3600 rpm, tandem-compound unit rated at 192 MVA.

A high initial response excitation system, the Althyrex system has been specified with a power system stabilizer for a number of steam-turbine generators in the Pacific southwest. The first unit, the cross compound Four Corners No. 4 unit, has been in service since July 1969. Its arrangement and performance will be described.

EXCITATION SYSTEM DESCRIPTION

The basic system concept of the Althyrex excitation system is illustrated in figure 1.

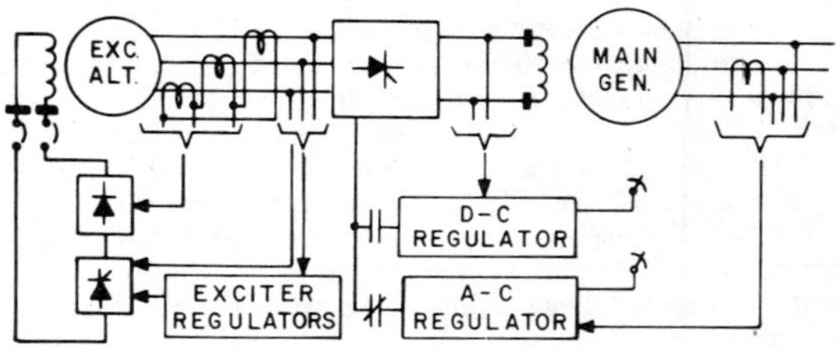

Figure 1 The Althyrex Excitation System

This system uses an ac alternator as a power source for excitation. AC voltage from this exciter alternator is phase controlled by the thyristors (SCR's) to furnish dc voltage for the main generator field. The exciter is self-excited. Exciter terminal voltage is maintained by a static voltage regulator.

Exciter and Controls

Basic excitation system components are illustrated by the detailed diagram, figure 2. The exciter alternator is a direct-coupled ac synchronous generator driven from the main generator rotor, with stationary armature and rotating field. It is air-cooled, with a closed ventilation circuit and an air-to-water heat exchanger located in the alternator frame.

Exciter field power is supplied from the armature through two redundant, parallel sources, each including:

1. A 3-phase full wave thyristor circuit supplied by an anode transformer from the exciter armature voltage.

2. A 3-phase full wave diode rectifier circuit supplied by current boost current transformers.

Either of the parallel bridge circuits can be disconnected for maintenance during the operation by an eight-pole switch. Either bridge will provide exciter field current without limiting system capability while the other bridge is out of service.

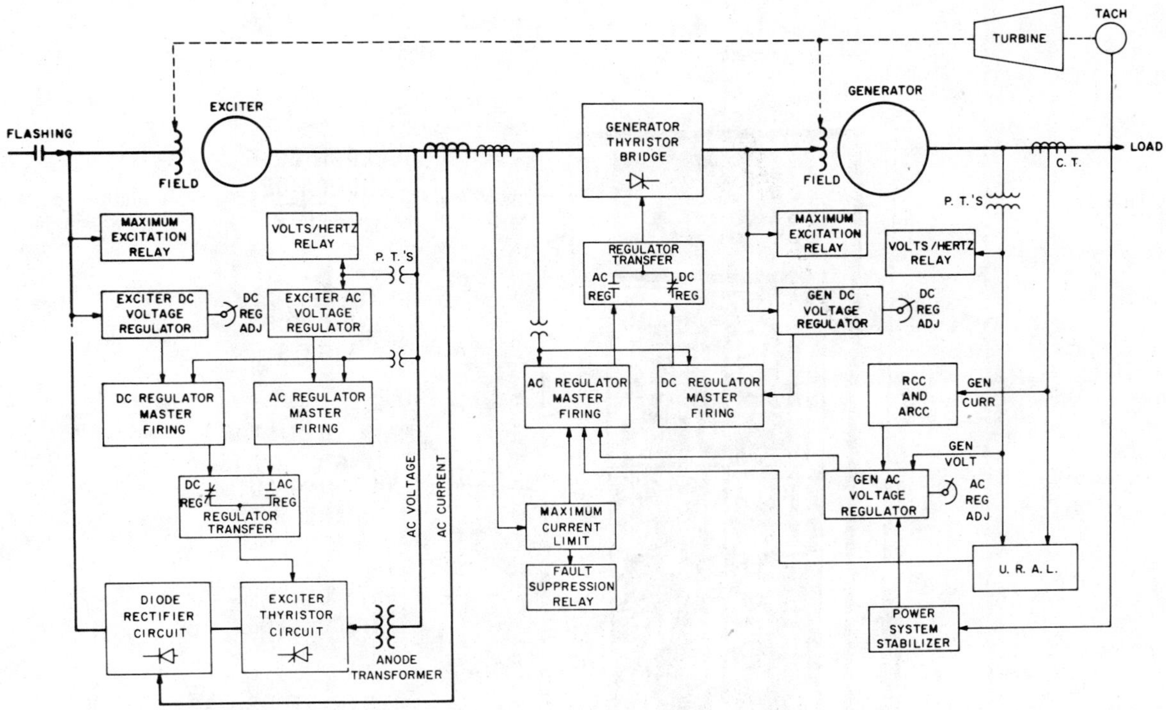

Figure 2 Detailed Diagram of Althyrex Excitation System

During normal operation, exciter field current control is obtained by phase-control of the exciter thyristor bridge circuit. During transient conditions, such as would occur during some system disturbances, the diode circuit may function to furnish additional exciter field current.

The exciter ac regulator is normally in service in order to provide a constant source voltage for the generator thyristor bridges. The exciter dc regulator, which maintains exciter field voltage, is used during startup or when the exciter ac regulator is not in service.

Generator Thyristor Bridge

The generator thyristor assembly consists of multiple bridges connected in parallel. For the cross-compound turbine generator set described, four parallel bridges are used on the low-pressure generator and six bridges are used on the high pressure generator. For each generator, any bridge can be removed from service, under load, by means of a 5-pole disconnect. With any bridge removed from service, the rating of the excitation system is maintained.

The assembly consists of thyristors mounted on heat sinks, which are in turn connected into parallel three-phase bridge circuits. Figure 3 illustrates a single thyristor mounted on a heat sink. The cooling water for the thyristors is furnished from the de-ionized water-pumping unit supplying the generator armature winding. The thyristor heat sinks, which also serve as part of the electrical circuit, are isolated from ground by the de-ionized water column and the insulating pipe assemblies.

When thyristors are used in parallel, as in the case of multiple bridges, the parallel

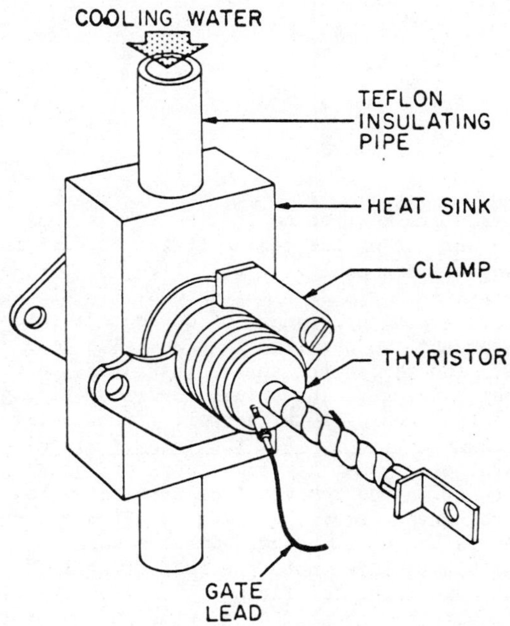

Figure 3 Thyristor and Heat Sink

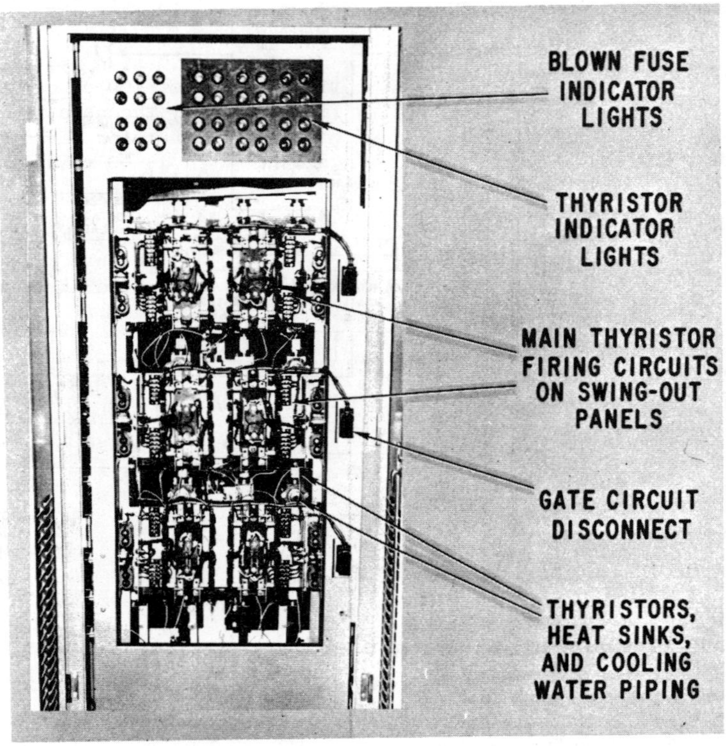

Figure 4 Generator Double Bridge Thyristor Assembly

thyristors must be controlled from a common source. The purpose of the master firing circuit is to generate a signal that will fire, or turn-on, each parallel thyristor simultaneously. This simultaneous firing provides good current division between the two parallel legs and also good voltage division across the series thyristors in the generator bridge. The actual thyristor firing is accomplished by a trigger circuit associated with each main bridge. All the trigger circuits are in parallel so that any individual bridge trigger circuit may be removed from service without disturbing the remaining bridges. Figure 4 shows one thyristor double bridge with the cabinet door open. Visible in this photo are indicating lights and the main thyristor firing circuits. The firing circuits are mounted on a swing-out panel in the foreground. The main thyristors, heat sinks, and cooling water piping are visible behind these panels.

The regulator transfer switch selects either the master firing circuit controlled by the dc regulator which holds a constant generator field voltage or the master firing circuit controlled by the ac regulator which holds a constant generator terminal voltage. A transfer voltmeter is used for matching signals to provide for smooth transfer between the two regulators.

Generator Excitation Regulators

The dc voltage regulator provides manual control by controlling firing of the thyristor bridges to hold constant generator field voltage.

If the excitation system is placed in service well below synchronous speed, as in the case of cross-compound generators, the dc regulator is used during startup to maintain constant generator terminal volts/hertz. The dc regulator is used also when the ac regulator is not available for service, or if manual control operation is desirable.

The inputs to the generator ac regulator include a signal from generator terminal voltage, and compensating signals from the reactive current compensator (RCC) and the active reactive current compensator (ARCC). If a power system stabilizer is used this signal also is fed into the ac regulator. The output of the generator ac regulator is a control signal to the corresponding master firing circuit.

The underexcited reactive ampere limit (URAL) provides a turn-on, take-over signal to the master firing circuit associated with the generator ac regulator if the underexcited reactive current increases beyond a desired value.

The input to the maximum current limit circuit is a signal from the exciter output current. The output is a turn-off, take-over signal to the master firing circuit controlled by the generator ac regulator.

The fault suppression circuit senses a collector ring fault by comparing signals from the generator field voltage and generator field current. The output is a signal that trips the master firing circuit power supply.

The reactive current compensator (RCC) provides a signal to modify the generator voltage input signal as required to achieve good paralleling of the generator with the power system. The active reactive current compensator (ARCC) provides a signal to modify the generator voltage input signal as required to compensate for line drip; thus it regulates voltage at some remote point in the power system.

The maximum excitation limit circuits will transfer regulator control from ac regulator to dc regulator if a higher than rated field voltage is maintained for a time in excess of the thermal capability of the field winding. Regulator transfer occurs after a preset time delay. A maximum excitation limit is furnished for both the exciter and generator.

Volts/hertz protective relay circuits may be provided to remove both the generator and exciter ac regulators from service at 90% speed by transferring control to the dc regulators. A dc regulator adjustor run back circuit may be included as a part of the volts/hertz protection on the generator.

EXCITATION SYSTEM PERFORMANCE

Phase controlled thyristor bridges supplied from a constant source voltage furnish a transient response capability that is not attainable with conventional excitation equipment. An added feature of thyristor bridges compared to diode bridges is that both positive and negative field voltages are available for forcing.

This excitation system is capable of a positive ceiling of 160 percent of nominal generator field voltage; the negative ceiling is 150 percent of nominal generator field voltage. Negative voltage can only be maintained transiently as the thyristor bridges will not allow negative current.

In order to illustrate the fast response capability of the thyristor exciter, figure 5 shows the response of the exciter voltage as

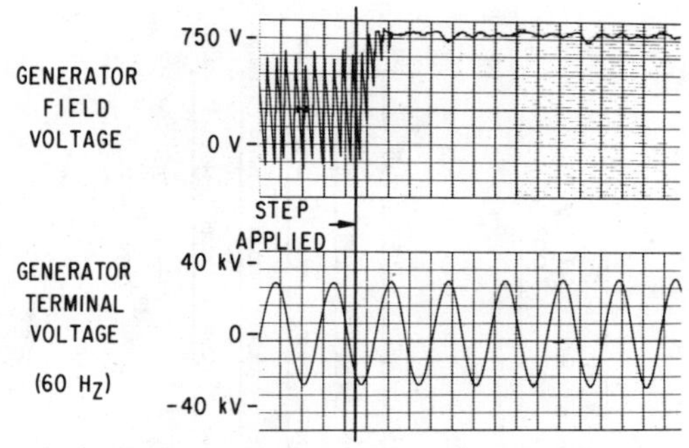

Figure 5 Response of Excitation Voltage Output

determined in factory tests. Initially the exciter was operating at a phased-back steady state condition to supply generator open-circuit rated voltage excitation. The generator ac voltage regulator was given a step of reference, and the generator field voltage changed to ceiling voltage in less than one cycle of the generator voltage.

CALCULATED PERFORMANCE UNDER LOAD

The performance features of the Althyrex excitation system that led to its selection for the Four Corners Station[1,2] are its high excitation system response ratio and short excitation system delays. These features provide benefits to both transient stability first swing reduction[4] and to damping of subsequent swings.[3] The degree of these benefits may be appraised from the results of the computer-simulated performance of the Four Corners No. 4 unit to a three phase stub fault at the high voltage bus, cleared in 3 cycles (0.05 seconds). The generators of the cross-compound unit were bussed at the terminals and, were assumed for this simulation to be connected through a transformer-transmission system with 40% impedance on the unit MVA base. The fault was assumed to occur on the high voltage terminals of a step-up transformer with 15% impedance. The initial unit loads were 90% power at unity power factor, expressed on the generator MVA rating.

The computer simulation used the FACE program[7] to provide detailed generator and excitation system models. The generator model included a representation of the solid rotor iron circuits and their shielding effects. The excitation system model included the effects of the gating circuit time delays and non-linearities. The gains and time constants of the representation shown in figure 10 were used in this computer simulation after adjusting for the inertias and field time constants of the high and low pressure units. However, the modeling uses a more detailed representation of the delays and non-linearities than those indicated in figure 10.

The results shown in figure 6 illustrate the generators operating with fixed excitation voltage. This case is presented as it provides a benchmark for performance with this combination of load, transmission system, and disturbance. The units were first swing stable and the rotor angle oscillations came to rest after 5 seconds.

In figure 7 excitation for the generators was supplied by the Althyrex excitation systems. The voltage regulator input was terminal voltage offset by 3% reactive current compensation; the power systems stabilizers were not connected. The excitation system steady-state and transient gains were adjusted as described in the appendix.[3]

The units were stable using both first swing and steady-state stability criteria. However, compared to the fixed excitation operation, figure 6, the rotor angle oscillations were not as well damped. The initial swings of the rotor angles were less, since the excitation acts to raise generator internal voltage. With constant excitation, the peaks of the angular swings of the high pressure and low pressure units, respectively, were 97° and 95°; with the excitation active, they were 95° and 93°.

Figure 8 illustrates the benefits of adding a power system stabilizing signal to the Althyrex excitation system. An increase in damping performance may be obtained with the application of power system stabilizing signals to a low time constant excitation system. The rotor angle oscillations were essentially damped out within 2 seconds from the time of the fault.

The lower two traces of figure 8 show the output of the power system stabilizing signals. A clamp circuit at the output of the transfer function limited the signal output to a range corresponding to ± 10 percent terminal voltage reference change. The limit circuit was included to prevent the stabilizing signal circuits from causing large voltage swings. The stabilizing signals extend into both upper and lower limits

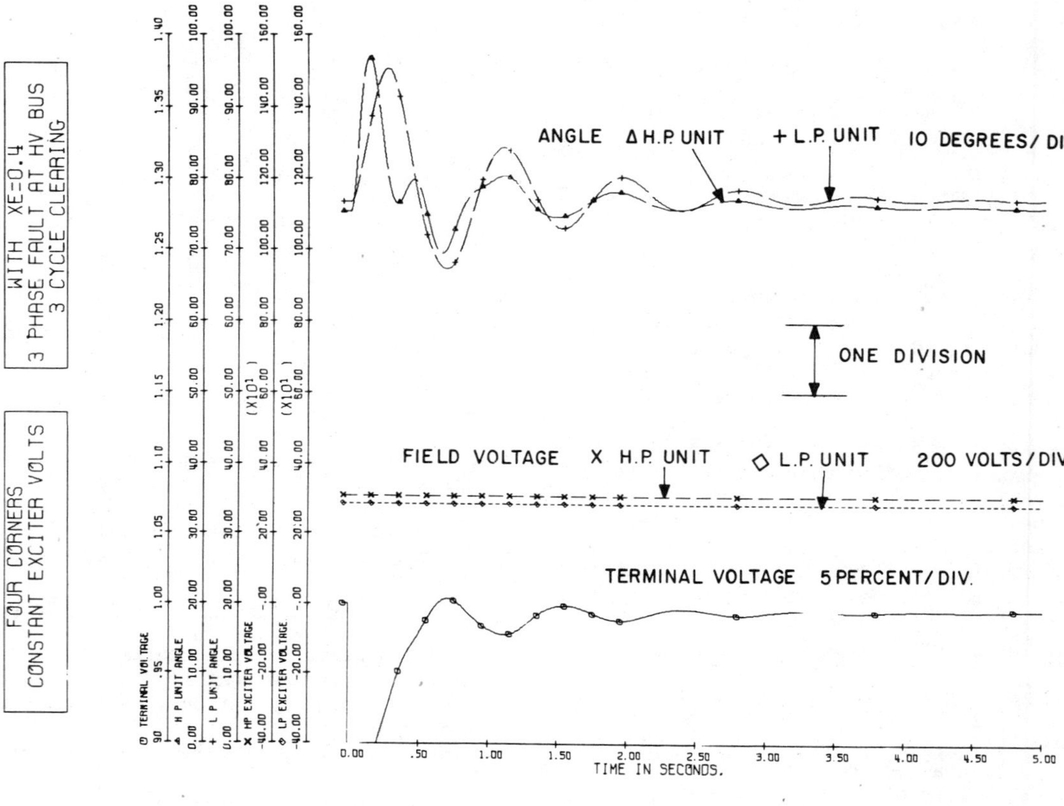

Figure 6 Computed Response of Unit with Constant Excitation

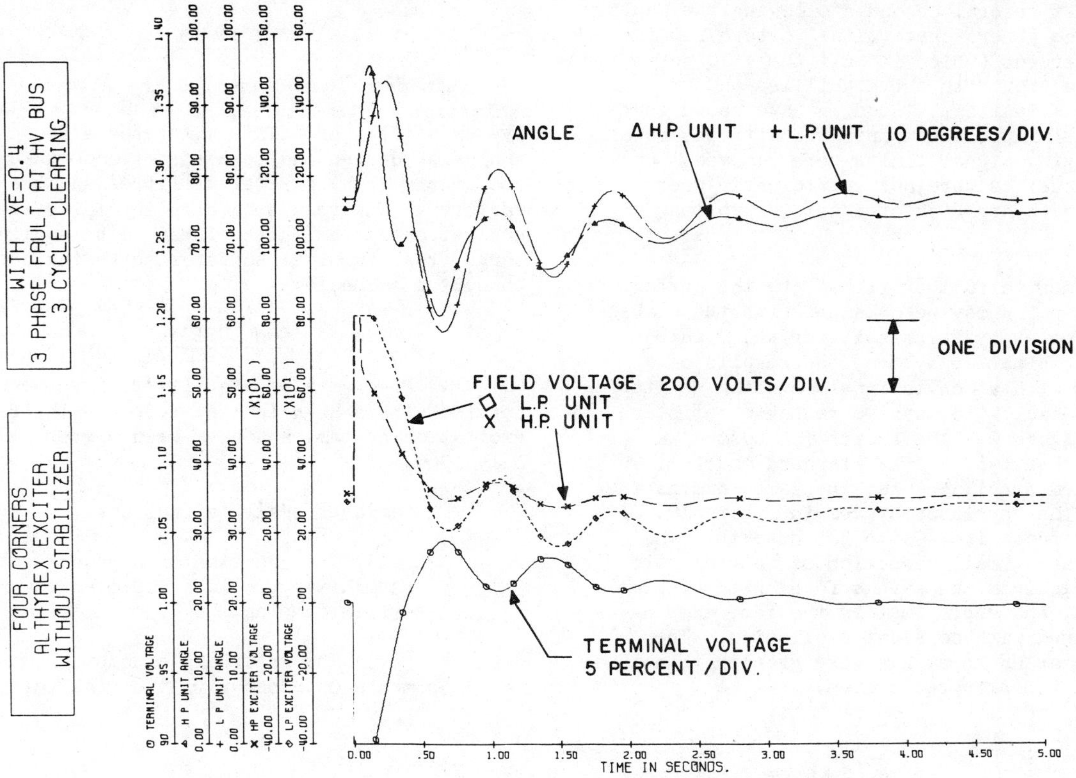

Figure 7 Computed Response of Unit with Althyrex

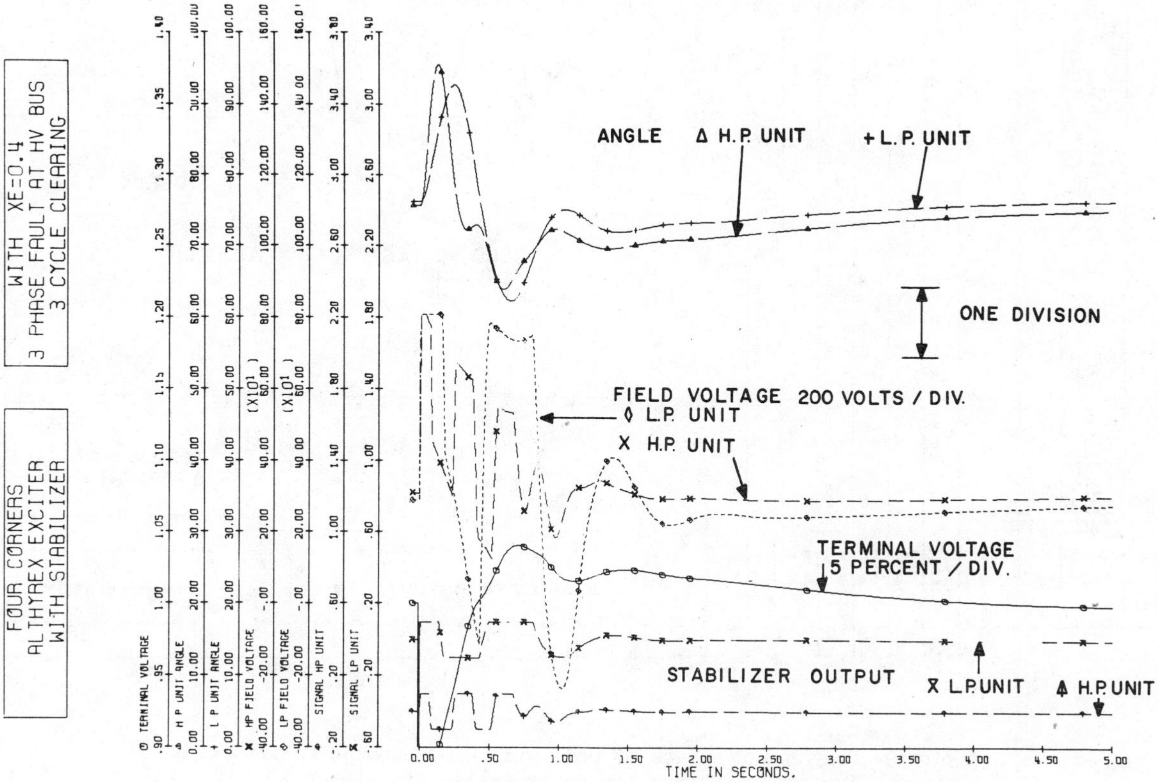

Figure 8 Computed Response of Unit with Althyrex with Power System Stabilizer

in the first several seconds following the fault. However, the generator terminal voltage remained within 5 percent (which is half the ± 10 percent signal band) for both the stabilized and unstabilized situations. Studies have shown that a very small increased damping benefit was obtained if the damping signal limits were removed, but that the benefits were not sufficiently great to justify the remote risk of excessive voltage variations.

In order to further illustrate the performance advantages which may be obtained with the combination of a power system stabilizer and a gated thyristor excitation system, the results of a simulation with a conventional rotating exciter with a 2.0 excitation system response ratio are given in figure 9. The excitation system was modeled with a type 1 IEEE standard model.[8] The power system stabilizer has the same form as that shown for the thyristor system in figure 10. Its gain was reduced from 96 to 30, however. The results show a small reduction of initial rotor angle swings from the swings found with constant excitation, and approximately the same damping as that obtained with constant excitation. The initial rotor angle swings were greater than those found with the Althyrex system.

SUMMARY

This paper has described the Althyrex excitation system and its speed derived power system stabilizer. This completes a sequence of analysis, design and test effort undertaken cooperatively by several utilities and the manufacturer. The resulting high initial response excitation system is the first to be applied to a large steam turbine-generator set using semiconductor technology.

CONCLUSIONS

1. Power system stabilizers have been applied with a high initial response thyristor excitation system, and have been operating since July 1969.

2. Computer analysis indicates:

(a.) The addition of a power system stabilizer provides improved power system damping.

(b.) The rotor angle damping improvements obtained with the high initial

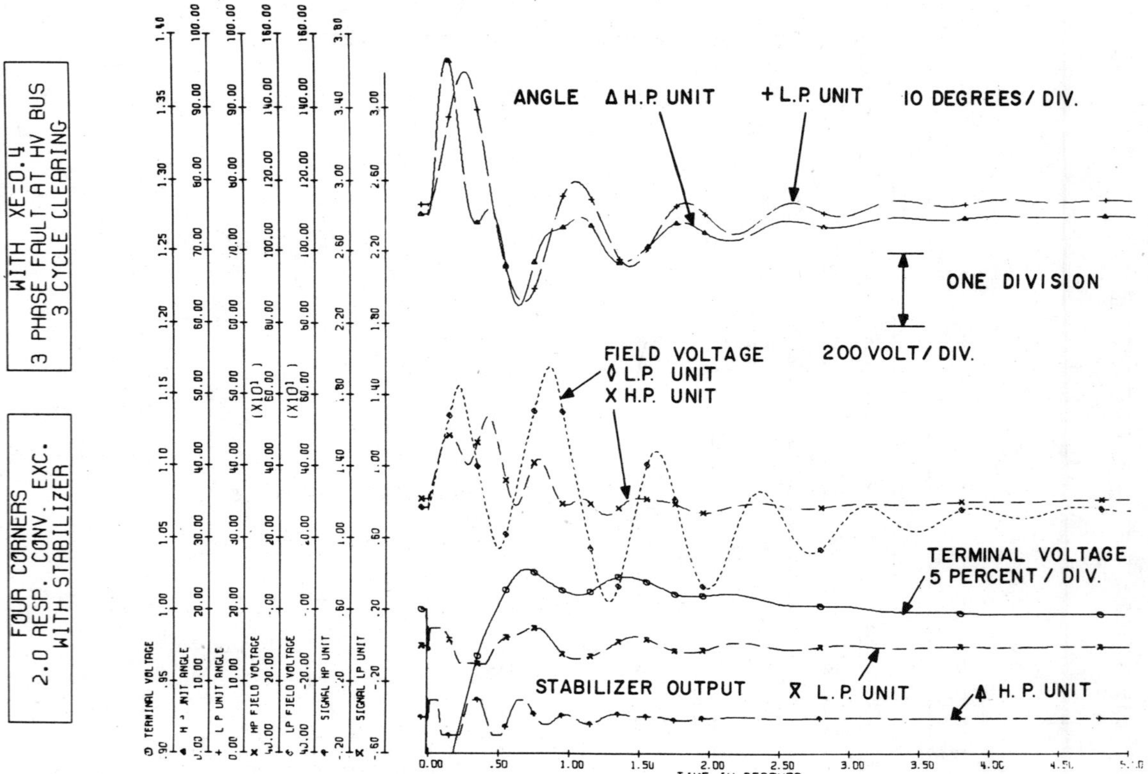

Figure 9 Computed Response of Unit with 2.0 Response Conventional Excitation System and Power System Stabilizer

(b.) ... exciter voltage response of the Althyrex excitation system were found to be greater than those obtained by applying a power system stabilizer to a conventional excitation system with the same excitation system response ratio.

(c.) The rotor angle damping with the Althyrex excitation system with the recommended voltage regulator adjustments was found to be acceptable with a moderately high (40%) equivalent system impedance.

3. Analysis of the first swing of the generators' rotors following faults also showed the benefits of high initial exciter voltage response in improving transient stability performance.

APPENDIX

Analytical Representation

Referring to figure 10 and the preceding detailed equipment description, the essential

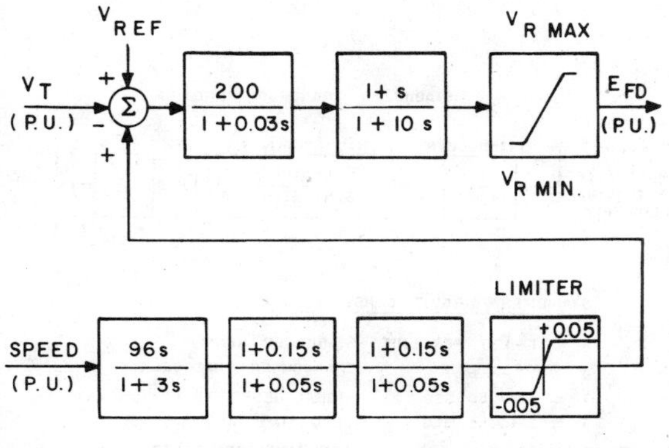

Figure 10 Simplified Representation of Excitation System and Power System Stabilizer

elements of performance that affect power system operation may be outlined:

1. The excitation power source was a shaft driven alternator whose armature voltage is essentially constant during the time periods considered in system stability studies.

2. The main generator excitation voltage was determined by phase-control of the main rectifier bridge. The inherent time delays and time constants associated with the regulator and firing circuits total about 0.01 second. However, the use of a single equivalent time constant 0.03 second seemed a reasonable value for use in large scale digital studies.

3. Voltage regulator gain constraints were necessary to allow satisfactory operation of the generator both on open circuit and on the power system. With steady state gain maintained at historically high values for voltage regulation, transient gain reduction was desirable.[3] The addition of a lag-lead network resulted in a well damped voltage regulating loop.

4. The addition of a supplementary signal to provide for damping of the generator rotor swings, i.e., a "power system stabilizer"[5] was provided. A tachometer mounted on the machine shaft was utilized as a signal source. The evolution of the stabilizing function is well documented.[1]

A simplified block diagram suitable for large scale power system studies shown in figure 10 represents these salient characteristics of the excitation system. By means of algebraic manipulations this can be converted into the Type 1 model of reference 8. The particular values shown would apply to a generator with an open circuit time constant (T'_{do}) of 5 second and a total inertia constant of 4 KW-sec/KVA.

REFERENCES

1. R. G. Farmer, R. H. Hartley, M. H. Kent, J. M. Wheeler, "Four Corners Project Stability Studies", IEEE Paper 68-CP-708-PWR, ASME/IEEE Power Generation Conference, San Francisco, Calif., Sept. 17, 1968.

2. E. A. Lohela, V. E. Dalke, "Auxiliary Power System for the Four Corners Plant 750 MW Units", IEEE Transactions on Power Apparatus and Systems, Vol. PAS-86, No. 12, December, 1967, pp. 1548-1554.

3. F. P. deMello, C. Concordia, "Concepts of Synchronous Machine Stability as Affected by Excitation Control", IEEE Transactions on Power Apparatus and Systems, Vol. PAS-88, No. 4, April, 1969, pp.316-329.

4. P. G. Brown, F. P. deMello, E. H. Lenfest, R. J. Mills, "Effects of Excitation, Turbine Energy Control, and Transmission on Transient Stability", IEEE Paper 70-TP-203-PWR.

5. IEEE Committee Report, "Proposed Excitation System Definitions for Synchronous Machines", IEEE Transactions on Power Apparatus and Systems, Vol. PAS-88, No. 8, August, 1969, pp. 1248-1258.

6. R. P. Schulz, M. Temoshok, R. G. Farmer, "Dynamic Stability Tests at Four Corners", 1969 American Power Conference, Proceedings, Vol. 31, pp. 872-882.

7. D. N. Ewart, R. P. Schulz, "FACE Multi-machine Power System Simulator Program", 1969 Power Industry Computer Applications Conference, Proceedings, #69-Cl-PWR, pp. 133-153.

8. IEEE Committee Report, "Computer Representation of Excitation Systems", IEEE Transactions on Power Apparatus and Systems, Vol. PAS-87, June, 1968, pp. 1460-1464.

DESIGN OF A POWER SYSTEM STABILIZER SENSING FREQUENCY DEVIATION

F. W. KEAY, Senior Member, IEEE and W. H. SOUTH, Member, IEEE

Westinghouse Electric Corporation
East Pittsburgh, Pennsylvania

ABSTRACT

A power system stabilizer specifically designed to the performance specifications of various utilities in the Western United States has been developed. The stabilizer utilizes synchronous machine terminal frequency as input. With this input signal, the stabilizing signal is derived using all static components and susceptibility to electrical noise is reduced. The design is modular with the sensing and signal conditioning elements implemented with all solid-state circuitry contained on draw-out printed circuit card assemblies. The design provides independent adjustment of gain and time constants with calibrated controls to allow direct settings from calculated data. Factory tests on the prototype verified its dynamic characteristics met or exceeded performance specifications. Field tests on a 422 MVA unit demonstrated improved damping of oscillations following switching of a tie line.

INTRODUCTION

The West Coast utilities during the winter of 1968-9 were faced with the problem of how to maximize the water runoff anticipated in the Spring from a record snow fall. Heavier than normal hydro electric power generation would have to be transmitted over existing lines. To get this power out without spilling excess water would require supplementary power system stabilizing equipments. Short delivery of these equipments were required to meet Spring thaws.

The function of these power system stabilizers is to provide a supplementary control signal input to a synchronous machine voltage regulator to improve power system dynamic performance. (1) Several different quantities, including electrical power, shaft speed, and frequency deviation, have been successfully used as the input to a power system stabilizer. A number of publications, (references 2-11) have described analytical studies and field tests in which improved damping has been achieved. This paper describes the design of a power system stabilizer utilizing a frequency signal in the manner reported by Schlief et al. (9,10)

DESIGN SPECIFICATIONS

The stabilizer described in this paper was specifically designed to meet the specifications of a number of electric utilities in the Western United States. These specifications were given in terms of end results and specific features and not in terms of hardware. All of the individual utilities specifications were quite similar with respect to the required dynamic characteristics, being principally based on the recommendations of a Technical Studies Task Force of the Western Systems Coordinating Committee. The transfer function specified is shown in Figure 1 along with the specified range of adjustment of circuit parameters.

In addition to the specifying dynamic characteristics, various users specifications also included the following additional requirements:

Paper 70 TP 561-PWR, recommended and approved by the Power Generation Committee of the IEEE Power Group for presentation at the IEEE Summer Power Meeting and EHV Conference, Los Angeles, Calif., July 12-17, 1970. Manuscript submitted March 13, 1970; made available for printing May 22, 1970.

Input Signal Transducer

Output not to be appreciably affected by changes in magnitude of input signal.

Output linear within 5% over the frequency range from 59-61 Hz.

Time constant less than or equal to .040 sec.

Signal Conditioning Circuitry

Noise level not more than 10% of the dynamic range.

Adjustment with calibrated dials relating the dial setting to the set parameter within ±10%.

Total noise filtering phase lag not more than 27 degrees at 1.33 Hz.

STABILIZER TRANSFER FUNCTION

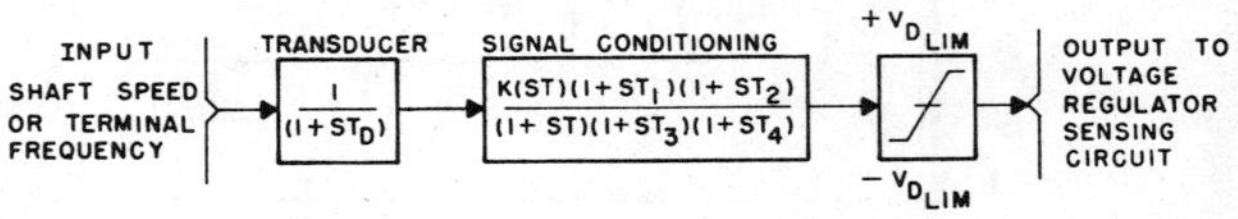

STABILIZER CIRCUIT CONSTANTS:

K	$= 1 - 100$	PER UNIT	STABILIZER GAIN
T_D	$\leq .04$	SEC.	TRANSDUCER TIME DELAY
T	$= .1$ TO 50	SEC.	SIGNAL RESET TIME CONSTANT
T_1 & T_2	$= .2$ TO 1.5	SEC.	LEAD TIME CONSTANTS
T_3 & T_4	$= .02$ TO .1	SEC.	LAG TIME CONSTANTS
$V_{D_{LIM}}$	$=$ (0–.10)	PER UNIT	SIGNAL LIMIT

Fig. 1. Power System Stabilizer Design Specifications

Optional Control Elements

Provision for accepting optional additional control functions including supplementary voltage control, reactive compensation, and a means for switching the stabilizing signal on and off.

DESIGN CHARACTERISTICS

Sensing Element

Most user specifications permitted the use of either shaft speed or terminal frequency as the input signal to the stabilizer. Although excellent results have been reported using a shaft speed signal (6), the author's company chose the terminal frequency signal because of the advantages it affords. With terminal frequency as input the stabilizing signal is derived utilizing all static circuitry. Problems inherent in mounting, aligning, and maintaining mechanical transducers are eliminated and susceptibility to electrical and mechanical noise is reduced. Any eccentric or lateral movement or tortional oscillations, in a shaft-mounted transducer results in excessive noise requiring special filtering. The all static circuitry is simpler and potentially more reliable.

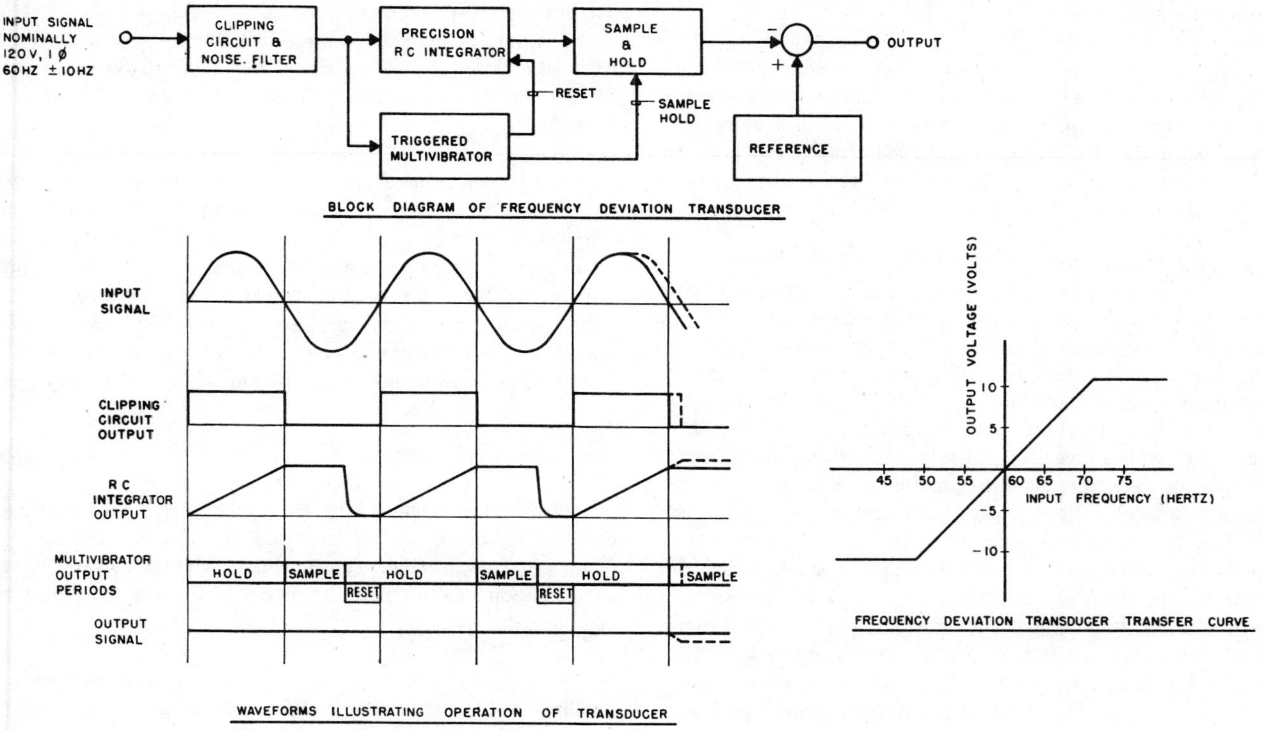

Fig. 2. Power System Stabilizer Frequency Deviation Transducer

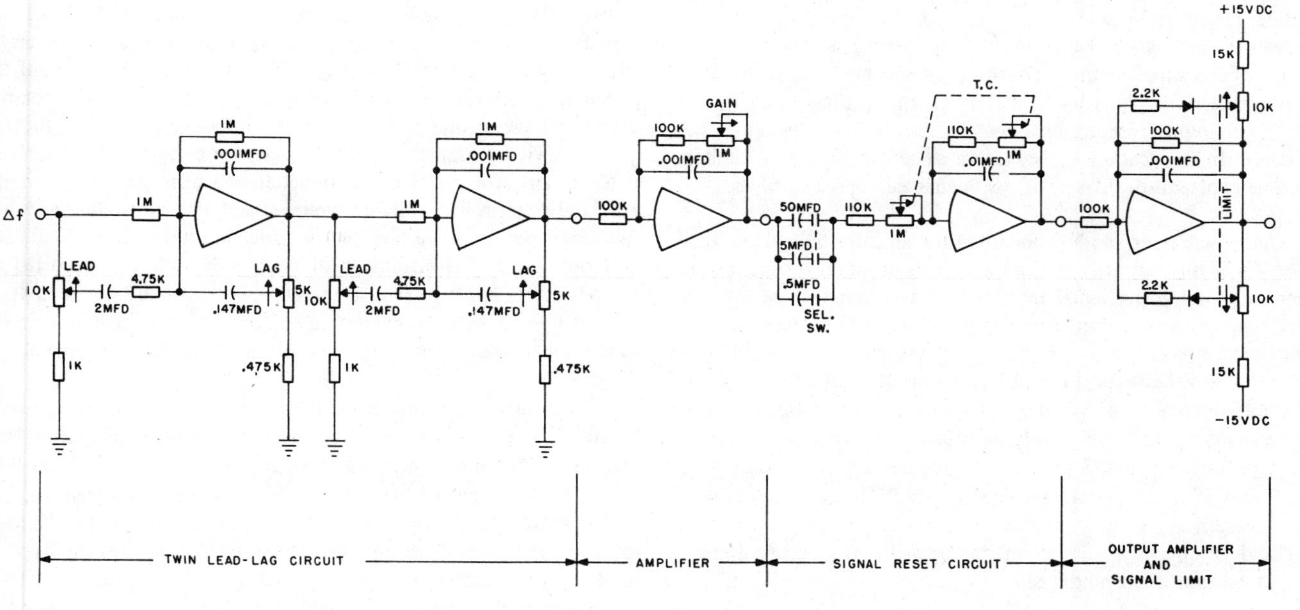

Fig. 3. Power System Stabilizer Signal Conditioning Elements

Frequency transducers with suitable characteristics were not found to be available commercially, it was, therfore, necessary to develop a novel transducer specifically for this application.

The block diagram and waveforms of Figure 2, illustrate the operation of the frequency deviation transducer which functions on a sample and hold principle. Each positive half cycle of input signal frequency, a ramp generator measures the positive period of the incoming signal and generates a d.c. voltage proportional to the period. This voltage is sampled and held for the next cycle. As long as the period of the incoming signal is constant, the output of the holding circuit is a constant d.c. voltage. This voltage is compared with a constant reference voltage. Should the frequency of the incoming signal change, the transducer output will be a polarized signal proportional to the magnitude of change in frequency. The output of the transducer is linear over the range of 50 to 70Hz with a gain of 1 volt/Hz. Its time delay is a transport delay of 1 cycle plus a filter time constant giving a total effective time constant of .03 seconds. The jitter in the output for a constant frequency input signal is less than .001%. Since the sensing circuitry functions as a zero crossing detector, its output is relatively unaffected by changes in magnitude of the incoming signal. For a 10% change in input signal magnitude, the output change is less than 25 mv.

Signal Conditioning Elements

Figure 3 shows the basic signal conditioning elements in elementary schematic form along with the principal component values. Power supply connections, amplifier stabilizing and balance adjustment circuitry are omitted for clarity. The desired transfer functions are readily realized with solid state operational amplifiers in the configurations shown in Figure 3.

The error signal from the frequency deviation transducer is applied to a twin-lead-lag circuit having four independently adjustable time constants. The lead time constants are adjustable from .18 to 2.0 seconds and the lag time constants are adjustable from .02 to .15 seconds. The lead time constants serve to compensate for the time delays inherent in excitation control systems. The lag time constants are necessary from practical noise considerations and should, in general, not be set less than 1/10 of the lead time constants. As shown in the Appendix, amplifier stabilizing and noise suppression circuitry introduces two small additional lags of .001 sec. and .01 sec., respectively. These lags contribute negligible phase shift (6 degrees at 1.33 Hz) in the range of frequencies of interest in power systems. Electronically varying the effective capacitance, rather than resistance, in the lead-lag signal conditioning element permits independent adjustment of circuit gain and time constants with calibrated controls. Precision resistors (± 1%) and capacitors (± 5%) are used to meet the requirement for calibrating the dial setting of adjustments within ± 10%.

All of the d.c. gain in the stabilizer is incorporated in a separate amplifier circuit. The gain of the amplifier is continuously adjustable over a ten to one range. Three input terminals provide a means of selecting ranges of gain from 0.1 to 1.0, 1.0 to 10, and 10-100 volts/volt respectively. The allowable gain setting is limited by the stability of the supplementary control loop and noise considerations. It is normally determined by field tests. In applications to excitation systems employing rotating exciters, the gain is normally set on the mid-range. In application to excitation systems, employing exciters with time constants of 0.1 sec. or less, higher gain may be utilized.

The signal reset module incorporates an adjustable long time constant which is normally set at 10 seconds or more. It functions to prevent the stabilizing signal from biasing terminal voltage for prolonger frequency excursions and also to wash-out the effects of drift in the preceeding circuitry. Since the time constant of the reset circuit is long, it has no effect on the phase or gain of the stabilizing signal over the frequency range of interest (normally .2 to 2 Hz). The reset time constant is also continuously adjustable over a ten to one range. A selector switch in the input circuit provides a means of selecting three ranges of time constant settings from 0.05 to .5, .5 to 5, and 5 to 50 seconds, respectively. Ganged potentiometers in the amplifier input and feedback circuits permit adjustment of the circuit time constant without changing the gain.

An output amplifier with unity gain serves to raise the stabilizer signal to a power level suitable for controlling the magnetic and/or rotating amplifiers used in existing excitation systems. The polarity of the signal is in the direction to raise excitation for terminal frequency higher than normal. The output amplifier also includes circuitry for limiting the magnitude of the stabilizing signal. As shown in Figure 3, biased diodes in the amplifier feedback circuitry provide for limiting the signal by clipping the output. The bias is adjustable providing a signal limit, when applied to existing regulators of Westinghouse manufacture, corresponding to 2% to 25% terminal voltage excursion.

To prevent the stabilizing signal from causing prolonged excitation forcing in the event of a component failure within the stabilizer equipment, the output of the stabilizer is monitored by a static level detector and timer module. This protective element provides an alarm and automatically disconnects the stabilizing signal from the regulator if the output of the stabilizer remains above or below a preselected level for a preselected time interval. Pickup of this device is adjustable from ± 1 to ± 15 V d.c. and the time delay is adjustable from .2 to 20 seconds.

A ± 15 volt regulated d.c. power supply is provided to operate all of the circuitry of the stabilizer. The input power for the power supply is 120 or 240 volts, 3 phase, 60 to 420 hertz and is usually taken from the 420 hertz permanent magnet generator that supplies the machine voltage regulator. In addition to the normal ± 15 volt fixed output voltage, an adjustable voltage is provided for test purposes. This voltage can be connected by means of a temporary jumper to simulate an input signal for any of the various modules.

Optional Control Elements

Provision is included in the stabilizer for incorporating the optional additional control functions shown in the Block Diagram, Figure 4. These include supplementary voltage control, reactance compensation, and an alternate means of limiting the signals effect upon terminal voltage.

Supplementary voltage control may be incorporated in the stabilizer by the addition of voltage deviation detector, No. 1, of Figure 4. When this module is applied, its sensing input is nominally 120 volts, three phase from the regulator potential transformers, but may, as shown in Figure 4, be derived from the output of a reactive compensator. The input signal is rectified, filtered, and compared to a constant d.c. voltage reference at the summing point of an operational amplifier. The output voltage error signal is applied to a lead-lag circuit in the signal conditioning elements which is duplicate of the circuit to which the frequency deviation signal is applied. The polarity of the voltage error signal is, however, opposite to that of the frequency deviation signal. Thus, with this configuration, the stabilizing signal is an error signal proportional to the difference between the ultimately controlled variable, voltage, and the controlling variable, frequency deviation, rather than a signal proportional to frequency deviation alone. The advantage of this configuration, as reported by Schleif, et al (10) is to reduce the dependence of the supplementary control upon the actual excitation system constants, and adjustment of the stabilizer compensating constants, is, therefore, less critical.

When used, the function of the reactance compensator shown in Figure 4 is to approximate synchronous machine quadrature axis reactance (Xq) and provide a sensing signal to the stabilizing circuits that is proportional to machine internal voltage. A frequency deviation signal derived from this voltage closely represents deviations in rotor velocity. Again, as described by Schlief, et al (10) this type of signal may be preferable in some, but not all, applications. Their conclusion is that this type of signal is preferable where the stability problem is primarily one of local, rather than interarea swings.

In application to older excitation systems, employing rotating amplifiers having slower response than magnetic or transistorized amplifiers, forcing capability of the stabilizing signal may be required to obtain improved system damping. In these applications, limiting may be accomplished by disconnecting the signal with a static switching circuit, rather than by limiting the output signal of the stabilizer. Though not shown in the schematic of Figure 3, this circuit is an integral part of the output amplifier.

The static switch functions to reduce the gain of the output amplifier to zero. It operates instantaneously when machine terminal voltage deviation exceeds a preselected level and resets within 0.1 seconds when the deviation drops below the selected level. The

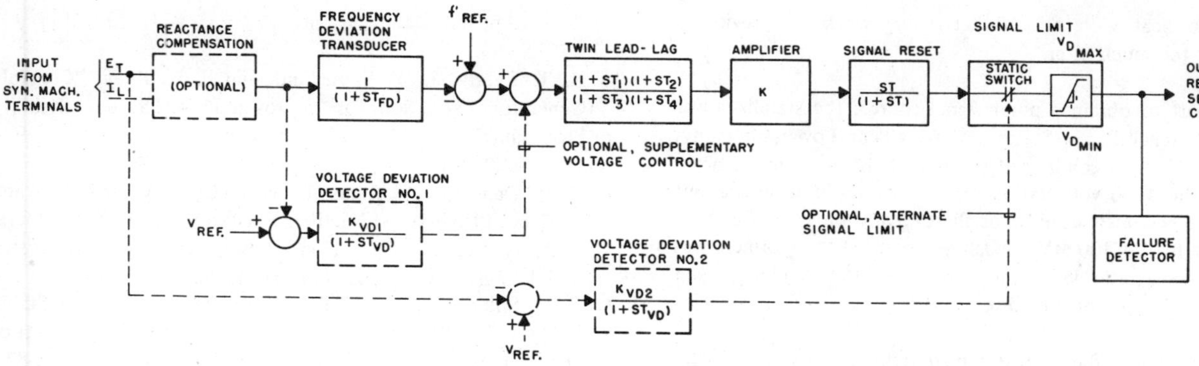

STABILIZER DESIGN CONSTANTS

T_{FD}	=	.03	SEC.	FREQUENCY DEVIATION TRANSDUCER TIME DELAY
T_1, T_2	=	.2 TO 2	SEC.	LEAD TIME CONSTANTS
T_3, T_4	=	.02 TO .15	SEC.	LAG TIME CONSTANTS
T	=	.05 TO 55	SEC.	SIGNAL RESET TIME CONSTANT
T_{VD}	=	.02	SEC.	VOLTAGE DEVIATION DETECTOR TIME DELAY
K	=	.1 TO 100	P. U.	STABILIZER GAIN
K_{VD1}	=	.02 TO .20	P. U.	OPTIONAL SUPPLEMENTARY VOLTAGE CONTROL GAIN
K_{VD2}	=	40 TO 400	P. U.	GAIN - ALTERNATE SIGNAL LIMITING
$V_{D_{MAX}}, V_{D_{MIN}}$	=	.02 TO .25	P. U.	LIMIT OF STABILIZING SIGNALS EFFECT ON VOLTAGE
COMPENSATION	=	0 TO 1.0	P. U.	OPTIONAL COMPENSATION - FOR SYN. MACH. X_Q

Fig. 4. Power System Stabilizer Design Characteristics

control signal for the static switching circuit, when used, is derived from a second voltage deviation detector. This detector is duplicate of the one described above except that it has higher gain and is always connected to sense machine terminal voltage.

PACKAGING

Figure 5 shows the physical configuration of the stabilizer equipment. The design is modular. Each circuit function shown in the block diagram, Figure 4, is incorporated on a single, draw-out, printed circuit assembly. Controls for adjusting the circuit parameters are mounted on the face plate brackets integral to each card, and are provided with calibrated dials and locking mechanism to assure tamper proof settings. For application to existing excitation systems, the cards are mounted in the rack assembly shown in Figure 5. Space is provided for incorporating the additional, optional, control elements described above.

PROTOTYPE TESTS AND SUBSEQUENT FIELD EXPERIENCE

Tests were run on the prototype frequency transducer to determine that it produced a signal proportional to frequency in the region of interest and that the noise and jitter in the output would be low enough such that with the high gain in the signal conditioning elements the final output signal did not have appreciable noise content. With the gains and time constants of the stabilizer set at their highest practical setting, the minimum signal to noise ratio allowable of the input signal is approximately 10^5. Difficulty was encountered in obtaining a signal source of 60 hertz that was free of jitter to within 1 part in 10^5 for testing.

The 60 hertz utility service at the Westinghouse, East Pittsburgh Plant, was first tried but was found to have a jitter in frequency of about 1 part in 10^5 due probably to random load switching throughout the plant. An audio frequency oscillator was tried, however, any mechanical vibration of the case of the oscillator or bench on which it was setting would show up as a frequency modulation of at best 1 part in 10^4. In addition to the mechanical noise problems, the oscillators

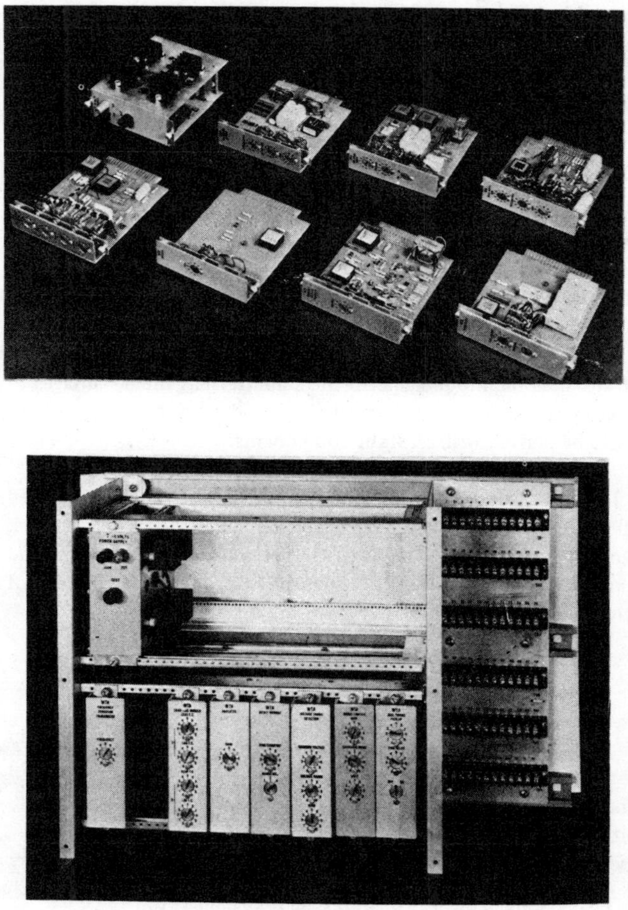

Fig. 5. Westinghouse Type F-7 Power System Stabilizer Equipment

would tend to beat with the 60 hertz line which as previously mentioned had too much jitter.

In an effort to obtain a purer signal source, the stabilizer was taken to the Mitchell Power Station, of West Penn Power Company, and connected to the secondary of the potential transformers. When connected to the 4160 volt station service bus, the noise was about the same as the best tests using an oscillator. Only when the stabilizer was connected to the 300 MVA machine potential transformers was the signal to noise ratio low enough to allow the maximum gain settings and still give acceptable noise in the output.

Complete dynamic data was obtained at the factory on a prototype model using a frequency response analyzer. This data confirmed the dynamic characteristics to be within specifications. Subsequently, in January, 1969, the prototype was field tested in cooperation with the Pacific Gas and Electric Co. at their Pittsburgh Plant. Recordings of MW swings following switching of a line showed improved damping with the stabilizer in service.

Thirty-four production units were delivered to Western Utilities in the first quarter of 1969. Subsequent operating experience revealed a design deficiency in the regulated power supplies in the stabilizer. The series droping resistors used in the power supply were initially mounted flush on the printed circuit. The heat dissipated in these resistors caused discoloration of the circuit boards and in three cases, resulted in failures of the power supply. The design was changed to correct this deficiency and all of the power supplies have been replaced with the revised design.

CONCLUSIONS

A power system stabilizer sensing synchronous machine terminal frequency deviation has been successfully developed. Extensive dynamic testing of a prototype confirmed that its characteristics met or exceeded the specifications established by various utilities in the Western United States.

Use of synchronous machine terminal frequency rather than shaft speed as the input signal to the stabilizer has the advantages of reduced susceptibility to electrical noise and permits the stabilizing signal to be derived with all static components.

Draw-out, modular, design of the stabilizer elements simplified installation testing and field servicing. The standardized building-block design also facilitates the addition of optional control elements and the application of the stabilizer elements to excitation systems of different designs.

REFERENCES

1. IEEE Committee Report, "Proposed Excitation System Definitions for Synchronous Machines," IEEE Transactions on Power Apparatus and Systems, Vol. PAS-88, pp 1248-1258, August, 1969.
2. H. M. Ellis, J. E. Hardy, A. L. Blythe, and J. W. Skooglund, "Dynamic Stability of the Peace River Transmission System," IEEE Trans. Power Apparatus and Systems, Vol. PAS-85, pp 586-600, June, 1966.
3. F. R. Schleif and J. H. White, "Damping for the Northwest-Southwest Tieline Oscillations–An Analog Study," IEEE Trans. Power Apparatus and Systems, Vol. PAS-85, pp 1239-1247, December, 1966.
4. F. R. Schleif, G. E. Martin, and R. R. Angell, "Damping of System Oscillations With a Hydro-Generating Unit," IEEE Trans. Power Apparatus and Systems, Vol. PAS-86, pp 438-442, April, 1967.
5. R. T. Byerly, J. W. Skooglund, and F. W. Keay, "Control of Generator Excitation for Improved Power System Stability," Proceedings of the American Power Conference, Vol. 29, pp 1011-1022, 1967.
6. P. L. Dandeno, A. Karas, K. R. McClymont, and W. Watson, "Effect of High-Speed Rectifier Excitation Systems on Generator Stability Limits" IEEE Trans. Power Apparatus and Systems, Vol. PAS-87, pp 190-201, January, 1968.
7. R. M. Shier and A. L. Blythe, "Field Tests of Dynamic Stability Using a Stabilizing Signal and Computer Program Verification," IEEE Trans. on Power Apparatus and Systems, Vol. PAS-87, pp 315-322, February, 1968.
8. O. W. Hanson, C. J. Goodwin, and P. L. Dandeno, "Influence of Excitation and Speed Control Parameters in Stabilizing Inter System Oscillations," IEEE Trans. Power Apparatus and Systems, Vol. PAS-87, pp 1306-1313, May, 1968.
9. F. R. Schleif, H. D. Humlins, G. E. Martin and E. E. Hattan, "Excitation Control to Improve Powerline Stability," IEEE Trans. Power Apparatus and Systems, Vol. PAS-87, pp 1426-1434, June, 1968.
10. F. R. Schleif, H. D. Hunkins, E. E. Hattan, and W. B. Gish, "Control of Rotating Exciters for Power System Damping; Pilot Applications and Experience," IEEE Trans. on Power Apparatus and Systems, Vol. PAS-88, pp 1259-1266, August, 1969.
11. F. P. DeMello and C. Concordia, "Concepts of Synchronous Machine Stability as Affected by Excitation Control," IEEE Transactions on Power Apparatus and Systems, Vol. PAS-88, pp 316-329, April, 1969.

APPENDIX

Transfer Function of Lead-Lag Circuit

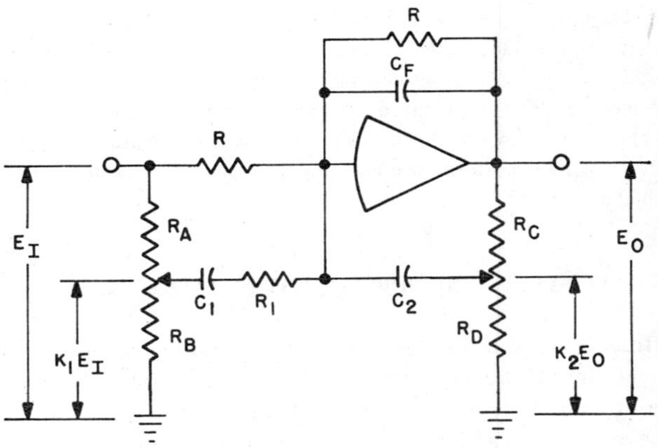

Assuming an Ideal Operational Amplifier and Negligible Potentiometer Loading:

S = Operator, d/dt

$$K_1 = \frac{R_B}{R_A + R_B}$$

$$K_2 = \frac{R_D}{R_C + R_D}$$

$$E_i \left[\frac{1}{R} + \frac{K_1}{R_1 + 1/SC_1}\right] = -E_o \left[\frac{1}{R} + \frac{1}{1/SC_F} + \frac{K_2}{1/SC_2}\right]$$

$$E_i \left[\frac{(1 + SR_1 C_1) + SK_1 RC_1}{R(1 + SR_1 C_1)}\right] = -E_o \left[\frac{1 + S(K_2 RC_2 + RC_F)}{R}\right]$$

$$\frac{E_o}{E_i} = \left[\frac{1 + S(K_1 RC_1 + R_1 C_1)}{(1 + SR_1 C_1)\,[1 + S(K_2 RC_2 + RC_F)]}\right]$$

$$\frac{E_o}{E_i} = \left[\frac{1 + S(T_A + T_B)}{(1 + ST_B)\,[1 + S(T_C + T_D)]}\right] \quad \text{where}$$

$T_A = K_1 RC_1$ – Lead Time Constant

$T_B = R_1 C_1$ – Noise Filter Time Constant

$T_C = K_2 RC_2$ – Lag Time Constant

$T_D = RC_F$ – Amplifier Stabilizing Circuit Time Constant

Referring to Figure 3, R = 1 Megohm, R_1 = 4.75 Kilohms, C_1 = 2 Microfarads, C_2 = .147 Microfarads and C_F = 1 Nanofarad. K_1 is adjustable from 1/11 to 1 and K_2 is adjustable from 1/8.67 to 1, therefore:

T_A = .182 to 2.0 Seconds

T_B = .0095 Seconds

T_C = .017 to .147 Seconds

T_D = .001 Seconds

Transfer Function of Signal Reset Circuit

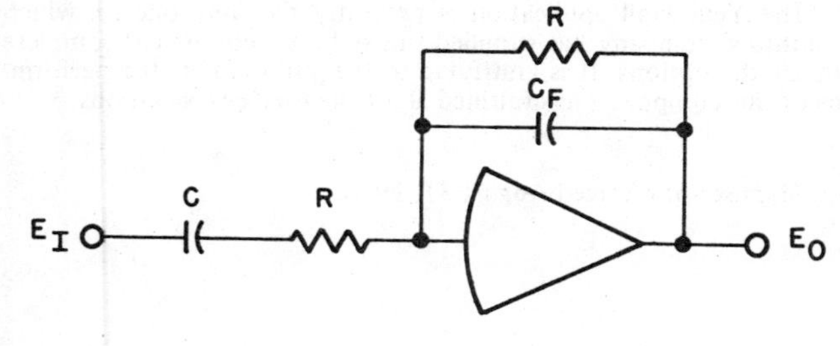

$$E_i \left[\frac{1}{R + 1/SC}\right] = -E_o \left[\frac{1}{R + 1/SC_F}\right]$$

$$\frac{E_o}{E_i} = -\left[\frac{SRC}{1 + SRC}\right]\left[\frac{1}{1 + SRC_F}\right]$$

$$\frac{E_o}{E_i} = -\left[\frac{(ST)}{(1 + ST)(1 + ST_F)}\right]$$

Again referring to Figure 3, C varies from 0.5 to 50 MF, C_F = .01 MF and R is adjustable from .11 to 1.11 Megohm, therefore, T is adjustable from .055 to 55.5 seconds and T_F varies from .0011 to .011 seconds.

Discussion

F. R. Schleif and **W. B. Gish** (Bureau of Reclamation, Denver, Colo. 80225): It has been quite gratifying to find this rather specialized control apparatus carefully designed to fulfill the wide range of specifications requirements even though many of the component functions, such as the frequency transducer, were not previously available.

All four units of our Yellowtail Powerplant near Hardin, Montana, have recently been individually equipped with this control. We believe this is the only application to date employing all the optional features. Even though influence of the special features had been explored with our pilot apparatus, special care was taken to check characteristics of these control units in our laboratory and finally to test performance of the systems in the field.

The frequency transducer proved to yield a clean, fast response signal. It does have some voltage sensitivity but happily this is of such polarity as to constitute a negative voltage feedback with the transfer function

$$\frac{0.0032\,\Delta E}{1 + 0.03S}.$$

Therefore it operates to very slightly augment the optional voltage feedback path we had specified so that in this respect the effect is not detrimental.

The voltage feedback circuitry was specified for the purpose of broadening the frequency range over which damping can be effectively improved.[10] Frequency response curves taken in the field on the compensated systems with and without the voltage feedback clearly showed its ability to maintain proper phasing to higher swing frequencies. However, the resulting effect on damping is more illustrative. This is shown by the transient response curves in Figure 6 of this discussion, which were produced by rejecting about 6 MW under each of several conditions. In Figure 6A it is to be seen that the natural local swing frequency is rather high at 80 cpm (1.33 Hz). As shown in Figure 6B, activation of the basic supplementary control with two stages of compensation does not bring about effective damping at this high swing frequency. However, with activation of the voltage feed-

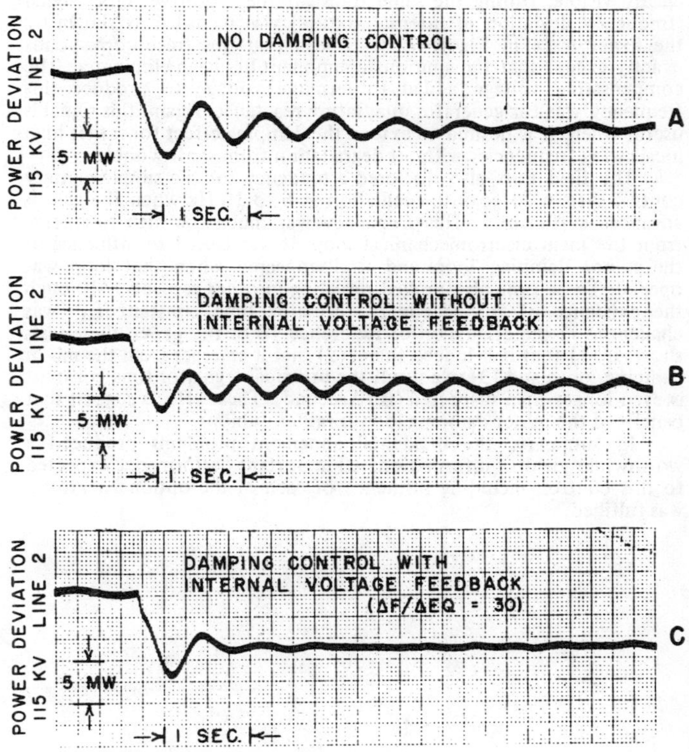

Fig. 6. Transient tests of damping control, Yellowtail power plant.

Manuscript received July 22, 1970.

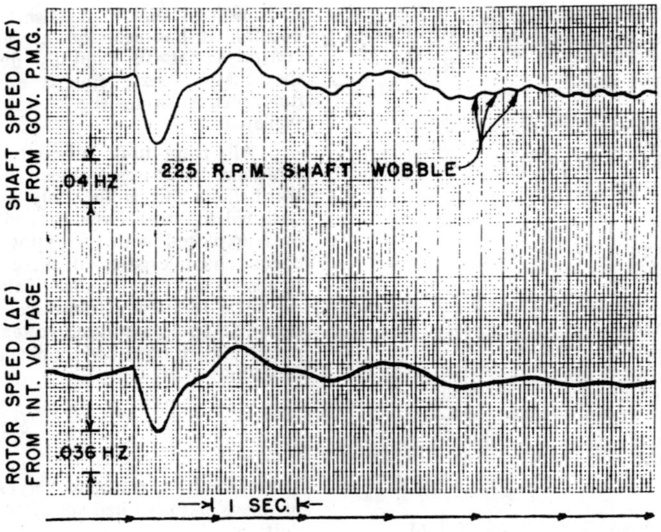

Fig. 7. Comparison of shaft speed and internal frequency deviation.

back circuit in Figure 6C, damping becomes effective. Thus this feature is useful where damping of high local swing frequencies is desired.

Another unique comparison yielded by the field tests at Yellowtail was effectiveness of the reactance compensation for yielding a high grade "rotor velocity" signal. Shaft speed (frequency deviation) from the shaft-mounted 60-Hz governor PMG is shown at the top in Figure 7. Substantial ripple from the shaft wobble at 225 rpm is clearly visible. During the same transient the "rotor velocity" signal (frequency deviation of internal voltage) was recorded as shown by the lower trace in Figure 7. It is seen that this trace corresponds exactly except that the spurious component has been eliminated. The compensating voltage added to terminal voltage to produce this frequency was for generator quadrature reactance. Again, this signal is useful where it is desired to extend the damping effect for rather high local swing frequencies without degradation from shaft wobble.

The field tests also disclosed a tendency for the supplementary control system to go into oscillation at 6 to 10 Hz when the gain is advanced sufficiently. This frequency was much too high to emerge from the main electromechanical loop. It was traced to influence of the minor damping loop and it disappeared when that loop was opened. These tests also proved revealing in other respects. Although the existence of the secondary regulator time constant T_A' was challenged in the discussion of Reference 10 of the paper, these tests showed: (1) that it is indeed there, (2) that it emerges from the minor damping loop, and (3) as a result the principal regulator time constant is even longer than suspected (0.12 second). These effects suggest that benefit of the minor damping loop is rather dubious.

In summary, the incidental alinement problems produced a worthwhile yield of information. All important performance expected to this control, including influence of each of the optional features, was fulfilled.

F. W. Keay and **W. H. South**: The authors thank Messrs. Schleif and Gish for their valuable discussion. Their oscillograms of field tests made with this equipment add to the paper. Figure 7 of the discussion is particularly valuable because it supports, with actual test data, the conclusions of both this paper and reference 10 that a compensated frequency derived signal is equivalent to a shaft-speed derived signal but without possessing the undesirable noise which is usually present in a shaft derived signal.

The voltage regulators to which these stabilizers are applied at Yellowtail do contain a rate feedback around the three stages of magnetic amplifiers which are used in the regulator. The discusser's conclusion is correct in that the effect of this damping loop is to increase the time constant. This damping loop was introduced early in the design of this type of regulator to insure against the possibility of encountering high frequency oscillations. Recent analysis of new tests conducted by the authors indicate that this damping loop is neither required nor desirable and we recommend that it be deleted in those applications where power system stabilizers are to be applied.

The Yellowtail application is presently the only one for which the authors' company has supplied this stabilizer equipment complete with all the options. It is gratifying to the authors that the performance of this equipment has fulfilled all of the user's expectations.

Manuscript received August 31, 1970.

A HIGH INITIAL RESPONSE BRUSHLESS EXCITATION SYSTEM

T. L. Dillman, Member IEEE, J. W. Skooglund, Senior Member, IEEE
F. W. Keay, Senior Member, IEEE, W. H. South, Member IEEE, and
C. Raczkowski, Senior Member, IEEE
Westinghouse Electric Corporation
East Pittsburgh, Pennsylvania

Abstract—A high initial response excitation system retaining the advantages of the absence of commutators, collectors and brushes while responding with the speed of an electronic exciter has been developed. Calculated performance of this system has been confirmed by factory tests. With the exciter loaded, the excitation system achieved 95% of ceiling voltage in 0.016 seconds following a 10% step change in the voltage regulator sensing signal. In addition to the high initial response capability, the excitation system also includes supplementary feedback control for improving the small signal dynamic performance characteristic of the exciter. With the addition of feedback, the effective exciter time constant is reduced by a factor of thirty compared to its value without feedback. The tests confirm that low time constants and high initial response can readily be attained in an excitation system which includes a brushless exciter with non-controlled rectifiers.

INTRODUCTION

Recent years have seen increased application of fast response, high performance excitation systems to improve the transient stability of synchronous machines. Many of these installations are at hydro plants where, typically, long transmission circuits connect the plant to the system and stability is an important design consideration. Electronic exciters have become nearly the standard for hydro installations. Today, solid-state, controlled rectifiers have replaced the mercury-arc rectifier for such applications.

The short time constants of electronic exciters have made it possible to apply very effective positive damping of machine oscillations with suitable supplemental signals. Such supplemental signals can provide positive damping with conventional rotating exciters, but more effective damping can be obtained with the faster responding, low time constant excitation systems.

The application of faster excitation systems to steam turbine generators has developed at a slower pace. Traditionally, steam turbine generators have derived excitation energy from the shaft with a direct-connected rotating exciter. The Westinghouse brushless excitation system employs a shaft driven alternator-rectifier exciter with rotating rectifiers directly connected to the generator field. This scheme retains the concept of a shaft power source, independent of power system disturbances, and provides improved reliability and reduced maintenance by eliminating all commutators and collector rings and associated brushes.

To provide faster response of this excitation system, an obvious approach is the replacement of the rotating diodes with controlled rectifiers. Such a system, al hough feasible, appears less desirable than the system described herein because of complexity and because an oversize alternator operating continuously at ceiling voltage would be required.

This paper describes an alternate scheme which retains the

Paper 71 TP 29-PWR, recommended and approved by the Power Generation Committee of the IEEE Power Engineering Society for presentation at the IEEE Winter Power Meeting, New York, N. Y., January 31-February 5, 1971. Manuscript submitted September 14, 1970; made available for printing November 11, 1970.

conventional rotating diodes of the brushless excitation system but provides fast response by means of high forcing and a reduction of the time constants of the rotating exciter and voltage regulator. This excitation system responds at a rate comparable to that of an electronic system and is defined by the proposed IEEE Excitation System Definitions for Synchronous Machines as a high initial response excitation system.[1] Such a system is defined as: "an excitation system having an excitation system voltage response time of 0.1 second or less–". Excitation system voltage response time is defined as: "the time in seconds for the excitation voltage to reach 95% of ceiling voltage under specified conditions."

This system retains the features of the conventional brushless excitation system while providing response times and performance of electronic systems.[2,3] The first such application will be put in service at the Pacific Gas and Electric Diablo Canyon nuclear plant. An exciter rated 6000 kW will provide excitation for a 1300 MVA generator.

Brushless Excitation System

The basic brushless excitation system is shown in Figure 1.

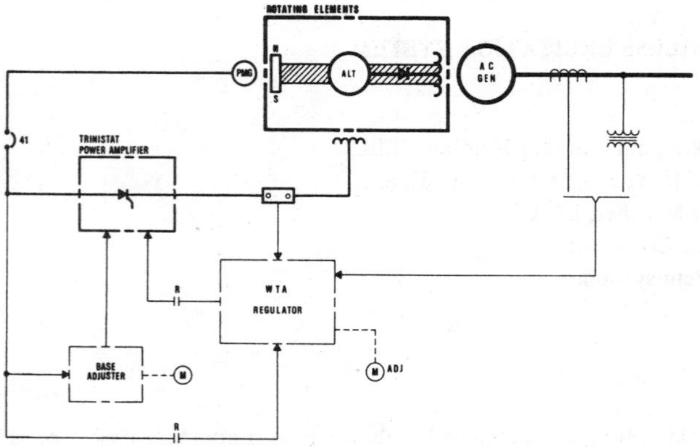

Fig. 1. Brushless Excitation System

This system includes an alternator rectifier main exciter and a permanent magnet generator pilot exciter both directly driven from the synchronous machine rotor. The main (brushless) exciter has a stationary field and a rotating armature which is directly connected through silicon rectifiers to the synchronous machine field without commutator, collectors and brushes. Field power for the brushless exciter is supplied from the permanent magnet generator, having rotating permanent magnets attached to the shaft and a stationary three phase armature. The a-c output of the PMG is converted to d-c by phase controlled, three phase, full wave, thyristor bridges.

The thyristor bridge assembly which supplies the exciter field is made up of a number of draw-out, drawer mounted, three phase, full converter bridges as shown in Figure 2.

The assembly contains one redundant bridge which may be removed under load for maintenance. In this particular application, the assembly consists of six drawers, which are forced cooled with main and reserve fans.

All of the thyristor bridges are controlled by a set of dual firing circuits operating in parallel as one master unit. The firing circuits are comprised of draw-out modules mounted in a draw-out drawer as shown in Figure 3. Two complete drawers provide 100% redundancy of the rectifier control circuits.

In this system, base excitation and manual control of synchronous machine voltage is provided by an adjustable d-c input setting to the thyristor gating circuits. This control signal is derived

Fig. 2. Thyristor Bridge Assembly

from the permanent magnet generator via a regulated d-c supply which also serves all of the regulator logic circuitry. The regulator controls excitation by supplying a buck-boost control signal which adds algebraically to the base setting. All of the regulator control elements utilize all solid-state circuitry with each control function contained on a draw-out printed circuit card module mounted within draw-out drawers similar to Figure 3.[4]

Features of the High Initial Response System

This brushless excitation system was designed to have some special characteristics to meet the high initial response requirement.

The alternator time constant was minimized by the following design features:

1. All alternator components that carry main pole flux were laminated.

2. The inductance of all windings was minimized by careful selection of copper size, number of turns, slot size, etc.

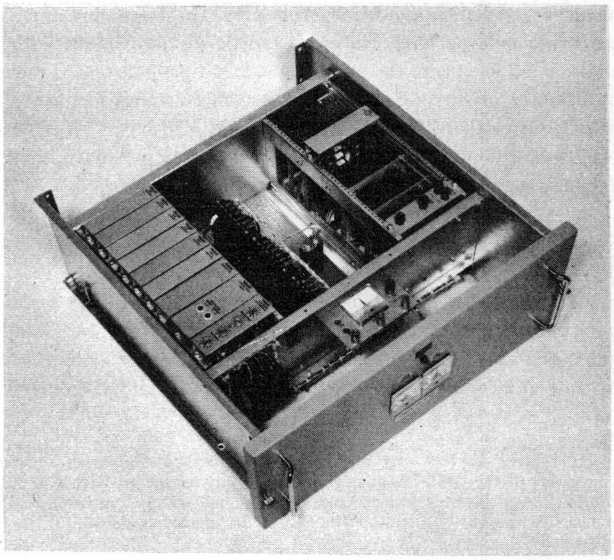

Fig. 3. Thyristor Firing Circuit Assembly

3. Damper windings in the pole face were eliminated.

4. The main pole through bolts were completely insulated from the main pole laminations.

A 0.185 ohm resistor was connected in series with the exciter field. The purpose of this resistor is to limit the exciter field current, in the event of a failure of the regulator field-forcing limiter to perform its limiting function. Physically, the resistor is located inside the exciter housing.

The pilot exciter was designed to have a kilowatt rating approximately 5 times that which would be required for an exciter of this size not having the high initial response feature. This higher pilot exciter rating is required to furnish sufficient forcing voltage to the exciter field. The exciter rating, and saturation curves are shown on Figure 4. The main exciter was 10% oversize.

Special features incorporated in the regulator include:

1. The use of a 24 phase rectifier in the sensing circuit to minimize ripple, reduce filtering requirements, and hence minimize sensing circuit time delay.

2. The addition of supplementary feedback control to reduce the effective time constant of the exciter and improve small signal dynamic performance.

3. The inclusion of an exciter field forcing limiter to protect the exciter from excessive values of field current.

4. The inclusion of a power system stabilizer sensing frequency deviation to provide positive damping of machine oscillations in this particularly critical application.

Factory Test Arrangement

Figure 5 shows the factory test set-up that was used during the combined regulator-exciter response tests. In general, the parameters of the test equipment were chosen to simulate in-service conditions as closely as possible.

The exciter was coupled to a standard brushless exciter test rig which basically consisted of a 7000 hp induction drive motor and a set of collector rings. The collector rings were used to transfer the direct current output of the exciter from the rotating circuit on the shaft to a set of stationary air core reactors. The reactors were connected to obtain a total load circuit resistance of 0.0575 ohms, which equals the generator field resistance at operating temperature. At this reactor setting, the measured load circuit inductance was 0.064 henrys, giving a load circuit time constant of 1.11 seconds. This value is less than the actual generator field time constant of 2.76 seconds but was the highest that could be obtained with available test equipment. This difference in time constants has negligible effect on the results since the exciter output current (I_0) did not change appreciably during the test interval of .05 seconds. The inertia of the entire rotating system was sufficient to maintain essentially constant speed during the response tests, even though an induction drive motor was used.

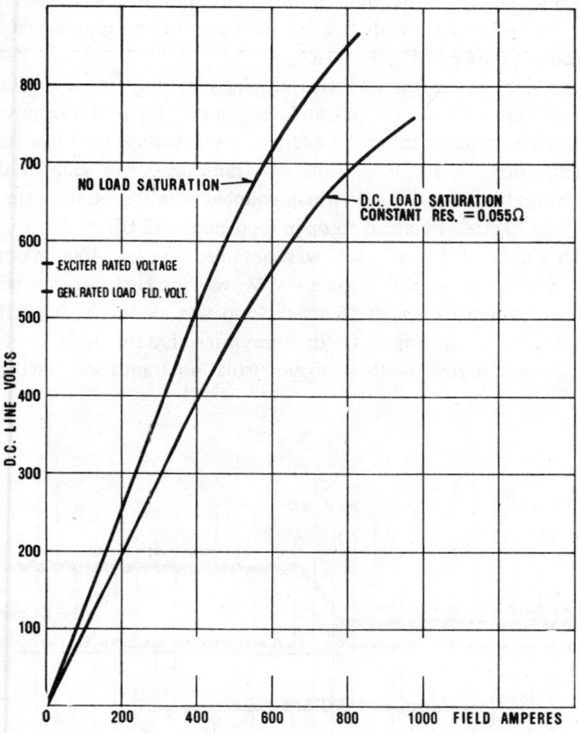

Fig. 4. 6000 kW Brushless Exciter Saturation Curve

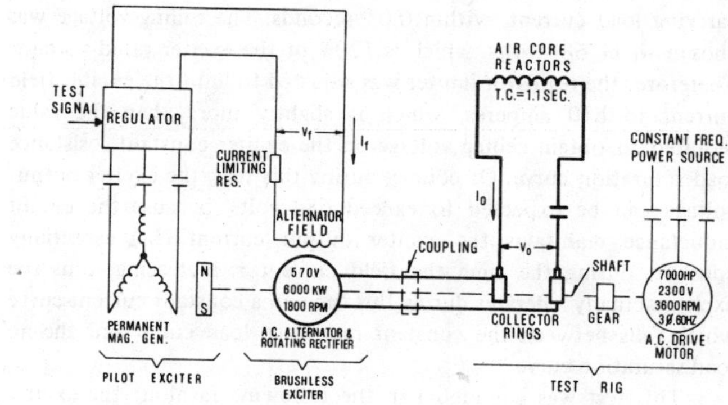

Fig. 5. Factory Test Set-Up

The parameters of the alternator field and the pilot exciter circuits and their connections to the regulator, were the same as they will be when the exciter is in service.

Factory Tests

Three types of response tests were taken on the excitation system and its components including: a response ratio test, a high initial response test, and frequency response tests on the exciter and on each element of the voltage regulator. The block diagram, Figure 6, shows the dynamic characteristics of the excitation system pertinent to the tests. All of the small time constants of the regulator elements are included in this detailed representation.

In addition to the basic voltage regulator, this excitation system includes reactive compensation, minimum excitation limiting, over-excitation limiting, volts/hertz limiting and a power system stabilizer. The intelligence for these additional control elements is derived from synchronous machine terminal quantities. Since they were inoperative during the tests, they are not shown in the block diagram.

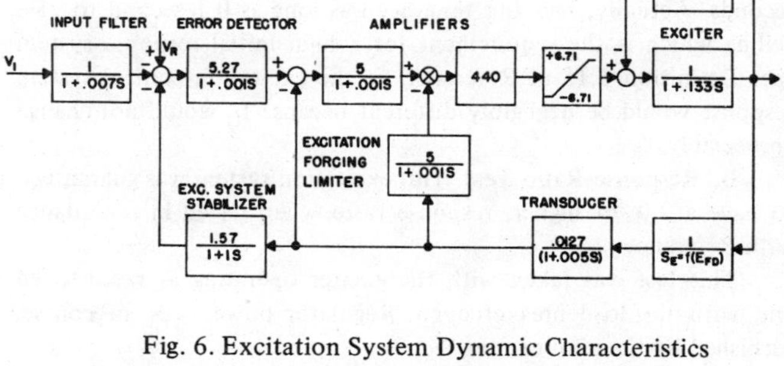

Fig. 6. Excitation System Dynamic Characteristics

A. High Initial Response Test Under Load. The excitation system was guaranteed to attain 95% of ceiling voltage, while carrying load current, within 0.05 seconds. The ceiling voltage was chosen to be 684 volts, which is 120% of the exciter rated voltage. Therefore, the regulator limiter was adjusted to limit the exciter field current to 810 amperes, which is slightly more than the value required to obtain ceiling voltage on the exciter constant resistance load saturation curve. Of course, during this test, the exciter output voltage can be expected to exceed 684 volts, because the circuit inductance maintains the exciter output current (I_O) essentially constant during the time the field current is increasing, thus the exciter actually operates during this time on a constant current curve which falls between the constant resistance load curve and the no load saturation curve.

This test was conducted in the following fashion: the exciter was driven unexcited at rated speed and with the load and pilot exciter breakers closed. The regulator was completely phased back. At time t_o the regulator, controlled by a cam operated timer, quickly forced the exciter field current to the value required to obtain 538 volts output on the constant resistance saturation curve. The exciter output current began to increase exponentially. It was necessary to initiate the response test before the exciter output attained the steady state condition due to test equipment limitations. Therefore, at time t_1 while the instantaneous exciter output voltage (V_O) and current (I_O) were at 578 volts and 7,350 amperes respectively, the timer caused the regulator to go to full forcing by decreasing the regulator sensing signal by 10%. At time t_2, the regulator limiter clamped the field current at 810 amperes, and at time t_3, the excitation was removed. Figure 7, which is not drawn to scale, illustrates this sequence.

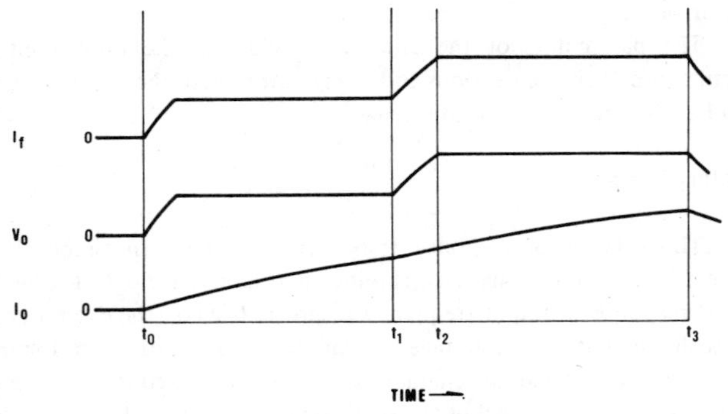

Fig. 7. Response Test Sequence

The time required to attain ceiling voltage, during load, under test conditions, was 0.032 seconds, as measured from the test curve shown on Figure 8. The time required to attain 95% of ceiling voltage during this test was 0.011 seconds. Had this test started at 538 instead of 578 volts, the authors calculate that the time to 95% of ceiling voltage would have been approximately 0.016 seconds. This compares very favorably with the guaranteed value of 0.05 seconds. Actually, had the time been as long as 0.1 second to 95% ceiling which is the requirement for a high initial response system (per Definition 3.15 of Reference 1), the excitation control system response would be negligibly different because I_O would not change appreciably.

B. Response Ratio Test. This excitation system was guaranteed to have a 2.0, or higher, response ratio when tested in accordance with Reference 1.

This test was taken with the exciter operating at rated speed and with the load breaker open. Regulator power was, of course, furnished by the pilot exciter.

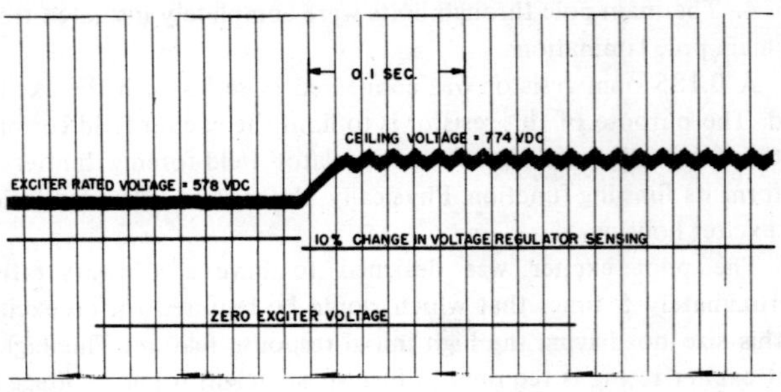

Fig. 8. Excitation System Voltage Time Response Under Load

The test was taken as follows:

1. The exciter field current (I_F) was adjusted, by means of the regulator base adjuster, to set the exciter output (V_O) at 538 volts, which is the rated load field voltage of the generator.

2. The regulator sensing circuit signal was instantaneously decreased by 10% causing the regulator power amplifier to go to the full forcing condition, with resultant exponential increases of field current and exciter output voltage. The regulator forcing limiter again limited the exciter field current to 810 amperes which produced a ceiling voltage of 852 volts on the no load saturation curve. The field voltage, field current, regulator error signal, exciter output voltage, and timing wave were recorded by a high response oscillograph. For clarity, Figure 9 shows a duplicate oscillograph showing only the exciter output voltage and regulator error signal.

The calculated exciter system voltage response ratio was 2.23, using this test curve and Definition 23 of Reference 1. A response ratio of 2.0 was guaranteed.

C. Frequency Response Tests. In addition to each element of the voltage regulator, frequency response tests were made on the exciter, both at no load and under load with and without supplementary feedback, with a frequency response analyzer. Also tests were made measuring the response of field current as well as exciter terminal voltage to verify their approximate equivalence. The results of two of these tests, with the exciter operating at approximately half load are shown in Figure 10-a.

The data of Figure 10-a was obtained by inserting a constant voltage, variable frequency, sinusoidal signal into the regulator power amplifier input and reading the exciter output voltage via a low time constant (.003 sec.) voltage isolation transducer. The gain of the power amplifier was 440 A/V which coupled with the exciter gain of 1 V/A (see Figure 4) yielded an open loop gain = 52 DB.

The data of Figure 10-b was obtained in the same manner except a stage of amplification (5 V/V) was inserted in series with the power amplifier and a feedback loop was closed from exciter field current to the input to this amplifier. Exciter field current feedback was derived with a current isolation transducer having a

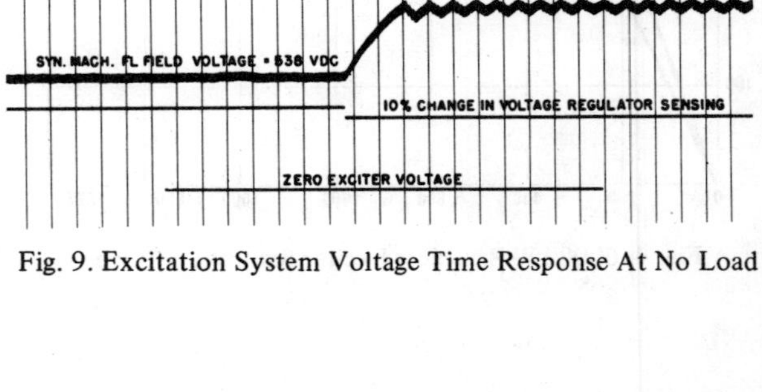

Fig. 9. Excitation System Voltage Time Response At No Load

gain of .0127 V/A and an equivalent time delay of approximately .005 seconds.

From Figure 10-b, it is evident that the addition of feedback resulted in a reduction in gain of 15 DB but also in a reduction of exciter time constant from .133 seconds to less than .01 second. That these results were to be expected is shown in the Appendix. The reduction in overall gain was compensated by an increase in the error detector amplifier gain.

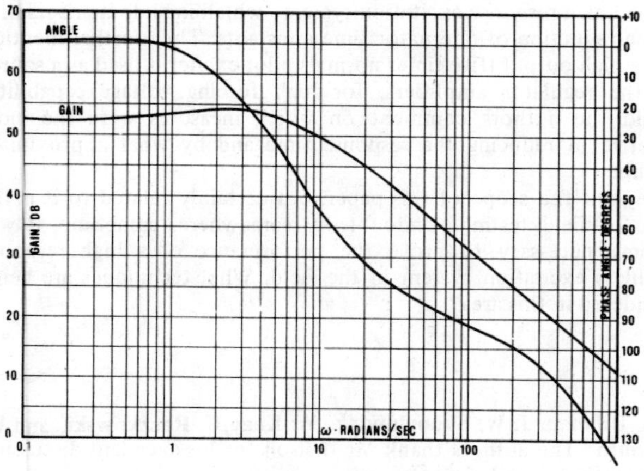

Fig. 10-a. Brushless Exciter Frequency Response Without Compensation

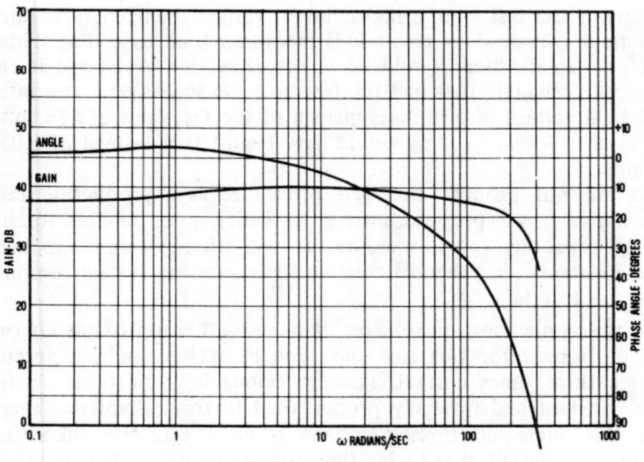

Fig. 10-b. Brushless Exciter Frequency Response With Compensation

CONCLUSIONS

By proper design, high initial response can be readily attained with an excitation system including a brushless exciter with non-controlled rectifiers.

A high initial response brushless excitation system retains the advantages of improved reliability and reduced maintenance by providing shaft derived excitation with the elimination of all commutators, collector rings and brushes.

The availability of modern solid-state controls affords the opportunity to modify the dynamic characteristics of excitation systems in a manner which was heretofore not possible. By the addition of feedback, with negligible time delay, the effective time constant of a rotating exciter, whether high initial response or not, may be reduced. This affords advantage in improved small signal dynamic performance and facilitates the application of supplementary power system stabilizing signals.

REFERENCES

1. IEEE Committee Report, "Proposed Excitation System Definitions for Synchronous Machines," IEEE Transactions on Power Apparatus and Systems, Vol. PAS-88, pp 1248-1258, August, 1969.
2. E. C. Whitney, D. B. Hoover, and P. O. Bobo, "An Electric Utility Brushless Excitation System," AIEE Transactions on Power Apparatus and Systems, Vol. PAS-78, pp 1821-1824, February, 1960.
3. E. H. Myers and P. O. Bobo, "Brushless Excitation System," Proc., 1966, SWIEEECO.
4. F. W. Keay and W. H. South, "A Solid State Regulator for Electric Utility Applications," IEEE 70 TP 80-PWR, July, 1970.
5. IEEE Committee Report, "Computer Representation of Excitation Systems," IEEE Trans. Power Apparatus and Systems, Vol. PAS-87, pp 1460-1464, June, 1968.
6. F. W. Keay and W. H. South, "Design of a Power System Stabilizer Sensing Frequency Deviation," IEEE 70 TP 561-PWR, May, 1970.

APPENDIX

Exciter Time Constant Compensation

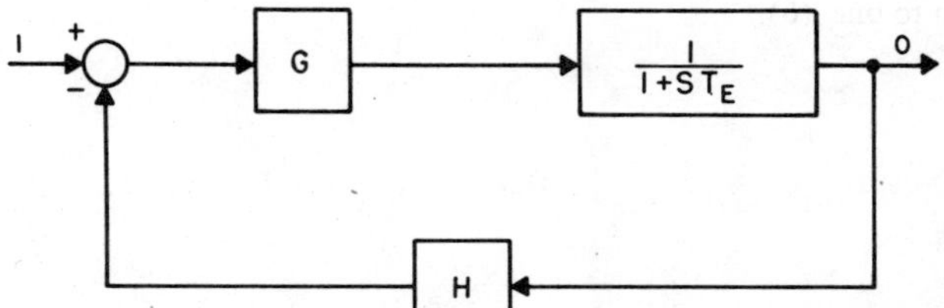

Fig. 11. Exciter With Time Constant Compensation

Assuming negligible time delay in the amplifier, G, and in the feedback transducer, H, the transfer function, without feedback, is:

$$\frac{O}{I} = \frac{G}{1 + ST_E}\text{, where } S = \text{operator } d/dt$$

With the addition of feedback, the transfer function becomes:

$$\frac{O}{I} = \frac{\dfrac{G}{1 + GH}}{1 + S\dfrac{T_E}{(1 + GH)}}$$

where both the gain and time constants have been reduced by the factor of (1 + GH).

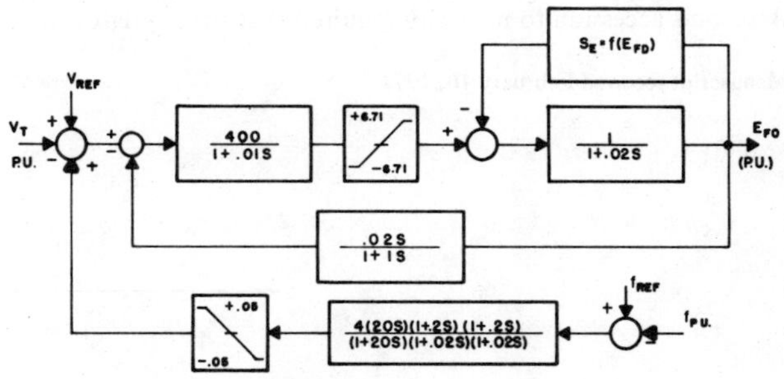

Fig. 12. Excitation System—IEEE Type 1 Representation

In the frequency response test of Figure 10-a, the amplifier gain (G) was 440 V/A and the exciter gain, as shown in Figure 4, was approximately 1 V/A, yielding a gain of 52 DB and a measured time constant = .133 second.

In the test of Figure 10-b, the amplifier gain (G) was increased to 2200 A/V and the feedback loop was closed, with H = .0127 A/V, yielding a closed loop gain of 76, an open loop gain of GH = 27.9, and a reduction of time constant to

$$T_E' = \frac{.133}{28.9} = 28.9 = .0046 \text{ second. This}$$

This agrees with the data of Figure 10-b. It should be noted, however, that in this frequency range, the small time constants in the transducers begin to have an effect.

Computer Representation of Excitation System (5)

Although the time constants in the excitation system total less than 15 ms, a simplified representation, with longer time constants, is adequate for use in system stability studies.

The constants for the excitation system stabilizer and power system stabilizer are typical. The gain and time constant of the excitation system stabilizer are both adjustable over a ten to one range. The gain of the power system stabilizer is adjustable over a range of 1000/1 and the time constants are adjustable over a range of ten to one. (6)

Discussion

H. S. Wilson (Detroit Edison Company, Detroit, Mich. 48226): The authors are to be complimented on their explicit account of factory testing of the brushless system as a high response system. In my discussion I wish to address myself to this aspect of the paper, comment on and ask questions about the procedures used, and raise other points of interest to the user.

As a user, I am most interested in two things when I purchase a piece of equipment: (1) does it meet the specifications upon which its price is based and (2) what are its actual characteristics so that I can accurately model its performance for system studies. To achieve these two goals, the test procedures must accurately measure the required parameters in such a manner that the results can be compared meaningfully to the appropriate minimum performance specifications.

The factory initial response test cited obviously required special test procedures and equipment next to impossible to duplicate in the field. Even in the factory certain compromises had to be made. For example, the load reactor time constant differed from the actual main generation field time constant. The error introduced by using reactors with a lower L/R ratio appears to be on the conservative side. That is, the time to 95% of exciter ceiling would be longer for the test case than one would expect if the actual field were the load. In the case cited, the performance of the equipment was well in excess of the guaranteed performance and definition requirements, however, this type error could prove troublesome in a marginal case.

A second accession to necessity required that the regulator forcing step be applied before the initial conditions had stabilized. This would appear to cause an error on the liberal side; namely, to predict a faster response than would be experienced in actual service. Would the authors please comment on the magnitude in the error introduced by the two test approximations?

The test procedure outlined appears to be a major undertaking. Does the authors' company propose this as a production test for high response brushless systems? If not, what alternate test method and analysis will be offered the user as documentation that the purchased equipment will meet the specified performance guarantees?

The authors have mentioned in the paper some of the "features" of the "high response excitation system" which include items relative to the reduction of alternator time constants. They further mention that a high output (five times normal) pilot exciter is used as a source for the regulator amplifiers, for high forcing voltage capability. Would the authors comment on which measures were the most effective in reducing the response time and by what approximate amounts.

Beyond the scope of this paper, but certainly related to it is the topic of field testing. Undoubtedly somewhere, sometime, it will become necessary to prove the performance of a high response brushless excitation system in the field. What techniques are being considered in this area?

Manuscript received February 16, 1971.

T. L. Dillman, J. W. Skooglund, F. W. Keay, C. Raczkowski, and W. H. South: The authors thank Mr. Wilson for his excellent discussion representing a user's point of view.

We agree with Mr. Wilson's conclusions regarding the compromises made in the test. As long as the time constant of the load is long with respect to the test interval, (ten times or more), the error introduced by the time constant of the test load being less than the actual load will be negligible.

Starting the test from a higher initial voltage than generator rated load field voltage does result in a measured time to ceiling that is shorter than the time would be in actual service. As stated in the paper, we calculate that had the test been started at generator rated load field voltage of 538 volts instead of the test value of 578 volts, the time to 95% of ceiling would have been 0.016 instead of 0.011 seconds.

We do not propose this test procedure as a production test. Alternatively, we propose testing to verify pilot exciter forcing voltage capability and to measure the excitation system component time constants, and then calculating the excitation system response using this test data.

In this application, the exciter time constant is approximately one half of normal for this size and type of exciter, and the forcing voltage three times normal. Time to ceiling is proportional to the time constant and inversely proportional to forcing voltage; therefore, the increased forcing voltage is more effective than time constant reduction in reducing the response time. The time constant compensation circuitry included in the regulator is only effective in reducing the exciter time constant for small signals, i.e., for operation within the linear range of the regulator. It has no effect in reducing the time to ceiling.

In field testing of a brushless excitation system, we recommend recording exciter field current which is approximately equivalent to exciter terminal voltage. Alternately, in most applications, it is possible to measure the exciter output voltage using a set of spring-loaded test probes. These probes can be inserted through ports in the exciter housing and lowered onto the diode wheel long enough to obtain a recording of exciter output voltage.

Manuscript received April 2, 1971.

A SIMPLIFIED ANALYSIS OF THE NO-LOAD STABILITY OF MECHANICAL-HYDRAULIC SPEED CONTROL SYSTEMS FOR STEAM TURBINES

M. A. Eggenberger

Turbine Control Systems Engineer

Large Steam Turbine-Generator Department

General Electric Company, Schenectady, N. Y.

INTRODUCTION

When parameters of a steam turbine speed-control system for a new unit are substantially different from past successful practice, it is necessary to evaluate the changes and their influence on the stability of the control system. In particular, it is advisable to analyze the no-load stability at synchronizing speed in order to make sure that the unit can easily be synchronized without the annoyance of hunting.

The subsequently described method for such a no-load stability analysis has been developed in order to make it possible to evaluate a set of given parameters with a minimum of work and, if necessary, to decide quickly what modifications must be made in order to obtain an acceptable system.

The presented method is based on the Bode frequency response theorem(1) and is using exclusively linearized equations whereby effects such as inertia of control components, compressibility, non-linear flow characteristics and similar effects have been neglected. However, this simplified method promises to be adequate for the intended purpose if a reasonably large phase margin is used.

It has been attempted to present this method so that personnel with little or no theoretical knowledge of control systems can learn to use the procedure purely as a routine.

DIMENSIONLESS REPRESENTATION

In order to make the analysis presented in this paper easily applicable to any unit without knowing every detail of the design of the control system and also for easy comparison of the conditions on different units, the variables of the control system have been put in non-dimensional form by referring the change of any variable to some characteristic magnitude of itself.

By handling the operation of non-dimensionalizing consistently it was possible to reduce the number of parameters to be used for the analysis to a minimum and the physical interpretation of the equations becomes more obvious.

The reference quantity to which the changes in variables shall be referred in the following analysis are either their rated values or the magnitude of that quantity between no load and full load of the unit.

The following procedure is used:

A speed change of the speed governor ΔS, (see Appendix I for definition of symbols), is referred to the rated speed S_r of the governor shaft such that

$$\frac{\Delta S}{S_r} = \sigma \qquad (1)$$

where σ (Sigma) is the relative speed change. For the calculation the following equation is used.

$$\Delta S = \frac{\Delta S}{S_r} S_r = \sigma S_r \qquad (2)$$

As another example, if X^* is the speed governor stroke corresponding to 5 percent speed change which is the normal steady-state regulation, the speed governor stroke changes are:

$$\Delta X = \frac{\Delta X}{X^*} X^* = \xi X^* \qquad (3)$$

where ξ (Xi) is the relative speed governor stroke.

The process of non-dimensionalizing other variables will be obvious from the analysis.

THE CONTROL SYSTEM

The entire control system is represented in block diagram form with the appropriate nomenclature in Fig. 1.

The speed signal is transformed by the speed governor into a position signal, which is subtracted from the reference input and the resulting error signal is fed into the speed relay, where it is amplified. A second-stage amplifier is provided by the servo-motor which operates the steam valves. The steam flow through these valves produces a torque which is opposed by the load torque and the very

(1)Numbers in parentheses refer to similarly numbered references in "Bibliography" at end of paper.

Reprinted with permission from *ASME Paper 60-WA-34.*

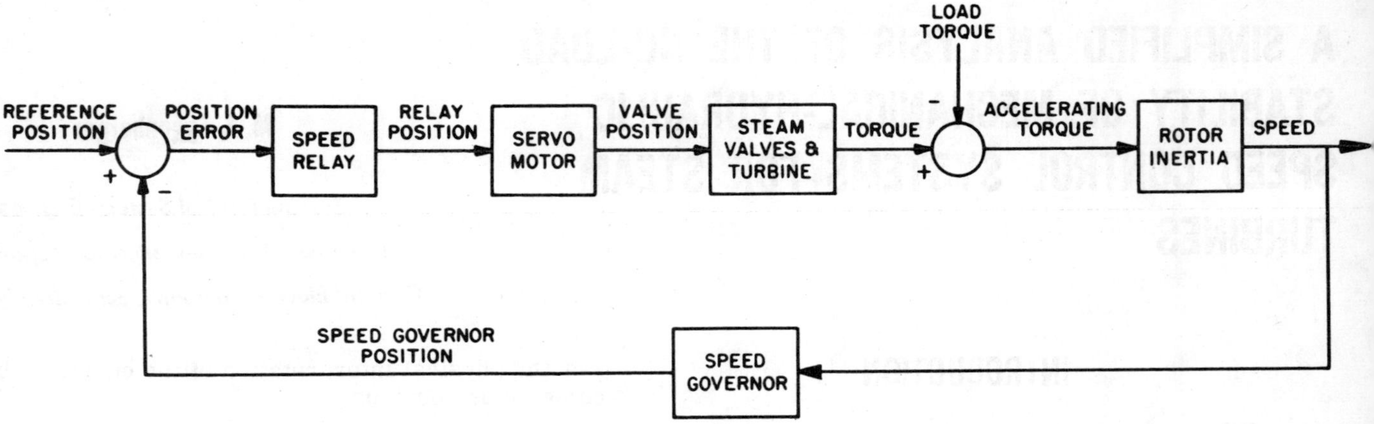

Fig. 1. Block diagram of steam turbine control system

small loss torque. The net remaining torque is used to accelerate the rotating mass of the shaft which feeds the speed signal back to the speed governor.

THE SPEED GOVERNOR

The speed governor is used to transform the speed signal into a position output by means of centrifugal forces opposed by a spring. It is shown schematically in Fig. 2.

A speed change ΔS of the input to the speed governor will produce a position change ΔX of the output of the speed governor or:

$$\Delta S\ K_g = \Delta X \quad (4)$$

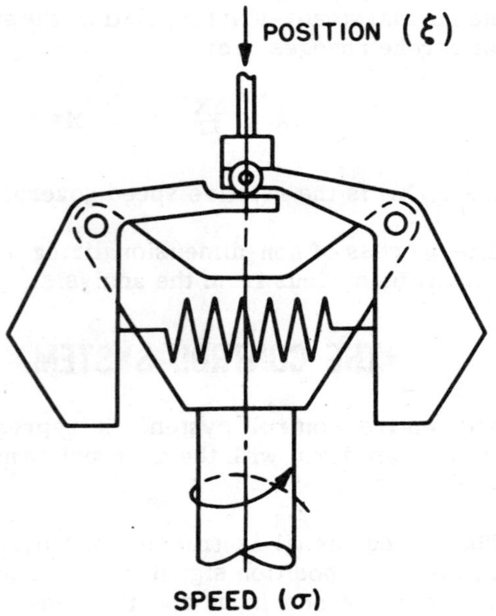

Fig. 2. Speed governor

where K_g is the quantity which relates speed and stroke on the particular speed governor under consideration.

Using the relative speed σ from Equation 1 and the relative stroke of the speed governor from Equation 2 we can write

$$\sigma\ S_r\ K_g = \xi\ X^*$$

or

$$\frac{\xi}{\sigma} = \frac{S_r\ K_g}{X^*} \quad (5)$$

letting

$$\frac{S_r\ K_g}{X^*} = C_g \quad (6)$$

we obtain

$$\frac{\xi}{\sigma} = C_g \quad (7)$$

which is the transfer function of the speed governor or the feedback transfer function of the system.

A customary value for C_g is 20 (5 percent regulation).

REFERENCE AND SPEED RELAY

By means of the arrangement shown in Fig. 3, the speed-governor stroke is compared to the reference input R which is the position of the synchronizing device and the resulting error signal ϵ (Epsilon) is amplified to obtain the output stroke of the speed relay.

The output of the speed relay is Y_1 and changes in this stroke are referred to the stroke Y^*_1 of the relay which is used to go from no load to full load of the unit:

$$\frac{\Delta Y_1}{Y^*_1} = \eta_1 \quad \text{(Eta sub 1)} \quad (8)$$

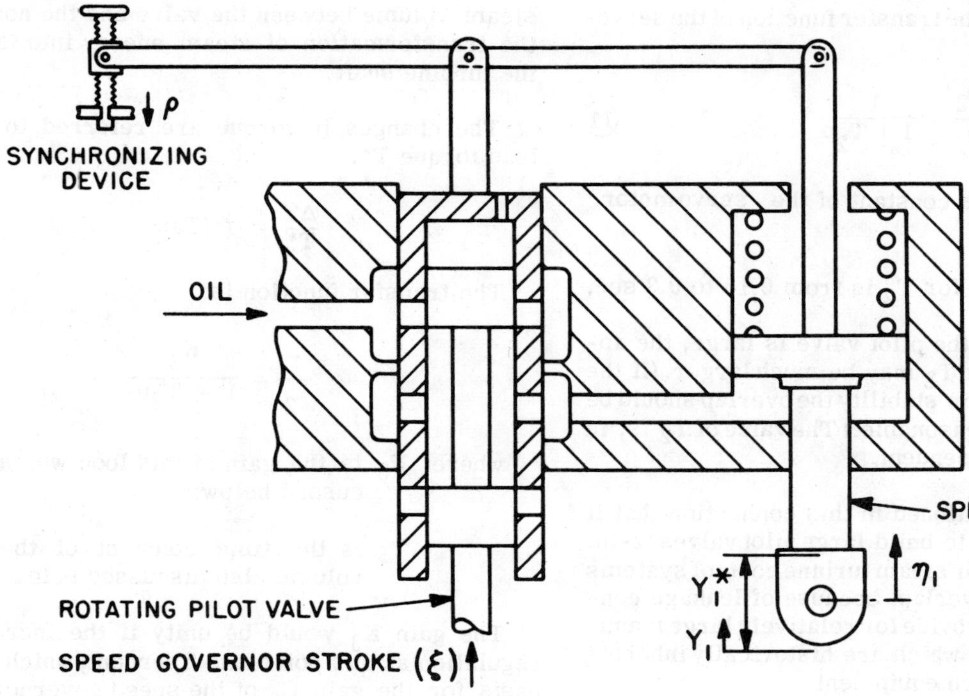

Fig. 3. Reference and speed relay

Note that the stroke Y^*_1 corresponds to the 5 percent speed change of the governor, which was used to define X*. In other words, a steady-state change of $\xi = 1$ will produce a steady-state change of $\eta_1 = 1$, therefore, the normalized gain of the speed relay is said to be unity.

The transfer function of the speed relay is:

$$\frac{\eta_1}{\xi} = \frac{1}{1 + T_1 p} \quad (9)$$

where: T_1 is the time constant of the speed relay. (See definition in appendix II.)

An average value for T_1 is:

$$T_1 = 0.008 \text{ to } 0.18 \text{ sec}$$

p is the operator ($p = \frac{d}{dt}$), the use of which will be described with the Bode diagram.

THE SERVOMOTOR

In order to obtain the large forces necessary to control the position of the steam valves, a second stage amplifier is used in form of a double-acting servomotor as shown in Fig. 4.

The output of the speed relay η_1 is now the input to the servomotor. Note that an interposed lever which may be used to amplify Y_1 will not change the value of η_1, because it has been referred to Y^*_1 which will be equally amplified by the lever ratio.

The output of the servomotor is Y_2 and an output change ΔY_2 will be referred to the active travel Y_2^* of the servomotor:

$$\frac{\Delta Y_2}{Y_2^*} = \eta_2 \text{ (Eta sub 2)} \quad (10)$$

For the same reason as for the speed relay, the normalized gain of the servomotor loop is unity.

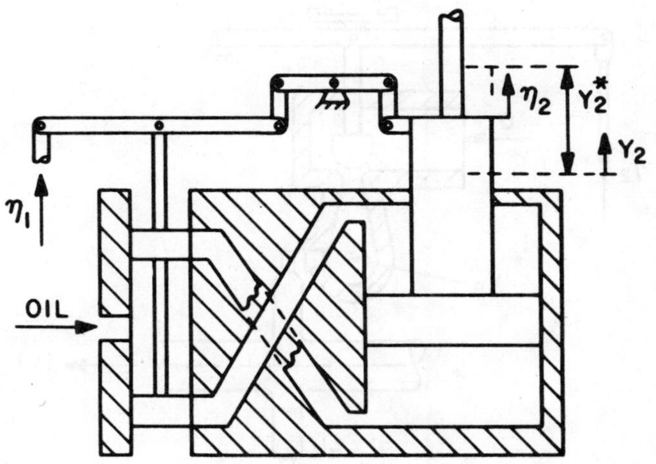

Fig. 4. Servomotor

(See Appendix II.) The transfer function of the servo-motor is:

$$\frac{\eta_2}{\eta_1} = \frac{1}{1 + T_2 p} \quad (11)$$

Where T2 is the time constant of the servomotor. (See Appendix II.)

An average value for T_2 is from 0.15 to 0.3 sec.

If the overlap of the pilot valve is large, the apparent time constant T_2 may be much larger. (If the system is marginal in stability the overlap should be made as small as reasonable.) The value of T_2 is, in effect, magnitude dependent.

It should be mentioned in this connection that it is common practice to build large pilot valves (2-in. to 5-in. diameter) for steam turbine control systems always with some overlap, because of leakage considerations and to provide for relatively large manufacturing tolerances which are historically inherent in the building of such equipment.

STEAM VALVES — TORQUE

The arrangement of the steam valve and the turbine rotor on which the driving torque is produced is shown in Fig. 5.

The input to the steam valve is the output η_2 of the servo-motor.

The transfer function relating η_2 with torque includes the characteristic of the control valves, the steam volume between the valve and the nozzle and the transformation of steam energy into torque on the turbine shaft.

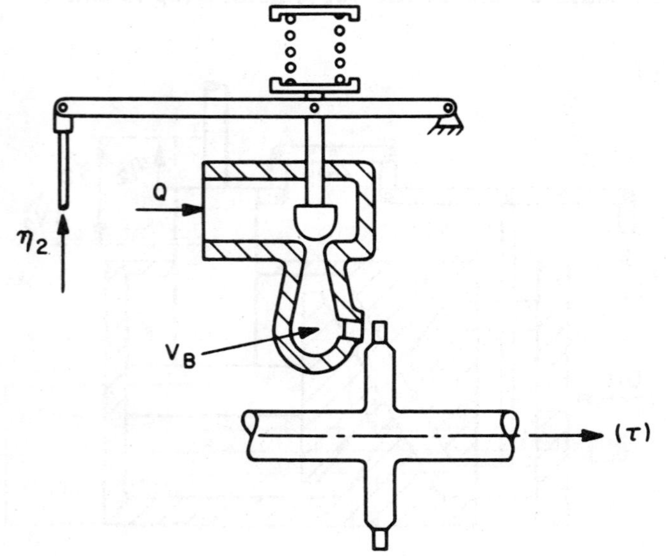

Fig. 5. Servomotor position, torque

The changes in torque are referred to the full load torque T*:

$$\frac{\Delta T}{T^*} = \tau \text{ (Tau)} \quad (12)$$

The transfer function is:

$$\frac{\tau}{\eta_2} = \frac{K_3}{1 + T_3 p} \quad (13)$$

where: K_3 is the gain of this loop which is discussed below:

T_3 is the time constant of the steam volume also discussed below:

The gain K_3 would be unity if the incremental regulation at this point was 5 percent, which was the basis for the gain C_g of the speed governor. Since this regulation varies from 2 to 12 percent over the range of the valve strokes and is usually made about 8 percent at the synchronizing position of the No. 1 valve, the gain K_3 must be corrected. (See Appendix II for definition of "regulation.")

$$K_3 = \frac{\delta_o}{\delta} \quad (14)$$

where: δ_o (Delta sub o) = over-all steady-state speed regulation (5 percent normally).

δ (Delta) = incremental speed regulation at the considered point (particularly no load).

Note that since δ_o and δ have the same units, they can be expressed either in per unit or in percent, the resulting K_3 will be the same:

For a normal unit δ_o = 5 percent and δ = 8 percent therefore:

$$K_3 = \frac{5}{8} \left(\text{or } \frac{0.05}{0.08}\right) = 0.625$$

K_3 together with C_g are parameters which can relatively easily be changed in order to obtain the desired phase margin.

The time constant T_3 represents the time it would take to fill the bowl volume V_B with steam of rated initial conditions at the full-load flow rate.

$$T_3 = \frac{V_B}{Q^*_v} \quad (15)$$

where: V_B = bowl volume [ft^3]

Q^* = full-load steam flow [lb/sec]

v = specific volume of steam at initial conditions [ft^3/lb.]

Average values for T_3 are:

T_3 = 0.05 to 0.4 sec depending on the location of the control valves.

On a reheat turbine there is a reheater of a substantial volume interposed between the high-pressure turbine and the reheat turbine. Its location with respect to the turbine stages has some influence on the response of the unit when a change in steam flow occurs. The corresponding transfer function for a reheat turbine should, therefore, be written as follows:(2)

$$\frac{\tau}{\eta_2} = \frac{K_3\,(1 + fT_R p)}{(1 + T_3 p)\,(1 + T_R p)} \quad (16)$$

where: f is the per unit fraction of the total power developed in the high-pressure section of the unit (0.2 to 0.3)

T_R is the time constant of the reheater.

The time constant T_R of the reheater is defined in analogy with T_3 as

$$T_R = \frac{V_R}{Q_R^* v_R} \quad (17)$$

where: V_R is the volume of the reheater and the cold and hot reheat line [ft^3]

Q_R^* is the full-load reheater steam flow [lb/sec]

v_R is the average specific volume of the steam in the reheater [ft^3/lb]

In order to calculate T_R it is necessary to take reasonable averages of the conditions which change appreciably across the reheater particularly the temperature.

An average value of T_R is from 3 to 11 sec.

THE EXTERNAL BRAKING TORQUE

The torque output of the turbine is opposed by the load which must be considered as the driving function of the system. In the particular case of no load there is no external torque on the unit and the only load of the unit will be the losses from bearing friction shaft pump power, windage and electrical losses.

TORQUE-SPEED

The elements involved are shown schematically in Fig. 6.

A change in net torque will accelerate the shaft at a rate determined by the rotor inertia. As the speed increases the loss torque on the shaft also increases with some power of the relative speed change, but since the losses are very small this feed-back torque is extremely small and can be neglected for this analysis. (See Appendix II.) The transfer function for the turbine shaft is, therefore:

$$\frac{\sigma}{\tau} = \frac{1}{T_4 p} \quad (18)$$

The characteristic time T_4 of a turbine is:

$$T_4 = \frac{(WR^2)\,(n_r)^2}{2.165\ KW^*\ 10^6}\ [\text{sec}] \quad (19)$$

where: WR^2 = inertia of the rotor [ft^2lb]

n_r = rated speed [rpm]

KW^* = maximum load [kw] corresponding to Y_1^*, Y_2^* and Q^*

T_4 is the time it would take the rotor to accelerate from standstill to rated speed if the rated torque T^* was applied as a step function at the time t = 0.

For some other considerations it is usually of advantage to know the so-called specific inertia of a turbine, which is

$$I_{sp} = \frac{(WR^2)}{KW^*}\left(\frac{n_r}{3600}\right)^2 \quad (20)$$

This is a convenient expression because its value turns out to be close to unity (0.9 to 2 for most units).

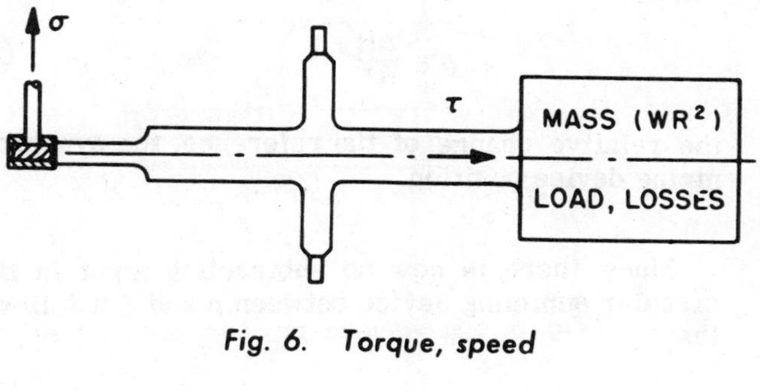

Fig. 6. Torque, speed

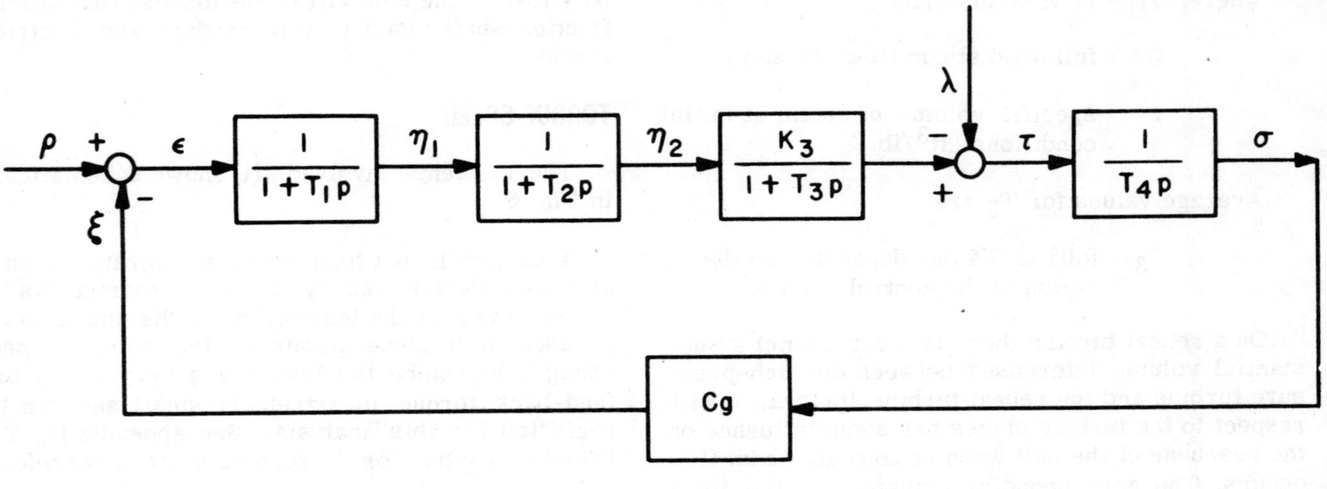

*Fig. 7. **Block diagram of straight condensing unit***

Knowing I_{sp}, the time T_4 is

$$T_4 = 5.98\, I_{sp} \;[\text{sec}] \quad (21)$$

STABILITY OF THE SYSTEM FOR A STRAIGHT CONDENSING UNIT

The block diagram of the control system can now be redrawn with the transfer functions inserted in the blocks; this is shown in Fig. 7.

It should be noted that the external load is usually the driving function of the system but in order to investigate the no load stability of the loop we will set the external load changes

$$\lambda = \frac{\Delta L}{L^*} = 0 \quad (22)$$

and determine the response to variations in the reference setting R.

In order to use the Bode method[1] by which the stability is determined from the open loop frequency response of a system, we cut the loop open at ξ and determine basically the response of $\frac{\xi}{\rho}$

where

$$\rho = \frac{\Delta R}{R^*} \quad (23)$$

the relative change of the reference, the synchronizing device position.

Since there is now no subtracting input in the circular summing device between ρ and ξ it follows that

$$\rho = \epsilon \quad (24)$$

or

$$\frac{\xi}{\rho} = \frac{\xi}{\epsilon} \quad (25)$$

The Bode method assumes that the input be changed sinusoidally at varying frequencies ω (Omega) and the limited-stability criterion[1] says that the system will be stable if the phase lag between the output ξ and the input ρ is less than 180 angular degrees at the point where the ratio

$$\frac{\xi}{\rho} = 1$$

The limited-stability criterion applies to this system.

Physically the process can be visualized by assuming a sinusoidal position change of the frequency ω to the synchronizing device while observing the magnitude and the phase angle of the motion of the speed governor output rod. The rotating pilot-valve would, hereby be imagined to be detached from the speed governor and held in a fixed position which corresponds to the concept of breaking the loop at ξ.

In order to draw the Bode diagram of the open loop, the following things must be done:

1) The complete open-loop transfer function must be written as

$$\frac{\xi}{\rho} = \frac{C_g K_3}{(1 + T_1 p)(1 + T_2 p)(1 + T_3 p)\, T_4 p} \quad (26)$$

2) The operator p is replaced by the term $j\omega$ where j is the imaginary number $\sqrt{-1}$ and ω is a frequency in radians/sec.

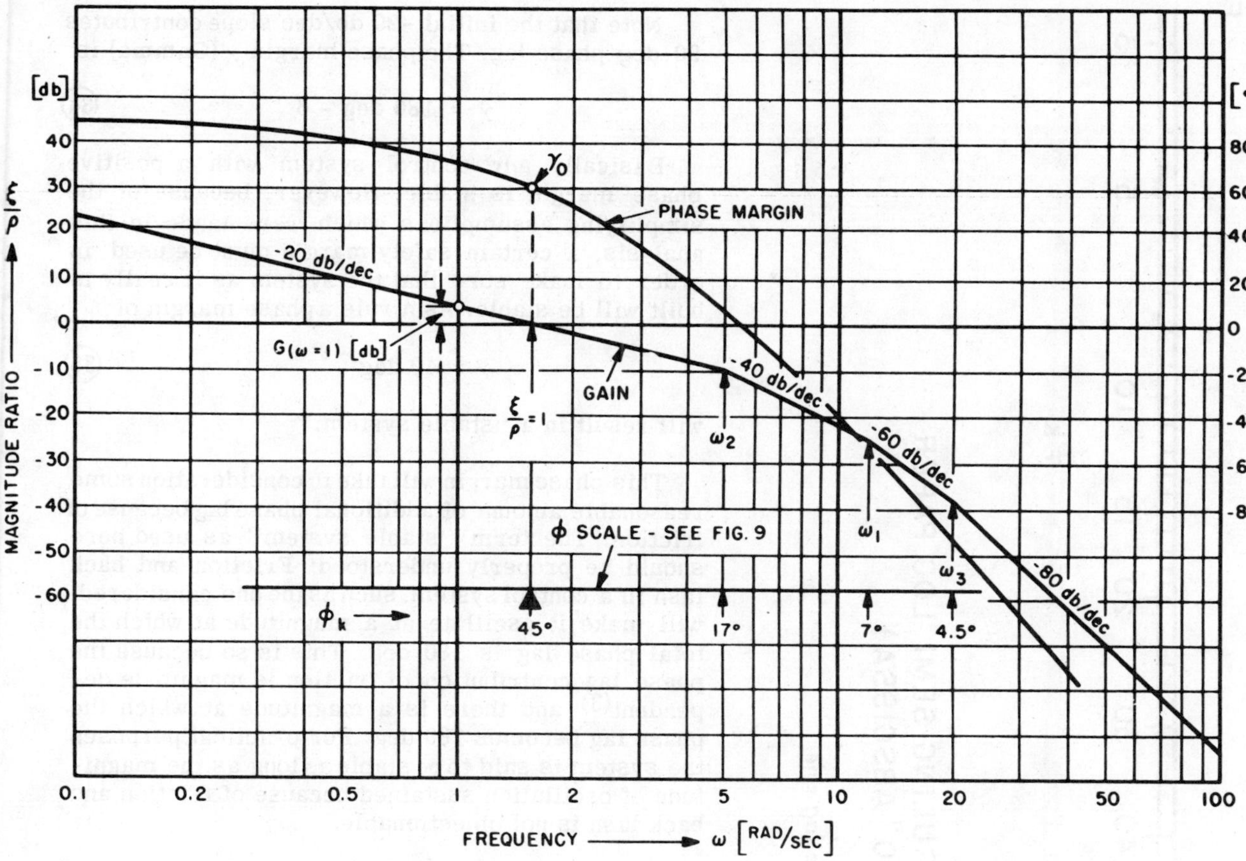

Fig. 8. Attenuation diagram for straight condensing unit

3) ω_1, ω_2, ω_3 must be determined as

$$\omega_1 = \frac{1}{T_1} \text{ [rad/sec]} \quad (27)$$

$$\omega_2 = \frac{1}{T_2} \text{ [rad/sec]} \quad (28)$$

$$\omega_3 = \frac{1}{T_3} \text{ [rad/sec]} \quad (29)$$

4) The system gain G must be determined as

$$G = \frac{C_g K_3}{T_4} \quad (30)$$

In order to use G in the Bode diagram we have to convert it to decibels

$$G_{[db]} = 20 \text{ Log}_{10} \left(\frac{C_g K_3}{T_4} \right) \quad (31)$$

The transfer function is now

$$\frac{\xi}{\rho} = \frac{G}{j\omega(1 + j\frac{\omega}{\omega_1})(1 + j\frac{\omega}{\omega_2})(1 + j\frac{\omega}{\omega_3})} \quad (32)$$

The term $\frac{G}{j\omega}$ represents a -20 db/decade attenuation line having the ordinate $G_{[db]}$ at the frequency $\omega = 1$.

The terms $\frac{1}{1 + j\frac{\omega}{\omega K}}$ represent an increase of the slope of -20 db/decade at the break point ω_K. The Bode diagram for a straight condensing unit is shown in Fig. 8.

The phase lag at the cross-over where $\rho = \xi$ or $\log(\frac{\xi}{\rho}) = 0$ (=0db) can be established by using the scale Fig. 9 by putting the 45-deg point vertically below the cross-over point and determining the phase lag contributions ϕ_K of each time constant: The total lag is

$$\phi = \Sigma\phi_K = 90 \text{ deg} + \phi_1 + \phi_2 + \phi_3 \quad (33)$$

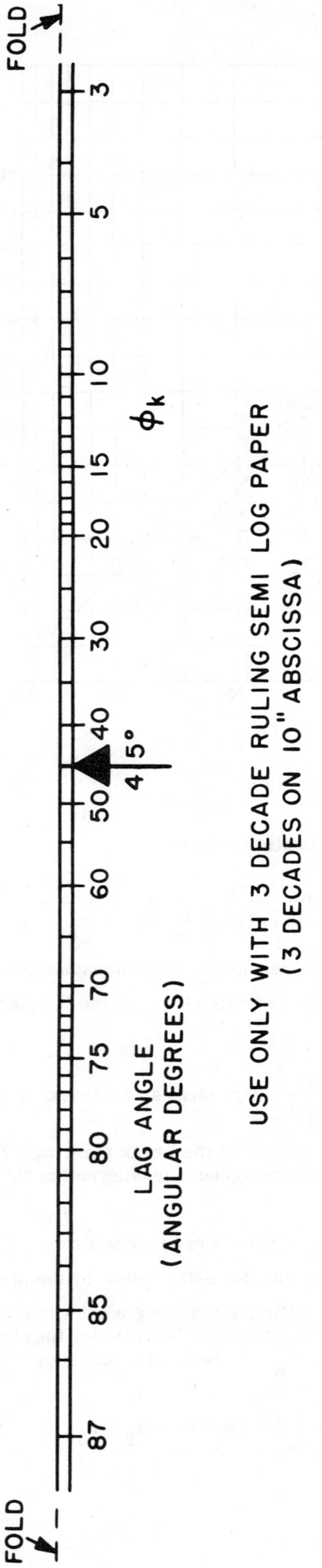

Fig. 9. Phase lag ruler

Note that the initial -20 db/dec slope contributes 90-deg phase lag. The phase margin γ (Gamma) is:

$$\gamma = 180 \text{ deg} - \phi \tag{34}$$

Basically any control system with a positive phase margin is stable. However, because of the simplifying assumptions which were made in this analysis, a certain safety margin must be used, in order to make sure that the system as it really is built will be stable. As a rule a phase margin of

$$\gamma \geq 40 \text{ deg} \tag{35}$$

will result in a "stable system."

This phase margin will take in consideration some reasonable amount of additional phase lag because of friction. The term "stable system" as used here should be properly understood: Friction and back lash in a control system, such as the one considered, will make it oscillate at a magnitude at which the total phase lag is 180 deg. This is so because the phase lag contribution of friction is magnitude dependent[3] and there is a magnitude at which the phase lag becomes 180 deg. For practical purposes the system is said to be stable as long as the magnitude of oscillation sustained because of friction and back lash is not objectionable.

If the system is within the above specified limits and turns out to oscillate at an objectionable magnitude, indications are that an excessive amount of friction is present which usually can be corrected and the system made satisfactorily stable.

STABILITY OF SYSTEM FOR A REHEAT UNIT

The simplified block diagram for a reheat unit is shown in Fig. 10.

The transfer function of the open loop for a reheat turbine is

$$\frac{\xi}{\rho} = \frac{C_g K_3 (1 + f T_R j\omega)}{T_4 j\omega (1 + T_1 j\omega)(1 + T_2 j\omega)(1 + T_3 j\omega)(1 + T_R j\omega)} \tag{36}$$

The attenuation diagram of this transfer function is drawn in a similar manner as outlined for the straight condensing unit using Equations 27 through 31 and

$$\omega_{R_1} = \frac{1}{T_R} \ [\text{rad/sec}] \tag{37}$$

$$\omega_{R_2} = \frac{1}{fT_R} \ [\text{rad/sec}] \tag{38}$$

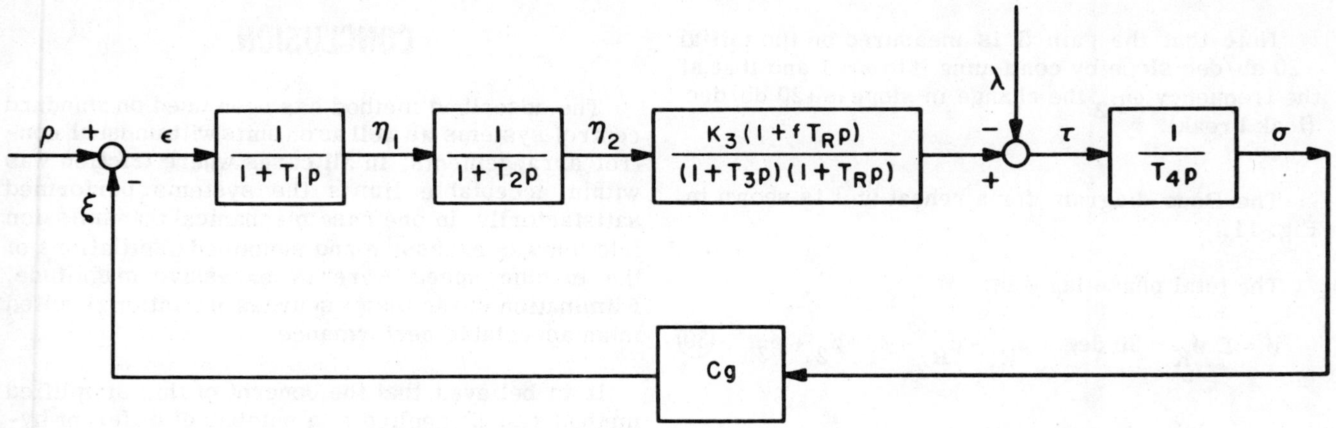

Fig. 10. Block diagram of reheat unit

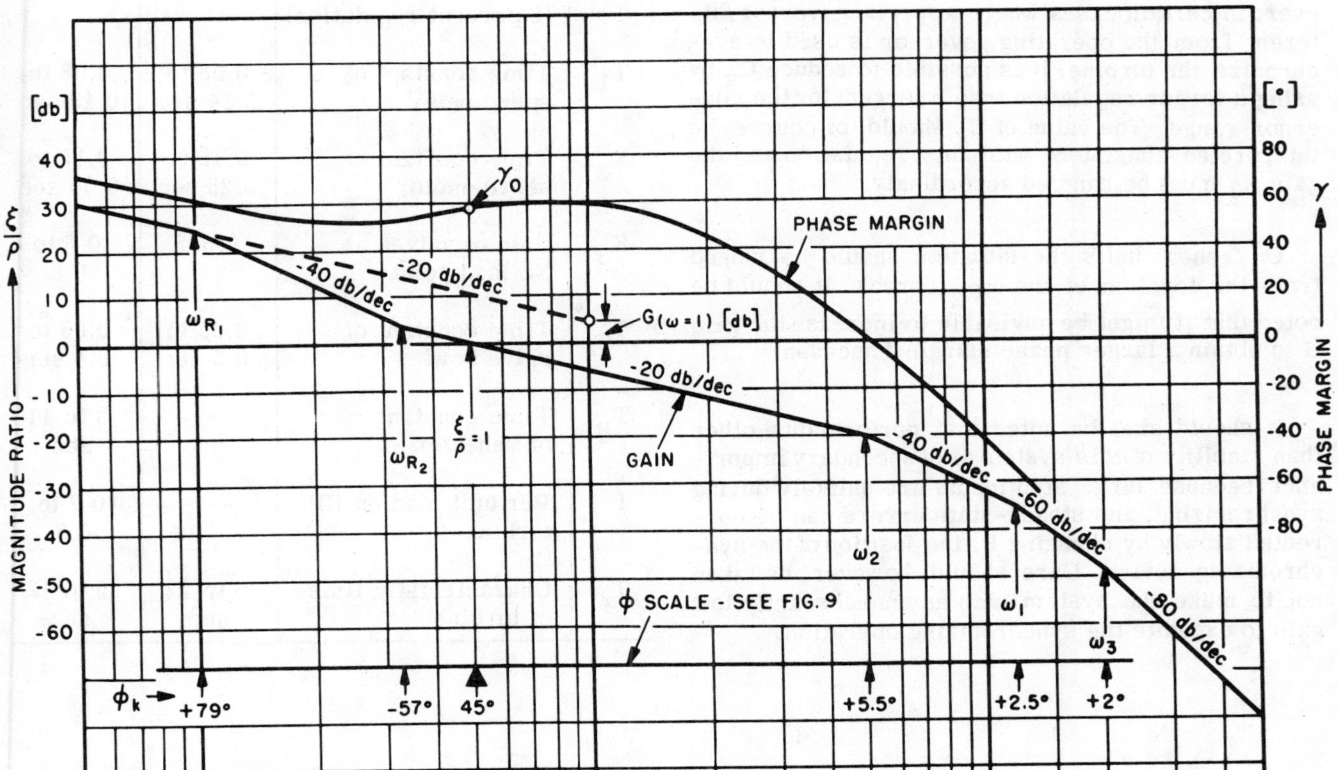

Fig. 11. Attenuation diagram for reheat unit

Note that the gain G is measured on the initial -20 db/dec slope by continuing it to $\omega = 1$ and that at the frequency ω_{R2} the change in slope is +20 db/dec (leak break).

The Bode diagram for a reheat unit is shown in Fig. 11.

The total phase lag ϕ is:

$$\phi = \Sigma\, \phi_K = 90 \text{ deg} + \phi_{R_1} - \phi_{R_2} + \phi_1 + \phi_2 + \phi_3 \quad (39)$$

The phase margin is again as given in Equation 34.

The criterion for the advisable magnitude of the phase margin is also the same as for the straight condensing unit. (See Equation 35.)

If the phase margin γ becomes less than 40 deg, the phase lag contribution of each break should be looked at, and a decision made as to whether or not it is practical to reduce the lag contributions of the particular device.

On straight condensing units a reduction of the gain G will reduce the phase lag ϕ and, therefore, increase the phase margin γ. The gain can, most practically, be decreased by a reduction of K_3. However, in certain cases where a special governor different from the operating governor is used to synchronize the turbine, it is possible to reduce C_g by using a larger regulation than 5 percent for the governor range. The value of C_g should, of course, be interpreted consistent with this regulation and the gain K_3 must be handled accordingly.

On reheat units the situation should be judged from the location of the ω_{R2}- break. It should be noted that it might be advisable to increase the gain G to obtain a larger phase margin if necessary.

It should also be noted that performance other than stability of this system is of secondary importance because large step inputs are unlikely during synchronizing, and steady-state errors can be corrected slowly by changing R, the position of the synchronizing device. Care should, however, be taken not to make the system with an unnecessarily low gain to expedite the synchronizing operation.

CONCLUSION

The described method has been used on standard control systems as well as on units with unusual control arrangements. In all cases where friction was within acceptable limits the systems performed satisfactorily. In one case mechanical transmission friction was excessive and sustained oscillations of the turbine speed were of excessive magnitude. Elimination of the major sources of friction resulted in an acceptable performance.

It is believed that the concept of this simplified method can be applied to a number of different hydraulic control systems thus making a quick evaluation of the important components a relatively easy operation without the knowledge of many details of the system.

TABLE I

TYPICAL VALUES OF CHARACTERISTIC CONSTANTS FOR LARGE STEAM TURBINE-GENERATORS

SYMBOL	NAME	NON-REHEAT TURBINE	REHEAT TURBINE
C_g	Normalized speed governor constant (5 percent regulation)	20	20
T_1	Time constant of speed relay	0.08 to 0.14 sec	0.08 to 0.18 sec
T_2	Time constant of servo-motor	0.15 to 0.25 sec	0.15 to 0.30 sec
K_3	Gain of valves at No load point	0.625	0.6 to 0.8
T_3	Time constant of valve bowl	0.05 to 0.3 sec	0.05 to 0.4 sec
T_R	Time constant of reheater	----	3 to 11 sec
f	Per unit load on HP turbine	----	0.2 to 0.3
T_4	Characteristic time of turbine	6 to 12 sec	5 to 12 sec

APPENDIX I

SYMBOLS AND DEFINITIONS

a	Input lever length on servomotor [in.]
a_m	Momentary angular velocity of turbine shaft [rad/sec]
Δa_m	Momentary angular velocity deviation [rad/sec]
a_r	Rated angular velocity of turbine shaft [rad/sec]
A_1	Oil flow cross section in pilot valve [$in.^2$]
A_2	Piston area of servomotor [$in.^2$]
b	Restoring lever length on servomotor [in.]
C_g	Normalized speed governor constant [dimensionless]
$d(\ )$	Symbol for differential
D	Diameter of round pilot valve ports [in.]
f	Per unit fraction of power of high-pressure turbine [dimensionless]
G	System gain (numeric) $\left[\frac{1}{sec}\right]$
$G_{(db)}$	System gain in decibels ($G_{(db)} = 20 \log_{10} G$) [db]
I_{sp}	Specific interia $\left[\frac{lbs\ ft^2}{kw}\right]$
j	$\sqrt{-1}$ imaginary unit
K_g	Speed governor constant [in./rpm]
K_3	Gain of valves at no-load point [dimensionless]
KW^*	Full-load power at generator terminals [kw]
L^*	Full-load power on turbine [kw]
ΔL	Change in load [kw]
L_n	Losses at speed n [kw]
L_{n_r}	Losses at rated speed [kw]
n	Speed of turbine shaft [rpm]
n_o	Speed at zero power output [rpm]
n_r	Rated speed of turbine shaft [rpm]
n'	Imaginary speed at full load [rpm]
N	Number of ports at one land in pilot valve bushing [dimensionless]

APPENDIX I (Cont'd)

SYMBOLS AND DEFINITIONS

p	Differential operator $\frac{d(\;)}{dt}$ $\left[\frac{1}{sec}\right]$
P_o	Initial oil pressure [psi]
P_1	Oil pressure under servomotor piston [psi]
Q^*	Full-load steam flow at control valves [lbs/sec]
$Q_R^{*'}$	Full-load steam flow at reheat turbine [lbs/sec]
Q_1	Oil flow through A_1 [$in.^3$/sec]
R	Position of speed reference [in.]
R^*	Change in speed reference position from no load to full load at constant speed [in.]
ΔR	Change in speed reference position [in.]
S_r	Rated speed of governor [rpm]
ΔS	Governor speed change [rpm]
t	Time [sec]
T_1	Speed relay time constant [sec]
T_2	Servo-motor time constant [sec]
T_3	Steam bowl time constant [sec]
T_4	Characteristic time of turbine [sec]
T_4'	Time constant of turbine shaft with loss torque [sec]
T_R	Reheater time constant [sec]
T_{R_1}	Reheater lag time constant [sec]
T_{R_2}	Reheater lead time constant [sec]
T^*	Full-load torque on turbine shaft [ft-lbs]
ΔT	Torque change on turbine shaft [ft-lbs]
T_L	Loss torque at speed n [ft-lbs]
T_{L_r}	Loss torque at rated speed n_r [ft-lbs]
ΔT_L	Change in loss torque [ft-lbs]
v	Specific volume of steam in bowl [ft^3/lb]

APPENDIX I (Cont'd)

SYMBOLS AND DEFINITIONS

v_R	Average specific steam volume in reheater [ft^3/lb]
V_B	Bowl volume [ft^3]
V_R	Reheater volume [ft^3]
W_p	Total port width of pilot valve [in.]
w_{p_2}	Width of a single port in pilot valve bushing [in.]
WR^2	Rotational inertia of turbine-generator rotor [lbs ft^2]
X	Speed governor stroke [in.]
ΔX	Speed governor stroke change [in.]
X^*	Speed governor stroke for 5 percent speed change [in.]
X_1	Pilot valve travel [in.]
Y_1	Speed relay stroke [in.]
ΔY_1	Change in speed relay stroke [in.]
Y_1^*	Speed relay stroke from no load to full load [in.]
Y_2	Servo-motor stroke [in.]
ΔY_2	Change in servomotor stroke [in.]
Y_2^*	Servomotor stroke from no load to full load [in.]
γ (gamma)	Phase margin [angular degrees]
γ_o (gamma sub zero)	Phase margin [angular degrees] at crossover point
δ (delta)	Incremental speed regulation [dimensionless]
δ_o (delta sub o)	Steady-state speed regulation [dimensionless]
ϵ (epsilon)	Speed error signal [dimensionless]
η_1 (eta sub 1)	Relative change of speed relay position [dimensionless]
η_2 (eta sub 2)	Relative change of servomotor position [dimensionless]
$\varkappa$ (kappa)	Exponent of loss function [dimensionless]
λ (lambda)	Relative load change [dimensionless]
ξ (xi)	Relative change of speed governor stroke [dimensionless]
ρ (rho)	Relative change of reference position [dimensionless]

APPENDIX I (Cont'd)

SYMBOLS AND DEFINITIONS

σ (sigma)	Relative speed change [dimensionless]
τ (tau)	Relative torque change on turbine shaft [dimensionless]
τ_L (tau sub L)	Relative change in loss torque [dimensionless]
ϕ_K (phi sub K)	Phase angle, general [angular degrees]
ϕ_1 (phi sub 1)	Phase lag produced by the presence of T_1 [angular degrees]
ϕ_2 (phi sub 2)	Phase lag produced by the presence of T_2 [angular degrees]
ϕ_3 (phi sub 3)	Phase lag produced by the presence of T_3 [angular degrees]
ϕ_{R_1} (phi sub R_1)	Phase lag produced by the presence of T_{R_1} [angular degrees]
ϕ_{R_2} (phi sub R_2)	Phase lead produced by the presence of T_{R_2} [angular degrees]
$\Sigma \phi_K$ (sum of ϕ_K)	Algebraic sum of all ϕ_K [angular degrees]
$\Phi(P_o, P_1)$	Oil flow coefficient for pilot valve [$\frac{\text{in.}}{\text{sec}}$]
ω (omega)	Angular frequency of sinusoidal excitation of the reference [$\frac{\text{rad.}}{\text{sec}}$]
ω_1 (omega sub 1)	Angular frequency corresponding to T_1 [rad/sec]
ω_2 (omega sub 2)	Angular frequency corresponding to T_2 [rad/sec]
ω_3 (omega sub 3)	Angular frequency corresponding to T_3 [rad/sec]
ω_{R_1} (omega sub R_1)	Angular frequency corresponding to T_{R_1} [rad/sec]
ω_{R_2} (omega sub R_2)	Angular frequency corresponding to T_{R_2} [rad/sec]
ω_K (omega sub K)	Angular frequency in general [rad/sec] corresponding to the time constant T_K

APPENDIX II

TRANSFER FUNCTION OF HYDRAULIC SERVO-MOTOR

A schematic diagram of a hydraulic servomotor on which the derivation of the transfer function shall be demonstrated is represented in Fig. 12.

For simplicity the oil connections are not crossed as in Fig. 4 and, correspondingly the feed back linkage has no reversing lever. (The crossing of ports and the reversing feed back lever are used in normal applications only because of safety consideration so that falling of the pilot valve would close the steam valves. For the analysis the two servomotors are equivalent.)

The oil flow cross section A_1 is given by the port width W_p of the pilot valve ports and the pilot valve lift X_1.

$$A_1 = W_p X_1 \qquad (40)$$

and the flow across this area is

$$Q_1 = A_1 \, \Phi \, (P_o, P_1) \qquad (41)$$

We assume now that $\Phi_{(P_o, P_1)}$ is constant so that the flow Q_1 is a linear function of X_1.

The motion of the piston which produces the stroke Y_2 is now given by the equation

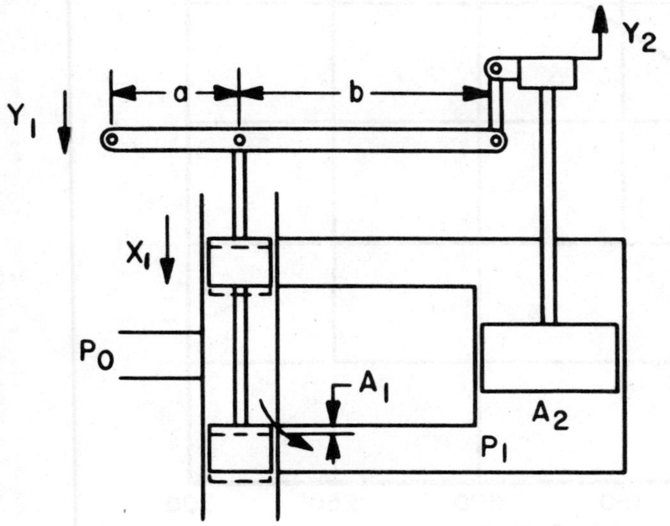

Fig. 12. Simple servomotor

$$A_2 \frac{dY_2}{dt} = Q_1 \qquad (42)$$

Substituting 40 and 41 yields

$$A_2 \frac{dY_2}{dt} = X_1 \, W_p \, \Phi \qquad (43)$$

Observing the operation of the feed-back lever the following relation is found for

$$X_1 = \Delta Y_1 \frac{b}{a+b} - \Delta Y_2 \frac{a}{a+b} \qquad (44)$$

letting

$$\frac{dY_2}{dt} = p \, \Delta Y_2$$

and

$$\frac{\Delta Y_1}{Y_1^*} = \eta_1 \quad \text{and} \quad \frac{\Delta Y_2}{Y_2^*} = \eta_2$$

and substituting in 43 yields

$$p \, \eta_2 Y_2^* A_2 = \frac{W_p \Phi}{a+b} (\eta_1 Y_1^* b - \eta_2 Y_2^* a) \qquad (45)$$

solving for $\frac{\eta_2}{\eta_1}$

$$\frac{\eta_2}{\eta_1} = \frac{\dfrac{W_p \Phi b \, Y_1^*}{(a+b) \, Y_2^*}}{p A_2 + a \dfrac{W_p \Phi}{a+b}}$$

or

$$\frac{\eta_2}{\eta_1} = \frac{\dfrac{b \, Y_1^*}{a \, Y_2^*}}{\left(1 + \dfrac{A_2}{\dfrac{a}{a+b} W_p \Phi} p\right)} \qquad (46)$$

from the feed-back lever it becomes obvious that

$$\frac{b}{a} = \frac{Y_2^*}{Y_1^*}$$

or

$$\frac{b\,Y_1^*}{a\,Y_2^*} = 1 \qquad (47)$$

so that Equation 46 becomes

$$\frac{\eta_2}{\eta_1} = \frac{1}{1 + T_2 p} \qquad (48)$$

where

$$T_2 = \frac{A_2}{\frac{a}{a+b} W_p \Phi} \qquad (49)$$

$\frac{a}{a+b}$ is the simple ratio of the stroke of the pilot valve to the stroke of the piston if the input point at Y_1 is held stationary. It can now be observed that any intermediate reversing lever cannot change this ratio as long as the ratio $\frac{Y_2^*}{Y_1^*}$ must stay constant.

The port width W_p of the pilot valve is, of course, the sum of the widths of all the ports if more than one port is used opposite the controlling land of the pilot valve.

If the ports are drilled, round holes of the diameter D, the width W_p of each port should be evaluated as

$$w_p = \frac{D}{3} \qquad (50)$$

and

$$W_p = N\,w_p \qquad (51)$$

where N is the number of ports at one land.

The flow coefficient Φ is given as a function of P_o in Fig. 13.

If this Φ is used for single acting servomotors it is assumed that the average operating pressure underneath the piston is

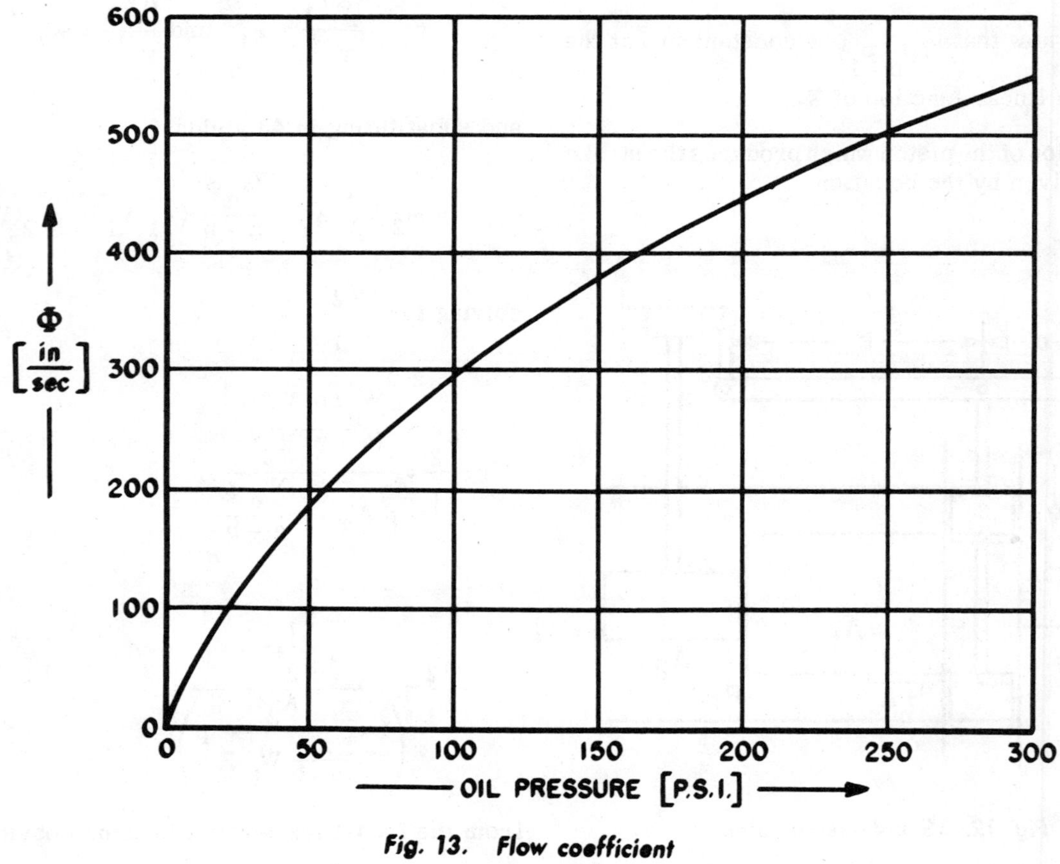

Fig. 13. Flow coefficient

$$P_1 \cong \frac{P_o}{2} \quad (52)$$

For double-acting piston, Φ is applicable if the upper piston area is approximately half the lower piston area which is somewhat like a design standard for steam turbines.

It should also be mentioned that the given values for Φ are for small changes such as considered for this stability analysis. For large strokes of the pilot valve, Φ becomes appreciably larger.

The dimension of Φ is $\left[\frac{\text{in.}}{\text{sec}}\right]$ or, stated in a more complete manner.

$$\Phi \left[\frac{(\text{in.})^3}{\text{sec (in. P.V. lift) (in. cut-off edge)}}\right] \quad (53)$$

LOSSES OF TURBINE

If it is desired to incorporate the loss feed back in the torque-speed loop (Equation 18) it is necessary to evaluate the rate of change of the losses with speed.

The losses are a combination of several components which do not respond to speed equally.[4] For an average steam turbine the loss function is:

$$L_n = L_{n_r}\left(\frac{n}{n_r}\right)^{\varkappa} \quad (54)$$

Where: L_n = losses at speed n [kw]

L_{n_r} = losses at rated speed n_r [kw]

n = actual speed [rpm]

n_r = rated speed [rpm]

$\varkappa$ = exponent of loss function (kappa)

an average exponent for the loss function is:

$$\varkappa = 2.7 \quad (55)$$

Expressed in terms of torque, Equation 54 becomes

$$a_m T_L = a_r T_{L_r}\left(\frac{a_m}{a_r}\right)^{\varkappa}$$

or

$$T_L = T_{L_r}\left(\frac{a_m}{a_r}\right)^{\varkappa - 1} \quad (56)$$

where a_m = momentary angular velocity [rad/sec]

T_L = loss torque at a_m [ft-lb]

a_r = rated angular velocity [rad/sec]

T_{L_r} = loss torque at a_r [ft-lb]

Expressed by means of differences:

$$T_{L_r} + \Delta T_L = T_{L_r}\left(\frac{a_r + \Delta a_m}{a_r}\right)^{\varkappa-1} \quad (57)$$

and recalling that

$$\frac{\Delta a_m}{a_r} = \sigma \quad (58)$$

$$T_{L_r} + \Delta T_L = T_{L_r}(1+\sigma)^{\varkappa-1} \quad (59)$$

Since σ is small compared to unity the first two terms of a Taylor series will sufficiently well approximate the value of $(1+\sigma)^{\varkappa-1}$:

$$(1+\sigma)^{\varkappa-1} \cong 1 + \sigma(\varkappa - 1) \quad (60)$$

Equation 59 becomes thus

$$\Delta T_L = T_{L_r}(\varkappa - 1)\,\sigma \quad (61)$$

with $\tau_L = \frac{\Delta T_L}{T^*}$ the loss torque change to speed change ratio becomes

$$\frac{\tau_L}{\sigma} = \frac{T_{L_r}}{T^*}(\varkappa - 1) \quad (62)$$

The torque speed loop thus looks as shown in Fig. 14.

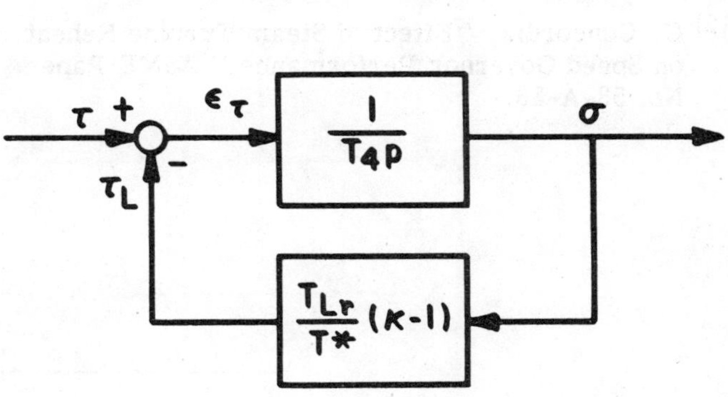

Fig. 14. Torque, speed loop with loss feedback

Assuming that the loss torque at rated speed is 1 percent of the full-load torque and using the average value of 2.7 for $\varkappa$, the feedback transfer function is

$$\frac{\tau_L}{\sigma} = 0.01\ (2.7 - 1) = 0.017 \quad (63)$$

The closed loop transfer function of this loop is now

$$\frac{\sigma}{\tau} = \frac{1}{T_4\, p\, (1 + \frac{1}{T_4 p}\, 0.017)}$$

or

$$\frac{\sigma}{\tau} = \frac{\frac{1}{0.017}}{1 + \frac{T_4}{0.017}\, p} \quad (64)$$

Since the time constant

$$T_4' = \frac{T_4}{0.017} \quad (65)$$

is obviously very large (300 to 400 sec) the break in the attenuation characteristic associated with T_4' at

$$\omega_4 = \frac{1}{T_4'} \quad (\cong 0.002 \text{ to } 0.003 \text{ rad/sec}) \quad (66)$$

is very far to the left and will for all practical purposes contribute 90-deg phase lag with respect to the cross-over point. It is, therefore, reasonable to neglect this feedback and write the transfer function of the turbine shaft as an integrator.

DEFINITION OF "REGULATION"

The *steady-state speed regulation* δ_o of a steam turbine is defined by the following expression

$$\delta_o = \frac{n_o - n}{n_r} \quad [\text{ Units }] \quad (67)$$

where n_o is the speed [rpm] at zero power output
n is the speed [rpm] at rated power output.
n_r is the rated speed [rpm]

with the reference position R remaining constant

The *incremental speed regulation* δ is defined by

$$\delta = \frac{n_o - n'}{n_r} \quad [\text{Units}] \quad (68)$$

where n_o and n_r are defined as above and

n' is a ficticious speed which would be necessary to run the turbine at rated output if the slope of the power increase at the particular point where the incremental regulation is measured was continued from no load to rated load with R = const.

ACKNOWLEDGEMENT

"The author wishes to express his appreciation to Harold Chestnut for his encouragement and guidance in writing this paper."

BIBLIOGRAPHY

(1) H. Chestnut and R. W. Mayer, "Servo Mechanisms and Regulating Systems," V. 1, John Wiley and Son, New York, N. Y.

(2) C. Concordia, "Effect of Steam Turbine Reheat on Speed Governor Performance," ASME Paper No. 58-A-36.

(3) H. Chestnut and R. W. Mayer, "Servo Mechanisms and Regulating Systems," V. 2, John Wiley and Son, New York, N. Y.

(4) M. A. Eggenberger, "Overspeed Protection of Reheat Turbine-Generators," ASME Paper No. 54-F-26.

Detailed Hydrogovernor Representation for System Stability Studies

D. G. RAMEY, MEMBER, IEEE, AND JOHN W. SKOOGLUND, SENIOR MEMBER, IEEE

Abstract—**In studying transient performance of power systems, large computer programs are now used with representation of individual unit governing systems included. Detailed representation of hydrogovernors is not usually available, but, rather, an equivalent transfer function is employed.**

A comparison is made of the transient performance of detailed versus equivalent hydrogovernor system representation. A procedure is suggested for the determination of the equivalent transfer function from plant data.

Introduction

THE INDUSTRY has received a number of papers describing tests and analytical procedures for optimizing adjustments of hydrogovernors to obtain fast, stable transient response [1]–[5]. Many of these studies represented hydro units in rather detailed form, and recent work by Schleif and Angell [1] used an analog computer in field tests on the individual unit to establish the best control system adjustments.

Unfortunately, for the users of large-scale system stability programs such detail of representation has not been available. The adequacy of available simplified representations has sometimes been in doubt.

The purpose of this study is to suggest a method of obtaining an equivalent hydrogoverning system transfer function from detailed plant data, and to then compare the performance of this equivalent with that of a more detailed representation.

Governor Transfer Function

Mechanical-Hydraulic Governor

A simplified schematic of a temporary-droop mechanical-hydraulic governor is shown in Fig. 1. In steady state the shaft-speed signal n_s is compared to the reference-speed setting n_r, modified by the permanent speed droop times gate position σz. Sometimes the permanent speed droop is obtained by using generator power output rather than gate position. The latter scheme is readily available on electric-hydraulic governors.

An imbalance between actual speed and the modified reference speed appears as a change in the input a to the pilot servo. When gate position is changing, a transient droop signal c is developed to oppose fast changes in gate position. In the mechanical-hydraulic governor illustrated, these signals are summed and transmitted through a system of floating levers from a mechanical motion to the operation of the pilot valve.

The derivation of the detailed and approximated transfer functions for the mechanical-hydraulic governing systems is shown in Appendix I. The results give the following:

Paper 69 C 2-PWR, recommended and approved by the Power System Engineering Committee of the IEEE Power Group for presentation at the IEEE PICA Conference, Denver, Colo., May 18–21, 1969. Manuscript submitted January 6, 1969; made available for printing June 20, 1969.

The authors are with Westinghouse Electric Corporation, East Pittsburgh, Pa. 15112.

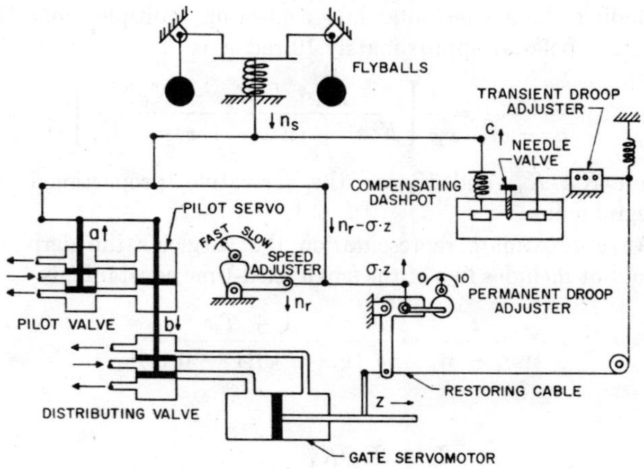

Fig. 1. Mechanical-hydraulic hydrogovernor.

detailed transfer functions

$$\frac{z}{n_r - n_s} = G_1$$

$$= \frac{\frac{1}{\sigma}(1 + T_r s)}{\frac{T_p T_r T_g s^3}{\sigma} + \frac{(T_p + T_r) T_g s^2}{\sigma} + \frac{[T_g + T_r(\sigma + \delta)]\, s}{\sigma} + 1}$$

where z is the deviation of gate position, and $n_r - n_s$ is the speed error;

approximate transfer functions

$$G_3 = \frac{\frac{1}{\sigma}(1 + T_r s)}{(1 + T_1 s)(1 + T_2 s)}$$

where

$$T_1 = \frac{T_g + T_r(\sigma + \delta)}{\sigma}$$

$$T_2 = \frac{T_r T_g}{T_g + T_r(\sigma + \delta)}$$

and s is the Laplace operator.

A comparison of the gain and the phase margin of these two transfer functions is made in Fig. 2. A comparison of computer simulation of detailed and approximate representations is given later.

Electric Governor

Since 1960 there has been increased usage of electric governors instead of the conventional mechanical-hydraulic hydrogovernors; their operation is functionally very similar. Speed sensing and permanent-droop compensation are performed electrically, and the output signal drives an electrical-mechanical transducer which operates a pilot valve and pilot-valve servomotor.

The electrical governor manufactured by the Woodward Governor Company has a three-term proportional-integral-derivative action [6].

A derivation of typical transfer functions is shown in Appendix I. It is convenient to express the transfer function in terms of proportional, integral, and derivative gains. A transfer function including these constants, but neglecting multiple small-time constant poles at approximately 10 rad/s, is

$$\frac{z}{n_r - n_s} = \frac{1}{\sigma}\left[\frac{K_d s^2 + K_p s + K_i{}'}{K_d s^2 + (K_p + 1/\sigma)s + K_i}\right]$$

where K_d, K_p, and K_i are the derivative, proportional, and integral gains.

An approximate representation that neglects the derivative gain, but includes two of the small fixed-time constants, is

$$\frac{z}{n_r - n_s} = \frac{1}{\sigma}\,\frac{1 + T_1 s}{(1 + T_2 s)(1 + 0.1s)^2}$$

where

$$T_1 = K_p/K_i$$

$$T_2 = (K_p\sigma + 1)/K_i\sigma.$$

This paper dces not illustrate a comparison of detailed versus approximate representation. The approximate representation is very nearly exact when using a very low derivative gain. However, a high derivative gain will produce excessive oscillations, possibly leading to instability. Therefore, it may be stated, at least qualitatively, that proper simulation of a hydro-unit with a large derivative gain setting requires detailed representation. From a review of published papers and their discussions, the authors conclude that derivative action is not normally recommended for plants strongly connected to an interconnection. Rather it seems that derivative action is most beneficial for isolated operation. Without derivative action, the transfer function of the electric governor is similar to that of the mechanical-hydraulic governor.

Turbine-Penstock Transfer Function

Appendix II shows a derivation of the turbine-penstock transfer function. Assuming an idealized turbine operating at full load, the classical transfer function results:

$$\frac{m}{z} = \frac{1 - T_w s}{1 + T_w s/2}.$$

A more general assumption is

$$\frac{m}{z} = a_{23}\,\frac{1 + (a_{11} - a_{13}a_{21}/a_{23})T_w s}{1 + a_{11}T_w s}$$

where a_{11} and a_{13} are partial derivatives of flow with respect to head and gate opening, and a_{21} and a_{23} are partial derivatives of torque with respect to head and gate opening. The partial derivatives are constants which depend upon the machine loading and represent the deviation from an idealized turbine.

Symbols

The following are deviations of variables from initial steady-state conditions:

- h turbine head (at turbine inlet in per unit)
- m turbine torque, per unit
- n turbine speed, per unit
- q turbine flow rate, per unit
- z gate position, per unit.

The following symbols represent the total magnitude of quantities:

- n_s shaft speed, per unit
- n_r reference speed, per unit
- Z gate opening, per unit
- Z_0 initial gate opening, per unit.

The following symbols are constants, some of which are dependent upon initial operating conditions:

- a_{11} partial derivative of turbine flow rate with respect to turbine head
- a_{12} partial derivative of turbine flow rate with respect to turbine speed
- a_{13} partial derivative of turbine flow rate with respect to gate position
- a_{21} partial derivative of turbine torque with respect to turbine head
- a_{22} partial derivative of turbine torque with respect to turbine speed
- a_{23} partial derivative of turbine torque with respect to gate position
- H inertia constant, kW·s/kVA
- T_c gate closing time, seconds
- T_g governor response time, or the inverse of governor open-loop gain
- T_m mechanical starting time, $=2H$
- T_p pilot valve time constant, seconds
- T_r dashpot time constant, seconds
- T_w water starting time, seconds
- β per unit speed deviation required to saturate the pilot valve
- δ temporary speed droop, per unit
- σ permanent speed droop, per unit
- D differential operator.

Comparison of Detailed Versus Equivalent

A comparison of gains and phase margins of the detailed and equivalent governor representations shown in Fig. 2 indicates that there should be relatively good agreement at oscillation frequencies up to at least 3 rad/s. To obtain a quantitative comparison, an analytical study of a simple one-machine infinite-bus system was made. Included in this study was a comparison of detailed versus classical turbine-penstock representation. The Westinghouse large-system stability program was used for this evaluation.

The conventional governor representation available in the program was used for the equivalent representation. New governor models were developed and added to the program to study in detail the mechanical-hydraulic and electric hydro-governors. A detailed representation of turbine-penstock was also added. These are illustrated in Fig. 3.

The governor and turbine penstock constants used for the evaluations are those given as typical values in Table I and Appendixes I and II. The generator constants used are

$$X_d = 1.01 \text{ pu} \qquad X_l = 0.15 \text{ pu}$$

$$X_d' = 0.31 \text{ pu} \qquad X_q = 0.64 \text{ pu}$$

$$T_{d0}' = 5.0 \text{ seconds} \qquad X_q' = 0.64 \text{ pu}$$

$$H = 6.0 \text{ kW·s/kVA}.$$

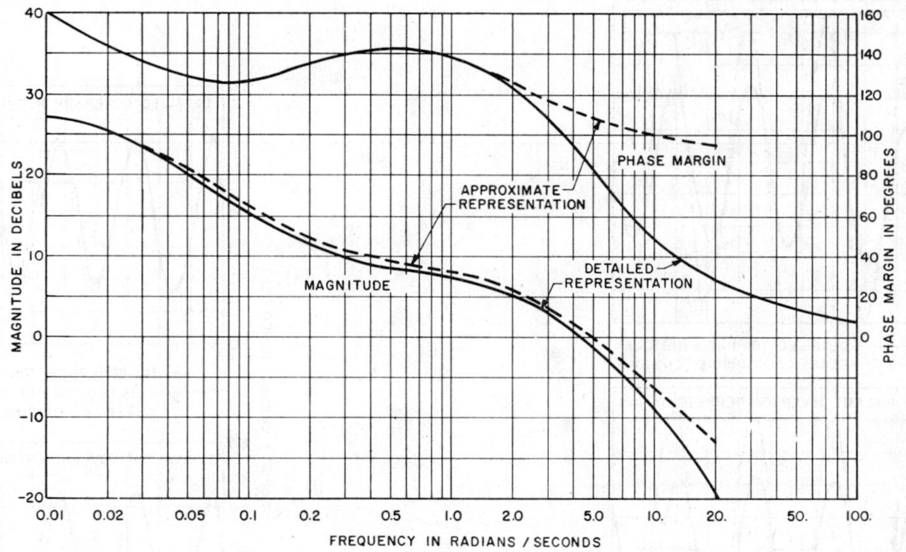

Fig. 2. Magnitude and phase-margin characteristics of mechanical-hydraulic hydrogovernor transfer function.

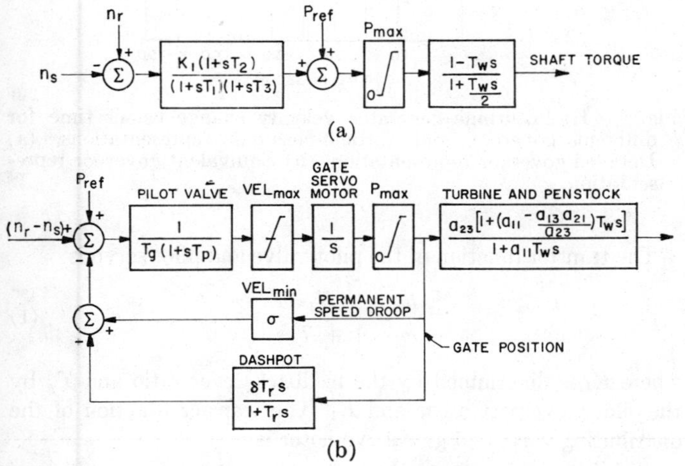

Fig. 3. Hydrogovernor–turbine representation for computer simulation. (a) Equivalent representation. (b) Detailed representation.

TABLE I

Typical Values and Range of Parameters

	Typical	Range
T_r	5.0	2.5 –25.0
T_g	0.2	0.2 – 0.4
T_w	1.0	0.5 – 5.0
T_m (= 2H)	8.0	6.0 –12.0
T_c	5.0	5.0 –10.0
σ	0.04	0.03– 0.06
δ	0.31	0.2 – 1.0
β	0.04	

Typically,

$$T_r = 5.0\,T_w$$

$$\delta = 2.5\,\frac{T_w}{T_m}$$

Velocity damping, generator damper circuits, and generator excitation system action were omitted from this study to isolate the effect of the governor.

With the machine constants fixed, the impedance from the generator bus to the infinite bus was varied to change the natural oscillation frequency. Frequencies up to 10 rad/s (1.6 Hz) were examined, and the largest difference between representations was found to be at approximately 5 rad/s (0.8 Hz). Figs. 4 and 5 illustrate the performance at a natural oscillation frequency of 5 rad/s. This frequency corresponds to a transformer reactance of 15 percent and a 25-percent external reactance to the infinite bus. Although the difference in phase margin of the different control system representations increases beyond 5 rad/s, gain is decreasing significantly, and there is little or no governor response at the higher frequencies.

In Fig. 4 turbine torque changes following a disturbance are shown for the different combinations of governor and turbine-penstock representation. There is very little difference in the transient response of the two governor representations; however, the classical turbine penstock shows a more oscillatory behavior than detailed. It may be concluded that the classical representation is conservative since it is somewhat less stable.

Fig. 5 shows a similar comparison of velocity changes. The difference in turbine-penstock representation is not nearly as noticeable as that shown in Fig. 4 because of the integrating effect of turbine–generator inertia. The response of a unit with no governor (constant mechanical input power) is shown for reference. This response illustrates the natural damping present with no speed governor action.

These figures show that for a normal stability study, the differences in representation are very slight. This is partly due to the small-frequency deviation and nearly zero average deviation following the transient disturbance. The only type of study where significant differences should be expected would be one in which there is a large low-frequency deviation, and where the limiting action of the pilot valve and gate is reached. Detailed representations are available for studying these problems.

Similar studies were made for the electric governor, but no figures are given here. Without derivative action the electric governor is very similar to the mechanical-hydraulic governor. Adding significant derivative gain causes an oscillatory buildup leading to instability. The authors conclude that only the detailed representation is valid for electric hydrogovernors with derivative gain.

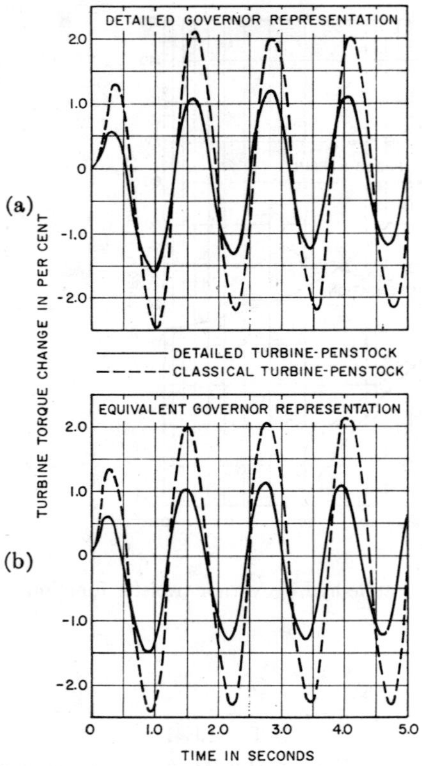

Fig. 4. Hydroturbine torque change versus time for different governor and turbine-penstock representations. (a) Detailed governor representation. (b) Equivalent governor representation.

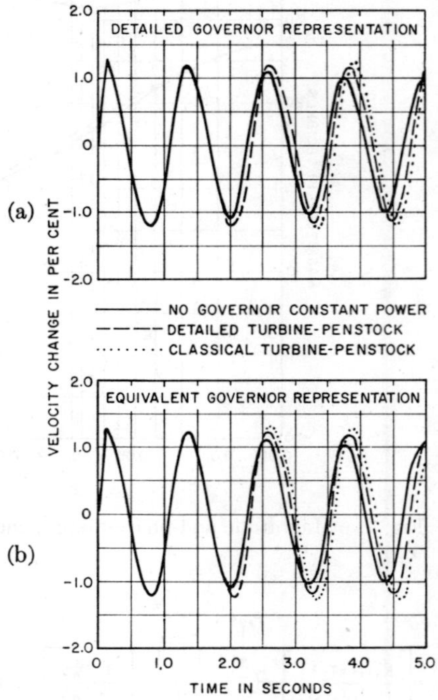

Fig. 5. Hydroturbine–generator velocity change versus time for different governor and turbine-penstock representations. (a) Detailed governor representation. (b) Equivalent governor representation.

Conclusions

1) A procedure is suggested for determining an equivalent representation of hydrogovernors in power system stability studies. The equivalent representation differs from the detailed primarily in the elimination of the pilot-valve time constant and the limiting of pilot-valve output (with no limit on gate velocity).

2) A comparison of a typical hydroturbine–generator response over a range of frequencies from dc (steady state) to approximately 100 rad/s shows good agreement between the detailed and approximate representations.

3) A significant difference in turbine torque changes following a disturbance was noted for detailed versus classical turbine-penstock representation. This was true for both detailed and equivalent governor representations.

4) Because of the integrating effect of turbine–generator inertia, less effect is noted in velocity than in torque.

5) Because the classical turbine-penstock representation is more oscillatory, it may be concluded that the use of the classical representation is a conservative assumption.

6) Detailed representations of hydrogovernors and turbine penstock are available on a large-scale system-stability program where such detail is desired.

Appendix I

Governor Transfer Function

Mechanical Hydraulic Governor

Fig. 1 shows a simplified schematic of a mechanical-hydraulic governor. The variables used in the derivation of the transfer functions are per-unit derivatives from initial steady-state values.

The transfer function of the pilot valve and pilot servo is

$$\frac{b}{a} = \frac{K_1}{1 + T_p s} \tag{1}$$

where K_1 is determined by the feedback lever ratio and T_p by the pilot-valve port areas and K_1. The transfer function of the distributing valve and gate servomotor is

$$\frac{z}{b} = \frac{K_2}{s}. \tag{2}$$

Combining (1) and (2)

$$\frac{z}{a} = \frac{K_1 K_2}{s(1 + T_p s)} = \frac{1}{T_g s(1 + T_p s)} \tag{3}$$

where $1/T_g = K_1 K_2$ is the governor open-loop gain. The pilot-valve feedback-level ratio and the port areas of the distributing valve and gate servo determine this gain.

T_g is equal to the time in seconds for a 1 pu change in frequency to produce a 1 pu change in gate opening. T_g is related to the gate closing time T_c by

$$T_g = \beta T_c$$

where T_c is the time required for full-gate travel with the gates moving at maximum velocity, typically 5 to 10 seconds, and β is the per-unit speed deviation required to saturate the pilot valve, typically 0.04 pu (2.4 Hz).

Assuming that the flow of dashpot fluid through the needle valve is proportional to dashpot pressure, the compensating dashpot transfer function is

$$\frac{c}{z} = \frac{\delta T_r s}{1 + T_r s} \tag{4}$$

where the temporary droop δ is determined by the selection of pivot-point for the lever connected to the input piston. The needle valve setting determines the reset time T_r.

Through the action of a system of floating levers, the reference speed, shaft speed, permanent droop, and temporary droop signals are added to produce the pilot-valve input signal:

$$a = n_r - n_s - \sigma z - \frac{\delta T_r s}{1 + T_r s}. \quad (5)$$

A combination of (3), (4), and (5) gives

$$\frac{z}{n_r - n_s} = G_1$$

$$= \frac{\frac{1}{\sigma}(1 + T_r s)}{\frac{T_p T_r T_g s^3}{\sigma} + \frac{(T_p + T_r)T_g s^2}{\sigma} + \frac{[T_g + T_r(\sigma + \delta)]s}{\sigma} + 1}. \quad (6)$$

If the pilot-valve time constant T_p is neglected, this transfer function reduces to

$$G_2 = \frac{\frac{1}{\sigma}(1 + T_r s)}{\frac{T_r T_g s^2}{\sigma} + \frac{[T_g + T_r(\sigma + \delta)]s}{\sigma} + 1}. \quad (7)$$

This may be closely approximated by the simpler transfer function

$$G_3 = \frac{(1/\sigma)(1 + T_r s)}{(1 + T_1 s)(1 + T_2 s)} \quad (8)$$

where

$$T_1 = \frac{T_g + T_r(\sigma + \delta)}{\sigma}$$

$$T_2 = \frac{T_r T_g}{T_g + T_r(\sigma + \delta)}.$$

Table I gives typical values and range of parameters for hydrogovernors. Substitution of these values in transfer function G_1 of (6) yields

$$G_1 = \frac{25(1 + 5s)}{2.5s^3 + 25.5s^2 + 48.75s + 1}.$$

Similarly, substitution in G_3 of (7) gives

$$G_3 = \frac{25(1 + 5s)}{(1 + 48.7s)(1 + 0.513s)}.$$

The magnitudes and phase margins of G_1 and G_3 are plotted in Fig. 2. The magnitudes agree very closely up to 10 rad/s (1.6 Hz), and the phase margins agree reasonably well up to 3 rad/s (0.48 Hz). At higher frequencies the approximate transfer function G_3 indicates a larger phase margin.

Electric Governor

Leum [6] describes the transfer function of the Woodward transistor electric governor as

$$z = \frac{1.44 \times 10^5 (s + \beta_1)(s + \alpha_1)[(n_r - n_s) - \sigma z]}{s(s + 1200\beta_1 k)(s + 21 + \alpha_1)(s + 10)^2}. \quad (9)$$

The permanent-droop feedback may be obtained from generator terminal power rather than from gate position to eliminate the nonlinearity of the gate-position versus power curve.

Assuming that the denominator time constants are small for typical oscillation frequencies of hydrounits, (9) may be simplified to

$$z = \frac{0.057}{ks}\left(\frac{s}{\alpha_1} + 1\right)\left(\frac{s}{\beta_1} + 1\right)[(n_r - n_s) - \sigma z] \quad (10)$$

rearranging terms

$$z = \frac{0.057}{k}\left[\left(\frac{1}{\alpha_1} + \frac{1}{\beta_1}\right) + \frac{1}{s} + \frac{s}{\alpha_1\beta_1}\right][(n_r - n_s) - \sigma z]. \quad (11)$$

The governing system can thus be described as a three-term controller: proportional, integral, and derivative. Expressed in a more general way

$$z = [K_p + K_i/s + K_d s][(n_r - n_s) - \sigma z] \quad (12)$$

where K_p, K_i, and K_d are the proportional, integral, and derivative gains; or

$$\frac{z}{n_r - n_s} = \frac{1}{\sigma}\left[\frac{K_d s^2 + K_p s + K_i}{K_d s^2 + (K_p + 1/\sigma)s + K_i}\right]. \quad (13)$$

Given the governor constants, (13) may be expressed in the form

$$\frac{z}{n_r - n_s} = \frac{1}{\sigma}\frac{(1 + T_1 s)(1 + T_2 s)}{(1 + T_3 s)(1 + T_4 s)}. \quad (14)$$

Since a double phase-lead term may not be available in the typical computer representation, a reduced approximation neglects the derivative constant K_d:

$$\frac{z}{n_r - n_s} = \frac{1}{\sigma}\frac{K_p s + K_i}{(K_p + 1/\sigma)s + K_i} \quad (15)$$

which may be expressed as

$$\frac{z}{n_r - n_s} = \frac{1}{\sigma}\frac{1 + T_1 s}{1 + T_2 s} \quad (16)$$

where

$$T_1 = \frac{K_p}{K_i}$$

$$T_2 = \frac{K_p\sigma + 1}{K_i\sigma}.$$

This approximation can be improved somewhat by adding the small time constants that were neglected in the derivation of (10). A value of 0.1 second is a good approximation. Since most programs have a capability of handling a total of three governor phase-lag terms,

$$\frac{z}{n_r - n_s} = \frac{1}{\sigma}\frac{1 + T_1 s}{(1 + T_2 s)(1 + 0.1s)^2}. \quad (17)$$

Appendix II

Turbine-Penstock Transfer Functions

Note that in the following equations the variables are deviations from initial steady-state conditions expressed in per unit.

The following expressions may be written for the turbine:

$$q = a_{11}h + a_{12}n + a_{13}z \quad (18)$$

$$m = a_{21}h + a_{22}n + a_{23}z. \quad (19)$$

The rate of change of flow deviation may be shown to be

$$Dq = \frac{-h}{T_w} \tag{20}$$

where

$$T_w = \frac{\Sigma\, lv}{gH_r} \tag{21}$$

with Σlv being the summation of length and velocity of sections in the water passage from draft tube intake to outlet, H_r the rated turbine head, and g the acceleration of gravity, and where

$$h = \frac{-a_{12}T_wDn - a_{13}T_wDz}{a_{11}T_wD + 1}. \tag{22}$$

Eliminating h in (19),

$$m = \frac{[(a_{11}a_{23} - a_{13}a_{21})T_wD + a_{23}]z}{1 + a_{11}T_wD} + \frac{[(a_{11}a_{22} - a_{12}a_{21})T_wD + a_{22}]n}{1 + a_{11}T_wD}. \tag{23}$$

Note that turbine torque changes are produced by two components, one related to gate changes and the other to speed changes. Since turbine speed changes are small, especially when operating connected to a system, the terms related to n will be neglected.

The turbine transfer function which relates a torque change in attaining a position change is

$$\frac{m}{z} = \frac{a_{23}[1 + (a_{11} - a_{13}a_{21}/a_{23})T_ws]}{1 + a_{11}T_ws}. \tag{24}$$

Typical values for a turbine at full load are

$$a_{11} = 0.58, \quad a_{21} = 1.40$$

$$a_{13} = 1.10, \quad a_{23} = 1.5.$$

For an ideal lossless turbine

$$a_{11} = 0.5Z_0, \quad a_{21} = 1.5Z_0$$

$$a_{13} = 1.0, \quad a_{23} = 1.0.$$

Substituting in (24)

$$\frac{m}{z} = \frac{1 - Z_0T_ws}{1 + 0.5Z_0T_ws}.$$

At full load $Z_0 = 1.0$ and

$$\frac{m}{z} = \frac{1 - T_ws}{1 + T_ws/2}$$

which is the classical turbine-penstock transfer function.

Acknowledgment

The authors wish to thank D. H. Berry of Ontario-Hydro for providing valuable guidance in the development of the equivalent transfer functions and supplying hydrosystem constants for the evaluation.

References

[1] F. R. Schleif and R. R. Angell, "Governor tests by simulated isolation of hydraulic turbine units," *IEEE Trans. Power Apparatus and Systems*, vol. PAS-87, pp. 1263–1269, May 1968.
[2] J. L. Woodward, "Hydraulic turbine transfer function for use in governing studies," *Proc. IEE* (London), vol. 115, March 1968.
[3] R. Oldenburger and J. Donelson, Jr., "Dynamic response of a hydroelectric plant," *AIEE Trans. (Power Apparatus and Systems)*, vol. 81, pp. 403–419, October 1962.
[4] L. M. Hovey, "Optimum adjustment of hydro governors on Manitoba hydro system," *AIEE Trans. (Power Apparatus and Systems)*, vol. 81, pp. 581–587, December 1962.
[5] J. M. Undrill and J. L. Woodward, "Nonlinear hydro governing model and improved calculation for determining temporary droop," *IEEE Trans. Power Apparatus and Systems*, vol. PAS-86, pp. 443–453, April 1967.
[6] M. Leum, "The development and field experience of a transistor electric governor for hydro turbines," *IEEE Trans. Power Apparatus and Systems*, vol. PAS-85, pp. 393–402, April 1966.
[7] D. G. Ramey and R. T. Byerly, "Dynamic simulation of interconnected systems," *1967 IEEE PICA Conf. Rec.* (Pittsburgh, Pa.), pp. 31–40.

Discussion

G. Quazza (Research Center on Automatic Controls, ENEL, Milan, Italy): The authors are to be commended for their attempt to improve upon existing programs for simulating transient response of power systems by including a detailed representation of hydrogovernors. I would like to comment, however, on the validity of the model for the water-supply system in the case of high-head hydroelectric stations.

The penstock-transfer function presented is actually only a medium-low-frequency approximation, with relevant errors in the high-frequency range. A more satisfactory description of penstock dynamics must account for the water-hammer phenomenon caused by the distributed character of the hydraulic line. Especially when the penstock, depending upon its embedding in concrete, is relatively elastic, the pressure waves, which are due to gate-opening variations and are reflected from the surge tank, travel at a relatively low speed. A typical value for the velocity of sound in these conditions may be 3000 ft/s so that the travel time for the pressure wave to return to the gate for a 3000-foot penstock may be 2 seconds. If the water-hammer theory, which dates back to the early 1900's (Allievi), is translated into the transfer function language, one obtains

$$\frac{m}{z} = a\,\frac{1 - j2\rho\,\tan(\omega\tau/2)}{1 + j\rho\,\tan(\omega\tau/2)} \tag{25}$$

where m is the turbine torque in per unit, z the gate opening in per unit, $a = \overline{m}/\overline{z}$ the steady-state ratio of turbine torque to gate opening around the given operating point, $\tau = 2l/c$ the time taken by the sound wave of speed c to travel through the penstock of length l twice, and $\rho = (\bar{q}/2\overline{H})(c/gA)$ a dimensionless factor (related to water velocity, with $\bar{q}$ the rate of flow of water, A the penstock cross section, $\overline{H}$ the head, and g the acceleration of gravity). Expression (25), where damping due to hydraulic friction has been neglected, yields the oscillating amplitude diagram versus frequency in Fig. 6(a) and the phase diagram in Fig. 6(b), showing that ang $m/z \to \infty$ for $\omega \to \infty$ [for comparison, see the dashed frequency response from (28)].

Resonant frequencies appear at

$$\omega_i \frac{\tau}{2} = (2i + 1)\frac{\pi}{2}, \quad i = 0,1,2,\cdots \tag{26}$$

where the relative amplitude $(1/a)|m/z|$ is 2, the phase minus $(2i + 1)\pi$ and antiresonant frequencies appear at

$$\omega_j \frac{\tau}{2} = j\pi, \quad j = 0,1,2,\cdots \tag{27}$$

where the relative amplitude is 1, the phase minus $2j\pi$.

In the preceding example $\tau = 2$ seconds; hence the first resonance occurs at $\omega_0 = 1.57$ rad/s. Here the phase lag is already 180°, while the classical nonminimum-phase approximation used by the authors never reaches this value. It is clear why the bandwidth of frequency

Manuscript received July 29, 1969.

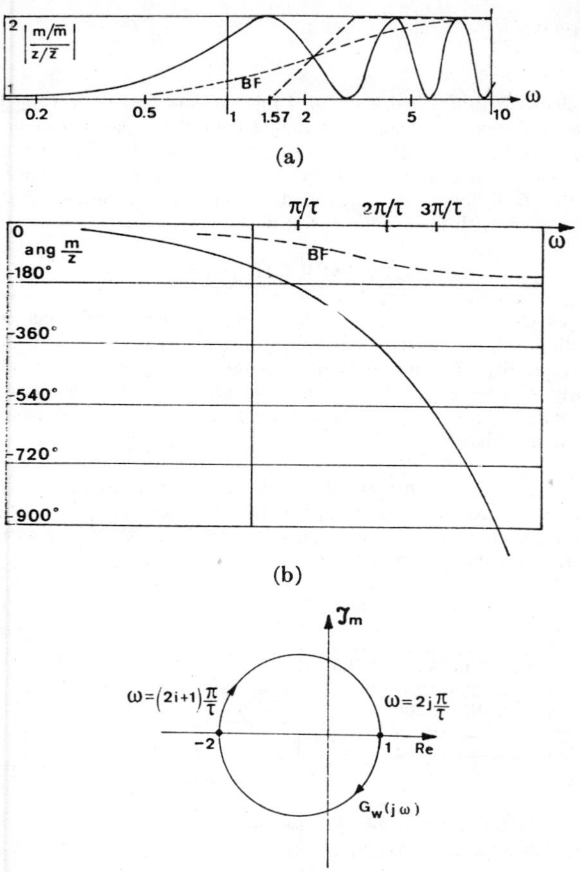

Fig. 6.

governors for high-head hydroelectric generators is kept low: the stability of the isolated generator calls for a crossover frequency not greater than $\sim\omega_0/3$. In our example, the frequency response diagrams in Fig. 2 should be corrected in the high-frequency range to account for Fig. 6.

Another relevant difference of (25) with respect to the first approximation

$$\frac{m}{z}\frac{1}{a} = \frac{1 - j\omega T_w}{1 + j\omega(T_w/2)} \tag{28}$$

[or equivalently, if $Z_0\tau_w$ replaces T_w, both being obtainable from (25) if $\tan(\omega\tau/2) = \sim\omega\tau/2$] lies in the fact that the resonant frequency ω_0 is independent of load, i.e., of operating flow rate while T_w is proportional to load, or at least closely related to it. Equation (28) may then generate the false impression that at low loads no relevant phase lags are to be expected in the useful frequency range.

The frequency response shown in Fig. 6 can be easily measured, at least for $\omega \leq 2\omega_0$ (by applying a sinusoidal input to the governor), as we have systematically done in the past several years on hydroelectric units used for load-frequency control. Hydraulic damping does not appreciably affect the frequency response for $\omega \leq \omega_0$. Expression (28) is not entirely adequate in the very-low-frequency range, as it does not account for the water-mass oscillations in the surge-tank reservoir. Expression (28) does not describe such oscillations since it is derived from sound-wave partial differential equations for a constant surge-tank level.

There would be no problem in completing the expression of the transfer function by including the preceding phenomenon [8]. It may be questionable, however, as to whether the added complication is worthwhile for small signal-control studies, the period of mass oscillations often being very long, say, 100 to 200 seconds or more, with no relevant influence upon water-hammer and governor response.

On the other hand, surge-tank water-level variations may become very important in the event of large load changes in the turbine. These variations are especially important if the unit is required to pick up full load at maximum rate just one half-period of the mass oscillation after a trip from full load to no load. There is a danger of disaster if the water level should decrease enough to uncover the penstock inlet. To avoid this danger, hydraulic plant designers usually include an additional wide chamber at the base of the surge tank. Except for the need to check the hydraulic design for this danger by suitable analog or digital computer simulation and by testing, reservoir-tank oscillations should not be of concern in the study of power system fast transients.

However, large signal operation requires a reasonably accurate description of governor and valve nonlinearities. First, high-head Pelton turbines have a double actuator on the water jet out of the penstock, i.e., on the the main distributor valve and the deflector. The action of the deflector (which regulates the jet, thus reducing turbine torque without water hammer) under a fast closing command is relatively rapid, say, 1.5 to 2 seconds closing time from full load to no load, while the distributor closing time may be as long as 30 to 40 seconds. Francis turbines have a synchronous discharge device, which is slower than the deflector but still much faster than the distributor. Since both distributor and deflector, or synchronous discharge, are rate saturated in opening with an opening time of 20 to 30 seconds, the need for close observation of the above nonlinearity is apparent.

Finally, consideration should be given to the relevant nonlinearities, including deadbands, in the distributor valve servoloop, and especially to the nonlinear relationship between turbine power, or torque, and water flow, or gate opening. This relationship is related to the basic expression of hydraulic power as the product of flow rate through the head and water speed proportional to the square root of the head.

In conclusion, if the digital computer program for power system transients is meant to cover in great detail several seconds of large perturbations, it is suggested that the basic equations for water-hammer pressure waves, with due regard to water compressibility, be directly introduced in the model, and that an adequate description of these nonlinearities be added.

REFERENCES

[8] G. Quazza and Saccomanno, "Considerazioni sul proporzionamento dei regolatori di frequenza per turbine idrauliche in relazione alle caratteristiche della rete," *Proc. 61st Annual Meeting of the Italian Electrical and Electronic Society* (Ancona, Italy, 1960).

[9] H. M. Paynter, "Surge and water hammer problems," Paper 2569, *ASCE Trans.*, vol. 118, 1953.

D. G. Ramey and **J. W. Skooglund**: The authors wish to thank Mr. Quazza for his valuable discussion. It clarifies the assumptions made in the turbine model and illustrates the limitations of the model.

The water starting time constant proposed in Fig. 6 is somewhat greater than that commonly encountered in our studies. As Mr. Quazza pointed out, this would normally result in governor time constants that would reduce the overall bandwidth of the speed-control system. This would lead to a response to the example disturbances similar to Figs. 4 and 5.

The authors agree with Mr. Quazza's comment that surge-tank reservoir water-mass oscillations and surge-tank water-level variations are normally of concern in the hydraulic design of the turbine rather than in control studies. Other nonlinear effects, such as deflectors or synchronous discharge values and deadbands in the distributor valve servoloop, are not difficult to include in the model. However, unless the hydrogovernor representation is a more significant factor in a study than that shown in the examples, the authors feel that the difficulty of determining accurate data and the computing effort required for these features are not justified.

One primary observation by the authors in this investigation was that the disturbances that have historically been studied with a power system stability program (i.e., faults, false tripping, etc.) result in large power flow and voltage disturbances, but in relatively small-frequency disturbances. With the current interest in longer simulation times it is desirable to review the assumptions not only of this model but of the other equipment in terms of the disturbances and characteristics of the system being studied. This discussion provides a useful summary of the more important assumptions in the derivation of the hydraulic turbine model.

Manuscript received August 21, 1969.

DYNAMIC MODELS FOR STEAM AND HYDRO TURBINES IN POWER SYSTEM STUDIES

IEEE COMMITTEE REPORT

ABSTRACT

Basic models for speed-governing systems and turbines in power system stability studies are presented. These models provide adequate representation for hydro, fossil-fired, and pressurized water reactor nuclear units in most stability analyses. Models for boiling water reactor nuclear units are to be presented at a later date. Typical parameters are given.

FOREWORD

The Dynamic System Performance Working Group of the Power System Engineering Committee established the Task Force on Overall Plant Response in 1968 to consider the effects of power plants on power system stability. The intent was to identify problems not already under investigation and to make recommendations regarding them to appropriate subcommittees and working groups.

In 1968, there were several computer programs which represented the effects of turbine speed control on system stability and many electric utilities had assembled related data files or were in the process of doing so. It appeared that a committee paper, similar to that on generator excitation[1], covering mathematical models, nomenclature, and typical data would be useful. After considering alternatives, the task force was expanded to include appropriate ASME representation with the assignment of writing this paper.

INTRODUCTION

This paper is concerned with the representation of steam and hydraulic turbines and their speed-governing systems for power system stability studies. The intent is to indicate the origin of the models currently in wide use, to bring about a greater understanding of when these models may and may not be used and to promote standardization of nomenclature.

In preparing this paper, the Task Force surveyed utilities and service centers currently using models of this general type. The model descriptions which follow are typical of those in use by the groups which responded to the survey.

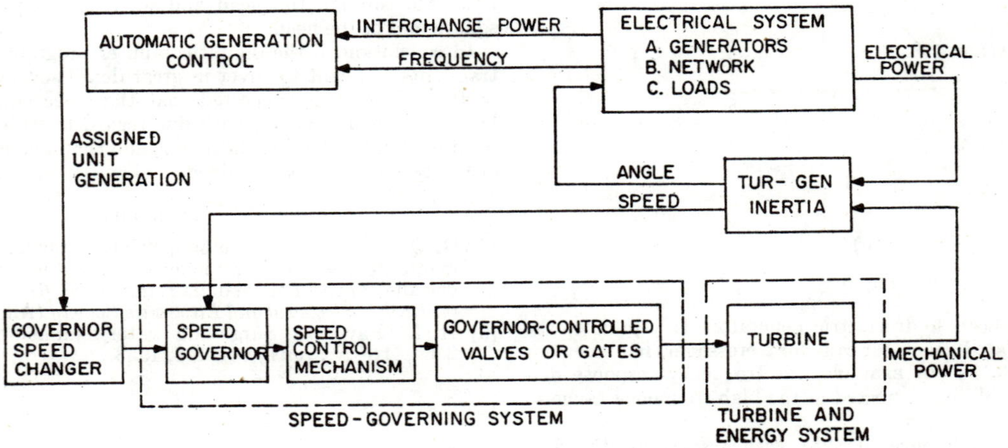

Fig. 1. Functional Block Diagram Showing Location of Speed-Governing System and Turbine Relative to Complete System.

Members of the Task Force on Overall Plant Response include: R. T. Byerly (Chairman), Ola Aanstad, D. H. Berry, R. D. Dunlop, D. N. Ewart, B. M. Fox, L. H. Johnson, and D. W. Tschappat. Contributing nonmembers include G. J. Bartok, F. P. DeMello, R. J. Mills, and C. C. Young.

Paper T 73 089-0, recommended and approved by the Power System Engineering Committee of the IEEE Power Engineering Society for presentation at the IEEE PES Winter Meeting, New York, N.Y., January 28-February 2, 1973. Manuscript submitted October 24, 1972; made available for printing December 14, 1972.

The diagram of Figure 1 includes functional blocks for the governor speed changer and for automatic generation control to show their relationship to the speed-governing system. The time span of most stability studies is short compared to the time required to make significant load changes, so that load changing equipment is not represented.

Four basic models are presented in this paper. These are:

1. Speed-governing systems for steam turbines.
2. Steam turbine systems.
3. Speed-governing systems for hydroturbines.
4. Hydroturbine systems.

For each category, a physical description of the equipment is given with one or more mathematical representations in block diagram form.

Reprinted from *IEEE Trans. Power App. Syst.*, Nov./Dec. 1973.

The paper includes general models applicable to most turbine and speed-governing system combinations. Typical constants are given.

The steam turbine models and associated speed-governing systems apply to conventional fossil-fired units. For purposes of stability studies, these also apply to turbines supplied by pressurized water reactors. It is the intent of the Task Force to present a standard model for turbines supplied by boiling water reactors at a later date. Further, it is assumed that all turbine control is accomplished by means of governor-controlled valves (i.e. excluding intercept valves and stop valves). Contingency valving, such as early valving or fast valving to obtain benefits for transient stability, is not specifically treated in this paper. It is suggested that power vs. time curves supplied by the turbine manufacturers be used in place of the turbine and speed-governing system models in cases where contingency valving is to be studied.

Finally, governor dead band is not shown in any of the models. In studies where the performance of the speed-governing system itself is the primary concern, it may be necessary to represent dead band, but it is not normally represented in large system studies.

SPEED-GOVERNING SYSTEMS FOR STEAM TURBINES

Mechanical-Hydraulic Control

A typical mechanical-hydraulic speed-governing system consists of a speed governor, a speed relay, a hydraulic servomotor, and governor-controlled valves which are functionally related as shown in Figure 2 (A). (See reference 2 for more detailed descriptions of mechanical-hydraulic controls.)

The block diagram of Figure 2 (B) shows an approximate nonlinear mathematical model. The speed governor produces a position which is assumed to be a linear, instantaneous indication of speed, and is represented by a gain K_G which is the reciprocal of regulation or droop. The signal, SR, is obtained from the GOVERNOR SPEED CHANGER of Figure 1, and is determined by the AUTOMATIC GENERATION CONTROL system. It represents a composite load and speed reference and is assumed constant over the interval of a stability study.

The speed relay is represented as an integrator with time constant T_{SR} and direct feedback. A nonlinear cam to compensate for valve nonlinearity is shown between the speed relay and the servomotor. The latter is represented by an integrator with time constant T_{SM} and direct feedback. The servomotor moves the valves and is physically large, particularly on large units. Rate limiting of the servomotor may occur for large, rapid speed deviations, and rate limits are shown at the input to the integrator representing the servomotor. Position limits are also indicated and may correspond to wide-open valves or the setting of a load limiter.

In power system studies, nonlinearities in the speed-control mechanism are normally neglected except for rate limits and the limits on valve position, as shown in Figure 2 (C). Typical parameters for mechanical-hydraulic systems are:

K_G = 20.0

T_{SR} = 0.1 seconds

T_{SM} = 0.2 to 0.3 seconds

$\dot{C}_{VOPEN}$ = 0.1 pu per second per valve

$\dot{C}_{VCLOSE}$ = 1.0 pu per second per valve

Electro-Hydraulic Control

An electro-hydraulic speed-control mechanism provides flexibility through the use of electronic circuits in place of mechanical components in the low-power portions.[3] The block diagram of Figure 3 (A) illustrates a typical configuration. The steam flow (or first stage pressure) feedback and the servomotor feedback loop provide for improved linearity over the mechanical-hydraulic system.

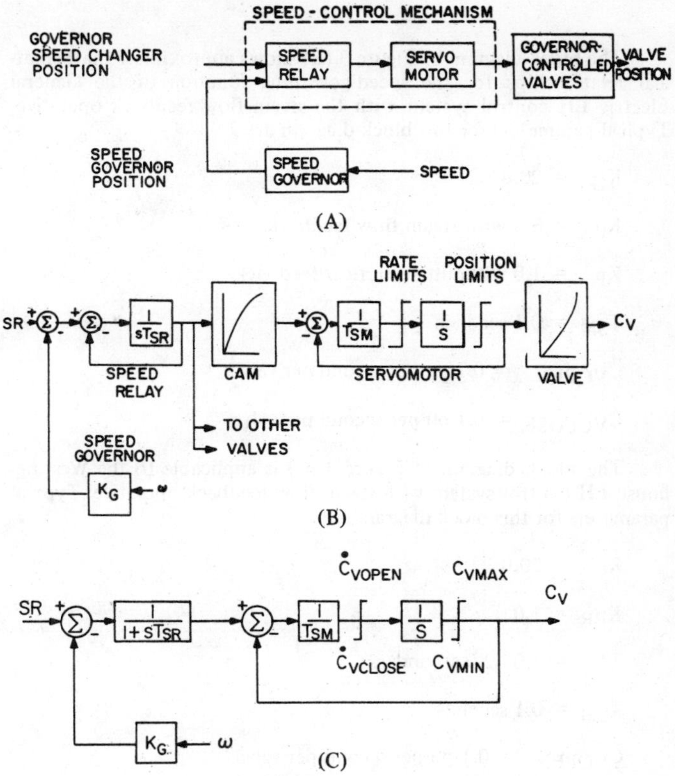

Fig. 2. Mechanical-Hydraulic Speed-Governing System For Steam Turbines. (A) Functional Block Diagram; (B) and (C) Approximate Mathematical Representations.

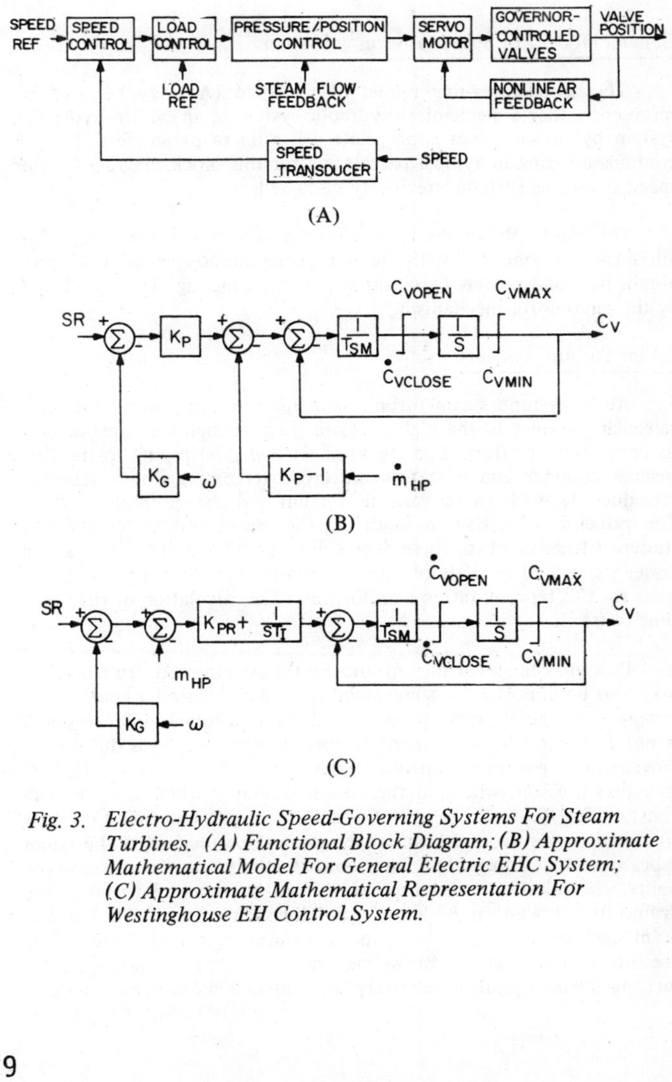

Fig. 3. Electro-Hydraulic Speed-Governing Systems For Steam Turbines. (A) Functional Block Diagram; (B) Approximate Mathematical Model For General Electric EHC System; (C) Approximate Mathematical Representation For Westinghouse EH Control System.

The block diagram of Figure 3 (B) shows approximate mathematical relationships for the speed-governing function of the General Electric EH control system with the steam flow feedback operative. Typical parameters for this block diagram are:

K_G = 20.0

K_P = 3.0 with steam flow feedback

K_P = 1.0 without steam flow feedback

T_{SM} = 0.1 seconds

$\dot{C}_{VOPEN}$ = 0.1 pu per second per valve

$\dot{C}_{VCLOSE}$ = 0.1 pu per second per valve

The block diagram of Figure 3 (C) is applicable to the Westinghouse EH control system with steam flow feedback operative. Typical parameters for this block diagram are:

K_G = 20.0

K_{PR} = 1.0

T_I = 1.0 - 2.0 seconds

T_{SM} = 0.1 seconds

$\dot{C}_{VOPEN}$ = 0.1 pu per second per valve

$\dot{C}_{VCLOSE}$ = 0.1 pu per second per valve

When steam flow feedback is not utilized with the Westinghouse EH system, the block diagram of Figure 4 (A) and the parameters of Table I are applicable.

General Model For Speed-Governing Systems

The speed-governing model of Figure 4 (A) may be used to represent either a mechanical-hydraulic system or an electro-hydraulic system by means of an appropriate selection of parameters. Table I contains a listing of typical parameters for this block diagram for the speed-governing systems previously discussed.

This model shows the load reference as an initial power P_O. This initial value is combined with the increments due to speed deviation to obtain the total power, P_{GV}, subject to the time lag, T_3, introduced by the servomotor mechanism.

Steam Turbine Systems

All compound steam turbine systems utilize governor-controlled valves at the inlet to the high pressure (or very high pressure) turbine to control steam flow. The steam chest and inlet piping to the first turbine cylinder and reheaters and crossover piping downstream all introduce delays between valve movement and change in steam flow. The principal objective in modeling the steam system for stability studies is to account for these delays. Flows into and out of any steam vessel are related by a simple time constant. Appendix I presents the basis for this representation and illustrates the calculation of the vessel time constant.

Pressure changes at the entrance to the governor-controlled valves may also be important in some stability studies.[11] Boiler controls are designed to regulate valve pressure, but the controlled boiler response is not fast enough to compensate for pressure variations due to the movement of governor-controlled valves. It can be assumed that there is a pressure somewhere in the steam generator which does remain constant for the time interval of a stability study, and that a pressure change dependent on the square of steam flow occurs in the boiler tubes from that constant pressure point to the valves. In a drum-type boiler, the point of constant pressure can be assumed to be at the drum. In a once-thru boiler no guidelines currently exist, but this point can be determined by tests or a simulation of the boiler. The pressure variation at the valves may be significant in simulating disturbances which result in relatively large, sustained frequency changes. It is not regarded as important when frequency variations are small and centered around the nominal system frequency.

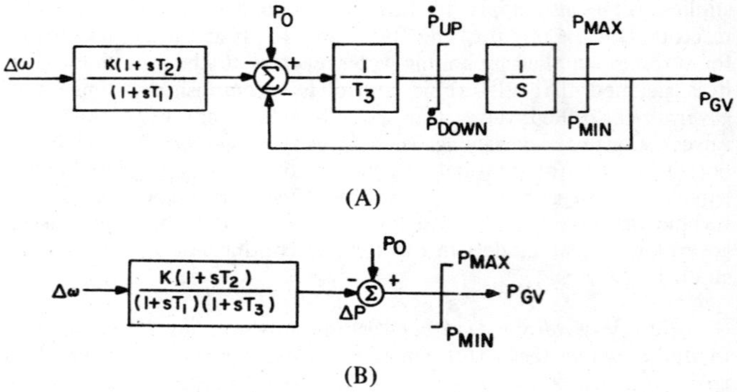

Fig. 4. General Models For Speed-Governing Systems. (A) Steam Turbine Systems; (B) Hydro Systems.

Table I. Speed Governing System Parameters For Use With The Block Diagram of Figure 4 (A)

System	Time Constants In Seconds		
	T_1	T_2	T_3
Mechanical-Hydraulic	0.2 - 0.3	0	0.1
General Electric EH With* Steam Feedback	0	0	0.025
General Electric EH Without Steam Feedback	0	0	0.1
Westinghouse EH With* Steam Feedback	2.8**	1.0**	0.15
Westinghouse EH Without Steam Feedback	0	0	0.1

Rate limits are nominally 0.1 pu per second except for mechanical-hydraulic system where $\dot{P}_{DOWN}$ is 1.0 pu per second.

Nominally, K = 100/(% Steady-State Speed Regulation)

See note 3 of Table V for further comment.

*Steam flow feedback includes the steam chest time constant T_{CH} which must be modified when Figure 4 (A) is used.

**These values may vary considerably from one unit to another.

Steam Chest and High Pressure Piping

Between the governor-controlled valves and the high pressure turbine is a steam bowl or chest as shown in Figure 5 (A). This vessel introduces a time delay between changes in valve steam flow and steam flow in the high pressure turbine. The mathematical model of Figure 5 (B) shows this effect as the time constant T_{CH}. In addition, this block diagram shows a method of accounting for boiler tube drop. The pressure P_{SG} is an internal boiler pressure assumed constant over the study interval and P_T is a variable pressure at the entrance to the governor-controlled valves. The parameter K_{PD} is a pipe-drop coefficient. The flow into the steam chest is:

$$\dot{m}_{CV} = P_{GV}\,(P_{SG} - K_{PD}\,\dot{m}_{CV}^{\,2})$$

where appropriate per unit variables are assumed.

The model of Figure 5 (C) shows a simpler method of approximating the effect of boiler tube drop by reducing the effective gain of the governor-controlled valves to the fraction F - (see note 3, Table V). If boiler tube drop is ignored, F is unity.

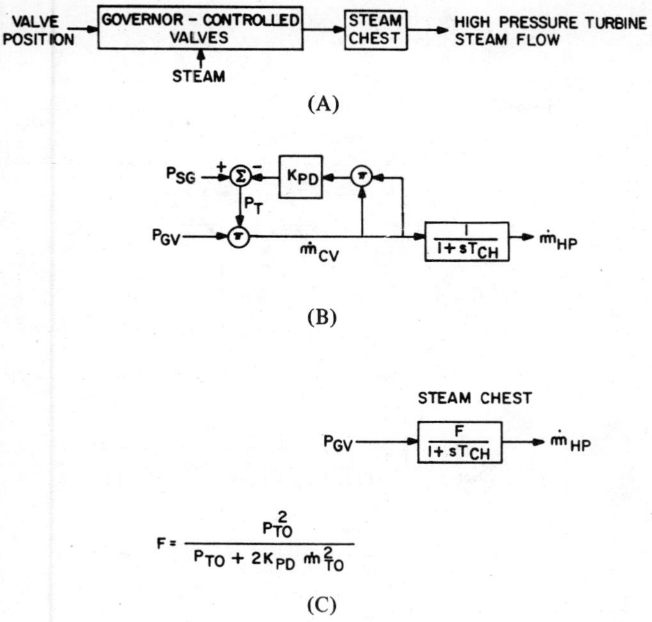

Fig. 5. Steam Chest and High Pressure Piping Representations. (A) Functional Block Diagram; (B) Approximate Nonlinear Model; (C) Linear Model.

Steam Configurations

Six common steam system configurations are shown in Figure 6. Corresponding mathematical models are shown in Figure 7. The time constants T_{CH}, T_{RH}, and T_{CO} represent delays due to the steam chest and inlet piping, reheaters, and crossover piping respectively. The fractions F_{VHP}, F_{HP} F_{IP}, and F_{LP} represent portions of the total turbine power developed in the various cylinders (see Table V and associated comments). Typical values for time delays and cylinder fractions are presented in Table II.

Extraction steam taken at various turbine stages to heat feedwater usually does not have significance for stability studies and is not shown in these models, although it is represented in some stability programs.[11]

General Steam System Model

Universal models are convenient for computer program development and as a basis for data files. All of the commonly encountered steam system configurations may be represented by the block diagram of Figure 8. Table III shows relationships between the parameters of Figure 8 and those of Figure 7.

Table II. Typical Parameters For Steam Turbine Models

Steam System Configuration	Typical Cylinder Fractions				Typical Time Constants			
	F_{VHP}	F_{HP}	F_{IP}	F_{LP}	T_{CH}	T_{RH} / T_{RH1}	T_{RH2}	T_{CO}
Nonreheat Fig. 7(A)	-	-	-	-	.2-.5	-	-	-
Tandem-Compound Single-Reheat Fig. 7(B)	-	.3	.4	.3	.1-.4	4-11	-	.3-.5
Tandem-Compound Double-Reheat Fig. 7(C)	.22	.22	.30	.26	.1-.4	4-11	4-11	.3-.5
Cross-Compound Single-Reheat Fig. 7(D)	-	.3	.3	.4	.1-.4	4-11	-	.3-.5
Cross-Compound Single-Reheat Fig. 7(E)	-	.25	.25	.5	.1-.4	4-11	-	.3-.5
Cross-Compound Double-Reheat Fig. 7(F)	.22	.22	.28	.28	.1-.4	4-11	4-11	.3-.5

SPEED-GOVERNING SYSTEMS FOR HYDROTURBINES

Mechanical-Hydraulic Control

The mechanical-hydraulic speed-governing system for a hydroturbine consists of a speed governor, a pilot valve and servomotor, a distributor valve and gate servomotor, and governor-controlled gates which are functionally related as shown in Figure 9 (A). The speed-governing requirements for hydroturbines are strongly influenced by the effects of water inertia, and the dashpot feedback of Figure 9 (A) is required to achieve stable performance.

The block diagram of Figure 9 (B) is an approximate nonlinear model for this speed-governing system. Note that the gate servomotor may be rate limited for large rapid speed excursions. However, transient droop feedback reduces the likelihood rate limiting in stability analysis. Position limits exist corresponding to the extremes of gate opening. Typical parameters for this model are shown in Table IV.

Table III. Interpretation of Parameters Used in General Model for Turbines

SYSTEM DESCRIPTION	Time Constants				Fractions							
	T_4	T_5	T_6	T_7	K_1	K_2	K_3	K_4	K_5	K_6	K_7	K_8
Nonreheat (Figure 7A)	T_{CH}	-	-	-	1	0	0	0	0	0	0	0
Tandem-Compound, Single-Reheat (Figure 7B)	T_{CH}	T_{RH}	T_{CO}	-	F_{HP}	0	F_{IP}	0	F_{LP}	0	0	0
Tandem-Compound, Double-Reheat (Figure 7C)	T_{CH}	T_{RH1}	T_{RH2}	T_{CO}	F_{VHP}	0	F_{HP}	0	F_{IP}	0	F_{LP}	0
Cross-Compound, Single-Reheat (Figure 7D)	T_{CH}	T_{RH}	T_{CO}	-	F_{HP}	0	0	F_{IP}	$F_{LP/2}$	$F_{LP/2}$	0	0
Cross-Compound, Single-Reheat (Figure 7E)	T_{CH}	T_{RH}	T_{CO}	-	F_{HP}	0	F_{IP}	0	0	F_{LP}	0	0
Cross-Compound, Double-Reheat (Figure 7F)	T_{CH}	T_{RH1}	T_{RH2}	T_{CO}	F_{VHP}	0	0	F_{HP}	$F_{IP/2}$	$F_{IP/2}$	$F_{LP/2}$	$F_{LP/2}$
Hydro (Figure 10B)	0	$T_{W/2}$	-	-	-2	0	3	0	0	0	0	0

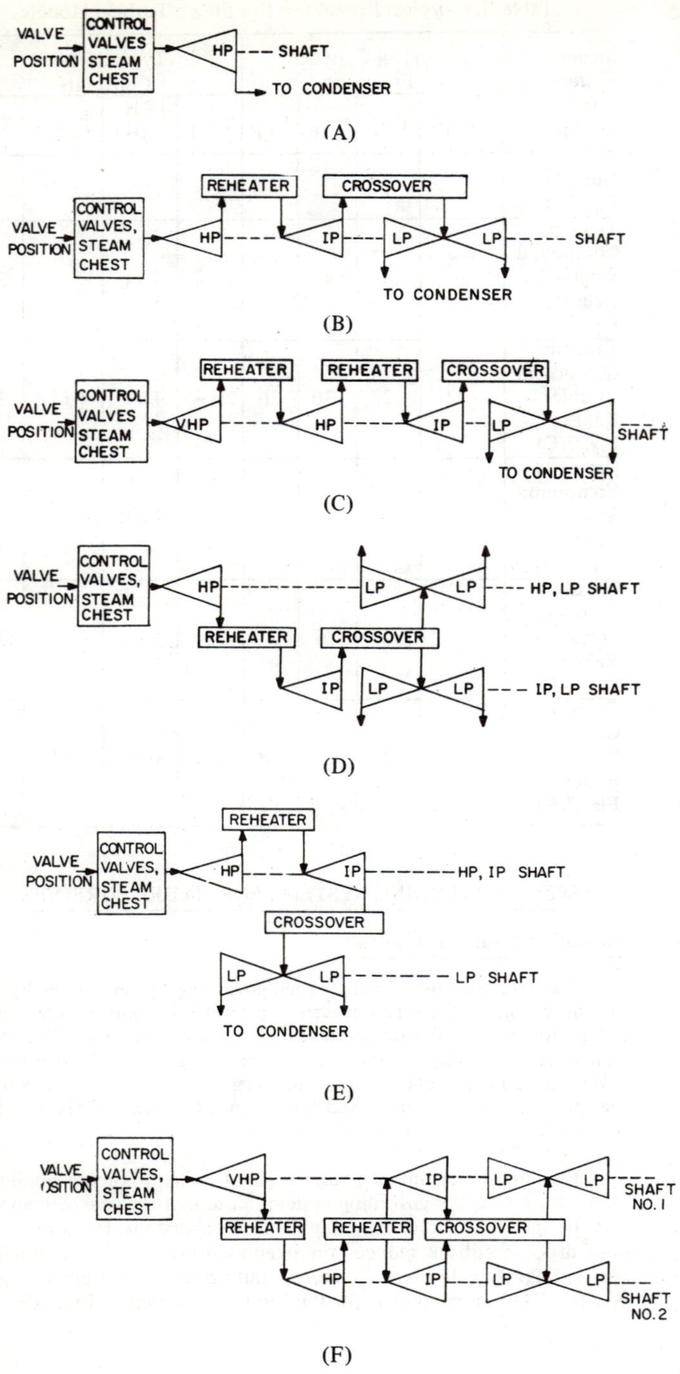

Fig. 6. Common Steam System Configurations

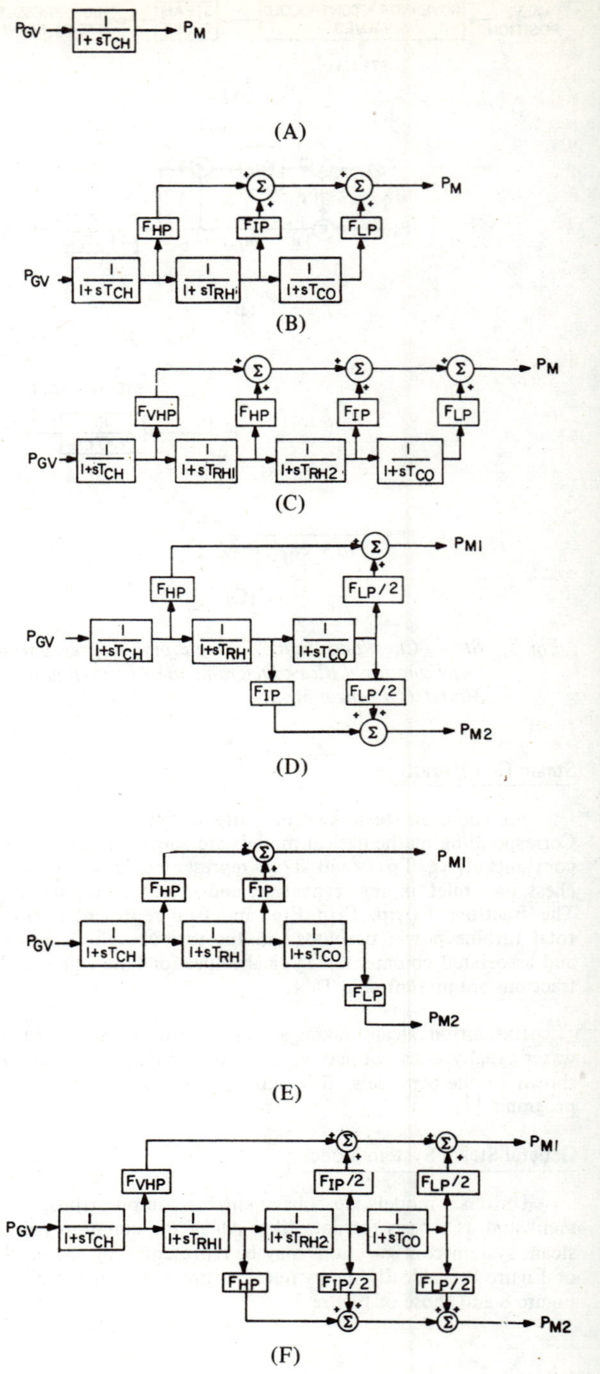

Fig. 7. Approximate Linear Models

A) Nonreheat
B) Tandem Compound, Single Reheat
C) Tandem Compound, Double Reheat
D) Cross Compound, Single Reheat
E) Cross Compound, Single Reheat
F) Cross Compound, Double Reheat

The simplified model of Figure 4 (B) is the one most often utilized for hydroturbine speed-governing systems in large system stability studies. This model may be derived from Figure 9 (B) by neglecting the time constant T_P and the gate servomotor limits. The assumption is also made that gate servomotor position and effective valve position are equal. Gate position limits are then imposed outside the feedback loops. Table IV also contains expressions for the parameters of Figure 4 (B) in terms of the parameters for 9 (B).

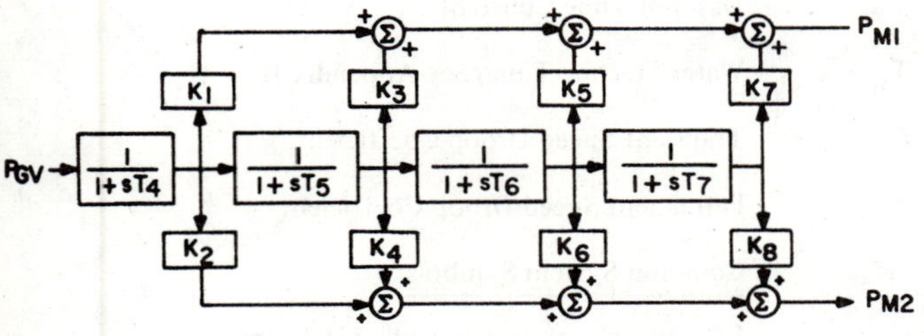

Fig. 8. General Model for Turbine Systems.

Table IV. Typical Parameters For Speed-Governing Systems For Hydroturbines

Parameter	Typical Value	Range
T_R	5.0	2.5 - 25.0
T_G	0.2	0.2 - 0.4
T_P	0.04	.03 - .05
δ	.3	0.2 - 1.0
σ	.05	0.03 - 0.06

Typically, $T_R = 5\ T_W$

$$\delta = 2.5\ T_W/2H$$

where H is the turbine-generator inertia constant on the machine MVA base.

Let $T_A = (\frac{1}{\sigma})\ T_R T_G$

$$T_B = (\frac{1}{\sigma})\left[(\sigma + \delta)\ T_R + T_G\right]$$

Then T_1 and T_3 of Figure 4 (B) are approximately

$$T_1, T_3 = \frac{T_B}{2} \pm \sqrt{\frac{T_B}{2}^2 - T_A}$$

Also in Figure 4 (B), $K = 1/\sigma$.

Electric-Hydraulic Control

Modern speed-governing systems for hydroturbines may involve electronic apparatus to perform the low-power functions associated with speed sensing and droop compensation. The electronic apparatus provides greater flexibility and improved performance in both dead band and dead time.[4] For interconnected system operation, however, the dynamic performance of the electric governor is necessarily adjusted to be essentially the same as that for the mechanical governor, so that a separate model is not needed.

Hydraulic Turbine

The transient characteristics of hydroturbines are determined by the dynamics of water flow in the penstock. The conversion of flow and head to power by the turbine involves only nondynamic relationships. The most precise models of water pressure and flow in the penstock are those which treat the travelling-wave phenomena. It is usually not necessary to use travelling-wave models for stability studies, although they are used regularly for detailed plant design studies. The block diagrams of Figure 10 contain the hydroturbine models most often used in system studies. These models are derived by various authors; for example, see Appendix II of reference 5.

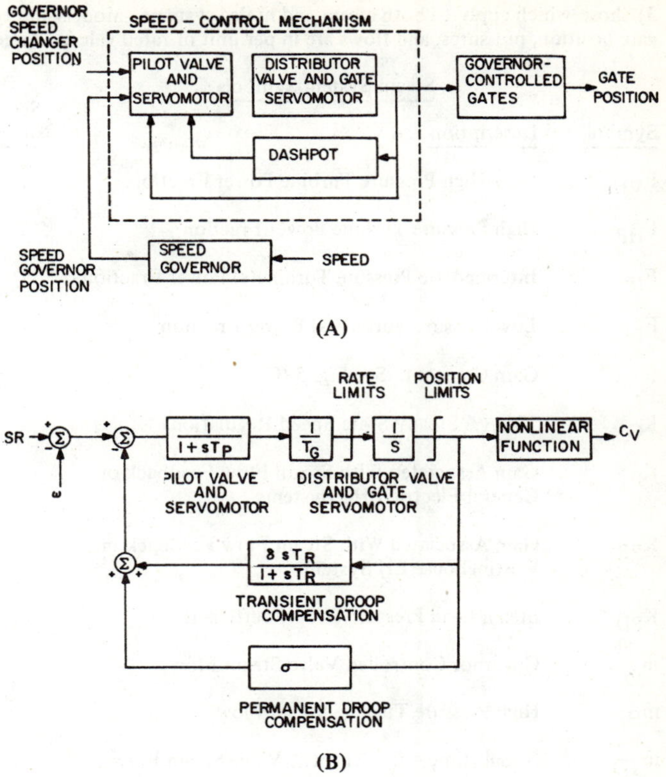

Fig. 9. Mechanical-Hydraulic Speed-Governing System for Hydroturbines. (A) Functional Block Diagram; (B) Approximate Nonlinear Model.

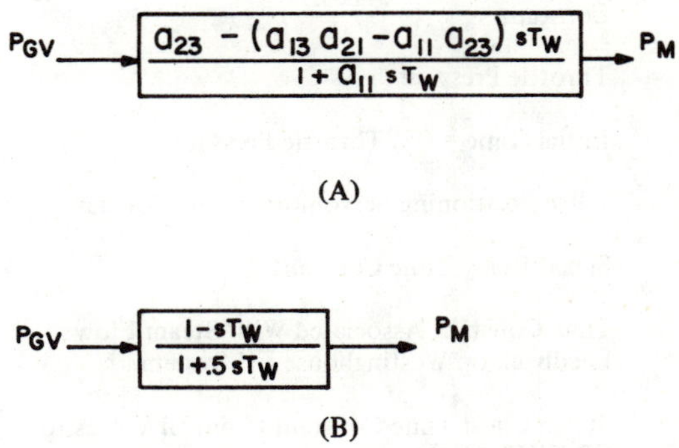

Fig. 10. Approximate Linear Models for Hydroturbines.

In Figure 10 (A), the time constant T_W is called the water starting time or water time constant. Appendix II presents a method for estimating this time constant from readily available parameters. The other model coefficients in Figure 10 (A) are associated with the turbine and are defined in Table V. For an ideal turbine, these coefficients are such that the model reduces to that shown in Figure 10 (B). The relationships of Figure 10 (B) can also be expressed in terms of the general model of Figure 8 through an appropriate choice of parameters. The last entry in Table III shows these parameters.

CONCLUSIONS

Basic models for speed-governing systems and turbines in power system stability studies are presented. These models provide adequate representation for hydro, fossil-fired, and pressurized water reactor nuclear units in most stability analyses.

Table V. Definitions for Model Symbols

This table contains three groups of symbols: 1) those which apply to steam systems only, 2) those which apply to hydro systems only, and 3) those which apply to both steam and hydro systems. Models and typical values are based on a per unit system in which power, speed, valve or gate position, pressures, and flows are in per unit of rated values. Time constants are in seconds.

Steam System Symbols

Symbol	Description	See Notes
F_{VHP}	Very High Pressure Turbine Power Fraction	1
F_{HP}	High Pressure Turbine Power Fraction	1
F_{IP}	Intermediate Pressure Turbine(s) Power Fraction	1
F_{LP}	Low Pressure Turbine(s) Power Fraction	1
F	Gain Constant. See Fig. 5 (C)	
K_G	100./(% Steady State Speed Regulation)	
K_P	Gain Associated With Steam Flow Feedback on General Electric EHC Systems	
K_{PR}	Gain Associated With Steam Flow Feedback on Westinghouse EH System	
K_{PD}	Steam Lead Pressure Drop Coefficient	
$\dot{m}_{cv}$	Governor-Controlled Valve Steam Flow	
$\dot{m}_{HP}$	High Pressure Turbine Steam Flow	
$\dot{m}_{TO}$	Initial (time = 0^-) Throttle Valve Steam Flow	
P_{M1}	Mechanical Power, Shaft 1	
P_{M2}	Mechanical Power, Shaft 2	
P_{SG}	Internal Boiler Steam Pressure, Assumed Constant	
P_T	Throttle Pressure	
P_{TO}	Initial (time = 0^-) Throttle Pressure	
T_{SM}	Valve Positioning Servomotor Time Constant	
T_{SR}	Speed Relay Time Constant	
T_I	Time Constant Associated With Steam Flow Feedback on Westinghouse EH System	
T_{CH}	Steam Chest Time Constant (Control Valves to HP (VHP)) Exhaust	4
$T_{RH}=$ T_{RH1}	Reheat Time Constant (HP VHP Exhaust to IP HP) Exhaust	4
T_{RH2}	Second Reheat Time Constant (HP Exhaust to IP Exhaust)	4
T_{CO}	Crossover Time Constant (IP Exhaust to LP Exhaust)	4

Hydraulic System Symbols

Symbol	Description	See Notes
a_{11}, a_{13}, a_{21}, a_{23}	Hydro Turbine Parameters	2
T_G	Gate Servomotor Time Constant	
T_P	Pilot Valve Time Constant	
T_R	Dashpot Time Constant	
T_W	Water Starting Time, See Appendix II	
δ	Transient Speed Droop Coefficient	
σ	Permanent Speed Droop Coefficient	

Common System Symbols

Symbol	Description	See Notes
C_V	Effective Governor-Controlled Valve or Gate Position	
Δ_{CV}	Effective Governor-Controlled Valve or Gate Incremental Position	
C_{VMAX}, C_{VMIN}	Valve or Gate Position Limits	
$\dot{C}_{VOPEN}$, $\dot{C}_{VCLOSE}$	Valve or Gate Servo Rate Limits	
$K_1 - K_7$	General Model Parameters See Figure 8 and Table III	
K	Total Effective Speed-Governing System Gain See Figure 5	3
P_M	Mechanical Power	
P_O	Initial (time = 0^-) Mechanical Power	
ΔP	Incremental Power Due to Valve or Gate Motion	
ΔP_{MAX}, ΔP_{MIN}	Incremental Power Limits Imposed by Gate or Valve Travel	
$\dot{P}_{UP}$, $\dot{P}_{DOWN}$	Limits on Rate of Change of Power Imposed by Control Valve Rate Limits	
P_{GV}	Power at Gate or Valve Outlet	
P_{MAX}, P_{MIN}	Power Limits Imposed by Valve or Gate Travel	
SR	Speed Reference	
T_1, T_2, T_3	See Figure 4 and Tables I and IV	
$T_4 - T_7$	General Model Time Constants See Figure 8 and Table III	
ω	Speed	
$\Delta\omega$	Speed Deviation	
s	Differential Operator	

Nomenclature Notes

1) Turbine power fractions are defined as the ratio of the steady-state mechanical power delivered by the particular turbine section (e.g., low pressure turbine) to the total steady-state mechanical power.

2) For small perturbations about a steady state condition the turbine may be represented by the following linearized equations:

$$q = a_{11}h + a_{12}n + a_{13}g$$

$$m = a_{21}h + a_{22}n + a_{23}g$$

where

q = p.u. deviation in flow

h = p.u. deviation in head

n = p.u. deviation in speed

g = p.u. deviation in gate position

m = p.u. deviation in torque

and

$$a_{11} = \partial q/\partial h, \quad a_{12} = \partial q/\partial n, \quad a_{13} = \partial q/\partial g$$

$$a_{12} = \partial m/\partial h, \quad a_{22} = \partial m/\partial n, \quad a_{23} = \partial m/\partial g$$

For the ideal turbine at rated speed, head and

$$a_{11} = 0.5 \quad a_{12} = 0 \quad a_{13} = 1$$

$$a_{21} = 1.5 \quad a_{22} = -1 \quad a_{23} = 1$$

3) The overall effective speed governing gain (see Figure 4) may be estimated with various degrees of complexity as follows:

a) For most stability studies, K is taken as 100./(%steady-state speed regulation).

b) Where there is need for more accurate representation and specific nonlinear characteristics are known, the steady-state incremental speed regulation may be used to compute K. When doing so, though, the user should check the range of applicability of the incremental regulation.

c) If the effects of steam lead pressure drops are not to be neglected then a gain correction of F (see Figure 5 (C) should be applied between P_{GV} and $\dot{m}_{HP}$. In this case $K = K_G F$.

d) Electrohydraulic control systems with steam flow feedback are treated as follows. For General Electric EHC Systems, use:

$$K = \frac{K_G K_P F}{1 + F(K_P - 1)}$$

For Westinghouse EH systems, use:

$$K = K_G F$$

4) See Appendix I for procedure of calculating steam time constants. The equivalent time constant should be calculated including all volumes in the section of steam path being represented.

5) No provision has been made in the speed-governing system models of Figures 2 and 3 to show separate speed and load references. In Figures 2 (B), 2 (C), 3 (B), and 3 (C), SR is necessarily a composite reference for both load and speed. In Figure 9 (B), it is a simple speed reference.

REFERENCES

1. IEEE Committee Report, "Computer Representation of Excitation Systems," IEEE Transactions on Power Apparatus and Systems, vol. PAS-87, No. 6 pp. 1460-1468, June 1968.

2. M. A. Eggenberger, "A Simplified Analysis of the No Load Stability of Mechanical-Hydraulic Speed Control System for Steam Turbines," ASME Paper 60-WA-34, December 1960.

3. M. Birnbaum and E. G. Noyes, "Electro Hydraulic Control for Improved Availability and Operation of Large Steam Turbines," ASME-IEEE National Power Conference, Albany, New York, September 19-23, 1965.

4. M. Leum, "The Development and Field Experience of a Transistor Electric Governor for Hydro Turbines," IEEE Transactions on Power Apparatus and Systems, vol. PAS-85, pp. 393-402, April 1966.

5. D. G. Ramey and J. W. Skooglund, "Detailed Hydro Governor Representation for System Stability Studies," Sixth PICA Conference Proceedings, pp. 490-501, May 1969, IEEE Transactions on Power Apparatus and Systems, vol. PAS-89, pp. 106-112, January 1970.

6. R. Oldenburger and J. Donelson, "Dynamic Responses of a Hydro Electric Plant," AIEE Transactions, part III, vol. 81, pp. 403-419, October 1962.

7. J. M. Undrill and J. L. Woodward, "Nonlinear Hydro Governing Model and Improved Calculation for Determining Temporary Droop," IEEE Transactions on Power Apparatus and Systems, vol. PAS-86, pp. 443-452, April 1967.

8. F. R. Schleif and A. B. Wilbor, "The Coordination of Hydraulic Turbine Governors for Power System Operation," IEEE Transactions on Power Apparatus and Systems, vol. PAS-85, pp. 750-756, July 1966.

9. J. L. Woodward, "Hydraulic-Turbine Transfer Function for Use in Governing Studies," Proc. IEE, vol. 115, No. 3, March 1968.

10. L. M. Hovey, "Optimum Adjustment of Governors on Manitoba Hydro System," AIEE TRANS, vol. PAS-81, pp. 581-587, 1962.

11. C. C. Young, "Equipment and System Modeling for Large-Scale Stability Studies," PICA, 1971.

12. E. I. Busby, J. W. Skooglund, and K. L. Williams, "System Response to Severe Contingencies Including Control Systems - Transmission Line Protection and Load Management," IEEE Conference Paper 69 CP65D-PWR, Summer Power Meeting, Dallas, June 22-27, 1969

13. IEEE Standard Definitions of Terms for Automatic Generation Control on Electric Power Systems, IEEE Transactions on Power Apparatus and Systems, vol. PAS-89, No. 6, July/August 1970.

Appendix I. Simplified Model of Steam Vessel

The continuity equation, with reference to Figure 11 is:

$$\frac{dW}{dt} = Q_{IN} - Q_{OUT} \tag{I-1}$$

where W is the weight of steam (lbs.) in the volume V (ft^3), t is time in seconds, and Q_{IN} and Q_{OUT} are flows (lbs/sec). Assuming the weight flow out of the vessel is proportional to pressure in the vessel,

$$Q_{OUT} = \frac{P}{P_O} Q_O \tag{I-2}$$

$$\frac{dQ_{OUT}}{dt} = \frac{Q_O}{P_O}\frac{dP}{dt} \tag{I-3}$$

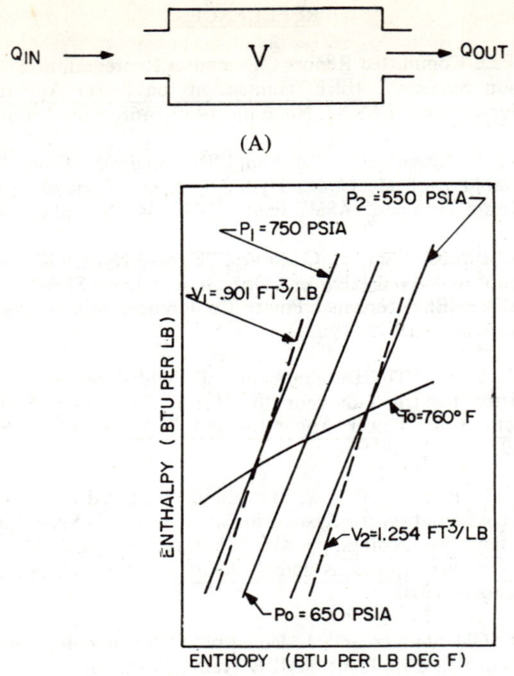

Fig. 11. Steam Vessel Time Constant.
(A) Steam Vessel
(B) Estimation of $\frac{\partial}{\partial P}\left(\frac{1}{v}\right)$

where P (psia) is a variable vessel pressure, P_O (psia) is the steady state vessel pressure, and Q_O(lbs/sec) is the steady state weight flow out of the vessel corresponding to pressure P_O. Assuming constant temperature in the vessel,

$$\frac{dW}{dt} = \frac{\partial W}{\partial P}\frac{dP}{dt} = V\frac{\partial}{\partial P}\left(\frac{1}{v}\right)\frac{dP}{dt} \tag{I-4}$$

where v is the specific volume (ft^3/lb) of steam in the vessel. Combining equations (I-1), (I-3), and (I-4),

$$Q_{IN} - Q_{OUT} = \frac{P_O}{Q_O} V \frac{\partial}{\partial P}\left(\frac{1}{v}\right)\frac{dQ_{OUT}}{dt} \tag{I-5}$$

Let

$$T = \frac{P_O}{Q_O} V \frac{\partial}{\partial P}\left(\frac{1}{v}\right) \tag{I-6}$$

Then

$$Q_{IN} - Q_{OUT} = T\frac{dQ_{OUT}}{dt} \tag{I-7}$$

or, in operational form,

$$\frac{Q_{OUT}}{Q_{IN}} = \frac{1}{1 + ST} \tag{I-8}$$

The following is an example calculation of the time constant T:

$Q_O = 900$ lbs/sec

$V = 200\ ft^3$

$P_O = 650$ psia

$T_O = 760^oF$

The derivative of the reciprocal of specific volume with respect to pressure at constant temperature is estimated from Figure 11 (B).

$v_1 = .901\ ft^3/lb$

$v_2 = 1.254\ ft^3/lb$

$p_1 = 750$ psia

$p_2 = 550$ psia

Then

$$T = \frac{(650)(200)}{(900)(750 - 550)}\left(\frac{1}{.901} - \frac{1}{1.254}\right)$$

T = 0.226 seconds.

Appendix II. Calculation of Water Time Constant

The water starting time, or water time constant, T_W, is associated with the acceleration time for water in the penstock between the turbine inlet and the forebay as shown in Figure 12, or between the turbine inlet and the surge tank if one exists. The basic equation for water time constant is:

$$T_W = \frac{LV}{H_T g} \tag{II-1}$$

where L is the length of the penstock in feet, V is water velocity in feet per second, H_T is the total head in feet, and g is the acceleration of gravity in feet per second squared. It is convenient to eliminate water velocity in terms of power generated by the plant.

$$P = \frac{VH_T Ae}{11.8} \tag{II-2}$$

where P is the generated electrical power in kilowatts, A is the average penstock area in square feet, and e is the product of turbine efficiency and generator efficiency. Thus,

$$T_W = \frac{11.8\ PL}{H_T^2\ Aeg} \tag{II-3}$$

or

$$T_W = \frac{0.366\ PL}{H_T^2\ Ae} \tag{II-4}$$

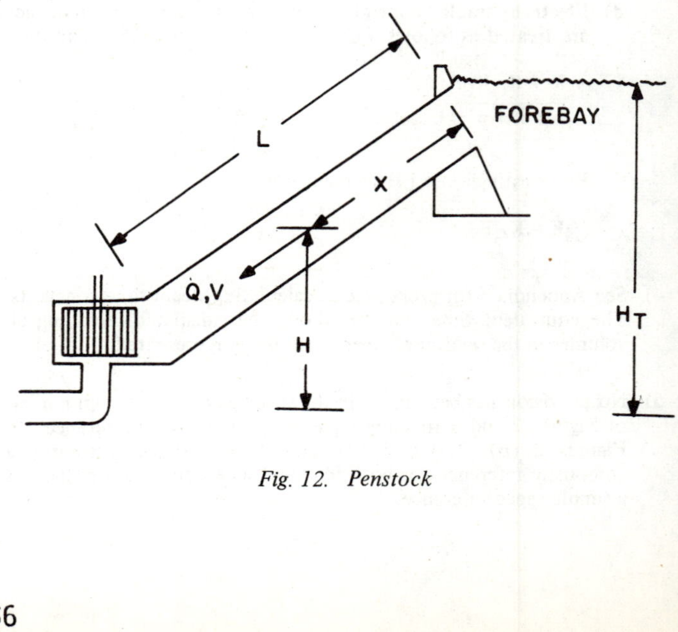

Fig. 12. Penstock

Discussion

Ronald P. Thompson (Pacific Gas and Electric Company, San Francisco, Calif. 94106): The Task Force has written a reference paper which will provide a valuable standard for modeling steam and hydraulic turbines and their speed-governing systems. Extensive work has been undertaken in recent years to model power plant response characteristics for stability studies. By acceptance of the models presented in this paper, the utility industry will have available, for the first time, an industry-wide accepted model which can add validity to test results and provide a means of comparison between proposed stability control systems utilizing the standard models.

Comments on the following points would be appreciated:

1. With the publication of this paper and the previously published paper on excitation system models[1], would the Task Force agree that IEEE standard power plant component models are now available to adequately represent plant responses for time periods not exceeding 10 seconds? Would the Task Force comment on the current status of the additional power plant component models (primarily boilers) necessary for representation of dynamic responses beyond 10 seconds?

2. Does the Task Force recommend use of the general speed governing and turbine systems presented in this paper for the longer term (greater than 10 seconds) dynamic stability time period? Would the models have to be simplified and approximations made for representation in the longer time period? If simplification is desirable, what plant parameters are significant enough to be retained in the turbine and speed governing models?

REFERENCE

1. IEEE Committee Report, "Computer Representation of Excitation Systems", IEEE PAS-87/No. 6, June 1968, pp. 1460-64.

Manuscript received February 8, 1973.

P. M. Anderson (Iowa State University, Ames, Iowa 50010): The IEEE Task Force on Overall Plant Response has performed a valuable service in publishing a paper which establishes a set of uniform models for steam and hydro turbine representation. Most of the turbine models now being used can be represented in the form presented in the paper and this creates a uniformity of representation which makes exchange of data easier. This also provides a standard model for manufacturers to use in specifying time constants for their equipment.

I am particularly fond of the general model representation presented in Figure 8 of the paper. This makes programming considerably easier than the specification of several different models for computer representation and should, in my view, be encouraged as a standard form for turbine representation.

One fault in the models presented by the task force is the omission of any mention of fast valving. Since fast valving is becoming an alternative which often requires study, it would seem appropriate to include in these models a provision for representing this alternative. This should be done in such a way that the control logic for the intercept valve can be specified by the user as this logic is not always the same.

It would also seem appropriate to include a provision for introduction of the boiler pressure at the governing valve. Since some studies require runs of over 10 seconds when boiler response may be a factor, inclusion of control valve representation will provide a means of adding this effect if one chooses to do so.

The inclusion of both the control valve and the intercept valve can be conveniently accomplished with multipliers as shown in Figures 1-3. Figures 1 and 2 are the same as Figure 6(B) of the paper but with control and intercept valve representation included. Figure 3 shows these valves added to the general model, Figure 8 of the paper. In the case of the control valve, the per unit pressure can be thought of as a signal which gain-adjusts the governor servomotor per unit stroke to give an output of per unit steam flow rate. The intercept valve inputs can be

Manuscript received February 16, 1973.

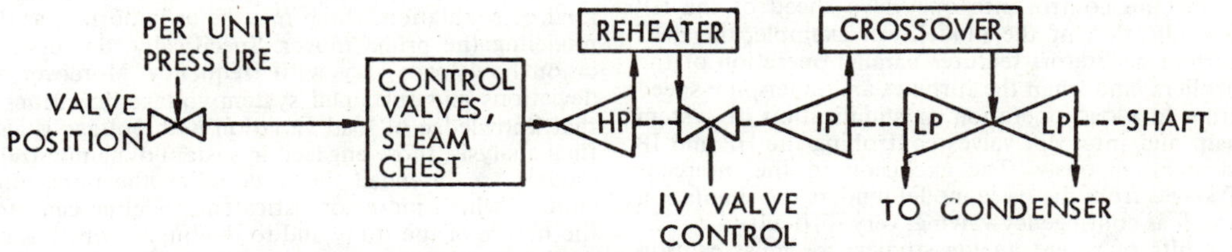

Fig. 1. Common Steam System Configurations for Tandem Compound, Single Reheat

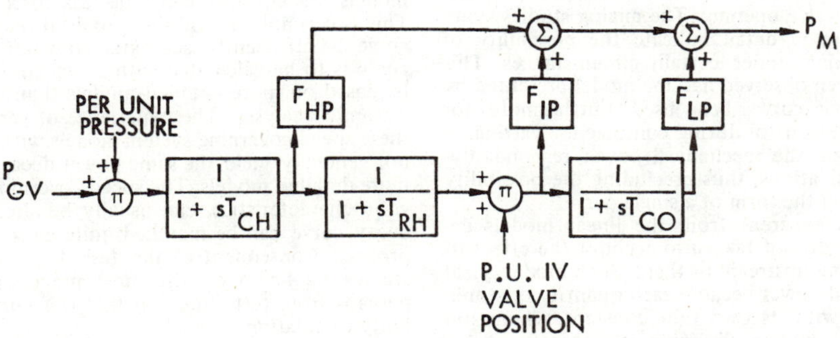

Fig. 2. Approximate Linear Models for Tandem Compound, Single Reheat

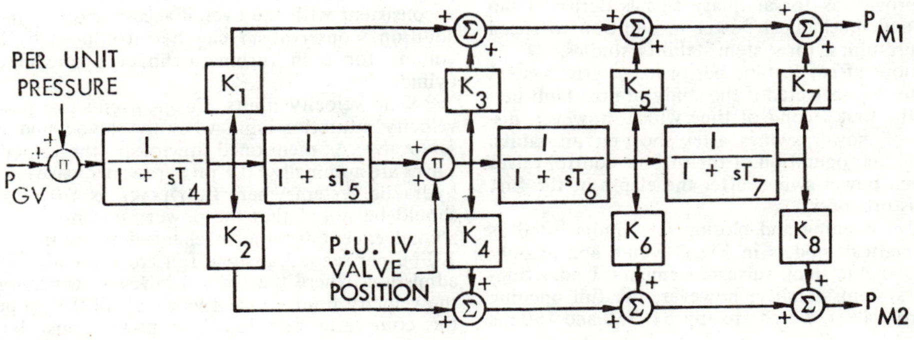

Fig. 3. General Model for Turbine Systems

thought of as the per unit steam available at the intercept valve (or intermediate stage input pressure) which is gain-adjusted by the per unit IV position to give an output signal of per unit intermediate pressure flow.

The manufacturers are beginning to suggest alternatives in intercept valve positioning, some of which involve complex logic to coordinate the control and intercept valves. These alternative control schemes are expensive and should be simulated, at least approximately, in order to make intelligent decisions regarding their purchase and later field adjustment. It is the intention of this discussion to suggest one way in which this might be accomplished. If these refinements are not required they may be either ignored or the separate input can be set to unity. If we can reach an agreement that such control and intercept valve representation is reasonable and desirable, then this makes future simulations easier for both the engineers making the study and the manufacturers who will be asked to supply data for the equipment. I would urge the Task Force to consider these alternatives in their models, which are otherwise a welcome addition to the technical literature.

J. Weissgerber (Allis-Chalmers Power Systems, Inc., West Allis, Wis. 53214): The authors are to be commended for attempting to establish a basic model of the governor system for use in digital stability studies. This effort should be appreciated by those appraising transmission system performance and the manufacturers of turbine-generators that have to respond to the data requests for such studies. To minimize the complexity of the model for digital studies, the authors had to resort to simplifications. The resultant limitations must be taken into account when applying the model.

Combining the turbine control into one valve ahead of the HP section restricts the application of the model. For example, the EHC system of A-CPSI turbine-generators features parallel operation of the speed and load controllers, and when the turbines are under pure speed control, namely during low-load operation or stabilization of system disturbances, the main and intercept valves controlling the HP and IP sections are stroked simultaneously. The exclusion of the intercept valve ahead of the IP stages from the basic model renders the simulation of such a control, as well as contingency valving, very difficult.

Based on the results of recent analog studies, we must caution against the authors' recommended use of power vs. time curves to simulate contingency valving. Such curves reflect idealized response of the control valves to being triggered by specific conditions into fast, total closure and time-delayed valve opening. The analog studies simulating the valve control system in detail indicate the possibility of multiple intercept valve stroking under certain circumstances. This phenomenon could not have been observed had the model presented by the authors and power vs. time curves been used. Furthermore, for turbines under continuous valve control during contingency valving, as employed in the A-CPSI design, the mechanical power responds dynamically to system power oscillations, thus precluding the possibility of defining the power response in the form of a single curve.

Another simplification is apparent from the linear models depicted in Figs. 7 and 8 which do not take into account the effect of turbine stages on the stages lying upstream of them. Such models yield too rapid changes of mechanical power because each quantity of stored energy changes in accordance with its own time constant, whereas in reality the steam flow and pressure in all stages are interdependent. Just considering one feedback loop around the HP section in the model can lead to deceptively favorable results. Another disadvantage of models that neglect this inter-stage coupling effect is that they make it difficult to incorporate intercept valves. The inclusion in the model of an intercept valve with provisions for auxiliary signals derived from comparison of electrical and mechanical power in addition to speed would greatly enhance its usefulness for system stability studies.

Steam pressure variations affect turbine performance after system disturbances. They can only be neglected if the study of post-fault performance is restricted to the first second of time which, however, disregards the later effects of power swings after short-circuit faults. Analog investigations reveal that operation of the turbine control valves may be required to dampen power swings after the elapse of the first second following serious disturbances.

The equally slow valve opening and closing rate limits listed as parameters for the mathematical models in Fig. 3 would appear only typical for load control involving small valve movements. Under transient conditions involving system stability, however, the full opening and closure times of control valves, which are approx. 750 and 150 ms respectively, would be operative. These translate into $\dot{C}_{VOPEN}$ = 1.3 pu/s and $\dot{C}_{VCLOSE}$ = 6.6 pu/s. An explanation of the assumptions made to arrive at the parameters that are indicative of slow valve action under conditions affecting system stability would be appreciated.

Manuscript received February 20, 1973

IEEE Task Force on Overall Plant Response, Dynamic System Performance Working Group, Power System Engineering Committee: The validity of these models is dependent not only on the time period being simulated but also on the nature of the system and the disturbance. For those studies in which the frequency deviations are small and the average frequency scarcely deviates from its nominal value, the principal effect of speed-governing is to alter the damping of power oscillations. Simulations of several cycles of dynamic oscillations using these models are justified even though the oscillation frequency may be low and the simulation period long, say up to thirty seconds or perhaps even longer. For very low frequency power oscillation, there may, however, be a need to simulate the automatic generation control system.

The situation is quite different when large sustained frequency deviations occur, such as when a subsystem becomes isolated from the interconnection with a resulting excess of load. In such cases, valves may open to their limits and remain there. Two questions pertinent to stability analysis arise when this occurs: 1) what immediate power is available at the governor valve position with open valves and 2) how does the available power decay with time. Detailed consideration of these questions is beyond the scope of this paper. The discussion related to Figure 5 suggests a means of calculating the immediate full open power at the governor valve but does not give typical data. Reference 11 has more detail on this subject and some comment on the deterioration of steam conditions with elapsed time. Within ten seconds, this phenomena need not be considered.

It is well to remember that there are other uncertainties in this kind of simulation which may be as important as those associated with modeling the prime mover. Specifically, one does not know with precision how loads vary with frequency. Moreover, sustained frequency deviations and sustained system voltage deviations usually co-exist, so that knowledge of load variation with voltage is also important. In the final analysis, those engaged in system dynamic studies—for which these models are intended—must consider the importance of increased accuracy, which more sophisticated modeling can provide, in relation to the nature of the study and to the impact which such sophistication has on the ultimate decisions being made.

The suggestion that fast-valving be simulated by representing governor and intercept valves as multipliers in the speed-governing models was considered by the task force as noted in the introduction. This is certainly a valid thing to do if the resultant turbine power decay curve is sufficiently accurate. However, if an accurate power decay curve is to be calculated with a high confidence level, it must in general be based on more detailed models than those justified for large system dynamic analyses. The experience of some task force members is that these speed-governing system models with valves added as multipliers do not generally yield the same power decay curves as those obtained with more detailed models. The speed-governing model parameters, including valve characteristics, can usually be altered so that an accurate power decay curve can be matched quite closely, but this is a trial and error process. Consequently, the task force recommends that fast-valving studies be based on the most precise power decay curve available, perhaps from test data, and that this curve be used directly in the stability calculation.

The effect of turbine stages on the stages upstream was well recognized by the task force. From the simulation point of view, the linearized models are much simpler to implement than non-linear ones which take interstage coupling into account. Considering the intended purposes of such models, the resultant response obtained in this manner is consistent with the overall accuracy of the models presented here. In addition, conservatism can be introduced by lumping the extraction volume for each turbine cylinder upstream of the blades for that cylinder.

The velocity limits, $\dot{C}_{V\ open}$ and $C_{V\ close}$, in Figure 3 represent velocity saturation limits when the servocylinder is under control of the servovalve. As mentioned previously, the velocity limits of the control valves are nominally ±0.1 pu per sec per valve except for the mechanical-hydraulic system where $\dot{C}_{V\ CLOSE}$ is 1.0 pu per second per valve. It should be noted that the power rate limits given in Table 1, ±0.1 per second except for the mechanical-hydraulic system where P_{DOWN} is 1.0 pu per second, assume full arc admission. In the case of partial arc admission, where the control valves are generally operated sequentially, the velocity limit on each valve is still 0.1 pu per second per valve, but the composite rate limit on power must be adjusted accordingly.

Manuscript received May 18, 1973

Further, if the error signal is large enough, more than one valve may be moving at a time. It is suggested that the user consult with the manufacturer on appropriate power rate limits to use, particularly under conditions of partial arc admission. During transient conditions involving other closing and reopening means, the $\dot{C}_V$ close could be as high as 10 pu/s and $\dot{C}_V$ open could be somewhat higher than .1 pu/s. However, this would be taken into account in computing the resultant power decay curve as recommended by the task force.

The task force has a backlog of work related to the representation in stability analyses of industrial gas turbines, aviation gas turbines, combined cycle plants, and BWR plants. It appears now that the continuous control of intercept valves should be added to that list.

Dynamic Modeling of Loads in Stability Studies

MAURICE H. KENT, MEMBER, IEEE, WAYNE R. SCHMUS, MEMBER, IEEE, FRANCIS A. McCRACKIN, MEMBER, IEEE, AND LUTHER M. WHEELER, SENIOR MEMBER, IEEE

***Abstract*—This paper emphasizes the need for a correct representation of electrical loads in stability studies. Several series of load-voltage tests performed on the Southern California Edison Company system are described. The test results are summarized and used as the basis for individual bus load representation.**

Paper 68 TP 705–PWR, recommended and approved by the Power Generation Committee of the IEEE Power Group for presentation at the ASME/IEEE Joint Power Generation Conference, San Francisco, Calif., September 15–19, 1968. Manuscript submitted June 3, 1968; made available for printing June 14, 1968.
The authors are with Southern California Edison Company, Los Angeles, Calif.

Introduction

THE NEED for transient stability studies on electric power systems has always been present. Historically, stability questions have been addressed principally to remote generating plants or synchronous motors connected radially to a relatively large network. A hydroelectric station feeding power into a metropolitan area is typical of this arrangement. In such a system, stability is known to be relatively insensitive to transient response of the load in the metropolitan area. The near-infinite bus characteristic of the receiving system results in a stabilized load which shows little direct effect on voltages at the sending

Reprinted from *IEEE Trans. Power App. Syst.*, May 1969.

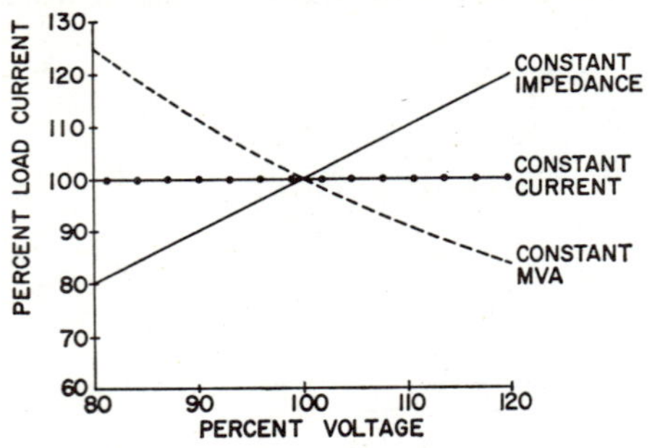

Fig. 1. Relationship between load current and bus voltage for simple load types.

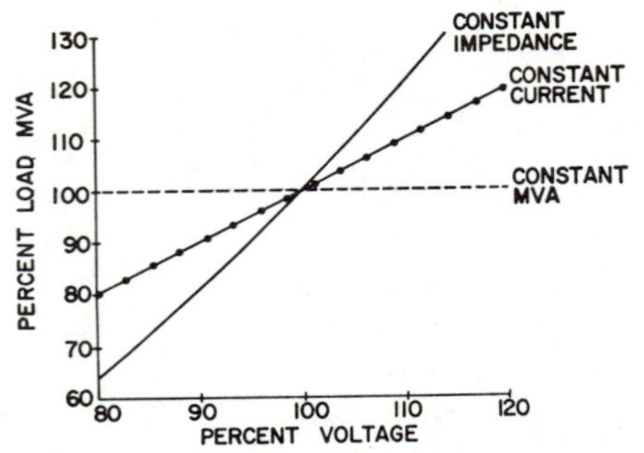

Fig. 2. Relationship between load MVA and bus voltage for simple load types.

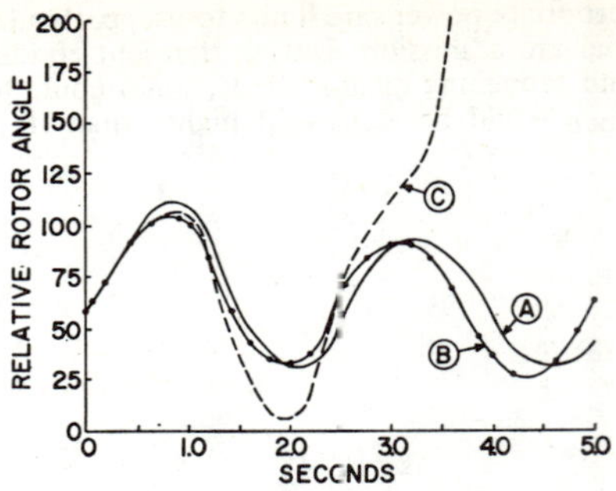

Fig. 3. Swing curve of remote generator relative to inertia center following fault and loss of generation. *A*—all constant impedance load; *B*—all constant current load; *C*—real power: 50-percent constant current load, 50-percent constant MVA load. Reactive power all constant current load.

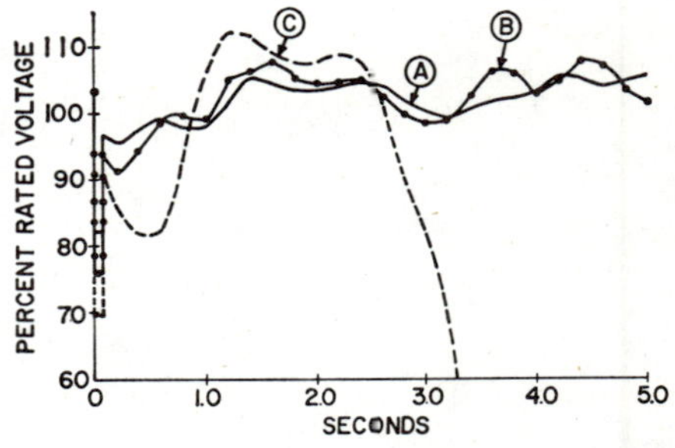

Fig. 4. Voltage at a principal transmission bus following fault and loss of generation. *A*—all constant impedance load; *B*—all constant current load; *C*—real power: 50-percent constant current load, 50-percent constant MVA load. Reactive power all constant current load.

or receiving ends of the transmission lines. Constant impedance representation has been used for the real and reactive components of loads in stability studies, partially as a carryover from ac network analyzer practice and to avoid the complication of nonlinear elements in network equations.

As a result of the rapid growth in system interconnections in the last decade and the tendency of load to spread away from metropolitan centers into urban areas, more stability studies are now concerned with transfer of bulk power through load-center networks. In this framework, load currents will directly influence and be influenced by the loading and voltages on transmission components. One such study contrasting constant MVA loads at all locations against constant current loads found that the stability limit for interarea power transfer could vary up to 50 percent.

To illustrate the sensitivity of a selected transmission system to load representation, a hypothetical system consisting of a single 500-kV line bridging power over several load centers connected by lower voltage lines was subjected to stability tests under three conditions of load representation. All loads were represented as 1) constant impedance; 2) constant current; or 3) 50-percent constant MVA, 50-percent constant current for real power, and 100-percent constant current for reactive power components. Plots of relative machine angle and voltage on an intermediate transmission bus, following a fault of five cycles on the 500-kV line, are shown in Figs. 1 and 2.

As can be seen from the plots, a change of 50 percent of the real power load from constant current to constant MVA representation was sufficient to cause instability.

Most modern transient stability programs have provision for the individual representation of bus loads as functions of bus voltage, bus frequency, or both. In order to make use of this provision, the response of each load bus or equivalent to changes in voltage and/or frequency must be predicted.

Load Behavior

The response of nearly all loads to voltage changes can be represented by some combination of constant impedance, constant current, and constant MVA devices. Actually, the constant current model is unnecessary as it is nearly equivalent to 50-percent constant impedance load combined with 50-percent constant MVA load. It has been found convenient to retain the constant current model as it is easily comprehended and is frequently used in the absence of more complete data. Figs. 3 and 4 show the relationships of load current and MVA with voltage for the three simple load types.

The constant MVA-type load representation is the most severe representation from the system stability point of view because of its effect in amplifying voltage oscillations: a drop in voltage will cause an increase in load current resulting in a further voltage drop. Conversely, constant impedance loads have a decided damping effect on voltage oscillations.

Individual loads may be viewed as being divided into two classes, static and rotating. The static class consisting of heating and lighting equipment will generally exhibit a constant near-unity power factor. The different characteristics of static loads tend to compensate each other, resulting in a composite effect of constant impedance load.

TABLE I
LOAD-VOLTAGE TEST INSTRUMENTS

Test Instruments	Description
Potential transformers	Existing in substations. 72 000/120-volt potential transformers.
Current transformers	Existing in substations. 66-kV circuit breaker bushing type metering CT's with ratios of 1200/5 or 2000/5.
Auxiliary current transformers	Knopp precision multirange current transformer, type P-800, range 5/5, 10/5, 20/5, 50/5, 100/5, Electrical Facilities, Inc.
Metering unit	
current auto balance transformer	General Electric, type JY-4. Available ratios are 5/5, 5/5.85, 5/7, 5/8.73, and 5/10.
phase shifting transformer	States Company, std. no. 402 three-phase three-wire, 120 volt.
ammeter	General Electric, type AB-18, 5-A ac, 0–6-A scale, 1 percent accuracy.
voltmeter	General Electric, type AB-18, 150-volt ac, 0–150-volt scale, 1 percent accuracy.
varmeter	General Electric, type AB-18, 5-A 120-volt ac, 250–0–250-var scale, 1 percent accuracy.
wattmeter	General Electric, type AB-18, 5-A, 120-volt ac 0–750-watt scale, 1 percent accuracy.
Transducer	Beckman four-channel Hall-effect transducer. 50-mV maximum dc output, zero supression. Response time: 40 ms, model 1790-68-0.
Dc amplifiers	Brush model 13-5335-00.
Oscillograph	Brush light beam oscillograph, model 16-2316-01. Frequency response: dc-2500 Hz. Chart speeds: 0.2, 0.4, 1, 2, 5, 10, 25, and 50 in/s.
Galvanometers	Brush galvanometer model 11-2111, type "O". Sensitivity 6.2 mA/in.

The rotating class of loads consists of synchronous motors and induction motors driving equipment with a variety of torque-speed characteristics. Motors will normally be constant MVA devices for moderate voltage changes, although the power factor may vary widely.

In a qualitative manner, the relative behavior of bulk power station loads may be predicted from knowledge of the type of connected load. Agricultural and industrial uses are generally motor loads and should be expected to have more constant MVA load than commercial and residential areas. Similarly, summer heavy loads resulting from air conditioning equipment will have more constant MVA load than winter heating and lighting demands.

Comparatively little is known about the behavior of loads for changing frequency. A commonly used guideline is the assumption of a linear one percent change in real power load for every percent change in frequency, although some sources have suggested a response several times this amount. Pertinent load testing for load-frequency response is difficult to accomplish because of several factors. Many utilities are unwilling to risk customer inconvenience or equipment damage associated with underfrequency; load variations due to frequency are difficult

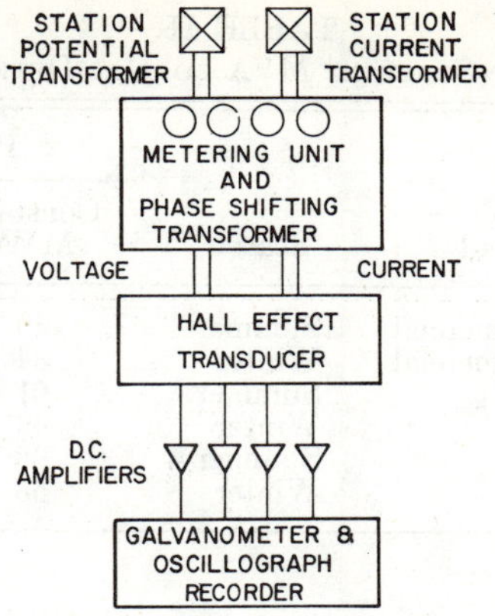

Fig. 5. Block diagram of load-voltage test instrumentation.

to separate from those caused by accompanying voltage changes; and the effects of load inertia must be considered in analyzing test results. For the traditional type of stability study, load variations due to frequency changes have been neglected. Studies are now being applied to load shedding applications and the effects of isolating areas following major disturbances. A better understanding of the behavior of loads with abnormal frequencies will be required to obtain meaningful results from such studies.

Load Tests

Recognizing the need for a factual basis for load representation, the Southern California Edison Company organized a field testing program in 1965 and 1966. The selected tests consisted of gradual and rapid voltage changes on 66-kV buses in three 220/66-kV stations, each having more than one half the connected substation load in a single category: residential–commercial, industrial, or agricultural. Rapid voltage changes were induced by dropping one of two or three parallel 220/66-kV transformers while gradual changes we e ontrolled by varying the excitation of a synchronous condenser on the 66-kV bus. To limit the risk of a customer interruption or equipment damage, all voltage changes were limited to 5 percent.

Tests were conducted at selected hours and days of the week during the general periods of summer peak, winter peak, and spring minimum to enable evaluation of load response for the usual conditions assumed in stability studies.

To measure the load and voltage changes, a portable test instrumentation board was constructed and moved to each test site. Descriptions of the instruments are included in Table I. A block diagram of the setup is shown in Fig. 5. Voltage and current signals from station metering transformers were fed into a metering unit to establish initial conditions and to verify test connections. A phase-shifting transformer was used to separate watt and var readings. From the metering unit, current and voltage were fed into a Hall-effect four-channel transducer equipped with suppressed zero capability. Dc signals from the transducer were amplified to drive a light beam oscillograph recorder.

Test Results

Oscillographs from the test recorder were replotted on scaled paper for analysis. Rapid voltage changes caused a transient of three- to six-cycles duration in the load response. No direct correlation was found between the magnitude and shape of the transient and other load characteristics, although each load

TABLE II
PERCENTAGE OF CONSTANT MVA LOAD AS DETERMINED BY TEST

Principal Type of Load	Season	Percentages Constant MVA	Constant Impedance
Residential–commercial	Summer	49	51
Residential–commercial	Winter	34	66
Industrial	Summer	61	39
Industrial	Winter	49	51
Agricultural	Summer	68	32
Agricultural	Winter	58	42

station tended to retain a distinct characteristic wave shape throughout all tests. Following the transient, loads generally stabilized to values predictable from the gradual voltage change tests. Since they follow abrupt voltage changes and have very short duration, load transients have not been considered in the final load model.

From each test, a factor $\Delta I/\Delta V$ was calculated by dividing the observed percentage change in load current by the final percentage change in bus voltage. The load-voltage relationship was assumed to be of the form

$$\mathrm{MVA}_1 = \mathrm{MVA}_0(A + (1 - A)V^2)$$

where

MVA_1 = load MVA at voltage V
MVA_0 = load MVA at initial voltage
V = ratio of voltage to initial voltage
A = fraction of load at initial voltage represented by constant MVA.

This equation is transformed in the Appendix to the following form which is used to calculate A, the fraction of MVA load:

$$A = (1 - \Delta I/\Delta V)/2.$$

Values of A obtained in this manner from the tests are presented in Table II.

The load-voltage tests have indicated the re ative proportions of constant MVA and constant impedance load for three representative load stations during various loading conditions. Although each station load consisted mainly of a particular load classification, the information does not lead immediately to a quantitative procedure for representing dissimilar loads.

If it were possible at the time of the test to learn the separate kilowatt demands of each load classification, the correct representation for 100-percent industrial load, 100-percent agricultural load, or 100-percent residential–commercial load could be calculated. These "pure" loads could then be recombined in particu ar proportions to represent any known system load.

As demand records are not generally available by classification, kWh sales records maintained by accounting personnel were substituted. In using kWh figures, the assumption is made that the daily load curves for all classifications of load are similar in shape, and the resulting load representations are precise only insofar as this is true. Although this assumption may be considered unrealistic in many cases, only major differences in daily load curve will be reflected as significant error in the final result, rendering the procedure useful for the normal composite load.

Sales records for the months of August and September 1965 are summarized below. Sales for street lighting are treated as commercial load.

	Residential–Commercial Area	Industrial Area	Agricultural Area
Percent sales to residential–commercial customers	68.4	46.7	29.3
Percent sales to industrial customers	30.5	53.3	11.7
Percent sales to agricultural customers	1.1	0	59.0
	100	100	100

From Table II, the following percentages of constant MVA load during summer peak are as follows:

Station Tested	Constant MVA (percent)
Residential–commercial	49
Industrial	61
Agricultural	68

Defining unknowns C, I, and A, equations expressing the fraction of constant MVA load in the composite loads tested can be established. Similar equations could be written expressing the fraction of constant impedance load.

C = fraction of constant MVA load in 100-percent commercial-residential use
I = fraction of constant MVA load in 100-percent industrial use
A = fraction of constant MVA load in 100-percent agricultural use

$$\begin{aligned} 0.684C + 0.305I + 0.011A &= 0.49 \\ 0.467C + 0.533I &= 0.61 \\ 0.293C + 0.117I + 0.59A &= 0.68. \end{aligned}$$

Solving the simultaneous equations yields the following percentages which may be used to represent any station for which kWh sales records are available.

Summer Loads	Constant MVA (percent)	Constant Impedance (percent)
100-percent residential–commercial load	32	68
100-percent industrial load	87	13
100-percent agricultural load	82	18

Sample Application

An estimate of the system average summer load response may be computed by substitution of total system kWh sales in the equations above. Sales figures for August and September of 1965 are as follows:

Classification	kWh (percent)
Commercial–residential	56.1
Industrial	37.7
Agricultural	6.2
Total	100.0

Substituting in the equation, the fraction of total summer load representable by constant MVA is derived.

$$(0.32)(0.561) + (0.87)(0.377) + (0.82)(0.062) = 0.56.$$

It is concluded that the average summer peak load is composed of 56-percent constant MVA and 44-percent constant impedance load. This method can be used to determine individual representations for each substation load by compiling sales records for the corresponding service area.

Areas for Additional Study

Although the measurements were taken to determine reactive power response to voltage changes, representation of vars has not been included in the analysis. The tests indicated that the var response was not correlated to factors measured in the tests. Power factors on the substations tested were usually corrected to near unity by widespread use of shunt capacitors, and the reaction of vars to voltage changes was apparently dependent upon the balance of actual var demand to correction capacitors as well as the actual load var response. As the load power factors were near unity, less importance has been placed upon the var representation and, generally, the same representation has been assumed for both real and reactive components of load.

In some systems, var representation could be as important as the MW representation to system stability and would require substantially more testing and analysis.

The increasing use of interconnections and underfrequency load shedding throughout the industry will require more attention to the response of load to changing frequency. Representative tests will be needed to guide in the formulation of a complete dynamic load model and ultimately to provide data for regional and nationwide stability studies.

Appendix

Transformation of load-voltage relationship is

$$\mathrm{MVA}_1 = \mathrm{MVA}_0(A + (1 - A)V^2) \tag{1}$$

where

MVA_1 = load MVA at voltage V
MVA_0 = load MVA at initial voltage
V = ratio of voltage to initial voltage
I = ratio of load current to initial load current
A = fraction of load at initial voltage represented by constant MVA

$\mathrm{MVA} = \mathrm{MVA}_1/\mathrm{MVA}_0 = IV$ by definition.

Dividing (1) by MVA_0 and substituting,

$$IV = A + (1 - A)V^2. \tag{2}$$

Dividing (2) by V and differentiating,

$$\frac{dI}{dV} = \frac{-A}{V^2} + 1 - A. \tag{3}$$

Assuming small voltage changes about normal values,

$$\frac{\Delta I}{\Delta V} = \frac{dI}{dV}$$

and

$$V^2 \cong 1.$$

Solving (3) for A,

$$A = \tfrac{1}{2}(1 - \Delta I/\Delta V). \tag{4}$$

References

[1] J. Baba, S. Hayashi, I. Yamada, H. Haneda, and F. Ishiguro, "Sensitivity analysis of power system stability," *1967 PICA Conf. Rec.* (Pittsburgh, Pa., May 15–17, 1967), pp. 48–68.

[2] H. A. Bauman, G. R. Hahn, and C. N. Metcalf, "The effect of frequency reduction on plant capacity and on system operation," *AIEE Trans. (Power Apparatus and Systems)*, vol. 73, pp. 1632–1637, 1954 (February 1955 sec.).

[3] R. T. Byerly and D. G. Ramey, "Dynamic simulation of interconnected systems," *1967 PICA Conf. Rec.* (Pittsburgh, Pa., May 15–17, 1967), pp. 31–40.

[4] S. B. Crary, *Power System Stability*, vol. 2. New York: Wiley, 1947.

[5] IEEE Joint Working Group on Load Characteristics, "The effect of frequency and voltage on power system load," presented at the IEEE Winter Power Meeting, New York, N. Y., January 30–February 4, 1966.

[6] A. H. El-Abiad and G. W. Stagg, *Computer Methods in Power System Analysis*. New York: McGraw-Hill, 1968.

[7] *Electrical Transmission and Distribution Reference Book*. East Pittsburgh, Pa.: Westinghouse Electric Corporation, 1950.

[8] R. M. Webler and C. C. Young, "A new digital computer program for predicting dynamic performance of electric power systems," *1967 PICA Conf. Rec.* (Pittsburgh, Pa., May 15–17, 1967), pp. 21–29.

Discussion

E. W. Kimbark (Bonneville Power Administration, Portland, Ore. 97208): Although load characteristics can have a considerable effect on power system stability, the information on load characteristics by field tests has been very limited until recently, and the amount of such information that has been published is still less [9]. Therefore the authors of this paper have done a notable service to the electric power industry by making the tests and by analyzing and publishing the results. A feature of the paper which enhances its usefulness is the determination of the composition of typical pure loads of three classes—agricultural, commercial–residential, and industrial—in summer and winter seasons.

The paper may give the impression that constant MVA loads are always detrimental to stability. Really, whether such loads are detrimental or beneficial depends on their location. Consider a power system consisting of a sending area and a receiving area, each having both generation and loads, the two areas being interconnected by tie lines. Any disturbance that threatens stability of the tie lines causes initial acceleration of the generators of the sending area and retardation of the generators of the receiving area. Consequently, the angular separation of the two groups of generators increases, and the voltages everywhere are depressed. Constant MVA loads in the sending area are more beneficial to stability than are constant impedance loads, for they maintain the loading on the sending-end generators, whereas constant impedance loads decrease with voltage and allow the generators to accelerate more. Constant MVA loads in the receiving area increase the retardation of the generators in that area and are detrimental to stability. Such loads along the intertie, where voltages dip more, are even more detrimental. Thus we can say that in the majority of cases constant MVA loads lower the stability limit.

The paper states that industrial loads are generally motor loads. This is true, but there are notable exceptions, such as 1) electric furnaces and 2) aluminum reduction plants and other electrolytic processes.

The effect of air conditioning on summer motor loads was mentioned. To this, pumping for irrigation is added in many parts of the western United States.

The three load-voltage characteristics provided in most digital stability programs can all be represented by one equation, as follows:

$$\frac{P}{P_n} = \left(\frac{V}{V_n}\right)^k$$

Manuscript received September 27, 1968.

wherein V_n is the initial value of voltage V and P_n is the initial value of the active or reactive power P; k is an exponent depending on the type of load.

For constant power (KP) load, $k = 0$. For constant current (KI) load, $k = 1$. For constant impedance (KZ) load, $k = 2$.

Exponent k describes the voltage sensitivity of the load. In a graph of P/P_n against V/V_n (Fig. 4), k is the slope of the curve at the normal operating condition (1, 1). That figure shows that two of the curves have constant slope, while the third has almost constant slope for voltage variation within ± 0.2 per unit.

Composite loads that are assumed to be mixtures of two or three of these standard types can be represented, within this limited voltage range, as straight lines with slopes between 0 and 2 or by the equation given above with k having some value between 0 and 2.

I propose that future stability programs provide for only one type of load specified by initial values of active and reactive power and by an exponent for the voltage sensitivity of each. This would simplify the input format of loads.

An aluminum reduction plant, for example, could be represented by

$$P + jQ = P_n(V/V_n)^{2.5} + jQ_n(V/V_n)^{2.7}.$$

If we were limited to the three standard types, the best approximation for the aluminum load would be to make both exponents equal to 2.0. Actually, the aluminum reduction plant, because of the counter EMF of the pot lines, has a somewhat greater voltage sensitivity than has a constant impedance load.

In the kind of disturbance with which most stability studies are concerned, loads are subjected to two kinds of voltage change: 1) abrupt changes caused by switching, chiefly by the application and removal of a short circuit; and 2) slow changes caused by angular swings. For slow changes of voltage, loads can be considered to be in a steady state, varying with the voltage according to laws already discussed. For fast changes in voltage there are transients, as mentioned in the paper, and these would be difficult to represent accurately. A practical question arises, however, in the treatment of constant power loads, especially of those near the fault, while the fault is on, for it may be impossible to solve the network under the assumed conditions. The simplest and most practical solution of this difficulty appears to be to change all or some loads to constant impedance type until the fault is cleared. This procedure appears reasonable if one remembers that an induction motor running at constant slip presents a constant impedance. The only question is, how much will the slip change in the 0.05 to 0.10 second while the fault is on?

The load characteristics usually assumed, and especially the constant power characteristics, can be valid only in some limited range of voltage, perhaps from 0.8 to 1.2 per unit of normal value. Studies have shown that voltages at some buses can dip to lower values, say 0.5 per unit, even in stable cases. Rather than assume that motor loads have a constant power characteristic down to so low a voltage, it would be more reasonable to assume that most such loads are shed at HV bus voltages around 0.75 to 0.80 per unit after a short time delay, and that many of these loads are not reconnected after the voltage rises again.

It may be argued that power systems should be designed so that load is not shed in this fashion except in unusual and extreme disturbances. With this contention I agree, but nevertheless maintain that engineers will be asked to reproduce by computation some of these extreme disturbances that do actually occur; therefore the computer program should be able to shed motor loads on low voltage. This is an argument for retaining the identity of constant power loads in the computer simulation.

For obvious reasons, the permissible change of voltage in staged tests of load characteristics is limited to much smaller ranges than occur during serious accidental disturbances. Therefore the only way that we can determine load characteristics over wide ranges of voltage and frequency is by providing instrumentation that will be connected and ready to record whenever a serious disturbance might occur.

References

[9] Five papers presented at meeting of Canadian Electrical Assoc., Toronto, Ont., Canada, April 17, 1968.

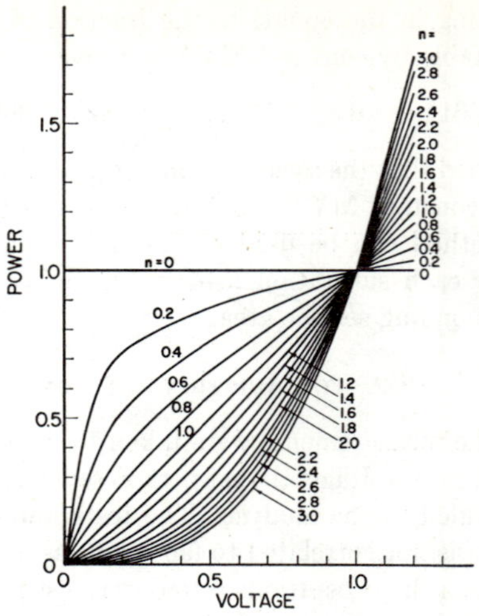

Fig. 6. Characteristic curves of $P = V^n$.

H. A. Peterson (University of Wisconsin, Madison, Wis. 53706): The need for reliable field measurements of load variations with variations of voltage has been felt for some time. It is especially significant, therefore, to find a paper which 1) presents results of such field measurements in one system and 2) derives from the basic measurements of $\Delta I/\Delta V$ information helpful in simulating loads during transient stability studies.

The experimental results are interpreted in terms of the three often used ways of load simulation: constant MVA, constant current, and constant impedance. Any one of these three is a particular case of the more general per unit relationship $P = V^n$. For example, if $n = 0$, the load is one of constant MVA; if $n = 1$, the load is one of constant current; and if $n = 2$, the load is one of constant impedance. Other noninteger values of n, including values greater than 2, could be used if experimental results so dictate. Furthermore, during large swings such as might precede system breakup leading to load shedding, voltages at certain load points might well be at or near zero at times. The expression $P = V^n$ always goes to zero when $V = 0$. The constant impedance and the constant current representation also go to zero when $V = 0$, while the constant MVA representation does not. Thus for large variations in voltage there might be some virtue in using the $P = V^n$ simulation, especially if it can be shown that at the same time dP/dV at $V = 1$ can be simulated as interpreted from the experimental results.

The general expression for $P = V^n$ is plotted in Fig. 6 with values of n as parameter over a range from 0 to 3. Curves for $n = 0$, 1, and 2 will immediately be identified with constant MVA, constant current, and constant impedance representations, respectively. For all values of n, $P = 0$ when $V = 0$, which is comforting. Of course at $V = 1$, $P = 1$ as it should be.

If

$$P = V^n \tag{5}$$

$$\frac{dP}{dV} = nV^{n-1} = n, \quad \text{if } V = 1. \tag{6}$$

Therefore, if we can get dP/dV from the experimental results, we will have n for the simulation expressed by (1).

$$P = VI \tag{7}$$

is quite valid and useful. Differentiation gives

$$\frac{dP}{dV} = I + V\frac{dI}{dV} \tag{8}$$

and at $I = 1$ and $V = 1$

$$\frac{dP}{dV} = 1 + \frac{dI}{dV}. \tag{9}$$

Manuscript received October 10, 1968.

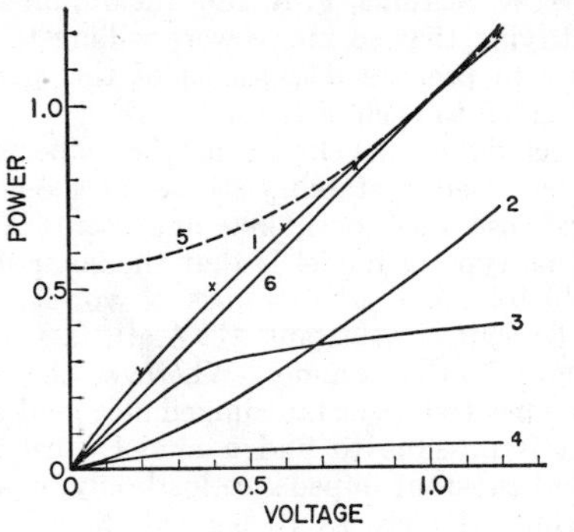

Fig. 7. Simulation of loads. 1—$P = V^{0.883}$; 2—$P = 0.561\ V^{1.36}$; 3—$P = 0.377\ V^{0.26}$; 4—$P = 0.062\ V^{0.36}$; 5—0.56 constant MVA, 0.44 constant impedance; 6—constant current; ×—sum of 2, 3, and 4.

The experimental results gave $\Delta I/\Delta V$ or, in the limit, dI/dV. Consequently,

$$n = 1 + \frac{\Delta I}{\Delta V} \tag{10}$$

and we have a straightforward way of utilizing the experimental results to give us the needed exponent n for simulation by (1).

Next, let us postulate that there is an n_1 which characterizes *pure* residential–commercial loads, and n_2 which characterizes *pure* industrial loads, and an n_3 which characterizes *pure* agricultural loads, and we wish to use the general V_n simulation. Then let us assume that we have three load locations in mind, identified by P_1, P_2, and P_3, which are *predominantly* residential–commercial, industrial, and agricultural, respectively. Thus these three per unit equations apply.

$$P_1 = a_1V^{n_1} + a_2V^{n_2} + a_3V^{n_3} \tag{11}$$

$$P_2 = a_4V^{n_1} + a_5V^{n_2} + a_6V^{n_3} \tag{12}$$

$$P_3 = a_7V^{n_1} + a_8V^{n_2} + a_9V^{n_3} \tag{13}$$

where $a_1 + a_2 + a_3 = a_4 + a_5 + a_6 = a_7 + a_8 + a_9 = 1$. Differentiation yields the following (at $V = 1$)

$$dP_1/dV = a_1n_1 + a_2n_2 + a_3n_3 \tag{14}$$

$$dP_2/dV = a_4n_1 + a_5n_2 + a_6n_3 \tag{15}$$

$$dP_3/dV = a_7n_1 + a_8n_2 + a_9n_3. \tag{16}$$

Measurements of $\Delta I/\Delta V$ at each of the three load locations enable us to identify, with the aid of (6), three values of $n = N_1$, N_2, and N_3, and we may write three equations:

$$a_1n_1 + a_2n_2 + a_3n_3 = N_1 \tag{17}$$

$$a_4n_1 + a_5n_2 + a_6n_3 = N_2 \tag{18}$$

$$a_7n_1 + a_8n_2 + a_9n_3 = N_3. \tag{19}$$

In the foregoing equations $a_1 \cdots a_9$ are fractional loads corresponding to that portion in each case with the associated n_1, n_2, or n_3. These coefficients are identified in the Section *Load Tests*, and for the $P = V^n$ simulation, we may write

$$0.684n_1 + 0.305n_2 + 0.011n_3 = 1.02 \tag{20}$$

$$0.467n_1 + 0.533n_2 = 0.78 \tag{21}$$

$$0.293n_1 + 0.117n_2 + 0.590n_3 = 0.64. \tag{22}$$

Solution of the equations yields 1.36, 0.264, and 0.353 for N_1, N_2, and N_3, respectively. Thus *pure* loads of the residential–commercial, industrial, and agricultural types would be represented by $V^{1.36}$, $V^{0.264}$, and $V^{0.353}$, respectively.

Any *composite* load, such as that described by the sales figures for August and September 1965, can now be represented by a single V^n term. Thus from the given data

Residential–commercial:	$a_1 = 0.561$,	$N_1 = 1.36$
Industrial:	$a_2 = 0.377$,	$N_2 = 0.26$
Agricultural:	$a_3 = 0.062$,	$N_3 = 0.36$

we can obtain

$$n = 0.561(1.36) + 0.377(0.26) + 0.062(0.36) = 0.883.$$

Comparison between the single term $P = V^{0.883}$ simulation and the three term $P = 0.561V^{1.36} + 0.377V^{0.26} + 0.062V^{0.36}$ simulation is shown in Fig. 7. Also shown is the simulation using the 0.56 constant MVA and 0.44 constant impedance simulation given in the paper.

From Fig. 7 it is clear that for a composite load, a single-term single-exponent simulation is essentially equivalent to a three-term three-exponent simulation, even though the three components are widely different from each other. Furthermore, while the two-component constant MVA constant impedance simulation and the V^n simulation are in close agreement in the vicinity of $V = 1$, at low voltages the disagreement is quite marked. As a final remark based on Fig. 7, the agreement between the constant current simulation and the V^n simulation for the composite load is strikingly good *in this case* throughout the voltage range.

G. W. Bills (North American Rockwell Company, Anaheim, Calif. 92803): Although power system engineers have been concerned with the effects of load characteristics on system stability for at least 40 years and many studies have been made on the subject, few electric utilities have made adequate tests to determine their own load characteristics. For this reason alone, this paper is significant because it describes a well-planned test to obtain load characteristics of a large electric system.

Understandably, the electric utilities are reluctant to subject their customers to very many planned disturbances for the purposes of analyzing transient behavior and not all utilities will make such tests. However, instrumentation and automatic recorders are now available which may make it feasible for utilities to obtain load characteristics without tests. By installing such equipment at important buses, whenever a system disturbance occurs, pertinent data can be recorded and even transmitted to a digital computer for analysis. In time, sufficient load data might be obtained to determine load characteristics, although as mentioned by the authors, separation of voltage and frequency effects is often difficult.

It is interesting that constant current loads can be represented by a combination of fixed impedance and constant MVA loads with adequate accuracy because this simplifies the load representation.

Does the Southern California Edison Company now employ this load representation in all its transient stability studies?

Perhaps this paper will initiate similar attempts by other electric utilities to determine their bus load characteristics and thus lead to more accurate stability studies and increased electric power system reliability.

Manuscript received September 30, 1968.

C. Concordia (General Electric Company, Schenectady, N. Y. 12305): We are extremely pleased that the authors have published their results. The characteristics of loads have always been recog-

Manuscript received September 25, 1968.

nized as important, and some tests have been made before, but the results of very few of them have been published. We hope that this paper will encourage others to publish whatever useful information on this subject they may have available. The information on the methods of testing used should also be useful in helping others to obtain consistent results.

The variation of reactive power with voltage is important, as the authors state, and it must be recognized that the increasing use of shunt capacitors has a tendency to change this characteristic from what may have been appropriate many years ago.

Testing the change of load with frequency is a much more difficult task. In connection with this, we have often thought that such tests might best be made on a small scale in a laboratory. Following up this thought, it seems to me that such laboratory tests might be made in schools as suitable, and indeed very desirable, thesis projects in the electric power field. Such laboratory work would be of considerable practical use to the electric power industry, and could, of course, be accompanied by theoretical studies to explain and generalize the results.

H. S. Worcester (State of California, Department of Water Resources, Sacramento, Calif. 95814): The authors are to be congratulated on identification of the need for improved data for representation of load-voltage response for stability studies. Due to insufficient computer capacity in the past little effort has been made to determine the true nature of the load for stability studies. The authors have established the importance of correct representation of load with voltage changes. The large amount of constant MVA load shown by the authors to exist on some of the stations implies that substantial inertia and damping may also exist at these stations.

It is suggested that the transient swings reported under the test results are due to a transient frequency change that developed from increasing the station reactance. A short time frequency reduction would be expected on the load side of the station as the power angle of the station increased due to the increase in station reactance. A slightly shorter transient frequency increase would be expected when the station reactance was returned to normal. It appears that the authors' test procedure, if modified to record frequency on the load side of the station, could provide data to determine the apparent inertia, damping, and, possibly, some load-frequency response information.

Manuscript received September 30, 1968.

M. H. Kent, W. R. Schmus, F. A. McCrackin, and **L. M. Wheeler:** It is very gratifying that so many were willing to contribute their time and effort to prepare discussions of this paper. Our sincere thanks are extended to each of them.

With the techniques developed in the paper, composite load models have been used in stability studies at the authors' company where load response was considered important. A well-recognized deficiency of this type of model is that the component of constant MVA load which, under all extremes of voltage, low voltage in particular, will maintain constant MVA. In fact, virtually no real loads will behave in this manner. Wherever the $\Delta I/\Delta V$ factor is calculated to be positive (constant impedance load exceeds constant MVA load), it is possible to find a model consisting of constant current load and constant impedance load only, which is equivalent for voltage swings as large as 10 percent. Such a tactic is highly desirable if a fault is to be studied in the vicinity of these loads.

An alternate load model suggested by both Mr. Peterson and Dr. Kimbark $[P = P_0(V/V_0)N]$ performs more realistically under voltage extremes for all loads except pure constant MVA load ($N = 0$) or load with a negative impedance component. Our tests indicate that these exceptions would be rare. We agree that the exponential model is superior to that presented in the paper for several reasons, not in the least of which is the substantial reduction in data storage requirement.

As there are some loads which will not correspond to exponents between zero and two (a constant Mvar load partially compensated by a shunt capacitor is an example), the flexibility should be retained to use values of N outside this range if necessary. Undoubtedly, a provision for undervoltage load shedding would be desirable for such a load.

Dr. Kimbark states that bus voltages as low as 0.5 per unit have been observed in stable cases. Our experience reaffirms this point, but it is important to note that following fault clearing, a voltage swing so severe as to trip off load represents a failure of system integrity to the affected customers, although in a technical sense, stability of generators was preserved.

Many have asked if additional load response tests are contemplated. For voltage response, additional tests could serve two purposes: to extend the test range beyond 5-percent change, and to determine the effect of system growth on load characteristics. Because of customer considerations, extending the test range is not practical. The alternate suggested by Mr. Bills and Dr. Kimbark of installing on-line monitoring of loads is probably the most practical to fulfill both the above objectives. Data obtained in this manner potentially can yield knowledge of frequency effects and severe voltage swings. However, the analysis problem is greatly complicated. It is expected that subsequent efforts to gather load data will employ the on-line method as well as the laboratory technique suggested by Mr. Concordia.

Manuscript received October 30, 1968.

STEADY-STATE VOLTAGE POWER CHARACTERISTICS FOR POWER SYSTEM LOADS

R. Barnett Adler Clifford C. Mosher
Drexel University, Philadelphia, Pa.

ABSTRACT

In an effort to assist the power utility engineer in his task of analytically representing system loads, steady-state data has been collected and summarized from three ordinary load types relating the real and reactive components of their power demand to the applied voltage. The load types are fluorescent lighting, small unit air conditioning, and a 150 hp induction motor.

The load tests were controlled and instrumented using the Edison Electric Institute "Intermediate Power AC/DC Transmission Model" facility. The data collected consisted of the simultaneous recording of the real and reactive power, and voltage, as voltage was varied in small steps covering a range of nominally 0.75 to 1.10 per unit.

Several tests covering this range were performed on each of the three load types. The data from each test were normalized using as bases the values of power recorded in the respective test at nominal voltage. The data from tests on the same load were then combined and smoothed into a single reasonably consistent data set. Analytical expressions were obtained for curves fitted to each smoothed data set summarizing the character of complex power variation as a function of voltage. Both smoothing and curve fitting procedures were structured around digital computer subroutines.

The experimental and reduced analytical results have been graphically presented to facilitate evaluation of the analytical expression used for each data set.

INTRODUCTION

In power system planning great care is taken to assure that the system operates in normal and abnormal modes reliably with maximum feasible economy. Studies using the digital computer dominate this area of engineering-planning.

The establishment of formulae to represent the complex power of a load as a function of voltage in the steady state remains a very uncertain but no less significant detail in the preparation of system studies. The load on a particular bus reacts to voltage change in a way associated with some function of voltage. A planning option, involving significant capital and with reliability ramifications, may be recommended or discarded based simply on the analytical expression chosen to represent this function of voltage.

This paper proposes to treat experimental results rather than to develop or amplify previous theory.[1,2] The goal is to obtain a quantitative representation for each of three load types: fluorescent lighting, small unit air conditioning, and a medium-size induction motor (150 hp).

This research has been confined to the steady-state voltage response. The results, however, may be of value in dynamic studies providing that one accepts the steady-state character as a good approximation.[3]

BACKGROUND

The real and reactive power required at a particular bus as a function of voltage depends on the types of connected load as well as the proportion of each type. One is able to state exactly the character of the load if one knows, the load composition and, second, the character of each "pure" load present in the composition.

Originally, to reduce the labor of manual calculations and simplify the circuitry of analog network analyzers, loads were represented as constant power, constant current or constant impedance. One recognizes that, since the nature of the load composition is generally uncertain, a more exact representation through more precise expression of the "pure" constituent loads appears to be an unwarranted complication. As a result, loads are yet described as having a nature similar to one or a portion of the three above characteristics. That is to say, the load is represented by a second order polynominal with positive or zero coefficients. (Constant power, constant current and constant impedance vary with the zero, first and second exponential powers of voltage, respectively).

In digitally assisted network studies, there is little significant economic advantage in these simpler (exponents 0, 1 and 2) expressions. The incorportation, mathematically, of inverse variation, negative coefficients or fractional exponents might better serve to approximate the response of a load to voltage change.

Reiterating, this paper attempts to establish experimentally the character of three "pure" loads. The portion of each load type tied to the bus in question, however, must be known for adequate load representation. Thus, by themselves, the expressions describing pure loads do not act immediately to increase the confidence with which one models a particular

Reprinted from *IEEE Paper 70 CP 706-PWR.*

load, but it serves as a necessary first step.

APPROACH

The data supporting the development of the steady-state voltage-power characteristic expressions for each load were obtained from a series of tests in which the load of interest was connected and the applied voltage varied from 1.0 per unit to approximately 0.75 per unit, and from 1.0 p.u. to 1.10 p.u., or in either case to the voltage at which the load reacted abnormally. The voltage was initially set at the nominal (1.0 p.u.) operating level for the load and then decreased in small steps to its lower limit. The nominal was then reset after which the voltage was increased in small steps to the upper limit. The real and reactive components of the power demand and the rms applied voltage were simultaneously recorded on a single oscillograph chart.

Test Facility

The load tests were performed using the laboratory facility of the Edison Electric Institute "Intermediate Power Level AC/DC Transmission Model."[4] This Model has been used principally for experimental studies on the effects of parallel ac and dc transmission. Its components include three DC motor driven alternators with a total capacity of 875 kVA, 500 kW, and 60 kVAR of independent static load units and two 150 hp induction motors loaded with variably excited eddy current brakes.

The Model features a well equipped instrumentation console to which, through a jackfield system, may be brought any electrical signal in the laboratory. The instrumentation includes four 3-wire 3-phase Watt transducers (two Esterline-Angus, Model 4956, and two Westinghouse, Model VP-840), and two Honeywell Visicorder, Model 1508, direct writing oscillographs. One of these light beam oscillographs provided an excellent means of recording simultaneously the three desired quantities.

The EEI Model was especially suitable for the load tests. In addition to providing the equipment for instrumentation, the power capacity of the Model was more than sufficient to support a significant block of load. Further the load under test was easily isolated affording maximum control over extraneous influences.

Test Loads

Included in the Model's apparatus are two rotating loads. Each is a General Electric, Type TS444DP, 440 volt, 3-phase, 60 Hz, 1770 rpm, 125 kW induction motor--coupled to a Kinatrol, Type EB482, water cooled, eddy current brake. Rotating Load No. 1 was directly switched to the Model and instrumented using provisions already installed. The motor was loaded to 85% rated capacity (95 kW motor demand at nominal 440 volts)--this loading chosen to represent an average.

The nominal 450 volt bus potential was stepped down through three 10 KVA two-winding transformers to the level dictated by the fluorescent lighting and unit air conditioning loads. Four circuits of the laboratory office lighting were available through a 120/208 volt power service panel. These provided a moderately balanced three phase fluorescent lighting load with nominally 1280 watts on each of two legs of a wye connection and 1600 watts on the third.

The small unit air conditioning load consisted of three York, Model LC-182-6A, 18200 BTU, nominally 3050 watt, 208/220 volt, single phase, window-type air conditioners. Each of these was connected phase to phase and supplied with a nominal operating voltage of 208 volts.

Test Arrangement

The voltage-load characteristic data were collected from the three loads types. A resistive load of approximately 12 kW was connected to the generator bus to stabilize the voltage level once it was adjusted by the appropriate generator "Voltage Control." (This Voltage Control served to vary the Model bus voltage through the range of interest--0.75 to 1.10 p.u.)

The 120/208 volt apparatus connected to the Model consisted of circuit breakers, three single phase transformers, and the necessary provisions for instrumentation. The circuit breakers along with three 208 volt receptacles for the air conditioners, a resistive voltage divider to obtain the desired 120 volt metering potential, and two 40 to 5 ampere current transformers were mounted for convenience on a temporary panel.

Three Standard Transformer, Type MG, 10 kVA, adjustable-tap, single phase transformers connected 456-120/208, delta-wye, provided 208 volts phase to phase for the air conditioners and 120 volts phase to neutral for the lights. A 456 volt Model bus potential proved to be a convenient central starting value, readily and adequately varied, as already mentioned, by the generator Voltage Control.

Instrumentation

Four Watt transducers were used to expedite the load testing--a pair for the induction motor, and a pair for the air conditioner and lighting loads. Reactive power for the nominally balanced motor load was monitored choosing two phase to neutral voltages orthogonal to the phase to phase voltages used in metering the real power. The resulting change in calibration was compensated by an adjustable level control at the input of the oscillograph.

The lighting and air conditioning loads were monitored using the two Esterline-Angus Watt transducers. Since the lighting and air conditioning loads were slightly unbalanced, one of the transducers was complemented by an Esterline-Angus, Model 6228, Var Adapter.

Employing "Hall Effect" semiconductors, the Watt transducers produce a dc output proportional to the ac watts measured. For a balanced three phase load, these transducers produce a dc output with insignificant ripple. The air conditioning and lighting loads, however, were somewhat unbalanced and necessitated output filtering. Low pass filters (20 hz cut-off) were placed between the Watt transducer outputs and the oscillograph inputs reducing ripple to an acceptable level.

Load Test Procedure

The Model was connected to render complete isolation of each load type. The load of interest was energized and the voltage varied down from nominal and then up from nominal, covering an approximate range 0.75 p.u. to 1.10 p.u. Care was taken during the first series of tests to assure that the initial voltage was nominal. In later tests nominal voltage was checked (and adjusted if necessary) at the start of both the voltage decreasing and voltage increasing sequences. This was advisable since, after some tests, there was question whether the real and reactive power had changed at nominal voltage during the test, or that the calibration of the oscillograph trace for nominal voltage had drifted. This is discussed below.

Data Management

Data collected from each load was normalized using the nominal voltage with corresponding real and reactive demands as bases. This served to reduce all data collected from the same load into a common form that was easily combined into a single block.

A computer program TABL was written to normalize the measurements obtained from the oscillographs. In the following discussion the word TEST is used to identify each set of data normalized with the appropriate nominal valves obtained from the oscillographs.

In some cases the data collected during both the voltage increasing and the voltage decreasing sequences were entered under the name of a single TEST. In other cases each sequence is listed under a separate TEST. This depended simply upon whether or not there existed a disparity in the nominal values recorded before the voltage increasing and the voltage decreasing test sequences.

The computer program TABL combines, in order of decreasing voltage, the normalized data for each of the real and reactive components of power for a particular load into a single collection. This collection was inspected visually for any values deviating significantly from the general trends. It was sufficient in most cases to normalize a particular voltage increasing and voltage decreasing test sequence using the nominal values of voltages and power obtained at the beginning, or voltage decreasing, part of the test. In other cases, it was necessary to normalize the data of the last part (voltage increasing) of a test using the most recently measured value of nominal voltage. If values of power inside the range 0.9 p.u. to 1.05 p.u. diverged, additional data treatment was necessary.

The additional treatment usually brought the data generally into line. Another correction, however, was required in a few of the TEST's. Since the values of complex power accompanying nominal applied voltage were used to normalize the remaining data of the TEST, any error in these values was propagated throughout the TEST. In these TEST's each value of real or reactive power demanded at nominal voltage which appeared obviously out of order was discarded and replaced by a value interpolated between adjacent points.

The causes of the variation in nominal values, especially between voltage increasing and voltage decreasing sequences within the same test, are explained as follows. In the motor load there is a thermal reaction to under-voltage. The reactive demand by the fluorescent lights tended to increase slightly upon restoration of nominal voltage following a voltage decrease. The same effect was not apparent after a voltage increase. Also at or near the lower voltage limit, the reactive power required by the lighting was at times observed to be discontinuous or changing independent of voltage.

A likely explanation of the variation in the values of complex power demanded at nominal applied voltage for the induction motor was that the eddy current brake displayed some instability. A 5 kW swing in the motor demand was noted once while voltage appeared constant. To minimize the possibility that excessive differences in data might pass un-noticed, two tests on the same load were never performed in succession but rather alternated with tests on the other loads. An additional feature of adherence to this procedure is that, with instrumentation of satisfactory accuracy, the precision of a specific data collection improves with the square root of the number of such test replications.5

With the above manipulations, the original data was brought into reasonable consistency. In the next step of data management, the ordered and normalized data collections, comprised of the simultaneous values of voltage, real and reactive power were transferred to data cards for subsequent adjustment by computer program ADJUST.

Program ADJUST uses a curve smoothing subroutine SQ13 available in the IBM "Scientific Subroutine Package"6 to compress the data collection for each load type into a more precise form. This subroutine smooths the data collection by evaluating for each point the least-squares first degree polynomial describing the point of interest, the point before and the point after, and then adjusting the point of interest to better conform to the polynomial. This operates on the collection from the second data point through to the second to the last data point.

Program ADJUST first applies this smoothing process to the ordered lists of the real and reactive components of power as functions of voltage provided by TABL, passing through each list (for instance, the reactive demand for fluorescent lighting) from nominal down to minimum voltage, and then from nominal up to maximum. The next step divides each of these arrays into two arrays--one consisting of the even numbered elements and the other the odd. Each of these is anchored with the nominal magnitude unity and smoothed. The four arrays are remeshed to reconstruct the order of the two original arrays. These two arrays (nominal down to minimum and nominal up to maximum) again are smoothed.

This cycle of smoothing, splitting, smoothing and recombining was repeated a second and final time after which it was noted that the data which had previously been precise remained little altered. The scattered data, especially near the extremes of 0.75 p.u. and 1.10 p.u., were satisfactorily compressed into a smooth curve. The reader is referred to the Calcomp plots of the original and smoothed data, Figures 1 and 3.

As a final operation program ADJUST punched into data cards, in order of increasing voltage, the normalized values of voltage with the corresponding original real and reactive power and their adjusted values. These were read by the subsequent programs, CURFIT and Calcomp plotting.

Smoothed Data to Analytical Expression

As a final step the smoothed data was summarized with analytical functions of voltage. Three forms were considered: 1) polynomials comprised of positive integer exponential powers up to the third of voltage; 2) similar polynomials including negative integer exponential powers up to the third of voltage; and, 3) a single positive fractional exponential power of voltage.

The objective was to describe the behavior of the load type with an expression similar to that of the second degree polynomial now used. A provision was made, however, to expand the form of the expression to include negative coefficients and higher orders of voltage raised to both positive and negative powers. A computer program CURFIT was written around a curve-fitting subroutine APMM to obtain these expressions.

APMM specifies a linear combination of linearly independent continuous real functions supplied by the user which constitutes the Chebyshev approximation of the given real function over the range described by the discrete data of interest.6

DISCUSSION AND COMMENT

Voltage Range

During the load tests the applied voltage was varied through the range 0.75 to 1.10 p.u. The motors were therefore held in a region of stable operation avoiding possible damage

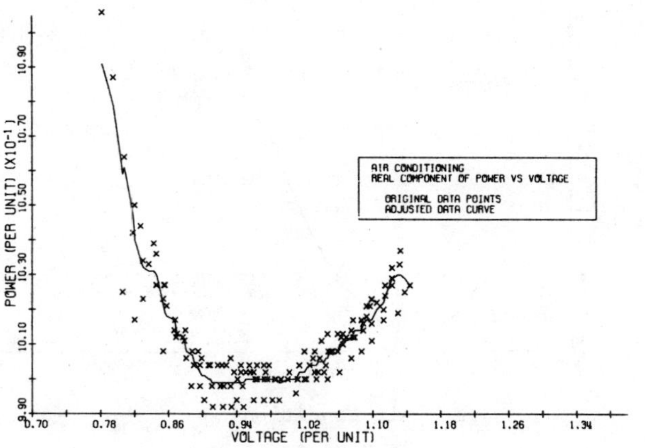

Air Conditioning

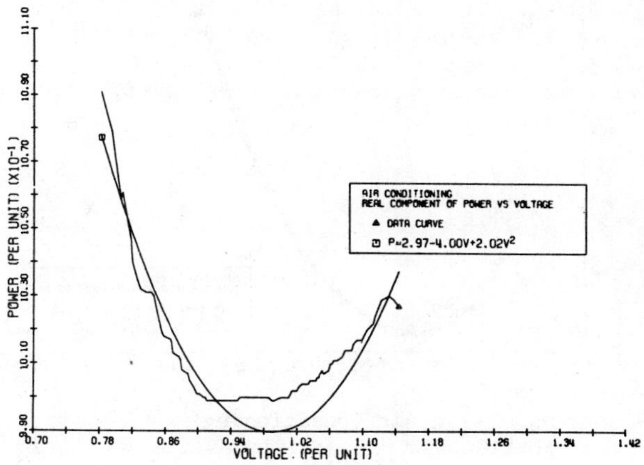

Air Conditioning

$P=2.97-4.00V+2.02V^2$

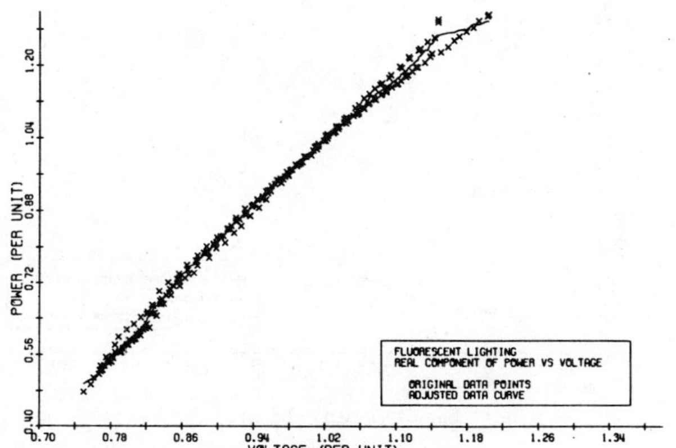

Fluorescent Lighting

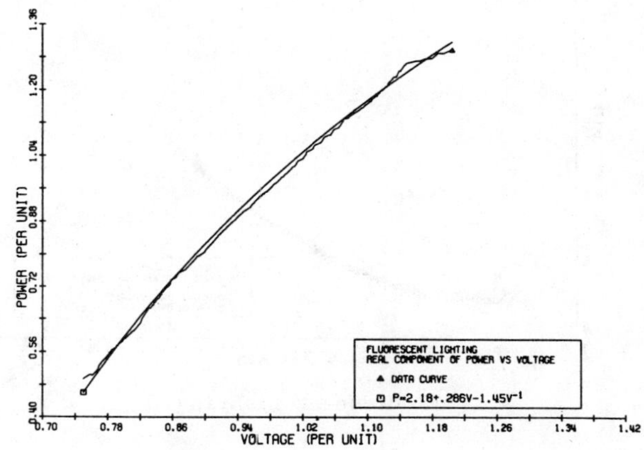

Fluorescent Lighting

$P=2.18+.286V-1.45V^{-1}$

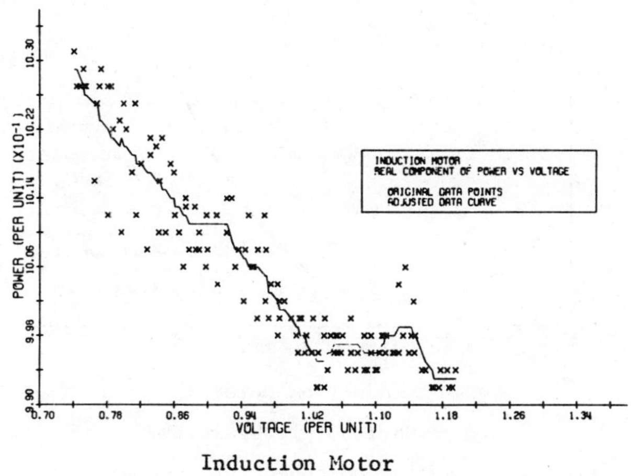

Induction Motor

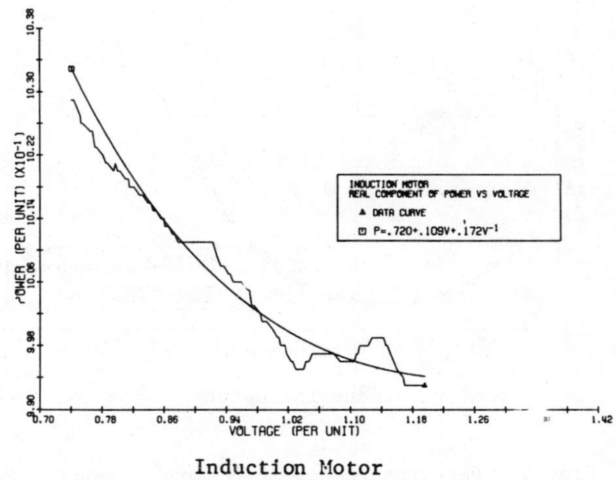

Induction Motor

$P=.720+.109V+.172V^{-1}$

Fig. 1. Real component of load versus applied voltage. Smoothed data curve superimposed on original data points.

Fig. 2. Reactive component of load versus applied voltage. Preferred analytical function of voltage and smoothed data curves.

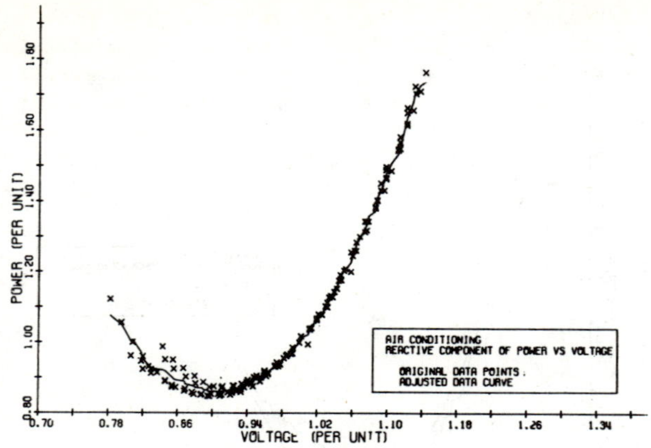

Air Conditioning

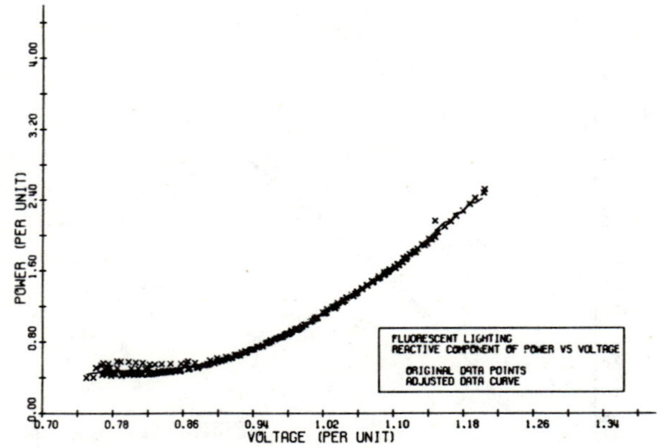

Fluorescent Lighting

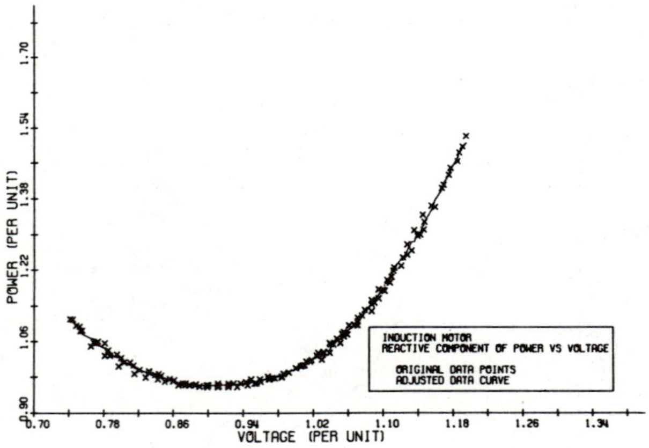

Induction Motor

Fig. 3. Reactive component of load versus applied voltage. Smoothed data curve superimposed on original data points.

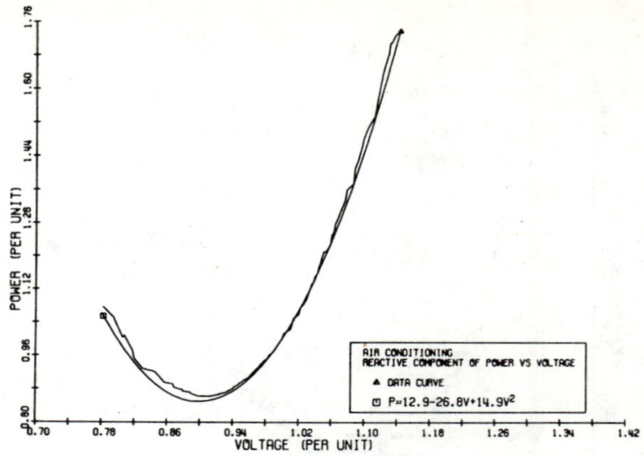

Air Conditioning
$P=12.9-26.8V+14.9V^2$

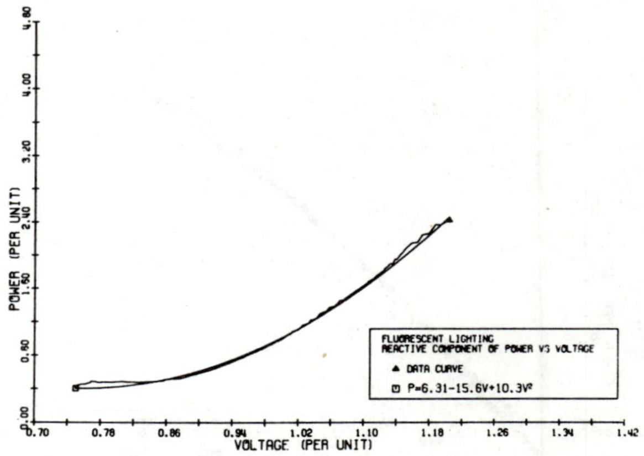

Fluorescent Lighting
$P=6.31-15.6V+10.3V^2$

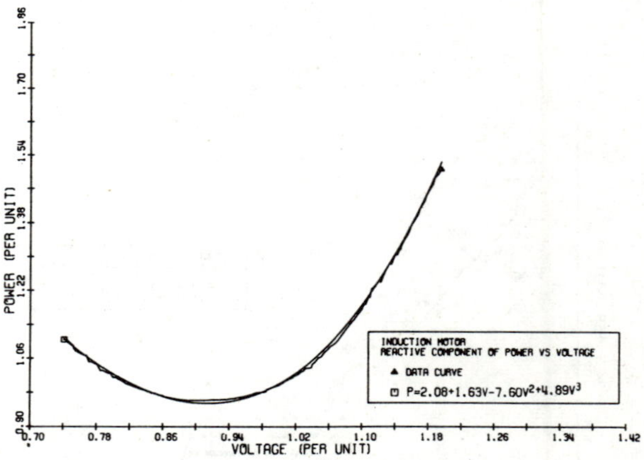

Induction Motor
$P=2.08+1.63V-7.60V^2+4.89V^3$

Fig. 4. Reactive component of load versus applied voltage. Preferred analytical function of voltage and smoothed data curves.

arising from thermal extremes. This range is justified since, for practical purposes, one can assume that the load on a particular circuit will be disconnected by relay action for a voltage outside its limits.

Data Consistency

In some of the test results for the fluorescent lighting and air conditioning loads, a .01 to .05 p.u. discrepancy was observed between the real or reactive demands at nominal voltage before the voltage was decreased to its lower limit and the same demands at the re-established nominal before the voltage was increased toward 1.10 p.u. Two additional tests were performed on each of these two load types to clarify this inconsistency and to justify the steps taken to rectify it. The first was to initiate the test with the voltage increasing; the second to begin with the voltage decreasing. Each time the test was completed only after allowing the load to operate for ten minutes at the re-established nominal voltage. In both instances, indentical values for power demand were recorded.

These tests justified the use of the most recent demands at nominal applied voltage as bases for normalization. These demands were either recorded at the commencement of, or at the time that nominal voltage was re-established midway through a test.

Care was taken to assure that the values used for normalizing were those associated with the nominal voltage. A voltage check was repeated before each series of voltage changes to establish more firmly the reference for the coincident merging of the separate test data into a single block.

The tests were repeated through five separate cycles in the order of: (1) induction motor, (2) lighting, and (3) air conditioning. Alternating load types served as a precaution against duplication of erroneous measurements.

CONCLUSION

Steady-state voltage-power data were collected from three load specimens, each chosen to manifest the typical behavior of a particular type of load. The character of this data, following the stated manipulations, was then summarized with analytical expressions. The expressions for steady-state power variation as a function of voltage may find application in several power system problems. These include a minimization of some uncertainties concerning load shedding, and load representation in stability or load flow problems.

In support of the data normalization and smoothing techniques, Calcomp plots were made to illustrate and compare the original data points and/with the associated adjusted data curve. These plots are found in Figures 1 and 3.

To assist in selecting the expression, generated by CURFIT, which best describes each adjusted data curve, Calcomp plots were also made for each expression which had potential to provide adequate description with coefficients limited to three significant figures, matching the precision of the original load test measurements. Each plot consisted of the function of voltage superimposed on its respective smoothed data curve.

After inspection of the Calcomp plots of each of these expressions, one expression was chosen for description of each of the real and reactive components of demand for each of the three loads.

Air Conditioning

Real: $P=2.97-4.00V+2.02V^2$

Reactive: $P=12.9-26.8V+14.9V^2$

Fluorescent Lighting

Real: $P=2.18+.286V-1.45V^{-1}$

Reactive: $P=6.31-15.6V+10.3V^2$

Induction Motor

Real: $P=.720+.109V+.172V^{-1}$

Reactive: $P=2.08+1.63V-7.60V^2+4.89V^3$

These expressions were selected for their precision and similarity to the presently used descriptions. The Calcomp plots of the above expressions are illustrated in Figures 2 and 4.

ACKNOWLEDGEMENT

The authors would like to express their appreciation to Mr. John Dougherty and staff of the Edison Electric Institute AC/DC Transmissions Model for their assistance and the Philadelphia Electric Company for advice, encouragement and computer time.

REFERENCES

1. M.H. Kent, W.R. Schmus and F.A. McCracken, "Dynamic Modeling of Loads in Stability Studies," IEEE TRANS. POWER APPARATUS AND SYSTEMS, vol. PAS-88, No. 5, May 1969.

2. V.A. Venikoo, Transient Phenomena in Electrical Power Systems, abridged translation by B. Alkins and D. Rutenberg. New York: Macmillan Co., Inc., 1964, pp.41-51.

3. J. Bubenko, T. Johansson and A. Olivegard, Load Representation in Stability Studies. Swedish State Power Board, 1965.

4. V. Caleca and J. J. Dougherty, "The Edison Electric Institute AC/DC Transmission Model," IEEE TRANS. POWER APPARATUS AND SYSTEMS, vol. PAS-87, No. 2, February, 1968.

5. H. Schenck, Theories of Engineering Experimentation, 2nd ed. San Francisco: McGraw Hill Book Co., Inc., 1968.

6. IBM, System 1360 Scientific Subroutine Package Programmer's Manual, (360A-CM-03X) Version III, Fourth ed. White Plains New York: IBM Technical Publication Dept. 1969.

BEHAVIOR OF LOADS DURING VOLTAGE DIPS ENCOUNTERED IN STABILITY STUDIES. FIELD AND LABORATORY TESTS

F. Iliceto
University of Rome,
Rome, Italy
Senior Member IEEE

A. Ceyhan
Turkish Electricity
Authority,
Ankara, Turkey

G. Ruckstuhl
CESI,
Milan, Italy

ABSTRACT

The lack of knowledge of the behavior of loads during the voltage depressions caused by faults in HV grids has been emphasized on many occasions. Field and laboratory tests were advocated for more realistic representation of loads in transient stability studies.

This paper describes and analyses a set of field tests performed on the 154 kV network of Turkey. Although they could not be extensive, to authors' knowledge they are the first reported tests with large voltage depressions. Laboratory tests carried out on some common electrical appliances are also reported.

Calculations are made of the power absorbed by electrochemical plants during a sudden drop in supply voltage. A fact previously underemphasized is the great importance of the selfinductance of the reduction cells circuit.

1. INTRODUCTION

Methods for stability studies of power systems have been remarkably improved in recent years with the use of large computers. The present state of the art limits the achievement of more accurate results due to insufficient knowledge of the dynamic behavior of some system components. These include:

1. Loads during the large voltage depressions caused by the faults in HV system;
2. Generator governors when studying long duration transients.

This article deals with the first problem.

Transmission system configurations are frequently encountered where important loads are supplied either by intermediate stations on long EHV lines, or by the receiving station located in areas where synchronous generators have small power ratings. In both cases the voltage at busbars feeding the load can be as low as 0.2-0.5 p.u. during the faults on EHV lines and as low as 0.50-0.75 p.u. during dead time before reclosure[1,3]. A load behaving as a constant impedance (P and Q proportional to V^2) would absorb much less power than a constant power load. Consequently the power flow in the transmission lines and the voltage drop itself at load busbars during dead time can be different to such an extent as to be the controlling factor of system stability[1].

In the past it has been common practice to represent the loads as constant impedances. Recently approaches were made for a more responsive simulation. The proposals put forth are based on the assumption that the absorbed power can be expressed as a function of the voltage alone, as could be measured in steady state conditions, i.e. when the transient caused by the sudden voltage depression is extinguished. Such functions P(V) and Q(V) are defined here as "static characteristics". In the literature the additional assumption has been generally accepted that the response of nearly all loads to voltage changes can be represented by some combination of constant impedance, constant current and constant MVA devices. Characteristics P(V) and Q(V) in a polynomial form, whose coefficients are calculated as empiric functions of the load type and composition have been proposed.

The above load modeling is elaborated upon and proved to be responsive by field measurements in the case of small sudden voltage drops[2]. In fact small voltage changes do not cause major modifications of the operating conditions of electrical appliances (motor speeds remain almost unaltered; discharge lamps do not quench, etc.). Therefore the transient conditions following the voltage drop have a much shorter duration than the voltage drop itself and may be neglected.

An extensive valuable set of field tests previously reported[2], had to be limited to a voltage drop of 5%. Field measurements with large voltage drops are lacking and the utilities are reluctant to perform such tests for obvious reasons. Nevertheless they have been recognised, along with laboratory tests, as an essential premise for a more accurate dynamic modeling of the loads. Encouragements to their implementation and publication of available results were made[3,4].

This paper presents the results of some field tests carried out on the 154 kV network of Turkey and a set of laboratory tests performed at Rome University on electrical appliances in common use. A simplified calculation of the transients of large electrochemical plants is also given. These showed that if the voltage depression is large (25 to 75%), the transient phenomena of synchronous and induction motors and of electrochemical plants have a duration of same order of the voltage depression in HV systems (0.5-1 s). Consequently the absorbed power varies widely with time during the voltage depression, even if voltage is kept constant. Besides, it showed that the transient conditions of induction motors are strongly influenced by several interacting factors: electrical and mechanical parameters of the machines and shape of voltage depression.

An attempt to synthesize experimental and analytical results has been made with the ambitious aim to contribute, in a practical way, to dynamic modeling of loads in stability studies. This will be dealt with in another paper. Still it is worthwhile outlining here how the simulation of families of induction motors will be approached. The basic work on this subject was already done when this paper was submitted.

The mathematical representation of induction motors in a reasonably simple but still accurate way is by far the major task, because: (1) great number of them are usually present, (2) large differences exist in the electrical and mechanical parameters and (3) motors are scattered in distribution networks at various voltage levels. Such difficulties are not encountered in synchronous motors

Paper T 72 169-6, recommended and approved by the Power System Engineering Committee of the IEEE Power Engineering Society for presentation at the IEEE Winter Meeting, New York, N.Y., January 30-February 4, 1972. Manuscript submitted September 23, 1971; made available for printing December 15, 1971.

Reprinted from *IEEE Trans. Power App. Syst.*, Nov./Dec. 1972.

and electrochemical plants which are few in number and have large rating: thus they can be represented as separate items with the appropriate differential equations, if an accurate simulation is required. On the other hand, the authors' tests seem to confirm that the representation of the common static loads, including non-linear components like the discharge lamps, is accurately performed with appropriate static characteristics.

A digital computer program based on Stanley's equations has been prepared for the analysis of the transients of induction motors. This was used to evaluate the influence exerted by the parameters of the motors and driven machines. It showed that the most important ones are the constant of inertia, H and the time constant of the rotor with open circuited stator, T'_{do}. The loading torque/speed characteristic, $T_l(n)$, was found important only in cases of small inertia. The results were summarized in a set of curve charts. A way for the dynamic modeling of a large family of induction motors would then be their groupment in a very few equivalent machines, each defined by the relevant H, T'_{do} and $T_l(n)$.

2. MEASURING PROCEDURES AND TREATMENT OF OSCILLOGRAPHIC RECORDS

Field and laboratory tests were carried out by causing a sudden three-phase symmetrical voltage depression of short duration. Voltages and currents at the load busses were recorded with light beam electromagnetic oscillographs, which in field tests were connected through the instrument transformers of the substation. Hall effect transducers were inserted to record the active and reactive powers. The available transducers had a time constant of 40 msec or greater and therefore it was known before the tests that they would not respond correctly during the fast variations of the input electrical parameters. In fact active and reactive power were calculated through the records of voltages and currents.

The oscillograms of currents and voltages, including those recorded without use of instrument transformers (see f.e. fig. 9) indicated that the unidirectional components may be important only in the currents and have short duration. Therefore they do not appreciably contribute to the power and can be neglected. Furthermore, they showed that the ac components can be approximated by a succession of half-sinusoidal waves, after the first cycle of the transient.

The above observations allow definition of the electrical parameters similar to a sinusoidal steady condition: the rms voltages and currents were deduced by reading the peak value of the pseudo half-sinusoids; the phase angles were calculated as the average of readings made between the corresponding maximum, minimum and zero values. The readings of the oscillograms were made with the help of a pencil follower and the data were elaborated with a computer.

3. FIELD TESTS IN THE 154 KV TURKISH NETWORK

The tests were carried out on industrial loads and on residential-commercial loads. The former was the load of the town of Izmit (90.000 inhabitants), near Istanbul, where important paper mills are located. The latter was the load of the areas of Adapazari, Eskisehir and Pasalar, south-east of Istanbul.

Large voltage drops were obtained by insertion of a long 154 kV line short-circuited at one terminal for 0.2-1 s. The small voltage drops were obtained through the deenergization of some hundreds of km of unloaded line.

Tests were performed at night, when the lines to be used as "shunt reactors" or "shunt capacitors" were available. To limit perturbances, loads were supplied by two generators of a hydro-power plant, previously separated from the national grid. The single-line diagram of the test circuit for the Izmit load is shown in fig.2.

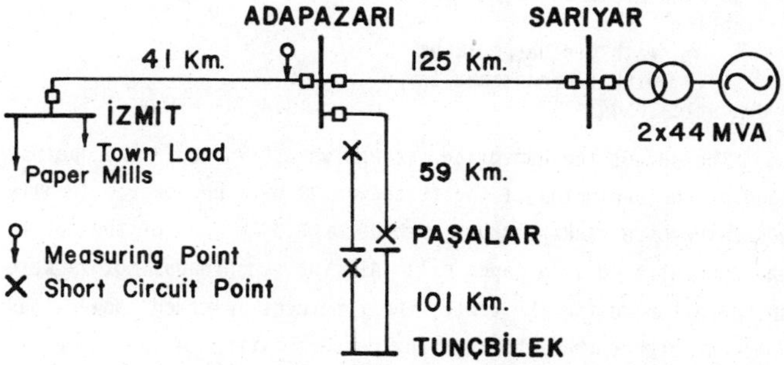

Fig. 2 - Connection diagram of the testing circuit for the load of Izmit town

System outages were not recorded during the tests, as voltage was automatically restored after the perturbance. A considerable part of the loads was however automatically disconnected by the customers protection relays, when the voltage dropped to 0.5 p.u. or less. This showed that the choice and setting of many customers'

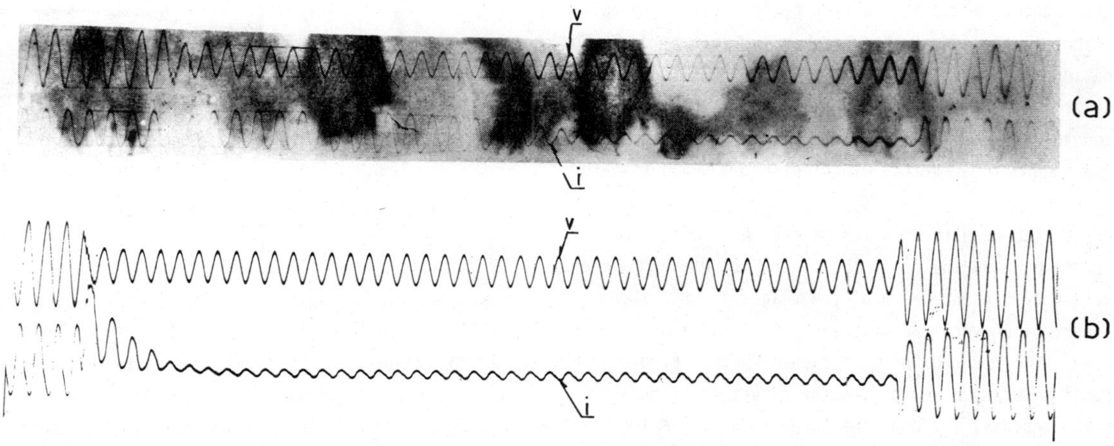

Fig. 1 - Oscillograms of voltage and current in one phase during two tests in the 154 kV Turkish grid
a) industrial load of Izmit; b) residential load of Adapazari

relays are not well coordinated with the protection and high speed restoration devices of the HV grid. Actually this situation is common in many countries. In fact the industrial customers should coordinate their protection in agreement with the power utilities, so that during faults in the HV grid the automatic load dropping is initiated only when the machinery or the production would otherwise be damaged. In many instances the operators of the HV grid have the same aim. In fact load dropping requires readjustment of power generation and voltage levels. It could also cause instability of those generators that are accelerating as a consequence of a line's fault, because dropping loads results in increased acceleration if the loads are connected to generator busbars.

The variations of frequency during the perturbances were small; they cannot be considered appreciably responsible for the recorded power demand change.

Tests on the industrial load

The Izmit connected load at night at the beginning of the tests was composed as follows, where reference is made to the nameplate rating of the motors:

- asynchronous motors 61.4%
- synchronous motors 17.9%
- electrochemical load (chlorine-soda) 3.5%
- lighting
  - with discharge lamps 2.3%
  - with incandescent lamps 2.5%
- domestic loads 12.4%

The sum of the nameplate ratings was 56 MW, but the actual load at the beginning of the tests was 27 MW. The power locally generated by a back-pressure turbine was 3.5 MW. Most of the motors was concentrated in a paper mill. All the synchronous motors were in operation during the tests; their percent absorbed power was somewhat higher than the relevant percent rating given above. It is worth recalling that the synchronous motors and back-pressure generators are commonly used in Europe; thus the above composition is typical of a highly industrialized European area.

The tests were carried out in the following cronological order, from 2 a.m. through 4 a.m.:

Fig. 3 relates to a voltage drop of about 10%. The reactive power Q absorbed from the 154 kV line dropped to about 50%. This should be attributed partly to the large increase of the reactive power generated by the synchronous machines in the load area and partly to the reduced demand of reactive power in the induction motors and other appliances, as a consequence of voltage drop. The same explanation justifies the current reduction in the 154 kV line. The slight increase of active power demand P during 0.4 s, may be attributed to the increase of power losses caused by the increase of the currents in the motors, distribution lines and transformers.

Fig.4 gives the results of a test where the voltage was reduced to about 0.5 p.u. during 0.7 s. The first part of the transient is characterized by the inversion of the reactive power flow, the increase of current above pre-disturbance value and a large demand of active power. The reversal of Q is caused by the motors which "feed the fault". During a few cycles the larger contribution is supplied by the residual field of the large induction motors. The increase of I above 1 p.u., in spite of the increase of PF, should be attributed to the large increase of the active component of the

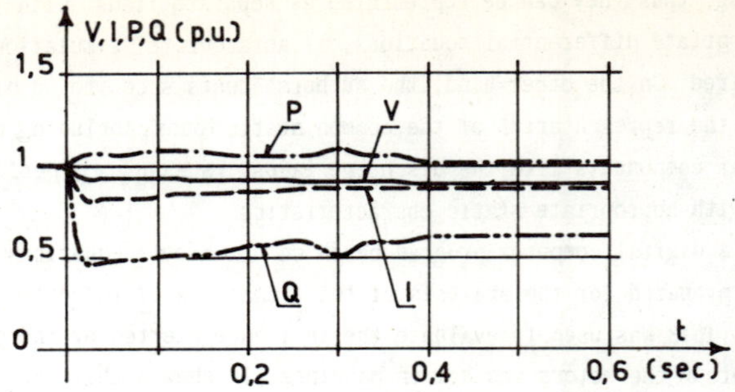

Fig. 3 - Test on the Izmit load. Initial conditions (1 p.u.):
V = 163 kV; I = 135 A; P = 23.5 MW; Q = 29.8 MVAR

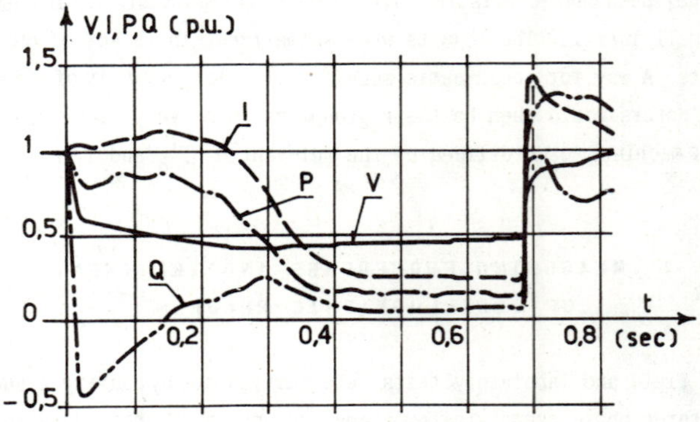

Fig. 4 - Test on the Izmit load. Initial conditions (1 p.u.):
V = 158 kV; I = 102 A; P = 18 MW; Q = 21.4 MVAR

current, as is evidenced by the fact that P is about 0.8 p.u. while V is only 0.5 p.u.

As shown by fig. 4 a considerable power decrease occurred some 0.3 s after the short circuit began. This was partly caused by the disconnection of many motors by under voltage relays. The load dropping was confirmed by reading of wattmeters made shortly after the tests.

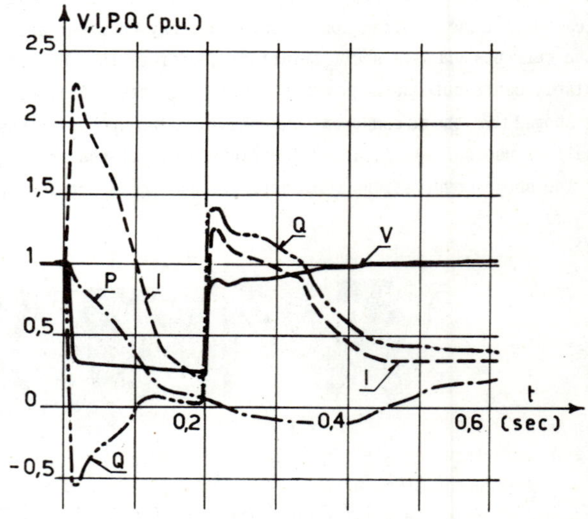

Fig. 5 - Test on the Izmit load. Initial conditions (1 p.u.):
V = 160 kV; I = 88 A; P = 15.5 MW; Q = 18.8 MVAR

The diagrams of fig.5 and 6 relate to voltage dips to 0.35 p.u. and 0.25 p.u., during some 0.2 s. The load composition and the absorbed power were different from prior tests because part of the loads had dropped. Besides, before the last test (fig.6) a standby back-pressure generator had been put in operation, thus dropping the power supplied from the 154 kV line.

The motors infed for some cycles a large current and reactive power into the 154 kV short-circuited line. This phenomenon appears amplified in fig. 6, owing to the more severe short circuit and to the contribution of the additional back-pressure generator.

The interpretation of other behaviors evidenced by fig.5 and 6 was not an easy task. The following explanations should therefore

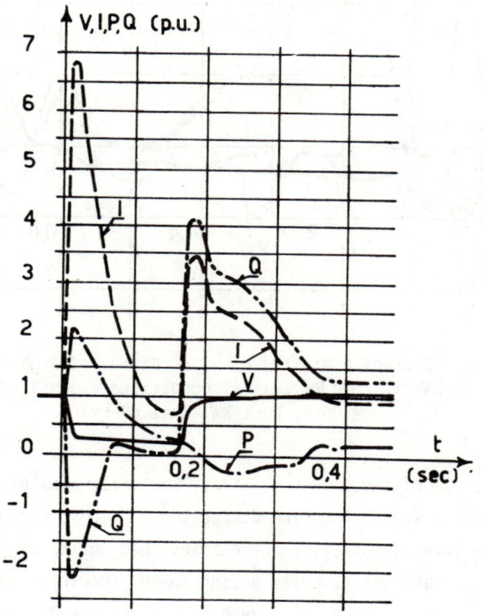

Fig. 6 - Test on the Izmit load. Initial conditions (1 p.u.):
V = 161 kV ; I = 69 A ; P = 11,8 MW; Q = 14.5 MVAR

be considered less reliable than the ones made for fig.3 and 4. The initial increase of active power demand in fig. 6 may be explained as follows: (1) the p.u. voltage on the 3 kV busbars of the paper mills was temporarily supported by the field of the motors and generators at a higher value than on the 154 kV feeding busbar and therefore many motors continued to absorb a large power, (2) the losses (in synchronous and asynchronous machines, cables, transformers, etc.) had increased manyfold because of the circulation of short circuit currents, (3) the power output of the back-pressure generators had decreased owing to the voltage reduction.

Both fig. 5 and 6 show slightly negative values of P for a short interval after fault clearing. The explanation may be that when the electrical transients of rotating machines extinguished, the voltage of the distribution systems collapsed, thus causing the disconnection of some motors and the extinction of the current of the discharge appliances and rectifiers. Besides, the back-pressure generators had accelerated during the voltage dip and the phase angle between their EMF and the system voltage had largely increased. So when the voltage was restored, the generators power output was temporarily much larger than their rated value and a little larger than the total power demand of the loads, which had strongly dropped for the reasons explained above.

Tests on residential-commercial loads

A set of tests was carried out on the load of the village of Pasalar and the town of Eskisehir. The night load was composed as follows:

- domestic loads and public lighting 70%
- asynchronous motors 30%

The results of one test are given in fig.7. The load area was supplied from the Adapazari station by a long 154 kV line operating almost unloaded (6 MW), so that the line itself generated more reactive power than the load demanded. As the measurements were effected at the sending end of the line, in Adapazari, the reactive power Q (fig. 7) was found negative, i.e. supplied to the 154kV system.

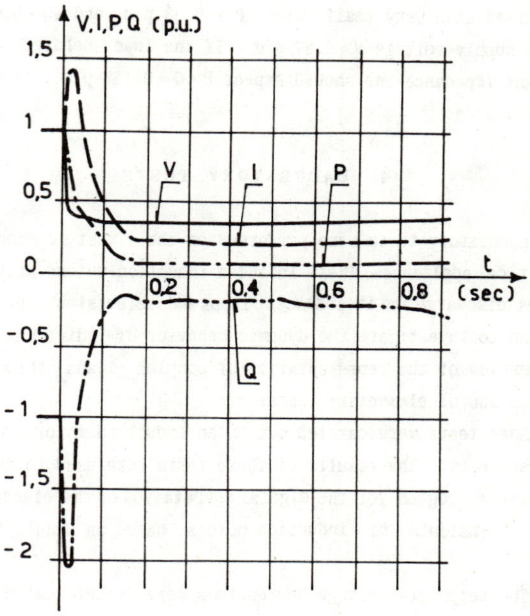

Fig. 7 - Test on the loads at Eskisehir and Pasalar. Initial conditions (1 p.u.): V = 154 kV; I = 22 A; P = 5.8 MW; Q = 1.12 MVAR (leading PF)

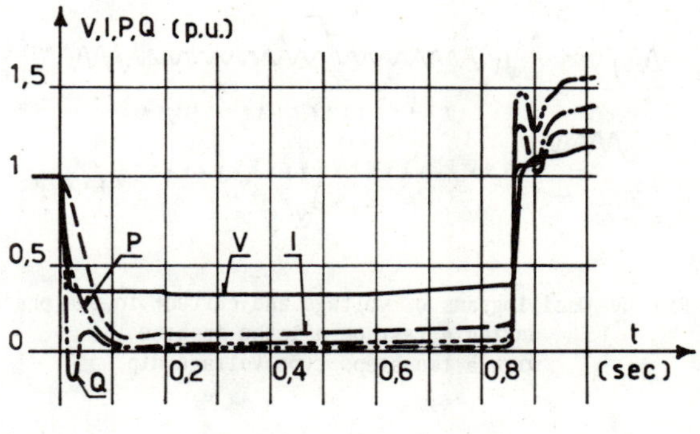

Fig. 8 - Test on the load of Adapazari. Initial conditions (1 p.u.):
V = 152 kV; I = 12.9 A; P = 2.45 MW; Q = 2.34 MVAR

The increase of I and Q during the first few cycles were caused by the residual magnetic flux of the motors. The transient condition following the voltage depression is of short duration, some 0.15 s. Thereafter an almost steady condition was recorded, with P = 0.08 p.u. and Q = 0.18 p.u. If the load had behaved like a constant impedance, the values consistent with V = 0.37 p.u. (fig. 7) would have been P = Q = 0.135 p.u. The larger reduction of P was caused by the quenching of the discharge lamps and by the automatic disconnection of part of the motors and other appliances, owing to operation of under voltage relays as well as to the release of contactors whose coils are energized with ac current.

Tests effected with different voltage depression showed a

similar pattern.

Fig. 8 gives the results of one of the tests carried out on the load of Adapazari, a little town. The night load was mainly domestic and street lighting. A small part was asynchronous motors feeding pumps.

The temporary reversal of the reactive power flow was due again to the induction motors. The duration of the initial transient condition was only some 0.1 s. Afterwards the absorbed power stabilized at a very small value: P = 0.03 p.u. and Q = 0.05 p.u. with a supply voltage V = 0.37 p.u. If the load behaved like a constant impedance one should expect P = Q = 0.135 p.u., i.e. much higher values.

4. LABORATORY TESTS

Laboratory tests were performed on the most common small electrical appliances. These included induction-motor driven machines, discharge and incandescent lamps and television sets. Their aim was to investigate the dynamic behavior of individual apparatus in view of the representation of complex loads through the composition of elementary loads.

Some tests were carried out on an induction motor coupled to a dynamo-brake. The results of these tests were used to check and optimize a program for the digital computation of the electromechanical transients of induction motors, based on Stanley's equations.

The temporary voltage depressions were symmetrical, either of the two-steps or single- step type. The two-steps type consisted of a three-phase reduction to 0.25-0.50 p.u. during 0.1 s, followed by a small reduction to 0.7-0.9 p.u. during some 0.5 s. This

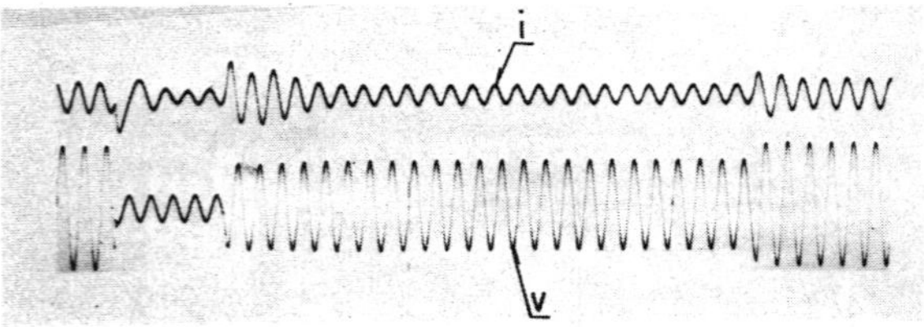

Fig. 9 - Oscillograms of voltage and current in one phase during a test on a lathe machine, with a two-steps type voltage dip

was intended to simulate the voltage transient caused by a three-phase fault cleared in 0.1 s, followed by a successful reclosure after some 0.5 s. For example, the oscillograms of fig. 9 show the record of a test on a lathe machine with a two-steps transient of type 0.25/0.75 p.u. The transients were obtained with a resistive divider switched in and off by contactors controlled by auxiliary and time relays.

Tests on induction motor driven machines

Only part of the tests performed will be described here. Of special interest are measurements on the machine tools. Their dynamic torque/speed characteristics are not known, so digital computation of the transients could not be correctly done.

Fig. 10 gives the transient of a milling machine during a two-steps voltage depression approximately of type 0.25/0.75 p.u. Such a machine, like other machine tools, has a limited inertia and a large braking torque at reduced speed. The large initial overcurrent and the reversal of reactive power flow were caused by the re-

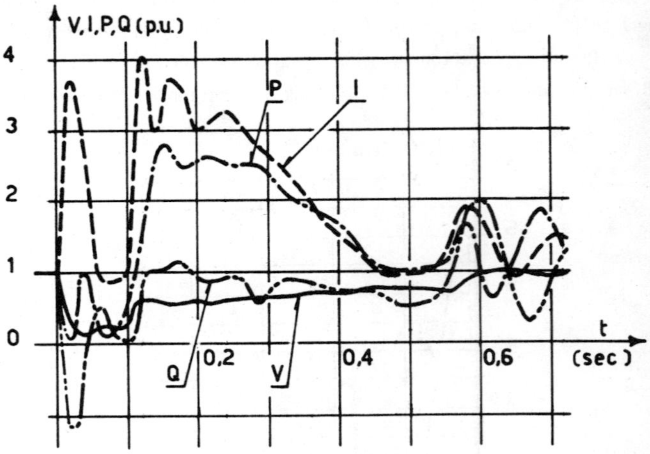

Fig. 10 - Test of a milling machine. Rating of 3-phase motor: P_n = 2,06 kW; V_n = 220 V. Initial conditions (1p.u.): V = 207 V; I = 4.05 A; P = 1 kW; Q = 1.07 kVAR

sidual field of the induction motor. The milling machine underwent an important speed reduction during the initial voltage dip as shown from the large re-starting current absorbed during the second step of voltage. The active power during speed recovery was about 2.5 times the steady value. At time 0.45 s a new steady condition was reached for V = 0.75 p.u., at nearly rated speed, with $P \simeq 1$ p.u., $I \simeq 1$ p.u. and $Q \simeq 0.6$ p.u. The reduction of Q was caused by the large decrease of the magnetizing power which overcompensated the increased consumption of reactive power in the series reactances of the motor. The latter was small because the current was only half rated value. At time 0.6 s the voltage recovered completely and the prior steady condition was reached through oscillations of limited amplitude.

Fig. 11 gives the results of a test in a small lathe, as were deduced from the oscillograms of fig. 9. The comments made above for the milling machine could partially be repeated. When the volt-

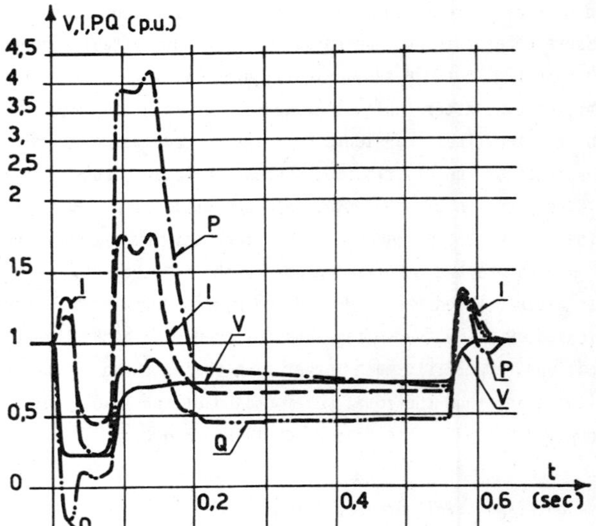

Fig. 11 - Test of a lathe machine. Rating of 3-phase motor: P_n = 0.368 kW; V_n = 220 V. Initial conditions (1 p.u.): V = 218 V; I = 1.55 A; P = 0.25 kW; Q = 0.53 kVAR age recovers to 0.75 p.u., the demand of active power reaches 4 p.u., while the current is only 1.70 p.u. The lathe did not slow as much as the milling machine and the new steady condition, with V = 0.75 p.u., was reached at time 0.2 s.

Fig. 12 relates to a reciprocating compressor operating at full load. The superimposed oscillations that the curves show are

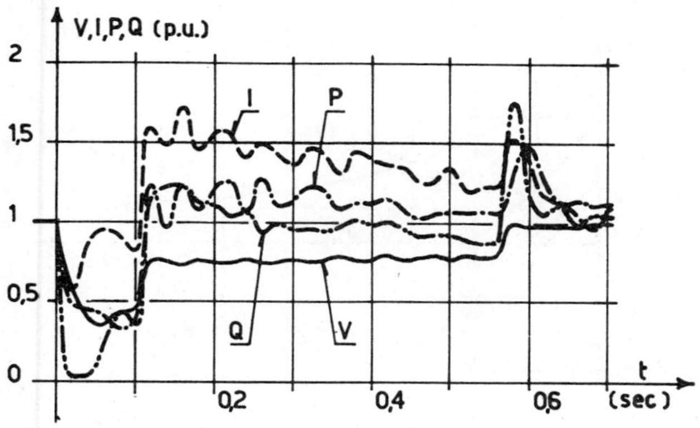

Fig. 12 - Test of a reciprocating compressor. Rating of 3-phase motor: P_n = 0.730 kW; V_n = 220 V. Initial conditions (1 p.u.): V = 220 V; I = 3.5 A; P = 1.02 kW; Q = 0.86 kVAR

caused by the alternating loading torque. At time 0.5 s the steady condition is reached for V = 0.75 p.u. The relevant active power is a little above 1 p.u., probably because of the increase of the losses.

Some general comments can be made about the machine tools:

1. They frequently operate at reduced load, much lower than the driving motors rating. Therefore, quite important voltage drops (say, 25% or so) would allow normal operation, with almost unaltered P and reduced Q, after an unconspicuous transient. At the occurrance of a large voltage dip which caused a speed drop, a large power is absorbed when voltage recovers because of the available margin, and speed is quickly regained. Fig. 10 and 11 illustrate these facts.
2. Voltage dips below 0.5-0.6 p.u. cause the release of ac energized contactors and disconnection of controlled motors. This was observed when testing the milling machine during the initial 0.1 s voltage drop to 0.25 p.u. Then an indipendent source was used for the auxiliary circuits and the results of fig. 10 were obtained.

Tests on discharge lamps

Analysis of the behavior of discharge lamps during sudden voltage depressions based on authors' tests show that the power demand is a function almost solely of the value of voltage. Therefore, the static characteristic defined in the introduction can be used for dynamic modeling.

The curves plotted in fig. 13, experimentally obtained, relate to a fluorescent tubular lamp and to a mercury vapor bulb lamp of standard type. The characteristics are nearly linear and therefore the active power may be expressed in the form:

$$P = P_n \frac{V - V_o}{V_n - V_o} \tag{1}$$

where P_n and V_n are rated power and voltage; V_o is the value of the voltage at which the linearized characteristic would intersect the axis of the abscissas. The formula (1) was found valid if:

V > 0.6-0.7 for fluorescent tubular lamps

V > 0.75-0.85 for mercury vapor bulb lamps

If the voltage drops below the above limits, even for a short while (0.1 s), the arc is quenched and current is drawn immediately to zero. The duration of the voltage drop has only a small

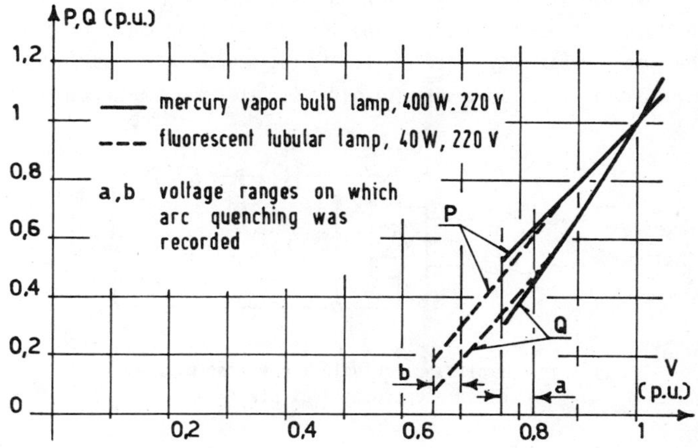

Fig. 13 - Test of a fluorescent tubular lamp, rated 40 W - 220 V, and a test of a mercury vapor bulb lamp for street lighting, rated 400 W - 220 V. Capacitors for PF correction were not connected

influence on the extinction value (10-15%). It is well known that, when the voltage is restored, the mercury vapor bulb lamps will not reignite during a few minutes.

The reactive power can be expressed in a form similar to equation (1).

Tests on incandescent lamps

The oscillographic records have evidenced an appreciable departure from typical behaviour of a constant resistance, owing to filament temperature decrease. The few tests performed showed that a better approximation is made with the formula:

$$P(t) = P_n \left(\frac{V(t)}{V_n} \right)^{\epsilon} \tag{2}$$

where the meaning of symbols is evident. The exponent ϵ decreases from the initial value, 2, to a final value that in the few tests carried out was found 1.5-1.6, appreciably independent from the amplitude of voltage reduction. The time interval during which ϵ decreases conincides with the thermal transient of the lamp; it remarkably depends on the amplitude of the voltage reduction. Values of some tenths of a second were recorded. As a rule of thumb, a linear decrease of ϵ with time can be assumed.

5. ANALYTICAL APPROACH OF THE TRANSIENTS OF THE ELECTROCHEMICAL PLANTS

In a steady condition an electrochemical plant is equivalent to a resistance R in series with a counter EMF E. The values of E and R are a function of the type of electrolysis and may be easily deduced from the treaties of electrochemistry. In practice the value of E ranges between 25% to 80% of applied dc voltage.

During the voltage depressions of short duration the values of R and E remain unaltered. The transient condition is however strongly dependent on the inductance L of the reduction cells circuit, as the time constant $\tau = L/R$ may be several tenths of a second in the large plants. So far this fact was not considered in the literature.

The simplified scheme of fig. 14 will be used here for the analysis of the transients. The inductance L is comprehensive of the ones, relatively very small, of the source and rectifier. The current i(t) is meaningful only if it is non-negative,

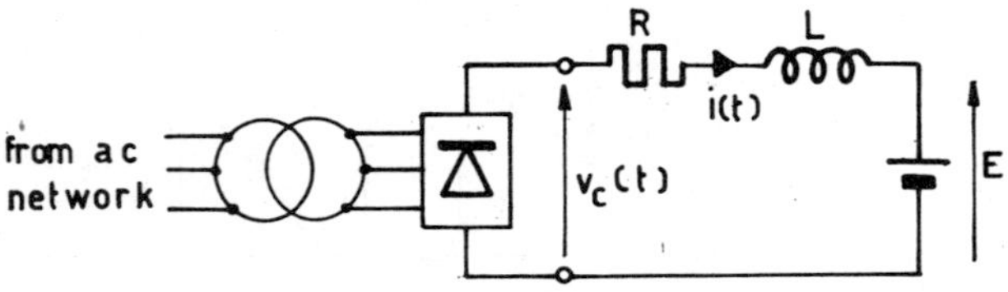

Fig. 14 - Simplified equivalent one-line diagram of an electrochemical plant

since the rectifier prevents negative current flow. The voltage $v_c(t)$ in the dc side of the rectifier can easily be calculated when the rms voltage on the ac side, $v_a(t)$, is known. The differential equation of the circuit then is:

$$L\frac{di}{dt} + Ri = v_c(t) - E \qquad (3)$$

If R is comprehensive of the equivalent resistance of converter plant, the active power demand from the ac system is:

$$P(t) = v_c(t)\ i(t) \qquad (4)$$

where non-negative values of P(t) only are meaningful.

The equations (3) and (4), along with the initial conditions, can be used for the representation of large electrochemical plants in transient stability computer studies.

If the voltage $v_a(t)$ is constant during the voltage depression and equal to a fraction K of the rated voltage V_a ($v_a(t) = V'_a = KV_a$), the dc voltage can be expressed as:

$$v_c(t) = V'_c = \rho h K V_a = \rho K V_c = \text{constant} \qquad (5)$$

where h is a constant of the rectifier and ρ is a factor accounting for the high-speed voltage control, if any ($\rho \geq 1$). If the rectifier has controlled ignition and α_o is the delay angle before the perturbance, at the occurrence of the voltage dip the delay angle will be drawn automatically to zero in the trial to support the voltage; then from theory of rectifiers it is found $\rho = 1/\cos\alpha_o$.

Let us call I and I' the dc currents circulating in steady condition when the ac voltage is V_a and $V'_a = KV_a$, i.e. $I = (V_c - E)/R$; $I' = (V'_c - E)/R = (K\rho V_c - E)/R$. Equation (3) is easily solved with the condition (5). The calculations, not reported here because commonplace, lead to the following expression of the active power (4):

$$P(t) = V'_c\, i(t) = \rho K V_c\left[(I - I')\, e^{-t/\tau} + I'\right] = \rho K V_c^2 \frac{1-K\rho}{R} e^{-t/\tau} + K\rho V_c \frac{K\rho V_c - E}{R} \qquad (6)$$

To give an idea of the importance played by the inductance L, the power absorbed by a chlorine-soda plant, after a reduction of the dc voltage to 75%, 50%, 25% respectively, was plotted in fig.15. The parameters of the plant, which is in operation, are specified in the figure's caption[5]. The value of the time constant $\tau = 0{,}45$s was determined through measurements and checked with calculations.

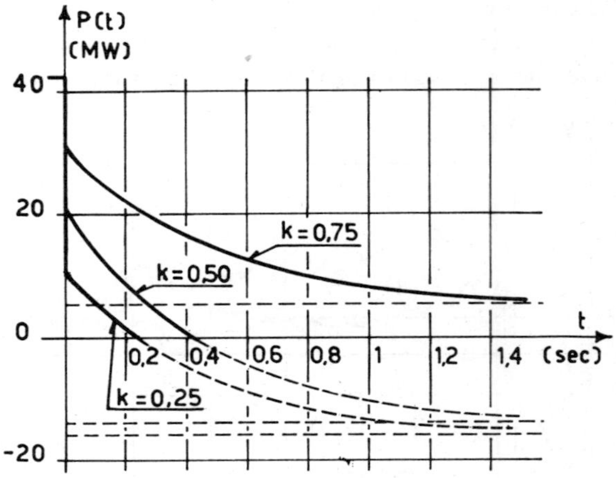

Fig. 15 - Active power demand versus time for a chlorine-soda reduction plant. Ratings: P = 43 MW; I = 220 kA; V_c = 195 V; number of cells in series 44; E = 136 V; R = 0,27 mΩ; L = 0,12 mH; τ = 0,45 s

The reactive power is easily calculated when the current and voltage are known. If the magnetizing power of the transformers is neglected, theory of rectifiers provides the following expression of the power factor, when the delay angle $\alpha = 0$, as supposed in the foregoing:

$$\cos\phi = 1 - \frac{\sqrt{2}}{2}\frac{IX}{V_a}$$

where X is the commutation reactance, I and V have the meaning above defined. Then the reactive power can be calculated at intervals of half a cycle with the formula:

$$Q(t) = P(t)\ \mathrm{tg}\,\phi(t) + q_o = P(t)\ \mathrm{tg}\ \mathrm{arc}\cos\left[1 - \frac{\sqrt{2}}{2}\frac{I(t)X}{V_a(t)}\right] + q_o \qquad (7)$$

The magnetizing power q_o of the transformers may be neglected or assumed proportional to the square of the voltage.

6. CONCLUSIONS

The experimental and analytical investigation carried out have emphasized some facts that can be summarized as follows:

1. On industrial loads including large motors, the transient phenomena caused by sudden voltage depressions have a duration of the same order of the voltage depression. The load dropping due to the operation of customers relays and to the release of ac energized contactors plays an essential role. The simulation of such complex loads in a lumped form with any characteristic of type P(V) + j Q(V), would unavoidably be far from reality.
2. The above conclusion is true also for the electrochemical plants, owing to the great value of the time constant of the equivalent circuit of the reduction cells. The large electrochemical plants may be represented in stability studies with a differential equation or with functions P(V,t) + j Q(V,t) deduced by said equation.
3. The residential loads composed mostly of so called static appliances, may be lumped in large blocks and represented with static characteristics of polynomial form, P(V) + j Q(V). To account for the extinction of the current in the discharge appliances (lamps, arc furnaces) and for the release of contactors, the coefficients of the polynomial functions should be modi-

fied when the voltage drops below critical values.

4. As outlined in the introduction, a method for the accurate simulation of complex industrial loads would be the representation as separate items of a few "equivalent induction motors" (with Stanley's equations), of the synchronous motors (with Park equations) and of the static load (with polynomial functions of voltage). An attempt for such a representation will be the subject of another paper.

ACKNOWLEDGEMENT

The authors are grateful to the Turkish Electricity Authority for having authorized the field tests on their 154 kV system.

REFERENCES

(1) F.Iliceto, G.Ruckstuhl, I.Tuncay, "Special features of the transient stability of the EHV Turkish system", paper presented to the IEEE Winter Power Meeting, New York, 1971.

(2) M.H.Kent, W.R.Schmus, F.A.McCrackin, L.M.Wheeler,"Dynamic modeling of loads in stability studies", IEEE Transactions on Power Apparatus and Systems, vol. PAS-88, pp.756-760,may 1969.

(3) E.W.Kimbark, Discussion to paper of ref.(2) above, vol.PAS-88, pp.760-761.

(4) C.Concordia, Discussion to paper of ref.(2) above,vol.PAS-88, pp. 762-763.

(5) P.L. Bertolozzi, S.Ferrario, "Gli impianti di conversione corrente alternata-corrente continua per elettrolisi", Rendiconti AEI, Milano, 1970.

Discussion

Edward W. Kimbark (Bonneville Power Administration, Portland, Ore. 97208): The representation of loads is generally recognized as a weak point in present stability studies. The paper significantly increases our knowledge of the characteristics of both individual and composite loads. It excels previous reports on this subject both in the magnitude of applied voltage dips and in the attention given to transient characteristics.

There are two kinds of voltage dips during most disturbances which threaten stability. One is caused by the short circuit; the other, by the subsequent relative angular motion of the synchronous machines. For brevity, these will be called the fault dip and the swing dip, respectively. Both are important in stability phenomena and in stability studies, but only the fault dip is specifically mentioned in the paper. The fault dip is important only in disturbances initiated by short circuits, whereas the swing dip is important in all severe disturbances, including those initiated by loss of generators, lines, or large loads.

The two kinds of dips differ in their characteristics. The fault dip is normally of short duration and has a sudden decrease of voltage, followed by a sudden increase. The swing dip is of longer duration and has a slower change of voltage. Because of their different characteristics, the two kinds of dips should be handled differently in stability calculations.

The transient characteristics of loads are prominent during and immediately after the fault dip. It does not necessarily follow from this, however, that the transients need be represented in detail. The total duration of the fault is normally so short that the angular motions of the machines during the fault are very small. Therefore its effect can be represented in computation more or less as an impulse, or the net effect on the acceleration of each machine can be averaged over the time of fault duration. To some extent, the effects of the transient due to occurrence of the fault may be offset by the effects of the transient due to clearing the fault. Also, it may be advisable to include in the special computation for the fault dip the braking effects from dc or negative-sequence components of current in the generators close to the fault.

During a stable swing, voltage at some loads may drop slowly as much as 50%, but the transients will be negligible and most of the loads can be adequately represented by their static characteristics—except for the disconnection of many motors by undervoltage relays or releases. Some of the disconnected motors will be automatically restarted but too late to affect stability.

Opinions differ on whether the disconnection of motor loads caused by undervoltage should be represented in stability computations. Obviously, it should be represented in studies made for comparison with staged tests or with recorded accidental disturbances. As to stability computations made for power system design, many persons claim that the system should be designed so that no loads will be disconnected during a disturbance resulting from a first contingency and that, accordingly, it is undesirable, or at least unnecessary, to represent disconnection of motors.

Manuscript received February 3, 1972.

R. H. Park (Main Street, Brewster, Mass. 02631): This paper represents a welcome addition to the limited amount of published data that is available in the area of how power system loads behave during severe voltage disturbances.

While there have been earlier papers, as see notably ref. (a) and its references, that brought out the point that what happens during line faults will greatly affect the way in which motor loads behave following fault clearance, due to the fact that there is a slowing down during the fault, with a consequent reduced input impedance when fault clearance takes place, there seem to have been far too few stability studies in which the actual behavior of motors has been fully factored into the calculations.

Among the results of studies that explicitly take the inertia of motor loads into account, what is shown in reference (a) is especially interesting, and it appears that too little consideration has been given to that paper.

How important is it to explicitly take into account the inertia of motor loads?

In a study that I carried out in 1928, I added the inertia of motors to that of nearby generators, but in those days faults that could seriously affect system stability occured on tie lines that were separated by relatively high impedances from load areas, and such areas received by far the bulk of their power from local generators.

However, as of today, it is becoming not uncommon for many load areas to be primarily supplied with power from remote generating stations and as brought out in the paper under discussion, as well as in reference (a), what takes place in these areas can depart greatly from what is commonly assumed in stability studies.

When it comes to the U.S., there are situations where the way in which motor loads affect stability can be especially important, as for example when it comes to cities which, in the summer, have a large amount of air conditioning load, yet receive the bulk of their power via high tension transmission.

REFERENCE

[a] B. J. Gevay and W. H. Schippel, "Transient stability of an isolated radial power network with varied load division", AIEE Trans. vol. 83, pp. 964, 970, Sept. 1964.

Manuscript received February 18, 1972.

G. J. Berg (University of Calgary, Calgary, Alta., Canada): To data, few large scale tests of the kind described in this paper have been reported. I would like to complement the authors on their success in carrying out system load response tests with large voltage depressions, and in publishing the results.

It is interesting to note that the P and Q patterns of the loads were determined from voltage and current recordings rather than by using Hall-effect transducers. Can the authors indicate the probable accuracy of their results, taking into account all phases of evaluation? Hall-effect transducers with unfiltered output are now available with rise time of the order of 1.0 millisecond. They should be suitable for recording large power swings. Small changes in P and Q due to sudden voltage drop appear difficult to record accurately by means of Hall-effect devices. In an effort to improve accuracy of measurements of active and reactive power, regardless of the magnitude of the changes, we have decided to use high speed sampling and data processing techniques by which the line current and voltage signals are transferred directly to computer memory for subsequent evaluation.

I agree with the author's comments regarding load modeling in general. With respect to induction motor load representation, the key question concerns the degree of approximation that can be tolerated. While comparative studies of simplified single induction motor load

Manuscript received February 18, 1972.

models have been made (1), work on representing a group of motors of mixed composition have only recently been reported (2,3). It is encouraging to learn that the authors have recognized mixed induction motor load model representation as a major task of load modeling, and that they are pursuing this problem further.

REFERENCES

[1] D. S. Brereton, D. G. Lewis, C. C. Young, "Representation of Induction Motor Loads During Power System Stability Studies", AIEE Trans. PAS-Vol. 76, 1957, pp. 451-61.

[2] M. Y. Akhtar, "Frequency Dependent Dynamic Representation of Induction Motor Loads", Proc. IEE, Vol. 115, 1968, pp. 166-69.

[3] G. J. Berg and A. K. Kar, "Model Representation of Power System Loads", 1971 PICA Conference Proceedings, pp. 153-62.

Francesco Iliceto, Argun Ceyhan, and **G. Ruchstuhl:** The authors wish to thank Professors E. W. Kimbark, R. H. Park and G. J. Berg for their discussions on well chosen points and for their interest and appreciation of the paper.

Professor Kimbark shrewdly suggests to consider two kinds of voltage dips: fault dips and swing dips. A voltage dip computed in the transient stability study of the Turkish system is reproduced in fig. 16

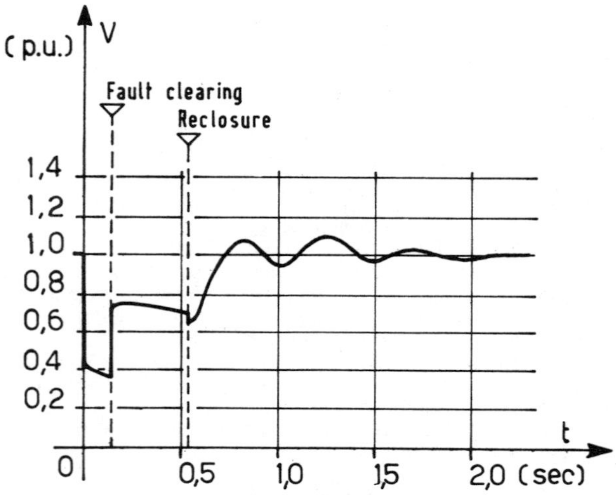

Fig. 16 – Computed voltage at 154 kV busbars in Izmir (Turkey), during a 3-phase fault at Seyitömer terminal of the Izmir-Seyitömer line, followed by successful reclosure

(from ref. (1), fig. 6). It represents the voltage at the Izmir 154 kV busbar after a 3-phase fault in the 380 kV line terminating at Izmir, followed by successful reclosure. During the fault and dead time the voltage is about 0.4 p.u. and 0.7 p.u., respectively. The swing dip is of of limited amplitude. The dead time dip appears of the same nature of the fault dip (an almost constant depressed voltage). Its duration is much longer than fault time, so that it turned out to be the controlling period for transient stability.

It is worth observing that a voltage behavior as shown in fig. 16 is typical of an initial radial system, as the present day EHV Turkish network. This justified the interest in performing the field measurements solely with fault dips, which on the other hand were the only possibility, taking into account the system operation limitations.

The swing dips play a more important role on stability of meshed grids. The swings usually have a damped sinusoidal shape around the new steady state value. The duration of the oscillations ranges from a few tenths of a second to a few seconds. It is easy to compute the behavior of individual electrical appliances during the swing dips. As an example, an induction motor of large rating (driving a boiler feed pump) was considered, having an open circuit rotor time constant T'_{do} = 6 s, a constant of inertia H = 1.25 s and a loading torque-speed characteristic of type $T_1 = kn^2$. The applied voltage was assumed to undergo a sinusoidal swing as represented in fig. 17. The computed P and Q absorbed by the motor were plotted versus voltage in fig. 17a-b.

Fig. 18 presents the calculated P and Q for the same motor versus time, during a voltage transient as represented in fig. 16

Manuscript received May 11, 1972.

The curves charts show that P and Q are not univocal functions of voltage also during a sinusoidal swing dip. A set of computer runs, not reported here for brevity, were performed making the motors parameters to vary in their practical range (H = 0.01÷6 s; T'_{do} = 0.1÷6 s ; $T_1 = kn^a$,

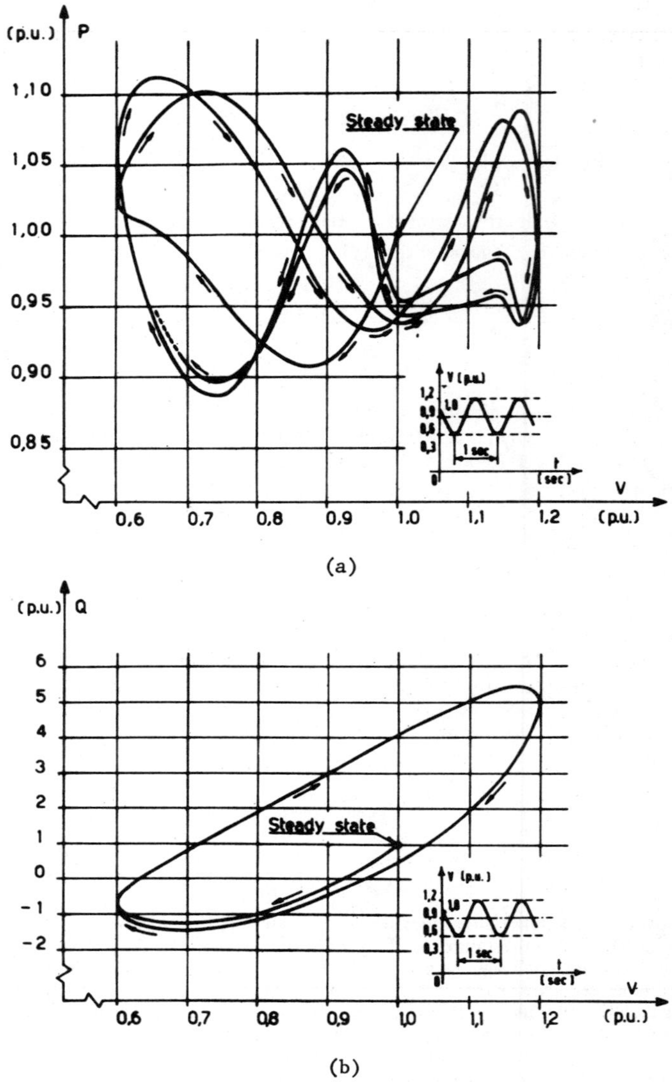

Fig. 17 – Dynamic behavior of a 6 MW-6 kV induction motor driving a boiler feed pump, during the voltage swing represented in the figure. Motor-pump parameters: T'_{do} = 6 s ; H = 1.25 s ; $T_1 = kn^2$. Active power P and reactive power Q are referred to steady state initial values (1 p.u.)

with a = -1÷2) and with different periods of oscillation of the applied voltage. This showed that the above conclusion is generally true for motors of medium and large rating; i.e. their simulation with static characteristics would give results that depart from reality. An alike conclusion is true for many electrochemical plants and rectifier driven dc motors.

Professor Park has pointed out the increasing interest for the correct simulation of motors in stability studies.

From a purely theoretical point of view, it is conceivable to represent individually all motors and other appliances. But this solution is obviously not practical, because of the enormous preparation and computation work. The acceptance of some kind of equivalents of complex loads is no doubt a necessity in stability studies. The essential point is determination of dynamic equivalents which are at same time reasonably simple and sufficiently accurate.

The paper given as reference in Professor's Park discussion describes a computer stability study of a power plant feeding through a transmission line a complex load. The latter was represented by three components in parallel: an induction motor, a synchronous motor and a constant impedance, each intended to simulate large families of motors and static appliances. Such a representation comes logically first to mind, because it appears a good compromise between the requirements of simplicity and accuracy. But so far it was not proved that large families of eterogeneous induction motors, supplied by systems of various rated voltages, can be simulated accurately by solely an equivalent motor. Moreover it was not justified how the parameters of such an equivalent could be determined. An alike remark could be repeated for synchronous motors.

It is worth observing here that the proper representation of so called static loads is not a simple task also. An increasing part are appliances fed through rectifiers (electrochemical plants, dc motors) and discharge appliances (discharge lamps, arc furnaces). As discussed in the paper, none of them is correctly represented by constant impedances.

To summarize, the authors believe that an essential premise for the correct simulation of loads in stability studies is the determination of proper dynamic equivalents of large families of omogeneous loads, above all of induction motors. Much work will be needed in this topic. The authors' paper is but one effort to present and explain some field and laboratory tests.

Prof. Berg has asked to indicate the probable accuracy of P and Q measurements during the voltage dips. It is obvious that the reading errors of voltage and current oscillographic records are important. Their percent influence is larger if the beam displacement of the oscillographer is little. Moreover, the accuracy of determination of P and Q is remarkably influenced by the reading and recording errors of phase differences between voltages and currents. Summing up all sources of errors (CT, PT, electromagnetic oscillographers, reading errors) the maximum error on active power was evaluated at 12%, when the PF is not less than 0.7. The error is usually kept within a few percents with proper calibration of instruments, if the light beam displacement of the electromagnetic oscillographer is sufficiently large and the recording paper speed is high. An alike comment is true for reactive power errors.

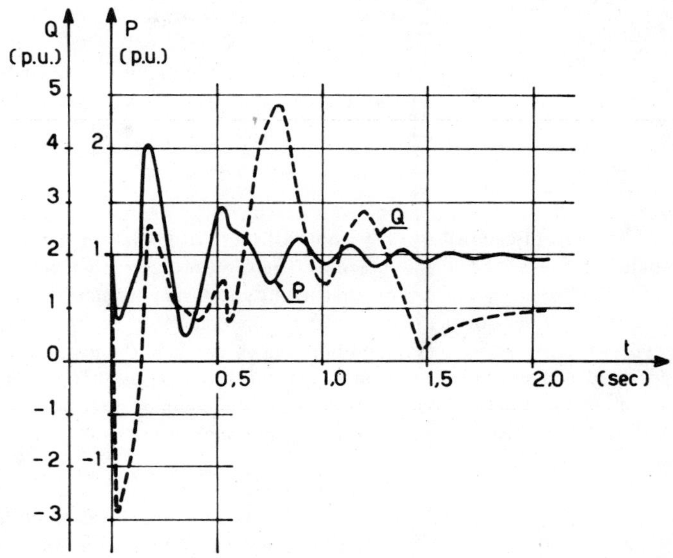

Fig. 18 – Dynamic behavior of the motor considered in fig. 17, with an applied voltage dip as in fig. 16

SYSTEM LOAD DYNAMICS – SIMULATION EFFECTS and DETERMINATION OF LOAD CONSTANTS

Computer Analysis of Power Systems Working Group
of the Computer and Analytical Methods Subcommittee–
Power System Engineering Committee

Abstract—Justification for using more realistic load models in stability studies than has been the practice heretofore is presented below. Examples are given which show that stability tests may be dramatically affected by load-voltage characteristics, or that more sophisticated models are necessary under conditions which preclude the usual composite load assumptions. In conclusion the paper states that system measurements are the most practical means of obtaining realistic load-voltage data for use in computer stability programs.

INTRODUCTION

This paper is divided into two basic parts. The first part deals with the effects of different dynamic load assumptions and modelling on the results of system stability studies. The second portion will deal with the methods of determining realistic and actual dynamic load data with particular reference to on site measurements of power system components. This work is devoted entirely to load-voltage relationships, and load-frequency characteristics will not be discussed in this particular presentation. Furthermore, attention is directed mainly to composite loads, while acknowledging that special industrial loads must be considered as separate dynamic models.

NECESSITY FOR KNOWLEDGE OF SYSTEM DYNAMIC LOAD CHARACTERISITICS

Special studies have been performed from time to time when it was obvious that the operational characteristics of large concentrated loads required special representation. The dynamic performance of large induction motor load concentrations were occasionally investigated prior to the advent of digital computers by means of analogue or network calculator studies. The induction motor models considered at least the primary and secondary leakage reactance, the motor-load inertia and some electrical load resistance which was dependent on slip. However, with these exceptions, virtually all stability studies in the 1935 to 1955 era which were performed on analogue or digital computers used a linear resistance and reactance connected in series or parallel combinations. Thus the effect of system load dynamics during stability studies was that Power and Reactive Power varied as the square of the voltage.

DEVELOPMENT IN LOAD REPRESENTATION TECHNIQUES

When load flow solutions first became available, with the advent of digital computers, the nodal iteration means of network solution became prevalent. Use of the electro-mechanical equations in stability studies, using driving point and transfer impedances as seen from the generators followed next. The first computer stability programs thus assumed in effect that bus loads were represented by constant impedances so that network driving point and transfer impedances as seen from the generators between switching times should remain constant.

As stability studies became more sophisticated, the solution of the entire network at each step of the stability calculation became prevalent, and algorithms were established which "wedded" or joined the generator and electrical equations to the electrical network. In the simplest representations however, the generator internal voltage merely became another system bus. The advantages of retaining the nodes in the electrical network were apparent since network voltage solutions produced much information at each calculation interval. Such programs as this permitted the current equations, which include the load current, to be modified at each node during every transient stability calculating interval. Present-day techniques permit constant power or constant current or constant impedance or a combination of all three characteristics. Alternate algorithms give the load characteristics as polynomial in terms of the voltage, and a quadratic function representation has usually been employed. Some comments on this latter approach are given in a later section.

It was found however, that such expressions for load power increased solution times and that the constant impedance assumptions gave the fastest answers. Furthermore, the extra core storage required from the additional programming and additional input data (for other than constant impedance representation) resulted in the tendency to still use constant impedance in stability studies. In many cases, constant impedance loads were used for convenience, usually because no system data was available.

Paper T 72 427-3, recommended and approved by the Power System Engineering Committee of the IEEE Power Engineering Society for presentation at the IEEE PES Summer Meeting, San Francisco, Calif., July 9-14, 1972. Manuscript submitted February 14, 1972; made available for printing April 20, 1972.

RESULTS OF VARYING POWER AND REACTIVE POWER DURING STABILITY STUDIES

As computer simulations expanded for load flow and transient stability studies, a natural tendency for power system analysts was to check the results of these studies with actual system conditions. During the 1960's as system interconnection grew, generating unit sizes also changed. The effect of sudden generator losses on interconnection power flows became of concern in many areas. In November of 1965 following the Northeast blackout it became necessary to simulate various other disturbances, both real and postulated. The determination of critical switching times, or the stability of large groups of generators, for assumed relay malfunctions, or for operation of circuit breaker failure protections, became widespread.

As noted above the assumption of constant impedance relative to the load representations were assumed to be acceptable due to the lack of any other information. Such assumptions were also thought to be pessimistic in most cases. One of the objectives of this working paper is to show the differences resulting from various load/voltage assumptions. Sensitivity to system simulation, particularly in dynamic conditions, both as to real and reactive power will now be discussed. Cases showing the results of different assumptions on steady state load flows will be presented in a following section.

One of the areas where the effect of load/voltage representations first became pronounced was in determination of transfer limits between pools or companies. In some instances where certain Metropolitan areas existed, the electrical utility serving them was, due to geographic or other physical constraints, interconnected over a well-defined electrical link to the "outside world" or remainder of the interconnected power systems. When generation was suddenly lost, the inertial swing was such that 80-95% of the generation loss was supplied from the interconnected system. In determining transfer limits from one pool to another, the

Reprinted from *IEEE Trans. Power App. Syst.*, Mar./Apr. 1973.

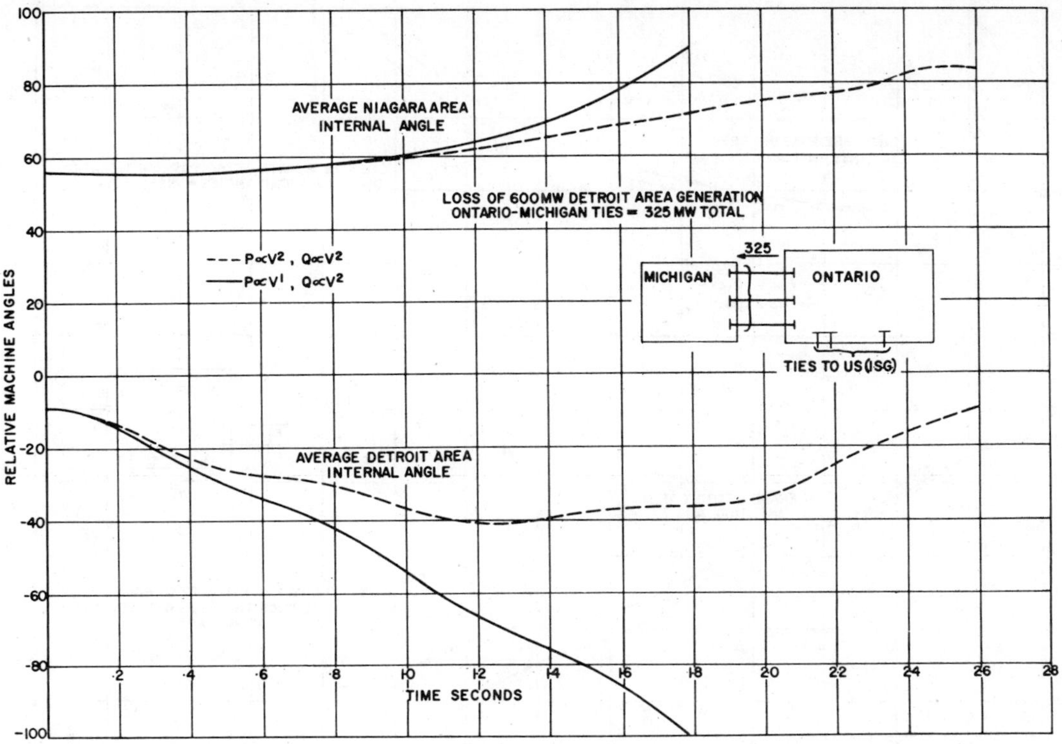

Fig. 1. Calculated loss of stability when power-voltage characteristic is changed.

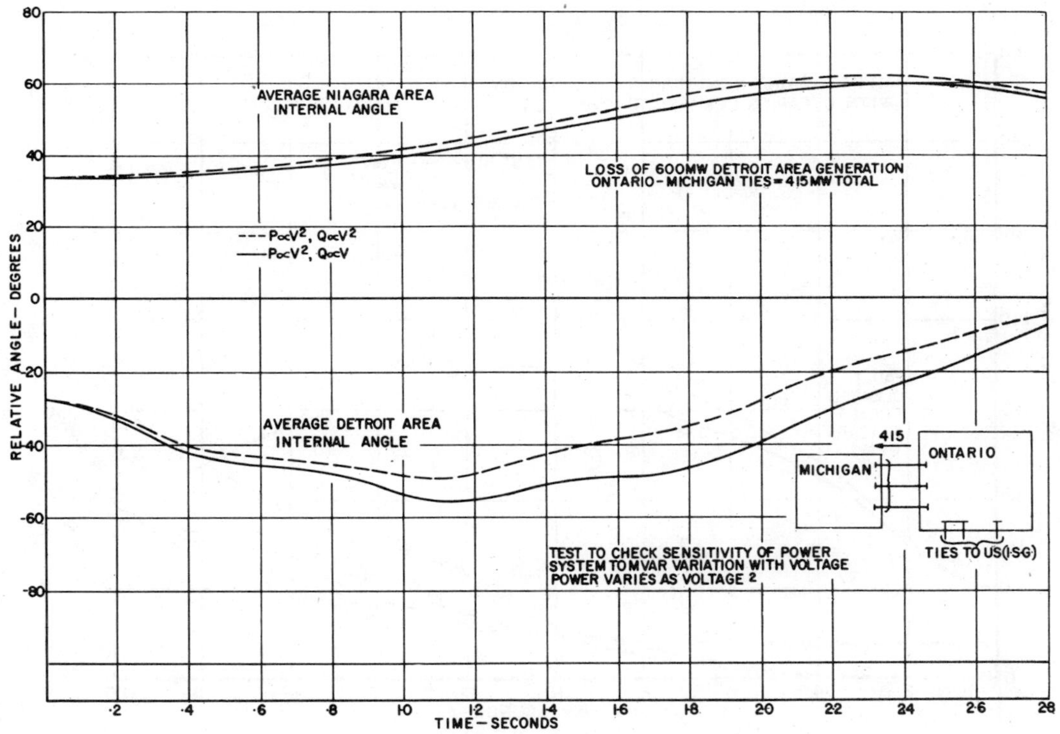

Fig. 2. Sensitivity of power system to reactive power representation – stable case.

effect of generation loss had also to be accounted for. One such case existed in practice for many years between the Detroit area generation in Michigan and the south and western 230 kv power network of Ontario. The effect of load representation throughout the entire region, including the Detroit area was found to be critical. Figure 1 shows the effects of an assumed sudden loss of 600 MW of generation in the Detroit area. The utility is importing 325 MW with the total utility load represented by about 5,000 MW. The resulting angular swings show the retardation of phase angle of one of the remaining generators in the system with respect to an important electrical bus in the outside interconnection. The assumption of constant impedance loads throughout the entire area indicates that the maximum angular retardation was about 100° at about 1.8″, and sufficient synchronizing power is exchanged that the angle is reduced to 95° at the end of 2.6″. The initial

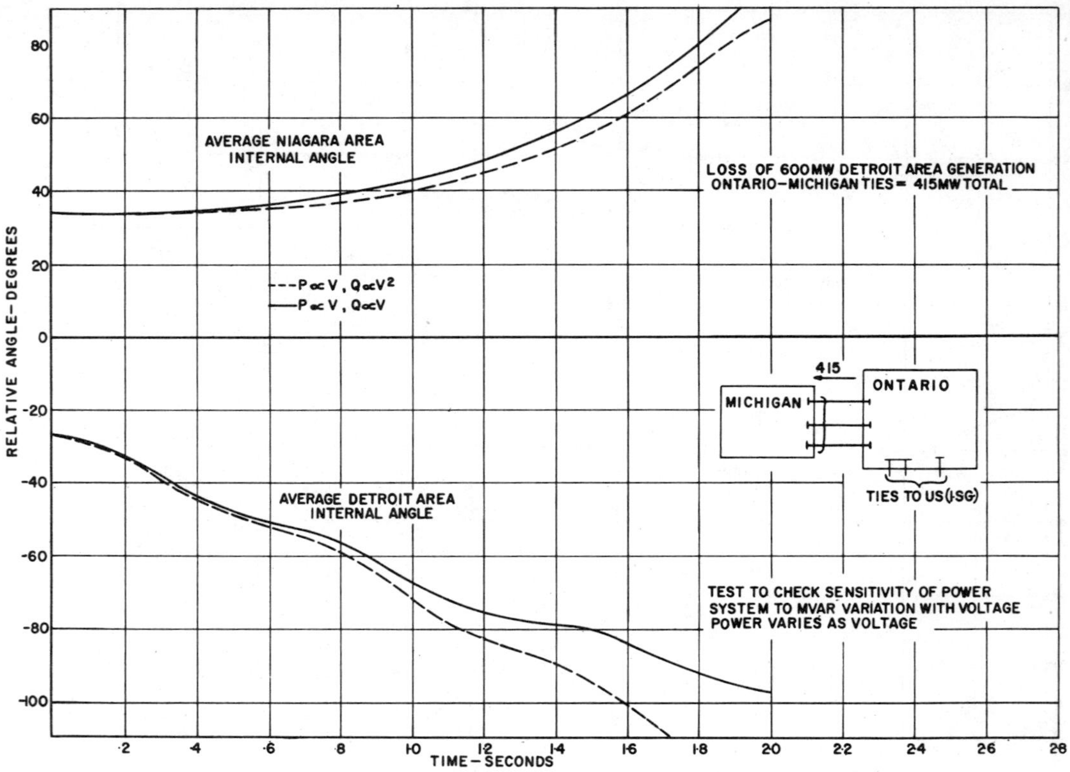

Fig. 3. Sensitivity of power system to reactive power representation – unstable case.

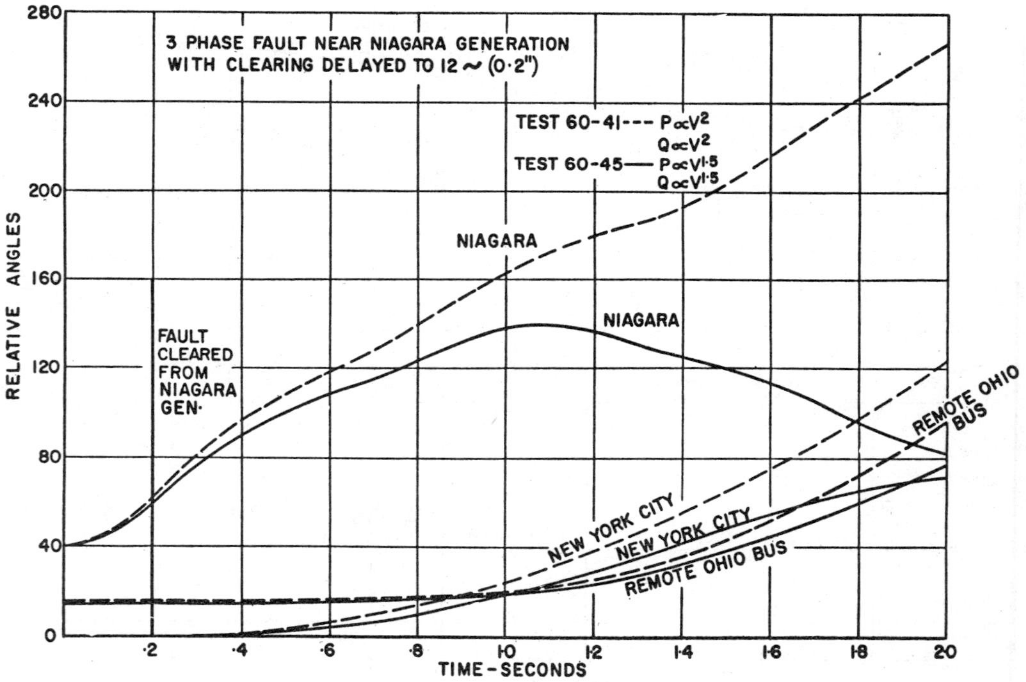

Fig. 4. Effect of change in load representation – fault near Niagara hydraulic generation.

displacement before the disturbance was 65°

When the load representation is changed to constant current ($P \propto V^{1.0}$), with no change to the reactive power representation, the receiving end utility is shown to lose synchronism with the remainder of the interconnection.

The next two figures show the sensitivity of a similar system to changes in the reactive power representation. (Although the system configuration is essentially the same as that studied in Figure 1, a different set of operating conditions was examined, and a different load flow formed the basis for the transient stability study, the basic results of which are plotted in Figures 2 and 3.) Initial transfers are slightly higher but the assumed generation loss is the same. A normally stable and a normally unstable case were examined. The solid curves show the effect of varying only the reactive component of the load representation.

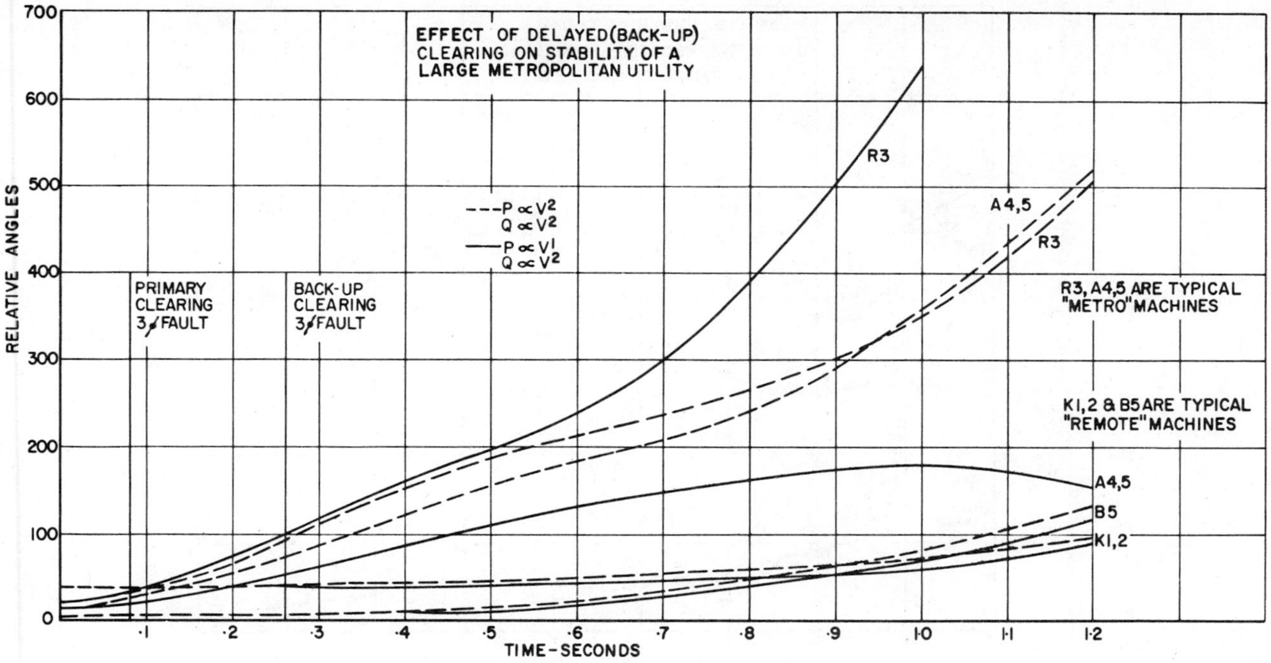

Fig. 5. Effect of change in load representation fault near Metropolitan thermal generation.

The maximum angular displacement at time t = 1.2 seconds for the case $P \propto V^2$, $Q \propto V^2$ is 93°. On the other hand allowing $Q \propto V^1$ shows an angular displacement of 98° at t = 1.2 seconds and a maximum angular displacement of 100° at a time of t = 1.8 seconds. It should also be noted that whilst the maximum angular displacement has increased slightly, the total acceleration or angular advance of the entire system is less.

For the unstable case in Figure 3 the opposite trend is apparent to that noted above, with the sending end system having advanced more while the receiving end system has been retarded a slightly less amount. We can generally conclude for this case that the effect of varying reactive power, while holding real power unchanged between computer tests, is definitely a second order effect and not significant for the usual range of voltage variations encountered in stability analysis (i.e. 80%-110%).

Another type of disturbance often under investigation is a slow or delayed clearing due to local breaker back-up operation. Two examples are presented here. The first compares the results of delayed clearing where power is being transmitted over a long distance from hydraulic plants out of a "generation-rich" area which, however, has a reasonable amount of "local" load. The generation is assumed cleared from the fault in this simulation in 12 cycles. In the first case all loads, including the local load, are assumed to have the characteristic $P \propto V^2$ and $Q \propto V^2$, and the generation near the fault becomes unstable. On the other hand, in the second instance when it is assumed that $P \propto V^{1.5}$ and $Q \propto V^{1.5}$ in all the loads, synchronism is retained and the generation "pulls back" or recovers. This is shown in Figure 4. It is the effect of the power-voltage relationship which makes this difference. The generation area voltage during the transient swing, (after the full removal) is between 85% and 95% of normal. Less electrical load is "seen" by the generation with a $P \propto V^2$ relationship, and the accelerating powers are sufficiently greater to cause loss of synchronism.

Figure 5 shows a similar situation where thermal generation, with local load and closely coupled to a Metro area, is faulted and cleared by breaker back-up protection. When P and Q are assumed to vary as voltage squared (constant impedance representation), all the closely coupled generation loses synchronism; in the second instance, with the same assumed fault clearing times and assuming that $P \propto V^{1.0}$, stability of the Metro group as a whole is maintained, with respect to the remote generation; only one unit, which is closest electrically to the fault, loses synchronism.

Once again, heuristic reasoning indicates, that after the fault is removed and the generation has advanced, the voltages in the area close to the generation continue to drop only until the peak of the generator angular swings. For a given voltage e.g. 80% of normal, more local load is "seen" by the generator if $P \propto V^1$, and hence with less accelerating power, the local generation eventually recovers, with the exception of the one unit (R3 in Figure 5).

MODELLING OF LOADS FOR STABILITY STUDIES

In order to achieve a voltage-MVA relationship other than a constant impedance one, (i.e. other than $P \propto V^2$ or $Q \propto V^2$) two basic approaches have been suggested. One is to use a voltage dependent polynomial. This usually takes the form $P = K + K_1 \cdot V^1 + K_2 \cdot V^2$, for Power (as an example). The second approach would be simply to state that $P \propto V^n$, where n might theoretically vary (in the extreme) between 0 and 3.

It may be argued that either method is acceptable during the range of voltages encountered during swings, particularly after any fault has been removed. However, the quadratic function with a constant term has been criticized because it does not pass through a zero P or Q point, for zero voltage. The Working Group is of the opinion that an exponential load model is preferred, with the exponent varying between some number near unity in the lower range to another number around 2 in the upper range. The number might be as high as 3 for the upper voltage-var relationship. The user should be cautioned against using exponents above 2, unless he feels assured that his own volt-Mvar characteristic is of this form. Another factor, influencing the volt-Mvar relationship particularly when composite characteristics are being used, is the amount of shunt capacitor power factor correction at the distribution level.

It is not the purpose of this paper to discuss the relative merit of network solution techniques during dynamic stability studies. Some opinions expressed by Working Group members were that a modified version of Newton's method was preferable to the Gauss-Seidel approach, particularly at the point in the time domain when angular swings were at their maximum. Others have continued to use the Gauss-Seidel techniques.

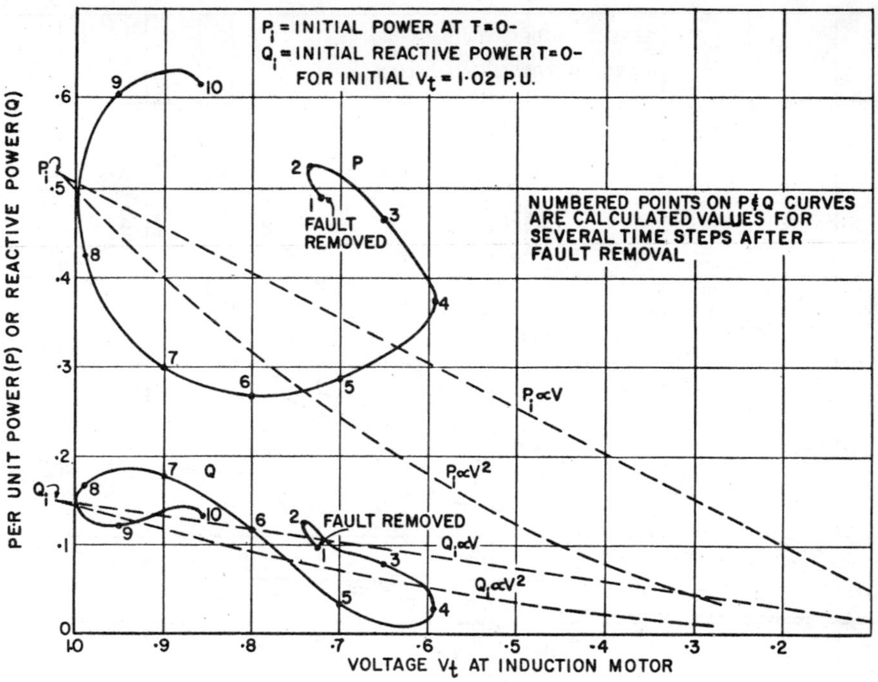

Fig. 6. Variation in P&QAS a function of voltage – induction motor load.

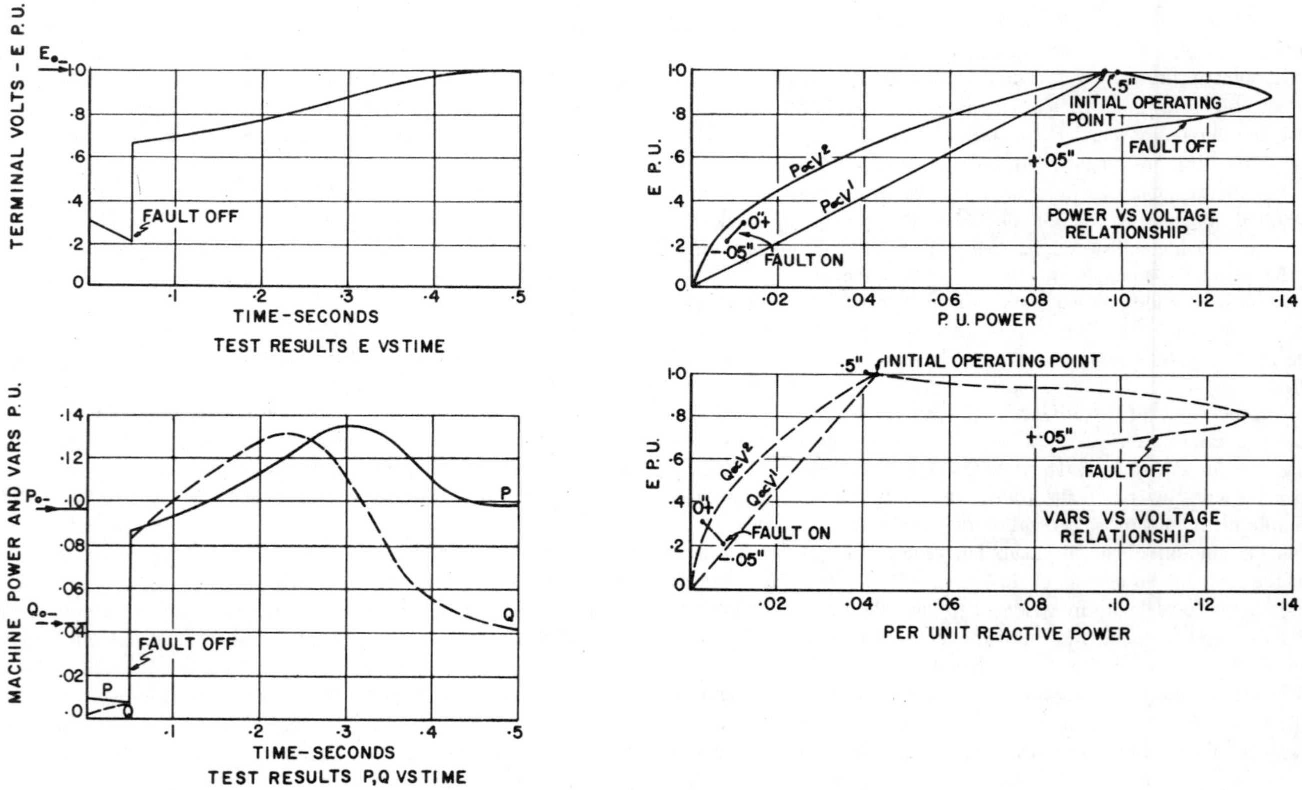

Fig. 7. 3ϕ fault at terminals of induction motor.

The point in which there is general agreement is that with constant impedance loads an exact solution can always be obtained with any method. It is the consensus that with constant power or constant current loads a solution may not exist. (See the following section on "Sensitivity of Load Representation During Faults" as well.) Any method may start tracking the "wrong" solution when non-linear load models are used. (This phenomena may also be present during steady state investigations, i.e., during Load Flow Analysis.)

All the above statements assume that the so-called dynamic characteristics of the load are as stated by the mathematical model; obviously in the case of induction motor representation, a specific model, including, in some cases the slip-torque characteristics of the driven load as well as the load-motor inertia must be accounted for.

As noted previously in this working paper induction motor models are available for analytical purposes and may be specifically used for stability studies of large groups of interconnected synchronous machines where loads are predominantly induction loads or for special purpose studies as indicated previously in the references.

Figure 6 shows a plot of the power and reactive power supplied to a particular load bus which was assumed to be composed predominantly of induction motors. The actual load mix is somewhat arbitrary – loads were considered to be 40% induction motor load, 10% synchronous motor load and 50% constant impedance load. The inertias of the induction loads and synchronous loads were 6.8 and 8.0 per unit of the MW rating of the loads and were chosen to be greater than normal in order to show a particular effect. The total load inertia in the system actually exceeded the total generator inertia and it is admitted that this situation rarely exists on a general basis for typical load mixes. In Figure 6, as the power and reactive power varied as a function of the terminal voltage, no single value expression such as a polynomial can be easily determined to adequately express the P, Q and V relationship and it is a multi-value function in this instance. Another example, Figure 7, shows the Power, Reactive Power, slip and voltage plotted as a function of time following a disturbance applied to a single induction motor. By corelating P, Q and E it can be seen that more than one value of P or Q exists for a particular voltage.

Each special load, whether real power or reactive power must be recognized when it becomes a significant part of the total system load. The modelling peculiar to that special load must then be simulated. Examples of this are rectifier loads supplying manufacturing processes or rapid transit loads. Another example occurs in representing non-linear reactor-capacitor compensating devices.

SENSITIVITY OF LOAD REPRESENTATION DURING FAULTS

Another area where load representation is important occurs during computer simulation of faults in the power system.

One computational situation is dependent on whether the type of load representation assumed permits solution of the network equations used during the network solution procedures. Constant power or constant current loads which are located near points of low voltage during assumed power system short circuits may be affected. Typical situations can be examined where, for example, the load would require current of 5 per unit at the normal voltage of 1 per unit. Should the bus voltage be depressed to such an extent during an assumed fault that the load current requirement (for an initial current pre-fault of 5 per unit) was now 25 per unit, it is easy to visualize a network which could not deliver such a current. Other less severe examples, such as constant current loads, also may lead to similar situations where iterations at a load bus, near a faulted bus, becomes excessive and the program halts because a solution does not exist. The absence of a network solution at low voltages may occur when using either the Newton-Raphson or the Gauss-Seidel method of solution.

At present many transient stability programs now change constant power or constant current loads to a constant impedance when the voltage falls below a certain minimum value. The loads are restored to their original values when the voltage is restored above the minimum.

In other actual system cases, loads near faulted buses are removed by relays operating on under voltage protection. This fact ought to be recognized when large load concentrations exist near generating station buses which are subjected to computer simulated faults, unless it is known that an adequate time delay exists in the protection system.

In summary it is evident and virtually axiomatic that a constant MVA or constant current characteristic would not exist over the whole spectrum of alternating current fundamental frequency power system voltages. This statement is particularly true at or near zero voltage and would certainly be reasonably valid over a range of voltages from zero up to at least 50% of normal.

EFFECT OF LOAD REPRESENTATION IN LOAD FLOWS

A practice has developed in power system analysis techniques wherein a post-transient load flow is utilized to determine the state of the system following a major disturbance. This is usually assumed to be the loss of a large block of generation, and the attendant flow of "inertial" power after such a generation loss is of considerable interest. The technique often employed is to increase all generation in the areas outside the "generator lost" area in proportion to their inertias or MVA ratings. This choice is somewhat arbitrary, but the MVA flows in this new post-transient load flow often give an indication of critical overloads.

When attempting to solve this second load flow, based on the loads assumed on the "pre-contingency" load flows, it is often found that the system will not solve. A solution will only exist if it is assumed that the Power and Reactive Power vary as some exponential power relative to the bus voltage. This assumption is definitely a practical one in this case, as the post-transient load flow can be considered to be a picture of the system in a dynamic state, anywhere from 5 to 30 seconds after the critical disturbance. With lower than normal voltages existing throughout most of the area in which the assumed loss of generation occurs, due to the heavy inflows of inertial power, power and reactive power could not be expected to remain at their pre-contingency values, except in cases where underload tap changing equipment has restored low tension bus voltage load delivery points to normal.

Figure 8 is a condensed power plot of a power system following the loss of 25% of its generation, and shows how two specific MVA-voltage assumptions in the post-transient state affect the answers obtained on critical tie line flows. The magnitude of voltage changes encountered at several major buses is in turn affected by such assumptions, as also shown in Fig. 8 by negative percentage numbers.

This type of analysis has also been used to determine capacitor requirements on extended intermediate and low voltage system, where capacitors are switched in during low voltage conditions following line or transformer outages.

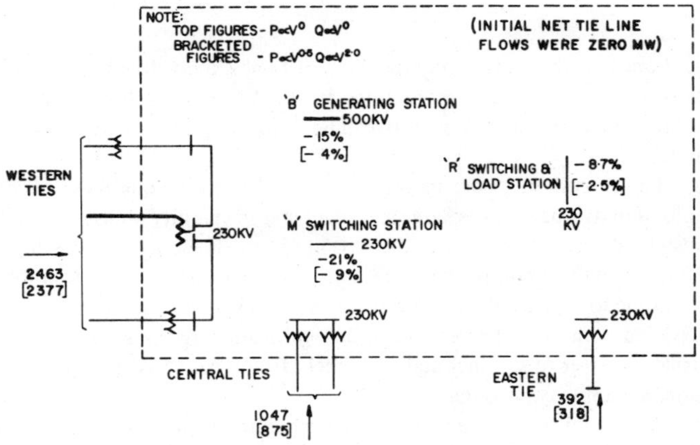

Fig. 8. Loss of entire plant output of 3960 MW near 'M' switching station.

METHOD OF DETERMINATION OF LOAD CHARACTERISTICS

There are alternative methods of establishing the characteristics of power system loads, that is how load varies with voltage or frequency. This paper is concerned only with variation with respect to voltage. Individual loads or types of load such as household appliances, tungsten or fluorescent lighting, can be analyzed from known parameters and real power and reactive power-voltage relationships established. As

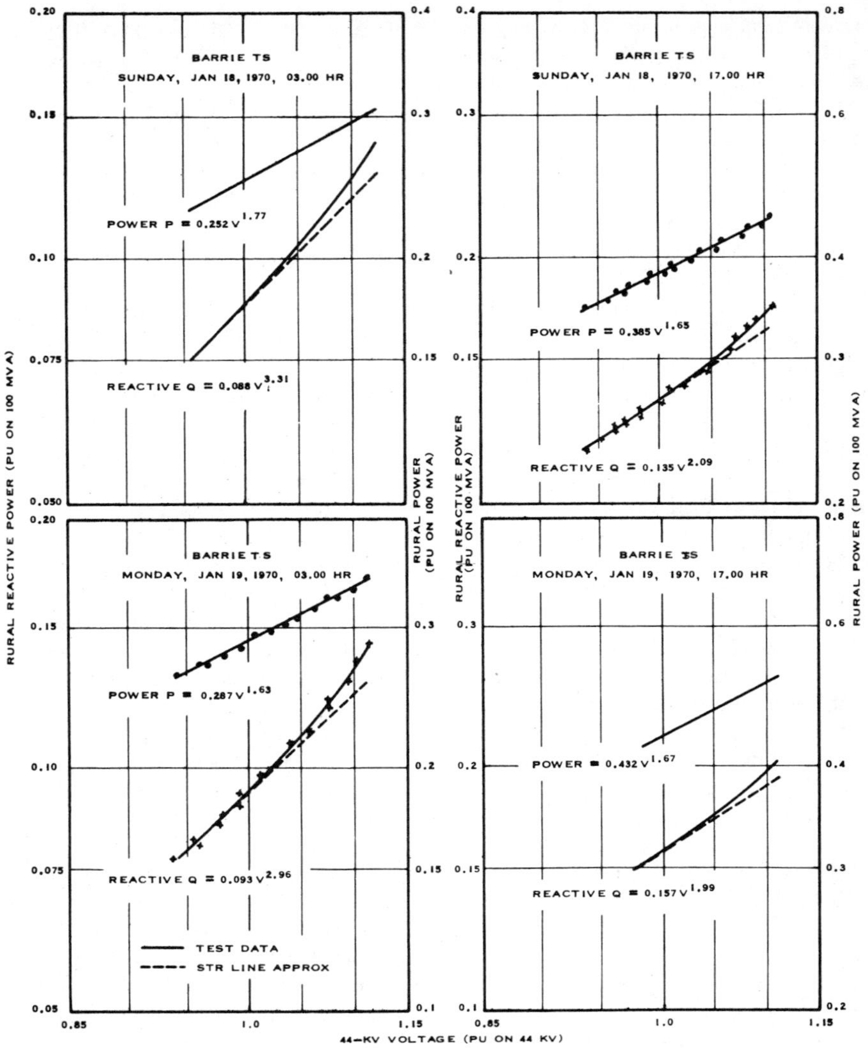

Fig. 9. Load-voltage curves for Barrie rural load.

mentioned in the section on Modelling of Loads, individual large motor loads can be specifically represented in stability studies and analytical methods or direct measurements can be utilized to develop satisfactory models.

Power system loads, in general, are a heterogeneous mass of all kinds of individual devices, large and small and in stability studies models must be devised to represent their composite behaviour. There will be variations in these composite characteristics, particularly if it is desirable or feasible to separate different categories such as residential commercial, heavy industrial, etc. There will also be significant variations, such as day to night, winter to summer and very likely from year to year, depending upon economic conditions.

The composite characteristics referred to above could be determined analytically by combining the effects of all types of devices. This, of course, would require some kind of detailed survey to determine the composition of loads supplied from various stations. An alternative to analytical and survey methods of determination of load characteristics is direct measurement at selected stations on power system networks.

MEASUREMENTS BY STAGED TESTS

Load characteristics can be obtained most directly by installation of suitable measuring equipment to monitor load and voltage and then varying the voltage and noting the consequent changes in load. The voltage can most conveniently be varied by changing taps at a transformer station. Such tests have been made at a number of stations and Figure 9 shows the results of such tests at Barrie Transformer Station.

TABLE I

BARRIE STATION LOAD-VOLTAGE TESTS
SUMMARY OF STEADY STATE dP/dV AND dQ/dV

Test # 1970	Station Total		Urban		Rural	
	P	Q*	P	Q*	P	Q*
1 Jan 18	1.62	3.2	1.58	4.09	1.77	3.31
3 Jan 18	1.61	2.45	1.64	2.80	1.65	2.09
5 Jan 19	1.63	3.32	1.57	3.88	1.64	2.96
7 Jan 19	1.64	2.44	1.57	3.25	1.67	1.99

*Represents straight line on log-log paper passing through low voltage points

Figure 9 shows the variations of real and reactive power for the rural area served by the Station, night and day on a Sunday and Monday in January, 1970. Straight line approximations of the measured data are also shown. The results for the entire station are summarized in Table 1. Similar results for other locations in Ontario are shown in Table II. These two Tables illustrate alternative methods of presenting the results.

TABLE II

SUMMARY OF TEST DATA AND RESULTS

Note: 100 MVA = 1.0 pu

Test Location	Peak Load MW-pu	Test Date	Test Time EST	$P=K_AV^{n_1}$ n_1	$Q=K_BV^{n_2}$ n_2
Rich	0.86	Jan 17	04:00	1.385	3.46
		Jan 4	14:10	1.40	3.06
		Jan 4	17:00	1.52	2.98
Ward	0.68	Jan 19	04:05	1.15	3.69
		Jan 18	14:55	1.36	3.04
		Jan 18	17:25	1.35	3.02
King	0.16	Mar 29	00:30	1.275	3.35
		Mar 28	15:37	1.345	3.15
		Mar 28	11:55	1.41	2.80
Kent	0.35	Mar 30	00:30	1.36	5.85
		Mar 29	15:00	1.22	4.275
		Mar 29	11:42	1.273	4.01
Wood	0.38	Jan 26	03:00	1.44	4.65
		Jan 25	15:01	1.26	3.28
		Jan 25	17:30	1.525	3.80
Galt	0.46	Jan 25	02:05	1.275	4.42
		Jan 24	15:00	1.370	3.566
		Jan 24	17:30	1.368	3.276

KA is initial Power
KB is initial Reactive Power

with the tap changers out of step. Switching out of one transformer then produces a step change in voltage to the load.

One of the principal disadvantages of the above methods of determination of the characteristics is that the permissible voltage variation is limited to about ± 7.5%. In practice during faults on a system, and for stability studies a much wider range is of interest.

MEASUREMENT BY CONTINUOUS MONITORING

Useful information on load characteristics has been obtained by installation of continuous monitoring equipment on selected loads, recording the output before, during and after system disturbances. Such a scheme requires the utilization of a memory-type device which can record pre-fault data as well as variations resulting from a disturbance. A schematic of such an arrangement used on two loads is shown in Figure 10.

The output was fed to a number of channels on an existing station oscillograph using magnetic tape recording. The number of channels required was kept to a minimum by continuously recording data on voltage, power and reactive levels to give pre-fault values and then, on initiation of a fault, switching these same channels to a higher gain level and recording changes only, for a pre-set period of time.

Figure 11 shows results obtained from a disturbance, in this case a fault on an adjacent circuit, causing a voltage dip on both 44 kV and 115 kV buses.

There is a marked difference in the characteristic obtained from the two feeders, for essentially a similar disturbance, as shown by the records. Feeder L6-7 is a 115 kV line to a heavy industrial load whereas 9M2 is a 44 kV city feeder and is largely commercial with some light industry and residential load. Although the voltage variations impressed on both loads were almost identical, there are large inertia oscillations

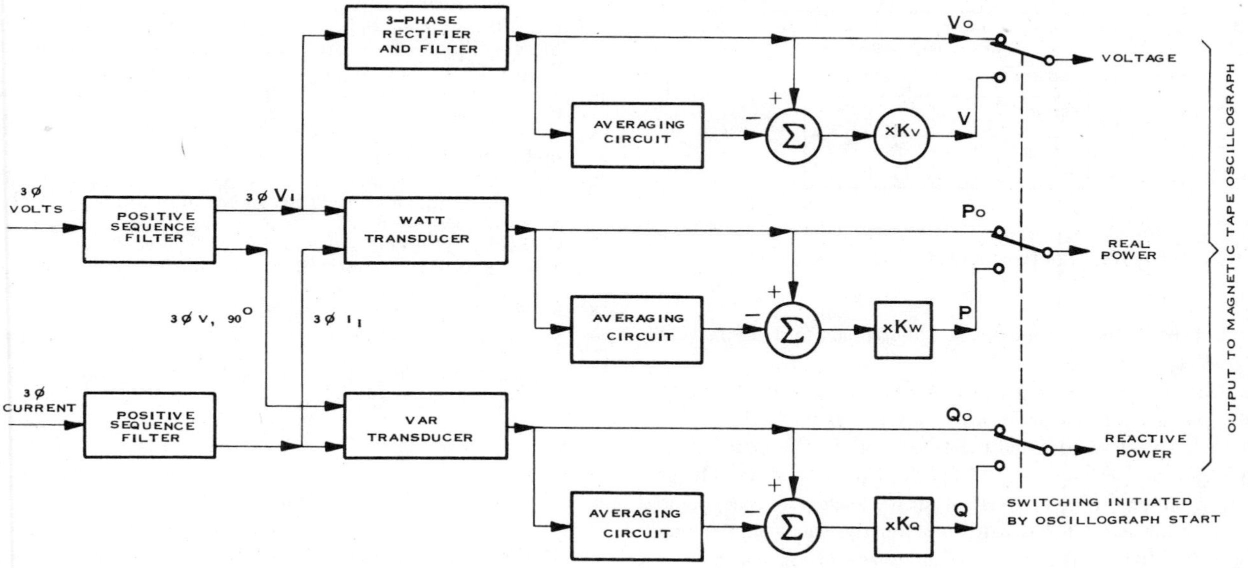

Fig. 10. Monitoring equipment for automatic recording of load/voltage characteristics on station type oscillograph.

In Table I values of dP/dV and dQ/dV are presented in per unit of nominal voltage power and reactive power levels. In Table II all values are in per unit of 100 MVA and exponential expressions of the relationship are presented.

Although transformer tap changing can only produce relatively slow changes in voltage it has been confirmed by other tests that the transient characteristics of most loads are not significantly different from those measured by using slow changes. One test, which has been used to verify this, is to supply a load through two parallel transformers

in the heavy industrial 115 kV load, whereas the 44 kV load follows the voltage variations very closely.

NATURE OF RESIDENTIAL LOADS

There has been some speculation on the nature of residential loads. In view of the large number of small motors utilized in domestic appliances it might be expected that some inertia effect would be present.

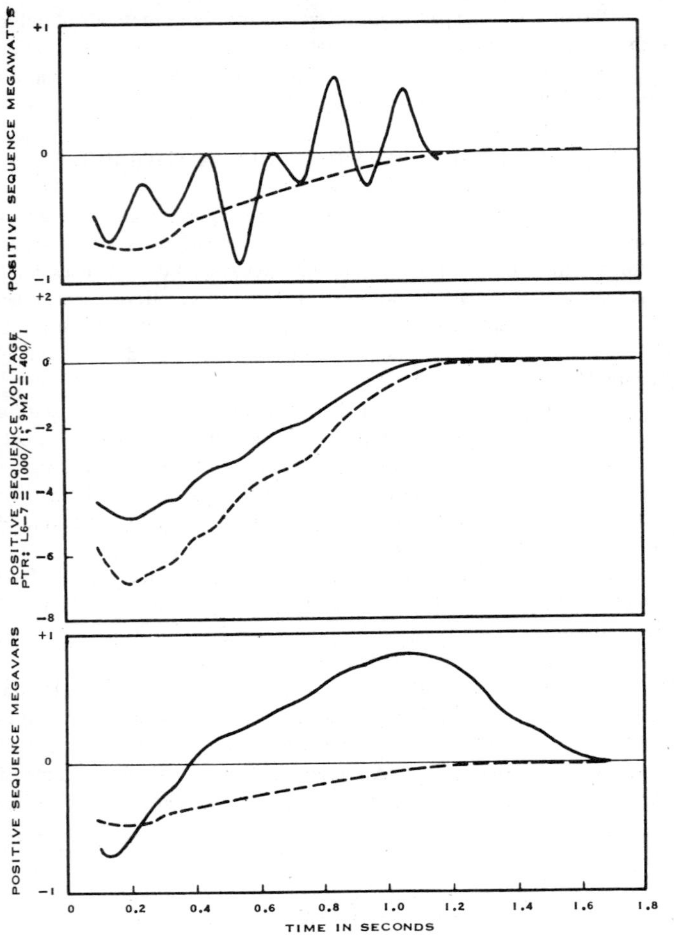

FDR	Vo KV	Po MW	Qo MX
—— L6−7	110	22.2	3.6^{T}
--- 9M2	44.2	10.4	3.0^{T}

Fig. 11. Load voltage characteristics – Martindale TS FC7 auto; May 28, 1969; 18:21:24 (Wednesday).

Tests have been made on a number of households to determine their characteristics.

In these tests typical households were supplied with step voltage changes brought about by switching to a variac supply, set at various levels, to give voltages above and below normal. By this means supply voltages as low as 50% were applied for short periods of time. It soon became apparent that the time constants of the loads were generally low enough to be negligible for two reasons. Firstly, small motors themselves have low time constants and secondly any major load such as a range or a water heater completely swamps the small motor loads.

Assuming a relationship $P \propto V^n$ with P and V both in per unit of nominal values, the above tests showed values of 'n' of 1.0 for rotating loads, rising rapidly to a maximum of 1.8 when significant amounts of resistive load, such as stoves and heating were added.

Available data shows that for incandescent lighting n = 1.6 and fluorescents, n = 1.0. Heating equipment shows a constant resistance characteristic with n = 2. The measurements taken on individual fluorescent lights for the Q vs V^n relationship show widely ranging values of n, and it is difficult to assign a common value of n to values of voltage above and below 100% voltage. This applies to most composite loads as well when voltage reactive power constants are being determined.

Some conclusions relating to the composite performance of residential loads can be drawn by examining the total kWhr per annum used by various appliances. Surveys in Ontario have shown that water heaters use 4,800 kWhr and electric ranges 1,100 kWhr per annum. A combination of primarily resistive loads such as water heater, range, TV set, kettle, radio and clothes dryer utilize 7,200 kWhr per year. Among appliances with motors, a frostless refrigerator freezer is one of the highest consumers, at 1,600 kWhr/annum and an oil furnace uses only 500 kWhr. Without the water heater the balance would be somewhat different.

In general it is valid to conclude that residential loads have negligible inertia time constants. A number of tests made on composite loads of varying character have confirmed that load inertia is generally negligible compared with generation inertia in the system. Only special individual loads have significant inertia effects, and such loads must be modelled as induction motors. or as composite groups of synchronous motors, when appropriate.

CONCLUSIONS AND RECOMMENDATIONS

The Working Group paper has pointed out the various effects of using different load-voltage characteristics on system dynamic simulation studies, and it has recommended that certain load voltage models be used depending on the composition or nature of the system loads. Finally the paper suggests, by citing examples of actual tests, that utilities should be encouraged to perform on-site measurements of composite load-voltage characteristics on their own power systems in order to obtain more realistic answers to their power system dynamic simulations

REFERENCES

[1] CIGRE Report 325 – 1950 – Report of Study Committee 13.

[2] Transient Stability Studies – I, Synchronous and Induction Machines, D. F. Shankle, C. M. Murphy, R. W. Lane, E. L. Harder, P. A. & S. Feb. 55, pp 1563-1577.

[3] Representation of Induction Motor Loads During Power System Stability Studies, D. S. Brereton, D. G. Lewis, C. C. Young, P. A. & S. Aug. 57, pp 451-56).

[4] J. A. Bubenko and T. Johansson, "Steady-State Voltage-Load Characteristics", Elteknik 1962, No. 5, p 77.

[5] IEEE Report 31 CP 66-64, "The Effect of Frequency and Voltage on Power System Load".

[6] M. H. Kent, W. R. Schmus, F. A. McCrackin, and L. M. Wheeler, "Dynamic Modeling of Loads in Stability Studies". IEEE Transactions Paper 68 TP 705 – PWR

[7] G. Funk, "The Voltage Dependence of Three-Phase Loads". Third Power Systems Computation Conference Proceedings, Rome, June 23-72, 1969, Vol. 1, Paper N.A. 7.

[8] Model Representation of Power System Loads – Conference Paper presented at PICA 7 Boston, May 71 by Berg and Kar (Paper V-A).

[9] F. Iliceto, A. Ceyhan, and G. Richstuhl, "Behaviour of Loads During Voltage Dips Encountered in Stability Studies Field and Laboratory Tests", Paper No. T 72 169-6.

ACKNOWLEDGMENTS

The Computer Analysis of Power Systems Working Group membership participating in the preparation of this report was as follows:-

Messrs: P. L. Dandeno Chairman
H. H. Brown
C. Dubé
D. W. Johnston
C. Kovlakas
W. A. Lewis
D. W. Olive
R. B. Shipley
J. W. Skooglund
G. W. Stagg
W. F. Tinney
C. C. Young
W. Watson

Discussion

R. H. Park (R. H. Park Company, Inc., Brewster, Mass. 02631): The writer would be interested to learn the duration of fault assumed in the carrying out of the calculations of Fig. 6.

References (a) (b) and (c) represent three interesting papers that deal with the effects of motor load on transient stability, while reference (d) presents a useful study of how motor loads can operate to influence processes of slow voltage collapse over a period of some seconds such as can occur following some types of system disturbance.

Reference (c) informs that the specific inertia of motors of two or less horsepower does not exceed 0.1 seconds, which implies chance of major slowdowns in load areas that experience a severe long duration voltage reduction such as can occur in the event of a stuck breaker line fault clearance situation.

When it comes to slow voltage collapse due to a suddenly occurring system dislocation a point not stressed in the paper but that is well brought out in reference (d) (cf fig. 5a) and that was also noted in reference (e) is that, as also in the case of transient stability, what takes place as regards dependence of load on voltage and time is influenced by the torque speed characteristic of the shaft load of the motor.

In the last paragraph of the paper the point is made that residential loads have negligible inertia time constants.

Here a point that applies is as to the context. When it comes to the frequency of intersystem swings the statement will usually be true, but when it comes to the transient behavior and braking effect of post fault loads, and the voltage and related load changes that occur as a result of sudden loss of generation or line trip-out, as brought out in the paper in discussion of the data of Fig. 6, inertia effects can have a major effect while also the L6-7 feeder data of Fig. 11 bring out the same point.

While, as brought out in the paper, in Ontario water heating loads and ranges account for the bulk of domestic load, the point applies that this would be less the case in areas where air-conditioning or water pumping for irrigation or a combination of the two represents a large element of the load.

In general, and as was noted in the writer's discussion of reference 9 the importance of motor loads as a feature effecting system stability depends in large measure on where the load is situated with respect to system generation and it therefore follows that in specific situations, and say, perhaps, by way of illustration in the Phoenix, Arizona, and nearby load areas, problems could arise due to a deficiency of local generation. This in turn might imply the desirability of providing a certain amount of synchronous condenser capacity rather than to rely on shunt capacitors as a means of power factor correction.

In the writer's view there not only is but will for some time continue to be a need to check on how typical loads behave when system conditions vary, and it would seem that the approach of measurement by continuous monitoring affords an excellent way to proceed, and that the power industry would be well advised to allocate R&D funds in this area of discovery.

Manuscript received July 24, 1972.

REFERENCES

[a] B. J. Gevay & W. H. Schippel – "Transient Stability of an Isolated Radial Power Network with Varied Load Division" – "IEEE Transactions," September 1964, pp 964-970.

[b] V. S. Ramsden, N. Zorbas, & R. R. Booth – "Prediction of induction-motor dynamic performance in power systems" – "PROC. IEE," Vol. 115, No. 4, April 1968, pp 511-518.

[c] M. Y. Akhtar – "Frequency-dependent dynamic representation of induction-motor loads" – "PROC. IEE," Vol. 115, No. 6, June 1968, pp 802-812.

[d] B. M. Weedy, & B. R. Cox – "Voltage stability of radial power links" – "PROC. IEE," Vol. 115, No. 4, April 1968, pp 528-536.

[e] R. H. Park & E. H. Bancker – "System Stability as a Design Problem" – "AIEE Transactions," Vol. 48, 1929, pp 170-193.

IEEE Computer Analysis of Power Systems Working Group: The Working Group thanks Mr. Park for his interest in our endeavours.

For the study results plotted in Figure 6, the actual fault clearing time was 10.5 cycles. Several tests were performed in order to assess the excursions and deviations of the MVA-Voltage plots from a constant current or a constant impedance relationship, and the fault clearing times ranged from 9 cycles to 15 cycles.

It is true as Mr. Park says that different values of motor inertia will make a difference in the results of studies. Furthermore, the presence of heave inertia in motor loads would invalidate the assumptions that composite loads could be represented simply as $P \propto V^a$, or $Q \propto V^b$, where 'a' for example, might vary from 0.8 to 2.0. If the above points were not made sufficiently clear in this Working Paper, we wish to emphasize them again. One suggestion that we might make is that an equivalent induction motor load should be capable of being represented on major load buses in large scale transient stability programs. Once again, the necessity for obtaining composite or average torque-slip characteristics, and average inertias is evident, as well as typical composite induction motor reactances and time constants.

The discusser's concluding paragraph certainly concurs with the opinions of the Working Group that continuous monitoring affords an excellent way to obtain actual system data. We do not wish either to underestimate the importance of staged tests where the station bus voltage is capable of being varied in a controlled manner from perhaps 15% below normal to 5% or 10% above normal. It appears that if industry cannot allocate R & D funds to such investigations, the Electric Research Council would be a logical centre for directing or following up the determination of load MVA-voltage characteristics. We trust the points brought out in this Working Group Paper have shown the need for such determinations.

Manuscript received September 12, 1972.

ELECTROMECHANICAL EQUIVALENTS FOR USE IN POWER SYSTEM STABILITY STUDIES

J.M. Undrill
General Electric Co.
Schenectady, N.Y.

J.A. Casazza
Public Service
Electric & Gas Co.
Newark, N.J.

E.M. Gulachenski
New England Electric System
Westboro, Mass.

L.K. Kirchmayer
General Electric Co.
Schenectady, N.Y.

ABSTRACT

This paper describes the development of dynamic power system equivalents for use in stability studies of interconnected systems. The problem of electromechanical equivalencing is formulated, and a method of constructing electromechanical equivalents is described. The performance of these equivalents has been evaluated by a series of computational tests. Theoretical material pertaining to electromechanical equivalencing is presented in a companion paper.

INTRODUCTION

This paper describes a research project directed towards the ultimate development of dynamic system equivalents which can be of use in stability and load flow studies of electric power systems. A new computational technique of constructing electromechanical equivalents for use in power system analysis is presented herein. The work reported here is the initial phase of a program whose ultimate objective is the construction of equivalents on-line in dispatch office computers.

It is widely recognized that the rapid execution of stability calculations in a dispatch office of an electric power system is a desirable objective in the development of operating techniques that will enhance the security and reliability of bulk power transmission systems. The ability to make rapid calculations of existing or projected system stability conditions enables a dispatcher to better anticipate periods of reduced stability margin, and to adjust system conditions to provide improved stability margins.

As power systems become increasingly interconnected, stability calculations become so complex that it is virtually impossible to represent an interconnected system in detail within the constraints of available memory systems in dispatch office computers. Nevertheless rapid calculation of stability conditions can be achieved by devising suitable equivalents that represent the effect of an interconnected system on the boundaries of any given system.

Computer programs are presently in use for the construction of electrical equivalent circuits of transmission networks. While the methods employed have developed little since their initial introduction for use with network analysers, the equivalent circuits produced by these programs are satisfactory for use in load flow studies and for the representation of low voltage subsystems having no major generators. They are at a disadvantage, however, when called on to represent bulk transmission networks interconnecting major generating plants outside the system under study, because they either require the retention of a node for each generator, or else they involve the empirical allocation of generator inertias in order to approximate inertia and dynamical effects as they appear at the boundaries of the system being equivalenced. This practice is not readily applied in an on-line dispatch office computer where the equivalent construction procedure must be repeated each time the configuration of the equivalenced system is changed.

The problem of electromechanical equivalent construction is formulated, and a method of constructing electromechanical equivalents is described in succeeding sections of this paper. The theoretical and analytical basis for this electromechanical equivalencing method are given in the companion paper.[1]

APPROACH TO EQUIVALENCING

The objective of this research project has been to develop a new computational technique of constructing electromechanical equivalents which are capable of representing the effects of machine inertias, amortisseur winding, voltage regulator and governor dynamics as they appear at the boundaries of the equivalenced system. The new technique recognizes that these electromechanical effects cannot be achieved by the current practices which are based on electrical equivalent circuits, and does not require explicit retention of individual generator models in the equivalent.

No attempt has been made in this paper to develop a procedure which could be used to generate a specific equivalent of any of the various extensive interconnections in the United States or Canada. Instead, the emphasis has been on the more important task of providing a consistent analytical basis for the subsequent development of procedures for constructing large scale equivalents.

ELECTRICAL AND ELECTROMECHANICAL EQUIVALENTS

The techniques presented in this paper make a distinction between the task of representing the performance of an electrical transmission network and the task of representing the electromechanical systems (synchronous machines, prime movers, and controls) that are interconnected by the transmission network. The electromechanical equivalents are not developed in terms of "equivalent circuits," "equivalent machines," or "equivalent loads," but as purely mathematical entities which can be incorporated into suitably modified transient stability programs.

It is necessary to distinguish between two types of equivalent; the electrical equivalent, and the electromechanical equivalent. The mathematical characteristics of these two types of equivalent are defined in the companion paper.[1]

The electrical equivalent is a set of algebraic equations representing the instantaneous relationship between power frequency currents and voltages at a set of selected points in the system being equivalenced. Electrical equivalent construction is based on the well known elimination formula

$$Y_{eq} = Y_1 - Y_2 Y_4^{-1} Y_3 \tag{1}$$

for the admittance matrix relation I_1 and E_1 when the current subvector I_2 is set to zero in the network equation

$$\begin{bmatrix} I_1 \\ I_2 \end{bmatrix} = \begin{bmatrix} Y_1 & Y_2 \\ Y_3 & Y_4 \end{bmatrix} \begin{bmatrix} E_1 \\ E_2 \end{bmatrix} \tag{2}$$

If I_1 and E_1 are currents and voltages at the selected set of nodes, and I_2 is forced to be zero by replacing the load currents by equivalent admittances to ground, then Y_{eq} is an electrical equivalent relating I_1 and E_1 according to

$$I_1 = Y_{eq} E_1 \tag{3}$$

The electromechanical equivalent is a set of differential and algebraic equations representing the relationship between power frequency currents and voltages at a selected set of nodes, with the dynamical effects of machines and controls at other nodes being implicitly included in the equivalent.

While electrical equivalent construction is well developed,[2,3] previous approaches[3,4,5] to the subject of electromechanical equivalent construction have been restricted to empirical treatment of inertia effects, and have presented no general theoretical basis for handling complete dynamical representations of machines and controls.

PROBLEM OF ELECTROMECHANICAL EQUIVALENCING

The problem of electromechanical equivalencing is, in essence, the problem of obtaining a compact and computationally efficient set of equations that specify the effect of the complete set of components of an electric power system on the performance of its boundary nodes where it is interconnected to other systems. An intimately

Paper 71 TP 137-PWR, recommended and approved by the Power System Engineering Committee and the IEEE Transmission and Distribution Committee of the IEEE Power Engineering Society for presentation at the IEEE Winter Power Meeting, New York, N. Y., January 31–February 5, 1971. Manuscript submitted September 14, 1970; made available for printing December 10, 1970.

Reprinted from *IEEE Trans. Power App. Syst.*, Sept./Oct. 1971.

related problem is the efficient incorporation of the electromechanical equivalent equations of one system into the transient simulation of another neighboring system in order to reflect its effect on the neighbor. The nomenclature and conventions that have been adopted for the purpose of constructing and using electromechanical equivalents follow.

Consider the task of studying a disturbance in a specific system, referred to as the *study system,* which is a part of a larger interconnection as shown in Figure 1. The study system, S, which need not be contiguous, will have a set of nodes at its boundaries with the remainder of the interconnection. The study system is surrounded by a number of *external systems* which have a significant effect on the current-voltage behavior of the boundary nodes of system S in response to the disturbance. A considerable amount of data is needed on the state of these external systems which usually are part of the region containing the study system. The external systems are, in turn, surrounded by the remainder of the interconnection, which has only a secondary effect on the study system, and for which only a limited amount of operational data is available in the study system.

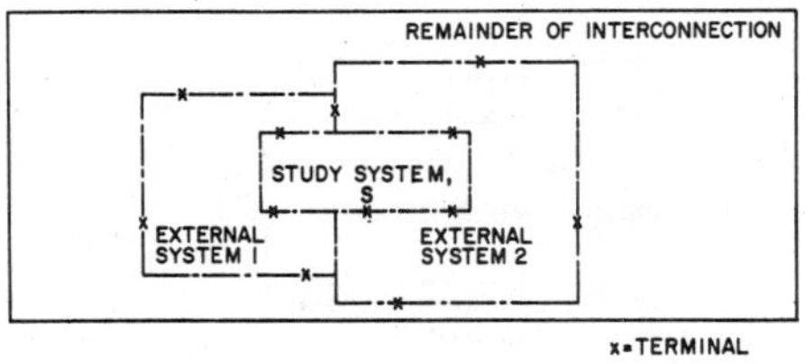

DEFINITIONS OF SUBSYSTEMS WITHIN THE INTERCONNECTED SYSTEM
FIG. 1

Each external system will have a set of nodes at its boundaries with the study system, with other external systems, and with the remainder of the interconnection. These nodes are referred to as *terminals* of the external system.

We consider, here, the construction of equivalents of the external systems. A basic set of assumptions is needed in order to make the problem mathematically tractable. The first assumption is that the terminals, T, of the study system are far enough removed from the site of the disturbance for the loads of the external system to be modelled by linearized current-voltage characteristics. This assumption allows the construction of an electrical equivalent of the external system for use in the construction of the electromechanical equivalent. The second assumption is that the disturbances propagated into the external system are sufficiently small to allow its generators to be modelled by linearized algebraic and differential equations.

These assumptions are believed to be reasonable since any part of the interconnection which would be expected to be disturbed to the extent of experiencing nonlinear behavior would normally be of specific interest in the stability study, and hence should be made a part of the study system.

Within the assumptions concerning linearity, each generating set in the external system may be modelled to any desired degree of detail, although the use of a high degree of detail will inevitably increase the dimensionality of the problem by increasing the order of the differential equations describing each generating set.

For this study, each generator is represented by the differential equations of its inertia, field winding, amortisseur windings, voltage regulator, and turbine governor. This relatively high degree of detail is used in order to demonstrate the principles of constructing electromechanical equivalents and to emphasize the generality of the formulation. It does not imply, however, that detailed modelling is a general requirement for the construction of successful equivalents.

CONSTRUCTION OF THE ELECTROMECHANICAL EQUIVALENTS

The fundamental philosophy of the electromechanical equivalencing method is:

i. The construction of a comprehensive linear differential equation describing the external system.
ii. Separation of natural modes of response by transforming the equations into a simplified canonical form.
iii. Reduction of the order of the canonical form equations by the deletion of selected natural modes.
iv. The integration of the reduced linear differential equation, in conjunction with the nonlinear differential equations of the study system, to give a simulation of the study system as it is affected by the external system.

The comprehensive linear differential equation takes the form:

$$\dot{y} = Ay + B\Delta V_T \tag{4}$$

$$\Delta I_T = \eta y' + \zeta \Delta V_T \tag{5}$$

where y is the vector of deviations of external system state variables from their equilibrium values, and ΔV_T is the vector of deviations of the terminal voltages from the equilibrium values that existed when the study and external systems were interconnected in the quiescent state. ΔI_T is the vector of deviations of the currents flowing from the study system into the external system, y' is a sub-vector of y.

Equations (4) and (5) are of high order and embody large non-sparse matrices. They are, therefore, processed by a modal analysis procedure[1] into the compact form

$$\dot{q} = \gamma q + \theta \Delta V_T \tag{6}$$

$$\Delta I_T = \eta a_1 + (\eta\beta_1 + \zeta)\,\Delta V_T \tag{7a}$$

$$s_R = a_2 q + \beta_2\,\Delta V_T \tag{7b}$$

where γ is a diagonal matrix and q is a transformed state vector which may be of lower order than the original state vector y. s_R is the rotor speed deviation of the reference generator. The use of the reduced order state vector, q, has the effect of reducing the size and storage requirement of the matrices in (6) and (7).

The efficiency of the equivalents in terms of computer core storage is, therefore, dependent upon the extent to which the order of q can be reduced below the order of y.

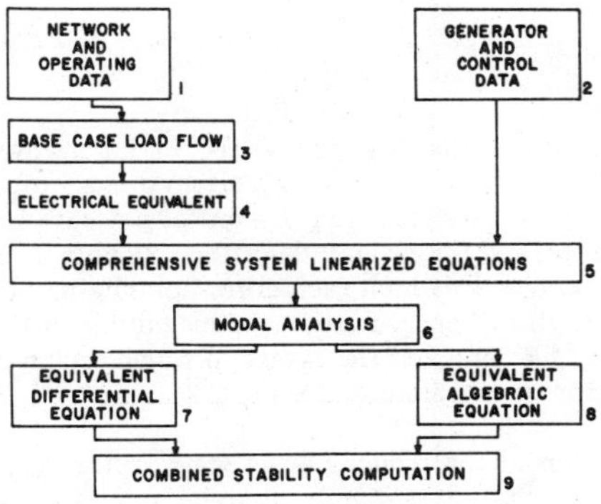

BASIC STAGES IN SYSTEM EQUIVALENCING
FIG. 2

The principal steps in the construction of the electromechanical equivalent are shown in Figure 2. Blocks one to four are standard data collection and computational functions and represent the fundamental data processing steps needed to obtain input data for electromechanical equivalencing. The analytical details of blocks five to nine are given in the companion paper.[1]

The load flow solution is needed to establish the steady-state values of all system state variables as required in the construction of the linear equations (4) and (5). No data on the study system is needed for the construction of the equivalent, but the conditions at the terminal buses of the external system must reflect the fact that the external system is, in actuality, connected to the study system. Hence it is usually necessary to specify the flows across the boundary at the terminal nodes as a part of the input data in block 1. This data can be acquired from a load flow study of the complete interconnection, or from telemetry in an on-line application.

The construction of the electrical equivalent (block 4) from the base case load flow is straightforward in concept but will, when the number of load buses is high, require carefully programmed computational procedures[6] to avoid excessive storage requirements. It should be noted that the electrical equivalent admittance matrix has an axis for each terminal and each generator of the external system, and is, in general, a fully populated matrix.

The comprehensive linearized equations (4), (5) of the external system are built up from the algebraic and differential equations of the individual synchronous machines and regulating elements. The elements of the system state vector y are the state variables of the individual differential equations describing each generating unit. For example, in equivalencing a system in which each generator is modelled by its field and amortisseur equations but has no voltage or speed regulating elements the state variables in y would be:

$$\Delta\psi_{fdi},\quad \Delta\psi_{kdi},\quad \Delta\psi_{kqi},\ s_i \qquad i = 1,2, ..., m$$

and

$$\Delta\delta_1, ..., \Delta\delta_{r-1},\quad \Delta\delta_{r+1}, ..., \Delta\delta_m$$

where m is the number of machines in the external system. The rotor angle of the r-th machine in the external system is assumed to be identically zero, and the direct and quadrature axes of this machine are taken as the angular reference for the combined simulation of the study and external systems. The reason for placing the reference axes in the external system is that the nonlinear study system equations are insensitive to the location of the references axes, while the linearized equations of the external system may, as discussed later, vary signficantly with differing choices of reference axis location.

The reduction of the comprehensive linear equations requires knowledge of the natural modes of the external system. This information is obtained by computing the eigenvalues and eigenvectors of (4) and using standard techniques[1] to transform (4) into the canonical form

$$\dot{p} = \Lambda p + \theta' \Delta V_T \tag{8}$$

$$y = R^{-1} p \tag{9}$$

where each element of the diagonal matrix Λ is an eigenvalue of (4). Since Λ is diagonal the real part of each eigenvalue is proportional to the rate at which the corresponding element of p decays to zero after a step disturbance, and the magnitude of the elements in each row of θ' indicates the extent to which each corresponding natural mode is excited by a step disturbance. Further the magnitude of the elements of each column of R' indicates the extent to which each natural mode contributes to the system state vector y.

The production of the reduced order state vector, q, is based on the fact that the deletion of rapidly decaying modes, insignificantly excited modes, or modes making negligible contribution to y, produces an acceptably small degradation in the accuracy with which the reduced equations (6), (7), reproduce the behaviour of the original equations (4), (5). The procedure detailed in the companion paper for the extraction of (6), (7) from (4), (5) is a modification and extension of established linear algebra techniques. This study is believed to be a more practically comprehensive evaluation of these techniques than has been described in the literature.

Techniques have been defined for combining the electromechanical equivalent equations (6), (7) with a variety of types of transient stability program for the execution of block nine of Figure 2. The results of this study were obtained with a stability program in which the study system loads are replaced by constant admittances to ground, the network is represented by an (electrical) equivalent admittance matrix, and the generator sub-transient impedances are appended on to the network in the manner described in Reference 7. The network solution at each integration step was obtained by combining (7a) with the study system network equation as described in the companion paper. Methods have also been defined to allow the incorporation of (6), (7) into transient stability programs using the nodal iterative, Newton Raphson, and factored admittance matrix methods in their network solutions.

PERFORMANCE OF ELECTROMECHANICAL EQUIVALENTS

Objectives

This section describes the results of a series of computational tests which were made to evaluate the performance of the electromechanical equivalents constructed by the method outlined in the preceding section. The principal objectives in these performance evaluations were:

i. To confirm the validity of constructing equivalents based on linearized modelling of the generators and their regulating elements.
ii. To confirm the validity of the modal analysis process for reducing the comprehensive system linear equations to a compact low-order form.

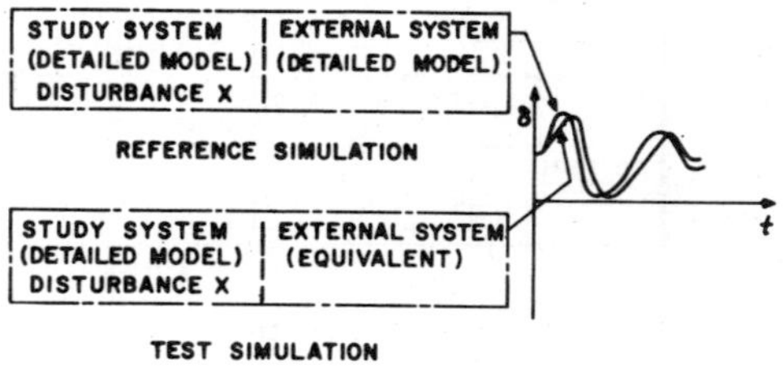

EQUIVALENT TESTING PROCEDURE
FIG. 3

Method of Testing

The method of performance evaluation is illustrated in Figure 3. All tests were made on the sample system whose network diagram and control system block diagrams are shown in Figures 4, 5, and 6. This system is generally typical of the 345kV system of the New England area; but, since it is intended as a basis for the comparison of one computational technique against another, no attempt was made to have it exactly reproduce any specific operating condition. The steps involved in the test procedure were as follows:

a. Obtain base case, faulted case, and post fault load flows for the complete system.
b. Execute a reference simulation of the complete system.
c. Partition the system and obtain base case load flows for each part to match the base case for the complete system.
d. Construct the equivalent of the external system.
e. Execute the combined solution of the study system and the equivalent using input conditions identical to those used in b.
f. Plot and compare transient responses from b. and e.

These tests covered three different external system designations as shown in Figure 4. Each equivalent was evaluated with respect to test disturbances at locations in the study system which caused major rotor swings on machines in the study system, but which were generally well removed from the terminals of the external system.

The sample system was chosen because it contains the representative features of bulk transmission system representations that would normally be used for dispatch office system security evaluations. The relatively small size of the sample system makes it a stringent test of equivalents because it is impossible to designate an external system that is so little affected by the test disturbance that it has no effect on the response of the system.

Two classes of test simulation were made with each combination of test disturbance and external system designation. The first class was intended to indicate the performance of the linearized model of the external system, before the linearized model is contracted into a low order equivalent by the rejection of natural modes. The second class was intended to indicate the effect on the linearized model of rejecting modes, and to evaluate various strategies for the rejection of modes.

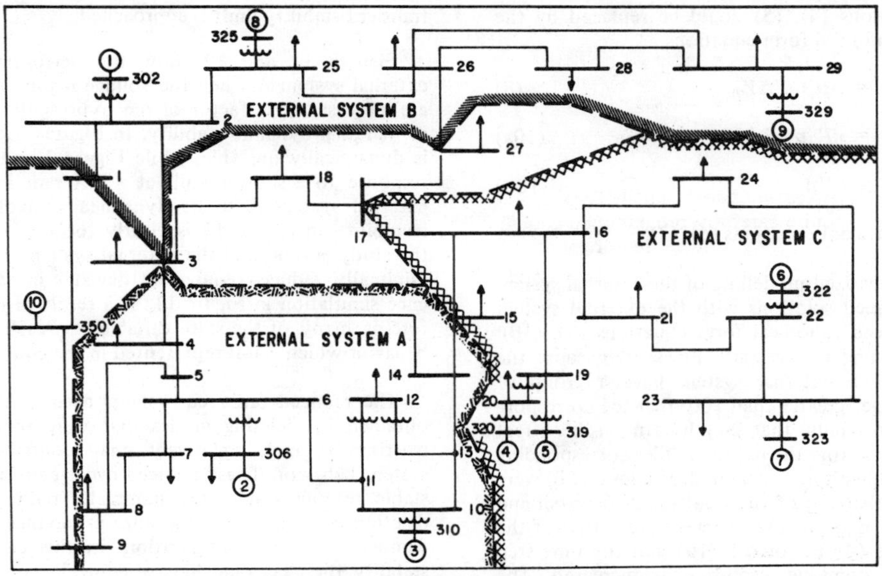

SAMPLE SYSTEM FOR EQUIVALENT TESTING
FIG. 4

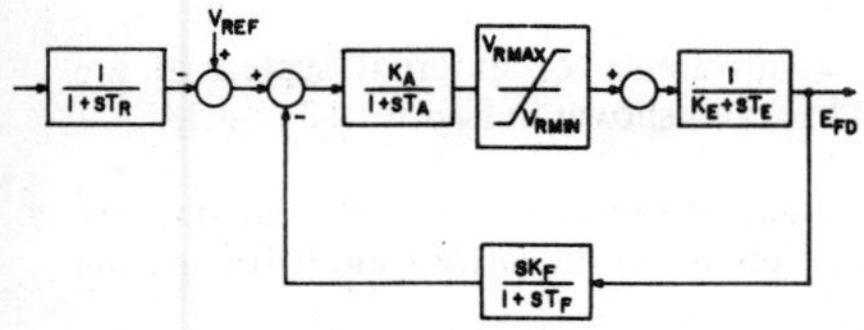

EXCITATION SYSTEM CONFIGURATION
FIG. 5

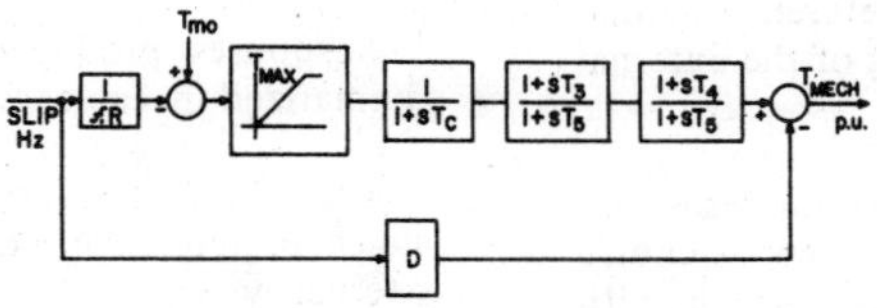

TURBINE GOVERNOR CONFIGURATION
FIG. 6

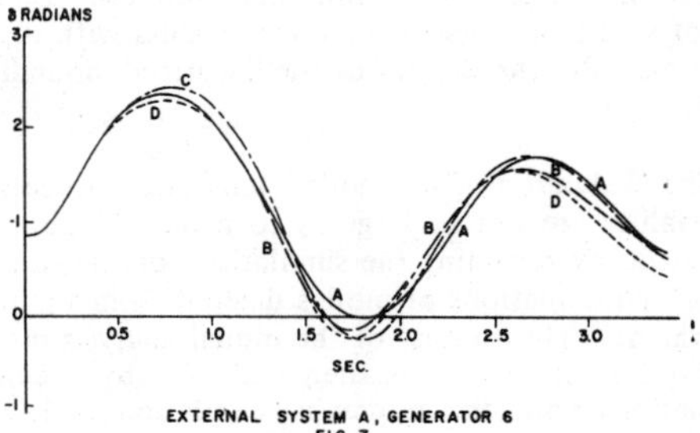

EXTERNAL SYSTEM A, GENERATOR 6
FIG. 7

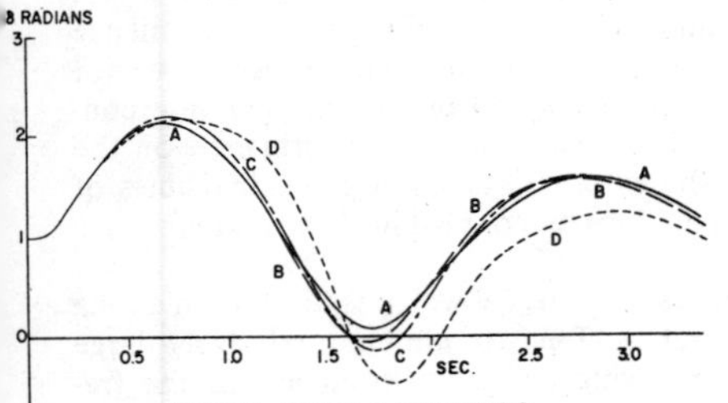

EXTERNAL SYSTEM A, GENERATOR 8
FIG. 8

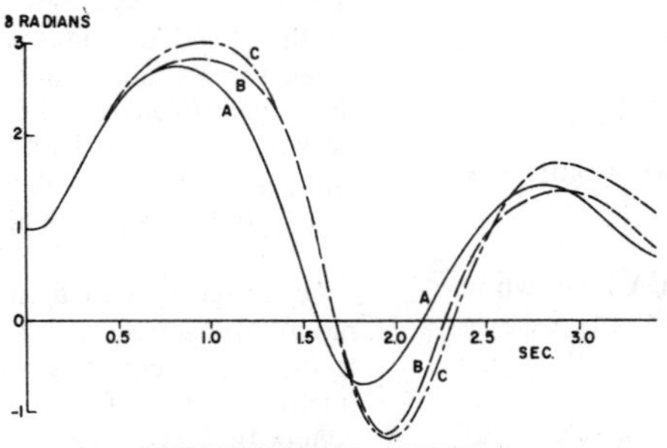

EXTERNAL SYSTEM A, GENERATOR 8
FIG. 9

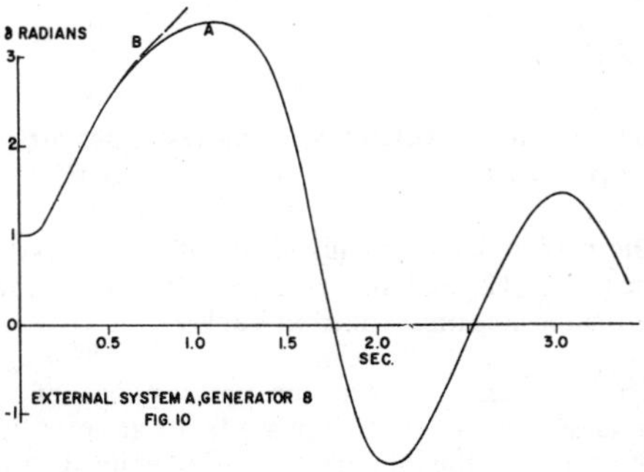

EXTERNAL SYSTEM A, GENERATOR 8
FIG. 10

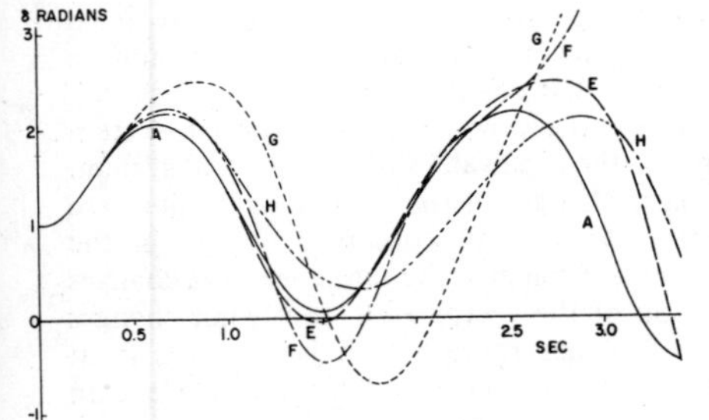

EXTERNAL SYSTEM A, GENERATOR 8
FIG. 11

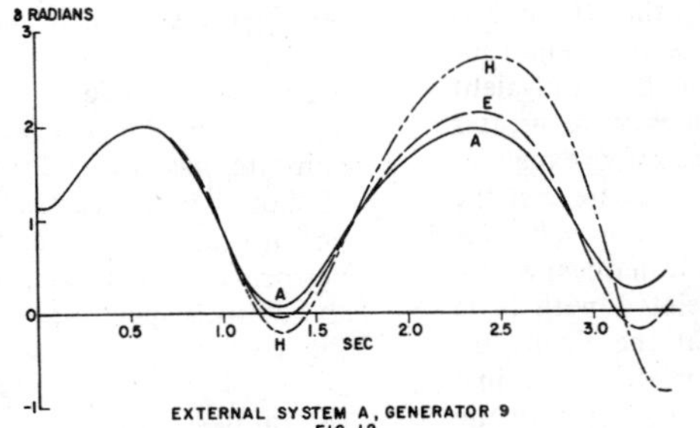

EXTERNAL SYSTEM A, GENERATOR 9
FIG. 12

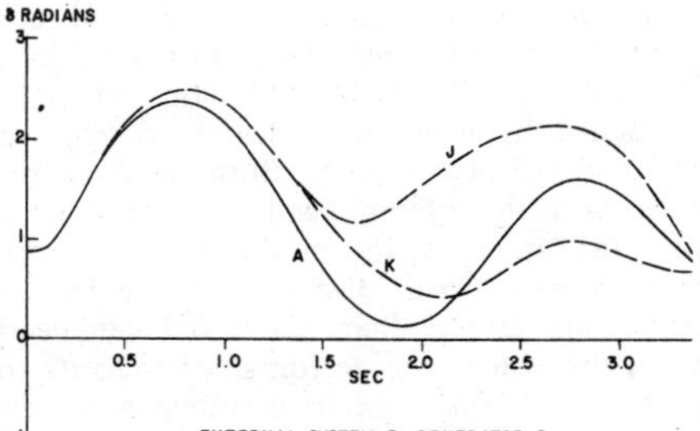

EXTERNAL SYSTEM B, GENERATOR 6
FIG. 13

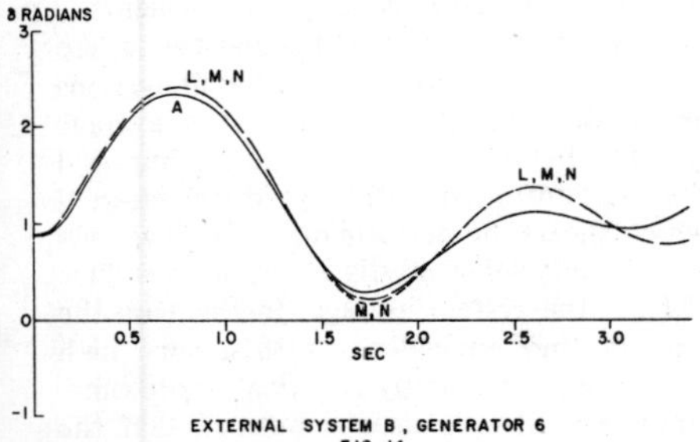

EXTERNAL SYSTEM B, GENERATOR 6
FIG. 14

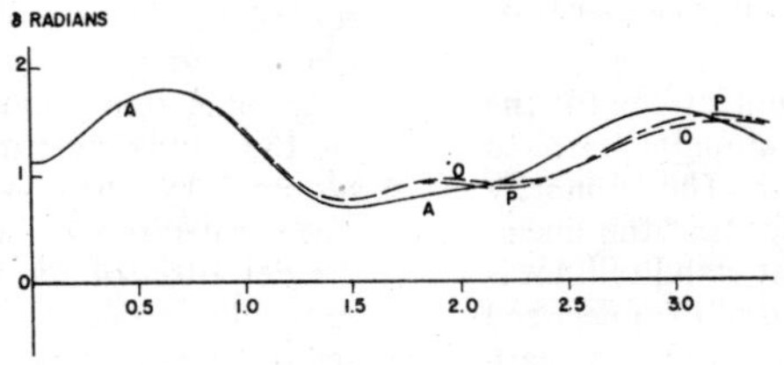

EXTERNAL SYSTEM C, GENERATOR 9
FIG. 15

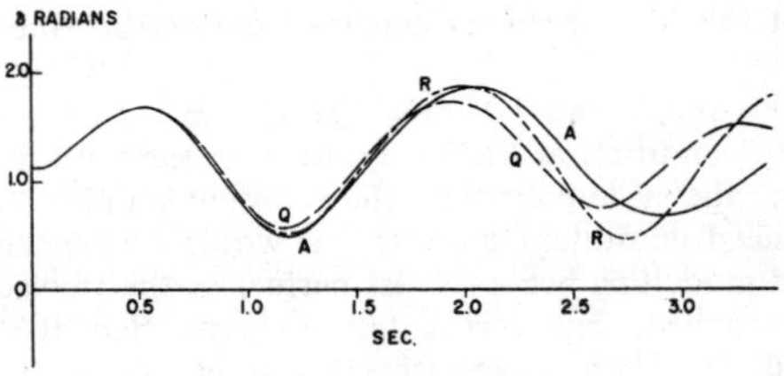

EXTERNAL SYSTEM C, GENERATOR 9
FIG. 16

For the purpose of simulating the response of the system the comprehensive linear equations (4), (5) could be replaced by the corresponding unreduced canonical form equations.

$$\dot{p} = \Lambda p + \theta' \Delta V_T \quad (8)$$

$$\Delta I_T = \eta R'' p + \zeta \Delta V_T \quad (10a)$$

$$s_R = R''' p \quad (10b)$$

or by the reduced order equivalent equations (6), (7).

A strict evaluation of linearized modelling of the external system would have required combined solutions with the external system represented by the unreduced canonical form equations (8), (10). This was both undesirable and unnecessary, however, because the differential equations of each external system have a group of eigenvalues which are so large (greater than 100) that the corresponding natural modes decay so rapidly that they have no visible effect on the dynamic response. Therefore the natural modes corresponding to large eigenvalues (with magnitudes greater than about 100) were deleted for all tests. The behaviour of the resulting differential and algebraic equations (6), (7) is negligibly different from that of the original linearized equations (4), (5) or (8), (10), but the time step that can be used in the combined solution is increased. The equivalents from which only these large eigenvalue modes have been deleted are referred to as "minimally reduced" equivalents. Comparison of simulations using these equivalents with the reference simulations indicates the validity of the linearized modelling of the external system.

The evaluation of the modal elimination process for modes whose eigenvalues are not so large as to make deletion an obvious choice was made by repeating the simulations of the test disturbances with various combinations of modes deleted from the minimally reduced equivalents. The success of the modal analysis method for reducing the equivalent order was then indicated by a comparison of these simulations with the reference simulation and with the results of using the minimally reduced equivalents.

Test Results

The reference simulations for all test cases are shown as curves A in Figures 7 to 16.

The performance of equivalents of external system A is shown in Figures 7 to 12, and the characteristics of the equivalents used in these figures are summarized in Tables I and II.

Figures 7 and 8 show the performance of the equivalents of external system A for three-phase faults at buses 22 and 2 respectively. In each case the minimally reduced equivalent produces excellent agreement with the reference simulation, indicating that the linearized modelling of the external system is valid in this case. The runs with 12 and 7 of the original 23 states retained in the equivalent were obtained by deleting modes whose eigenvalues were greater than 10 and 0.5 respectively. Both of these runs show excellent agreement with the reference simulation; and hence indicate that the reduction of order by the modal analysis procedure is valid for modes with quite small eigenvalues. The fact that modes whose eigenvalues are greater than about 0.5 can be deleted with little effect on the equivalent performance suggests that the significant energy storage effects in external system A are associated with time constants of two seconds or longer; and this is consistent with the inertia constants of the machines in the external system. The run with only 4 of the 23 states retained required the deletion of modes whose eigenvalues are smaller than 0.5, and Figures 7 and 8 show that this has somewhat degraded the performance of the equivalent.

Figures 9 and 10 show the system response simulations for the same condition as Figure 8, but with the fault duration increased to bring the system close to the transient stability limit. The minimally reduced equivalent indicates, as would be expected, that the linearized modelling becomes less perfect as the transient stability limit is approached; but fortunately suggests that the use of linearized modelling gives conservative (less stable) simulation results. Comparison of traces C in Figures 7, 8, and 9 suggests that the reduced equivalents give results which become more conservative as the transient stability limit is approached.

Figures 11 and 12 show the performance of equivalents of external system A when the voltage regulator gains of the machines are increased by a factor of ten to bring the system into a condition of marginal dynamic stability. In Figure 11 the reference simulation is dynamically unstable, while Figure 12 shows a marginally stable response to a shorter fault at a different location. Table II shows external system A to be dynamically stable, and so the dynamic instability in Figure 11 is clearly the result of interaction between the study system and the external system. The simulation using the minimally reduced equivalent deviates increasingly from the reference simulation in Figure 11, as a result of the growing disturbance forcing many of the state variables to reach the range of non-linear behavior which is not represented in the equivalent.

The reduced order equivalents used in Figures 11 and 12 were obtained by deleting modes that were indicated by the θ and a matrixes to make relatively small contributions to the external system behavior. These reduced order equivalents give generally less stable responses than the minimally reduced equivalent, with the deletion of modes having a more pronounced effect than in the well damped standard configuration. In this case of marginal dynamic stability the electromechanical equivalents continue to give satisfactory and slightly conservative results with respect to first swing transient stability.

Responses produced by equivalents of external system B are summarized in Tables III and IV and shown in Figures 13 and 14.

Table IV and Figure 13 illustrate the importance of a satisfactory choice of reference axis location in the linearized modelling of the external system.

It is assumed in the linearized modelling of the external system that the currents and voltages at all points in the external system have the same frequency, and the location of the reference axes on the rotor of a selected machine forces the frequency to correspond to the speed of this machine. This assumption is no inconvenience when there is a machine in the external system whose speed does not differ significantly from the average speed of the complete interconnection, but can lead to differences in the performance of the equivalent in cases where the external system contains machines of relatively low inertia which are loosely coupled into the system.

External system B includes generator 9 which is located on a long radial transmission line and is free to undergo relatively large frequency excursions while having a smaller influence on the frequency of currents and voltages within the system. It is apparent that generator 9 is a poor choice of reference location for external system B, while generator 1 should be a better site by virtue of having closer electrical coupling to the network as a whole. This is demonstrated by Figure 13 which shows the performance of minimally reduced equivalents having reference axes located at generator 1 and at generator 9. Table IV indicates that changing the reference location, and hence the definition of the frequency of external system currents, has changed some of the eigenvalues of the external system. It should be noted that these changes in the system eigenvalues are not necessarily the major reason for the observed change in the performance of the equivalents. Changing the reference axes changes the defined speed of rotation of the vector V_{To} and hence changes the vector of terminal voltage deviations which is obtained from $\Delta V_T = V_T - V_{To}$. This also changes the effect of the equivalent on the study system.

Because the performance of the electromechanical equivalents is sensitive to the location of the reference axes it follows that these equivalents will be most reliable when applied to external systems whose frequency does not deviate significantly from the average interconnection frequency when disturbances of interest are imposed on the study system. This is confirmed with regard to external system B by Figure 14, which shows the performance of equivalents when generator 9 is assigned a very large inertia in order to reduce the deviation of its speed from the system average. In this case the minimally reduced equivalents, and equivalents with a very high degree of reduction, give good and essentially identical approximations to the reference simulation. It will also be noticed that the

TABLE I

Test Runs – External System Designation A.

Figure	Disturbance	Curve	Equivalent Description	Minimally Reduced	System Condition
7	0.075 second fault at bus 22. Open line 21–22	B C D	19 of 23 modes ($\lambda < 100$) 7 of 23 modes ($\lambda < 0.5$) 4 of 23 modes	Yes No No	Standard
8	0.075 second fault at bus 2 Open line 2–25	B B C D	19 of 23 modes ($\lambda < 100$) 12 of 23 modes ($\lambda < 10$) 7 of 23 modes ($\lambda < 0.5$) 4 of 23 modes	Yes No No No	Standard
9	As Fig. 8 but 0.125 second fault	B C	19 of 23 modes ($\lambda < 100$) 7 of 23 modes ($\lambda < 0.5$)	Yes No	Standard
10	As Fig. 8 but 0.15 second fault	B	19 of 23 modes ($\lambda < 100$)	Yes	Standard
11	As Fig. 8	E F G H	19 of 23 modes ($\lambda < 100$) 6 of 23 modes 9 of 23 modes 12 of 23 modes	Yes No No No	All Voltage regulator gains multiplied by 10.
12	0.05 second fault at bus 29. Open line 26–29	E H	19 of 23 modes ($\lambda < 100$) 12 of 23 modes	Yes No	As Fig. 11.

TABLE II

Modes Retained in Equivalents of External System A

STANDARD CONFIGURATION					HIGH REGULATOR GAIN CONFIGURATION				
Eigenvalues	Minimally Reduced, B	Reduced B	Reduced C	Reduced D	Eigenvalues	Minimally Reduced, E	Reduced F	Reduced G	Reduced H
-494					-494				
-1000					-1000				
-708					-708				
-833					-833				
-17.82	+				-10.99 ±j16.38	+			
-15 ±j0.53	+				-15.88 ±j0.54	+	+		
-15.76	+				-8.82 ±j8.50	+			+
-2.47 ±j5.74	+				-2.47 ±j5.74	+		+	+
-10	+				-10	+			
-6.65	+	+			-6.64	+	+		+
-4.21	+	+			-2.44	+	+	+	+
-2.44	+	+			-2.22	+	+	+	
-2.22	+	+			-0.42	+	+	+	
-1.90	+	+			-0.18 ±j0.13	+	+		+
-0.42	+	+	+	+	-0.019	+		+	+
-0.18 ±j0.13	+	+	+		-0.10	+		+	+
-0.018	+	+	+	+	-0.00085	+			+
-0.10	+	+	+		-0.0028	+			+
-0.00039	+	+	+	+					
-0.0015	+	+	+	+					

TABLE III

Test Runs – External System Designations B and C

Figure, External System	Disturbance	Curve	Equivalent Description	Minimally Reduced	System Condition
13, B	0.075 second fault at bus 22. Open line 21–22				No regulator governor action in external system
		J	11 of 13 modes ($\lambda < 100$)	Yes	Reference at generator 9
		K	11 of 13 modes ($\lambda < 100$)	Yes	Reference at generator 1
14, B	As for Fig. 13	L	11 of 13 modes ($\lambda < 100$)	Yes	No regulator or governor action in external system.
		M	5 of 13 modes ($\lambda < 5$)	No	Reference at generator 1.
		N	2 of 13 modes ($\lambda < 1.5$)	No	Generator 9 inertia constant, $T_m = 200$.
15, C	0.05 second fault at bus 29.	O	15 of 19 modes ($\lambda < 100$)	Yes	No regulator or governor action in external system C.
	Open line 26–29	P	7 of 19 modes	No	
16, C	0.025 second fault at bus 29.	Q	40 of 47 modes ($\lambda < 100$)	Yes	Standard, except voltage regulator gains multiplied by 10.
	Open line 26–29	R	19 of 47 modes ($\lambda < 10$)	No	

TABLE IV

Modes Retained in Equivalents of External System B

STANDARD CONFIGURATION			INERTIA OF T_m = 200 AT GENERATOR 9, REFERENCE AT GENERATOR 1			
Eigenvalues Ref. at 9	Eigenvalues Ref. at 1	Minimally Reduced, J,K	Eigenvalues	Minimally Reduced, L	Reduced M	Reduced N
-2887	-2887		-2887			
-562	-562		-562			
-556	-556		-556			
-30.78	-30.78	+	-30.78	+		
-1.7 ±j8.10	-13.48	+	-13.3 ±j0.51	+		
-3.86 ±j7.83	-4.08 ±j8.00	+	-4.08 ±j8.00	+		
-13.73	-10.27	+	-8.38	+		
-11.04	-8.25	+	-0.74 ±j0.17	+	+	+
-8.63	-2.64 ±j2.62	+	-1.58	+	+	
-1.59	-1.58	+	-3.13	+	+	
-3.28	-2.44	+	-0.061	+	+	+
-0.20	-0.071	+				

change in inertia of generator 9 has made a major change in the damping in the response, confirming that the external system B does have a significant effect on the response of the corresponding study system.

The results for external system C are summarized in Table III and displayed in Figures 15 and 16. Figure 15 shows, for the standard configuration (but with control action blocked in the external system), that the minimally reduced equivalent gives reasonable agreement with the reference simulation, but exaggerates a response component with a period of about 0.75 seconds. The reduced equivalent gives very close agreement with the minimally reduced equivalent, indicating that the order reduction procedure is satisfactory even though the linear analysis gives the observed deviation from the reference simulation. Figure 16 shows the performance for equivalents under the same conditions as for Figure 15 but with control action included in the external system and all voltage regulator gains increased by a factor of ten. The system is slightly dynamically unstable and, as with external system B (when dynamically unstable), the minimally reduced equivalent indicates an increasing deviation between the reference simulation and the simulation using linearized analysis. The reduced order equivalent shows an increasing deviation from the minimally reduced equivalent, indicating, as do Figures 11 and 12, that stability of the external system alone does not guarantee that the reduced order equivalents will give decreasing deviations from the reference response.

The increasing error of the reduced order equivalents when a stable external system is a part of a dynamically unstable interconnection results from the fact that the decoupled canonical form equations (8) or (6) become re-coupled when incorporated into the equation of the study system, and that the eigenvalues of the external system alone are not identical to eigenvalues of the interconnected system. Evidently the differences between eigenvalues of the isolated and interconnected external system cause no problem when there is ample dynamic stability margin: but cause difficulty in cases with marginal dynamic stability.

The test results presented here indicate that the use of linearized modelling of the external system is generally satisfactory but requires the location of the reference axes at a suitably chosen generator on the external system. The reduced order equivalents are indicated to be satisfactory and generally pessimistic with respect to first swing transient stability; and to be satisfactory with respect to dynamic stability as long as the system has an adequate overall margin of dynamic stability. The electromechanical equivalents would, therefore, appear to be best suited to studies of the operation of systems which are known from detailed design studies to have good overall damping. They should, apparently, be used with caution when applied to detailed design studies involving questions of damping of marginally damped systems.

It is emphasized, finally, that these tests describe the performance of equivalents in the stringent environment of a small system model. More general information on the performance of equivalents constructed by the method described here will become available as they are tried and evaluated in studies of larger and different systems.

CONCLUSIONS

This paper has described the approach taken in an extensive analytical and computational study of power system equivalencing. The objective of this study was to set out a sound analytical basis for the construction of power system equivalents which would, in addition to representing the classical generator inertia effect, be capable of representing any dynamic effect, such as voltage regulator or governor action, in situations where this is required.

The analytical basis for equivalent construction summarized here assumes linearity in the system being equivalenced. A comprehensive linear differential equation of the system is used as the starting point for the construction of compact equivalents by a modal analysis procedure. The performance evaluations described in this paper have confirmed the validity of this approach within the limitations imposed by the assumption of linearity, and have also shown the linearity assumption to be reasonable, even in a relatively small system where all parts of the system respond significantly to any normal short circuit type disturbance.

The applicability of the method presented here to large scale systems depends upon the degree of detail used in the generator modelling. High order modelling of all generators, as used here to prove the applicability of the method to the general case including control action, results in about five generators producing matrices approaching the convenient upper size limit of about one hundredth order. The use of the classical inertia and constant voltage behind reactance model for the majority of generators in a system should allow the application of the method to systems with up to fifty generators within the limitation to approximately one hundredth order matrices.

The equivalencing method described above offers promise of being a significant aid in system planning and operating work. It is the authors' belief that initial applications of these equivalencing methods will be in planning studies. This should provide ample opportunity for testing the adequacy of the equivalencing on different systems under different types of disturbances. Hopefully, suggestions from those who will try the equivalencing methods will be forthcoming and lead to still further improvements in methods. After sufficient confidence has been built up in the equivalencing, the next step will be the actual application of equivalents on dispatching computer systems.

REFERENCES

(1) J.M. Undrill, A.E. Turner, "Construction of Power System Electromechanical Equivalents by Modal Analysis," (Companion Paper).

(2) J.B. Ward, "Equivalent Circuits for Power Flow Studies," *Trans. A.I.E.E.*, vol. 68, pp. 373-383, 1949.

(3) W.T. Brown, W.J. Cloues, "Combination Load Flow and Stability Equivalents for Power System Representation on A.C. Network Analysers," *Trans. A.I.E.E.*, vol. 74, pp. 782-787, 1955.

(4) H.E. Brown, R.B. Shipley, D. Coleman, R.E. Neid, "A Study of Stability Equivalents," *I.E.E.E. Trans. on Power Apparatus and Systems*, vol. PAS-88, pp. 200-207, 1969.

(5) A. Chang, M.M. Adibi, "Power System Dynamic Equivalents," *I.E.E.E. PICA Conference Record*, pp. 174-183, 1969.

(6) H.H. Happ, J.M. Undrill, Discussion of Reference 4, *I.E.E.E. Trans. on Power Apparatus and Systems*, vol. PAS-88, pp. 206-207, 1969.

(7) J.M. Undrill, "Structure in the Computation of Power System Nonlinear Dynamical Response," *I.E.E.E. Trans. on Power Apparatus and Systems*, vol. PAS-88, pp. 1-6, 1969.

ACKNOWLEDGEMENT

The authors acknowledge the valuable contribution of the members of the Electric Research Council Steering Committee on System Equivalents for the development of the subject matter of this paper. This project is one of six related research efforts being sponsored by the Electric Research Council under its Bulk Power System Security program. It is supported by the Edison Electric Institute, and by contributions from the Tennessee Valley Authority and the United States Department of the Interior, and is being performed by the General Electric Company under the guidance of a Steering Committee composed of representatives of the supporting agencies. In addition to J.A. Casazza and E.M. Gulachenski, the members of the steering committee were G. Heins of Consumers Power Company, A.H. Schmidt, Jr. of the U.S. Bureau of Reclamation, and P. Van Olinda of Public Service Electric and Gas Company.

Combined Discussion[1,2]

A. Kuppurajulu and **S. Elangovan** (Indian Institute of Technology, Madras, India): The authors are to be congratulated for their interesting contribution to the dynamic stability analysis of power systems. Although the discussors have shown the applicability of model reduction methods to single machine and interconnected power systems in [8], the test results for large interconnected systems were not given. It is gratifying that the authors in their companion paper [1] have illustrated the utility of these methods by considering large interconnected systems.

The idea of dividing the complete system into study system, external system and interconnected system, and representing the external system by its linearized reduced equivalent is certainly new and is based on sound reasoning. This is well supported by the appended test results.

It is interesting to note that transient processes in power systems are characterized by a wide variety of modes which last from a few microseconds to a few minutes. Thus while the complete representation of the system is definitely more accurate, it is not practicable because of the high order of the equations. However, if the interest is confined to a detailed study of the system behavior only during a certain interval of the whole period of response of the system, then the model reduction techniques presented in [13] may be applicable. Thus it is possible to derive different low order models applicable for different time periods of response. The authors' comments on this aspect are welcome.

[13] A. Kuppurajulu and S. Elangovan, "System analysis by simplified models," in IEEE Trans. Automatic Control, Vol. AC-15, pp. 234-237, April 1970.

Manuscript received March 1, 1971.

[1] J. M. Undrill and A. E. Turner, "Construction of power system electromechanical equivalents by modal analysis," this issue, pp. 2049-2059.

[2] J. M. Undrill, J. A. Casazza, E. M. Gulachenski, and L. K. Kirchmayer, "Electromechanical equivalents for use in power system stability studies," this issue, pp. 2060-2071.

J. E. Van Ness (Northwestern University, Evanston, Ill. 60201): This well written paper describes one method of developing a lower order electromechanical equivalent for a large system by eliminating certain modes of response of the total system. We have been looking at the same general problem and had at one time considered the basic method described here. Its major difficulty, as it has been applied in this paper, is that the new variables obtained usually do not correspond to specific variables in the original system. If the variables are transformed back to those originally defined, interconnections will develop in the reduced system which have no counterpart in the actual physical system. Because of this limitation, we chose to try a different path in our attempt to solve this problem, but have not yet been successful.

By defining a study system and an external system and then applying the reduction only to the external system, the authors of this paper have removed most of this type of difficulty. With their formulation the study system retains all of its physically significant variables, while the less important external system can be reduced, thus simplifying the problem. This does still leave some disadvantages. No reductions are possible within the parts of the network included in the study system, and the external system loses all of its internal identity.

It has been my experience that some modes of response can be identified as coming from the interaction of certain pairs of machines. This type of phenomena is discussed with respect to Fig. 3 in the paper where both machines are in the external system. Have the authors given any thought to modes of response for the total system which might come from interactions between a machine in the study system and one in the external system? In a sense, an eigenvalue of this type is a closed loop eigenvalue, while the reduction is being made on an open loop basis. I would expect the method described in this paper to be very good for eigenvalues which arise almost entirely within the external system, but not represent as well eigenvalues depending on interactions between the study system and the external system. It would be interesting to compare all of the eigenvalues of the total system before and after the reduction.

Manuscript received February 10, 1971.

I would like to point out two modifications in the techniques used that might simplify the procedures. In Table I and in the text it is stated that the eigenvalues and eigenvectors are carried in complex form. With a slight modification[1,2] it is possible to make all of the matrices real without increasing their size. A pair of complex conjugate eigenvalues are replaced by a 2 x 2 block, and two adjacent columns containing complex conjugate eigenvectors are replaced by two real columns, one with the real part and the other with the imaginary part of the eigenvectors. We have used this method in all of our work for many years, and it greatly simplifies the computer programs.

Under the section titled Computational Considerations, it is stated that the reduction procedure is not valid in the case of equal eigenvalues. Equal eigenvalues which are diagonalizable, or to state it another way, which occur in Jordan blocks of order 1, will have distinct eigenvectors and can be handled by this method as if they were unequal. In particular, identical eigenvalues which come from non-interacting sections of the external system will be diagonalizable and, therefore, can be handled directly by the reduction method described in this paper. An example where this was done for a different type of problem is given in Ref. 2.

[1] K. Ogata, *State Space Analysis of Control Systems,* Prentice-Hall, Inc., 1967, Englewood Cliffs, N. Y., pp. 143-145.

[2] F. P. Imad and J. E. Van Ness, "Finding the Stability and Sensitivity of Large Sampled Systems," *IEEE Trans. on Automatic Control,* Vol. AC-12, pp. 442-445, August 1967.

Lawrence Michaels (Electronic Associates, Inc., West Long Branch, N. J. 07764): The authors have presented a method of equivalencing which is based on the well known control theory principle of reduction of a linear system to canonical form. Once the canonical form has been obtained, certain natural system modes are deleted; primarily those which are not significantly excited from the terminals of the external system, or which do not appear at the terminals. This analysis is closely related to the study of controllability and observability of linear systems.

It appears that the following problems present themselves if this equivalencing technique is to be applied in a practical case:

a) The dynamic equations of the system to be equivalenced must be linear. Thus, the general non-linear system equations must be linearized about the present system operating point. This requires that a load flow be performed before a system equivalent may be determined, as was indicated by the authors.

The fact that the system equivalent varies with system load is perhaps the greatest weakness of this method, unless the form of the equivalent can be determined uniquely, with the parameters of the equivalent given as explicit functions of the load flow solution. The authors have not indicated if this is the case. It seems that computation requirements would preclude any one-line use of the equivalencing technique, if the equivalent must be completely recalculated as system load changes.

b) Engineering judgment is required to determine which natural modes of the system should be deleted. It is not clear just what criterion one should use to make this determination. For example, the authors claim that it is reasonable to delete modes whose eigenvalues are large, because these modes would decay rapidly, and thus would not contribute significantly to the system response. However, if a system had a set of eigenvalues whose values were: 50, 35, 25, 10, 5, 1, .1, it is not clear just where the cutoff point should be, since the eigenvalues are not naturally grouped.

c) Choice of the location of the boundary between the study system and the external system should be made such that the external system operates linearly in response to disturbances in the study system. This choice is also dictated by engineering judgment, and thus is subject to the same problems as described above in connection with the deletion of selected modes. In addition, many transient stability simulations of the complete system may be required to help determine the location of the study system/external system boundary, with the resultant consumption of large blocks of computation time.

d) The fact that the linearized equations of the external system vary with the choice of the system reference axis is another problem which must be dealt with via engineering judgment. A reference machine must be chosen which will not respond severely to disturbances in the study system. An important question which remains unanswered

Manuscript received February 22, 1971.

is how the choice of the reference axis affects the accuracy of the resultant equivalent system model.

e) Finally, there is a question about the ultimate savings realized from the utilization of an equivalent determined by this method. It is assumed that less computer time will be required for transient stability simulations which make use of an equivalent electromechanical model. However, the computer time required to determine the equivalent has not been specified, and if this calculation must be repeated frequently (such as when the system load changes), the total computer time used may turn out to be equal to, or greater than, that required without the use of the equivalent.

Robert F. Silva (Northeast Utilities, Hartford, Conn. 06101): A method for constructing power system equivalents which represent electromechanical effects has been desired for as long as power systems have been interconnected. The authors are to be commended for developing a technique which if properly implemented in a compact low order form shows considerable promise in off-line as well as on-line equivalencing applications.

The method described in this paper has been successfully applied at Northeast Utilities using a nodal iterative stability program modified to accept electromechanical equivalents based on classical machine representation with control action represented by a simple damping term proportional to generator slip. The results of a sample test case are shown on Figure 17 and 18. In Figure 17, the complete test system was used, whereas in Figure 18, the case was repeated with external system A represented by the equivalencing method. It is significant that the two machines in the external area represented 22% of the system generation which if not properly equivalenced would have resulted in a radical change in the response of the system to the disturbance. While maintaining accuracy, the elapsed computer time printed at the top of each print out shows a reduction of 26% when solving the case with the reduced system. There are many aspects in the use of this method which remain to be tested with larger systems and different types of disturbances, but initial experience indicates that the method is valid and has features worth developing.

A method for determining which modes can be eliminated beyond a "minimally reduced" equivalent without degrading the accuracy of the method in my opinion needs to be more clearly defined. From the discussion of test results in the paper, it appears that experimentation with disturbances which cause a system to be marginally stable is required. Also, selecting the reference machine appears to be a cut-and-try procedure. If this approach is required for each external area whenever a series of stability cases are studied, then reduction beyond a "minimally reduced" equivalent does not appear to be worthwhile. Do the authors have any comments which may suggest a procedure for this?

In Figure 2 of this paper, a base case load flow is performed within the equivalents construction program. This requires that an external area be contiguous. If this step was eliminated by entering network and operating data in load flow balanced form, could not an external area be non-contiguous?

If the area external to the study system can be non-contiguous, then the makings of a completely automated stability program with "minimally reduced" equivalencing is apparent. In other words, the equivalents construction program could be incorporated as a routine within the stability program which acts on a full system load flow that has the study area defined by busses to be retained. The areas are then separated maintaining load flow balanced form and the external area passed to the equivalents construction routine which is looped through twice. The first time all modes are retained. The second time modes with eigenvalues greater than 100 are eliminated. The stability cases are then performed in the study area with the "minimally reduced" equivalent. Would the authors please comment on any pitfalls or ramifications with this procedure which is an attempt to eliminate several manual links in the existing process?

Manuscript received February 12, 1971.

G. I. Stillman (Power Authority of the State of New York, New York, N. Y. 10019): The authors have chosen to use state-space techniques, eqs. (4) and (5) of the paper, to model detailed external systems (including the generator and turbine control loops) as shown in Fig. 1, in order to obtain equivalents of the external systems. The state space equations are then de-coupled by matrix diagonalization to obtain equations (6) and (7), which are generally simpler in form and which are characterized by a set of fictitious natural frequencies or modes. These modes are analyzed, and those which judgment, based on experience, indicates to be of minor significance can be discarded, yielding a reduced set of simplified algebraic and differential equations which constitute, respectively, the desired electrical and electromechanical equivalents. The process of diagonalizing equations which "are of high order and embody large non-sparse matrices" is a time consuming one on a computer.

Manuscript received February 19, 1971.

It would appear from the authors' assumptions concerning boundaries of the study system ("the terminals T . . . are *far enough* removed from the site of the disturbance . . ." – my emphasis, G.I.S.) that rederivation of equivalents of the external systems may be required, for example, when the site of the disturbance for which the stability of the study system is being tested changes, or when it is desired to assess the stability of the study system for disturbances after major EHV transmission or generation plant has been forced out within an adjacent external system. Rederived equivalents would, of course, be required for similar circumstances with previously published methods (Ref. 4 of the paper). The authors offer no proof that their method consumes less computer time than previously published approaches to the equivalents problem. It would be helpful if they could offer a computer-time comparison of their method with the method of Ref. 4 starting with similar initial representations of the external systems.

The authors' method provides the capability of lumping the effects of turbine and generator control loops and inertias into the external system equivalent much more readily than with other published methods. If the study system boundaries are "far enough" away from the site of the disturbance, however, the added accuracy provided by this capability is not clear. Here too, it would be helpful if the authors could provide results comparing their method, including external system control details, with, say the method of Ref. 4, excluding such control details, to show (a) that their method yields results of greater accuracy for the equivalents, or (b) that their method permits the study system boundaries to be located significantly closer to the site of the disturbance.

I would appreciate some clarification of the authors' *Test Results*. If I understand figure 4 correctly, buses 9, 10, 1, 3, 18, 17 and 27 fall wholly or partly within the "study system" as defined in figure 1, while the remaining buses of fig. 4 are in external systems A, B, or C. Furthermore, to check the validity of an equivalent for an external system, the disturbance must be taken within the study system, as is correctly shown in fig. 3. Why, then, do the results of figs. 7 and 8, for example, "show the performance of the equivalents of external system A for three-phase faults at buses 22 and 2 respectively" when these buses are not in the study system but in external systems C and A respectively? Would the authors' kindly clarify this point?

The authors state the assumptions concerning the boundaries of the study system succinctly: they must be *far enough* away from the site of disturbance to allow linearized representation of external system loads and so that the effects of the disturbance on the external system are *small enough* to permit linearization of generator models. The questions of what is "far enough" and "small enough" have always plagued personnel engaged in studying stability. These questions have usually been resolved by the use of "engineering judgment". But one can visualize a more precise approach to the resolution of these questions: one could perform a series of stability runs on a detailed interconnected system starting with the boundaries of the study system arbitrarily far away and moving them successively closer electrically to the disturbance site, for a given fault duration, run by run. When certain key variables in the *external* system (such as angular deviation or speed deviation of generators, for example) exceed certain predetermined limits, which, for the moment, are arbitrary, then the boundary can be considered to be established; the precision of the study system boundary for a given fault location, duration and type is a function of the "arbitrary" limits one has chosen for the above-mentioned key variables in the external system. The procedure just described is, of course, tedious and impractical, but, it seems to me, that if the authors were to turn the power of the state-space approach on this procedure, suitable state transformations would result in the *limits* of the key variables generating a surface in a transform hyperspace; the *actual* variables for given disturbances would then be points in this hyperspace, and, depending on whether they fall within the limiting surface or outside of it, would determine whether key variable limits have been exceeded or if they have not been exceeded, what the margin is. It seems to me that the state-space approach offers the possibility of getting away from engineering judgment to determine what is "far enough" and "small enough". A real contribution to the solution of stability problems could be made by providing an efficient procedure for determining the boundary of the study system. I should appreciate the authors' comments on the above, and, if they see some merit in it, I would suggest that they consider it in their ongoing research project.

My compliments to the authors for a clear and concise paper in a fundamental and difficult area of solving stability problems.

R. Bruce Shipley (Tennessee Valley Authority, Chattanooga, Tenn. 37401): The authors and their sponsors should be commended for the considerable effort that has gone into this paper. It is a contribution to a field of endeavor that has been dormant for too long. Present inadequacies of digital computers, both from the standpoint of hardware and software, force the analyst to make approximations. If he elects to represent machines in detail, he must study small systems and hope they adequately represent the system in question. If he elects to represent the large system, then he must abbreviate machine representation and hope the results will be adequate. In the past making an equivalent has required judgment. Accordingly, the same person making an equivalent of some system conditions might not arrive at the same equivalent twice in succession. Unfortunately, the method presented in the present paper may not completely eliminate this dilemma since it also requires judgment in the reduction of variables. Can this be reduced to a procedure where two different analysts would arrive at the same equivalent for the same system condition? I think this should be a goal.

Further comments on the following conditions might be helpful. When a machine represented in an equivalent hits a limit—voltage regulator limit, field current reduction due to protective action, etc.— is this ignored due to linear assumptions? The suggested use of the equivalent in planning studies requires clarification on whether this means long-range system planning or short-range operational planning. In long-range system planning the analyst may not even know the transmission configuration accurately, not to mention details of generator characteristics, control characteristics, and possibly more important load characteristics. Therefore, elegant analytical techniques may be of limited value to him. In short-range planning the load characteristics may still be missing. Other problems may also be present depending upon how the dispatcher may use the computer. If it is to be used in a preventative mode, he may assume three-phase faults at various locations to evaluate whether or not he can prevent instability by shifting generator schedules. This could result in undue expenses for fuel since such faults are not likely to occur; and if they do, they may evolve in such a manner that the generator has considerable damping, invalidating results that do not properly consider it. If the dispatcher takes into account many different fault locations and different sequences of how phase faults may evolve, he possibly has an impossible chore.

The above discussion indicates complete immediate success in the determination of equivalents should not be expected. It is an indication that analytical development should continue. An adaptive means of determining system characteristics directly from the power system could be an attractive approach, but it probably should be taken a step further to include the probable response of the system.

Adaptive developments, not necessarily using conventional analog or digital equipment, can conceivably be applied to prevent system instability after the disturbance has occurred. This would prevent expensive generator rescheduling as used in preventive measures. Tools for this application are available in adaptive analytical methods, fast valving, switching of circuit parameters, etc. In speculating, the computer techniques discussed in this paper should be developed to provide a means for analyzing the effectiveness of other measures that would actually be used on the power system to improve operations.

Manuscript received February 8, 1971.

J. M. Undrill, L. K. Kirchmayer, A. E. Turner, E. M. Gulachenski, and **J. A. Casazza:** We thank the discussors for their interest in this work and their comments on the papers. Our principal objective in the study was to gain an improved understanding of the equivalent construction problem by replacing the conventional inertia-lumping approach to electromechanical equivalencing with a method based firmly in the mathematics of dynamical systems. We have been encouraged by our progress, but are aware, as are the discussors, of many questions requiring further study concerning this complex subject.

It is encouraging to learn that Prof. Van Ness is pursuing alternative approaches to the equivalent construction problem, because the problem of constructing power system equivalents will require a continuing research effort in order to provide methods and programs to satisfy the needs of the industry.

Prof. Van Ness notes correctly that our method will modify the natural modes that are associated with oscillations between pairs of machines where one is in the study system and one is in the external system. We believe that this opening of closed loops is partly responsible for the poor performance of the equivalents in Fig. 11 of our paper TP 137. We note, however, that the tests described in this paper suggest that this is only a problem when the system is in a condition of very low damping and marginal dynamic stability. The stable cases, and cases that we have run using classical constant voltage behind reactance generator models, have given no problems of a type that could be associated with this loop-opening effect. This question deserves further investigation.

Manuscript received March 31, 1971.

Our programs did not take advantage of the storage saving method used in Refs. 1 and 2 of Prof. Van Ness's discussion, but this approach can certainly be taken where programming considerations indicate its desirability.

It would have been better for the paper to have stated that the order reduction procedure *may not* be valid in cases with equal eigenvalues, and we thank Prof. Van Ness for pointing this out.

We are pleased to acknowledge the comments of Profs. Kupperajulu and Elangovan, and will look into their suggestion described in the reference given in their discussion.

Mr. Silva has been successful in using our electromechanical equivalent technique in conjunction with a large-scale nodal-iterative type stability program, and we will be most interested in the results of his continuing tests on equivalents produced by our method.

Mr. Shipley and Mr. Stillman discuss the power system equivalencing problem on a general basis. Messrs. Stillman and Michaels point out that careful engineering judgment is needed to select the external system boundaries before an equivalent construction program can be used. At present, the selection of external system boundaries appears to require experimentation with complete stability solutions if all the questions concerning validity of small disturbance assumptions are to be answered explicitly in each case. This places equivalents in the role of reducing the computing cost of making repetitive studies of a given large system, such as the system under consideration by a power pool control center. However, we have been encouraged to find in our test runs that there has been little difficulty in selecting external boundaries in such a manner that our linearity assumptions are satisfactory. We have found no need in practice to make large numbers of stability simulations of the complete system in order to locate satisfactory study system boundaries.

We agree with Mr. Stillman that it may not always be necessary to represent control loop dynamics in equivalents where the equivalent represents machines that are subject to small disturbances. However, we would point out that the small disturbance behavior of control loops does become important in studies concerned with dynamic stability when the control loops are being relied upon to provide stabilizing and damping action. Further, the collected small disturbance effects at many units of an external system can have a large effect on a study system in cases such as loss of generation where the compensating action must come from the external system, even though the effect on any one external system element may be very small.

In answer to Messrs. Stillman's and Michaels' questions on computer times, the construction of equivalents by our method takes more time than the use of the algorithm given in Ref. 4 of paper TP 137 because that algorithm is essentially an electrical equivalent construction technique. In fact, block 4 of our equivalent construction technique (Fig. 2) uses essentially the same algorithm as Ref. 4 to construct the electrical equivalent of the external system. Mr. Silva's discussion gives an indication of the effect of using our equivalent on the time for the stability computation. We would expect the equivalents constructed by the method of Ref. 4 to have essentially the same effect on stability computation times as our equivalents; but we have not made tests to check this.

In clarification of Fig. 4 of the paper, only one of the three designated external areas was considered to be the external system in each run. Hence, when external area A was represented by an equivalent (i.e., Figs. 7-12), all other buses including those in areas B and C were in the study system, and so on.

Our equivalent construction method does not represent nonlinear effects, such as voltage regulator limits, in the external system, but it is perfectly compatible with detailed treatment of these effects in the study system simulation. Mr. Michaels is correct in stating that our equivalent, being a linearization, varies as system load is varied. Our experience with broad spectrum of power system simulations suggests that this should not significantly limit its practical utility.

Several discussors raise the question of automation of the mode selection process so that it becomes independent of the judgment of the user. The mode selection process was not automated in our studies because one of our main requirements was to learn the effects of the different selection strategies on the performance of the equivalents. Subsequent experience with equivalents based on "constant-voltage-behind reactance" generator modeling indicates that automatic mode selection should be possible once criteria have been selected for the maximum natural frequency and the maximum natural damping factor of importance in the representation of a given system. Once the criteria have been set, all modes of natural frequency or damping factor higher than the maximum values of importance would be deleted by program decision. The one real eigenvalue mode associated with classical modeling would always be retained. The cut-off values of natural frequency and damping would depend upon system characteristics and would have to be set after due consideration of the general system characteristics. For example, the maximum natural frequency of importance would be lower for the loosely coupled systems of the Western United States than for closely coupled metropolitan systems.

With regard to Mr. Silva's suggested procedure to eliminate manual links, we suggest by-passing the step retaining all modes and proceeding directly to the minimally reduced equivalent. We would point out, however, that the minimally reduced equivalent differs from the case with all modes retained only when the external system has short time constant elements. In the case of constant-voltage-behind-reactance modeling, the minimally reduced equivalent normally has no modes deleted.

We believe that the "conventional equivalent method" referred to by Messrs. Byerly and Skooglund corresponds to the use of an electrical equivalent to reduce the size of the network representation of the external system without reducing the number of individual machines that are represented. Our prime objective was to start from the electrical equivalent used in the "conventional approach" and produce a new equivalent with the individual machine representations eliminated and with only the terminal nodes having axes in the resultant algebraic equation. The reduced network described by Messrs. Byerly and Skooglund is exactly the electrical equivalent used in our construction of the electromechanical equivalents for external system A. Since our method handles control dynamics, its performance may be superior to that of equivalents where generators are lumped by simple additon of inertias and paralleling of reactances.

The fact that the numbers of modes retained in our equivalents of external system A correspond to the numbers of equations cited by the discussors is coincidental. In fact, when using classical generator modeling, we have been able to reduce the number of retained modes to one, both for external system A, and for another external system which contained machines 2-7 and 10.

The stability program used for our study represented all loads as constant impedances to ground, and hence gave reference simulations which correspond to the use of the conventional equivalent described by Messrs. Byerly and Skooglund. The figures of the paper, therefore, may be interpreted as indicating the difference between the performance of our equivalent and the performance of the conventional equivalent with all machine axes retained.

The storage advantage of our electrical equivalent over the "conventional equivalent" increases as the number of machines increases and as the number of retained natural modes is decreased. The use of multiple equivalents requires only a minor programming change to account for the fact that one set of reference axes is used for all equivalents. We see no real difficulty in choosing a suitable reference in very large systems. Dr. Michaels' question d. concerning reference selection is answered explicitly by Figs. 13 and 14 of our paper TP 137. The choice of location for the reference axes is not affected by the number of equivalents in use; and the reference machine should still be one whose speed remains close to the mean electrical speed of the interconnection. Noncontiguous external areas cause no problem since they can be represented by separate equivalents. The engineering effort needed to prepare data for our equivalent construction program is comparable to that needed to assemble input for a load flow and stability run. We have usually made two program runs for each case; one with all modes retained to give data on the natural modes, and one with the desired selection of modes to store the equivalent on magnetic tape.

Means of determining equivalents directly from measured system currents and voltages as mentioned by Mr. Shipley were explored as part of the ERC research project on power system equivalents, and this work has been reported to the Electric Research Council.

In conclusion, we are pleased by the interest expressed in this subject and would like to encourage further experimentation and research in this area.

Part II

Dynamic Stability

The stability characteristics of large electric power systems is too varied to permit a simple classification of all behavior, and it is not possible to separate all analyses into simple categories. Nevertheless, the terms "transient stability" and "dynamic stability" are widely used to distinguish between two kinds of studies. These terms are not precisely defined nor universally used, but they are helpful in the organization of this book.

Transient stability studies are limited to relatively short time intervals, typically 1 s or less. They are most often used to determine the stability of a single unit or plant during the initial period of high stress immediately following a nearby fault. Such studies may require the representation of a large system, but primarily to secure accuracy rather than to obtain total system response. Traditionally, transient stability studies have utilized classical generator models, but modern practice is to represent generators, excitation systems, and speed-governing systems in detail. Reprints and a supplementary bibliography related to transient stability are presented in Part III.

Dynamic stability studies cover longer real-time intervals, perhaps 5–10 s, occasionally up to 30 s. Attention may be focused on a single unit or plant, but frequently these studies are made to obtain large system response. Dynamic models for generator flux, excitation systems, speed-governing systems, and other apparatus are essential.

There are various reasons for conducting dynamic stability studies. For example, it may be desirable to simulate the breakup of a large system, such as occurred in the Northeast Power Failure of 1965. Studies made subsequent to that event simulated the cascade circuit interruptions which occurred in the first 7 s and were made to help understand more fully what had happened. A detailed report of these studies is contained in the third reference following the Introduction to this book. The results of this type of study are usually too complicated to publish except as lengthy reports, and there is essentially no periodical literature dealing with them.

Dynamic stability studies are also made to determine the consequences of suddenly having a significant excess of either load or generation in an isolated power system. Such a system might be normally isolated and lose load or generation, or it might normally be part of a larger system and become separated. Papers dealing with situations of this type and related topics are covered in Part V.

Much of the effort devoted to dynamic stability analysis is concentrated on the problems associated with undamped or poorly damped oscillations, and the papers in this section pertain to that subject. Oscillations may occur between one machine or plant and the rest of the system, or between large machine groups. Spontaneous oscillations, or those which are initiated by minor disturbances, are not common to all large power systems, but they have occurred in parts of the western and midwestern United States and in parts of Canada. This kind of stability behavior constitutes a severe threat to system security and creates very difficult operating problems. Conditions for oscillations between major subsystems may depend on many variable factors, such as load levels, generator loadings, and circuit outages, and are difficult to predict with precision. They are virtually certain to occur, however, whenever operating conditions conducive to them exist.

The only effective way to deal with spontaneous oscillations is to alter the inherent system characteristics which cause them. Experience has been that it is usually possible to eliminate or significantly reduce the incidence of spontaneous oscillations through the use of power system stabilizers. These devices act on generator field voltages through the excitation systems. Power system stabilizers are also effective in damping oscillations subsequent to disturbances.

One interesting facet of spontaneous oscillation is that it does not initially involve nonlinear behavior in any essential way. Therefore, linearized mathematical models and linear analytical methods may be used to study the problem. This is particularly important as a means for examining the sensitivity of system damping to various factors which influence it. Several papers are reprinted here which use linear methods to improve comprehension of dynamic stability phenomena.

This section also includes a number of papers describing large-scale dynamic stability computer programs. Computer simulation is the only effective means presently available for studying any of the system problems described in the preceding paragraphs. The same simulation programs are applicable to transient stability studies.

A group of four papers is included primarily to document some circumstances under which undamped or poorly damped oscillations have occurred. Finally, seven papers on power system stabilizers are reprinted covering installation, testing, and field experience.

Again, there are many excellent papers on these topics, and space is simply not available to reprint them. The following bibliography contains some of them.

Supplementary Bibliography

Concepts and Studies Based on Linear Analysis

[1] J. E. Van Ness, "Response of large power systems to cyclic load variations," *IEEE Trans. Power App. Syst.*, vol. PAS-85, pp. 723–727, July 1966.

[2] J. E. Van Ness and W. F. Goddard, "Formation of the coefficient matrix of a large dynamic system," *IEEE Trans. Power App. Syst.*, vol. PAS-87, pp. 80–83, Jan. 1968.

[3] R. Kasturi and P. Doraraju, "Sensitivity analysis of power systems," *IEEE Trans. Power App. Syst.*, vol. PAS-88, pp. 1521–1529, Oct. 1969.

[4] —," Relative dynamic stability regions of power systems," *IEEE Trans. Power App. Syst.*, vol. PAS-89, pp. 966–974, May/June 1970.

[5] J. A. Pinnello and J. E. Van Ness, "Dynamic response of a large power system to a cyclic load produced by a nuclear accelerator," *IEEE Trans. Power App. Syst.*, vol. PAS-90, pp. 1856–1862, July/Aug. 1971.

[6] M. K. El-Sherbiny and D. M. Mehta, "Dynamic system stability. Part II. Design realistic compensating network for excitation control using digital computer," IEEE paper C 73-221-9.

Nonlinear Analytical Methods

[7] R. B. Shipley, N. Sato, D. W. Coleman, and C. F. Watts, "Direct calculation of power system stability using the impedance matrix," *IEEE Trans. Power App. Syst.*, vol. PAS-85, pp. 777–782, July 1966.

[8] K. Prabhashankar and W. Janischewsyj, "Digital simulation of multimachine power systems for stability studies," *IEEE Trans. Power App. Syst.*, vol. PAS-87, pp. 73–80, Jan. 1968.

[9] J. M. Undrill, "Dynamic stability calculations for an arbitrary number of interconnected synchronous machines," *IEEE Trans. Power App. Syst.*, vol. PAS-87, pp. 835–844, Mar. 1968.

[10] A. Colombo, F. Redaelli, G. Ruckstuhl, and A. Vian, "Determination of the dynamic response of electrical systems by means of digital program," *IEEE Trans. Power App. Syst.*, vol. PAS-87, pp. 1411–1418, June 1968.

[11] J. M. Undrill, "Structure in the computation of power-system nonlinear dynamical response," *IEEE Trans. Power App. Syst.*, vol. PAS-88, pp. 1–5, Jan. 1969.

[12] J. E. Van Ness and F. B. Kern, Jr., "Use of the exponential of the system matrix to solve the transient stability problem," *IEEE Trans. Power App. Syst.*, vol. PAS-89, pp. 83–88, Jan. 1970.

[13] K. N. Stanton and S. N. Talukdar, "New integration algorithms for transient stability studies," *IEEE Trans. Power App. Syst.*, vol. PAS-89, pp. 985–991, May/June 1970.

[14] J. D. Morgan and D. W. Shaw, "A hybrid computer solution of the power system transient stability problem," IEEE paper 71 CP 600-PWR.

[15] A. Kuppurajulu and S. Elangoran, "Simplified power system models for dynamic stability studies," *IEEE Trans. Power App. Syst.*, vol. PAS-90, pp. 11–23, Jan./Feb. 1971.

[16] S. N. Talukdar, "Iterative multistep methods for transient stability studies," *IEEE Trans. Power App. Syst.*, vol. PAS-90, pp. 96–102, Jan./Feb. 1971.

[17] D. G. Ramey, R. T. Byerly, and D. E. Sherman, "The application of transfer admittances to the analysis of power system stability studies," *IEEE Trans. Power App. Syst.*, vol. PAS-90, pp. 993–999, May/June 1971.

[18] J. Nanda, "Analysis of steady state stability of a two machine system by the D-decomposition technique," *IEEE Trans. Power App. Syst.*, vol. PAS-90, pp. 1848–1855, July/Aug. 1971.

[19] V. A. Venikov and V. A. Stroev, "Power System stability as affected by automatic control of generators—Some methods of analysis and synthesis," *IEEE Trans. Power App. Syst.*, vol. PAS-90, pp. 2483–2494, Nov./Dec. 1971.

[20] K. R. Padiyar and R. S. Ramshaw, "Dynamic analysis of multimachine power systems," *IEEE Trans. Power App. Syst.*, vol. PAS-91, pp. 526–535, Mar./Apr. 1972.

[21] W. Janischewskyj and P. Kundur, "Simulation of the non-linear dynamic response of interconnected synchronous machines. Part I. Machine modelling and machine-network interconnection equations," *IEEE Trans. Power App. Syst.*, vol. PAS-91, pp. 2064–2065, Sept./Oct. 1972.

[22] P. L. Dandeno and P. Kundur, "Simulation of non-linear dynamic response of interconnected synchronous machines. Part II. Network solution procedures and comparisons of particular computational methods," *IEEE Trans. Power App. Syst.*, vol. PAS-91, pp. 2069–2076, Sept./Oct. 1972.

Studies, Field Tests, and Experience

[23] D. N. Ewart, G. L. Langren, M. Temoshok, and W. W. Walker, "Stability studies on a 532 MVA cross-compound turbine generation set," *IEEE Trans. Power App. Syst.*, vol. PAS-84, pp. 526–528, Apr. 1965.

[24] F. R. Schleif and J. H. White, "Damping for the Northwest-Southwest tieline oscillations—An analog study," *IEEE Trans. Power App. Syst.*, vol. PAS-85, pp. 1239–1247, Dec. 1966.

[25] J. S. Hooper, G. E. Adams, and J. C. Conder, "Damping of system oscillations with steam electric generating units," IEEE paper 31 CP 66-176.

[26] F. R. Schleif, G. E. Martin, and R. R. Angell, "Damping of system oscillations with a hydrogenerating unit," *IEEE Trans. Power App. Syst.*, vol. PAS-86, pp. 438–442, Apr. 1967.

[27] W. A. Mittelstadt, "Four methods of power system damping," *IEEE Trans. Power App. Syst.*, vol. PAS-87, pp. 1323–1329, May 1968.

[28] A. M. El-Serafi, "Optimization of excitation systems of synchronous machines for maximum possible capacitive loading," IEEE paper 70-CP-202-PWR.

[29] M. K. El-Sherbiny and A.-A. Fouad, "Digital analysis of excitation control for interconnected power systems," *IEEE Trans. Power App. Syst.*, vol. PAS-90, pp. 441–447, Mar./Apr. 1971.

[30] A. Klopfenstein, "Experience with system stabilizing excitation controls on the generation of the Southern California Edison Company," *IEEE Trans. Power App. Syst.*, vol. PAS-90, pp. 698–706, Mar./Apr. 1971.

[31] E. J. Warch, F. R. Schleif, W. B. Gish, and J. R. Church, "Alignment and modeling of Hanford excitation control for system damping," *IEEE Trans. Power App. Syst.*, vol. PAS-90, pp. 714–724, Mar./Apr. 1971.

[32] J. L. Dineley and A. J. Morris, "The Dynamic interaction of closely-coupled synchronous generators," *IEEE Trans. Power App. Syst.*, vol. PAS-90. pp. 2158–2166, Sept./Oct. 1971.

[33] H. A. M. Moussa and Y.-N. Yu, "Improving power system damping through supplementary governor control," IEEE paper C72 470-3.

[34] R. G. Farmer and L. L. Daviet, II, "Case study of system frequency fluctuations," IEEE paper C 72 604-7.

[35] D. R. Hayes and G. E. Craythorn, "Modeling and testing of valley steam plant supplemental excitation control system," *IEEE Trans. Power App. Syst.*, vol. PAS-92, pp. 464–470, Mar./Apr. 1973.

Concepts of Synchronous Machine Stability as Affected by Excitation Control

FRANCISCO P. DEMELLO, SENIOR MEMBER, IEEE, AND CHARLES CONCORDIA, FELLOW, IEEE

Abstract—**The phenomena of stability of synchronous machines under small perturbations is explored by examining the case of a single machine connected to an infinite bus through external reactance.**

The analysis develops insights into effects of thyristor-type excitation systems and establishes understanding of the stabilizing requirements for such systems. These stabilizing requirements include the voltage regulator gain parameters as well as the transfer function characteristics for a machine speed derived signal superposed on the voltage regulator reference for providing damping of machine oscillations.

Introduction

THE PHENOMENON of stability of synchronous machine operation has received a great deal of attention in the past and will receive increasing attention in the future. As economies in system design are achieved with larger unit sizes and higher per unit reactance generating and transmission equipment designs, more emphasis and reliance is being placed on controls to provide the required compensating effects with which to offset the reductions in stability margins inherent from these trends in equipment design [1].

Concurrent with these trends are improvements in calculating methods and computing capability which permit predicting complex dynamic effects [2]–[5], providing the means for designing control equipment with the proper characteristics.

Among several aspects of stability of synchronous machine operation, an important one is the mode of small perturbation stability referred to as steady-state, dynamic or conditional stability. Increasing attention has been focused recently on the effects of excitation control on the damping of oscillations which characterize the phenomena of stability. In particular, it has been found useful and practical to incorporate transient stabilizing signals derived from speed, terminal frequency, or power [6], [7] superposed on the normal voltage error signal of voltage regulators to provide for additional damping to these oscillations.

This paper deals with an analysis of the phenomena of stability of synchronous machines under small perturbations by examining the case of a single machine connected to a large system through external impedance. The object of this analysis is to develop insights into effects of excitation systems and to establish an understanding of the stabilizing requirements for such systems. By examining a wide range of system and machine parameters and probing into causes and effects, a logical set of guide rules and concepts is developed to explain the nature of the problem and to arrive at a set of recommendations for stabilizing through excitation control. In order to limit the scope of this paper, the excitation system investigated is one which could be characterized by a small time constant of between 0.03 and 0.05 second which is typical of thyristor-type systems.

Paper 68 TP 129–PWR, recommended and approved by the Power Generation Committee of the IEEE Power Group for presentation at the IEEE Winter Power Meeting, New York, N. Y., January 28–February 2, 1968. Manuscript submitted September 18, 1967; made available for printing November 29, 1967.

The authors are with the General Electric Company, Schenectady, N. Y.

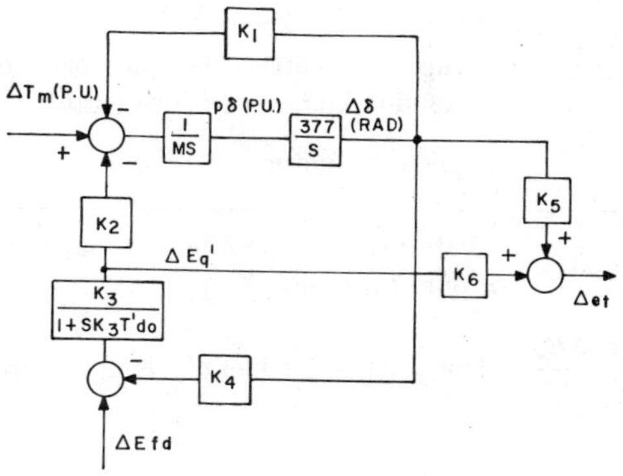

Fig. 1. Linearized small perturbation relations of a single generator supplying an infinite bus through external impedance.

The method of analysis of course can easily be extended to excitation systems with different dynamic characteristics. It is believed that from this analysis covering a wide range of conditions for the single machine case, one can project recommendations for stabilizing machines in multimachine systems.

Dynamics of a Single Machine Connected to a Large System—Fundamental Concepts

Block Diagram Relations

The phenomena of stability and damping of synchronous machines for the mode of small perturbations can be examined with the aid of block diagrams relating the pertinent variables of electrical torque, speed, angle, terminal voltage, field voltage, and flux linkages. The relations in the block diagrams discussed in this paper apply to a 2-axis machine representation with a field circuit in the direct axis but without amortisseur effects. Although a more rigorous representation should include amortisseur or solid iron eddy current effects in both axes, this simpler representation is sufficient to establish the basic effects and develop concepts.

The basic phenomenon in question is the stability of the torque-angle loop, i.e., the behavior of the rotor angle and speed following a small disturbance such as a mechanical torque disturbance. Although the whole subject can be and has been explored by various stability analysis techniques such as Routh's criterion [8], eigenvalue analyses [4], etc., considerable value is found in analyzing the phenomena in the light of elementary servomechanism and frequency response theory, thereby developing an insight into the basic elements that cause various effects.

We will attempt to relate the familiar concepts of small perturbation stability of a single machine supplying an infinite bus through external impedance to the elements and relations shown in Fig. 1. These relations and block diagram have been treated previously in [9]. The parameters in these relations are listed in the Appendix. They are derived by small perturbation

Reprinted from *IEEE Trans. Power App. Syst.*, Apr. 1969.

analysis on the fundamental synchronous machine equations, and hence are functions of machine and system impedances as well as operating point.

The parameters of Fig. 1 are defined as follows:

$K_1 = \left.\dfrac{\Delta T_e}{\Delta\delta}\right\vert_{E_q'}$	change in electrical torque for a change in rotor angle with constant flux linkages in the d axis
$K_2 = \left.\dfrac{\Delta T_e}{\Delta E_q'}\right\vert_{\delta}$	change in electrical torque for a change in d-axis flux linkages with constant rotor angle
K_3	impedance factor
$K_3 = \dfrac{X_d'+X_e}{X_d+X_e}$	for the case where the external impedance is a pure reactance X_e
$K_4 = \dfrac{1}{K_3}\dfrac{\Delta E_q'}{\Delta\delta}$	demagnetizing effect of a change in rotor angle
$K_5 = \left.\dfrac{\Delta e_t}{\Delta\delta}\right\vert_{E_q'}$	change in terminal voltage with change in rotor angle for constant E_q'
$K_6 = \left.\dfrac{\Delta e_t}{\Delta E_q'}\right\vert_{\delta}$	change in terminal voltage with change in E_q' for constant rotor angle
T_{do}'	field open circuit time constant
$T_{dz}' = K_3 T_{do}'$	effective field time constant under load.

It is important to recognize that, with the exception of K_3 which is only a function of the ratio of impedances, all other parameters change with loading, making the dynamic behavior of the machine quite different at different operating points. Since these parameters change in rather complex manner, it is difficult to reach general conclusions based on parameter values for one operating point only.

Performance with Constant d-Axis Flux Linkages

Returning to Fig. 1, let us first look at the portion shown in Fig. 2, which is pertinent for the condition of constant flux linkages in the d axis.

This portion shows a torque-speed-angle loop which gives rise to a natural period of oscillation $\omega_n = \sqrt{377\, K_1/M}$ rad/s. Neglecting damping due to electrical load characteristics, we note that this torque-speed-angle loop is a pure oscillator with zero damping much like a spring-mass system. The lighter the machine (smaller M) and the higher the transient synchronizing power coefficient K_1, the higher is the oscillation frequency. From the relations in the Appendix, we note that K_1 is decreased by increasing system and machine reactances. It is also affected by machine loading in a somewhat peculiar manner due to transient saliency effects.

In most practical cases K_1 is positive giving rise to oscillations. It is possible in very unusual cases of very long ties and relative high loading on these ties to have K_1 go negative. In these cases, even if we were to maintain constant flux linkages in the d axis, the machine would be unstable in a real root stability sense, i.e., the machine angle would run away monotonically and exponentially with time. Although with special stabilizing in the control of excitation, it is possible to render some of these systems stable, they are primarily of academic rather than practical interest.

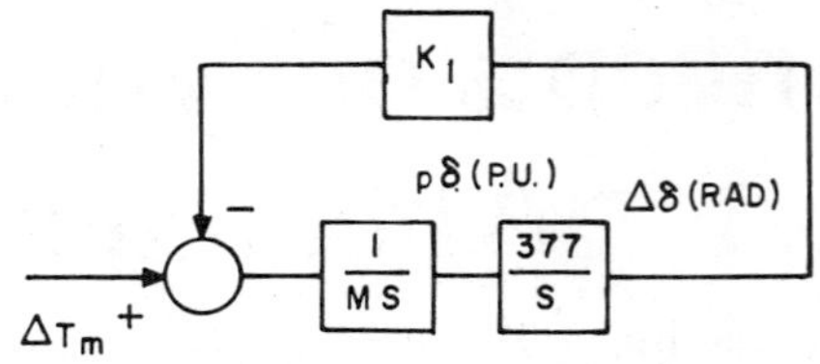

Fig. 2. Linearized torque-angle relationships for the condition of the constant flux linkages in the d axis.

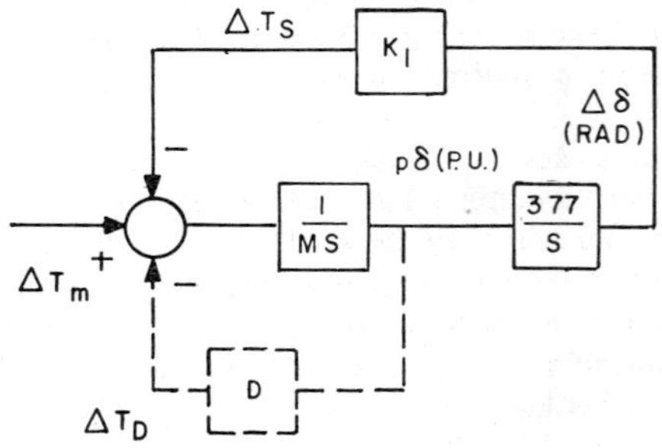

Fig. 3. Torque-angle loop defining synchronizing torques ΔT_s and damping torques ΔT_D.

Damping Torques, Synchronizing Torques—Damping Ratio

Before introducing the effects of field losses and excitation control, it is desirable to develop a few simple concepts relating the behavior of the second-order system of Fig. 3 which introduces a damping term D.

In this system, damping is provided by developing a negative torque proportional to and in time phase with speed $p\delta$. The resulting characteristic equation is

$$S^2 + (D/M)\ S + (K_1\, 377/M) = 0 \tag{1}$$

giving rise to damped oscillations with frequency $\omega_n\sqrt{1-\zeta^2}$ and damping ratio ζ where

$$\omega_n = \sqrt{K_1\, 377}, \quad \zeta = \tfrac{1}{2}D/\sqrt{K_1 M\, 377}.$$

One form of instability is approached as the damping ratio goes to zero. For normal values of damping ratios, the frequency of oscillation is not materially different from the value for zero damping ratio. For conceivable ranges of inertia, impedances, and loading values, this frequency of oscillation will be in the neighborhood of 0.5 to 2 Hz with the possibility of extreme values of up to 4 Hz at the high end and 0.1 Hz at the low end.

At this point it is appropriate to introduce the concept of synchronizing and damping torques. At any given oscillation frequency, braking torques are developed in phase with the machine rotor angle and in phase with the machine rotor speed. The former are termed synchronizing torques and the latter damping torques. The torque oscillations developed by any particular means can be broken down into these components for an insight as to their effect on stability.

Stability can be endangered by a lack of either or both synchronizing and damping torques. The traditional stability criterion with which industry is most acquainted concerns the tests for positive synchronizing torques which determine whether or not forces will be set up to restore the rotor angle of the machine following an arbitrarily small displacement of this angle. No correspondingly simple guide rules have been developed to test for the possibility of another form of instability which would arise for cases where damping torques became negative. This latter form of instability is becoming increasingly important as machine stable operation is more than ever dependent on automatic control of excitation.

Performance with Constant Field Voltage

Introducing now the effects of field losses, we refer to Fig. 4 which adds the armature reaction terms expressed as a demagnetizing influence with increasing rotor angle through the block K_4.

Examine now the contribution of torque due to this effect shown by the dashed lines in Fig. 4.

The effect of this component of torque can be described by the expression

$$\left.\frac{\Delta T}{\Delta\delta}\right|_{\text{due to }\Delta E_q'} = -\frac{K_2K_4K_3}{1+ST_{do}'K_3}. \quad (2)$$

These coefficients are always positive, hence we note that at steady state or zero oscillation frequencies, this demagnetizing component of torque is $-(K_2K_4K_3)\Delta\delta$ which is opposed in sign to the transient synchronizing component discussed in the previous section, $+K_1\Delta\delta$.

The familiar steady-state stability criterion with constant field voltage defines the stability limit as the condition for which the steady-state synchronizing power coefficient $K_1 - K_2K_4K_3$ is zero.

We note that at very high oscillation frequencies, $\omega \gg 1/K_3T_{do}'$ (note that $1/K_3T_{do}' = 1/T_{dz}'$), the phase angle of the torque component in question given by (2) is plus 90 degrees which means that at these high frequencies, these torques are almost entirely damping torques. The magnitude of the torques, of course, gets attenuated with increasing frequency. For typical machine and system parameters resulting in oscillation frequencies of about 1 Hz, the field would only contribute a damping ratio of between 0.03 and 0.05.

In terms of time domain behavior, the series of traces in Fig. 5 help describe the significance of the various effects discussed so far. The plots are the time response of machine angle following a small step change in mechanical torque.

Fig. 5(a) shows the case of constant E_q' and $K_1 > 0$ establishing the characteristic oscillations with zero damping. As mentioned before, damping due to load characteristics is neglected in these discussions.

Fig. 5(b) shows the effect of the field demagnetizing resulting in the same type of oscillations with slight damping superposed on a monotonic drift as the machine establishes new steady-state flux level conditions. This case is for $K_1 - K_2K_4K_3 > 0$, i.e., the machine has steady-state stability or stability under close manual control in the traditional sense (positive zero frequency synchronizing power coefficient).

Fig. 5(c) illustrates the angle response for the case where the machine is operating with constant field voltage beyond the steady-state limit. Here we continue to see the oscillations; however, they are superimposed on a monotonically unstable component. For such an operating point, stable operation can only be achieved by superimposing the effects of excitation control which cancel out the unstable monotonic component. Operation in this region under regulator control has been referred to as operation with dynamic stability or conditional stability. Providing stability by excitation control requires cancelling out the monotonic instability without deteriorating damping or giving rise to negative damping. These effects will be discussed in the next section.

Finally, Fig. 5(d) shows the behavior for the rare case where K_1 is very small or negative and the system manifests instability without oscillations even under assumption of constant direct axis flux linkages E_q'.

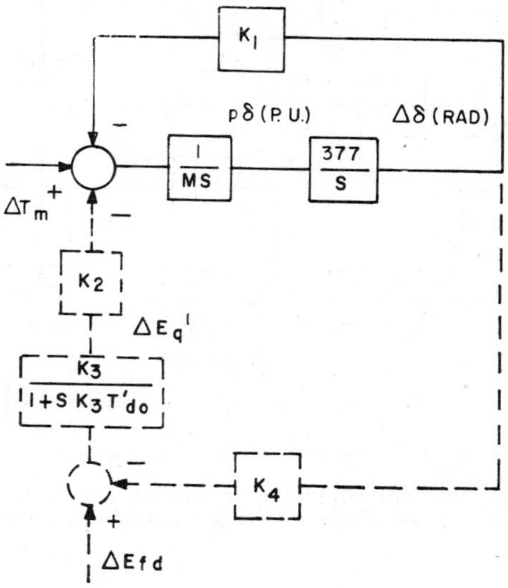

Fig. 4. Torque-angle loop including direct axis field effects.

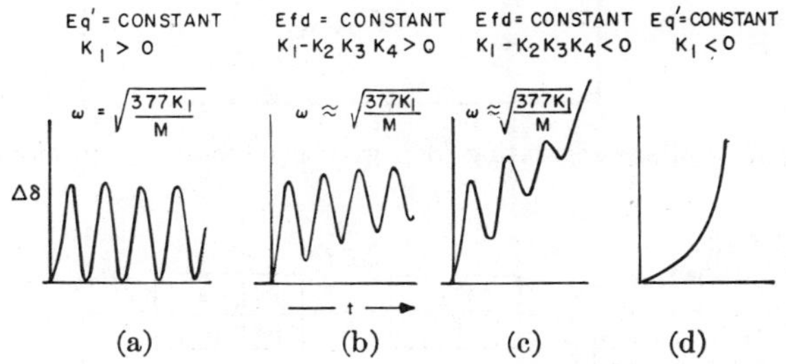

Fig. 5. Typical responses of rotor angle following small step increases in mechanical torque.

Effects of Voltage Regulators

Block Diagram Relations

The effect of the voltage regulator on damping and synchronizing torques will now be analyzed by referring to the block diagram of Fig. 6 where the additional necessary relations between E_q' and e_t, δ and e_t, and E_{fd} and e_t have been included.

In this treatment, we are considering a simple voltage regulator-excitation system, i.e.,

$$\Delta E_{fd}/\Delta e_t = -K_\epsilon/(1+ST_\epsilon) \quad (3)$$

with a gain K_ϵ and small time constant T_ϵ as might be typical of thyristor-type exciters.

By means of small perturbation analysis, the terminal voltage deviations Δe_t can be related to changes in angle $\Delta\delta$ through the proportionality factor K_5 and to changes in flux linkages $\Delta E_q'$ by the proportionality factor K_6.

From the relations in the Appendix, it can be seen that K_6 is always positive although its magnitude gets reduced with loading (large machine internal angles) and with small external impedances. The constant K_5, on the other hand, can have either sign and considerable range in magnitude depending on the impedances and the operating condition. This variation in the parameter K_5 has considerable bearing on the effects on damping of voltage regulators. Before developing insights into these effects, let us analyze regulator gain constraints.

Regulator Gain Constraints

One important criterion of voltage regulator performance is its operation with the machine on open circuit for which case

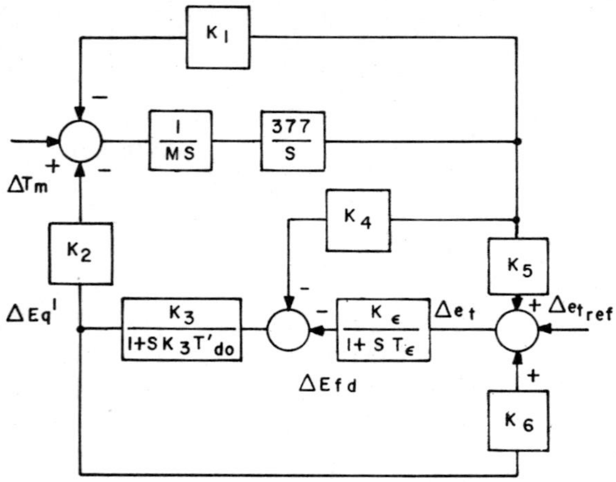

Fig. 6. Single machine supplying an infinite bus through external impedance including effects of voltage regulator-excitation system.

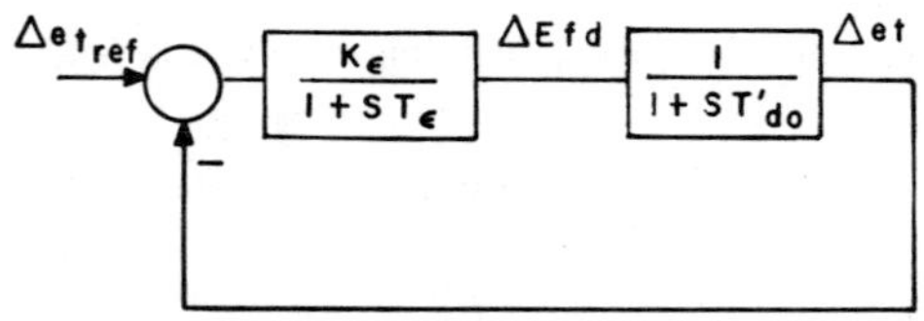

Fig. 7. Voltage regulating loop with machine on open circuit.

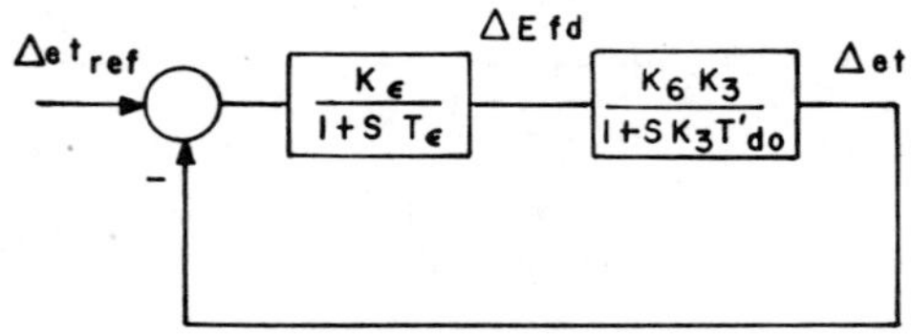

Fig. 8. Voltage regulating loop with machine connected to system assuming no changes in rotor angle.

machine angle has no significance. The regulator-machine loop is therefore described by Fig. 7.

For well-damped performance of this regulating loop [10], it is desirable to maintain the crossover frequency less than $1/2T_\epsilon$ which would mean that the gain K_ϵ should be approximately less than $T_{do}'/2T_\epsilon$. For an exciter time constant of 0.05 second and open-circuit field time constant T_{do}' of 5 seconds, we note that it would be desirable to hold the regulator gain to less than 50 due to these open-circuit performance considerations.

It should be noted that the gain we speak of can be interpreted also as the transient regulator gain. Since it may be desirable, from a steady-state regulation or droop consideration, to have considerably higher regulator steady-state or dc gain, the desired transient gain reduction can be obtained with a series lag-lead network $(1 + T_1s)/(1 + T_2s)$ where

$$T_1/T_2 = \text{(transient gain)}/\text{(steady-state gain)}$$

or alternately with rate feedback from exciter voltage.

By selecting $1/T_1$ as considerably lower than the crossover frequency, the effects of concern are not significantly different if the assumption is made that the regulator is represented by $(K_{\epsilon 1})/(1 + ST_\epsilon)$ rather than by

$$[K_{\epsilon 2}(1 + ST_1)]/[(1 + ST_2)(1 + ST_\epsilon)]$$

where $K_{\epsilon 1} = K_{\epsilon 2}\, T_1/T_2$ is the transient gain.

One more point of interest is in order. The regulating loop, as described in Fig. 7, also has significance with the machine under load. This loop by itself describes the closed-loop effects of voltage as affected by flux changes only, i.e., with no changes in angle. Under loaded conditions, this loop is given by the block diagram of Fig. 8. The machine function $\Delta e_t/\Delta E_{fd}$ is now

$$\frac{\Delta e_t}{\Delta E_{fd}} = \frac{K_6 K_3}{1 + ST_{do}'K_3} \tag{4}$$

which can also be expressed as

$$\frac{\Delta e_t}{\Delta E_{fd}} = \frac{K_6(T_{dz}'/T_{do}')}{1 + ST_{dz}'}. \tag{5}$$

Disregarding the effect of K_6 which can vary from about 1 to a considerably lower value (occasionally for very high external reactances combined with heavy shunt resistance loads K_6 can be slightly greater than the open-circuit value of 1), we note that the effect of the shorter field time constant under load T_{dz}' is offset by the lower gain T_{dz}'/T_{do}' so that the closed-loop response of this voltage component regulating loop under load conditions is not materially different from the closed-loop response on open circuit.

As a first approximation using familiar frequency response analysis techniques [10], this closed-loop response between the terminal voltage component due to flux changes and change in voltage reference can be described by two time constants in series, one equal to $T_{do}'/K_\epsilon K_6$ and the other equal to T_ϵ.

$$\left.\frac{\Delta e_t}{\Delta e_{t\,\text{ref}}}\right|_{\delta\ \text{constant}} \approx \frac{K_\epsilon K_6}{(1/K_3 + K_\epsilon K_6)[1 + S(T_{do}'/K_\epsilon K_6)](1 + ST_\epsilon)}$$

$$\approx \frac{1}{[1 + S(T_{do}'/K_\epsilon K_6)](1 + ST_\epsilon)} \quad \text{for large } K_\epsilon. \tag{6}$$

Similarly

$$\left.\frac{\Delta E_q'}{\Delta e_{t\,\text{ref}}}\right|_{\delta\ \text{constant}} \approx \frac{K_\epsilon}{(1/K_3 + K_\epsilon K_6)[1 + S(T_{do}'/K_\epsilon K_6)](1 + ST_\epsilon)}$$

$$\approx \frac{1}{K_6[1 + S(T_{do}'/K_\epsilon K_6)](1 + ST_\epsilon)} \quad \text{for large } K_\epsilon. \tag{7}$$

Voltage Regulator Effects on Machine Stability

Comparing Fig. 6 with Fig. 4, we note first that, with the voltage regulator in operation, the demagnetizing component $K_4\Delta\delta$ causes a greatly attenuated effect relative to that which it had with no voltage regulator. That is, if we compare the expressions for change in electrical torque for a change in angle due to this demagnetizing action, we have

$$\frac{\Delta T}{\Delta \delta} = \frac{-K_4 K_2 K_3}{1 + ST_{do}'K_3} \tag{8}$$

in the case of no voltage regulator, and we previously described its effects in terms of field damping and negative synchronizing torques.

For the case with voltage regulator, this particular component of torque becomes approximately

$$\frac{\Delta T}{\Delta \delta} \approx \frac{-K_4 K_2}{K_\epsilon K_6[1 + S(T_{do}'/K_\epsilon K_6)]}. \tag{9}$$

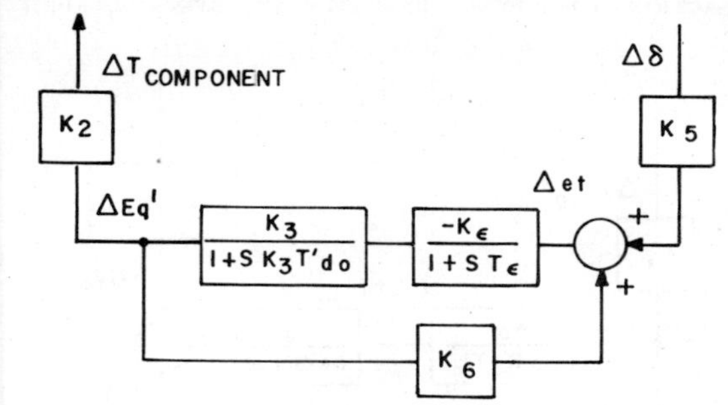

Fig. 9. Component of torque produced by voltage regulator action response to voltage deviations produced by rotor-angle deviations.

When comparing (8) and (9), we note that at low frequencies (9) is considerably smaller than (8) by the factor $1/K_\epsilon K_6 K_3$, while at very high frequencies, the two expressions approach each other. The significance of these facts is that the steady-state negative component of synchronizing power coefficient is practically eliminated with a voltage regulator with fairly high gain. Meanwhile, since the effective time constant has been reduced from T_{dz}' to $T_{do}'/K_\epsilon K_6$, the damping component of torque due to this effect is correspondingly reduced, i.e., the phase lag approaches 90 degrees for considerably higher oscillation frequencies.

Now that we have disposed of the component of torque produced by the K_4 branch in Fig. 6 as rather negligible and contributing only slightly to damping, let us analyze the contribution coming through the branch K_5 accounting for the effect of angle on terminal voltage. This terminal voltage in turn affects the flux and torque through the closed voltage regulator loop.

The component in question is described by the block diagram of Fig. 9.

The exact expression for this component of torque due to a change in angle and its effect on voltage is

$$\frac{\Delta T}{\Delta\delta} = -\frac{K_2 K_\epsilon K_5}{(1/K_3 + K_6 K_\epsilon) + S(T_\epsilon/K_3 + T_{do}') + S^2 T_{do}' T_\epsilon}. \tag{10}$$

Using the approximate expressions for the closed loop which apply for the usual range of constants in question, this becomes

$$\left.\frac{\Delta T}{\Delta\delta}\right| \approx \frac{-K_2 K_5}{K_6[1 + S(T_{do}'/K_\epsilon K_6)](1 + ST_\epsilon)}. \tag{11}$$

Finally, including the previous component produced by the K_4 branch, the demagnetizing effect of angle, we have the complete expression for changes in torque due to changes in angle via their effect on E_q'

$$\left.\frac{\Delta T}{\Delta\delta}\right|_{\text{due to } \Delta E_q'} = \frac{-K_2\{[K_\epsilon K_5 + K_4] + ST_\epsilon K_4\}}{1/K_3 + K_\epsilon K_6 + S(T_\epsilon/K_3 + T_{do}') + S^2 T_{do}' T_\epsilon}. \tag{12}$$

Examine now the synchronizing and damping torque contributions from this expression for different oscillation frequencies.

Synchronizing Torques

This torque is the component in time phase with angle. Neglecting the demagnetizing effect through K_4, i.e., using the simpler expression (10), we note that this synchronizing torque component T_S at oscillation frequency ω is

$$\frac{-K_2 K_\epsilon K_5}{1/K_3 + K_6 K_\epsilon - \omega^2 T_{do}' T_\epsilon}\Delta\delta. \tag{13}$$

For very low frequencies

$$T_S = \frac{-K_2 K_\epsilon K_5 \Delta\delta}{1/K_3 + K_6 K_\epsilon}$$

$$\approx \frac{-K_2 K_5 \Delta\delta}{K_6} \quad \text{for high values of } K_\epsilon.$$

We note that with K_5 positive this synchronizing component is negative. This is of no particular concern as for those cases where K_5 is positive (low to medium external impedance and low to medium loadings) K_1 is usually high so that the net synchronizing component $K_1 - K_2 K_5/K_6$ is still significantly greater than zero.

For K_5 negative, which occurs for moderate to high system transfer impedances, and heavy loadings, we note that the synchronizing torque component is positive at low frequencies. This is beneficial in those cases where the transient synchronizing coefficient K_1 is low or negative, or where the steady-state synchronizing coefficient without regulator $K_1 - K_2 K_4 K_3$ is negative.

As oscillation frequencies increase and ω^2 approaches $(1/K_3 + K_6 K_\epsilon)/(T_{do}' T_\epsilon)$, we note from (13) that the synchronizing torque becomes infinite and for even larger frequencies it reverses sign. Actually before this can happen, the oscillation frequency becomes very high and instability develops because of negative damping.

Damping Torques

From (11), we see that the damping torque component T_D due to voltage regulator action is

$$\frac{K_2 K_\epsilon K_5 (T_\epsilon/K_3 + T_{do}')\omega}{(1/K_3 + K_6 K_\epsilon - \omega^2 T_{do}' T_\epsilon)^2 + (T_\epsilon/K_3 + T_{do}')^2 \omega^2}\Delta\delta. \tag{14}$$

This component gives positive damping whenever K_5 is positive but for a large number of cases where K_5 is negative, it contributes negative damping. Further, we note that, with K_5 negative, increasing K_ϵ increases the magnitude of negative damping causing instability. However, at the other extreme with zero regulator gain, we only get the small amount of field damping described before.

We thus have a conflicting problem. In those cases where K_5 is negative and which are generally the cases involving stability problems, a voltage regulator is of major help in providing synchronizing torques and curing that part of the stability problem. However, in so doing it destroys the natural damping of the machine which is small to start with. The recourse has been to have just enough regulator gain to provide synchronizing power coefficient without cancelling all of the inherent machine damping.

This can be a satisfactory solution in most cases; however, there can be instances where stability is provided by the regulator with very poor damping, making operation extremely oscillatory. In some special cases of very long lines requiring operation near the line limit, the solution is to have a fairly high regulator gain to provide the necessary synchronizing power coefficient. In these cases, one effective way to solve the damping problem is to provide a special stabilizing signal derived from machine speed, terminal frequency, or power.

Recognizing that the problem of damping can be solved by auxiliary means removes some of the constraints placed on regulator gain. However, as was noted before, there remain the regulator gain constraints explained in the Section *Regulator Gain Constraints*. Since there are fixed relationships between the variables, speed, terminal frequency, and power which can be used to relate the requirements of one signal in terms of the requirements of another, we will confine our analysis to use of signals from machine rotor speed. It is recognized that the use of any particular signal in preference to another will depend on hardware and implementation considerations outside the scope of this paper.

Stabilizing Signals from Speed

Phase and Magnitude Relations

The stabilizing signal requirements to provide damping through transient manipulation of the voltage reference in response to speed can be derived as follows. Fig. 10 shows the elements of the system in question relating the effect of speed through the stabilizing function $G(S)$ through the voltage regulator loops affecting $\Delta E_q'$, which produces a component of torque ΔT_{sig}.

The question is: what should be the nature of $G(S)$ to provide adequate damping over a wide range of machine and system parameters and operating conditions?

The over-all function between the component of torque produced by the stabilizing signal and speed is

$$\frac{\Delta T_{\text{sig}}}{p\delta} = \frac{G(S)K_2K_\epsilon}{(1/K_3 + K_\epsilon K_6) + S(T_\epsilon/K_3 + T_{do}') + S^2T_{do}'T_\epsilon}. \tag{15}$$

For the usual range of constants, this function can be approximated by

$$\frac{\Delta T_{\text{sig}}}{p\delta} \approx \frac{G(S)K_2K_\epsilon}{(1/K_3 + K_\epsilon K_6)[1 + S(T_{do}'/K_\epsilon K_6)](1 + ST_\epsilon)}. \tag{16}$$

For large values of K_ϵ it can be further approximated by

$$\frac{\Delta T_{\text{sig}}}{p\delta} = \frac{K_2}{K_6}\frac{G(S)}{[1 + S(T_{do}'/K_6K_\epsilon)](1 + ST_\epsilon)}. \tag{17}$$

If we wish $\Delta T_{\text{sig}}/p\delta$ to provide pure damping throughout the frequency range, then $G(S)$ ideally should be a pure lead function with two zeros, i.e., $G(S) = K(1 + ST_{do}'/K_6K_\epsilon)(1 + ST_\epsilon)$. Such a function is not physically realizable and further we do not wish to have a permanent steady-state effect of speed on the voltage reference. Hence, the stabilizing function should be some compromise function which provides damping over the spectrum of expected frequencies of oscillation, i.e., a function with enough phase lead to compensate a significant part of the phase lag contributed by the machine and regulator.

Reverting back to the concepts of synchronizing and damping components of torque, we note the following. If at a given frequency of oscillation ω, $G(j\omega)$ has more lead angle than the lag angle of the machine function

$$\frac{\Delta T}{\Delta e_{t\,\text{ref}}} = \frac{K_2K_\epsilon}{(1/K_3 + K_SK_6) + j\omega(T_\epsilon/K_3 + T_{do}') - \omega^2T_{do}'T_\epsilon}$$

which for large K_ϵ is approximately equal to

$$\frac{K_2}{K_G(1 + j\omega T_{do}'/K_6K_\epsilon)(1 + j\omega T_\epsilon)}$$

(this difference being less than 90°), then in addition to the damping component, we have a negative synchronizing component, the magnitude of these two components being related to the sine and cosine of this difference in angle. That is, let the lead angle of $G(S)$ be θ at a given frequency of oscillation ω, and the lag angle of the machine regulator function $\Delta T/\Delta e_{\text{ref}}$ be γ at this same frequency. Then the relative magnitude of the synchronizing torque T_S to damping torque T_D at that frequency produced by the stabilizing signal is

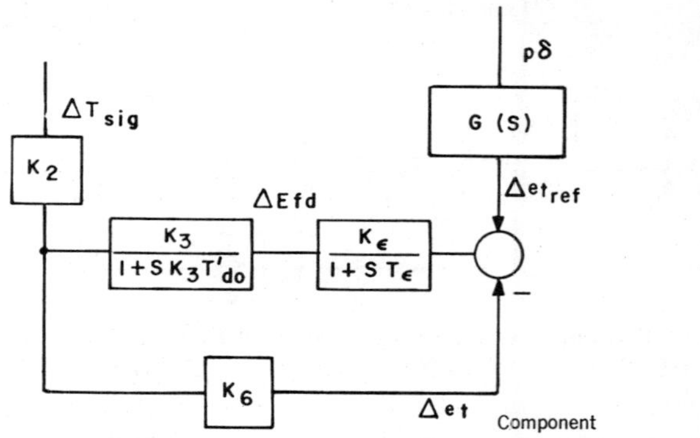

Fig. 10. Component of torque produced by voltage regulator action in response to a speed-derived signal.

$$\frac{T_S}{T_D} = \frac{\sin(\gamma - \theta)}{\cos(\gamma - \theta)}. \tag{18}$$

Since it is very difficult to synthesize a function which will provide exact angle cancellation for all conditions, i.e., having $\theta = \gamma$ at all points, and further since the stability problem is often helped by providing some synchronizing component, we come to the following conclusions relating to the nature of the stabilizing signal function.

1) At low frequencies of oscillation which are indicative of weak synchronizing torques, one should strive to hold θ less than γ, or if this is not possible, at least keep θ close to γ. At these low frequencies, the machine regulator function $\Delta T_e/\Delta e_t$ has small lagging phase angles, hence the stabilizing function should correspondingly have small leading phase angles. As was mentioned before, a medium-high-regulator gain provides synchronizing torques through the regulator by virtue of the effect of angle on terminal voltage K_5. Hence, what should be attempted at the low-frequency range is to provide damping without appreciably hurting the synchronizing torque.

2) At the high extreme of the spectrum of frequencies of oscillation, there is no particular danger to subtracting from the synchronizing component. As a matter of fact, since the machine regulator function's phase angle becomes increasingly lagging with increasing frequency whereas, because of hardware limitations, the stabilizing function ceases to provide increasing phase leads at these higher frequencies, it is natural to expect θ to be less than γ. However, in this range of frequencies (γ-θ) should not be so large as to add significantly to the synchronizing component with its attendant effect on increasing the frequency of oscillation. As long as the difference between the machine regulator lag and the phase lead for the stabilizing function is within ±30° for the frequency spectrum of concern, acceptable results are obtained.

Stabilizing Signal Constraints

There are several constraints that have to be observed when prescribing a stabilizing signal design that will be physically realizable and desirable from other considerations. These considerations and constraints are as follows.

1) The signal should not produce a steady-state offset of voltage reference with frequency hence it must approach a pure rate signal at very low frequencies.

2) Lead functions can only be realized with lead-lag pairs, and there is a limit as to how small the lag time constant can be. It is conservative to assume that this lag time constant cannot be smaller than 0.05 second.

Search for Universally Applicable Stabilizing Function

General Approach

The insights that have been developed into the problem of providing damping from stabilizing signals may convince one that there is no way of developing a universally applicable stabilizing function that would be optimal for a wide range of machine and system parameters, and operating conditions. However, even if not optimal, it would be highly desirable to have a signal transfer function which would provide damping in those cases where it was badly needed without deteriorating damping in those instances where there was no particular need for providing supplementary damping.

The single machine against infinite bus analysis outlined in this paper has been used to explore a variety of conditions of a single machine with different parameters and loading conditions. For each of these conditions there are any number of functions which provide adequate damping simply by having the right range of magnitude and phase at the particular frequency of oscillation of concern.

By examining the phase and magnitude of these functions as related to the frequency of oscillation that was pertinent for the particular case, we arrive at bands of phase angles and magnitudes as functions of frequency of oscillation. It turns out that this band falls within ranges for which a general function with a rather well defined range of parameters serves reasonably well.

Heuristic reasoning permits casting the results of this single machine analysis to the general multimachine case. It is known that in a multimachine situation, there is a wide band of modes of oscillation ranging from those between tightly coupled light machines to those between groups of machines separated by long ties. It is reasonable then to expect that to be effective in the general power system environment, the stabilizing signal on any given machine must accommodate this wide range of possibilities. Further, from a practical standpoint, it is essential that the range of adjustment not be critical since systems change from month to month as do operating conditions.

Range of Single Machine Cases Studied

The parameters K_1, K_2, K_3, K_4, K_5, and K_6, defined in the Appendix, were calculated for a number of cases.

Two system configurations were studied. One was described by a single machine supplying an infinite bus through an external impedance, Fig. 11 (a). Table I describes all the different combinations of parameters and loading explored with this system configuration.

The machine constants that were used are

	Hydro	Steam
X_d	1.14	1.6
X_d'	0.24	0.32
X_q	0.66	1.55
T_{do}'	12	6

The second configuration attempting to simulate the condition of a system supplying local load connected to a large system through a weak tie is shown on Fig. 11 (b).

Table II lists the different combinations of parameters explored for this configuration.

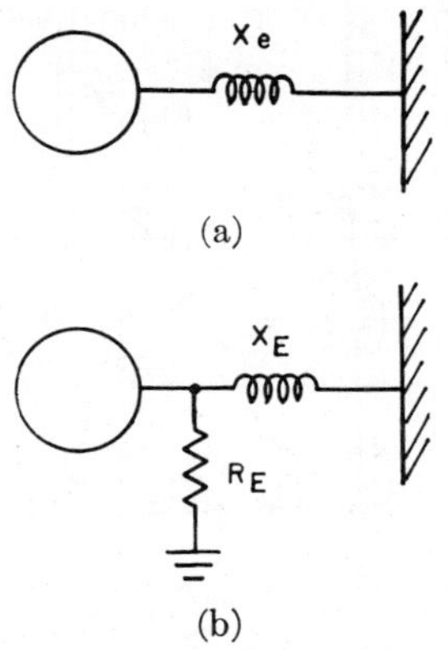

Fig. 11. External impedance system configurations for single machine against infinite bus.

TABLE I
SINGLE MACHINE AGAINST INFINITE BUS THROUGH EXTERNAL REACTANCE—TERMINAL VOLTAGE = 1.0

Machine Type	Inertia H	External Reactance X_e	Machine Loading $P + jQ$
Hydro, steam	1.5, 5	0.1, 0.4, 0.7, 1.0	$0.1 + j0$, $0.5 + j0$, $1 + j0.5$, $1 + j0$, $1 - j0.5$

TABLE II

Machine Type	Inertia H	External Impedance X_E	R_E	Machine Loading $P + jQ$
Steam, hydro	1.5, 5.0	1.0	1.0	$1.0 + j0$, $1.5 + j0.3$, $1.5 - j0.3$, $0.5 + j0.3$, $0.5 - j0.3$
		5.0	1.0	$1.0 + j0$, $1.2 + j0.1$, $1.2 - j0.1$, $0.8 + j0.3$, $0.8 - j0.3$

Guide Rules for Stabilizing Function Phase and Magnitude Requirements

As developed in the Section *Phase and Magnitude Relations,* the lead phase angle of the signal $G(S)$ at a particular oscillation frequency ω required to provide pure damping at that frequency would be equal to the phase lag of the expression

$$\frac{\Delta T}{\Delta e_{t\,\mathrm{ref}}} = \frac{K_2 K_\epsilon}{1/K_3 + K_\epsilon K_6 + S(T_\epsilon/K_3 + T_{do}') + S^2 T_{do}' T_\epsilon}. \tag{19}$$

The phase angle of (19) as well as the reciprocal of its magnitude A for the frequency of oscillation $\omega = \sqrt{K_1 377/M}$ radians was calculated for the various cases listed above using the expressions for the coefficients K_1 to K_6 in the Appendix.

As far as a criterion for the required magnitude of the signal is concerned, we could arbitrarily select a magnitude of this signal that would result in a damping ratio of 0.5.

Referring to the torque-angle characteristic equation

$$S^2 + (D/M)S + (K_1 377/M) = 0 \tag{1}$$

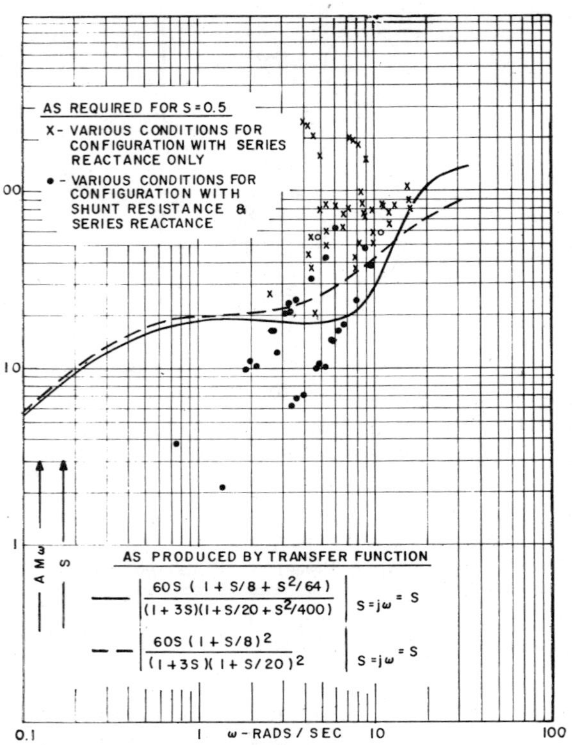

Fig. 12. Plot of magnitude of stabilizing functions versus frequency of oscillation. Scatter points are values of $AM\omega$ for a variety of steam machine system parameters and loading conditions as listed in Tables I and II.

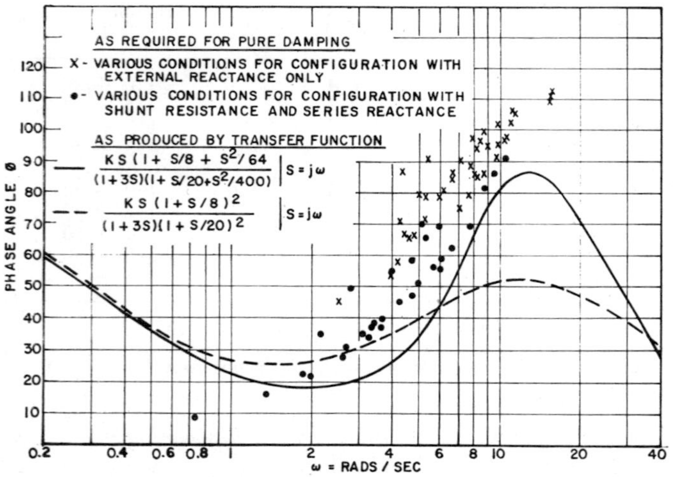

Fig. 13. Phase angle versus frequency of oscillation. Scatter plot points are phase angles required to cancel the lagging phase angles of the machine function $(\Delta E_q')/(\Delta e_{t\ \mathrm{ref}})$ for a variety of steam machine system parameters and loading conditions as listed in Tables I and II.

where D describes the gain of the damping term $\Delta T_D/p\delta$, we note that $\omega_n = \sqrt{K_1 377/M}$ and $\zeta = \frac{1}{2}D/\sqrt{377K_1M}$.

For a constant damping ratio, say $\zeta = 0.5$, we have $D = \sqrt{377\,K_1M} = M\sqrt{K_1377/M} = M\omega$.

The magnitude of the signal that would be required would then be $A\sqrt{377K_1M}$ or $AM\omega$ where A is the reciprocal of the magnitude of the function $\Delta T/\Delta e_{t\ \mathrm{ref}}$ for $\omega = \sqrt{K_1 377/M}$.

The voltage regulator transient K gain $K_{2\epsilon}$ was taken at a good compromise value of $4T_{do}'$.

Fig. 12 is a scatter plot of the desired magnitude $AM\omega$ of the signal versus frequency of oscillation ω as dictated by the above criterion. Points for cases with the system configuration of Fig. 11(a) are marked with X, whereas those with the system configuration of Fig. 11(b) are marked with dots.

A composite plot of phase angles dictated by the above idealized criterion for all cases incorporating the two configurations is shown on Fig. 13. Again X are points corresponding to cases with the configuration of Fig. 11(a), whereas dots correspond to those for the configuration of Fig. 11(b).

It is interesting to note the ranges of the scatter plots for the two different configurations. Most stability problems are characterized by oscillations at the low end of the spectrum and at this end, the stabilizing function needs little phase lead. This confirms the results of several studies which were concerned with long distance transmission problems.

On Figs. 12 and 13 we have superposed plots of magnitude and phase, respectively, for two stabilizing functions discussed in the next Section.

Stabilizing Functions Studied—Analog Computer Results

From the general shape of the scatter plots of the ideal stabilizing function by the above criterion, and guided by the understanding of the many effects developed in previous sections, several stabilizing functions were tried on the system of Fig. 1 simulated on an analog computer.

These analog computer studies confirmed with time domain results the general findings of the frequency response analysis. In particular, these results were valuable in obtaining a calibration on how far from the ideal values of $AM\omega$ and ϕ the magnitude and phase angle of a stabilizing function could be without adverse effects.

The nature of the scatter plots shows that ideally the function should have little phase angle in the frequency range of 1 to 2 rad/s. and then have a rapidly rising phase angle as the frequency of oscillation increases to about 15 rad/s.

The magnitude and phase angle of the function

$$\frac{\Delta e_{t\ \mathrm{ref}}}{p\delta} = \frac{60S(1 + S/8 + S^2/64)}{(1 + 3S)(1 + S/20 + S^2/400)}$$

are shown in Figs. 12 and 13, respectively. The function in question is made up of a rate term with long washout of 3 seconds in series with a complex pair of zeros ($\omega_n = 8$, $\zeta = 0.5$) and a complex pair of poles ($\omega_d = 20$, $\zeta = 0.5$).

In Figs. 12 and 13 we also have plotted the magnitude and phase of the function

$$\frac{\Delta e_{t\ \mathrm{ref}}}{p\delta} = \frac{60S(1 + S/8)^2}{(1 + 3S)(1 + S/20)^2}$$

which is similar to the previous function except for a damping ratio $\zeta = 1.0$ which turns the zeros and poles to real values.

Figs. 14 and 15 are sample results of the analog study illustrating some significant aspects of the problem. The traces are deviations in speed for a small mechanical torque disturbance. Where pertinent, values of $AM\omega \angle\phi$, as ideally desired by the above criterion, have been listed, and noted alongside are the actual phase and magnitude values of the stabilizing functions $S \angle\theta$ at frequency ω. Parameters K_1 through K_6 have also been listed.

Among these parameters describing machine dynamics, the most significant ones are $\omega = \sqrt{K_1 377/M}$, K_5, and K_6. ω indi-

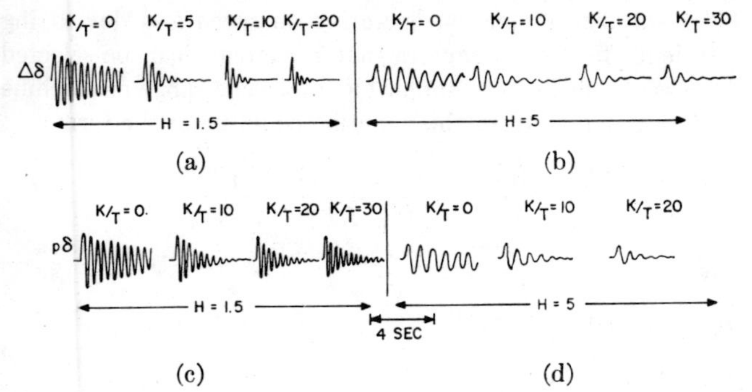

Stabilizing Function

For (a) and (b):

$$\frac{KS(1 + S/8 + S^2/64)}{(1 + TS)(1 + S/20 + S^2/400)}.$$

For (c) and (d) $T = 3$ seconds:

$$\frac{KS(1 + S/8)^2}{(1 + TS)(1 + S/10 + S^2/400)^2}$$

$$X_e = 0.1, \quad P + jQ = 1.0 + j0.5,$$

$$K_1 = 1.01, \quad K_2 = 1.149, \quad K_3 = 2.78,$$

$$K_4 = 1.47, \quad K_5 = -0.097, \quad K_6 = 0.419.$$

Fig. 14. Speed deviations of a steam unit following a small step decrease in mechanical torque. Effect of stabilizing with speed-derived signal. Conditions: regulator-exciter gain $K_\epsilon = 25$, $T_\epsilon = 0.05$ second.

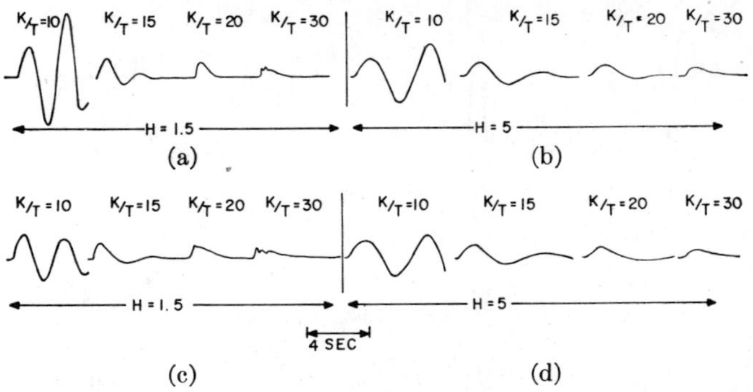

Stabilizing Function

For (a) and (b):

$$\frac{KS(1 + S/8 + S^2/64)}{(1 + TS)(1 + S/20 + S^2/400)}.$$

For (c) and (d):

$$\frac{KS(1 + S/8)^2}{(1 + TS)(1 + S/20)^2}$$

$$T = 3 \text{ seconds}$$

$$R_E = 1.0, \quad X_E = 5.0, \quad P + jQ = 1.2 + j0.1,$$

$$K_1 = -0.191, \quad K_2 = 2.35, \quad K_3 = 2.227,$$

$$K_4 = 0.046, \quad K_5 = -0.136, \quad K_6 = 1.062.$$

Fig. 15. Speed deviation of a steam unit following a small step decrease in mechanical torque. Effect of stabilizing with speed-derived signal. Configuration of Fig. 11(b). Conditions: regulator-exciter gain $K_\epsilon = 25$, $T_\epsilon = 0.05$ second.

cates the frequency of oscillation of concern. K_5 by its sign and magnitude indicates the extent to which the voltage regulator per se helps or hinders damping. Finally, K_6, as explained in the Section *Voltage Regulator Effects on Machine Stability*, influences the phase lag contributed by the voltage regulator loop.

Fig. 14(a) and (b) shows the effects of stabilizing with various strengths of the complex function

$$\frac{KS(1 + S/8 + S^2/64)}{(1 + 3S)(1 + S/20 + S^2/400)}$$

for the case of a fully loaded overexcited steam unit supplying a stiff system ($X_e = 0.1$). Cases are shown for inertia constants of $H = 1.5$ and $H = 5.0$. All pertinent constants are listed in the captions.

Fig. 14(c) and (d) contains similar results for the case where the stabilizing function is composed of real poles and zeros

$$\frac{KS(1 + S/8)^2}{(1 + 3S)(1 + S/20)^2}.$$

Table III lists the values of ω, $AM\omega \angle \phi$, and $S \angle \theta$ for the various cases of Fig. 14.

These values show that for the light inertia cases [Fig. 14(a) and (c)], the first of the stabilizing functions is more effective than the second, and this is confirmed by the analog computer results.

Fig. 15 contains results of similar nature for the configuration of local load with high tie reactance ($R_E = 1$, $X_E = 5$) with a relatively high tie loading of $0.2 + j0.1$.

The two stabilizing functions are almost equally effective. This rather extreme loading case shows that it is possible to stabilize with a voltage regulator and supplementary speed signal even though the constant flux linkage synchronizing coefficient K_1 is negative.

TABLE III

	ω	$AM\omega \angle \phi$	$S \angle \theta$ for $K = 60$
Fig. 14(a)	11.27	81.4 ∠98°	50 ∠86°
Fig. 14(b)	6.17	81.8 ∠80.3°	18 ∠44.5°
Fig. 14(c)	11.27	81.4 ∠98°	50 ∠52°
Fig. 14(d)	6.17	81.8 ∠80.3°	29 ∠44°

In order to illustrate the effect of voltage regulator gain on damping, we have included in Fig. 16(a) and (b) results of angle deviations for a step change in mechanical torque for the case where no stabilizing is used. Fig. 16(c) is for a case where the parameter K_5 is positive, leading to some benefit in damping with voltage regulator gain, whereas Fig. 16(b) shows the adverse effects of voltage regulator gain on damping for the case where K_5 is negative.

Conclusion

The small perturbation stability characteristics of a single machine supplying an infinite bus through external impedance have been explored by means of frequency response analyses giving insights into effects of machine and system parameters, voltage regulator gain, and stabilizing functions derived from speed and working through the voltage reference of the voltage regulator.

The study has explored a variety of machine loadings, machine inertias, and system external impedances with a determination of the oscillation and damping characteristics of voltage or speed following a small disturbance in mechanical torque. An attempt has been made to develop some unifying concepts that explain the stability phenomena of concern, and to predict desirable phase and magnitude characteristics of stabilizing functions.

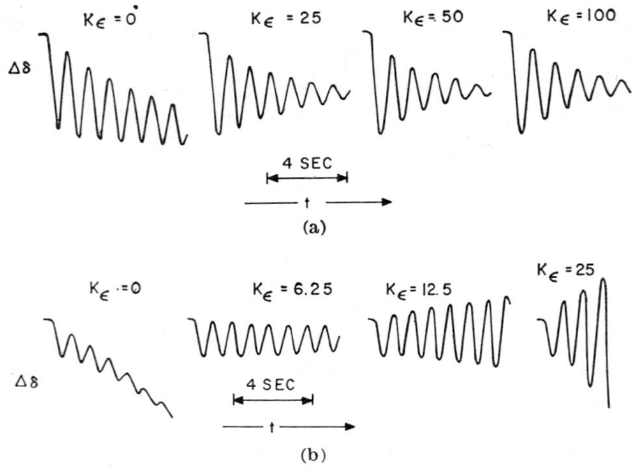

For (a):

$X_e = 0.4,\quad P + jQ = 0.5 + j0,\quad H = 5,$

$K_1 = 1.05,\quad K_2 = 1.07,\quad K_3 = 2.78,$

$K_4 = 1.37,\quad K_5 = 0.0544,\quad K_6 = 0.439.$

For (b):

$X_e = 0.4,\quad P + jQ = 1.0 - j0.5,\quad H = 5.$

$K_1 = 1.18,\quad K_2 = 1.73,\quad K_3 = 2.78,$

$K_4 = 2.21,\quad K_5 = -0.255,\quad K_6 = 0.080.$

Fig. 16. Angle deviations of a steam unit following a small step decrease in mechanical torque. Effect of voltage regulator-exciter gain K_ϵ at two different loading conditions.

It is evident that for every combination of machine and system parameters, and loading conditions, there are a variety of stabilizing signal transfer functions operating on speed which give essentially the right phase and magnitude relation for the frequency of concern. It is a challenge, however, to find a universal function which would be adequate for the whole spectrum of possibilities. By examining this spectrum of possibilities over a credible band of machine and system parameter values, as well as loading conditions, it appears that a fair compromise function can be recommended that will be almost universally applicable.

In the general case of a machine in a power system, there are many modes of oscillation between machines and groups of machines; these modes can change from day to day with loading conditions and from year to year with changes in power system configuration. Accordingly, the search for a universally applicable stabilizing function has a great deal of merit.

For a thyristor-type excitation systems characterized by a time constant in the neighborhood of or smaller than 0.05 second, certain general conclusions and recommendations can be made to cover the majority of applications. These are as follows.

1) A transient voltage regulator gain of about 25 pu ΔE_{fd}/pu. Δe_t is a good compromise for a machine with an open-circuit time constant $T_{do}' = 6$. An examination of the block diagram relations and the constraints set by voltage regulator loop damping on open circuit convinces us that the most direct way of compensating for differences in machine open-circuit time constant is to vary the transient voltage regulator gain in direct proportion to this time constant.

2) The transition from steady-state gain to transient gain can be obtained with rate feedback from exciter voltage or with a lag-lead network $(1 + T_2S)/(1 + T_1S)$ in series with the regulator where $T_2/T_1 =$ (transient gain)/(steady-state gain) and T_2 should be larger than 1 second.

3) A stabilizing function operating on speed and developing a transient offset of voltage regulator reference can be selected to be close to universally applicable for a wide range of machine and system parameters. This function could be of the form

$$\frac{\Delta e_{t\ \mathrm{ref}}(\mathrm{pu})}{p\delta(\mathrm{pu})} = \frac{KS(1 + ST_1)^2}{(1 + ST)(1 + ST_2)^2}$$

where

T	2 to 4
T_1	0.1 to 0.2
T_2	0.05
K/T	10 to 40.

This function usually provides a significant amount of damping to the rotor angle oscillations. It should not be concluded, however, that other functions are not equally or better suited for particular cases, or that stabilizing is mandatory in all cases. Normal damping forces provided by load characteristics and amortisseurs are entirely adequate in many instances.

4) There appears to be some correlation between the strength of the signal required and the inertia of the machine. From examination of several sets of runs where the only difference was inertia, we conclude that the strength of the signal should vary approximately as the square root of inertia.

5) The need for stabilizing increases with increasing dependence on the excitation to provide synchronizing power through a high response excitation system.

These conclusions and recommendations have been checked out in detailed digital multimachine simulations. Among several problems being presently studied are the dynamics under large disturbances and design requirements of signal limiting. Also being studied are the relative merits of stabilizing with terminal frequency or with a signal derived from electrical power instead of speed.

Appendix

For a machine connected to an infinite bus with voltage E through an external impedance $r_e + jX_e$, Fig. 11(a), assuming as negligible amortisseur effects, armature resistance, armature $p\psi$ terms, and saturation, the following relationships apply.

$$e_t^2 = e_d^2 + e_q^2$$

$$-e_d = \psi_q = -X_q i_q$$

$$e_q = \psi_d = E_q' - X_d' i_d$$

$$E_q = E_q' + (X_q - X_d') i_d$$

$$T_e = E_q i_q$$

$$i_d = [E_q - E\cos\delta]\{[X_e + X_q]/[r_e^2 + (X_e + X_q)^2]\} - E\sin\delta\{r_e/[r_e^2 + (X_e + X_q)^2]\}$$

$$i_q = [E_q - E\cos\delta]\{r_e/[r_e^2 + (X_e + X_q)^2]\} + E\sin\delta\{[X_e + X_q]/[r_e^2 + (X_e + X_q)^2]\}$$

$$E_q' = X_{ad} I_{fd} - (X_d - X_d') i_d$$

$$T_{do}'(dE_q'/dt) = E_{fd} - X_{ad} I_{fd}$$

$$T_m - T_e = M[d(p\delta)/dt].$$

Expressing all equations in small oscillation form and preserving the basic variables Δe_t, $\Delta E_q'$, and $\Delta\delta$, one can derive the following relations.

$$\Delta e_t = K_5 \Delta\delta + K_6 \Delta E_q'$$

$$\Delta E_q' = \frac{K_3 \Delta E_{fd}}{1 + S T_{do}' K_3} - \frac{K_3 K_4 \Delta\delta}{1 + S T_{do}' K_3}$$

$$\Delta T_e = K_1 \Delta\delta + K_2 \Delta E_q'$$

where

$$K_1 = \frac{E_{qo} E_o}{A} [r_e \sin\delta_o + (X_e + X_d') \cos\delta_o]$$

$$+ \frac{i_{qo} E_o}{A} [(X_q - X_d')(X_e + X_q) \sin\delta_o$$

$$- r_e (X_q - X_d') \cos\delta_o]$$

$$K_2 = \left[\frac{r_e E_{qo}}{A} + i_{qo}\left(1 + \frac{(X_e + X_q)(X_q - X_d')}{A}\right)\right]$$

$$K_3 = \left[1 + \frac{(X_e + X_q)(X_d - X_d')}{A}\right]^{-1}$$

$$K_4 = \frac{E_o (X_d - X_d')}{A} [(X_e + X_q) \sin\delta_o - r_e \cos\delta_o]$$

$$K_5 = \frac{e_{do}}{e_{to}} X_q \left[\frac{r_e E_o \sin\delta_o + (X_e + X_d') E_o \cos\delta_o}{A}\right]$$

$$+ \frac{e_{qo}}{e_{to}} X_d' \left[\frac{r_e E_o \cos\delta_o - (X_e + X_q) E_o \sin\delta_o}{A}\right]$$

$$K_6 = \frac{e_{qo}}{e_{to}} \left[1 - \frac{X_d' (X_e + X_q)}{A}\right] + \frac{e_{do}}{e_{to}} X_q \frac{r_e}{A}$$

$$A = [r_e^2 + (X_e + X_d')(X_q + X_e)].$$

The steady-state operating values of δ_o, E_{qo}, E_o, e_{do}, and e_{qo} are derived from the standard machine vector diagram. Expressed as a function of steady-state terminal voltage e_{to}, and steady-state real and reactive load currents I_{po} and I_{qo}, these values are

$$E_{qo} = \sqrt{(e_{to} + I_{qo} X_q)^2 + (I_{po} X_q)^2}$$

$$E_o = \sqrt{(e_{to} - I_{po} r_e - I_{qo} X_e)^2 + (I_{po} X_e - I_{qo} r_e)^2}$$

$$\sin\delta_o = \frac{e_{to} I_{po} (X_q + X_e) - r_e X_q (I_{po}^2 + I_{qo}^2)}{E_{qo} E_o}$$

$$\cos\delta_o = \frac{e_{to}(e_{to} - I_{qo}(X_q - X_e) - I_{po} r_e)}{E_{qo} E_o} - \frac{X_e X_q (I_{po}^2 + I_{qo}^2)}{E_{qo} E_o}$$

$$i_{qo} = [I_{po}(e_{to} + I_q X_q) - I_{qo} I_{po} X_q] / E_{qo}$$

$$i_{do} = [I_{po}^2 X_q + I_{qo}(e_{to} + I_{qo} X_q)] / E_{qo}$$

$$e_{qo} = [(e_{to} + I_{qo} X_q)/E_{qo}] e_{to}$$

$$e_{do} = i_{qo} X_q.$$

For the special case of zero external resistance, the expressions become

$$K_1 = \frac{X_q - X_d'}{X_e + X_d'} i_{qo} E_o \sin\delta_o + \frac{E_{qo} E_o \cos\delta_o}{X_e + X_q}$$

$$K_2 = \frac{E_o \sin\delta_o}{X_e + X_d'}$$

$$K_3 = \frac{X_d' + X_e}{X_d + X_e}$$

$$K_4 = \frac{X_d - X_d'}{X_e + X_d'} E_o \sin\delta_o$$

$$K_5 = \frac{X_q}{X_e + X_q} \frac{e_{do}}{e_{to}} E_o \cos\delta - \frac{X_d'}{X_e + X_d'} \frac{e_{qo}}{e_{to}} E_o \sin\delta_o$$

$$K_6 = \frac{X_e}{X_e + X_d'} \frac{e_{qo}}{e_{to}}.$$

The configuration of Fig. 11(b) is converted to the form of Fig. 11(a) by making a Thevenin equivalent where

$$r_e = R_E X_E^2 / (R_E^2 + X_E^2)$$

and

$$X_e = R_E^2 X_E / (R_E^2 + X_E^2).$$

Nomenclature

All quantities in per unit on machine base.

i_d, i_q	armature current, direct and quadrature axis components
e_d, e_q	armature voltage, direct and quadrature axis components
e_t	terminal voltage
E_q'	voltage proportional to direct axis flux linkages
E_{fd}	generator field voltage (one per unit is the value for 1 per unit terminal voltage on the air gap line, open circuit)
X_e, X_E	equivalent system reactances
R_E	system shunt resistance
S	Laplace operator, or magnitude of stabilizing signal
δ	angle between quadrature axis and infinite bus
$p\delta$	per unit speed deviation from synchronous
T_e	electrical torque
H	inertia constant, seconds
M	inertia coefficient = $2H$, seconds
D	damping coefficient
E	infinite bus voltage.

Subscript o means steady-state value.
Prefix Δ indicates small change.

References

[1] F. P. deMello, D. N. Ewart, and M. Temoshok, "Stability of synchronous machines as affected by excitation systems, machine and system parameters," *Proc. American Power Conf.*, vol. XXVII, pp. 1150–1159, 1965.

[2] D. N. Ewart and F. P. deMello, "A digital computer program for the automatic determination of dynamic stability limits," *IEEE Trans. Power Apparatus and Systems*, vol. PAS-86, pp. 867–875, July 1967.

[3] D. N. Ewart and F. P. deMello, "FACE, a digital dynamic analysis program," presented at the Power Industry Computer Application Conference, Pittsburgh, Pa., May 15–17, 1967.

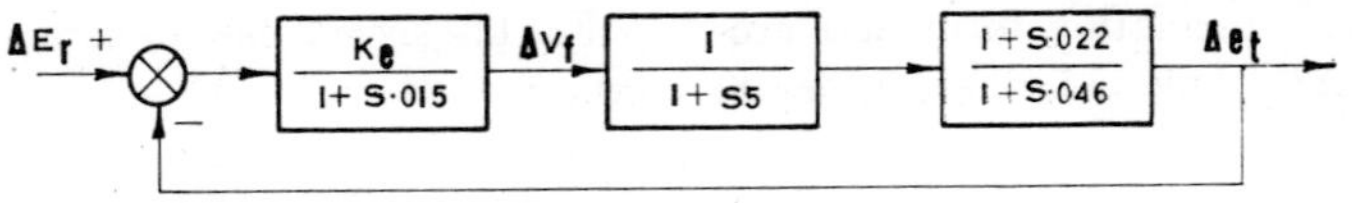

Fig. 17. Typical transfer function for hydro generator on open circuit including amortisseur effects.

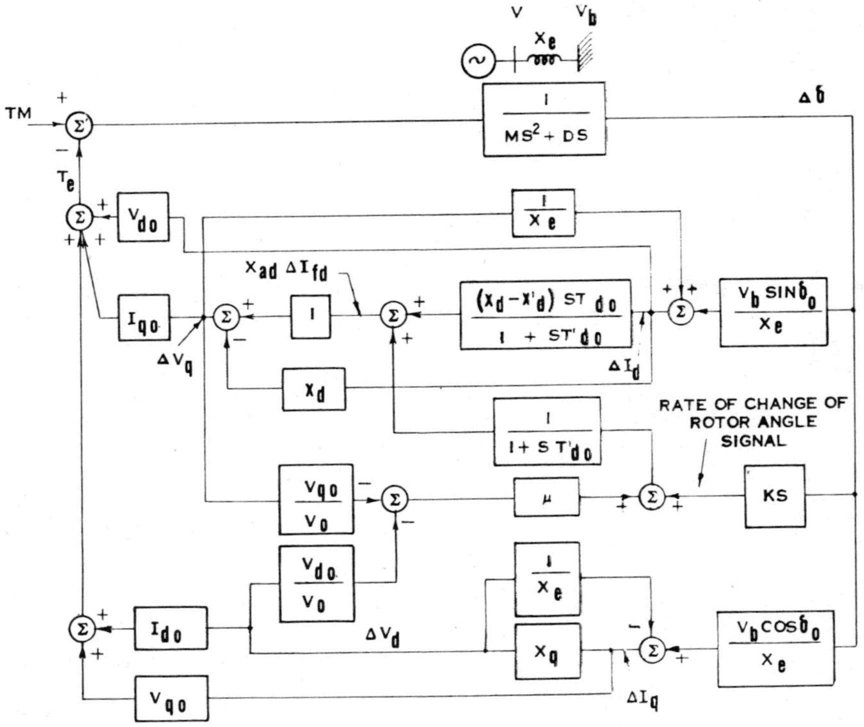

Fig. 18. Synchronous machine and excitation system.

[4] J. M. Undrill, "Dynamic stability calculations for an arbitrary number of interconnected synchronous machines," *IEEE Trans. Power Apparatus and Systems*, vol. PAS-87, pp. 835–844, March 1968.

[5] M. S. Dyrkacz, C. C. Young, and F. J. Maginniss, "A digital transient stability program including the effects of regulator, exciter, and governor response," *AIEE Trans. Power Apparatus and Systems*), vol. 79, pp. 1245–1257, 1960 (February 1961 sec.).

[6] P. L. Dandeno, A. N. Karas, K. R. McClymont, and W. Watson, "Effect of high-speed rectifier excitation systems on generator stability limits," *IEEE Trans. Power Apparatus and Systems*, vol. PAS-87, pp. 190–201, January 1968.

[7] F. R. Schleif, H. D. Hunkins, G. E. Martin, and E. E. Hattan, "Excitation control to improve power line stability," *IEEE Trans. Power Apparatus and Systems*, vol. PAS-87, pp. 1426–1434, June 1968.

[8] C. Concordia, "Steady-state stability of synchronous machines as affected by voltage regulator characteristics," *AIEE Trans.*, vol. 63, pp. 215–220, May 1944.

[9] W. G. Heffron and R. A. Phillips, "Effect of modern amplidyne voltage regulators on underexcited operation of large turbine generators," *AIEE Trans.* (*Power Apparatus and Systems*), vol. 71, pp. 692–697, August 1952.

[10] H. Chestnut and R. W. Mayer, *Servomechanisms and Regulating System Design*, vol. 1. New York: Wiley, 1951.

Discussion

G. Manchur, K. McClymont, and **W. Watson** (The Hydro Electric Power Commission of Ontario, Toronto, Canada): The authors are to be congratulated for their paper which, with careful study, can give an excellent insight into transfer function requirements for static excitation systems.

Manuscript received February 16, 1968.

In our experience, the technique of developing linear transfer functions for a synchronous machine connected through reactance to an infinite bus, and the study of these functions to determine the requirements of the excitation system, should be the first step in synthesizing excitation system transfer functions. This technique was used in developing the speed stabilizing signal described in [6] using block diagrams similar to those described by the authors.

We question whether the gain K_E should be limited to less than $T_{do}'/2T_E$ as suggested by the authors. It appears that this recommendation is based on achieving a heavily damped system having a damping factor of 0.707 for the two time constant system considered. In practice, we think the amortisseur windings should also be considered in establishing limits of gain for the controlled-rectifier excitation system.

The effect of amortisseurs is to reduce the permissible value of voltage regulator gain to maintain the same relative stability of the closed-loop system. For example for the system shown in Fig. 17, in order to maintain a phase margin of 65°, which the relationship $K_E = T_{do}'/2T_E$ would give neglecting amortisseurs, the gain must be reduced to approximately $K_E = 0.5(T_{do}'/2T_E)$ when including amortisseur effect. However, if the design criterion for the control loop were chosen to require 30° phase margin, a commonly accepted value, instead of 65°, the permissible value of voltage regulator gain, including amortisseur effects would be $K_E = 4(T_{do}'/2T_E)$, an increase in gain of eight times. We have used gains in the range $T_{do}'/2T_E$ to $4(T_{do}'/2T_E)$ on a number of hydro machines and the higher values have not created any difficulties.

Equation (17) relates the torque component ΔT_{sig} to generator speed variations. It shows that the requirement for phase lead in the function $G(S)$ will be minimized if the excitation system gain K_E is high. For example, with K_E equal to 200, K_6 equal to 0.5, and T_{do}', 5 seconds, the time constant T_{do}'/K_6K_E is only 0.05 second and phase lead would not be required on the usual hydro generator. This is in accordance with our experience since phase lead of the stabilizing signal has not been required on any of our hydro generators. On the other hand for thermal units, closely coupled to the system, K_6 will be significantly smaller (about 0.2) so that phase lead networks will usually be required for optimum system performance. The use of high gain and a small time constant T_E where possible would appear to be superior to the use of phase lead, since the optimum amount of phase lead required varies with system conditions and as noted by the authors, it is essential that the excitation system perform satisfactorily over a wide range of system conditions.

The block diagrams in the paper demonstrate why all successful stabilizing signals developed to date use a speed or frequency deviation signal or derive a speed signal from generator power variations.

In our work we have used block diagrams based on the same small perturbation analysis presented by the authors but having a different form as shown in Fig. 18. This diagram shows the relationship for balanced conditions of all the currents and voltages which can be measured on a synchronous machine. By including saturation and amortisseur windings, this model has provided a basis for comparing calculated transfer functions with those measured in field tests using frequency response techniques.

Manuscript received February 9, 1968.

M. H. Kent (Southern California Edison Company, Los Angeles, Calif. 90053): In their introduction, the authors state that "... more emphasis and reliance is being placed on controls to provide the required compensating effects with which to offset the reductions in stability margins inherent from these trends in equipment designs." It should be pointed out that the concurrent improvements in calculating methods would be tending to emphasize response of control equipment regardless of the trends in equipment designs, and that a blind reliance on assumed margins built into equipment should never substitute for a more complete understanding of the problem. This paper clearly and significantly contributes to the needed understanding of system damping in the area of small perturbation stability.

In striving to derive a universal stabilizing signal, the analysis should include a large perturbation mode of stability since saturation effects and limits on regulator voltage, field voltage, and stabilizing signals can produce surprising results.

In view of the important fundamental concept involved in the application of supplemental excitation control as given by the authors, it may be helpful to examine this concept a little more qualitatively. The response of a generator and its excitation system in a manner such that it contributes negative damping can be illustrated as in Fig. 19(a). This figure shows the vector diagram of a generator that has experienced a decrease in electrical power output, P_e. The generator rotor position δ will change directly in time-phase with P_e, and the generator terminal voltage V_t will increase. Due to the rotor position change there will result a velocity, ω (the derivative of δ with respect to time), accompanied by an induced field voltage, $E_{fd\ \mathrm{induced}}$, and an induced field current will lag this voltage by something less than 90 degrees. Accelerating power P_a of the machine will be directly proportional to the second derivative of δ with respect to time, as given by the basic machine equations in the paper. Thus, the time-phase relationships of the major machine parameters are established, and the effect of automatic excitation control will now be discussed.

The positive change in terminal voltage V_t will result in a negative error signal V_{error} being applied to the automatic voltage regulator. This error signal will act through the regulator and exciter to produce a change in field voltage to buck the increased terminal voltage. The change in field current I_{fdv} due to the change in field voltage will lag by nearly 90 degrees because of the large main field time constant. The two field current vectors, I_{fdv} and $I_{fd\ \mathrm{induced}}$, can be added using superposition to obtain the net change in field current. The net field current vector can now be resolved into components along the rotor-position δ and rotor-speed ω axes. Since torque is directly proportional to the net field current, it can be seen that there is a torque component (the synchronizing component) in phase with rotor position, but that the component reflected onto the axis of rotor speed (the damping component) is negative.

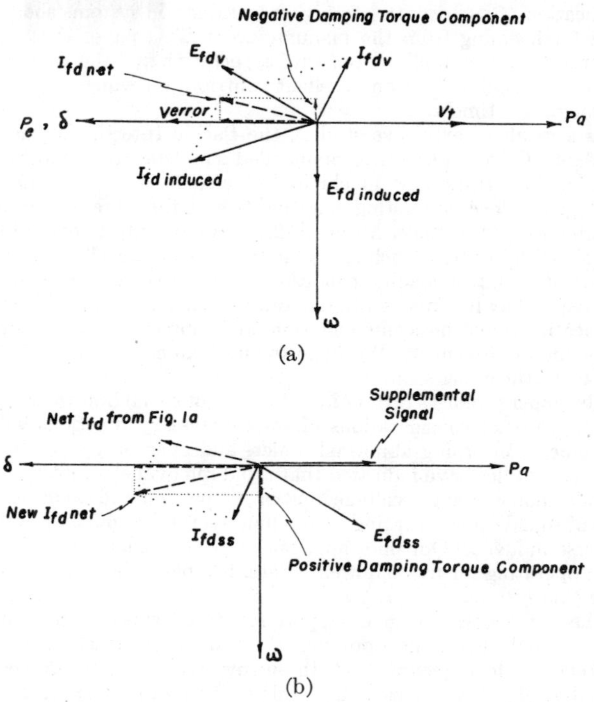

Fig. 19. (a) Negative damping. (b) Positive damping using supplemental control.

The effect of a selectively chosen supplemental regulator control signal is shown in Fig. 19(b). In this figure, the result of adding a supplemental signal is superimposed on Fig. 19(a). For this illustration, the supplemental signal is chosen so that it is in time-phase with accelerating power, and since it acts through the voltage regulator and its associated time delay, a change in field voltage will follow which in turn will produce a change in field current, I_{fdss}. Adding the net field current vector obtained in Fig. 19(a) to that resulting from the supplemental signal produces a new net field current vector. Resolving this vector into components along the rotor-position and rotor-speed axes, shows that the damping torque component has become positive. Thus, supplementary regulator controls can eliminate negative damping emanating from the excitation system and introduce positive damping.

As noted in the paper, the time-phase relationships shown in Fig. 19 are determined by the frequency of oscillation and the time constants involved, and the supplemental signal itself could originate from speed or frequency or from accelerating power. Because different systems oscillate at different frequencies and have different time constants, it would ordinarily not be desirable to have the supplemental signal in exact time-phase with accelerating power, but the illustration shows the principle. It can be seen in Fig. 19(b) that an improperly chosen supplemental signal can result in significant reduction in synchronizing torque, which would not be desirable, while maximizing damping torque.

The authors show that under many system circumstances, excitation systems can be a major source of negative damping on system oscillations. Since the circumstances required for this to occur are not especially unique, perhaps consideration should be given to including stabilizing signals as an integral part of all conventional and high-speed excitation systems.

Manuscript received February 13, 1968.

F. R. Schleif and **W. B. Gish**, (U. S. Bureau of Reclamation, Denver, Col.): This is a most welcome exploration and documentation of the principles pertinent to the practical application of supplementary excitation control. Preliminary recommendations concerning application of supplementary control to rotating exciters were proposed on the basis of explorations described in [7] with the remark that application of such control to higher speed static systems should be soon forthcoming from the manufacturers. This paper contributes generously to the fulfillment of that expectation and the authors are to be congratulated on an excellent contribution which is not only thorough but timely.

As a result of extensive studies, the Pacific Intertie Engineering Guidance Committee has recommended supplementary control not only on important new machines but also to a large number of strategically located existing machines having the more conventional rotating exciter systems. Although these systems must contend with additional limitations such as nonlinearity, lower speed of response, and relationship to loading conditions, some of the guides offered in this paper may be extended to the rotating exciter systems. However, applications of such recommendations are complex and there is much room for improvement. We hope further extensions into this field will be forthcoming soon.

The paper promises further investigation of signal limiting requirements. Further considerations of these necessary functions will be welcome. Concerning abnormal under- and overvoltage conditions, our own studies have indicated that rapid removal of supplementary signals upon crossing a voltage limit of ± 5 percent and rapid restoration of signal upon returning within limits serve this purpose without adverse influence. Our pilot application at Glen Canyon, which has been operating for five months to date, has performed satisfactorily with this system.

The paper does not give support for its conclusion which states "The signal should not produce a steady-state offset of voltage reference with frequency ...." In our own study of the problem as stated in [7], it is not clear that this conclusion is warranted in all cases. It has been proposed [11] that a substantial benefit can be accomplished by reducing voltage as a function of frequency under disaster conditions. The frequency supplementary signal would automatically accomplish this type of action where desirable, such as near load centers. On the other hand, generating stations serving long transmission lines may not benefit by this action. The point of this remark is that reset in the supplementary signal need not be mandatory. It may be advantageous to make it optionally available.

References

[11] K. L. Hicks, "Disaster control coordination for large interconnected systems," *IEEE Spectrum*, vol. 4, pp. 52–55, November 1967.

F. P. deMello and **C. Concordia**: We wish to thank all of the discussers for their comments, which add considerably to the value of this paper.

In regard to the various points raised by Mr. Manchur, Mr. McClymont, and Mr. Watson, we would comment as follows.

The optimum value of the gain K_E is a matter of judgment and the value of $T_{do}'/2T_E$ suggested in the paper was considered a good compromise considering its effect on performance under small and large perturbations. It is quite true that from the point of view of stability under the linear small perturbation mode, a considerably larger value of K_E is perfectly satisfactory. Also, it is true that with this larger value of K_E, the effective response of flux to voltage reference has less phase lag and therefore the lead requirements for the stabilizing signal are correspondingly less. However, the higher K_E, the more likely is the mode of operation where the exciter hits limits and, under these conditions, the effective phase lag of flux to signal can be drastically different and considerably greater than that under the linear mode of operation. Use of describing function methods of analysis of the mode of operation where the exciter is hitting limits will show that in such a mode, the stabilizing signal will need considerably greater phase lead to produce damping.

Manuscript received March 15, 1968.

An effective means of preserving the beneficial damping characteristics of the stabilizing signal under large disturbance conditions, such as occur following fault conditions, as well as under the small perturbation mode, is to have moderate values of K_E which do not cause the exciter to continually hit limits under typical large disturbance conditions, thereby avoiding the drastic change in phase angle characteristics under these conditions.

We do appreciate the need for more complete machine representations including effects of amortisseurs, saturation, etc. As a matter of fact, the concepts and recommendations developed in the paper have been checked out on very complete digital representations of machines in multimachine situations. For purposes of clarity, however, it was felt that the basic effects and concepts should be established with the simplest possible 2-axis representation, excluding amortisseur effects.

We appreciate Mr. Schleif and Mr. Gish's comments on their studies for the Pacific Intertie system and agree that much is yet to be learned in terms of application requirements considering the nonlinear aspects of the problem. Our own investigations of large disturbance conditions have confirmed the general recommendations contained in the paper, particularly as regards effects of maintaining moderate regulator transient gains as explained above.

In regard to the value of a permanent offset of voltage with respect to speed, although this may be desirable from the point of view of balancing generation with load, it should be noted that the gains of voltage offset per unit of frequency deviation are in the order of 20 times what could be tolerated on the basis of a constant volts per cycle criterion. As long as the stabilizing signal effect is limited to a safe excursion in terminal voltage, we agree that there may be cases where a permanent offset voltage with frequency may be desirable.

It should be noted that this offset would be more effective at the load utilization points than at the generator terminals, since reduction of generator voltage would also cause a reduction of transmission voltage and intertie capacities.

Mr. Schleif and Mr. Gish suggest removing the stabilizing signal when the voltage deviation exceeds a certain limit, rather than limiting the signal itself. We do not think this is a good idea, and indeed there may well be cases where removing the signal might drive the voltage even farther away from normal. On the other hand, when the signal itself is simply limited, it is at least always contributing toward damping regardless of the magnitude of the voltage deviation.

We appreciate Mr. Kent's endorsement of the need for calculating methods for a more complete understanding of the problem. His qualitative description of the phenomena of damping and synchronizing powers as affected by voltage regulation is an interesting addition to this subject. However, we must point out that it neglects the interaction of the two signals, i.e., the voltage error vector in Fig. 19(a) should be modified by voltage regulator action.

BLOCK DIAGRAMS AND TORQUE-ANGLE LOOP ANALYSIS OF SYNCHRONOUS MACHINES

I. Nagy
Research Institute for Automation, Hungarian Academy of Sciences
Budapest, Hungary

Abstract—Block diagrams of synchronous machines supplying an infinite bus or lumped parameter load are developed for large and small-signals. This representation includes the effects of saliency, the amortisseurs, variable speed operation, positive and negative sequence components of the infinite bus voltage and an approximate treatment of the saturation. The paper illustrates the application of the block diagram technique. First, the electrical torques proportional to the load angle and their contribution to the overall transfer function of the torque-angle loop are investigated by the elementary theory of servomechanisms. Second, transfer functions are derived for excitation system analysis. Finally self-excitation of the machine connected to a capacitive load is described.

INTRODUCTION

The block diagram of synchronous machines can be a useful and comprehensive aid to conventional calculations in dynamic analysis; e.g. in the dynamic investigation of excitation control or for setting up computer programs. By giving quantitive and causal relations among principal variables, it supports the mathematical calculations and offers an insight into the mode of action of the machine. The block diagram promotes unified treatment of various dynamic problems. By applying virtually the same computer program once set up a number of different problems can be solved.

The first part of the paper shows the block diagrams of a machine connected to a large system through external impedance or to a lumped parameter load. In setting up the block diagrams, the object was a more general approach to the problem than previously applied[1,2,3,4]. Block diagrams were developed for large and small perturbations, and they included the effects of saliency, amortisseurs, and variable speed. These results apply, with some restrictions, also to infinite bus voltage having negative sequence components. An approximate treatment of saturation is mentioned as well.

After intermediate steps, the paper arrives at a small signal block diagram (Fig. 3) which can be particularly helpful for application of the elementary theory of feedback control systems. Its use offers certain rewards with its simplicity in this field as well. In this block diagram, the transfer functions are in a form conventional in control systems and in a relatively simple relationship to each other while their time constants and gains can be calculated from well-known machine characteristics. Consequently, setting up an analogue computer program can be more favourable on the basis of block diagrams of Figs. 1 or 3 rather than on the basis of fundamental equations listed in Appendix II.

The second part involves three applications of the block diagrams. The first example treats the so-called torque-angle loop established by the electrical torque developed by the variation of the load angle. The second example discusses the transfer function e/e_f of small signal representing the relationship between the field voltage and the terminal voltage for a generator connected to an infinite bus and its approximations. The third example deals with the transfer function e/e_f and self-excitation of a generator connected to a balanced capacitive load.

Complex vectors with superposed bar "-" and without suffix "s" are Park vectors expressed in the d-q rotor coordinate system. For instance, the definition for the phase voltage vector $\overline{E}$

$$\overline{E} = \frac{2}{3}(E_a + \overline{a}E_b + \overline{a}^2E_c)e^{-j\theta} = \overline{E}_s e^{-j\theta} \tag{1}$$

where E_a, E_b and E_c are the instantaneous values of phase voltages in the a, b, and c stator phases, respectively, $\overline{a} = \exp j2\pi/3$ and $\overline{E}_s$ is Park vector expressed in the stator coordinate system[5]. Park vector $\overline{E}$ may be transformed to the stator coordinate system by multiplying by $\exp j\theta$. When the phase voltages have positive and negative sequence components of rated frequency only and the rotor turns at synchronous speed, $\overline{E}_{1s} = E_1 \cdot \exp j(t + \beta_1)$ and $\overline{E}_{2s} = E_2 \cdot \exp j(\beta_2 - t)$, respectively. Amplitudes E_1 and E_2 are constant. Park vector $\overline{E}_{1s}$ is at rest relative to the rotor, while $\overline{E}_{2s}$ turns in the opposite direction at twice synchronous speed.

BLOCK DIAGRAMS

The fundamental equations expressing the voltage and torque relations of a generator supplying an infinite bus through external impedance may be found in Appendix II. The block diagram built up from the fundamental equations for large signals is shown by solid lines in Fig. 1. Here the variables are time functions and s = d/dt.

The block diagram deviates from the fundamental equations in two respects: In (A4) and (A5) the components $s\delta\psi_d$ and $s\delta\psi_q$ proportional to speed variations were neglected. In general, $\overline{E}_N$ contains both positive and negative sequence components. In this case E_{Nd} and E_{Nq} can be obtained as follows: $E_{Nd} = E_{N1}\sin\delta + E_{N2}\cos(2t + \gamma_2)$ and $E_{Nq} = E_{N1}\cos\delta - E_{N2}\sin(2t + \gamma_2)$, where $\gamma_2 = \vartheta - \beta_2$. However, in Fig. 1, $\overline{E}_{N2} = 0$ was assumed.

The superposition principle may be applied i.e. the influence of E_{N1} and E_{N2} could be examined separately with the following restrictions: The mechanical transient phenomenon should be approximately neglected (δ = const.) and the components of signals E_d, E_q and I_d, I_q produced one by one by the input signals E_{N1}, E_{N2} and E_f are first to be summed up algebraically and vector $E\angle\varphi_E$, and $I\angle\varphi_I$ may be calculated only then.

Having determined the d-q components, the Park vector in rotor and stator coordinate system for rotor turning at synchronous speed is e.g. $\overline{E} = E_d + jE_q$ and $\overline{E}_s = \overline{E}\exp j(t + \vartheta)$. Instantaneous values of phase quantities in a certain winding can then be obtained by projecting Park vector $\overline{E}_s$ on the axis of corresponding phase winding: $E_a(t) = \mathrm{Re}(\overline{E}_s)$, $E_b(t) = 0.5[-\mathrm{Re}(\overline{E}_s) + \sqrt{3}\mathrm{Im}(\overline{E}_s)]$ and $E_c(t) = -0.5[\mathrm{Re}(\overline{E}_s) + \sqrt{3}\mathrm{Im}(\overline{E}_s)]$.

On the basis of the block diagram, an analog computer program may be set up. In programming it is preferable to take advantage of the fact that denominators of G(s), F(s) and $X_d(s)$ are identical.

By omitting from the block diagram the blocks serving for calculation of $E\angle\varphi_E$ and $I\angle\varphi_I$, and taking the load angle δ as approximately constant, ensuring thus linear relationship among the signals, the variables may be regarded also as Laplace transformed quantities. In such cases, "s" is the independent variable and, in place of the time functions of the negative sequence components, the

Paper 70 TP 801-PWR, recommended and approved by the Power Generation Committee of the IEEE Power Group for presentation at the 1970 Joint ASME-IEEE Power Generation Conference, Pittsburgh, Pa., September 27-30, 1970. Manuscript submitted April 24, 1970; made available for printing June 29, 1970.

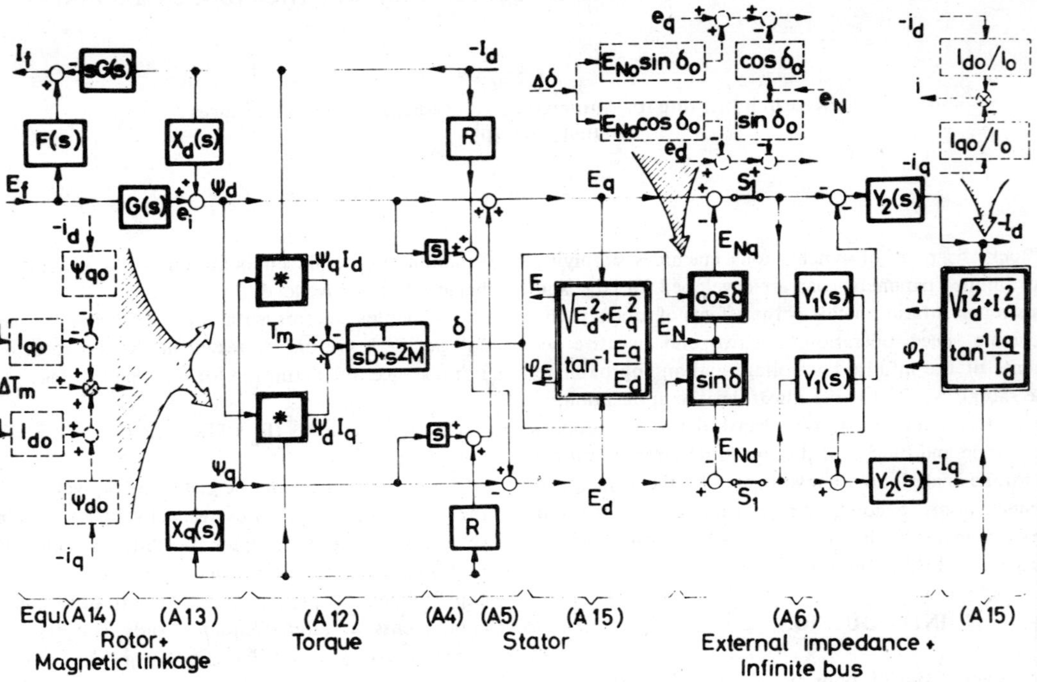

Fig. 1. Block diagram for large signals (solid lines) and for small signals (replace the double enframed blocks by the corresponding sets of dotted line) of synchronous machine supplying an infinite bus through external impedance.

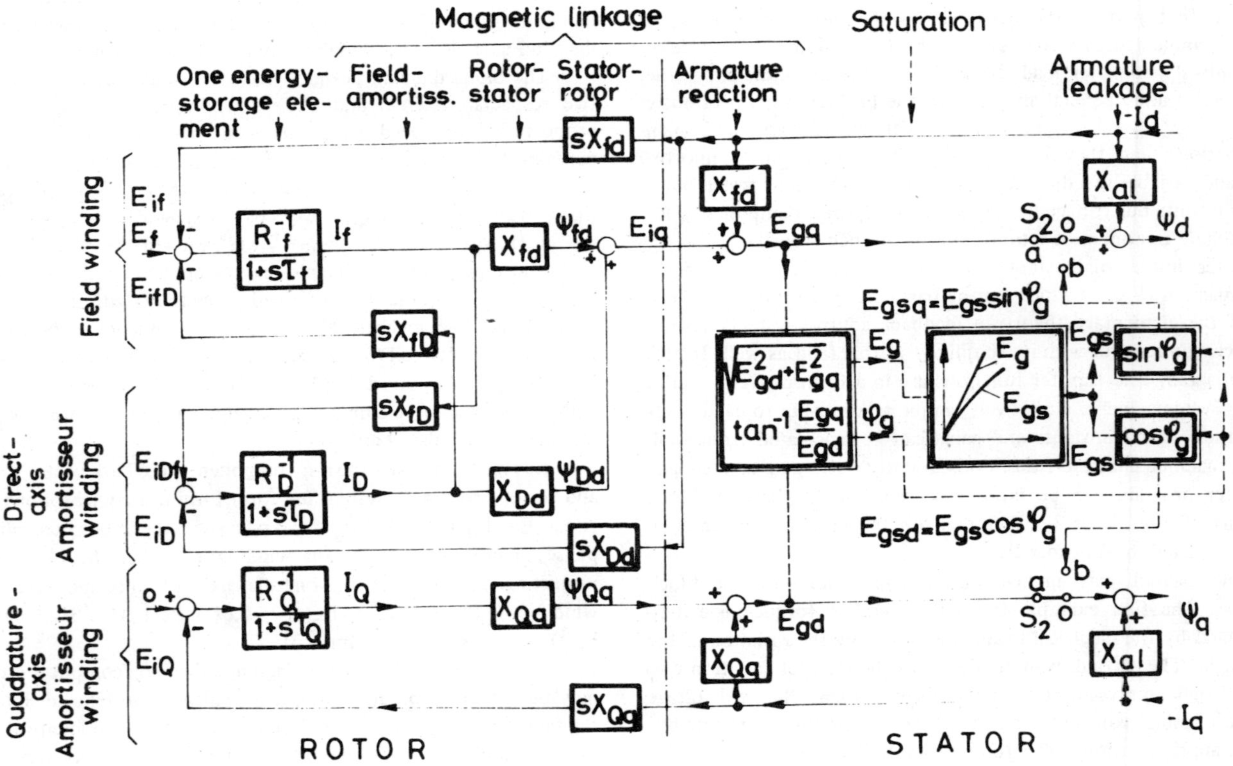

Fig. 2. Detailed sketch of a block diagram part of Fig. 1 built up on basis of (A13) and (A14) by taking into account approximately the saturation effect as well.

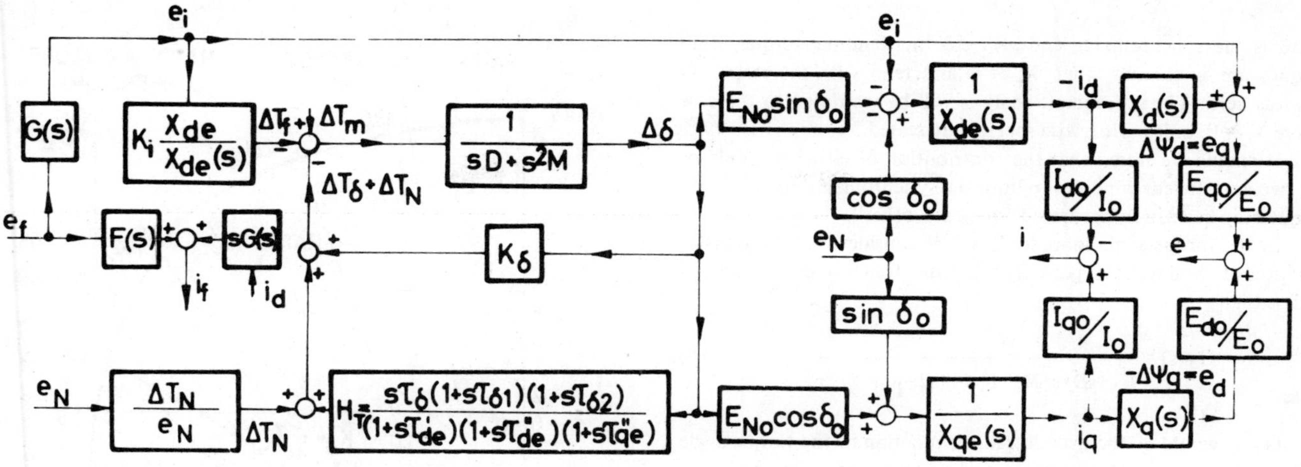

Fig. 3. Rearrangement of block diagram of Fig. 1 for small signals.

Laplace transformed values should be introduced:

$$L\left[\bar{E}_{N2}\right] = L\left[\bar{E}_{N2m}\right] e^{-j2t} = \frac{\bar{E}_{N2m}}{s+2j};$$

$$L\left[\bar{E}_{N2d}\right] = \mathrm{Re}\left[\bar{E}_{N2m}\right]\frac{s}{s^2+4};\ L\left[\bar{E}_{N2q}\right] = \mathrm{Im}\left[\bar{E}_{N2m}\right]\frac{-2}{s^2+4}$$

The transfer matrix [W(s)] providing the relationship among output and input signals is

$$\begin{bmatrix} I_f \\ E_i \\ \Psi d,q \\ E_{d,q} \\ I_{d,q} \end{bmatrix} = \begin{bmatrix} W_{11}(s) & W_{12}(s) \\ & \\ & \\ W_{81}(s) & W_{82}(s) \end{bmatrix} \begin{bmatrix} E_f \\ E_N \end{bmatrix} = [W(s)] \begin{bmatrix} E_f \\ E_N \end{bmatrix}$$

The transfer functions $W_{11}(s) \ldots W_{82}(s)$ may be derived from the block diagram.

At no-load, the switch S_1 should be opened and the block diagram takes a simple form. Substituting $E_N = 0$ into Fig. 1 the block diagram of a synchronous generator connected to a lumped R-L or R-C load is obtained. Some modifications are to be carried out here. Introducing ω in place of δ, calculating with the intact terms $\omega\psi_d$ and $\omega\psi_q$, especially in the case of capacitive load and taking into consideration the transfer function $P(s) = T_m/\omega$, the result may be a fairly good representation of the generator. P(s) depends on the prime mover and its speed regulator.

By replacing the part of the block diagram of Fig. 1 built up to represent (A13) and (A14) by the one shown in Fig. 2 which was drawn directly on the basis of the operating principle of the machine, a deeper insight into the dynamic processes of the machine may be gained. It is also rewarding from instructional aspects. The two corresponding block diagram parts seen from outside are completely equivalent apart from the dotted paths accounting for saturation effect. The dotted paths serve to include approximately the reducing effect of saturation on stator flux linkages and induced voltages. When taking saturation into consideration, switch S_2 is in position "o-b".

Block diagrams for small signals

By being restricted to small perturbations around some initial condition, the block diagram of the solid line shown in Fig. 1 may be linearized. After linearization, the frequency method may be applied to the dynamic analysis of the system including the mechanical transient phenomena. Linearization should be carried out on the nonlinear operational blocks of Fig. 1 drawn by double lines. The rest of the block diagram remains unchanged. The linearized version of the double enframed nonlinear blocks are shown in Fig. 1 as well and are drawn by dotted lines. For simplicity, the linearized version of the block determining the absolute value of voltage E is omitted. Replacing the nonlinear blocks by inserting the linearized parts into their places, a block diagram for small signals is obtained which can also be built up on the basis of relations derived in Appendix II. Variables may be regarded either as functions of time or as Laplace transformed quantities here.

By rearranging Fig. 1 to the form of Fig. 3 (Appendix III), the resulting block diagram and its transfer functions are well suited for studying the transient phenomenon with the technique of elementary servomechanism theory. The time constants and gains are related only to machine characteristics usually known. In Fig. 3 R = $R_e = 0$ and $s\Delta\psi_d = s\Delta\psi_q = 0$.

Electrical torque deviation ΔT_e is due to a change in quantities δ, E_N and E_f or E_i. Transfer functions between input signals $\Delta\delta$, e_N and e_f and respective electrical torque components ΔT_δ, ΔT_N and ΔT_f as output signals are given by (A20), (A23) and (A26). The relations described here between small changes may also be read in steady-state from the vector diagram of the synchronous machine.

TORQUE-ANGLE LOOP ANALYSIS

The condition of the torque-angle loop illustrated in Fig. 3 affects the stability of the synchronous machine to a high degree. Knowledge of the overall frequency response of the torque-angle loop $G_o(j\nu) = \Delta\delta/\Delta T_f$ may be necessary for stability analysis. A time-saving, approximative method will be shown for estimating the frequency response $|G_o|$. By using the asymptotic amplitude Bode diagrams, approximative plots of $|G_o(\nu)|$ can be obtained for various operating conditions of the machine in a short time and what is more important, the influence of single time constants and their variation on the plots of $|G_o|$ can be quickly learned.

According to Fig. 3 $\Delta T_\delta = [K_\delta + H_T(s)]\Delta\delta$, where $K_\delta\Delta\delta$ is the change in torque in steady-state caused by the field current and saliency, while $H_T(s)\Delta\delta$ is the change in torque created by the transient field current and amortisseur winding currents. ΔT_δ may be written

$$\Delta T_\delta(j\nu) = [K_s(\nu) + j\nu K_d(\nu)]\Delta\delta \tag{3}$$

$K_s\Delta\delta$ is the synchronizing and $\nu K_d\ \Delta\delta$ the damping torque. Both torques are maximum when $X_e = 0$ and tend with increasing X_e towards zero. The subsequent analysis will be restricted to the case where $X_e = 0$, unless otherwise specifically stated.

It should be noted that the contribution of either the field or the two amortisseur windings to both the synchronizing and damping torque may be entirely separated from each other[6].

In the torque-angle loop, let $H_T(s)$ be considered as the transfer function of the feedback path. The transfer function of the forward path is then

$$G_T(s) = \frac{1}{K_\delta+sD+s^2M} = \frac{1}{K_\delta(1+s2\xi_T\tau_T + s^2\tau_T^2)} \qquad (4)$$

$G_T(s)$ would be the overall transfer function of the torque-angle loop if there were no amortisseurs and no transient currents flowing through the field winding. The overall transfer function is

$$G_o(s) = \frac{G_T}{1 + H_TG_T} \cong \begin{cases} G_T & \text{if } |H_T^{-1}| >> |G_T| \\ H_T^{-1} & \text{if } |H_T^{-1}| << |G_T| \end{cases} \qquad (5)$$

G_o is approximately equal to either $|G_T|$ or $1/|H_T|$. The approximation is poor only when and where $|G_T| \cong 1/|H_T|$. Some general comments on plotting the asymptotic Bode diagrams of $|G_T|$ and $1/|H_T|$ given below in connection with an example will assist in drawing them simply and quickly.

The exact frequency response G_o can be determined from G_T and H_T, e.g. by the Nichols' diagram used in the analysis of servomechanisms or by calculation.

Amplitude approximations of G_T and H_T^{-1}, along with the resonance peaks of G_T are shown in Fig. 4 for different operating conditions of a nonsalient pole machine in the range $0° < \delta < 90°$. The phase angle φ_s increases here from 0° to 180°. A few points on the exact curve of $|G_o|$ determined by calculation are also marked. In Table I the numerical values of the parameters necessary for plotting the Bode diagrams are listed. Based on the transfer functions G_T, H_T and on the example as well, the following conclusions may be drawn:

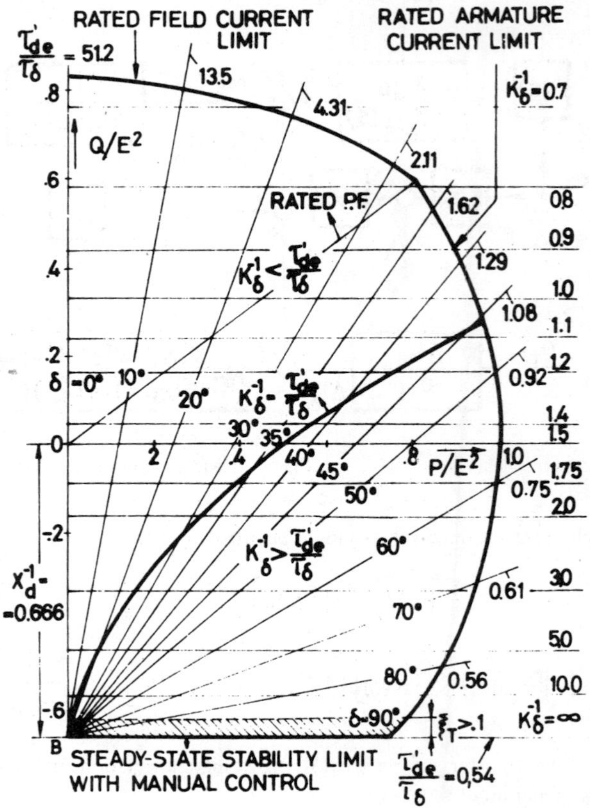

Fig. 5. Straight lines K_δ^{-1} = const. and τ'_{de}/τ_δ = const. as well as limit curve $K_\delta^{-1} = \tau'_{de}/\tau_\delta$ in working domain of generator in power chart.

a. The asymptote of $|G_T|$ at high frequencies is the same in all operating conditions and has a slope of - 40 db/decade. The asymptote at low frequencies has a slope of 0 db/decade and its magnitude is $|G_T(0)| = K_\delta^{-1}$.

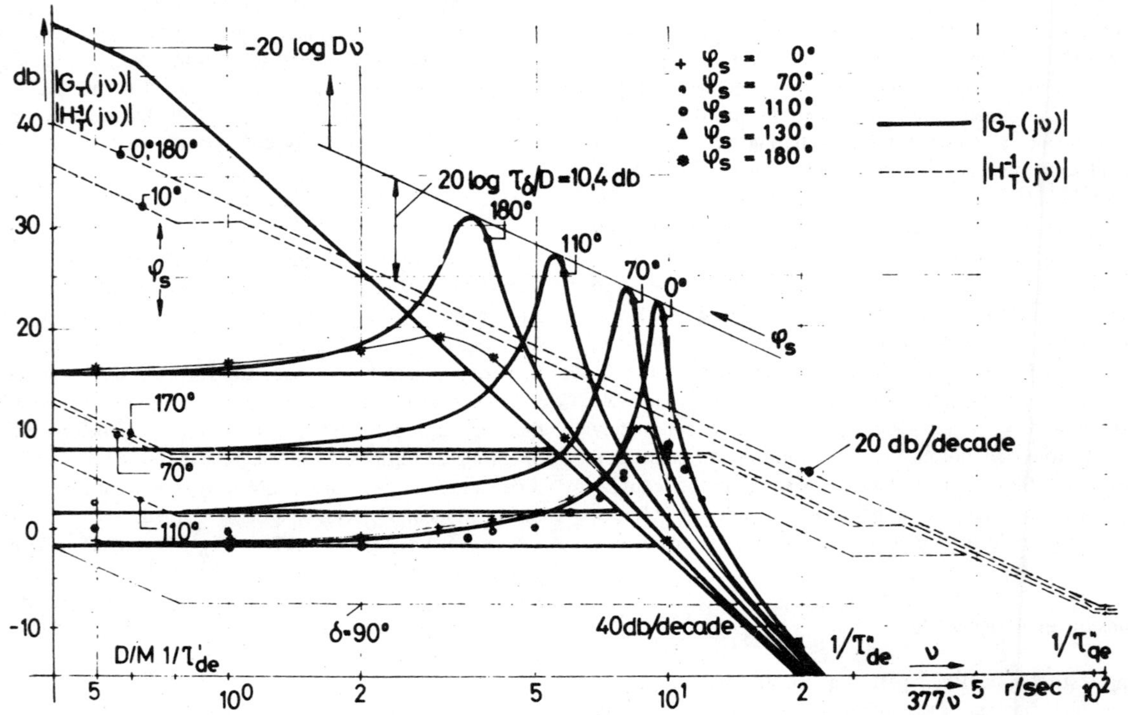

Fig. 4. Amplitude Bode diagrams of torque-angle loop for different φ_s angles. Data: $I_o = .5$ $E_o = 1$, $X_d = X_q = 1.5$, $X'_d = .4$; $X_e = 0$; $X''_d = .3$; $X''_q = .4$; $H = 2.5$; $D = 3$; $\tau'_{do} = 1885$; $\tau''_{do} = \tau_Q = 18.85$

TABLE I

Time in seconds. $I_o = .5$

φ_s^o	δ^o	ξ_T	τ_T	τ_δ	$\tau_{\delta 1}$	$\tau_{\delta 2}$	K_δ^{-1}
0	0	.032	.107	.026	1.33	.0375	.87
10	4.5	.032	.107	.041	.857	.037	.86
70	29.5	.038	.126	.061	.08	.027	1.2
110	43.5	.055	.18	1.18	.062	.019	2.52
130	47.5	.058	.19	1.35	.059	.018	2.85
170	28.5	.084	.279	.58	.083	.028	5.92
180	0	.085	.282	.026	1.33	.038	6.0

$$K_\delta = \frac{EE_i}{X_d}\cos\delta = \frac{E^2}{X_d} + EI\cos\varphi_s = \frac{E^2}{X_d} + Q \tag{6}$$

and assuming E = const. and I = const. K_δ^{-1} is minimum at $\varphi = 0$ and increasing monotonically with φ_s. $|G_T(0)|$ reaches its maximum value at $\varphi_s = 180°$, when δ remains smaller than 90°. At larger current I, load angle δ becomes 90° before φ_s had reached 180°. In the case of $\delta = 90°$ the transfer function is

$$G_T(s) = \frac{1}{sD}\frac{1}{1 + sM/D};\ K_\delta = 0 \tag{7}$$

b. Reducing current I, $G_T(0)$ varies less and less with φ_s, its lowest value being larger and its highest value smaller.

c. The resonance peak of $|G_T|$ is

$$Q_p = \frac{1}{2\xi_T}\frac{1}{\sqrt{1-\xi_T^2}} \cong \frac{1}{2\xi_T} \tag{8}$$

The approximation is good since $\xi_T < 0.1$ in the predominant range of δ (Table I and Fig. 5.).

d. The values τ_δ, $\tau_{\delta 1}$, $\tau_{\delta 2}$ depend on the operating condition of the machine, but for a given machine they are solely determined by angle δ when E_N = const. |equ. (A21a)|.

e. As long as (8) is a good approximation, the points of resonance peaks Q_p lie on a straight line $Q_p(\nu) = -20 \log D\nu$. With decreasing K_δ the resonance peak Q_p is shifting left and shrinking.

f. In the frequency range interesting from the viewpoint of G_o, the torque proportional to $\Delta\delta$ is, for small δ;

$$H_T(s)\Delta\delta \cong \frac{s\tau_\delta(1 + s\tau_{\delta 1})}{1 + \tau'_{de}s}\Delta\delta \tag{9}$$

This torque is produced together by the d,q amortisseurs and the field winding. The smaller δ is, the greater is the part of the torque produced by the q amortisseur. In case of $\delta = 0$ the whole torque is created by the q amortisseur $[H_T = s\tau_\delta/(1 + s\tau''_{qe})]$. On the other hand, with increasing δ, an increasing part of the torque $H_T\Delta\delta$ still influencing the overall frequency response originates from the field winding.

g. For larger angles δ(over 30°∿40°), the values of $\tau_{\delta 1}$ and $\tau_{\delta 2}$ are small (Table I). Now the torque $H_T\Delta\delta$ in the interesting frequency range is approximately

$$H_T(s)\Delta\delta \cong \frac{\tau_\delta s}{1 + \tau'_{de}s}\Delta\delta \tag{10}$$

In producing this torque, apart from the field winding, the two amortisseurs also take part through τ_δ. In our example $\tau_\delta = 2.45\sin^2\delta_o + .026\cos^2\delta_o$ in secs. and with no amortisseurs $\tau_\delta = 2.44\sin^2\delta_o$ in secs. At $\delta = 20°$, $\tau_\delta = .309$ sec. and with no amortisseurs $\tau_\delta = .285$ sec. When, however, $\delta_o = 40°$ then $\tau_\delta = 1.03$ sec, while with no amortisseurs $\tau_\delta = 1.01$ sec. Apparently, neglecting the amortisseurs in (10) is a good assumption in the range $\delta_o > 30° \sim 40°$. (The quality of approximation deteriorates towards higher frequencies even for large angle δ.)

h. Change in operating conditions of the machine modifies the relative position of $|G_T|$ and $|H_T^{-1}|$. The amplitude $|H_T^{-1}| = \tau'_{de}/\tau_\delta$ calculated from the asymptotic approximation at frequency $\nu = 1/\tau'_{de}$ is being reduced by increasing δ, since now τ_δ is becoming greater, as well [equ. (A21a)]. On the other hand, $|G_T(0)|$, determined by the reactive power Q, increases with decreasing Q [equ. (6)].

Fig. 5 shows the straight lines τ'_{de}/τ_δ = const. and K_δ^{-1} = const. in the operating range of a nonsalient pole synchronous generator on the power chart. The operating range is bounded by the rated field current limit, the rated stator current limit and by the static stability limit. Curve $K_\delta^{-1} = \tau'_{de}/\tau_\delta$ divides the operating range into two parts. The farther the operating point of the machine is from this curve, the more accurate is the approximation (5) in the corresponding frequency ranges; in the example, in most part of the underexcited range $K_\delta^{-1} > \tau'_{de}/\tau_\delta$. This means that with increasing δ the frequency range where $|G_T| >> |H_T^{-1}|$, i.e. where $|G_O| \cong |H_T^{-1}|$, widens.

TRANSFER FUNCTION e/e_f

In excitation control of synchronous machines it is useful to know the relation between small changes of field voltage e_f and the absolute value of the terminal voltage Park vector e. As it was previously shown, in a significant part of the operating range and especially in the underexcited range, the presence of amortisseurs has only little effect on the overall frequency response G_o and they can be neglected, especially at lower frequencies. Transfer function e/e_f then takes a relatively simple form, as from Fig. 1.

$$\frac{e}{e_f} = \frac{(X_{df}/R_f)K_e(1 + s2\xi_e\tau_e + s^2\tau_e^2)}{1 + sK_\delta^{-1}(D + K'_\delta\tau'_{de}) + s^2K_\delta^{-1}(D\tau'_{de} + M) + s^3K_\delta^{-1}M\tau'_{de}} \tag{11}$$

where

$$K_e = A_e + \frac{E_{No}^2\sin^2\delta_o}{K_\delta X_{de}}\left[\frac{X_eX_dE_{io}}{X_{de}^2E_o} + \frac{E_{No}\cos\delta_o}{E_o}\left(\frac{X_d^2}{X_{de}^2} - \frac{X_q^2}{X_{qe}^2}\right)\right] \tag{12}$$

$$2\xi_e\tau_e = \frac{A_e}{K_eK_\delta}D;\ \tau_e^2 = \frac{A_e}{K_eK_\delta}M;\ A_e = \frac{X_e}{X_{de}}\ \frac{E_{qo}}{E_o} \tag{13}$$

Using the same numerical data as in Fig. 4 but now $X_e = .2$.Table II shows the values of constants in transfer function (11) for a few initial conditions.

TABLE II

$\tau = K_\delta^{-1}(D + K'_\delta\tau'_{de})$, $b = K_\delta^{-1}(D\tau'_{de} + M)$, $c = K_\delta^{-1}M\tau'_{de}$ . . . Time is measured in sec.

I	φ_s^o	δ^o	K_e	τ_e	ξ_e	τ	b	c
.5	0	0	.117	.115	.035	1.76	.029	.025
.5	30	16	.123	.116	.034	1.9	.03	.026
.5	70	36	.142	.097	.029	2.66	.041	.035
.5	110	49	.174	.08	.024	5.26	.084	.07
.5	150	50	.18	.076	.023	8.11	.13	.11
.5	180	0	.117	.115	.035	1.81	.17	.14
1	0	0	.117	.115	.035	1.77	.023	.02
1	30	25	.13	.107	.033	2.0	.025	.021
1	70	55	.198	.076	.023	3.46	.04	.035
1	110	81	.71	.027	.008	15.8	.17	.15

For synchronous condensers ($\delta = 0°$, $K_\delta' = K_\delta$) equ. (11) is simplified to

$$e/e_f = (X_{df}X_e/R_fX_{de})/(1 + s\tau'_{de}) \quad (14)$$

The denominator of (11) may be approximately expressed as the product of a first degree and a second degree term

$$D(s) = (1 + s\tau)\,[1 + K_\delta^{-1}(M + D\tau'_{de})\frac{s^2}{1 + s\tau} + K_\delta^{-1}M\tau'_{de}\,\frac{s^3}{1 + s\tau}]$$

where

$$\tau = K_\delta^{-1}\,(D + K'_\delta\tau'_{de}) \cong (K'_\delta/K_\delta)\,\tau'_{de} \quad (15)$$

Up to the corner frequency $\nu = 1/\tau$ the second degree term is approximately unity, while above it $(1 + s\tau) \cong s\tau$ and then

$$s^2/(1 + s\tau) \cong s/\tau \text{ and } s^3/(1 + s\tau) \cong s^2/\tau \text{ and}$$

$$D(s) \cong (1 + s\tau)\left\{1 + [D + M/\tau'_{de})/K'_\delta]\,s + (M/K'_\delta)\,s^2\right\} \quad (16)$$

It is a good approximation with the exception of the coefficient of s^2 which is always larger than the exact value. Using another approximation

$$D(s) \cong (1 + s\tau)(1 + sD/K'_\delta + s^2M/K'_\delta) \quad (17)$$

the coefficient of s^2, that is, $\tau D/K_\delta' + M/K_\delta'$ is always smaller or at most equals the exact value, since $K'_\delta \geqslant K_\delta$. Approximation(16) (or 17) gives a value larger (or smaller) than the exact value for damping ratio in the second degree term, or, in an extreme case, the exact value.

Transfer function e/e_f is approximately

$$\frac{e}{e_f} = \frac{X_{df}}{R_f}K_e\,\frac{1 + s(A_eD/K_eK_\delta) + s^2(A_eM/K_eK_\delta)}{(1 + s\tau)[1 + sD/K'_\delta + s^2M/K'_\delta]} \quad (18)$$

or at low frequencies

$$e/e_f = (X_{df}/R_f)K_e/(1 + s\tau) \quad (19)$$

SYNCHRONOUS GENERATOR CONNECTED TO SEPARATE LOAD

The block diagram of a generator running at constant speed and supplying a load of concentrated R-L or R-C parameters could be directly plotted on the basis of Fig. 1 (Fig. 6). In the simplest case, when the amortisseurs are neglected, and $R = 0$; $s\Delta\psi_d = 0$; $s\Delta\psi_q = 0$ as well as $\tau_{eL} = 0; \tau_{eC} = 0$, the transfer function e/e_f from Fig. 6 is:

$$e/e_f = K_e/(1 + s\tau'_{dz}) \quad (20)$$

where

$$K_e = \frac{X_{fd}}{R_f}\,\frac{E_{do}R_eX_q + E_{qo}\,(R_e^2 + X_eX_{qe})}{R_e^2 + X_{de}X_{qe}} \quad (21)$$

and

$$\tau'_{dz} = (R_e^2 + X'_{de}X_{qe})\,/\,(R_e^2 + X_{de}X_{qe}) = N_o/D_o \quad (22)$$

In the case of a capacitive load, self-excitation may arise ($X_e = X_{ec}$). Were the mathematical model of (20) exact, it could be stated that self-excitation arises when $\tau'_{dz} < 0$. This inequality will be satisfied when $N_o < 0$ or $D_o < 0$. $N_o = 0$ and $D_o = 0$ indicate the stable and unstable ranges in plane $R_e - X_e$ (Fig. 7). When $R_e \geqslant \frac{1}{2}(X_q - X'_d)$, no self-excitation develops.

At pure capacitive load and at high frequencies compared to the subtransient time constants τ''_{do} and τ''_{Q}, voltage components $s\Delta\psi_d$ and $s\Delta\psi_q$ can no longer be neglected but, in turn, in such cases

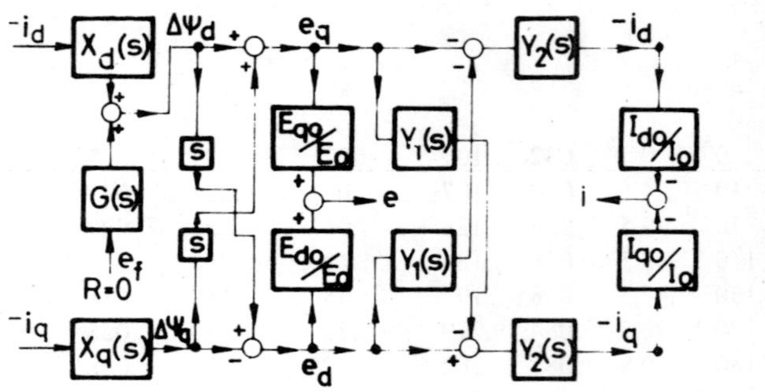

Fig. 6. Block diagram of a synchronous generator connected to separate load.

$$X_d(s) \cong X''_d \text{ and } X_q(s) \cong X''_q \quad (23)$$

Let e vary due to some external signal by Δe_q. From Fig. 6

$$i_d = \frac{C_e^2X''_q\,(1 + s^2) - C_e}{sD_1}\,\Delta e_q;\; i_q = \frac{C_e^2\,X''_d(1 + s^2) + C_e}{D_1}\,\Delta e_q \quad (24)$$

where

$$D_1 = 1 + (s^2 - 1)C_e(X''_d + X''_q) + (s^2 + 1)^2\,C_e^2\,X''_dX''_q$$

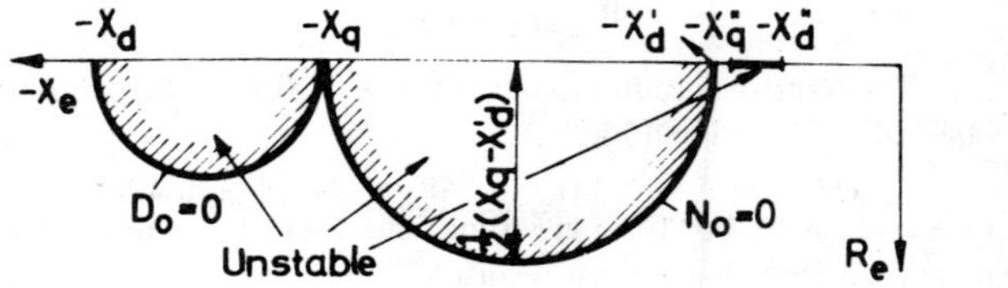

Fig. 7. Stable and unstable ranges of synchronous generator in the load impedance plane.

Due to neglecting the resistances, the current has undamped sinusoidal oscillating components of frequencies ω_a and ω_b.

$$\omega^2_{a,b} = 1 + \frac{\frac{1}{2}(X''_d + X''_q) \pm [2C_eX''_dX''_q(X''_d + X''_q) + 25(X''_d - X''_q)^2]^{\frac{1}{2}}}{C_e\,X''_d\,X''_q},$$

At least one of the frequencies is greater than the rated frequency. At frequencies ω_a and ω_b, (23) and therefore (24) are good approximations. The term independent of frequency in denominator D_1 is: $1 - C_e(X''_d + X''_q) + C^2\,X''_d\,X''_q$. It is negative in the range of capacitive loads $-X''_d > X_e > -X''_q$ and here i_d and i_q increase beyond all limits; i.e. self-excitation develops (Fig. 7). In reality the presence of damping resistances R and/or R_e can prevent self-excitation.[7]

CONCLUSIONS

The block diagrams for large and small signals are useful means as they help in the application of the elementary theory of feedback control to the investigation of the dynamic behaviour of synchronous machines. They promote unified treatment of related problems, assist the setup procedure of computer programs and give a deeper insight into causal relations among signals and into the basic processes. The adaptability of block diagrams 1, 3 and 6 is supported by the fact that they have only generally-known parameters time constants, synchronous, transient and subtransient reactances.

The overall frequency response of the torque-angle loop may

quickly be determined by the amplitude asymptotic Bode diagrams of G_T and H_T^{-1} for various working conditions of the machine. Since the asymptote locus is drawn by straight lines whose slopes change at each corner frequency, a freehand sketch is often sufficient. An easily surveyed picture may be obtained to estimate the influence of the variation of time constants on the overall frequency response.

The transfer function giving the relation between terminal voltage and field voltage is approximately one-energy-storage element with time constant $\tau'_{de} K'_\delta / K_\delta$, at low frequencies.

APPENDIX I Nomenclature

Capital letters denote large excursions. Δ quantities or lower case denote small excursions around an initial operating point. Time is measured in radians, unless otherwise specifically noted. Rotor quantities are referred to the stator and normalized in the same way as stator quantities.

D	combined prime mover and generator damping factor
E	machine terminal voltage
E_i	machine no-load voltage
I	armature line current
$j = \sqrt{-1}$	imaginary unit
K_d, K_s	damping and synchronizing torque coefficients, respectively
$K_\delta = K_s$	when $s = 0$
$M = 4\pi f H$	where H is per unit inertia coefficient
P, Q	active and reactive power, respectively
R, R_e	armature and external resistances, respectively
s	time derivative operator $\frac{d}{dt}$ or independent variable in the domain of Laplace transformation
t	time in radians
T_m	prime mover torque
$T_e = \psi_d I_q - \psi_q I_d$	electrical torque
$X_e = L_e$	external inductive reactance
$X_{ec} = -1/C_e$	external capacitive reactance
X_d, X'_d, X''_d	direct axis reactances (synchronous, transient, subtransient)

The symbols above denote per unit quantities.

Greek letters

δ	load angle between the q axis and the positive sequence infinite bus voltage vector
ϑ	initial angle between stator and rotor coordinate system
θ	angular position of the d axis to the real axis of the stator coordinate system
$\tau_{eC} = R_e C_e$;	
$\tau_{eL} = L_e/R_e$	
τ_f	field circuit time constant
$\tau'_{do}; \tau''_{do}$	d axis no-load time constants
$\tau'_d; \tau''_d$	d axis short circuit time constants
$\tau'_{de}; \tau''_{de}$	d axis short circuit time constants with external reactance
$\tau_D; \tau_Q = \tau''_{qo}$	amortisseur time constants in d and q axis, respectively
τ''_q	q axis short circuit time constant
τ_{qe}	q axis short circuit time constant with external reactance
φ	angle between vector $\bar{E}$ and $\bar{I}$
$\varphi_s = \varphi + 90°$	
ψ	flux linkage
$\omega = s\theta = 1 + s\delta$	

$\mu_D = 1 - X_{fD}X_{Dd}/X_D X_{fD}$; $\sigma_{fd} = 1 - X_{fd}^2 / X_d X_f$; $\sigma_{fD} = 1 - X_{fD}^2/X_f X_D$;

$\sigma_{Dd} = 1 - X_{Dd}^2/X_d X_D$; $\sigma_{Qq} = 1 - X_{Qq}^2 / X_q X_Q$

Subscripts

aℓ	armature leakage
d,q or D,Q	direct and quadrature field or amortisseur axis, respectively
f	field winding
g	air gap
i	induced voltage
N	infinite bus
o	initial state
s	stator coordinate system
1,2	positive and negative sequence system, respectively

Functions

$$X_d(s) = X_d \frac{1 + s(\sigma_{fd}\tau_f + \sigma_{Dd}\tau_D) + s^2 \sigma_{fD}\tau_D\tau_f X''_d / X_d}{1 + s(\tau_f + \tau_D) + s^2\sigma_{fD}\tau_D\tau_f}$$

$$= X_d \frac{(1 + s\tau'_d)(1 + s\tau''_d)}{(1 + s\tau'_{do})(1 + s\tau''_{do})}$$

$$X_{de}(s) = X_d(s) + X_e = (X_d + X_e)\frac{(1 + s\tau'_{de})(1 + s\tau''_{de})}{(1 + s\tau'_{do})(1 + s\tau''_{do})}$$

$$X_q(s) = X_q \frac{1 + s\sigma_{Qq}\tau_Q}{1 + s\tau_Q}; \quad X_{qe}(s) = X_q(s) + X_e = (X_q + X_e)\frac{1 + s\tau''_{qe}}{1 + s\tau_Q}$$

$$\tau'_{de} = \tau'_{do}\frac{X'_d + X_e}{X_d + X_e}; \quad \tau''_{de} = \tau''_{do}\frac{X''_d + X_e}{X'_d + X_e}; \quad \tau''_{qe} = \tau_Q \frac{X''_q + X_e}{X_q + X_e}$$

$$G(s) = \frac{(X_{fd}/R_f)(1 + s\mu_D\tau_D)}{(1 + s\tau'_{do})(1 + s\tau''_{do})}; \qquad F(s) = \frac{(1/R_f)(1 + s\tau_D)}{(1 + s\tau'_{do})(1 + s\tau''_{do})}$$

$$Y_1(s) = (R_e/X_e)(1 + s\tau_{eL}); \qquad Y_2(s) = X_e/Det(s);$$

$$Y_{1c}(s) = \frac{R_e}{X_{ec}}\left[1 + \left(\frac{X_{ec}}{R_e}\right)^2 s\tau_{ec}(1 + s\tau_{ec})\right]; \quad Y_{2c}(s) = X_{ec}/Det_c(s)$$

APPENDIX II Basic equations

Voltage equations in rotor coordinate system with Park vectors: Armature

$$\bar{E} = -R\bar{I} + (s + j\omega)\bar{\psi} \tag{A1}$$

Between terminal and infinite bus

$$\bar{E} - \bar{E}_N = R_e\bar{I} + (s + j\omega)X_e\bar{I} \tag{A2}$$

For generator supplying capacitive load

$$\bar{E} = R_e\bar{I} - \frac{X_{ec}}{s + j\omega}\bar{I} \tag{A3}$$

Rewriting them into d, q component equations

$$E_d = -RI_d + s\psi_d - \omega\psi_q \tag{A4}$$

$$E_q = -RI_q + s\psi_q + \omega\psi_d \tag{A5}$$

$$\begin{bmatrix} I_d \\ I_q \end{bmatrix} = \frac{X_e}{Det} \begin{bmatrix} Y_1(s) & 1 \\ -1 & Y_1(s) \end{bmatrix} \begin{bmatrix} E_d - E_{Nd} \\ E_q - E_{Nq} \end{bmatrix} \quad (A6)$$

further from (A3)

$$\begin{bmatrix} I_d \\ I_q \end{bmatrix} = \frac{X_{ec}}{Det_c} \begin{bmatrix} Y_{1c}(s) & 1 \\ -1 & Y_{1c}(s) \end{bmatrix} \begin{bmatrix} E_d \\ E_q \end{bmatrix} \quad (A7)$$

where

$$Det = R_e^2 (1 + s\tau_{eL})^2 + X_e^2;\ Det_c = R_e^2 + X_{ec}^2 (1 + s\tau_{ec})^2$$

Voltage equations for rotor windings and matrix equation giving the relations between flux linkages and currents are

$$E_f = R_f I_f + s\psi_f \quad (A8)$$

$$O = R_D I_D + s\psi_D \quad (A9)$$

$$0 = R_Q I_Q + s\psi_Q \quad (A10)$$

$$\begin{bmatrix} \psi_d \\ \psi_q \\ \psi_f \\ \psi_D \\ \psi_Q \end{bmatrix} = \begin{bmatrix} -X_d & 0 & X_{fd} & X_{Dd} & 0 \\ 0 & -X_q & 0 & 0 & X_{Qq} \\ -X_{fd} & 0 & X_f & X_{fD} & 0 \\ -X_{Dd} & 0 & X_{fD} & X_D & 0 \\ 0 & -X_{Qq} & 0 & 0 & X_Q \end{bmatrix} \begin{bmatrix} I_d \\ I_q \\ I_f \\ I_D \\ I_Q \end{bmatrix} \quad (A11)$$

The rotor torque relation

$$T_m - \psi_d I_q + \psi_q I_d = Ms^2\theta + D(s\theta - 1) = (s^2M + sD)\delta \quad (A12)$$

where $s\theta = 1 + s\delta$. From (A8) . . . (A11)

$$\begin{bmatrix} \psi_d \\ \psi_q \end{bmatrix} = \begin{bmatrix} -X_d(s) & 0 \\ 0 & -X_q(s) \end{bmatrix} \begin{bmatrix} I_d \\ I_q \end{bmatrix} + \begin{bmatrix} G(s) \\ 0 \end{bmatrix} E_f \quad (A13)$$

$$I_f = F(s)\, E_f + sG(s)\, I_d \quad (A14)$$

From d,q components the vector, e.g. in case of E_d and E_q is: $\bar{E} = E\exp j\varphi_E$ where

$$E^2 = E_d^2(t) + E_q^2(t) \text{ and } \varphi_E = \tan^{-1}[E_q(t)/E_d(t)] \quad (A15)$$

Expressing equations (A4) . . . (A7) and (A12) . . . (A15) in small oscillation form, one obtains

$$e_d = - Ri_d + s\Delta\psi_d - \Delta\psi_q \quad (A4a)$$

$$e_q = - Ri_q + s\Delta\psi_q + \Delta\psi_d \quad (A5a)$$

$$\begin{bmatrix} i_d \\ i_q \end{bmatrix} = \frac{X_e}{Det} \begin{bmatrix} Y_1(s) & 1 \\ -1 & Y_1(s) \end{bmatrix} \begin{bmatrix} e_d - e_{Nd} \\ e_q - e_{Nq} \end{bmatrix} \quad (A6a)$$

$$\begin{bmatrix} i_d \\ i_q \end{bmatrix} = \frac{X_{ec}}{Det_c} \begin{bmatrix} Y_{1c}(s) & 1 \\ -1 & Y_{1c}(s) \end{bmatrix} \begin{bmatrix} e_d \\ e_q \end{bmatrix} \quad (A7a)$$

$$\Delta T_m - \psi_{do} i_q - I_{qo}\, \Delta\psi_d + \psi_{qo} i_d + I_{do}\Delta\psi_q = (s^2M + sD)\Delta\delta + \quad (A12a)$$

$$\begin{bmatrix} \Delta\psi_d \\ \Delta\psi_q \end{bmatrix} = \begin{bmatrix} -X_d(s) & 0 \\ 0 & -X_q(s) \end{bmatrix} \begin{bmatrix} i_d \\ i_q \end{bmatrix} + \begin{bmatrix} G(s) \\ 0 \end{bmatrix} e_f \quad (A13a)$$

$$i_f = F(s)\, e_f + sG(s)\, i_d \quad (A14a)$$

$$e = (E_{do}/E_o)e_d + (E_{qo}/E_o)e_q \quad (A15a)$$

In (A4a) and (A5a) the voltage component proportional to $s\Delta\delta$ has been neglected. In (A4a) . . . (A15a) the variables may be regarded as Laplace transformed quantities.

APPENDIX III.

Rearrangement of small signal block diagram

Small excursion in electrical torque

$$\Delta T_e = \psi_{do} i_q + I_{qo}\Delta\psi_d - \psi_{qo} i_d - I_{do}\Delta\psi_q \quad (A16)$$

The constants are to be determined from (A4) . . . (A6) and (A13), or from the vector diagram ($R = R_e = O$): $\psi_{do} = E_{qo}$; $\psi_{qo} = -E_{do}$

$$\psi_{do} = \frac{X_e\, E_{io}}{X_{de}} + \frac{X_d\, E_{No}\cos\delta_o}{X_{de}};\ \psi_{qo} = -\frac{X_q E_{No}\sin\delta_o}{X_{qe}} \quad (A17)$$

$$I_{do}X_e = E_{qo} - E_{No}\cos\delta_o;\ I_{qo}X_e = E_{No}\sin\delta_o - E_{do} \quad (A18)$$

Torque ΔT_e is caused by three quantities, namely the change in δ, E_N and E_f: $\Delta T_e = \Delta T_\delta + \Delta T_N + \Delta T_f$. Be the superposition terms calculated one by one from Fig. 1.

First let $\Delta\delta \neq O$ and $e_N = e_f = O$. From the upper and lower negative feedback loops

$$i_d = \frac{E_{No}\sin\delta_o}{X_{de}(s)}\Delta\delta \text{ and } i_q = \frac{E_{No}\cos\delta_o}{X_{qe}(s)}\Delta\delta \quad (A19)$$

By substituting (A13a), (A17) . . . (A19) into (A16) and by using the values of $X_{de}(s)$ and $X_{qe}(s)$

$$\frac{\Delta T_\delta}{\Delta\delta} = K_\delta + \frac{s\tau_\delta\,(1 + s\tau_{\delta 1})\,(1 + s\tau_{\delta 2})}{(1 + s\tau'_{de})\,(1 + s\tau''_{de})\,(1 + s\tau''_{qe})} = K_\delta + H_T(s) \quad (A20)$$

where

$$K_\delta = \frac{E_{io}E_{No}}{X_{de}}\cos\delta_o + \frac{X_d - X_q}{X_{de}X_{qe}} E_{No}^2 \cos^2\delta_o \quad (A21)$$

$$\left.\begin{aligned} &\tau_\delta = a_1\tau_1 + a_2;\ \tau_\delta(\tau_{\delta 1} + \tau_{\delta 2}) = a_1(\tau_2^2 + \tau''_{qe}\tau_1) + a_2(\tau'_{de} + \tau''_{de}); \\ &\tau_\delta\tau_{\delta 1}\tau_{\delta 2} = a_1\tau''_{qe}\tau_2^2 + a_2\tau'_{de}\tau''_{de} \\ &\tau_1 = \tau'_{do} + \tau''_{do} - \tau'_{de} - \tau''_{de};\ \tau_2^2 = \tau'_{do}\tau''_{do} - \tau'_{de}\tau''_{de} \end{aligned}\right\} \quad (A21a)$$

$$a_1 = \frac{E_{No}^2 \sin^2\delta_o}{X_{de}};\ a_2 = \frac{E_{No}^2\cos^2\delta_o}{X_{qe}}\ (\tau_Q - \tau''_{qe})$$

Secondly let $e_N \neq 0$ and $\Delta\delta = e_f = 0$, now

$$i_d = -\frac{\cos\delta_o}{X_{de}(s)} e_N \text{ and } i_q = \frac{\sin\delta_o}{X_{qe}(s)} e_N \quad (A22)$$

By a similar procedure as used for deriving (A20)

$$\frac{\Delta T_N}{e_N} = K_N - \frac{s\tau_N\,(1 + s\tau_{N1})\,(1 + s\tau_{N2})}{(1 + s\tau'_{de})\,(1 + s\tau''_{de})\,(1 + s\tau''_{qe})} \quad (A23)$$

where

$$K_N = \frac{E_{io}\sin\delta_o}{X_{de}} + \frac{X_d - X_q}{X_{de}X_{qe}} E_{No}\sin 2\delta_o \quad (A24)$$

$$
\left.
\begin{aligned}
\tau_N &= \frac{E_{No}}{2}\sin 2\delta_o\left(\frac{\tau_1}{X_{de}} - \frac{\tau_Q - \tau''_{qe}}{X_{qe}}\right);\ \tau_N(\tau_{N1}+\tau_{N2}) = \\
&= \frac{E_{No}}{2}\sin 2\delta_o\left[\frac{\tau_2^2 + \tau''_{qe}\tau_1}{X_{de}} - \frac{(\tau_Q - \tau''_{qe})(\tau'_{de}+\tau''_{de})}{X_{qe}}\right] \\
\tau_N\tau_{N1}\tau_{N2} &= \frac{E_{No}}{2}\sin 2\delta_o\left[\frac{\tau''_{qe}\tau_2^2}{X_{de}} - \frac{(\tau_Q - \tau''_{qe})\tau'_{de}\tau''_{de}}{X_{qe}}\right]
\end{aligned}
\right\} \quad \text{(A24a)}
$$

Finally let $e_f \neq 0$ and $\Delta\delta = e_N = 0$, then

$$
i_d = \frac{G(s)}{X_{de}(s)}\, e_f \text{ and } i_q = 0 \qquad \text{(A25)}
$$

Similarly to the above

$$
\Delta T_f = \frac{E_{No}\sin\delta_o}{X_{de}(s)}\, G(s) e_f = K_i \frac{X_{de}}{X_{de}(s)}\, e_i \qquad \text{(A26)}
$$

Neglecting amortisseurs, the high frequency values of transfer functions (A20), (A23) and (A25) will be $(s \to \infty)$

$$
K'_\delta = \frac{\Delta T_\delta}{\Delta\delta} = \frac{E_{No}^2}{X'_{de}}\left(\frac{E'_{qo}}{E_{No}}\cos\delta_o - \frac{X_q - X'_d}{X_{qe}}\cos 2\delta_o\right) \qquad \text{(A27)}
$$

$$
K'_N = \frac{\Delta T_N}{e_N} = \frac{E'_{qo}}{X'_{de}}\sin\delta_o + \frac{X'_d - X_q}{X'_{de}X_{qe}}\, E_{No}\sin 2\delta_o \qquad \text{(A28)}
$$

$$
K'_i = \frac{\Delta T_f}{e_i} = \frac{E_{No}\sin\delta_o}{X'_{de}} \qquad \text{(A29)}
$$

The expression for K'_δ, K'_N, and K'_i may be directly determined from equations for K_δ, K_N, and K_i, by writing E'_{qo} and X'_{de} in place of E_i and X_{de}, respectively. Constants (K) are the respective derivatives of the static and dynamic torque-load angle characteristics.

REFERENCES

1. W. G. Heffron, R. A. Phillips, "Effect of a modern amplidyne voltage regulator of underexcited operation of large turbine generators" AIEE TRANS. vol. 71, pt. III, pp. 692-697, August 1952.
2. M. Riaz, "Analog computer representations of synchronous generators in voltage-regulation studies", AIEE TRANS. vol. 75, pt. III, pp. 1178-1184. December 1956.
3. I. M. Canay, "Block diagrams and transfer functions of synchronous machine", AIEE TRANS. vol. 85, pp. 952-959, September 1966.
4. F. P. Demello, C. Concordia, "Concepts of synchronous machine stability as affected by excitation control", IEEE TRANS. POWER APPARATUS AND SYSTEMS, vol. 88, pp. 316-329, April 1969.
5. K. P. Kovács, I. Rácz, "Dynamic behaviour of a.c. machines" in German, Budapest: Academy Publisher, 1959.
6. R. V. Shepherd, "Synchronizing and damping torque coefficients of synchronous machines", AIEE TRANS. vol. 80, pt. III, pp. 180-187, June 1961.
7. Mo-Shing Chen, W. C. Duesterhoeft, Jr., "Transients in synchronous machines with balanced capacitive loads", IEEE TRANS. vol. 83, pp. 1242-1245, December 1964.

DYNAMIC SYSTEM STABILITY
PART I – INVESTIGATION OF THE EFFECT OF DIFFERENT LOADING AND EXCITATION SYSTEMS

Mohamed K. El-Sherbiny, Member IEEE
Gannon College, Erie, Pa.

Dinesh M. Mehta
Gannon College, Erie, Pa.

Abstract–Analysis of the phenomena of stability of synchronous machines under small perturbations as examined by Demello and Concordia[1], has been extended to the following:

First, the change of the parameters K_1 to K_6 of the block diagram describing the system (Fig. 1) has been investigated for different loading and power factors. From this a set of curves has been plotted. Such curves can be used to reach a general conclusion about the variations of such parameters by changing the operating condition, which is an important factor in dynamic stability studies.

Second, the dynamic stability of the system, as affected by different excitation, has been analyzed. The systems investigated are thyristor-type and type-1 excitation systems. Their effect on dynamic stability has been compared. The results obtained by mathematical analysis have been checked by both digital and analog computer. From this and analysis a set of rules was developed which helps in understanding the stability requirements.

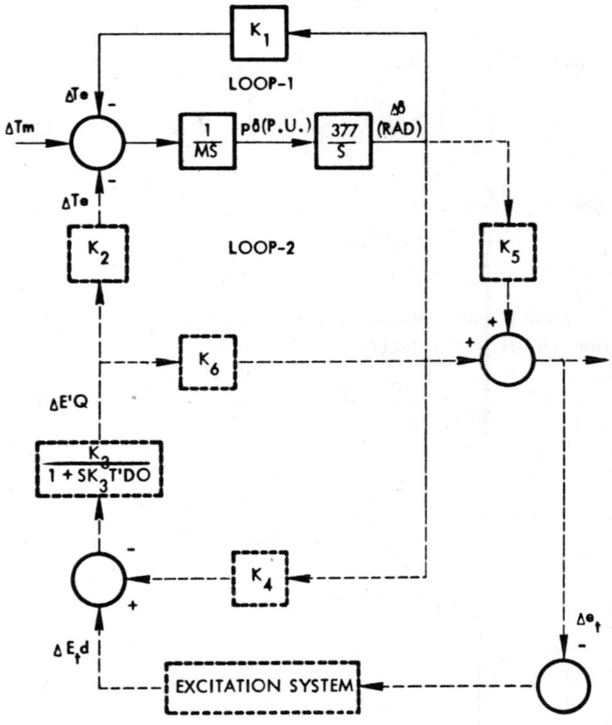

Fig. 1. Block diagram of system with voltage regulator

INTRODUCTION

Demello and Concordia[1] analyzed the phenomena of stability of synchronous machines under small perturbation. They investigated the case of a single machine connected to a large system through external impedance. The object of their analysis was to develop insights into effects of excitation systems and to establish an understanding of the stabilizing requirements for such system. However, the scope of their paper was limited to a simple excitation system which could be characterized by a small time constant of between 0.03 and 0.05 sec, which is typical of a transistor-type system.

Their method of analysis is extended to the realistic excitation system recommended by the IEEE Excitation Committee[2]. It has been found that the dynamic behavior of the machine is quite different at different operating points. Hence, a wide range of system operating conditions has been investigated which enables explanation of the nature of the dynamic system stability as affected by different excitation systems. Also, the analysis presented in this paper explains and proves the adverse effect of the excitation system, which is verified by experiments and digital results in only some of the literature[3-6].

BLOCK DIAGRAM AND PARAMETERS

Figure 1 shows the block diagram of a single machine supplying an infinite bus through external impedance. The relations shown have been treated previously[7] by the concept of small perturbation on the fundamental synchronous machine equations. The relations in the block diagrams discussed in this paper apply to a two-axis machine representation with a field circuit in the direct axis but without amortisseur effects. The parameters K_1 to K_6 are functions of machines and system impedances as well as the operating points. Demello and Concordia[1] found that the concept of synchronous machine stability depends on these parameters, since both synchronizing and damping torques are functions of those parameters, as will be discussed later. Much effort has been spent studying the behavior of such parameters for different operating points. The equations for these parameters are shown in Appendi JA, as given previously[1].

Paper T 73 220–1, recommended and approved by the Power Generation Committee of the IEEE Power Engineering Society for presentation at the IEEE PES Winter Meeting, New York, N.Y., January 28–February 2, 1973. Manuscript submitted September 14, 1972; made available for printing January 9, 1973.

Model 1-A

This is the case for one machine against infinite bus and without local load as shown in Fig. 2a, in this case r_e usually $<< x_e$ and the approximate expression can be used. A wide range of operating conditions has been considered and the values of K have been plotted as shown in Figs. 3a through e, for K_1, K_2, K_4 . . . K_6 respectively. K_3 is excluded because it is independent of loading. To explain the nature of such curves, the vector diagram of the system is shown in Fig. 4, where the constant power P and reactive power Q lines are shown. In this case note that the power angle δ is always between 0 and 90 degrees (for lagging power factor). By increasing P and/or Q, all the parameters are positive except for K_5. From Fig. 4 it can be seen that by increasing P (Q constant), the power angle will increase causing the expression for K_5 in Appendix A to be negative.

In this case, one can conclude from these figures that all parameters, K_1 to K_6, are positive for the entire loading range; only K_6 changes its positive sign for higher loading and/or lower power factor. This conclusion has been proven for any case such that no local load and $r_e << x_e$.

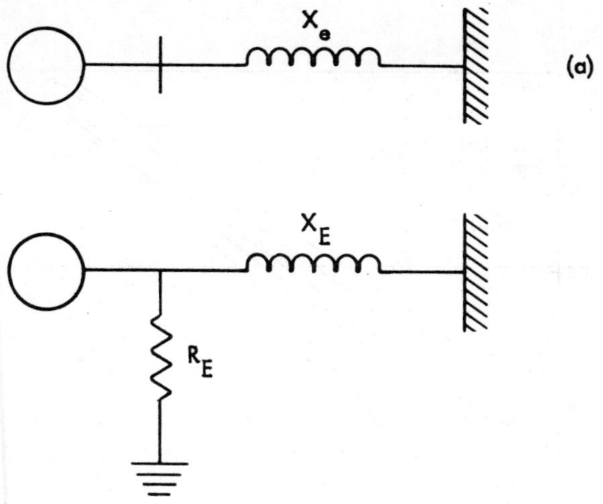

Fig. 2. External impedance system configurations for single machine against infinite bus.

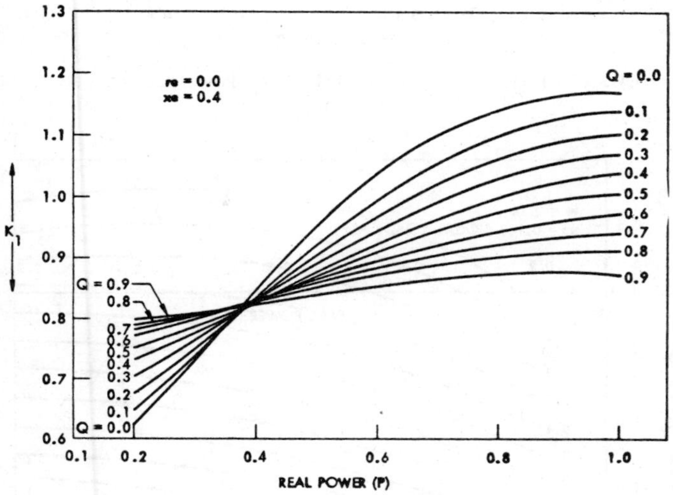

(a) K_1 vs P (Real Power) and Q (Reactive Power) as parameter

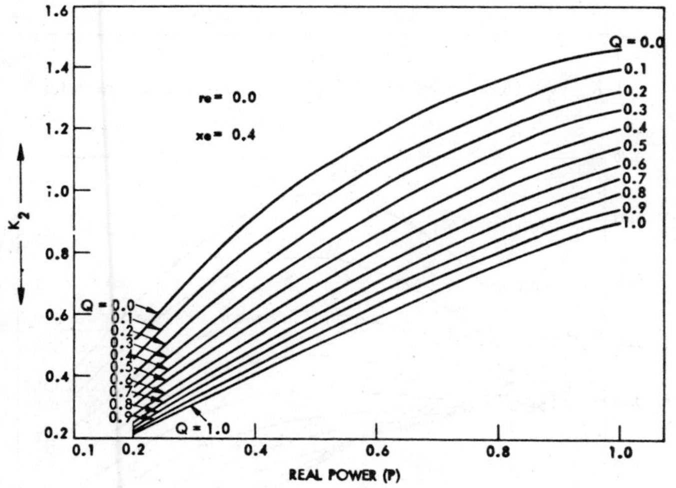

(b) K_2 vs P (Real Power) and Q (Reactive Power) as parameter

Model 2

This is the case where local load is considered as shown in Fig. 2b. This model can be converted to the form of Model 1 by making the Thevenin equivalent where:

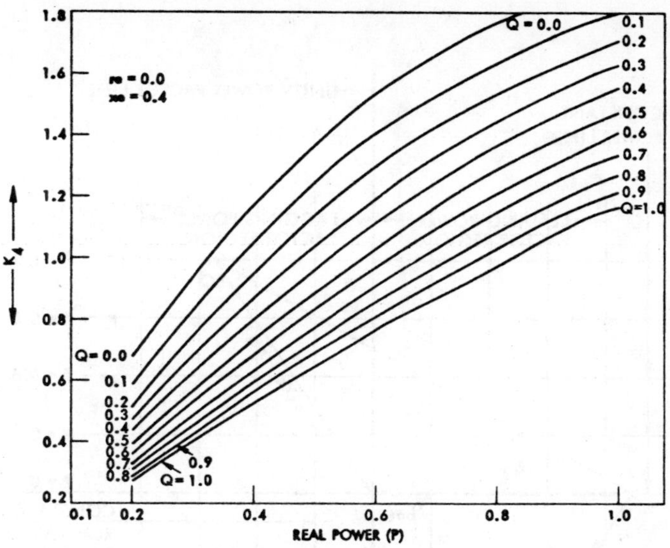

(c) K_4 vs P (Real Power) and Q (Reactive Power) as parameter

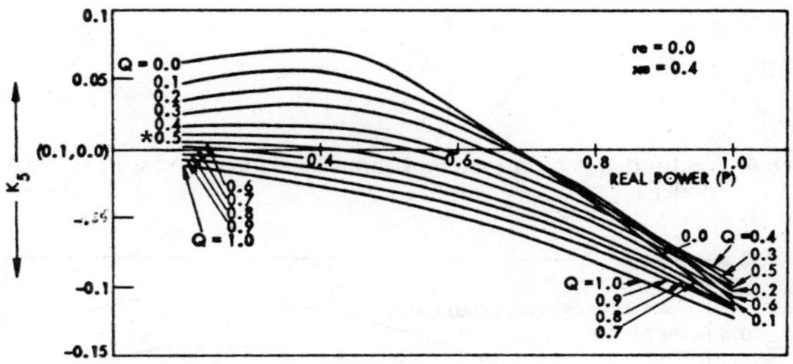

(d) K_5 vs P (Real Power) and Q (Reactive Power) as parameter

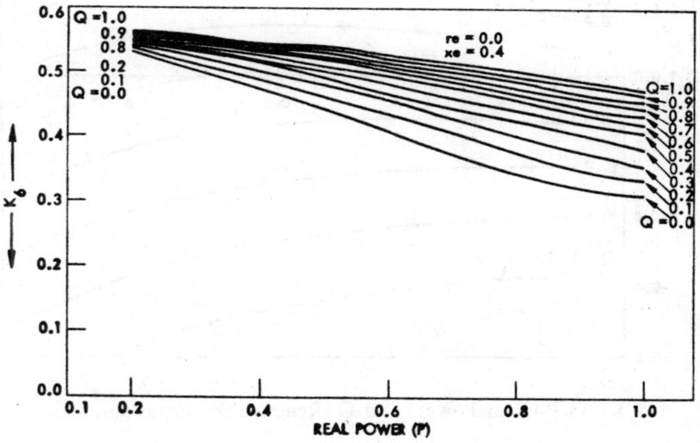

(e) K_6 vs P (Real Power) and Q (Reactive Power) as parameter

Fig. 3. Variation of parameter $K_1 \ldots K_6$ with different operating conditions (P and Q) for Model 1 ($r_e << x_e$).

$$r_e = \frac{R_E \cdot X_E^2}{R_E^2 + X_E^2}$$

$$x_e = \frac{R_E^2 \cdot X_E}{R_E^2 + X_E^2}$$

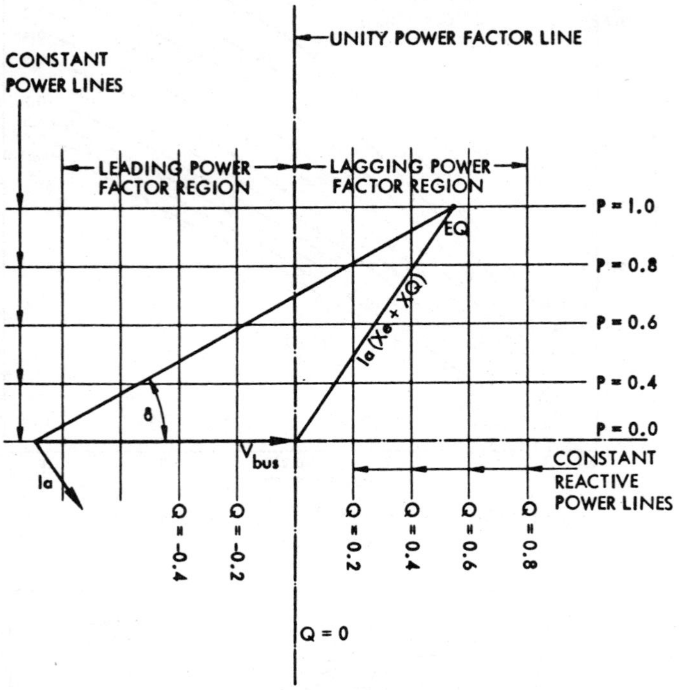

Fig. 4. Vector diagram for system without local load, $r_e << x_e$ ($r_e = 0$) Model 1

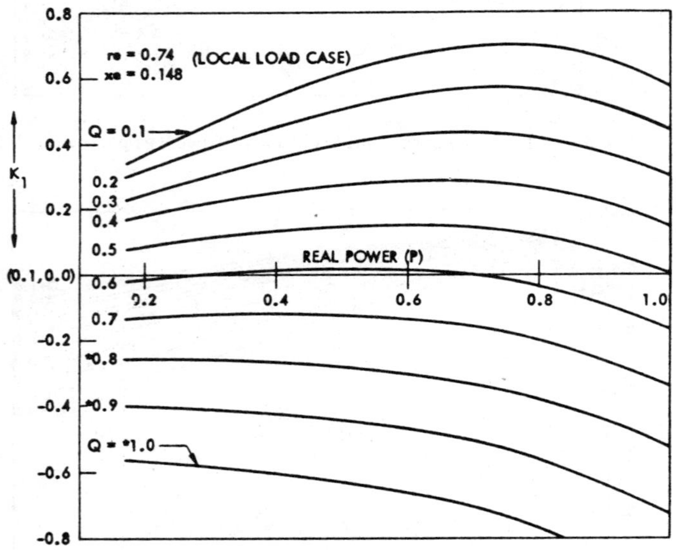

(a) K_1 vs P (Real Power) and Q (Reactive Power) as parameter

Figures 5a through e show how K_1 to K_6 vary as machine loading P and Q vary. Different results are obtained, in this case for local load, where $r_e > x_e$, and the vector diagram of Fig. 4 becomes that shown in Fig. 6. The torque angle δ in this case can vary anywhere between $-90 < \delta < 90$, or the rotor angle δ is positive (δ_A) for higher loading and power factor and negative (δ_B) for lower power and/or power factor. Hence, from Fig. 6 one can interpret the results obtained in the curves of Figs. 5a through e. The conclusion in this case is that, K_2, K_5 and K_6 are always positive. K_1 and K_4 can change signs from positive to negative as Q increases and/or power factor decreases. This conclusion has been proved for the cases where there is a local load and $r_e > x_e$. Cases studied for this case are:

$$r_e = 0.3,\ 0.6,\ 0.9,\ 1.2$$

$$x_e = 0.4,\ 0.6,\ 0.8$$

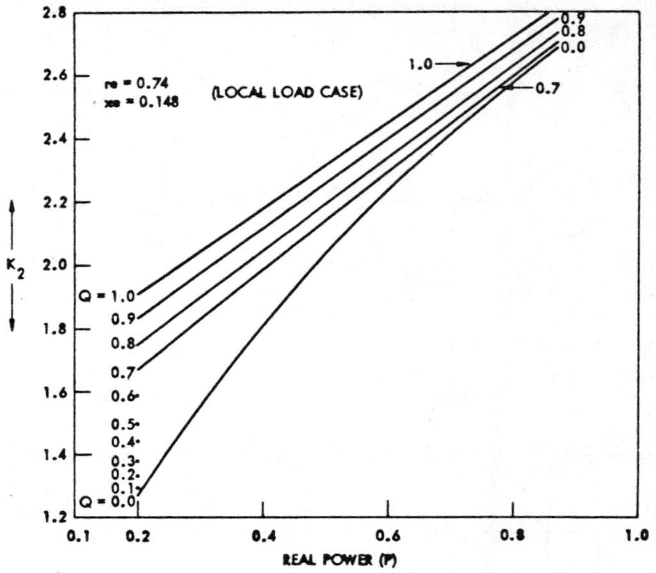

(b) K_2 vs P (Real Power) and Q (Reactive Power) as parameter

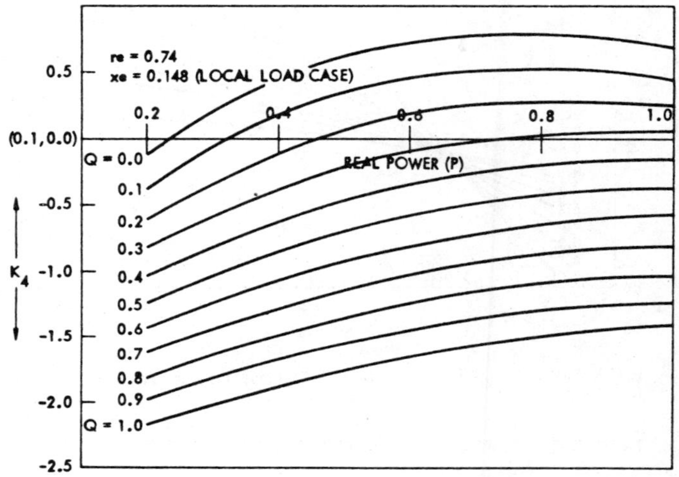

(c) K_4 vs P (Real Power) and Q (Reactive Power) as parameter

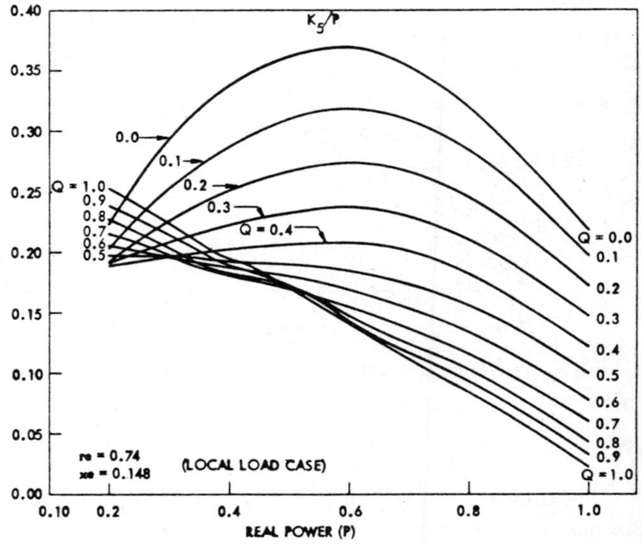

(d) K_5 vs (Real Power) and Q (Reactive Power) as parameter

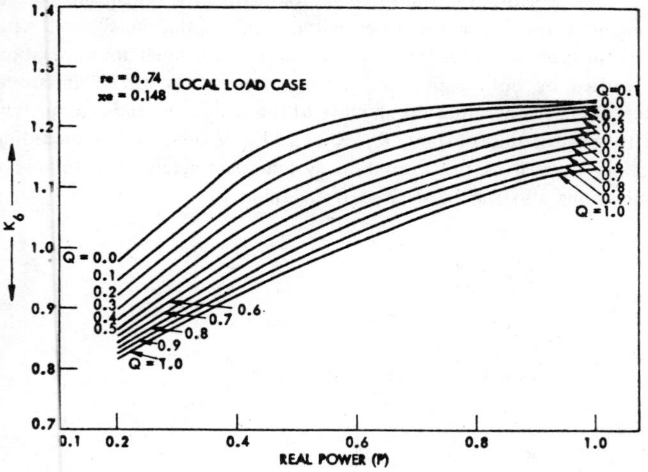

(e) K_6 vs P (Real Power) and Q (Reactive Power) as parameter

Fig. 5. Variation of parameters $K_1 \ldots K_6$ with different operating condition P and Q) for Model 2 ($r_e \not> x_e$)

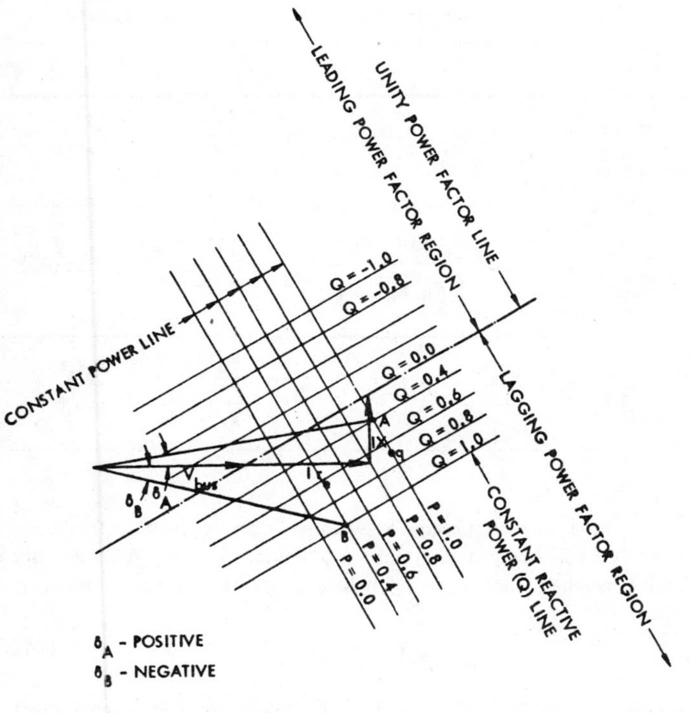

Fig. 6. Vector diagram for system with local load $r_e >> x_e$

CONCEPT OF MACHINE STABILITY

The concept used by Demello and Concordia of synchronizing and damping torques will be used in this paper. At any given oscillation frequency, breaking torques are developed in phase with machine rotor angle δ (synchronizing torques) and in phase with machine rotor speed Pδ (damping torques). The torque developed by any particular means can be broken down into these components for stability study. Sufficient and necessary conditions for the system to be stable is that both synchronizing and damping torques be positive. Positive synchronizing torques assure restoring of the rotor angle of the machine following an arbitrary small displacement of this angle. Positive damping is necessary to damp out any oscillation due to any perturbation.

EFFECT OF EXCITATION SYSTEM ON MACHINE STABILITY

Figure 1 shows the block diagram of a synchronous machine and excitation system. The block diagram $G_E(s)$ in Fig. 1 represents the excitation system of any kind. In this paper the effect of two excitation systems has been discussed and compared with the case of no excitation system. Namely, the thyristor and the type-1 excitation system.

From Fig. 1, the transfer function between the change in torque and rotor angle is:

$$\frac{\Delta T_e}{\Delta\delta} = \frac{-K_2 \cdot G_3 (K_4 + K_5 \cdot G_e)}{1 + G_e \cdot G_3 \cdot K_6} \tag{1}$$

where

$$G_3 = \frac{K_3}{1 + ST} \qquad T = K_3 \times T'_{do} = T_{dz} \tag{2}$$

G_e = The transfer function of the excitation system under study.

Thyristor-Type Excitation System

This case has been discussed by Demello and Concordia[1]. It is mentioned here for the purpose of comparison.

In Eq. (1) the term G_e is

$$G_e = \frac{\Delta E_{fd}}{\Delta e_t} = \frac{-K_E}{1 + ST_E} \ . \tag{3}$$

Substituting from Eq. (3) into Eq. (1), then for any frequency of oscillation substitute S by jω, and break the resultant torque into two components. The synchronizing torque (in phase with angle) becomes

$$\Delta T_s = \frac{-K_2/K_6 \left(\frac{K_4}{K_E} + K_5\right)}{1 + \omega^2 \cdot T_{eQ}^2} \cdot \Delta\delta \tag{4}$$

where $T_{eQ} = T_{do}/K_6 K_E$. Normally the term K_4/K_E is very small in comparison with K_5; hence, Eq. (4) reduced to

$$\Delta T_s \cong \frac{-K_2 \cdot K_5/K_6}{1 + \omega^2 \cdot T_{eQ}^2} \cdot \Delta\delta \ . \tag{5}$$

Similarly, the net damping torque (in phase with speed)

$$\Delta T_d = \frac{\omega \cdot K_2 \cdot T_{eQ} \cdot /K_6 \cdot \left[\frac{K_4}{K} + K_5\right]}{1 + \omega^2 \cdot T_{eQ}^2} \cdot j\Delta\delta \tag{6}$$

$$\cong \frac{T_{eQ} \cdot K_2 \cdot K_5/K_6}{1 + \omega^2 \cdot T_{eQ}^2} \cdot j\omega\Delta\delta \tag{7}$$

When the effect of the excitation system is neglected, or there is no voltage regulator, ΔT_s and ΔT_d become

$$\Delta T_s = \frac{-K_2 \cdot K_3 \cdot K_4}{1 + \omega^2 \cdot T^2} \cdot \Delta\delta \tag{8}$$

$$\Delta T_d = \frac{T \cdot K_2 \cdot K_3 \cdot K_4}{1 + \omega^2 \cdot T^2} \cdot j\omega\Delta\delta \tag{9}$$

where $T = T_{do} \times K_3$ and the total torque (ΔT_{tot}) is the vectorial sum of synchronizing and damping torques.

Type-1 Excitation System

The block diagram of a type-1 excitation system, given by the Excitation System Committee[3], is used as shown in Fig. 7. From this

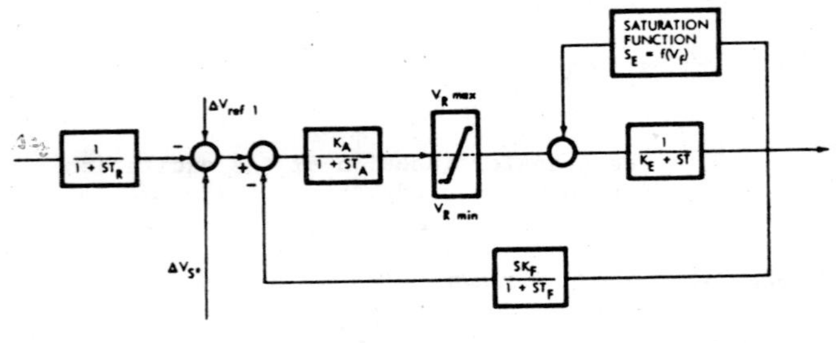

Fig. 7. Type-1 excitation system

the expression G_e, given in Eq. (1) with T_r neglected, is

$$G_e(s) = \frac{-K_A \cdot (1 + ST_F)}{(1 + ST_A)(1 + ST_F)(s_E + K_E + ST_E) + SK_A \cdot K_F}$$

Substitute $G_e(s)$ in Eq. (1) and follow the same analysis in finding the synchronizing and damping torques as discussed above. The corresponding expressions for Eqs. (8) and (9) are:

$$\Delta T_s = -\left(\frac{A_o \cdot B_o + A_1 B_1}{B_o^2 + B_1^2}\right) \cdot \Delta\delta \tag{10}$$

$$\Delta T_d = -\left(\frac{A_1 B_o - A_o B_1}{B_o^2 + B_1^2}\right) \left(\frac{1}{\omega}\right) (j\omega\Delta\delta) \tag{11}$$

where

$$A_o = a_o - a_2 \cdot \omega^2 \tag{12}$$

$$A_1 = a_1\omega - a_3 \cdot \omega^3 \tag{13}$$

$$B_o = b_o - b_2 \cdot \omega^2 + b_4 \cdot \omega^4 \tag{14}$$

$$B_1 = b_1\omega - b_3 \cdot \omega^3 \tag{15}$$

where a and b are constant and functions of a synchronous machine and excitation system as listed in Appendix B. The expressions as given in Eqs. (10) and (11) are not clear enough for investigating the concept of stability.

Table 1 shows the results of approximating the synchronizing and damping torques for the cases without an excitation system, with thyristor-type and with type-1 excitation system. Such approximation is verified by both digital and analog computer as will be discussed later. Table 1 shows that the stability of the system depends completely on the sign of the constant, $K_1, K_2, \ldots K_6$ discussed in the preceding section, as well as the excitation system parameters. The following conclusions also can be derived from Table 1.

where

$$T = T_{do}' \cdot K_3$$

$$b_3 = b_3$$

$$F = K_a^2 \cdot K_2K_3K_4/(b_o^2 + b_1^2 \cdot \omega^2)$$

$$M = \frac{K_F}{K_4} - K_3 \cdot K_6 \cdot \frac{K_F}{K_5}$$

Table 1. Net synchronizing and damping torques for synchronous machine without excitation system with thyristor and type-1

	Lower frequency $\omega < 1$		Higher frequency $\omega >> 1$	
	K_s	K_d	K_s (net)	K_D
Without excitation system	$K_1 - K_2 \cdot K_3 \cdot K_4$	$TK_2 \cdot K_3 \cdot K_4$	$K_1 - \frac{K_2K_3K_4}{\omega^2T^2}$	$\frac{K_2K_3K_4}{\omega^2T}$
With thyristor type	$K_1 - \frac{K_2K_5}{K_6}$	$\frac{T\ K_2\ K_5}{K_E\ K_6^2\ K_3}$	$K_1 - \frac{K_2K_3^2K_5K_6 \cdot K_E^2}{\omega^2T^2}$	$\frac{K_2K_3K_5K_E}{\omega^2\ T}$
With type-1	$K_1 - \frac{K_2K_5}{K_6}$	$M.F.K_5$	$K_1 - \frac{K_2 \cdot K_3 \cdot K_4 \cdot b_3}{T_A \cdot T_F \cdot T_E \cdot T^2 \cdot \omega^2}$	$\frac{K_2K_3K_4}{\omega^2T}$

System without Voltage Regulator. At zero oscillation frequency (that is, at a steady-state condition), the actual steady-state stability limit is reached when the net synchronizing torque is zero. This occurs when

$$(K_1 - K_2K_3K_4)\ \Delta\delta = 0, \tag{16}$$

or the quantity $(K_1 - K_2K_3K_4)$ is the synchronizing power coefficient described by $dP/d\delta$ in some of the literature[8]. This quantity decreases as the power increases up to the actual steady-state stability limit, as shown by the curves in Figs. 8a and 8b for Models 1 and 2, respectively. It is interesting to note that even in the very unusual cases ($Q >> P$) where K_1 is negative, as shown in Fig. 5a, K_4 also is negative which causes the synchronizing power coefficient to be positive, as shown in Fig. 8b. However in this case the machine would be unstable at a higher frequency. The synchronizing power coefficient,

$$K_s = K_1 - K_2K_3K_4/1 + \omega^2T^2,$$

will be dominated by K_1.

It is clear also that the field circuit contributes a positive damping which did not concern us in our discussion of steady-state stability. In Fig. 9a, the system is stable since the net synchronizing coefficient K_s is positive, and vice versa in Fig. 9b.

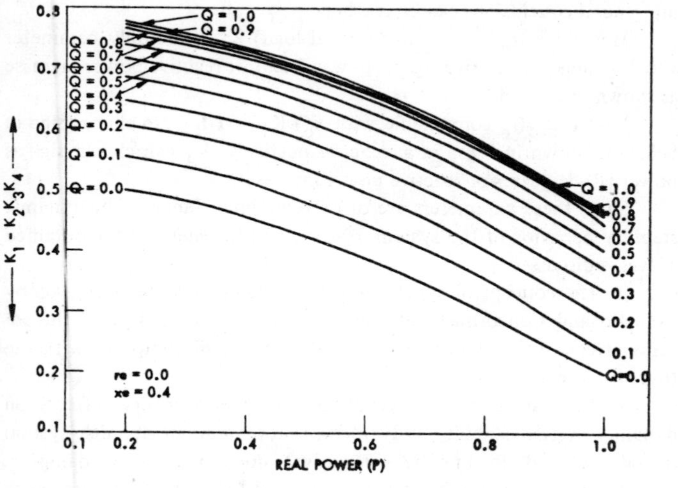

(a) $K_1 - K_2K_3K_4$ vs P (Real Power) and Q (Reactive Power) as parameter

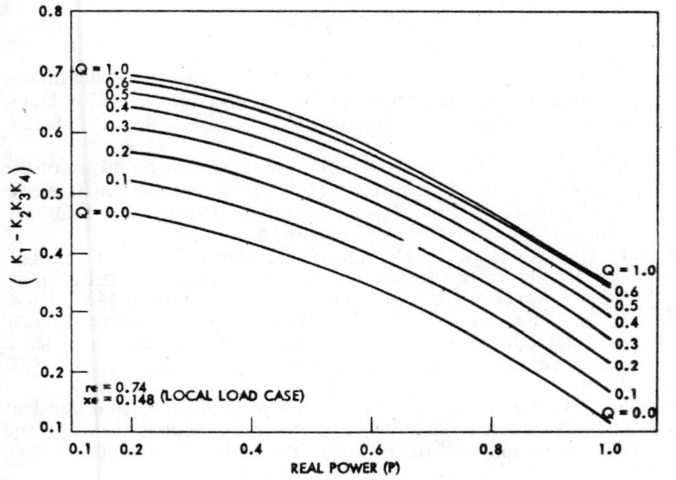

(b) $K_1 - K_2K_3K_4$ vs P (Real Power) and Q (Reactive Power) as parameter

Fig. 8. Net synchronizing power coefficient for different operating conditions

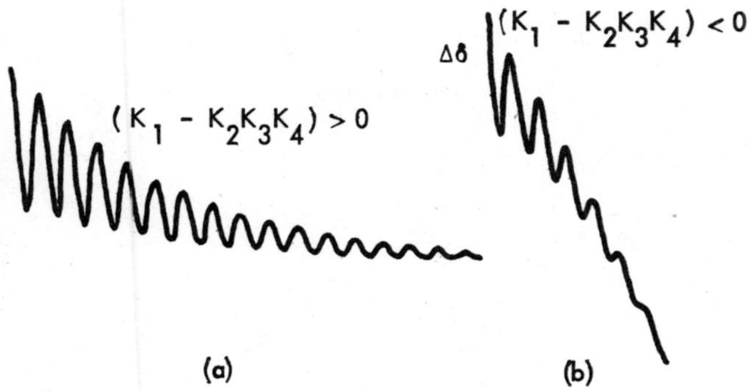

Fig. 9. System without any excitation system

System with Thyristor-Type Excitation System

Discussion of the effect of such excitation system was discussed by Demello and Concordia[1]. Only the adverse effects will be stressed in this part.

a. If K_5 is positive, which is the case for low to medium loading and external impedance, the margin of stability will decrease as the net synchronizing torque decreases. But the damping torque in this case is positive.

b. If K_5 is negative, which is the case for moderate and high loading and external impedance, instability will occur due to negative damping, as shown in Fig. 10 a and b, respectively.

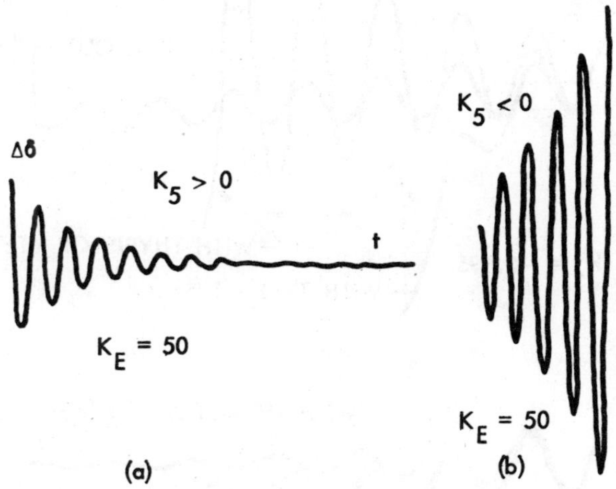

Fig. 10. With thyristor-type excitation system

System with Type-1 Excitation System

From Table 1 at lower frequencies ($\omega < 1$), the synchronizing and damping torques are:

$$\Delta T_s \cong \left(K_1 - \frac{K_2 \cdot K_5}{K_6}\right) \Delta\delta \tag{17}$$

$$\Delta T_d \cong F \cdot M \cdot K \,(j\omega\Delta\delta) \tag{18}$$

where

$$F = K_A^{\,2}\, K_2K_3K_4/b_o + b_1^{\,2}\omega^2 \tag{19}$$

$$M = \frac{K_F}{K_4} - \frac{K_3 \cdot K_6 \cdot K_F}{K_5} \tag{20}$$

It is clear that the stability will depend also on the sign of the constant $K_1 \ldots K_6$. For the data of Model 1 discussed in the preceding section, we can make the following conclusions:

a. For the cases where K_5 is positive, the synchronizing torque in Eq. (17) is reduced. M in Eq. (20) is almost negative for the practical range discussed in Fig. 3. This will cause instability of the system due to negative damping. This is contradictory to the case for the system with a thyristor-type excitation system.

b. For the cases when K_5 is negative, M is positive and the damping torque contributed to the system is negative, as shown in Eq. (18), which causes instability.

This analysis is verified for many cases by the digital and analog computer. Comparison of the three cases is shown in Figs. 11a and b. In Fig. 11a ($K_5 < 0$) the adverse effect discussed above is shown. Both thyristor-type and type-1 excitation systems cause instability. Due to the thyristor-type system the instability is caused due to negative damping only, while the instability caused by the type-1 excitation system is caused by negative damping at low frequency, as well as synchronizing torque at high frequency. In Fig. 11b ($K_5 > 0$) the contradicting effect of the thyristor-type and type-1 excitation systems is clear. In this case the system's margin of stability is improved by the thyristor-type due to positive damping, while the type-1 excitation system with negative damping causes the instability.

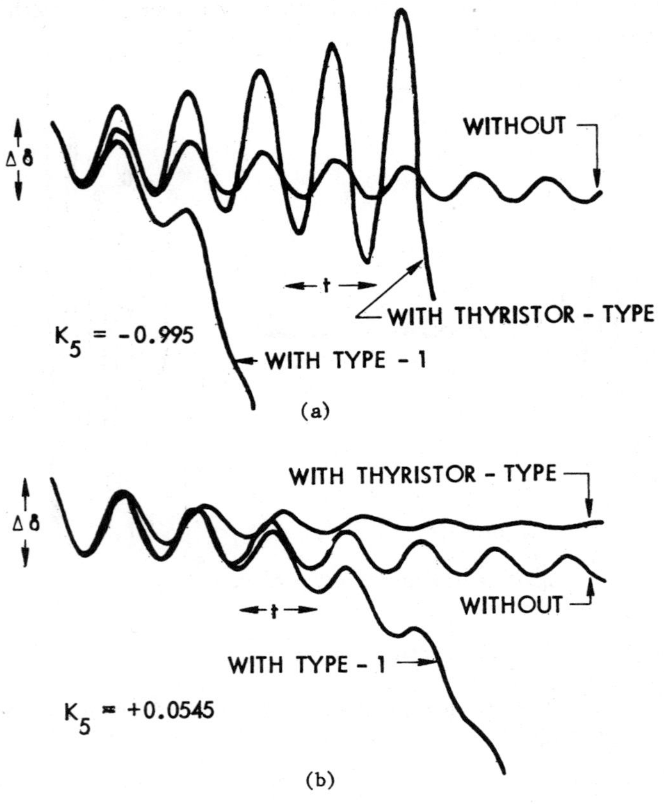

Fig. 11. Comparison between cases without excitation system, with thyristor-type, and type-1 excitation system

DIGITAL AND ANALOG ANALYSIS

The results obtained in this paper are obtained by digital computer program. The program has been designed to calculate the block diagram constants $K_1 \ldots K_6$, and to compare the effects of thyristor-type and type-1 excitation systems on dynamic system stability. This is done by calculating the synchronizing and damping torque for each case at all range of operating frequency (0.2 to 20 Hz). This program is designed in a general manner to handle any similar system.

The cross check of the above digital results has been made by simulating the system on a TR-48 analog computer. The results were tallied and found consistent with the mathematical analysis described. Appendix C includes the data of the system, used in the study.

WORK IN PROGRESS

Work is under way to compensate for the adverse effects of the type-1 excitation system and/or to improve the system stability margin by introducing a stabilizing scheme. This program will be extended to design and optimize such schemes based on optimizing the synchronizing and damping torques rather than using the trial and error method.

CONCLUSION

1. A digital computer program has been used to plot a set of curves for the block diagram (Fig. 1) parameters K_1 to K_6. This set of curves has been plotted for the variation of K_1 to K_6, as the operating condition (P and Q) varies, for any given system. From this the following general conclusions can be reached:

Model 1: ($r_e << x_e$ or no local load) K_5 is the only parameter which changes it positive sign as power P and/or reactive power increase as shown in Fig. 3d.

Model 2: (r_e comparable with x_e) K_1 and K_4 can change sign in behavior shown in Figs. 5a and 5c, respectively. K_1 can change sign as power (P) decreases or reactive power increases.

Since these parameters are of extreme importance in the dynamic stability behavior of the system, reasoning of the shape of the obtained curves is discussed.

2. The concept of synchronizing and damping torques as given by Demello and Concordia[1] has been used to analyze and prove the adverse effect of the type-1 excitation system which compared with the thyristor type.

3. The thyristor-type excitation system has adverse effects on dynamic system stability only in cases for moderate to high system impedance and loading (K_5 negative) due to negative damping.

4. In all cases the excitation system of type-1 has been proven to have adverse effects on system stability due to negative damping.

REFERENCES

1. F. P. Demello and C. Concordia, "Concepts of synchronous machine stability as affected by excitation control," IEEE Trans. (Power Apparatus and Systems), Vol. 88, No. 4, pp. 316-327, April 1969.
2. IEEE Working Group of excitation Systems, Subcommittee of the Power Generating Committee, "Computer representation of excitation systems," IEEE Trans. (Power Apparatus and Systems), Vol. 87, pp. 1460-1464, June 1968.
3. F. R. Schleif, H. D. Huukins, G. E. Martin, and E. E. Hattan, "Excitation control to improve power line stability," IEEE Trans. (Power Apparatus and Systems), Vol. 87, pp. 1426-1434, 1968.
4. R. M. Shier and A. L. Blythe, "Field tests of dynamic stability using a stabilizing signal and computer program verification," IEEE Trans. (Power Apparatus and Systems), Vol. 87, pp. 315-322, 1968.
5. M. S. Dyrkacz, C. C. Young and F. J. Maginniss, "Digital transient stability program including the effect of regulator, exciter and governor response," IEEE Trans. (Power Apparatus and Systems), Vol. 78, pp. 1673-1677, February 1960.
6. M. K. El-Sherbiny and A. A. Fouad, "Digital analysis of excitation control for interconnected power systems," IEEE Trans. (Power Apparatus and Systems), Vol. 90, pp. 441-448, March/April 1971.
7. W. G. Heffron and R. A. Phillips, "Effect of modern amplidyne voltage regulator on underexcited operation of large turbine generators," AIEE Trans. (Power Apparatus and Systems), Vol. 71, pp. 692-697, August 1952.
8. E. W. Kimbark, *Power system stability,* Vols. 1 and 3, New York: John Wiley and Sons, Inc., 1948.

APPENDIX A

Equations of Parameter K

$$\Delta e_t = K_5 \Delta\delta + K_6 \Delta E_q'$$

$$\Delta E_q' = \frac{K_3 \Delta E_{fd}}{1 + ST_{do}' K_3} - \frac{K_3 K_4 \Delta\delta}{1 + ST_{do}' K_3}$$

$$\Delta T_e = K_1 \Delta\delta + K_2 \Delta E_q'$$

where

$$K_1 = \frac{E_{qo} E_o}{A} [r_e \sin \delta_o + (X_e + X_d') \cos \delta_o]$$

$$+\frac{i_{qo}E_o}{A}[(X_q - X_d')(X_e + X_q)\sin\delta_o - r_e(X_q - X_d')\cos\delta_o] \quad \text{(A-1)}$$

$$K_2 = \left[\frac{r_eE_{qo}}{A} + i_{qo}\left(1 + \frac{(X_e + X_q)(X_q - X_d')}{A}\right)\right] \quad \text{(A-2)}$$

$$K_3 = \left[1 + \frac{(X_e + X_q)(X_d - X_d')}{A}\right]^{-1} \quad \text{(A-3)}$$

$$K_4 = \frac{E_o(X_d - X_d')}{A}[(X_e + X_q)\sin\delta_o - r_e\cos\delta_o]$$

$$K_5 = \frac{e_{do}}{e_{to}}X_q\left[\frac{r_eE_o\sin\delta_o + (X_e + X_d')E_o\cos\delta_o}{A}\right] \quad \text{(A-4)}$$

$$+\frac{e_{qo}}{e_{to}}X_d'\left[\frac{r_eE_o\cos\delta_o - (X_e + X_q)E_o\sin\delta_o}{A}\right] \quad \text{(A-5)}$$

$$K_6 = \frac{e_{qo}}{e_{to}}\left[1 - \frac{X_d'(X_e + X_q)}{A}\right] + \frac{e_{do}}{e_{to}}X_q\frac{r_e}{A} \quad \text{(A-6)}$$

$$A = [r_e^2 + (X_e + X_d')(X_q + X_e)] .$$

The steady-state operating values of δ_6, E_{qo}, e_{do}, and e_{qo} are derived from the standard machine vector diagram. For the special case of zero external resistance, the expressions become

$$K_1 = \frac{X_q - X_d'}{X_e + X_d'}i_{qo}E_o\sin\delta_o + \frac{E_{qo}E_o\cos\delta_o}{X_e + X_q} \quad \text{(A-7)}$$

$$K_2 = \frac{E_o\sin\delta_o}{X_e + X_d'} \quad \text{(A-8)}$$

$$K_3 = \frac{X_d' + X_e}{X_d + X_e} \quad \text{(A-9)}$$

$$K_4 = \frac{X_d - X_d'}{X_e + X_d'}E_o\sin\delta_o \quad \text{(A-10)}$$

$$K_5 = \frac{X_q}{X_e + X_q}\frac{e_{do}}{e_{to}}E_o\cos\delta - \frac{X_d'}{X_e + X_d'}\frac{e_{qo}}{e_{to}}E_o\sin\delta_o \quad \text{(A-11)}$$

$$K_6 = \frac{X_e}{X_e + X_d'}\frac{e_{qo}}{e_{to}} \quad \text{(A-12)}$$

APPENDIX B

The constants of Eqs. (13) – (15) are

$$a_c = K_2K_3K_4M + K_2K_3K_5K_A$$

$$a_1 = K_2K_3K_5T_AT_F + K_2K_3K_4K_AK_F + K_2K_3K_4T_E + K_2K_3K_4M(T_A + T_F)$$

$$a_2 = K_2K_3K_4MT_AT_F + K_2K_3K_4T_E(T_A + T_F)$$

$$a_3 = K_2K_3K_4T_AT_FT_E$$

$$b_o = M + K_3K_6K_A$$

$$b_1 = T_E + M\cdot(T_A + T_F) + MT + K_3K_6K_AT_F + K_A\cdot K_F$$

$$b_2 = T\cdot T_E + MT(T_A + T_F) + M\,T_AT_F + T_E(T_A + T_F) + T\cdot K_A\cdot K_F$$

$$b_3 = T_AT_FT_E + TMT_AT_F + TT_E(T_A + T_F)$$

$$b_4 = T\,T_AT_FT_E$$

where

$$M = K_E + S_E$$

APPENDIX C

1. Machine constants for tie line (in per unit on machine base)

System without local load				With local load			
Macine Constants							
x_d	= 1.6	x_Q	= 1.55	x_d	= 1.7	x_Q	= 1.62
x_d'	= 0.32	T_{do}	= 6.0	x_d'	= 0.2	T_{do}	= 3.9 sec
Tie-Line							
r_e	= 0.0	x_e	= 0.4	r_e	= 0.74	x_e	= 0.148
				These are Thevenin equivalent values r_e = 1, x_e = 5.			

2. Excitation systems

Thyristor-Type				Type-1			
K_E	= 50	T_E	= 0.05	K_A	= 50	K_E	= -0.17
				T_A	= 0.05	T_E	= 0.95
				K_F	= 0.04	S_E	= 0.95
				T_F	= 1.0		

3. Loading

Real power (P) = 0.2 . . . 1.0
Reactiver power (Q) = 0.1 . . . 1.0

NOMENCLATURE

All quantities in per unit on machine base.

i_d, i_q armature current, direct and quadrature axis components

e_d, e_q armature voltage, direct and quadrature axis components

e_t	terminal voltage
E_q	voltage proportional to direct axis flux linkages
E_{fd}	generator field voltage (one per unit is the value for 1 per unit terminal voltage on the air gap line, open circuit)
X_e, X_E	equivalent system reactances
R_E	system shunt resistance
S	Laplace operator, or magnitude of stabilizing signal
δ	angle between quadrature axis and infinite bus
$p\delta$	per unit speed deviation from synchronous
T_e	electrical torque
H	inertia constant, sec
M	inertia coefficient = 2H, sec
D	damping coefficient
E	infinite bus voltage

Subscript o means steady-state value.

Prefix Δ indicates small change.

Excitation System Type-1

T_R	regulator input filter time constant
K_A	regulator gain
T_A	regulator amplifier time constant
K_F	regulator stabilizing circuit gain
T_F	regulator stabilizing circuit time constant
K_E	exciter constant related to self-excited field
T_E	exciter time constant
S_E	exciter saturation function

Thyristor Type

K_E	= regulator gain
T_E	= regulator time constant

Loading

P, Q	real and reactive power, respectively
T_s, T_D	synchronizing and damping torques, respectively
K_s, K_D	coefficient of synchronizing and damping torques respectively

ACKNOWLEDGMENT

The authors wish to acknowledge Gannon College for the financial support of this project. Also to Mr. E. P. Paulochik, Pennsylvania Electric Company for providing the necessary data and to Mr. M. Crosby and J. Brazel of Lord Manufacturing Company at Erie, Pennsylvania, for providing analog computer facilities.

Discussion

F. P. deMello (Power Technologies, Inc., Schenectady, N.Y. 12301): I wish to congratulate the authors for their useful contributions in extending a method of analysis presented in Reference 1 and in deriving useful quantitative results which further explore and explain damping characteristics of generating units as affected by loading, system impedances, and excitation system characteristics.

I would just like to state that the mathematically simple excitation system characterized by a single, short time constant chosen in Reference 1 was indeed helpful, by virtue of its simplicity, in permitting analysis and insight into the problem. However, it should not be assumed that this simple system is unrealistic as the authors seem to imply. Thyristor type excitation systems are becoming increasingly popular especially in the case of hydro units.

In addition to extending the method of analysis to the more complex excitation systems, as reported by the authors, it would seem worthwhile to include improvements in representation of the generators, including amortisseur and saturation effects. It is hoped that future contributions will be made to assess these effects.

Manuscript received March 22, 1973.

Mohamed K. El-Sherbiny and **Dinesh M. Mehta** – We wish to express our appreciation for Mr. DeMello's remarks, which added considerably to the value of this paper.

We strongly agree with Mr. DeMello that it is the simplicity of the excitation system which is characterized by a single, short-time constant chosen in Reference 1 which permits the analysis in Reference 1 as well as of this paper. Moreover, he added to the value of this analysis when he mentioned that such thyristor type excitation systems are becoming increasingly popular for the case of hydro units. The main reason that we did not emphasize such a system is because we were trying to remain with one type of excitation system as recommended by the Excitation System Committee shown in Reference 2.

We do emphasize the need for more complete machine representation including the effects of amortisseurs, saturation, etc. However, neglecting the effect of amortisseur windings leads to a conservative design, since the damping contributed by such windings will be positive. For the purpose of including the effect of saturation, it was felt that the saturation curve ($v = f(i_f)$) of the machine should be simulated by a mathematical expression. A sum of exponential curves of the form:

$$V = a_o + \sum_{i=1}^{4} a_i \epsilon^{-\alpha_i i_f}$$

has been found to simulate efficiently the open circuit characteristics from the origin, through the knee of the curve to the saturation region with an accuracy better than 4% over the whole range of the curve. α and a are constants for a given curve as discussed in Reference 9. Work is now in progress to investigate the effect of saturation as simulated by the above equation on the block diagram of the machine, shown in Figure 1.

REFERENCES

9. M. K. El-Sherbiny, "Representation of Magnetizing Characteristics by Sum of Exponentials". I.E.E.E. Transaction of Magnetic, March, 1973.

Manuscript received April 17, 1973.

New Techniques for the Calculation of Dynamic Stability

D. W. OLIVE

***Abstract*—A 100-machine transient stability program has been developed which represents the effects of machines much more completely than has been available previously. Both *d*- and *q*-axis quantities are fully represented and saturation effects in both axes are incorporated. The speed-governing and voltage-regulating systems available are extensive and complete, thus allowing a very close approximation to any physical system. Facilities for nonlinear loads and induction motor transients are also included.**

The theory and assumptions underlying the machine, regulator, and governor representations, and the computational techniques used, are discussed. Emphasis is given to damping corrections and the iterative procedures necessitated by the choice of representation.

The capabilities and limitations of this program are discussed briefly.

Paper approved and recommended by the Transmission and Distribution Committee of the IEEE Power Group for presentation at the PICA Conference, Clearwater, Fla., May 19–21, 1965. Manuscript submitted January 25, 1965; made available for printing August 11, 1965.

The author is with the Westinghouse Electric Corp., East Pittsburgh, Pa.

ONE OF THE FIRST applications of the high-speed digital computer was to the analysis of power system stability. The tedious calculations necessary in any network calculator study made this an ideal type of study for digital methods. In 1954 such a program was developed at the Westinghouse Electric Corporation [1]–[3]. The procedure adopted for computations was the standard classical procedure. In the ensuing years the digital computer has been used more intensively for power system problems with most emphasis being placed on the economic problems associated with the operation and expansion of a power system or of pooled systems. Concurrently, technological advances have greatly increased the complexity of decision making by increasing the number of possible choices and decreasing the incremental costs associated with any choice.

This desire to accurately determine the economics of operation or expansion has again focused attention upon

Reprinted from *IEEE Trans. Power App. Syst.*, July 1966.

system analysis methods. The old procedures are no longer sufficiently accurate to furnish the precise figures necessary for economic and engineering evaluations.

The analysis programs have then evolved into design programs also. They must be used not only to determine whether system operation will be successful but also to differentiate the degree of success possible with alternate plans. It is only in this way that dollar differentials may be assigned. The *optimistic* calculation of ten years ago is inadequate to give these figures because the degree of optimism within the calculations is largely unknown.

It is for this reason that the development of a new digital program for the accurate determination of the stability of large power systems was undertaken.

Philosophy of Development

When the accent is placed upon *accurate*, the first question that immediately arises concerns the degree of accuracy. It is clearly impractical, if not impossible, to include a complete analysis of all the lines, loads, transformers, machines, and other devices. Obviously the passion for perfection in representation and program capability must be weighed against the urge for economy in data collection and in ultimate running cost. Therefore, since the behavior of the machine is of primary importance in any study, it was first decided to concentrate upon the representation of the machine and its regulation and governing systems. The power system was to be represented much as it had been previously—as a collection of constant impedances operating in the steady state over any specified time interval. This then eliminates the inductive and capacitive effects resulting from voltage and current level changes in the network, saturation effects in transformers, and changes in frequency because of the swinging machines. This system representation was later amended to include facilities for nonlinear loads.

The philosophy was to provide a program which would be a reliable yardstick against which the changes made in a small area of the system could be measured. It was fully realized that the system representation itself was not an extremely accurate one and that this would be compounded further when calculations were made using system data that was accurate to, at most, three significant figures. This system, however, will be a reasonably close approximation to the true system and will furnish the yardstick against which changes in a truly accurate representation of the machine or in the region near the machine may be gauged.

This paper presents the treatment accorded to the machine and its associated regulation and governing systems. This includes the equations and the way in which they are used in the solution of the problem. No further justification will be attempted as to the magnitude of the effect of this detail in the representation upon the results of a study. (References [4] and [5] contain information pertaining to this.) The representation possible in this program contains not only the recognized important effects but also effects which have not been considered of sufficient importance to warrant inclusion prior to this time. This result has been achieved with a negligible increase in program running time and no decrease in size of system represented.

Even so, it has been necessary to approximate some effects (such as damping and saturation), to neglect others (such as harmonic phenomena), and to average others (such as damping and certain calculation procedures). This need has been dictated for a variety of reasons.

1) The effect is truly negligible.

2) The effect is extremely difficult to calculate in that more time would be required than is warranted.

3) The data for the calculation or for its ultimate utilization is not available.

4) The restriction of input data to a reasonable amount and to save storage space.

5) The reduction or elimination of the need for predictive and corrective techniques and iteration requirements in calculations.

A New Digital Program

For reference purposes, the capabilities available in this program and discussed in this paper are listed as follows:

1) No restrictions whatsoever are placed upon any machine impedances. Both the direct and quadrature axes impedance and time constants are used. Flux decay and voltage-current relations are possible including subtransient, transient, and steady-state regions.

2) Machine saturation is included.

3 Voltage regulator effects are included.

4) Speed governor effects are included.

5) Induction motor behavior, including rotor transients, is included.

6) Synchronous condensers may be treated conventionally or as zero-inertia, zero-power output devices.

7) Phase shifter representation is possible.

8) Loads may be represented as any combination of constant impedance, constant power, or constant current and power-factor.

9) Machine positive-sequence damping and braking because of stator direct currents and negative sequence currents is included.

10) Systems of 100 generators, 500 busses, and 1000 lines may be studied.

In addition, certain features exist with regard to input calculations and output. This is a partial list of the most salient features.

1) The representation of a machine is completely selective; that is, all features may be included or excluded at will.

2) Automatic program terminations exist for those cases when one synchronous machine exceeds a certain velocity or if two machines differ in angular position by 2π radians.

3) Swing curves may be plotted.

4) Line currents and bus voltages, line flows and imped-

ances seen may be printed for any line in the system. No restriction exists as to whether the line is an off-nominal tap transformer or a pi-section line.

5) No restriction exists on the number of any type of line or load element (within the total allowed). All lines could be pi-lines or off-nominal transformers. Further, negative impedances or loads are completely acceptable.

6) Switching operations may be performed on as many as ten system lines during a study.

7) The effects of small system changes (steady-state stability) may be investigated by ramping or stepping power input into selected machines or changing selected machine angles.

8) Various machine voltages, currents, powers, saturation, and damping factors may be printed at each interval. In addition, voltage regulator input-output information may also be printed.

Network Considerations

The desire to achieve high speed in the computations dictated that matrix procedures be used. At the inception of this project the optimum matrix procedure, if such exists, was by no means apparent. Therefore, several different matrix methods were investigated. The overriding consideration was that of minimum running time, and each method was evaluated on the following points:

1) Ease of initial matrix construction, including possibilities for inputting data or rearranging data to minimize this and later steps

2) Ease of first reductions—matrix or algebraic manipulations to the matrix size that allowed switching operations to be included easily

3) Ease of network solution—that procedure to arrive at the machine outputs (this would involve a final reduction or matrix inversion)

4) Whether the form of the solution appears to be of material aid in the reduction of iterations required or in the calculations to be done

5) Other considerations, such as ease of reconstructing network voltages and currents, storage requirements and tape manipulation.

Of these, the fourth point was the most difficult to evaluate since iterative procedures were uncertain. Yet, it was also the most important, for while the other procedures were done only a few times, the iteration would require obtaining a solution, correcting voltages or currents, and repeating the solution hundreds and possibly thousands of times.

Of the methods investigated the Y formulation was chosen; first, because it appeared to be equal to or superior to the other formulations on a total time basis, and, secondly, because the original stability program used this method and the programming and logic had been extensively tested.

The matrix procedures used will be discussed very briefly in this paper as these procedures have been extensively detailed [2], [3].

The basic matrix manipulation employs a Y matrix. The input data is internally ordered in such a fashion that the original construction yields a matrix of the form

$$\begin{bmatrix} I_G \\ I_F = 0 \\ I_V = 0 \\ I_N = 0 \end{bmatrix} = \begin{bmatrix} Y_{GG} & Y_{GF} & Y_{GV} & Y_{GN} \\ Y_{GF} & Y_{FF} & Y_{FV} & Y_{FN} \\ Y_{GV} & Y_{FV} & Y_{VV} & Y_{VN} \\ Y_{GN} & Y_{FN} & Y_{VN} & Y_{NN} \end{bmatrix} \begin{bmatrix} E_G \\ E_F \\ E_V \\ E_N \end{bmatrix}.$$

The subscripts indicate

G busses to which controlled sources are connected
F busses involved in switching operations
V busses on which voltages are to be measured
N all other busses.

The initial reduction eliminates the last row and column by a series of single row and column substitutions (denoted here in matrix form).

Since $I_N = 0$

$$E_N = -Y_{NN}^{-1}(Y_{GN}E_G + Y_{FN}E_F + Y_{VN}E_V).$$

No information need be saved in this reduction since information concerning these busses is not desired.

The second reduction proceeds exactly as the first but now information necessary to reconstruct the voltages E_V are saved since voltages will be calculated for these busses.

The third and final reduction is made after the desired switching operations have been performed. The procedure is as before both as to the reduction and as to the saving of information.

The resulting matrix equation

$$I_G = Y_{GG}'\ E_G$$

is the direct solution of the network. The classical approach to transient stability uses the assumption that the voltages E_G remain constant over a switching time. The more detailed representation of the machine used in the new program will, in general, not allow this assumption. When saturation and saliency are present no suitable voltage meeting this assumption can be found. It is thus necessary to continuously change these voltages such that they are consistent with known constants or with the constraints imposed because of terminal conditions. This modification of the E_G matrix is the iteration process and will be discussed in that section.

Synchronous Machines

The development of the equations used to represent a synchronous machine proceeds from the representation of the machine as a linear collection of circuits. This representation is shown in Fig. 1. F represents the field winding and winding G is a fictitious winding inserted in the model to account for quadrature axis transient behavior. Windings A, B, and C are the three phases of the armature winding.

The development of the equations follows procedures given previously [6], [7] and has been extensively detailed [8]. The assumptions made in the development are that

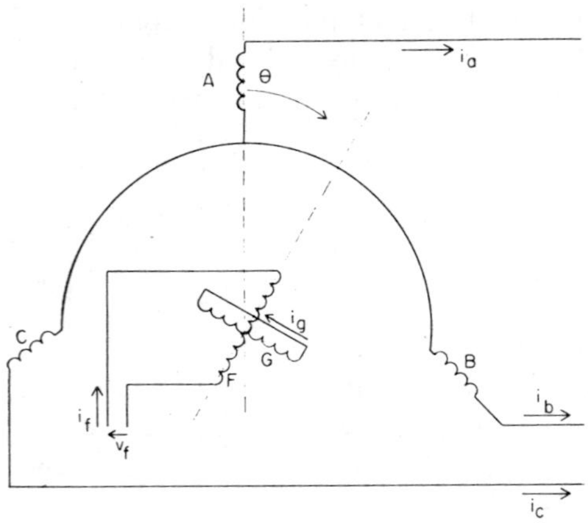

Fig. 1. Schematic diagram of synchronous machine.

1) all inductances are independent of current
2) superposition applies
3) all self and mutual inductances may be represented as constants plus a simple sine variation of the rotor angle Θ
4) distributed windings may be adequately represented as lumped windings
5) higher order space and time harmonics are negligible.

The equations (in per unit) used to describe machine transient behavior are

$$V_d = -R_aI_d + X_qI_q + E_d \tag{1}$$

$$V_q = -R_aI_q - X_dI_d + E_q \tag{2}$$

$$E_q' = -(X_d - X_d')I_d + E_q \tag{3}$$

$$E_d' = (X_q - X_q')I_q + E_d \tag{4}$$

$$dE_q' = \frac{1}{T_{d0}'}(V_f - E_q) \tag{5}$$

$$dE_d' = -\frac{1}{T_{q0}'}E_d. \tag{6}$$

Equations (3)–(6) are exact equations in that they are derived by defining quantities and rewriting true equations. Equations (1) and (2) are derived assuming that the rate of change of the d and q axis stator fluxes are negligible over the interval being calculated. It may be noted that the elimination of any rotor winding will reduce the number of equations by two. The elimination or neglecting of the G winding will eliminate (4) and (6) and set E_d to zero in all other equations.

Machine Saturation

A nonlinear relation between flux and current will, if solved rigorously, eliminate the use of superposition. Since this makes any large scale problem virtually insolvable, the accepted solution is to modify the linear relations so as to retain linearity while approximating saturation effects. A preferred procedure is not known at present for, although many ways have been advanced, the comparisons between methods and with the actual machine have not been made with regard to the transient operation of a machine.

The procedure adopted for use here is based upon a correction of the flux linkages within the machine. It is approximate in that leakage flux has been assumed known for stator circuits only and has been assumed zero for rotor circuits. This tends to indicate a higher degree of rotor saturation than would actually exist. The method does have the advantage that it reduces to the method for saturation correction in the AIEE test code for synchronous machines [9].

Saturation corrections involve three fundamental assumptions. These are

1) that the reluctance of the direct and quadrature axes paths is due only to differences in the length of the air gap in each axis,
2) that the degree of saturation in an axis is proportional to the components of the voltage behind Potier (or leakage) reactance. This saturation level exists for all rotor and stator circuits in that axis, and
3) that the distortion of any air gap flux waves does not change the unsaturated inductance values or destroy the sinusoidal variations assumed for rotor and stator inductances.

The saturation factor at any specified voltage for the direct or quadrature axis S_d or S_q is defined as

$$\frac{\text{Field current to give that voltage on the magnetization curve}}{\text{Field current to give that voltage on the air gap line}} - 1.$$

The direct axis saturation factor may be obtained directly from the machine open circuit curve corresponding to the value of the quadrature axis component of the Potier voltage, or at

$$V_{qp} = V_q + R_aI_q + I_dX_p \tag{7}$$

where

$$S_d = f(V_{qp})$$

and, as a consequence of the first assumption the value of the quadrature axis saturation factor may also be obtained from the same open-circuit curve by multiplying the value found at the quadrature axis component of Potier voltage by the ratio of X_q/X_d.

$$V_{dp} = V_d + R_aId - X_pI_q \tag{8}$$

where

$$S_q = \frac{X_q}{X_d}\cdot f(V_{dp}).$$

It is assumed that the linkages of the d and q stator axes are composed of two parts, one a result of mutual flux and the other leakage flux and that the former is subject to saturation and the latter is not. Then the machine per unit equations are modified to

$$V_q = -R_ai_q - \left[\frac{X_d - X_p}{1 + S_d} + X_p\right]I_d + E_q \tag{9}$$

$$V_d = -R_a I_d + \left[\frac{X_q - X_p}{1 + S_q} + X_p\right] I_q + E_d \quad (10)$$

$$E_q' = -\frac{(X_d - X_d')}{1 + S_d} I_d + E_q \quad (11)$$

$$E_d' = \frac{X_q - X_q'}{1 + S_q} I_q + E_d \quad (12)$$

$$\frac{dE_q'}{dt} = \frac{1}{T_{d0}'}(V_f - (1 + S_d)E_q) \quad (13)$$

$$\frac{dE_d'}{dt} = -\frac{1}{T_{q0}'}(1 + S_q)E_d \quad (14)$$

These equations may be used to draw a phasor diagram representing machine behavior at some specific time. This diagram appears in Fig. 2.

Equations (9)–(14) reduce to (1)–(6) if saturation is neglected; that is, if $S_d = S_q = 0$. From the definitions E_d' and E_q' are proportional to the components of the rotor linkages. As such, by the theorem of constant flux linkages in closed circuits, they may be assumed to remain constant over any switching time. Unfortunately, it is impossible to find a constant voltage behind a constant impedance for use with the matrix. This requires the definition of another voltage for use with the matrix and iterative techniques to insure the correct relationship between the matrix voltage and the above constant voltages. The voltage selected for use is that behind transient direct-axis reactance and projections of this voltage on the d, q axes are then

$$E_{dm} = R_a I_d - X_d' I_q + V_d \quad (15)$$

$$E_{qm} = R_a I_q + X_d' I_d + V_q \quad (16)$$

or in terms of the constant voltages

$$E_{dm} = E_d' + \left[\frac{X_q' - X_p}{1 + S_q} + (X_p - X_d')\right] I_q \quad (17)$$

$$E_{qm} = E_q' - \left[\frac{X_d' - X_p}{1 + S_d} + (X_p - X_d')\right] I_d. \quad (18)$$

If $S_d = S_q = 0$ and $X_d' = X_q'$ the dependency on current vanishes and no iteration is necessary. When iteration is required the equations to be satisfied are (17) and (18).

A similar development may be made when subtransient circuits (damper windings) are included. Because the subtransient period is so short the effect of the subtransient circuits on a conventional stability study is slight. However, since some specialized studies of excitation systems or machines do require this, facilities have been included to represent some of the machines completely, including damper winding effects.

Machine Damping

It is well known that certain extraneous torques will exist within a machine operating in a transient state. The following list contains most torques of interest:

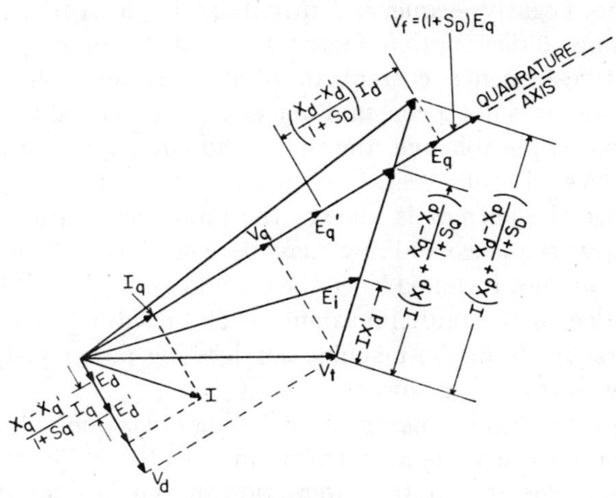

Fig. 2. Phasor diagram—saturated synchronous machine in transient state.

1) Those resulting from fundamental positive-sequence flux changes in the machine

2) Those resulting from the interaction of the damper windings with any negative-sequence fields present (negative-sequence braking)

3) Those resulting from the interaction of the damper windings with the field produced by any armature direct currents (dc braking)

4) Those resulting from induction motor/generator action of dampers and the positive-sequence field (positive-sequence damping)

5) Those resulting from the interaction of rotor circuits with the field produced by the low-frequency currents obtained as all machines slowly swing on the system

6) Those resulting from the higher harmonic fields obtained with unbalanced operation or as a consequence of abrupt system changes.

The first of these has been automatically included in the formulation of the machine equations. All of the others have been eliminated by the assumptions used in formulating the system and the machine. When the stator linkage change was assumed negligible over an interval and when the system was formulated on the basis of positive-sequence impedances, the remaining torques were automatically excluded.

Since the reinsertion of these torques is desirable, approximations have been made and the results entered in the calculations. The general procedure is as follows.

Negative-Sequence Braking: The presence of negative-sequence currents in the stator circuits creates a negative-sequence field in the air gap. The interaction of this field and the damper windings creates a torque which always tends to slow the machine down. The analysis is made based upon induction motor analysis and assuming a known value of negative-sequence current.

The derivation of this component of damping P_{Dn} has been reported in the literature [8]. The equation used is

$$P_{Dn} = I_2^2(R_2 - R_1). \quad (19)$$

The negative-sequence current I_2 is determined by defining a distribution factor as the ratio of generator negative-sequence current to positive-sequence current. This factor is not a constant and so an average value must be used applicable over the time period in which negative-sequence currents exist.

Once the current is known, the power associated with negative-sequence braking may be computed. (Resistive losses are not included since they would already have been included in the fault impedance.) The resulting power is subtracted from the machine accelerating power yielding a new accelerating power.

Direct-Current Braking: A switching operation will produce direct currents in the armature windings. The stator resistive losses and the interaction of the field of these currents and the rotor windings produce a braking torque on the machine. An induction motor equivalent circuit may be used to analyze this situation also [8].

The following equation has been corrected for the resistive losses because of the direct currents also. The braking power attributed to direct currents P_{Dd} is

$$P_{Dd} = \frac{i_{dc}^2}{2}(4R_2 - 3R_1). \tag{20}$$

After the average value of the direct current in the armature windings i_{dc} has been calculated, this equation may be used to find the braking power and to correct the accelerating power of the machine.

Positive-Sequence Damping: As a machine swings on the system, the positive-sequence field in the air gap interacts with the damper windings to produce an asynchronous torque. This swinging produces other positive-sequence currents which should be added to those obtained by the solution of the machine equations previously given (9)–(14). That is, the currents and voltages produced because of this induction motor/generator action are of positive-sequence and should be included in the analysis. It is assumed, however, that the components are small in comparison and so negligible. Since the torque produced is not negligible, the torque is included in the analysis. This torque, or power, is calculated based upon the interaction of this machine to an equivalent machine with the reactance to this machine and the machine voltage being calculated at each step. Since the torque is proportional to the rate of change of the direct-axis angle, a damping factor is calculated rather than a damping power. The damping factor is utilized directly in the solution of the swing equation.

The equation shown has been derived previously under the assumption of one machine to infinite bus operation [8], [10], [11].

The equation is slightly modified from the equation as derived and is given as

$$P_D = \left[\frac{V_{dB}^2 T_{d0}''(X_d' - X_d'')}{(X_d' + X_e)^2} + \frac{V_{qB}^2 T_{q0}''(X_q' - X_q'')}{(X_q' + X_e)^2}\right]\frac{d\delta}{dt} \tag{21}$$

where

P_D = Power associated with damping torque.

For use in the calculations this is represented as

$$P_D = K_D \frac{d\delta}{dt}$$

the factor K_D is utilized directly in the solution of the swing equation.

Synchronous Condensers

One of the difficulties associated with the accurate representation of condensers in a study is that the low inertia constants can produce high-amplitude swings at relatively high frequencies. While such a representation is possible, it may be desirable to choose one of slightly less accuracy which will eliminate this difficulty. This added representation requires considerably less input data also. In this case the condenser is represented as a inertialess, lossless device which must operate at a terminal power-factor of zero. Therefore only d axis impedances are of interest. The iteration procedure varies the position of the direct axis such that the phasor positions of the internal voltage and the terminal voltage are the same. No restriction is placed on automatic voltage regulation capabilities if this alternative representation is used.

Induction Machines

The representation of an induction machine is sometimes desirable in a stability study. This situation may arise if appreciable system loads are known to be induction motor loads and representation accuracy is of importance or if the slip variation of an induction motor load is crucial and must be known.

Many procedures have been advanced for calculating induction motor transients. These range from the use of the steady-state equivalent circuit through approximate equations which should give the asymptotic behavior to exact equations derived from the circuit analysis of the machine. The equations used here may be derived by matrix methods [12], or by a conventional d, q analysis of the induction machine similar to that done for the synchronous machine. As a consequence of balanced rotor and stator circuits the four equations in d, q coordinates may be combined into the following two phasor equations:

$$V = -(R_a + jX')I_s + E' \tag{22}$$

$$\frac{dE'}{dt} = -\frac{1}{T_{d0}}[E' + j(X - X')I_S] - j\omega sE' \tag{23}$$

where

V = terminal voltage
R_a = stator resistance
X' = blocked rotor reactance
T_{d0} = rotor time constant
X = open-circuit reactance
ω = stator electrical angular velocity
s = slip, per unit

E' = rotor voltage proportional to rotor flux linkages
I_s = stator current, out of machine.

Two other equations are necessary

$$T = Re\{E'I_s\} \tag{24}$$

$$\frac{ds}{dt} = \frac{1}{2GH}(T - T_L) \tag{25}$$

where

T = developed torque, synchronous watts
H = inertia constant, MJ/MVA
T_L = load torque, synchronous watts (assumed constant)
G = machine rating, MW.

[The equations may be converted to per unit by elimination of G in (25).]

The major assumption is, as it was for synchronous machines, that of assuming negligible changes in stator flux over a calculation interval.

The literature contains numerous references to the difficulties of calculating induction motor transients by digital means [1], [13], [14]. Although most of the objections could be overcome by using smaller calculation intervals, the reduction is far beyond that necessary for accurate synchronous machine calculations. It is for this reason that the use of the steady-state equivalent circuit has been used extensively or approximate methods used for large-scale system analysis.

This program uses the equations given here and to compensate for the need of small calculation intervals converts the rest of the system into an impedance to a bus, reduces the time interval by a factor of up to 1/40 (depending upon the amount of change in the linkage voltage) and performs a series of calculations to bring the induction motor up to the next system calculation time. Nonetheless, some doubt may be raised concerning the results if the slip ever exceeds 0.10.

Other Representations

Nonlinear loads which may be expressed in the following form may be included

$$P = C_1|V| + C_2$$

$$Q = C_3|V| + C_4.$$

These may be recognized as a combination of constant real and reactive power (C_2,C_4) and of constant current, constant power-factor (C_1,C_3). There is no restriction on the constants—they may be zero, positive, or negative.

This facility has been included even though it may be argued, and quite validly, that little enough is known of load response to a series of steady-state changes let alone to the unspecified changes encountered in a transient disturbance. The capability is available if desired.

Facilities are included for the representation of phase shifters also. These are capable of representing a pure phase shift between two busses.

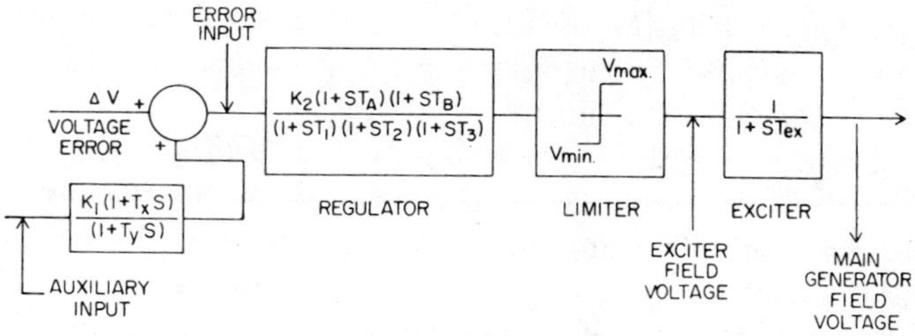

Fig. 3. Block diagram of excitation system.

Excitation Systems

The excitation system has long been recognized as crucial to the operation of a system in the period immediately following a system disturbance. During the course of developing this program it was found to be just as important in that period when the system is slowly swinging back to an equilibrium state. The representation chosen for the voltage regulator is dictated by both cases with the latter case being primarily responsible for the complexity.

The excitation system is shown in the block diagram of Fig. 3. The transfer functions are expressed in terms of the complex frequency, or Laplace transform variable. In all cases the substitution of $p(= d/dt)$ for S may be made if so desired.

The input to the regulator may consist of two signals if so desired. The primary signal is that resulting from an error in the terminal voltage and is equal to reference voltage minus true terminal voltage. This signal is summed with an auxiliary input signal from any one of four sources (at present). These are machine velocity (deviation from synchronous), machine acceleration, rate of change of main machine field current, and rate of change of machine output power.

The auxiliary input capability was inserted for the investigation of the *Stability* type regulators. The choice of signals was dictated by a study made during the program development. The block in the auxiliary input is inserted to control the relative strength of auxiliary and error voltage input, to shift the phase angle of the auxiliary signal for maximum effectiveness and to filter out undesirable signals.

The block representing the voltage regulator itself is complex and yet is about the minimum necessary for an accurate representation of the regulator for slow transients. This block is actually the reduction of a more complicated feedback system into an equivalent block, or transfer function with retention of the more important time constants. Capability exists in this block for a third-order transfer function. Whether all of this capability is used depends on the physical system represented and the data available.

The output of this block is then limited. This limit may represent a physical limitation on the regulator output—as for example saturation in a magnetic amplifier or a machine or it may be a compromise between this and saturation limitations on the exciter.

The next block represents the exciter as a simple delay function. The output of this block is then the field voltage, or more accurately the deviation from reference field voltage of the main machine.

A per unit system is used throughout and all gains in the excitation system are arbitrarily placed in the regulator section. This simplifies insofar as possible the input data preparation.

Several other possibilities exist in the representation of the excitation system. These are of more use in specific excitation system investigations rather than extensive system stability studies and will only be mentioned.

1) Exciter saturation may be included
2) *droop* in the exciter may be included
3) representation more suitable to rectified ac or multified dc exciters may be chosen
4) ceiling voltages may be varied as a function of line currents and terminal voltage
5) a representation for rheostatic (direct-acting) regulators is available.

The determination of the change in main field voltage requires the solution of a relatively complicated system, and this system must be solved so as to create minimum inaccuracies in the result. The assumption of machine and system variables constant over a calculation interval dictates that care must be taken to minimize any further errors.

The calculation procedures used in the solution of the excitation system are therefore chosen to eliminate any errors other than the unavoidable (without extensive predictive or extrapolative techniques) initial input.

Governing System

The speed governing system is shown in block diagram form in Fig. 4. The input is the rate of change of the angle of the machine direct axis (in rad/s). This signal is applied to a block representing the control and/or activator system. This is normally a simple lag network but for hydro systems a lag-lead network may be desirable to compensate for water-hammer effects. The next block is a simple lag network representing the valve action. The output of this block is then limited indicating that no more change can be called for than the energy system is capable of supplying. The last block is the energy system transfer function. The form of the transfer function in this block may be inferred from test data or derived. It is equally valid for steam [15] or hydro systems. For a steam system the parameters may be defined as

T_3 = conventional steam system time constant
T_4 = reheat system time constant
K_2 = 1 − pu steam up stream from reheat.

For nonreheat units the numerator term cancels a denominator term leaving only a simple lag function.

For hydro systems no simple correspondence may be given. The two time constants are sufficient to depict operation and K_2 becomes negative to simulate water-hammer effects.

Those comments concerning the solution of the regulator system apply to the solution of the speed governing system. The same procedures are used here as were discussed at that time although the precautions are perhaps not as important here since a governing system is considerably slower acting than that of a voltage regulator. The input information is, as a consequence, easier to obtain in that it is less critical and errors in specification are not as important. The important constants are those of the energy system block and the gain in the control block. The latter parameter is simply related to the speed regulation of the unit.

Calculation Procedure

The procedures used in performing, or calculating a study may best be shown by referring to the symbolic block diagram of Fig. 5. This is a very simple representation of a program that currently consists of a 12 link chain.

Links 1, 2, and part of 3 are input links. All data is read in, initial conditions are calculated, innumerable checks for consistency and completeness are made, and data is reordered for optimum speed in network reductions. The rest of link 3 constructs the original matrix, converting all constant impedance bus loads to admittance and constructing necessary tables. Link 4 performs the initial matrix reduction. Link 5 completes the matrix reduction, first eliminating busses on which current is desired and then reducing this matrix to only those controlled voltage busses as many times as necessary to represent all system conditions for the study in question. Link 6 calculates time zero quantities, reference quantities, and impedance relations. Link 7 performs all calculations for swing curves and voltage variations. Link 8 is an output link for the machine angle tabulation. Links 9, 10, and 11 are input, calculation, and output for line current and bus voltage information. Link 12 plots the swing curves.

Link 7 is that part of the program of interest here and the procedure will be discussed further. This link is entered at time 0+ and the currents into all controlled busses are determined from the pertinent matrix. An iteration subroutine is now entered which utilizes this current, the matrix voltage and impedance data against the constraints imposed on that bus. If the value found does not agree with the constraints, a change is made in the matrix voltage, and the currents is calculated for the next bus. If agreement is found, or if it is known that iteration is not required, the next bus is checked immediately. The entire list will be rechecked if any bus voltage is modified in passing down the list. When the iteration is completed, the changes in synchronous machine voltages and power inputs are determined and the damping powers in the machine. The new direct axis angle is then computed and the new internal voltages. For the induction machines the new internal voltage and the new slip are computed.

New values of matrix voltages are computed for all sources and after checks as to whether the study is completed or whether a new matrix is necessary, the program either returns or continues.

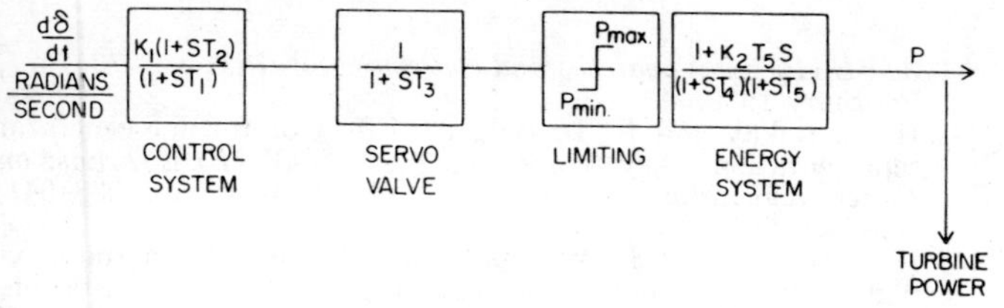

Fig. 4. Block diagram of speed governing system.

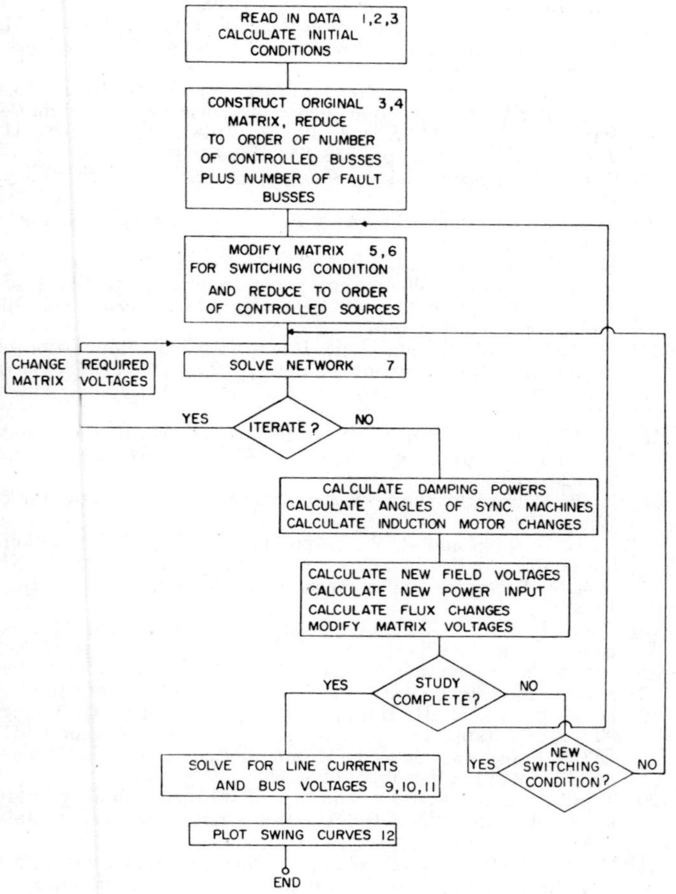

Fig. 5. Symbolic flow diagram—dynamic stability program (numbers in blocks indicate link numbers).

The swing equation solved is

$$P_a = \frac{2H}{\omega}\frac{d^2\delta}{dt^2} + \frac{d\delta}{dt}D. \tag{26}$$

The D constant is a turbine-generator-system damping factor, and it is in fact not a true constant. This factor is determined not only by the viscous damping coefficient of the turbine-generator but also by the initial loading on the generator and the characteristics of the system which dictate how much load the machine will pick up as a function of $d\delta/dt$. Since it can vary widely during the course of a study, an average or pessimistic value must be used.

The equation actually solved is modified from (26) by the addition of the positive-sequence damping factor discussed previously as an additional term of the left side multiplying $d\delta/dt$.

The procedure used for solution is the classical step-by-step method [16]. The small increase in accuracy which might result from more elaborate projective-corrective techniques or more elaborate expansion methods is counterbalanced by more data storage requirements and longer calculation times.

Iteration Procedures

The iteration procedures, for the reasons that have been discussed previously, have been the subject of continuous and rigorous refinement. They are now considered to be very efficient, reliable, and accurate and are sufficiently general for use in any problem encountered. Predictive techniques are not used and maximum effectiveness is made of the natural damping properties of the system.

The iterative procedure passes through one bus at a time. The system is then solved insofar as this bus only is concerned. A check is then made against the constraints imposed for that bus.

1) *Synchronous Machines:* Two voltages are checked. One is derived from the matrix voltage, the other from the flux linkage voltages. The tolerance is 0.5 percent.

2) *Synchronous Condensers:* The angle of internal voltage and terminal voltage must agree to within 0.0004 radian or 0.1 percent, whichever is greater.

3) *Nonlinear Loads:* Apparent power of controlled generator and apparent power based on terminal voltage and the equations must agree within 0.5 percent.

4) *Phase Shifters:* Voltages on the busses are defined according to phase shift. The current components must agree within 0.5 percent (corrected for phase shift).

If the tolerance is not met the matrix voltage is corrected (in the case of phase shifters, both voltages are corrected).

This procedure is so efficient that the program could be almost called noniterative. The number of times through the bus list will seldom exceed two when no disturbances occur and slow swings are the rule, and will seldom exceed four times through for severe switching conditions.

An increase or a decrease in tolerance does not materially affect the number of iterations either. The convergence is extremely rapid and has no overshoot as far as can be determined.

The number of iterations appears to be virtually independent of the size of swing allowed in the interval. This is also a function of the method used to correct the matrix voltages for the start of the next calculation.

The iteration procedure has been remarkably effective in all recent studies. It has not failed even when stability was obviously lost.

Conclusion

The capabilities discussed here and included in the program all give detectable changes in the behavior of a system. Cases may be found where the inclusion of any factor discussed is the difference between instability and stability for the system.

Further investigation appears to be warranted in many areas. A few of these include

1) the area of saturation in machines

2) the use and determination of equivalents in system analysis. Along with the use of regulators and governors as well as more complicated computer dispatch systems, this area requires much work

3) load characteristics. These can be shown to be the difference between stability and instability. Accurate knowledge of these is very desirable.

Nomenclature

V_d, V_q	projections of terminal voltage on machine axes
I_d, I_q	projections of stator current on machine axes
X_d, X_d', X_d''	steady-state, transient, and subtransient direct axis reactance
X_q, X_q', X_q''	steady-state, transient, and subtransient quadrature axis reactance
R_a, R_1	stator (positive-sequence) resistance
R_2	negative-sequence resistance
T_{do}', T_{qo}'	direct and quadrature axes open-circuit transient-time constants
T_{do}'', T_{qo}''	direct and quadrature axes open-circuit subtransient-time constants
E_q	internal machine voltage proportional to field current
E_d	internal machine voltage proportional to quadrature axis rotor current
E_q'	internal machine voltage proportional to field linkages
E_d'	internal machine voltage proportional to quadrature axis linkages
X_p	Potier (leakage)reactance
V_f	field voltage
I_2	negative-sequence stator current
V_{dp}, V_{qp}	projections of the voltage behind Potier reactance on machine axes
E_{dm}, E_{qm}	projections of voltage behind transient direct axis reactance on machine axes
PD_n	negative-sequence decelerating power
V_{db}, V_{qb}	equivalent infinite bus voltage projections on machine axes
X_e	reactance between the machine and an equivalent infinite bus
δ	angular position of direct axis, radians
P_a	accelerating power
H	inertia constant
ω	base frequency.

References

[1] D. F. Shankle, C. M. Murphy, R. W. Long, and E. L. Harder, "Transient stability studies, I—Synchronous and induction machines," *Trans. AIEE (Power Apparatus and Systems)*, vol. 73, pp. 1563–1580, February 1955.

[2] C. M. Lane, R. W. Long, and J. N. Powers, "Transient stability studies, II—Automatic digital computation," *Trans. AIEE (Power Apparatus and Systems)*, vol. 77, pp. 1291–1295, February 1959.

[3] L. J. Rindt, R. W. Long, and R. T. Byerly, "Transient stability studies, III. Improved computational techniques," *Trans. AIEE (Power Apparatus and Systems)*, vol. 78, pp. 1673–1677, February 1960.

[4] H. E. Lokay and R. L. Bolger, "Effect of turbine-generator representation in system stability studies," *IEEE Trans. on Power Apparatus and Systems,* vol. PAS-84, pp. 933–941, October 1965.

[5] H. E. Lokay and J. W. Skooglund, "Power system stability-digital analysis showing effects of generator representation, types of voltage regulators and speed governing systems," presented at the PICA Meeting, Clearwater, Fla., May 19–21, 1965.

[6] W. A. Lewis, *The Principles of Synchronous Machines.* Chicago, Ill.: Illinois Institute of Technology, 1949.

[7] C. Concordia, *Synchronous Machines.* New York: Wiley, 1951.

[8] E. W. Kimbark, *Power System Stability,* vol. 3. New York: Wiley, 1956.

[9] *AIEE Test Code for Synchronous Machines,* no. 503, June 1945.

[10] R. H. Park, "Two reaction theory of synchronous machines—Part I," *Trans. AIEE,* vol. 48, pp. 716–727, July 1929; II, *Trans. AIEE,* vol. 52, pp. 352–354, June 1933.

[11] S. B. Crary, *Power System Stability,* vol. 2. New York: Wiley, 1947.

[12] G. Kron, *Equivalent Circuits of Electric Machinery.* New York: Wiley, 1951.

[13] D. S. Brereton, D. G. Lewis, and C. C. Young, "Representation of induction-motor loads during power system stability studies," *Trans. AIEE (Power Apparatus and Systems),* vol. 76, pp. 451–461, August 1957.

[14] J. L. Gabbard, Jr., and J. E. Rowe, "Digital computation of induction motor transient stability," *Trans. AIEE (Power Apparatus and Systems),* vol. 76, pp. 970–977, December 1957.

[15] C. Concordia, "Effect of steam-turbine reheat on speed-governor performance," *Trans. ASME,* vol. 81, ser. A, pp. 201–207, 1959.

[16] E. W. Kimbark, *Power System Stability,* vol. 1, New York: Wiley, 1948.

[17] D. B. Breedon and R. W. Ferguson, "Fundamental equations for analog studies of synchronous machines," *Trans. AIEE (Power Apparatus and Systems),* vol. 75, pp. 297–306, June 1956.

[18] M. S. Dyrkacz, C. C. Young, and F. J. Maginniss, "Digital transient stability program including the effects of regulator, exciter, and governor response," *Trans. AIEE (Power Apparatus and Systems),* vol. 79, pp. 1245–1254, February 1961.

[19] H. E. Brown, H. H. Happ, C. E. Person, and C. C. Young, "Transient stability solution by an impedance matrix method," *IEEE Trans. on Power Apparatus and Systems,* vol. PAS-84, pp. 1204–1214, December 1965.

[20] T. Shibataki, "Digital calculation of transient stability study in power systems," *Mitsubiski Denki Lab Repts,* vol. 5, July 1964.

[21] G. Shadeshift, "Digital solution of power-system steady state and transient stability," presented at the Conference on Digital Computation for Electrical Power Systems, Queen Mary College, London, England, September 16–20, 1963.

[22] R. B. Shipley, D. W. Coleman, C. F. Watts, and N. Sato, "Direct calculation of power system stability using the impedance matrix," this issue, pp. 777–782.

Discussion

Victor A. Kinitsky (Drexel Institute of Technology, Philadelphia, Pa.): This paper deals effectively with the difficult problem of a correct representation of various elements of a power system by calculating the dynamic stability. The author should be complimented for his achievement. I would like to submit two comments to clarify the assumptions made.

For the representation of a synchronous machine it is assumed in the paper that all self- and mutual-inductances may be represented as constants plus a simple sine variation of the rotor angle θ. Actually, the self- and mutual-inductances of the armature

Manuscript received June 3, 1965.

winding consist of a constant term and a second harmonic variation of the rotor angle. The self-inductances of rotor windings have only constant terms, and the mutual-inductances between the rotor and stator windings consist of variable terms only, which variate with the cosine of the rotor angle.

Further, it is assumed in the paper that the leakage flux of a synchronous machine is not subject to saturation. The leakage flux consists of the slot, end-winding, tooth-top, and differential leakage. Of these components the end-winding leakage flux is passing through the air only, but the other three components of the leakage flux have their magnetic path partially through the iron, therefore, they are influenced by the iron saturation.

D. W. Olive: The comments of Dr. Kinitsky serve to clarify some of the assumptions. These omissions, made in the interest of brevity, are warranted in the light of references to the original paper.

Manuscript received November 15, 1965.

The inductance variations are as given by Dr. Kinitsky or as in Lewis [6] and Concordia [7]. It might be added that physical variations are actually infinite series but that all terms involving angular variations of more than twice the rotor angle are negligible in determining fundamental frequency behavior.

The assumption of a leakage flux not subject to saturation is questionable for, as Dr. Kinitsky has said, components of the leakage flux do have iron paths. This may be reasonably justified in that the large airgaps involved are effective in preserving a near linear relationship between current and flux regardless of the degree of saturation of the iron portion of the path.

I reiterate that the area of saturation requires considerable investigation since saturation has been shown to be important in both machine and excitation system action. The procedure outlined is distinguished chiefly by its simplicity and yet should yield results suitable for a first order correction. A more precise representation will be more complex as to data requirements and calculation procedures.

DYNAMIC SIMULATION OF INTERCONNECTED SYSTEMS

D. G. Ramey
Advanced Development Engineer
Electric Utility Engineering
Westinghouse Electric Corporation
East Pittsburgh, Pennsylvania

R. T. Byerly
Advanced Development Engineer
Electric Utility Engineering
Westinghouse Electric Corporation
East Pittsburgh, Pennsylvania

INTRODUCTION

Stability studies involving interconnected companies have become common. It is also common to simulate transient behavior for several cycles of the dominant modes of oscillation, and a number of real system studies have indicated loss of synchronism after the system had survived the first swing. Computer programs to conduct this type of study employ dynamic mathematical models of prime movers, governors, generators, and voltage regulators. Experience with the application of such a program to large system stability studies over the past two years provides the basis for the extended computer capabilities discussed in this paper and provided in a new program. The following paragraphs set forth briefly the most significant of these new capabilities.

First, there is the need to handle larger systems. Many studies already press present programs to the limits, and the tendency is to study even larger interconnections. Increased program capacity does not eliminate the need or desirability for network and generating plant equivalents, but it does place fewer restrictions on the exercise of judgment with regard to such equivalents.

It is also very desirable to simulate protective relaying for the transmission network, including the automatic opening and reclosing of breakers. The same end result can be achieved by successive computer runs, repeating calculations where analysis indicates switching should have occurred. This is expensive, especially in terms of engineering analysis, and can require considerable elapsed time where cascaded line outages are simulated.

A third requirement of a new program is to provide for underfrequency and undervoltage load shedding studies. For this type of study, it seems desirable to estimate individual bus frequencies, since severe oscillations may result in transient frequency deviations differing in magnitude and phase across the system. With individual bus transient frequency deviations computed, there is also the possibility of modifying network parameters and loads as functions of these frequencies. The use of individual bus frequencies provides a better basis for representing load damping than the use of rotor speeds.

A number of studies have been made reporting the use of special damping signals fed into governors or voltage regulators.[1, 2, 3, 4, 5] Some utility companies now predict the widespread use of such methods for improving stability. Thus there is the necessity to provide for the introduction of any of several signals into any machine governor or voltage regulator.

Finally, any new stability program should anticipate the evolution of this art and be organized so that it can be adapted as quickly as possible to changing demands.

GENERAL DESCRIPTION

This program is dimensioned for 1000 real system buses and 1500 real system lines. It permits up to 500 synchronous and induction machines which may be grouped in any way on the real system buses. Each of the synchronous machines may have four rotor circuits, two on the direct axis and two on the quadrature axis. Each synchronous machine may also have an automatic voltage excitation system, a governor and prime mover system, and may have auxiliary signals input to either the voltage regulator or governor.

The real and reactive power variations with voltage magnitude at any bus may be specified as a general quadratic function. Frequency deviation is computed at each system bus, and real and reactive power may be varied as a function of that frequency. Load shedding on either underfrequency or undervoltage is permitted at 200 buses.

Any line may be a simple series line, a pi-line, or a transformer. Impedance or overcurrent relays may be used for automatic breaker operations at 200 locations.

The user may specify up to 30 system conditions involving faults and associated breaker operations. Up to 200 additional switching operations may occur automatically.

GENERATING PLANT REPRESENTATION

Synchronous Machines

A maximum of 500 synchronous machines may be simulated by the program with no limitations on the number of machines that can have subtransient circuits or saliency. To facilitate the preparation of data for such a large number of machines, the user has only to input the machine constants, and the program automatically selects the machine representation to be used. The basic machine representation is the standard two-axis model which is derived in

Reprinted with permission from *PICA Conf. Proc.*, 1967.

various references.[6, 7] Six variations of this model, depending on the time constants and reactances which are read in, are recognized by the program. They are:

1. Transient and subtransient circuits on both axes

2. Transient and subtransient circuits on the direct axis; either a transient or subtransient circuit on the quadrature axis

3. Transient circuits on both axes

4. Transient and subtransient circuits on the direct axis; constant flux linkage on the quadrature axis

5. Transient circuit on the direct axis; constant flux linkage on the quadrature axis

6. Classical; constant voltage behind transient reactance

The program selects the most complete representation possible based on input data and checks the consistency of the data for that representation. Any extra reactances will be ignored.

The problem of determining an equivalent voltage behind a fixed impedance for load balance when salient machines are represented has received considerable attention.[10, 11] This problem can be avoided by representing the internal-terminal branch by a non-symmetrical 2 x 2 matrix as shown in Appendix I. The advantage of this representation is that the internal voltage can be the voltage which is proportional to the flux linkages on each axis and can be assumed to remain constant for the solution of the load balance equations.

Saturation effects in synchronous machines during transient conditions are quite complex if treated precisely. Fortunately, various studies have indicated that precise treatment is not warranted in stability studies. However, it does seem desirable to approximate the effects of saturation. Accordingly, machine saturation is represented in this program by changing the air gap flux linkage on the direct and quadrature axes. This change is such that:

$$\gamma'_{ad} = \gamma_{ad}/(1. + s_d)$$

$$\gamma'_{aq} = \gamma_{aq}/(1. + s_q)$$

where

$$S_d = A\gamma_{ad}^{\ B}$$

$$S_q = A\gamma_{aq}^{\ B}$$

and the primed variables are saturated values.

In these equations the constants A and B are determined from the saturation factors at 1.0 and 1.2 per unit voltage. These factors are defined as:

$$\frac{\text{Field current on the saturation curve}}{\text{Field current on the air gap line}} - 1$$

The final assumption that is made is that saturation is a slowly varying effect and that the equations need be modified only at the end of a calculation interval for this effect rather than iteratively as the other dynamic equations are being solved. All dynamic equations are solved by established numerical integration techniques. There are many references available for methods of numerical integration. The authors found reference 8 to be particularly useful.

Voltage Regulator and Exciter Representation

Modern excitation systems are manufactured from a variety of equipment in many different arrangements. Fortunately, almost all of these are equivalent to one of several general mathematical models. As a result of the efforts of a Working Group of IEEE*, a report will soon be available which presents general mathematical models, matches the products of several manufacturers to those general models, and recommends typical constants for stability calculations. Three of these general models are described here and are in agreement with the preliminary considerations of the Working Group. Another model is for a static exciter combining both voltage and current sources which will be programmed after the Working Group publishes its final report. If their final recommendations relative to the three general block diagrams presented here differ significantly, programmed models will be altered where possible.

Prior to the presentation of the general mathematical models for excitation systems, two computational restrictions will be discussed. These restrictions apply not only to excitation systems, but to governors, prime movers, and auxiliary signal transfer functions as well.

The first restriction is a limitation on the minimum time constant, either pole or zero, whose effect will be calculated. This has been set at .02 seconds. Smaller time constants require smaller integration step sizes, and their effects do not significantly alter results. Where smaller time constants are entered as data, they will either be set to zero, or a diagnostic will be printed and calculation halted after the initialization portion of the program.

*The working group on computer representation of excitation systems of the Excitation Systems Subcommittee of the Power Generation Committee will tentatively publish its report at the Summer Meeting this year.

The second restriction is one of physical realizability. Transfer functions which are physically realizable lead to differential equations which can be solved by various widely used numerical integration procedures. It should be noted that physical realizability checks on transfer functions are made after very small time constants have been set to zero. Failure to pass these checks halts the calculation after initialization with suitable diagnostic messages.

The block diagram of Figure 1 corresponds to what will be called a type 1 excitation system. It is intended for all rotating exciter systems in which damping is derived after the exciter delay. Exciters without a shunt-connected field are obtained by setting K_E to 1. An exciter with a shunt-connected field has K_E set to a negative constant. A variety of models is actually available from this one general model because the program permits T_R, T_A, K_E, and K_F to be set to zero. Saturation will be ignored in the calculations if no saturation data is entered. The time constant T_E must be greater than .02 seconds, and T_F must be greater than .02 seconds if K_F is not zero.

The block diagram of Figure 2 depicts a type 2 excitation system. It is intended for rotating exciter systems with damping feedback from the regulator terminals. Again, the program permits T_R, T_A, and K_E to be zero, and saturation will be ignored in the absence of input data. However, damping feedback is always assumed to be present, and T_{F1} and T_{F2} must be greater than .02 seconds. The exciter time constant, T_E, must also be greater than .02 seconds.

Rheostatic control of d-c exciters is represented by the block diagram of Figure 3.

Governor and Prime Mover Representation

All single shaft governor and prime mover characteristics are obtained from the block diagram of Figure 4. Any of the poles or zeros of the gate or valve servo or of the steam or water supply system may be zero, and as already mentioned, will be set to zero by the program if they are less than .02 seconds. Because of the limiter between servo and energy system transfer functions, each of them must be realizable. If not, calculation is terminated with appropriate diagnostic messages. Cross-compound governor representation is given by the block diagram of Figure 5.

Auxiliary Inputs to Voltage Regulators and Governors

Entry points for auxiliary signals to voltage regulators and governors are shown on all their block diagrams. Auxiliary inputs are computed using the block diagram of Figure 6. The previously stated band width restrictions still apply and realizability is checked in the initializing programs. Input to the auxiliary transfer function is any of several computed variables. Presently programmed inputs are:

1. Machine slip
2. Accelerating power
3. Electrical power
4. Terminal voltage
5. Bus frequency deviation

LOAD REPRESENTATION

In general, loads are sensitive to voltage magnitude and frequency, whether they are assumed to be motors or general loads. Voltage magnitudes at buses are computed as transients, and sensitivity to voltage is normally considered in stability studies at present. Where sensitivity to frequency is taken into account, it is usually done on the basis of an average of generator rotor speeds or on the basis of some nearby generator rotor speed.

For this program, load shedding studies based on either voltage magnitude or frequency deviations were anticipated, and the calculation of an approximate transient frequency at each bus seemed very desirable. This stems from the fact that magnitudes and time phases of frequency deviations are not the same at all buses any more than are generator rotor speeds. The latter are, of course, integrals of rotor accelerations, and there is no question of their availability or reliability within the general assumptions pertaining to stability studies. Bus frequencies are not so readily available. Rotor speed deviation from normal speed could be obtained at least approximately by differentiating rotor phase angles, and this suggests that approximate bus frequencies can be obtained by differentiating bus phase angle functions. Some discussion supporting this notion is presented in Appendix II. Carrying out this calculation in the program requires saving past values of bus angles and using some approximate differentiating formula. In this specific program, two past angles are saved so that frequency deviations are based on three angle values.

Unless otherwise specified, loads are assumed to be represented by constant impedances. At any bus, however, the load may be represented as

$$P = a + b|v| + c/v/^2$$

$$Q = d + e|v| + f/v/^2$$

by reading in the data required to separate the load into these components. This requires unique data for each bus so represented.

Loads may also be made sensitive to bus frequency deviation.

$$P' = P(1 + k_1 p\theta)$$

$$Q' = Q(1 + k_2 p\theta)$$

where ($p\theta$) is the estimated bus frequency deviation and k_1 and k_2 are two data constants applied at all buses.

At up to 200 buses, abrupt changes in load are permitted either on the basis of frequency or voltage. This requires reading in the percentage changes in P and Q and the voltage level or frequency deviation at which the change is to occur.

Motor loads may be specified at any bus by supplying machine data. Synchronous motors are represented in just as much detail as generators, including excitation systems and auxiliary inputs to the voltage regulator. Two models for induction motors are programmed, one representing only mechanical differential equations, and the other representing rotor transients as well. Load buses may also be d-c line terminals. A limit of three two-ended d-c lines is presently programmed.

AUTOMATIC LINE SWITCHING

The stability program which the authors have used over the past two years[10] required the user to provide input data for all the fault and switching conditions for a study. A considerable amount of difficulty has been experienced in predicting which lines in the system should be switched as a result of power swings following a disturbance and at what times these lines should be switched. To help overcome this difficulty, this program provides for relaying functions at up to 200 locations in the system which will simulate either impedance or overcurrent type relays. These relaying functions are intended for use primarily during swings when the positive-sequence simulation is assumed to represent real network conditions, so the switching associated with fault conditions will continue to be input data to the program. Provision is also made to include some carrier functions and out-of-step blocking functions by making decisions based on the simulation of two or three relays. A description of the models used will help clarify these applications.

Impedance Relay Model

The impedance relay is characterized by three circles in the R-X plane as shown in Figure 7. Each circle represents a zone of protection and is represented by a radius and center. At each time interval the program calculates the apparent impedance of the line from its positive-sequence voltage and current. When this impedance enters one of the zones, a timer picks up and records the amount of time spent in that zone. If this timer reaches the preset maximum time for the zone, a breaker operation is initiated, and the breaker is tripped after enough additional time has passed to account for the time required for breaker operation. There is also an option for the user to specify the amount of time from the line opening until the breaker is reclosed. If the impedance moves out of any zone before the trip time has elapsed, the program assumes an instantaneous reset. An additional option allows the program to simulate directional elements by specifying that a portion of the R-X plane, such as the portion to the left of line A-B in Figure 7 is outside the impedance zones. There are no limitations on the location of impedance circle centers, and drop-out may be specified as a fraction of pick up.

Overcurrent Relay Model

The basic overcurrent relaying function is illustrated by Figure 8. Here the time axis represents the time required for the relay contacts to close for various values of current. The technique used is to check the value of line current on the monitored line at each calculation interval, find the corresponding trip time from the relay characteristic, and accumulate relay travel by the following formula.

$$X_{n+1} = X_n + \frac{\Delta t}{T_n}$$

Here X_n represents the fraction of the closing distance traveled by the nth time interval, T_n is the total time required to close the contacts with the nth current value, and Δt is the time interval during which the current is assumed constant. If X_n reaches 1.0 the contacts have closed and breaker action is initiated. When the current falls below the pick-up value, a negative value is assigned to T_n which is independent of current, and the relay begins to reset at a linear rate.

The input information required for the overcurrent relay will be relay type, time dial setting, current transformer ratio and tap setting, and nominal bus voltage. The program then makes use of routines developed for a relay coordination program[9] to calculate the time current curve of Figure 8. Additional information may be input to describe directional elements if they are desired. In addition, a positive-sequence current multiplier may be input to approximate the actual currents seen by the relay during asymmetrical faults.

Carrier Functions and Out-of-Step Blocking

Logic has been included which will allow the combination of two or three impedance type relays to control a breaker operation. These relays may be at the same location or they may be on different lines. Carrier functions may be simulated by making the breaker operation dependent on the apparent impedance being in a given zone at all the controlling locations. Out-of-step blocking may be performed by simulating two relays at the same location and making the breaker operation dependent on the length of time the impedance has been in a given zone for one of the relays before it enters the corresponding zone for the other.

DIGITAL COMPUTER UTILIZATION

The general requirements for this program present several difficulties in computer usage. Most apparent are the difficulties associated with system size. Even though the program is intended to operate on computers with 64, 000 words of central memory or more, there is too much data to consider storing it entirely in central memory at one time when the maximum size capability is employed. Hence, memory priorities need to be established. The highest priority for central memory allocation is given to the data which must be used in iterative calculations: specifically, the data pertinent to obtaining bus voltages and certain data which must be modified each calculation interval. Central memory which is not required for data used in iterative type calculations is available for synchronous machine data, regulator or governor data, relay data, or other data for dynamic devices. Small problems will thus fit entirely in central memory. When the available central memory is exhausted the remaining data is placed on back-up storage in blocked record form. The decision to employ back-up memory is under program control. The block records are designed so that the central processor time that is required to update the record each calculation interval is as nearly equal to the time required to read it into central memory as possible. With the buffered input-output capabilities of present digital computers this minimizes the increase in time associated with the use of back-up storage.

The amount of time required for the solution of a stability problem is inversely proportional to the time interval between load balances and directly proportional number of iterations required to achieve a load balance as well as system size. The program minimizes the number of iterations required for a load balance by making use of the fact that in the absence of switching, the real and imaginary components of voltage at load buses do not change rapidly. If a history of the voltages at a bus is kept, a simple extrapolation technique may be used to predict the next voltage. This prediction is then used as a starting point for the load balance equations. The accuracy of the prediction depends on the amount of previous data that can be stored so there is a trade-off between computer memory requirements and reduction in iteration time. The authors' present experience indicates that two previous values of voltage are sufficient to achieve a significant reduction in the number of iterations required for load balance.

The relatively slow change in bus voltages indicates that load balances do not need to be calculated with as small a step size as that required for accurate representation of dynamic devices with shorter time constants. The program uses separate step sizes for load balance calculations and numerical integration. For the interval between load balances, the system beyond the terminal bus of a dynamic device may be represented either by a characteristic impedance which is assumed constant over the interval or by a varying voltage which can be extrapolated from the previous history. Input data to the program includes the desired time interval for load balances and the program automatically calculates the time interval for each dynamic device to be small relative to the smallest time constant for the device. All of this is an attempt to account for relatively rapidly changing variables, such as currents in damper windings, without always balancing the system buses at extremely small time intervals. Some experience will be required to see how effective this may be.

On large programs of this type, the organization of information so that the input data is reasonable and the output can be interpreted is essential. This program will accept system load flow information from either input cards or magnetic tape and it provides for three restart options that permit additional stability problems to be run from a load flow with a minimum of additional input data. These options are:

A. Continue previous simulation with possible changes in the prescribed line switching and fault conditions.

B. Rerun previous simulation with same machines and initialization but different prescribed faults and switching conditions.

C. Change the representation of selected generating stations, and loads, reinitialize and rerun the previous simulation (switching conditions and faults may also be changed with this option).

All restart conditions require only the additional input data to make the desired changes in the system.

Output from the program can be obtained for any number of the machines, buses, or lines in the system and consists of:

A. Tabulated machine variables: includes turbine power, electrical power, accelerating power, velocity deviation, rotor angle, field voltage, field current, terminal voltage magnitude and angle, and generator current and angle. These are summarized over the entire transient time for a given machine. All variables are available for plotting. Standard plot will include accelerating power and angle.

B. Swing curves: a maximum of five machines per plot with any machine in the system as reference. A tabulation of the reference machine angles precedes the plot.

C. Tabulation of bus voltages: the magnitude and angle of voltage for any bus in the system.

D. Summary of line currents, power flows, and apparent impedances on any line: these line flows are also tabulated separately for each line.

Other information describing the system initialization, the fault conditions, prescribed breaker operations, and automatic relay operations is also printed.

CONCLUSION

The development of this stability program provides a means of studying large interconnected power systems and an analytical tool for investigating the adequacy of system representation as well. It is designed to provide as much detail for representation as present knowledge justifies, and is adaptable so that mathematical models of present devices may be replaced with improved models, or new devices may be added.

REFERENCES

1. Dynamic Stability of the Peace River Transmission System, H. M. Ellis, J. E. Hardy, A. L. Blythe, J. W. Skooglund. IEEE Transactions on Power Apparatus and Systems, vol. PAS-85, no. 6, June 1966, pp. 586-600.

2. Damping for the Northwest-Southwest Tieline Oscillations--An Analog Study, F. R. Schleif, J. H. White. IEEE Transactions on Power Apparatus and Systems, paper no. 31 TP 66-13.

3. Damping of System Oscillations with a Hydro-generating Unit, F. R. Schleif, G. E. Martin, R. R. Angell. IEEE Transactions on Power Apparatus and Systems, paper no. 31 TP 66-12.

4. Excitation Control to Improve Powerline Stability, F. R. Schleif, H. D. Hunkins, G. E. Martin, E. E. Hattan. IEEE paper no. 31 PP 67-51.

5. Effect of High-Speed Rectifier Excitation Systems on Generator Stability Limits, P. L. Dandeno, A. N. Karas, K. R. McClymont, W. Watson. IEEE paper no. 31 PP 67-20.

6. Electrical Transmission and Distribution Reference Book, Westinghouse Electric Corporation, Pittsburgh, Pennsylvania, fourth edition, 1950.

7. Electromechanical Energy Conversion, D. C. White, H. H. Woodson. John Wiley & Sons, Inc., New York, New York, 1959.

8. Linear Differential Operators, Cornelius Lanczos. Van Nostrand, 1961.

9. Digital Computer Protective Device Coordination Program I--General Program Description, R. E. Albrecht, M. J. Nisja, W. E. Feero, G. D. Rockefeller, C. L. Wagner. IEEE Transactions on Power Apparatus and Systems, vol. 83, no. 4, April 1964, pp. 402-410.

10. New Techniques for the Calculation of Dynamic Stability, D. W. Olive. IEEE Transactions on Power Apparatus and Systems, vol. PAS-85, no. 7, July 1966, pp. 767-777.

11. Direct Calculation of Power System Stability Using the Impedance Matrix, R. B. Shipley, N. Sato, D. W. Coleman, C. F. Watts. IEEE Transactions on Power Apparatus and Systems, vol. PAS-85, no. 7, July 1966, pp. 777-782.

APPENDIX I

SALIENCY BY DIRECT CALCULATION

The basic approach to the calculation of salient branches between the internal-terminal buses of a machine is the generalized transformation which transforms a vector from one set of coordinates to another co-planar set of coordinates. This transformation is illustrated by Figure 9 which shows two sets of co-planar coordinates, the R-I coordinates, and the D-Q coordinates, separated by angle δ, and a vector $\overline{A}$. The transformation of $\overline{A}$ from the D-Q coordinates to the R-I coordinates is:

$$\begin{bmatrix} A_R \\ A_I \end{bmatrix} = \begin{bmatrix} \cos\delta & \sin\delta \\ \sin\delta & -\cos\delta \end{bmatrix} \begin{bmatrix} A_Q \\ A_D \end{bmatrix}$$

The internal voltage and terminal voltage of a machine are related by the equations:

$$E_Q = V_Q + I_Q R + I_D X_D$$

$$E_D = V_D + I_D R - I_Q X_Q$$

where R is the armature resistance and X_D and X_Q are the appropriate armature reactances.

Then:

$$\begin{bmatrix} E_Q - V_Q \\ E_D - V_D \end{bmatrix} = \begin{bmatrix} R & X_D \\ -X_Q & R \end{bmatrix} \begin{bmatrix} I_Q \\ I_D \end{bmatrix}$$

Transforming these equations to the real and imaginary axis gives:

$$\begin{bmatrix} \cos\delta & \sin\delta \\ \cos\delta & -\cos\delta \end{bmatrix} \begin{bmatrix} E_R - V_R \\ E_I - V_I \end{bmatrix} = \begin{bmatrix} R & X_D \\ -X_Q & R \end{bmatrix} \begin{bmatrix} \cos\delta & \sin\delta \\ \sin\delta & -\cos\delta \end{bmatrix} \begin{bmatrix} I_R \\ I_I \end{bmatrix}$$

This equation may be reduced to

$$\begin{bmatrix} G_1 & B_1 \\ B_2 & G_2 \end{bmatrix} \begin{bmatrix} E_R - V_R \\ E_I - V_I \end{bmatrix} = \begin{bmatrix} I_R \\ I_I \end{bmatrix}$$

where:

$$G_1 = \frac{R + (X_Q - X_D)\sin\delta\cos\delta}{R^2 + X_D X_Q}$$

$$G_2 = \frac{R + (X_D - X_Q)\sin\delta\cos\delta}{R^2 + X_D X_Q}$$

$$B_1 = \frac{X_Q \sin^2 + X_D \cos^2}{R^2 + X_D X_Q}$$

$$B_2 = -\frac{X_Q \cos^2 + X_D \cos^2}{R^2 + X_D X_Q}$$

The assumptions which are made are the usual assumptions that the angle δ and the internal voltage components E_R and E_I do not change instantaneously so they are assumed to be constant for the load balance equations.

APPENDIX II

FREQUENCY VARIATIONS

Frequency-sensitive effects appear to have an increasingly important place in stability calculations. For example, future studies will almost certainly deal with underfrequency load shedding. Generator rotor speeds are not all the same during transient conditions, and this results in internal generated voltages whose frequencies are not all the same. Bus voltages will intuitively have different frequencies as well. The general way in which different bus frequencies come to exist may be illustrated by a simple situation. For the voltage divider of Figure 10, let

$$e_1 = E_1 \sin\omega t \qquad \text{(II-1)}$$

$$e_2 = E_2 \sin(\omega + \epsilon)t \qquad \text{(II-2)}$$

$$e_3 = ne_1 + (1 - n)e_2 \qquad \text{(II-3)}$$

where n is a fraction representing the tap setting. If n = 1, the frequency observed at the tap is ω. If n = 0, the frequency at the tap is $(\omega + \epsilon)$. For intermediate tap settings, the waveform is not sinusoidal, but the period can be determined. To have a specific example, let $E_1 = E_2 = 2$ and

$n = \frac{1}{2}$.

$$\text{Then } e_3 = \sin\omega t + \sin(\omega + \epsilon)t \qquad \text{(II-4)}$$

The zeros of e_3 require that

$$\sin(\omega t + \epsilon t) = -\sin(\omega t) \qquad \text{(II-5)}$$

which is satisfied if

$$\omega t + \epsilon t = 2k\pi + \alpha \qquad \text{(II-6)}$$

and

$$\omega t = 2k\pi - \alpha \qquad \text{(II-7)}$$

Adding (II-6) and (II-7) and dividing by 2,

$$(\omega + \epsilon/2)t = 2k\pi \qquad \text{(II-8)}$$

Equation (II-8) gives the alternate values of t which cause e_3 to be zero, and implies a frequency

$$\omega' = \omega + \epsilon/2$$

at a tap half way between the two ends of the divider.

This simple situation may be somewhat more loosely described by a sequence of phasor diagrams which imply that e_1 and e_2 have the same frequency, strictly speaking, but show ever-increasing phase displacement.

The difference in frequencies can be computed by dividing incremental phase displacement by incremental time. Now if the tap voltage is constructed on each phasor diagram, ignoring the logical difficulties which stem from the fact that it is not even sinusoidal, then its apparent frequency difference is obtained by dividing incremental angles by incremental times.

This notion is extended to stability calculations, where time sequences of phasor diagrams for the entire system are being calculated. Frequency deviations from normal system frequency are implied, and are computed by dividing angle increments by time increments.

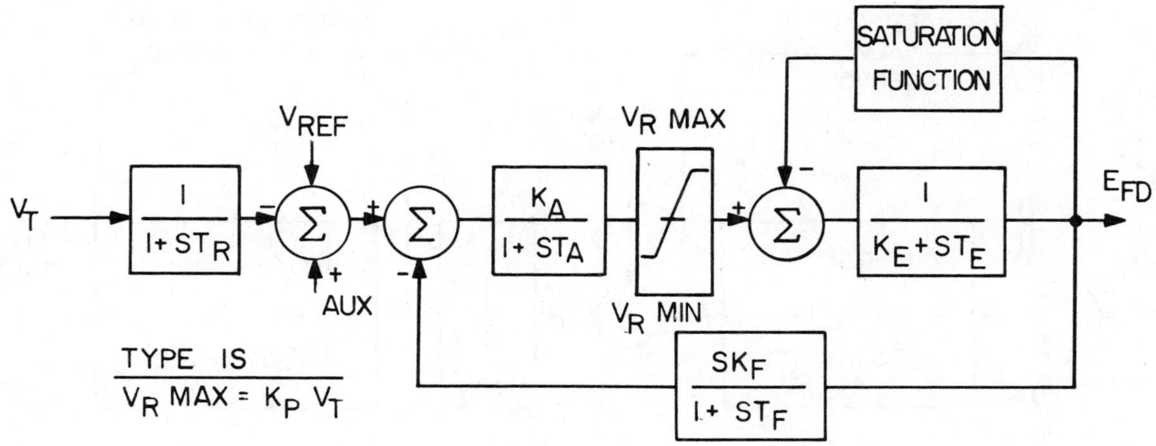

FIG. 1 TYPE 1 EXCITATION SYSTEM

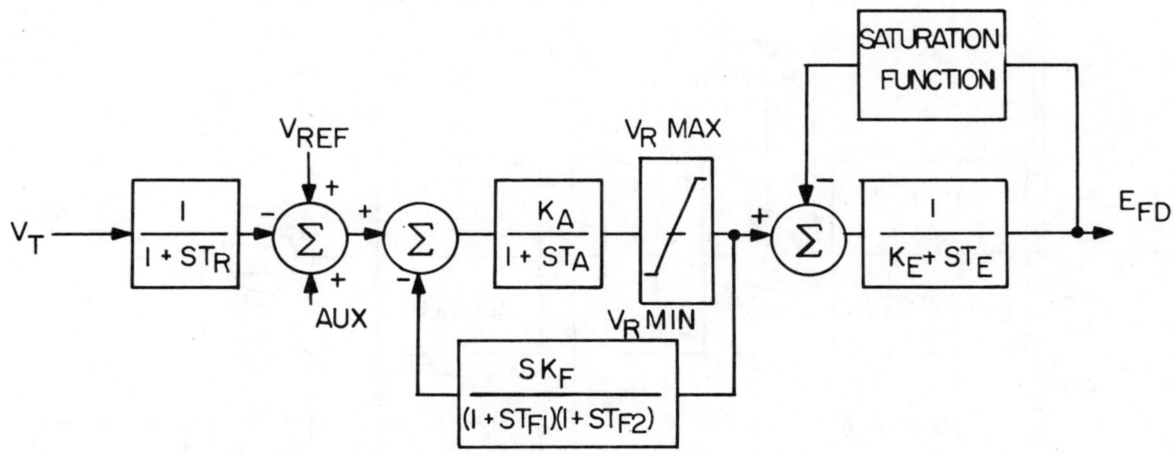

FIG. 2 TYPE 2 EXCITATION SYSTEM

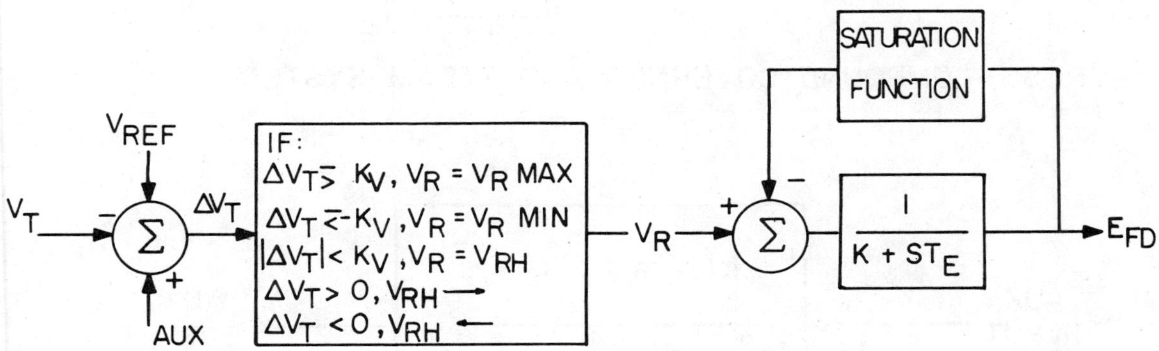

NOTE:

V_{RH} LIMITED BETWEEN V_R MIN AND V_R MAX

TIME CONSTANT RHEOSTAT TRAVEL = TRH

FIG. 3 RHEOSTATIC CONTROL FOR DC EXCITERS

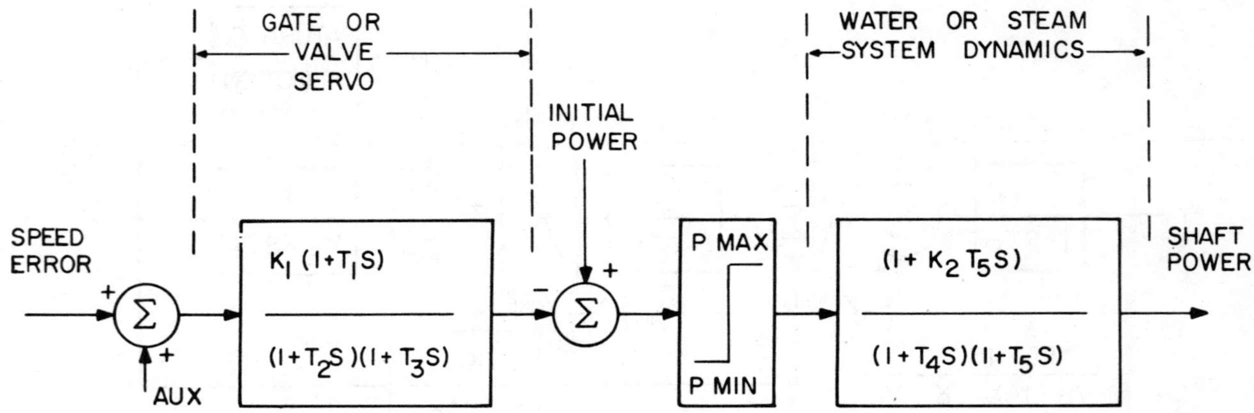

FIG. 4 SINGLE-SHAFT GOVERNOR AND ENERGY SYSTEM

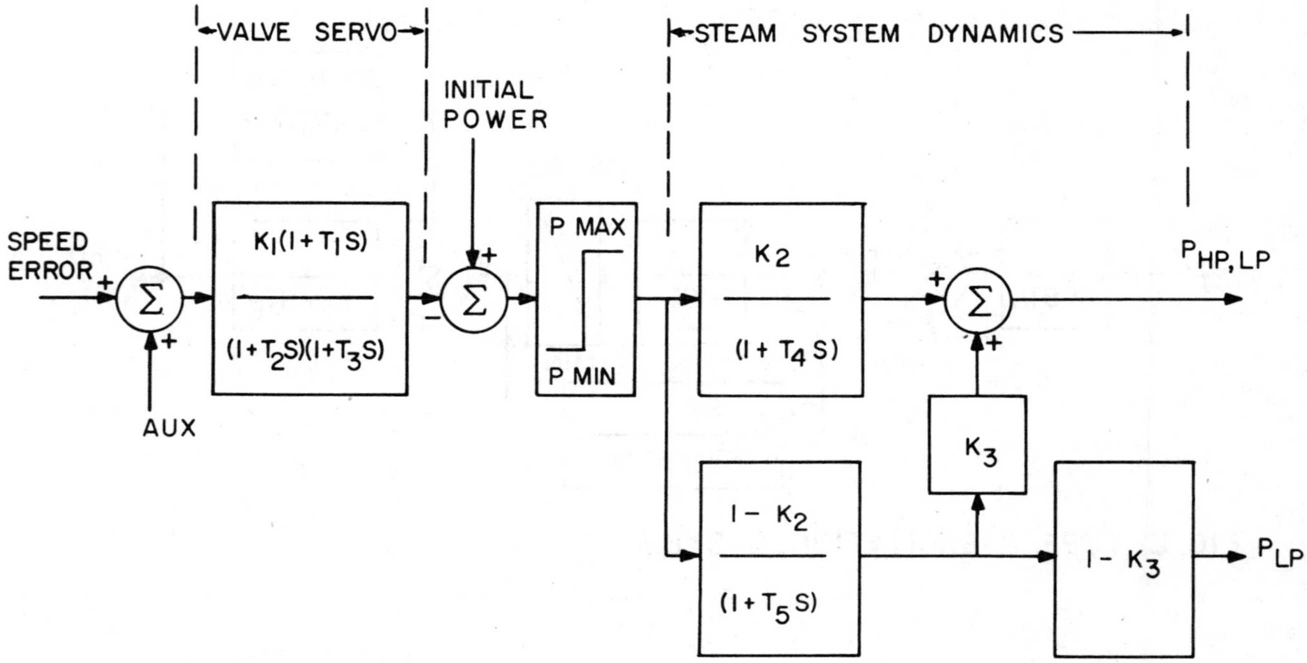

FIG. 5 CROSS-COMPOUND GOVERNOR AND STEAM SYSTEM

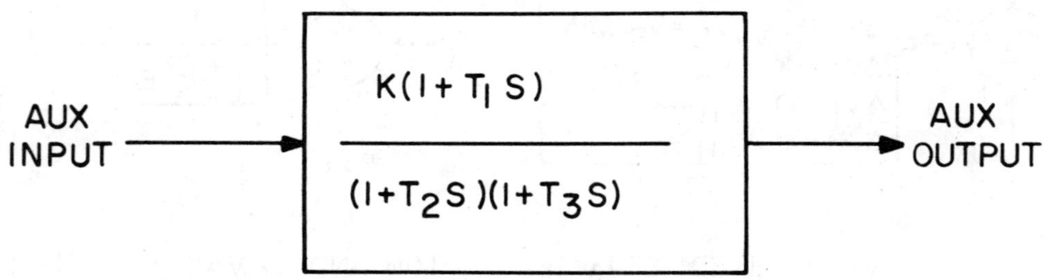

FIG. 6 AUXILIARY TRANSFER FUNCTION FOR VOLTAGE REGULATOR OR GOVERNOR

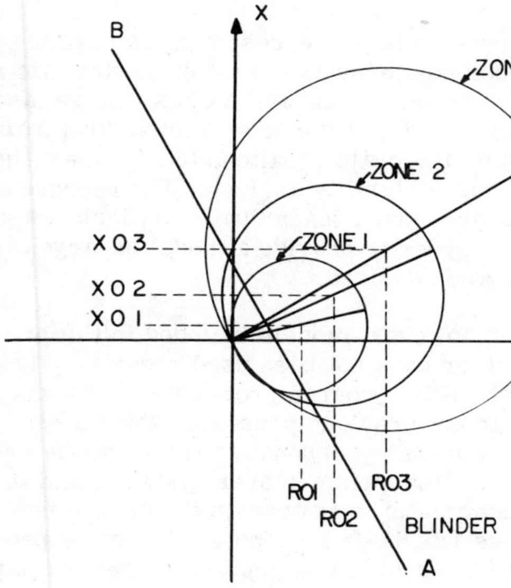

FIG. 7 IMPEDANCE RELAY CHARACTERISTIC

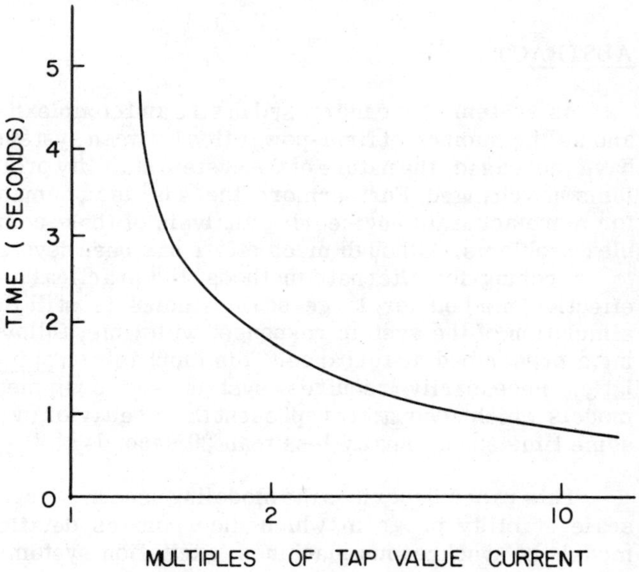

FIG. 8 OVERCURRENT RELAY CHARACTERISTIC

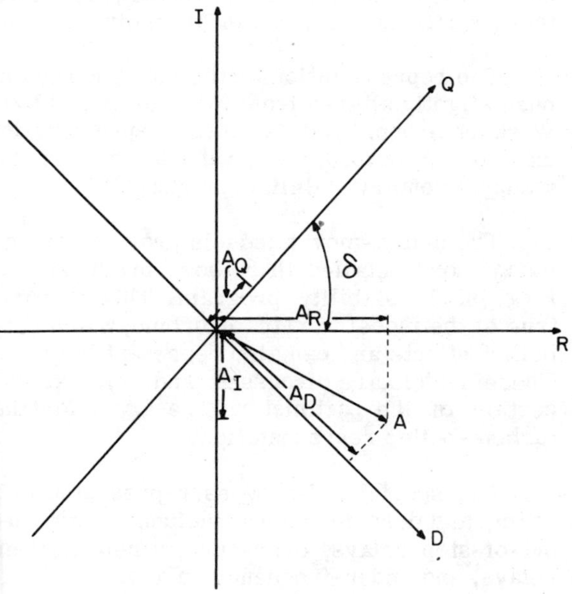

FIG. 9 ROTATION OF AXES IN A PLANE

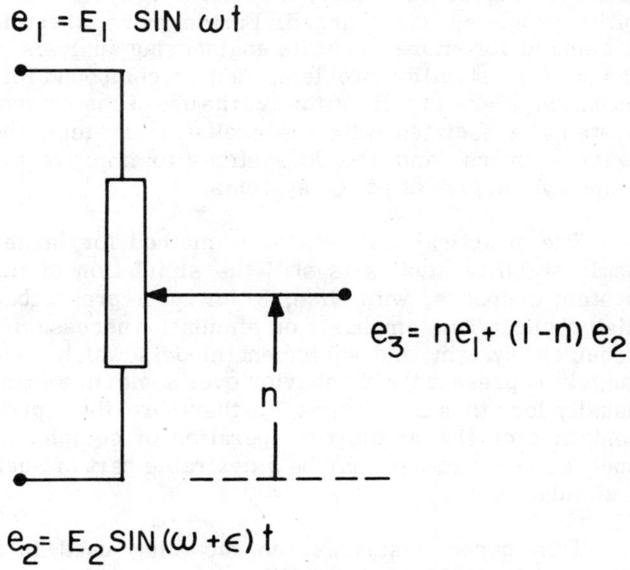

FIG. 10 VOLTAGE DIVIDER

EQUIPMENT AND SYSTEM MODELING FOR LARGE-SCALE STABILITY STUDIES

C. C. Young
General Electric Co.
Schenectady, N. Y.

ABSTRACT

As systems have increased in size and complexity, and as the number of firm-power ties between systems have increased, the nature of the system stability problems has changed. Furthermore, there is also a demand for more accurate engineering analysis of these complex problems. Although much effort has been devoted to searching for alternate methods, the practical and effective method for large-scale studies is still the simulation of the system response, with time, following a prescribed disturbance. This emphasis on simulation necessarily requires system and equipment models which adequately represent their behavior over some time span, usually less than 20 seconds.

This paper describes the modeling used in a large-scale stability program which incorporates detailed models of synchronous machines, excitation systems, prime-movers, loads, relays, series capacitor protection devices, and DC systems and controls. The equations or block diagrams for the representation of the equipment are presented for those models which either have not been used previously in a large-scale program or which have been altered from that previously used.

I. INTRODUCTION

Power systems have grown in both size and complexity, and, consequently, the nature of system stability problems has changed. Furthermore, there is a demand for more accurate engineering analysis of the system stability problem. And to compound this situation, there is a trend toward the use of the control systems associated with the excitation system, the prime-movers, and the dc systems to improve the stability margins of power systems.

The practical and effective method for large-scale stability studies is still the simulation of the system response, with time, following a prescribed disturbance. This emphasis on simulation necessarily requires system and equipment models which adequately represent their behavior over some time span, usually less than 20 seconds. Furthermore, the representation of the automatic operation of equipment, such as relay action, may be a desirable part of such a simulation.

This paper describes the modeling used in a large-scale (2000 bus) stability program, which incorporates detailed models of synchronous machines, excitation systems, prime-movers, loads, relays, series capacitor protection devices, and dc systems and controls. Some of these models have been previously used and described. These include the representation of composite system loads[1] and the representation of dc systems and their controls.[1,2] The principal changes in these representations have been to make the composite system load characteristic a function of bus frequency as well as bus voltage, and the addition of a model of the dc system and controls which uses only algebraic equations to represent the equipment during a stability analysis. The representation of induction motor loads[3] has been included in this program in order to provide a model of large industrial loads when desired.

There are, however, models included in this program which either have not been used previously in a large-scale stability program, or which have been altered from those previously used. In this category are the models of the synchronous machines, the excitation systems, the prime-mover systems, and the simulation of automatic relay operation. The representation of relays has been incorporated in other programs,[4,5] but the specific assumptions or details may be different.

The representation of synchronous machines in stability programs should be as simple as possible to minimize computer costs. But, at the same time, there are situations where an accurate representation is required. Furthermore, there may be uncertainty as to whether a simpler model will be adequate. For these reasons, models of the synchronous machine have been included which involve the representation of between one and three rotor-iron (or amortisseur) circuits. These models are discussed in Section II and their equations are given in Appendix I.

The representations of the excitation systems have been slightly altered from those proposed by the IEEE Working Group,[6] and the current representations being used are described, along with the model provided for a supplementary stabilizing signal.

The prime-mover models provided are generally much more detailed than those previously used in a large-scale stability program. This is particularly true of the model of a steam turbine, which can include boiler effects and can also represent boiler controls. These models are discussed, and "typical" values for certain of the parameters are given for the steam turbine-boiler representation.

The special relay types represented in the program, and described here, include impedance relays, out-of-step relays, over-current relays, over-power relays, and under-frequency relays.

II. SYNCHRONOUS MACHINE MODELS

During any given system stability study, it is common to have synchronous machines in the system being represented by models of varying detail. For example, the machines near the disturbance may be represented

Paper 71 C 26-PWR-V-B, recommended and approved by the Power System Engineering Committee of the IEEE Power Society for presentation at the PICA Conference, Boston, Mass., May 24-26, 1971. Manuscript submitted January 6, 1971; made available for printing June 9, 1971.

Reprinted from *IEEE Trans. Power App. Syst.*, Jan./Feb. 1972.

by a very detailed model, whereas machines remote to the disturbance may be represented by the simplest model. Furthermore, the machines at intermediate locations may use still another model. Therefore, one requirement for a sophisticated stability program is that it have models available to the user which encompass the range of representation required.[7]

In the development of the models used in this program, it was decided to identify three basic models, and it was noted that by the proper selection of the numerical values used in these models, one could represent two additional models. The detailed equations of the models are described in Appendix I. In the order of increasing detail they are:

1. The classical transient stability model, which uses a constant voltage magnitude behind transient reactance (MODEL I). When represented by a Norton Equivalent, the transient reactance (X_d') is represented by a reactance to ground at the machine's terminal, and a current is injected at this node equal to ($\tilde{e}'/X_d'$) where $\tilde{e}'$ is the complex voltage behind X_d'.

2. The second model (MODEL II) allows the representation of the stator circuits and two rotor circuits. The rotor circuits are the field circuit in the direct axis and an iron circuit in the quadrature axis. This representation can be used to directly include a substantial portion of the damping effects which arise in a steam-turbine generator due to the presence of rotor iron currents.

By setting the quadrature-axis transient reactance (X_q') equal to the quadrature-axis synchronous reactance (X_q) in the equations, this model may also be used to represent the stator circuits and only the field circuit in the rotor. This particular representation has been widely used in stability studies, and is the simplest model which includes the direct representation of the effects of the excitation system behavior.

When represented by a Norton Equivalent, this model uses a reactance to ground equal to the transient reactance (X_d') and an injected current equal to ($\tilde{e}'/X_d'$) where $\tilde{e}'$ is the complex voltage behind X_d'. In this case, the magnitude and the angle of $\tilde{e}'$ is a function of both the known internal conditions of the machine and the terminal conditions. Therefore, an iterative procedure is necessary to solve for the proper value of $\tilde{e}'$.

3. The most detailed model (MODE I, III) represents the stator circuits and four rotor circuits. The rotor circuits are the field circuit, one amortisseur circuit in the direct axis, and two amortisseur circuits in the quadrature axis. In the case of a steam-turbine generator, these amortisseur circuits represent the rotor iron current effects.

This is the most detailed model that can be used where standard machine data are supplied by a manufacturer. Studies now underway suggest that this degree of detail of representation is sufficient for an accurate representation in a stability program.

The second amortisseur circuit in the quadrature axis is intended to represent a rotor iron circuit in a steam-turbine generator. In the case of a laminated rotor machine (such as a hydraulic turbine generator or synchronous motor), data for only one amortisseur circuit in the quadrature axis may be available. The representation for this type of machine may be made by setting the quadrature-axis transient reactance (X_q') equal to the quadrature-axis synchronous reactance (X_q).

When represented by a Norton Equivalent, this model uses a reactance to ground equal to the subtransient reactance (X_d'') and an injected current equal to ($\tilde{e}''/X_d''$) where $\tilde{e}''$ is the voltage behind X_d''. The magnitude and the angle of $\tilde{e}''$ is known from the solution of the machine's differential equations, and no iteration is necessary for this model.

In terms of the numbers of the various machine types that may be needed in a stability program, it is common for all of the machines represented in a study to be either MODEL I or MODEL II; whereas, it is likely that only a few MODEL III machines will be used in a study. Therefore, it is a common practice in stability programs to limit the number of MODEL III machines to some relatively small number, such as 25 machines.

The amount of machine data required increases as the detail of representation increases, and, generally, the computational costs also increase (particularly for MODEL III machines). But at the same time, the more detailed representations presumably permit a more accurate evaluation of system behavior and stability limits. Studies have been made in an attempt to define the effects of added detail in the representation (sometimes with conflicting conclusions), but there is a continuing need to define the simplest model which will meet the needs of system planners and operators in a given situation.

III. EXCITATION SYSTEMS

The basic representations for various types of excitation systems was proposed by an IEEE Working Group,[6] and its recommendations have been adopted for many stability programs. Generally, these have been satisfactory and flexible enough to represent a wide range of excitation system types. However, experience has indicated the need to modify the representation for Type I and to add the time constants T_{A1} and T_{A2} shown in Figure 1.

There are two principal reasons for making this change in the representation: (1) The analysis of some actual excitation systems has indicated that a more accurate representation of the excitation system behavior might require these additional constants. Therefore, where such tests have been made, this altered representation may be desirable. (2) In high initial response excitation systems, a reduction in transient gain may be physically attained by the use of series functions rather than by feedback functions. Therefore, to represent this type of equipment, the use of the function including T_{A1} and T_{A2} is desirable. One can, in effect, simulate the gain reduction by using only the feedback function, but this is undesirable if it can be avoided.

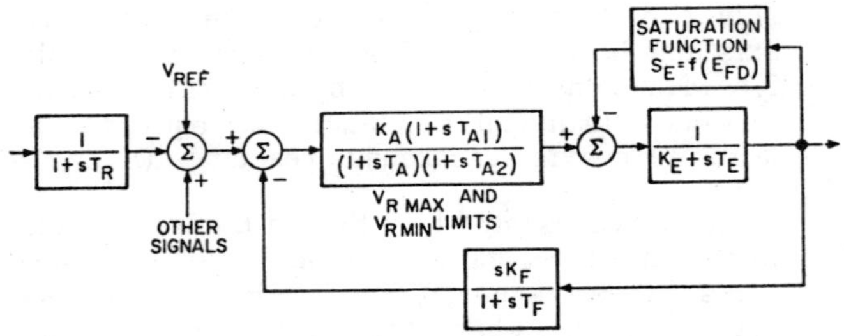

Fig. 1. Block diagram of excitation system

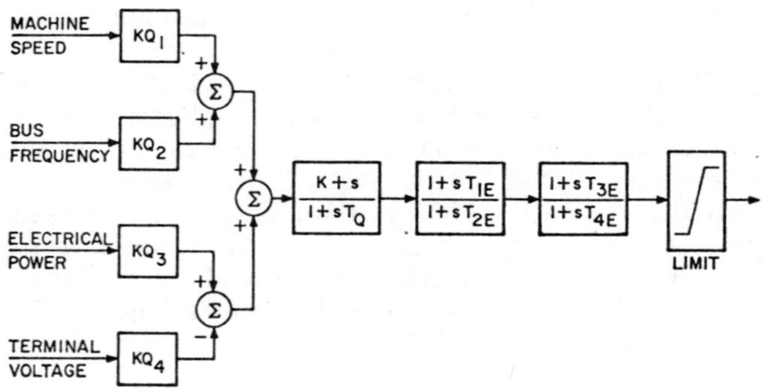

Fig. 2. Block diagram of stabilizer

The representation of supplementary stabilizing has not been standardized. Therefore, it is necessary to use as general a representation as possible. The representation shown in Figure 2 is intended to represent a general function, and inputs from machine speed, bus frequency, electrical power, and terminal voltage may be used separately or in combination.

IV. PRIME-MOVERS

The prime-mover models used in the program are more elaborate than those used in many large-scale stability programs. The use of such detailed models was prompted by two considerations: (1) In some studies, particularly those that involve the loss of generation or load, the response of the prime-movers have an important effect upon tie-line loading and frequency variations.[8] The effects which are now more directly modeled have been approximated before by empirical methods. (2) Surprisingly enough, there are times when it is more difficult to establish the proper data to be used with a simple model than with a more detailed model. Obviously, this is true only up to a certain point, but it has been noted that the manufacturers who must supply this data are more likely to be familiar with a representation where many of the important elements of the prime mover control system are separately identified. If only a simple representation is allowed, either the manufacturer must attempt to fit his transfer functions to the model or the program user must attempt it. Therefore, it seems that there may be actually an advantage in providing a certain amount of detail in a prime-mover representation.

The two principal types of prime-movers that will be discussed are the hydraulic turbine (and controls) and the steam turbine (and controls).

1. Hydraulic Turbine

The model of the hydraulic turbine and its controls are shown in Figure 3. If there is a surge tank associated with the hydraulic turbine, its representation is shown in Figure 4. These representations have been in use in the FACE Multi-Machine Power System Simulator Program,[9] and they include the effect of the surge tank and water-hammer effects in the penstock on either side of the tank, as well as speed and head effects on the turbine runner. The model is non-linear and can be used for a variety of studies. Each synchronous machine being represented by the program may have a hydraulic turbine represented with this degree of detail.

2. Steam Turbine

The representation of the steam turbine and its controls are shown in Figure 5. The model represents the effects of boiler pressure changes due to any changes in the steam flow which may occur in the turbine. If desired, a simplified representation of the combustion controls is available which includes a throttle pressure regulator control (Figure 6). Each synchronous machine being represented may have a steam-turbine represented with this degree of detail.

All values used with this model are in per-unit of the turbine rating. For cross-compound units, the combined rating of the high-pressure, intermediate-pressure, and low-pressure turbines is the base.

Governor dead-band may be represented, and a distinction can be made between a mechanical-hydraulic control system and an electro-hydraulic control system (as applied to General Electric turbines.)

The speed relay, valve, and steam bowl effects are separately represented by linearized models.[10] The limit on the rate-of-change of valve position may be represented as well as the limit representing the maximum valve position. When a change in the high-pressure turbine flow occurs, the effect upon the high-pressure turbine throttle pressure due to the change in the pressure drop and the boiler pressure may be modeled.

A reheater (or pipe volume) between the high and intermediate turbine and a reheater (or pipe volume) between the intermediate and low pressure turbines may be separately modeled. The intercept value position for each reheater is identified, but in this program these are maintained at a constant value. The functions $\frac{1 + sT_4}{1 + sT_5}$ and $\frac{1 + sT_6}{1 + sT_7}$ represent the effects of feed-water heaters.

The combustion control representation includes a limit before the representation of the combustion elements. This is a limit which may exist in a given control scheme. In any case, it also affects the maximum power that the boiler may produce even after a long

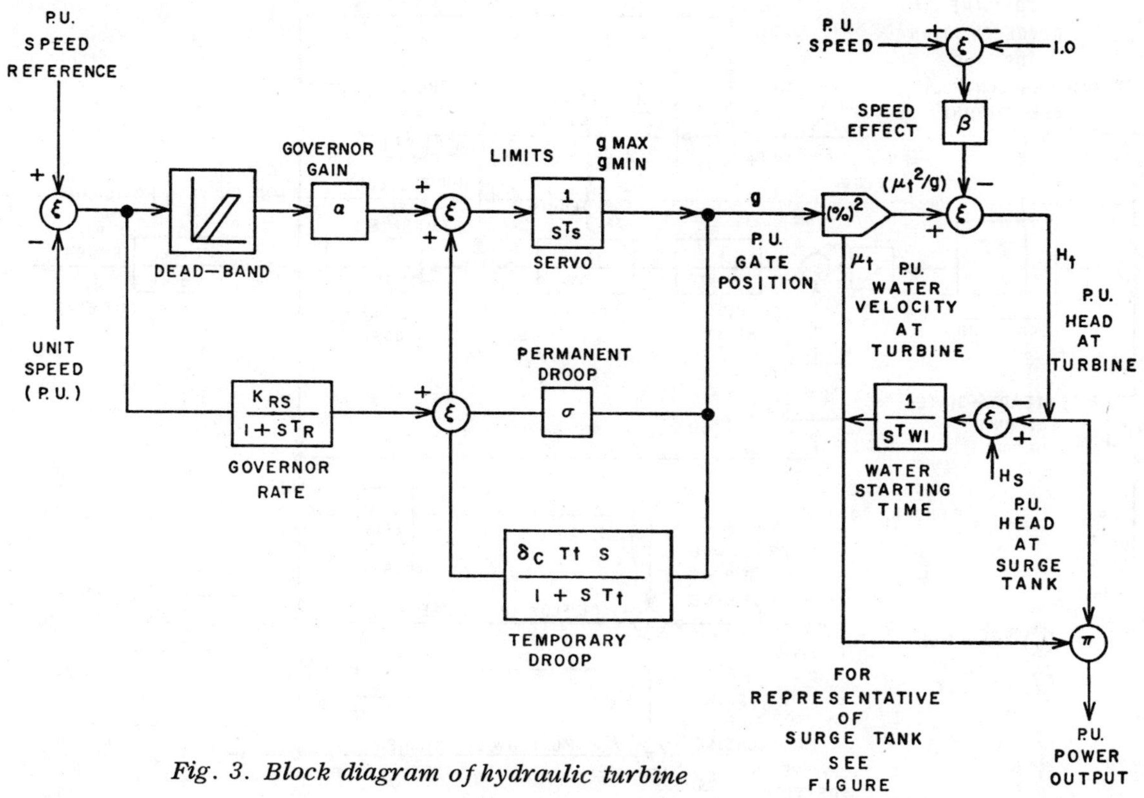

Fig. 3. Block diagram of hydraulic turbine

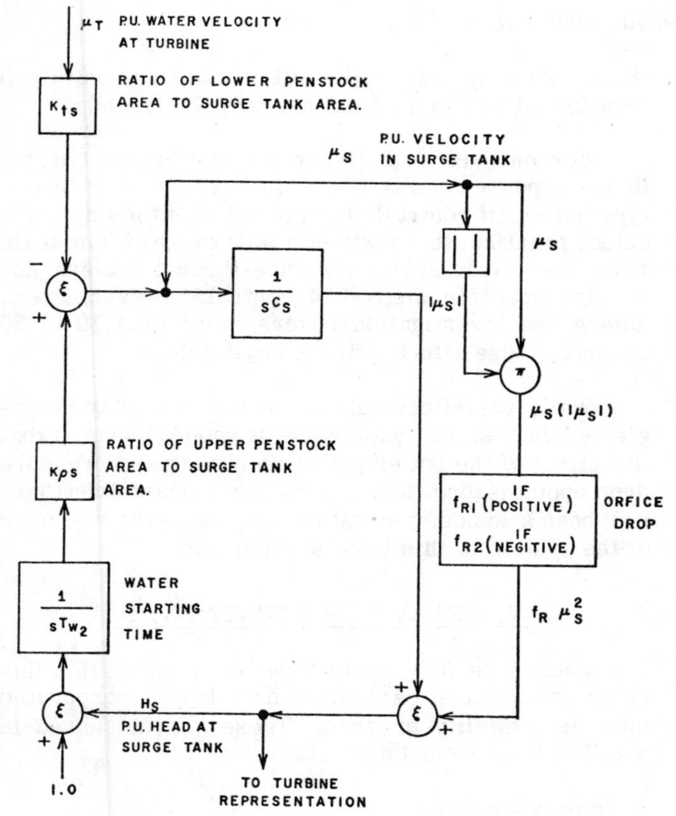

Fig. 4. Block diagram of surge tank representation

period of time. Therefore, if such a limit does not exist in the control, the value used should be the maximum available boiler output in per-unit of the combined turbine rating. This value would ordinarily neglect any effects which could not occur without manual intervention.

The values for the representation of the turbine are either obtained from the manufacturer for specific machines, or from typical values provided by the manufacturer. Since reheat time-constants depend upon the equipment used outside of the turbine, these are usually computed by the user.

In order to provide guidance to a user as to the typical numbers that might be used in the representation of the boiler and feed-water heater effects, the responses of a variety of boiler types were examined, and the following constants appear to represent average responses (see Figures 5 and 6 for the meaning of the symbols used):

K_2 = 0.08
T_D = 7 seconds for gas and oil-fired boilers
T_D = 25 seconds for fast responding coal-fired units
T_D = 60 seconds for slow responding coal-fired units
T_8 = 5 seconds for gas and oil-fired boilers
T_8 = 30 seconds for coal-fired units
T_4 = 29.4 seconds
T_5 = 30.0 seconds
T_6 = 29.4 seconds
T_7 = 30 seconds

For units less than 300 MW:

$$K_3 = 0.5 \times 10^{-2}$$

For units more than 300 MW:

$$K_3 = 0.8 \times 10^{-2}$$

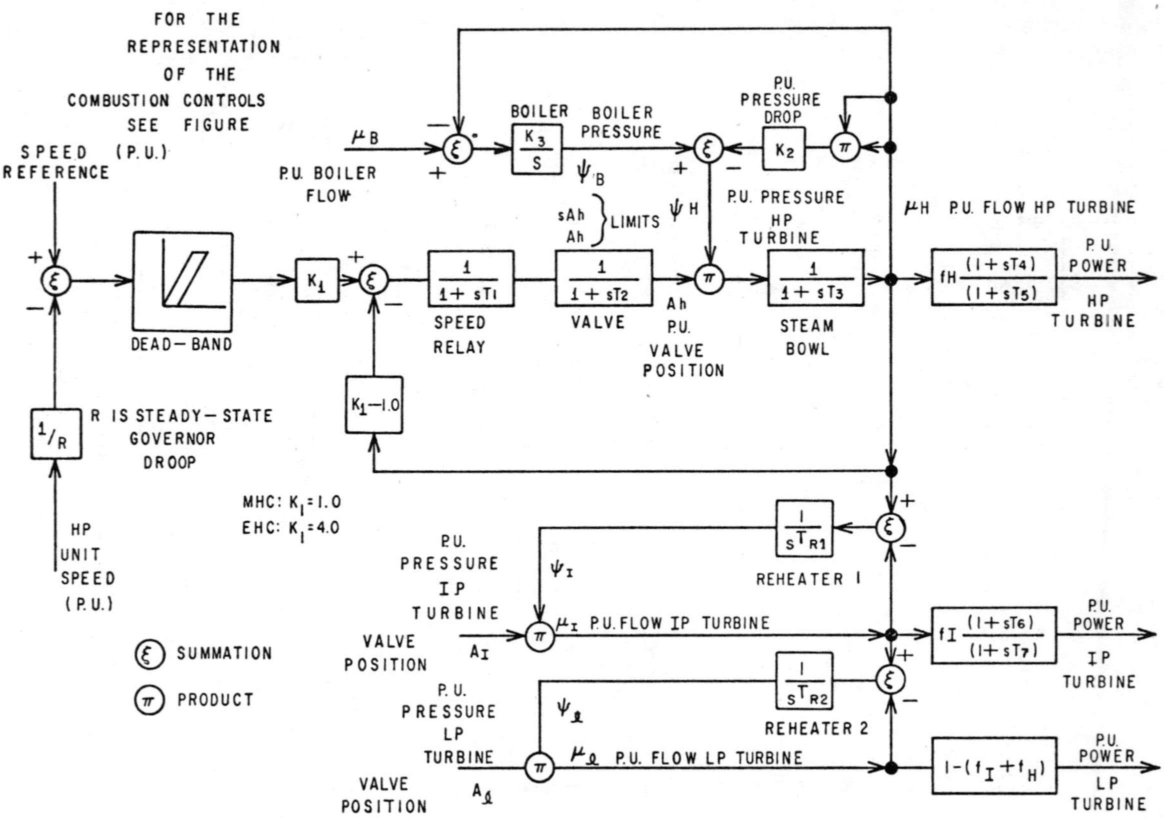

Fig. 5. Block diagram of steam turbine and boiler representation

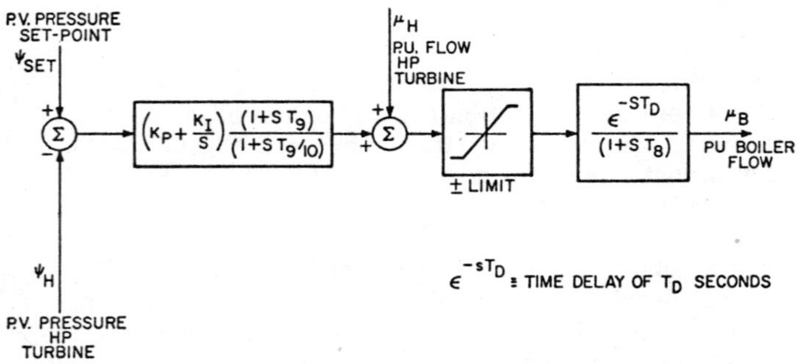

Fig. 6. Block diagram of boiler control

The pressure control settings are tabulated as follows:

	K_P	K_I	T_9
Oil, gas > 300 MW	3.33	.045	6
Oil, gas < 300 MW	5.0	.05	6
Fast coal > 300 MW	1.67	.02	36
Fast coal < 300 MW	3.33	.022	24
Slow coal > 300 MW	1.05	.0105	60
Slow coal < 300 MW	2.20	.011	36

Although the values given are considered to be typical, it should be recognized that there could be significant variations from these for a unit in a station.

There have been three types of responses included, one for gas and oil-fired boilers, one for fast coal and one for slow coal. However, since the effects of combustion control will not be felt for anything less than T_D seconds, a study which is concerned with a time period of less than 10 seconds will not be affected by these controls, and only the oil or gas fired units will be affected for time spans less than 25 seconds.

For once-through boilers, it has been noted that their responses can be quite similar to those of drum-type boilers if controlled in the boiler-follow mode. In actual practice, there will be a wide range of long term response characteristics of once-through boilers, depending upon the method of control; however, again, unless the investigation covers more than 10 to 20 seconds, these effects will be negligible.

While the effects of boiler controls may be neglected in a stability analysis, it should be noted that the effect of the boiler pressure change and pressure drop upon the throttle pressure is important when there has been a loss of generation and where the response of the system to this loss is important.

V. RELAY REPRESENTATIVES

There are four basic types of transmission line relay characteristics[11] which have been incorporated into the stability program. These are discussed in detail in the succeeding sections.

1. Impedance Relays

The impedance relay characteristic is in the form of a circle on the impedance plane (Figure 7). This circle is defined by its center coordinates (R_c, X_c) and its radius (r_1), and the relay is identified, by input, as being located at one end of a particular transmission line. The blocking relay characteristic is assumed to be a concentric circle, with a specified radius (r_2).

The apparent line impedance (Z_ℓ) is calculated at each time step, and when the impedance enters the outer circle, but not the inner circle, the blocking logic is initiated. If the impedance Z_ℓ stays in the outer circle for three cycles or more before entering the inner circle, tripping is blocked. If the impedance Z_ℓ leaves the outer circle, the relay is assumed to reset instantaneously.

If the impedance enters the interior circle, and if tripping is not blocked, the logic is initiated to trip the line. When Z_ℓ leaves the interior circle, if the time inside was less than one cycle, the relay is assumed to reset instantaneously. If the time was more than one cycle, the tripping is continued.

The choice of three cycles for the blocking relay and one cycle for the operation of the tripping relay is arbitrary and can be changed for all relays of this type.

The user specifies, as part of input, the total line tripping time (relay plus breaker time) and the dead-time between the tripping point and the time when line reclosing will be done. Single-shot reclosure is allowed.

This same relay may trip several additional lines, each with its own trip time and dead-time for reclosure.

2. Directional Over-current Relays

The directional element of an over-current relay is represented as shown in Figure 8. The angle of

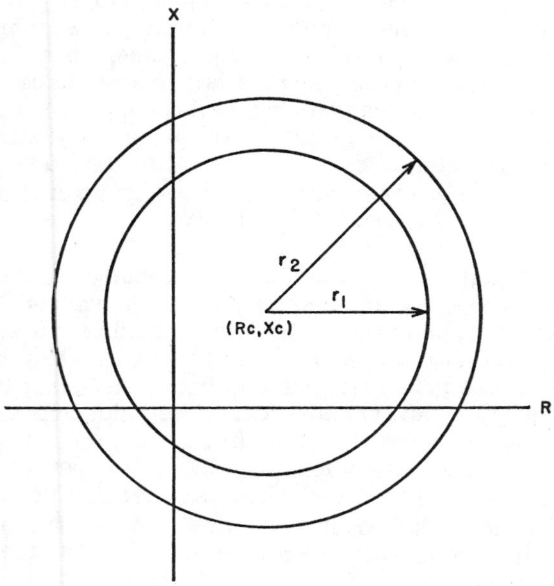

Fig. 7. Impedance relay characteristic

maximum torque (θ) is an input value. When the line current I_ℓ is in the tripping region, the overcurrent relay is allowed to operate.

When I_ℓ is out of the tripping region, the relay is reset at a fixed rate (also given by input), or is held at the reset position if it has not moved.

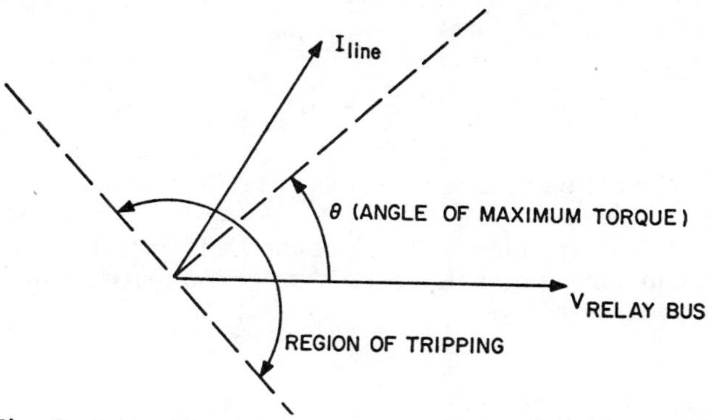

Fig. 8. Directional over-current relay characteristic

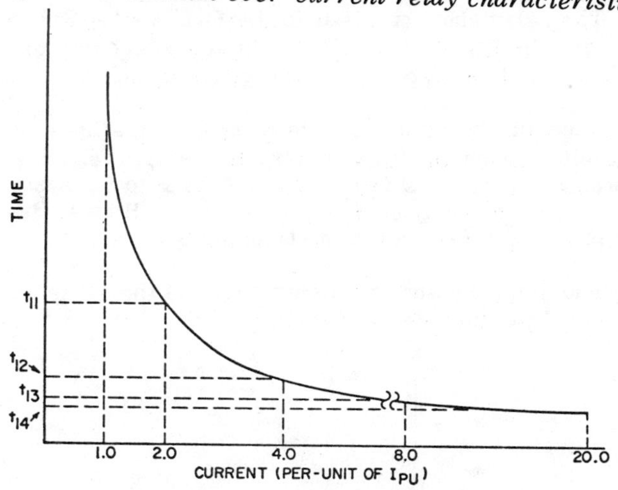

Fig. 9. Over-current relay time characteristic

The relay time-current characteristic (Figure 9) is curve fitted by a third-order polynominal. The specific time-dial setting for the relay is used to formulate this curve. The pick-up current (in per-unit on the system base) is given as input data, together with four relay operating times as functions of specific multiples of the pick-up current.

The following logic is used:

(1) If the current is in the proper direction, the relay operating time (t_c) is calculated from the polynominal fit of the current-time characteristic. If the line current is greater than 1.5 times the pick-up current, the change in the per-unit position of the relay (D_R) is:

$$\Delta D_R = \frac{\Delta t}{t_c}$$

$$D_R = D_R + \Delta D_R$$

When $D_R = 1.0$, line tripping is initiated, and the line is removed at a fixed time later (equal to the breaker time, and read in as part of the input).

If the direction of the current goes outside of the tripping region of the directional unit, or if the line current becomes less than 1.5 times the pick-up current, then the relay begins to reset:

$$\Delta D_R = \frac{-\Delta t}{t_{RESET}}$$

$$D_R = D_R + \Delta D_R$$

The action is hated when D_R reaches zero.

The tripping and relcosing logic for the line and for remote lines is identical to that for impedance relays.

3. Out-of-Step Relays

The relay characteristic for the Out-of-Step Relay is shown in Figure 10. The slopes and intercepts (on the R axis) of the two curves are given as input.

Once the starting zone is identified, the logic of the unit causes the line to trip if the apparent line impedance (Z_ℓ) goes from zones A to B to C. Also, if the impedance goes from zones C to B to A, the line is tripped. Otherwise, no tripping occurs.

The tripping and reclosing logic is the same as described for impedance relays.

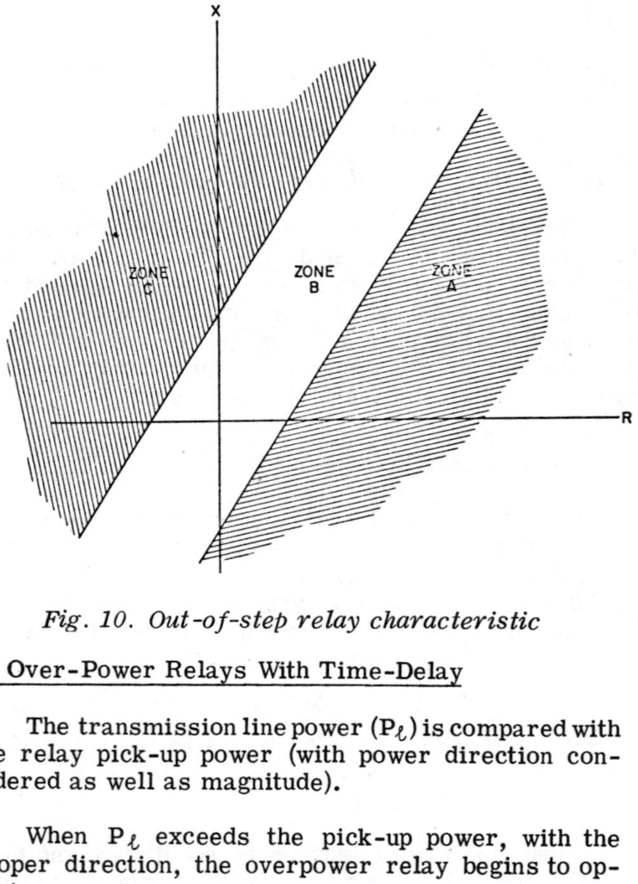

Fig. 10. Out-of-step relay characteristic

4. Over-Power Relays With Time-Delay

The transmission line power (P_ℓ) is compared with the relay pick-up power (with power direction considered as well as magnitude).

When P_ℓ exceeds the pick-up power, with the proper direction, the overpower relay begins to operate:

$$\Delta D_R = \frac{\Delta t}{t_{pu}}$$

where t_{pu} is the relay operating time (assumed constant).

$$D_R = D_R + \Delta D_R$$

and line tripping is initiated when $D_R = 1.0$.

Whenever P_ℓ is less than the pick-up power or reverses sign:

$$\Delta D_R = \frac{-\Delta t}{t_{RESET}}$$

The tripping logic is the same as for impedance relays.

For all relay types, the relay can be used to initiate the tripping of a number of remote lines, where each line has its individual breaker time and reclosing time.

In addition to the representation of conventional transmission line relays, the representation of the automatic operation of the series capacitor protective equipment can be useful. It is necessary to provide a current setting which will operate the protective gap in terms of the fundamental frequency symmetrical RMS current. When the line current exceeds this value, the shorting of the series capacitor is simulated. As part of input, the time between the gap closing and reinsertion is provided, and it is used to determine the time at which the capacitor is reinserted. Any number of removals and reinsertions is allowed.

5. Under-Frequency Relays

The representation of under-frequeny relays provides for the modeling of either a static type relay or an induction cup type relay. In either case, when the relay at a bus is represented, the bus frequency is calculated, and the load can be shed in three steps. Each of 200 buses may have an under-frequency relay with different characteristics, and different load shedding schedules. The breaker operating time is identified separately from the relay operating time.

For a static relay, the input data required is the relay trip time, the breaker trip time, the values of the three frequencies at which load shedding will be done, and the amount of bus load to be shed at each frequency. The principal consideration here is that if the frequency recovers, the correct decision will be made either to reset the relay (if the time since the relay began to operate is less than the relay operating time) or to proceed to trip the load as scheduled (if the time since the relay began to operate is more than the relay operating time but less than the combined breaker and relay time).

The representation of an induction cup type of under-frequency relay is more complicated. The representation assumes that there is a unique relay characteristic which describes the relation between the rate-of-change of frequency and the relay operating time. This characteristic is represented in the program by two straight-line segments. One extends from a point, where the maximum rate of frequency (MAXR) and minimum time (MINT) are defined, down to a "break point" with BRR and BRT defining the point. The second segment extends from the break point to the point that defines the minimum rate of frequency (MINR) and maximum time (MAXT).

When the bus frequency drops below a particular relay setting, the time (ST) is stored. Then, at subsequent times, where the frequency continues to be less than the set point, the average rate of change of frequency is calculated:

$$AVR = \frac{\text{RELAY SETTING} - \text{FREQUENCY}}{\text{TIME} - \text{ST}}$$

If AVR is greater than MAXR, then the shed time is the sum of MINT and the breaker time. If AVR is less than MINR, the shed time is the sum of MAXT and the breaker time. If the value of AVR is between MAXR and MINR, then the appropriate straight line segment must be identified, and the value of the relay operating time calculated. This time is added to the breaker time in order to calculate the shed time, and the shed time is recalculated at each time step.

If the bus frequency rises above the set point, the relay is assumed to reset instantaneously, unless the time since the relay began to operate is more than the relay operating time but less than the combined breaker and relay time. In that case, the load is tripped at the scheduled time.

REFERENCES

(1) C. C. Young, R. M. Webler, "A new stability program for predicting the dynamic performance of electric power systems," Proceedings of the American Power Conference, vol. XXIX, 1967, pp. 1126-1138.

(2) G. D. Breuer, J. F. Luini, C. C. Young, "Studies of large ac/dc systems on the digital computer," IEEE Transactions (PA&S), vol. PAS-85, no. 11, November 1966, pp. 1107-1116.

(3) D. C. Brereton, D. G. Lewis, C. C. Young, "Representation of induction motor loads during power system stability studies." AIEE Trans. Pt. III (PAS), vol. 76, August 1957, pp. 451-461.

(4) D. G. Ramey, R. T. Byerly, "Dynamic simulation of interconnected systems," PICA Conference Record, 1967, pp. 31-40.

(5) G. W. Stagg and A. H. El-Abiad, Computer Methods in Power System Analysis, McGraw-Hill Book Company, New York, 1968.

(6) IEEE Committee Report, "Computer representation of excitation systems," IEEE Trans., vol. PAS-87, June 1968, pp. 1460-1464.

(7) C. C. Young, "Modern concepts of power system dynamics – the synchronous machine," IEEE Special Publication 70M62-PWR.

(8) C. Concordia, F. P. DeMello, L. K. Kirchmayer, R. P. Schulz, "Effect of prime-mover response and governing characteristics on system dynamic performance," American Power Conference, April, 1966.

(9) D. N. Ewart, R. P. Schulz, "FACE multi-machine power system simulator program," PICA Conference Record, 1969.

(10) M. A. Eggenberger, "Introduction to the basic elements of control systems for large steam turbine-generators," General Electric Publication GET-3096A.

(11) C. R. Mason, The Art and Science of Protective Relaying, John Wiley & Sons, Inc., New York, 1956.

APPENDIX

MACHINE EQUATIONS

Each of the mathematical models used to represent a synchronous machine in a stability study consists of a set of simultaneous algebraic equations and a set of simultaneous differential equations. The phasor diagram associated with each representation is a way of describing the algebraic equations which relate the machine's internal and terminal conditions. In some cases, additional algebraic equations may exist. In all cases, the differential equations are expressed separately from the phasor diagram.

The equations which are not represented by the phasor diagram are described in this appendix for each of the representations considered.

A. MODEL I

Model I is the "Classical" stability representation for the synchronous machine (Figure 11). The only differential equation to be solved is the acceleration equation.

B. MODEL II

Model II is the representation of the machine which includes the field effects and represents the effect of a single iron circuit in the quadrature axis. In addition to the equations implied in the phasor diagram (Figure 12) other equations are:

$$\tilde{\bar{e}}' = \{\bar{e}'_q + j\,[\bar{e}'_d - (X'_q - X'_d)\,\bar{i}_q]\}\epsilon^{j\delta}$$

It should be noted that $\tilde{\bar{e}}'$ is a phasor, but $\bar{e}'_q$, $\bar{e}'_d$, and $\bar{i}_q$ are the magnitudes of these variables.

$$\frac{d\bar{e}'_q}{dt} = \frac{1}{T'_{do}}\,(\bar{E}_{fd} - \bar{E}_I)$$

$$\frac{d\bar{e}'_d}{dt} = \frac{1}{T'_{qo}}(-\bar{E}_{Ikq})$$

$$\bar{E}_{Ikq} = \bar{e}'_d + (\bar{X}_q - \bar{X}'_q)\,\bar{i}_q$$

$$\bar{E}_I = \bar{e}'_q + (\bar{X}_d - \bar{X}'_d)\,\bar{i}_d + \Delta\bar{E}_I$$

$$\frac{d^2\delta}{dt^2} = \frac{180\,f}{H}(\bar{T}_m - \bar{T}_e) - \bar{K}\frac{d\delta}{dt}$$

$$\Delta\bar{E}_I = f \text{ (index of saturation)}$$

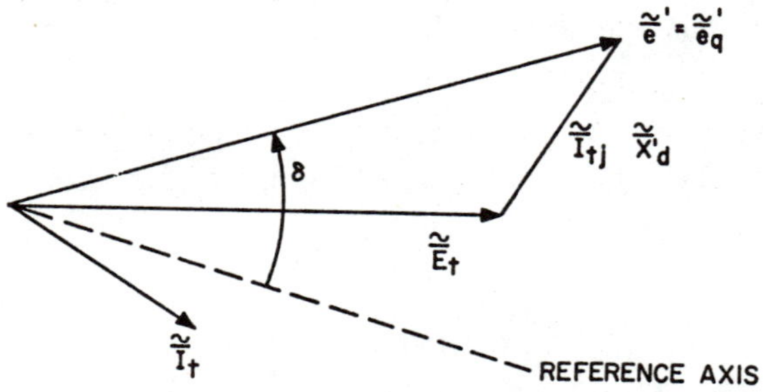

Fig. 11. Model I phasor diagram

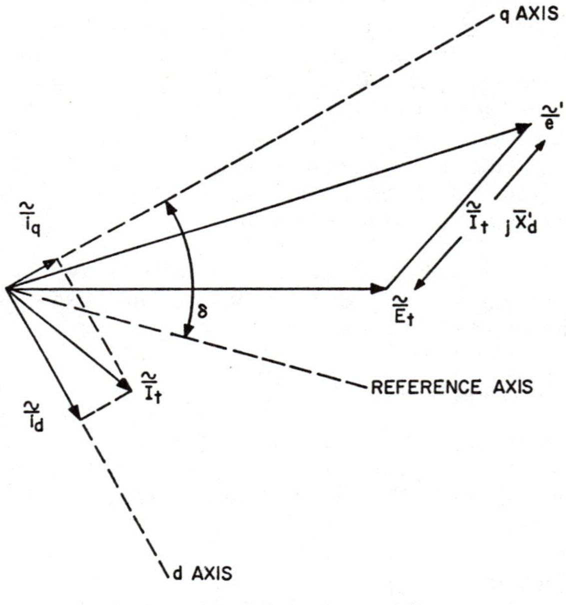

Fig. 12. Model II phasor diagram

C. MODEL III

Model IV is the representation of the machine which includes the field effects and the effects of amortisseur circuits in both axes. The equations given here (in addition to those of the phasor diagram (Figure 13)) are for the case where there is one amortisseur circuit in the direct axis and two amortisseur circuits in the quadrature axis, and $\bar{X}''_d$ equals $\bar{X}''_q$:

$$\bar{\psi}''_d = \bar{\psi}_{kd} + \left\{\frac{\bar{X}''_d - \bar{X}_1}{\bar{X}'_d - \bar{X}_1}\right\}(\bar{e}'_q - \bar{\psi}_{kd})$$

$$\bar{\psi}''_q = \bar{\psi}_{kq} + \left\{\frac{\bar{X}''_d - \bar{X}_\ell}{\bar{X}'_q - \bar{X}_\ell}\right\}(\bar{e}'_d - \bar{\psi}_{kg})$$

$$\frac{d\bar{e}'_q}{dt} = \frac{1}{T'_{do}}(\bar{E}_{fd} - \bar{E}_I)$$

$$\frac{d\bar{\psi}_{kd}}{dt} = -\left\{\frac{(\bar{X}''_d - \bar{X}_1)^2}{(\bar{X}'_d - \bar{X}''_d)\,T''_{do}}\right\}(\bar{i}_{kd})$$

$$\bar{E}_I = \bar{e}'_q + (\bar{X}_d - \bar{X}'_d)(\bar{i}_d - \bar{i}_{kd}) + \Delta\bar{E}_I$$

$$\bar{i}_{kd} = \frac{(\bar{X}'_d - \bar{X}''_d)}{(X'_d - X_1)^2}\left\{\bar{\psi}_{kd} - \bar{e}'_q + (\bar{X}'_d - \bar{X}_1)\,\bar{i}_d\right\}$$

$$\frac{d\bar{e}'_d}{dt} = \frac{1}{T'_{qo}}(-\bar{E}_{Iq})$$

$$\frac{d\bar{\psi}_{kq}}{dt} = \frac{1}{T''_{qo}}(-\bar{E}_{Ikq})$$

$$\bar{E}_{Iq} = \bar{e}'_d + (\bar{X}_q - \bar{X}'_q)\,\bar{i}_q - \frac{(\bar{X}_q - \bar{X}'_q)(\bar{X}'_q - \bar{X}''_d)}{(\bar{X}'_q - \bar{X}_\ell)^2}\,E_{Ikq}$$

$$\bar{E}_{Ikq} = \bar{\psi}_{kq} - \bar{e}'_d + (\bar{X}'_q - \bar{X}_\ell)\,\bar{i}_q$$

$$\frac{d^2\delta}{dt^2} = \frac{180\,f}{H}(\bar{T}_m - \bar{T}_e) - K\frac{d\delta}{dt}$$

$$\Delta\bar{E}_I = f \text{ (index of saturation)}$$

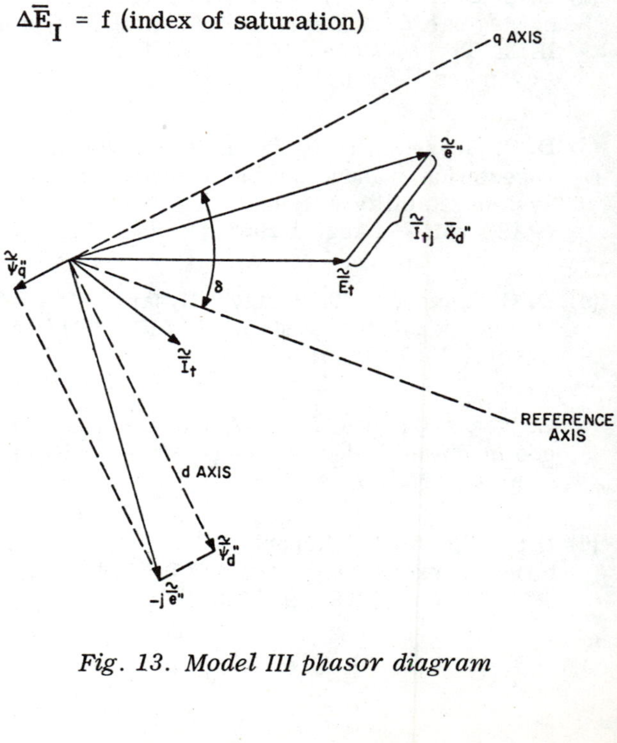

Fig. 13. Model III phasor diagram

APPENDIX II

SYMBOLS

- δ — Angle between the q axis and the reference axis (degrees)
- t — Time (seconds)
- f — Base frequency (hertz)
- H — Inertia constant (MW-seconds/MVA)
- K — Damping coefficient
- $\overline{T}_m$ — Mechanical torque (per-unit)
- $\overline{T}_e$ — Electrical torque (per-unit)
- $\overline{X}_1$ — Stator leakage reactance (per-unit)
- $\overline{X}''_d$ — Direct-axis subtransient reactance (per-unit)
- $\overline{X}'_d$ — Direct-axis transient reactance (per-unit)
- $\overline{X}_d$ — Direct-axis synchronous reactance (per-unit)
- $\overline{X}''_q$ — Quadrature-axis subtransient reactance (per-unit)
- $\overline{X}'_q$ — Quadrature-axis transient reactance (per-unit)
- $\overline{X}_q$ — Quadrature-axis synchronous reactance (per-unit)
- $\overline{X}_p$ — Potier reactance (per-unit)
- T''_{do} — Direct-axis subtransient open-circuit time-constant (seconds)
- T''_{qo} — Quadrature-axis subtransient open-circuit time-constant (seconds)
- T'_{do} — Direct-axis transient open-circuit time-constant (seconds)
- T'_{qo} — Quadrature-axis transient open-circuit time-constant (seconds)
- $\overline{i}_q$ — Quadrature-axis component of stator current (per-unit)
- $\overline{i}_d$ — Direct-axis component of stator current (per-unit)
- $\overline{i}_{kd}$ — Direct-axis amortisseur current (per-unit)
- $\overline{E}_{Ikq}$ — Quadrature-axis amortisseur current (per-unit)
- $\overline{E}_I$ — Field current (per-unit)
- $\overline{E}_{Iq}$ — Quadrature-axis iron current
- $\Delta\overline{E}_I$ — Correct to field current for saturation (per-unit)
- $\tilde{\overline{e}}'$ — Phasor voltage behind transient reactance (per-unit)
- $\tilde{\overline{e}}''$ — Phasor voltage behind subtransient reactance (per-unit)
- $\tilde{\overline{e}}'_q$ — Field flux-linkages (per-unit)
- $\tilde{\overline{e}}'_d$ — Quadrature-axis iron-circuit flux-linkages (per-unit)
- $\overline{E}_{fd}$ — Field voltage (per-unit)
- $\overline{E}_p$ — Potier voltage (per-unit)
- $\overline{\psi}_{kd}$ — Direct-axis amortisseur flux-linkages (per-unit)
- $\overline{\psi}_{kq}$ — Quadrature-axis amortisseur flux-linkages (per-unit)
- $\overline{\psi}''_d$ — Direct-axis component of rotor subtransient flux-linkages (per-unit)
- $\overline{\psi}''_q$ — Quadrature-axis component of rotor subtransient flux-linkages (per-unit)

NOTE: Wherever a variable is a phasor, it may appear in two forms. For example, the per-unit field current may appear as a phasor, where the symbol $\tilde{\overline{E}}_I$ is used to designate this form; or, its per-unit magnitude may be used in an equation, where the symbol $\overline{E}_I$ is used to designate this form.

Discussion

K. R. Padiyar and **R. S. Ramshaw** (University of Waterloo, Waterloo, Ont., Canada): In the section II of the paper, the author has presented three different models of the synchronous machine. It is of interest to note that the number of models given in this paper, is one less than the number presented earlier by the author in reference [7] of this paper. This shows that there is no basic difference between the various models presented. Actually, all these models can be derived from one common framework which does not put any restriction on the number of rotor circuits to be considered in the analysis. Present methods of analysis consider not more than two rotor circuits per axis and the different models apply for the particular cases of specified number of rotor circuits.

We have shown the general nature of the machine model (applicable to any arbitrary number of rotor circuits) in our paper presented in PICA-7 conference[1]. The 'dynamic saliency' is also taken into account in this approach.

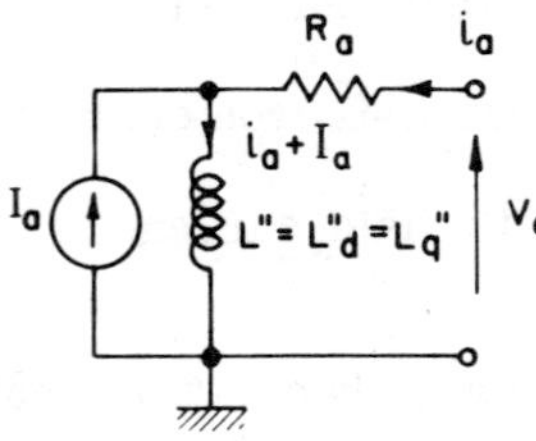

Fig. 1.

The author states that the Norton's equivalent circuit of the machine can be derived from the voltage source model given in the paper. This is not strictly correct, since the current source model shown in Fig. (1) and derived in reference [1] is the exact representation, while the voltage source model of the paper is not. The Thevenin's equivalent of the current source of Fig. (1) is given by

$$e_{Th} = L''p\, I_a$$

where

$$I_a = I_d \cos\theta + I_q \sin\theta,\ I_d = \underline{c}_1^t\, \underline{\psi}_{rd},\ I_q = \underline{c}_2^t\, \underline{\psi}_{rq} + \xi i_q$$

$\underline{c}_1$ and $\underline{c}_2$ are constant vectors, $\underline{\psi}_{rd}$ and $\underline{\psi}_{rq}$ are respectively rotor direct axis and quadrature axis flux linkages.

Differentiating I_a, we obtain

$$e_{Th} = L''\left(\frac{dI_d}{dt}\cos\theta + \frac{dI_q}{dt}\sin\theta\right) + L''\,\dot{\theta}(-I_d \sin\theta + I_q \cos\theta)$$

Ignoring the first term in comparison with the second term in the above equation and ignoring the variation of θ, we get the voltage source model of the paper.

Hence the voltage source model is inexact in comparison with the current source model and the correct conversion of the voltage source into a current source model

$$\left(I_N = \frac{1}{L''p}\, e\right)$$

will yield wrong results. However the transformation of the sources presented in the paper is valid only for steady-state conditions of the network at constant (nominal) frequency.

It would be interesting to know the integration method used by the author. The time-step for Runge-Kutta routines is determined by the smallest time constant of the system. Thus, the step size may be determined as much from the small regulator time constants as the damper circuit time constants.

Finally, we would like to compliment the author for this excellent work in the field of system modelling, which still remains a difficult task.

REFERENCE

[1] K. R. Padiyar and R. S. Ramshaw, "A Generalized Synchronous Machine Model for Large Scale Power System Dynamic Studies", Conference Proceedings, PICA-7, pp. 261-268, May 1971.

Manuscript received June 7, 1971.

C. C. Young: As system problems, the methods of analysis, and the characteristics of computing devices change, it is important to periodically examine the different models that one might use for system equipment, and to select those that are best suited for a particular situation. Therefore, I am always interested in discussions and technical papers on the subject of synchronous machine modeling for power system studies.

Although I have been directly involved in the modeling of synchronous machines in system studies for 20 years, I am continually amazed at just how many different ways one can write Park's equations. Professor Ramshaw and Mr. Padiyar have developed an interesting form of these equations that uses an injected current plus an inductance to ground to represent a machine. In comparing the discussor's model with other models which have been developed, ther are two important facts to be recognized: First, if all of the various models start with Park's equations and then make the same simplifying assumptions, and if the mathematic process of manipulating the resulting equations are rigorous, then the models developed will all be equally accurate, regardless of how different they may be in their form. Considering the assumptions implicit in Park's equations, none of these models should be considered "exact", although they may be entirely adequate for a particular type of analysis.

Second, because of the practical problems of obtaining data and because of the computing costs that are involved in the stability analysis of a large power system, it is very important to avoid superfluous detail in the modeling of equipment which adds to the cost without contributing significantly to the accuracy of the results. Experience has indicated that it is not desirable and is not necessary to attempt to represent all of the synchronous machines in a system by a single, detailed model. Only a relatively few machines may actually require an accurate simulation. It is this situation that led us to define three distinct models of varying complexity.

The discussors seem to be confused about the differences between their model and the models described in the paper. Comparing the assumptions used in deriving these models (7) with the assumptions used by the discussors (12), they are identical except for two differences: (1) the discussors represent stator transients (fundamental and harmonic currents and fluxes arising from the presence of the $p\varphi d$, $p\varphi q$ terms in the stator voltage equations), and (2) the discussors represent the effect of speed upon the generated terminal voltage of of the machine. While there are some types of studies which require that one or both of these effects be represented, it has been found that these effects do not have to be directly represented in large-scale stability studies. (13,14) In fact, since the system connected to the machines will very rarely have its transients or speed effects represented, the representation of machine stator transients is usually inconsistent, and is therefore not as accurate a simulation as one might suppose. In any case, if the stator transients are represented, the size of the time step required for numerical integration is considerably smaller than that required by models where these effects are omitted. Therefore, the cost may become prohibitive for large system studies.

If the stator transients and speed effects are neglected, the machine stator equations become algebraic equations, and can be represented by phasors, as are the system equations. This leads to the forms shown in the paper. The Norton Equivalent discussed in the paper is an exact equivalent of the voltage form.

In the stability program, the methods used for numerical integration are predictor-corrector methods, and dual time steps are allowed where it is appropriate.

REFERENCES

[13] K. Prabhashankar, W. Janischewskyj, "Digital Simulation of Power System Dynamic Performance, Part I," IEEE Conference Paper 68CP94-PWR.

[14] D. W. Ewart, F. P. De Mello, "A Digital Computer Program for the Automatic Determination of Dynamic Stability Limits," IEEE Trans., Vol PAS 86, July, 1967, pp. 867-875.

Manuscript received July 8, 1971.

FAST TRANSIENT STABILITY SOLUTIONS

H. W. Dommel N. Sato

Bonneville Power Administration
Portland, Oregon

ABSTRACT

Techniques are described for improving the speed of large transient stability studies without sacrificing accuracy. A fast iterative method for solving the algebraic network equations, including the effect of generator saliency, is explained. A new technique for solving the differential equations with the implicit trapezoidal rule of integration is introduced. These two techniques can be combined into one simultaneous solution, thereby eliminating the problem of interface error between the differential and algebraic equation solutions of the traditional approach.

INTRODUCTION

Transient stability studies now require considerable computer time, especially for solving large systems. This has motivated a search for faster solution techniques. Two possibilities for improving the speed of transient stability studies are: (a) Reduction of the total system to a smaller one, which could be solved faster, and (b) improvements in the numerical solution techniques.

Reduction techniques produce equivalents which are normally only approximations. Therefore, in spite of recent progress in finding equivalents, the user must have a "feeling" for the problem when equivalents are used. It should also be noted that solutions based on sparsity techniques become faster through reduction only to a certain point because of fill-in caused by the reduction process[1]. Judicious use of equivalents can make stability solutions faster; however, this paper is concerned with the second possibility, improved numerical solution techniques.

Improvements in numerical solution techniques can speed up the solution without placing any burden of judgment on the user. This paper describes experiments carried out at Bonneville Power Administration (BPA) over the last few years with the objective of obtaining faster solutions for the steady-state equations as well as for the differential equations which arise in stability studies. Speeding up the solution of the steady-state equations has been accomplished by applying sparsity techniques and solution methods which require fewer iterations. Speeding up the solution of the differential equations has been achieved with the implicit trapezoidal rule of integration, which has already been used successfully for the solution of switching transients at BPA[2]. This method is numerically stable and accurate enough. It should be noted that explicit techniques, including the Runge-Kutta methods, require higher accuracy primarily to insure numerical stability. As a consequence, such techniques also require longer solution times. It is expected that implementation of these ideas into BPA's production program will make the solution about five times faster.

Brackets are used to indicate vector and matrix quantities. Parentheses usually indicate a functional relationship or the value for time t. A super dot is used to indicate derivatives with respect to time.

STRUCTURE OF TRANSIENT STABILITY EQUATIONS

The initial conditions for a stability study are determined by a steady-state power flow solution. Thereafter, two sets of equations must be solved simultaneously as a function of time, namely a system of steady-state equations (which describe the steady-state behavior of the network including steady-state models of loads and the algebraic equations of synchronous machines),

$$\begin{aligned} &g_1(x_1, \ldots x_n, y_1, \ldots y_m) = 0 \\ &\cdots\cdots\cdots\cdots \\ &g_n(x_1, \ldots x_n, y_1, \ldots y_m) = 0 \\ &\text{or } [g([x],[y])] = 0 \end{aligned} \tag{1}$$

and a system of differential equations (which describe the dynamic behavior of the machines and their control circuits),

$$\begin{aligned} &\dot{y}_1 = f_1(x_1, \ldots x_n, y_1, \ldots y_m, t) \\ &\cdots\cdots\cdots\cdots \qquad \text{or } [\dot{y}] = [f([x],[y],t)] \\ &\dot{y}_m = f_m(x_1, \ldots x_n, y_1, \ldots y_m, t) \end{aligned} \tag{2}$$

The structure of Eq. (1) will change at certain moments in time due to fault initialization, fault clearing, line switching, etc. Such changes, which require re-solutions without advancing the time, produce discontinuities in the value of the vector [x]. No discontinuities can appear in [y].

CLASSIFICATION OF SOLUTION METHODS

On a digital computer, the values of [x] and [y] are computed at discrete points in time, thereby producing a sequence of snapshot pictures at intervals Δt. The sampling rate $1/\Delta t$ determines the upper limit of the frequencies which will be adequately represented in [x] and [y]. As a consequence of the discretization, interpolation assumptions must normally be made on [x] to define the state between discrete points. In most cases linear interpolation is adequate.

The methods for solving Eqs. (1) and (2) simultaneously fall roughly into three categories: (a) Alternating solution of Eqs. (1) and (2), (b) elimination of [x], and (c) algebraization of Eq. (2) by implicit integration. The distinction is not always clear. As an example, the numerical solution of differential equations is in itself based on an algebraization process.

(a) Alternating Solution Method

This technique is probably used in most existing stability programs. The algorithm is roughly as follows:

1. Establish initial conditions by solving Eq. (1) at t = 0.
2. Predict how [x] behaves over the next time interval by using zero, first or second order extrapolation.
3. Solve Eq. (2) for [y(t)] by any numerical technique, using the polynomial obtained in step 2 as an interpolation formula.
4. Solve Eq. (1) for [x(t)] by any of the well-known power flow solution techniques.

Paper T72 137-3, recommended and approved by the Power System Engineering Committee of the IEEE Power Engineering Society for presentation at the IEEE Winter Meeting, New York, N.Y., January 30-February 4, 1972. Manuscript submitted September 20, 1971; made available for printing November 19, 1971.

Reprinted from *IEEE Trans. Power App. Syst.*, July/Aug. 1972.

5. Use the computed value $[x(t)]$ to correct the prediction over the interval from $t\text{-}\Delta t$ to t.

6. Solve Eq. (2) for $[y(t)]$ by any numerical technique, using the polynomial obtained in step 5 as an interpolation formula.

7. Check differences of two successive solutions for $[y(t)]$. If the differences are negligible proceed to step 8. Otherwise go to step 4.

8. Advance by one time step and loop back to step 2.

Several variations are possible. In the version described above, the steady-state equations (1) must be solved once, most of the time, since the differences in step 7 are normally sufficiently small on the first pass. The prediction and correction process of the alternating solution method should not be confused with predictor-corrector methods for solving differential equations; if such methods are used, steps 3 and 6 would also involve another kind of prediction-correction process in itself.

It is essential to include the check of step 7 to avoid so-called "interface" errors. If this is not done, a careful assessment should be made about the acceptability of the accumulated errors. With the interface loop closed, the quality of the prediction (step 2) will determine how often the loop must be executed and, in effect, how long the study will take, but it will not degrade the accuracy of the solution because of corrections initiated in step 7. This is not true for open-loop solutions, where the quality of the prediction will have a decisive influence on the accuracy.

(b) Elimination of $[x]$

This approach is used for small disturbance stability studies[3]. For small variations $[\Delta x]$, $[\Delta y]$ around an operating point, Eq. (1) can be linearized,

$$\left[\frac{\partial g}{\partial x}\right][\Delta x] + \left[\frac{\partial g}{\partial y}\right][\Delta y] = 0$$

with the Jacobian matrices $\left[\frac{\partial g}{\partial x}\right], \left[\frac{\partial g}{\partial y}\right]$ being evaluated at the given solution point. Eq. (2) can also be linearized around the operating point,

$$[\Delta\dot{y}] = [A][\Delta y] + [B][\Delta x]$$

Both linear equations combined become

$$\begin{bmatrix} \left[\frac{\partial g}{\partial x}\right] & \left[\frac{\partial g}{\partial y}\right] \\ [B] & [A] \end{bmatrix} \begin{bmatrix} [\Delta x] \\ [\Delta y] \end{bmatrix} = \begin{bmatrix} 0 \\ [\Delta\dot{y}] \end{bmatrix}$$

and after elimination of $[x]$ by triangularization,

$$\begin{bmatrix} \diagdown & \\ 0 & [A]_{reduced} \end{bmatrix} \begin{bmatrix} [\Delta x] \\ [\Delta y] \end{bmatrix} = \begin{bmatrix} 0 \\ [\Delta\dot{y}] \end{bmatrix}$$

or

$$[\Delta\dot{y}] = [A]_{reduced} \cdot [\Delta y] \tag{3}$$

The eigenvalues of $[A]_{reduced}$ give the damping (real part) and natural frequencies (imaginary part) of all possible modes of oscillations without having to make any assumption about the specific (small) disturbance. The system will be stable as long as the real parts of all eigenvalues are negative (first method of Lyapunov).

(c) Algebraization by Implicit Integration

The step-by-step solution of Eq. (2) is best expressed in integral form,

$$[y(t)] = [y(t-\Delta t)] + \int_{t-\Delta t}^{t} [f([y],[x],\tau)]\, d\tau$$

Implicit integration techniques use interpolation functions for the expressions under the integral. Interpolation means that the functions must pass through the yet unknown points at time t, which must therefore be expressed as variables. In general, the solution may require iterations. If the differential equations are linear, however, then a direct solution becomes possible.

Let the differential equations (2) be linear of the form

$$[\dot{y}] = [A][y] + [B][x] \tag{4a}$$

Most differential equations are linear in stability studies. There are some nonlinearities, however, but their inclusion poses no serious problem, as shown later for the excitation system. Eq. (4a) can be rewritten as a step-by-step integration,

$$[y(t)] = [y(t-\Delta t)] + [A]\int_{t-\Delta t}^{t}[y]d\tau + [B]\int_{t-\Delta t}^{t}[x]d\tau \tag{4b}$$

The simplest implicit integration scheme is the trapezoidal rule of integration.* It is based on the assumption that $[x]$ and $[y]$ vary linearly over the interval from $t\text{-}\Delta t$ to t. Then Eq. (4b) becomes

$$[y(t)] = [y(t-\Delta t)] + \frac{\Delta t}{2}[A]\Big[[y(t-\Delta t)] + [y(t)]\Big] + \frac{\Delta t}{2}[B]\Big[[x(t-\Delta t)] + [x(t)]\Big]$$

which is simply a system of linear algebraic equations

$$\left[[U] - \frac{\Delta t}{2}[A]\right][y(t)] - \frac{\Delta t}{2}[B][x(t)] = [c(t-\Delta t)] \tag{5a}$$

The vector on the right-hand side is known from the values already computed at the preceding time step:

$$[c(t-\Delta t)] = \frac{\Delta t}{2}[B][x(t-\Delta t)] + \left[[U] + \frac{\Delta t}{2}[A]\right][y(t-\Delta t)] \tag{5b}$$

$[U]$ denotes the unit or identity matrix. Eq. (5a) can now be solved directly with Eq. (1) as one complete system of steady-state equations.

NUMERICAL STABILITY OF THE TRAPEZOIDAL RULE OF INTEGRATION

The trapezoidal rule of integration is numerically stable. To illustrate the problem of numerical stability, it will be assumed that a fast oscillation in a control circuit produces "ripples" of very small magnitudes which do not have any influence on the overall behavior of the circuit. Such a mode of oscillation could be described by

$$\ddot{y} + y = 0, \text{ with } y(0) = 0,\ \dot{y}(0) = 10^{-4} \tag{6a}$$

The exact solution for this ripple is $y = 10^{-4}\sin(t)$; its amplitude of 10^{-4} will be considered as very small by definition. Eq. (6a) must be rewritten as a system of first-order differential equations in order to apply any of the numerical solution techniques:

$$\begin{bmatrix} \dot{y}_1 \\ \dot{y}_2 \end{bmatrix} = \begin{bmatrix} 0 & 1 \\ -1 & 0 \end{bmatrix} \begin{bmatrix} y_1 \\ y_2 \end{bmatrix} \text{ with } \begin{matrix} y_1 = y \\ y_2 = \dot{y} \end{matrix} \tag{6b}$$

The exact step-by-step solution is

$$\begin{bmatrix} y_1(t) \\ y_2(t) \end{bmatrix} = e^{[A]\Delta t} \begin{bmatrix} y_1(t-\Delta t) \\ y_2(t-\Delta t) \end{bmatrix} \quad \text{whereby } [A] = \begin{bmatrix} 0 & 1 \\ -1 & 0 \end{bmatrix}$$

with the transition matrix $e^{[A]\Delta t}$. Application of the trapezoidal rule to Eq. (6b) gives

$$\begin{bmatrix} y_1(t) \\ y_2(t) \end{bmatrix} = \frac{1}{1 + \frac{\Delta t^2}{4}} \begin{bmatrix} 1 - \frac{\Delta t^2}{4} & \Delta t \\ -\Delta t & 1 - \frac{\Delta t^2}{4} \end{bmatrix} \begin{bmatrix} y_1(t-\Delta t) \\ y_2(t-\Delta t) \end{bmatrix} \tag{7}$$

* There are also higher-order generalizations of the trapezoidal rule[4].

It can be shown that $y_1^2(t) + y_2^2(t) = y_1^2(t-\Delta t) + y_2^2(t-\Delta t)$ in Eq. (7) for any choice of Δt. Therefore, if the process is started with the correct initial condition $y_1^2(0) + y_2^2(0) = 10^{-8}$, the solution for y will always lie between -10^{-4} and $+10^{-4}$, even for step-widths which are much larger than one cycle of the oscillation. In other words, the trapezoidal rule "cuts across" oscillations which are very fast but of negligible amplitude, without any danger of numerical instability.

Explicit integration techniques, which include Runge-Kutta methods, are inherently unstable. They require a step-width tailored to the highest frequency or smallest time constant (rule of thumb: $\Delta t \leq \frac{1}{5} T_{min}$) even though this mode may produce only negligible ripples, with the overall behavior determined by large time constants, as in so-called "stiff-systems."

Applying the conventional fourth-order Runge-Kutta method to Eq. (6b) is identical to a fourth-order Taylor series expansion of the transition matrix in case of linear systems,

$$e^{[A]\Delta t} \cong [U] + \Delta t[A] + \frac{\Delta t^2}{2}[A]^2 + \frac{\Delta t^3}{6}[A]^3 + \frac{\Delta t^4}{24}[A]^4 .$$

Adding the 5 terms gives

$$\begin{bmatrix} y_1(t) \\ y_2(t) \end{bmatrix} = \begin{bmatrix} 1 - \frac{\Delta t^2}{2} + \frac{\Delta t^4}{24} & \Delta t - \frac{\Delta t^3}{6} \\ -\Delta t + \frac{\Delta t^3}{6} & 1 - \frac{\Delta t^2}{2} + \frac{\Delta t^4}{24} \end{bmatrix} \begin{bmatrix} y_1(t-\Delta t) \\ y_2(t-\Delta t) \end{bmatrix} \tag{8}$$

Plotting the curves with a reasonably small Δt, e.g., 6 samples/cycle, reveals that the Runge-Kutta method of Eq. (8) is more accurate at first than the trapezoidal rule, but tends to lose the amplitude later on (Fig. 1).

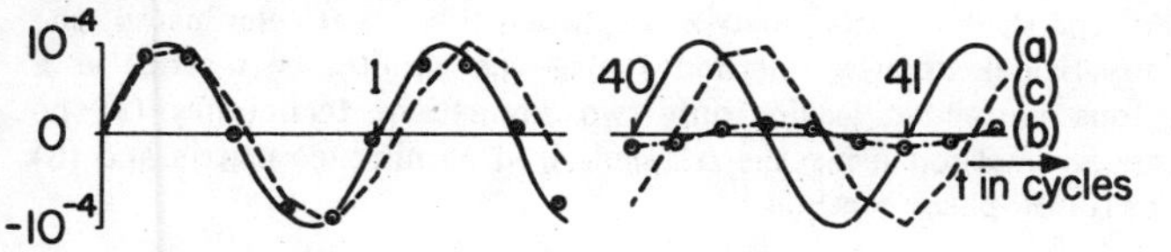

Fig. 1. Numerical solution of $\ddot{y} + y = 0$
(a) Exact, (b) Runge-Kutta,
(c) trapezoidal rule

This is not serious since the ripple is assumed to be unimportant in the first place. If the step-width is increased, however, to $\Delta t > \frac{\sqrt{2}}{\pi}$ cycles ($\Delta t > 0.4502$ cycles), then the amplitude will eventually grow to infinity. This is illustrated in table I for $\Delta t = 1$ cycle.

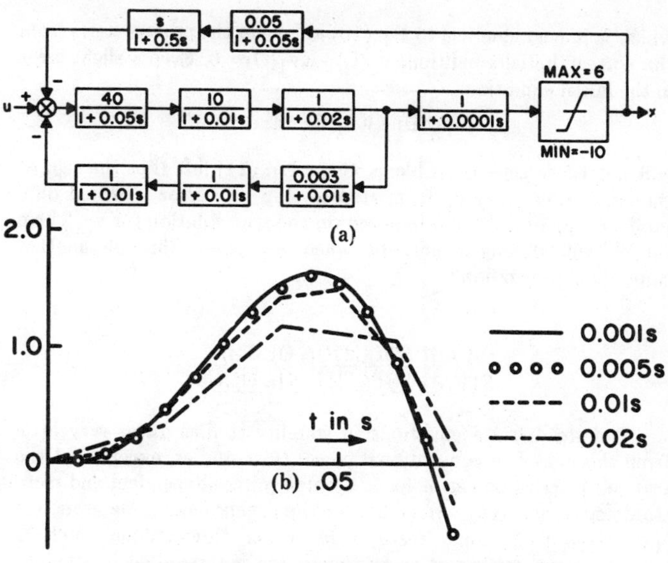

Fig. 2. Test example for trapezoidal rule.
(a) Control circuit. (b) Solution for step-function input u = 0.2.

The plots of Fig. (2b) were obtained with the trapezoidal rule, using different step-widths. The results are acceptable with a step-width as large as 0.01 s, which is 500 times larger than the approximate step-width $\frac{1}{5}T_{min}$ required for the Runge-Kutta method.

Block diagrams with small time constants are often described by equations of the form

$$T\frac{dy_2}{dt} = Ky_1 - y_2, \tag{9}$$

as in Fig. (2a). If T is very small, then the block is basically a multiplier,

$$y_2 = Ky_1 \quad \text{if } T \cong 0 \tag{10}$$

If explicit integration techniques are used in such cases, then it becomes necessary to set small time constants equal to zero to avoid numerical instability. This is bothersome to the user because the problem manifests itself quite often only after the algorithm has failed to give a solution. On the other hand, the programmer hesitates to build the decision into the program because it is difficult to draw the line between small and not so small time constants for all possible situations. This problem does not arise with the trapezoidal rule since the limit case $T \to 0$ turns out to be identical to the algebraic equation (10). Applying the trapezoidal rule to Eq. (9) gives

$$(1 + \frac{2T}{\Delta t})\, y_2(t) - Ky_1(t) = -(1 - \frac{2T}{\Delta t})y_2(t-\Delta t) + Ky_1(t-\Delta t)$$

and for $T \ll \Delta t$

$$y_2(t) - Ky_1(t) = -\left(y_2(t-\Delta t) - Ky_1(t-\Delta t)\right), \tag{11}$$

Table I. Numerical solution of Eq. (6a) with Δt = 1 cycle

t in cycles	1	2	3	4	5	6
exact	0	0	0	0	0	0
trapezoidal rule	0.58 10^{-4}	-0.94 10^{-4}	0.96 10^{-4}	-0.63 10^{-4}	0.06 10^{-4}	0.53 10^{-4}
Runge-Kutta	-0.004	-0.32	-18	-590	6800	2,600,000

VERY SMALL TIME CONSTANTS IN THE TRAPEZOIDAL RULE

The fact that the trapezoidal rule of integration is stable even if the step-width is much larger than the smallest time constant is of practical importance for stability studies. Fig. (2a) shows a hypothetical control circuit with a smallest time constant of 0.0001 s, and a step function input, which is regarded as the most severe test.

which is indeed identical to Eq. (10) as long as the process starts from the correct initial conditions $y_2(0) - Ky_1(0) = 0$. Even a slight error in the initial conditions,

$$y_2(0) - Ky_1(0) = \epsilon$$

will not cause serious problems. Since Eq. (11) just flips the sign of the expression $y_2 - Ky_1$ from step to step, the error ϵ would only produce ripples $\pm\epsilon$ superimposed on the true solution for y_2. These ripples will disappear entirely when y_2 passes through another trapezoidal integration.

FAST SOLUTION OF THE STEADY-STATE EQUATIONS

The steady-state equations in stability studies differ very little from those used in conventional power flow studies, except that the real and reactive power of loads become voltage dependent and more steady-state equations are added for the synchronous generators. On the other hand, some features in power flow studies, such as automatic tap setting of transformers, are not required in stability studies.

Generator Model

A generator model with three armature windings, one field winding and one hypothetical g-coil (which represents deeper flowing eddy currents in steam generators), is adequate to represent transient effects (subtransient effects ignored). The additional steady-state equations for this model are *):

$$\begin{bmatrix} I_q \\ I_d \end{bmatrix} = \frac{1}{R_a^2 + X'_d X'_q} \begin{bmatrix} R_a & X'_d \\ -X'_q & R_a \end{bmatrix} \begin{bmatrix} E'_q - V_q \\ E'_d - V_d \end{bmatrix} \qquad (12)$$

with the quantities in the q, d-reference frame related to the phasor quantities $\bar{I}, \bar{V}, \bar{E}'$ of the network solution by

$$I_q + jI_d = \bar{I} \cdot e^{j\delta}, \text{ with } \bar{I} = I_{real} + jI_{imag}, \text{ analogous for } \bar{V}, \bar{E}' \qquad (13)$$

The angle δ measures the rotor position of the generator relative to the synchronously rotating reference frame, which is implied in phasor solutions of the network. The parameters δ, E'_q, E'_d are part of the vector [y], whose value is determined from the solution of the differential equations, whereas I_q, I_d, V_q, V_d are part of the vector [x]. In the alternating solution method, δ, E'_q, E'_d are treated as known quantities during the solution of the steady-state equations.

Eq. (12) is easy to handle if saliency is ignored, that is, if $X'_q = X'_d$. In this case, the 2 equations can be written as one equation with phasor quantities,

$$\bar{I} = \frac{1}{R_a + jX'_d} (\bar{E}' - \bar{V}) \text{ for } X'_q = X'_d \qquad (14)$$

which is simply a known voltage $\bar{E}'$ behind transient impedance $R_a + jX'_d$. In effect, each internal node with the voltage $\bar{E}'$ is a slack node.

The inclusion of Eq. (12) in the network solution becomes more complicated if saliency is considered, that is, if $X'_q \neq X'_d$. When Eq. (12) is transformed to the reference frame of the network solution,

$$\begin{bmatrix} I_{real} \\ I_{imag} \end{bmatrix} = [M(t)] \begin{bmatrix} E'_{real} - V_{real} \\ E'_{imag} - V_{imag} \end{bmatrix}, \qquad (15)$$

$$\text{with } [M(t)] = \frac{1}{R_a^2 + X'_d X'_q} \begin{bmatrix} \cos\delta & -\sin\delta \\ \sin\delta & \cos\delta \end{bmatrix} \begin{bmatrix} R_a & X'_d \\ -X'_q & R_a \end{bmatrix} \begin{bmatrix} \cos\delta & \sin\delta \\ -\sin\delta & \cos\delta \end{bmatrix}$$

then two complications in Eq. (15) as compared to Eq. (14) become apparent:

1. It is no longer possible to combine the two equations (15) into one phasor equation, and

2. the matrix [M(t)] is now a function of time, with δ entering into it.

If subtransient effects are to be included, then the model must have two more windings "kd" and "kq" in the direct and quadrature axis to represent damper bars or hypothetical coils for surface eddy currents. This adds 2 more differential equations, but the algebraic equations can again be reduced to the form of Eq. (12) except that the primed quantities are now replaced by double primed quantities,

$$\begin{bmatrix} I_q \\ I_d \end{bmatrix} = \frac{1}{R_a^2 + X''_d X''_q} \begin{bmatrix} R_a & X''_d \\ -X''_q & R_a \end{bmatrix} \begin{bmatrix} E''_q & -V_q \\ E''_d & -V_d \end{bmatrix} \qquad (16)$$

All arguments for the model with transient effects are, therefore, also applicable to the model including subtransient effects.

The complete system of steady-state equations is solved by adopting well-known techniques for fast power flow solutions to the stability problem. The improvement in the solution of systems of linear equations by optimally ordered triangular factorization with the sparsity of the matrix exploited [5,6] has eliminated the Gauss-Seidel iterative method (using the admittance matrix) as a serious contender, leaving only two competitive techniques: (a) the iterative method using the triangularized admittance matrix and (b) Newton-Raphson method.

(a) Iterations with the Triangularized Admittance Matrix

First, the inclusion of Eq. (12) or (16) must be resolved. The simplest way would be to use the terminal current $I_{real} + jI_{imag}$ as a given node current and adjust it iteratively as improved answers are obtained for $V_{real} + jV_{imag}$. Experiments have shown, however, that this simple technique fails to converge quite frequently. Convergence can be improved considerably by creating a fictitious slack node behind an impedance; the value of the slack node voltage will then depend more or less on $V_{real} + jV_{imag}$ and must, therefore, be adjusted iteratively. Using $R_a + jX'_d$ as the impedance to the fictitious slack node improves the situation, but in some cases convergence is still difficult.

It was found that the convergence becomes very fast if the admittance to the fictitious slack node is set to

$$\bar{Y}_{fictitious} = \frac{R_a - j\frac{1}{2}(X'_d + X'_q)}{R_a^2 + X'_d X'_q}$$

Two or three iterations are normally sufficient to adjust tne fictitious slack node voltage, or--if an equivalent current source is used in parallel with the shunt admittance $\bar{Y}_{fictitious}$ instead of a voltage source, as in BPA's stability program (Fig. 3)--to adjust the fictitious current source.

*) The newly recommended position for the quadrature axis lagging 90° behind the direct axis is adopted.[7]

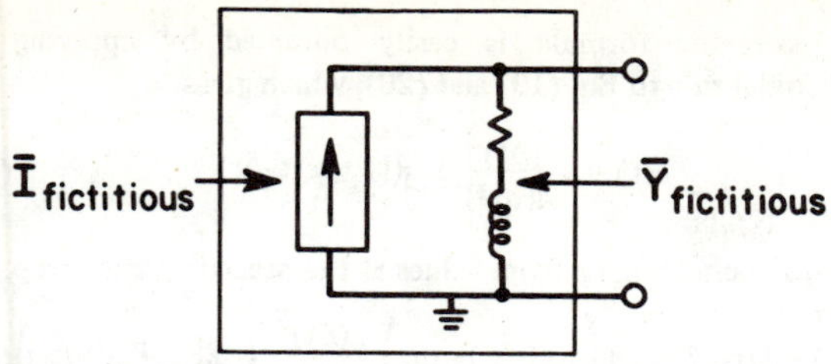

Fig. 3. Equivalent circuit for generator.

The value of the current source is determined from

$$\bar{I}_{fictitious} = \frac{R_a - j\frac{1}{2}(X'_d + X'_q)}{R_a^2 + X'_d X'_q} \cdot \bar{E}' + \bar{I}_{salient} \tag{17a}$$

with the first part being known and the second part being adjusted iteratively:

$$\bar{I}_{salient} = -j\frac{1}{2} \frac{X'_q - X'_d}{R_a^2 + X'_d X'_q} (\bar{E}'^* - \bar{V}^*)\, e^{j2\delta} \tag{17b}$$

(all quantities are phasors in the network solution reference frame; "*" denotes conjugate complex). As an initial guess for the terminal voltages, the values obtained at the preceding time step are used, except that their angle is advanced by the same amount by which the internal angle δ has changed from $t\text{-}\Delta t$ to t. Omitting this rotation would double the number of required iteration steps.

Techniques for avoiding the iterations due to saliency altogether are hardly worthwhile, because these 2 or 3 iterations do not add to the number of iterations which are normally required for non-impedance load models anyhow. Also, gains in one place would probably be offset somewhere else (example: If two real equations with two real variables were used on terminal nodes of generators, then Eq. (15) could be used without any need for iterations. However, the admittance matrix would then have to be partly triangularized in each time step because of the time-dependence of [M]).

The nodes can now be subdivided into two sets,

subset 1 of all "external" nodes (terminals of generators, loads, passive nodes) and,
subset 2 of all "internal" nodes (true slack nodes if $X'_q = X'_d$, fictitious slack nodes otherwise).

The nodal admittance matrix is partitioned accordingly,

$$[\bar{Y}] = \begin{bmatrix} [\bar{Y}_{11}] & [\bar{Y}_{12}] \\ [\bar{Y}_{21}] & [\bar{Y}_{22}] \end{bmatrix}$$

With $[\bar{V}_2]$ known (slack bus voltages), $[\bar{V}_1]$ has to be found by solving the system of linear equations,

$$[\bar{Y}_{11}][\bar{V}_1] = [\bar{I}_1] - [\bar{Y}_{12}][\bar{V}_2] \tag{18}$$

All other quantities can then be computed from $[\bar{V}_1]$ and $[\bar{V}_2]$. Eq. (18) is solved by triangular factorization, using optimal ordering and sparsity techniques. The matrix $[\bar{Y}_{11}]$ must only be triangularized initially and whenever network changes (short-circuit, line switchings) take place. Otherwise, it remains constant for all iteration steps as well as over the time steps. With $[\bar{Y}_{11}]$ already triangularized, Eq. (18) is solved by a "repeat" solution, in which the triangularization process is extended to the right-hand side only. Repeat solutions are about five times faster than triangularization of the matrix. Typical solution times on a CDC 6400 for a 1190-node case are:

Triangular factorization of the matrix	8.6 s
repeat solution	1.6 s

The term $-[\bar{Y}_{12}][\bar{V}_2]$ in Eq. (18) is identical with the generator currents from Eq. (17a), with $\bar{I}_{salient} = 0$ if $X'_q = X'_d$. The vector $[\bar{I}_1]$ represents the current injections on the external nodes, including load nodes. Loads are converted to equivalent impedances which give the correct power at the nominal voltages of the initial power flow solution; their values enter into the diagonal elements of $[\bar{Y}_{11}]$. If all loads are represented as impedances then Eq. (18) would be non-iterative with $[\bar{I}_1] = 0$, except for iterations due to saliency. If loads are not modeled as constant impedances, then $[\bar{I}_1]$ is adjusted iteratively to account for the difference between quadratic voltage dependence of the impedance and the specified voltage dependence. Converting the loads to impedances at nominal voltage reduces the number of iterations by one half as compared to representing the loads entirely in $[\bar{I}_1]$. The average number of iterations depends very much on the load representations; 5 to 7 iterations are typical figures. Unrealistic load representations may increase the number of iterations appreciably. The user should have good reason, therefore, to represent loads which differ appreciably from constant impedance.

(b) Newton-Raphson Method

The Newton-Raphson method applied to Eq. (18) has already been explained in detail elsewhere[8,9]; therefore, the equations will not be repeated here. The most promising approach with the Newton-Raphson method is the version which uses the current equation form and rectangular coordinates for the variables[9], with the following modification: Re-use the old Jacobian matrix from the preceding time step for the first iteration step (this requires saving the lower as well as the upper triangular matrix, though) and compute a new Jacobian matrix for the next iteration step. These two steps (one repeat solution plus one complete solution) usually solve the steady-state equations with sufficient accuracy. The version using the current equation and rectangular coordinates has the advantage of being non-iterative for nodes with constant impedance loads[9]. Saliency does not pose any problem because 2N real equations and variables are used in the Newton-Raphson method anyhow, instead of N complex equations and variables, and because the matrix must be triangularized anew in each time step. In this case, Eq. (15) is easy to handle because it is linear in the unknown variables V_{real} and V_{imag}.

(c) Comparison of the Two Methods

The iterative solution using the triangularized admittance matrix and the Newton-Raphson method are closely competitive. At this time, the former method appears to be faster and preferable for the following reasons:

1. The Newton-Raphson method requires one re-triangularization in each time step, whereas the admittance matrix must only be re-triangularized when network changes take place. Tests indicate that the solution with the triangularized admittance matrix is faster if it converges in less than 5 to 10 iterations.

2. The Newton-Raphson method requires more storage, up to three times as much when the lower triangular matrix is saved.

In the second argument it was assumed that the matrix $[\bar{Y}_{11}]$ is symmetric; the lower triangular matrix is not needed then. The comparison becomes less advantageous for the triangularized admittance matrix if the network contains phase shifting transformers. In

* Loads which cannot be modeled as constant impedance.

such cases, those columns of the lower triangular matrix which are affected by the phase-shifting transformers must also be stored.

It is entirely possible that the problem formulation will change in future stability studies in such a way that the Newton-Raphson method will become preferable. One situation which is difficult to handle with the triangularized admittance matrix, but is very easy with the Newton-Raphson method, is the node type where real power and voltage magnitude are specified. Another situation which might be easier with the Newton-Raphson method is frequency-dependence of machines, of loads, and line impedances.

SIMULTANEOUS SOLUTION OF STEADY-STATE AND DIFFERENTIAL EQUATIONS

The feasibility of speeding up stability studies by using the trapezoidal rule of implicit integration has been demonstrated at BPA with a number of test cases. These experiments will be explained in order of increasing complexity.

Swing Equation

The only differential equations in "classical" stability studies are the swing equations for each generator:

$$(\omega J)\frac{d\omega}{dt} = P_m - P_{el} \qquad (19)$$

and

$$\frac{d\delta}{dt} = \omega - \omega_s \qquad (20)$$

with J = moment of inertia,

ωJ = angular momentum,

ω = speed (ω_s = synchronous speed) on electrical side,

δ = rotor position relative to synchronously rotating reference frame,

P_m = shaft power input,

P_{el} = electrical power output, corrected for electrical losses.

For simplicity, damping terms will be neglected in Eq. (19), but their inclusion poses no serious problem. It is also customary to set $\omega J \cong \omega_s J$, which is permissible as long as the speed deviates very little from the synchronous speed. In other cases, ωJ should be changed as the speed changes.

Since classical stability studies have been performed for many years, long before digital computers became available, it is worthwile to look at the techniques which were used for hand calculations. The most frequently used procedure has been a predictor formula[10]:

$$\delta(t) = 2\delta(t\text{-}\Delta t) - \delta(t\text{-}2\Delta t) + \frac{(\Delta t)^2}{\omega(t\text{-}\Delta t)J}\left(P_m - P_{el}(t\text{-}\Delta t)\right) \qquad (21)$$

Eq. (21) can be derived by making two assumptions:

1. Integrate Eq. (19) from midpoint $t\text{-}\frac{3}{2}\Delta t$ to midpoint $t\text{-}\frac{1}{2}\Delta t$, assuming that $\frac{P}{\omega}$ varies linearly in this interval, which gives
$$\omega(t-\tfrac{1}{2}\Delta t) = \omega(t-\tfrac{3}{2}\Delta t) + \frac{\Delta t}{\omega(t-\Delta t)J}\left(P_m - P_{el}(t\text{-}\Delta t)\right)$$
2. Integrate Eq. (20) from $t\text{-}\Delta t$ to t, assuming that ω varies linearly in this interval with its average value being $\omega(t-\frac{1}{2}\Delta t)$.

Eq. (21) is sufficiently accurate for normal step-widths of 1 to 5 cycles. There is no reason, therefore, why it should not be used in computer programs, especially if a corrector formula is added as a safeguard.

The corrector formula is easily obtained by applying the trapezoidal rule to Eq. (19) and (20), which gives

$$\delta(t) = -\frac{(\Delta t)^2}{4(\omega J)} P_{el}(t) + \alpha(t\text{-}\Delta t) \qquad (22a)$$

with $\alpha(t\text{-}\Delta t)$ being known from values at the preceding time step,

$$\alpha(t\text{-}\Delta t) = \delta(t\text{-}\Delta t) + \Delta t\left(\omega(t\text{-}\Delta t) - \omega_s\right) + \frac{(\Delta t)^2}{4(\omega J)}\left(2P_m - P_{el}(t\text{-}\Delta t)\right) \qquad (22b)$$

In Eq. (22) it is assumed that P_m is constant, but the equation can easily be changed if P_m changes as a function of time or if the torque T_m is constant.

Eq. (22a) must be used after discontinuities where Eq. (21) is no longer valid and should be used at other times to avoid interface errors. It is merged into the solution of the steady-state equations as follows:

1. Predict δ with Eq. (21) (at times of discontinuities assume that δ will not change over the next interval).
2. At the end of each steady-state iteration step (or in the back-substitution) compute P_{el} and insert it into Eq. (22a) to get a corrected angle. Correct the angle only if the change is noticeable, to avoid unnecessary computations of $\sin\delta$ and $\cos\delta$.
3. Update α after the steady-state solution has been completed.

A more direct approach could be used with the Newton-Raphson method, where Eq. (22a) could be added as an additional equation and δ as an additional variable. The size of the Jacobian matrix would not increase if δ is eliminated as soon as the elimination hits the rows of the terminal node, that is, if the technique is used which was described for phase shifting angles in reference 9.

Except at times of discontinuities, the corrections are normally quite small, and in such cases, Eq. (22a) serves only as a check. If iterations are required anyhow because of saliency or non-impedance loads, then the corrector formula Eq. (22) will not add to the total number of iterations.

Differential Equations of the Generator

For a generator model with transient effects only, there are two differential equations,

$$\frac{dE'_q}{dt} = -\frac{1}{T'_{do}}\left(E'_q - (X_d - X'_d)I_d - E_{FD}\right) \qquad (23a)$$

$$\frac{dE'_d}{dt} = -\frac{1}{T'_{qo}}\left(E'_d + (X_q - X'_q)I_q\right) \qquad (23b)$$

(E_{FD} = voltage applied to terminals of field winding). Application of the trapezoidal rule transforms Eq. (23) into

$$E'_q(t) = a_d(X_d - X'_d)I_d(t) + F_q \qquad (24a)$$

$$E'_d(t) = -a_q(X_q - X'_q)I_q(t) + F_d \qquad (24b)$$

with F_q and F_d being known from values at the preceding time step at $t\text{-}\Delta t$ and from $E_{FD}(t)$ (which is also known if the exciter is represented as a simple voltage source),

$$F_q = E'_q(t\text{-}\Delta t) + a_d\Big(E_{FD}(t\text{-}\Delta t) + E_{FD}(t) - 2E'_q(t\text{-}\Delta t) + (X_d - X'_d)I_d(t\text{-}\Delta t)\Big) \qquad (25a)$$

$$F_d = E'_d(t-\Delta t) - a_q\left(2E'_d(t-\Delta t) + (X_q - X'_q)\, I_q\, (t-\Delta t)\right) \qquad (25b)$$

with $a_d = \dfrac{\Delta t}{\Delta t + 2T'_{do}}, \quad a_q = \dfrac{\Delta t}{\Delta t + 2T'_{qo}}$

Eq. (24) inserted into Eq. (12) gives the complete description of all algebraic and differential equations, reduced to an equivalent of two algebraic equations:

$$\begin{bmatrix} I_q \\ I_d \end{bmatrix} = \frac{1}{R_a^2 + X'_{dmod} X'_{qmod}} \begin{bmatrix} R_a & X'_{dmod} \\ -X'_{qmod} & R_a \end{bmatrix} \begin{bmatrix} F_q & -V_q \\ F_d & -V_d \end{bmatrix} \qquad (26)$$

with the modified reactances

$$X'_{dmod} = X'_d + a_d\,(X_d - X'_d)$$

$$X'_{qmod} = X'_q + a_q\,(X_q - X'_q)$$

For normal time constants and step-widths in the order of a few cycles, the modified transient reactances differ only by a few percent from the original transient reactances. The only approximation which had to be made is linear interpolation between $t-\Delta t$ and t for the quantities E'_q, E'_d, I_q, I_d, E_{FD}. It is significant that Eq. (26) has the same structure as Eq. (12). As a consequence, Eq. (26) can simply replace Eq. (12) with no need to change the solution algorithm which was already developed for Eq. (12). The differential equations (23) will be solved implicitly without requiring any extra work, except that F_q and F_d have to be updated from step to step. At times of discontinuities with no advance in time, Eq. (12) must be used in place of Eq. (26) for the second "post-change" solution, or, if it is desirable to keep the impedances unchanged, Eq. (26) may be used iteratively with the following values for F_q and F_d:

$$F_q = E'_q - a_d\,(X_d - X'_d)\, I_d$$

$$F_d = E'_d + a_q\,(X_q - X'_q)\, I_q$$

with E'_q, E'_d determined from the "pre-change" solution and I_q, I_d being obtained iteratively within the "post-change" solution.

The same ideas should be applicable to generator models which include subtransient effects. In this case, large step-widths (greater than 1-2 cycles) may not be able to represent the fast decay of the currents in the kd- and kq-windings after discontinuities very accurately, but the subsequent oscillations and the over-all damping effect should be fairly accurate. Tests are planned to verify this conjecture.

Excitation System

Inclusion of the excitation system was tried with a 9-bus test example. It contained three generators with their swing equations. Two of them were "classical" generator representations, described by Eq. (14), and one was a fully represented generator, described by Eq. (12) and (23), with an exciter model of type 1 as defined in Ref. 11. The smallest time constant in the exciter model was $T_R = 0.06$ s.

First, the differential equations of the four exciter blocks are transformed into algebraic equations with the trapezoidal rule. These can then be reduced to one linear algebraic equation of the form,

$$E_{FD}(t) = E_o - b\, V_T\,(t) \qquad (27)$$

and one inequality $E_{LOW} \leq E_{FD}\,(t) \leq E_{HIGH}$ (28)

The saturation is taken into account by linearizing the feedback $S_E E_{FD}$ around the solution point at the preceding time step at $t-\Delta t$, which gives an equation of the form

$$S_E E_{FD} = k_1 E_{FD} - k_2 \qquad (29)$$

with k_1 and k_2 being known. A simpler way would be to use a piecewise linear saturation curve; in this case, k_1 and k_2 would not have to be reevaluated in each time step but only when a new region of the piecewise linear curve is entered. For the 9-bus test example, a linearized curve with only 2 slopes, one for the unsaturated and one for the saturated region, gave results which were close enough for practical purposes to those obtained with the detailed saturation curve.

The parameter b in Eq. (27) is recomputed whenever k_1 and k_2 change, or whenever the regulator output reaches its limit,

$$b = \frac{K_A/a_R}{a_A a_E + K_F K_A/(T_F + \frac{\Delta t}{2})} \qquad (30)$$

with $a_R = \dfrac{2T_R}{\Delta t} + 1, \; a_A = \dfrac{2T_A}{\Delta t} + 1, \; a_E = \dfrac{2T_E}{\Delta t} + K_E + k_1$

The value E_o is computed from a linear combination of the values of five state variables (including E_{FD} and V_T) at the preceding time step $t-\Delta t$.

The limits E_{LOW} and E_{HIGH} are dynamic limits and not the steady-state limits of the exciter. They can easily be computed in each time step by assuming that the regulator output V_R goes from its state at $t-\Delta t$ to its upper or lower limit at time t,

$$E_{HIGH} = (b_E + V_{R\,MAX})/a_E$$

$$E_{LOW} = (b_E + V_{R\,MIN})/a_E$$

with $b_E = E_{FD}(t-\Delta t)\left(\dfrac{2T_E}{\Delta t} - K_E - k_1\right) + V_R(t-\Delta t) + 2k_2.$

Once the regulator output reaches its limit, the parameter K_A is temporarily set to zero in computing b from Eq. (30), and V_R is set to its limit. Thereafter, the derivative of the regulator output is used to check when to back off the limits again, in which case K_A is restored to its original value.

The inclusion of the excitation system in the solution of the steady-state equations is now very simple:

1. Update the state variables of the exciter at time $t-\Delta t$; compute E_o, b, E_{HIGH}, E_{LOW} and an estimate for $E_{FD}(t)$ with Eq. (27) by assuming that V_T will not change from $t-\Delta t$ to t.

2. At the end of each steady-state iteration, compute $E_{FD}(t)$ from Eq. (27), observing the constraints of Eq. (28), and correct F_q as defined in Eq. (25a) accordingly.

The iterative adjustments of $E_{FD}(t)$ and F_q will not add to the 2-3 iterations required for saliency or to additional iterations caused by non-impedance loads.

It cannot be emphasized strongly enough that the few extra statements in the steady-state solution (two expressions for finding E_{FD} and correcting F_q and two IF-statements for checking the limits of E_{FD}) will implicitly bring about the solution of the differential equations without adding more iterations than are required for saliency and non-impedance loads, and without producing interface errors. The updating formulas are also simple enough to be very fast (35 FORTRAN statements in the test program).

Fig. 4 shows the field voltage E_{FD} using the described technique with step-widths of 1 and 5 cycles (fault applied at t=O, fault cleared and line removed at t=5 cycles). The larger step-width produces acceptable results without any danger of numerical instability.

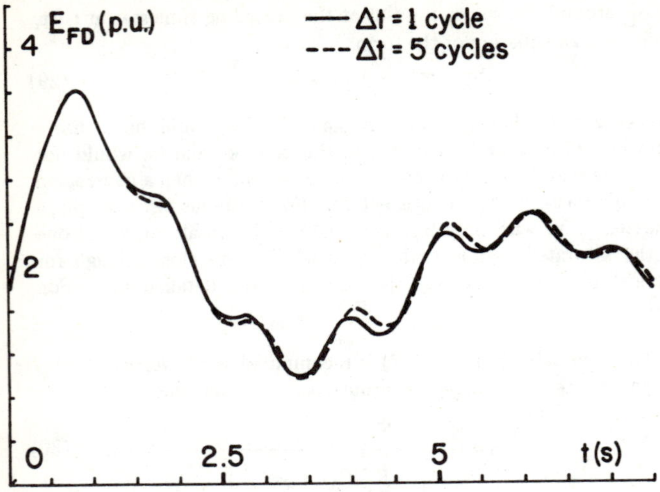

Fig. 4. E_{FD} as function of time.

A more direct approach would be possible with the Newton-Raphson method where an additional equation (27) and an additional variable E_{FD} could be added directly in the same fashion as indicated for the swing equation.

CONCLUSION

In describing the authors' experience in solving transient stability problems, this paper covers two methods in particular which drastically reduce solution times without sacrificing accuracy: (1) Efficient solution of the steady-state network equations using optimally ordered triangular factorization, and special techniques to reduce the number of iterations due to generator saliency and non-impedance loads and (2) Solution of the differential equations with the trapezoidal rule of integration, which is very fast and numerically stable. It also eliminates the interface problem.

ACKNOWLEDGEMENT

The authors are indebted to F. D. Byrnes, Bonneville Power Administration, for his many contributions, notably for his technique for handling saliency as described in the paper.

REFERENCES

1 W. F. Tinney and N. M. Peterson, "Steady-state security monitoring," Brown-Boveri Symposium, Real-Time Control of Electric Power Systems, Baden, Switzerland, Sept. 27-28, 1971.

2 H. W. Dommel, "Digital computer solution of electromagnetic transients in single and multiphase networks," IEEE Trans. Power Apparatus and Systems, vol. PAS-88, pp. 388-399, April 1969.

3 J. A. Pinnello and J. E. VanNess, "Dynamic response of a large power system to a cyclic load produced by a nuclear accelerator," IEEE Trans. Power Apparatus and Systems, vol. PAS-90, pp. 1856-1862, July/August 1971.

4 E. J. Davison, "A high-order Crank-Nicholson technique for solving differential equations," The Computer Journal, vol. 10, no. 2, pp. 195-197, August 1967.

5 N. Sato and W. F. Tinney, "Techniques for exploiting the sparsity of the network admittance matrix," IEEE Trans. Power Apparatus and Systems, vol. 82, pp. 944-950, December 1963.

6 W. F. Tinney and J. W. Walker, "Direct solutions of sparse network equations by optimally ordered triangular factorization," Proc. IEEE, vol. 55, pp. 1801-1809, November 1967.

7 IEEE Committee Report, "Recommended phasor diagram for synchronous machines," IEEE Trans. Power Apparatus and Systems, vol. PAS-88, pp. 1593-1610, November 1969.

8 W. F. Tinney and C. E. Hart, "Power flow solution by Newton's method," IEEE Trans. Power Apparatus and Systems, vol. PAS-86, pp. 1449-1460, November 1967.

9 H. W. Dommel, W. F. Tinney, and W. L. Powell, "Further developments in Newton's method for power system applications," IEEE Paper No. 70 CP 161 - PWR, presented at the IEEE Winter Power Meeting, New York, Jan. 25-30, 1970.

10 O. G. C. Dahl, Electric Power Circuits, vol. II, Power System Stability. New York: Mc-Graw Hill, 1938, pp. 391-401.

11 IEEE Committee Report, "Computer representation of excitation systems," IEEE Trans. Power Apparatus and Systems, vol. PAS-87, pp. 1460-1464, June 1968.

Dynamic Stability of the Peace River Transmission System

H. M. ELLIS, SENIOR MEMBER, IEEE, J. E. HARDY, MEMBER, IEEE, A. L. BLYTHE, MEMBER, IEEE, J. W. SKOOGLUND, SENIOR MEMBER, IEEE

Abstract—**The transient and dynamic stability of the Peace River 500-kV transmission system was studied with a new digital computer program. The detailed representation included direct and quadrature machine axes, saturation, excitation, and governing systems. Stability is maintained following faults by employing a braking resistor and additional stabilizing signal to the generator excitation system.**

Introduction

THE PEACE RIVER Hydro-Electric Project of the British Columbia Hydro and Power Authority (B. C. Hydro) is now under construction with the first stage consisting of three units scheduled for completion in the autumn of 1968. The ultimate system will consist of ten units with a total output of 2300 MW and two 500-kV transmission lines covering a distance of 580 miles to Vancouver, B. C.

Portage Mountain Development

The Peace River Project involves 2300 megawatts in the main site at Portage Mountain Dam on the Peace River, near the head of the Peace River Canyon some 55 air miles west of Fort St. John. A later stage visualizes a run-of-river plant approximately ten miles downstream, referred to as Site No. 1, involving an additional 600 MW.

The Portage Mountain Dam will be an earth-fill structure of 60 million cubic yards with a crest length of 6700 feet and a height of approximately 600 feet. The reservoir created will have an area of 680 square miles forming a 225-mile long lake with a storage capacity of 62 million acre feet.

The powerhouse will be underground with an installation of ten 227-MW, 95-percent power-factor generators, each direct connected to a single runner Francis turbine rated at 310 000 hp, 150 r/min.

General System Description

The bulk of the population, and the greatest use for power, is in the lower British Columbia area, and while very rapid development is taking place in the Central Interior and Northern parts of the Province, the initial need is to transmit this Peace River power southward to the Lower Fraser Valley and Greater Vancouver Area involving a transmission distance of 580 miles.

An EHV transmission voltage of 500 kV was selected for this purpose. Higher voltages up to 735 kV were found to be uneconomical because of increased carrying charges in the early years, and a lower voltage, 345 kV, resulted in increased capital costs with additional problems resulting from the limited availability of rights-of-way through the mountainous terrain of B. C. for the number of circuits that would be required.

It should also be pointed out that there are additional hydro sites in the northern part of the Province, including the lower Peace River, which could produce several thousand megawatts of capacity. Additional EHV transmission lines will be required for these sites at a later date.

For the initial 10-unit development at Peace River producing 2300 MW, a 2-line 500-kV transmission system will be used. With two intermediate switching stations, the line sections are approximately 200 miles in length. The main transmission system is shown in Fig. 1. A more complete description of the Peace River EHV Transmission System is presented by Steede et al. [15].

Substation Characteristics

The switching station at the generating plant has a modified breaker-and-a-third scheme. This scheme limits loss of generation to two units for breaker failure on switching out a faulted line or for a 500-kV bus fault.

The two intermediate switching and load substations, located at Prince George and Kelly Lake, have a modified double bus arrangement. With this type of switching arrangement, only one line section is disconnected for breaker failure on switching a faulted line or for a 500-kV bus fault. The 500/230-kV transformer banks at these intermediate stations are rated 300 MVA each, based on load and switched reactor requirements.

The two receiving stations in the Lower Mainland, Ingledow and Meridian, form the backbone of the Metropolitan network. The 500/230-kV auto-transformers at these stations will be rated 600 MVA per bank. The total land requirement for each of these stations is 15 to 20 acres.

The initial stage of the transmission system, scheduled for in-service in September 1968 will include three generators together with a single 500-kV line. High-speed reclosing together with dynamic braking resistors will be

Paper 31 TP 65-813 recommended and approved by the Transmission and Distribution Committee of the Power Group for presentation at the IEEE EHV-AC Transmission Conference, Richmond, Va., October 4–6, 1965. Manuscript submitted July 5, 1965; made available for printing August 20, 1965.

H. M. Ellis and J. E. Hardy are with the British Columbia Hydro and Power Authority, Vancouver, B. C., Canada.
A. L. Blythe is with the Canadian Westinghouse Company, Ltd., Hamilton, Ont., Canada.
J. W. Skooglund is with the Westinghouse Electric Corporation, East Pittsburgh, Pa.

Reprinted from *IEEE Trans. Power App. Syst.*, June 1966.

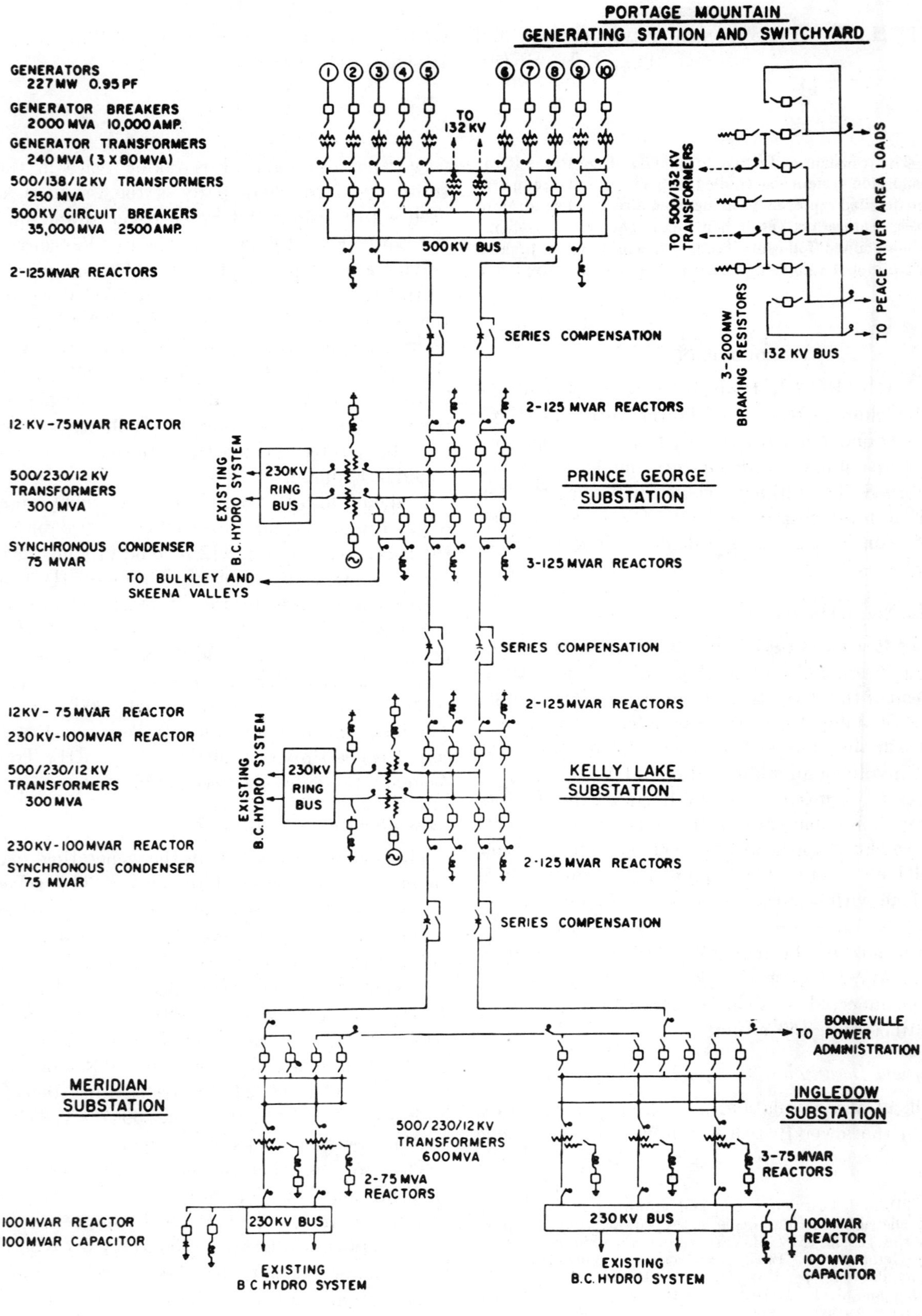

Fig. 1. Peace River 500-kV transmission system—1-line diagram.

utilized throughout the stage development to provide optimum transmission system stability. It is anticipated that the second line will be completed in 1970.

Series capacitor compensation will be gradually built up as the transmission loading increases. When the 2300-MW stage is reached, capacitor banks, installed at the midpoint of the line sections, will have installed capacities of 350 to 700 MVAR.

System Studied

For the purposes of the study, the major receiving stations at Ingledow and Meridian could not be considered an infinite bus and therefore, to obtain as much accuracy as possible, a detailed representation of the B. C. Hydro system and its ties to the Northwest Power Pool (BPA) was used. The system representation used in the study is illustrated in Fig. 2, which shows hydro generating stations at Portage Mountain, Bridge River No. 1, Bridge River No. 2, and Cheakamus, a thermal generating station at Burrard, equivalent generating station representations for the Vancouver Island system, the Lower Mainland 60-kV system, and the BPA system. All major transmission ties between these plants were included so that an accurate representation of system impedances between generating stations was obtained.

Preliminary transient stability studies on the ac network analyzer (using constant internal voltage behind transient reactance) indicated that only marginal stability would be obtained when the 500-kV lines were 50 percent series-capacitor compensated. Since compensation beyond 50 percent was considered undesirable because of relaying difficulties, it was decided to investigate what improvement in stability margin could be obtained by using high-speed excitation, optimizing machine characteristics, and employing special devices such as braking resistors and excitation forcing signals.

An initial stage of development was also studied with two units at Portage Mountain and a single 500-kV line to Kelly Lake, connected into the existing B. C. Hydro transmission system. For this stage there is no Meridian bus and no 500-kV tie between Ingledow and BPA. With only a single line from Peace River, a permanent fault would isolate the plant, and the output cannot be considered firm. However, it is highly desirable to maintain the plant output through a temporary fault condition if possible. For this reason, studies were conducted to determine the feasibility of 3-pole and single-pole reclosing on the initial system.

Criterion of Stability

The objective was to design a system which would be dynamically stable with one 500-kV line section out of service and transiently stable for a permanent double-line-to-ground (DLG) fault on one circuit cleared in six cycles, with high-speed unsuccessful reclosure into the fault. Thus, stability of the system for any one fault was not to be dependent upon successful reclosure. However, since the lines pass through an area of relatively low isokeraunic level, they are designed without overhead ground wires (except for short lengths adjacent to the stations). With no ground wires, the probability of successive faults during a given storm is relatively high, hence it is necessary to restore a faulted circuit to service as soon as possible.

This severe stability criterion resulted in the need for dynamic braking resistors to maintain stability for both the ultimate and initial systems. The switching of these resistors, the representation of series-capacitor gap flashover and reinsertion, the detailed representation of a relatively large number of machines and transmission lines all contributed to the complexity of the study. A simplifying assumption of a 3-phase fault for five cycles rather than a DLG fault for six cycles was made for most of the cases studied. Only a comprehensive digital computer program is capable of representing the detail required for this study. The main features of such a digital stability program are described in the following section.

Study Techniques

System Representation

The digital computer program used for this study includes a very detailed system representation of up to 100 generators and 1000 lines. This permitted the representation of the B. C. Hydro high-voltage lines without using equivalents. A detailed discussion of the computer program will not be given here but is included in the work by Olive [1].

Transmission circuits were considered to be constant impedance pi sections for the longer lines. System loads are also considered constant impedance unless treated as *nonconforming*. Except for a few cases, the constant impedance load representation was used.

Series capacitors were included as negative reactance paralleled by a low-impedance branch when simulating gap flashover. Since the digital program does not use an iterative solution for the network flow, negative impedances and branches of very low impedance may be employed to simulate series capacitors and various switching operations. In this investigation the series capacitors were reinserted five cycles after fault clearing.

The eleven synchronous machines were represented in different degrees of complexity, the most detailed including: direct and quadrature axes, flux linkage decay in both axes, saturation in both axes, positive sequence damping, mechanical damping, and detailed excitation and governor systems. Inertia constants were used for all units including the BPA system. No infinite bus was used.

A block diagram illustrating the excitation system representation is given in Fig. 3. The K's are gains and the T's, time lag (denominator) or phase lead (numerator). The original excitation system configuration with feedback is converted to an equivalent forward transfer function. Typical constants shown are for the electronic exciter of the Peace River generators. An additional exciter stabilizing signal of Δw (velocity deviation) was used to improve the

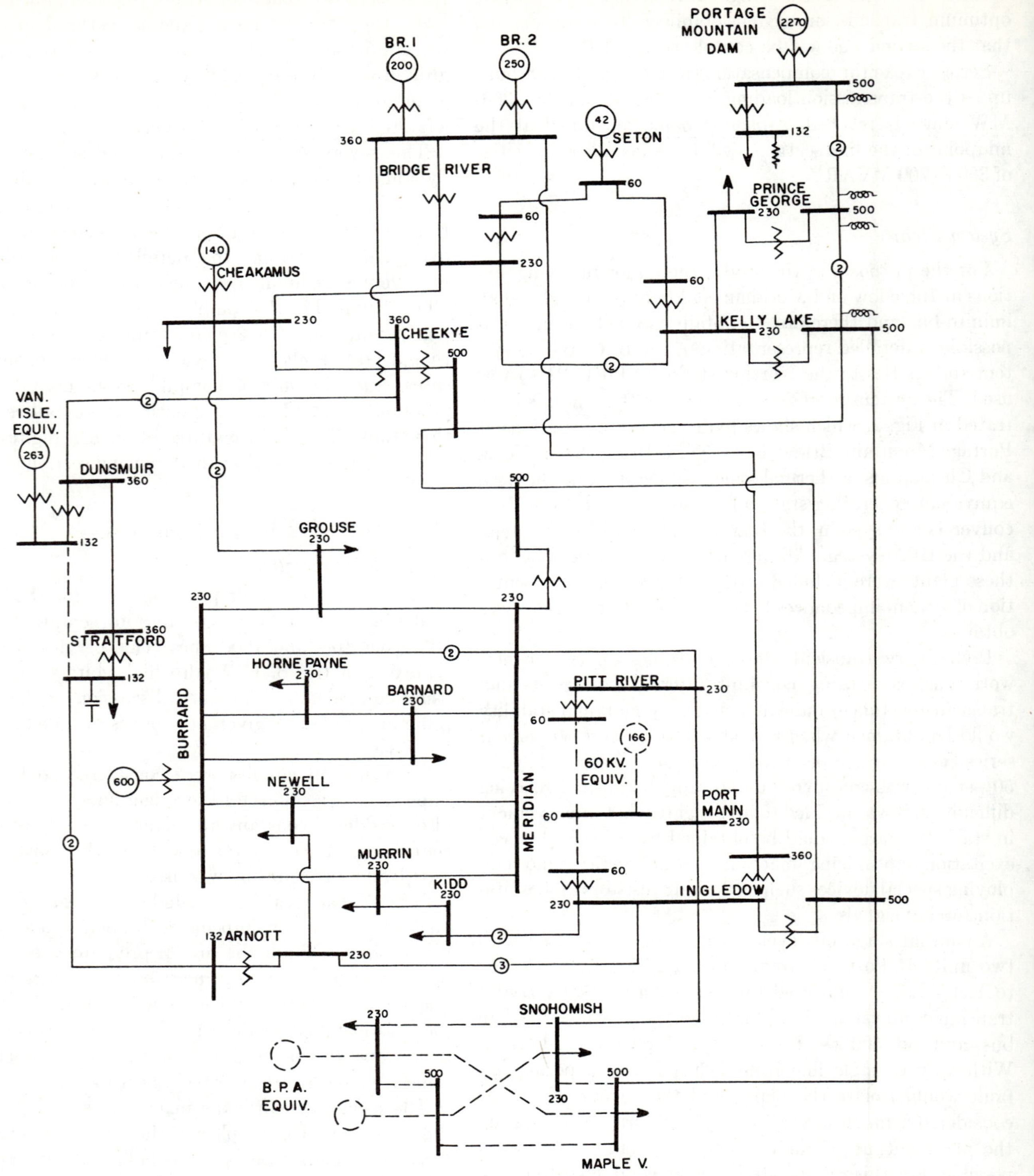

Fig. 2. B. C. Hydro system—ultimate Peace River development.

system damping in the oscillatory period following a disturbance. A velocity deviation above synchronous speed produces an increase in exciter voltage, conversely deviation below synchronous speed lowers exciter voltage.

A similar description of the governor-hydraulic system is shown in Fig. 4. Three different sets of constants are given. In the early part of the study, water inertia was accidentally not represented. Results of those cases when shown are identified and should be used only for relative, rather than absolute comparison.

Excitation Stabilizing Signal

It was established early in the study that the stability of the Peace River transmission system could be considerably improved by use of a static excitation system. Conventional excitation systems, even with very high response ratios and ceiling voltage, would not provide the necessary stabilizing effect so the decision was made to use static excitation. The fast response and high ceiling available from this type of excitation system, when activated by a

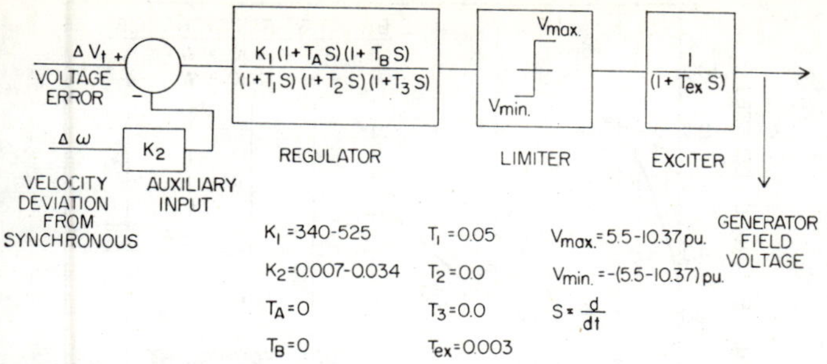

Fig. 3. Block diagram of excitation system representation for Portage Mountain generators. Static exciter and stabilizing input.

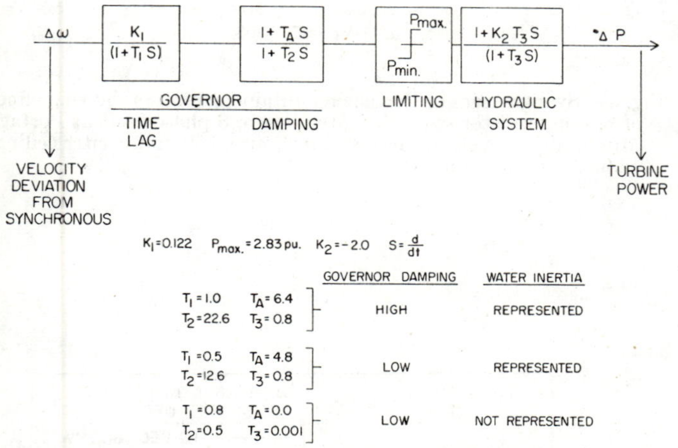

Fig. 4. Block diagram of governor-hydraulic system representation for Portage Mountain generators.

voltage error signal only, improve stability on the first cycle of transient swing but do not provide the damping necessary to ensure stability on subsequent cycles of the swing. Hence, an investigation was made to find a suitable signal for stabilizing these subsequent swings. This signal should act on the excitation in the right direction and at the right time to create greater stability and should have some reasonable proportionality to the magnitude of the swing. It should reset to zero in the steady state so that normal voltage regulation is maintained. It should be relatively easy to obtain in actual practice. The following possible stabilizing signals were considered:

1) rate of change of terminal voltage, pE_a
2) speed error, Δw
3) acceleration, pw
4) rate of change of field current, pI_{fd}
5) subtransient direct axis field current, i_{sd}
6) subtransient quadrature axis field current, i_{sq}
7) rate of change of armature current, pI_a.

To evaluate these signals most rapidly, a simplified system representation was set up on an electronic differential analyzer. The Portage Mountain generation was set up as one equivalent machine with full 2-axis representation with saturation. Excitation was fully represented, but fixed input shaft power was assumed.

The generator was connected through a transformer and a single line to the load concentrated at the Ingledow bus. The BPA system was represented as an infinite bus connected by a line of equivalent impedance to Ingledow. A load change of approximately ten percent of Peace River output was used to initiate a system swing.

Several of the signals considered showed a stabilizing influence and the most promising of these were pI_{fd} and Δw. The pI_{fd} signal exhibited signs of instability in the excitation system. This was to be expected since this signal forms a closed positive feedback loop in the excitation system restrained from instability only by the interaction of the generator armature and field. Before the gain of the pI_{fd} signal could be raised sufficiently to provide adequate generator damping, the instability had to be controlled by introducing a time constant. The signal thus became $pI_{fd}/\text{-}1 + pT$. While this permitted the gain to be raised, the signal was phase shifted such that it was not at all times acting in the correct direction to improve damping.

The Δw signal gave particularly good results, indicating that it could produce any degree of damping desired.

The digital computer transient stability program was modified to use either the Δw or pI_{fd} signal. Runs were made simulating the simplified system used on the differential analyzer and good agreement obtained for similar cases.

The complete B. C. Hydro system, as previously described, was set up and all further investigations were made digitally on the complete system.

The limitations of the pI_{fd} signal were again shown in the digital simulation, so this signal was abandoned.

The Δw signal again proved to be ideally suited to the purpose and various degrees of sensitivity were tested. It was apparent that the stronger this signal became, the better the system was stabilized. The strongest signal used in the study required a speed measuring device with a sensitivity of approximately 0.01 percent.

In the results that follow, reference is made to speed error signals in terms of 2.5 Δw, 5 Δw, 12.96 Δw. The strength of these signals is compared with that of the normal voltage error signal in Fig. 5.

Braking Resistor Application and Control

The braking resistor was switched on the 138-kV bus at Portage Mountain Switchyard through two 150-MVA, 500/138-kV step-down transformers, which also feed local area load at 138 kV. Consideration was given to switching individual braking resistors on the terminals of each of the generators, and while this would have improved the braking effect, the switching cost would have been considerably higher and the control would have become more complicated. One major objective of the study was to find a suitable control scheme for switching the resistor to suit a variety of system conditions including the following:

1) Variation in plant output—A single size of braking resistor would have considerably greater effect on braking five machines than ten machines.

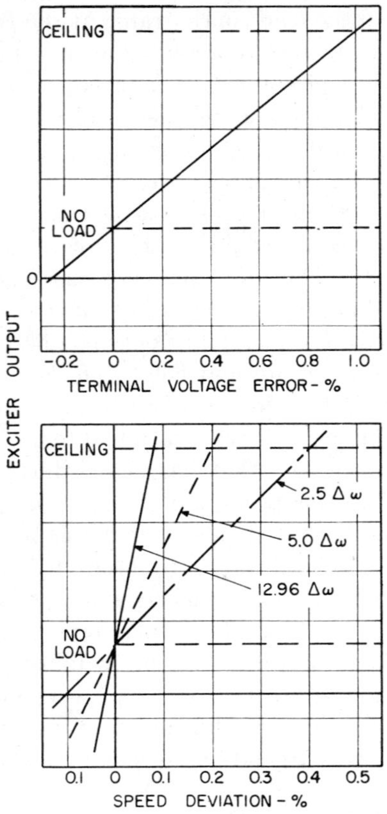

Fig. 5. Comparison of signals applied to voltage regulator.

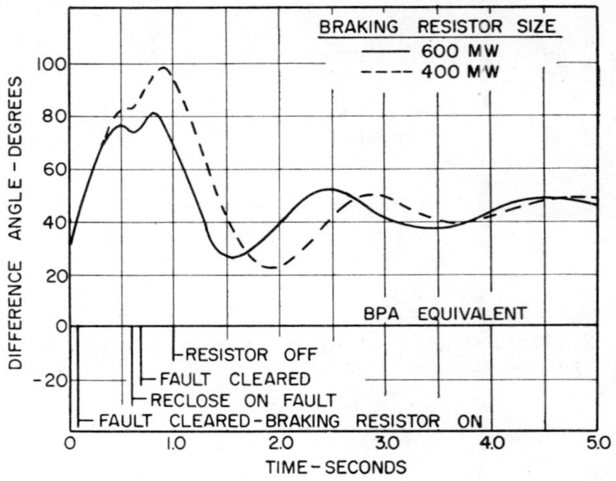

Fig. 6. Swing curves of Portage mountain generators showing effect of braking resistor size. Ultimate system, 3-phase fault at Portage Mountain, 35-cycle unsuccessful reclosure, 750-volt exciter ceiling, 12.96 Δω, no water inertia.

2) The type and location of the fault—The accelerating power for a given number of machines would vary considerably depending upon the type and location of the fault.

3) Variation in reclosing time for temporary and permanent faults.

Digital Stability Analysis and Results

Ultimate System

Braking Resistor Size and Timing: Of the schemes studied, the most effective single means of improving transient stability is application of a braking resistor. The resistor was switched on at the time of fault clearing and then off at some fixed time later. A range of resistor size between 400 MW and 600 MW was established for the 10-machine case, and a comparison of the effects of these two sizes is illustrated in Fig. 6. The swing curves of the Portage Mountain generators are plotted with the BPA system as reference, since that system and its interconnection to western United States is very nearly an infinite bus to Peace River transients. As expected, the 600-MW resistor produces less angular swing during the first swing cycle, however, this results in a larger swing in the second cycle, indicating that the size of resistor has an upper as well as a lower limit for any given number of machines connected. Variation of the time of removal of the resistor is illustrated in Fig. 7. A duration of one second is nearly

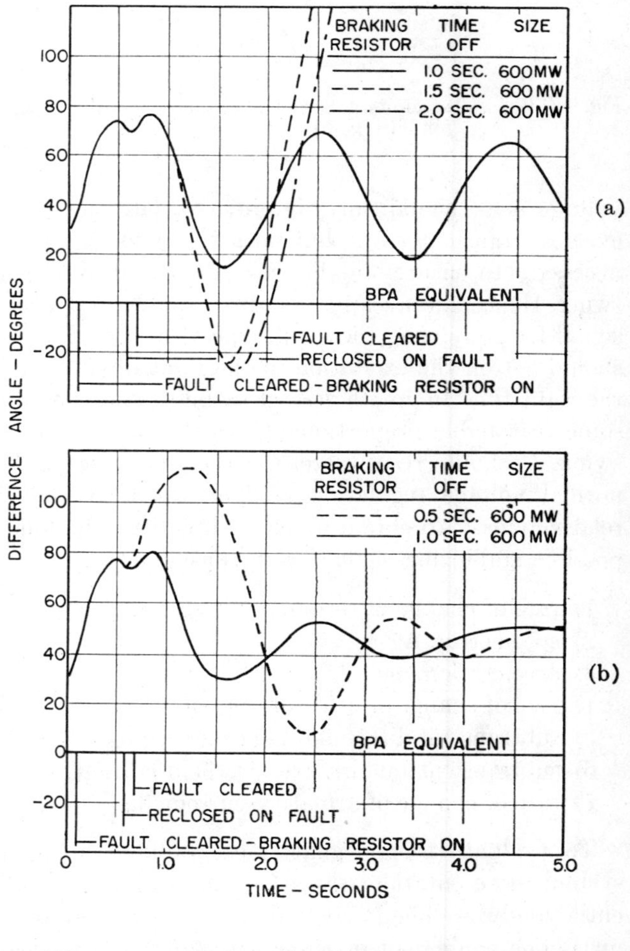

Fig. 7. Swing curves of Portage Mountain generators showing effect of braking resistor timing. Ultimate system, 35-cycle unsuccessful reclosure, 750-volt exciter ceiling, no water inertia. (a) 3-phase fault at Portage Mountain, 2.5Δω. (b) 3-phase fault at Prince George, 12.96Δω.

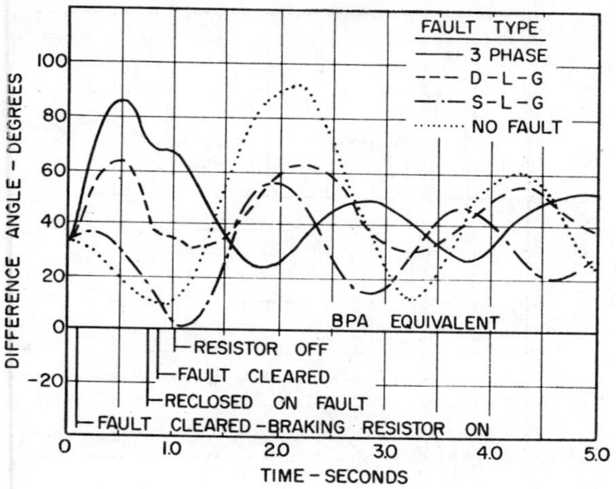

Fig. 8. Swing curves of Portage Mountain generators showing effect of fault type. Ultimate system, faults at Prince George, 45-cycle unsuccessful reclosure, 600-MW braking resistor for one second, 750-volt exciter ceiling, $5\Delta\omega$.

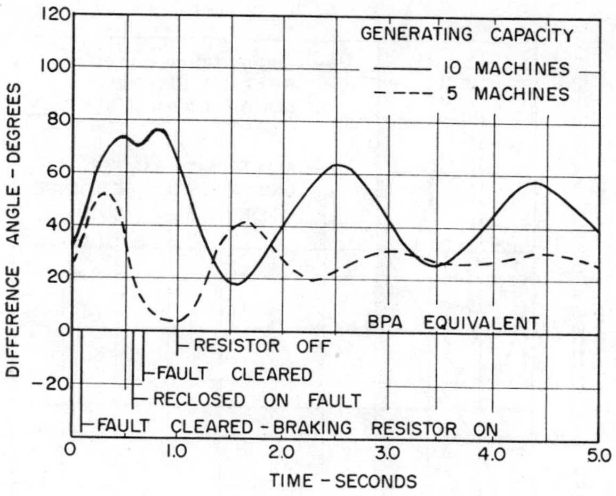

Fig. 9. Swing curves of Portage Mountain generators showing effect of generating capacity. Ultimate system, 3-phase fault at Portage Mountain, 35-cycle unsuccessful reclosure, 600-MW braking resistor for one second, 750-volt exciter ceiling, $2.5\Delta\omega$, no water inertia.

optimum, and any longer time causes too large a negative oscillation and instability in the second cycle.

Fault Type: Having established the range of resistor size between 400 and 600 MW and the time of removal between 0.5 and 1.0 second, the next step was to determine the effect of the type of fault. Using a 600-MW resistor applied for 1.0 second, single-line-to-ground (SLG) and DLG faults were compared with 3-phase faults at a given location, all cases being stable (Fig. 8). Applying the resistor with no fault (such as might occur through inadvertent closing of the resistor breaker) produces a severe disturbance. The system is stable, although in this case the second cycle shows the greatest angular swing.

Generating Capacity: It was established that a fixed resistance-time control could be used to provide transient stability for ten machines at full output. A similar series of tests was performed using the same resistor–time combination when only five machines were operating at full output. The resulting angular swing is less severe than for the 10-machine case as shown in Fig. 9.

Exciter Ceiling Voltage: A range of exciter ceiling voltages from 5.5 per unit (750 volts) to 10.3 per unit (1400 volts) was considered for the electronic exciter. As shown in Fig. 10, field voltage rises quickly and remains at ceiling because of the low voltage during and after the fault and through action of the stabilizing input signal. The magnitude of ceiling is only significant during the first swing cycle because field voltage is below even the lowest ceiling thereafter. Ceiling voltage has a significant effect on the magnitude of initial angle swing because the high ceilings can increase field flux in the 0.5 second time to the first peak. The range of ceiling voltage is approximately 2.75 to 5.1 times full-load excitation. Figure 10 indicates a tendency

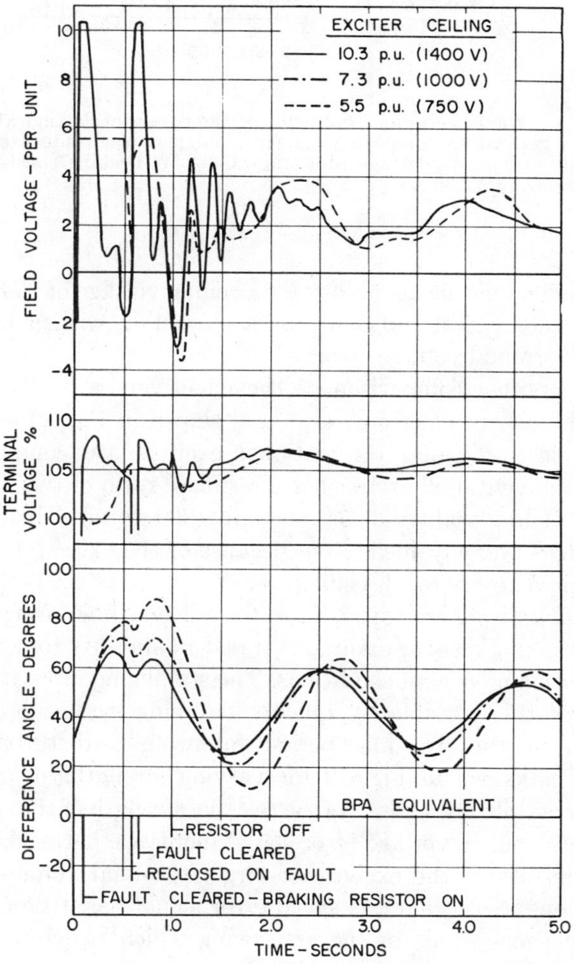

Fig. 10. Angle, terminal, and field voltage curves of Portage Mountain generators showing effect of exciter ceiling voltage, 3-phase fault at Portage Mountain, 35-cycle unsuccessful reclosure, 600-MW braking resistor for one second, $5\Delta\omega$.

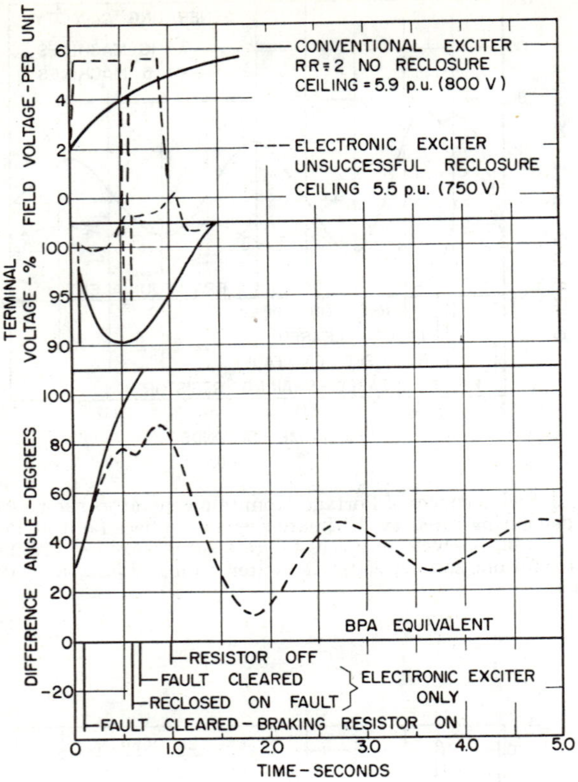

Fig. 11. Angle, terminal, and field voltage curves of Portage Mountain generators comparing conventional and electronic exciters. 3-phase fault at Portage Mountain, 600-MW braking resistor for one second, 12.96 $\Delta\omega$.

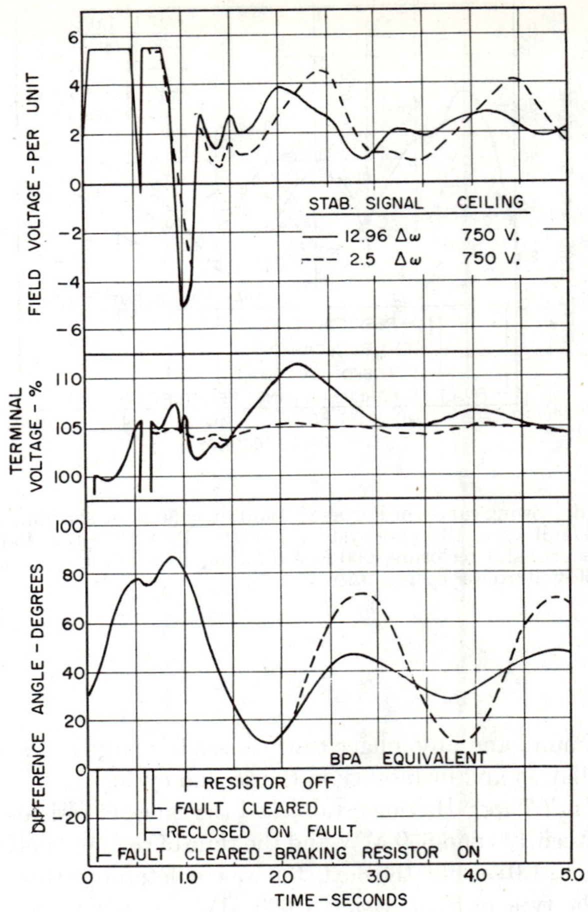

Fig. 12. Angle, terminal, and field voltage curves of Portage Mountain generators showing effect of stabilizing input signal. Ultimate system, 3-phase fault at Portage Mountain, 35-cycle unsuccessful reclosure, 600-MW braking resistor for one second.

toward regulator instability for a ceiling voltage of 10.3 per unit and a gain sufficient to reach ceiling with a 1-percent terminal voltage error.

A graphic comparison of the effectiveness of the low time constant electronic exciter is shown in Fig. 11 where the electronic and conventional exciters are compared. The conventional exciter has a response ratio of two and a high ceiling voltage of 5.9 per unit (800 volts). This high ceiling is largely ineffective because of the excessive time required to approach ceiling.

Stabilizing Input Signal: A fixed load–time control of the braking resistor maintains transient stability for a wide variation of system conditions. The stabilizing input signal makes this possible by quickly damping system oscillations in the dynamic period following a disturbance. This is shown in Fig. 12 for varying strengths of speed error stabilizing signal. As the relative strength of the speed error signal to voltage error signal increases, better damping results at the expense of larger terminal voltage excursions. Note that the speed error signal has little effect on the magnitude of the first swing which is determined primarily by the ceiling voltage.

Reclosing Time: Generally, the stability of a system is improved as reclosing time is reduced, however, the unsuccessful reclosure criterion of stability used in this study reverses the relationship. By delaying the reclosure to suit the natural period of oscillation of the system, the effect of reclosing on a permanent fault is far less severe. This is indicated in Fig. 13, which shows the effect of reclosing times of 25, 35, and 45 cycles.

Generator Reactance and Inertia: As expected, the initial angular swing is inversely proportional to the turbine-generator inertia constant as shown in Fig. 14. Similarly the initial swing angle is proportional to the generator transient reactance.

Governor-Hydraulic System Characteristics: Before it was practical to represent excitation and governing systems in stability studies, input power was held constant and the general opinion was that this gave conservative results. Any governing action would reduce angular swing and improve stability. This is true for thermal plants, but the water-inertia effect of the water column feeding the hydraulic turbine makes governor-hydraulic system representation more significant. Any change in gate position causes an accompanying pressure change, and at high loads, a rapid gate closure causes a temporary power in-

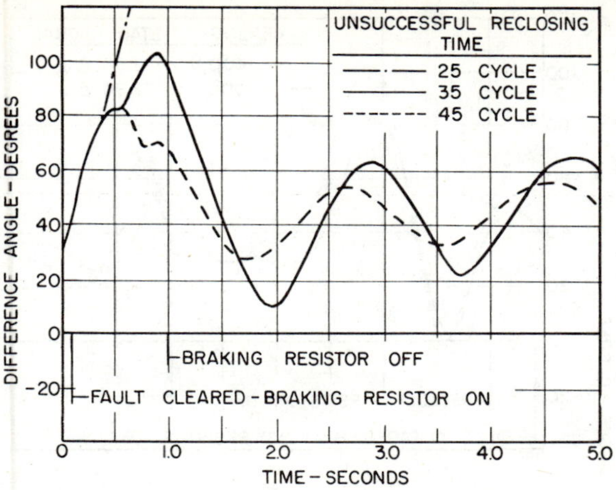

Fig. 13. Swing curves of Portage Mountain generators showing effect of reclosing time. Ultimate system, 3-phase fault at Portage Mountain, 600-MW braking resistor for one second, 750-volt exciter ceiling, $2.5\Delta\omega$, no water inertia.

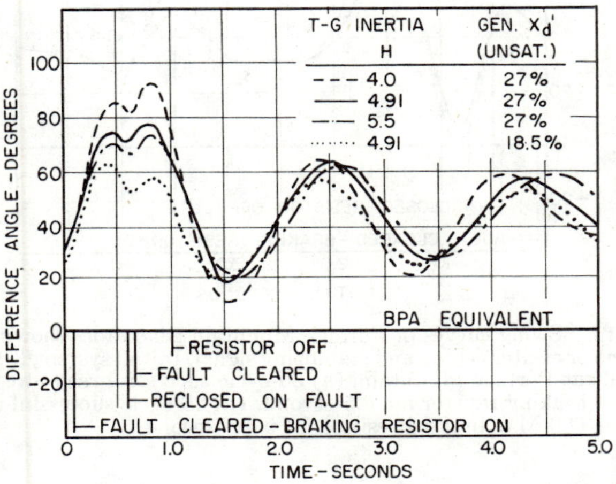

Fig. 14. Swing curves of Portage Mountain generators showing effect of generator inertia and reactance. Ultimate system, 3-phase fault at Portage Mountain, 35-cycle unsuccessful reclosure, 600-MW braking resistor for one second, 750-volt exciter ceiling, $2.5\Delta\omega$, no water inertia.

crease before decreasing. This leads to the classic transfer function of the hydraulic system.

$$\frac{1 - T_W S}{1 + T_W S/2}$$

where T_W = water starting time. For the Portage Mountain plant $T_W = 1.6$ seconds.

Figure 15 illustrates the significance of this form of response. As mentioned previously, some of the early cases had no water inertia represented. This produced the lower (initial swing) swing curve and input power responded directly to gate changes. When water inertia was added, the initial angle swing increased and the machines experienced an oscillatory buildup to ultimate pull out. Input power oscillations also increased in amplitude.

By increasing the governor damping and compensation, this oscillatory condition was eliminated and stability

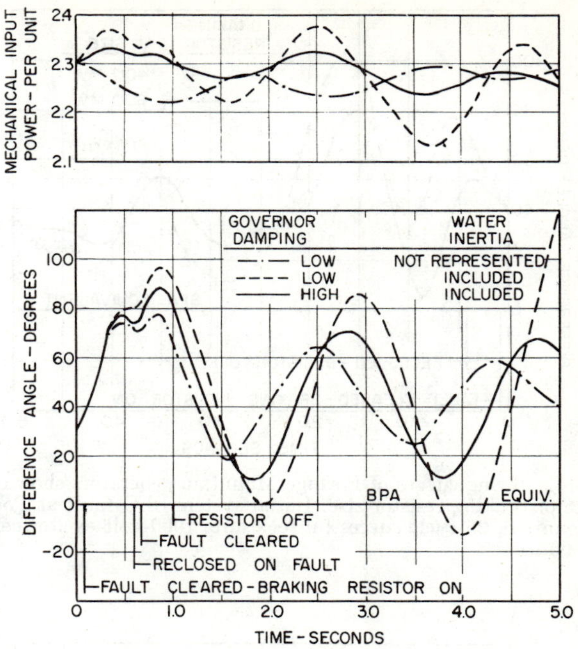

Fig. 15. Angle and input power curves of Portage Mountain generators showing effect of governor damping and water inertia. Ultimate system, 3-phase fault at Portage Mountain, 35-cycle unsuccessful reclosure, 600-MW braking resistor for one second, 750-volt exciter ceiling, $2.5\Delta\omega$.

maintained. Fortunately, electric governors will be used on the Portage Mountain turbines, and a wide range of adjustment is available to permit adequate damping of hydraulic system oscillations.

Initial System

A similar set of studies was conducted for the initial 450-MW single-line system, the main difference in these studies is the fact that a 500-kV line outage, Peace River to Prince George, results in a complete separation of the Portage Mountain plant from the rest of the system. For successful 3-pole reclosing, the braking resistor must limit the machine angle without the aid of any synchronizing power. While single-pole reclosing was studied and found to be superior, the ability to achieve satisfactory arc extinction on a 200-mile section of 500-kV line remains in doubt. It is, therefore, planned to make provision for single-pole reclosing facilities so that they can be readily installed if further investigation indicates that successful arc extinction can be achieved.

The effect of braking resistor size is shown in Fig. 16. In this case it is evident that a large resistor is more advantageous because it reduces the initial angle excursion. Figure 17 shows that the removal of the resistor must be initiated before an excessive downswing is produced, otherwise instability will occur on the second cycle. The resistor size-timing combination is much more critical in this initial stage than in the ultimate stage. Figure 18 illustrates that, while a 600-MW resistor for 0.6 seconds will maintain stability for 3-phase and DLG faults, this combination is too severe for SLG faults. In Fig. 16, stability is

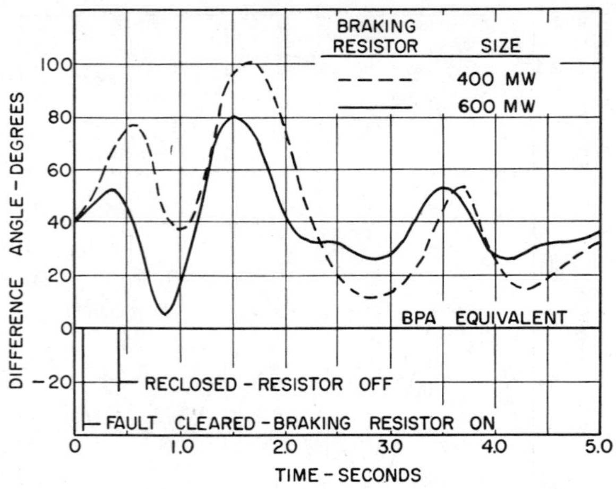

Fig. 16. Swing curves of Portage Mountain generators showing effect of braking resistor size. Initial system SLG fault at Portage Mountain, 25-cycle successful reclosure, 1000-volt exciter ceiling, 12.96Δω.

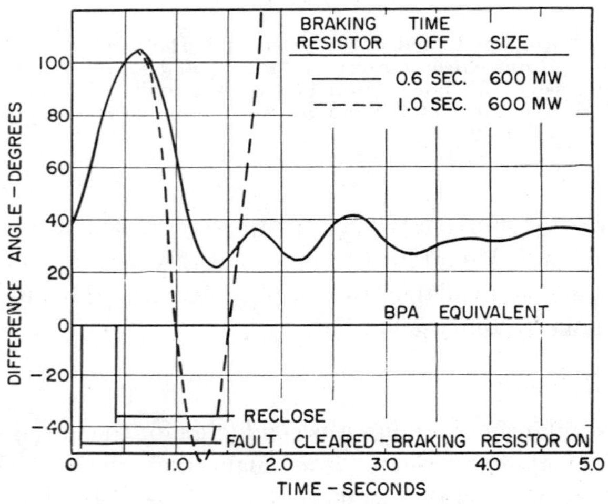

Fig. 17. Swing curves of Portage Mountain generators showing effect of braking resistor timing. Initial system, 3-phase fault at Portage Mountain, 25-cycle successful reclosure, 1000-volt exciter ceiling, 12.96Δω.

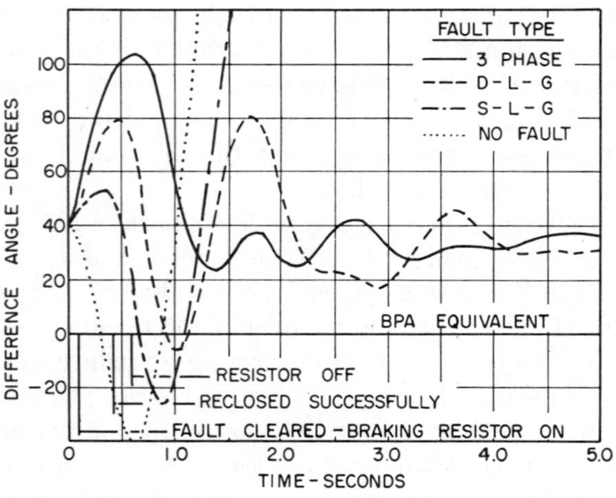

Fig. 18. Swing curves of Portage Mountain generators showing effect of fault type. Initial system, fault at Portage Mountain, 25-cycle successful reclosure, 600-MW braking resistor for 0.6 second, 1000-volt exciter ceiling, 12.96Δω.

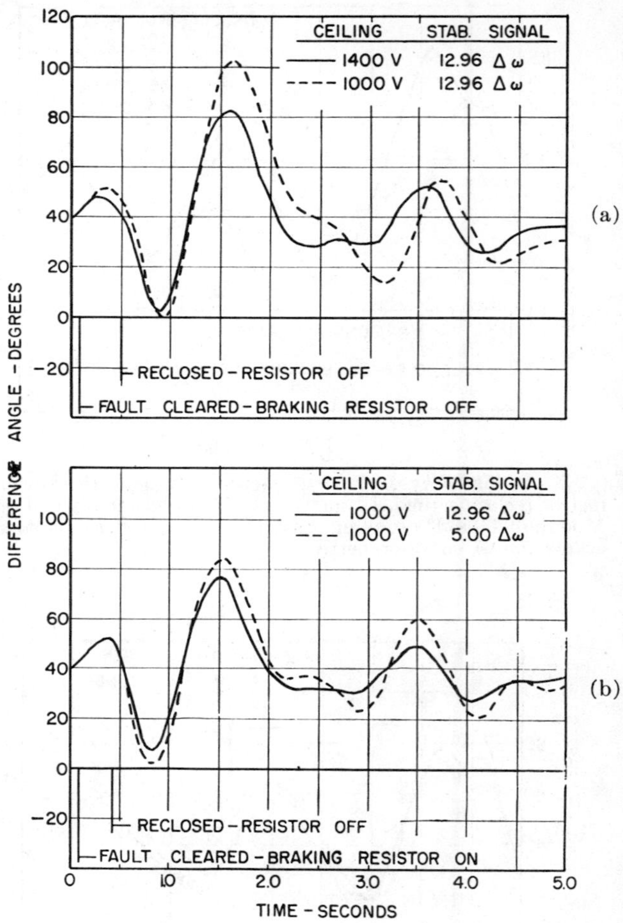

Fig. 19. Swing curves of Portage Mountain generators showing effects of exciter ceiling and stabilizing signal. Initial system, 3-phase fault at Portage Mountain. (a) 30-cycle successful reclosure, 600-MW braking resistor for 0.5 second. (b) 25-cycle successful reclosure, 600-MW braking resistor for 0.417 second.

maintained for SLG faults by switching the resistor off at 0.417 seconds (25 cycles). Since SLG faults are much more common than phase faults, this faster time is preferable, even if it means instability for the 3-phase fault case.

Because of the critical resistor–time combination, a high ceiling voltage and a sensitive speed error stabilizing signal are particularly helpful during this initial stage. Figure 19(a) shows the improvement obtained with a higher ceiling voltage, and Fig. 19(b) shows that obtained from a more sensitive speed error signal.

It is obvious that, for the initial system where successful reclosure is essential to continued operation of the plant, the reclosing time should be as fast as possible, consistent with arc deionization time.

Practical Application of Computer Results: Having determined that the system transient stability can be improved significantly by applying a fixed resistor–time control scheme for the braking resistor in conjunction with a speed error stabilizing signal, it remains to determine whether a practical excitation stabilizing signal can be obtained in practice. A series of field tests performed on the Clowhom machine of the B. C. Hydro system to determine

TABLE I

INCREMENTAL RELATIVE COST OF ALTERNATIVE METHODS OF PRODUCING A 1° REDUCTION IN INITIAL TRANSIENT STABILITY SWING ANGLE

Method of Reducing Swing Angle	Incremental Relative Cost
Increase in series compensation (based on 50 percent series compensation)	1.0
Decrease in machine transformer reactance (based on $X_t = 0.15\ pu$)	0.5 to 1.0
Increase in inertia constant (based on $H = 4.6$)	0.8 to 1.0
Decrease in machine reactance (based on $X'd = .25\ pu$)	0.4 to 0.5
Increase in exciter ceiling voltage from 5.0 *pu* to 7.5 *pu* (using solid state exciter)	0.3 to 0.5
Increase in braking resistor size from two 200-MW units to three 200-MW units	0.2 to 0.4

a practical means of obtaining a speed error signal (or equivalent signal) have been completed and further tests are to be done. The results of these tests will be the subject of a future paper. The results show that a stabilizing signal produces significantly improved damping of machine oscillations.

The somewhat critical resistor–time combination for the initial system, plus the need for reliability of resistor application, leads to the conclusion that the resistor should be split into three or four parts, each with a switching breaker so that part or all of the resistor can be switched, if necessary, depending upon the severity of the fault or the plant output.

The methods of improving system transient stability included in this study were evaluated relative to inserting more series capacitors in the lines as the loading increases. The cost of alternate methods of improving stability relative to the present worth of installing additional series capacitors is given in Table I. Analysis of Table I indicates that improvement of system stability can be achieved more economically by optimizing machine characteristics, than by installing additional series capacitors. Thus, when specifications were issued for the generators, a quantitative evaluation was placed upon a machine with a lower transient reactance than specified, the maximum value specified being a 0.25 per unit X_d' saturated, a *normal least cost* value. This method of specifying machine reactance was preferred to that where a fixed low reactance would be specified, because the latter method would require an intimate knowledge of design costs. Similarly a quantitative evaluation was placed upon a machine with a higher inertia than that specified, with the specified minimum value being 3.9 kW-s/kVA, a normal least cost value. This value was also the minimum desirable for hydraulic design considerations. The machine which had the lowest evaluated cost of all machines bid was one with X_d' of 0.176 per unit and inertia constant of 4.63 kW-s/kVA, characteristics which should be ideally suited to maintain the stability of the Peace River Transmission System.

CONCLUSION

A braking resistor is very effective in maintaining transient stability following a severe fault disturbance. For the double-circuit system, a fixed resistor size and time duration were suitable for all fault types. However, it is planned to split the resistor into three parts for finer control.

High ceiling voltages provided by an electronic exciter are also effective in improving transient stability. A conventional exciter, response ratio = 2, is ineffective in maintaining stability.

A stabilizing input to the voltage regulator increases damping of machine oscillations and improves system stability following the initial transient period. A stabilizing signal of velocity deviation from synchronous was found to be effective in the digital study. Governor and hydraulic-system constants have a significant effect on system stability. Stability is maintained with normal generator reactance and inertia when 50-percent series compensation is used.

A new digital stability program has been used to evaluate system performance including control system response.

REFERENCES

[1] D. W. Olive, "New techniques for the calculation of dynamic stability," presented at the IEEE PICA Conference, Clearwater, Fla., May 19–21, 1965.

[2] E. W. Kimbark, *Power System Stability.* New York: Wiley, 1956.

[3] D. B. Breedon and R. W. Ferguson, "Fundamental equations for analogue studies of synchronous machines," *Trans. AIEE (Power Apparatus and Systems)*, vol. 75, pp. 297–306, June 1956.

[4] Croft and Hartley, "Improving transient stability by use of dynamic braking II," presented at the AIEE Fall General Meeting, Detroit, Mich., October 15–20, 1961.

[5] Gornshtein and Luginskii, "The use of repeated electrical braking and unloading to improve the stability of power systems," *Electric Technology USSR*, vol. 2, p. 292, 1962.

[6] Trofimenko, "The stability of a hydro-electric generator with electric braking," *Electric Technology USSR*, vol. 1, p. 70, 1962.

[7] Easton, Fitzpatrick, and Parton, "The performance of continuously acting voltage regulators with additional rotor angle control," presented at the 1960 CIGRE, Paris, France, Paper 309.

[8] Venikov, Hertzenberg, Kostenko, Neumann, Savalov, and Sokolov, "Boost regulation on electric power systems in the USSR," presented at the 1960 CIGRE, Paris, France, Paper 325.

[9] Kostenko, discussions of papers 309, 325, 331, presented at the CIGRE 1960, Paris, France.

[10] IEEE Working Group on Arc Deionization Times of the General Systems Subcommittee, "Arc deionization times on high speed three pole reclosing," *IEEE Trans. on Power Apparatus and Systems*, vol. 825, pp. 236–253, 1963.

[11] H. Schiott, "Optimum setting of water turbine governors," *Trans. Soc. of Instrument Technology*, pp. 1–8, November 1959.

[12] H. E. Lokay and R. L. Bolger, "Effect of turbine generator representation in system stability studies," *IEEE Trans. on Power Apparatus and Systems*, PAS-84, pp. 933–942, October 1965.

[13] A. R. Scott, "Optimization of generator and excitation system parameters for Grand Rapids development," *Trans. Canadian Electrical Association*, vol. 2, pt. I, 1963.

[14] H. E. Lokay and J. W. Skooglund, "Power system stability-digital analysis showing effect of generator representation, types of voltage regulators and speed governor systems," presented at the IEEE PICA Conference, Clearwater, Fla., May 19–21, 1965.

[15] J. H. Steede, P. J. Croft, and H. M. Ellis, "Peace River EHV transmission system," presented at the Winter Power Meeting, New York, N. Y., January 31–February 5, 1965.

Discussion

E. W. Kimbark and **M. L. Shelton** (Bonneville Power Administration, Portland, Ore.): We, and others, have been studying the stability of the planned 500-kV ac interties between the Pacific Northwest and the Pacific Southwest, and it is therefore interesting to consider both similarities and differences between that project and the one under discussion.

In both projects there are two long parallel 500-kV lines, and in both projects careful study is required to make the system stable when these lines are loaded to two gigawatts or more. In both projects series compensation is required.

An important difference between them is that the Peace River system is essentially a *radial system* (that is, substantially the entire output of the sending-end generators will be transmitted to the receiving-end system) whereas the Pacific intertie is an *interregional tie*, and the power transmitted will be only about 10 percent of the output of the sending-end generators, the remainder being consumed by loads in the same region. Therefore, the ratio of inertia constant to power transmitted is greater in the Pacific interties than in the Peace River system, and the period of swing of the Pacific interties is longer—about 3 seconds, as compared to 1.8 seconds in the Peace River system.

With the shorter period of swing, fast switching is more important than it is with the longer period. However, we note that the Peace River study assumed 6-cycle fault clearing, whereas the Pacific intertie study assumed 3-cycle clearing at the near end and 4-cycle at the far end of a line section. Inasmuch as 2-cycle 500-kV circuit breakers are available from several manufacturers, we wonder why faster clearing than six cycles was not assumed. Likewise, in the Peace River study, series capacitors were assumed to be inserted in five cycles after fault clearing, whereas we believe that more rapid reinsertion (approximately 1 cycle) will be feasible if required.

Another difference between radial and interregional transmission is that in the radial case it is feasible to depart from standard characteristics of generators and excitation systems, for example, to use generators of lower than usual reactance and higher than usual inertia, and to use excitation systems with higher than usual ceiling voltages, faster than usual response, and regulator input signals, such as rate of change of voltage or angle, in addition to the usual voltage signal. In an interregional system, such changes are less feasible because they would involve hundreds of machines, most of which were installed before the date of the study. Thus, in a study of interregional lines, more attention is likely to be given to the transmission system.

Our studies have shown that switched series capacitors are an effective means of improving stability. We plan to have in some line sections 50 percent normal compensation with an additional 25 percent inserted immediately after a fault occurs. Was this possibility considered in the Peace River study?

Braking resistors can be effective on either type of system. We have studied application of braking resistors to the Pacific interties and have found them to improve stability if the power and application time are appropriately chosen. One second seemed to be a good application time and more effective than 1.5 seconds with equal energy dissipation. In some cases, the resistor made the system survive the first swing, but synchronism was lost on the second swing after disconnection of the resistor.

In general, we have found generator dropping just as effective as resistor braking and less expensive because the cost of the resistor and of a circuit breaker for it are saved. However, we have found that generator dropping near the sending end of the intertie lines (John Day plant) is less effective than generator dropping back in the system (for example, at Grand Coulee). The reason is probably that generator dropping near the sending end lets the sending-end voltage drop more. Perhaps, for the same reason, generator dropping would be less effective than braking on a radial system. Has generator dropping been considered for the Peace River system?

The representation of the BPA system as an infinite bus in the Peace River study should give somewhat more pessimistic results than representing it with a finite inertia, according to our experience with similar cases.

Manuscript received October 13, 1965.

W. A. Morgan (Bechtel Corporation, San Francisco, Calif.): The authors have presented a practical solution to a complex stability problem. Also, they have refuted a long held assumption that it is conservative to ignore governor action on hydro-electric generating units. It is significant that the initial, single, 580-mile, 500-kV transmission-line system is being designed on the basis of successful breaker reclosure because of the expected frequency of transient lightning flashovers with the line not provided with overhead lines, whereas the ultimate, 2-circuit, 2-intermediate switching station system is being designed to withstand reclosing on a solid fault for greater reliability of the 2300-MW capacity generating source.

While a great variety of stability parameters are discussed with emphasis properly given to the newer stability techniques, perhaps in rebuttal it may be explained how extensive the effect of fault clearing times and fault location were studied. A previous paper dealing with the same problem as that confronting the authors presented the results of studies of a hydro plant at the end of a long, 230-kV line, connected to a relatively high enertia system.[1] This paper showed that the power limit of such a system is quite sensitive to the type and location of the fault along the line and to the time of fault clearing and reclosure. Figure 20 taken from Morgan's paper,[1] illustrates these effects.

It would appear worth evaluating the cost of clearing times faster than the 5 and 6 cycles mentioned in the paper. With 2-cycle breakers and 1-cycle maximum operating time static line protective relays available, it would appear that 3- or 4-cycle clearing times could be considered. Actually, because of the importance of clearing time on stability limit, there should be no reason to limit the study by using

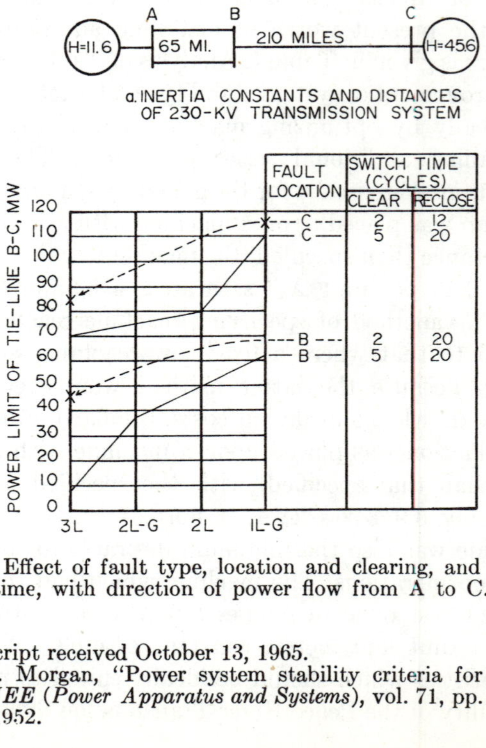

Fig. 20. Effect of fault type, location and clearing, and reclosing time, with direction of power flow from A to C.

Manuscript received October 13, 1965.

[1] W. A. Morgan, "Power system stability criteria for design," *Trans. AIEE* (*Power Apparatus and Systems*), vol. 71, pp. 499–503, January 1952.

only 1-cycle increments of alternative clearing times. With the digital computer, the tripping times can be varied in 0.001 second intervals.

Another refinement is the relative clearing times of each end of a faulted line. For a fault near one terminal, with either carrier-current or microwave relaying channels, the terminal nearest the fault should clear before the terminal at the far end of the line, and these different times have an effect on stability and can be represented by the digital computer. In a recent study this discusser made, a change of 0.002 second, near end, and 0.006 second, far end, tripping times made the difference between instability and stability.

Faults of various types located at varying distances from the generating end bus might require smaller braking resistors than 600 MW. With the total resistor split into 200-MW elements, perhaps instantaneous current relay elements in each phase would distinguish between the types of faults, and also the location and severity. Have the authors considered the severity of the fault and the possibility of developing a flexible amount of braking resistance as an alternative to the duration of time the resistor is applied?

P. L. Dandeno and **K. R. McClymont** (The Hydro-Electric Power Commission of Ontario, Toronto, Canada): The authors of this paper are to be commended for their interesting presentation and for the sound solutions they have found to the formidable stability problems associated with incorporating the Peace River generation into the B.C. Hydro system.

During 1961 and 1962 Ontario Hydro, in association with Canadian and European manufacturers, developed a stabilizing signal for use on rectifier excitation systems which would permit improved generator stability limits, by taking full advantage of the nearly instantaneous response of rectifier type exciters. In 1963 rectifier exciters employing a speed deviation stabilizing signal were placed in service on three Ontario Hydro hydraulic generators on the Moose River system. At the present time Ontario Hydro has five such excitation systems in service and eight on order. In view of our early adoption of controlled rectifier excitation systems for the purpose of improving generator stability, we are pleased to see other North American utilities now adopting these systems.

We would like to reinforce the authors comments on the importance of this type of excitation system as a means of improving damping on remote hydraulic stations. Generators at such remote locations tend towards instability and are poorly damped because of the absence of that damping contributed by loads and because damping from generator amortisseur windings and from turbines tends to be small. However, with the use of a proper stabilizing signal, large damping forces can be obtained from a controlled rectifier excitation system. This is illustrated in Fig. 21 which indicates the comparative stability performance of a group of generation. The value of $K\omega$ shown in the figure indicates the per unit value of error signal (in terms of per unit field voltage) for a speed change in cycles per second. The dashed curve (Test 1) shows the undamped angular swing (δ) of the generators subsequent to a disturbance ($K\omega = 0$). The solid curves show, for a speed stabilizing signal ($K\omega = 30$) the angular swing (δ) of the generators, and the change in excitation system output voltage (Efd). Also shown is the time variation of Q_N, the stabilizing signal itself.

The controlled rectifier exciters which Ontario Hydro has purchased obtain their excitation power from a transformer connected to the generator terminals. While this arrangement is a disadvantage when excitation ceiling voltages are reduced during the fault period, it eliminates the need for a rotating ac source of supply and provides a very economic excitation system. It appears from Figures 10–12 that B.C. Hydro intends to employ a separate rotating excitation supply. We would appreciate the authors comments on the need in their case for the more expensive system.

We are interested in the authors comments on exciter ceiling voltage. We have selected ceiling voltages in the range of 4.4 to 5.5 per unit for all our purchases for hydro generators. While high ceiling voltages can be obtained at relatively small cost we found the gain in stability by increasing ceiling voltages tended to fall off very quickly beyond a certain point, because of generator saturation. High ceiling voltages can also increase dynamic overvoltages in some conditions and we are, therefore, anxious not to increase such voltages any more than necessary. Could the authors indicate the ceiling voltage they have selected, and their reasons for such a choice?

The question of dynamic braking has come to the fore in several recent designs for long distance transmission. There is an alternative to dynamic braking which is much more direct and simpler to execute—that of generation rejection. When the problems of stability in long distance transmission involve generator angular swings against a relatively stiff bus, deliberate generation rejection for transient faults will have little effect on overall system frequency. Generation rejection, in general, requires that the ties within a system or with interconnected systems be strong enough to stand the loss of a generator, so this should not generally be a problem. In addition to assisting the transient stability problem, the dynamic stability of the system, with one line section permanently out of service, would also be improved after generation rejection. Have the authors considered this as an alternative to dynamic braking?

We understand from the authors comments that electric governors will be used on the Peace River system, as they permit a wide range of adjustments. Ontario Hydro has used both mechanical and electric governors on their hydraulic generators, and we believe that at some locations the electric type has advantages over the mechanical type. We have found that the mechanical governors provide an ample range of adjustment for the type of performance required for power system stability. We, therefore, hope the authors will provide more specific details as to any expected stability performance improvements they hope to achieve by the use of electric governors.

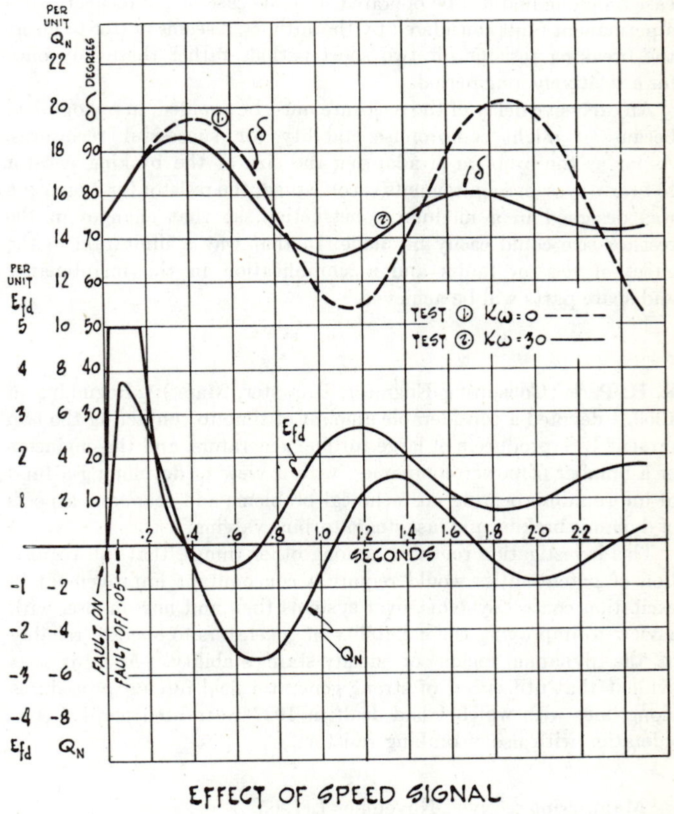

Fig. 21.

Manuscript received October 25, 1965.

G. Catenacci (CESI, Milan, Italy):[1] The authors are to be commended for their very extensive work on a matter of primary interest to many transmission engineers. The deep investigations on the influence of the excitation systems, voltage regulator characteristics, governor effects, hydraulic system characteristic, and dynamic breaking possibilities are extremely interesting.

It may be useful to report the results obtained during the study of a similar system for Argentina. Anticipating some of the results which will be presented in a paper for the 1966 CIGRE meeting, the system studied consists of two 500-kV lines to transmit, at the initial stage of development, 1100 MW from El Chocon and Cerros Colorados power stations up to Buenos Aires through a distance of 650 miles. Two intermediate switching stations are provided. No series compensation is necessary at the first stage for steady-state stability. A special arrangement of the intermediate switching stations allows the future addition of series compensation. To obtain the needed high degree of transient stability in the first stage, conclusions similar to the authors' were obtained:

1) Static excitation will be used for the important improvement in transient stability and in dynamic overvoltages it permits.
2) Careful design of voltage regulator stabilization is of primary importance.
3) Dynamic breaking resistors of 600 MW will be used since they permit important increases in the transient stability limits at the lowest cost.

During the extensive studies carried out on the ac network analyzer and on the E.d.F. "microréseau" it was found that an insertion time of 0.3 second is optimum. Much shorter times will give instability on the first swing, while much longer times will do so on the third swing. Fixed size and timing of the braking resistor, namely 600 MW and 0.3 second, were proven suitable for any system condition and fault type and position. Excessive breaking is not produced even with a reduced number of generators because of the natural reduction of the power absorbed by the resistor as a consequence of the natural supply voltage drop. Very strong braking is useful when only a few generators are in service, because in that case only one line will be operated. For the case of the reclosing into a permanent fault considered by the authors, it seems better to insert the breaking resistor for two short periods rather than only once for a relatively long period.

An excessive delay of the reclosure must be avoided, in my opinion, because it might compromise stability for successful reclosures. As far as the problem of adapting the size of the braking resistor to future system requirements is concerned, the resistor for Argentina was designed in a modular construction, so that changes in the resistor size could easily be made. In that way a limitation in the effect of resistor faults and a simplification in the maintenance and spare parts will be achieved.

R. H. Park (Consulting Engineer, Brewster, Mass.):[2] Beginning in 1959, I devoted a considerable amount of time to contacting the two present U.S. producers of large turbine generators, and the engineers of a number of power companies, with a view to developing a fund of information, bearing on technical problems and economic aspects of dynamic braking and fast steam turbine valving.

This investigation revealed, among other things, that full realization of potentialities would require a concomitant improvement in excitation control systems, over systems then and now in use, with a view to improving the capability of generators to operate reliably in the dynamic region of steady-state stability. Also, it was evident that utilization of strong generator field forcing procedures along lines with which I had dealt in 1927 warranted parallel consideration with use of braking resistors.

Admittedly, there can always be a question as to when and how fast to move ahead into employment of new areas of technology. Accordingly, what is needed to bring matters to a head is a detailed economic evaluation.

In the power industry today, latest computational techniques, using improved methods of generator and system representation, offer an opportunity of quantitatively evaluating economic aspects of relatively complex system design alternatives on a reliable basis. The authors have put these techniques to good use and in the process have clarified long standing questions as to economic value of dynamic braking and advanced excitation systems with results which, as it appears, confirm earlier findings of USSR engineers.

In general, from the data shown, it should be possible to derive considerable advantage from provision of automatic means to control extent and duration of braking load application, not only in relation to fault severity and prefault conditions, but also in relation to the occurrence or nonoccurrence of a refault.

In the matter of the braking of generators located in generating stations remote from load areas, the point of what in detail happens to load area generators has been raised at various times. In this connection the assumption of constant impedance types of load could act to mask the significance of the effect and, in the case of certain types of load areas, could possibly lead to errors that ought not to be ignored.

Any information that the authors might be prepared to furnish on this point as applying to the British Columbia Hydro power system would be of interest.

When and if receiving system effects are important, they are capable of being compensated for by

1) Automatic fast boosting of driving power of spinning reserve generator prime movers, where feasible.
2) Automatic shedding of preselected receiving system loads.

Also, in relation to utility of dynamic braking, where steam generated power is involved, it is possible to advantageously supplement and, in some cases, avoid need of the use of braking resistors, by employment of fast turbine valving as a means, not only to prevent transient instability, but also to allow effecting a fast step reduction in prime mover driving power. This avoids the need for generator trip outs and also offering a way of dealing with problems of steady-state stability.

Accordingly, it is a corollary of the findings reported in the paper that engineers engaged in planning the construction of mine-mouth steam plants or system interconnections, involving export of steam generated power, can increase transmission capability and reduce transmission system costs through employment of dynamic braking and fast valving.

Another corollary of the paper, which, though already touched on, warrants emphasis, is the indication that the relatively low-ceiling excitation systems, which the power industry has up to now mainly elected to use, do not represent a favorable design choice in application to generators that will be located at plants remote from load areas.

This should not be attributed to lack of manufacturers' capability to produce more advanced equipment, but, rather, to the fact that the power industry has at best only recently realized that it may be desirable to set up higher standards.

In this connection and barring what is presented in the paper under discussion, there has been a lack of quantitative economic investigations sponsored by power companies or their associations, bearing on what types of generator excitation and excitation control system characteristics ought to be specified.

It seems likely that suitably planned quantitative studies would reveal, not only that the bulk of new machines should be furnished with more advanced systems than are currently being supplied, but also that reconstruction of many existing systems would be desirable.

Also it appears that power companies should carry out more tests, designed to prove out the performance capabilities of excitation and excitation control systems under unusual and critical conditions. In the last several years, tests have been carried out that have re-

[1] Manuscript received November 1, 1965.
[2] Manuscript received December 22, 1965; final revision March 2, 1966.

vealed deficiencies of present systems. Again, in several areas, sustained transmission-line power oscillations have been reported which, it would seem, could have been prevented with use of favorably designed excitation systems, such as are planned for use in the Peace River installation.

REFERENCES

[1] R. H. Park, discussion, and R. E. Doherty, closure of Doherty, R. E., "Excitation systems, their influence on short circuits and maximum power," *Trans. AIEE*, vol. 47, pp. 970–974, July 1928.

[2] L. W. Thompson and P. J. Walton, "Automatic regulation of synchronous condensers equipped with superspeed excitation," *Trans AIEE*. vol 49, pp. 315–318, January 1930.

[3] R. C. Buell, R. J. Caughey, E. M. Hunter, and V. M. Marquis, "Governor performance during system disturbances," *Trans. AIEE*, vol. 50, pp. 354–369, March 1931.

[4] V. A. Venikov, *Transient Phenomena in Electrical Power Systems*. New York: MacMillan, 1964.

[5] R. H. Park, "Means for maintaining stability of power transmission systems notwithstanding a fault," U.S. patent 3234397, February 1966.

H. M. Ellis, J. E. Hardy, A. L. Blythe, and **J. W. Skooglund:** We wish to thank the various discussers for their contribution to our paper. Mr. Park's comments reflect his interest in the subject of dynamic braking. The discussers covered similar topics so we will attempt to answer their questions together.

W. A. Morgan, E. W. Kimbark, and M. L. Shelton questioned the use of the 5-cycle clearing time for 3-phase faults used in the study. They suggested that reductions in fault clearing times are available, down to three cycles, which would improve the stability limit of the system. We agree that faster clearing times can be achieved at relatively little extra cost. We are not so certain, however, that these faster clearing times can be achieved consistently when breakers have been standing in extremely cold ambient temperatures and when pilot channels have not recently been finely tuned. We therefore decided to allow some margin in clearing times. If experience shows that faster clearing times are achieved consistently, a lower percentage of series capacitor compensation will be installed in the later stages. Some studies using faster clearing times were carried out to check the effect of the braking resistor under fast clearing conditions.

A number of questions were raised by E. W. Kimbark and M. L. Shelton on the use of series compensation. They suggest that fast reinsertion and the insertion of additional compensation in the unfaulted line could assist in improving stability. Five-cycle reinsertion time of series capacitors was used assuming bypass and reinsertion switching following excessive fault currents. The 1-cycle reinsertion can only be obtained by use of self-extinguishing bypass gaps. We feel that the longer time of reinsertion allows us a choice in the type of installation. In our study the largest system swings occur for a fault on the Peace River 500-kV bus and no series capacitor installation flashes over for this fault. Therefore the 1-cycle reinsertion would not help stability for the worst fault location but could certainly help for other fault locations.

We did look at switching from 33-percent compensation to 50-percent compensation after the fault had cleared but found that this did not show any improvement over operating with 50-percent compensation from the start. However, this was done for a fault at the Peace River 500 kV bus, where no capacitor flashover took place in either case, thus one could hardly expect to get improvement by switching in this case. Another approach, particularly where flashover on external faults is marginal, is to eliminate flashovers on external faults entirely by increasing the voltage rating of the capacitor bank. In marginal cases this could be less expensive than providing rapid reinsertion equipment.

Manuscript received November 22, 1965.

All of the discussers showed considerable interest in the use of dynamic braking resistors to obtain transient stability. It should be pointed out that the braking resistors are applied only to maintain the transient stability during the first swing of the Peace River generators. They should be removed immediately following the angular recovery of the Peace generators. Following this swing the system can be damped by the fast acting excitation system using appropriate forcing signals.

W. A. Morgan and R. H. Park suggested that the amount of braking resistor to be applied might be chosen according to fault location and/or fault magnitude. These possibilities have been studied and various methods suggest themselves. One possible application method being considered is to insert the braking resistor in blocks of 200 MW according to the amount of accelerating power measured during a fault. The total time of application would remain the same for each fault.

R. H. Park mentions possible system oscillations being caused by the use of braking resistors. We agree that such oscillations certainly can occur if the resistor is kept on too long. The braking resistor must be removed in time to allow the Peace River generators to swing in step with the rest of the system. If the resistor is kept on too long, Peace River generators may swing excessively and cause instability with the rest of the system.

The suggestion that generator rejection might have been considered rather than the use of braking resistors is advanced by Kimbark and Shelton and by Dandeno and McClymont. Generation rejection was not considered as a substitute for the braking resistor. In the early stages of the development both generator rejection and dynamic braking will be required. In the ultimate stage generator rejection could be considered as a means of reducing the amount of series capacitors required. However, with machines of this size, we are somewhat reluctant to subject them to full load rejection overspeed any more than is absolutely necessary.

P. L. Dandeno and K. R. McClymont present a number of comments on the excitation system. The Peace River generator excitation will be supplied from the generator terminals and the exciters will have a ceiling voltage of 7.5 per unit. Higher ceiling voltages could have been obtained but studies showed possible exciter instability at higher ceiling voltages and high gains. The first swing is considerably reduced for the higher ceiling voltages as illustrated in Fig. 10, however subsequent damping shows little effect of higher ceiling voltages.

R H. Park proposes the use of fast valving as a means of stabilizing a system. This proposal requires changing the input power. Governor action could certainly assist stability through fast valving on a steam turbine but in a hydro system because of the water inertia the exact opposite is true. Fast governor action may produce instability as shown in Fig. 15 of the paper. The electric governors used for the Peace River machines will not improve transient stability, but will allow ease in selection of damping, and insertion of load control signals.

It is gratifying to hear that G. Catenacci is designing on a similar basis to that used in our system. While we agree that, theoretically, insertion of the braking resistor twice for unsuccessful reclosure and once for successful reclosure would be preferable, we doubt that any greater degree of reliability would be achieved. Reclosure on a permanent fault would come at a time when the braking resistor is off and a few cycles will be required before the resistor can be switched back on. The system could be in a more unstable situation for this second bump than it was for the first one and the delay in getting the resistor back on may be more critical. We would prefer to keep the switching sequence as simple as possible, consistent with satisfactory results.

Since carrying out the studies described in the paper we have carried out field tests on static excitation with various control signals. These field tests and later digital studies proved that a power signal, equivalent to the speed signal, can be used in damping the system oscillations following transient fault recovery. We plan to report some of this work in future papers.

Influence of Excitation and Speed Control Parameters in Stabilizing Intersystem Oscillations

OSCAR W. HANSON, MEMBER, IEEE, C. J. GOODWIN, AND PAUL L. DANDENO,
SENIOR MEMBER, IEEE

Abstract—**Intersystem power oscillations have occurred between the interconnected utility systems of Saskatchewan, Manitoba, and Ontario (West System). Such power swings have been most evident when the power transmission angles on the interconnections or on certain internal 138-kV and 115-kV transmission have been of relatively great magnitude. The oscillations have been controlled or reduced by adjustment of the excitation of speed control parameters on various groups of machines throughout the three interconnected systems. The analytical techniques utilized to determine the adjustments will be of considerable benefit in planning future extensions to the interconnections and facilities within the power systems in the three provinces.**

Introduction

TIE LINE oscillations have been experienced on the Saskatchewan-Manitoba-Ontario West interconnection in varying degrees of severity since Manitoba and Ontario became interconnected in 1956, followed by Manitoba's interconnection to Saskatchewan in 1960.

Fig. 1 is a single line diagram of the system, and Table I provides the salient details of the larger generating stations. In general, the interconnection is a 2700 MW system serving approximately 200 000 square miles, with major load centers located at Port Arthur, Fort William, Winnipeg, Regina, and Saskatoon. When one compares this to, say, the power system serving Metropolitan New York—a 7500 MW system serving an area less than 1000 square miles—it is reasonable to state that this interconnection is a relatively "loosely connected" system.

Historical Notes

Commencing in 1956 and until 1960, only Manitoba and Ontario West were interconnected, and transfers of up to 50 MW took place in either direction on the tie line without trouble. By 1960 the Manitoba load was 600 MW, while the Ontario West system load was just over 450 MW.

In 1960, an interconnection was also established between Manitoba and Saskatchewan. Under normal transfer conditions the three systems operated in parallel successfully. It was found that frequency and tie line power bandwidths were considerably reduced in Manitoba and Ontario when connected with Saskatchewan. However, severe disturbances in any of the three areas caused the Saskatchewan-Manitoba tie line to relay out.

In the spring of 1962, tie line power and frequency oscillations due to dynamic instability also caused the Saskatchewan-Manitoba tie line to relay out,[1] and the factors appearing to have a bearing on the matter were considered to be as follows:

1) Heavier east to west power transfers in Manitoba had come about due to increasing loads in the Brandon area and to power interchange to Saskatchewan. Transmission extensions had not kept pace with these requirements.

2) The stabilizing effect of the Manitoba thermal plants, particularly at Brandon, was lost when operation was cut back after winter peaking operation.

3) Oscillations were probably aggravated with the hydroelectric units in the eastern end of the interconnection, both in Manitoba and Ontario, interacting with the higher inertial and more responsive thermal-electric units in the west.

4) The long distance (1100 miles) between the most remote generating sources (Saskatoon in Saskatchewan and the Nipigon River plants in Northern Ontario) resulted in a low equivalent synchronizing power coefficient between these groups in particular.

In the fall of 1962 the problem disappeared once again as load increased and generating patterns changed. The oscillations might not have returned in 1963 because Manitoba's transmission deficiency between Winnipeg and Brandon was removed when a new 230-kV line was placed in service between these two cities. However, in April, 1963, Saskatchewan's Squaw Rapids plant was commissioned and connected to the end of the loosely tied interconnection, increasing the distance between the most remote plants from 1100 to 1300 miles.

The main oscillation problem discussed in this paper, including cyclic changes in voltage, frequency, and power, began with this extension of the system.

Nature of the Oscillations

Electromechanical oscillations can develop between any groups of machines on an interconnected system. The frequency of the oscillations depends on the electrical length of the tie lines between the machines and on the inertia of the machines themselves.[2] The frequency of oscillation in the Saskatchewan-Manitoba-Ontario West System was approximately 0.35 Hz unless either tie line opened when it increased to 0.45 Hz on the remaining two areas.

The oscillations were sinusoidal in nature, as illustrated in Fig. 2. In the period 1962–1965, the tie line power and voltage oscillations were usually measurable and sometimes objectionable. They often continued to grow until the Saskatchewan-Manitoba tie line was tripped by impedance relays. Fig. 3 shows typical system recording meter charts for such an occasion.

Paper 31 TP 67-21, recommended and approved by the Power Generation Committee of the IEEE Power Group for presentation at the IEEE Winter Power Meeting, New York, N. Y., January 29–February 3, 1967. Manuscript submitted November 2, 1966; made available for printing August 15, 1967.

O. W. Hanson is with the Saskatchewan Power Corporation, Scarth-Victoria Regina, Sask., Canada.

C. J. Goodwin is with Manitoba Hydro, Winnipeg, Man., Canada.

P. L. Dandeno is with the Hydro-Electric Power Commission of Ontario, Toronto, Ont., Canada.

Reprinted from *IEEE Trans. Power App. Syst.*, May 1968.

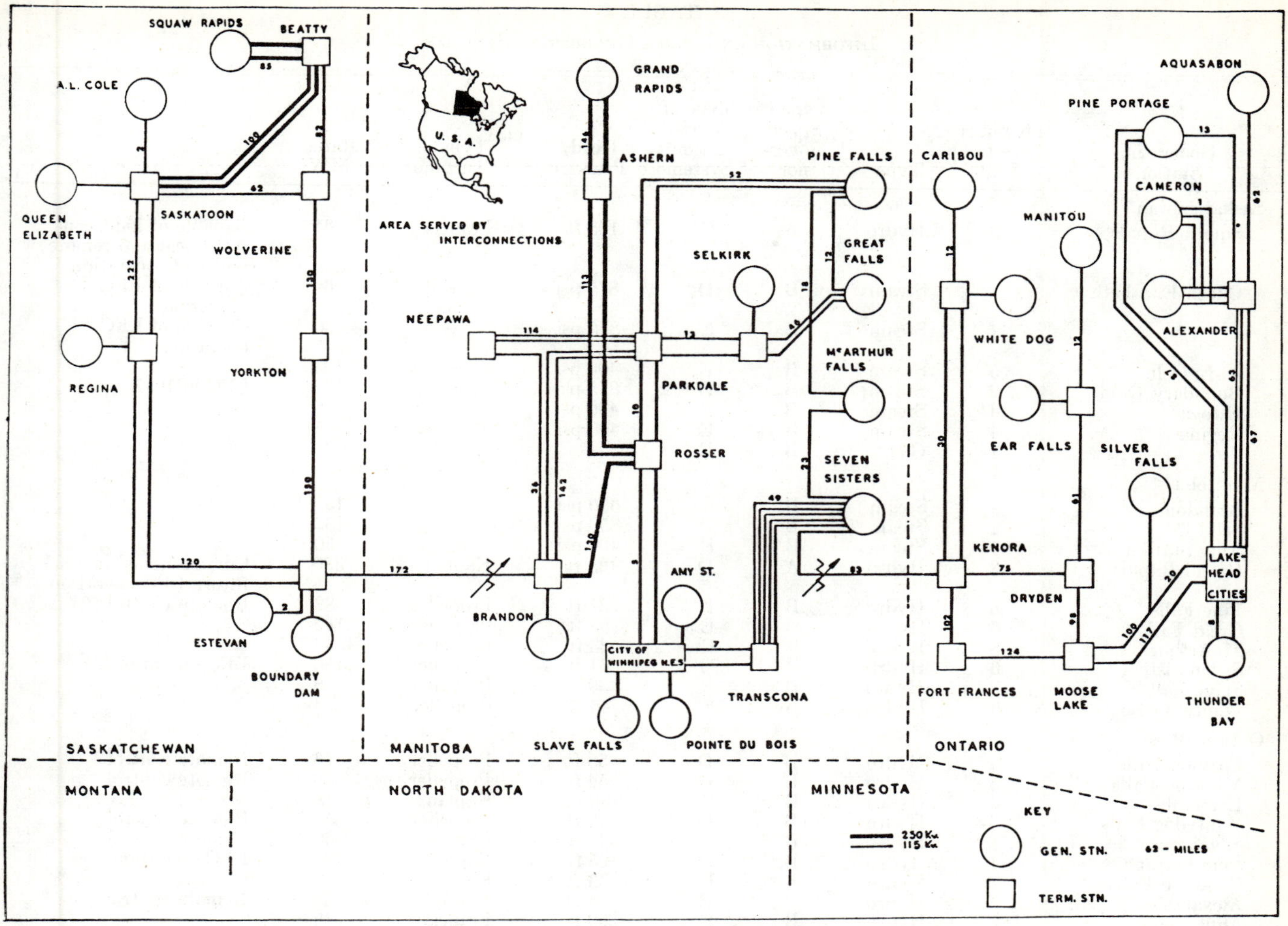

Fig. 1. System diagram (June 1965).

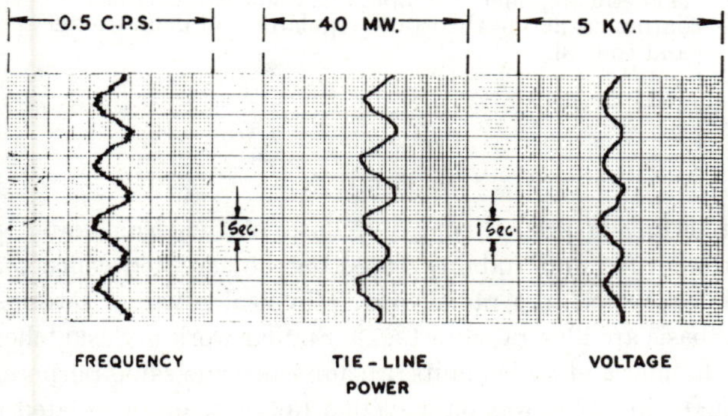

Fig. 2. Typical oscillation pattern.

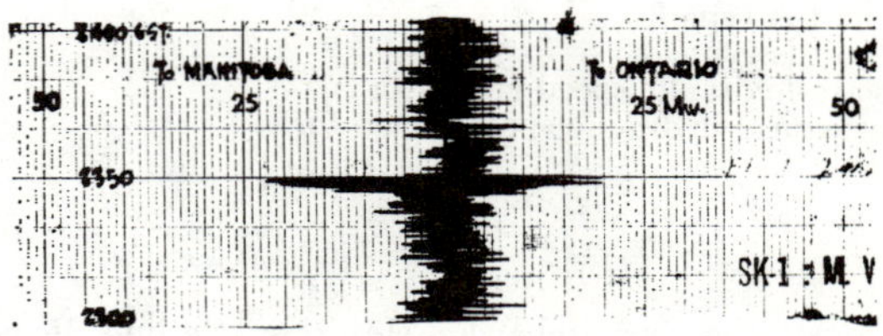

(a)

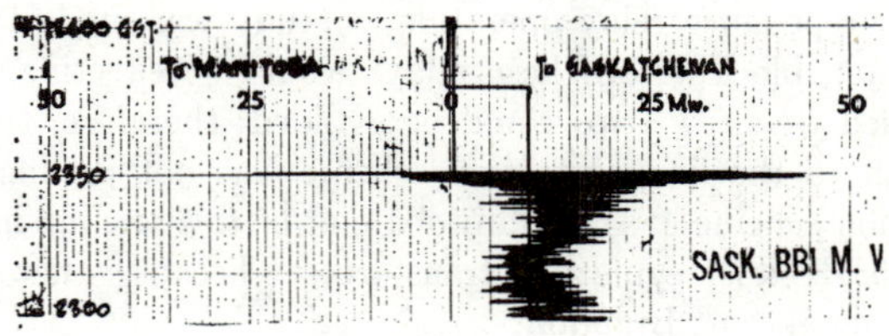

(b)

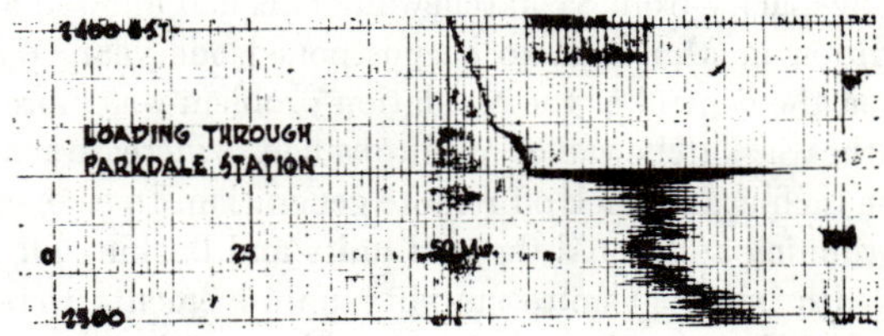

(c)

Fig. 3. Oscillations building up and systems pulling out of step. (a) Manitoba—Ontario tie line. (b) Saskatchewan—Manitoba line. (c) Power flow west from Parkdale Station in Manitoba.

Initial Investigations into the Problem

In the period 1962–1965, efforts were made to locate the cause of the oscillations. These efforts were mainly directed toward finding a predominant cause; in fact, it was possible that the cause lay in many areas. At the beginning it was believed that interim solutions would suffice, either because planned system improvements such as the Winnipeg-Brandon 230-kV line would make a permanent solution, or because of changing tie line transfers.

Power transfers between Manitoba and Saskatchewan were generally minor in 1962, but Manitoba was buying substantial economy power from Ontario. The easiest solution at this time

TABLE I

INFORMATION ON LARGER GENERATING STATIONS

Generating Stations	Number of Units	Type	Type of Speed Gover-nor	Type of Exci-tation System	Head/ Pressure	Type of Turbine	Rated MW	
Saskatchewan								
Squaw Rapids	6	Hydro	A	C	105 ft	Francis	201	Capable of joint control (local and remote) and LFC operation
Queen Elizabeth	1	Steam	B	D	875 psi	—	66	Capable of LFC operation
	1	Steam	B	E	875 psi	—	66	Capable of LFC operation
A. L. Cole	5	Steam	B	E	400 psi	—	105	
Boundary Dam	2	Steam	B	D	875 psi	—	132	LFC in 1967
Estevan	4	Steam	B	E	420 psi	—	70	
Regina	4	Steam	B	E	400 psi	—	65	
	1	Gas	B	E	—	—	30	
Manitoba								
Brandon	4	Steam	B	C	600 psi	—	132	
Selkirk	2	Steam	B	D	850 psi	—	132	
Amy Station	4	Steam	B	E	400 psi	—	50	
Grand Rapids	3	Hydro	A	D	125 ft	Kaplan	330	Commissioned in 1965 future LFC operation
Pine Falls	6	Hydro	B	D	37 ft	Propeller	82	Remote control
Great Falls	6	Hydro	B	E	58 ft	Propeller	132	
McArthur	8	Hydro	B	E	23 ft	Propeller	56	
Seven Sisters	6	Hydro	B	E	61 ft	Propeller	150	LFC operation
Slave Falls	8	Hydro	B	F	30 ft	Propeller	68	
Pointe du Bois	16	Hydro	B	F	46 ft	Propeller	72	
Ontario West								
Caribou Falls	3	Hydro	B	D	58 ft	Propeller	76	Remote control
Manitou Falls	5	Hydro	B	D	54 ft	Propeller	69	Remote control
Ear Falls	4	Hydro	B	F*	36 ft	Kaplan	18	
Whitedog Falls	3	Hydro	B	D	50 ft	Propeller	51	Remote control
Silver Falls	1	Hydro	B	D	330 ft	Francis	46	Remote control
Pine Portage	4	Hydro	B	D	105 ft	Francis	130	LFC operation
Cameron Falls	7	Hydro	B	E†	73 ft	Francis	72	
Alexander	5	Hydro	B	E†	58 ft	Francis	62	Remote control
Aquasabon	2	Hydro	B	E	290 ft	Francis	46	
Thunder Bay	1	Steam	B	D	1450 psi	—	100	

All steam units are non-reheat.
* One unit is type E.
† One unit is type C.
A—electro-hydraulic speed generator
B—mechanical-hydraulic speed generator
C—magnetic amplifiers and main exciter
D—magnetic amplifier + amplidyne and main exciter
E—continuous acting rheostatic regulator and main exciter
F—hand control.

was to open the Saskatchewan-Manitoba tie line. Because instability usually caused the Saskatchewan-Manitoba tie line to trip by impedance relay operation, a study was made to see if out-of-step blocking on these relays would be effective. It was concluded that few trips could be avoided; therefore, the complication to the line protection scheme was not justified. The protection scheme at the Manitoba end of this tie line included the 115/138-kV regulating transformer, because there is no 138-kV breaker at Brandon.

During 1963–1964 Manitoba had an agreement for the purchase of capacity from Saskatchewan. This had allowed a capacity addition in Manitoba to be postponed one year. With this condition the control of the oscillation problem was more costly, since either Saskatchewan or Manitoba had to increase generation near the tie line and back off remote generation. In each case this meant running additional thermal units and backing off run-of-river hydro units. Definite stability improvements were noted when transmission voltage levels were kept as high as possible, even at light load periods.

Prior to 1962, Manitoba Hydro had begun a series of tests to ensure that the hydro governor parameters were known and that the settings (speed droops and dashpot time constants) were optimum for unit stability based on the accepted concept of a unit connected to a radial load.[3] Typical governor settings on this basis are illustrated in Fig. 4. Similar work in Saskatchewan on thermal and hydro units accomplished the same purpose, although this work was on a system rather than an isolated unit basis.[4] Fig. 5, for example, shows the effect of various "service damping" settings of the Squaw Rapids Generating Station electric governors on the magnitude of power oscillations on the Saskatchewan-Manitoba tie line. Measurements on the Ontario West system indicated that all unit governors operated in a stable manner.

In summarizing these investigations, it was initially felt that damping forces on the interconnected systems were small and perhaps negative at times. The preliminary analysis of such negative damping forces considered the importance of small synchronizing power coefficients due to the large electrical distances between major groups of generation. The settings of the speed governors in the Saskatchewan and Manitoba systems in particular were closely examined. When it became apparent that the problem had not been completely resolved, the possibility of

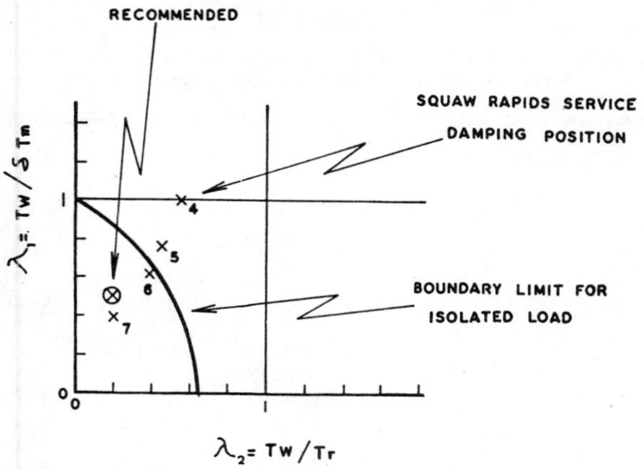

Fig. 4. Governor settings. Settings outside the boundary result in instability based on generator supplying an isolated load.

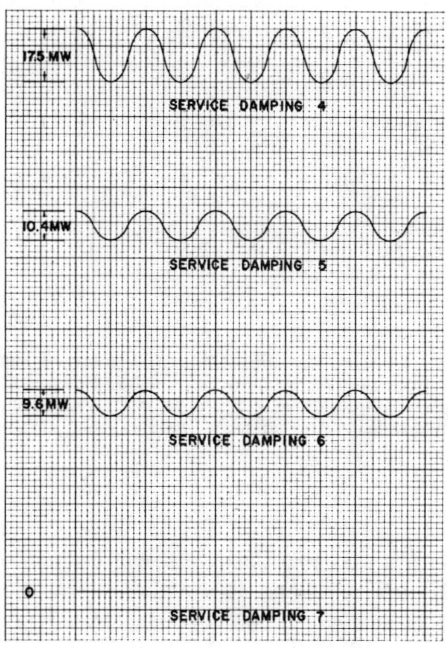

Fig. 5. The effect of the damping settings on the governors at Squaw Rapids on tie-line oscillations.

negative damping forces due to unsuitable voltage regulator settings was considered, and analytical efforts were then concentrated in this direction.

Analytical Studies of this Type of Problem

For many years it has been evident that a change in the forward gain of the voltage regulator and excitation system of a synchronous machine can result in higher generator stability limits.[5] Generally speaking, obtaining typical excitation system voltage response ratios of 1.0 per unit with large direct-connected slow-speed exciters requires relatively high gains in the overall excitation system.

Analytical studies and special field tests with generation connected radially over long lines have shown that reduction in main stage amplification by approximately one half for typical amplidyne or magamp voltage regulators did result in noticeable improvement in stability limits under dynamic (small oscillation) conditions.

The analytical studies also showed for the particular test conditions that reduction of the main stage gain to zero resulted in a very low stability limit.

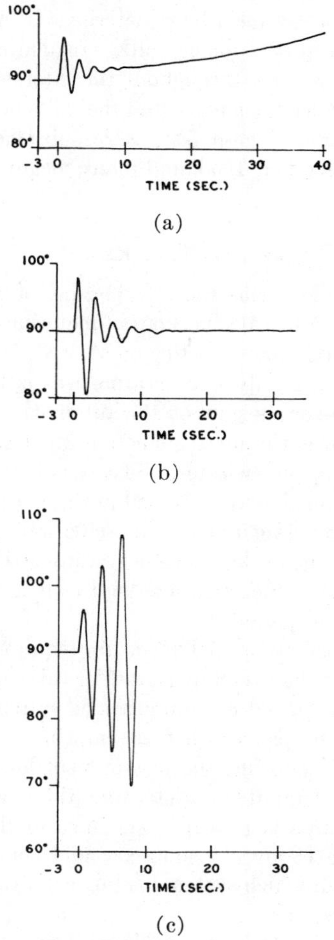

Fig. 6. Effect of excitation system gain. Plot shows rotor angles after disturbance is applied. Initial rotor angle is 90°. (a) Zero gain. (b) Medium gain. (c) High gain.

Fig. 6 shows the output from some analog computer runs performed at Ontario Hydro to assess the effect of excitation systems on the dynamic (small oscillation) characteristics of a single generator connected through a high impedance to an infinite bus. A change in torque of the impulse type was made. In the first curve the gain of the excitation system is zero, in the second curve a medium gain was used, and in the third curve relatively high gain was simulated. The latter gain is required to give an excitation system voltage response ratio of just over 1.0 per unit for a five percent error signal. These computer runs were made with the generator loaded to give a rotor angle of 90° to the infinite bus.

These studies subsequently led to a recommendation to conduct field tests on the combined systems just referred to, with a view to determining the effect of reducing gains on several of the principal generating stations in the interconnection. These stations had excitation systems that were controlled by continuously acting voltage regulators with either magnetic amplifiers or rotating amplifiers (amplidynes) in their main stage amplification. Due to the configuration of parts of the interconnected system and the methods of operation desired at the time, it was decided to accept any decrease in transient stability limits due to this lowered gain and take advantage of the higher dynamic limits.

After a discussion with various interested parties of the proposed studies and of other probable causes of the intersystem oscillations, a test program was planned for which each provincial utility set up special recording metering. It was proposed to determine the effect of reducing voltage regulator gain settings at various locations *in turn* throughout the interconnected systems and finally to reduce the gains on all the voltage regulators on all systems at one time. These tests were subsequently performed in June and September, 1965, and many meaningful conclusions were obtained.

June 1965 Test Results

Tests were made during the working day of Tuesday, June 1, 1965, following a full-scale checkout of metering and communications and some preliminary testing on May 31, 1965.

For each test a group of generators was either put on hand control of voltage or the gain on the automatic voltage regulator was reduced. The entire interconnection was treated in this way. At the beginning of each test, Saskatchewan generation was increased in the north and decreased in the south until the heavy north-south power transfer induced oscillations of nearly 15 MW peak-to-peak on the Saskatchewan-Manitoba tie line. When this point was reached, generation was kept constant while recorders were run for a short period.

The need to set up special recording equipment throughout the system cannot be overemphasized for this type of testing, nor can the need for a good communications system and an overall chief-of-test be neglected when attempting to coordinate the actual test runs. Following the system tests, load flow cases were run on a digital computer to determine the power angles in the system for comparison with the measured oscillations of each test. Table II gives a summary and comparison of three particular tests in which system loading conditions and transfers were reasonably similar.

The following conclusions were drawn from the tests in particular and from a series of about a dozen similarly instrumented tests:

1) Reduced gain on automatic voltage regulators at many plants improved stability. Where the gains at several plants were reduced together, greater benefits were noted. Certain plants throughout the interconnected system showed marked effects.

2) Removal of automatic voltage regulators from service is detrimental to stability. In other words, even regulators with too great a gain are better than none.

Following the tests, gains on automatic voltage regulators were reduced where possible at several plants in which the effect was obviously beneficial. This led to the conclusion that oscillations on the three systems are properly damped and no longer lead to instability. A tolerable voltage oscillation level instead of tie line power oscillations may now be used as a guide for determining when a tie line shall be opened or when power transfers must be changed.

Due to changes in system load during the test period, comparisons to a "base case" became inconclusive for the Manitoba area. Accordingly, further system tests, scheduled for September, 1965, were arranged to include governor setting changes also.

September 1965 Test Results

Similar methods to those of June 1, 1965, were used for the system tests made on Tuesday, September 14, 1965. The following conclusions were obtained.

1) Reduction of automatic voltage regulator gain at the Manitoba plants also improved stability.

TABLE II

Comparison of Voltage, Frequency, and Power Oscillations with Different Voltage Regulator Gains

	Test 1A Voltage Regulator Gains "As Found"	Test 7 Voltage Regulator Gains Reduced by Approximately 50 Percent	Test 6B Voltage Regulator Gains Reduced at Certain Plants and Certain Voltage Regulators Out of Service
Voltage Oscillations, Percent			
Squaw Rapids	1.8	0.5	2.6
Yorkton	3.3	1.9	3.75
Winnipeg	0.6	0.4	0.70
Port Arthur	0.73	0.52	0.83
Frequency Oscillations, Hz			
Squaw Rapids	0.19	0.11	0.21
Port Arthur	0.09	0.07	0.14
Tie-Line Power Oscillations			
Saskatchewan-Manitoba Tie Line	14.5	13.5	20.
Manitoba-Ontario Tie Line	9.6	6.8	12.8

Note: Tests 1A and 6B had the same relative angular spread across the interconnected systems. In Test 7, Squaw Rapids output was at its highest, and the relative angle of that plant with respect to the point of interconnection with tie line BB1 had increased by 7°.

2) The effect of changing the permanent droop in several of the Manitoba Hydro Winnipeg River plants was ascertained by varying the permanent droop settings from 5 percent to 3.5 percent. Power oscillations on the tie lines from Manitoba to Saskatchewan and from Manitoba to Ontario were reduced by as much as 20 percent peak-to-peak. Voltage oscillations in the systems external to Manitoba Hydro were also reduced, principally in the Saskatchewan system. Although the effect of reducing the permanent droop in the governing system is to increase the gain and thus make the governor more sensitive, this effect is more than offset by a considerable increase in the predominant time constant of the governing system. It might be concluded, then, that making the Winnipeg River plants more slow to respond by reducing the permanent droop removed or minimized one of the perturbing forces on the interconnected systems.

3) Oscillations were significantly reduced all over the systems when the gate limits were put on at Squaw Rapids. Despite the fact that the plant has conventional stable governor settings, its position at the end of a long electrical system required that under dynamic conditions it should take no part in system governing. Fig. 7 shows the effect of this gate limiting on the oscillations on the Saskatchewan-Manitoba tie line. The explanation for this effect appears to be that the application of the gate limit is equivalent to reducing the forward gain of the governor to a very small number.

4) Voltage oscillations were reduced when synchronous condensers are in operation near the electrical center of the systems.

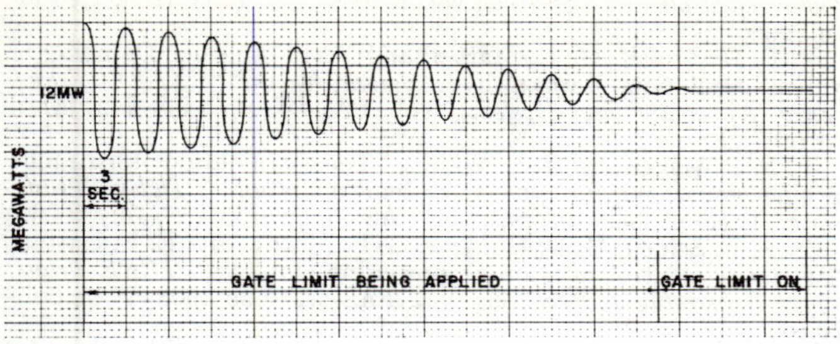

Fig. 7. The effect of wicket gate limiting at Squaw Rapids on tie-line oscillations.

This was simulated by running two thermal units at Brandon with steam valves closed and measuring voltage oscillations at various locations. In this case the principal reason for the improvement is the location of these synchronous units at approximately the electrical center between the Saskatchewan and Manitoba systems.

Present State of the Systems

Tie lines have not tripped due to dynamic instability since the June, 1965, tests and subsequent to reduction of gains on automatic voltage regulators throughout the interconnection. Tie line tripouts due to heavier disturbances (remote faults, etc.) have been fewer than before and are not of great concern. It is now possible to operate important system internal lines at heavier loadings without causing instability, and the three areas have continued to operate for most of the time as an interconnected group.

The voltage oscillations, while still present on a much reduced scale, cause only occasional operating restrictions. These include the running of thermal units solely for voltage support, the necessity of live line maintenance, or, on very isolated occasions, the opening of one of the two tie lines, when key system internal circuits are out of service.

Future Work

Work is being directed to minimizing the effects of the voltage oscillation problem and to ensuring that the dynamic stability problem remains under control.

Measurement of excitation system and governor parameters, in suitable form for computer analysis of dynamic stability, is continuing. Frequency response measurements have yielded the best information for analytical purposes. Transient responses have been made to step input functions and the results duplicated, using the block diagrams of transfer functions obtained through frequency response tests, by simulation on an analog computer.

Analytical work to find future problem areas in advance and to suggest solutions is being done. This work now plays a more predominant role in the analysis and planning for each system when considering additions which are expected to affect the dynamic stability of the interconnection. For instance, the Coteau Creek Hydroelectric Station in Saskatchewan, which has very long penstocks, will have very slow governor action except over the range of the Vernier stabilizer.[6]

In 1967, Manitoba Hydro will convert the conventional rotating main exciter excitation systems on two of the three units at Grand Rapids to static systems using silicon controlled rectifiers (SCRs) to increase the transient stability limits. Studies leading to this decision included simulating the effect of a speed stabilizing signal in the automatic voltage regulator for the controlled rectifier system.[7]

In specifying the excitation systems for new 150-MW units at Boundary Dam, the Saskatchewan Power Corporation called for high-speed, high-ceiling units and has bought static SCR systems, with provision for a stabilizing signal combined with the input to the automatic voltage regulator.

In specifying a 100-MW unit extension at Brandon, Manitoba Hydro called for the provision to run it as a synchronous condenser. Both of these facilities have been purchased. In addition, one unit of the existing four is being converted to synchronous condenser operation.

In Manitoba, the Kelsey plant and its associated hoisting, smelting, and refining loads will be connected to the interconnection at Grand Rapids.[8] This extension will require close scrutiny from an analytical viewpoint to ensure that problems of dynamic instability can be kept under control.

Conclusions

Probably the most significant conclusion reached by the three interconnected utilities was that dynamic instability problems must be approached from an overall system basis.

The parameters of synchronous machine control equipment in generating stations can have a considerable influence on overall system performance; they should not be adjusted purely for local requirements. Excitation control and speed control elements must definitely be set to satisfy the system from an overall performance viewpoint.

Field test techniques must be developed to ensure that the control devices on new units are properly set up to satisfy the system as a whole. At the present time installed units are periodically checked to ensure provision of the best performance possible.

Further field work must be done to supply valid input data to the powerful computer programs now used to design future extensions to interconnected systems. The computer study results will be no better than the input data used. This should ensure that the best possible control devices are purchased.

By conducting coordinated interprovincial tests using special recording equipment and then carefully analyzing the results, new settings for automatic voltage regulators in particular and some speed governors as well were implemented which have minimized the dynamic instability problem on the Saskatchewan-Manitoba-Ontario West interconnection.

References

[1] C. J. Goodwin, "Stability problems on the western Ontario-Manitoba-Saskatchewan Interconnection," Canadian Electrical Association, October 1963.

[2] S. B. Crary, *Power System Stability*, vol. 2. New York: Wiley, 1947, pp. 26, 284.

[3] L. M. Hovey, "Optimum adjustment of hydro governors on Manitoba Hydro system," *AIEE Trans.* (*Power Apparatus and Systems*), vol. 81, pp. 581–597, December 1962.

[4] O. W. Hanson, "Power system tests to optimize controls," *Trans. Engrg. Inst. Can.*, December 1964.

[5] C. Concordia, "Steady state stability of synchronous machines as affected by voltage regulator characteristics, *AIEE Trans.*, vol. 63, pp. 215–220, May 1944.

[6] L. Allas and E. R. Smith, "The Coteau Creek project," Canadian Electrical Association, March 1966.

[7] P. L. Dandeno, A. N. Karas, K. R. McClymont, and W. Watson, "Effect of high-speed rectifier excitation systems on generator stability limits," *IEEE Trans. Power Apparatus and Systems*, vol. PAS-87, pp. 190–201, January 1968.

[8] L. M. Hovey, "The Kelsey hydro-electric development on the Nelson River, Manitoba," presented at the AIEE Summer General Meeting, Buffalo, N. Y., June 22–27, 1958.

Discussion

F. R. Schleif (Bureau of Reclamation, Denver, Colo.): The authors have described an important problem of system electromechanical oscillations, analysis of sources, verification tests, and relief measures. Their experiences seem to have closely paralleled certain experience[9],[10] of power systems in the Western states. This opportunity to compare and consolidate analyses and relief measures is especially welcome.

Recognition by the authors of possible contribution to the system oscillations from many different sources seems entirely proper and well supported. Thus, the conclusion that dynamic instability problems should be approached from an overall system basis is correct in that each dynamic element should be considered. However, the conclusion that the various parameters should be set for overall system performance rather than purely for local requirements seems to improperly imply that a conflict is inherent. This interpretation is probably more general than the authors intended. Nevertheless, coordination is presently important enough to interconnected operation as to warrant some discussion. Our studies and experience in coordinating governors of the Bureau of Reclamation plants indicate that occasionally system resonance may impose an additional requirement on coordination, but system operation does not dispense with the primary need for stability on an isolated basis.

The behavior illustrated for Squaw Rapids has attracted our interest and study in this connection. With parameters approximating the plant and system by analog, the behavior the authors describe is reproduced. The appearance of a difference in frequency of system oscillation between the conditions of Fig. 5 and those of Fig. 7 suggests rather sharp resonance at the latter conditions. The analog confirms this to be possible and further, that the governor/penstock parameters can be such as to contribute to the oscillation in a manner similar to Fig. 10 of Schleif and Wilbor.[11] The study also indicates that a reduction of gain in the loop between the pilot and main servos is one way to suppress contribution to the resonance. Normal governing action might then continue with only slightly compromised characteristics.

The study also indicates that contribution to the resonance is strongly dependent upon the penstock parameter T_w which, in effect, varies with load. This introduces the interesting possibility that contribution to the resonance is aggravated by the joint control function in keeping all units responding in unison. A mitigating influence would then be to distribute load nonuniformly among the Squaw Rapids units.

The authors' identification of Squaw Rapids as a contributor to certain system oscillations and their expedient for suppressing the contribution are both incontestable and obviously practical. The comments offered regarding other measures are in the interest of clarifying or reconciling the problem of governor coordination. Similarly, coordination of other dynamic components of the system is a part of the problem.

Benefits to system damping obtainable through reduction of gain of voltage regulators is consistent with the frequency response analysis. The Bureau's recent studies[12] have offered an explanation of the process by which rapid voltage regulation contributes negative damping. These studies and tests have also shown that it is possible, by inserting a properly phased frequency-derived supplementary signal, to correct the phase lag without reducing gain so as to achieve good positive damping. As the authors indicate, the static exciter systems yield most readily to this refinement. This type of exciter would therefore be a logical choice for new equipment where powerful damping is needed. However, the rotating exciter systems, many of which are in existence, need not be replaced for this reason alone because they, too, can yield strong damping if they are supplemented with proper frequency-derived signals.

The authors are to be congratulated for presentation of a valuable analysis of a problem vital to our growing power systems.

References

[9] F. R. Schleif and J. H. White, "Damping for the Northwest–Southwest tieline oscillations—An analog study," *IEEE Trans. Power Apparatus and Systems*, vol. PAS-85, pp. 1239–1244, December 1966.

[10] F. R. Schleif, G. E. Martin, and R. R. Angell, "Damping of system oscillations with a hydrogenerating unit," *IEEE Trans. Power Apparatus and Systems*, vol. PAS-86, pp. 438–442, April 1967.

[11] F. R. Schleif and A. B. Wilbor, "The coordination of hydraulic turbine governors for power system operation," *IEEE Trans. Power Apparatus and Systems*, vol. PAS-85, pp. 750–758, July 1966.

[12] F. R. Schleif, H. D. Hunkins, G. E. Martin, and E. E. Hattan, "Excitation control to improve powerline stability," *IEEE Trans. Power Apparatus and Systems*, vol. PAS-87, June 1968 (to be published).

Manuscript received February 23, 1967.

Y. N. Yu, K. Vongsuriya, and **G. E. Dawson** (The University of British Columbia, Vancouver, B.C., Canada): The authors are to be congratulated for their important contribution to the dynamic studies of interconnected systems. We believe that the authors' recommendation on the adjustment of system parameters based on the overall system performance is a sound one which should serve as a guide for future works in this area. The problem of analytical determination of optimum system parameters for a system of this size eventually leads to complicated procedure in digital computation. Moreover, valid mathematical system models and system data are required to make the analytical studies meaningful. The authors did not elaborate on the analytical technique being used. It would be of interest to know whether and how well the technique of parameter sensitivity analysis can be applied to this problem for assessing the influence of key parameters on the overall system performance.

References

[13] R. Tomovic, *Sensitivity Analysis of Dynamic Systems.* New York: McGraw-Hill, 1963.

Manuscript received March 2, 1967.

A. R. Scott (Canadian General Electric, Ltd., Peterborough, Ont., Canada): This paper, together with Dandeno *et al.*,[7] further serves to underline the importance of regulating system performance in the operation and design of large power systems. In a sense, the authors were fortunate that the interconnection which modified the constraints of the previous system arrangements was primarily an economic interchange and could be dispensed with from time to time until a satisfactory solution was found.

The problem has been defined as inadequate total system damping, and both excitation system and speed governoring system adjustments have been made.

While the proof of system performance improvement by field test is the best possible substantiation, the authors clearly recognize the limitations of field tests to handle future system operating conditions and even, one suspects, to economically investigate a full range of present operating conditions. In addition, system tests can only rarely explore the behavior following severe disturbances because of the risks to customer supplies.

The voltage regulator changes have improved damping at the expense of transient response. Have the authors explored the reduced transient stability margins which result from these changes to to determine whether the best compromise has been reached?

The problem of conducting sufficiently detailed dynamic performance studies on computers is recognized to be more limited by the availability of accurate data on the adjustment of specific regulating systems than by the limitations in stability program rep-

Manuscript received March 2, 1967.

resentation, and the authors are to be commended for taking steps to obtain such control data for their systems from actual field measurements and other sources.

C. Concordia (General Electric Company, Schenectady, N. Y.): During a rather long association with the electric power industry, we can recall at least three other cases of system oscillations which seem very similar to that reported in this paper. All of these had one condition in common: at least one of the systems involved consisted predominantly of hydraulic turbine-driven generators. We believe that this is not a coincidence, but rather that it is related to the transient characteristics of the hydraulic system (the water inertia) and to the characteristics of the usual turbine speed governors which are designed to accommodate these hydraulic characteristics. It can easily be demonstrated that, for most reasonable values of the period of the oscillation, hydraulic turbine speed governor action will result in negative damping, quite independent of whether or not the criterion of stability for isolated operation is satisfied. We think this was clearly demonstrated in the paper by the fact that good performance was obtained by blocking the governors at Squaw Rapids. It is, of course, impossible to block all governors. However, it should be recognized that it would be very helpful to go to, say, much greater transient droops and longer reset times than would be indicated by the criterion for an isolated machine. That is, since the damping contributed is practically sure to be negative, we might as well make it as small as possible.

This is not at all to say that the instability must be cured solely by operation on the speed governors. Instability is a system problem, and it is not very helpful to lay the sole blame on a particular element of the system. It is really a question of which is the least expensive and most expedient change to make in order to eliminate the instability. The authors have shown that a very simple and effective way is to reduce the gain of the generator excitation system control, as can be concluded from Concordia.[5]

Manuscript received February 24, 1967.

O. W. Hanson, C. J. Goodwin, and P. L. Dandeno: The authors are particularly indebted to Mr. Schleif for his interest in the Squaw Rapids hydro governor problem. As a result of his analog simulation study, he has verified our concern for setting system controls without taking the system natural frequencies into consideration. He also suggests that cognizance must be taken of the stability problem when supplying isolated loads. These are two valid comments, but the question still arises as to which criteria must hold if there is a conflict. In today's large interconnected power systems, units will rarely be running isolated except when synchronizing or desynchronizing. The most probable situation would be following a disturbance which split the system. On this basis, if there is a conflict, it seems reasonable to suggest the general system criteria should prevail. Provision could be made for other fail-safe means to handle the occasional isolated load problem.

Mr. Schleif's comment relative to the strong effect on system resonance of the "variable" penstock parameter T_w is most interesting. Before the tests and before changes to the control settings were made, the three interconnected systems experienced the most severe dynamic stability problems when Squaw Rapids approached full output (largest T_w). This also coincided with the largest power angles in the Saskatchewan portion of the interconnection. Applying the Squaw Rapids gate limit at full station output to overcome the governor contribution to the dynamic problem is fortunately not an operating handicap. However, it is significant that T_w is not constant at all generating levels, and this particular point provides further insight into the problem of hydraulic turbine speed control.

Manuscript received June 12, 1967.

The interest and work in progress by the staff members at the Bureau of Reclamation indicates that they are also quite concerned with similar power system control problems. Major efforts will be required to get the best performance possible out of existing equipment in any given power system. Further studies of an increasingly complex nature will be necessary to ensure that the parameters of control devices now in the planning stage take cognizance of the "dynamic" as well as the "transient" stability problem.

We certainly agree with Yu, Vongsuriya, and Dawson when they suggest that the search for optimum system control parameters will eventually require digital computer computation. Whether the technique of parameter sensitivity analysis is the best approach is very difficult to say at this time. The "overall" system approach obviously requires more study, and we hope it will serve as a guide for future work at the University of British Columbia.

Mr. Scott's comment on control changes to improve damping at the expense of reduced transient stability margins is a question most worthy of discussion. Up until very recently, the available control data were insufficient to properly simulate the systems to ascertain the best compromise between "dynamic" and "transient" stability conditions. The question still remains, how does one know when the best compromise has been reached for all future patterns of generation and all fault conditions? Since the transient condition may happen once a year, while the dynamic condition is occurring the remainder of the time, the decision to reduce the transient stability limit was accepted when much improved dynamic operation resulted from reduced control system gains.

The long-term solution to this problem is to design and install future "system" control equipment, which performs equally well under all conditions, and modify presently installed facilities with auxiliary feedback signals to get the most out of this existing equipment.

The authors appreciate Mr. Concordia's remarks, particularly his point regarding the influence of hydraulic governors in contributing to negative system damping. One major experience with system oscillations to which he alluded concerned the southern portion of the Ontario Hydro 230-kV system which is not connected with Manitoba or Saskatchewan. Interconnections existed with a closely-knit thermal system immediately to the west of southern Ontario, as well as to an all-hydro system to the east. Oscillations occurred when Ontario Hydro was tied to both systems at the same time. Opening the interconnections to the west or to the east caused the oscillations to stop. It was shown by system tests that the oscillations occurred partly because of low or negative synchronizing power between the generators in the western and the eastern interconnected systems. It should be noted, of course, that the thermal generators in the western system were as much as 800 miles from some major groups of generation in the eastern interconnected system, and that power flows were predominantly in one direction through the eastern interconnected system and through the Ontario Hydro system as well, with a consequent "build up" of power angle.

In retrospect, it is now evident that some study or system tests would also have shown that the governors in the all-hydro system were contributing to negative damping. The same remarks could be made about the excitation system gains in the all-hydro system.

The authors wish to give due credit to Mr. Concordia for his early work in this field.[5] It is surprising and regrettable that almost 20 years elapsed before power system analysts in the electric utility industry put the results of Mr. Concordia's paper to general use. We are particularly grateful for his advice in connection with the problems discussed in the authors' paper. Fortunately, these concepts of matching synchronous machine control system parameters to meet the requirements of interconnected operation are now widely accepted and recognized. Large-scale digital computer programs are also available[14],[15] which should permit a reasonably detailed analysis to be made of the effect of varying important synchronous machine control parameters on a system-wide basis.

References

[14] R. M. Webler and C. C. Young, "A new digital computer program for predicting dynamic performance of electric power systems," presented at the 5th PICA Conf., Pittsburgh, Pa., May 1967.

[15] D. G. Ramey and R. T. Byerly, "Digital simulation of interconnected systems," presented at the 5th PICA Conf., Pittsburgh, Pa., May 1967.

FOUR CORNERS PROJECT STABILITY STUDIES

R. G. Farmer R. H. Hartley
Arizona Public Service Company
Phoenix, Arizona

M. H. Kent L. M. Wheeler
Southern California Edison Company
Los Angeles, California

ABSTRACT

Four Corners Project stability studies employed two types of digital computer programs. A program with the classical representation for generators and loads was used to establish the preliminary system requirements. A program with more accurate representation was used to make detailed studies. Braking resistor, generator dropping, fast valving, fast load runback, and combinations of these were studied as aids to system stability.

INTRODUCTION

The Four Corners Project is a further development in coal-fired steam generation in the Southwest. The present project consists of two 755-MW (nameplate) generators in the Northwest of New Mexico near the boundary intersections of the four states of Colorado, Utah, New Mexico, and Arizona. The maximum capability of each new unit is 795 MW. The two new units are owned on a joint basis by six participants with the following shares:

Arizona Public Service Co.	15%	239 MW
El Paso Electric Company	7	111
Public Service Co. of N. Mex.	13	207
Salt River Project	10	159
Southern California Edison Co.	48	763
Tucson Gas & Electric Co.	7	111
	100%	1590 MW

The two units are an addition to the existing Arizona Public Service Company installation at Four Corners of three generators having a combined nameplate capacity of 580 MW. The first new unit is scheduled for the spring of 1969 and the second unit for the spring of 1970.

The Four Corners Project is the first coal-fired generation in which a southern California utility has participated. The plant is over 600 miles from the Los Angeles load center, and as such, it has led to a significant expansion in 500 kV transmission facilities in the area. Feasibility economics have dictated that plant output to California, 763 MW, be carried on a single 500 kV line for a distance of approximately 400 miles. With a projected 9,000-10,000 MW system in 1969-70, it was anticipated that 750 MW generators would have been the next unit size beyond the 500 MW machines recently placed in service in the Los Angeles basin. Thus, the single 500 kV line, with its relatively low estimated outage rate compared to the generators, is viewed as an equivalent 750-MW unit from California.

The Four Corners Project has a very profound effect on the Arizona power system. From a power flow viewpoint, approximately 450 MW of additional Four Corners power will be transmitted to the Phoenix-Tucson area in 1970. Studies show that this can be accomplished by increasing the capability of the 345 kV transmission lines north of Phoenix.

From a stability viewpoint, the stored energy in the two new units will amount to approximately 50% of the stored energy of all the existing Arizona machines. With one 500 kV line from Four Corners toward California, a major part of the Arizona power system acts as a relatively high impedance parallel path to the 500 kV line. This path consists of two 345 kV lines from Four Corners to Pinnacle Peak (Phoenix), a 230 kV system around Phoenix to the Liberty Substation, and a 345 kV line from Liberty to Mead Substation, near Hoover Dam. In parallel with the Four Corners-Pinnacle Peak lines is a path consisting of the 230 kV connection between Four Corners, Shiprock and Glen Canyon, and the two Glen Canyon-Pinnacle Peak 345 kV lines. A diagram of the geographic location of the related facilities is shown in Figure 1.

Two major areas of concern to the Arizona power system are emphasized by the Four Corners Project:

1. Control of power flow under varying loading conditions.

2. Stability of the Arizona system within itself and with respect to California, New Mexico, Colorado, and Utah.

The two problems are closely related because system stability is affected by the power flow on key lines at the time that a major disturbance might occur.

INITIAL STUDIES

Several joint studies have been performed to determine system parameters such as series compensation, shunt reactors, excitation system characteristics, etc. This paper discusses selected studies performed to establish transmission system and plant requirements for stability.

System Studied

The initial system studies, beyond those which were performed for the feasibility analysis, were generally based on projected 1970 summer peak loading conditions for maximum system

Reprinted from *IEEE Paper 68 CP 708-PWR.*

stress. The system studied includes New Mexico, Arizona and Southern California, and was represented for computer studies by a 250-bus system with 50 generators. The major objective of this series of studies was to determine an operable system within the framework of the two new 755-MW units at Four Corners (in addition to the existing three units), two 345 kV lines to Phoenix, two new 345 kV lines to Albuquerque, and the single 500 kV line to California. The power transfer was to be about 760 MW to California, 320 MW to Albuquerque and El Paso, and 1030 MW (including existing-plant output) to Phoenix and Tucson.

Purpose

As an integral part of the initial studies, it was necessary to determine the critical contingencies which might require more definitive study; to obtain a closer approximation for series capacitor compensation in the major transmission lines than was assumed in the feasibility study; and to investigate alternate methods of connecting the new generators to the system.

Study Methods

Digital computer programs were used as the tools for both load flow and stability analyses. For this phase of the study, the stability program used was limited to classical representation of the system, i.e., constant voltage with a fixed reactance and inertia for generators, and constant impedance simulation of load. This type of system model limits transient stability studies to the first-swing criteria, but it was considered to be an economical and expedient approach.

Results and Discussion

Results of the initial studies led to four basic conclusions:

1. The most critical single-contingency for transient stability is a permanent three-phase fault at Four Corners, normally cleared in 5 cycles by opening the 500 kV Four Corners-Eldorado line. This is the basic contingency referred to throughout this paper.

2. Loss of the Four Corners-Eldorado line requires tripping one of the 755-MW units at Four Corners in order to maintain system stability. This conclusion amounted to confirmation of the conceptual design determined in the feasibility study.

3. It would be necessary to install series compensation in the Four Corners-Eldorado 500 kV line and the Four Corners-Pinnacle Peak 345 kV lines to provide for the desired division of power flow and for steady-state and transient stability.

4. The preferred method of connecting the new 755-MW units to the system would be to transform the output of one directly to 500 kV and the other to 345 kV.

Loss of the 500 kV Four Corners-Eldorado line presents a special situation because there is not excess transmission capacity through the Phoenix system to allow transfer of 760 MW to California through it. Thus, to maintain synchronism between the systems, generation was dropped at Four Corners in an amount approximately equal to the 500 kV line loading prior to its loss. After loss of the 500 kV line, the tie lines between Arizona and California that remain intact consist of two 161 kV lines, one 230 kV line, and one 345 kV line, and it has not been considered prudent to design the Four Corners system in a manner that would jeopardize these ties for a single contingency.

Series capacitor compensation for the system based on steady-state stability was studied using the digital computer program and a pulsing technique whereby the mechanical power input to all the Four Corners units was increased by ten percent for one-tenth of a second. Steady-state stability was then determined by the degree of damping of the subsequent oscillations of the generators. As shown in the swing curves of Figure 2, the results indicated that about 40% series compensation in the 500 kV line would be needed for steady-state stability, and Figure 3 shows that some additional compensation in the 345 kV Arizona system would be necessary.

In order to gain the desired loading of the 500 kV line, it was found that more series compensation, about 50% total, would have to be installed in it than was required for steady-state stability, and of course, the actual amount would be somewhat dependent on the compensation placed in the 345 kV lines.

The original Project concept of installing the two 755-MW generators on a 500 kV bus at Four Corners with tie transformation to an existing 230 kV bus was studied with two alternatives to obtain the best system performance. As shown in Figure 4a, the existing Four Corners Plant is connected such that switching of the 345 kV lines to Phoenix is accomplished on the 230 kV side of the line autotransformers. Figure 4b shows the modifications to the existing plant as proposed by the original concept. This arrangement was based on the premise that most of the output of the new units would be transmitted to California. It was a modification of this premise that led to the alternatives shown in Figure 4c, with one of the new units on the 230 kV bus; and in Figure 4d, in which one of the units is placed on a new 345 kV bus. The Alternate shown in Figure 4d was ultimately chosen, since it provides the lowest equivalent impedance for 830 MW to the 345 kV systems and 760 MW to the 500 kV system, and as shown in Figure 5, it is the only alternate which would be adequate for transient stability upon loss of the 500 kV line (and dropping one unit) unless more extensive system changes were made or more remedial actions were taken.

STUDIES WITH DETAILED REPRESENTATION

The initial studies were very helpful in establishing a starting point for stability studies with detailed machine and load representation. This representation was required to study:

a) The effect of the SCR excitation system to be furnished with the new Four Corners units.

b) System and machine damping.

c) The effect of generation and load changes over a 5- to 10-second period.

Computer Stability Program

The stability program used for the detailed study was the General Electric E-ST6 program.[1] This program includes the following features:

a) variable flux linkages
b) generator saturation
c) generator saliency
d) exciter saturation
e) voltage regulator
f) regulator and excitation system time constants
g) feedback of slip as a supplementary signal source
h) governor representation
i) frequency damping
j) loads represented as constant impedance, constant current, or constant mva.

Load Representation

Loads were represented as constant current for the real power portion of the load, and constant impedance for the reactive portion. This representation was based on reports of various load-voltage staged tests and the reasonable results obtained by other study groups that have used a similar representation.

A comparison of results using constant current and constant impedance load representation for the critical Four Corners stability case is shown in Figure 6. From this it can be seen that the load representation has little effect on the first swing, but the difference on damping of subsequent swings is quite significant.

Damping

Damping due to frequency change is a system characteristic which is well-known to operating personnel.[2] With everything else in the system constant, an increase in frequency produces a reduction in effective mechanical torque input to generators and causes an increase in electrical loads on the system.

The particular stability program used for these studies did not have provision to change load as a function of frequency, but it did allow direct simulation of the mechanical torque-frequency relationship. The magnitude of this factor used in the studies allowed a 2% change in mechanical torque for every 1% change in frequency (or 2 per unit) for each generator.

The effect of damping due to frequency change is illustrated in Figure 7, which compares results for the critical stability case using 1- and 2-per-unit factors. Very limited experience in comparing computer-program results with field-test results indicates that the 2-per-unit factor is conservative.

Check of Initial Studies

The classical representation studies indicated that the critical stability contingency would be a three-phase fault on the Four Corners-Eldorado line near Four Corners, and that stability would be maintained by normally clearing the faulted line and tripping a 755-MW unit. This contingency was re-studied using the E-ST6 program to verify results of the initial studies and to determine any differences in results due to changed representation.

The new study showed a condition of apparent dynamic instability, and it was found that one way the condition could be overcome was by reducing the gain of the excitation system regulators on the generators at the U.S. Bureau of Reclamation's Glen Canyon hydro plant. The Glen Canyon plant is an integral part of the Arizona power system, and its regulators are of the modern rotating-amplifier type with high ASA response ratio. Swing curve results for reducing the regulator gain at Glen Canyon to about one-fourth the original value used are shown in Figure 8. These results demonstrate the phenomenon of negative damping originating from excitation systems with high gains and relatively large time constants. The Bureau of Reclamation has studied the characteristics of the Glen Canyon system and has designed a supplemental excitation control device which eliminates the negative damping feature with no sacrifice in voltage sensitivity.[3]

The Four Corners studies were continued using reduced regulator gain at Glen Canyon, and the results of the "classical" studies were confirmed. A comparison of results from the two methods of study is shown in Figure 9.

Series Compensation

The next phase of the study concentrated on optimizing series compensation in the 345 kV lines in Arizona. Since the Bureau of Reclamation's Glen Canyon network was involved, a joint study was performed which resulted in plans to install 60% series compensation in the Glen Canyon-Pinnacle Peak lines. This alternative eliminated the need for additional compensation in the Four Corners-Pinnacle Peak lines and it provides increased stability margin for the critical Four Corners contingency, as shown by the swing curves in Figure 10. This arrangement results in a savings of approximately 150 MVAR of capacitors with mutual benefits. The cost of the added 345 kV series compensation will be shared by the Arizona Public Service Company and the Bureau of Reclamation.

METHODS TO MINIMIZE UNIT DROPPING

The new Four Corners generators will have super-critical boilers and cross-compound reheat turbines which are very sensitive to rapid and extreme temperature changes. Thus, dependence on a deliberate unit-trip scheme for system stability subsequent to loss of the 500 kV line has the disadvantage of causing the unit that is tripped to be removed from service during an estimated six-to-nine-hour restart period.

In an effort to determine a satisfactory method for minimizing the use of the unit-trip mode of operation, several possibilities have been investigated. Generally, the possibilities have involved use of (a) fast-control and intercept-valve manipulation with valve reset to either half-load or full-load position, (b) fast turbine-load runback to half-load, and (c) dynamic braking resistors. Four specific methods which have been studied in conjunction with the 500 kV Four Corners fault are:

1. Immediately initiate fast valving of both 755-MW units, automatically reclose the 500 kV line, then if the fault is permanent, trip one of the units.

2. Apply a braking resistor, automatically reclose the 500 kV line, then if the fault is permanent, trip one of the units.

3. Initiate fast valving of both 755-MW units and apply a braking resistor. Reset valves to about half-load position.

4. Utilize fast turbine-load runback to half-load and apply a braking resistor.

Method 1

Method 1 uses a fast-control and intercept-valve closure scheme devised by the turbine manufacturer which causes a significant decrease in mechanical input power within 0.3 seconds. The valves must be reopened about 1.0 second after being closed. Figure 11 shows turbine power plotted versus time for fast valving with valve reset to full power, as furnished by the turbine manufacturer.

It was anticipated that the application of fast valving might reduce accelerating power enough to allow one attempt at line reclosure for the 500 kV fault, and if the fault were permanent a generator could then be dropped. It is estimated that 3 out of 4 faults on the 500 kV line will be temporary, so successful reclose would permit all units to remain in service at full power for most line faults. The minimum reclose time for the 500 kV line is about 40 cycles as limited by deionization time at the fault point.

Results of the fast valving study plotted in Figure 12 show that first swing stability could not be obtained for the permanent three-phase 500 kV fault. Continuing the study by using the fast valving scheme with a permanent single line-to-ground fault instead of a three-phase fault showed that first swing stability could be obtained, but that the large negative swing created instability on the second swing, Figure 12.

Finally, the permanent single line-to-ground fault case was restudied with added series compensation in the 345 kV system over the amount planned, and stability was obtained as shown in Figure 12.

For Method 1 to be successful, it would be necessary to:

a) Discriminate between single line-to-ground faults and all other types.

b) Trip a 755-MW unit without delay for faults other than single line-to-ground.

c) Install additional series capacitors in the 345 kV network.

d) Reclose the 500 kV line across an angle of approximately 90 degrees.

Method 1 has not been adopted since it involves uneconomical capital expenditures for additional capacitors, and it has serious technical shortcomings.

Method 2

Method 2, use of a braking resistor, was considered because the basic problem associated with loss of the 500 kV line is an excess of about 760 MW of generation at Four Corners. Although resistors alone would not eliminate the necessity for tripping generation in the case of permanent faults, they could be effective in minimizing the number of unit-trips required due to line outages.

The plant arrangement offers alternate voltage levels for application of a braking resistor. The 500 kV bus or the 230 kV bus have been considered in this case, and although a resistor would be more effective on the 500 kV bus, switching equipment is more readily available for 230 kV. Thus, most Method 2 studies have been based on the assumption that a resistor would be switched at the 230 kV level.

Using a line-test scheme at the Eldorado end of the 500 kV line so that the Four Corners plant is not subjected to a second shock due to reclose on permanent faults, the swing curve of Figure 13 shows that use of an 800-MW resistor to absorb energy at the plant would allow time to test the line for reclosure. For permanent faults, one of the new generators would, of course, have to be tripped after testing the line. It is possible that a slightly smaller resistor would serve the purpose, but Figure 13 shows that a 400-MW size would not provide for even first-swing stability. Again, the possibility of discriminating between single line-to-ground faults and other types could result in a smaller resistor, but such a scheme would require more complicated protection and would introduce additional chance of incorrect action of critical relays.

Method 3

Method 3 combines modifications of Methods 1 and 2. Using fast valving on the 755-MW units but with valves reset to half-power position instead of full-power, the final Four Corners output would be about the same as if one unit had been dropped and both units would be available for full output when the Four Corners-Eldorado line is returned to service. Estimated valving curves for reset to half power have been furnished by the manufacturer and are shown in Figure 14.

The studies of Method 1 showed that first-swing stability could not be obtained with fast valving alone for other than single line-to-ground faults on the 500 kV line. Method 3 includes the use of a braking resistor for about one-half second to obtain first-swing stability for two line-to-ground faults, but as shown in Figure 15, stability is lost on the second swing when the resistor size is 400 MW. Instability on the second swing was attributed to the "overshoot" of mechanical power when the valves were reset. From Figure 14 it can be seen that the mechanical power on one unit is 460 MW after three seconds, or about 60% of full load. The case was rerun with a revised valving curve, also shown in Figure 14, which approaches half power asymptotically with no overshoot, and the result was a stable system.

Fast valving with controlled reset of the valves is attractive in principle, but the variation of mechanical power cannot be assured until testing is performed on the actual units. Therefore, no further consideration will be given to Method 3 until the testing has been accomplished.

Method 4

The fourth method of substituting for the unit-trip mode of operation involves use of a braking resistor in conjunction with fast load runback on the two 755-MW units. For fast load runback, it was assumed in the studies that turbine power could be reduced linearly from full-load to half-load within about thirty seconds using normal governor controls. The braking resistor used with this scheme would require a rating for 20- or 30-second duty, and based on data for other such installations, it is estimated that the resistance would approximately double within the duty period.

For Method 4, the swing curves of Figure 16 show that the system would be stable for at least seven seconds using an 800-MW resistor following a five-cycle three-phase 500 kV fault at Four Corners, but the system would be unstable using a 700-MW resistor. The swing curves do not indicate a substantial degree of margin for the 800-MW resistor case, and in view of the significant transient to be expected upon removal of the resistor and the lack of assurance of the satisfactory operation of the load runback feature, this method has not been chosen as an alternate to the unit-trip mode.

CONCLUSIONS

A classical stability program has value in obtaining results of a preliminary nature. A more comprehensive program is required to study the effects of damping, excitation systems, and governor systems.

The classical-type stability studies were instrumental in establishing the final plant arrangement which includes a new 755-MW unit on a new 345 kV bus and one on the new 500 kV bus. The minimum compensation needs were also defined by these studies.

For this study it was found that load representation had a much greater effect on damping than it had on the magnitude of first swings. This study also showed that the difference between 1.0 per unit damping due to frequency and 2.0 per unit damping was slight.

The experience gained from this study with negative damping effects emanating from excitation systems has emphasized the need for more concentration on this subject by the industry.

This study shows, as have others,[4, 5] that a braking resistor can be effective in the reduction of accelerating torque. The most promising application for the Four Corners Project system is in conjunction with fast valving. Fast valving is a relatively new system stability aid that shows much promise in this study. As such, further studies for this application will be based upon field test results. The control equipment for fast valving is being supplied with the Four Corners turbines which will allow testing of the fast valving feature.

Fast load runback for this system necessarily requires a braking resistor with a relatively long time rating (20 to 30 seconds). This increases the effective resistor size required and complicates the problem of resistor removal.

System control and relaying is being designed so that unit tripping will be initiated whenever a 500 kV fault or line opening occurs. This control can be easily modified to incorporate the stability aids discussed above if later studies show them to be practical.

Further studies are being conducted to investigate many other contingencies and operating conditions not discussed in this paper.

REFERENCES

1. "The effect of frequency and voltage on power system load," IEEE Working Group on Load Characteristics, System Planning and System Operations Subcommittee of the IEEE Power System Engineering Committee, IEEE Winter Power Meeting, New York, 1966.

2. R. M. Webler and C. C. Young, "A new digital computer program for predicting dynamic performance of electric power systems," Power Industry Computer Applications Conference, 5th Conference Record, pp. 21 - 29, May 15 - 17, 1967.

3. F. R. Schleif, H. D. Hunkins, G. E. Martin and E. E. Hattan, "Excitation control to improve powerline stability," IEEE TRANS. ON POWER APPARATUS AND SYSTEMS, Vol. PAS - 87, June 1968.

4. H. M. Ellis, J. E. Hardy, A. L. Blythe and J. W. Skooglund, "Dynamic stability of the Peace River transmission system," IEEE TRANS. ON POWER APPARATUS AND SYSTEMS, Vol. PAS - 85, pp. 586 - 600, June 1966.

5. W. H. Croft and R. H. Hartley, "Improving transient stability by use of dynamic braking," AIEE TRANSACTIONS, Vol. 81, Part III, pp. 17 - 26, 1962.

6. F. P. deMello, D. N. Ewart, M. Temoshok and M. A. Eggenberger, "Turbine energy controls aid in power system performance," presented at American Power Conference, Chicago, Illinois, April 28, 1966.

7. E. W. Kimbark, Power System Stability, New York, Wiley, 1956.

8. S. B. Crary, Power System Stability, Vol. II, New York, Wiley, 1947.

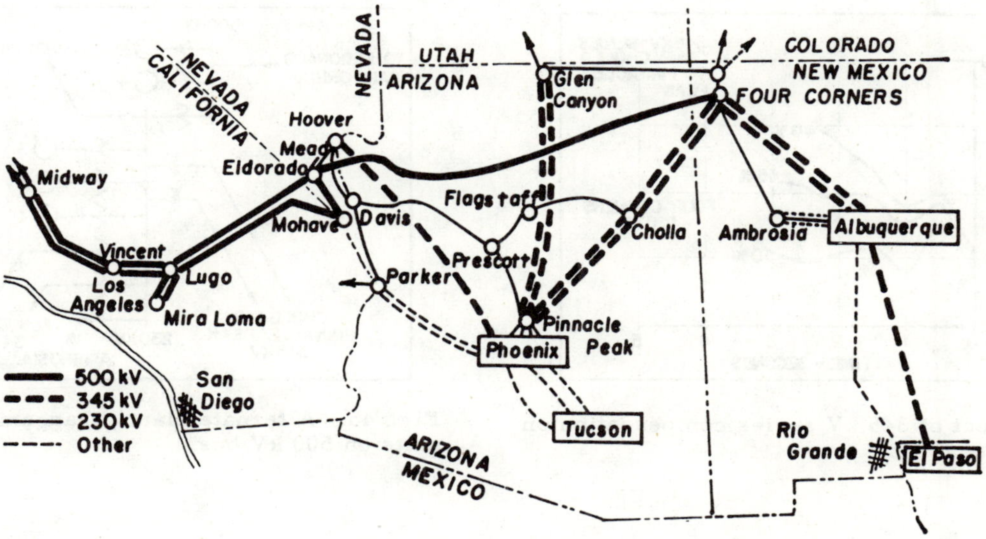

Fig. 1. Four Corners and regional transmission network.

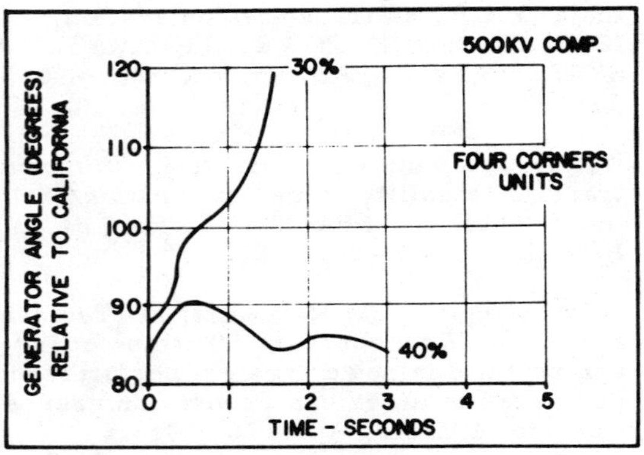

Fig. 2. Effect of 500 kV series compensation on steady state stability.

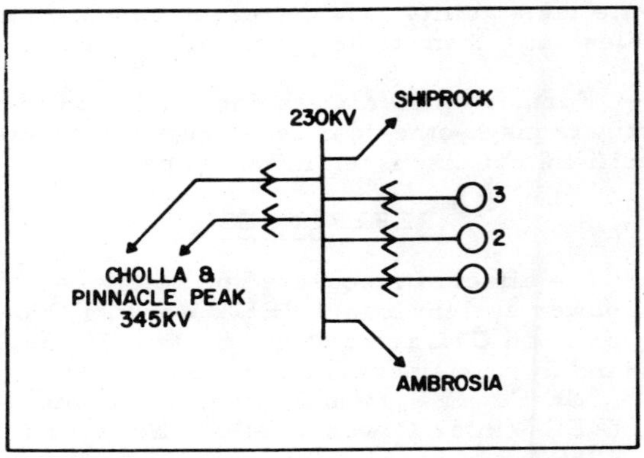

Fig. 4a. Existing Four Corners arrangement.

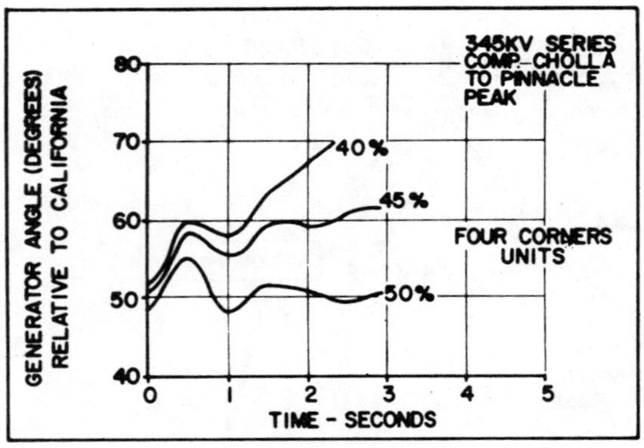

Fig. 3. Effect of 345 kV series compensation on steady state stability.

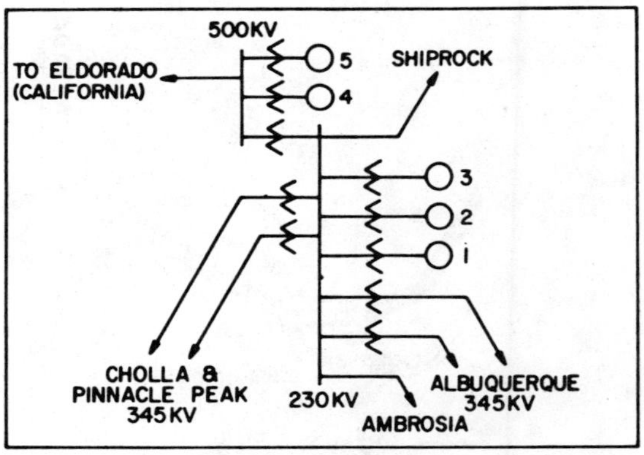

Fig. 4b. Alternate plant arrangement with new units on 500 kV bus.

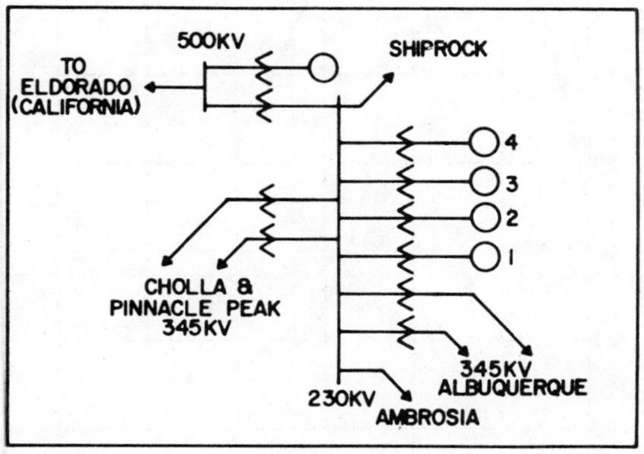

Fig. 4c. Alternate plant arrangement with one new unit on 230 kV bus and one on 500 kV bus.

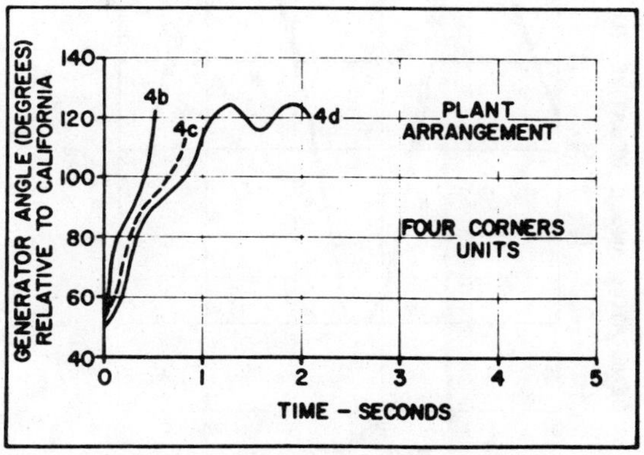

Fig. 5. Effect of plant arrangement on transient stability for 5 cycle, 3-phase 500 kV fault.

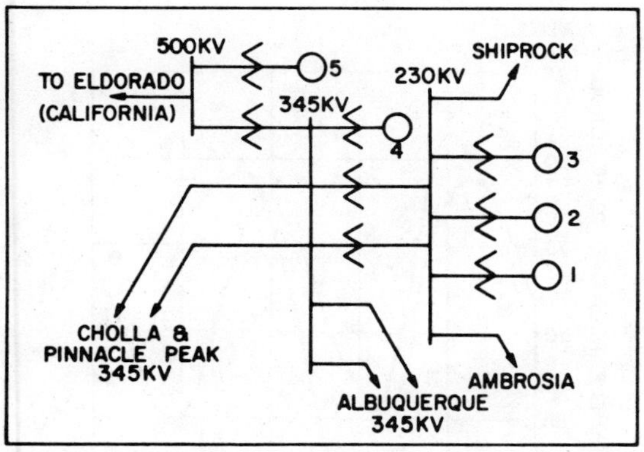

Fig. 4d. Final arrangement based on stability study.

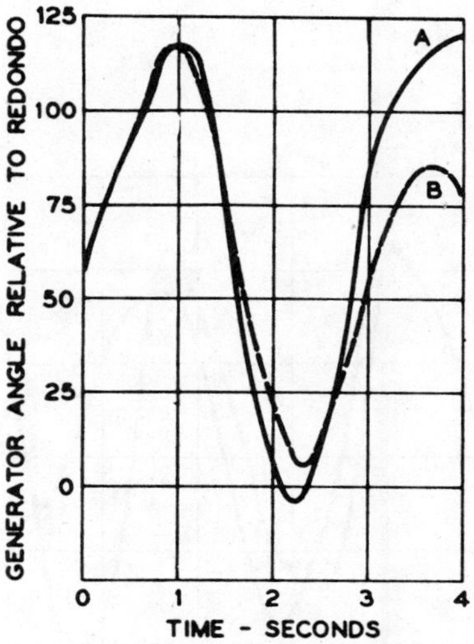

Fig. 6. Four Corners Unit 4 swing curves for the critical contingency and Four Corners Unit 5 dropped at 5 cycles. (A) Loads represented as constant current. (B) Loads represented as constant impedance.

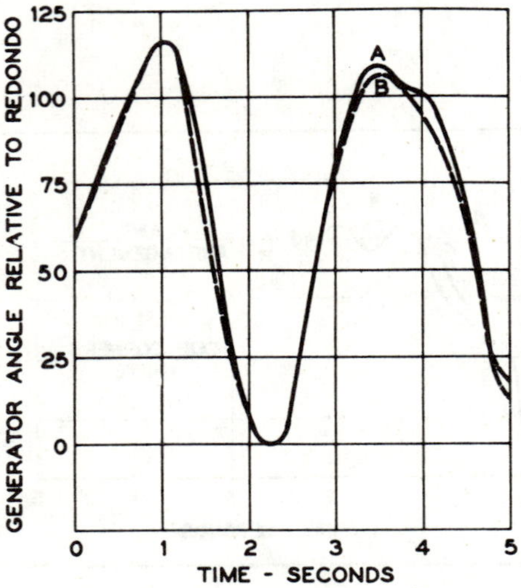

Fig. 7. Four Corners Unit 4 swing curves for the critical contingency and Four Corners Unit 5 dropped at 5 cycles. (A) 1-per-unit frequency damping. (B) 2-per-unit frequency damping.

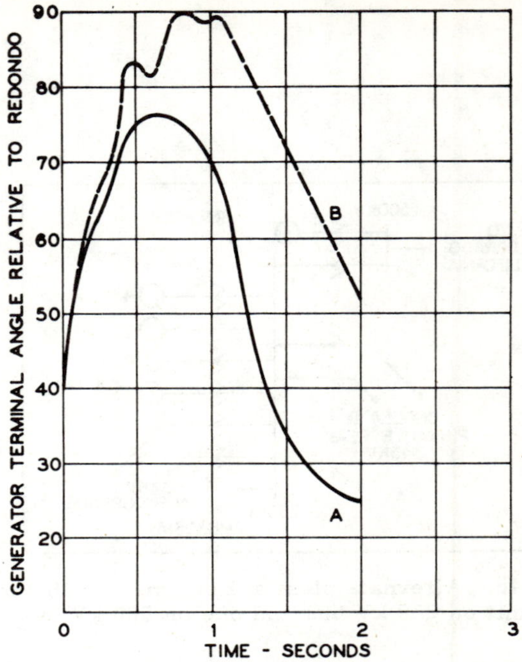

Fig. 9. Four Corners Unit 4 terminal swing curves for the critical contingency and Four Corners Unit 5 dropped at 5 cycles. (A) E-ST6 program. (B) Classical stability program.

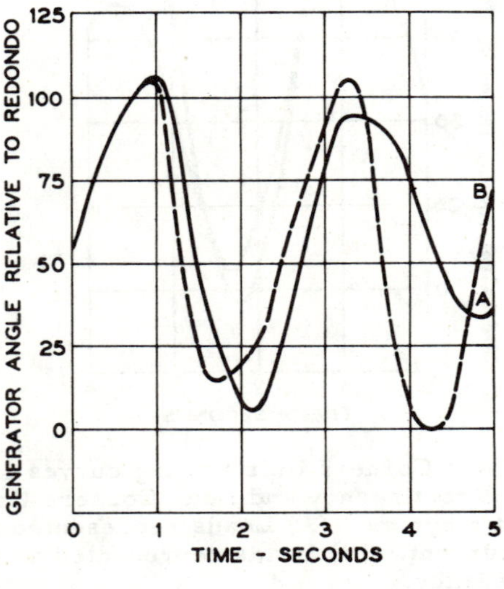

Fig. 8. Four Corners Unit 4 swing curves for the critical contingency and Four Corners Unit 5 dropped at 5 cycles. (A) Glen Canyon regulator gain one-fourth of normal and no supplementary speed signal. (B) Glen Canyon regulator gain normal and no supplementary speed signal.

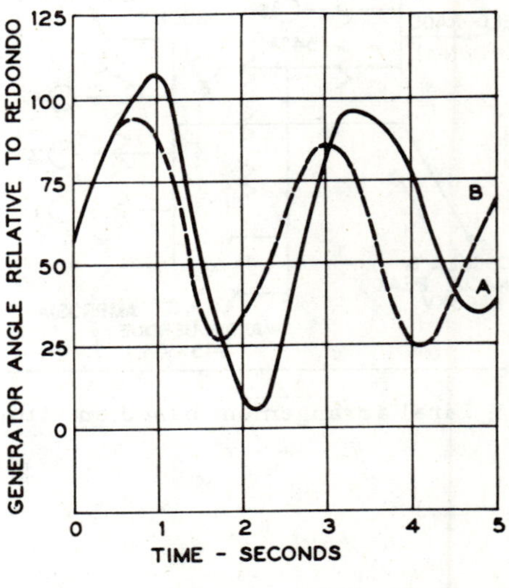

Fig. 10. Four Corners Unit 4 swing curves for the critical contingency and Four Corners Unit 5 dropped at 5 cycles. (A) Series compensation added to the Four Corners-Pinnacle Peak 345 kV lines only. (B) Series compensation added to the Glen Canyon-Pinnacle Peak 345 kV lines only.

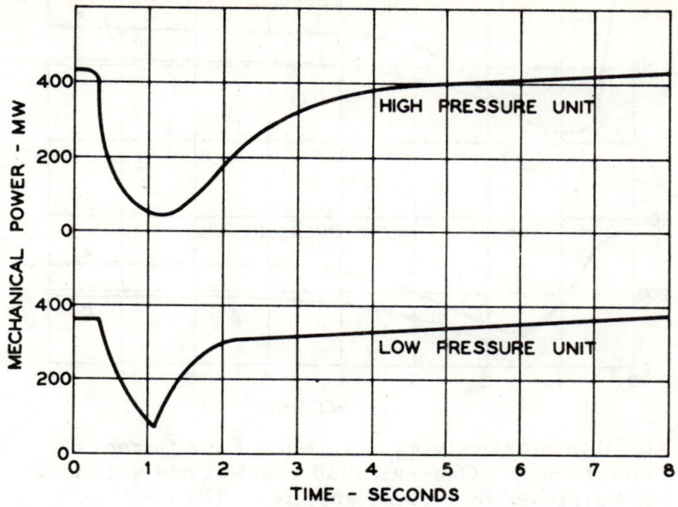

Fig. 11. Fast valving curves for Four Corners Units 4 and 5. Full load reset at one second.

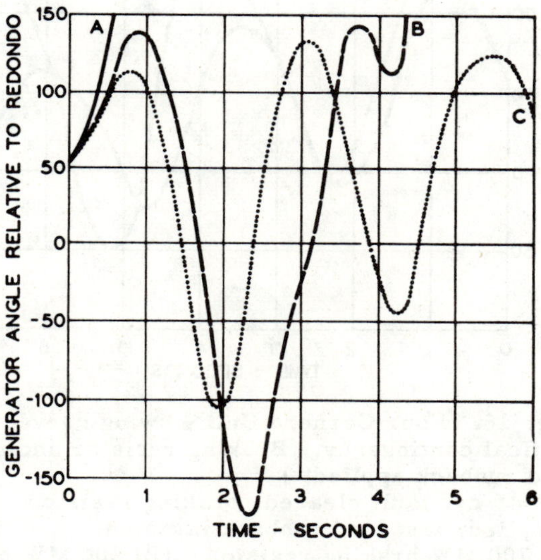

Fig. 12. Four Corners Unit 4 swing curves for fast valving with full load reset at one second applied to Four Corners Units 4 and 5.
T = 5 cy, fault cleared and valving initiated.
T = 45 cy, reclosed on fault.
T = 50 cy, fault cleared.
T = 53 cy, drop Four Corners Unit 5.
(A) Critical contingency. (B) Permanent SLG 500 kV fault near Four Corners. (C) Permanent SLG 500 kV fault near Four Corners with increased series compensation in the Arizona Triangle.

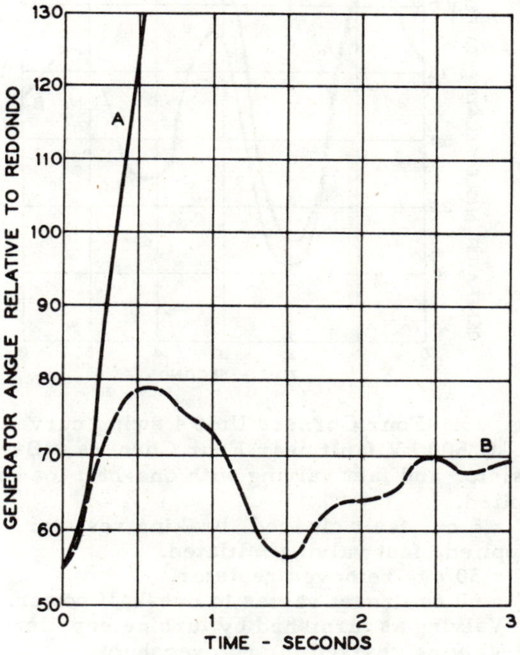

Fig. 13. Four Corners Unit 4 swing curves for critical contingency and braking resistor applied.
T = 5 cy, fault cleared and braking resistor applied.
T = 45 cy, reclosed line from Eldorado only.
T = 50 cy, fault cleared.
T = 53 cy, dropped Four Corners Unit 5 and removed resistor.
(A) 400 MW braking resistor. (B) 800 MW braking resistor.

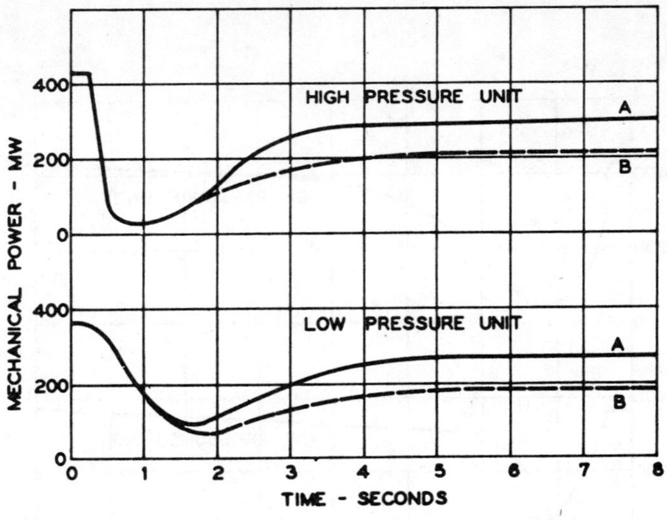

Fig. 14. Fast valving curves for Four Corners Units 4 and 5. One-half load reset at one second. (A) Furnished by turbine supplier. (B) Revised for no overshoot.

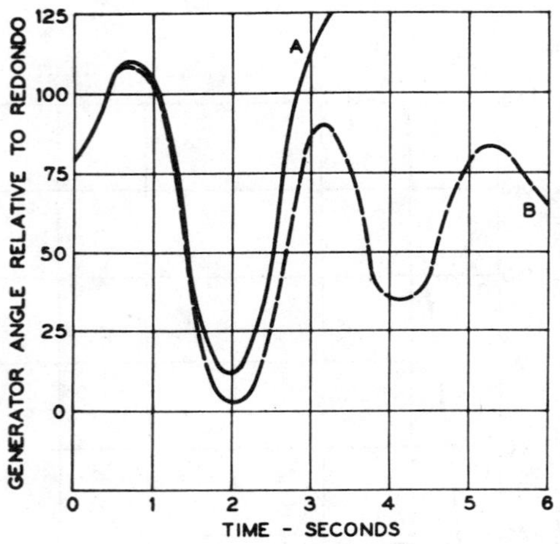

Fig. 15. Four Corners Unit 4 swing curves for 2 L-G 500 kV fault near Four Corners. Braking resistor and fast valving with one-half load reset applied.

T = 5 cy, fault cleared, braking resistor applied, fast valving initiated.
T = 30 cy, removed resistor.
T = 60 cy, reset valves to one-half power.
(A) Valving as furnished by turbine supplier.
(B) Valving revised for no overshoot.

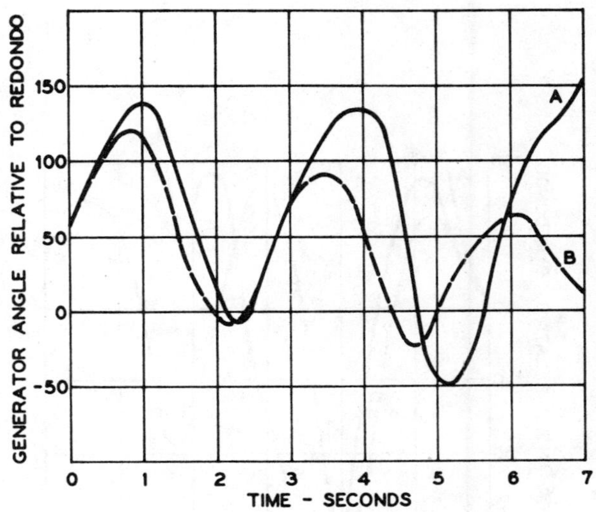

Fig. 16. Four Corners Unit 4 swing curves for critical contingency. Braking resistor and fast load runback applied.

T = 5 cy, fault cleared, braking resistor applied, fast load runback initiated.
(A) 700 MW braking resistor. (B) 800 MW braking resistor.

Damping of Power Oscillations in Salient-Pole Machines with Static Exciters

RICHARD T. BYERLY, SENIOR MEMBER, IEEE, F. W. KEAY, SENIOR MEMBER, IEEE, AND JOHN W. SKOOGLUND, SENIOR MEMBER, IEEE

***Abstract*—It is generally accepted that a high-performance excitation system may reduce the effective internal damping of a synchronous machine. Alternately, studies and field tests indicate that a fast-acting excitation system offers the best opportunity for increased damping through the use of an auxiliary signal into the voltage regulator. The application of static exciters to water-wheel generators requires an awareness of these possible changes in machine damping. The use of an auxiliary damping signal into the voltage regulator may become especially important for machines that are inherently lightly damped.**

The measurement of electrical power, using current and potential transformers with a Hall generator watt transducer, provides a useful signal source to achieve improved damping. Studies showing the effectiveness of this damping signal are presented. These studies include observation of the effects of nonlinear generator damping characteristics that result from saturation of rotor iron in the vicinity of the damper bars. Simulated performance compares favorably with field observations.

Introduction

DAMPING OCCURS in synchronous machines because of currents induced in the rotor by oscillations. Since rotor currents are also affected by changes in field voltage, it follows that damping may be altered by the excitation system. Generally speaking, the effect of a high-performance excitation system is to reduce damping, although reduction in damping is not always significant. This is indicated by the large number of machines in service with high-performance excitation systems exhibiting satisfactory damping characteristics. In certain situations, however, combined characteristics of load, generator, excitation system, and prime mover result in significantly reduced damping. There are instances of sustained oscillations that cease when the voltage regulator is taken out of service.

Fortunately, the excitation system can be made to increase damping rather than reduce it. This requires an additional input signal to the voltage regulator. A number of published reports document both studies and field tests in which improved damping has been achieved. The successful use of a frequency signal with conventional exciters to improve system damping has been reported in [2], [3], [6]. Ontario-Hydro [5], [7] has reported the use of a shaft speed signal for stabilization of a hydrounit with static exciter.

The work reported here goes beyond a study previously reported [4]. It deals with a pumped hydromachine well damped for large oscillations but lightly damped for low-level oscillations due to the damper bar design. The combination of low inherent damping and static excitation results in power oscillations that have been observed in operation and reproduced by computer simulation. Good damping characteristics are achieved by means of a supplementary signal into the voltage regulator derived from electrical power. The model of the machine used in simulation is unique because it represents local saturation in the vicinity of the damper bars to match more closely actual machine damping characteristics.

Salina Pumped-Storage Project

Early in 1968 the Salina Pumped-Storage Project of the Grand River Dam Authority was placed in service. Initially, this project consisted of three generator/motors employing static excitation systems. The Salina generator/motors have the following characteristics: generator—48 000 kVA, 0.9 power factor (PF); motor—64 000 hp, 1.0 PF; speed—171.4 r/min. These machines are designed for across-the-line starting at rated voltage and for reduced-frequency or partial-synchronous starting with the driving machine initially operating at one-half speed. The Salina Plant is located approximately six miles from the Markham Ferry Plant, which has an installed capacity of four 30 000-kVA generators. The short-circuit capacity at Markham Ferry is 2000 MVA minimum at present and will be as high at 7000 MVA maximum in the future. The impedance from the Salina unit terminals to Markham Ferry is approximately 32 percent (100-MVA base).

Primarily because of the starting requirement, the damper bars for these units are constructed of steel and embedded in iron. This type of damper produces relatively small damping torque at low levels of rotor oscillation.

When the first machine went into service as a pump motor, low-level sustained oscillations occurred with the voltage regulator in service, but with no supplementary signal. These oscillations did not occur when the excitation control was switched to manual. The design and construction of the excitation system provided for an auxiliary damping signal as well as internal rate feedback from the exciter output to the regulator input. Experience indicated that an auxiliary damping signal derived from electrical power measurement would stabilize the machine, and this was subsequently shown to be the case.

An analytical study of this situation was undertaken primarily to establish values for adjustable parameters in the power damping circuit. Existing computer programs were available for such a study, but several aspects of this problem encouraged development of a special program. The oscillations were localized at the Salina station and extensive system representation did not seem to be required. This circumstance made a special program feasible. The flexibility of a special program with regard to machine and excitation system models was desirable. A special program served the immediate need as well as the secondary purpose of model and technique development.

Paper 69 TP 735-PWR, recommended and approved by the Power Generation Committee of the IEEE Power Group for presentation at the IEEE Summer Power Meeting, Dallas, Tex., June 22–27, 1969. Manuscript submitted February 17, 1969; made available for printing April 30, 1969.

The authors are with Westinghouse Electric Corporation, East Pittsburgh, Pa. 15112.

Reprinted from *IEEE Trans. Power App. Syst.*, July/Aug. 1970.

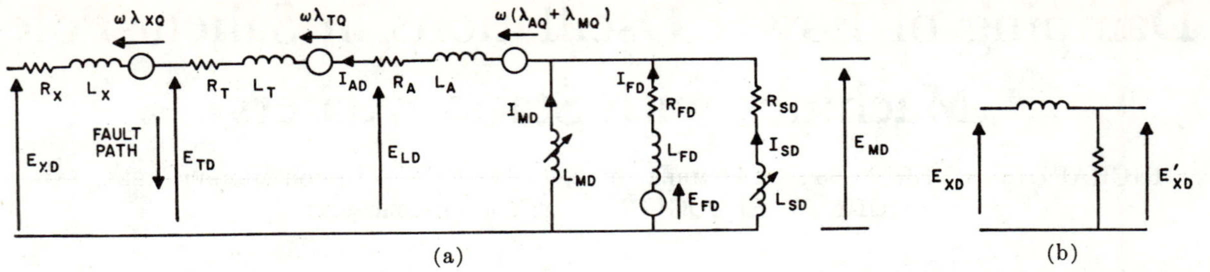

Fig. 1. Direct-axis equivalent circuit including transformer and external system impedances. (b) Schematic representation of infinite bus to improve system damping characteristics after fault clearing.

Analytical Study

Oscillations had occurred spontaneously in operation without the advent of any major disturbance. Nevertheless, the behavior of the machine subsequent to nearby faults is of primary importance, and simulation of faults provides the best means of judging performance.

The program developed has the capability of representing a single generator with its excitation and governing systems. The generator is delivering power to an infinite bus. The function of the program is the computation of the transient behavior of the generator and its controls, assuming that the generator unit transformer high-side terminals are faulted and then cleared.

Generator Mathematical Model

The equivalent circuit for the direct axis of the generator is shown in Fig. 1. A similar circuit may be drawn for the quadrature axis, with the subscript D replaced by Q and the speed-induced voltages reversed. Also, on the quadrature axis, there is no rotor current path corresponding to the direct-axis field circuit since the rotor is laminated. The equivalent circuit includes direct-axis projections of the infinite bus voltage E_{XD}, the transformer high-side voltage E_{TD}, and the generator terminal voltage E_{LD}. Representation of the latter two voltages requires separating the speed-induced voltages into component parts as indicated.

No derivation of this equivalent circuit is given here; for a more complete discussion of equivalent circuits for synchronous machines, any of several textbooks may be recommended [9].

The mutual inductance L_{MD} and the subtransient leakage reactance L_{SD} are saturable. All other inductances on the direct axis are assumed to be linear. On the quadrature axis, saturation of the mutual inductance is ignored, but the subtransient leakage inductance is treated as saturable. The mutual inductance saturation and the subtransient leakage inductance saturations are independent. The direct-axis mutual inductance is saturated in a manner that would cause the model to have approximately the same no-load steady-state saturation characteristic as the actual machine. The subtransient leakage inductances are saturated to simulate the nonlinear damping characteristics of embedded bars.

Fig. 2 shows a pole cross section with embedded damper bars. This drawing is not intended to show the specific construction of this unit but is useful in illustrating how nonlinear damping occurs. For low levels of damper bar current, the iron between the bar and the air gap will be unsaturated, with correspondingly high incremental inductance, which limits incremental current and damping. When the bar current is forced to higher values, the bridge of iron above the bar saturates, with consequent lower incremental inductance and larger incremental current and damping. The generator will have a variable damping characteristic, being lightly damped for small oscillations, and more heavily damped for large oscillations.

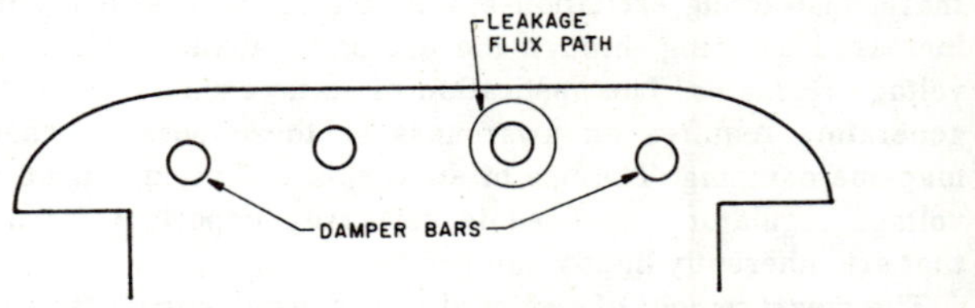

Fig. 2. Field pole cross section showing embedded damper bars.

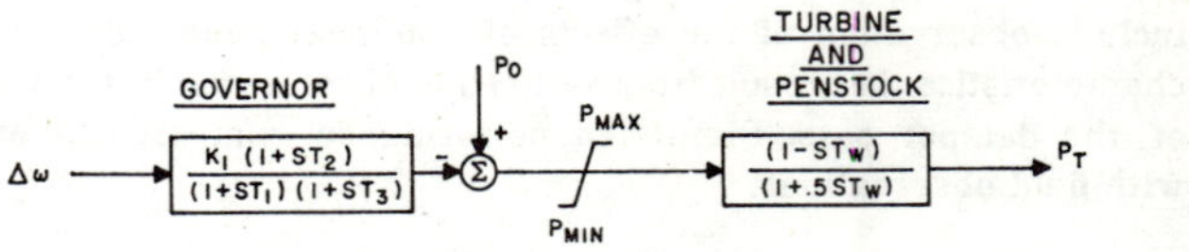

Fig. 3. Block diagram of governor–turbine penstock system.

TABLE I
Governing System Constants

Symbol	Description	Value
K_1	Low-frequency gain	20.0
T_1	Governor time constant	42.0 seconds
T_2	Governor phase lead	4.80 seconds
T_3	Governor time constant	0.45 second
T_w	Water time constant	1.20 seconds

Precise representation of saturation phenomena is not possible within the framework of the lumped-parameter equivalent circuit shown. However, treating the direct-axis mutual inductance and the subtransient leakage inductance as saturable does give a reasonable model for this study.

To facilitate representation of these two independent saturations, the equations for the machine equivalent circuits are written in terms of branch currents and flux linkages rather than loop currents and flux linkages. All the branch flux linkages are determined from the numerical solution of differential equations. Thus the 60-Hz currents in the rotor circuits and the consequent torques are not neglected. Transformer coupling from rotor to stator coils is also simulated.

Governor and Energy System Model

The mathematical model for the speed governor and energy systems is that conventionally used for power system stability studies. Fig. 3 illustrates the form of the transfer functions with the numerical values used in this study.

The dashpot transfer function and permanent speed droop feedback loops have been combined into an equivalent forward

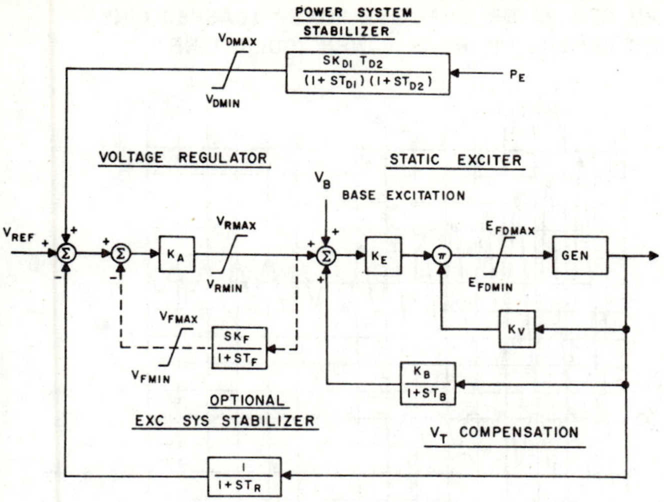

Fig. 4. Block diagram of excitation system.

TABLE II
EXCITATION SYSTEM CONSTANTS

Symbol	Description	Value
K_A	Regulator gain	200
T_R	Transducer time constant	0.05 second
$V_{R\,\max}$	Regulator output limit	5.53
$V_{R\,\min}$	Regulator output limit	−5.53
K_F	Damping gain	0.05
T_F	Damping time constant	1.0 second
$V_{F\,\max}$	Damping circuit limit	5.53
$V_{F\,\min}$	Damping circuit limit	−5.53
K_E	Exciter gain	1.0
$E_{FD\,\max}$	Exciter output limit	3.98
$E_{FD\,\min}$	Exciter output limit	0
V_B	Base excitation	2.72
K_B	Compensation gain	−1.11
T_B	Compensation time constant	0.01
K_{D1}	Power damping gain	0.1
T_{D1}	Damping time constant	0.05
T_{D2}	Damping time constant	1.0
$V_{D\,\max}$	Damping output limit	0.108
$V_{D\,\min}$	Damping output limit	−0.108

transfer function with a gain, a single-lead function, and two-phase lag functions. The representation of the water column and hydraulic turbine is the conventional $(1 - ST_w)/(1 + 0.5\,ST_w)$ where T_w is the water time constant. Table I lists constants used for this study.

Excitation System Representation

The block diagram of the excitation system is shown in Fig. 4. This representation is similar to the Type 1S of [10] modified to include the dynamics of the manual control and to show the excitation system rate feedback loop as it is actually applied in the physical system.

In this static excitation system, base excitation and manual control are obtained with an adjustable dc-signal input to the exciter. The regulator controls exciter output by supplying a buck-boost control signal about the base value. The base excitation input setting is a dc voltage that is the sum of a fixed voltage V_B and a voltage that is proportional to exciter supply voltage V_T.

The purpose of the variable signal is to compensate for the effect of supply voltage variations on the exciter output voltage for ±30-percent variation in supply voltage. When the exciter is connected to the synchronous machine the manual control circuit actually constitutes a second low-gain voltage regulator. In Fig. 4, K_B is the gain of the compensation circuit and T_B is the time constant of the signal filter.

Solid-state operational amplifiers, with negligible time delays, are used in the excitation system control modules. Therefore, the only significant delay in the system is the time constant T_R of the filter in the regulator sensing circuit.

Excitation system damping, where required, is obtained by feedback of the rate of change of the dc-input voltage to the exciter firing circuits. There is negligible time delay in the exciter, and this voltage has the practical advantage of being a low-level ripple-free signal.

In the power system stabilizing circuit a signal proportional to electrical power is derived from a single-phase Hall watt transducer, amplified, filtered (time constant T_{D1}), and applied to a derivative circuit with a reset time constant T_{D2}. The gain K_{D1} is adjustable from 0.03 to 0.3 in terms of per-unit terminal voltage/per-unit power. The reset time constant is adjustable from 1 to 5 seconds. To prevent the stabilizing signal from causing excessive terminal voltage excursions on load rejection and to retain the benefit of regulator forcing for large disturbances, the magnitude of the damping signal is limited. The limit level ($V_{D\,\max}$ and $V_{D\,\min}$) is adjustable over a range corresponding to 0.10-pu terminal voltage error. Constants for this specific regulator are listed in Table II.

Analysis of Performance

Evaluation of generator and excitation system performance is based on a rather large number of computer runs, only a few of which can be presented here. These have been selected to show the effects of damper circuit saturation, of the excitation system without auxiliary inputs, and of the power damping signal.

To facilitate comparison, each computer run represents the response of the machine to a six-cycle three-phase fault at the high side of the unit transformer. The initial load is rated generator power at unity power factor. Operation as a pump motor and at a variety of loads has also been studied. The damping of oscillations is not significantly changed for these other circumstances, although it should be noted that the polarity of the auxiliary damping signal must be reversed as the change is made from generating to motoring.

Damper Winding Saturation

Typical study output is illustrated in the curves of Fig. 5, which are reproduced from a cathode-ray tube plotter under computer control. The data of Fig. 5 are based upon an assumption of linear damper reactance equal to the saturated value; that is, a subtransient reactance that is of a low magnitude and constant. Fig. 5(a) shows the generator real and reactive power and total current for the first 0.14 second. This interval includes the duration of the fault and 40 milliseconds subsequent to fault clearing. Fig. 5(c) shows generator field voltage and current on the same expanded time scale as Fig. 5(a). Note that the plot of generator field current shows an induced 60-Hz component.

The response over this 140-ms interval is shown to emphasize the nature of the mathematical model of the generator. The 60-Hz component of field current [dashed line in Fig. 5(c)] exists because the stator flux linkages and consequently the stator currents are determined from the stator differential

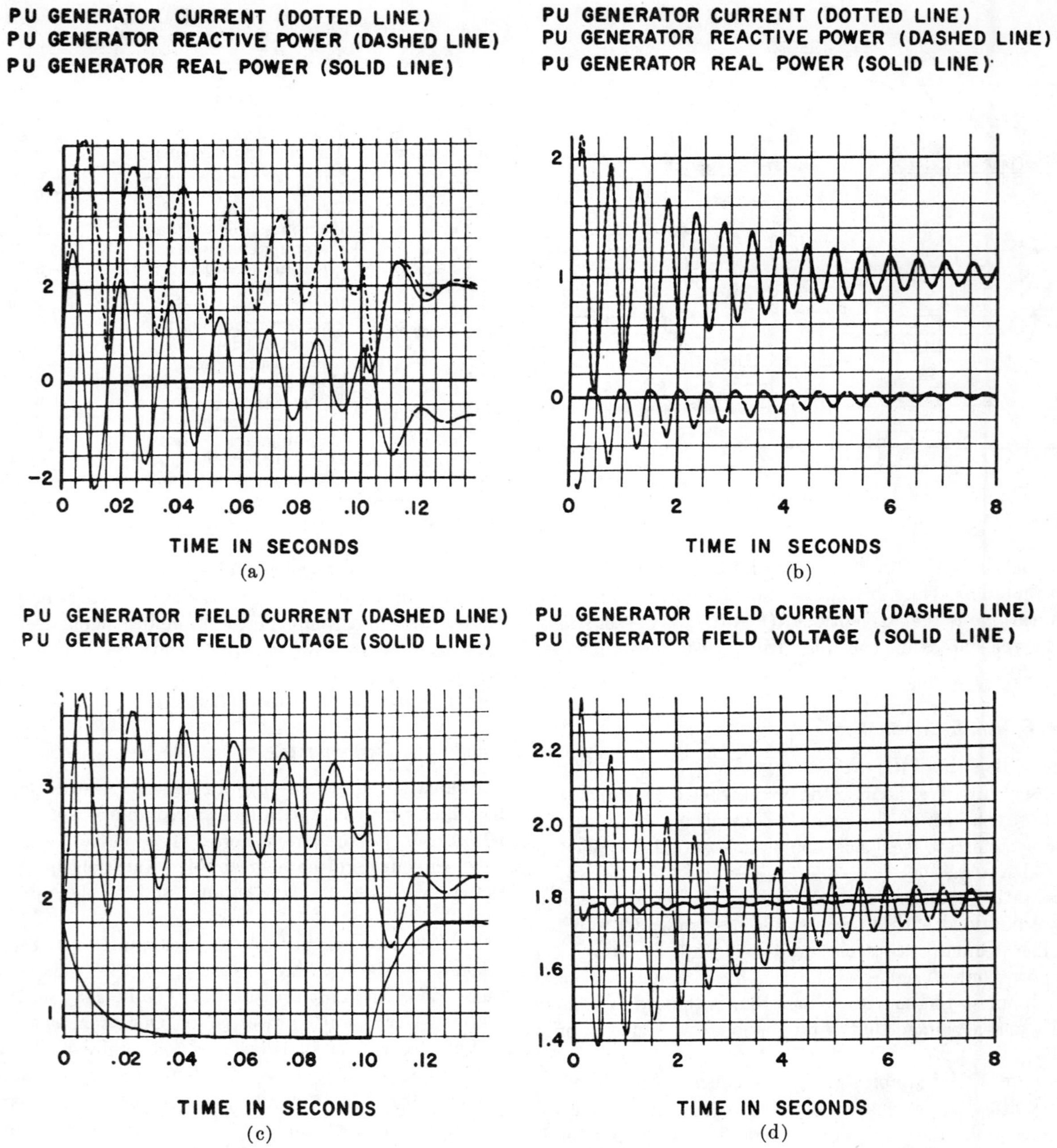

Fig. 5. Machine response to six-cycle three-phase fault with low-reactance linear dampers; manual excitation and constant turbine torque.

equations. This contrasts with machine models generally used for short-circuit and stability analysis. The normal procedure is to assume that stator currents can change instantaneously, which leads to the concept of an internal generator voltage and the algebraic determination of stator equivalent D- and Q-axis currents. The damping changes of interest are in themselves small, and it was decided to simulate the stator differential equations rather than make the normal assumptions. The damper currents and stator equivalent currents also contain 60-Hz components. This component damps rapidly after the fault is cleared because of effects external to the machine (see Appendix).

For comparison, Fig. 6 shows the unit response for conditions similar to those of Fig. 5 except that the damper winding reactance is assumed to be a saturable quantity. The value of subtransient reactance varies from a high value when unsaturated to a value equal to that of Fig. 5 when well saturated. Fig. 6(b) and (c) are similar to those illustrated previously. Fig. 6(a) shows high- and low-side voltage, while Fig. 6(d) plots unit velocity deviation from synchronous. Following the clearing of the fault, damper bar current is not high enough to effectively saturate the iron above the bar, and this effectively limits damper bar incremental current and hence damping torque. The saturation characteristics selected for the damper leakage flux path result in a somewhat pessimistic damping characteristic for the model. The actual machine has more inherent damping at the lower level of oscillation. The machine characteristics shown in Fig. 6 are used in the remaining response curves.

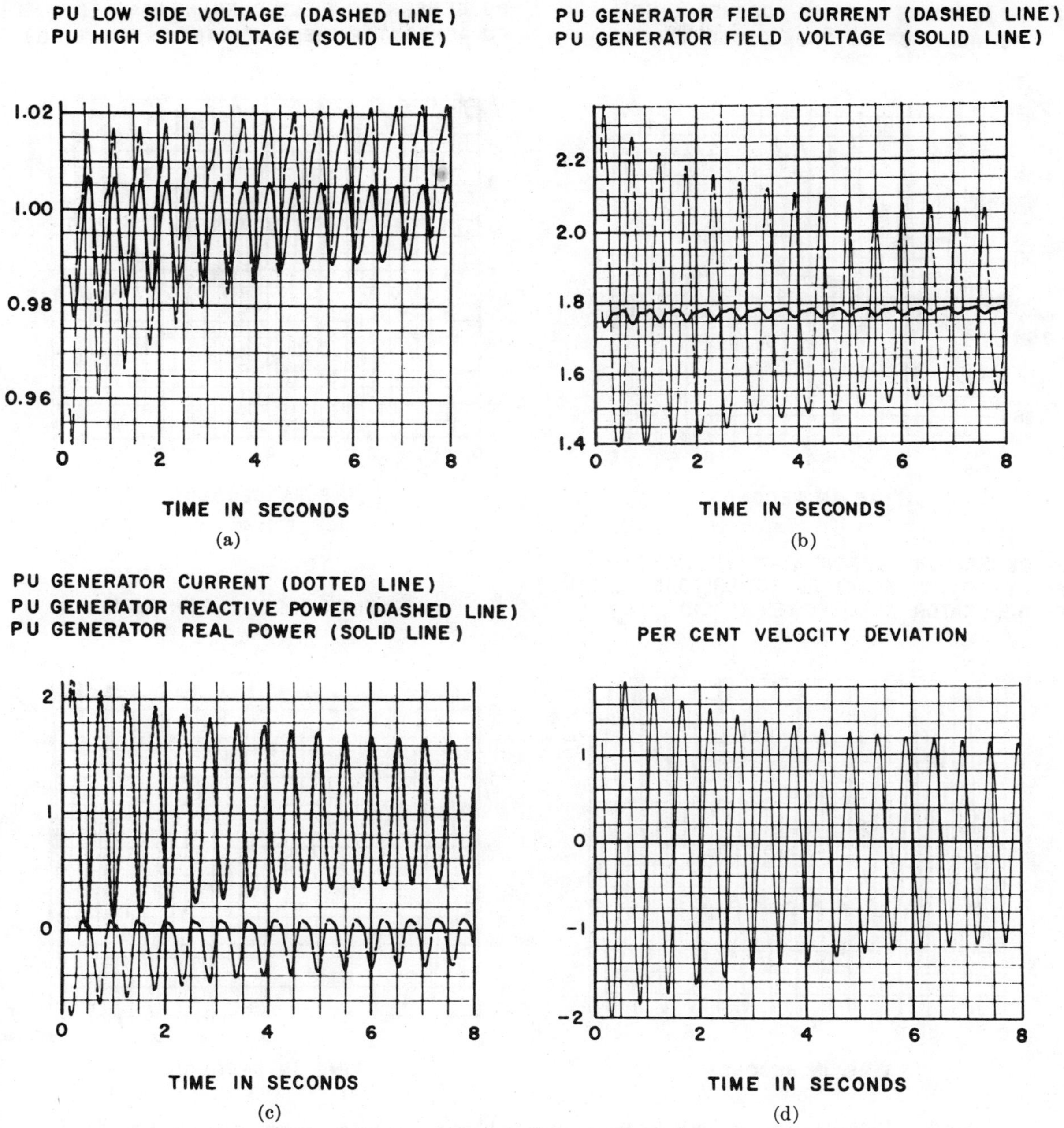

Fig. 6. Machine response to six-cycle three-phase fault with nonlinear dampers; manual excitation and constant turbine torque.

Effect of Control Systems

The effect of governor response and water inertia on damping is slight, and response curves for governor control without automatic excitation are not shown. The curves of Fig. 7 are for a generator with governor control and automatic excitation. Most of the change from Fig. 6 is due to the action of the excitation system. Machine oscillations continue to build up with evident negative damping. This excitation system model did not have an internal rate feedback loop (excitation system stabilizer). A normal static excitation system, regulating voltage alone with no power damping signal, would have internal stabilization. The data of Fig. 7 are included in the paper because this is the system that is presently in service with the power stabilizing signal.

A static excitation system with a rate feedback loop is illustrated in Fig. 8. Oscillations persist for an extended time and, although there is a small decrement in successive oscillations, performance is not satisfactory. The negative damping of the high-gain static excitation system produces these extended oscillations, and it is likely that a limit cycle condition will be reached where oscillations continue indefinitely.

Fig. 9 illustrates the performance of the excitation system that is now in service. This includes a static excitation system with no internal rate feedback loop but with a power damping signal obtained from generator terminal real power. The curves show good damping of machine oscillations, and at the end of 6 seconds machine velocity has essentially stabilized. Note that the excitation system oscillates between the maximum and minimum

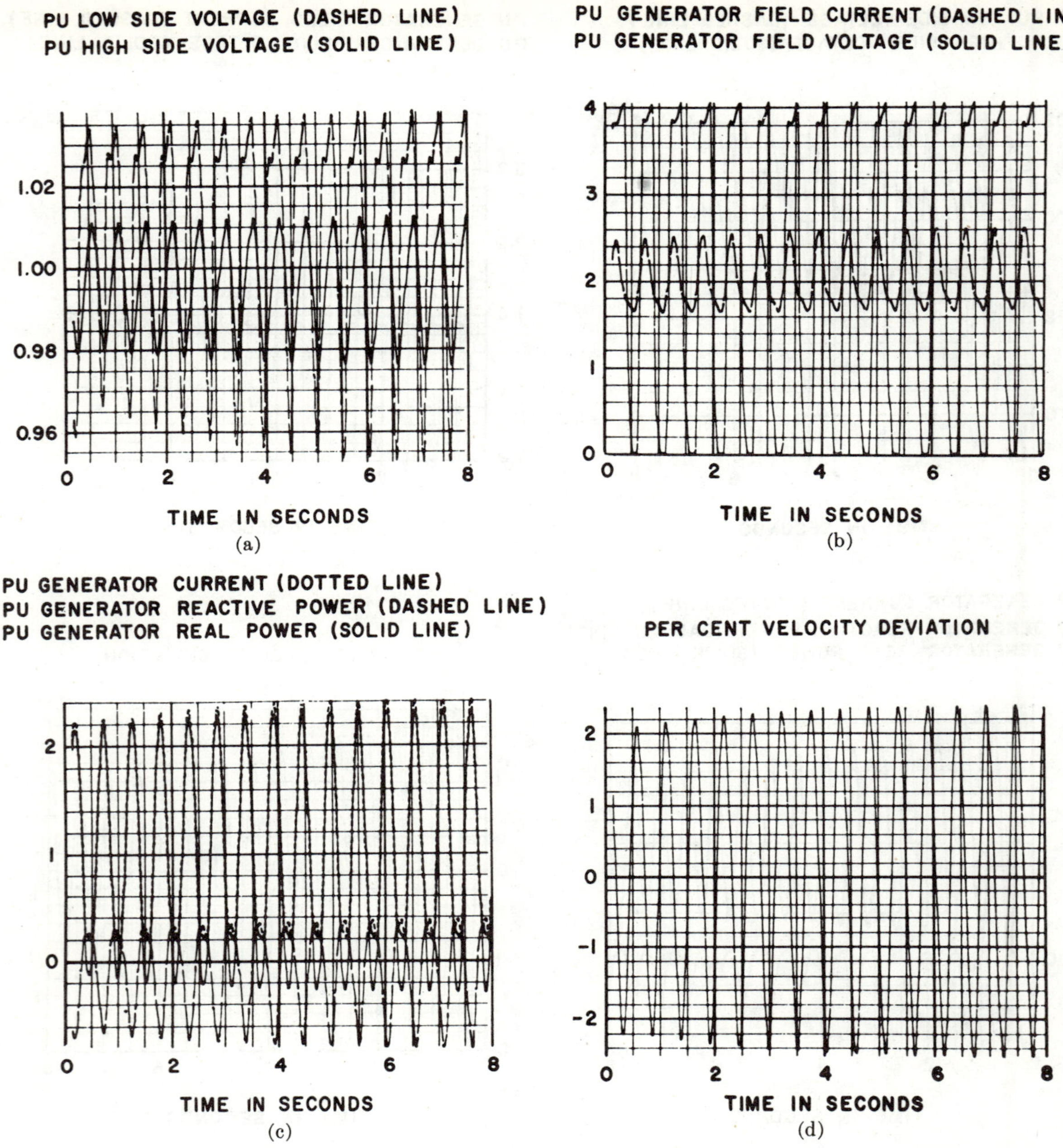

Fig. 7. Machine response to six-cycle three-phase fault with nonlinear dampers, governor control, and automatic excitation without rate feedback or power damping.

values for approximately 4 seconds before coming off limits and finally settling out in about 7.5 seconds. The damping of machine oscillations is considerably better than that for nonlinear dampers with no excitation system (shown in Fig. 6) and is approximately the same as that for the theoretical example of linear damper windings.

Fig. 10 illustrates the performance when both power damping and an internal rate feedback loop are employed. The performance is very similar to that of power damping alone, but with slightly poorer damping in Fig. 10. Since the internal rate feedback loop reduces the transient excitation system gain by approximately 10/1 at the oscillation frequency, the gain of the power damping signal was increased by ten to compensate for this gain reduction.

Evaluation of Damping

In the performance of a study of power system damping such as the one described, it is difficult to make a quantitative evaluation of the magnitude of damping present. The authors have examined a number of different functions derived from such quantities as torque, angle, and velocity and have chosen a few for illustration in Figs. 11 and 12. It is not suggested that the functions shown are the best to illustrate damping, but they do show some of the gross effects. Fig. 11 shows the performance of the static excitation system with rate feedback stabilization. Fig. 12 includes power damping.

Figs. 11(a) and 12(a) show electrical torque as a function of machine shaft velocity. The component of torque in phase with

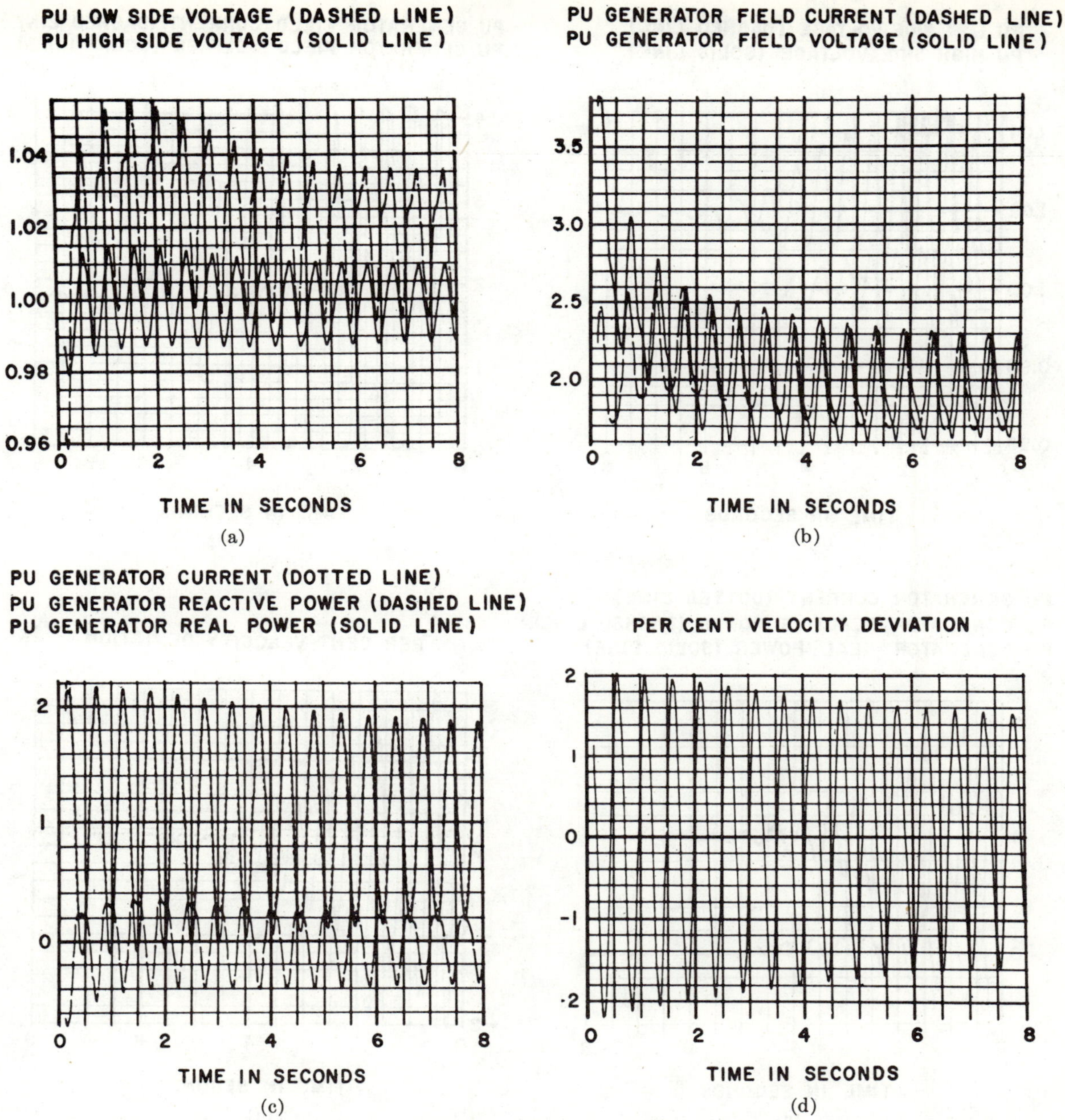

Fig. 8. Machine response to six-cycle three-phase fault with nonlinear dampers, governor control, and automatic excitation with rate feedback but no power damping.

velocity is commonly called damping torque. Since the shape is very nearly elliptical there is only a small damping torque compared to the total torque. Note, however, the graphic indication of the decrement of oscillations. Fig. 12(a) shows a rapidly closing spiral settling in 1.0-pu torque and zero velocity deviation. In Fig. 11(a) the decrement is not as large and there is some appearance of diminishing decrement such as would occur if a limit cycle were reached. Oscillations would then continue indefinitely. An unstable case has an ever-increasing spiral up to the point of instability.

A plot of electrical torque as a function of angle deviation is shown in Figs. 11(c) and 12(c). The component of torque in phase with angle deviation is called synchronizing torque. Note that in both cases most of the torque is synchronizing since there is very nearly a linear relation between torque and angle deviation.

The integral of per-unit damping energy is illustrated in Figs. 11(b) and 12(b). The oscillatory case—Fig. 11(b)—shows some positive damping initially but then the magnitude drops to a low value and is slightly negative at 8 seconds. Values are highly oscillatory with positive or negative half-cycles essentially equal in size. The well-damped case—Fig. 12(b)—shows a rapid build-up in damping energy which settles out at a value of approximately 0.0008 pu.

Similar plots of the integral of synchronizing energy are shown in Figs. 11(d) and 12(d). Since the oscillations of Fig. 11(d) have not settled out, the integral of synchronizing energy continues to increase.

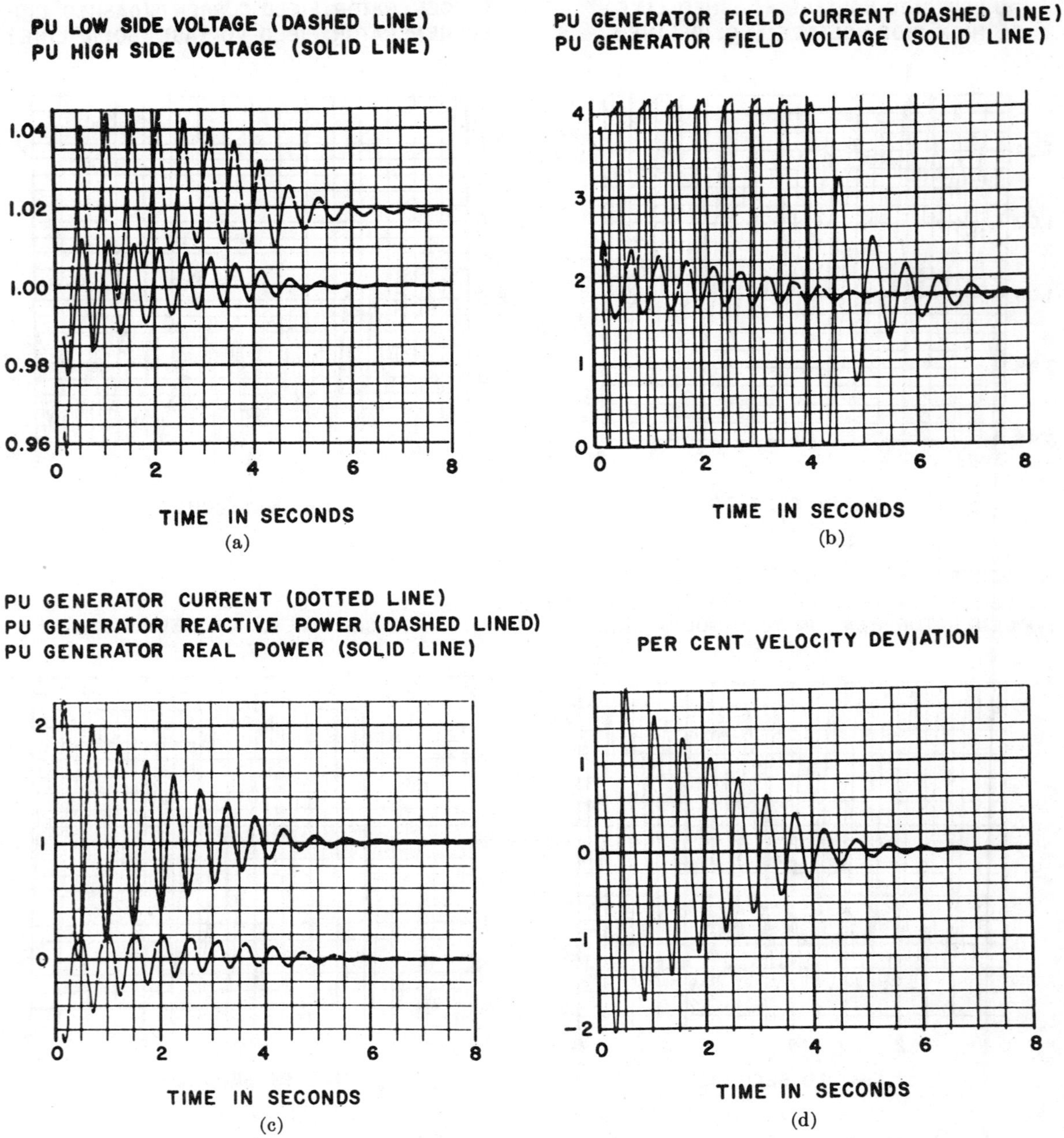

Fig. 9. Machine response to six-cycle three-phase fault with nonlinear dampers, governor control, and automatic excitation with power damping but without rate feedback.

Conclusions

The application of a static exciter to a salient-pole machine has the potential for providing improved damping of dynamic oscillations. This is achieved through the use of supplementary signals added to the normal voltage regulator error signal. In this study no attempt was made to examine alternate stabilizing signals, but a terminal power signal was shown to be very effective.

A static exciter with no stabilizing signal and no internal rate feedback loop is highly oscillatory and, for this application, unsatisfactory. A similar system (no stabilizing signal) but with an internal rate feedback loop shows extended oscillation with a small positive decrement.

It may be concluded that static excitation systems show improved performance with a stabilizing signal, but may be unstable without such a signal, especially if no internal rate feedback loop is used. In the general application of static excitation systems an internal rate feedback loop should be available for use either in operation continuously or activated when the stabilizing signal is out of service.

The design of machine damper windings has a significant effect upon the machine's damping characteristic. Embedded damper windings are relatively ineffective in damping small oscillations.

The machine model developed for these studies provides a more detailed representation than has been generally used for digital simulation. In addition to the simulation of the saturation of damper windings, this model also provides for more precise determination of machine currents on which torque, and hence damping, depend.

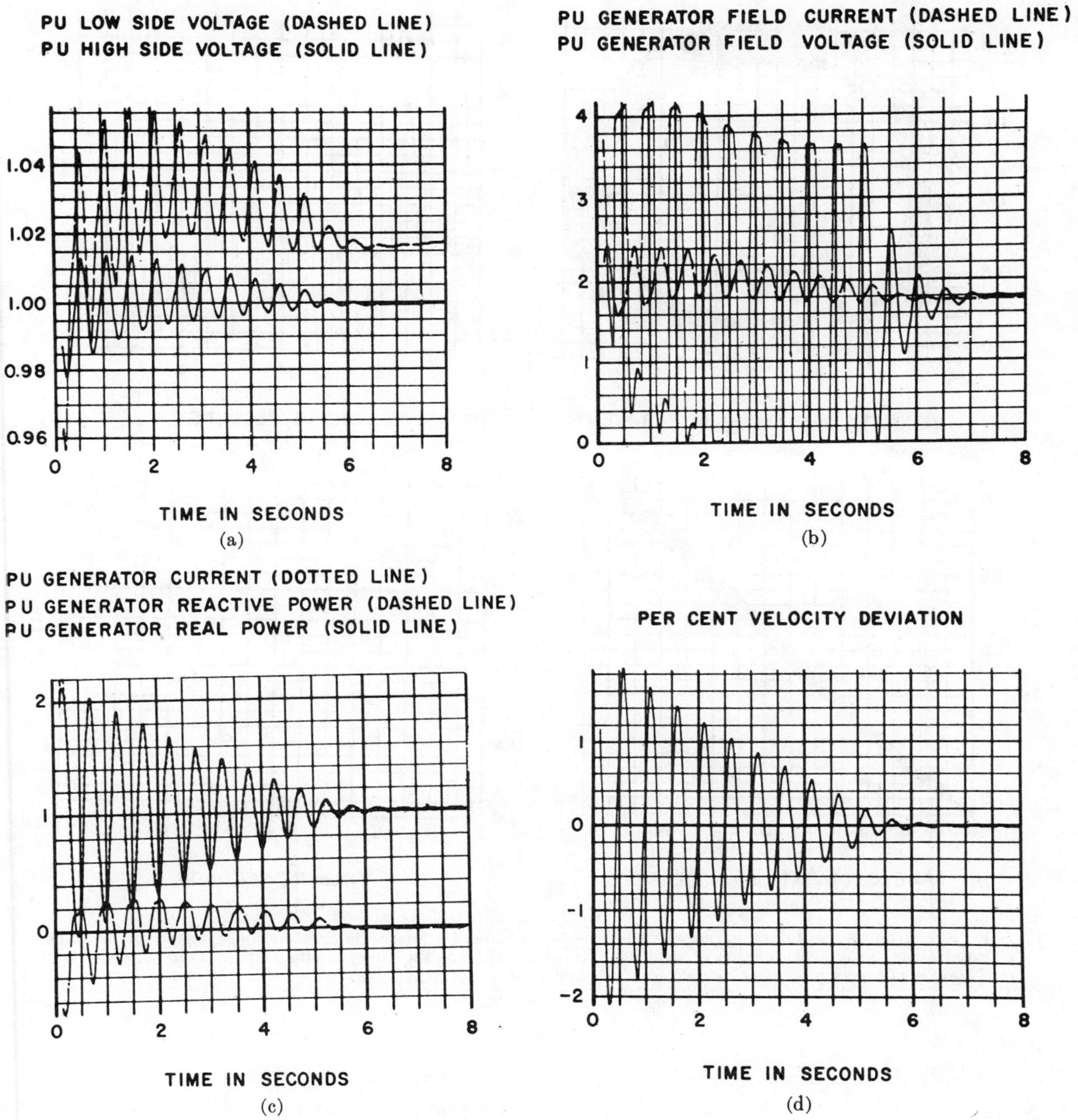

Fig. 10. Machine response to six-cycle three-phase fault with nonlinear dampers, governor control, and automatic excitation with power damping and rate feedback.

Appendix

Generator Mathematical Model

Calculation of generator flux linkages is based on differential equations written for the direct-axis equivalent circuit of Fig. 1 and a corresponding circuit for the quadrature axis not shown. Direct-axis equivalent circuit symbols are defined in the next section. Branch flux linkages are selected as the variables to be calculated to simplify the treatment of saturation. For the direct axis, two saturation functions are needed: S_{MD} for the mutual inductance L_{MD}, and S_{SD} for the damper circuit leakage inductance L_{SD}. On the quadrature axis, only the saturation of the damper leakage reactance is assumed to be significant. Saturations on the direct axis are calculated from

$$S_{MD} = A_M|\lambda_{MD}|^{BM}$$

and

$$S_{SD} = A_S|\lambda_{SD}|^{BS}$$

where subscripts MD and SD are branch flux linkages, and A_M, A_S, B_M, and B_S are constants. The quadrature-axis saturation function is calculated similarly.

The decision to compute branch flux linkages rather than loop flux linkages requires the voltage E_{MD} to be known, and the corresponding voltage E_{MQ} on the quadrature axis. These voltages may be determined at any moment in terms of the stator terminal voltages, the field voltage, the branch flux linkages, and the saturation characteristics. An equation for E_{MD} may be obtained by

$$I_{AD} = I_{MD} + I_{FD} + I_{SD}.$$

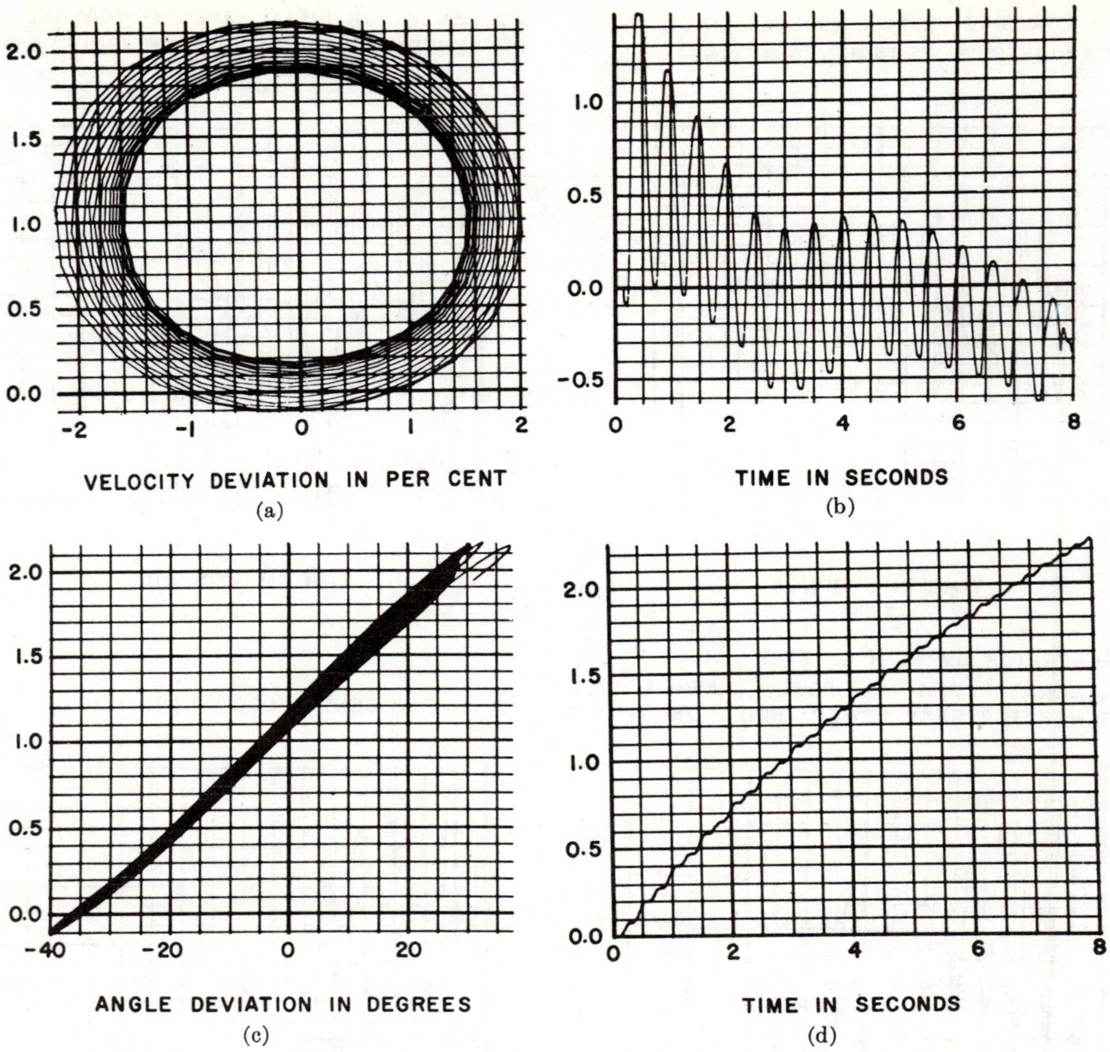

Fig. 11. Synchronizing and damping torque and energy functions for poorly damped response, no power damping. (a) Pu electrical torque as function of velocity deviation. (b) Damping energy in pu seconds $\times 10^{-4}$. (c) Pu electrical torque as function of angle deviation. (d) Synchronizing energy in pu seconds.

Differentiating,

$$pI_{AD} = pI_{MD} + pI_{FD} + pI_{SD}.$$

These current derivatives may be expressed in terms of E_{MD} and the variables listed above, yielding an equation for E_{MD}. There is no assurance that the use of branch flux linkages will result in a more economical computation. The same saturation functions could be represented if the equations were formulated in terms of loop flux linkages with two fewer differential equations and no need to calculate E_{MD} and E_{MQ}. However, iteration of these terms would be needed to obtain the saturation functions.

The equivalent circuits are initialized from known terminal conditions for the generator. The fault is applied by setting E_{TD} and E_{TQ} to zero. As shown in Fig. 5, this abrupt change in terminal voltage results in the 60-Hz components of currents in all branches of the equivalent circuits. These current components are the result of the speed voltage coupling between the two axes and will appear in response to any abrupt terminal voltage change. The first trial runs with this model simply restored the connection of the machine to the infinite bus to simulate fault clearing. This resulted in additional 60-Hz currents, which were damped at the same rate as those that occurred during the fault. The persistence of these currents after the fault was cleared seemed unrealistic, and the infinite bus was changed to an "almost infinite" bus as shown in Fig. 1(b). This can be visualized schematically by replacing E_{XD} of Fig. 1(a) with E_{XD}' and the circuitry of Fig. 1(b). The parameters of Fig. 1(b) may be chosen to damp the 60-Hz components after fault clearing with almost no damping effect at swing frequencies.

Definitions

R_X	external resistance
L_X	external inductance
R_T	transformer resistance
L_T	transformer inductance
R_A	stator resistance
L_{MD}	mutual inductance
R_{FD}	field resistance
L_{FD}	field leakage inductance

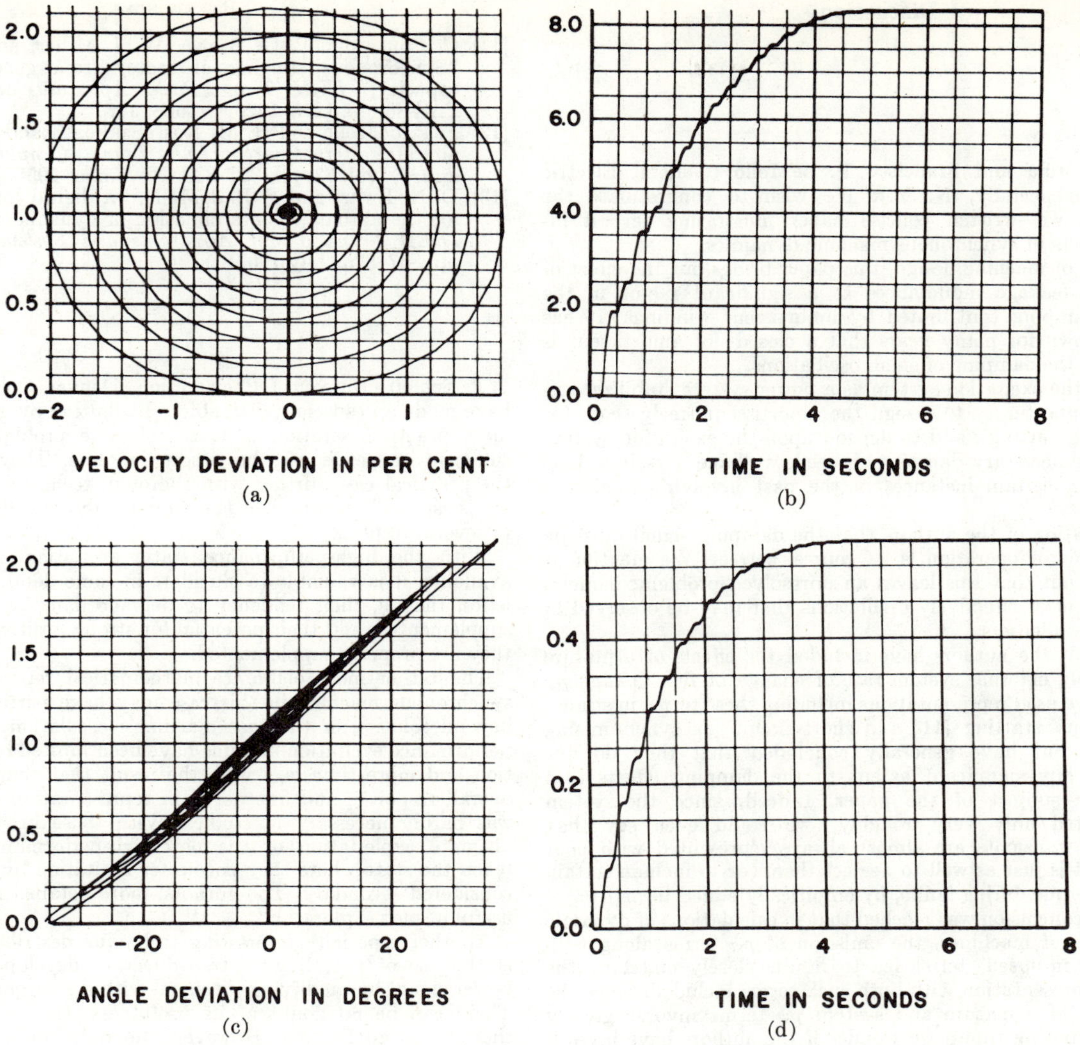

Fig. 12. Synchronizing and damping torque and energy functions for well-damped response, with power damping. (a) Pu electrical torque as function of velocity deviation. (b) Damping energy in pu seconds $\times 10^{-4}$. (c) Pu electrical torque as function of angle deviation. (d) Synchronizing energy in pu seconds.

E_{FD}	field voltage
R_{SD}	damper resistance
L_{SD}	damper leakage inductance
E_{MD}	mutual voltage
E_{XD}	infinite bus voltage
E_{TD}	transformer high-side voltage
E_{LD}	generator terminal voltage
λ_{XQ}	quadrature-axis external flux linkage
λ_{TQ}	quadrature-axis transformer flux linkage
λ_{AQ}	quadrature-axis stator leakage flux linkage
λ_{MQ}	quadrature-axis mutual flux linkage
ω	frequency, rad/s
L_A	stator leakage inductance.

References

[1] H. M. Ellis, J. E. Hardy, A. L. Blythe, and J. W. Skooglund, "Dynamic stability of the Peace River transmission system," *IEEE Trans. Power Apparatus and Systems,* vol. PAS-85, pp. 586–600, June 1966.

[2] F. R. Schleif and J. H. White, "Damping for the Northwest–Southwest tieline oscillations—an analog study," *IEEE Trans. Power Apparatus and Systems,* vol. PAS-85, pp. 1239–1247, December 1966.

[3] F. R. Schleif, G. E. Martin, and R. R. Angell, "Damping of system oscillations with a hydrogenerating unit," *IEEE Trans. Power Apparatus and Systems,* vol. PAS-86, pp. 438–442, April 1967.

[4] R. T. Byerly, J. W. Skooglund, and F. W. Keay, "Control of generator excitation for improved power system stability," *Proc. American Power Conf.*, vol. 29, pp. 1011–1022, 1967.

[5] O. W. Hanson, C. J. Goodwin, and P. L. Dandeno, "Influence of excitation and speed control parameters in stabilizing intersystem oscillations," *IEEE Trans. Power Apparatus and Systems*, vol. PAS-87, pp. 1306–1313, May 1968.

[6] F. R. Schleif, H. D. Hunkins, G. E. Martin, and E. E. Hattan, "Excitation control to improve powerline stability," *IEEE Trans. Power Apparatus and Systems,* vol. PAS-87, pp. 1426–1434, June 1968.

[7] P. L. Dandeno, A. N. Karas, K. R. McClymont, and W. Watson, "Effect of high-speed rectifier excitation systems on generator stability limits," *IEEE Trans. Power Apparatus and Systems*, vol. PAS-87, pp. 190–201, January 1968.

[8] F. P. deMello and C. Concordia, "Concepts of synchronous machine stability as affected by excitation control," *IEEE Trans. Power Apparatus and Systems*, vol. PAS-88, pp. 316–329, April 1969.

[9] D. C. White and H. H. Woodson, *Electromechanical Energy Conversion.* New York: Wiley, 1959, ch. 9.

[10] IEEE Committee Report, "Computer representation of excitation system," *IEEE Trans. Power Apparatus and Systems,* vol. PAS-87, pp. 1460–1464, June 1968.

Discussion

Charles Concordia and **Francisco P. deMello** (General Electric Company, Schenectady, N. Y.): We wish to congratulate the authors on a well-written concise paper illustrating several interesting aspects of synchronous machine dynamics.

In the area of machine design, this paper brings out the effect of saturation of leakage inductances as a significant factor in the amount of damping contributed by amortisseur windings. It has been well known for many years that a closed-slot amortisseur is ineffective for the damping of small oscillations.

The use of the excitation system as a power system stabilizer can provide the opportunity to design the amortisseur freely (e.g., for the purpose of starting), and to depend upon the excitation system to provide the necessary damping. Indeed, we have ourselves done exactly this in certain instances in the past involving industrial applications.

The observation of the authors that the damping signal must be reversed for motor operation is, of course, correct. We mention it only to point out that this leaves an unresolved problem; namely, how to damp most effectively oscillations that may be observed in synchronous condensers.

We note that the authors have included the effects of armature and presumably external system rate of change of flux linkage $p\varphi$ terms. We have used representations including these terms in studies of synchronous starting [11] and short-circuit or synchronizing torques [12], but have generally concluded that they do not contribute to any significant extent to the damping effects that are the main subject of the paper. Indeed, since the system is approximated only very roughly, we could even say that the armature transients are almost always represented with poor accuracy; so it is just as well to neglect them (or to include certain effects, e.g., torque during faults, by empirically added terms).

In [13] the conclusion was reached that in calculations of dynamic stability limits of machines, the omission of $p\varphi$ terms along with $p\delta$ (variations in speed) terms led to results closely matching the more exact representation with both such terms included. Since the representation of armature and system $p\varphi$ terms involve greatly expanded computing times, we wonder if the authors have investigated this matter and, in particular, if they have made comparisons of results with and without inclusion of such terms.

We agree with the authors that the present tendency toward the use of external stabilizing signals gives the opportunity to have both high response and good damping in the same apparatus.

The study demonstrates the effectiveness of a power-derived stabilizing signal. With the justifiable assumption that changes in mechanical torque are negligible, there is a very predictable relation between electrical power and speed, which was the signal used for stabilizing machine rotor angle oscillations in [8].

From the recommendations in [8], confirmed with several subsequent studies, we conclude that the parameters shown for the power system stabilizer function are not necessarily the optimum values. In this respect we question the authors' observation in discussing Fig. 10 that reduction of the transient excitation system gain justified an equivalent increase of the stabilizing signal gain. As pointed out in [8], the effect of variation in voltage regulator gain is mainly in changing the bandwidth or phase lag of the closed-voltage regulator loop. Accordingly, we believe that better results could have been achieved with the proper choice of another lead lag pair in the stabilizer function, rather than a mere increase in signal gain.

With respect to nomenclature, we prefer to call the damper an external stabilizer to distinguish it from the internal feedback stabilizers usually incorporated in the voltage regulating loop. The power system stabilizer is the whole excitation system, including the external stabilizer, which may be said to convert the excitation system into a power system stabilizer.

Manuscript received July 7, 1969.

References

[11] C. Concordia, P. G. Brown, W. J. Miller, and L. Wuosmaa, "Synchronous starting of motor from generator of small capacity," *IEEE Trans. Power Apparatus and Systems*, vol. PAS-86, pp. 1215–1226, October 1967.

[12] F. P. deMello and H. M. Rustebakke, "Random paralleling of aircraft alternators," *AIEE Trans.* (*Applications and Industry*), vol. 77, pp. 322–330, November 1958.

[13] D. N. Ewart and F. P. deMello, "A digital computer program for the automatic determination of dynamic stability limits," *IEEE Trans. Power Apparatus and Systems*, vol. PAS-86, pp. 867–875, July 1967.

F. R. Schleif (Bureau of Reclamation, Denver, Colo.): The authors have made an especially valuable contribution by going beyond the mere practical solution of a troublesome problem to an extent sufficient to get it off the production line. They have exploited the practical case further with thorough technical analyses, which act is especially welcome at this time to advance the state of the art in several respects.

With the high-performance static excitation systems recently available, it is valuable to elucidate in some detail their characteristics, that is, their tendency to degrade damping if they are unsupplemented, and their potential for strong positive damping when they are properly supplemented.

The rather more elaborate mathematical representation of the synchronous machine in this case may have pertinent application beyond solution of the case machine's oscillation problem. Lower performance excitation systems have been limited by exciter saturation and more time lags with the result that they dominated the overall response, and fine detail in representation of the generator was hardly necessary in studies. With those limitations relieved, ultimate exploitation of the new high-performance systems may reach the state where they encounter limitation by effects formerly considered secondary. The authors' more detailed model should be useful in such exploration.

Another especially interesting technique described in this paper is the use of a Hall watt transducer to develop an accelerating power signal by modifying its signal with a 1-second reset function. There can be no doubt of its usefulness; their work has proved that it has application. However, the range of applicability may be worth comment. Some of our recent work has demonstrated that the beneficial influence obtainable through stabilizing signals to the excitation system is greatly enhanced when a voltage-sensitive load (constant impedance) is influenced by the generator voltage. This effect is not confined to a transient interval but may be sustained as long as the voltage is influenced. Consequently, a substantial aid to governing becomes possible, but when this is to be done it is necessary to employ a more sustained type of signal, such as that obtainable from frequency deviation or shaft speed.

The authors are to be commended for an interesting and revealing contribution.

Manuscript received June 30, 1969.

Otto J. M. Smith (Electronics Research Laboratory, University of California, Berkeley, Calif.): Senior students at the University of California have implemented an excitation control on a 15-kVA machine similar to that studied in this paper, except that the exciter was rotary, not silicon-controlled rectifier (SCR). The open-loop mechanical frequency was 2 Hz. The closed-loop frequency of excitation instability at excessive gain was 5 Hz. With suitable noise rejection filters following the power transducer and a stabilizing filter designed by root-locus techniques to reduce the 5-Hz response, the mechanical transient decrement was about 75 percent per

Manuscript received July 14, 1969.

period. The authors seem to have achieved about 20-percent decrement per period.

Digital computer studies by Richard Oswald simulating 300-MW machine with an 8-second field time constant and utilizing excitation feedback from power, velocity, angle, field current, and terminal voltage, achieved a one-half cycle mechanical swing in 1 second, followed by a negligible 4-second monotonic transient to the target state. The mechanical oscillations were essentially removed in the one-half period of 1 second. The five state variable design for the five-loop feedback was obtained by simulating an optimal control algorithm, and then adaptively improving the algorithm coefficients using engineering judgment on the resulting responses.

Could the authors explain why the derivative of power was used as a control signal in Fig. 4? Could the authors explain why the velocity and power transient envelope time constants were approximately 4 seconds with an SCR exciter as given in this paper, whereas the theoretical ability of this type of control should be about 1 second.

Richard T. Byerly, F. W. Keay, and John W. Skooglund: The authors wish to express their appreciation to Mr. Concordia, Mr. deMello, Mr. Schleif, and Prof. Smith for their discussions.

Mr. Concordia and Mr. deMello agree with the need to reverse the polarity of the auxiliary damping signal between generating and motoring modes. They state that this leaves unresolved the means for effective damping of oscillations in synchronous condensers. Perhaps the problem of damping in condensers is more fundamental than this statement implies. Any synchronous machine receiving or delivering zero power in its steady state will have its stator current acting entirely on the direct axis. Phase displacements from equilibrium will produce relatively large variations in quadrature-axis currents, but extremely small variations in the direct-axis currents. The interaction of varying field voltages with these small direct-axis-induced currents represents neither a sink nor a source of energy at the oscillation frequency. Studies made by the authors show no measurable negative damping effect due to normal voltage regulator action on a synchronous condenser. By the same token, various damping schemes, including signals that switch polarity between the generating and motoring phase displacements, have produced no significant improvement in inherent machine damping.

Mr. Concordia and Mr. deMello express the opinion that the solution of the stator flux linkage differential equations increases computer costs while having a minor influence on damping. In general this is true, and in large system studies it is definitely quite costly to solve these differential equations. In this instance, however, simulation costs were low and there was no motivation to do anything other than solve the equivalent circuit equations without modification. Since other programs available to the authors do not permit representation of nonlinear dampers, no critical comparison of the effects of solving the stator flux linkage equations has been made.

Mr. Concordia and Mr. deMello suggest that better results, presumably more rapid damping, could have been achieved. This may be true. It is important to note, however, that the results achieved were good in several respects. Once the voltage regulator enters the linear operating range, there is a decrement of about 57 percent per period (not 20 percent as Prof. Smith states). This is certainly acceptable. The voltage excursions at the low-side bus resulting from the low-frequency characteristic of the auxiliary signal transfer function are within acceptable limits. The field voltage remains at ceiling through the first positive angle swing after the fault in spite of the tendency of the auxiliary damping signal to cause a prior reduction of that voltage. Finally, all these characteristics are relatively insensitive to parameter variations.

Mr. Concordia and Mr. deMello offer some suggestions for nomenclature, and certainly there is a need to work out and adopt industry standards for this relatively new but already widespread control application. This matter should be brought to the attention of appropriate institutional groups.

The authors thank Mr. Schleif for his very interesting comments regarding the use of sustained signal sources such as velocity deviation to not only provide transient damping, but to improve frequency regulation as well.

Prof. Smith asks why the derivative of power is used as a control signal in Fig. 4. The purpose of the signal transfer function in Fig. 4 is to obtain from P_E alone a variation that resembles the accelerating power on the unit. Accelerating power could be obtained by measuring both turbine power and electrical power. However, turbine power is a more difficult measurement, and the accelerating power is principally the negative variation of electrical power since governor action is slight. The purpose of the auxiliary signal transfer function in Fig. 4 is to pass a band of frequencies with essentially no phase shift to approximate accelerating power. The derivative action takes place below the dominate frequencies of oscillation.

Prof. Smith asks why the transients persist for approximately 4 seconds whereas the theoretical ability of this type of control should be about 1 second. Perhaps it is because the theoretical response to which he refers does not concern itself with regulator or exciter ceilings. Note for example in Fig. 10 that the exciter output is limited until about 5 seconds. In Fig. 9, the exciter is limited until about 4.5 seconds.

Manuscript received September 8, 1969.

Field Tests of Dynamic Stability Using a Stabilizing Signal and Computer Program Verification

RICHARD M. SHIER, MEMBER, IEEE, AND ALLAN L. BLYTHE, MEMBER, IEEE

Abstract—**Stability studies were made[1] on a digital computer of the British Columbia Hydro and Power Authority Peace River System which indicated the advisability of using high-speed excitation with a stabilizing feedback. Consideration of the novelty of this technique led to a decision to conduct field tests to confirm that the equipment would behave as forecast. In addition, these tests could provide confirmation of the assumptions and techniques employed in the use of the digital computer program. The tests were made on a small hydroelectric station connected to the main system by a relatively weak line. Single phase to ground faults were applied to the station 132-kV bus. The results of these tests and of computer simulations are shown.**

Introduction

A STUDY was made in 1964 of the stability problems expected on the British Columbia Hydro and Power Authority (BCHPA) Peace River development and general solutions to these problems were found.[1] These studies made use of the Westinghouse Electric Corporation's digital computer program.[2] It was established that a high-speed excitation system with a stabilizing speed error feedback could contribute significantly to system stability.

To confirm that an adequate stabilizing signal could be obtained in practice and that confidence could be placed in the assumptions and techniques used with the digital computer program, it was considered desirable to carry out field tests. Accordingly, agreement was reached to conduct these tests on the Clowhom Falls plant of BCHPA.

Power System

Clowhom Falls generating station is located approximately 40 miles northwest of Vancouver, British Columbia, at the head of a long inlet leading from the Strait of Georgia. It comprises a single hydro generator rated 30 MW, 31.58 MVA, 13 800 volts with subtransient reactance of 0.354 per unit and inertia of 2.55 J/VA (H constant). The transmission system is shown in Fig. 1, with the principal power flows in MW and impedances in percent on 100 MVA. Under the test conditions, the Powell River plant of MacMillan, Bloedel, and Powell River Limited has a load consisting of four 5000-hp synchronous motors and 7.0 MW of other load. The nearest other generating station is Cheakamus, rated 140 MW, 160 MVA. The rest of the system is shown as an infinite bus, connected close to Cheakamus.

To initiate reasonably consistent dynamic disturbances a temporary set-up was made in the outdoor substation of a single-pole, motor-operated disconnect switch and a power fuse, connecting

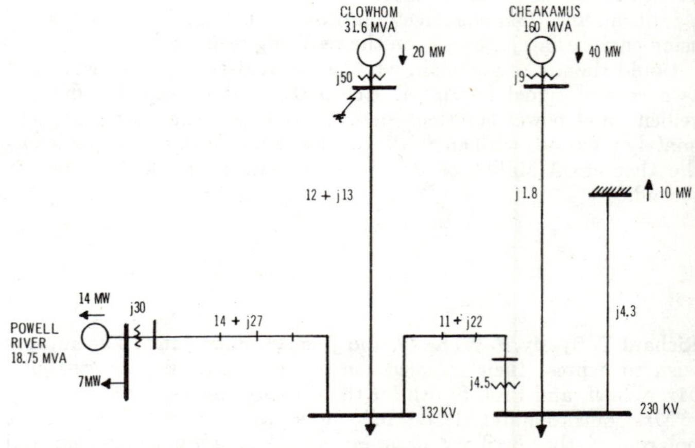

Fig. 1. Power system showing major power flows in MW and impedances in percent on 100 MVA.

one phase of the 132-kV bus to ground. Closing the switch caused a single phase to ground fault lasting approximately 16 cycles (0.267 seconds).

Excitation System

The normal excitation system of the generator was a rotating exciter controlled by a type WMA magnetic amplifier voltage regulator. The high-speed excitation system was a modified version of an industrial type of thyristor power supply with a solid-state voltage regulator. This was strictly a temporary installation and the excitation system was equipped with only a minimum of protective devices. The thyristor power supply and gating controls were supplied from a transformer bank connected to the line side of the generator breaker and was, therefore, subject to voltage variations during system transients. The ceiling voltage of the static exciter was 5.08 per unit as compared to 3.6 per unit for the rotating exciter. The voltage regulator was adjusted so that a 1.3 percent drop in terminal voltage would cause an increase in exciter voltage from that at generator rated full load to ceiling value.

As a precautionary measure, an automatic transfer scheme was arranged to cause transfer to the rotating exciter should any trouble arise in the static exciter. The open-circuit voltage of the rotating exciter was manually controlled to match that of the static exciter at all times. Also, the static exciter was equipped with a current-limiting control which would prevent continuous operation at currents in excess of 1000 amperes in order to protect the thyristors and the generator field winding from accidental overcurrents.

Recordings were taken on a 16-channel Brush 2300 recording oscillograph.

Stabilizing Feedback

The Peace River stability studies had indicated that the ideal stabilizing feedback was speed error. The results of those studies indicated that any speed measuring device should be capable of detecting a speed change of 0.01 percent.

Paper 31 TP 66–415, recommended and approved by the Transmission and Distribution Committee of the IEEE Power Group for presentation at the IEEE Summer Power Meeting, New Orleans, La., July 10–15, 1966. Manuscript submitted April 11, 1966, made available for printing October 4, 1967.

R. M. Shier is with the British Columbia Hydro and Power Authority, Vancouver, B. C., Canada.

A. L. Blythe, deceased, was with Canadian Westinghouse Company, Ltd., Hamilton, Ont., Canada.

Reprinted from *IEEE Trans. Power App. Syst.*, Feb. 1968.

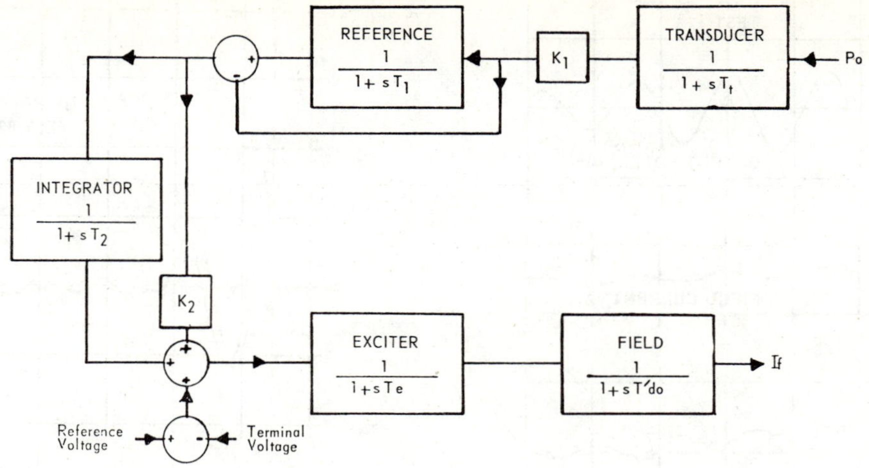

Fig. 2. Feedback signal block diagram.

A pulse-counting technique of speed measurement using a toothed wheel was experimented with and good results were obtained in the factory. This device was mounted as closely as possible to the generator thrust bearing and development continued. Difficulty was experienced due to the various modes of vibration of the generator shaft. While most of these could be readily compensated, the effects of torsional vibrations were difficult to eliminate.

Results were obtained by this method which did show a reasonable degree of damping. However, to produce a reasonably "clean" signal, filtering was necessary. This introduced a time delay which reduced its capability. Before further improvement of this method could be made, another approach was found which was purely electrical in nature and consequently much simpler to apply to any machine. The success of this method caused discontinuance of the pulse-counting technique.

The success of the speed feedback in theoretical studies is dependent on the following reasoning.

The machine swings are a consequence of the instantaneous speed of the machine oscillating around that of the system. If a control can be applied to reduce the speed differences, the transient will be quickly damped out. A high-speed excitation system is capable of rapidly varying the field current and hence the power output during a system swing, thus producing acceleration or deceleration of the machine and so controlling the speed. A negative feedback of speed error can thus act to control speed and provide transient damping.

It was reasoned that the speed of the machine could also be considered to be stored rotational energy. If the stored energy in the rotor is controlled, so is the speed. The logical feedback is then the error in stored energy. This may be measured from the changes in output and input energy.

During any transient, the delays in the input energy system are such that the input power varies by only small amounts. For instance, on the Peace River studies, the most severe transient produced swings of input power of only about two percent. On the other hand, output power would swing almost 100 percent. For transient performance then, the input power variations could be ignored and change in output energy would be a reasonably close measurement of change in rotor energy.

The method of deriving this signal is shown in the block diagram, Fig. 2. A watt transducer is used to measure output power. Operational amplifiers are used to convert this to the final signal. The power signal is fed to a time constant circuit, the *reference*, whose output will be the power level under steady-state conditions. The instantaneous power signal is subtracted from the reference value to produce a signal which is the negative of the instantaneous power error from pretransient conditions. This signal is then approximately time-integrated by means of a time constant circuit to produce a result proportional to the error in energy. This resultant signal is the main stability signal used to control the excitation.

An additional component can be added to the feedback signal by adding in a portion of the power error signal. This permits control of the phasing of the signal to compensate for delays in the watt transducer and exciter.

The stability signal is actually added to the error signal from the voltage regulator and the resultant controls the exciter. The stability signal is strong enough to override the voltage signal during a transient. It will be noted that, except during the fault duration, the stability and voltage signals are in opposition, as the voltage signal will always resist the efforts of the stability signal to change the terminal voltage from the set value.

In the steady-state condition, the instantaneous power signal and the reference are of equal value so the stability signal is zero and control reverts to the voltage regulator.

A limiting circuit is applied to the stability signal so that under no circumstances can the stability signal cause undesirably large voltage excursions. In the field tests, this limit was set at a value equivalent to a 12 percent voltage error.

The quantities K_2 and K_1 in the figures represent multipliers (or gains). The numerical values given to K_1 are the gains from generator per unit power to per unit terminal voltage. Thus, for K_1 equal to 6.0, an energy error of 0.01 per unit·second would produce a signal to the exciter numerically equal to that produced by the voltage regulator for a terminal voltage error of 0.06 per unit. The value of K_2 is a pure number.

A limited number of tests were made as it was not desirable to deliberately fault the system more than absolutely necessary. No attempt could be made to optimize all the factors in the feedback. The reference time constant T_1 was chosen to be 10 seconds. Based upon the observed frequency of oscillation, it was considered that integration would be obtained closely enough by using a time constant circuit T_2 in place of true integration, choosing T_2 to be 1 second. This had the advantage of limiting any build-up due to stray pick-up and noise. Neither of these two time constants was varied during the tests, although studies made since have indicated that these would not be the optimum values.

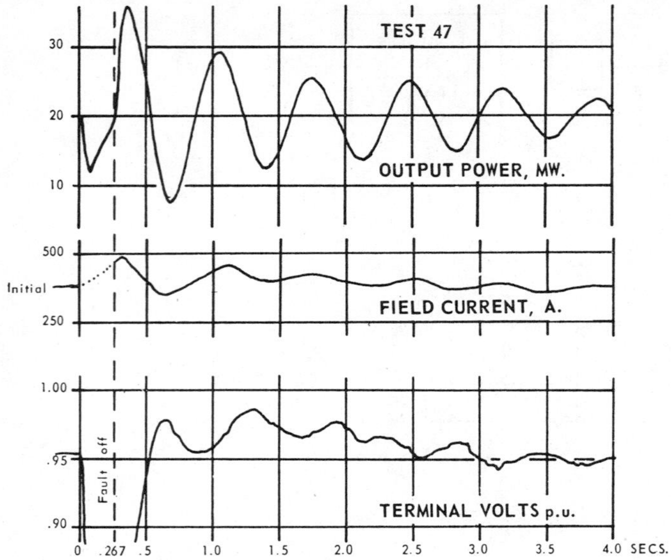

Fig. 3. Field test result with rotating exciter and magnetic amplifier regulator. Generator output 20 MW.

The two gains were varied somewhat, although optimization could not be attempted. Control theory has since been applied to the optimization of the signal parameters and this is described generally in the Appendix.

Discussion of Test Results

To provide a basis of comparison, one test was made with the rotating exciter, Fig. 3. The system is naturally damped, but the oscillations take about 10 seconds to disappear.

With the static exciter and feedback, Fig. 4 shows the transient to be essentially controlled in less than 1.0 second after the fault is removed.

The extremely high speed of the static excitation can be seen in its movement from positive to negative ceiling in less than 0.1 second. This forces the field current to change rapidly, yet the actual current values reached are not excessive and in the short time involved would not cause overheating of the field windings.

The voltage swings are relatively minor and, in fact, are little worse than with the rotating exciter. The major difference is in the phasing of the voltage swings which now occur at the right time to produce the desired system damping.

Records were not taken at Powell River as this would have been difficult to co-ordinate. However, the subsequent computer simulation indicated that the small oscillations persisting after 1.5 seconds were attributable to oscillations of the Powell River machines which took some time to damp out.

In this test K_1 was 6 and K_2 0.08. Fig. 5 shows the results for K_1 of 6 and K_2 of 0.04. The damping is not quite as good. However, it should be noted that the decrease in K_2 causes not only a change in phasing but also a decrease in signal strength. The phasing of K_2 at 0.04 could possibly be superior to that of K_2 at 0.08 if the value of K_1 were adjusted to give the same overall gain. Careful analysis must be made in optimizing the value of K_2.

Fig. 6 shows results for K_1 of 12 and K_2 of 0.04. This is not quite as good a result as that of Fig. 5. However, improvement could hardly be expected as K_1 of 6 already makes full use of the ceiling values of excitation and the higher gain can only marginally affect the timing. Evidently, forcing too far beyond the ceiling capability actually spoils the timing somewhat.

Fig. 7 is comparable to Fig. 5 except that the steady-state generation is 30 MW instead of 20 MW. As might be expected, the transient is more difficult to damp because of the ceiling on the exciter voltage. Similarly, the 10-MW case of Fig. 8 is easier to control.

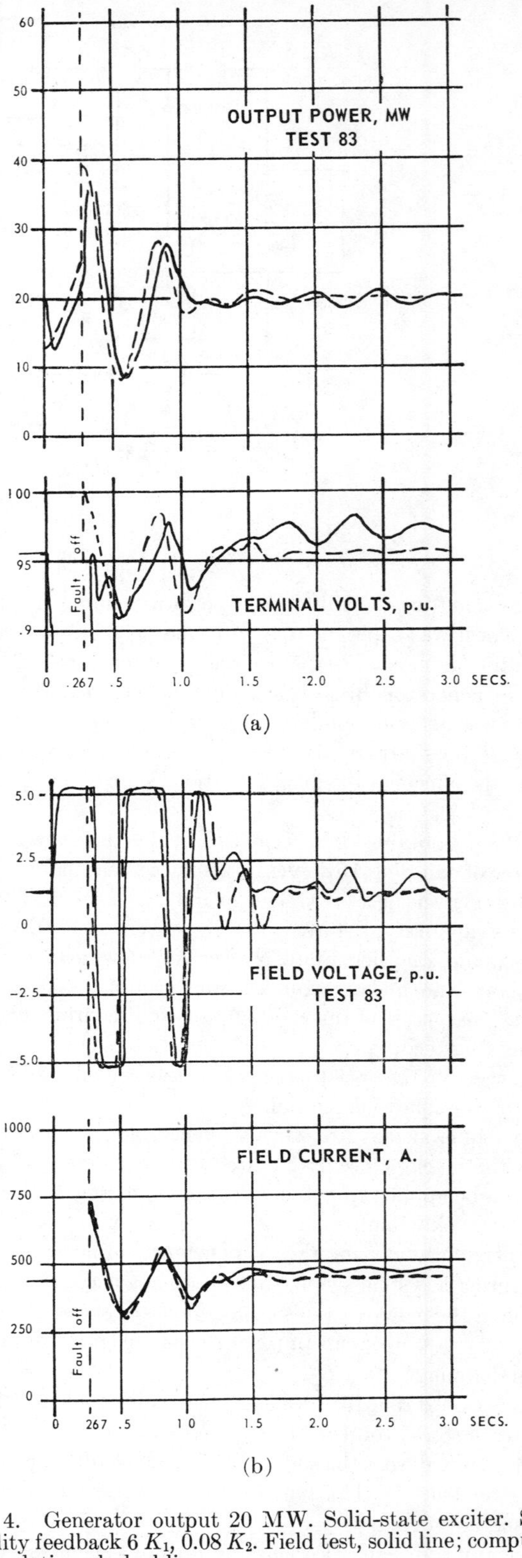

Fig. 4. Generator output 20 MW. Solid-state exciter. Stability feedback 6 K_1, 0.08 K_2. Field test, solid line; computer simulation, dashed line.

Generally speaking, the gains used in the tests were unfortunately all too high to provide sufficient variation in the results. It would have been easier to compare results and draw conclusions if the excitation was not being forced to ceiling so hard. Nevertheless, the purposes of the field tests were fulfilled.

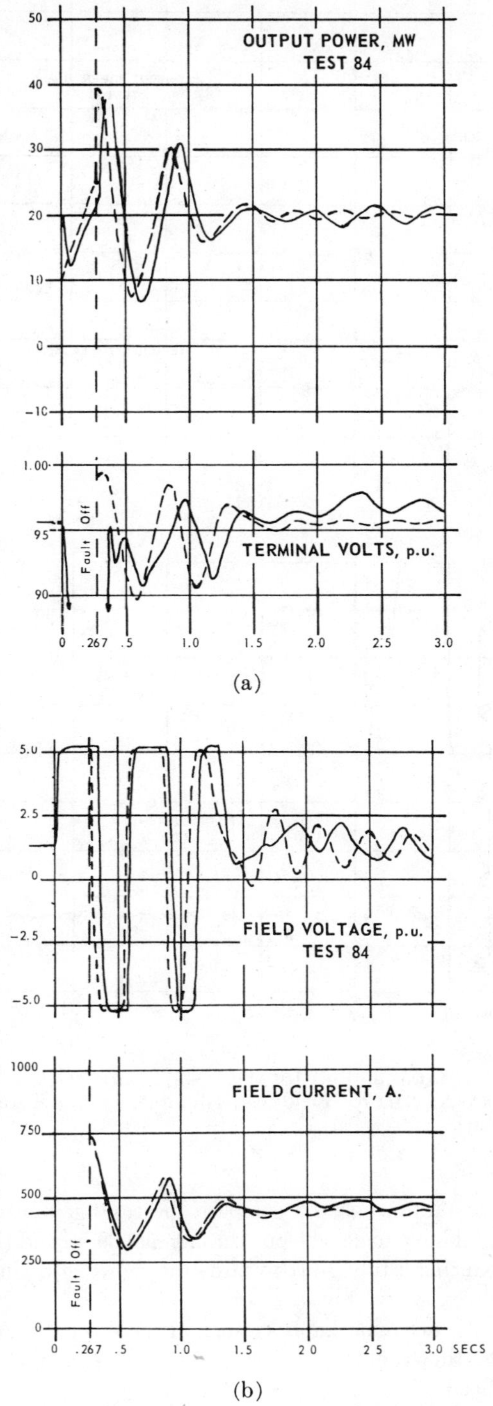

Fig. 5. Generator output 20 MW. Solid-state exciter. Stability feedback 6 K_1, 0.04 K_2. Field test, solid line; computer simulation, dashed line.

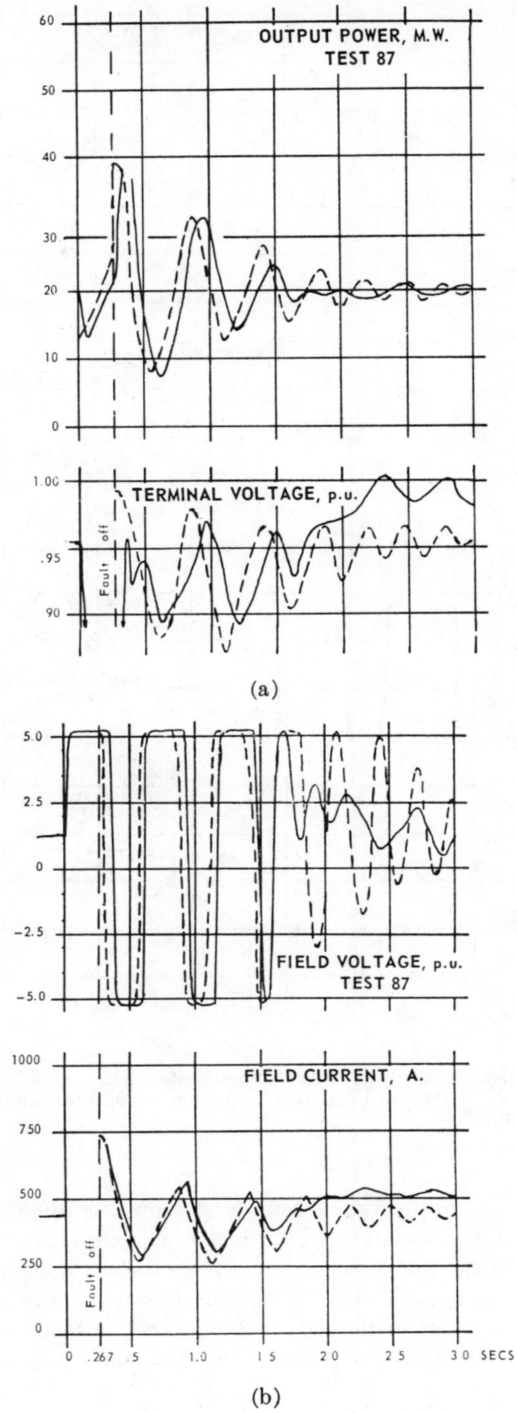

Fig. 6. Generator output 20 MW. Solid-state exciter. Stability feedback 12 K_1, 0.04 K_2. Field test, solid line; computer simulation, dashed line.

Incidentally, the discontinuities in some of the power traces were due to their going off the edge of the recording paper. Fortunately, this does not seriously affect the usefulness of the results.

In each of the tests, the terminal voltage is elevated around the 2.5-second mark. This was due to an offset which developed in the stability signal. This decayed as the integrating time constant took effect. It was evidently due to partial saturation or nonlinearity of the operational amplifiers. While it is not a serious effect, this would not appear in a production model as it is easily designed out by the use of better circuits and amplifiers.

Computer Simulation

The tests shown in Figs. 4–8 inclusive were simulated by means of the Westinghouse Electric Corporation's Stability Program on the IBM 7094 digital computer at East Pittsburgh, Pa.

The transmission system was represented as shown in Fig. 1, with the addition of line charging, transformer off-nominal taps, and estimates of the smaller loads.

Clowhom was fully represented with direct and quadrature axes, internal damping, saturation, voltage regulator with stability feedback, exciter, and governor action.

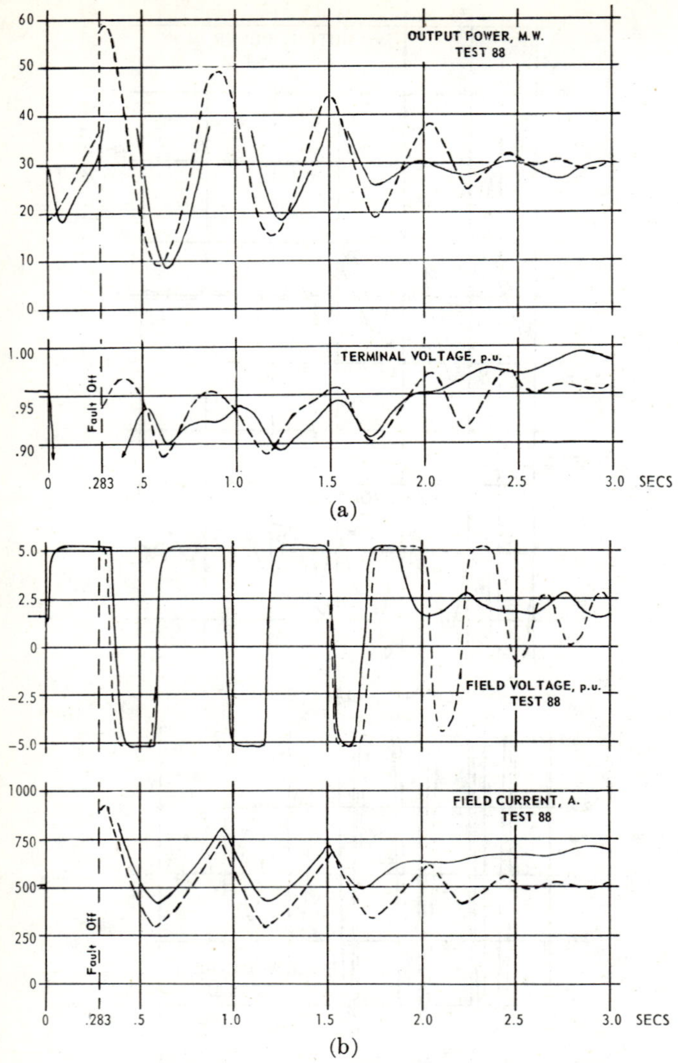

Fig. 7. Generator output 30 MW. Solid-state exciter. Stability feedback 6 K_1, 0.04 K_2. Field test, solid line; computer simulation, dashed line.

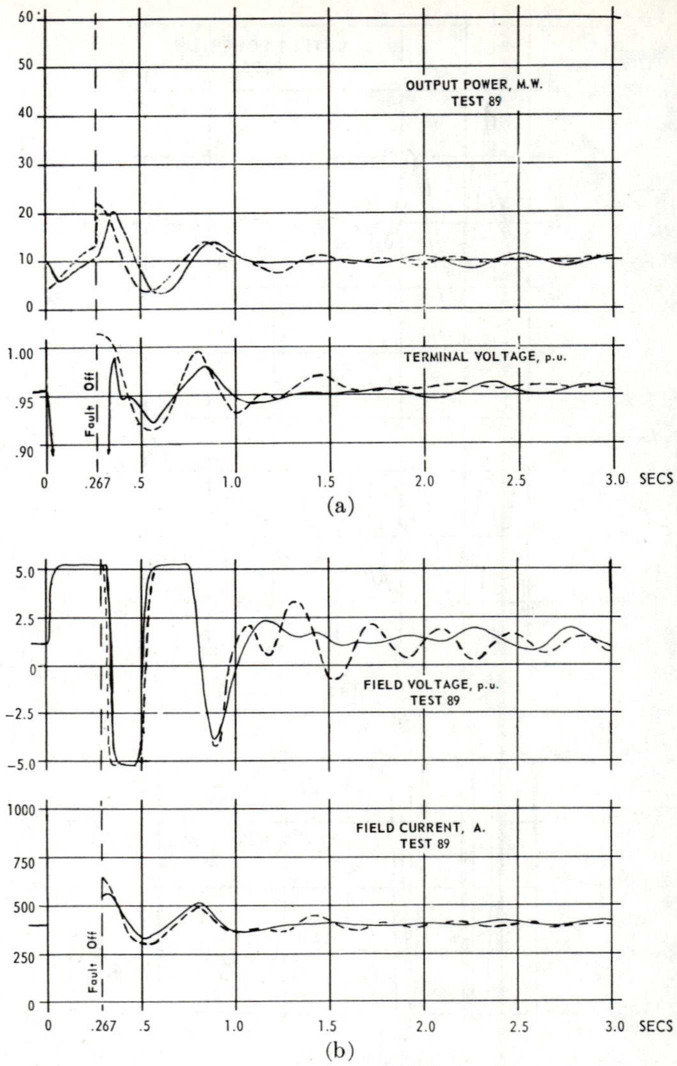

Fig. 8. Generator output 10 MW. Solid-state exciter. Stability feedback 6 K_1, 0.04 K_2. Field test, solid line; computer simulation, dashed line.

The computer program is capable of simulating the effect of terminal voltage variations on the exciter capability. However, input limitations would then prevent the inclusion of other more important factors. The voltage variation has comparatively little effect on the exciter performance except during the fault and even then the overall effect is minor. Constant supply voltage was assumed.

Cheakamus was represented as fully as Clowhom except that internal damping was omitted. As Cheakamus swings only slightly with respect to the infinite bus, damping was unimportant.

Powell River motors were represented as an equivalent generator with negative power input. This included 2-axis representation, no saturation, fixed excitation, and fixed shaft power. The additional 7-MW load at Powell, while principally induction motors, was represented as fixed impedance load, as also were all other minor loads.

The digital program uses a step-by-step procedure and the calculation interval was chosen as 0.02 second. Results were printed out every 0.1 second. Hence, any variations occurring in less than 0.1 second do not show in the curves.

The transducer used in the field tests to record instantaneous power had a time constant of 0.09 second. The computer values are truly instantaneous. To compare the results, the computer values would have to also be put through a 0.09-second time constant, producing a time shift towards the right and reducing the peaks slightly.

The results are remarkably close, and there is every reason to believe that they could be even closer if the following factors could be introduced:

1) Closer knowledge of the actual load values instead of estimates.

2) Representation of loads, particularly at Powell River, as nonconforming and/or as induction motors. This would require a detailed knowledge of the load characteristics, which was not readily available.

3) Adjustment of the damping factor D on each of the machines. This factor allows for such things as mechanical turbine damping and load damping due to variations in frequency and voltage.

4) Representation of the variations in excitation supply voltage due to the excitation being supplied from the generator terminals.

5) A representation of the feedback circuit which would incorporate the limitations in the equipment used, i.e., amplifier nonlinearity and noise pick-up. This is clearly impractical.

Conclusions

1) The field tests demonstrated that a practical stabilizing feedback has been devised using simple, easily maintained, electrical devices.

2) The digital computer simulation techniques are entirely adequate not only for system studies, but also to further study and optimize the stability feedback design.

Appendix

Using the symbols indicated in Fig. 2 and

I_f field current

s Laplace operator

the feedback circuit can be represented as a forward transfer function if the voltage error signal is ignored:

$$I_f(s) = \frac{1}{1+sT_t} K_1 \left(\frac{1}{1+sT_1} - 1\right)\left(K_2 + \frac{1}{1+sT_2}\right) \times \frac{1}{1+sT_e}\frac{1}{1+sT'_{do}} P_o(s)$$

$$= -\frac{s(1+s\,K_2T_2/(1+K_2))\,K_1\,(1+K_2)T_1}{(1+sT_t)\,(1+sT_1)\,(1+sT_2)\,(1+sT_e)\,(1+sT'_{do})} \times P_o(s).$$

The overall gain necessary depends on the inertia constant of the generator, the ceiling voltage of the exciter, and a number of other factors associated with the transmisssion system. The time constants to use can be derived approximately (and they are not critical) provided the period of oscillation of the generator is known.

It is advantageous to the normal operation of the generator that the time constants T_1 and T_2 be as small as possible, so that the signal returns to zero as quickly as possible after a change in load.

Consider the transfer function of a time constant circuit $1/(1+sT)$. If T is very large, the transfer function becomes $1/sT$, pure integration $(1/s)$ with low gain $(1/T)$. If T is large compared to the period of oscillation, the output will be almost constant at the average value of the input. As T is reduced, the function becomes less close to pure integration and the gain increases.

For the integration circuit, T_2 must not be made too small, otherwise unsatisfactory integration is obtained. For the reference, T_1 should be fairly large. However, if it is reduced, the effect will be to reduce the power error signal (which can be corrected by increasing the value of K_1) and to change its phasing (which can be corrected by using K_2).

In studies for the Peace River system it has been found that T_1 can be made as small as one half the period of oscillation and T_2 about 1.5 times the period.

If T_1 and T_2 have been selected, the optimum value of K_2 can be calculated. The response of a transfer function to a sinusoidal forcing function can be calculated by substituting $j\omega$ for s in the transfer function, where ω is the frequency of oscillation in rad/s. The transfer function then becomes

$$I_f(j\omega) = \frac{-j\omega\,(1+j\omega K_2\,T_2/\,(1+K_2))\,K_1\,(1+K_2)\,T_1}{(1+j\omega T_t)\,(1+j\omega T_1)\,(1+j\omega T_2)\,(1+j\omega T_e)\,(1+j\omega T'_{do})} \times P_o(j\omega).$$

The angle of lag of the vector I_f behind the vector P_o is then calculated as in normal vectorial analysis to be

$$\arctan(\omega T_t) + \arctan(\omega T_1) + \arctan(\omega T_2) + \arctan(\omega T_e) + \arctan(\omega T_{do}') + 90° - \arctan\frac{(\omega K_2T_2)}{(1+K_2)}.$$

The instantaneous power output of the generator is directly proportional to the internal voltage, that is, to the field current. Immediately after the transient, the power output must be increased to provide the greatest deceleration to the generator rotor. Ideally, the field current should be exactly in phase with the power output. We can derive

$$\arctan\frac{(\omega K_2T_2)}{(1+K_2)} = \arctan(\omega T_t) + \arctan(\omega T_1) + \arctan(\omega T_2) + \arctan(\omega T_e) + \arctan(\omega T_{do}') + 90°.$$

From this equation, the optimum value of K_2 can be readily obtained. It is not necessary that the phasing be exact. This is fortunate, in that the period of oscillation is partly dependent on the transmission system and may, therefore, vary over a small range because of normal switching operations.

References

[1] H. M. Ellis, J. E. Hardy, A. L. Blythe, and J. W. Skooglund, "Dynamic stability of the Peace River transmission system," *IEEE Trans. Power Apparatus and Systems*, vol. PAS-85, pp. 586–600, June 1966.

[2] D. W. Olive, "New techniques for the calculation of dynamic stability," *Proc. PICA Conf.* (May 1965), pp. 212–229.

[3] E. W. Kimbark, *Power System Stability*. New York: Wiley, 1956.

[4] D. B. Breedon and R. W. Ferguson, "Fundamental equations for analogue studies of synchronous machines," *AIEE Trans. (Power Apparatus and Systems)*, vol. 75, pp. 297–306, June 1956.

[5] A. R. Scott, "Optimization of generator and excitation system parameters for Grand Rapids development," *Trans. Canad. Elect. Assoc.*, vol. 2, pt. 1, Paper 63-A-27, 1963.

[6] H. E. Lokay and J. W. Skooglund, "Power system stability—digital analysis showing effect of generator representation, types of voltage regulators and speed governor systems," *Proc. PICA Conf.* (May 1965), pp. 230–241.

[7] Easton, Fitzpatrick, and Parton, "The performance of continuously acting regulators with additional rotor angle control," *CIGRE*, Paris, France, Paper 309, 1960.

[8] Grabbe, Ramo, and Wooldridge, *Handbook of Automation, Computation and Control*, vol. 1. New York: Wiley, 1958.

Discussion

G. A. Jones and J. F. Luini (Pacific Gas and Electric Company, San Francisco, Calif.): It is always valuable to have material presented that compares the results of an analytic study and field tests. The fact that the investigation is concerned with a novel method of increasing system stability makes this presentation even more interesting.

The system considered is essentially a radial one, and it is worthwhile to note that when studying radial systems with respect to an infinite bus, more realistic results are obtained than if the system was a more interconnected one. Such results tend to be somewhat optimistic.

The tests were made with an adjacent line supply for the solid-state exciter, and the digital studies assumed a constant supply. If all disturbances are considered, the variations in supply greatly influence the performance of the system. How will the actual installation at the Peace River Development be supplied?

It is more appropriate to attribute the machine oscillations to the accelerating power rather than as the consequence of the machine's instantaneous speed with respect to the system. Similarly,

Manuscript received September 29, 1966.

it is control of the transfer of energy to and from the machine that produces the return to a steady-state condition.

If the change in machine speed is to be derived from the negative integral of the change in output power, some justification should be given. This is seen easily as follows.

Consider the kinetic energy KE of the rotor with respect to the powers and speed of the machine. The variables with 0 subscripts denote the steady-state or initial condition. The change in rotor KE is

$$\Delta KE = KE - KE_0 = I\omega^2 - I\omega_0^2 \qquad \omega = \omega_0 + \Delta\omega$$

$$= \int (\Delta P_{\text{in}} - \Delta P_{\text{out}})dt$$

$$= I(2\Delta\omega\,\omega_0 + \Delta\omega^2)$$

but

$$\Delta\omega^2 \ll 2\Delta\omega\omega_0, \text{ and } \Delta P_{\text{in}} \ll \Delta P_{\text{out}}$$

$$\therefore\ 2I\Delta\omega\omega_0 \cong -\int \Delta P_{\text{out}}dt$$

so

$$\Delta\omega \cong -\frac{1}{2I\omega_0}\int \Delta P_{\text{out}}dt.$$

Thus a signal derived from the negative integral of the change in output power is proportional to the change in machine speed.

It now remains to resolve the following problem. A signal is derived that is proportional to the change in speed and is added to the terminal voltage error as shown in Fig. 2. This then disagrees with the sign convention shown in Fig. 3 of Ellis *et al.*[1] To further confuse matters, such feedback is stated to be negative in the immediate paper. The addition of a signal *as shown* in Fig. 2 is positive feedback and is the correct choice. (For example, to stabilize an overspeeding machine, change the field current to reduce the mechanical energy stored, and increase the electrical energy output. This is achieved by increasing the field current. Thus, an increase in $\Delta\omega$ implies an increase in I_f and is consistent with Fig. 2.) A comment on this is surely in order.

Once a controller has been chosen, its parameters may be "optimized" with respect to various disturbances. However, this is somewhat superficial, since true optimization grows out of the development of a controller, or optimal policy that will optimize some predetermined criterion (e.g., finding the optimal policy to minimize the functional J, where $J = \int_{t_1}^{\infty} \Delta P_{\text{out}}^2 dt$, and t_1 is the time that the second cycle of power oscillation begins; such a policy would minimize output power oscillations following the first cycle). The use of a linear controller to drive the field voltage to limits can often cause loss of control as well as the suboptimal performance exhibited in the paper. In truth the optimal policy for the system is nonlinear.[9]

Results obtained by analytic optimization of a system with respect to an infinite bus should be used with care for they may not be satisfactory for situations which deviate from the radial representation. This is emphasized by Ewart and deMello[10] in their discussion of a technique for analyzing synchronous machine capabilities.

Perhaps the authors would explain how one might "optimize" a more interconnected system with no infinite bus representation. Such systems have the characteristic that the optimum performance of the machines as a system is not identical to the optimum performances of the individual machines with respect to the system. This can be related to the problem of determining the period of oscillation of a machine when it is influenced by the performance of the excitation system which in turn is "optimized" by knowledge of the period of oscillation of the machine. A lengthy analysis could result for a system with many machines.

It would be interesting to know how the single line to ground fault was simulated on the digital computer representation.

References

[9] G. A. Jones, "Transient stability of a synchronous generator under conditions of bang-bang excitation scheduling," *IEEE Trans. Power Apparatus and Systems*, vol. PAS-84, pp. 114–121, February 1965.

[10] D. N. Ewart and F. P. deMello, "A digital computer program for the automatic determination of dynamic stability limits," *IEEE Trans. Power Apparatus and Systems*, vol. PAS-86, pp. 867–875, July 1967.

Manuscript received September 6, 1966.

R. W. Haywood (Atomic Energy of Canada Ltd., Manitoba Hydro, Winnipeg, Canada): The paper describes an ingenious method of obtaining a stabilizing signal at a minimum cost. It appears that the signal and its application as described in the paper would perform well for this type of case, but might be detrimental to system performance in other circumstances.

Sagging of the system frequency due to loss of a large block of generation is a condition which might warrant further detailed study. What would the excitation system do under this condition? It would appear that the tendency would be for the exciter to go to full boost, while the machine was slowing down, even though it was slowing down in synchronism with the rest of the system. If the wicket gates were not at the full open position, then the shaft power would be increasing, thus adding to the rate of power increase seen by the transducer. Under these conditions would the exciter go to full boost and remain there during the protracted period of system deceleration, thus overheating the field windings and causing loss of the machine when it is most needed on the system, or are the time constants in the reference circuit such that the exciter would not go to full boost with a rate of frequency decline that would allow the system to remain stable?

Analysis of this would likely require a digital computer program that treats inductances and capacitances as such, with their impedances being dependent on frequency, and also permits good prime mover generator and load representations over long periods of changing conditions.

Manuscript received September 6, 1966.

R. M. Shier and **A. L. Blythe**: We would like to thank the discussers for their interest and contributions to the paper. We will reply to Mr. Jones and Mr. Luini first.

We agree that it is easier to obtain realistic results on a radial system connected to an infinite bus, but we would ascribe this to the smaller amount of data necessary and the greater care that can be taken to assure the accuracy of these data.

The variation in supply voltage to the excitation does not necessarily have much effect on performance. If the ac supply voltage to the thyristors is sufficiently above the controlled ceiling voltage of the exciter, a considerable variation in ac voltage will not affect the ceiling capability. Hence, except for the brief period of fault duration, the exciter performance may be unaffected.

On the Peace River system, the exciter will be fed from transformers connected to the generator terminals, but the design is such that ceiling voltage can be produced with only 85 percent normal voltage available. Computer simulations indicate that system voltage variations affect the first swing by only a degree or two of generator angle on 6-cycle 3-phase fault at Peace 500-kV.

The justification for the signal can be based entirely on energy, as stated in the paper. The derivation in terms of speed was used for comparison with earlier studies but we did not feel it was necessary to include this derivation in the paper.

The confusion in signs is noted. The sign convention of Fig. 3 of Ellis *et al.*[1] is based on the terminal voltage producing a positive signal when it is above the set value. Hence, a positive signal would force voltage down. With this convention, the speed signal must be subtracted. In this paper, we avoided the sign convention problem by showing the derivation of the voltage signal. This gives the opposite sign to that of the Ellis *et al.* and hence the stabilizing signal is added.

When the error in a variable is measured, manipulated, and fed back in such manner as to reduce the error, we have negative feedback. The fact that it may control an intermediate variable in a positive manner does not make it positive feedback. The energy error (or speed error) is fed back to reduce the error in energy (or speed) and the direction in which excitation is changed is only incidental.

As demonstrated in the paper, the ceiling voltage of the exciter creates a nonlinearity which places a limit on the gain which can be used. In our opinion, the optimum gain is that which, on the occurrence of the most severe system disturbances, causes the excitation

Manuscript received September 6, 1966.

voltage to barely touch ceiling on the second positive peak. With this value of gain selected, the phase angle of the signal can be adjusted to its best value.

It is agreed that there can be no true optimization because the normal variations in an interconnected system as to loads, generators connected, lines open, etc., will cause variations in the period of oscillation and in the effectiveness of the excitation in transiently controlling output power. However, it is possible to choose settings which will provide the best control over a range of foreseeable system conditions and it is in this sense that the term "optimizing" is used in the paper. For a large system this can only be done by a number of computer simulations, using the general appreciation obtained from simpler studies to guide one's efforts.

With respect to the simulation of line to ground faults, the computer program simulates only the positive-sequence network, but permits the inclusion of a fault impedance. This fault impedance was made equal to the sum of the negative and zero-sequence driving point impedances at the fault point, thus giving the correct positive sequence values during the fault.

With respect to Mr. Haywood's discussion, we would point out that the stabilizing feedback acts in such manner as to resist all changes in output power. Under the situation of a loss of a large block of generation, the particular generator would see a large increase in output power. The feedback would reduce excitation to reject this increase. The result would be a drop in system voltage with a corresponding reduction in load. The long term effects would depend upon the response of governors over the whole system.

The paper describes only the tests made at Clowhom and the computer verifications, but additional studies have been carried out on the computer and a modification added to the feedback.

When a scheduled change in loading is made, the feedback resists the change in output power. With increasing load, this would cause decreasing excitation and the generator will move to a higher power angle in order to output the power. Undesirable var flows and possible instability are a result.

The governor manufacturers make provision for a gate-position potentiometer. We now use this potentiometer to give us a signal roughly equal to input power. This is subtracted from the output power signal and the resultant input to the feedback circuit in place of the output power signal. The "reference" now becomes equal to only the generator losses and measurement errors.

This modification has no effect on transient performance as the water hammer effect is greater than the gate movement. It does completely overcome the effect of load changes.

In the problem raised by Mr. Haywood, it would also prevent any further decrease in voltage as a result of governor action.

Excitation Control to Improve Powerline Stability

FERBER R. SCHLEIF, SENIOR MEMBER, IEEE, HARVEY D. HUNKINS, SENIOR MEMBER, IEEE, GLENN E. MARTIN, MEMBER, IEEE, AND ELLIS E. HATTAN

***Abstract*—The influence of generator voltage regulators upon powerline stability is closely scrutinized by analog techniques. Contrary to common concepts, it is shown that high-speed control of excitation to maintain constant terminal voltage actually hampers rather than aids damping and powerline stability. It is further shown that damping and stability are greatly improved by supplementing excitation control with a specially derived function of frequency deviation. Results of the studies are verified in actual performance obtained in field tests.**

Introduction

THE comparatively recent programing of digital computers for transient stability studies of very large power system interconnections has led to a number of interesting developments, one the subject of this paper. With the computer capacity to include the detail of exciters and voltage regulator characteristics and to detail system behavior over a period of many seconds, stability curves began to show divergence with machine groups pulling out of step after several oscillations rather than on the first swing after a disturbance.[1]–[3] In short, insufficient damping often became evident. Both the influence of voltage regulators or inadequate representation of natural system damping characteristics are possible contributors. This paper is concerned only with the former.

During 1965 studies of oscillations of about six cycles per minute experienced between the Pacific Northwest and Southwest over a very long 230-kV tie, it was found that special supplementary signals derived from tieline power or frequency differences could be applied to selected machine governors to produce damping torques.[4],[5] These studies also showed that, when further tielines were built, the higher frequencies of oscillation expected could exceed the frequency response range of special governor control, especially for those frequencies of oscillations associated with dynamic stability following severe disturbances.

Results of dynamic stability studies on Peace River[1] showed that a speed error signal applied to machines with static exciters would produce damping. The use of special signals derived instead from bus frequency deviations to control voltage regulators of the rotating amplifier type is the principal subject of this paper.

Since system-wide stability studies for up to ten seconds real time are complex and costly of computer time, exploration of the potential of special control was made on a simplified system basis. The supplementary control under consideration had not been employed before and since it could affect a large number of machines of the Corps of Engineers and the Bureau of Reclamation, significant study was pursued in those organizations. This paper describes the studies made by the Bureau of Reclamation with consultation and check runs by the Corps of Engineers and an actual field test confirming results of the study. Both the analog study and system test involved the Bureau of Reclamation's Glen Canyon Powerplant near the Arizona–Utah border.

Paper 31 TP 67-51, recommended and approved by the Power System Engineering Committee of the IEEE Power Group for presentation at the IEEE Winter Power Meeting, New York, N. Y., January 29–February 3, 1967. Manuscripts submitted October 26, 1966; made available for printing January 10, 1968.

The authors are with the U. S. Bureau of Reclamation, Denver, Colo. 80225.

Procedure

As a first step, the Glen Canyon generators and excitation systems were simulated on the analog computer. Response of the generator voltage to sudden changes of reactive load was studied with the regulator 1) as originally designed, 2) supplemented with a dv/dt signal without other change of the excitation system, and 3) with the supplementary dv/dt signal and the original internal stabilizing circuit optimized. This performance was studied as essentially isolated from the system. Then the simulation was expanded to include representation of an equivalent tieline to a remote infinite bus in order to evaluate the influence of the excitation and control system on tieline stability and damping. A block diagram of the excitation system and simplified representation of the generator and tieline are shown in Fig. 1.

Isolated Generator and Excitation System

Performance of the regulator, exciter, and generator without the tieline is shown in Fig. 2. It should be noted that:

a) In the original condition, response is satisfactorily damped. Response varies with the size of step and direction owing to the regulator limits and difference in forcing range in the buck and boost directions, Fig. 2(a) and (b).

b) When a supplementary dv/dt signal is added that is not optimized, the response is faster but is underdamped, Fig. 2(c).

c) Fast, well-damped response can be obtained by shortening the time constant of the stabilizing circuit and optimizing the stabilizer and dv/dt gains, Fig. 2(d).

d) A 0.05-per-unit step of voltage is sufficient to sacrifice an appreciable part of the initial derivator excursion against the regulator limit and the subsequent recovery is retarded by derivator action rather than accelerated. Fig. 2(e) shows the result for a 0.1-per-unit step.

Regulator, Generator, and Tieline to Infinite Bus

The representation employed for the generator and tieline shown in Fig. 1 is considerably simplified, a procedure required to come within capacity of the computer available at the time of the study. Nevertheless, the performance of the analog proved satisfactory in the most important respects. A close check was obtained between results of this analog and those of another incorporating full detail.

Tieline power was represented as linearly rather than sinusoidally related to angle, this quantity being further modified in proportion to terminal voltage to constitute the relation $P - KE_TE_S/X_S$. The performance thus may be considered sufficiently representative of small swings around any chosen operating point to permit comparison of the effect of various control measures on stability and damping. The total generator reactance is repre-

Reprinted from *IEEE Trans. Power App. Syst.*, June 1968.

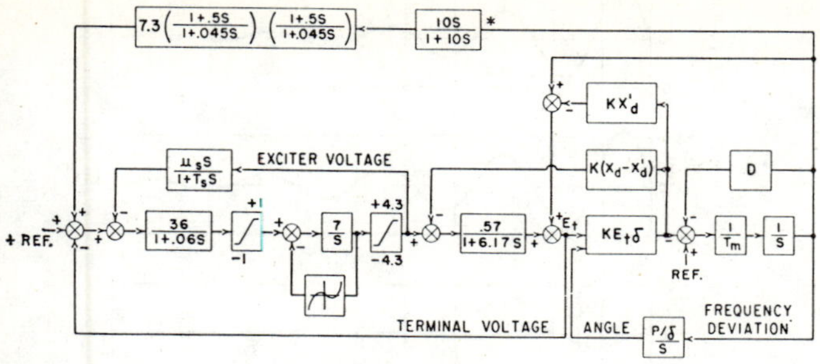

Fig. 1. Block diagram of Glen Canyon generator, excitation system, supplementary control, and equivalent tieline to infinite bus. Asterisk indicates function that may be used for reset of frequency signal (not used in results shown).

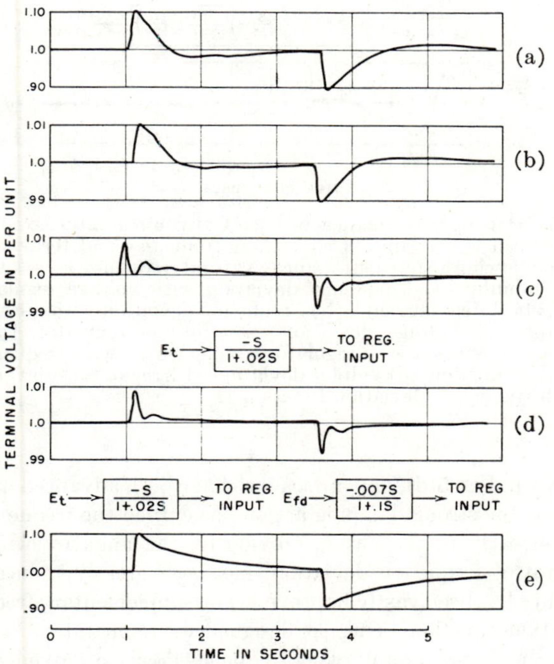

Fig. 2. Analog performance for Glen Cayon regulator, exciter, and generator without tieline in controlling terminal voltage. (a) Normal regulator response to 0.1-p.u. voltage step. (b) Normal regulator response to 0.01-p.u. voltage step. (c) Regulator and *dv/dt* supplementary control with constants *not* optimized; 0.01-p.u. voltage step. (d) Regulator and *dv/dt* supplementary control optimized (stabilizer function altered as shown); 0.01-p.u. voltage step. (e) Regulator and *dv/dt* supplementary control optimized as for (d); 0.10-p.u. voltage step.

sented as X_d, of which the X_d–X'_d portion is influenced by the open circuit time constant T'_{d0}.

Since the power factor angle is not represented as such in the simplified simulation used in this study, the reactance drops subtract directly from internal voltage rather than vectorially. Thus, the inclusion of the full amount of the IX_d drop would closely simulate the effect of zero power factor load. To simulate the effect of unity power factor load, a coefficient of 0.34 is used with the reactance of the Glen Canyon generators.

$$(\sqrt{1 + X_d^2} - 1)/X_d = 0.34.$$

The equivalence of the simplified simulation to one detailing the respective internal reactances in the direct and quadrature axes may be explained as follows. In Fig. 3, solid lines show the quantities represented in the simplified simulation. In dashed lines, the diagram is completed to show the classical phasor dia-

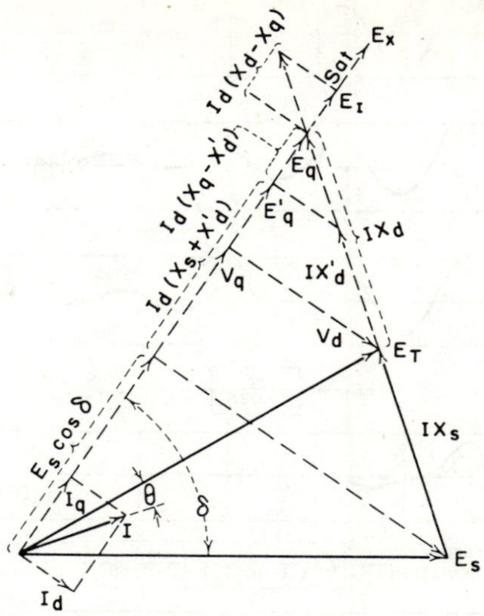

Fig. 3. Phasor diagram of synchronous machine and tieline to system. I—load current; I_d—direct axis component of load current; I_q—quadrature axis component of load current; X_s—system reactance; X_d—generator synchronous reactance; X'_d—generator transient reactance; X_q—generator quadrature axis reactance; E—system internal voltage (at infinite bus); E_q—internal voltage behind X_d; E'_q—internal voltage behind X'_d; E_T—terminal voltage; θ—power factor angle; δ—angle between internal voltages.

gram for the synchronous machine in which the internal reactances and voltage drops are identified. These internal voltage drops serve to relate terminal voltage E_T to the internal voltages E_I and E_q.

To represent the entire power/angle curve the voltages E_q, E_s, and the angle δ must be considered. Power transfer from E_q to E_s is affected by the magnitude of E_q and the angle δ (assuming E_s, the internal voltage of the system or infinite bus, to remain constant). Power transfer from E_T to E_s is similarly affected by the magnitude of E_T. For small deviations around the angle δ, E_T, and E_q respond similarly to changes in excitation. Hence, the simplified simulation is sufficient for examining the influence of various excitation controls upon E_T and thus upon the damping of small swings. This simulation would not be adequate for determination of stability limit, but that is not necessary to the examination of damping characteristics.

Representation of Swing Frequency and Natural Damping

The principal natural swing frequency used in these studies of the effectiveness of supplemental controls for improving damping for the Glen Canyon generators was 50 cycles per minute (0.83 Hz). This was achieved by adjusting tieline reactance. This swing frequency is approximately that of the loaded Glen Canyon generators operating over their transmission lines. It compares reasonably well with the natural frequency indicated in a recent system study where the frequency had not been appreciably shifted by the supplemental controls. However, checks are included of the effect of various supplemental controls over a wide range of swing frequencies.

A fixed amount of natural damping, exclusive of the influence of the excitation system, is represented in the analog. This natural damping factor has been held constant in all studies so that the effects of various controls can be readily compared. Although the amount of natural damping is somewhat immaterial when used in this manner, the damping factor of 6.9 was selected to make results compare with observed performance. (Subsequent analysis of tests made at Glen Canyon in October 1965 indicates the damping factor for the amortisseurs to be 7.3 at 50 cycles per minute.)

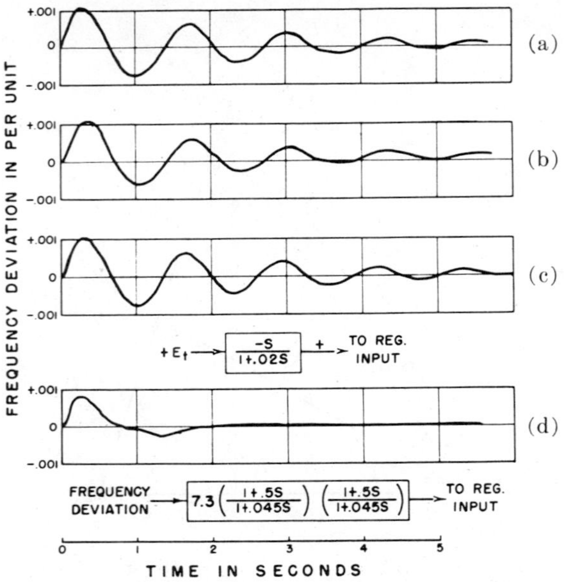

Fig. 4. Analog performance for Glen Canyon regulator, exciter, and generator with tieline. Natural swing frequency of system F_n = 50 cycles per minute. (a) With normal regulator. (b) Voltage regulator off (constant field current). (c) Regulator with dv/dt supplementary control. (d) Regulator with frequency deviation function.

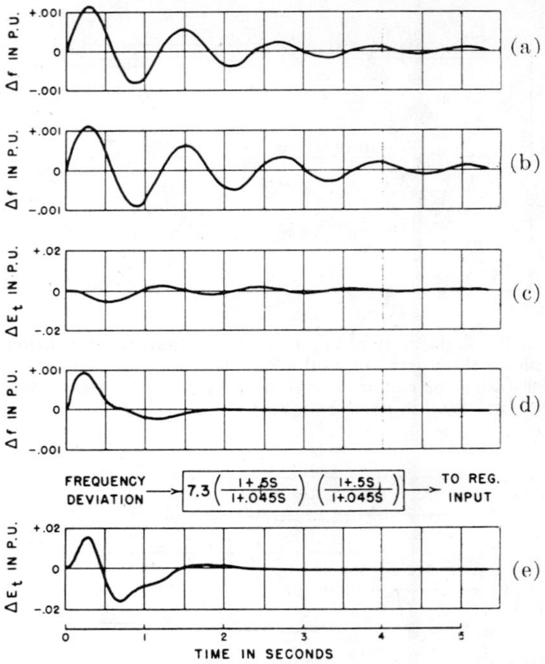

Fig. 5. Analog performance for Glen Canyon regulator, exciter, and generator with tieline in controlling frequency and terminal voltage following a step disturbance. Natural frequency F_n = 50 cycles per minute. (a) Frequency deviation with voltage regulator off (constant field current). (b) Frequency deviation with normal regulator. (c) Voltage deviation with normal regulator. (d) Frequency deviation with regulator supplemented with frequency deviation function. (e) Voltage deviation with regulator supplemented with frequency deviation function.

Results

The most important comparisons of results of this study are shown in Fig. 4. The damping of system swings, recorded from the computer in terms of frequency deviation, is shown for the normal voltage regulator in Fig. 4(a); with the voltage regulator on manual control (constant field current) in Fig. 4(b); with excitation control supplemented with a dv/dt signal, Fig. 4(c); and finally with excitation control supplemented with the frequency deviation function developed in these studies, Fig. 4(d).

It will be seen that the damping rate with the normal voltage regulator is modest. With the voltage regulator on manual (field current constant) damping of system swings is appreciably stronger. (The reason for this will be discussed subsequently.) With excitation control supplemented with a dv/dt signal, damping is slightly better at certain swing frequencies than with the normal voltage regulator, but it is still not as strong as when no automatic voltage regulation is used. When excitation control is supplemented with a frequency deviation signal of the form indicated in Fig. 4(d), damping is very strong.

The action of the frequency deviation function for supplementary excitation control to achieve strong damping of system swings is shown in Fig. 5. It will be noted that with the normal voltage regulator, the voltage deviations shown in Fig. 5(c) lag the frequency deviations, Fig. 5(b), by essentially 90 degrees, corresponding closely with the angular swings (integral of frequency deviations) and power flow. With the frequency deviation function supplementing control of excitation, deviations of terminal voltage, Fig. 5(e), have been changed in sign and advanced by nearly 90 degrees where they are nearly in phase with the frequency deviations of Fig. 5(b). Damping has now become very strong, as shown by comparing the frequency deviations in Fig. 5(d) with those in Fig. 5(b).

This damping is as nearly optimum as it is practical to attempt. It should be noted that the terminal voltage deviation, Fig. 5(e), has been shifted to be in phase with the unmodified frequency deviation, Fig. 5(b). The maximum of the remaining frequency deviation, Fig. 5(d), only appears to have been advanced in phase because the strong damping action has curbed the frequency deviation early in the swing. Obviously, the angular swing, the integral of frequency deviation, has been greatly reduced and stability has been vastly improved. The supplementary frequency deviation function being so designed as to maintain response properly phased for any swing frequency down to zero has not induced noticeable secondary swings. It is recognized, however, that with saturation effects in the exciter and generator, the control system is nonlinear and phase correction cannot be perfect for all signal levels.

Fig. 5 also shows, by comparison of the voltage deviations in Fig. 5(e) and (c), why a supplementary signal of rate-of-change of voltage could not produce such a pronounced effect. It would have eliminated its own source of signal without yet having produced the voltage deviation in the opposite direction of which the frequency deviation function is capable.

Reset can be introduced into the frequency deviation function, as indicated in Fig. 1, if it should be considered necessary. This would avoid biasing terminal voltage with prolonged offset in frequency. However, this would sacrifice proper phasing at low swing frequencies and could lead to aggravating rather than suppressing slow swings. Since prolonged offset in frequency is uncommon and tends to be relieved by the voltage bias when it does occur, it is not conclusive that the reset is desirable.

Fig. 6 shows a comparison of performance of the simplified system for the unmodified excitation system, (a), for control of the excitation system supplemented with a dv/dt function, (b), and for control of the excitation system supplemented with the frequency deviation function derived by these studies, (c), as best suited to the characteristics of this particular machine and excitation system. It will be noted that with the frequency deviation

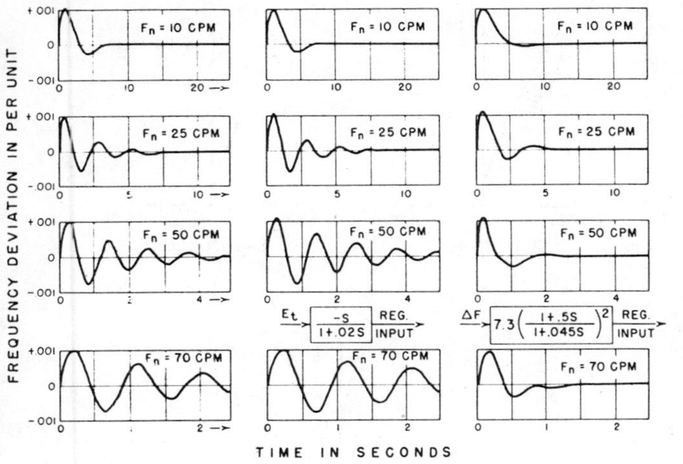

(a) (b) (c)

Fig. 6. Analog performance of Glen Canyon regulator, exciter, and generator with tieline producing different natural frequencies F_n in cycles per minute. Response shown is to step disturbance. (a) Normal voltage regulator. (b) Regulator with dv/dt supplementary control. (c) Regulator supplemented with frequeny deviation function.

function damping is made strong over a wide range of system swing frequencies.

These studies also included exploration of tieline power as a source of supplementary control. This was shown to have greater capabilities than the use of dv/dt. However, the capabilities were not as good as for the frequency deviation function shown and application would be more complex. Consequently, details of this method are not included here.

Discussion

Since it may not be immediately evident how the results of this study reconcile with former conceptions of how voltage regulator performance influences stability, some discussion of the influences seems appropriate.

The principal mechanism by which voltage control can influence damping of tieline swings may be illustrated by the phasor diagrams of Fig. 7. If the terminal voltage phasor E_T should be increased in magnitude during the time that the angle δ between terminal voltage and receiving end voltage E_r is increasing, power transfer across the line will be increased. Acceleration of local inertia will be diminished by the increase of power transferred from it across the tieline, and change of the angle δ will be opposed. Conversely, a reduction of terminal voltage (similarly internal voltage) while the angle δ is decreasing will oppose the decrease and the swing is damped. In the production of positive damping, the tip of phasor E_T will describe a counterclockwise spiral which converges. If the influence is opposite, the damping is said to be negative and the locus is a clockwise spiral. If there is no change in magnitude of terminal voltage, its influence upon damping is zero and there is no spiraling of the locus in the phasor diagram.

The principal source of natural damping is the retarding influence of the field time constant T'_{d0} upon the armature reaction. With no automatic regulation of terminal voltage as tieline power and angle increase, terminal voltage tends to decrease due to the combined effect of armature reaction and other impedance drops, Fig. 7(a). However, in the transient state terminal voltage is retarded in reaching the steady-state locus by the influence of the field time constant upon armature reaction. The result is that terminal voltage remains higher than the steady-state locus during increasing angle and lower than steady-state locus during recovery swing of angle. Thus, the locus of terminal voltage is a diminishing counterclockwise spiral which indicates positive damping.

If terminal voltage should be rigidly and instantaneously regulated to a constant value during swings of power and angle, Fig. 7(b), the locus would be a simple arc of a circle with no spiraling, either counterclockwise or clockwise, and there would be no contribution, either positive or negative, from voltage to damping. This is the principal explanation of the decrease of damping between Fig. 5(a) with no automatic voltage regulation and Fig. 5(b) with the normal voltage regulator in service.

Considering the inertia and tieline representation, an additional small but direct damping influence is the inherent tendency for the unregulated terminal voltage (and therefore tieline power and its decelerating torque) to vary directly with speed (frequency). Due to the function of integration by inertia upon accelerating torque to obtain frequency, the frequency deviation lags this torque by 90 degrees of the swing cycle. At the middle of the angular swing, frequency and frequency deviation would be at their maximum. The unregulated voltage would also be at its maximum. Therefore, the influence of this voltage component upon tieline power reaches its maximum, exerting maximum retarding influence at the center of the angular swing. Previous studies have demonstrated that this phase relation produces the maximum damping influence. However, high-speed voltage regulating action tends to cancel this beneficial influence.

In Fig. 7(c), by increasing terminal voltage above normal during increasing angle and below normal during decreasing angle, damping is made strong and the counterclockwise spiral diminishes rapidly to a steady-state point. To help clarify the influence of voltage upon damping as explained with Fig. 7, the phase plane plots in Fig. 8 between terminal voltage and tieline angle were recorded from the analog computer. They correspond directly with the loci of terminal voltage in the phasor diagrams of Fig. 7.

The foregoing is not necessarily in conflict with earlier concepts that stability is held or lost with the first swing and that stability is improved by action of the voltage regulator. Before voltage regulators and excitation ceilings reached the state of development presently available, there was sufficient lag of excitation behind voltage change to produce an effective damping influence. With the high speeds and high ceilings for field forcing presently available, the phase of response to terminal voltage can be sufficiently advanced to cancel the damping produced by the effect of the field time constant in retarding response of terminal voltage to load and to cancel the damping influence of speed upon voltage. These effects are worth serious consideration in connection with the tendency in recent studies, considering high-speed voltage control, to show loss of stability after a number of oscillations.

It may be noted that the improvement in damping through use of dv/dt is rather small compared with the improvement possible through use of the complex frequency feedback function proposed. This is largely a matter of proper phasing for the control signal, but it must also be noted that by using the frequency signal as a control source, terminal voltage may be biased above and below normal as best aids stability.

The foregoing should not be construed to indicate that high-speed excitation control is invariably disadvantageous. It merely suggests that very rigid control to maintain constant terminal voltage would not help stability as much as biasing that response as a function of frequency deviation.

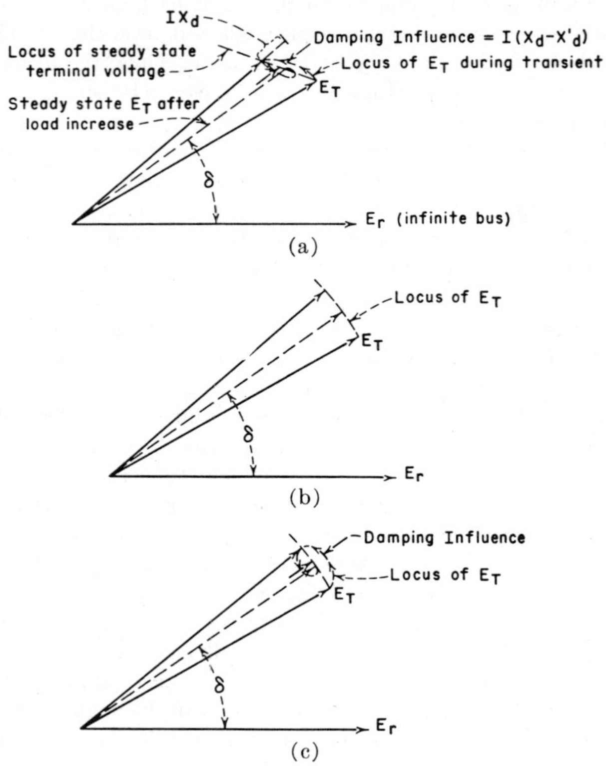

Fig. 7. Phasor diagrams showing influence of terminal voltage on damping for single machine on tieline to infinite bus. (a) Terminal voltage unregulated. (b) Terminal voltage regulated to constant value. (c) Terminal voltage biased by frequency deviation function.

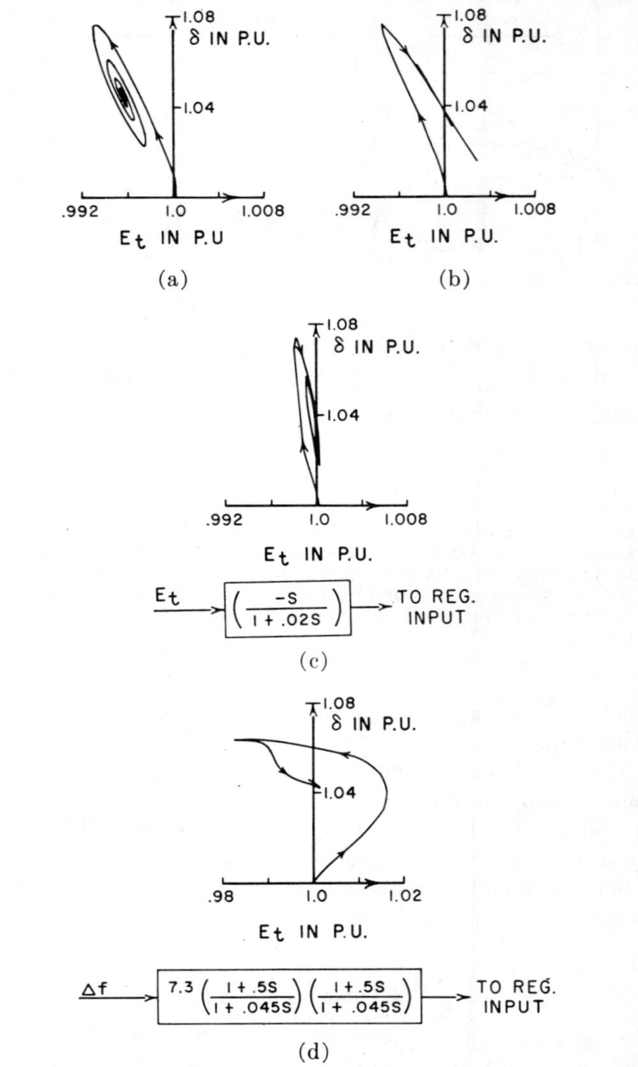

Fig. 8. Phase plane plots of angle δ vs. terminal voltage, both in per unit, for a system natural swing frequency of 50 cycles per minute and for an initial frequency deviation of 0.001 per unit. (a) With voltage regulator off (constant field current). (b) With normal voltage regulator. (c) With voltage regulator supplemented with *dv/dt* control. (d) With voltage regulator supplemented with frequency deviation function.

It is, of course, necessary to recognize that in producing voltage response as a function of frequency, the problem of overvoltage upon load rejection could be aggravated unless appropriate limits are employed. There appears to be adequate margin in which to insert fixed limits for this influence.

Verification by Field Tests

The study phase indicated considerable potential for producing damping torques but, to the authors' knowledge, no such control had ever been attempted with a rotating-amplifier-type excitation and control system. Therefore, field tests were considered worthwhile, not only to verify the study results but to explore the practicality of application. Arrangements were made to conduct the field tests at Glen Canyon Powerplant which had been analyzed in the studies.

A small analog computer was chosen as the principal tool. It was used to operate upon the input frequency deviation signal in the same manner as during the studies but with the actual machine and power system closing the control loop instead of the analog representation. In addition, the analog computer circuitry was set up to provide protection against abnormal voltage and frequency. The field test setup is shown in Fig. 9.

Insertion of the supplementary control signals for the tests proved to be easily accomplished. The signals were inserted into extra input windings of the magnetic amplifiers of the voltage regulator systems. As many as seven machines available at the time were teamed up in response to a single control signal for certain tests. Fewer machines were controlled simultaneously for other tests as necessary with obtainable power system configurations.

The first step undertaken in the field tests was to run frequency response curves of the machine and excitation systems to determine how the actual characteristics compared with the typical data supplied for the studies. It was found that the phase lag characteristics among the machines tested were quite consistent but that response was appreciably slower than indicated by the typical data. A plot of the phase lag characteristics from the data supplied for the studies and from the actual tests is shown in Fig. 10.

While it appeared likely that readjustments were possible to make the actual performance closely match the typical data, it was considered preferable for test purposes to leave the voltage regulator adjustments exactly as they were and to readjust the supplementary control functions to compensate for the additional phase lag. Fortunately, the analog computer employed in the field tests afforded flexibility. Also, the studies indicated that during regular operation, without a supplementary control signal, stability and damping would be better with the slower response of the excitation system.

Fig. 9. Analog computer, recorders, and other apparatus in use during field tests at Glen Canyon Powerplant.

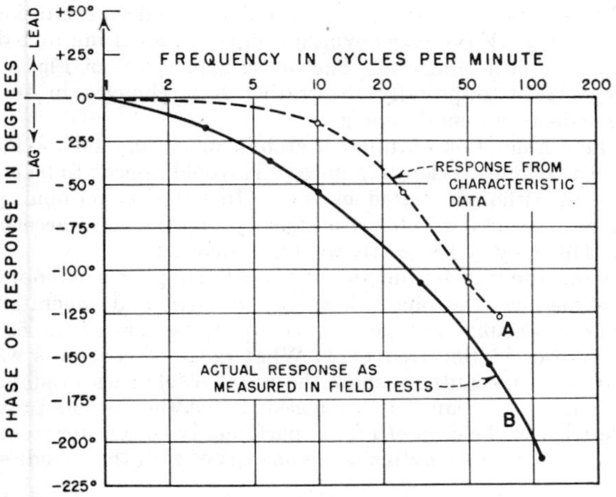

Fig. 10. Plot of phase lag characteristics for Glen Canyon generator. *A*—obtained from analog study using typical data provided; *B*—actual response of Glen Canyon Unit 1 as measured. (Other units similar.)

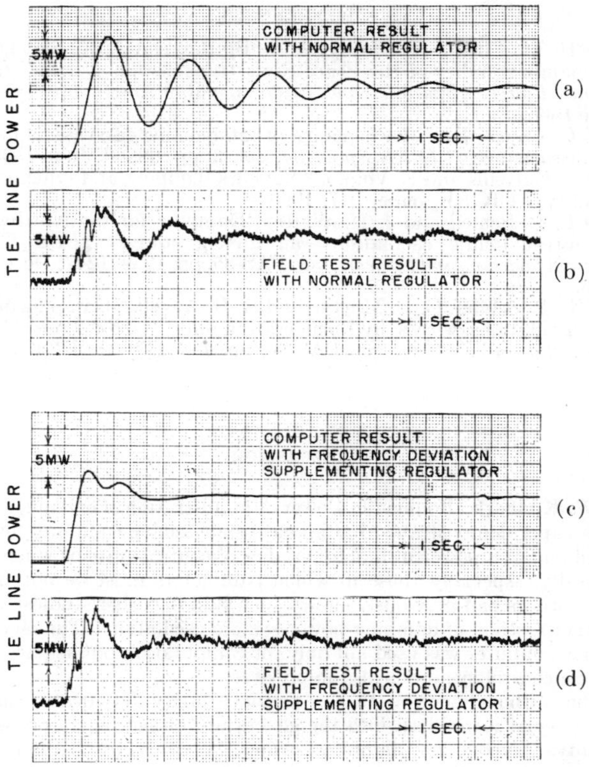

Fig. 11. Transient response of tieline power from Glen Canyon generator group to step disturbance measured during field tests compared with response indicated by analog. (a) Analog computer result with normal regulator. (b) Actual response from generator group under normal regulator control. (c) Analog computer result with voltage regulation supplemented with frequency deviation function. (d) Actual response from generator group with voltage regulation supplemented with frequency deviation function.

Trial field test runs before adjusting the control function to compensate for the slower exciter system response yielded little improvement in stability, but introduced many secondary higher frequency swings in a manner which had become familiar from observing performance with imperfect compensation on the computer. This verified the importance of proper phasing of the control signal and the value of using a control function sufficiently complex to maintain proper phasing over a wide range of swing frequencies. Otherwise, the control could suppress one swing frequency and simultaneously aggravate others.

After synthesizing a control function to accurately complement the phase lag characteristics of the excitation systems, the results shown in Fig. 11 were obtained. To clarify the comparison, the power swing transients recorded during the field tests are shown below the corresponding computer readout obtained during the study. Fig. 11(a) shows the computer result and Fig. 11(b) the field test result with the excitor controlling in its normal manner. Fig. 11(c) shows the computer result and Fig. 11(d) the field test result with the excitation control supplemented with the frequency deviation function.

The operation recorded in Fig. 11 followed the opening of a slightly unbalanced power system loop suddenly leaving four Glen Canyon machines (rated 125 MVA each) delivering 245 MW over one 238-mile, 345-kV line radially into Pinnacle Peak Substation near Phoenix, Ariz.

Conclusions

1) The damping of power system swings may be hampered, rather than aided, by very fast response of the excitation systems to maintain constant terminal voltage. The fast control of terminal voltage can help to limit the first swing but without adequate damping, stability would be lost on subsequent swings.

2) Use of rate-of-change of voltage (dv/dt) to supplement the control of excitation can produce either positive or negative damping, depending upon the relative gains and time constants. The maximum positive damping it can yield is much less than that obtainable from a frequency deviation function.

3) A control signal offering the most promise for improving stability is a complex function of frequency amounting to frequency deviation for slow rates of swing and rate-of-change of frequency for high rates of swing. To achieve efficient damping at very high swing frequencies, use of the second derivative of frequency may also be needed.

4) At nodal points in the power system near the electromechanical center of the swing, the integral of tieline power deviation may be used to replace or supplement the frequency signal. This detail would not ordinarily be worthwhile.

5) Conclusions of the study phase were verified by actual performance in special field tests.

Acknowledgment

Valuable assistance, consultation, and independent checking of results were rendered by R. L. Krahn of the U. S. Corps of Engineers, Portland, Ore.

References

[1] H. M. Ellis, J. E. Hardy, A. L. Blythe, and J. W. Skooglund, "Dynamic stability of the Peace River transmission system," *IEEE Trans. Power Apparatus and Systems*, vol. PAS-85, pp. 586–600, June 1966.

[2] C. Concordia, "Performance of interconnected systems following disturbances," *IEEE Spectrum*, vol. 2, pp. 68–80, June 1965.

[3] J. Kattelus and E. Voipio, "Network damping and synchronous stability," CIGRE, paper 317, 1966.

[4] F. R. Schleif and J. H. White, "Damping for the Northwest–Southwest tieline oscillations—an analog study," *IEEE Trans. Power Apparatus and Systems*, vol. PAS-85, pp. 1239–1247, December 1966.

[5] F. R. Schleif, G. E. Martin, and R. R. Angell, "Damping of system oscillations with a hydrogenerating unit," *IEEE Trans. Power Apparatus and Systems*, vol. PAS-86, pp. 438–442, April 1967.

Discussion

E. W. Kimbark (Bonneville Power Administration, Portland, Ore.): This paper demonstrates that an electronic analog computer, even a small one, is a powerful tool for studies of control systems when used skillfully, especially those in which parameters must be optimized. The authors had previously shown such skill in earlier studies[4],[5] of another method of damping power-system oscillations. The fact that such a computer was used not only for computation but as part of an on-line control system is also impressive.

The authors mention that a previous study[1] indicated that a speed signal introduced into the generator voltage regulators would be advantageous for damping swings. A study by Althammer[6] gives similar indications. Attempts made to obtain a speed signal directly from the shaft of the machine encountered difficulties from vibration of the shaft.[7] Two methods of obtaining a satisfactory speed signal from electrical quantities have recently been described in the literature. Both methods have been tested in the field. The one described in the present paper utilizes deviations of the frequency from its normal value. A method described earlier by Shier and Blythe,[7] utilizes the changes of electrical power output; this signal is proportional to the acceleration of the generating unit and, by ingrating it electrically, one obtains a speed signal. I wonder whether the authors would comment on the relative advantages of the two methods.

The value of the paper would be enhanced if the circuitry used for obtaining the frequency-deviation signal were described or if a reference describing it were given.

The papers advocating use of a supplementary speed signal for damping swings apply the method to generators feeding power into an infinite bus or large power system. Instead, we will consider a synchronous motor receiving power from an infinite bus and swinging with respect to that bus. It seems to me that, to achieve damping of the swings of the motor, we would have to increase the synchronizing power by raising excitation and hence voltage while the motor is swinging away from the bus and has a speed below normal, and decrease the synchronizing power by lowering excitation and hence voltage while the motor is swinging toward the bus and has a speed above normal. Thus, the speed signal would have to be applied to the voltage regulator of the motor with the sign opposite to the correct one for application to the voltage regulator of the generator. Do the authors agree? If so, how should the speed signal be applied to generators at the receiving end of a long intertie? Should it be in the same or opposite way from that used for the sending-end generators? If different, would the signal at the receiving end contribute to damping of intertie swings or only to local swings between the machines at the receiving end?

Similar considerations apply to the use of change of tie-line power as a supplementary signal. Here, however, one could readily use changes in the *absolute* instead of the *algebraic* value of power.

Fig. 6 shows that a supplementary speed signal can produce effective damping of swings having frequencies from 10 to 70 cycles per minute (0.17 to 1.2 Hz). What is the lower limit of frequency at which this method is effective? What is the range of frequencies of swing at which a supplementary signal to speed governors is effective? And what is the range in which damper windings are effective?

Manuscript received January 30, 1967.

Comparison of Fig. 11(a) and (b) shows that the inherent damping of power system swings is greater than that indicated by computer results. This result is reassuring.

References

[6] P. Althammer, "Questions of stability of E.H.V. transmission systems," *Brown Bov. Rev.*, vol. 51, pp. 10–20, January/February 1964.

[7] R. M. Shier and A. L. Blythe, "Field tests of dynamic stability using a stabilizing signal and computer program verification," *IEEE Trans. Power Apparatus and Systems*, vol. PAS-87, pp. 315-322, February 1968.

J. W. Skooglund and **R. T. Byerly** (Westinghouse Electric Corporation, East Pittsburgh, Pa.): The authors have presented a well-organized and timely paper illustrating the effectiveness of an excitation system in increasing system damping. They show that this increased damping can be achieved with an auxiliary stabilizing signal of ΔF (frequency deviation). The fact that a modern excitation system with only ΔV (voltage deviation) input does not improve damping is also well documented. The phase plane plots of Fig. 8 are a unique way of graphically illustrating how changes in terminal voltage affects system damping.

In the examination of different stabilizing signals did the authors consider machine accelerating power? It would appear to be equivalent to ΔF although shifted in phase. In practice, terminal power would be measured to obtain a signal proportional to accelerating power. This may be less costly to obtain than ΔF.

How was the transfer function of the ΔF damping signal obtained? We assume that the phase delay between ΔF and generator field current (or terminal voltage) was chosen to be zero at the 50 cycle per minute oscillation frequency. What range of constants was examined? Can the authors show the effect of changes in stabilizing signal gain or constants? It is suggested that the authors provide a complete list of the data of Fig. 1 (particularly the exciter stabilizing loop) so that others may make a comparison with their studies.

The authors discuss damping factors of 6.9 used in the analytical studies and 7.3 from the field tests. What are the dimensions of these constants?

We disagree somewhat with the wording of the authors' first conclusion. For severe disturbances, high excitation system ceiling voltage and fast response will, we believe, always reduce the magnitude of the first transient swing and, if the system has sufficient natural damping, subsequent performance will be satisfactory. It is agreed that high-performance excitation systems are more in need of additional stabilizing signals. They also can more readily provide additional system damping with proper stabilizing signals. With additional stabilizing systems the system can be stabilized because of positive excitation system damping, rather than being stable despite regulator action.

The authors should be complimented on a complete presentation in which the results of field data are compared with their analytical study. Was the effect of different gains and phasing studied in the field tests? This could be very important in establishing requirements for equipment that is to operate for long periods with a minimum of attention.

Could the authors describe how the phase lag characteristics of the Glen Canyon generators were obtained in their field tests? It would be helpful if they added to the plot of Fig. 10 the overall unit response including the ΔF stabilizing signal.

Manuscript received February 15, 1967.

R. L. Krahn (Corps of Engineers, Portland, Ore.): As the authors mentioned, this discusser made similar studies, simultaneously but independently, using the analog computer facilities of the Bonneville Power Administration. Because of the larger capacity of the BPA computer, however, the synchronous machine model used was ex-

Manuscript received February 17, 1967.

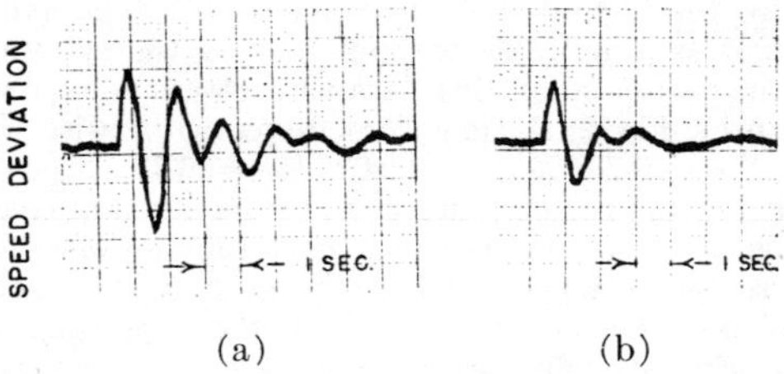

Fig. 12. Performance of detailed analog of regulator, exciter, and generator with tieline including saturation direct and quadrature axis, and power angle nonlinearity. (a) With normal regulator. (b) Regulator supplemented with frequency deviation function.

panded to include refinements such as saturation, both direct and quadrature axis, and the power angle nonlinearity. Nevertheless, when similar data were used, results were substantially the same. For example, Fig. 12 shows responses of the more complicated model both with and without a special supplementary speed signal. Fig. 12(a) should be compared with Fig. 4(a) and Fig. 12(b) with Fig. 4(d) to note the similarity of results. This check strengthens the validity of the simplification and again demonstrates the ingenuity of Mr. Schleif and his associates in reducing a complicated problem to manageable proportions.

When large system studies first indicated that excitation systems, which by themselves were well stabilized, were contributing appreciable negative damping to system oscillations, the addition of a dv/dt input signal to the voltage regulator was investigated with interest. Study results confirmed that this could produce a significant increase in damping under certain conditions. As illustrated in the paper, however, the improvement, at least for small oscillations, is theoretically limited to the condition of constant terminal voltage. While such a perfect voltage regulating system would have less damping than that without a regulator, it would nevertheless reduce the angular swing, thereby increasing transient stability. The addition of a suitability controlled supplementary input signal appears to allow damping to be increased while retaining the improvement in transient stability provided by high-speed automatic voltage regulation.

The Glen Canyon field tests are significant in that, in addition to demonstrating operation on actual equipment, they also indicate the feasibility of using a single supplementary signal device for a number of generating units. This could be important in reducing the cost and simplifying the adjustment of such equipment in multiple unit stations.

The authors deserve a great deal of praise for their logical and complete development of a practical method to provide additional positive damping to power systems. Their cooperation in exchanging ideas and study results has been greatly appreciated.

Manuscript received February 24, 1967.

K. R. McClymont (Hydro-Electric Power Commission of Ontario, Toronto, Ont., Canada): The authors are to be congratulated for their excellent work in demonstrating that special signals can be used with conventional excitation systems to obtain a major increase in system damping.

It is interesting to compare the frequency signal which the authors use with a generator speed signal on a controlled rectifier excitation system, described in Dandeno *et al.*[8] From the standpoint of its effect on damping, one would not expect that there would be much difference between a speed and a frequency signal if there is a large reactance between the generator or group of generators and the main system. Our investigations showed that very effective damping was obtained if the exciter voltage varies in phase with the generator speed variations. The ratio of exciter voltage in per unit to speed deviation in electrical radians per second, i.e., the gain, should be in excess of five and preferably at least ten to obtain rapid damping. (The authors appear to use a per unit system in which rated frequency is one per unit per second. For this per unit system the gain should be 377 times larger.) If we assume the objective is to produce a variation in exciter voltage proportional to and in phase with frequency deviations, then the frequency signal should be put into a network which cancels the time delays of the regulator and exciter. In the linear range, the transfer functions of the regulator, exciter, and exciter feedback in Fig. 1 can be replaced by a single forward transfer function which relates regulator input and exciter voltage. Using the regulator and exciter constants of Fig. 1 and typical values for the feedback constants, this single transfer function is

$$\frac{54\,(s+1)}{s(0.14s+1)^2}.$$

There are a number of networks which can be devised to effectively cancel out the time constants of the excitation system so that exciter voltage varies in phase with frequency deviations. An obvious choice would be to use a network having the transfer function

$$\frac{70s(0.14s+1)^2}{s+1}.$$

This network might be difficult to realize in practice. With the transfer functions used by the authors, the gain and phase shift from the frequency input to the exciter voltage output was calculated to be approximately as follows:

Frequency, cycles per minute	Gain, p.u./rad/s	Angle, degrees
10	1.7	−10
25	2.6	+30
50	4.9	+30
70	6.4	+10

These figures indicate that the damping provided by the network used is relatively small at the lower frequencies of 10 and 25 cycles per minute. Examination of Fig. 6 tends to verify this.

Do the authors believe there would be any advantage to increasing the gain at these frequencies of 10 and 25 cycles per minute? Have they encountered any practical limitations in the gains and time constants they can employ in their frequency processing network?

The authors' comments on the effect of fast response excitation system in reducing system damping are well taken. We found it desirable to reduce the excitation system gain at several locations on our system to obtain improved damping.[9]

The authors state, "Since prolonged offset in frequency is uncommon and would tend to be relieved by the voltage bias when it does occur, it is not conclusive that the reset is desirable." In our appliction we took a somewhat different point of view. For certain unusual conditions the generators and load could be separated from the main system and the frequency of this isolated system departs from 60 Hz. Under these conditions we thought it undesirable to have a sustained deviation in voltage. On the other hand, we had no evidence that there was any requirement for increased damping at frequencies much below 10 cycles per minute.

REFERENCES

[8] P. L. Dandeno, A. N. Karas, K. R. McClymont, and W. Watson, "Effect of high-speed rectifier excitation systems on generator stability limits," *IEEE Trans. Power Apparatus and Systems*, vol. PAS-87, pp. 190–201, January 1968.

[9] O. W. Hanson, C. J. Goodwin, and P. L. Dandeno, "Influence of excitation and speed control parameters in stabilizing intersystem oscillations," *IEEE Trans. Power Apparatus and Systems*, vol. PAS-87, pp. 1306–1313, May 1968.

F. R. Schleif, H. D. Hunkins, G. E. Martin, and E. E. Hattan: We thank the discussers for their gracious assessment of our efforts in this rather complicated problem.

Dr. Kimbark's contributions add much toward properly completing the picture. Though our attentions were focused mainly on achieving a practical approach, it was not our intention to disregard any of the preceding theoretical recognitions of the problem. Althammer[6] constitutes a theoretical treatment of a static excitation system

Manuscript received March 27, 1967.

hypothetically supplemented with rate of change of load angle. Though not quite the same, many similarities are evident and the background is valuable.

Both the discussions of Dr. Kimbark and Mr. Skooglund and Mr. Byerly inquired whether the integral of accelerating power was considered as a signal source. It was indeed considered and even field tested for comparison. We concluded that for our particular application the advantages were somewhat outweighed by the limitations. It has the virtues of basis on speed difference rather than deviation and slightly simpler hardware for a single unit, but the reset should closely match the prime mover response and this is complicated to achieve on a hydro unit. Application gang fashion to a number of units on a bus becomes more complicated. This signal source inherently incorporates reset, which we were not seeking for reasons to be mentioned later.

The frequency deviation signal source we did use had been obtained commercially for other test purposes but yielded sufficient sensitivity and response: 5 volts per hertz deviation and 90-percent response in less than 85 ms. (It should be noted that the time constant would thus be less than 40 ms. approximately.)

Dr. Kimbark is correct that the polarity of the supplementary signal for damping with a synchronous motor is opposite to that for a generator. Nevertheless, another series of our studies indicated that damping by generator excitation control on the basis of a frequency derived supplementary signal is always correct regardless of direction of flow of tieline power. The explanation seems to be that for tieline power to flow toward a generating station, that station must have a still greater local load to which energy transfer is increased by raising excitation. A signal from change in the absolute value of tieline power would be applicable to a synchronous condenser.

Effectiveness of damping by excitation control of generators, we conclude, is limited at the lower frequencies, varying approximately as the inertia available and the square of the swing frequency. Though more effective at lower frequencies than the amortisseurs, this damping would be hard put to reach down to the frequency range of prime mover control.

Mr. Skooglund and Mr. Byerly also demonstrate keen comprehension by their remarks and searching questions. Their question concerning accelerating power as a signal source has been answered earlier. The transfer function of the damping signal was obtained by synthesizing a practical analog computer circuit which would best compensate for the phase lags in the complete system. The net phase lag was not only compensated to zero for 50 cycles per minute but from zero to the highest frequency practicable. The gain was also made as high as practicable without making the system unstable. Values of the exciter stabilizing parameters were not shown because for certain tests, Fig. 2(d), they were altered. Otherwise, $\mu_s = 0.032$, $T_s = 1.0$, $X_d = 0.76$, $X'_d = 0.24$, $T'_{d0} = 6.17$, $T_m = 6.9$. The damping factors are per unit change in torque per unit change in frequency. The Glen Canyon phase lag characteristics were obtained by frequency response tests, including the generator, at normal load. The net phase lag characteristics with the supplementary signal would be nearly a straight horizontal line within 15 degrees of zero on Fig. 10 up to 60 cycles per minute above which the lag increases.

Not only are Mr. Krahn's comments and data contributions valuable to the paper, his advice and consultation were exceedingly helpful throughout the study. In addition to corroborating our results in Fig. 12, it was particularly interesting that when transfer functions with reset were used for the supplementary signal, low frequency oscillations remained.

Mr. McClymont's comments are especially valuable because of the excellent work he and his associates have recently performed with static exciter systems.[8],[10] His analysis of the linearized problem demonstrates that much can be done by this means. Though the function $70s(0.14s + 1)^2/(s + 1)$ is impossible to realize by the means we have been using, the approximation of $70s(0.14s + 1)^2/(s + 1)(0.02s + 1)^2$ could be achieved. However, our objective has been to compensate to zero net phase lag the terminal voltage rather than the exciter voltage. We wonder if this may explain the seemingly excess compensation of phase angle and the rising gain in his analysis.

Increasing the gain should improve damping at the lower frequencies but this is limited by inability to keep the response in phase beyond a certain upper limit (and lower limit if reset is used). Since phase shift must inevitably reach 180 degrees, the function gain must be limited by necessity to keep the loop gain less than one at this phase shift, otherwise instability will result. This was demonstrated at Glen Canyon by cautiously increasing the gain until oscillations at approximately 2 Hz appeared. This corresponded well with the computer results.

We are most grateful to the discussers for their contributions to understanding and controlling this problem.

References

[10] K. R. McClymont, G. Manchur, R. J. Ross, and R. J. Wilson, "Experience with high-speed rectifier excitation systems," this issue, pp. 1464–1470.

Effect of High-Speed Rectifier Excitation Systems on Generator Stability Limits

PAUL L. DANDENO, SENIOR MEMBER, IEEE, ALEX N. KARAS, MEMBER, IEEE, KENNETH R. McCLYMONT, SENIOR MEMBER, IEEE, AND WILFRED WATSON, SENIOR MEMBER, IEEE

***Abstract*—A device to provide a signal proportional to changes in generator speed has been developed by Ontario Hydro. Field tests have been performed to demonstrate that use of this signal with a high-speed rectifier excitation system results in dynamic stability limits which approximate those which can be obtained with a zero reactance generator. Transient stability limits are greater than those obtainable with high-response conventional excitation systems. Results from the field tests and a computer study are compared.**

INTRODUCTION

WITHIN the last six years, Ontario Hydro has installed 550 MW of hydraulic generation in four plants on the Moose River System in northern Ontario. This generation is incorporated into the main system over a 440-mile, 500-kV transmission line which extends to the load center in the Toronto area. Switching and step-down transformation are installed in the Sudbury area near the middle of the line. This 500-kV transmission system has been described in a previous paper.[1]

The remote location of this generation with the consequent high cost of transmission facilities focussed attention on methods of increasing transient and dynamic stability limits. Hydraulic generators have very poor inherent damping forces since the damping obtained from the amortisseur windings is quite small in the usual frequency range for natural oscillations. Also, because of water inertia, the speed governors usually cannot contribute significant damping. The major damping forces are those obtained from the prime mover, the excitation system, and the load. In view of the remoteness of this generation from the load center, there was concern about the possibility that damping forces would be inadequate when the transmission system was heavily loaded. Severe oscillations at other locations on the Ontario Hydro system with similar conditions have been experienced.

Preliminary investigations on an analog computer in 1961 indicated that a very significant improvement in stability limits could be obtained with a high-speed rectifier exciter with a voltage regulator and supplementary derivative-type control signal. These results were also reinforced by the favorable experience reported by the U.S.S.R. and by one other article in the technical literature.[2] It was therefore decided to purchase seven rectifier excitation systems for those generators on the Moose River system for which conventional rotating exciters had not been ordered.[3]

After placing the order for rectifier excitation systems, investigations were continued to determine the supplementary control signal required to obtain optimum generator performance. Further investigation showed that a control signal proportional to change in generator speed would provide the best generator performance for the system. Considerable effort was spent in trying out other signals since it was known that it would be difficult to obtain a signal proportional to speed. However, none of the other signals investigated gave the desired performance.

Since a device for obtaining this signal was not commercially available, it was decided to construct one. A device was designed and placed in service on three generators in 1963. While it gave very good results, it suffered from two major disadvantages, viz.,

1) Gain was limited by an inner loop stability consideration.

2) The device employed rotating components which are subject to wear.

It was therefore decided to improve the design of the equipment to eliminate the two disadvantages listed. This was done and the new device placed in service on one generator early in 1966. It performed as anticipated, and three more were placed in service in the summer of 1966.

This paper describes this improved speed signal device, and presents the results of tests to determine dynamic stability limits and of tests to measure the transient performance. Results from field tests are compared with results from computer studies.

SPEED STABILIZING SIGNAL

Method of Speed Sensing

The speed signal must be extremely sensitive to small variations in speed, since a speed deviation of only about 0.15 percent should be sufficient to drive the exciter to its ceiling. When dealing with such high sensitivity, considerable care must be exercised to insure that "noise" will not cause unnecessary swings in the exciter voltage. Noise at low frequencies such as that corresponding to the speed of the shaft cannot be removed by conventional electric filters, and its elimination must be inherent in the method for obtaining the speed signal. Various methods of measuring the speed of rotation of the generator rotor were considered. The large shaft diameter and relatively large bearing clearances of a hydraulic generator considerably complicate the problem. The four generators on which the equipment has now been installed each have a shaft diameter of 38 inches at the guide bearing, and rotate at 100 r/min. Lateral movements of the shaft of 0.030 inch are typical at points close to the generator guide bearing. It was decided that a flexible coupling, used to drive a speed sensor, would be unsuitable for the sensitivity required, and therefore it would be necessary to measure the speed of rotation of the generator shaft itself. As access to the shaft at the top of the rotor was difficult, due to the presence of slip rings and

Paper 31 TP 67-20, recommended and approved by the Power Generation Committee of the IEEE Power Group for presentation at the IEEE Winter Power Meeting, New York, N. Y., January 29–February 3, 1967. Manuscript submitted November 1, 1966; made available for printing November 29, 1966.

P. L. Dandeno, K. R. McClymont, and W. Watson are with the Hydro-Electric Power Commission of Ontario, Toronto, Ont., Canada.

A. N. Karas is with the National Energy Board, Ottawa, Ont., Canada. He was formerly with the Hydro-Electric Power Commission of Ontario, Canada.

Reprinted from *IEEE Trans. Power App. Syst.*, Jan. 1968.

the governor speed sensor, a location below the generator guide bearing was chosen as the most suitable.

Fig. 1 shows the equipment installed on one of the generators. A steel gear wheel with 600 teeth was mounted on the shaft and magnetic pickups supported from the generator lower bracket were positioned around it. Movement of the gear teeth across the face of each pickup generates an alternating voltage in the coil. The output of the pickups is at 1000 Hz at rated speed of the generator. A number of problems were overcome in the design of the pickup arrangements:

1) The apparent variation in rotational speed as seen by one pickup, resulting from lateral movement of the generator shaft, is eliminated by summing the outputs from several pickups spaced around the shaft.

2) The effects of inaccuracies in spacing and profile of the gear teeth are minimized by using a number of pickups and by designing each pickup to sense the passage of a number of gear teeth; for example, in the equipment developed, each pickup senses seven adjacent teeth and there are five pickups. This gives the average position of 35 teeth at any instant.

3) To permit assembly around the shaft, the gear wheel is necessarily made in at least two pieces. As each joint between the pieces passes a pickup, some discontinuity in the signal results.

Fig. 1. Stabilizing signal equipment mounted on a generator.

Since the joints are diametrically opposite and there are five pickups, a relatively small disturbance occurs in the signal 10 times per revolution (i.e., at 16.67 Hz).

Signal Processing Circuits

Fig. 2 is a block diagram of the signal circuits. The signals from the five magnetic pickups are summed, differentiated, and clipped in two amplifier stages. This has the effect of eliminating low-frequency components arising from the joints in the gear wheel and from the lateral movement of the generator shaft. The frequency of the resultant signal is then converted to a dc signal by a commercially available transducer. The latter uses a saturating core transformer which gives a constant volt-second output for each input 1/2 cycle. The output of the saturating core transformer is rectified, giving unidirectional pulses having a frequency equal to twice the input frequency. Each of these pulses has the same average value so that the mean level of the output is proportional to the number of pulses and therefore to the input frequency. The output of the frequency transducer is filtered to remove the 2000-Hz component and amplified in two capacitor-coupled operational amplifier stages. The first capacitor coupling has a time constant of the order of 10 seconds and eliminates the steady dc value. The second stage coupling incorporates a "washout" feature which, after a time, reduces the output to zero for sustained changes in speed. The washout time constant which has been used to date is 1.6 seconds.

The output level of the final stage is designed to match the level of the particular voltage regulator circuit with which the stabilizer is being used. An output of from 2 to 5 V/rad/s has been found suitable for all the applications made to date. The maximum output of the amplifier used is 10 volts. The signal is connected through a suitable resistor directly to the summing junction of the operational amplifier normally used in a controlled-rectifier excitation system.

Generator Voltage Limit

To facilitate synchronizing, the stabilizing signal is connected into the regulator only when the generator breaker is closed. To prevent excessive terminal voltage due to an increase in speed resulting from loss of load, a limit circuit is provided which prevents the generator voltage from exceeding about 115 percent of its rated value.

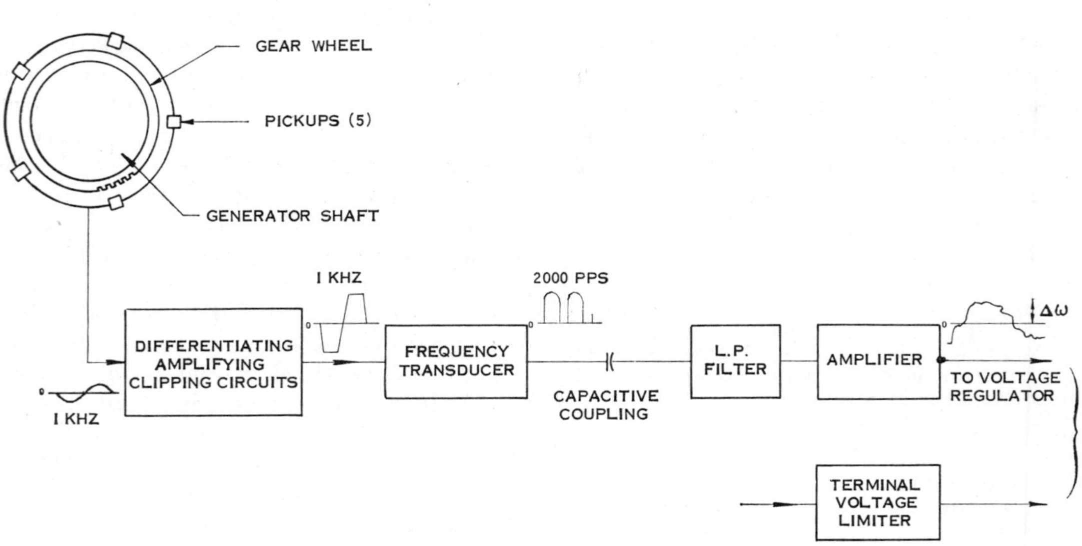

Fig. 2. Block diagram of stabilizing signal equipment.

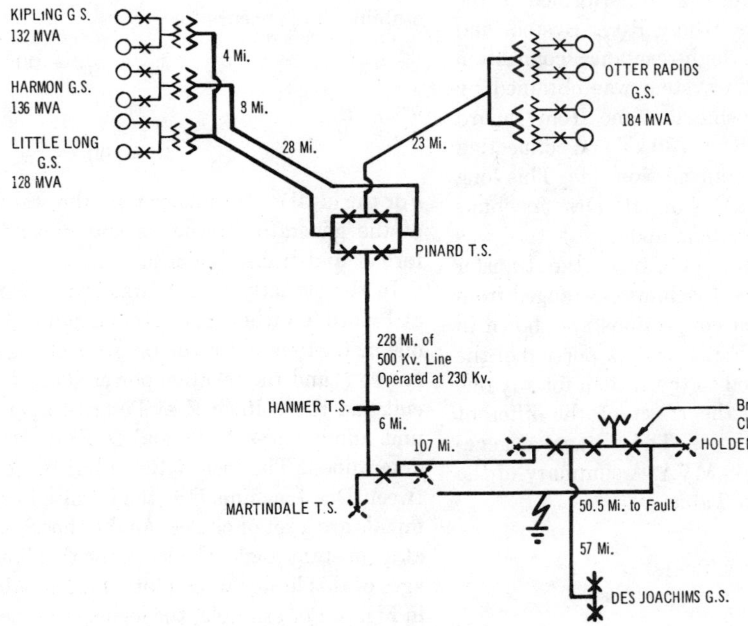

Fig. 3. Single line diagram of system tested.

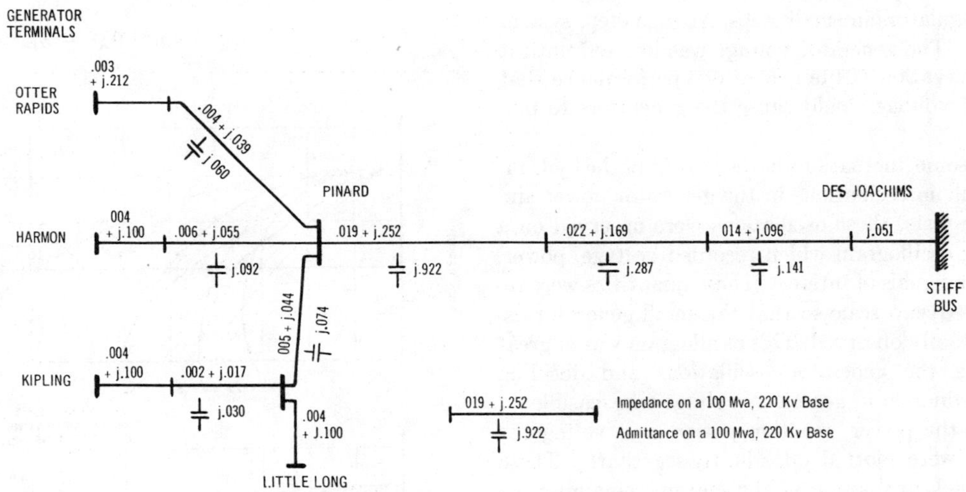

Fig. 4. Impedance diagram of system tested.

TABLE I

SUMMARY OF GENERATOR CHARACTERISTICS

Station	Number of Generators	Rating of Generators, MVA	Type of Exciter	Generator Reactances Per Unit on Rating: *Xd* (unsaturated)	*Xq*	*X'd*
Otter Rapids GS	3	46	rotating	0.68	0.44	0.20
	1	46	rectifier	0.68	0.44	0.20
Little Long GS	2	64	rectifier	0.63	0.40	0.24
Harmon GS	2	68	rectifier	0.70	0.45	0.23
Kipling GS	2	66	rectifier	0.69	0.45	0.24

System Tested

Dynamic and transient stability tests were performed in the Fall of 1966 using generators on the Moose River System and special transmission connections. A high-reactance connection between the generators and the main system was obtained by operating the 230-mile 500-kV transmission line from Pinard to Hanmer Transformer Stations (TS) at 230 kV and connecting it in series with 165 miles of 230-kV transmission line. This long line was terminated on the 230-kV bus at Des Joachims Generating Station (GS). For the system under test this was a fairly stiff bus. On the generation MVA base, the transfer reactance between Pinard TS and Des Joachims GS ranged from 105 to 126 percent. The transmission connections are shown in Fig. 3 and an impedance diagram in Fig. 4. Only a portion of the Moose River generation was connected to the system for any one test, as this facilitated investigating the effect of the different types of excitation system characteristics. The connected generating capacity ranged from 224 to 268 MVA. A summary of the generator characteristics is shown in Table I.

Dynamic Stability Tests

Test Procedure

The dynamic stability tests were performed with four generators on the Moose River. These generators were initially loaded close to their rating with normal generator voltage. Then the generator voltage was lowered by changing the setting on the generator voltage regulators in small steps. At each step, system readings were taken. The generator voltage was lowered until it was decided by observation of the generator's performance that further reduction of voltage would cause the generators to pull out of step.

There are always some fluctuations in the output of the hydraulic turbine that result in oscillations in the generator power and voltage. During the tests, these oscillations were observed on a 6-pen direct writing oscillogram which recorded voltage, power, frequency, and other signals of interest. These quantities were recorded on a depressed-zero scale so that the small generator oscillations could be readily observed. This oscillogram was of great value in evaluating the generator oscillations and deciding whether a further reduction in generator voltage was feasible.

At each reading, the power, reactive power, and voltage as read at Pinard TS were plotted on admittance charts. These charts permitted quick evaluation of the margin remaining between an operating point and the theoretical stability limit.

Theoretical Limits—Admittance Charts

To simplify the analysis of stability limits it is convenient to assume the four generators used in the test are paralleled at the generator low-voltage terminals and to represent the generators by a single equivalent generator. The output of the equivalent generator can be expressed by

$$P_1 = Y_{11}E_1^2 \sin\alpha_{11} + Y_{12}E_1E_2 \sin(\delta_{12} - \alpha_{12})$$

where all quantities are in percent on a convenient base and P_1 is the generator power output.

$Y_{11}\underline{/\theta_{11}}$ is the driving point admittance seen from the generator terminal looking towards the infinite bus and $\theta_{11} = -90° + \alpha_{11}$.

$Y_{12}\underline{/\theta_{12}}$ is transfer admittance between the generator terminals and the infinite bus and $\theta_{12} = -90° + \alpha_{12}$.

$E_1\underline{/\delta_1}$ is the voltage of the generators, $E_2\underline{/\delta_2}$ is the voltage of the infinite bus, and $\delta_{12} = \delta_1 - \delta_2$.

An ideal excitation system has the ability to effectively hold the generator terminal voltage constant. With E_1 constant the stability limit occurs for

$$\delta_{12} - \alpha_{12} = 90°$$

or

$$\delta_{12} = 90° + \alpha_{12}.$$

For the ideal excitation system, the stability limit is independent of the generator constants and dependent only on the transformer and transmission line reactances.

In the stability tests it was convenient to use meter readings at Pinard TS where the total output of the plants is available on one meter to monitor the progress of the tests. By converting the power P and the reactive power Q on the Pinard to Hanmer circuit and the voltage V at Pinard to per unit quantities, the per unit admittances P/V^2 and Q/V^2 as seen at Pinard TS can be determined. The loci of this admittance for constant angles between Des Joachims GS (the infinite bus) and the generator terminals are a set of circles. As the theoretical stability limit occurs at a constant angle, the locus for this limit and for fixed percentages of this limit can be plotted on an admittance chart as shown in Fig. 5. For example, the locus for 90 percent in Fig. 5 is defined as the locus for the condition

$$\sin(\delta_{12} - \alpha_{12}) = 0.9$$

or

$$\delta_{12} = \sin^{-1} 0.9 + \alpha_{12}.$$

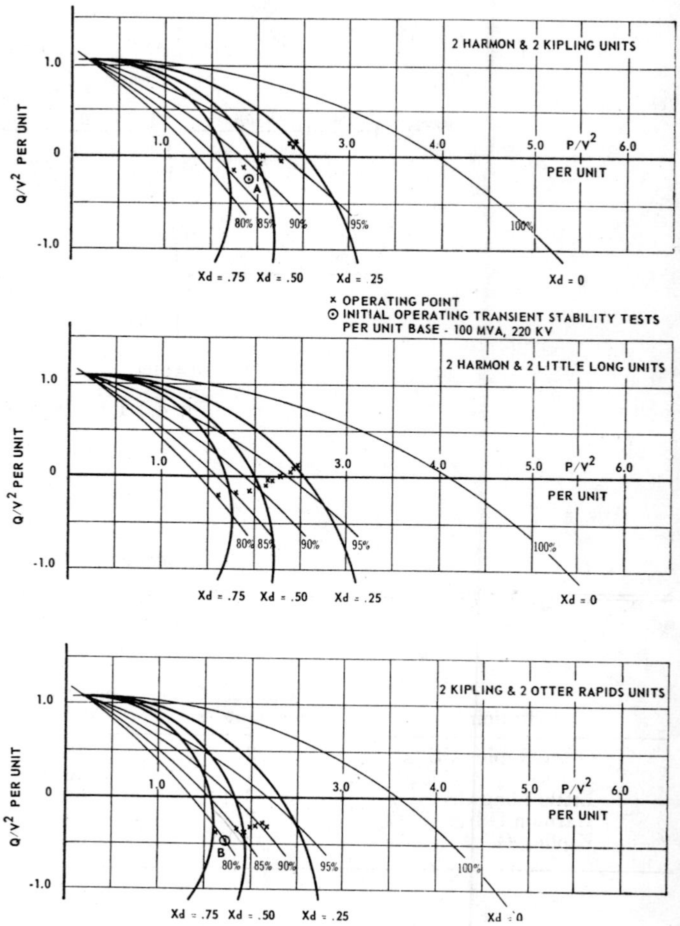

Fig. 5. Admittance charts showing operating points.

Also, the loci of the theoretical stability limits for generators having effective reactances other than zero, i.e., the voltage regulator is not ideal, can be plotted on the same chart. These charts provide a simple, quick way of evaluating test results.

Test Results

As shown in Table II, stability limits were obtained for three combinations of generators. The test results for these different combinations are plotted in Fig. 5. With rectifier excitation systems and speed stabilizing signals, the stability limit obtained was approximately 97 percent of the limit obtainable with the ideal regulator which corresponds to the theoretical stability limit of an unregulated generator having a synchronous reactance of 25 percent on the generator base. In the test using two generators with rotating-amplifier voltage regulators and two generators with rectifier exciters and speed stabilizing, the stability limit was equal to 92 percent of the limit obtainable with an ideal regulator or equal to the theoretical stability limit obtainable with a generator having a synchronous reactance of 40 percent on the generator base.

Figs. 6 and 7 show the oscillations which occurred at operating points considered to be practical stability limits. The decrease in magnitude of oscillations is evident after the voltage was raised to avoid pulling out of step. In Fig. 6 the four generators being tested have rectifier exciters and speed stabilizers. In this figure the long period of the oscillations and the small magnitude of the power swings contrast with the results obtained in Fig. 7. In Fig. 6 the period of the oscillations decreases as the generators move away from the stability limit. Fig. 8 demonstrates the effectiveness of the speed stabilizing signal. While operating with three generators at a point well below the previously determined dynamic stability limit, the speed stabilizing signals were removed. Generator oscillations built up as shown in Fig. 8 until the speed signal was reinserted on one unit.

Transient Stability Tests

Test Procedure

The same transmission line connections as used for the dynamic stability tests were used for the transient stability tests. A double line to ground fault was applied near the midpoint of a

TABLE II

Summary of Steady-State Stability Tests

Generator Excitation System Combinations	Practical Stability Limit as Percent of Theoretical Stability Limit	Reactance of Unregulated Generator Having Same Stability Limit (Per Unit on Generator Base)
Two Harmon and two Kipling units all with rectifier exciters and speed stabilizing	97	0.25
Two Harmon and two Little Long units all with rectifier exciters and speed stabilizing	97	0.25
Two Kipling units with rectifier exciters and speed stabilizing; two Otter Rapids units with rotating-amplifier voltage regulators	92	0.38

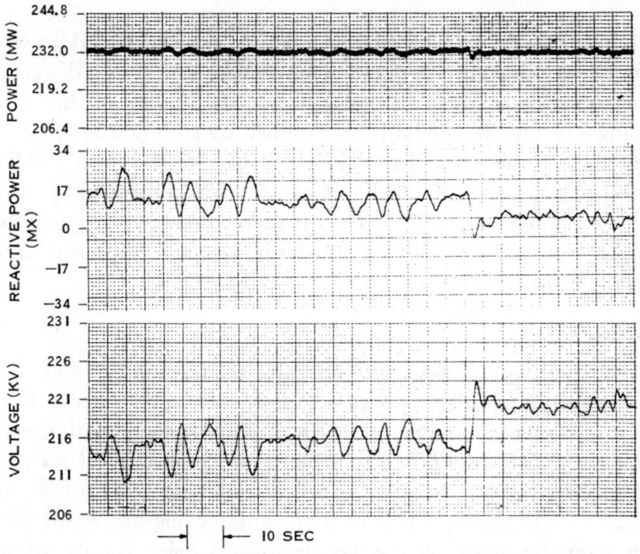

Fig. 6. Oscillogram of operation at dynamic stability limit, two Kipling and two Harmon units.

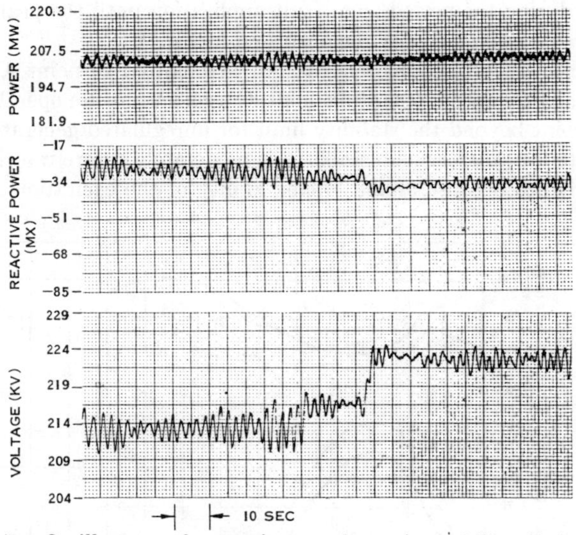

Fig. 7. Oscillogram of operation at dynamic stability limit, two Kipling and two Otter Rapid units.

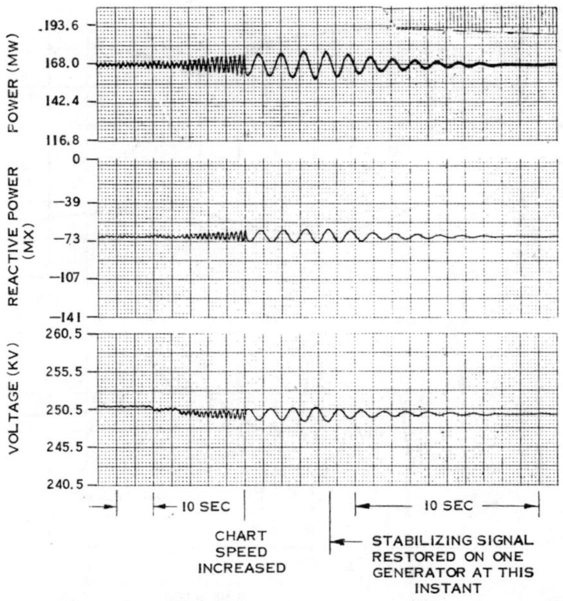

Fig. 8. Oscillogram showing oscillations with speed signal removed and then restored, two Kipling and one Harmon units.

110-mile 230-kV transmission line which was connected to the radial transmission at a point about 15 percent of the distance from the infinite bus to the generation. The location of the fault is shown in Fig. 3. The fault was applied by closing a breaker to energize the line which had previously been rigged with a solid double line to ground fault. Fault severity was controlled by controlling the opening time of the breaker. Direct writing oscillographs at Pinard TS, Kipling GS, and Harmon GS recorded important quantities.

Test Results

Test results obtained with two units at Harmon GS and two units at Kipling GS in service are plotted as solid lines in Fig. 9. All of these units have rectifier excitation systems with speed stabilizing. The double line to ground fault was applied for approximately 0.2 second. The excellent damping of this system can be seen in the time plot of the angular swing at Pinard TS. The maximum angle on the second upward swing is only a few degrees above the initial value. Also, the ability of the excitation system to increase generator terminal voltage above its initial value during the period from fault application until the peak of the first angular swing is demonstrated in Fig. 9.

The initial conditions for this test are plotted as point A on Fig. 5. At the initial operating point, the generators are operating at a point beyond the stability limit for unregulated generators. Despite this, they remain stable for a moderately severe transient disturbance because of the forcing ability of the excitation systems. The test arrangement was simulated on a computer substituting "conventional" excitation systems with rotating exciters for the rectifier exciters. Computer results indicated that the system was unstable for excitation system response ratios up to at least 2.0 with conventional excitation systems.

Test results obtained with two units at Kipling GS and two units at Otter Rapids GS "in-service" are plotted as solid line in Fig. 10. The units at Kipling GS have rectifier excitation systems with speed stabilization and the units at Otter Rapids GS have rotating exciters and voltage regulators using rotating amplifiers. The system is adequately damped but damping is considerably less than that obtained when all units have rectifier excitation systems.

Initial conditions for this test are plotted as point B in Fig. 5. At the initial operating point, the generators are close to the stability limit for unregulated generators.

Computer Simulation

The transient tests were simulated on a digital computer to investigate the validity of the computer representation. A transient stability program similar to that described elsewhere[4] was used with some additions to the excitation system representation as described in the following. The main features of this program for simulation of these tests results are:

1) Generator saliency and transient effects due to the generator field are correctly represented; however, generator and exciter saturation and effects arising from amortisseur windings are neglected.

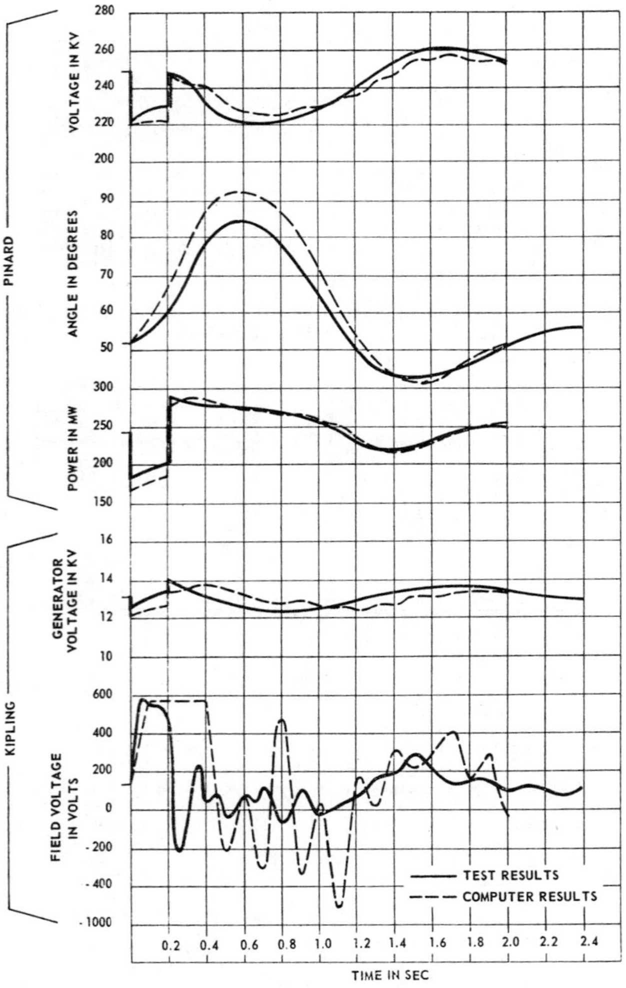

Fig. 9. Time plots for transient stability test, two Kipling and two Harmon units.

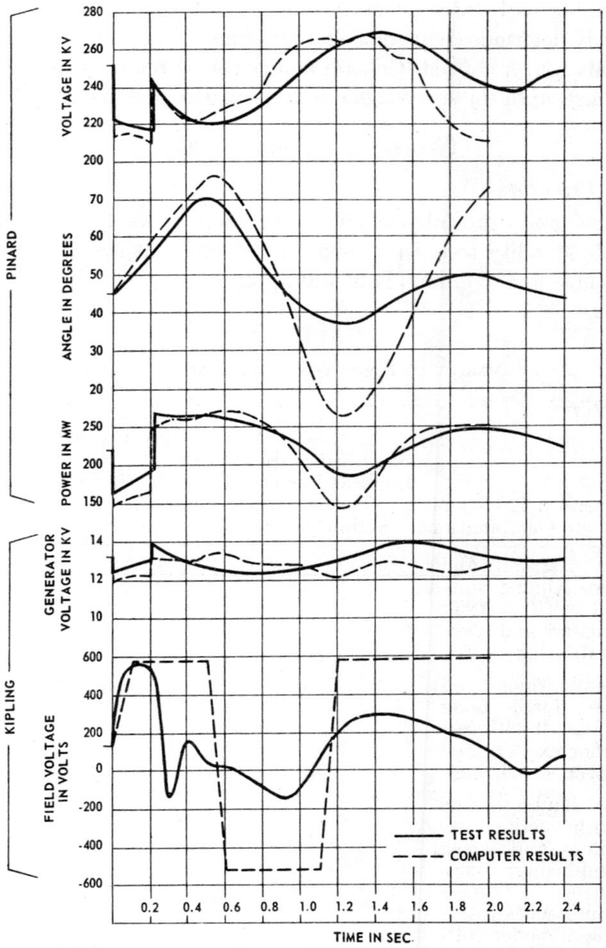

Fig. 10. Time plots for transient stability test, two Kipling and two Otter Rapids units.

2) Conventional rotating exciters and voltage regulators are represented as single time constant systems. The parameters required by the computer are time constant, gain, and exciter ceiling voltage. Accurate representation of controlled rectifier exciters is possible with the single time constant representation. An addition was made to the program to include a speed stabilizing signal.

Computer results are plotted as dashed lines in Figs. 9 and 10 and may be compared with tests results. From Fig. 9 it can be seen that the calculated swing is larger than the swing obtained in the field tests. It is thought this is largely due to neglecting subtransient effects in the computer program, particularly during the fault. Actual voltages and power transfers during the fault period are significantly higher than the calculated values. Aside from the magnitude of the first swing, agreement between calculated and test results is very close.

In Fig. 10 also the calculated swing is greater than the actual swing. For this test, actual damping was much higher than that indicated by the computer. This difference is largely due to the simplified representation of the conventional excitation system at Otter Rapids. The voltage regulator provides considerably more damping than that obtainable in the simplified representation. In the immediate future modifications to the existing transient stability program are expected to be made which will give a more accurate representation of the conventional voltage regulators.

Conclusion

A device for producing a signal proportional to small changes in generator speed has been successfully developed. Field tests have confirmed that using this speed stabilizing signal in conjunction with a high-speed rectifier excitation system increases generator stability limits. If high-speed rectifier excitation systems with speed stabilizers are used on remote hydro-electric generators connected into large systems, the dynamic stability limits obtained are approximately equal to that which could be obtained by holding generator voltage constant at every instant (zero reactance generator). Transient stability limits are greater than those obtainable with high-response conventional excitation systems.

A comparison of field tests with computer results shows the computer gives pessimistic results. This conservatism is believed to be primarily because generator subtransient effects are neglected in the computer program. Damping was accurately represented by the computer simulation of the system using the rectifier exciter and speed stabilizing signals. Computer simulation of a system using both rectifier exciters with speed stabilizing and rotating exciters with rotating-amplifier voltage regulators was inadequate, the actual damping being much higher than the computed damping. It is thought that this difference between test and computer results is due to the simplified representation of the conventional excitation systems. Steps are being taken to improve the excitation system representation.

References

[1] K. R. McClymont, W. Watson, and G. Nesgos, "Overvoltages on Ontario Hydro's 500 KV system and their effect on system design," *IEEE Trans. Power Apparatus and Systems* (*Supplement*), vol. 82, pp. 459–475, 1963.

[2] A. S. Aldred and G. Shackshaft, "A frequency response method for the predetermination of synchronous-machine stability," *Proc. IEE*, pt. 3, pp. 2–10, March 1960.

[3] K. R. McClymont, G. Manchur, R. J. Ross, and R. J. Wilson, "Experience with high-speed rectifier excitation systems," presented at the IEEE Winter Power Meeting, New York, N. Y., January 29–February 3, 1967.

[4] M. S. Dyrkacz, C. C. Young, and F. J. Maginniss, "A digital transient stability program including the effects of regulator, exciter, and governor response," *Trans. AIEE* (*Power Apparatus and Systems*), vol. 79, pp. 1245–1257, 1960 (February 1961 sect.).

Discussion

G. I. Stillman (American Electric Power Service Corporation, New York, N. Y.): This stimulating paper provides field test data showing the effectiveness of the latest development in excitation systems—the controlled solid-state rectifier excitation system. Although the paper is concerned with application of SCR systems to hydro-electric units, the conclusions are undoubtedly applicable to thermal units as well.

In the section entitled *Generator Voltage Limit*, the authors state that under loss of load, "a limit circuit is provided which prevents generator voltage from exceeding about 115 percent of its rated value." The controlled rectifiers are supplied by step-down transformers supplied from the generator terminals and, consequently, the voltage supplied to the rectifiers would also be limited to about 115 percent of rated value. Yet, in a related paper,[3] it is stated that the excitation system was specified to be capable of withstanding fundamental frequency voltages up to 170 percent of rated with crests up to 200 percent of rated because of possible self-excitation of the unit. I would appreciate it if the authors would clear up this point.

In the section that discusses the results of the transient stability tests, the authors state that point A of Fig. 5, the initial operating point, shows that the Harmon and Kipling units "are operating at a point beyond the stability limit for unregulated generators." This, of course, is true when one notes that Xd for each unit as given in Table I is 0.7 per unit on the generator base. Normally, however, one does not use Xd when considering transient stability. Upon consideration of the derivation of the curves shown in Fig. 5, it is clear that the loci labelled Xd could just as easily have been labeled Xd' with the same numerical value as given in Fig. 5 applying to each locus. With this in mind, and obtaining from Table I values of 0.23 and 0.24 per unit for Xd' for the Harmon and Kipling units, respectively, it becomes clear from Fig. 5 that point A is well within the transient stability limits (based on constant voltage behind transient reactance) of the machines.

In their discussion of the method of speed sensing, the authors mention that various methods of measuring speed of rotation were considered. Were stroboscopic means considered—either by use of a stroboscope or by counting light pulses from a light beam shining on a disc with equally spaced slits? Were there any particular drawbacks to this method?

Manuscript received February 13, 1967.

A. Blythe (Canadian Westinghouse Company, Ltd., Hamilton, Ont., Canada): The authors are to be congratulated on their development of an adequate speed error signal. This is no mean feat, as it involves the measurement of speed errors as small as 2 rad/s or about 0.01 of one percent of rated speed. It would be interesting if the authors could supply data as to the sensitivity, noise bandwidth, and time delay of the system.

The ideal stabilizing signal is not the true speed error of the generator, but rather the difference between the generator speed and the average speed of the system. It is suggested that the "washout" feature in the speed sensing device has an effect which causes the signal to approximate the difference between the instantaneous and average speeds of the generator. As the generator's average speed will

Manuscript received February 13, 1967.

be very close to that of the system, this effect is a desirable feature. This indicates the choice of time constant for the washout feature can be made on a logical basis in that it should be close to the speed time constant of the system, which is twice the equivalent inertia constant (2.0 H). Experience has indicated a value closer to H to be reasonable and, for a predominantly hydro-electric system, this would be of the order of 3 seconds, rather than the 1.6 seconds used.

When the governor speed setting is changed to meet changes in load, the generator moves to a new load angle. This requires a small, but definite, speed deviation. Did the authors experience any undesirable effects on generator terminal voltage due to the action of the speed signal under these circumstances, and have they introduced any means of correction?

This discussor was involved in the development of a stabilizing signal with British Columbia Hydro and Power Authority.[5] Field tests were conducted at a small hydro-electric plant. Results very similar to those shown in Fig. 8 were obtained with a generator loading of about 80 percent. However, no oscillations occurred when the generator was operated at 110 percent loading with its normal rotating exciter and magnetic amplifier. In other words, dynamic instability was introduced by going to a static exciter, and was then controlled by the stabilizing signal. It is not suggested that this is the case in the tests described in this paper. It is, rather, a word of warning that static exciters may cause instability if used without an adequate stabilizing signal.

While very good damping is exhibited in Figs. 9 and 10, the field voltage swings appear to lead the angular swings by about 120° of a swing cycle, instead of the ideal 90°. This may be due to the low value of the "washout" time constant used, which could have an effect of this nature, and it is suggested that a larger value might further improve the results.

Noting that the speed sensing equipment is situated in a position subject to severe vibration and difficult of access, would the authors care to comment on any anticipated maintenance problems?

References

[5] R. M. Shier and A. L. Blythe, "Field tests of dynamic stability using a stabilizing signal and computer program verification," presented at the IEEE Summer Power Meeting, New Orleans, La., July 10–15, 1966.

J. W. Skooglund (Westinghouse Electric Corporation, East Pittsburgh, Pa.): The authors are to be congratulated on their pioneering effort in the utilization of the excitation system to improve system damping and stability. Although the authors' efforts, as reported, were limited to hydro-electric generators, the principles developed are also applicable to thermal units. As additional interconnections are completed throughout Canada and the United States, additional damping of thermal units may be required to reduce tie-line oscillations.

A shaft speed stabilizing signal ($\Delta\omega$) has been shown to be very effective in improving damping and dynamic stability. What other signals were investigated and what was the basis for choosing $\Delta\omega$ over the others?

What is the time constant or phase delay of the stabilizing signal from the input of shaft speed to the output applied to the voltage regulator? It would be helpful if the transfer functions, including time constants and gains, of the entire excitation system were shown indicating how the $\Delta\omega$ signal supplements the regulator voltage damping circuits.

Is it planned to apply the stabilizing signal to those units with rotating exciters? High ceiling, rectifier excitation systems certainly offer advantages in improving transient stability. However, a low time constant excitation system is not required to achieve a significant improvement in system damping with an auxiliary stabilizing signal. The required stabilizing signal transfer function is, admittedly, more complex, but achievable with practical circuitry.

Manuscript received February 15, 1967.

R. L. Krahn (Corps of Engineers, Portland, Ore.): This paper is timely in that it deals largely with the problem of system damping which has recently been receiving great attention as the result of digital computer simulation of large systems. Such studies are indicating the poor damping forces that may exist as system designs become more refined to attempting to obtain optimum economics.

The use of high-speed excitation systems has long been recognized as a relatively inexpensive method of increasing stability limits in many cases. Static excitation systems appear to offer the practical ultimate in high-speed performance, thereby providing maximum assistance to transient stability limits. Unfortunately, the very high speeds and gains that give them this capability also result in poor system damping. Special stabilizing methods, such as the supplementary speed signal discribed by the authors, therefore become necessary.

With full appreciation of the difficulties in obtaining a signal from the machine shaft, I would like to ask why this was considered necessary rather than using the terminal frequency? The two quantities are related (thought not the same) and the latter is much easier to obtain. The use of terminal frequency would also allow a single stabilizing network to be applied simultaneously to more than one generator in a multiunit station.

To obtain maximum damping effect from the stabilizer, our studies show that the change of internal voltage due to excitation system action should always be in phase with the speed deviations. For low-frequency oscillations where the phase lags are small, the stabilizer input into the regulator can therefore be in phase with the speed deviation. For higher frequency oscillations, however, the signal should be advanced in phase to compensate for the phase lag through the field time constant. Ideally, the stabilizing network should be designed to automatically adjust the phasing to match the oscillation frequency. If not, significant damping can still be achieved with a static excitation system by tuning the stabilizer to a particular dominant frequency. It is assumed that the latter method was used in the equipment described in the paper. When attempting to apply speed stabilization to systems with rotating exciters, however, the use of automatic phasing is believed more necessary to avoid aggravating some "untuned" frequencies.

The field demonstration of the improvements in stability from the stabilized static excitation systems is particularly impressive. Actual proof of the benefits must be gratifying when one has risked so much in a pioneering effort.

My congratulations are extended to the authors for such a bold approach to system design. The increasing application of such static excitation systems will attest to their foresight and resourcefulness.

Manuscript received February 17, 1967.

F. R. Schleif and **H. D. Hunkins** (Bureau of Reclamation, Denver, Colo.): The excellent work described in this paper is especially interesting to us because of some parallel aspects in our respective efforts and some differences. The purposes of our efforts have been similar, but the apparatus we deal with differs considerably. Although the rectifier-type excitation systems would be regarded better candidates for special controls in new units, there are many units with rotating exciters and regulators already in existence or nearing completion in the various power systems now interconnected in the West. In this regard, the rotating-type exciter systems have come under study by the Bureau of Reclamation[6] and Corps of Engineers. Despite the slower response of these systems, results with supplemental signals are surprisingly similar to the rectifier-type systems, but they require more persuasive signals and signal-conditioning networks.

The sophisticated system for speed sensing described in the paper is impressive and it seems certain to serve many applications. Notwithstanding the disappointment of the authors with other signals, we have found terminal frequency to be quite successful also as a source for supplementary excitation control, even where successive

Manuscript received February 23, 1967.

derivatives of the signal are employed. Although the shaft speed signal and the terminal frequency signal may have some subtle differences, perhaps the differences can be used to advantage. Referring to Fig. 3 of Schleif *et al.*,[6] the application of a load increment will cause an immediate increase in lag of terminal voltage behind internal voltage. Thus the terminal frequency deviation signal may slightly precede the shaft speed signal.

The mention in this paper of poor damping characteristics of hydraulic generators interests us, particularly since it differs from our own experience. We have found damping torques from the amortisseurs, which are proportional to slip or rate of change, to be quite appreciable toward the higher end of the range of dynamic frequencies, that is from 50 cycles/min on up. Their influence may be limited to some extent by system impedance; however, more of their benefit may be offset by the adverse influence of such high-speed excitation systems as the rectifier type if they operate only to maintain constant terminal voltage. An explanation of the mechanism by which fast voltage control diminishes the inherent stability has been offered.[6]

Since part of the damping torque by excitation control comes from the acceleration or deceleration of machine inertia as the generator internal angle changes in response to excitation, influence available from this control is strongest for the higher swing frequencies, in general diminishing below about 25 cycles/min.

We have found the damping torques obtainable from hydraulic turbines and governors to be quite good up to above 3 cycles/min, depending upon water inertia. With special control[7] this can be extended up to about 12 cycles/min. Thus, substantial aids to damping and dynamic stability at the upper and lower ends of the swing frequency range have become available, there remaining for the moment some limitation at intermediate frequencies on supplementary damping torques which can be generated. With such ingenuity as that demonstrated by the authors, perhaps the deficiency can be filled soon.

The authors are to be commended for a significant, straightforward contribution to power system stability, bolstered by the demonstration of techniques and hardware.

References

[6] F. R. Schleif, H. D. Hunkins, G. E. Martin, and E. E. Hattan, "Excitation control to improve powerline stability," presented at the IEEE Winter Power Meeting, New York, N. Y., January 29–February 3, 1967.

[7] F. R. Schleif, G. E. Martin, and R. R. Angell, "Damping of system oscillations with a hydrogenerating unit," *IEEE Trans. Power Apparatus and Systems*, vol. PAS-86, pp. 438–442, April 1967.

C. Concordia (General Electric Company, Schenectady, N. Y.): This paper has made a very worthwhile and timely contribution by demonstrating that it is completely practical to employ shaft speed as a stabilizing signal for generator excitation systems. While it has been recognized for a long time that shaft speed was a very effective stabilizer, it was also recognized that there might be many practical difficulties in constructing a dependable, accurate, and noise-free sensor. I have just a few comments.

1) Although the introductory paragraph of the paper states that "stability limits which approximate those which can be obtained with a zero reactance generator" were obtained, it is pointed out in the discussion of the test results that the limit actually corresponded to the theoretical stability limit of an unregulated generator having a synchronous reactance of 25 percent on the generator base. (In this case the stability limit was approximately 97 percent of that which would have been obtained with a zero reactance generator.) This shows, first, that the stability limit was not that of a zero reactance generator but rather that of a generator as represented approximately by its transient reactance, and, second, that expressing the power limit as a percentage of the limit which would be obtained with a zero reactance generator is a very insensitive way of evaluating the relative performance of different regulating systems. It seems to us much better to use an equivalent generator reactance. Our previous studies have indicated that it is in practice very difficult to reduce the equivalent reactance below the transient reactance, and this seems to be confirmed by the authors' experience.

2) With reference to the authors' discussion of comparisons with test, we will mention that the component of generator torque caused by the direct current, which may or may not exist during the fault, is very important in determining the magnitude of the generator initial exceleration. For a double line to ground fault this dc component can vary all the way from zero to an amount equal to that for a 3-phase fault with simultaneous closing. The corresponding torque can vary from zero to about 50 percent. In making stability calculations, it is our usual practice to assume a rather small value for this component of torque in order to produce a fault which can be more severe than that which would be observed in a staged test. This is also true for a 3-phase fault, where a staged test with simultaneous closing will result in a much milder disturbance than might occur in real life.

3) The higher damping observed in tests is ascribed by the authors to "the simplified representation of the conventional excitation systems." We are somewhat surprised at this, since it has been our experience that simplifying the representation of a control system almost always results in a more stable system.

4) In common with several other recent papers, the present one refers to the damping torque produced by the amortisseur as a function of the frequency of oscillation. Although amortisseur damping does vary with frequency, it is much more a function of the armature circuit impedance. The relation with frequency is largely indirect since natural frequency is also affected by impedance.

5) Although the authors have been very successful in using shaft speed as an effective stabilizer, we should point out that a great many signals can be used as stabilizers if they are used in conjunction with appropriately chosen transfer functions, since, if the generator is oscillating, practically all quantities are oscillating.

Manuscript received February 24, 1967.

C. J. Goodwin (Manitoba Hydro, Winnipeg, Manitoba, Canada): The authors have made a strong case for the need to have effective stabilizing signals for these high-speed rectifier excitation systems, the signal proportional to change in generator speed being found the most effective. What other signals were investigated on the analog computer? Did any other types of signal approach the expected performance of the change of speed signal?

In introducing the speed signal to the excitation controller a "washout" time constant of 1.6 seconds is used. Will the authors state whether this value was determined during commissioning or during the simulation studies?

Manuscript received February 24, 1967.

H. Glavitsch (Brown Boveri and Company, Baden, Switzerland): The authors are to be highly commended for their excellent work out in the field, on the computer, and in presenting the material. An insight into the problems of the system of northern Ontario at an early stage of the development has lead us to a more extended study of the theoretical background of the stability of synchronous machines as affected by signals other than voltage error. A few remarks on this problem may be appropriate.

A rectifier excitation system allows the field voltage to change from negative to positive ceiling or vice versa with practically no delay. Having this in mind and knowing that the field voltage can affect the motion of the rotor of the synchronous machine, a simplified system consisting of a synchronous generator connected to an infinite bus was studied with the intention of finding the optimum control of the rotor motion.[8] The results of this study showed that the field voltage has to be switched in a bang-bang fashion and the switching instants can be derived for the model under consideration from slip and angular acceleration of the rotor in addition to the rotor angle in

Manuscript received February 24 1967.

order to obtain optimum control. For the linearized model, switching surfaces could be devised and programmed for the computer study. Since, however, the exact optimum requires a highly sophisticated control system which is hard to justify by the results, a control system was designed which consists of a regular electronic voltage regulator and a device which derives a slip and an acceleration signal from the active power measured at the terminals of the synchronous machine. The last mentioned device employs power transducers and conventional analog computer circuitry. These signals are fed over adjustable gains together with the voltage signal to the grid control set of the rectifier. With this realistic control system at hand the results of the optimum control system mentioned were used as a standard of perfection to adjust the gains in such a way as to give responses as close as possible to the desired standard. The results of the computer study indicate that such a control system can yield an almost optimum response whereby the settings are independent of the load of the synchronous generator. This is still true when armortisseur windings and saliency of the generator are taken into account. These results also indicate that the acceleration signal is desirable for the improvement of generator stability.

References

[8] H. Glavitsch, "Möglichkeiten der Verbesserung der Stabilität und der Spannungsregelung von Synchronmaschinen mit Hilfe der Gleichrichtererregung," to be published *Elektrotechnik u. Maschinenbau (Austria)*, vol. 85, no. 2, 1968.

A. R. Scott (Canadian General Electric, Ltd., Peterborough, Ontario Canada): Computer simulation is an important tool in determining probable methods which may be employed to improve the performance of control systems of various types. Recognizing the usefulness of a particular parameter which can be easily obtained in a noise-free state from the computer simulation and achieving equivalent results in the field are entirely different problems. The authors have reported an achievement demonstrating considerable engineering skills and patience in the successful results which they report.

The paper underlines the growing importance of viewing a power system as a large interacting control system and demonstrates the very substantial improvements in dynamic performance derived from well-designed excitation systems. The voltage regulation function of these excitation systems clearly becomes of secondary significance in this situation. Indeed the damping action depends on the ability of the stabilizing signal to override the voltage regulator and to produce terminal voltage modulation in a different phase position to that which would accompany angle swings. The excitation system thus improves upon a constant terminal voltage mode.

The importance of excitation system action in the real system underlies the need for adequate control system representation in system studies, especially when the adequacy of total system damping in the proposed design is being evaluated. The authors have noted a situation with their own program where the approximate representations available for conventional excitation systems did not provide a sufficiently close agreement in predicting actual damping performance of the system.

The behavior of a speed stabilizing signal under overspeed conditions causes some concern, especially when load rejection occurs with some portion of the transmission system remaining connected to the generators. It would seem preferable to disable the speed signal at say, two percent above the predisturbance speed. With the level of gain quoted in the paper this would in effect limit the voltage rise to about 15 percent, but, more important, when further increase in speed took place the terminal voltage would rapidly be restored to rated value. This would be a special benefit at higher speeds where the Ferranti effect becomes of increasing importance.

Finally, will the authors state whether the instability noted in the computer tests of "conventional" system was first swing or subsequent swing instability?

Manuscript received March 2, 1967.

J. E. Hardy, D. G. McFarlane, and **R. M. Shier** (British Columbia Hydro and Power Authority, Vancouver, B. C., Canada): The authors are to be complimented on their excellent contribution in the area of excitation control to improve generator stability limits.

In our opinion, a speed-derived signal introduced into a high-speed excitation system is certainly the most effective signal for damping transient swings. However as experienced by the authors, many difficulties are inherent in obtaining this type of signal. The British Columbia Hydro and Power Authority considered the use of a speed signal for their Peace River excitation, and numerous computer and field tests were performed in an attempt to develop a suitable signal. It was found that a signal derived from accelerating power could be used to provide the necessary damping.[5]

An optical device to measure machine speed, under development by the British Columbia Research Council, has been tested on a B.C. Hydro 120 r/min generator. Pairs of mirrors are mounted directly around the generator shaft. The time interval between receipt of light beams from the two mirrors in the pair is measured to provide an electrical signal corresponding to generator speed. The system is inherently insensitive to translational motion of the shaft and can detect speed changes of less than 0.01 percent of synchronous speed.

In applying the damping signal to an excitation system, the gain employed for voltage error and speed deviation usually interact with each other. What values of voltage error gain are being used by the Hydro-Electric Power Commission? How was the speed deviation of 0.15 percent to drive excitors to ceiling arrived at? Is this the maximum value obtainable or did studies indicate that further sensitivity was unnecessary or undesirable? Finally, with regard to this damping signal, has Hydro-Electric compared their system with other derived signals?

The field results shown in Figs. 9 and 10 show angles between Pinard and Des Joachims. Was this angle measured or calculated? Was there any evidence of swings occurring between the various generating stations during the tests?

The speed signal can be of great help in damping out system oscillations following a transient recovery; however, it can also produce other problems. One such problem noted in the paper is that of load rejection. The generator speed will increase and the excitation system through its speed signal will force voltage higher. It is stated that an upper limit of 115 percent of rated is provided. Is any provision made for removal of the signal for such conditions? Also, have the authors experienced any voltage variations during normal governor action for load changes?

This paper represents a significant contribution in the field of stability and provides practical evidence of what can be accomplished. Hopefully B.C. Hydro can contribute such evidence of practical application in the near future at our Peace River generation plant.

Manuscript received March 17, 1967.

P. L. Dandeno, A. N. Karas, K. R. McClymont, and **W. Watson:** We are very gratified by the interest and knowledge displayed by the discussors and thank them for their valuable contributions.

We have arranged our replies to the questions by subject matter since a number of the questions cover overlapping areas.

Choice of Stabilizing Signal

The use of signals derived from terminal voltage, armature current, power, reactive power, and field current, in addition to speed, were investigated on a computer representation of a simple two or three machine system. Originally we thought that a number of these quantities could be used to stabilize the generator by choosing appropriate lead or lag networks. It was considered essential that the stabilizing signal provide damping which would be relatively independent of the impedance external to the generator. With this requirement, none of the other quantities appeared to offer the same

Manuscript received April 17, 1967.

promise as signals derived from generator speed. Mr. Hardy, Mr. McFarlane and Mr. Shier, Mr. Skooglund, and Mr. Blythe report favorable experience with a speed signal derived from generator power. We also considered deriving a signal proportional to generator speed from a measurement of generator power as we had developed a device to provide such a signal for use in our power swing relay.[9] We did not pursue this method because we thought it would not provide a good indication of speed variation near the steady-state stability limit where power variations are small and it requires considerable filtering to remove 120-cycle components under fault conditions.

Mr. Glavitsch reports that optimum control is obtained with slip and angular acceleration signals. We acknowledge his assistance with our analytical studies and look forward to reading the forthcoming article he refers to.

Mr. Schleif, Mr. Hunkins, and Mr. Krahn have obtained good results with a frequency signal measured at the generator terminals. One would expect that for many applications the effect of this signal would be almost the same as that of frequency signal derived from the generator shaft. Possible advantages of measuring speed rather than terminal frequency are

1) the speed signal starts out as a high-frequency signal which is easily filtered without introducing significant time constants
2) the speed signal is isolated from electrical transients which presumably might in some instances give an erroneous signal
3) the frequency deviation at the shaft is normally greater than at the terminals.

It appears to us that a speed signal, a frequency signal, or speed signal derived from accelerating power will give very similar results, and the choice between signals should be decided on an individual application basis.

Mr. Krahn suggests that it is necessary to tune the stabilizer to a particular dominant frequency and Mr. Schleif and Mr. Hunkins suggest that the damping produced by the speed stabilized excitation system is effective over a limited frequency range. It appears that these conclusions are based on their valuable work with rotating exciters and that controlled rectifier excitation systems have different characteristics. Our analysis has shown that with a controlled rectifier excitation system having generator voltage and speed as inputs, the damping is effective over the entire frequency range of significance from a power-stability standpoint. The effectiveness at the low-frequency end is controlled by the choice of the washout time constant as discussed next.

Excitation System Transfer Functions and Computer Representation

A transfer function for the static excitation system is shown in Fig. 11. The gain of the speed deviation signal was determined from computer studies which showed that a gain of this magnitude was desirable. The exact value of gain is not extremely critical and satisfactory results would be obtained at twice this gain.

The computer representation of the rotating exciter excitation system used in obtaining Fig. 10 is shown in Fig. 12. A revision to our computer program now permits a more accurate representation of the rotating exciter excitation system. The transient swing obtained with the improved representation and the transfer function are shown in Figs. 12 and 13.

Washout Time Constant

In reference to Mr. Goodwin's comments on the magnitude of the washout time constant, we found that from a stability standpoint, the washout time should be as long as practical. However, for this application we selected a short washout time constant so that if the generators become separated from the system with a sustained frequency deviation, the voltage regulator would quickly resume control.

For normal system connections, the oscillation frequency of the generator will be high enough that the washout feature will not have a significant effect. As noted in the paper, as the generator approaches the dynamic stability limit, the oscillation frequency decreases and approaches zero at the limit. Therefore, for the tests a longer washout time or ideally no washout would likely have increased the dynamic stability limits.

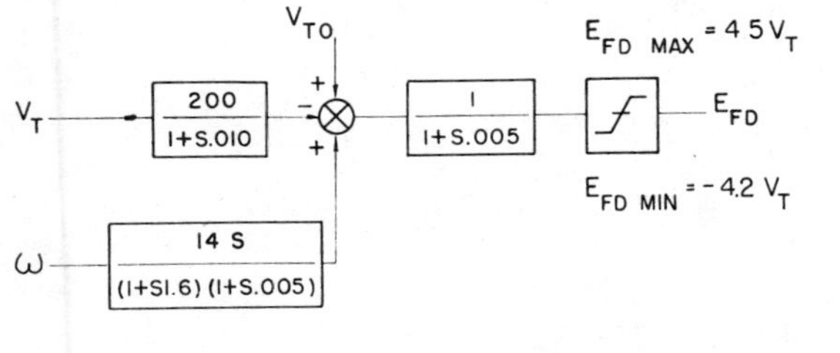

Fig. 11. Transfer function of typical controlled rectifier excitation system.

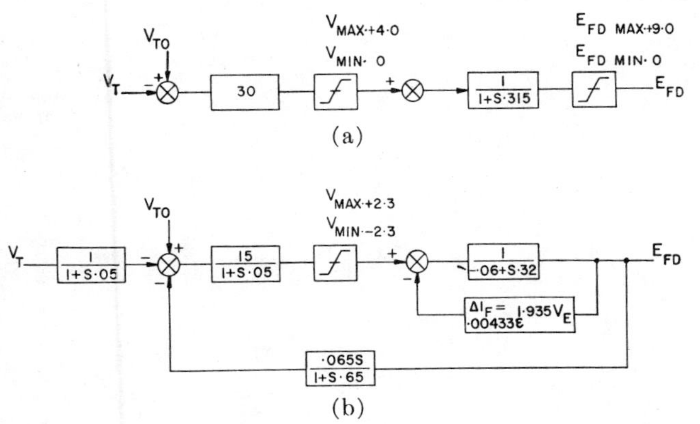

Fig. 12. Computer representation of an Otter Rapids rotating amplifier voltage regulator and rotating exciter. (a) Representation corresponding to Fig. 10. (b) Representation corresponding to Fig. 13.

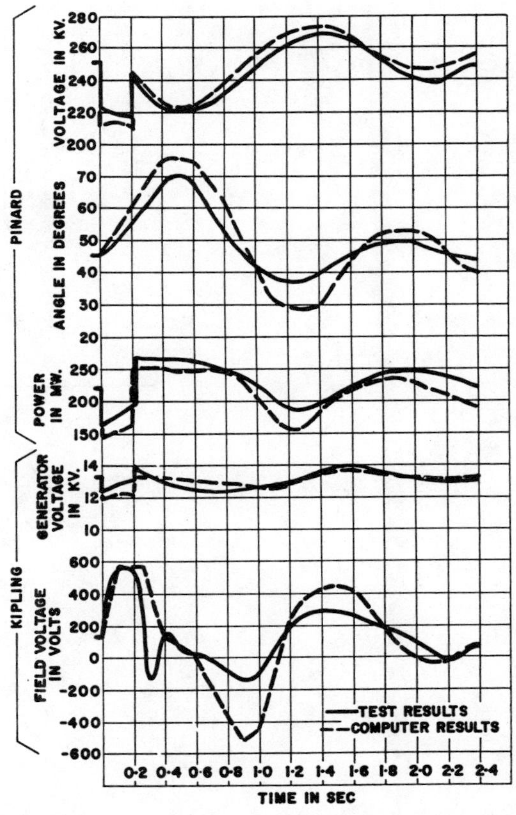

Fig. 13. Time plot for transient stability test—improved computer representation of Otter Rapids excitation system, two Kipling and two Otter Rapids units.

Test Results

Mr. Concordia's early and classic work on the affect of excitation systems on stability limit is well known.[10] He, Mr. Scott, and their colleagues have provided valuable assistance and guidance, particularly in our early analytical investigations in this area.

In the paper, operating limits are expressed in terms of the theoretical limits for unregulated generators having a specified synchronous reactance. It would not of course be practical to operate an unregulated generator at these theoretical limits and we are not proposing the use of equivalent reactances in analytical investigations involving the determination of system facilities. Expressing the operating limit as a percent of the limit obtainable with a zero reactance machine shows the power margin between this latter limit and the operating limit.

Our computer studies show the limit obtainable with a controlled rectifier, speed-stabilized excitation system would correspond to the theoretical limit of a zero-reactance generator. We believe that the main reason this was not demonstrated in practice was due to the effect of the washout time constant, as discussed. It is perhaps significant too that the power margin between the lowest operating point and the zero reactance stability limit was only about 7 MW.

The solid curves for the Pinard-Des Joachims angles shown in Fig. 10 were obtained by measuring the angle between the Pinard voltage and a reference voltage derived from a stable 60-cycle oscillator. Therefore, the change in angle was measured but the initial value of the angle was calculated. There was no evidence of swings between generating stations during the tests.

Effect of Normal Load Changes

We have not experienced any undesirable voltage variations as a result of governor action for load change.

Overvoltages

Mr. Stillman asked for an explanation of the difference between the limit setting and withstand voltage. The generator voltage limit prevents the voltage from exceeding about 115 percent on a sustained basis when such control is within the capability of the excitation system. The withstand value of 170 percent of rated voltage was specified to ensure that the excitation equipment would not fail if exposed to this high overvoltage due to self-excitation or for short periods before the regulator could reduce the voltage.

As an additional protection against excessive voltage under load rejection conditions and to eliminate excessive exciter voltage variations at partial loads where water effects may cause high-frequency rotor oscillations, the speed signal is removed when gate position is below about 75 percent.

Other Methods of Speed Measurement

Stroboscopic methods of speed measurement were considered but were eliminated primarily because their output is affected by lateral movements of the shaft. The optical device described by Mr. Hardy, Mr. McFarlane, and Mr. Shier for measuring shaft speed independent of translation motion is interesting as it provides another way of producing a speed signal.

The discussors have made many important points which do not require any comments on our part. We are grateful to the discussors for their contributions.

REFERENCES

[9] R. D. Brown and K. R. McClymont, "A power swing relay for predicting generation instability," *IEEE Trans. Power Apparatus and Systems*, vol. PAS-84, pp. 219–224, March 1965.

[10] C. Concordia, "Steady-state stability of synchronous machines as affected by voltage-regulator characteristics," *Trans. AIEE*, vol. 63, pp. 215–220, May 1944.

IMPROVED STABILITY WITH LOW TIME CONSTANT ROTATING EXCITER

H. R. Perry J. F. Luini J.C. Coulter
Pacific Gas and Electric Company
San Francisco, California

Abstract—This paper discusses the improved stability performance offered by a low rise time, high response excitation system to be installed on Unit No. 1 at the Diablo Canyon Site. Such an excitation system is provided by a low time constant rotating exciter with minor loop current feedback.

INTRODUCTION

In the early 1960's several factors were present in the northern and central California area that led to the decision by the Pacific Gas and Electric Company to install an extensive 500 kV transmission system that would overlay the 230 kV system in the area. The principal factors were (1) the economics of higher voltage transmission for a system that would soon reach a 10-million kilowatt peak and the increasing difficulty of acquiring transmission line rights-of-way, (2) the economics of going to 750 megawatt and larger generator size, and (3) the participation in the construction of the Pacific Northwest-Pacific Southwest Interconnection.

This decision brought with it the need to give more study to the problem of system stability. This was not surprising since the growth of the 230 kV system over the recent decades had resulted in a general improvement in stability. Long 230 kV transmission lines were being drastically shortened by the construction of intermediate substations with less of the remote hydro generation being transmitted long distances to the San Francisco Bay Area load center. Also, the newer and larger thermal units were being located nearer load and there was general improvement in breaker clearing times. As a result of these developments, periodic stability studies made during the 1950's and early 1960's showed the system becoming more stable with time.

These pre-EHV stability investigations were made on an AC Network Calculator and later by digital computer using the "classical" solution method. These studies were run only for the first swing duration. The excitation system on most of the post-World War II machines was an 0.5 ASA response rotating d.c. generator controlled by a continuous acting voltage regulator. The primary objective at that time was to maintain desired steady-state voltage levels with little emphasis placed on the dynamic performance of the excitation system.

Early EHV stability studies, including the new large generators and the Pacific Intertie, were made using the classical method. Subsequent studies, as early as 1965, represented generator saliency and crude approximations of load characteristics, excitation and governor systems. The results of studies run for more than the first swing gave reasons for concern over the proper representation of the system dynamics—particularly the load behavior and excitation system. The overall results of these and more refined studies are outside the scope of this paper.

This paper will discuss the reasons for selecting a high response excitation system for Unit No. 1 at the Diablo Canyon Site. This is a nuclear unit scheduled for operation in 1973. The generator is rated 1300 MVA, .90 PF, 25 kV. A second unit of similar size is planned at the site for 1974 operation.

The paper illustrates that the low time constant rotating exciter with minor loop current feedback is a practical way of providing fast rise times. We further demonstrate that conventional rate feedback combined with a power system stabilizer is very successful in damping oscillations while maintaining high regulator gain in the feed forward loop.

Paper 71 TP 16-PWR, recommended and approved by the Power Generation Committee of the IEEE Power Engineering Society for presentation at the IEEE Winter Power Meeting, New York, N. Y., January 31-February 5, 1971. Manuscript submitted September 24, 1970; made available for printing October 30, 1970.

REGULATOR-EXCITER DESIGN REQUIREMENTS

With the construction of the Pacific Intertie, large centers of rotating mass were connected producing an interconnected system with a natural oscillation frequency of 1/3 Hertz, often modulated by higher local frequencies. This low frequency of oscillation required that studies be run for a longer time which disclosed the inadequacies of the then-current excitation system design. Not only were ceiling voltages too low, but field voltage corrections lagged 90° to 130° behind the desired corrections. This resulted in negative damping causing an increase in the magnitude of subsequent system oscillations. Subsequent simulation, design, and installation of power system stabilizers (supplementary excitation control) with frequency or shaft speed deviation input provided the necessary phase advance at a nominal investment[1]. In order to obtain the desired higher ceiling voltages, it would only be practical to purchase 2.0 ASA response exciters on generators ordered after the 1965 studies.

It is easily understood that maximum damping requires maximum volt-seconds at the right time to control generator field flux. The power system stabilizer provides the right time, and an exciter with a one or two-cycle ceiling voltage rise time will produce more volt-seconds. Others in the western United States have demonstrated that an SCR excitation system is also a very practical way of meeting the above rise times[2].

Based on the results of stability studies previously mentioned, the excitation system for Unit No. 1 at the Diablo Canyon Site was specified as a 2.0 ASA response with a rise time of one to two cycles.

The manufacturer provided preliminary design data that showed the maximum time constants for the regulator and exciter would be .02 and .015 seconds respectively. The ceiling voltage was estimated to be 4.45 p.u. and the nominal regulator gain to be 400. It was determined that the above four parameters would provide an excitation system that would satisfy the specified requirements. The selected exciter was to be an a.c. alternator with rotating rectifiers. Rate feedback would be taken from the alternator field current. Figure 1 shows in block diagram form the design data for the regulator-exciter system.

REGULATOR-EXCITER TESTS

Prior to shipping, extensive frequency response tests were performed by manufacturer's engineers on all excitation system components. Since most components are of solid state design, the frequency response corner frequencies are greater than 100 radians/sec. and therefore have a negligible effect on the dynamic performance. Consequently, regulator sensing filter, voltage error detector,

Reprinted from *IEEE Trans. Power App. Syst.*, Sept./Oct. 1971.

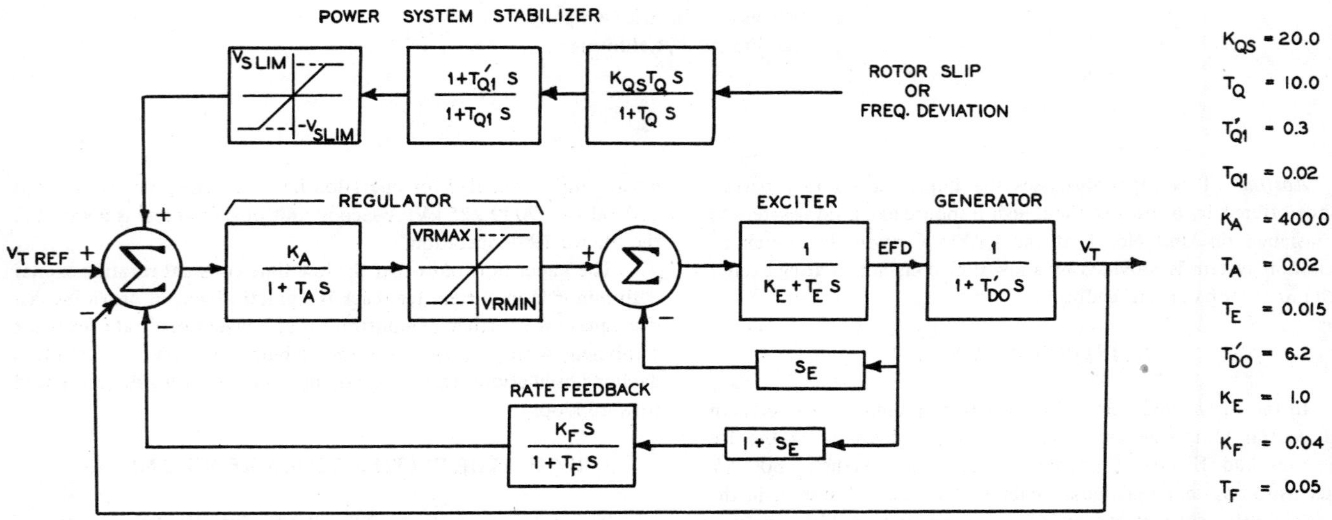

Fig. 1. Excitation and power system stabilizer block diagram with tabulation of design values.

power amplifier and exciter test results will be emphasized over other components.

Figure 2 lists the ratings for the PMG, brushless rotating rectifier exciter, and the a.c. generator. The solid state regulator provides the firing pulse information to modulate the PMG output to the a.c. alternator field. The alternator a.c. output is rectified by diodes mounted on the shaft and fed directly to the generator field.

Figure 3 is the plotted frequency response test data for the regulator sensing filter. The corner frequency indicates a time constant of .0067 seconds. Although the frequency response curves for the error detector amplifier are not shown, they indicate time constants of .0005 seconds at a minimum gain of .32 volt/volt, and .005 seconds at a maximum gain of 36 volt/volt. In addition, the time constant for the power amplifier is .0007 seconds with a gain of 112 volts/volt. If the regulator time constant can be approximated by the simple sum of the above components, it has a minumum value of .0079 seconds and a maximum value of .0124 seconds. These values are well within the estimated regulator time constant of .02 seconds of Figure 1.

The exciter design is unique because it employs current feedback to reduce the effective time constant of the alternator. Since the a.c. alternator output voltage is inaccessible (rotating rectifier is directly connected to the generator field) its field current is used as an approximation of its output. The field current signal is obtained from a current shunt. This current signal is conditioned by the time constant compensator and then fed back into the power amplifier. This compensation is equivalent to negative feedback around the alternator with a gain setting of about 10. The net effect is to produce an effective exciter time constant which is 10 percent of the actual time constant.

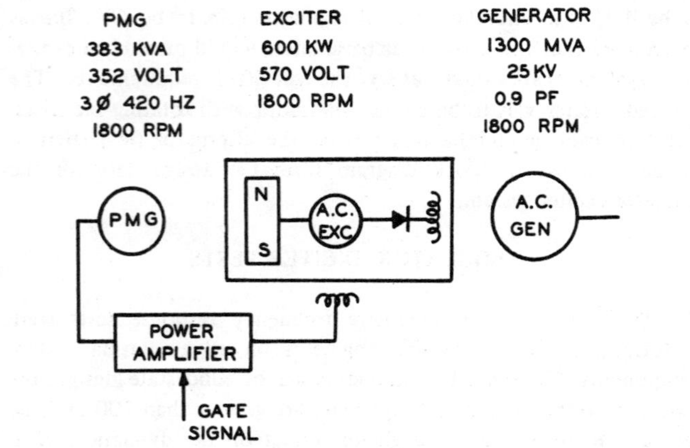

Fig. 2. Schematic of rotating exciter, PMG, generator, and regulator.

Figure 4 compares exciter frequency response tests with and without the time constant compensator. As desired, the compensated time constant is less than 10 percent of the uncompensated value of .125 seconds. Figure 5 illustrates the exciter response to a step input. As shown, the desired rise time of less than two cycles has been accomplished. The actual ASA response was calculated from Figure 5 to be 2.23 and the ceiling voltage was 5.0 p.u. (base voltage = 153 volts).

DISCUSSION

Presently, there are three conventional means of damping an excitation system. They are power system stabilizer, regulator gain reduction, and rate feedback[4]. A very effective damping approach is to maintain higher regulator gain, substantial rate feedback, and a power system stabilizer.

The high regulator gain provides high initial response. The rate feedback provides damping at the cost of closed loop gain and phase lag, and the stabilizer settings make up the lost gain and phase lag.

The lost gain and phase can be explained by considering the block diagram of Figure 1. If rate feedback is not used, the closed loop portion represented by V_T/V_{TREF} can be expressed as

$$\frac{K_A}{(1+T_AS)(1+T_ES)(1+T'_{do}S)+K_A} \qquad \text{Eq. (1)}$$

where S is the Laplacian operator. If rate feedback is used,

$$V_T/V_{TREF} = \frac{K_A(1+T_FS)}{K_A(1+T_FS)+[K_AK_FS+(1+T_ES)(1+T_AS)(1+T_FS)](1+T'_{do}S)} \qquad \text{Eq. (2)}$$

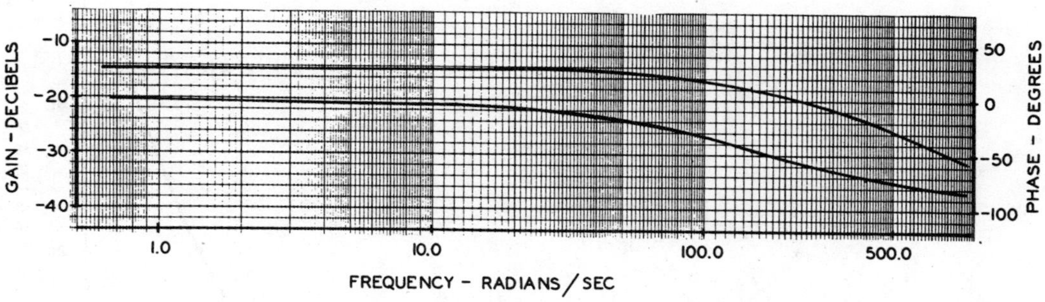

Fig. 3. Frequency response of the regulator error sensing filter.

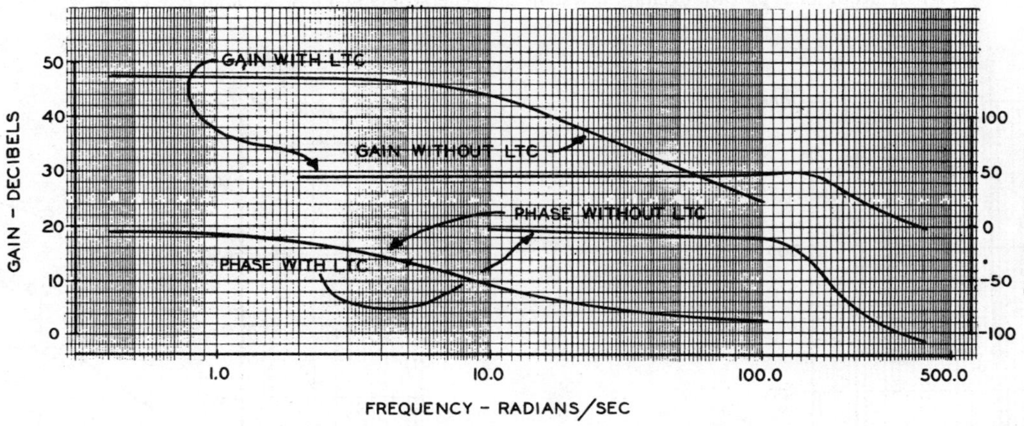

Fig. 4. Frequency response tests for exciter, with and without low time constant compensation.

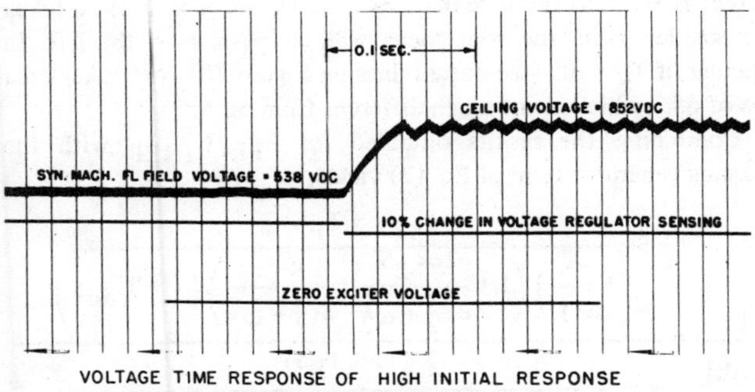

Fig. 5. Exciter ASA response.

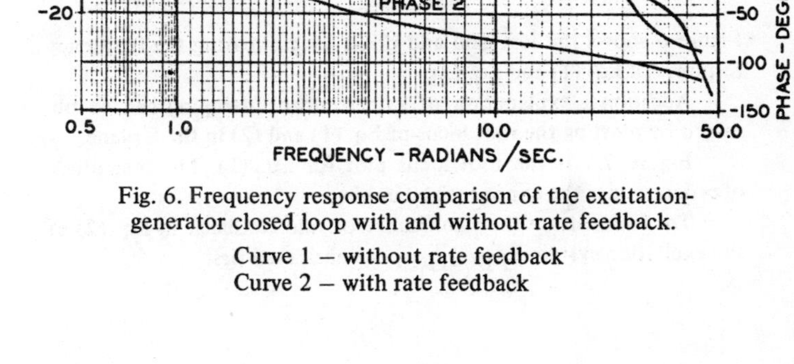

Fig. 6. Frequency response comparison of the excitation-generator closed loop with and without rate feedback.

Curve 1 – without rate feedback
Curve 2 – with rate feedback

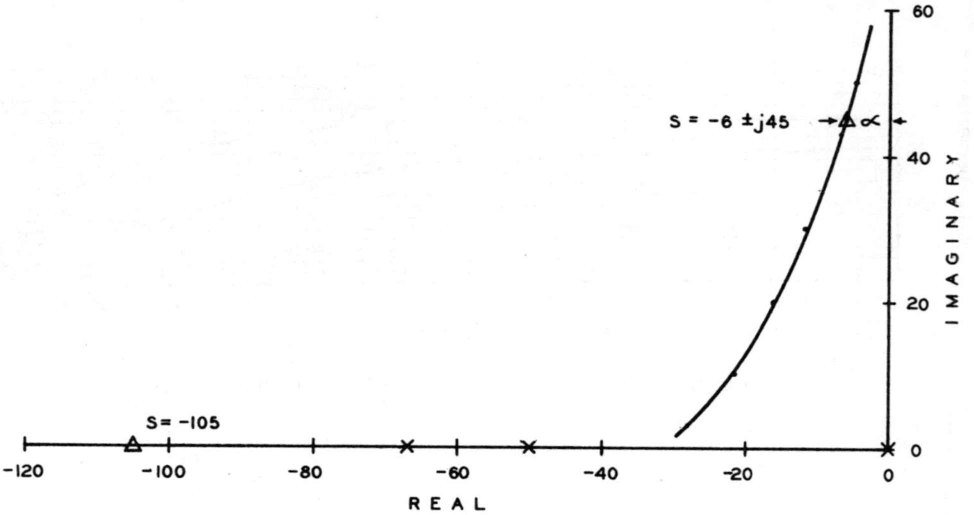

Fig. 7A. Root locus plot for excitation system and generator closed loop without rate feedback.

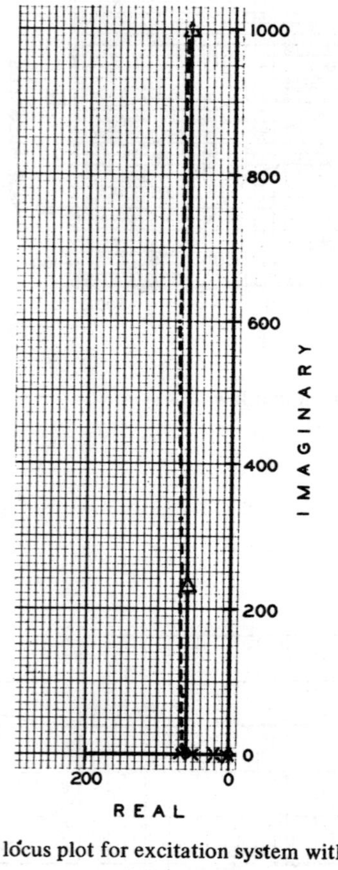

Fig. 7B. Root locus plot for excitation system with rate feedback.

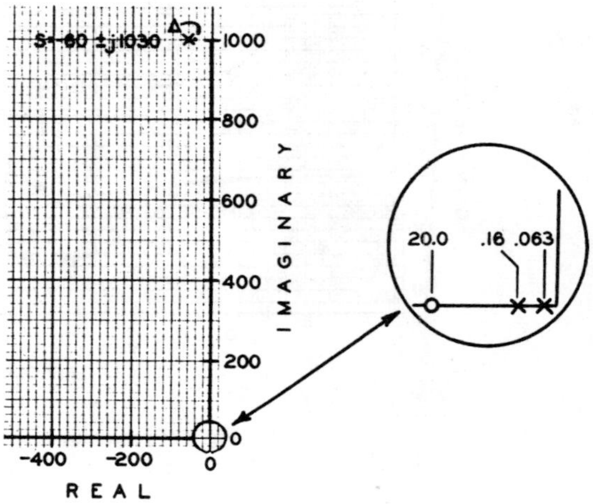

Fig. 7C. Root locus plot for excitation system and generator closed loop with rate feedback.

Figure 6 shows the lost gain and phase by comparing the frequency response of Eq. (1) and Eq. (2).

Additional explanation of rate feedback damping can be obtained by plotting the root locus of Eq. (1) and (2) in the S plane.

Figure 7A is the root locus plot for Eq. (1). The magnitude of α determines the exponential damping.

To demonstrate the significance of rate feedback in Eq. (2) of the excitation system E_{FD}/V_{TREF} is considered first.

$$E_{FD}/V_{TREF} = \frac{K_A(1+T_FS)}{K_FK_AS+(1+T_AS)(1+T_ES)(1+T_FS)} \qquad \text{Eq. (3)}$$

Figure 7B is the root locus plot for Eq. (3) for a rate feedback time constant $T_F = 1.0$. Notice here that α has a value of 58.5. If T_F takes on a smaller value the root locus will be pushed to the left. In particular, if $T_F = .05$ (see dotted lines on Figure 7B) α takes on a value of 68.3 which is not much different from 58.5.

Combining the results obtained for E_{FD}/V_{TREF} with the remaining generator term of Eq. (2) yields,

$$\frac{V_T}{V_{TREF}} = \frac{\dfrac{K_A(1+T_FS)}{\left(1+\frac{S}{\alpha_1}\right)\left(1+\frac{2\alpha_2S}{\alpha_2^2+\omega_2^2}+\frac{S^2}{\alpha_2^2+\omega_2^2}\right)\left(1+T'_{do}S\right)}}{1+\dfrac{K_A(1+T_FS)}{\left(1+\frac{S}{\alpha_1}\right)\left(1+\frac{2\alpha_2S}{\alpha_2^2+\omega_2^2}+\frac{S^2}{\alpha_2^2+\omega_2^2}\right)\left(1+T'_{do}S\right)}} \qquad \text{Eq. (4)}$$

which is equivalent to Eq. (2).

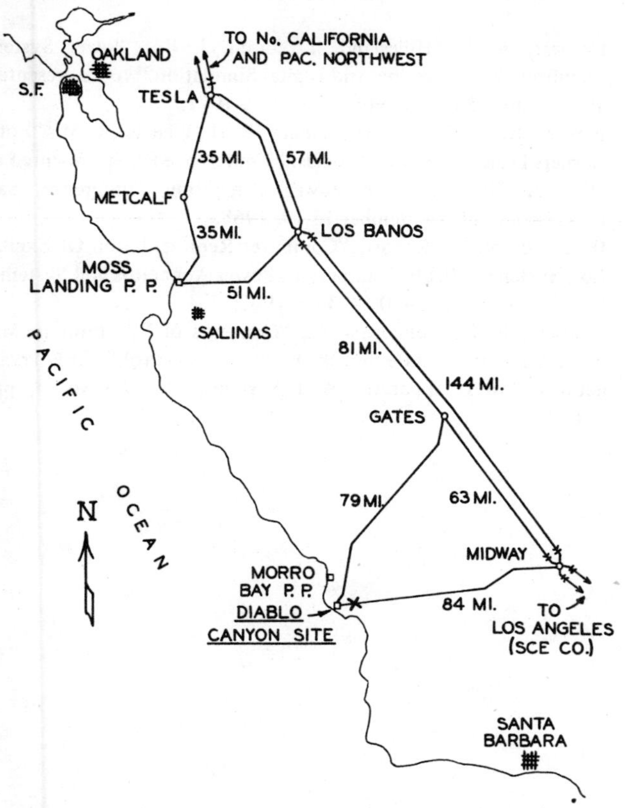

Fig.8. One line diagram of Diablo Canyon Site with fault location.

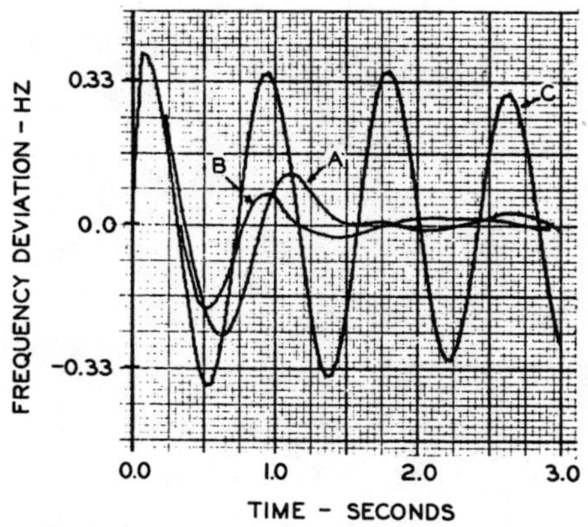

Fig. 9. Frequency deviation comparison for a 4-cycle fault on the Midway circuit adjacent to the 500 kV bus at the Diablo Canyon Site.
Curve A – 2.0 ASA conventional excitation system.
Curve B – low time constant excitation system with rate feedback.
Curve C – low time constant excitation system without rate feedback.

Figure 7C is the root locus plot for Eq. (4), which includes rate feedback. As can be seen, the dominant α's are much greater than those of Eq. (1) without rate feedback.

Damping can also be changed by varying the rate feedback gain K_F while leaving T_F fixed. Actually, a combination of both parameter changes would probably be more effective, as can be seen from the dotted line of Figure 7B.

Compensation for the detrimental effects of increased rate feedback can be obtained by appropriate power system stabilizer settings. Referring again to Figure 6, for T_F = .05 seconds the appropriate lead setting is at the 45° phase lag frequency, i.e., T'_{Q1}, (Figure 1) = 1/3.3 = .300 seconds. An acceptable lag setting, T_{Q1}, is

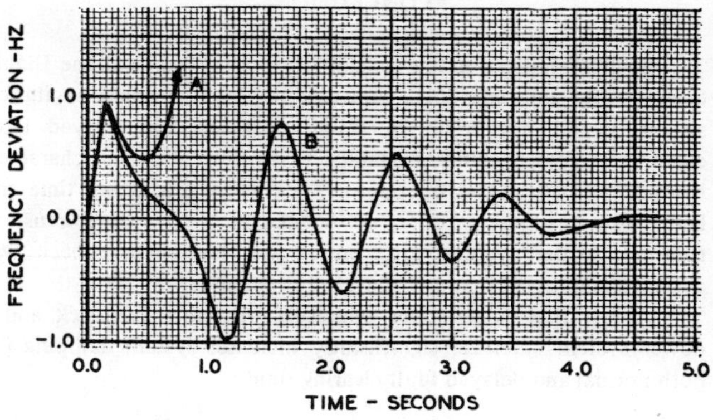

Fig. 10. Frequency deviation comparison for a 9.6-cycle fault on the Midway circuit adjacent to the 500 kV bus at the Diablo Canyon Site.
Curve A – 2.0 ASA conventional excitation system.
Curve B – low time constant excitation system with rate feedback.

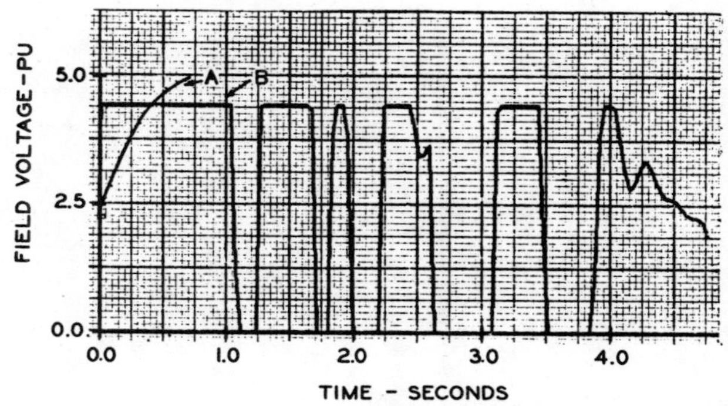

Fig. 11. Field voltage comparison for a 9.6-cycle fault on the Midway circuit adjacent to the 500 kV bus at the Diablo Canyon Site.
Curve A – 2.0 ASA conventional excitation system.
Curve B – low time constant excitation system with rate feedback.

.02 seconds. It should be emphasized that only one lead-lag stage is necessary with a high initial response exciter due to the low exciter time constant.

Of the many disturbances that could occur, a three-phase fault on the Midway circuit adjacent to the 500 kV bus at the Diablo Canyon Site was selected to demonstrate the results of the above concepts on system damping. Figure 8 shows a one-line diagram for the Diablo Canyon area with the fault location. Two fault conditions are compared; (1) normal 4-cycle fault clearing and line removal to Midway, (2) 9.6-cycle fault clearing and line removal to Midway. The 9.6-cycle fault represents backup fault clearing. For each fault condition a conventional 2.0 ASA excitation system is compared to the low time constant excitation system with rate feedback. Both systems include a power system stabilizer.

Figure 9 compares generator frequency deviation for the 4-cycle fault. Curve A is for the 2.0 conventional system, Curve B is for the low time constant system with rate feedback, and Curve C is for the low time constant system without rate feedback. Figures 10 and 11 compare generator frequency deviation and field voltage for the 9.6-cycle fault.

A comparison of Curves B and C on Figure 9 shows the importance of rate feedback and Curve B shows improved damping over Curve A. A more significant advantage of the low time constant–rate feedback system is shown in Figures 10 and 11. The difference amounts to a stable or unstable system and can be attributed to the fast rise time of the field voltage.

CONCLUSIONS

System studies for the installation of Unit No. 1 at the Diablo Canyon Site showed that a conventional high response excitation system would not result in stable operation for delayed fault clearing. Analytical studies enabled us to determine the characteristics needed for stable operation. These were a low rise time and high ASA response exciter. Subsequently a low time constant rotating exciter system with minor loop current feedback was designed and tested that met these requirements.

A low rise time exciter operating with rate feedback and a power system stabilizer significantly increases system damping for both normal and delayed fault clearing time.

REFERENCES

1. Gerhart, A. D., Hillesland, T., Luini, J. F., "Power System Stabilizer: Field Testing And Digital Simulation," to be presented at the Winter Power Meeting, February 1971.
2. Farmer, R. G., Kent, M. H., Hartley, R. H., Wheeler, L. M., "Four Corners Project Stability Studies," Conference Paper presented at 1968 ASME/IEEE Joint Power Generation Conference, San Francisco, Calif., September 15-19, 1968.
3. IEEE Committee Report, "Computer Representation Of Excitation Systems," IEEE Transactions Power Apparatus and Systems, Vol. 87/No. 6, pp. 1460-64, June 1968.
4. DeMello, F. P., Concordia, C., "Concepts of Synchronous Machine Stability As Affected By Excitation Control," IEEE Transactions Power Apparatus And Systems, Vol. 88/No. 4, pp. 316-29.

POWER SYSTEM STABILIZER: FIELD TESTING AND DIGITAL SIMULATION

A. D. Gerhart — San Francisco, California
T. Hillesland, Jr. — Emeryville, California
J. F. Luini — San Francisco, California
M. L. Rockfield, Jr. — Emeryville, California

Pacific Gas and Electric Company

Abstract—A field test program is described which was used to evaluate and then to adjust the 33 power system stabilizers installed on the Pacific Gas and Electric Company's system. Also included is a comparison of the results of these tests with those of a digital computer simulation.

INTRODUCTION

Many studies made in recent years have shown that instability of Pacific interconnected transmission and generating systems can occur if there is insufficient system damping. From these studies and field tests, it has been found that improved system damping can be achieved by adding special supplementary or stabilizing controls to the voltage regulators at selected plants.[1] The function of these power system stabilizers is to sense shaft speed or frequency deviations, condition the signal thus obtained, and then feed it into the voltage regulator as a supplemental input.

In 1966 the Pacific Intertie System Technical Studies Task Force conducted studies with the aid of P G and E's stability program. Following the subsequent Task Force recommendation P G and E prepared specifications and placed orders with two manufacturers for 33 power system stabilizers. These were installed at eight plants – five thermal, three hydroelectric – for 25 generating units with a combined capacity of 5,242 megawatts. Unit capacities range from 40 MW for a hydro unit to 750 MW for a cross-compound steam unit.

Since the stabilizers represented a new development for each of the two equipment manufacturers, it was considered desirable to field test a prototype unit from each manufacturer. This was done to check compliance with specifications and to check equipment performance. Each prototype test was conducted with the respective design engineers on location.

In addition, an extensive test program was undertaken to test each stabilizer and associated excitation control system (regulator, exciter and generator) before the stabilizer was placed in operation. This was done to obtain optimal adjustment of the equipment and to obtain data for computer modeling.

It is the purpose of this paper to describe the tests made, procedures and instruments used, and the results obtained.

POWER SYSTEM STABILIZERS AND VOLTAGE REGULATORS

The signal conditioning part of the power system stabilizers obtained from each manufacturer is essentially the same. Functionally, it includes a reset stage, two lead-lag stages, and a limiter. However, the manufacturers' design concepts differed somewhat for the input transducers. One manufacturer furnished frequency deviation transducers for the relatively slow speed hydroelectric machines, and speed-sensing transducers (using a shaft-mounted three phase tachometer as the signal source) for the steam units. The other manufacturer supplied frequency deviation transducers for both hydroelectric and steam units. The transfer function for these stabilizers is illustrated in the text.

Each of the generating units to which a power system stabilizer was added has a main exciter of the dc generator-commutator type and a continuously acting voltage regulator. The regulators are of three basic designs: (1) Impedance type, with a rotating amplifier output stage; (2) Magnetic amplifier type with a transistorized front-end and a rotating amplifier output; (3) Magnetic amplifier type, with a buck-boost magnetic amplifier output. Of the six generators described in this paper two with regulators of the second design have an American National Standard Institute voltage response ratio of 2.0. All other excitation systems have a response ratio of 0.5.

Simplified diagrams of these regulators showing inputs for the power system stabilizers are given in Figures 1-3.

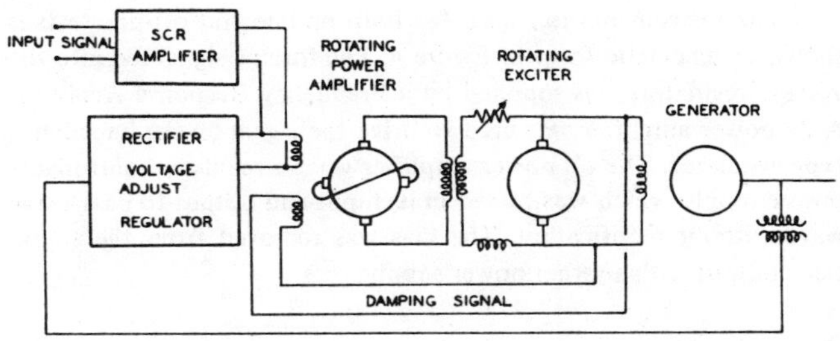

Fig. 1. Impedance regulator with rotating amplifier output.

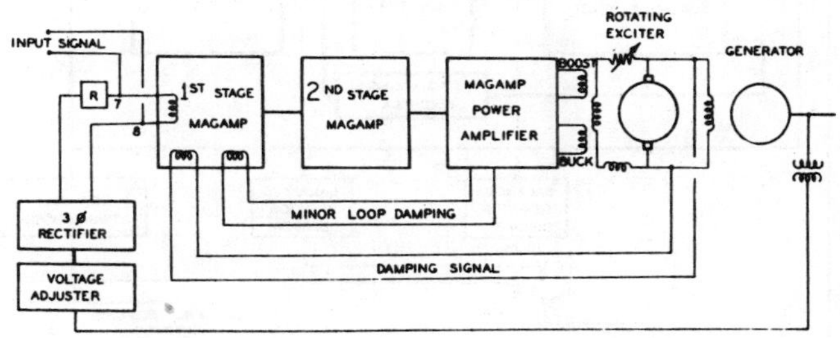

Fig. 2. Magnetic amplifier regulator with magnetic amplifier output.

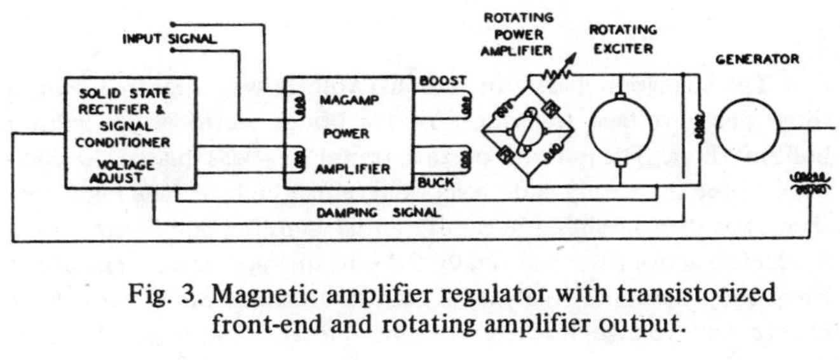

Fig. 3. Magnetic amplifier regulator with transistorized front-end and rotating amplifier output.

Paper 71 TP 77-PWR recommended and approved by the Power Generation Committee of the IEEE Power Engineering Society for presentation at the IEEE Winter Power Meeting, New York, N. Y., January 31-February 5, 1971. Manuscript submitted September 17, 1970; made available for printing December 3, 1970.

Reprinted from *IEEE Trans. Power App. Syst.*, Sept./Oct. 1971.

TEST DESCRIPTION, RESULTS AND INSTRUMENTATION

One of the first tests made was to determine the damping of the voltage regulator itself because of its effect on the overall performance of the excitation control system. With the machine at rated speed and voltage, with no load, a step signal was applied to the voltage regulator while monitoring generator terminal voltage on a chart recorder. Results showed that overshoot ranged from 1.6 percent to 65 percent. With one exception, attempts to improve upon this wide range of damping were unsuccessful since the adjustment range needed was not available.

Frequency response tests were made from 0.01 hertz to 5.0 hertz of the uncompensated excitation control system, system stabilizer alone, and control system plus stabilizer. Stabilizer lead-lag settings were determined from Bode plots of the fequency response data. These settings were also used for computer modeling.

The test method used on cross-compound units was to treat each machine as a single unit. The high pressure unit would be checked for gain and frequency response while its corresponding low pressure unit voltage regulator would be in the "test" position. The opposite conditions would exist when the low pressure unit was tested. Voltage regulators for other units paralleled on the same bus would be kept in the "test" position whenever possible. The cross-compound units were left connected on the low side during testing since they could only be separated with great difficulty. The control gain was set by raising the HP and LP control gains together.

The instrumentation used for both on-line and off-line tests is shown in schematic form in Figure 4. The forcing signal fed into the voltage regulators was supplied by a Frequency Response Analyzer. A dc power amplifier was used to drive the signal on the impedance type regulator. The dc power amplifier was in reality a controllable power supply which was biased in its input and output to pass a sine wave without rectification. The bias was removed from the amplifier's output with another power supply.

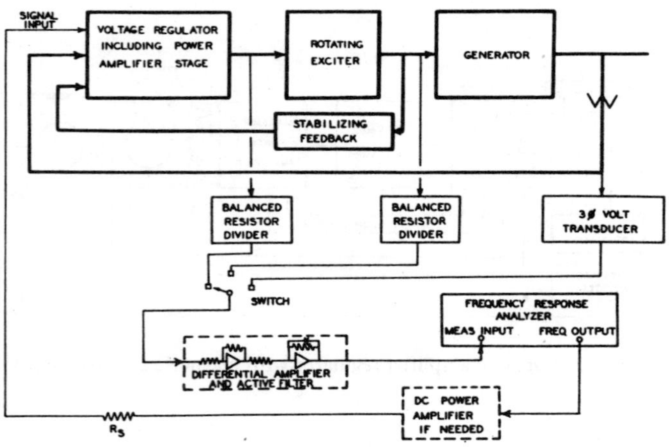

Fig. 4. Test instrumentation schematic

The change in generator terminal voltage was measured using a three phase voltage transducer of the bridge rectifier type with a built-in filter. The output of this transducer was biased to compensate for the steady-state level of terminal voltage. The remaining signal was then amplified by a differential amplifier and filtered using a selective active filter and finally fed into the measuring input of the Frequency Response Analyzer. The measurements of exciter voltage change and voltage regulator output (either rotating amplifier or power mag-amp output) were made in a similar manner except that balanced resistor dividers were used in place of the voltage transducer. The Frequency Response Analyzer would display digitally the change in amplitude of the return signal at the fundamental frequency of the forcing signal as well as its phase shift with respect to the forcing signal. This instrument saved a significant amount of data gathering and reduction time during the testing period in addition to providing a higher degree of accuracy than could be achieved by other measurement methods.

Typical results given in Figures 5 to 9 provide a frequency response comparison of the exciter-generator closed loop for the following conditions: calculated off-line using manufacturers' typical data, test off-line, test on-line, and test on-line with power system stabilizer.

These figures clearly indicate a difference in break frequencies between off-line and on-line tests. This difference can be directly attributed to loading effect of the system on the unit. In all cases the lead time constants were set for the latter break frequencies. In addition, all lag time constants were set to the minimum dial setting of .02 seconds to provide compensation over a fairly wide range of frequency.

Figure 10 illustrates the IEEE standard excitation model along with the lead-lag stages of the power system stabilizer. The lead time constants in the two lead-lag stages were set equal. They provide 90

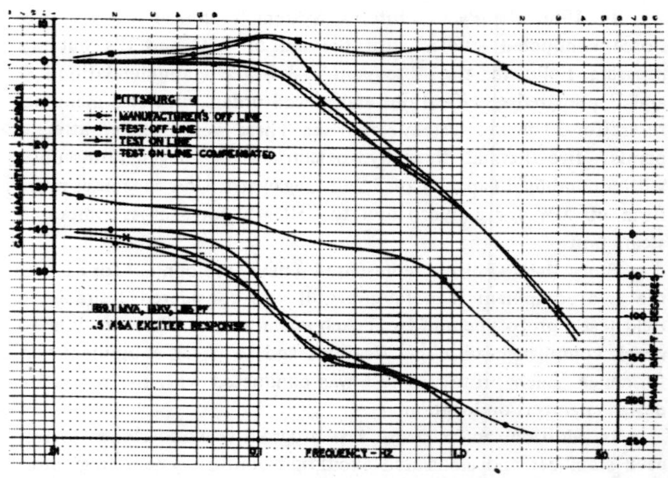

Fig. 5. Frequency response, impedance regulator with rotating amplifier output. Pittsburg 4, tandem compound, 169.1MVA.

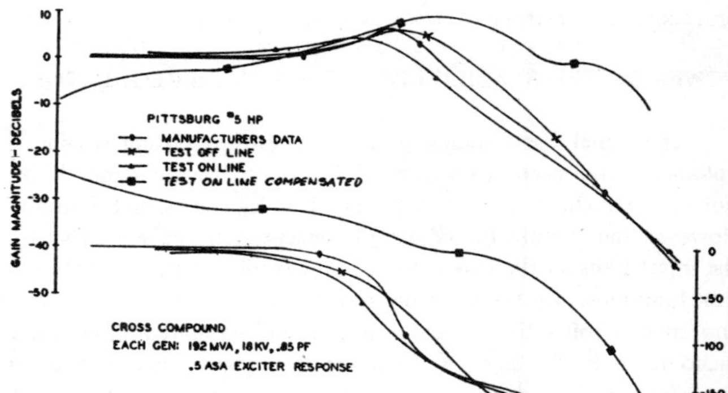

Fig. 6. Frequency response, magnetic amplifier regulator with magnetic amplifier output. Pittsburg 5 HP, cross compound, 192MVA each generator.

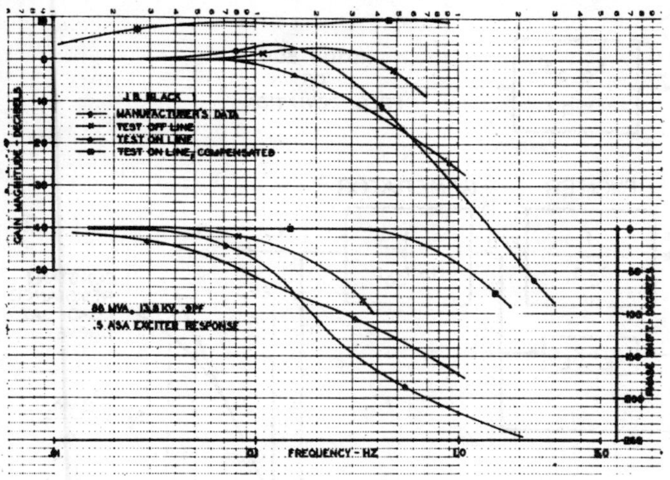

Fig. 7. Frequency response, magnetic amplifier regulator with transistorized front-end and rotating amplifier output. J.B. Black 1, hydro-electric, 86MVA.

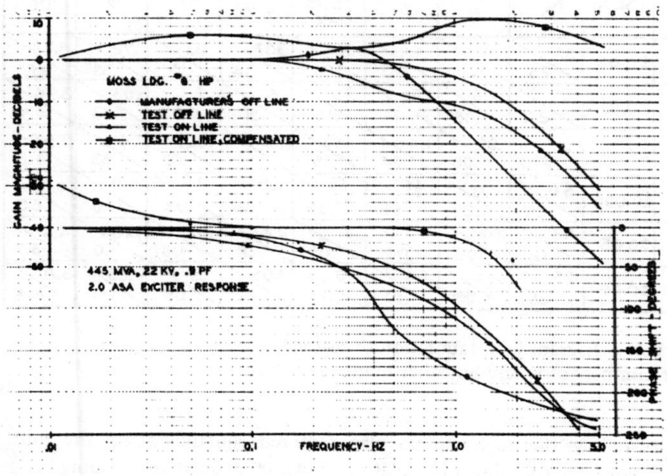

Fig. 8. Frequency response, magnetic amplifier regulator with transistorized front-end and rotating amplifier output. Moss Landing 6 HP, cross compound, 445MVA.

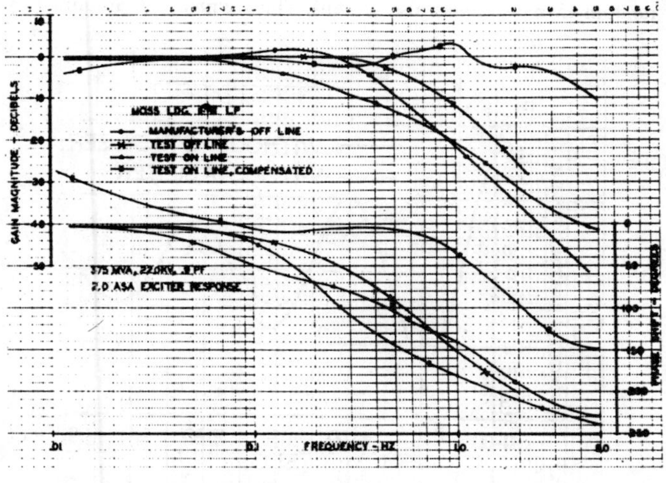

Fig. 9. Frequency response, magnetic amplifier regulator with transistorized front-end and rotating amplifier output. Moss Landing 6 LP, cross compound, 375MVA.

degrees phase advance (45 degrees contribution from each stage) at the 90 degree phase lag frequency. This approach is satisfactory provided the closed loop response exhibits a dominant second order characteristic over the frequencies of interest. Since the Pacific Intertie exhibits a natural frequency of .3 Hertz the foregoing assumption is a good one.

Depending on unit response, the ratio of lead to lag time constants ranged from 10:1 to 65:1. This has probably caused some control gains to be lower than they would be if the control was merely adjusted for good compensation at lower frequencies.

The stabilizer compensated curves of Figures 5 to 9 illustrate the net gain and phase with the control open loop and gain set to 1.0. As a final test of the basic test program, the maximum usable gain setting of the power system stabilizer was determined. With the test machine at or near rated load, the stabilizer gain was first increased until a small sustained oscillation was seen on the monitoring meter. Then, for a final setting, the gain was decreased about one third to give a control loop gain margin of about 10 db. With few exceptions, the final gain settings had values less than one per unit.

For this test to have the most value, it is desirable to have the test machine connected to an isolated transmission line. However, in some cases this was not possible because of dispatching schedules and the test was made with the machine tied solidly to the system.

On several occasions it was possible to make comparative switching tests with and without the power system stabilizer. Often the stabilizer would significantly improve the damping even though the frequencies of oscillation were several times higher than the .3 hertz of the Pacific Intertie.

The practical field application of power system stabilizers does not require in itself a detailed knowledge of the generator and excitation components within the closed loop. Frequency response testing reveals the various equivalent break frequencies and gains of the equivalent closed loop. However, from an overall stability viewpoint, component parameters must be known in order to justify and use component simulations. Additional off-line frequency response monitoring was performed to gather data on the generator and excitation systems.

In many instances system stability analysis must be performed without available test information. Fortunately the manufacturers have provided so-called typical excitation data and in some cases actual test values. Consequently, it is important to compare test and typical data to develop a measure of confidence where test results are unavailable. Because of its relative importance, a test value for the direct-axis open circuit time constant is also desirable. Among other things, Figure 10 provides a tabulation of typical test values that were found.

Bode plots for generator, exciter, regulator, and damping feedback were plotted from steady-state and frequency response tabulations of gain and phase. Conventional interpretation of the Bode plots yields the values for T'do, K_A, T_A, K_F, and T_F. However, the break frequency on the exciter diagram does not yield the T_E of Figure 10 directly. An additional calculation is needed for this constant. For example, the break frequency for Moss Landing 6LP corresponds to T_{EQ} = 5.3 sec. and Vex/V_c = 4.29 volts/volt. Therefore, T_E = 5.3/4.29 = 1.24 seconds. (See Appendix).

To obtain the total gain of regulator and exciter the exciter gain must be converted to units of P.U./volt. The base voltage for the exciter is 133.7 volts. The exciter gain is 4.29/133.7 = 0.032 P.U./volt. Since the regulator gain is 9756 volts/P.U. the total gain is 312.

One test consisted of opening a circuit breaker which isolated Pittsburg No. 5 on a radial line. The test simulation was attempted with the P G and E digital stability program which models generator saliency, saturation, non-conforming loads, and governors in addition to excitation and stabilizer control.

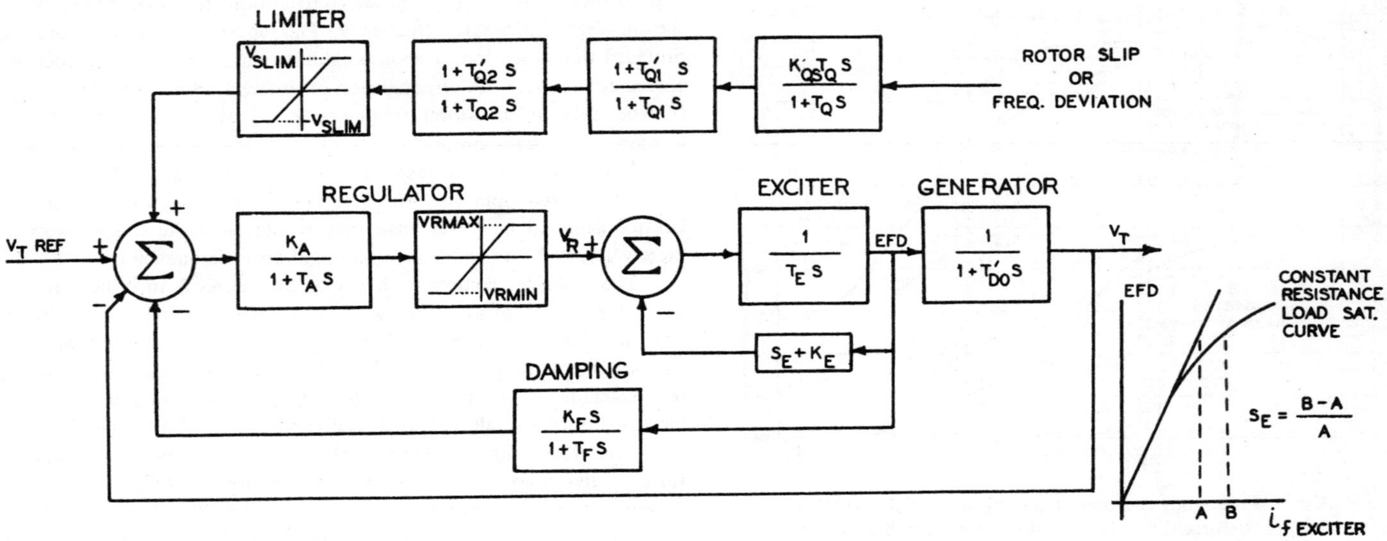

* K_A & T_E DIVIDED BY VRMAX TO NORMALIZE REGULATOR LIMITER TO ±1.0

UNIT NAME	REGULATOR SCHEMATIC	K_{QS}	T_Q	T'_{Q1}	T_{Q1}	T'_{Q2}	T_{Q2}	VSLIM	K_A	T_A	T_E	$S_{E .75\,MAX}$	S_{EMAX}	K_F	T_F	T'_{DO}
PITTSBURG #4	* FIG 1	1.12	10.0	1.33	0.02	1.33	0.02	±0.05	38 / 25	.3 / .2	.86 / .63	.18 / .09	307 / .34	.03 / .10	.72 / .35	5.7 / 9.2
PITTSBURG #5 HP	FIG 2	0.3	10.0	0.57	0.02	0.57	0.02	±0.10	— / 270	.06 / .06	1.28 / .94	.15 / .15	.51 / .51	.2 / .03	.53 / .33	3.94 / 4.5
PITTSBURG #5 LP	FIG 2	0.3	10.0	0.665	0.02	0.665	0.02	±0.10	— / 270	.06 / .06	1.44 / .94	.15 / .15	.51 / .51	.02 / .03	.4 / .33	3.54 / 4.5
J B BLACK #1	* FIG 3	3.15	10.0	0.758	0.02	0.758	0.02	±0.095	46 / 25	.06 / .2	.46 / .65	.45 / .09	1.5 / .35	.10 / .08	1.0 / 1.0	4.0 / 6.6
MOSS LND #6 HP	* FIG 3	1.17	10.0	0.228	0.02	0.228	0.02	±0.05	62 / 12	.053 / .06	.03 / .09	.81 / .09	2.43 / .2	.07 / .06	1.88 / 1.0	4.8 / 3.7
MOSS LND #6 LP	* FIG 3	1.17	10.0	0.45	0.02	0.45	0.02	±0.05	25 / 28	.05 / .06	.04 / .10	1.0 / .09	3.05 / .22	.09 / .03	2,25 / 1.0	8.0 / 8.4

Fig. 10. Excitation and power system stabilizer block diagram with tabulation of test values (upper numbers) and manufacturer's data (lower numbers or where no other numbers given).

Figure 11 compares the field tests with and without the stabilizer. Figure 12 compares the simulation with and without the stabilizer.

FIELD EXPERIENCE

Installation of the three types of stabilizers was basically similar except that difficulties were encountered mounting the tachometers on some units since existing equipment had to be relocated. Installing the tachometers also required shutting down the unit. Frequency transducer installation could be made at any loading.

The frequency transducers also had less noise than the tachometers in their outputs. The tachometers first exhibited an output variation at the fundamental frequency of the unit's speed apparently due to eccentricity. This was discovered in prototype tests with the manufacturer and corrected by adding a Twin-T filter tuned to reject the variation. Additional noise at about 13 Hertz, but more random in nature, was also present. This frequency is amplified by

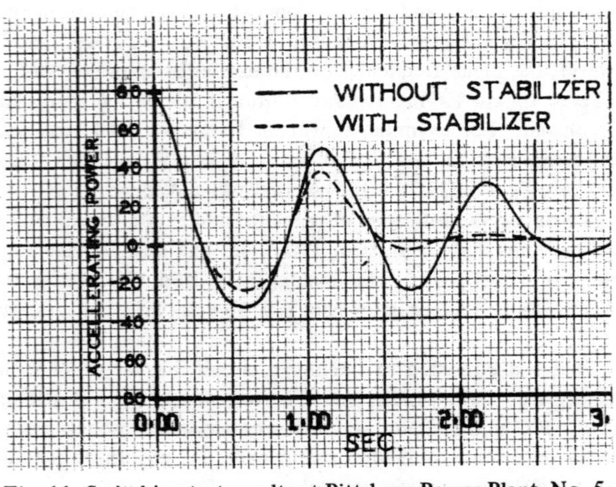

Fig. 11. Switching test results at Pittsburg Power Plant, No. 5.

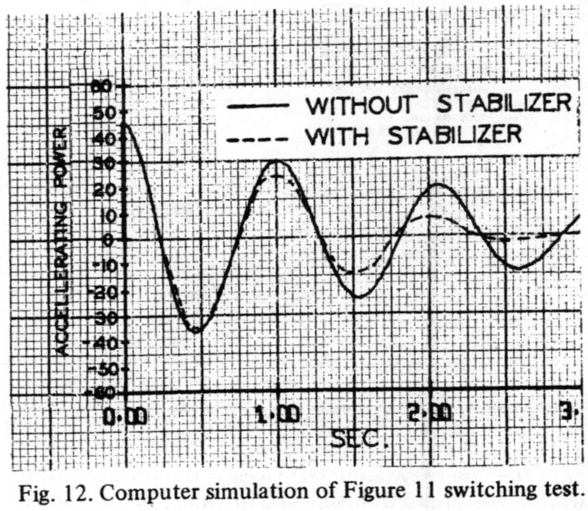

Fig. 12. Computer simulation of Figure 11 switching test.

the lead circuits in the control and is quite significant in the final stabilizer control output. At Moss Landing tachometer noise limited gain to a low value. Replacement of tachometers by frequency transducers permitted an order of magnitude gain increase.

Due to the high current level required to control the impedance type regulator a thyristor-type power amplifier was used between the normal stabilizer output and the rotating amplifier's control field.

Excitation system testing led to the discovery of one regulator with badly unbalanced power mag-amps which were subsequently replaced. The testing program also showed that regulator sensitivities vary over a wide range even though they may be of the same make and model. For example, the impedance type regulators had outputs ranging from 10 ma to 85 ma/1% terminal voltage error. In most instances the sensitivity was readjusted to approximately 30 ma thereby conforming to stabilizer output design.

The regulators with mag-amp inputs experienced an off-set when the power system stabilizer was switched on. This was due to a loading effect caused by the control output impedance on the regulator input. This problem was corrected by incorporating a switched resistor on the regulator input so that it would always see a constant impedance. The regulator gain and/or range width would usually require readjustment after the installation of the stabilizer.

A protection scheme was developed to supplement the protection in the control which cuts itself off from the regulator if certain components fail. The scheme requires the unit to be paralleled before the control is switched on automatically to the regulator.

The control output was set to limit at ± 10 percent change in terminal voltage at most plants. In one type of regulator the control output was not capable of driving the regulator to the ± 10 percent range required. In this case the limits were set at about 6 percent.

Numerous component failures were discovered during the initial testing and installation of the power system stabilizers. In addition, several failures have occurred since the first controls were installed early in 1969. Several frequency deviation transducers, power supplies and meter relays have failed. Missing capacitors in the reset circuit and cold solder joints were also encountered. In one incident the control was feeding a regulator which was referenced to ground. The control output was designed to be isolated from ground but a potentiometer shorted to ground in the power supply during service and caused a large offset to appear on the control output which resulted in driving one regulator of a cross-compound machine into saturation. The regulator tripped out, and could not be put back in service until the leads from the power system stabilizer were lifted because only one lead of the control had switch contacts. This problem was corrected by switching both leads on the control output by an auxiliary relay. These experiences illustrate the need for a careful check out and test program to put the controls in service.

The installation of some controls on a concrete wall which supported the turbine-generators resulted in some erratic operations of the control apparently due to vibration causing a relay in the power supply to occasionally drop out. This was corrected by shock mounting the controls.

CONCLUSIONS

1. An instrumentation system was developed to properly test and adjust power system stabilizers. This same apparatus was used to study excitation system characteristics on several typical generating units.

2. Excitation system data provided by manufacturers are generally accurate, but regulator sensitivities were found to vary widely.

3. The switching tests and digital simulations showed that power system stabilizers are effective in damping system oscillations.

4. The test program exposed many kinds of component failures and hardware problems including improperly operating voltage regulators.

5. The adjustment range in voltage regulators is generally not adequate to significantly reduce overshoot. Greater damping in regulators would allow the power system stabilizer to operate more effectively.

APPENDIX

Shunt Field Excitation

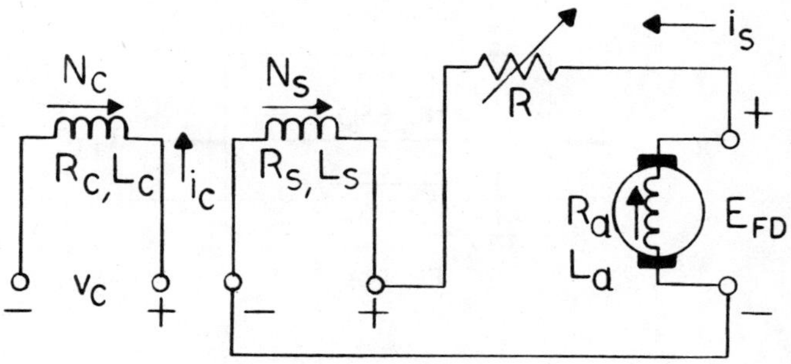

Fig. 13.

Assume: The same mutual flux links all windings, leakage inductance is negligible, R_a and L_a are negligible.

The mutuals are: $L_{cs} = \frac{N_s}{N_c} L_M$, $L_{sc} = \frac{N_c}{N_s} L_M$,

$$G_s = \frac{N_s}{N_c} G_c,\; G_c = \frac{N_c}{N_s} G_s$$

$$V_c = R_c i_c + L_M\, p\, i_c + \frac{N_s}{N_c} L_M\, p\, i_s \qquad (1)$$

$$E_{FD} = \frac{N_c}{N_s} L_M\, p\, i_c + (R + R_s)\, i_s + L_M\, p\, i_s \qquad (2)$$

$$E_{FD} = G_c\, w\, i_c + G_s\, w\, i_s \qquad (3)$$

Solving Eq. (1) and (2) for i_c and i_s:

$$i_c = \frac{V_c}{R_c} - \frac{(L_M\, p\, i_c + \frac{N_s}{N_c} L_M\, p\, i_s)}{R_c} \qquad (4)$$

$$i_s = \frac{E_{FD}}{R + R_s} - \frac{(L_M\, p\, i_s + \frac{N_c}{N_s} L_M\, p\, i_c)}{R + R_s} \qquad (5)$$

Substituting Eqs. (4) and (5) into Eq. (3) yields;

$$E_{FD} = \frac{G_c\, w\, V_c}{R_c} + \frac{G_s\, w\, E_{FD}}{R + R_s} - w\,(T_c + T_s)\, p\, (G_s i_s + G_c i_c) \qquad (6)$$

But: $E_{FD} = w\,(G_s i_s + G_c i_c)$

Therefore:

$$E_{FD} = \frac{G_c w}{R_c} V_c + \frac{G_s w}{R + R_s} E_{FD} - (T_c + T_s)\, p\, E_{FD} \qquad (7)$$

or: $$\frac{E_{FD}}{V_c} = \frac{G_c w/R_c}{\left[1 - \frac{G_s w}{R + R_s}\right]\left[\frac{(T_c + T_s)\,p}{1 - \frac{G_s w}{R + R_s}} + 1\right]} \qquad (8)$$

Eq. (8) holds for Figures 1 and 2. For Figure 3 $G_c w/R_c$ is replaced by $G_s w/R + R_s$.

If an equivalent block diagram is made for the exciter portion of Figure 10, equation (8) can be directly related to it as follows:

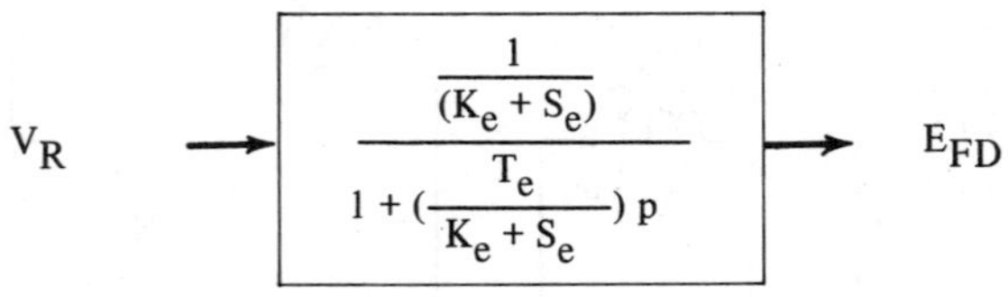

Fig. 14A.

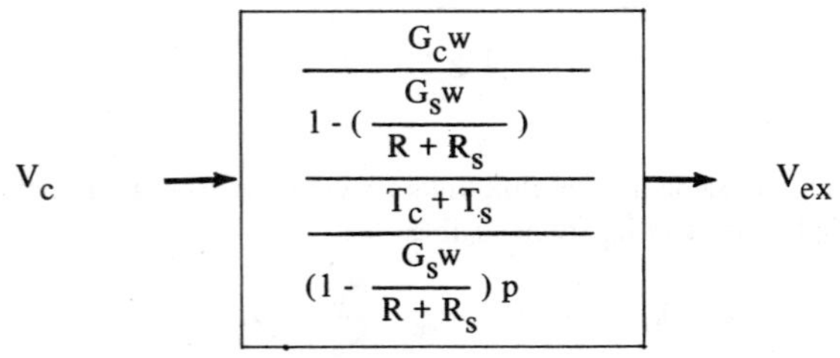

Fig. 14B.

At no load $S_e = 0$, and $K_e = 1 \cdot \frac{G_s w}{R + R_s}$.

The equivalent time constant can be defined as $\dfrac{T_c + T_s}{1 - \frac{G_s w}{R + R_s}}$,

and the equivalent gain as $\dfrac{G_c w}{\left(1 - \frac{G_s w}{R + R_s}\right)}$.

The equivalent gain is measured as $\frac{V_{ex}}{V_c}, \frac{\text{Volts}}{\text{Volts}}$

E_{FD} - Exciter armature voltage (P.U.)
$G_c w$ - Exciter armature voltage/control field current
$G_s w$ - Exciter armature voltage/shunt field current
i_c - Control field current (amps)
i_s - Shunt field current (amps)
L_a - Exciter armature inductance (henries)
L_{cs} - Mutual inductance between the control and shunt winding (henries)
L_M - Self inductance of either the control or shunt winding (henries)
L_{sc} - Mutual inductance between the shunt and control winding (henries)
N_c - Control winding turns
N_s - Shunt field turns
p - d/dt operator
R - Rheostat resistance setting (ohms)
R_a - Exciter armature resistance (ohms)
R_c - Control field resistance (ohms)
R_s - Shunt field resistance (ohms)
T_c - Control winding time constant (L_M/R_c)
T_s - Shunt field time constant ($L_M/R + R_s$)
V_c - Control field voltage (volts)
V_{ex} - Exciter armature voltage (volts)
V_R - Regulator output voltage (P.U.)
w - Exciter shaft velocity

REFERENCES

1. Schleif, F. R., Hunkins, H. D., Martin, G. E., Hattan, E. E., "Excitation Control to Improve Powerline Stability," IEEE Transactions on Power Apparatus and Systems, PAS-87/No. 6, June 1968, pp. 1426-34.
2. IEEE Committee Report, "Computer Representation of Excitation Systems," IEEE Transactions on Power Apparatus and Systems, PAS-87/No. 6, June 1968, pp. 1460-64.

Discussion

Glenn R. Meloy (U. S. Corps of Engineers, Portland, Oreg.): The authors have made a valuable contribution toward shedding some light on the procedure for installation and adjustment of power system stabilizers. The tabulation of data and comparison with manufacturers' "cook book" data is extremely interesting and does a lot toward taking some of the mystery out of this subject. Have the authors made any computer studies, based on their test data, to optimize the bandwidth of the stabilizer to provide maximum damping? In addition to the increased noise associated with the increased bandwidth, the selection of the minimum lag time constant establishes the lead time constants to accomplish compensation at the 90 degree phase lag frequency. It would appear that manipulation of the poles and zeros of the stabilizer transfer function to coordinate with the associated system might optimize the damping effort provided by the stabilizer.

With the exception of JB Black #1, the tabulation lists gains in the order of one. Would the authors care to amplify their comments on increased gain by use of frequency transducers as opposed to tachometer inputs?

Manuscript received February 22, 1971.

A. D. Gerhart, T. Hillesland, Jr., J. F. Luini, and M. L. Rockfield, Jr.: The authors wish to thank Mr. Meloy for his discussion. We have run several computer studies evaluating the system response to various disturbances with and without the stabilizers, but no studies to compare stabilizer bandwidth with increased gain. Our original aim in adjusting the compensation circuits of the stabilizers was to correct for excitation system phase shift over as broad a frequency range as possible. This approach resulted in a minimum of compensated system phase lag at the critical intertie frequency of 0.3 Hz. Recent field studies on Pittsburg 4 unit showed that higher control gains could be achieved when the control bandwidth was reduced by moving the lag time constants closer to the lead settings. For example, the maximum gain without oscillation on that unit was 3.5 per unit. When the poles were moved from 0.02 seconds to 0.1 seconds the gain could be increased to 8 per unit. However, the phase shift at 0.3 Hz increased from -22 degrees to -40 degrees.

Replacement of the tachometer with a frequency transducer allowed substantial increases in gain. At Moss Landing, we were able to increase the gain from 0.26 to 1.17 per unit by using a frequency transducer due principally to a reduction in noise. The recent tests at Pittsburg 4 also included a comparison between tachometer and transducer. With a frequency transducer and using the same settings as given previously, the gains were 9.0 (at 0.02 seconds lag setting) and 29 (at 0.1 seconds). However, for this unit, noise was not a limiting factor.

Manuscript received March 26, 1971.

EXPERIENCE WITH SUPPLEMENTARY DAMPING SIGNALS FOR GENERATOR STATIC EXCITATION SYSTEMS

Wilfred Watson, Gerald Manchur
The Hydro-Electric Power Commission of Ontario
Toronto, Canada

Abstract – Virtually all hydraulic generating units installed in Ontario Hydro since 1964 have been equipped with static excitation systems, necessitating supplementary stabilizing signals to suppress power oscillations. This paper reviews the experience gained with alternative types of stabilizing systems at these stations and discusses advantages and disadvantages of each, problems in meeting accuracy requirements, permissible levels of noise content in the signals and the problems in application of stabilizing signals to thermal machines.

Generator excitation is increasingly being supplied from rectifiers, fed either from rotating ac exciters, from the generator terminals, or from the station service bus[1-4]. The principal advantage of these systems lies in the negligible time constant of thyristor control of excitation with consequent possible improvement in transient stability.

The use of high-speed excitation systems has been found to result in decreased damping of generators in a power system, to an extent that steady-state stability may be lost even at normal full load operation. The damping, however, can be increased by the introduction of a supplementary signal to the excitation control[2-5]. In fact, a rectifier type exciter with supplementary stabilizing signal can be adjusted to give essentially the ultimate in steady-state stability limit, equivalent to a very low generator internal impedance[3].

NATURE OF THE STABILIZING SIGNALS

Analysis has shown that an effective damping signal will be a combination of generator slip speed and acceleration. However, such a signal may be obtained by measurement of various electrical quantities on the machine and applying appropriate transfer functions, as alternatives to direct measurement of the mechanical quantities.

In practice the following methods have been used and may be considered analytically equivalent.

(1) Rate of change of phase of generator internal voltage relative to phase of average system voltage ($p\delta$).

(2) Difference between generator speed and average system speed ($\Delta\omega$).

(3) Integral of generator accelerating power ($\Delta P/\rho$)[4].

Methods (2) and (3) require that the average system speed be constant. Practical power systems vary in speed to some degree, but satisfactory results can be obtained by applying the stabilizing signal through a transfer function of the form $pT/(1 + pT)$ thus giving a "washout" or reset effect. The value of the reset time constant T does not appear to be critical provided it does not introduce significant phase shift at frequencies of interest or that such phase shift is compensated elsewhere. A value of $T = 1.5$ seconds results in satisfactory operation.

The gain used with the stabilizing signal is dependent upon the damping considered to be adequate or to be useful. The signal may also be used with a transient gain reduction in the exciter as shown in Figure 1(A)[2] or without it as in 1(B)[3]. Whereas the parameters shown in Figure 1(A) will result in significant damping and no doubt adequate steady-state stability margin, those of Figure 1(B) will give very heavy damping of generator oscillations and even higher stability margin. It has been the practice in Ontario Hydro to favour high gains and heavy damping.

Paper T 72 227-2, recommended and approved by the Power Generation Committee of the IEEE Power Engineering Society for presentation at the IEEE Winter Meeting, New York, N. Y., January 30-February 4, 1972. Manuscript submitted September 15, 1972; made available for printing November 30, 1971.

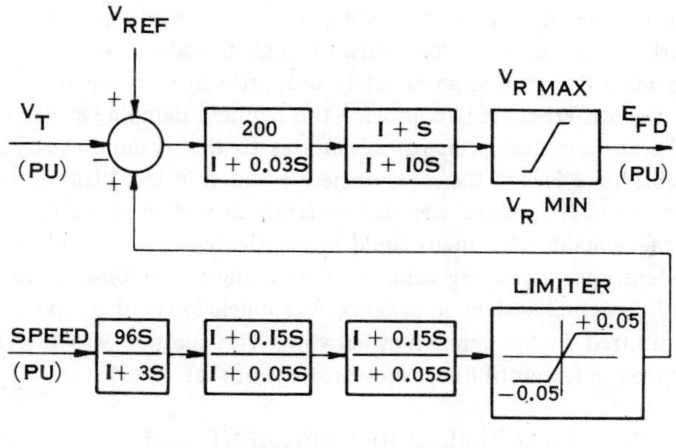

FIGURE 1(A) SIMPLIFIED REPRESENTATION OF EXCITATION SYSTEM AND POWER SYSTEM STABILIZER

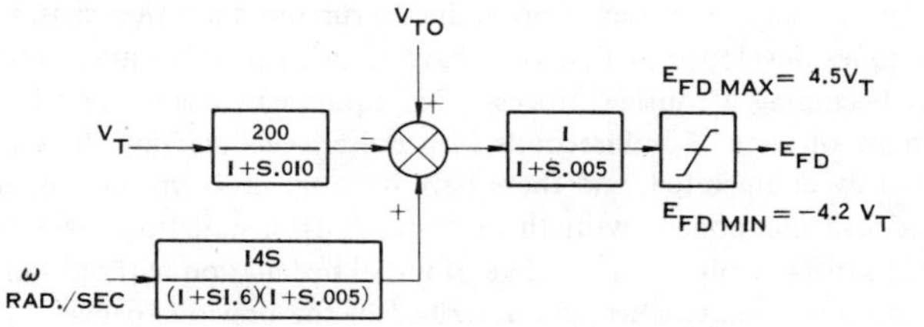

FIGURE 1(B) TRANSFER FUNCTION OF TYPICAL CONTROLLED RECTIFIER EXCITATION SYSTEM

When designing equipment to provide the stabilizing signal, certain rather rigorous requirements must be met. For example, for the parameters of Figure 1(B) utilizing a shaft speed measurement, a change in speed of 0.01% will result in a change of field voltage of 33%. With this order of sensitivity, "noise" in the stabilizing signal must be kept to a minimum. Moreover only a minimum of filtering is permissible since in application it is required to phase lead the signal, ie, add a proportion of rate of change (acceleration) to the speed signal to obtain the required damping effect. This differentiation increases the noise problem.

The most important criterion relating to noise is that it shall not significantly influence either the active or reactive power output of the generator. The tolerable noise level is dependent upon the frequency. For noise frequencies below about 5 Hz the level must be very low (<0.02%) since significant changes in field current can be produced by low frequency changes in field voltage. A frequency corresponding to shaft rotational speed and resulting from shaft run-out is likely to be the most important noise component in this range. At higher frequencies (>10 Hz) higher amplitudes can be tolerated. Torsional vibrations generated in hydraulic turbines at frequencies of 10-12 Hz have resulted in typical field variations of up to 150 volts peak-to-peak, without any deleterious effects after years of operation. For thermal units such a condition cannot be tolerated for reasons which will be discussed later.

It would seem to be important to keep noise at "synchronous" frequencies (60 Hz and harmonics) at a low level, since such com-

Reprinted from *IEEE Trans. Power App. Syst.*, Jan./Feb. 1973.

ponents in the thyristor control would cause undesirable unbalance between the banks in different phases.

ANGLE SIGNALS (pδ)

The first static exciters utilized in Ontario Hydro were installed in 1964 and were equipped with a stabilizing signal derived from phase angle measurement between generator internal voltage (E_q) and remote system voltage (E_s). Both these voltages were obtained from simulating networks at the generator terminals. An angle transducer was developed to provide a dc signal approximately proportional to phase angle. This signal was differentiated to provide the required damping signal (pδ).

There were two principal advantages to this system. Firstly, unavoidable time lags in the measurement resulted in less than optimum performance. The second was that variation in system impedance seen from the generator terminals could in practice vary over a wide range, dependent upon operating conditions, resulting in instability of the device when actual system impedance was much lower than the preset value utilized in the simulating network. This method was therefore abandoned in favour of direct measurement of shaft speed.

SHAFT SPEED MEASUREMENT (Δω)

The techniques used to derive a stabilizing signal from shaft speed of a hydraulic generator are described in reference[3]. Four such installations were put into service in 1966 and six have been made subsequently. Although during development difficulties were experienced in making adjustments to minimize noise due to run-out and other causes, the techniques developed at that time have resulted in subsequent installations becoming a routine process. The equipment has an overall time constant of about 5 milliseconds with noise levels meeting the criteria previously enumerated, and there have been no problems in applying a phase lead component with this signal. All 10 installations have functioned satisfactorily from the time of initial installation without failures and without maintenance. As described in the previous paper[3], it has remained the practice to disconnect the stabilizing signal by a pallet switch at gate positions below about 70 per cent, to prevent excessive modulation of the field voltage by torsional vibration generated in the turbine at part-gate positions.

INTEGRAL OF POWER CHANGE (ΔP/ρ)

Immediately after the installation of the initial delta-omega equipment in 1966, it was necessary to install stabilizing signals on two 75-MVA generators at Mountain Chute Generating Station. These generators were already under construction and physical arrangements were such that it was impossible to incorporate a speed measuring device of the type which had been developed. It was decided, therefore, to utilize a signal based upon the integral of accelerating power.

It can be shown that, as the speed of a machine on a power system varies by small amounts from the mean, the variation in speed can be expressed as,

$$\Delta\omega = \frac{1}{M}\int (P_m - P_e)dt, \qquad (1)$$

where $M = 2H$

H = inertia constant

P_m = mechanical power input

P_e = electrical power output

For a hydraulic unit running with constant gate position and constant head, mechanical power input P_m is sometimes assumed to be constant; and hence

$$\Delta\omega = -\frac{1}{M}\int \Delta P_e \, dt$$

where ΔP_e = changes in electrical power from the mean.

Such an arrangement functions satisfactorily except when it is required to change the loading on the machine by changing the gate position. It is then necessary to disconnect the stabilizing signal to prevent excessive excursions of field current and reactive power during real power changes on the generator. In many situations on the Ontario Hydro system, disconnection of the stabilizing signal would result in immediate loss of stability of the machine. Where generators are utilized for load control the stabilizing signals might be disconnected most of the time. In the case of the Mountain Chute generators there is a requirement for the units to operate frequently as synchronous condensers or at maximum efficiency loading, with maximum power output available as spinning reserve. This reserve must be available at the maximum rate at which the gates can be opened. For the foregoing reasons, in providing a stabilizing signal for these units the more complete simulation represented by equation (1) was necessary. The arrangements adopted are shown in the block diagram of Figure 2.

A signal obtained from a linear gate position transducer is fed to a function generator adjusted to represent the non-linear gate/power relationship for average plant head. In the following transfer function, representing water inertia effects, the value of water starting time constant T_w is adjusted to correspond to average plant head and efficiency gate opening. Electrical power is measured with instantaneous hall watt transducers, compensated to provide negligible power factor errors. The integrating and proportional amplifiers provide the desired speed and acceleration signals through a conventional washout circuit.

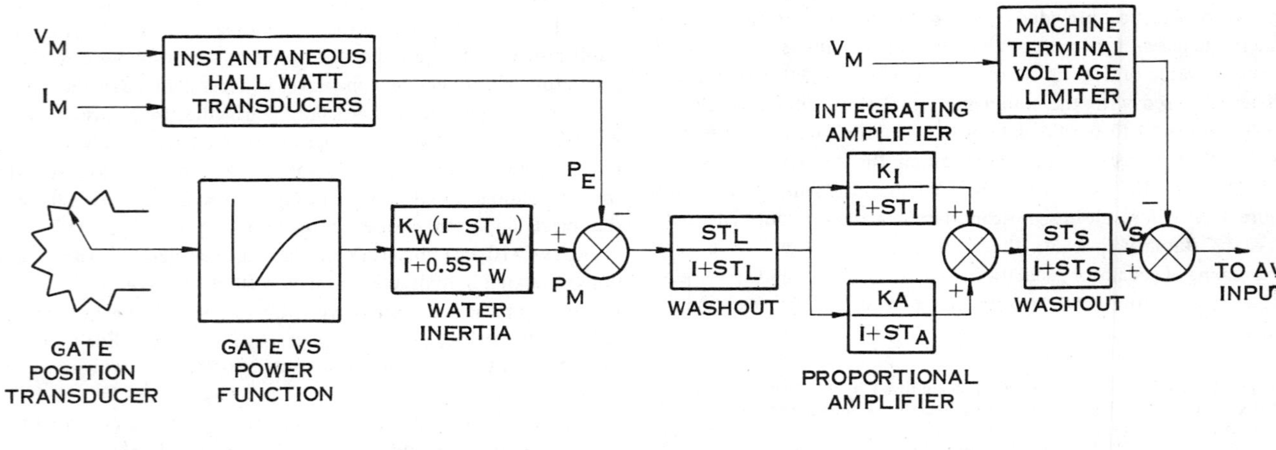

FIGURE 2 SIMPLIFIED ARRANGEMENT OF INTEGRATED POWER STABILIZER

It should be noted that with this arrangement it was necessary to cut off the stabilizing signals at loadings below 60% gate. There were significant variations in electrical power output of the machine at about 50% gate without gate movement. These were attributed to water hammer in the penstock. The simulation shown in Figure 2 gave satisfactory operation in service, with gate velocities corresponding to zero to 100% in 30 seconds with acceptable disturbance in reactive output (<20 MVAR). It should be noted, however, that the simulation is only correct for constant head. For a plant with significant variations in head it might be necessary to introduce another variable in the computation.

APPLICATION OF STABILIZING SIGNAL TO THERMAL UNITS

In December, 1969 an attempt was made to commission the first stabilizing signal on a 500 MW thermal unit at the Lambton Generating Station of Ontario Hydro. This signal was based on generator shaft speed measurement and utilized the techniques previously successfully applied on hydraulic units. A signal with a very low noise level (<0.005%) was thereby achieved. The sensitivity available with this low noise level made it possible to detect torsional vibrations in the rotational masses and it was discovered that introduction of the stabilizing signal to the exciter control resulted in a build-up of such torsional vibrations. This has proved to be a difficult problem which has now been resolved and is the subject of a separate paper[6].

STABILIZING SIGNAL DERIVED FROM TERMINAL FREQUENCY

Among the many sources of information available for a stabilizing signal, frequency of generator terminal voltage has often been advocated. Simulating networks can also be used to combine terminal voltage and current to produce quadrature axis voltage whose frequency more closely corresponds to shaft speed. In either case it becomes necessary to produce a dc control signal from a 60 Hz quantity with the conflicting requirements of low ripple amplitude, low time constant and high sensitivity. Keay and South[5] have described a frequency transducer developed for this purpose. It may be noted that this device has an inherent time constant of 30 milliseconds. This equipment has principally been applied on rotating exciter systems where longer time constants and higher noise levels are tolerable. When applied with rectifier exciters it may be necessary to limit the usable gain because of noise and thereby limit the degree of damping which can be achieved.

When applying such a device on a thermal unit, it must be assumed that the output signal closely approximates shaft speed. For this reason the problems associated with torsional vibrations previously described must be anticipated.

CONCLUSIONS

Experience with alternative stabilizing systems has shown that each have their own peculiar development and application problems to overcome to achieve desired performance. Thus, whereas systems using direct measurement of shaft speed place emphasis on noise content in the signal, other systems using indirect measurements of speed place emphasis on the accuracy requirements of transducers and simulating networks to faithfully reproduce the desired speed signal.

A rational comparison of the various systems in use can only be made in relative terms since there is no absolute "figure of merit" by which to measure their performance. The authors have found that a stabilizing system based on direct measurement of shaft speed has the advantage of being virtually independent of system configurations and operating procedures and thus approaches the universally applicable stabilizing system sought after.

REFERENCES

[1] K. R. McClymont, G. Manchur, R. J. Ross, and R. J. Wilson, "Experience with High-Speed Rectifier Excitation Systems", IEEE Trans Power Apparatus and Systems, Vol PAS 87, pp 1464-1469, June 1968.

[2] M. L. Crenshaw, W. J. Miller, R. P. Schulz, and M. Temoshok, "Althyrex Excitation System with Power System Stabilizer", IEEE CP70-563-PWR, presented at Summer Power Meeting 1970.

[3] P. L. Dandeno, A. N. Karas, K. R. McClymont, and W. Watson, "Effect of High-Speed Rectifier Excitation Systems on Generator Stability Limits", IEEE Trans Power Apparatus and Systems, Vol PAS 87, pp 190-196, January 1968.

[4] R. M. Shier, A. L. Blythe, "Field Tests of Dynamic Stability Using a Stabilizing Signal and Computer Program Verification", IEEE Trans on Power Apparatus and Systems, Vol PAS 87, pp 315-322, February 1968.

[5] F. W. Keay, W. H. South, "Design of a Power System Stabilizer Sensing Frequency Deviation", IEEE Transaction on Power Apparatus and Systems, Vol PAS 90, pp 707-712, March/April 1971.

[6] W. Watson, M. Coultes, "Static Exciter Stabilizing Signals on Large Generators – Mechanical Problems". Paper to be presented at IEEE Winter Power Meeting, 1971.

Discussion

F. P. deMello (Power Technologies, Inc., Schenectady, N.Y. 12301): Industry owes a debt of thanks to Ontario Hydro and their engineers for the many outstanding contributions that they have shared with the technical community.

This paper by Messrs. Watson and Manchur is another example of a series of worthwhile contributions in the area of application of supplementary stabilizing in excitation systems.

The discussion of the various alternate methods of developing stabilizing signals opens opportunities for exchanges of ideas which are of benefit to those who are considering the use of stabilizers. In this spirit, the following comments are offered:

Under the section "Nature of the Stabilizing Signals" the authors discuss the merits of the system of Fig. (1B) versus that in Fig. (1A) which was discussed in Ref. 1. Basically the authors prefer to keep the very high gain in the voltage regulator loop as opposed to the recommendations in Ref. 1 that a transient gain reduction be used as in Fig. 1A, to yield more moderate regulator gains.

In addition to the arguments raised in favor of the scheme of Fig. 1A offered in the closure of Ref. 1, we believe that the following further considerations should be noted:

(1) A desirable feature from a reliability standpoint should be that the excitation system be inherently stable even without the signal. A very high gain regulator can be marginally stable or even unstable should loss of the signal occur.

(2) As has been reported in a companion paper entitled "Static Exciter Stabilizing Signals on Large Generators – Mechanical Problems" a very high gain excitation system loop keeps the flux-torque speed loop active at frequencies much higher than those of interest causing instabilities at shaft resonant frequencies.

The remedy is to cut off the action of the signal at the higher frequency bandwidth which could be done at the front end with a filter on the signal, but in our opinion is better done (from an energy and noise effect criterion) by limiting the bandwidth of the voltage and excitation closed loop which is accomplished with the transient gain reduction.

Incidentally, in Fig. 1B it would seem that the value of gain 200 should be in the block for exciter and regulator rather than in the block which connects V_T to the reference summing point.

One final comment relates to the discussion of the correctness of the model used to simulate mechanical power changes in Fig. 2. It should be noted that the simulation is approximate even if no significant head variations exist since the effective starting time T_W varies with operating point.

REFERENCE

[1] "Concepts of Synchronous Machine Stability as Affected by Excitation Control" by F. P. deMello and C. Concordia. Discussions and Closure. *IEEE Transactions on Power Apparatus and Systems,* PAS-88, No. 4, April, 1969, pp. 316-329.

Manuscript received February 7, 1972.

C. E. J. Bowler and **M. L. Crenshaw** (General Electric Company, Schenectady, N.Y.): The use of generator excitation controls to improve power system performance has become a matter of increasing importance. The application of thyristor excitation systems to achieve high initial response and high ceiling voltages for large steam turbine generators being installed in the Southwestern U.S. is a dramatic example.[1] Units totaling some 3770 MW are presently in service and 3000 MW additional are presently on order utilizing these excitation systems.

The system requirements and operating philosophies in the United States differ somewhat from those in Canada. To complement the authors' valuable experience report, we would like to briefly summarize some of our findings.

The short time delays and high ceiling voltages associated with this excitation equipment permit improvements in transient stability. The use of additional signals from speed or frequency (power system stabilizer) has been used in the U.S. principally to provide additional damping of low frequency, intersystem oscillations having periods of several seconds.

In order to prevent the generation of negative damping torques and reduction in steady-state stability should the power system stabilizer signal not be available for some reason, these applications have utilized the transient gain reduction referred to by the authors. In most instances, somewhat less gain reduction than the 10:1 shown in figure 1(A) of the paper would be adequate for this purpose. Transient gain reduction, while providing a heavily damped voltage response, also increases gain margin in the region of frequencies of interest. The additional regulator loop phase lag can be compensated for by adjustment of the power system stabilizer as shown in figure 1. If the stabilizer

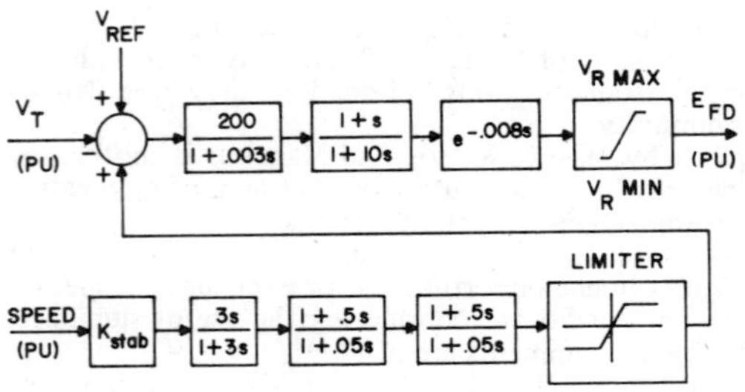

Fig. 1. Excitation System With Power System Stabilizer – With Transient Gain Reduction.

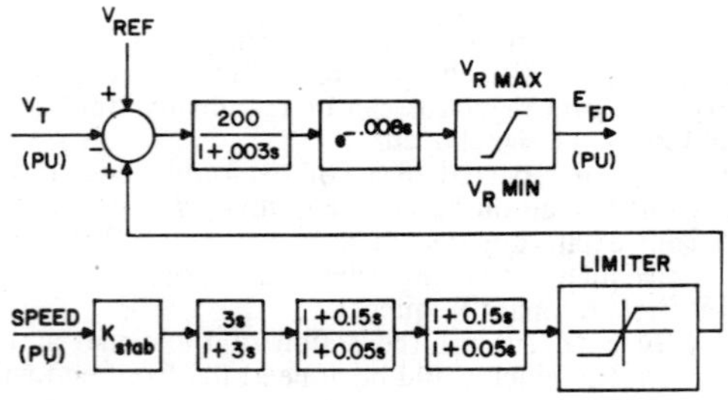

Fig. 2. Excitation System With Power System Stabilizer – Without Transient Gain Reduction.

gain is modified to account for the change in regulator gain, the damping for "local" oscillation modes can be made identical (with or without transient gain reduction). A root locus plot, figure 3, shows the essentially similar trajectories for these conditions.

Transducer design for low noise is of particular concern when the power system stabilizer signal obtained is used to modulate the output of a very fast excitation system. Spurious noise can occur for various reasons; principally, (1) shaft run out effects with shaft speed sensors, (2) unbalanced phases and rectifier harmonics in tachometer systems and, (3) switching harmonics in sample and hold systems.

Manuscript received February 22, 1972.

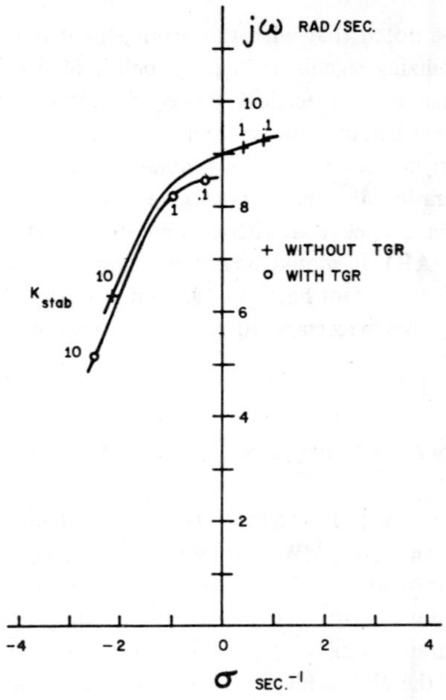

Fig. 3. Root Locus Plot With Variation Of Stabilizer Gain – With And Without Transient Gain Reduction.

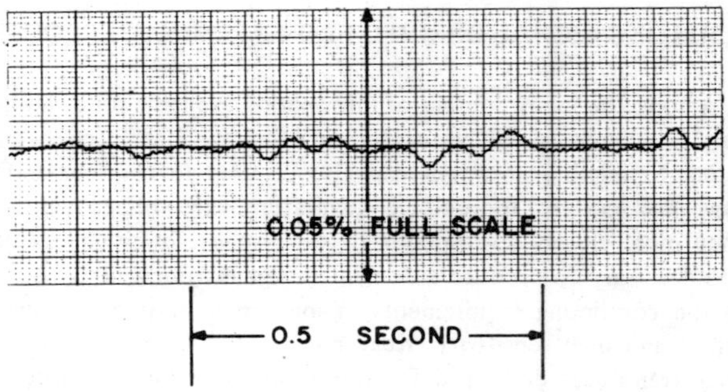

Fig. 4. Power System Stabilizer Output Under Normal System Operating Conditions.

Initial application of thyristor exciters to steam turbine generators at the Four Corners and Mohave Stations utilized signals from 6 pole 3 phase tachogenerators. Operating experience with these units has been good. Spurious noise signals at multiples of shaft speed are of concern but can be removed by filtering.

Present applications of thyristor exciters are being supplied with frequency derived from terminal voltage. Recent tests were conducted on a stabilizer of this type, and a recording of the stabilizer output is shown in figure 4. A clear indication of freedom from noise and high resolution is seen by noting the amplitude of the 1st torsional resonance and the very low magnitude 120 Hz noise.

The fact that torsional frequency modulation of field voltage can cause problems raises the question of stabilizer bandwidth. The stabilizer is designed to maintain dynamic stability of the power system by supplying phase lead in the torque angle loop. The highest frequency associated with the natural frequency generator system mode is of the order of 1 to 2 Hz. The maximum allowable time constant of the sensor is limited by the requirement that the stabilizing signal reflect adequate damping at this highest frequency. A reasonable sensor time constant can be compensated by additional lead in the shaping function. A lag of 30 msec can be easily tolerated without adverse effect on performance.

The authors indicate a preference for their type of speed transducer as approaching the sought for universal stabilizer source. At this stage of development it probably is an excellent system, but with the high reliability of electronic components it seems less flexible with regard to application to different units as compared with a frequency signal derived from voltage. With the latter method, noise components are of predictable frequency making it applicable to all types of excitation systems and generating units.

REFERENCE

[1] M. L. Crenshaw, R. G. Farmer, R. P. Schulz and M. Temoshok, "System Design Considerations and Operating Experience With High Performance Thyristor Excitation Systems", Proc. Amer. Power Conf., Vol. 32, p. 1169-1177 (1971).

Wilfred Watson and **Gerald Manchur**: The authors wish to thank the discussers for their comments and contributions which add to the experience value of the paper. They also recognize their many fine contributions to the industry in the field of synchronous machines and excitation system control.

Several points favouring the use of transient gain reduction in the excitation system forward loop are brought up in both discussions, so these will be dealt with jointly. Traditionally, in the application of continuously acting voltage regulators and excitation systems to synchronous machines, there have been conflicting requirements of high gain, fast response for improvement of transient stability, and moderate gain, more damped response for improvement of steady state stability. With the advent of static excitation systems and supplementary damping signals, the problems of steady state stability have been effectively overcome and it would appear that the above conflicting requirements could finally be met. The use of transient gain reduction, however, by virtue of its effect in slowing down the excitation system response, would seem to be tending in the wrong direction. Perhaps a more appropriate direction might be to supplement the present stabilizing systems with additional control which would react more favourably on the excitation system during non-linear operation when the exciter is hitting against its limits.

Steady state operation with a reduced exciter gain and no stabilizing signal may not be possible or practical in some quite representative situations. This is illustrated in Figure 1 which is a plot of the steady state stability limits of a 2000 megawatt thermal plant on the Ontario Hydro system. Curve 1 is for the case with no stabilizing signal, and it is evident that operation at any value of exciter gain would not be realistic, for full load output from the plant. By comparison, curve 2 shows the improvement obtained by modifying the voltage regulator transfer function to be more representative of a conventional excitation system, while curve 3 indicates the large improvement realizable with supplementary speed stabilizing. These results were obtained by means of an eigenvalue analysis of the machines and system. It has been our practice to modify the arrangement of an excitation system to give the performance shown in curve 2 whenever a stabilizing signal is not available. The point of using transient gain reduction to reduce the available energy at the higher torsional frequencies of a turboalternator is well taken. This could still be accomplished if the gain reduction is made effective at frequencies above the local generator mode oscillation. The margin of stability would be improved by the consequent reduction in gain. However, we have not considered to date that gain reduction alone is an acceptable solution for prevention of instability of torsional mechanical modes of oscillation.

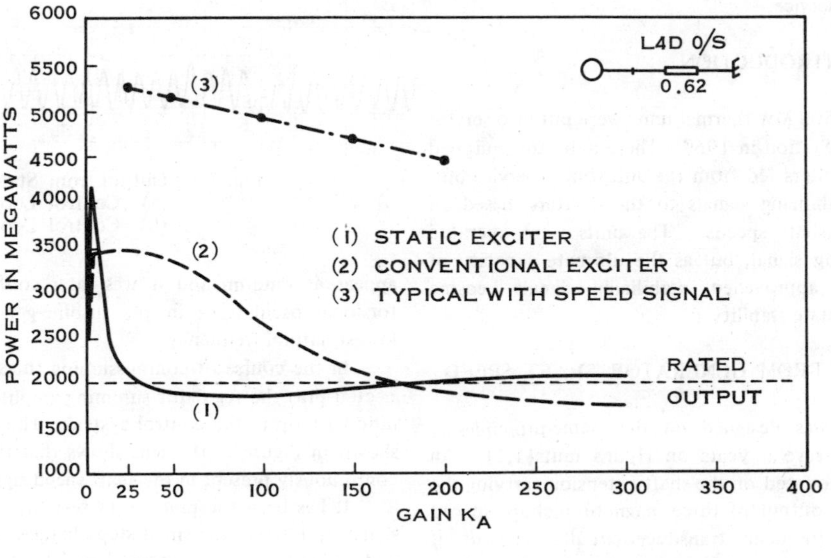

Fig. 1. Effects of Exciter Gain on Steady State Stability Limits.

Manuscript received April 25, 1972.

STATIC EXCITER STABILIZING SIGNALS ON LARGE GENERATORS - MECHANICAL PROBLEMS

Wilfred Watson, Murray E. Coultes
The Hydro-Electric Power Commission of Ontario
Toronto, Canada

Abstract – Torsional vibrations have been observed on large thermal units with both static exciters, and stabilizing signals based on shaft speed measurement. Studies of transfer functions of round rotor machines and mechanical modes of torsional oscillation of a 500 MW turbo-generator, have led to a solution based on derivation of the stabilizing signal from a speed measurement close to the node of the first torsional mode. New arrangements have been designed and installed on four units now in service.

INTRODUCTION

Ontario Hydro's first 500 MW thermal units were put into service at the Lambton Generating Station in 1969. These units are equipped with thyristor controlled exciters fed from the unit station service bus. Provision was made for stabilizing signals to the exciters, based on measurement of generator shaft speeds. The units were operated initially without a stabilizing signal, but as the ultimate capacity of four 500 MW units was approached, stabilizing signals became essential to maintain steady-state stability.

STABILIZING SIGNAL FROM GENERATOR SHAFT SPEED

The stabilizing signal was designed on the same principles as those used successfully for several years on Hydro units[1,2]. An 80 tooth gear wheel was mounted on the shaft extension carrying the sliprings, and the combined output of three magnetic pickups spaced around the wheel was fed to frequency transducer circuits. The output was a dc signal proportional to changes in speed ($\Delta\omega$) together with a component of acceleration ($p\Delta\omega$) provided by a phase-lead adjustment. A transfer function block diagram of the arrangement is shown in Figure 1. Figure 2(a) shows a record of the speed variations indicated by the output of the stabilizer (phase lead=0) with the generator operating on the line at full load. The high sensitivity (0.042 per cent full scale) should be noted, thus demonstrating that a signal with extremely low noise level can be obtained with little difficulty, from shaft speed measurement on a thermal unit. The high sensitivity of the record makes it possible to distinguish some small speed oscillations which have frequency of approximately 16 Hz. These appear and disappear at random and it was later confirmed that they result from torsional oscillations in the turbine-generator rotating mass, at its lowest natural frequency.

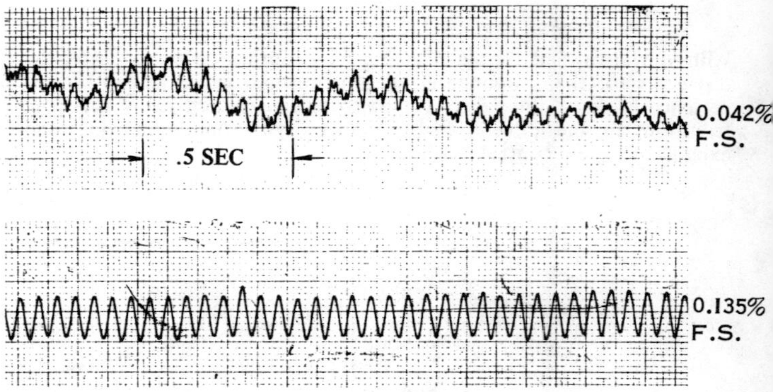

Fig. 2. Output from Stabilizing Circuit
(a) Control Loop Open
(b) Control Loop Closed

In the course of commissioning the stabilizer, the output was connected into the regulator summing amplifier circuit, thus closing an additional loop in the control system. The stabilizer output was then as shown in Figure 2(b). This shows that the 16 Hz component became continuously present in the shaft speed signal.

It has been the practice to test the effectiveness of the stabilizing signal by introducing small step changes to the voltage regulator input and observing the resulting damped oscillation of generator power, speed and other variables as illustrated in Figure 3(a). The gain and phase lead of the stabilizer are then adjusted to obtain the required damping. The result of this process is illustrated in Figures 3(a), 3(b), and 3(c). However, the most significant effect apparent on the records of Figure 3(c) is the continuously increasing amplitude of the torsional oscillation at 16 Hz which continued until the stabilizing signal was removed. It was evident that a regenerative loop was formed by introduction of the stabilizing signal to the exciter and it was feared that continued operation would build up excessive mechanical torques in the machine at the torsional natural frequency.

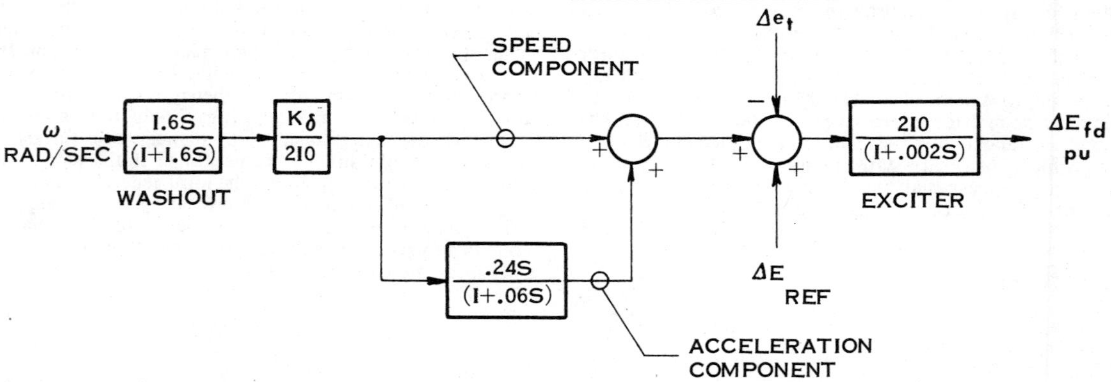

Fig. 1. Block Diagram of Original Stabilizing Signal

It was then decided that it was required to obtain a better understanding of the observed phenomena by analysis and, if necessary, by simulation. At this stage there was insufficient data for further analysis and investigations were initiated to provide this data in two main areas:

Paper T 72-228-0, recommended and approved by the Power Generation Committee of the IEEE Power Engineering Society for presentation at the IEEE Winter Meeting, New York, N.Y., January 30-February 4, 1972. Manuscript submitted Spetember 16, 1971; made available for printing November 30, 1971.

Reprinted from *IEEE Trans. Power App. Syst.*, Jan./Feb. 1973.

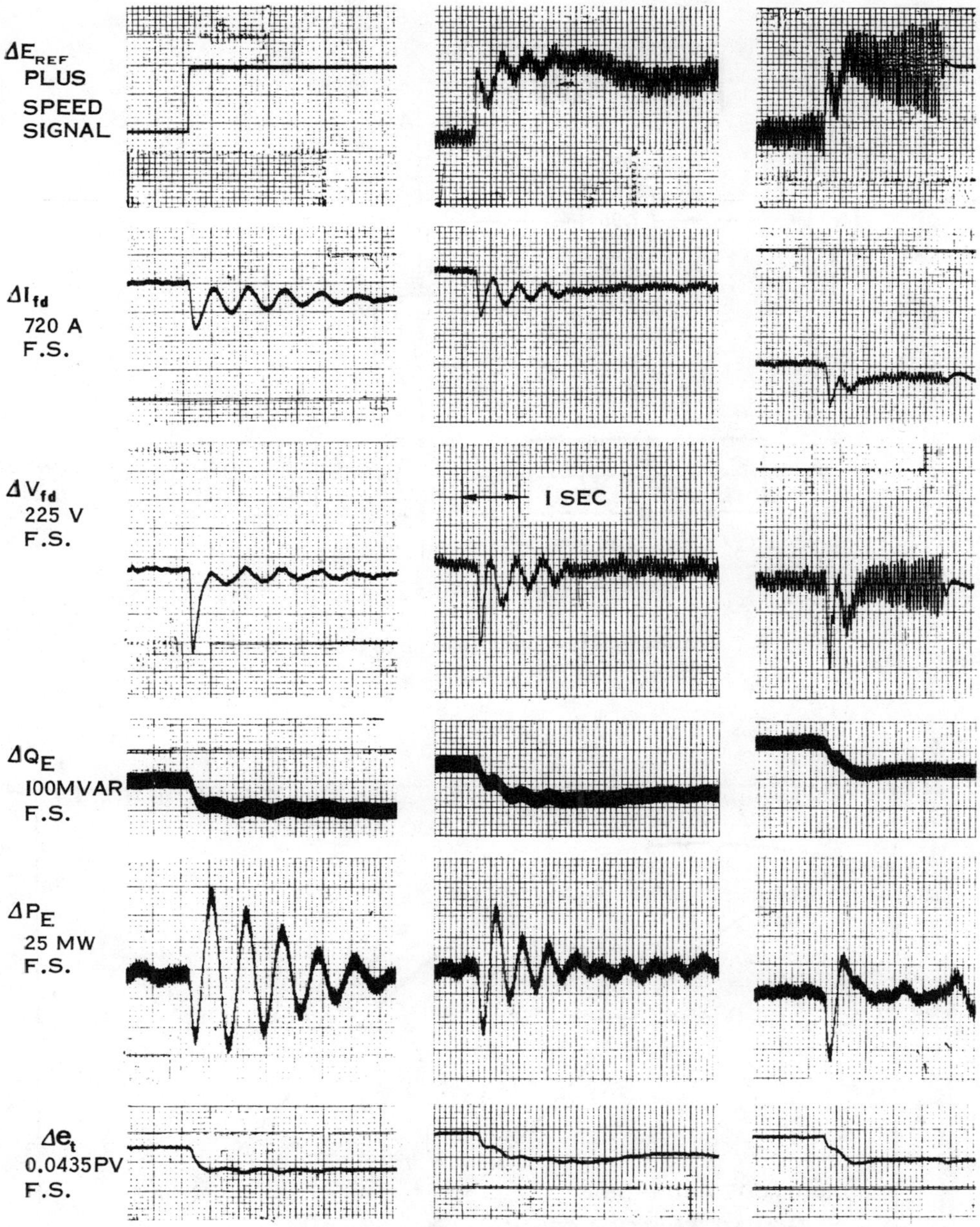

Fig. 3. Generator Response to Small Changes in Regulator Reference Voltage
(a) No Stabilizing Signal to Regulator
(b) Speed Signal Added to Regulator Input
(c) Speed and Acceleration Added to Regulator Input

a. development of a generator model adequate to represent the relationship between torque output and field excitation input for small perturbations of the field at frequencies up to 16 Hz.

b. modelling of the turbine-generator mechanical characteristics related to torsional oscillations.

DEVELOPMENT OF GENERATOR REPRESENTATION

In the course of development of excitation stabilizing signals for synchronous machines a small-signal generator model based on Park's equations and on frequency response tests on a hydraulic generator had been used. A similar model had been derived for the Lambton thermal machine utilizing manufacturer's data, including saturation effects. With this model it was not found possible to duplicate the behaviour of the unit with respect to torques and oscillations at frequencies around 16 Hz. Basically, it was not apparent why a signal applied to the field would effectively damp out oscillations of the generator against the system at 1.6 Hz but would not damp oscillations of the same generator against its turbine at a frequency of 16 Hz. Clearly, there was a phase shift in the actual system which was not duplicated in the model. This was substantiated by observation on site that the magnitude and phase relationships between field current and field voltage perturbations did not correspond with the simulation. An extensive test program to establish an adequate model of the machine was therefore undertaken and is described in a separate paper [3]. This diagram shows the relationship between generator shaft torque and control input signals to the exciter for the generator operating on full load with closed loop regulator control. Included also is a lead-lag block normally incorporated to provide zero phase shift between a speed signal input ($\Delta\omega$) and torque thereby producing optimum damping at the normal oscillation frequency of 1.67 Hz between the generator and the system. The Bode diagram in Figure 5 shows that this same signal produces a phase shift of -134° at 16 Hz and thereby provides an explanation for the negative damping observed at this frequency.

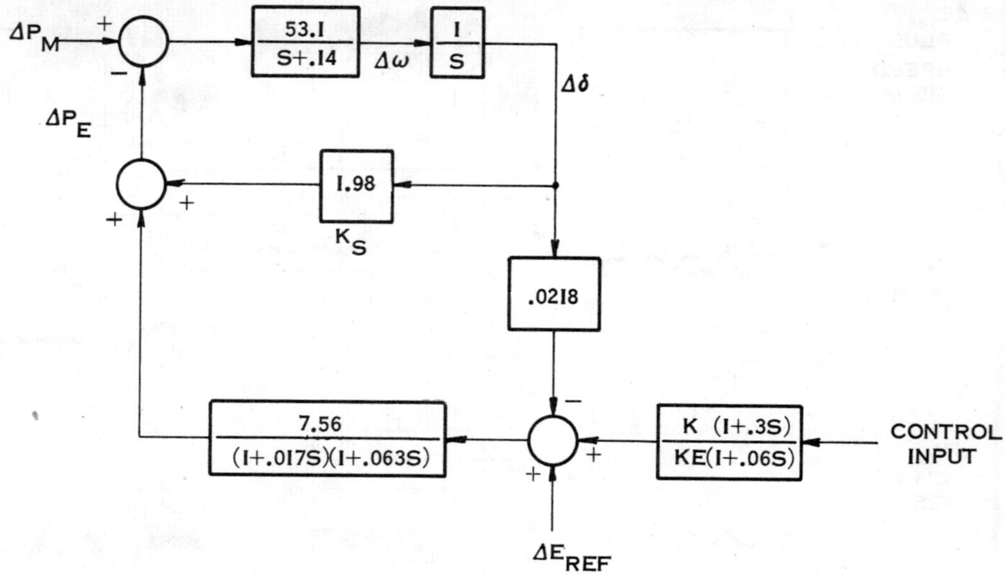

Fig. 4. Transfer Function Block Diagram of Generator with Exciter, Full Load Observation

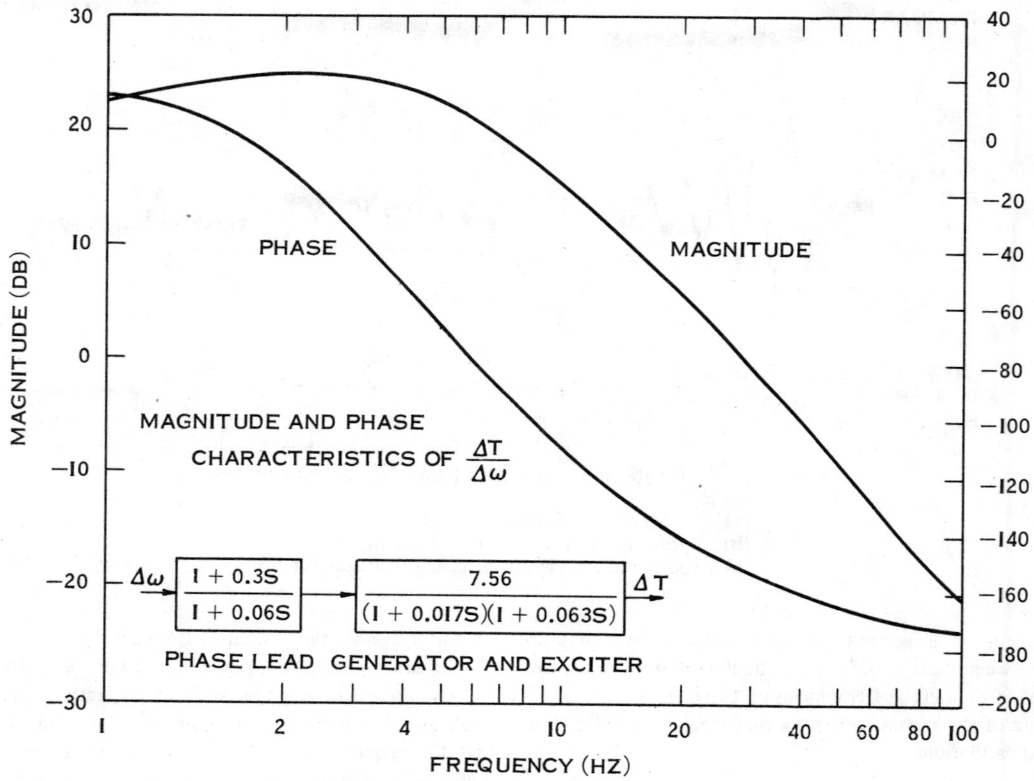

Fig. 5. Bode Diagram for Model of Figure 4

REPRESENTATION OF TORSIONAL OSCILLATIONS

The turbo-generators at Lambton each comprise five principal rotating masses, namely the turbine high pressure, intermediate pressure and two low pressure sections and the generator rotor. For the purpose of analysis these were considered as concentrated masses, which were coupled by shafts of known torsional elasticity. Data on the inertia of the masses and stiffness of the shafts were readily supplied by the manufacturer. The generator itself was represented as coupled to an infinite power system through the synchronizing power coefficient. Differential equations were written for the mechanical system composed of the five rotating masses. The eigenvalues of the resultant matrix gave the five natural torsional frequencies of the system. Evaluation of the eigenvectors gave the relative amplitudes of

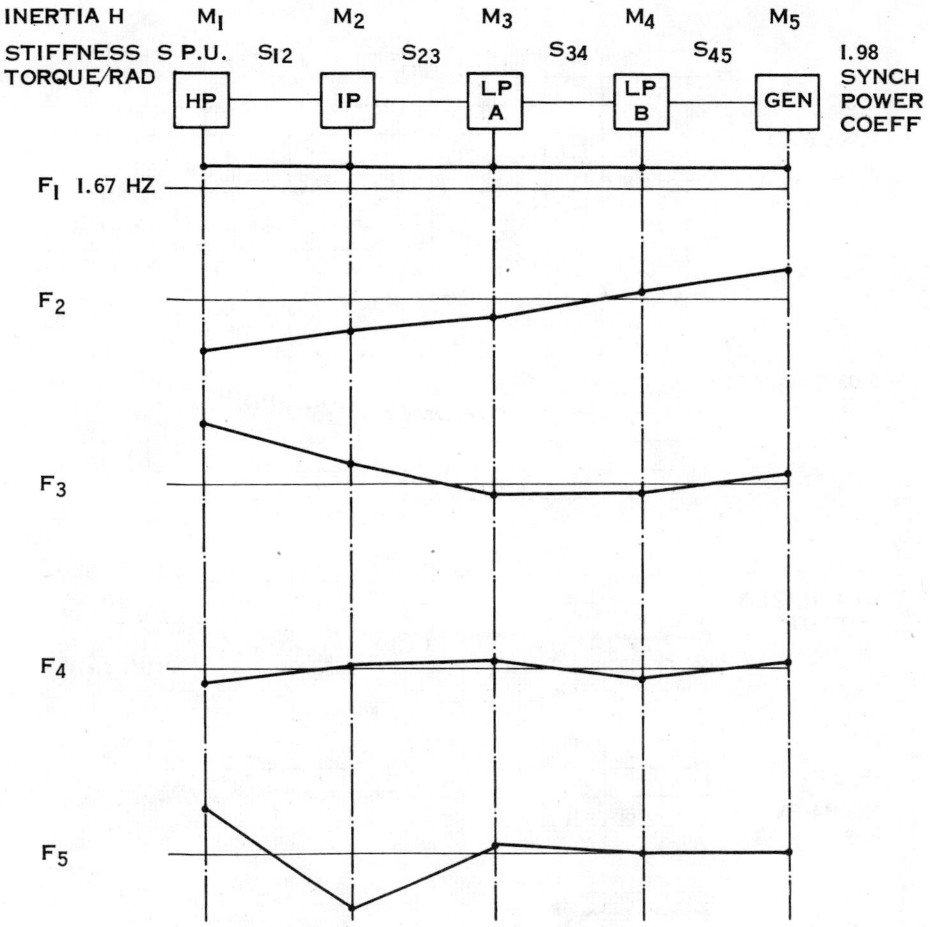

Fig. 6. Modes of Torsional Oscillation of Turbine-Generator

torsional oscillation of the masses for each of the five modes. Figure 6 is a plot showing the natural modes of torsional oscillation. The lowest frequency, 1.67 Hz, is the normal oscillation frequency of the unit against the system. The remaining four modes at frequencies of from 16 Hz upwards are torsional modes of oscillation of the turbine-generator masses against each other. The lowest torsional frequency corresponds, within the limits of measurement, with that of the oscillations excited by application of the stabilizing signal to the exciter.

SIMULATION OF FIELD CONDITIONS

The generator model derived from field measurements and the model of the mechanical system were set up on an analog computer to simulate the conditions observed in the field. It was necessary to introduce components to represent damping of the mechanical torsional oscillations. In the absence of other data a damping component was introduced on each turbine section equal to the per unit mechanical power of that section. In fact, at full load, approximately 1/4 of the power is produced by each of the HP, IP and 2 LP sections so this is equivalent to an assumption of turbine damping equal to total per unit power output distributed equally among the four sections.

The adequacy of the simulation is apparent by comparison of the record obtained from the analog simulation shown in Figure 7 with the field results of Figure 3(c).

POSSIBLE SOLUTIONS TO THE PROBLEM

It is required to develop an exciter control signal based on shaft speed, which will provide satisfactory damping of generator oscillations against the system at 1.67 Hz, and will not excite torsional resonances in the shaft which occur at frequencies of 16 Hz and above, even though there is a phase lag in the generator of 134° or more at the latter frequencies. It would be possible to provide electrical filters in the stabilizing signal circuitry which would achieve the required performance and this was verified by site tests at an early stage in the investigation. However, this solution alone was not considered acceptable to date, on the grounds of reliability. It would make the possibility of damage to a number of major generators over a long period of time, dependent on the reliability of a number of electronic components. Should any of these fail there would be the possibility of exciting torsional vibrations in the shaft to a level where serious damage could result. None of the existing protective devices would detect the condition and the development of a new type of protection would introduce yet another set of problems.

The use of a stabilizing signal based on changes in electrical power was considered. It can be shown that such a signal would not have the same tendency to excite torsional oscillations as a shaft speed signal. However, for satisfactory operation, it is also necessary to provide a mechanical power signal [2] with a high sensitivity and this could be very difficult to obtain on a thermal unit.

It was realized early in the development that a signal derived from a speed measurement at the high-pressure end of the unit would not excite torsional vibrations at 16 Hz. As shown in Fig. 6, oscillations at this frequency are in opposite phase at this point to those at the generator, so that use of such a signal would actually damp oscillations at this frequency. However, although oscillations at the higher torsional frequencies have never been observed in the field, it is important to avoid the possibility of exciting them, and as shown in Figure 6, a measurement at the HP turbine end might excite oscillations

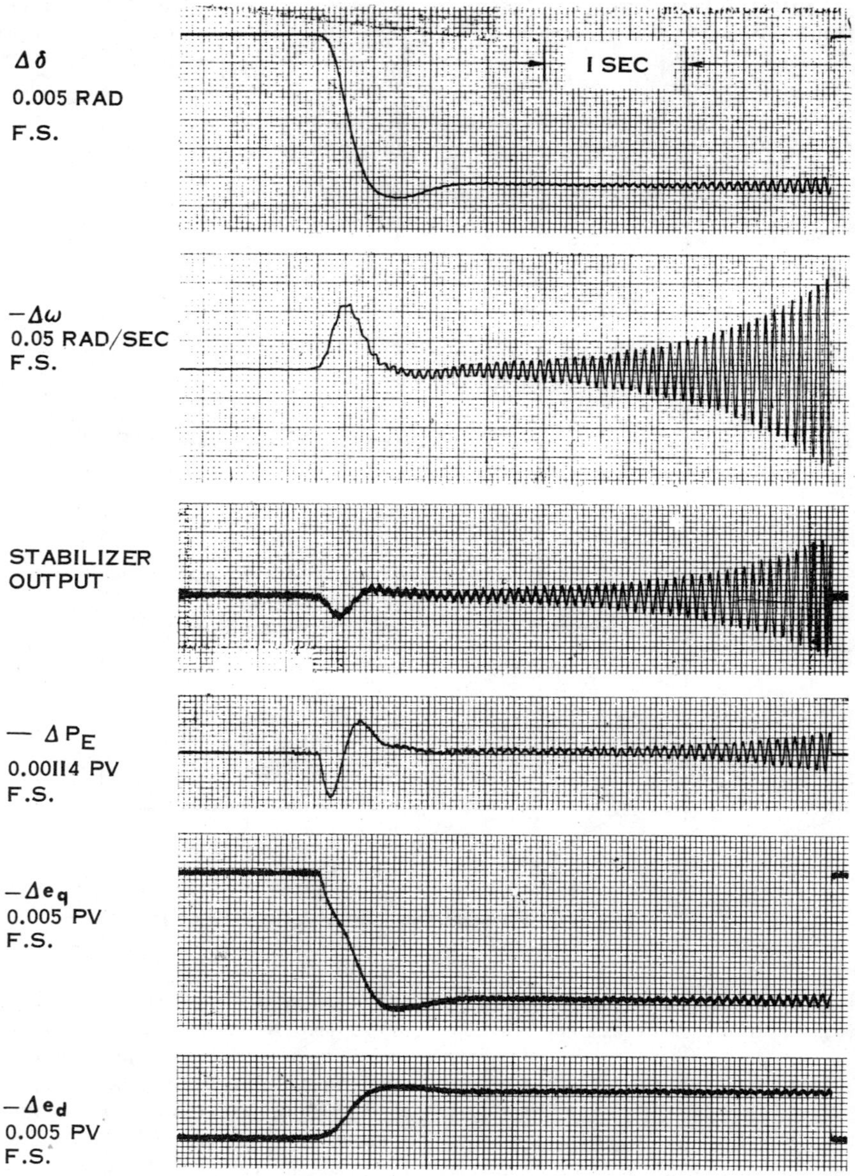

Fig. 7. Behavior of Analog Simulation of Generator

at the next higher frequency F_3. Consideration therefore was given to measuring shaft speed at a location which would not result in excitation of oscillations at any torsional frequency. A point between the two LP sections appeared to meet the requirement. In the first three torsional modes all the amplitudes at this point are negative and in the fouth mode the amplitude is very small. It was decided, therefore, to install speed measuring equipment at this location.

STABILIZING SIGNAL INSTALLED AT LAMBTON GS

The following considerations were taken into account in designing the stabilizing signal equipment now in service at Lambton:

1. The speed pickups should be at or close to a node at the first torsional frequency (16 Hz). Any departure from the node position should be in the negative (away from the generator) direction.
2. Higher torsional frequencies measured at this point on the shaft should be of such magnitude and/or phase that they are not excited by introduction of the stabilizing signal.
3. The exact position of the nodes is not well-defined and there is, in any case, a remote possibility that at some time in the future the stabilizing signal on a unit may be accidentally reversed.

It was decided therefore to incorporate a filter with a notch at the lowest torsional frequency and substantial attenuation at all the torsional frequencies.

The block diagram arrangement of Figure 8 incorporates the features outlined above. Extensive analytical studies were made with this arrangement and it was proved to operate satisfactorily under all normal conditions. Studies were also made with the stabilizing signal reversed. The location of the speed measurement between the two LP sections was varied to allow for any uncertainties about the node position for the first torsional mode. No problems relating to excitation of torsional vibration were discovered.

The speed signal is derived from 9 steel teeth spaced around the shaft at an accessible point between the two LP sections. The teeth are 1/8" x 1/8" x 5". They were accurately spaced around the shaft and cemented in position and then secured by two steel bands. Two magnetic pickups spaced 180° apart in a horizontal plane have their outputs summed to give, at rated speed, 1080 pulses per second, to the transducer circuits. The use of two pickups eliminates any apparent speed changes due to movement of the shaft in a vertical plane. The output of the equipment is proportional to speed change plus a component of acceleration (phase lead) as required.

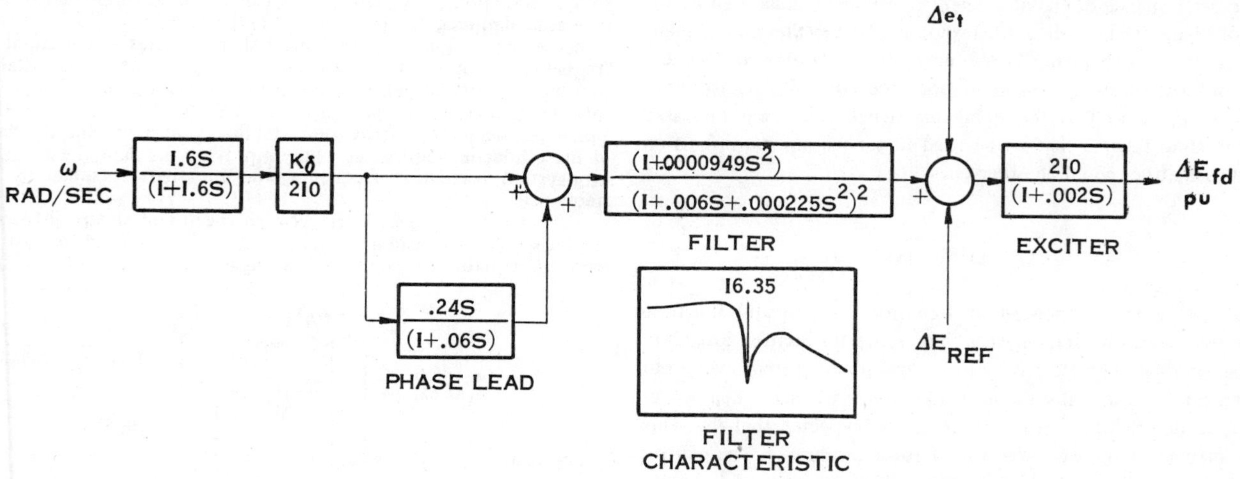

Fig. 8. Block Diagram of New Stabilizing Signal

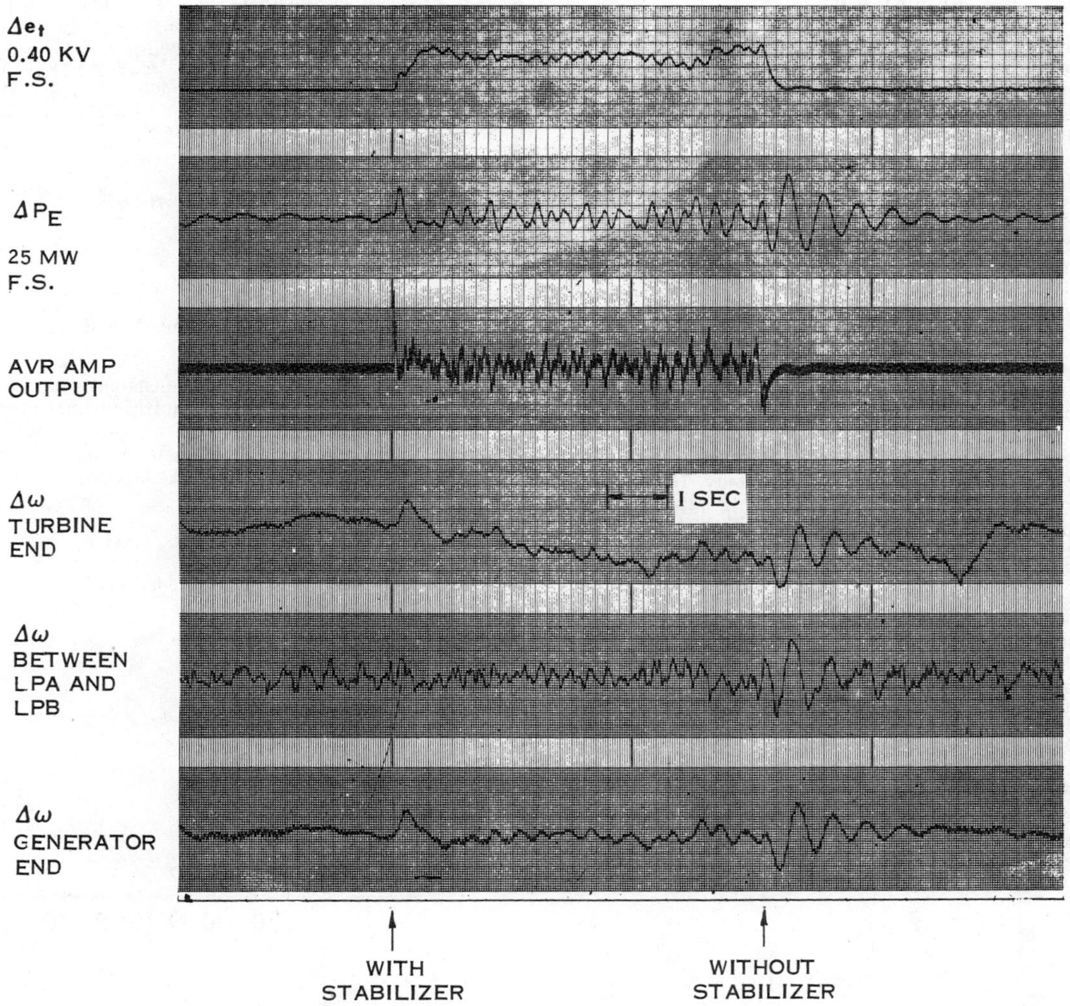

Fig. 9. General Response to Small Change in Reference Voltage With New Stabilizer

The performance of the equipment in service is illustrated by the records of Figure 9 which show the response of the exciter and machine power output to a step change in reference voltage. During commissioning, the outputs of speed sensors at both the HP and generator ends were recorded, as well as the stabilizing signal. Gain was increased to a value almost twice that to be used in service and no effects on torsional oscillations could be observed.

CONCLUSIONS

The problems experienced at Lambton dramatically illustrate the fact that static exciters have a very powerful control capability which can be used very effectively to control power system swings but can also produce potentially harmful side effects when not kept under the required degree of control. It should be expected that the same problem may occur on any large round rotor machines. The control strategies developed have solved the immediate problem and are expected to be applicable to other units in the future which have a similar mechanical configuration. However, the equipment must be designed to match the characteristics of the particular units. It would be desirable to have a general solution applicable regardless of the turbo-generator design. Such a solution has not yet been developed.

REFERENCES

[1] P.L. Dandeno, A.N. Karas, K.R. McClymont, W. Watson. "Effect of High Speed Rectifier Excitation Systems on Generator Stability Limits". IEEE Trans Power Apparatus and Systems, Vol. PAS 87, pp. 190-196, January 1968.

[2] W. Watson, G. Manchur, "Experience with Supplementary Damping Signals for Generator Static Excitation Systems". Paper to be presented at IEEE Winter Power Meeting, 1972.

[3] G. Manchur, D.C. Lee, M.E. Coultes, J.D. Griffin, W. Watson, "Generator Models Established by Frequency Response Tests on a 555 MVA Machine". Paper to be presented at IEEE Winter Power Meeting, 1972.

Discussion

F.P. de Mello (Power Technologies, Inc., Schenectady, N.Y. 12301): The authors are to be commended for documenting experiences and problems in the application of stabilizing signals with high-speed static excitation systems.

These comments amplify on some points also made in discussing a companion paper entitled "Experience With Supplementary Damping Signals for Generator Static Excitation Systems."

The amplification of resonant shaft vibrations of 16 Hz could possibly be due to too high a gain at these frequencies between speed change and torque change produced via the stabilizing signal and excitation system.

Were the effects of a transient gain reduction in the regulator considered or tried as one way of avoiding the problem?

Ref: "Concepts of Synchronous Machine Stability as Affected by Excitation Control" by F.P de Mello and C. Concordia. Discussions and Closure. IEEE Transactions on Power Apparatus and Systems, PAS-88, No. 4, April, 1969, pp. 316-329.

Manuscript received February 7, 1972.

C.E.J. Bowler and N.W. Simons (General Electric Company, Schenectady, N.Y. 12345): The authors have demonstrated that a power system stabilizer can significantly decrease the damping of shaft torsional oscillations. A stabilizer transfer function designed to provide increased damping for oscillations in the vicinity of 1.67 Hz resulted in decreased damping for the torsional frequencies which are at higher frequencies. In most applications to date in the United States the power system stabilizer is designed to increase the damping of system intertie oscillations in the vicinity of 0.3 Hz. Recognition of the torsional problem means that these stabilizers must provide the damping at these intertie frequencies while improving the damping at the generator-system oscillation frequency and the shaft torsional oscillation frequencies.

That a 134° phase shift between speed and electrical torque can produce torsional instability, as stated by the authors, can be shown from the equations of motion of a single torsional system. In Figure 1,

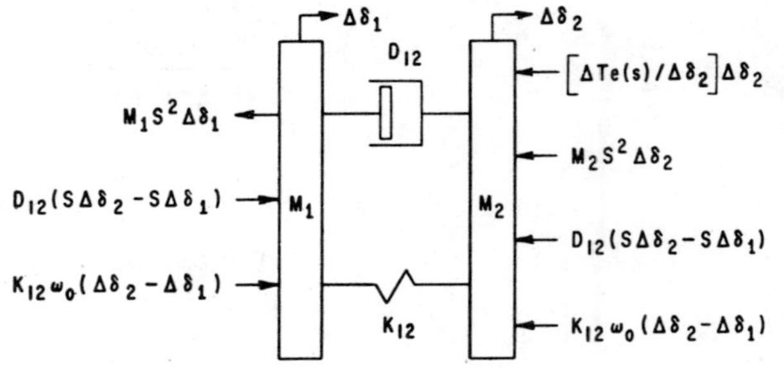

Fig. 1 Simple Torsional System

M_n represents the nth rotor inertia, T_e is the electrical torque, D_{12} is the damping coefficient, and K_{12} is the shaft stiffness. Letting $\delta_2 - \delta_1 = \Delta\delta$ and noting that $\delta_2 + \delta_1 = 0$ from the mode shape, the equation of motion is:

$$0 = S^2\Delta\delta + \frac{D_{12}}{Meq} S\Delta\delta + \frac{K_{12}\omega_0}{Meq}\Delta\delta + \frac{\Delta T_e}{\Delta\delta_2}(s)\frac{\Delta\delta}{2M_2}$$

where $\frac{1}{Meq} = \frac{1}{M_1} + \frac{1}{M_2}$. Letting

$$\frac{\Delta T_e}{\Delta\delta}(s)\,\Delta\delta = (T_s(s) + sT_D(s))\,\Delta\delta$$

the equation of motion becomes

$$0 = S^2\Delta\delta + \left[\frac{D_{12}}{Meq} + \frac{T_D(s)}{2M_2}\right] S\Delta\delta + \left[\frac{\omega_o K_{12}}{Meq} + \frac{T_S(s)}{2M_2}\right]\Delta\delta$$

From this equation it is seen that instability results if the damping torque, $T_D(s)$, is negative and of sufficient magnitude to overcome the inherent mechanical damping. Figure 2 is a polor plot of $\Delta T_e(s)/\Delta\omega$ ($S\Delta\omega = \Delta\omega$) as shown in Figure 5 of the paper. Note that $T_D(s)$ is negative when $\Delta T_e(s)$ lags $\Delta\omega$ by more than 90° but less than 270°.

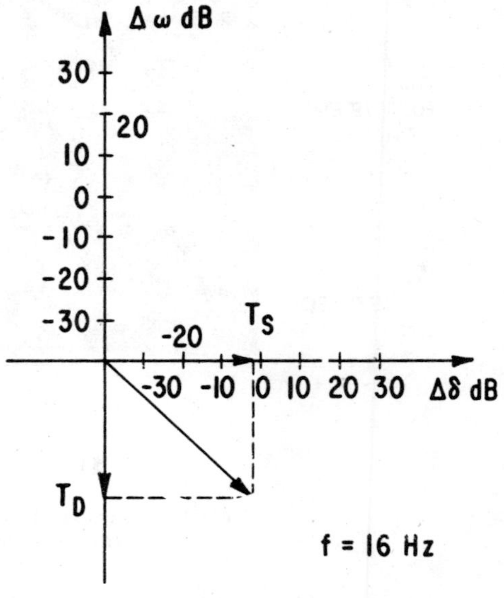

Fig. 2 Polar Diagram of $\Delta T_e/\Delta\omega$

Manuscript received February 22, 1972

The solution developed by the authors for the Lambton generators appears to be adequate. Location of the speed sensor at the shaft's first torsional node should eliminate the possibility that this torsional will be made unstable. Since it is unlikely that the sensor would be placed exactly at the node, locating it toward the turbine end of the machine will increase the damping of the first torsional. As noted by the authors, this location also provides additional damping for the second and third torsional frequencies; the fourth torsional damping is decreased. The plot of the natural modes of torsional oscillation in the paper shows that the relative magnitude of the fourth torsional is small at a point on the shaft corresponding to the node of the first torsional; however, the absolute magnitude of angle or speed change at this point can be quite large at this frequency. Our studies have shown, however, that there is no problem with the higher torsional frequencies on the Lambton machines.

As pointed out by the authors, their solution is not general, and might not be applicable to some future units. It is likely that some future units would not have an accessible location on the shaft that would be acceptable for the shaft's first torsional frequency as well as the higher torsional frequencies.

Our own studies to date have indicated two requirements which in conjunction with the power system stabilizer could result in an unstable torsional oscillation.

1. The generator's shaft torsional frequencies must be in the range where the stabilizer has some effect. Large tandem compound units tend to have lower shaft torsional frequencies.

2. The excitation system must be very fast, with characteristics of a thyristor system, in order for the stabilizer to be effective in the range of the torsional critical frequencies. The excitation systems for the Lambton units satisfy this criterion as does the General Electric Althyrex System.

To better understand this problem and to evaluate the possibility of damage to a generator because of the stabilizer, tests were run on a large tandem compound unit, rated over 850 MVA, during October and November 1971. The generator uses our Althyrex excitation system, and the stabilizer uses frequency as its input signal.

The series of tests were planned to observe the excitation control system while placing the power system stabilizer in service, and during transients designed to stimulate the system and turbine-generator shaft. In order to better understand the duty on the turbine-generator shaft and to obtain more accurate information on the torsional oscillations, strain gages were installed at two locations on the shaft. One set was located between the two low pressure turbines at the approximate node of the first torsional. This location is the point of maximum stress for this torsional. The second set of strain gages was located between the generator and alternator. This unit's second torsional has a high magnitude at this point, while the point between the two low pressure turbines is near a node of the second torsional. Generator terminal voltage, generator field voltage, speed, and several stabilizer signals were also recorded.
Besides being displayed on a chart recorder, many of these signals were recorded on magnetic tape for later analysis.

During the tests all of the unit's torsional frequencies were observed. Also, although addition of the stabilizer did decrease the damping of the first torsional oscillations, it did not cause negative damping. Figure 3 shows a typical strain signal recorded. Several methods of stimulating the shaft were used, from steps of reference in the voltage regulator to operation of system breakers. In general, each method tended to stimulate the shaft in a different way.

Figure 4 demonstrates the increased damping of the generator-system oscillation with the power system stabilizer. These particular tests were initiated by switching out a transmission line. Besides the generator-system oscillation at 1.4 Hz, the higher frequency first torsional is also apparent.

Although it was found that the torsionals were stable at all usable stabilizer gains, a special filter was inserted into the stabilizer so as to increase rather than decrease the damping. Thus, an electrical device is used in this application to maintain the original level of torsional

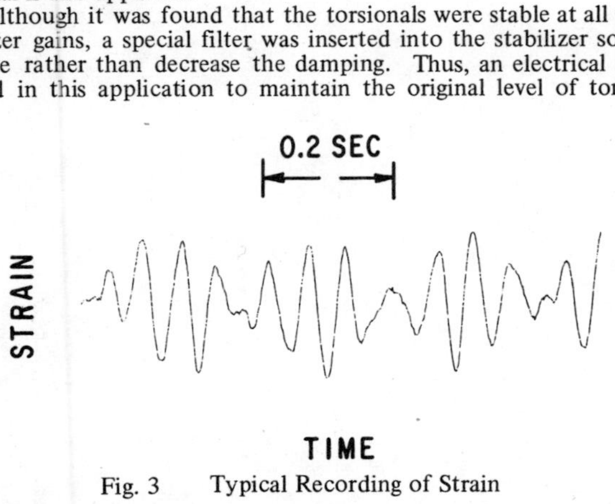

Fig. 3 Typical Recording of Strain

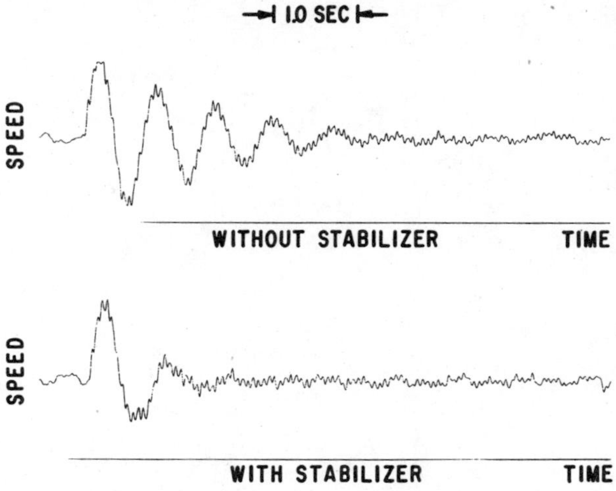

Fig. 4 Speed Recording on Line Switching

damping. In general, the reliability of such electrical equipment has been excellent, and we expect to apply similar devices in the future. Also, because of the consequences of an undetected torsional instability, a protective device to detect unstable torsional oscillations may be desirable.

The need to consider torsional instability where stabilizers are used with high initial respones excitation systems is clear based on these two reported experiences. More analytical studies will be required together with tests to verify the conclusions.

Wilfred Watson and Murray E. Coultes: The authors thank Messrs. Bowler and Simons for their contribution. The analysis presented by the discussers parallels that made by the authors and confirms the results of the tests made by ourselves and by the discussers. It can be concluded therefore that the nature of the problem is now well understood and that solutions can be devised accordingly.

The adequacy of any given method of deriving a stabilizing signal may be evaluated with respect to the following criteria:

1. Physical arrangements are such that the input inherently does not include components likely to excite torsional oscillations.

2. The signals are processed electrically in such a manner that components at torsional mechanical frequencies will add to the damping rather than subtracting from it.

3. The gain and phase of the overall transfer functions at the torsional frequencies are such that instability at these frequencies will not occur.

The solution adopted at Lambton meets the first criterion above, by virtue of the mechanical arrangements and also criterion 3 by means of electrical filtering. We feel that the conservatism inherent in this solution was justified since this was the first experience of the problem. The solution adopted by the discussers essentially meets criteria 2 and 3 above in that electrical arrangements modify the signal and gains are such that even without these arrangements no serious problem would result. Once again this may be considered a conservative solution. It may be that satisfactory applications can be made which satisfy only criterion 3. However, such applications are difficult to evaluate in the light of experience to date. Performance depends firstly upon the reliability of electronic components which may be used to provide attenuation at torsional frequencies. Secondly, the margins of safety are dependent upon the inherent damping of the mechanical oscillations and it is not known at present to what extent this varies with time on individual machines, between similar machines, or between different designs of machines.

The use of transient gain reduction as suggested by Mr. de Mello fall into the category covered by criterion 3, in that increased margins of stability could be obtained. We have not considered to date that solutions meeting only criterion 3 are satisfactory, preferring instead an arrangement which is inherently stable, whereupon the additional margins obtainable through gain reductions are of secondary importance.

No doubt as more experience is accumulated in this field, various methods of applying stabilizing signals will be developed with increasing confidence in the satisfactory performance of the equipment.

Manuscript received April 25, 1972

Part III

Transient Stability

Transient stability analysis is primarily concerned with the effects of transmission-line faults on generator synchronism. During a fault, electrical power from nearby generators is reduced, perhaps drastically, while power from machines somewhat removed from the fault may be scarcely changed. The resultant differences in acceleration produce speed differences over the time interval of the fault, and it is important to clear faults quickly to limit these speed differences and the associated changes in angle differences. Clearing the fault necessarily removes one or more transmission elements from service and at least temporarily weakens the transmission system. This change in the transmission system also implies that the generator angle differences which existed prior to the fault no longer represent equilibrium conditions. If speed differences and accumulated angle differences at the time of fault clearing are sufficiently bounded, and if transfer impedances between generators subsequent to fault clearing are sufficiently low, the accelerated machines will pick up load due to their advanced angular positions, slow down, and eventually a new synchronous equilibrium will be established. Loss of synchronism, if it occurs, will usually be evident within 1 s of the initial disturbance.

Faults on heavily loaded circuits are more likely to cause instability than faults on lightly loaded lines because they tend to produce more extreme accelerations during the fault and because clearing the fault results in a weaker transmission network. Three-phase faults produce greater accelerations than those involving one or two conductors, and they have a greater demagnetizing effect on nearby generators. Faults not cleared by primary relaying produce greater accelerations during the fault, greater demagnetization of nearby generators, and a weaker transmission system after the fault since fault clearing may remove several transmission elements from service.

It is not feasible to obtain the same standards of transient stability performance for all systems at all times. Transmission distance is the most critical factor in limiting performance goals for system planning studies. However, the availability of generation sites and transmission right-of-way, constraints on capital expenditures, and environmental restrictions all influence the selection of performance standards. In North America, system design criteria cover a wide range, from three-phase faults cleared by backup relays to single-line-to-ground faults with single-pole breaker operation and successful reclosure.

Transient stability studies are done as part of planning for virtually all new generators. Such studies are carried out routinely and specific study procedures and results are seldom published. It would be wrong to infer from this that these studies are necessarily simple. To be confident of accuracy, they frequently involve the representation of large systems of from one to three hundred generators along with corresponding networks and loads. System response is inherently nonlinear and may be extremely sensitive to factors tending to cause loss of synchronism. Extrapolation of study results from one situation to another is not reliable, so that a variety of operating conditions and contingencies must be thoroughly examined.

One means of improving the stability characteristics of long transmission lines is the use of series capacitors to reduce the effective transfer reactance between generators. A group of five papers related to this subject are reprinted in this section. The first of these is a fundamental application paper covering economics, performance improvements, and certain limitations and restrictions to be observed. The next two papers both pertain to the use of switched series capacitors and could well have been included in Part II as dynamic stability papers. The desirability of collecting these papers on series capacitors in one location prevailed, however. The final two papers on series capacitors discuss the problems of subsynchronous resonance in series compensated circuits.

A group of four papers on single-pole switching is included. Two of these describe methods for arc suppression to reduce dead time for single-pole reclosing. The remaining two in this group describe analytical methods for simulating the effects of single-pole operations and the design and testing associated with a particular project.

Finally, four papers dealing with fast load control are reprinted. Principal interest here centers on fast valving as a means of reducing steam turbine power quickly enough to prevent loss of synchronism. Other control methods, such as resistance braking, generator dropping and load dropping are also discussed. The final paper in this group is a report of design studies on a specific plant.

Supplementary Bibliography

Transient Stability

[1] J. Baba, S. Hayashi, I. Yamada, H. Haneda, and F. Ishiguro, "Sensitivity analysis of power system stability," in *1967 PICA Conf. Rec.*, pp. 47–68.

[2] W. A. Mittelstadt, "A method of improving power system transient stability using controllable parameters," *IEEE Trans. Power App. Syst.*, vol. PAS-89, pp. 23–27, Jan. 1970.

[3] P. G. Brown, F. P. DeMello, E. H. Lenfest, and R. J. Mills, "Effects of excitation, turbine energy control, and transmission on transient stability," *IEEE Trans. Power App. Syst.*, vol. PAS-89, pp. 1247–1252, July/Aug. 1970.

[4] C. N. Kashkari, "Effects of feed-back signals on transient stability—Phase plane approach," *IEEE Trans. Power App. Syst.*, vol. PAS-90, pp. 2228–2232, Sept/Oct. 1971.

[5] A. C. Sullivan and F. J. Evans, "Some model experiments in fast valving to improve transient stability," IEEE paper C 72 242-1.

[6] J. W. Balance and S. Goldberg, "Subsynchronous resonance in series compensated transmission lines," presented at the 1973 Winter Meeting of the IEEE PES.

[7] S. Goldberg and W. R. Schmus, "Subsynchronous resonance and torsional stresses in turbine-generators shafts," presented at the 1973 Winter Meeting of the IEEE PES.

The Use of Series Capacitors to Obtain Maximum EHV Transmission Capability

G. D. Breuer, Senior Member IEEE *H. M. Rustebakke*, Senior Member IEEE

R. A. Gibley, Member IEEE *H. O. Simmons, Jr.*, Senior Member IEEE

Summary: Series compensation can substantially increase the transmission capability of EHV (extra-high-voltage) systems, permitting a higher loading per right of way. Several of the possible circuit arrangements that can be used for insertion of series capacitors into a transmission system are cited in this paper. Certain technical problems of using series capacitors are analyzed. The effects of fault switching time on stability limit for series-compensated circuits and of delaying reinsertion of series capacitors after fault clearing are indicated for particular configurations. The relative economics of various degrees of series compensation and number of intermediate switching stations for 500 kv and distances up to 400 miles are also shown.

The growing electrical load and the limited availability of rights of way often require maximum utilization of each right of way in transmitting energy. Also technical and economic considerations may tend to dictate very high energy transmission per circuit. Increased capacity per circuit has been obtained traditionally by higher transmission voltages, higher speed breakers, intermediate switching stations, and reduced reactances of generators and transformers. High-speed excitation systems, also common today, tend to increase transient stability limits. More recently, employment of bundled phase conductors has reduced the series line impedance, also increasing stability limits. This paper assumes these techniques are used, and further increase in possible loading is secured by direct series compensation.

In addition to these techniques, high-speed reclosing will augment transient stability limits of transmission systems. The reclosing characteristics of series-compensated lines are unlike those of uncompensated lines. This difference is discussed and some curves, showing the stability limit as a function of reclosing time, are presented.

Series capacitors can be utilized in several possible circuit arrangements. The purpose of the compensation discussed in this paper is to obtain very high right-of-way loading and, since one arrangement proved to give higher capability than most others, this circuitry was studied in some detail.

The maximum power capability of a transmission system may be restricted by thermal considerations, transient or steady-state stability. When series compensation is used, the possibility exists of a sustained overvoltage to ground at the series capacitor terminals. This overvoltage can be the power-limiting criterion at a high degree of compensation. Curves demonstrating these characteristics are presented and analyzed.

Technical Assumptions

The circuit studied is represented in Fig. 1. This is a 2-circuit 500-kv transmission system with various combinations of intermediate switching stations and different degrees of series compensation. Three circuits were also considered for short distances. Conductor size and line configuration assumed for this study consist of 1,431-MCM (thousand-circular-mil) dual conductors in 18-inch bundles

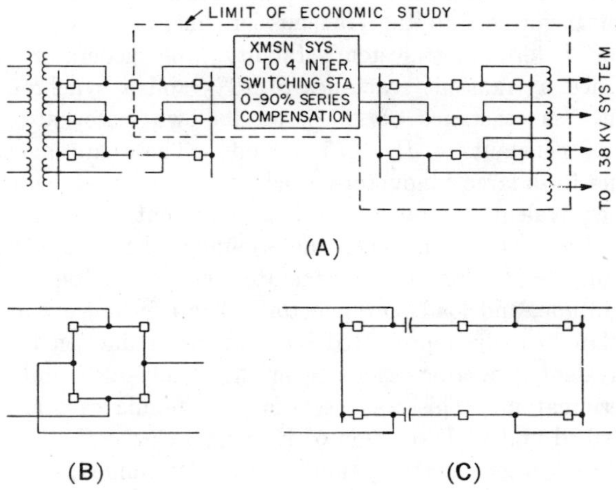

Fig. 1. Circuit studied

A—Limit of economic evaluation
B—Intermediate switching station arrangement without series compensation
C—Same arrangement as B, but with series compensation

Paper 64-61, recommended by the IEEE Transmission and Distribution Committee and approved by the IEEE Technical Operations Committee for presentation at the IEEE Winter Power Meeting, New York, N. Y., February 2–7, 1964. Manuscript submitted November 1, 1963; made available for printing April 9, 1964.

G. D. Breuer, H. M. Rustebakke, R. A. Gibley, and H. O. Simmons, Jr., are with the General Electric Company, Schenectady, N. Y.

Thanks are extended to A. J. Bingham, T. N. Hand, and C. E. Orsak for making network analyzer studies of the systems studied and C. Concordia for his consultation. They wish to especially thank Miss C. A. Skitko for not only programming but also running the economic calculations.

Reprinted from *IEEE Trans. Power App. Syst.*, Nov. 1964.

Table I. List of Intermediate Switching Stations and Degree of Series Compensation Investigated in Study

Miles	Number of Intermediate Switching Stations	% Series Compensation
100	0	0
		50
	1	0
		50
200	1	50
		70
	2	0
		50
		70
	3	0
		50
	4	0
300	1	90
	2	50
		70
	3	0
		50
		70
	4	0
400	3	50
	3	70

with a phase separation of 39.5 feet, flat spacing. The surge impedance loading (SIL) of this line, using the usual definition, is 810 mw (megawatts) per circuit or 1,620 for the two circuits studied, based on an operating voltage of 500 kv.

A minimum standard step-up transformer impedance of 17% on its own base was used in the stability study. The receiving-end voltage transformation was accomplished by an autotransformer with a minimum standard impedance of 11% on its own base. The receiving-end 138-kv system was assumed to have a short-circuit capability of 15,000,000 kva (kilovolt-amperes), and an infinite bus was considered to exist behind the system impedance. The circuit breakers were arranged in a breaker-and-a-half scheme to connect the autotransformers to the transmission line.

For determining system stability, the generator was assigned a transient reactance of 22% and a synchronous direct-axis reactance of 157% on its own base, also an inertia constant of $H = 3.75$ seconds. These are average values for a large conductor-cooled, steam turbine-generator. An equivalent reactance, used to represent the effect of generator saturation when determining the steady-state stability,[1,2] is a function of generator terminal voltage, load magnitude, and load power factor. The effect of a voltage regulator can be represented by a further reduction in the equivalent generator reactance in the steady-state stability determination. This was used for a 400-mile case and is described under "Discussion of Economic Curves."

The transient stability limits were determined, using a 3-cps (cycles per second) switching time. In addition to switching time required to remove the faulted line, the series capacitor may be momentarily bypassed by a protective gap connected across the capacitors. The protective gap is usually set to break down at 250% of the design voltage of the series capacitor although it would be possible, but not economical, to design the capacitor with sufficient voltage rating to withstand fault currents. In order for this capacitor to be effective during a transient swing, the gap must be deionized after fault-clearing. Also, it is necessary to prevent another protective gap breakdown during system swings where a current of no more than twice the normal current would be expected in most of the configurations.

Whenever the line current is high enough to cause the protective gap to break down, such as during a line fault, the gap must be deionized before restoring the series capacitors to service. Special equipment can be used to deionize the gap at each current zero so the capacitor will be available immediately upon fault clearing. This is called instantaneous reinsertion, and if it is not used, the deionizing time extends beyond fault clearing; thus, the series capacitors are inoperative immediately after fault clearing. The sequence of events is listed here:

1. Prefault, the system is operating normally.
2. There is a 3-phase fault on a line section adjacent to the generator high-voltage bus.
3. The fault is cleared in 3 cycles.
4. Series capacitors are immediately returned to service (instantaneous reinsertion) or remain short circuited for a few cycles after fault clearing (delayed reinsertion).
5. Thus, three cycles (instantaneous), or longer with delayed reinsertion, after the fault initiation, the system is operating with one line section and one capacitor bank out of service and all remaining series capacitors in service. The series capacitor switching arrangement may modify this statement, as explained in the next paragraph.

The word "remaining" in point 5 can be understood by referring to Fig. 1(C), which shows how the series capacitors were located in the switching stations for the switching scheme studied in greatest detail. Studying this figure shows that one of the parallel series capacitors is removed from service with any one of the four lines switched off. Other arrangements which might not switch any of the series capacitors out of service were investigated, but the setup shown in Fig. 1(C) was selected for detailed study since the transient stability limit was much higher than for other arrangements.

Not all possible combinations of switching stations and compensation were considered, since some were obviously impractical. The combinations found most practical for a given distance are shown in Table I.

Performance requirements were established with the practical operation of a power system in mind. The sending-end high-voltage bus of the transmission system was kept at rated voltage—100%. The receiving-end bus was kept between 95% and 100% for all normal load conditions. The generator was operated in such a manner that it would never "take reactive"—that is, operate underexcited—at heavy load. Also, when the line required reactive, the sending-end power factor would never drop below 95% overexcited. Necessary receiving system voltage support was obtained by locating shunt capacitors on the 138-kv receiving bus. Shunt reactors were employed for overvoltage control and power-factor correction.

The sustained fundamental frequency series capacitor terminal voltages to ground are usually unequal under normal operating conditions, and this voltage difference tends to increase under heavy load. Since this difference is principally a function of the direction of var (reactive volt-ampere) flow, some voltage control can be exercised by a shunt capacitor or reactor connected to the capacitor terminals. This was done with the maximum voltage limit set at 110% under emergency conditions. Emergency conditions, for the purpose of this study, include maintaining firm transmission with a single line section outage, or with the transmission system energized from one end. Transient overvoltages were not investigated and are beyond the scope of this paper.

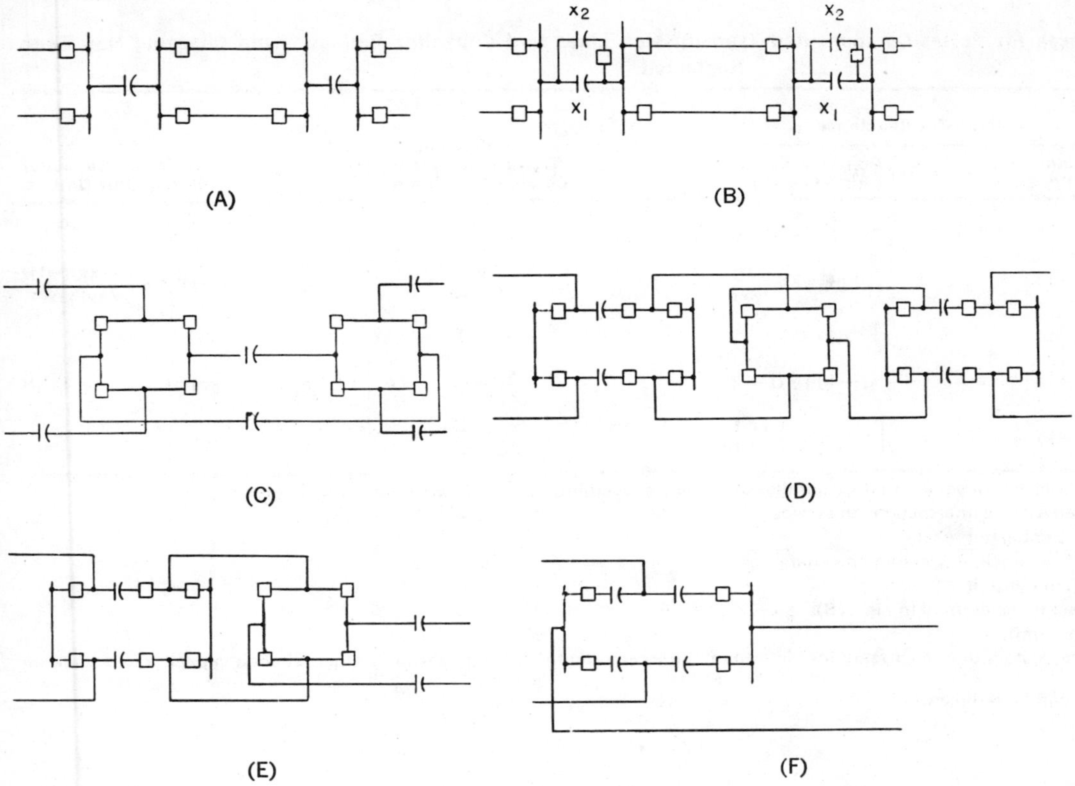

Fig. 2. Several series capacitor arrangements

A—Fixed series capacitors located in intermediate switching station bus
B—Switched series capacitors in same location as in A
C—Series capacitors in each line section and switched with line
D—Series capacitors incorporated into a breaker-and-a-half scheme in intermediate switching station bus. This is satisfactory for an odd number of intermediate switching stations
E—Same arrangement as in D, but showing last intermediate switching station for an even number of stations
F—Split series capacitors, equivalent to D but with fewer breaker positions and higher vulnerability

Series Capacitor Arrangements

Several arrangements of series capacitors could be made in transmission circuits. Usually the purpose of the capacitor is to increase the stability limit by reducing transmission line transfer impedance. To be most effective in increasing the stability limit, this reduction in transfer impedance must be effective with a line section switched out, as well as with all circuits normal. Not all capacitor arrangements, for a given degree of series compensation, have the same postswitching transfer reactance. Further, the required capacitor rating for equal compensation depends upon the scheme used. Thus, economics enter the decision, not only of *whether* to use series capacitors, but *how*.

Series capacitors can be located in line sections and switched as part of the line, or located in switching stations. Fig. 2 shows several possible arrangements.[3] A summary of characteristics, neglecting line charging, is given in Table II. Column 4 of Table II lists the change in transfer reactance as a line is switched out of service. Circuit A shows an increase of $x/2$ ohms, which is the same change as experienced by an uncompensated line. Circuit B can be made to present no change or even reduce the transfer reactance when a line is switched out of service by proper selection of x_1 and x_2. Circuits D and E will offer low transfer reactance changes for moderate compensation.[4] The term $(1-2k)$ for scheme D indicates that no change in transfer reactance is experienced at 50% series compensation. For a higher degree of compensation, the postswitching reactance will be less than the preswitching reactance. Caution must be exercised in using these arrangements with high compensation since overcompensation can occur with a line section switched out of service. It is this characteristic that gives circuits D, E, or F of Fig. 2 a higher stability limit than the other circuits. The percentage of compensation has been defined, neglecting the line charging.

The circuit in Fig. 2(B) requires a very high-speed switch, which can disconnect the auxiliary capacitor quickly enough to be effective in increasing the transient stability limit. Although both configurations, A and B, minimize the number of breaker positions required, these bus arrangements present a definite disadvantage from the standpoint of system reliability. A bus fault, or failure to clear a line fault with first line protection (breaker or relaying), will result in complete loss of power transfer ability. To increase the reliability, more circuit breakers must be added. The configuration of Fig. 2(C), which has a ring bus, permits failure of line or capacitors and, even in the event of a breaker failure, leaves a through circuit intact. The breaker-and-a-half arrangement of Fig. 2(D) likewise can isolate any faulted line or bus and only in the case of failure of the breaker located between incoming and outgoing lines are two lines removed, but a through circuit remains intact. This arrangement requires two extra circuit breakers, but the increase in reliability is significant.

From the standpoint of the total series capacitor rating requirements, circuit A of Fig. 2 appears to have the advantage, as indicated in column 5 of Table II. However, it has the highest postswitching transfer reactance (column 3), hence, the lowest stability limit. Since the highest stability limit occurs with circuit D, which additionally represents a bus arrangement with high reliability, this system was studied in detail. The total series capacitor bank rating is high in this scheme, but it is possible to rationalize a reduced rating, as considered later. When any of the line sections of configuration D are switched off, one series capacitor unit is removed from the active circuit. The remaining capacitor unit in this station must then carry the entire current. Thus, the high series capacitor rating is required only to accommodate the entire load current during line outages. This is also true for configuration C. Since, for 2-circuit transmission, the current through the series capacitor will double under this condition, the capacitor

Table II. Transfer Reactances for Series-Compensated Transmission Lines and Capacitor Ratings; Line Charging Has Been Neglected

	2	3	4	5
	Transfer Reactance			
Arrangement from Fig. 2	All Lines IN Service	One Line Section Switched OUT	Change in Transfer Reactance Column 3—Column 2	Total Series Capacitor Rating, Per Unit
A	$\frac{n}{2}x(1-k)$	$\frac{n}{2}x(1-k)+\frac{x}{2}$	$\frac{x}{2}$	$\frac{kn}{2}I^2x$
B	$\frac{n}{2}x(1-k)$	$\frac{x}{2}[1+n(1-k)]-\frac{x_1^2}{x_1+x_2}$	$\frac{x}{2}-\frac{x_1^2}{x_1+x_2}$	$(n-1)I^2x_1+\frac{I^2x_1^2x_2(n-1)}{(x_1+x_2)^2}$*
C	$\frac{n}{2}x(1-k)$	$\frac{n+1}{2}x(1-k)$	$\frac{x}{2}(1-k)$	$2knI^2x$†
D	$\frac{n}{2}x(1-k)$	$\frac{n}{2}x(1-k)+\frac{x}{2}(1-2k)$	$\frac{x}{2}(1-2k)$	$2knI^2c$†
E	$\frac{n}{2}x(1-k)$	$\frac{x}{2}\left[n+1-kn\frac{n+3}{n+1}\right]$	$\frac{x}{2}\left[1-\frac{2kn}{n+1}\right]$	$2knI^2x$†

n=number of equal-length line sections in series, that is, number of equally spaced intermediate switching stations, plus 1.
k=degree of series compensation with all line sections in service.

$$k=\frac{\text{total circuit series capacitor reactance}}{\text{total uncompensated circuit reactance, neglecting line charging}}.$$

x=reactance of one line section, in per unit.
x_1 and x_2 are series capacitor reactances defined in Fig. 2(B).
I=total transmitted current, in per unit.
* This includes more switched capacitors than necessary for more than one intermediate switching station. One switched capacitor can serve for all adjacent line sections.
† Text describes how this rating can be reduced.

rating or size will have to be increased by a factor of 4.

One possibility to consider in determining the bank ratings is utilization of the 150% capacitor voltage capability for 5 minutes. Thus a capacitor, adequate for an emergency at twice-normal voltage, could be rated at two-thirds this emergency voltage and be satisfactory for at least 5 minutes. The kvar (kilovar) rating of this unit would be $(1.33)^2$ or 1.78 times normal, compared to the rating of 4 times normal required if the capacitor rating is selected for continuous operation under emergency conditions. If the load is not at design peak, the capacitor will be satisfactory for longer than 5 minutes, perhaps continuously. If the load at the time of a permanent line fault is no more than one-half of the design limit, a capacitor suitable for rated load under normal conditions would be sufficient. Note that a successful high-speed reclosure would cause capacitor overloading only during the reclosing interval. On the other hand, if a capacitor is overloaded because a contingency has removed one capacitor, there would be at least 5 minutes for taking corrective action of any form that would unload the capacitors—reclosing, reducing line flow, reconnecting the idle capacitor in parallel with the overloaded unit, etc.

The scheme shown in Fig. 2(E) is the same as in Fig. 2(D), but showing how an even number of switching stations can be used. Fig. 2(D) is for an odd number of switching stations. Fig. 2(F) shows an alternate capacitor arrangement that can be used in circuits D and E.

Table III shows a tabulation for some of the rows or arrangements of Table II for $n=4$ and $k=0.6$. The condition for row B is the selection of x_1 and x_2 to give no change in reactance when one line and a capacitor (x_2) are switched out of service. Row D will be 100% compensated after switching with $k=5/6$ or $83^1/_3$% compensation. In general, this will occur when $k=(n+1)/(n+2)$. If k is greater than this, the compensation, after switching, will reach into the terminal reactance such as the transformers.

Series Capacitor Terminal-to-Ground Overvoltages

The series capacitor voltage (IX_c) between its terminals is the phasor difference of the voltages to ground from either terminal. Thus, the magnitude of these voltages depends upon the direction and magnitude of the current flow with respect to line-to-ground voltage. The greatest concern is when one terminal voltage exceeds specified limits; this can happen when there is a high degree of series compensation and a heavy load.

Table III. Specific Case for Three Intermediate Switching Stations and 60% Series Compensation (n=4, k=0.6)

	Transfer Reactance			
Arrange-ment from Fig. 2	All Lines IN Service	One Line Section Switched OUT	Change in Transfer Reactance Column 3—Column 2	Total Series Capacitor Rating, Per Unit
A	0.8x	1.3x	0.5x	1.2 I^2x
B	0.8x	0.8x	0	3.367 I^2x
C	0.8x	x	0.2x	4.8 I^2x
D	0.8x	0.7x	−0.1x	4.8 I^2x

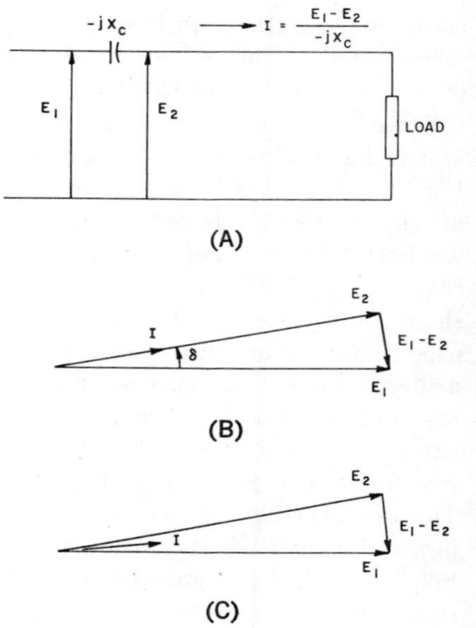

Fig. 3

A—Series capacitor in a load circuit
B—Phasor diagram for A with unity power-factor load
C—Phasor diagram for A with lagging power-factor load

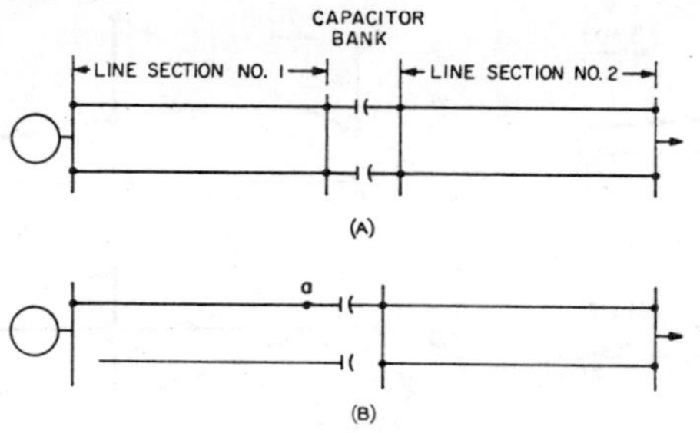

Fig. 4

A—Two-circuit, one intermediate switching station transmission with series capacitors connected as in Fig. 2(D)
B—Same as A except that a line section and a series capacitor have been switched out of service

Fig. 3(A) shows a circuit containing a series capacitor. The phasor diagrams for two load conditions are also shown. The power flow is into the load or from E_1 to E_2 through the series capacitor. The phasor diagram of Fig. 3(B) is shown for a resistive load; that is, no reactive is delivered to the load from E_2. The series capacitor will generate I^2X_c vars and they flow from the voltage E_1 toward the power source. Note that E_1 is greater than E_2.

Fig. 3(C) differs from 3(B) only in that the voltage E_2 has been increased, which causes the current phasor to lie between the two phasor voltages. Thus, reactive is delivered to both right and left from the series capacitor. Had E_2 been increased until I was in phase with E_1, the entire series capacitor I^2X output would be delivered to E_2, the higher voltage. Power flows toward the leading voltage through a series capacitor, and reactive tends to flow toward the higher voltage. Thus, with E_1 fixed, adding a shunt reactor at E_2 will cause E_2 to increase in magnitude. Likewise a shunt capacitor at E_2 will reduce that voltage.

Under most conditions, the series capacitor terminal voltages will be within voltage limits when all line sections are in service. However, when using capacitor arrangements to obtain maximum transmission capability, voltage problems may exist with a line section out of service. When a line section is switched out for Fig. 2(D), the remaining section will carry the entire current, which will approximately quadruple the reactive (I^2X) used by that line section, or double the I^2X required per twin circuit right of way. This reactive requirement increase will tend to be supplied by the series capacitor, increasing the capacitor terminal voltage at location α as shown in Fig. 4(B). The overvoltage to ground can impose a power limitation when it exceeds the design or operating voltage limits. The series capacitor can be broken into smaller units and distributed over the line. This will permit a higher power flow for the voltage-limited case and may be an economical means of extending the capability.

The nature of some series capacitor overvoltages can be studied in Fig. 5. These curves were constructed to show the performance of a 200-mile 500-kv line taking 3,000 mw from a generating station with a line section out of service as shown in Fig. 4(B). The sending-end voltage is held at 1.00 per unit in all cases. The curves of Fig. 5(A) are for 30% series compensation. The top two curves show the line-to-ground voltage and its angle as a function of position on the line. The intermediate points were considered to be of minor interest so they were not determined. The voltage is not necessarily as shown, linear with distance, but will tend to be below the straight line. This loading and compensation show that the depressed capacitor voltage can be increased by raising the receiving-end voltage as in the third curve. The fourth curve shows the depressed voltage raised by a 300-mvar (megavar) shunt capacitor bank on the high-voltage terminal. The fifth or bottom curve shows the series capacitor terminal voltages to ground, equalized by using a 600-mvar capacitor and a 275-mvar reactor.

The curves of Fig. 5(B) are similar to 5(A) except that they are for 50% series compensation. The first curve now shows an overvoltage that is brought within limits by reducing the receiver voltage, third curve; a 190-mvar reactor, fourth curve; and a 200-mvar capacitor, fifth or bottom curve. The reactor connected to the low-voltage terminal had very little effect on that voltage, but it reduced the opposite or high-voltage terminal. A similar effect is noted in the last curve, although here the capacitor caused some reduction in the high voltage.

The curves of Fig. 5(C) are for 70% series compensation. Nearly 24% overvoltage is displayed in the top curve, showing that the overvoltage is a function of the degree of series compensation. The third curve demonstrates the overvoltage reduction caused by lowering the receiving-end voltage. Note that the voltage on that side of the series capacitor has increased, while on the other terminal, it has decreased. If the receiving-end voltage were lowered farther, that side of the series capacitor would have an overvoltage of more than 110%. In light of the previous discussion, the series capacitor would then be supplying most of its generated reactive toward the receiving end, which now takes reactive from the line due to its depressed voltage.

The fourth curve displays the effect of a 240-mvar reactor connected on the receiving-end terminal as shown. The reactor is exercising only enough voltage control at 70% compensation to bring the voltage within the 110% limit. The bottom curve shows the effect of a shunt reactor that is too large (610 mvar). This reactor has shifted the overvoltage from the source-side terminal to the load-side terminal. The top curve indicates that a shunt capacitor will not be a satisfactory voltage corrective device since both terminals are experiencing high voltage, particularly the sending-end side. A capacitor on the high-voltage terminal has tended to raise the low voltage while a reactor on the low-voltage terminal lowers the high voltage.

Capability of a Series-Compensated Line

The transmission capability of an EHV line will be determined by: (1) conductor thermal limits, (2) transient stability, (3) steady-state stability, or (4) overvoltage at the capacitor terminals of a series-compensated line. The thermal limit of conductors will be reached when the current (heating) exceeds the safe value, usually experienced only on short lines. Transient stability is usually the limitation experienced on long lines and steady-state on short lines. However, as series compensation is added to a line, it appears electrically to be shorter, since the series reactance has been reduced. Thus, the steady-state stability criterion can limit the capability of long, highly compensated lines.

The effect of series compensation on the stability limit criterion is displayed in Fig. 6. The curves in this figure show transient and steady-state stability limits for a 200-mile 500-kv 2-circuit transmission system with the terminations, etc., as cited under "Technical Assumptions." The two upper curves have been determined for a series capacitor

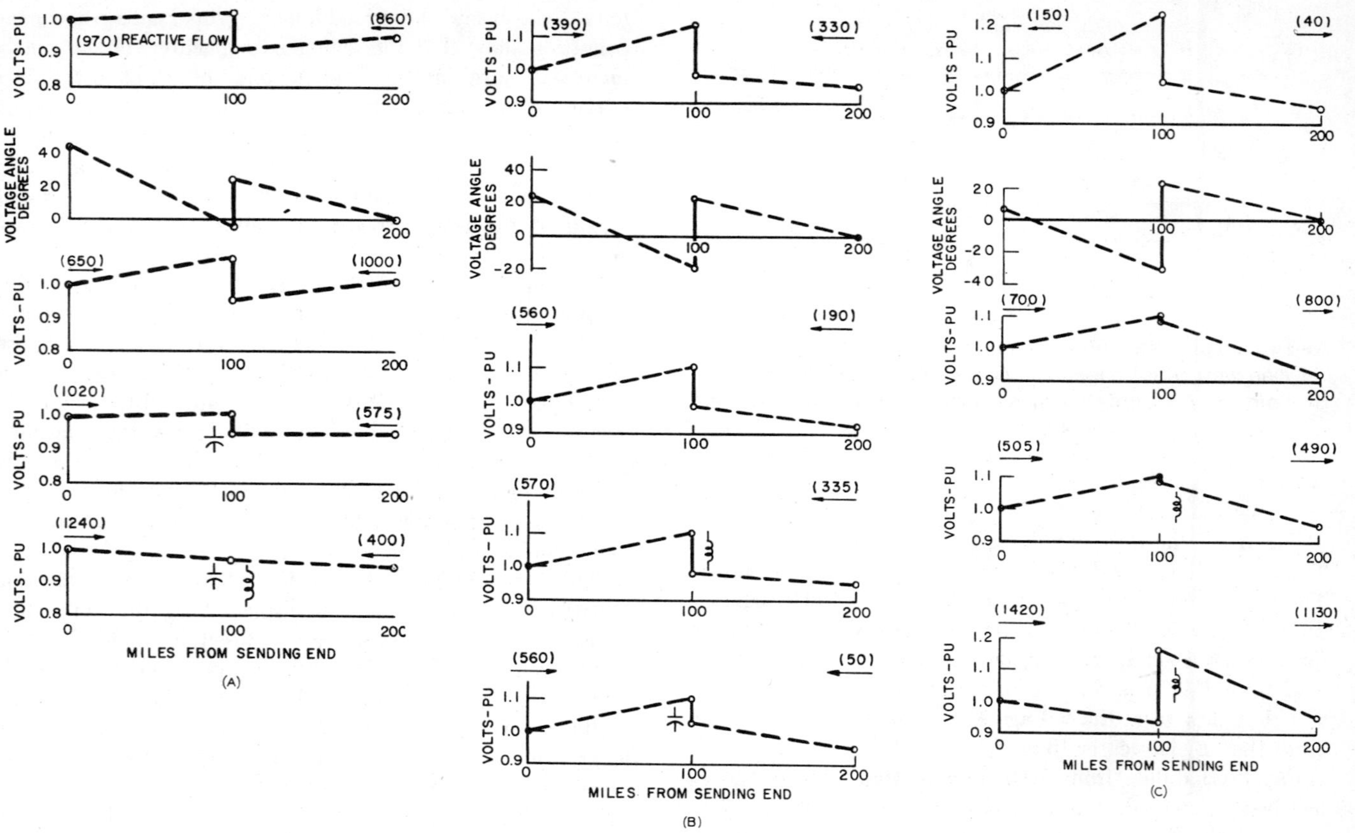

Fig. 5. Voltage and angle profiles for 2-circuit 500-kv 200-mile transmission circuit with line section switched out of service as in Fig. 4(B). In every case, the line receives 3,000 mw at rated voltage

A—30% series compensation B—50% series compensation C—70% series compensation

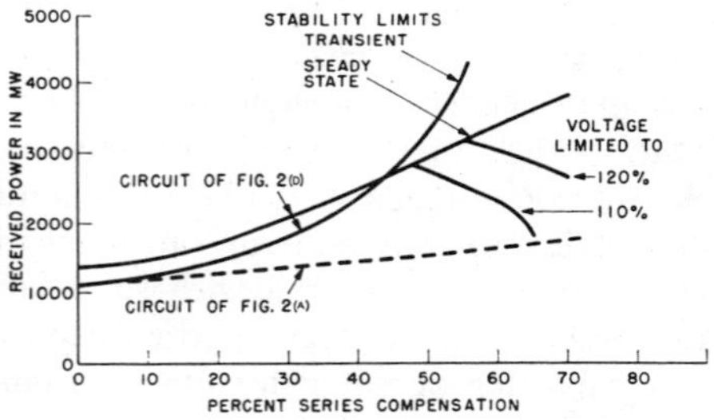

Fig. 6 (left). Power limits for 500-kv 200-mile 2-circuit transmission system with one intermediate switching station as a function of series compensation

arrangement as in Fig. 2(D), while the lower curve is for the capacitor arrangement of Fig. 2(A).

The steady-state stability curve cited in the previous paragraph was made for a 2-circuit configuration with lines switched to give the higher transfer reactance, or the lower stability; that is, below about 50% series compensation, the limit was determined with a line section out of service, and with all line sections in service for higher compensations. Since the system must be stable for both conditions, only the lower limit was determined.

The solid curves in Fig. 6 increase with additional series compensation but at different rates. The transient curve is the power limitation up to a series compensation of about 45%. Beyond this point, the steady-state criterion will limit the capability of this transmission system. The steady-state curve is terminated at 70% compensation but, at this point, one capacitor terminal voltage to ground is 148% or 427 kv. Since it is unlikely that such a voltage could be tolerated, steps must be taken to reduce the overvoltage to acceptable limits. This can be accomplished by reducing the power transmitted to lower values. Two branches from the steady-state stability curve are shown. On one branch, the voltage is limited to 120% by a reduction in transmitted power, and on the other, the voltage is held at 110%. This shows that, with no corrective action such as shunt reactors, there is an optimum compensation to obtain the maximum transmission capability, keeping the voltages within design limits.

These curves demonstrate 3 of the 4 transmission capability limitations cited previously. The maximum capability for this system is 2,800 mw at about 47% series compensation when the voltages are to be kept within 110% of nominal. If 120% voltage is permitted, the maximum capability is 3,150 mw at 55% series compensation.

Higher powers than 2,800 mw can be attained and the 110% voltage criterion met if shunt reactors are used. The reactor should be connected to the terminal of the capacitor having lower voltage, which normally will be on the opposite side of the series capacitor from the switched line. A reactor could be switched to either terminal at high speed as the need arises. A 490-mvar reactor so switched will raise the capability from the 110% voltage curve at 55% series compensation to the steady-state stability limit curve of 3,150 mw.

This switching will limit the steady-state overvoltages to 110%. However, there are some complications in so switching a reactor. If one were wise enough to know which terminal needed the reactor, and permanently connected it to this terminal, no further complications would arise. Without this foresight, a very-high-speed switching arrangement to connect the reactor to the proper capacitor terminal would be needed. The cost of the switching scheme and reactor could be evaluated to determine if it is justified for an additional 650 mw. This overvoltage problem is quite difficult to control by switched reactors when high compensation is used, particularly on extra long lines. The equipment and line exposed to the overvoltage could be adequately insulated to withstand the overvoltages, or the series capacitor might be broken into smaller units as cited earlier.

The dashed curve of Fig. 6 shows the stability limit, using the series capacitors in the bus as shown in Fig. 2(A). A comparison of the two schemes can be made by transmitting 2,000 mw or 1,500 mw. For the 2,000-mw case, only the D circuit can safely carry it. The A circuit appears to approach 2,000 mw at 100% compensation but many problems would be encountered at this compensation; see the Appendix. To transmit 1,500 mw requires 22% compensation for the circuit of Fig. 2(D) and 46% compensation for the circuit of Fig. 2(A). From Table II, the capacitor rating for circuit A is 0.46 I^2X and circuit D requires 0.88 I^2X. Converting the difference into series capacitor rating on this example shows circuit D requires about 240 mvar more than A. If the 150% voltage rating is utilized for short times, the circuit D required rating is 0.39 I^2X or lower than for circuit A. Thus, at 1,500 mw, either circuit arrangement could be used, but D would be more expensive unless short-time voltage capability were used.

Effect of Switching and Reinsertion Time

As the switching time is reduced, the transient stability limit tends to increase, since the power disturbing influence is more quickly removed. The usual curve for 3-phase generator-end high-voltage bus faults is shown as the bottom curve of Fig. 7. This curve is for one switching station and no series compensation on a 2-circuit 500-kv 200-mile transmission system. It has the characteristically small slope for very low switching times.

The addition of 50% series compensation gives a substantial increase in the stability limit for any switching time up to 6 cycles (on a 60-cps basis) as shown. If the series capacitor is available when the faulted circuit is interrupted (instantaneous reinsertion), the top curve applies, while the intermediate curve represents a 5-cycle delay in inserting the capacitors. The two curves for series compensation are terminated at the steady-state stability limit of 2,950 mw.

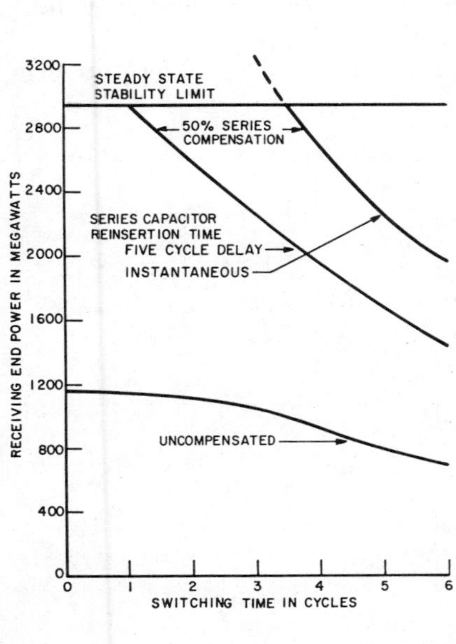

Fig. 7. Stability limits as function of switching time for a 500-kv 200-mile 2-circuit transmission system with one intermediate switching station. Comparison of uncompensated circuits and 50% series compensation shown; also displayed is effect of capacitor reinsertion time versus switching time

Fig. 8. Transient stability limit as function of series capacitor reinsertion time for a 500-kv 200-mile 2-circuit transmission system with one intermediate switching station using 3-cycle switching

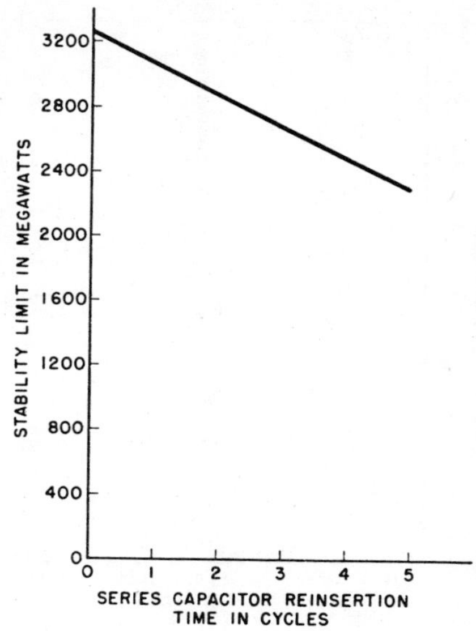

The curves of Fig. 7 for series compensation continue to rise as the switching time is reduced and do not flatten out as for the uncompensated case. This can be explained from the circuitry displayed in Fig. 4(B). Here it is apparent that the transfer impedance—neglecting line charging—is unchanged for 50% series compensation when a line is switched out of service. Without series compensation, the postswitching impedance of a twin-circuit line is always higher by one-half the switched line impedance.

The curves of Fig. 7 demonstrate the effectiveness of series compensation at 200 miles. The capacitors have increased the transmission capability substantially over the uncompensated value. This increase is for 50% compensation with more gain to be realized with higher compensation. Much higher compensation will not necessarily be desirable since operating problems will appear at some value of high compensation.

The steady-state stability limit as shown in Fig. 7 will not permit the transient limit to be realized at a switching time of 3 cycles and instantaneous reinsertion of the series capacitor. As mentioned previously, the steady-state stability was determined without evaluating the increase due to the beneficial action of the voltage regulator. This may permit the transient limit to be used; if so, the effect of varying series capacitor reinsertion time can be investigated. Fig. 8 shows such a curve; it indicates that loss in stability limit with reinsertion time is linear with a slope of about 195-mw loss in stability limit per cycle of reinsertion time. This is for a specific line—200 miles, 500 kv, twin circuit, etc.,—and would not necessarily apply to other configurations or distances. This should, however, demonstrate the nature of the phenomenon.

High-Speed Reclosing with Series Compensation

When series capacitors are used in the arrangement of Fig. 2(D), the change in transfer impedance when switching a line is $x/2\ (1-2k)$, as shown in column 4 of Table II. This 50% compensation causes no change in transfer impedance when switching a line section into or out of service. In view of this, a line that has 50% series compensation in this switching arrangement does not need high-speed reclosing. This stability limit is substantially independent of the reclosing time, as shown by the top curve in Fig. 9. If the compensation is less than 50%, high-speed reclosing

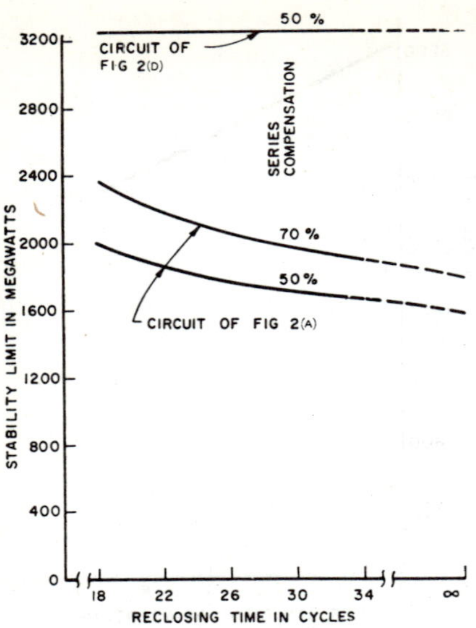

Fig. 9. Transient stability limit as a function of reclosing time for a 500-kv 200-mile 2-circuit transmission system with one intermediate switching station

will show some advantage, depending upon the degree of series compensation. However, high-speed reclosing will reduce the stability limit when the compensation exceeds 50%. Even when the stability limit is unaffected by high-speed reclosing, it is advantageous to reclose from the standpoint of reliability.

High-speed reclosing shows a gain in stability limit for the bus capacitor arrangement of Fig. 2(A). Curves for 50% and 70% series compensation, using the Fig. 2(A) scheme, are shown in Fig. 9.

Economic Assumptions

The generators, step-up transformers, and associated breakers were not included in the economic portion of this study. It was considered that the station arrangement would not affect the economic choice of transmission system configurations. However, three breakers were included on the sending end, since they are associated with the two transmission lines in a breaker-and-a-half scheme (see Fig. 1).

The average cost of a 500-kv dual Plover 1,431-MCM line was assumed to be $115,000 per mile. This included the right of way and erection costs. Installed switching station costs and associated equipment were evaluated at $413,000 per breaker position.

The receiving-end autotransformers (1,550 kv/450-kv basic impulse insulation level) were limited to 500-mva output. The transformer costs were $383,650 at 100 mva; $513,650 at 200 mva; $642,850 at 370 mva; and $726,830 at 500 mva. A multiplier of 1.2 was applied to the unit price to obtain the installed cost of the transformer.

Series capacitors were evaluated at $7.00/kvar. This cost included supporting structure, protective equipment, equipment for instantaneous reinsertion, and installation costs, as well as the capacitor units themselves. The switched shunt capacitors were installed for $5.00/kvar, and switched shunt reactors at $3.50/kva. The losses were evaluated at $150/kw capacity charge with a 2.5 mills/kwh (kw-hour) energy charge.

The economic curves shown in this paper are based upon 60% load factor, 46% loss factor, and 15% carrying charge. Other values are shown for 200-mile transmission for comparison. These values are 12% carrying charge, 80% load factor, 70% loss factor, and installed series capacitor cost of either $6.00 or $8.00/kvar.

Economic Calculations Program

The economic calculations were made with an IBM (International Business Machines Corporation) *704* data processing system. The computer program consists of a load flow that will yield all the usual load flow information including required ratings of any series capacitors and of a set of economic calculations. The economic calculations can be made to include a large portion of the system, that is, up to 50 lines. Lines, transformers, series and shunt capacitors, shunt reactors, and switching stations can be represented in the economic computations. Also, several different unit costs can be used for each type of equipment and for each quantity, such as annual carrying charge, load factor, etc. The output will give the costs for each combination of unit costs and other quantities.

The output from the program includes all of the input, load flow results, losses in the area under economic study, series capacitors' ratings, and investments in each component, such as lines, transformers, etc. Further, the annual cost, as well as unit energy cost for each combination of inputs, is in the output. The unit energy cost is based on the energy delivered by the area under economic study.

Calculations can include an evaluation of the change in losses in the system not under economic study when comparing alternatives. As the area under economic study increases in size, this program becomes an even greater time saver than it is for a single element, such as one transmission line. It also yields results with great ease and speed when varying parameters.

Discussion of Economic Curves

Curves that terminate at the thermal limit will end in a circle, followed by two dashes to indicate a higher stability limit. Curves terminated at the steady-state stability limit have an *s* to indicate this limit; those without symbols terminate at the transient stability limit. The transient limit was determined, using instantaneous reinsertion of the series capacitors after fault interruption.

Fig. 10 shows the transmission costs at 100 miles for various combinations of equipment. These curves, and all other energy cost curves, terminate at a stability limit or an arbitrary thermal limit. The thermal limit was taken as 2,900 mw, which represents a conductor temperature rise of approximately 75 C (degrees centigrade) over a 40 C ambient in a 2-foot-per-second wind, assuming a 0.5 emissivity. In cases where curves were terminated at this thermal limit, it is to be understood that the stability limit has not yet been reached.

The curves in Fig. 10 demonstrate the need for more than

Fig. 10. A 500-kv 2-circuit transmission for 100 miles. *Not including step-up transformer or step-up transformer breaker costs common to each scheme. (The statement after the asterisk applies also to Figs. 11–15)

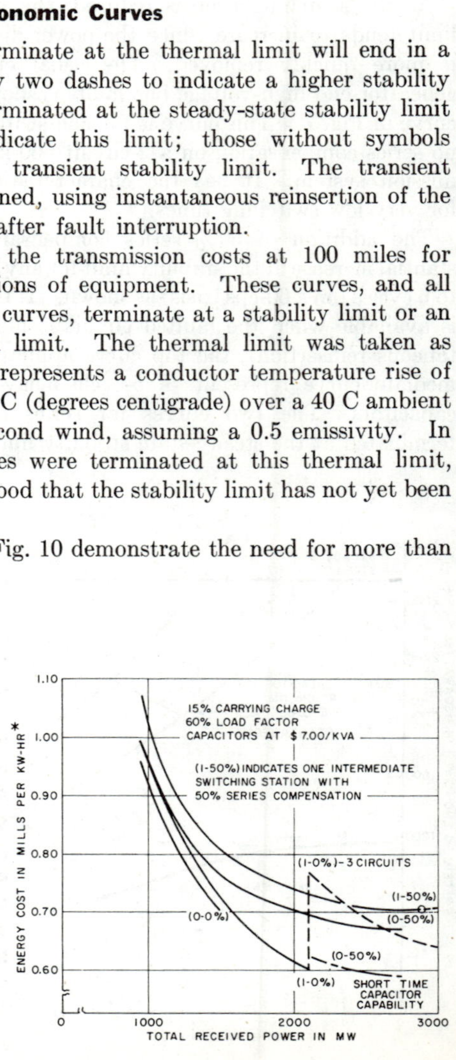

one intermediate switching station for loads in excess of 2,100 mw. Another circuit will extend the capability with favorable economics for loads of 2,600 mw. Series compensation will also extend the power capability as shown by the two curves (0–50%) and (1–50%). The latter arrangement will extend the capability beyond the arbitrary thermal limit. The (0–50%) configuration has a transient stability limit of 2,750 mw.

The cost curves are all shown for series capacitor ratings sufficient to carry the indicated power during a single line section permanent outage. As noted in the section "Series Capacitor Arrangements," it may be feasible to utilize the short-time overload capability of the capacitors. When this is done, the capacitor size is reduced in the ratio of 16/9 to 4 or 4 to 9. This reduction in the capacitor size could noticeably lower the energy cost curves, particularly at high loads. A section of the curve for 0–50%, using the short-time overload capacitor capability, is shown in Fig. 10 as "Short-Time Capacitor Capability."

Fig. 11 shows the importance of series capacitors in obtaining favorable capability and economics for transmitting heavy loads 200 miles. One intermediate switching station with 50% series compensation gives a capability approximately twice the uncompensated limit, even using four intermediate switching stations. Increasing the number of stations or compensation over the 1–50% will give no technical or economic advantage within the framework of this study. Again the economic effect of reducing the series capacitor size to its short-time overload capability is shown. The 2–50% curve terminates in steady-state stability at about 2,500 mw, while the 1–50% curve has a stability limit beyond the thermal limit. This apparent incongruity of a lower limit with more switching stations can be understood by computing the postswitching transfer impedance. Neglecting the effect of charging the ratio of postswitching to preswitching transfer impedance for the 1–50% case is unity, but for the 2–50% case is 7 to 6.

In contrast with the economics of 100-mile transmission, the addition of a third circuit without series compensation yields no economic advantage in 200-mile transmission. No curve is shown for three circuits at 200 miles. The third circuit with three intermediate switching stations would extend the stability limit from 1,460 mw to approximately 2,180 mw but at a higher unit energy cost than for the curves shown. The possibility exists that three circuits with series compensation may be economically attractive for extremely high loads.

Fig. 12 shows unit cost curves for 300-mile transmission. Even at this distance, series compensation can extend the stability limit beyond the arbitrary thermal limit selected in this study. The curves indicate the need for higher compensation at this distance. The curve for 2–70% has a much higher limit than that for 2–50%. Apparently, some intermediate compensation will be optimum for loads beyond the limit of that for three switching stations and 50% series compensation. The economy of using the short-time series capacitor overload rating is substantial in this case as shown by the broken curve for 2–70%.

The circuit shown in Fig. 2(A) was cited previously as having a lower stability limit than Fig. 2(D), as shown on the curve 3–70% for 300 miles in Fig. 12. The delayed reinsertion (5 cycles) stability limit, using the circuit of Fig. 2(A), is marked *D*, and its instantaneous reinsertion limit is marked *I*.

Fig. 13 shows economic curves for 400-mile 500-kv 2-

Fig. 11. A 500-kv 2-circuit transmission for 200 miles

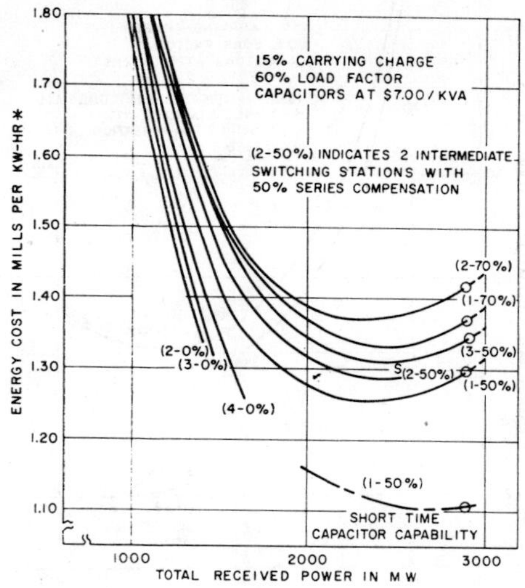

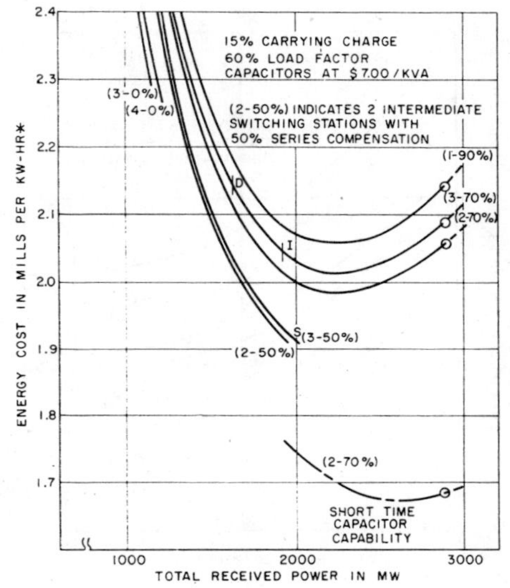

Fig. 12. Same as Fig. 11 except for 300 miles

circuit transmission. The 3–70% curve was limited by the steady-state criterion to less than 2,300 mw when machine saturation alone was considered. The equivalent machine reactance for steady-state calculations can be reduced by using a voltage regulator to some value between the equivalent reactance using saturation only (X_{eq}), and transient reactance (X_d')[2]. It was determined that a reactance of 0.6 ($X_{eq}+X_d'$) would extend the steady-state stability to 3,000 mw. A well-designed and properly adjusted voltage regulator is capable of this performance. The transient stability limit is above 3,000 mw.

Using the series capacitor arrangement of Fig. 2(D) resulted in overvoltages with a line section switched out of service and delivering 2,900 mw. To reduce this voltage, capacitors were redistributed, placing some in the line sections. This resulted in a combination of Figs. 2(C) and 2(D). The voltages were brought within 110%, using a line-to-station capacitor reactance ratio of 0.375 to 1.

The economic advantage of the short-time capacitor capability becomes greater with distance. At 400 miles, it shows a saving of more than 0.4 mill/kwh at 2,500 mw. At 300 miles, the saving was slightly greater than 0.3 mill/kwh; at 200 miles, it was 0.15; and it was 0.08 at 100 miles, as may be noted in Figs. 10 through 12.

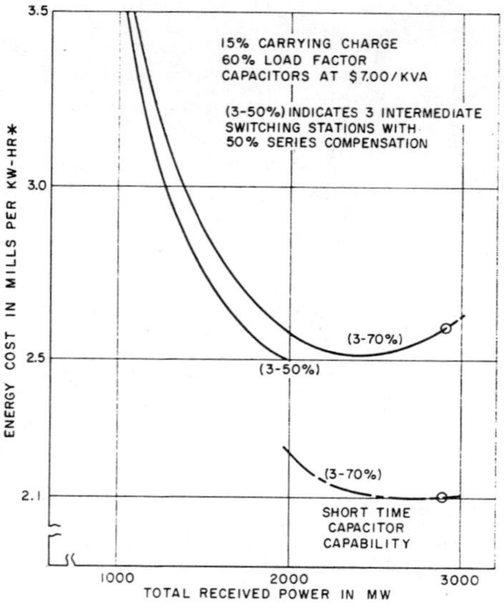

Fig. 13. Same as Figs. 11 and 12 except for 400 miles

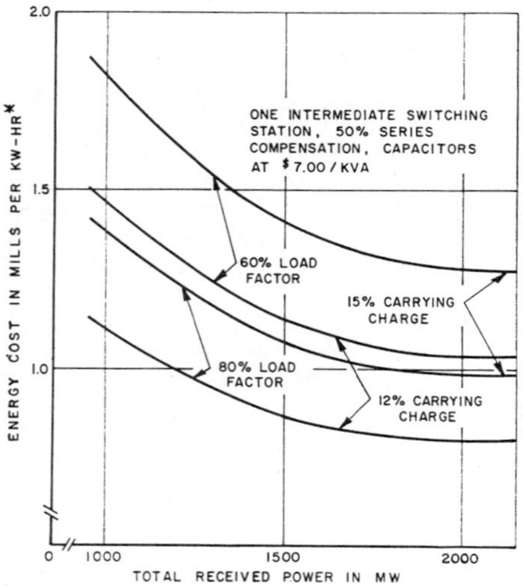

Fig. 14. Effect of changing load factor and carrying charge for 500-kv 2-circuit transmission for 200 miles

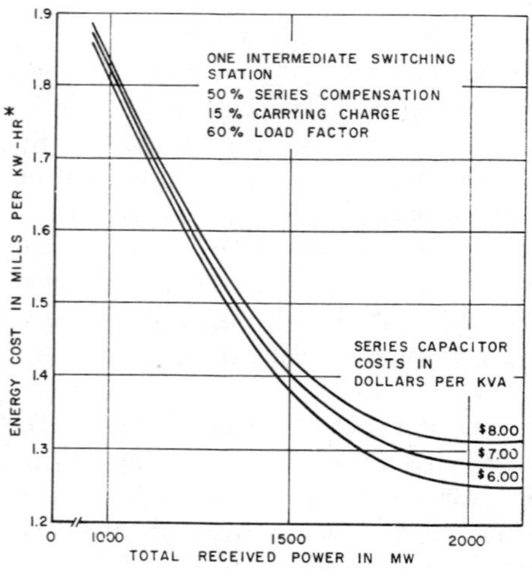

Fig. 15. Effect of series capacitor costs for 500-kv 2-circuit transmission for 200 miles

Figs. 14 and 15 illustrate the effect of varying carrying charge, load factor, and series capacitor costs for a fixed transmission distance and configuration of 200 miles, one intermediate switching station, and 50% compensation. Extrapolation or interpolation can be made from the curves of these figures to obtain a quick estimate of energy costs for other load factors and/or carrying charges. Fig. 15 shows the effect of different series capacitor costs on the cost of energy as a function of load. These curves tend to separate at higher loads since the required series capacitor rating tends to increase as the square of the transmitted power.

Conclusions

1. Substantial increases in transmission capability can be obtained by using series capacitors. This is usually the most economical approach, and permits a substantial increase in transmission capability.
2. Series capacitors can be economically employed at transmission distances of 200 miles or more. Under certain conditions, these capacitors are economically attractive for distances as low as 100 miles.
3. In series-compensated circuits, modern 2-cycle circuit-breaker interrupting times contribute significantly to higher transmission capability.
4. Application of series capacitors within a breaker-and-a-half switching scheme at a station results in the highest transmission capability and greatest reliability.
5. The reinsertion time of a series capacitor has a marked influence on the stability limit of a transmission system.
6. High-speed reclosing is unnecessary when using the breaker-and-a-half arrangement for connecting series capacitors into a transmission system, unless the degree of compensation is small.
7. Slow-speed reclosing can be advantageous in maintaining high reliability and for utilizing safely and economically the short-time overload capability of the series capacitors. For other series capacitor arrangements, where the transfer impedance increases with a line section switched out of service, high-speed reclosing can be effective in increasing the stability limit.
8. The transmission capability of a series-compensated circuit can be limited by excessive fundamental frequency voltages at the terminals of a series capacitor bank. Distributing the series capacitors in small units throughout the line will reduce the overvoltage. Or it can be controlled by switched shunt reactors.

Appendix. Maximum Steady-State Power Characteristics

This approximate upper bound of transmission capability is definitely not the point of maximum efficiency or even acceptable efficiency for power transmission. The approximation—that the impedance rating of the series capacitor can neutralize the same number of series inductive ohms—is used, and line charging is neglected. Curves displaying the results are included.

The nomenclature used in this appendix is:

$E_S = e_S e^{j\phi S}$, sending-end voltage

$E_R = e_R e^{j\phi R}$, receiving end voltage

$\phi = \phi_S - \phi_R$, angle by which E_S leads E_R

$Z = ze^{j\lambda}$, equivalent impedance between E_S and E_R

$= r + jx$

When $\phi = \lambda$, the condition for maximum power as a function of voltage angle is satisfied. The compensation for maximum power transmission is found in the relationship

$$x = r\sqrt{4\frac{e_R^2}{e_S^2} - 1} \tag{1}$$

The positive square root is used since overcompensation is not desired. If the terminal voltages are equal, the net reactance for maximum delivered power is

$$x = r\sqrt{3} \tag{2}$$

The maximum possible power can be determined as

$$P_{R\max} = \frac{e_S^2}{4r} \tag{3}$$

Equation 3 indicates that, with the compensation such that the net reactance x is correct for maximum power transfer according to equation 1, the maximum power is then only a function of sending-end voltage and the circuit resistance.

The required receiving-end reactive for conditions of maximum received power is given by equation 4.

$$Q_R(\text{for } P_{R\max}) = -\frac{x}{r} P_{R\max} \tag{4}$$

Equation 4 indicates that the reactive required at the receiver for maximum power will be prohibitive.

For moderate compensation, the maximum power will be realized with highest permissible voltages. As the compensation is carried to excessive values of near 100%, the receiving-end voltage should be depressed. At 100% compensation, $x=0$, and the receiver voltage for maximum power is

$$e_R = e_{S/2} \tag{5}$$

This is identical with the well-known relationship for d-c power systems where, for maximum power, the load resistance is equal to the source resistance; that is, load voltage is one-half the source voltage.

The foregoing indicates that, when a circuit is completely compensated between fixed voltages, only resistance remains, and the transmission of power is accomplished by voltage magnitude relationships rather than phase angles. A discussion of stability for this case shows that the normal phase angle change, caused by application of load, will result in a reactive rather than a power flow, so synchronizing torque has been lost. Another observation might be that load division between parallel circuits is affected by series compensation. To minimize the system loss, the resistance to reactance ratios of parallel lines should be equal.

The assumption made for the curves of Fig. 16 are equivalent to a line with 100% shunt compensation and series capacitors distributed throughout the line. Then the charging current is cancelled and the line reactance is effectively reduced in each small unit of length. The results shown in these curves would be modified by the charging current of long lines. However, the charging current is small compared with Q_R.

Fig. 16 shows several curves for a single-conductor transmission line that have been constructed from the preceding equations. The conductor size was not changed at heavy loading in order to obtain maximum power characteristics with less effort. These characteristics are only for a qualitative understanding of the steady-state boundary characteristics.

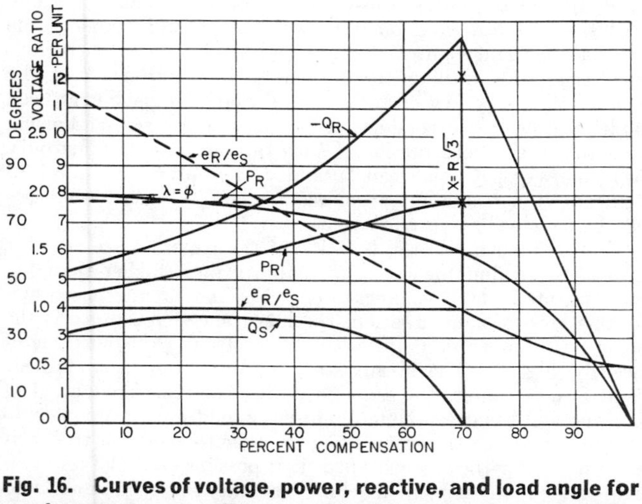

Fig. 16. Curves of voltage, power, reactive, and load angle for maximum power conditions as a function of series compensation

In constructing the curves of Fig. 16, the sender voltage was held at its maximum operating voltage. The receiver voltage was determined by the voltage ratio for maximum power transfer as a function of voltage angle as given by equation 6.

$$\frac{e_R}{e_S} = \frac{x^2 - r^2}{2rx} \sin\phi + \cos\phi \tag{6}$$

However, the receiver voltage was not permitted to exceed its operating voltage limit, which was assumed to be the same as for the sending end. The maximum power for fixed and equal terminal voltage occurs at the point given by equation 2, $x = r\sqrt{3}$.

However, if unlimited adjustment of the receiver voltage magnitude were permitted at a fixed sender voltage, e_R would follow the dashed curve; then the value of maximum power would be independent of compensation and also would be shown as a dashed line.

The curve for reactive at the receiver terminal shows a very large reactive required. This abnormal ratio of Q_R/P_R will be greatly reduced at lower values of power. Information can be obtained about the heavily compensated line characteristics if the receiver voltage is permitted to drop and the power and reactive are recalculated. The points plotted on Fig. 16 with x marks are for $e_R = 0.95e_S$. Note that the reactive is reduced much more rapidly than the power, which is nearly unaffected.

A summary of the conclusions, indicated by this steady-state derivation and resulting curves, includes:

1. There would be a maximum power that could be transmitted, independent of compensation, if the sending-end voltage were fixed and the receiving-end terminal voltage varied to satisfy equation 6.
2. Restricting voltages to their practical operating limits, but satisfying equation 6, when applicable, gives a maximum possible power transfer at a given degree of compensation.
3. With terminal voltage magnitudes fixed, there is a maximum power capability at a definite value of series compensation.
4. The compensation indicated by equation 1 should not be exceeded. In applying this equation, note that reactance x is the total between the internal voltages, and not just the line reactance.
5. The compensation of Fig. 16 indicates that the maximum power is attained at nearly 70% series compensation. This value will be a function of the over-all x/r ratio for the transmission line and terminal equipment between the fixed voltages.
6. The power curve of Fig. 16 is an upper bound on power transmission.
7. The reactive required by the line is excessive at the very high power flows and high compensation. Lowering the voltage ratio e_R/e_S reduced the receiving reactive required without noticeably reducing the power delivered. Thus, at high compensation, reactive problems can be expected, and modification of the terminal voltage ratio may yield desirable results.
8. With a series-compensated line paralleling a low-voltage network, the desired control of power and reactive flow in the parallel systems may require phase shifters and voltage control if the degree of compensation is great. No difficulties would be expected with moderate compensation. The x/r ratio for the parallel circuits should be equal for minimum loss. Thus, it is possible that the low-voltage circuits should be series-compensated.

References

1. EQUIVALENT REACTANCE OF SYNCHRONOUS MACHINES, **S. B. Crary, L. A. March, L. P. Shildneck.** *AIEE Transactions*, vol. 53, Jan. 1934, pp. 124–32; *discussion*, Apr. 1934, pp. 603–07.

2. STEADY-STATE STABILITY OF SYNCHRONOUS MACHINES AS AFFECTED BY VOLTAGE REGULATOR CHARACTERISTICS, **C. Concordia.** *Ibid.*, vol. 63, May 1944, pp. 215–20; *discussion*, pp. 489–90.

3. LOCATION OF SERIES CAPACITORS IN HIGH-VOLTAGE TRANSMISSION SYSTEMS, **S. B. Crary, L. E. Saline.** *Ibid.*, vol. 72, pt. III (*Power Apparatus and Systems*), Dec. 1953, pp. 1140–51.

4. THE BREED PLANT TRANSMISSION SYSTEM: CONCEPTUAL,

FEASIBILITY, AND DESIGN STUDIES, **C. F. DeSieno, A. Hauspurg.** *Ibid.*, pt. III, vol. 80 (Feb. 1962 section), 1961, pp. 910–19.

5. SERIES SYNCHRONOUS CONDENSERS FOR GENERATION OF VOLTAGE CONSUMED BY LINE INDUCTANCE, **T. H. Morgan.** *Ibid.*, vol. 49, Jan. 1930, pp. 162–65.

6. ANALYSIS OF SERIES CAPACITOR APPLICATION PROBLEMS, **J. W. Butler, C. Concordia.** *Ibid.*, vol. 56, Aug. 1937, pp. 975–88; *discussion*, vol. 57, Feb. 1938, pp. 110–11.

7. SERIES CAPACITORS FOR TRANSMISSION CIRCUITS, **E. C. Starr, R. D. Evans.** *Ibid.*, vol. 61, 1942, pp. 963–73; *discussion*, p. 1045.

8. STEADY-STATE AND TRANSIENT STABILITY ANALYSIS OF SERIES CAPACITOR IN LONG TRANSMISSION LINES, **J. W. Butler, J. E. Paul, T. W. Schroeder.** *Ibid.*, vol. 62, 1943, pp. 58–65; *discussion*, p. 379.

9. ECONOMICS OF LONG DISTANCE A-C POWER TRANSMISSION, **S. B. Crary, I. B. Johnson.** *Ibid.*, vol. 66, 1947, pp. 1092–1102.

10. ECONOMIC ASPECTS OF SERIES CAPACITORS IN HIGH-VOLTAGE TRANSMISSION, **T. S. Seymour, E. C. Starr.** *Ibid.*, vol. 70, pt. II, 1951, pp. 1663–70.

11. AN ECONOMIC STUDY OF HIGH-VOLTAGE TRANSMISSION, **J. M. Henderson, A. J. Wood.** *Ibid.*, vol. 75, pt. III, 1956, pp. 695–706.

12. SERIES CAPACITORS FOR HIGH TENSION LINES, **A. A. Johnson, J. E. Barkle, D. J. Povejsil.** CIGRE, Paris, France, vol. III, art. 311, 1952.

13. SELF-EXTINGUISHING GAPS IN LARGE SERIES CAPACITOR STATIONS, **L. Ahlgren, B. Grundmark.** CIGRE, vol. III, art. 317, 1956.

14. THE USE OF SERIES CAPACITORS ON HIGH-VOLTAGE TRANSMISSION SYSTEMS, **S. Lalander, L. Norlin.** CIGRE, vol. III, art. 330, 1958.

15. REIHENKONDENSATOREN FUR HOCHSTSPANNUNGSNETZE (book), **Hans Zimmermann.** *AEG-Mitteilungen*, AEG, Berlin-Grunewald, Germany, vol. 48, Aug./Sept. 1958.

16. PROTECTION OF CIRCUITS WITH SERIES CAPACITORS, **John Berdy.** *IEEE Transactions*, pt. III, vol. 81 (Feb. 1963 section), pp. 929–35.

Discussion

Robert H. Park (Consulting Engineer, Brewster, Mass.): As made clear in the paper by Breuer, Rustebakke, Gibley, and Simmons, some of the curves of power can limit application in the steady state. For example, Figs. 6, 7, and 11 are based on a limit computed with neglect of regulators, whereas others, particularly Figs. 8 and 13, take the stabilizing ability of regulators into account. Fig. 13 is based on the assumption that voltage regulators can be relied on in the steady state to reduce synchronous reactance to a value 0.6 $(x_{deq}+x_d')$, while also, relative to this assumption, there is the statement that a well-designed and properly adjusted voltage regulator is capable of this performance.

Two comments are offered in this connection:

1. If the statement is true, why is it in order to bother with calculating a so-called "limit," based on neglect of regulators, the point being that a limit so based can be viewed as overconservative. This would seem inappropriate in an economic study in which all other factors are realistically evaluated.

2. In 1928, C. A. Nickle developed a vibrating contact-type regulator, which acted to reduce x_d to $x_d'+0.05\ (x_d-x_d')$. See R. E. Doherty's closing discussion of Nickle and Carothers' paper on "Automatic Voltage Regulators" in *AIEE Transactions*, July 1928, p. 974. Is it not true that the value 0.6 $(x_{eq}+x_d')$ is also overconservative? Or, if this is not so, isn't it true that today's regulators, which generally utilize or are able to utilize solid-state circuitry, are unable to accomplish as much as C. A. Nickle's regulator with its vibrating contacts? Admittedly, a regulator using vibrating contact could hardly be expected to have stabilizing ability. However, with modern regulators, this limitation no longer applies.

Then, with the benefit of solid-state technology, the power industry should have at its disposal—and doubtless does—a choice of regulators that will, at least, do what was found feasible with vibrating contacts 36 years ago. Moreover, this feasibility was accompanied by only a relatively slow rate of exciter response.

It appears that, up to now, the power industry has not taken much interest in how well voltage regulators can improve steady-state stability and that, consequently, voltage regulating systems have been designed, produced, sold, and used without a naturally warranted concern as to their steady-state stabilizing capabilities.

This paper serves to point up the economic importance of this feature. It implies that purchasers of generating equipment may advisedly take the initiative and press for complete information on stabilizing qualities.

Since apparently voltage regulators can provide stabilizing action—not only equaling but exceeding the criterion of converting saturated synchronous reactance to transient reactance—it is in order that computational studies include some consideration of how these capabilities vary with variation in degree of effective reactance compensation. The range of values $1/2(x_d\text{sat}+x_d')$, x_d' and $x_d'/2$ could be selected, for example, in order to allow evaluation of the worth of favorable stabilizing performance.

With determination of economic worth and initiation of commercial demand, manufacturers would be stimulated to do everything possible to develop and offer for sale regulators and regulating systems incorporating optimum properties.

E. C. Starr (Bonneville Power Administration, Portland, Ore.): This paper is a timely, interesting, and valuable aid to understanding and appreciating the important role the series capacitor plays in achieving good economy in long-distance power transmission. Large blocks of power, long distances, and EHV are always assumed to be closely related.

Recognizing the potential gains and problems involved in going to ever higher voltages for transmission of greater blocks of power over long distances, attention should be called to the latent capability existing at our present HV and EHV levels. HV is employed basically to obtain low equivalent transfer impedance in the network: low reactance to give high power capability and low resistance to minimize losses.

Normal steps in EHV—that is, from 345 to 500, 500 to 700, or 700 to 1,000 kv—will increase basic line capability by a factor of approximately two. The same increase, or more, can be effected by series compensation and with excellent economy if the line is constructed with conductors, preferably bundled, with adequate cross section for the higher currents involved.

Success at EHV of 500 to 1,000 kv or more will obviously depend upon insulation reliability and definite control of switching-type voltage surges. Such control, to a limited degree, can be effected by opening and closing resistors in circuit breakers and, to a better degree ultimately, by opening resistors and synchronized closing. The latter, though difficult of accomplishment at EHV, can and will be achieved through the application of techniques now available to us.

In the meantime, and at all foreseeable stages, the series capacitor, in combination with shunt reactors where necessary, should be studied and considered for application as a very economical and useful building block in increasing transmission capability. Certainly for all very long distance a-c transmission, where half-wave tuning is impractical, the series capacitor must be employed to achieve most economical transmission. Along with this, it is almost axiomatic that bundled conductors of large cross section and high conductivity go hand in hand with compensation. Concerning the former, I have often heard the remark that very few, if any, transmission lines have ever been built with adequate conductivity to serve the decade following that of their construction. With appropriate conductivity provided, compensation can be added at a later date to afford substantial increases in transmission capability. Such increases can be obtained at very low unit costs.

With reference to Figs. 6 and 7 of the paper, it seems a bit unrealistic to find the transient stability limit higher than the steady state at 50% compensation. For the circuit chosen and at this level of compensation, the line transfer reactance is the same postfault as prefault, with the postfault resistance significantly higher. The steady-state loading prefault and postfault are, of course, the same as during the transient period of faulting and isolating the faulted section and must exist prior to and following the transient swing in order to establish transient stability. For the cases studied, it is possible that the relationship assumed between machine saturated synchronous and transient reactances was responsible for the apparent inconsistency. Also, it is well known that high-speed regulation and excitation can contribute substantially to increasing steady-state limits.

The higher degrees of compensation require special relay application techniques to identify the faulted line section if the protective gaps do not fire. These techniques are available and must be considered together with possible reclosing overvoltages and resonance phenomena.

G. D. Breuer, H. M. Rustebakke, R. A. Gibley, and H. O. Simmons, Jr.: The discussions by Mr. Park and Mr. Starr contribute much to the understanding of this subject and are appreciated.

Mr. Starr refers to Fig. 6 and the curve of transient stability exceeding steady-state stability. This curve was intended to demonstrate characteristics but was not clearly presented. We hope to clarify the significance of these curves. First, the term "transient" stability should have been "first swing transient" stability. The system must be stable, not only after the first swing but also on the subsequent swings and in the steady state, to meet the usual definition of transient stability which will certainly preclude the transient ever exceeding the steady-state stability. Perhaps the first-swing transient curve should not have been extended beyond the crossover point, but it seemed informative to show this. The lower curve will always be the capability limitation.

Both discussers cite the excitation system as a possible source of improved steady-state performance. Mr. Park discusses the effect of voltage regulators in considerable detail.

The voltage regulator has three effects pertinent to this discussion. It will control the steady-state voltage, affect transient stability, and affect steady state.[1,2] Good performance in each area can be designed into a voltage regulator (excitation system), but adjustments cannot always be made on an existing regulator that will improve performance in the three categories simultaneously.

The operation of a power system at a power level exceeding the constant field current limit makes the stable operation of this system dependent on the proper operation of the control system. Users usually have not chosen to do this in the past. This philosophy can be changed if desirable. If this is done, it will result in a change in emphasis in future designs and in readjustments of existing equipment. The curves for this paper were prepared with past practice in mind where machine saturation was recognized in determining the steady-state stability curves, but the potential benefits of the voltage regulator were not used. However, when the curves for Fig. 13 (400 miles) were prepared, the steady-state stability limit proved to be quite low relative to the first-swing transient limit. The series compensation could have been increased, but the decision was made to utilize, in what the authors thought was a reasonably conservative way, the potential beneficial effects of a voltage regulator. The user would need to decide which alternative to choose.

The characteristics of increasing transmission capability by using a particular type of voltage regulator can be learned from reference 2. This type of regulator has the characteristic of very close voltage control in the steady state, and also can be adjusted to improve steady-state stability by properly choosing the "separately excited amplification factor" or "transient gain." Fig. 2 of this reference shows a typical effect of this amplification factor or gain on the steady-state power limit. At low gain, the maximum stable operating angle is 90 degrees, as expected. The angle can be increased with higher amplification factors, say 1 to 20. But, for amplification factors of more than about 35 to 40, there is a reduction in steady-state power limit. Note that the stronger stabilizer can give higher steady-state stability limits. However, the response of the system would be reduced and it would tend to lower the transient limit.

Mr. Park mentions that the voltage regulating system should be capable of operation such that the equivalent machine reactance is less than the machine transient reactance X_d'. This is true; the equivalent machine reactance could theoretically be made negative if the voltage on the transformer high-voltage terminals is held constant. Refer to Fig. 3 of reference 3. The intersections of curves E and A show operation at 116 degrees; this calculation showed to be with an equivalent machine direct axis reactance of X_d'. Fig. 5 of this same paper[3] shows the capability decreasing as the voltage regulator amplification factor is increased. If damping torque is introduced, this curve reverses direction and increases with greater amplification factors. It is in this latter region of high gain that the equivalent reactance can be less than X_d' if damping is introduced, either directly or by proper stabilizing means.[4]

References

1. Steady-State Stability of Synchronous Machines as Affected by Voltage Regulator Characteristics, **C. Concordia.** *AIEE Transactions,* vol. 63, May 1944, pp. 215–20, 480.

2. Effect of Buck-Boost Voltage Regulator on Steady-State Power Limit, **C. Concordia.** *Ibid.*, pt. II, vol. 69, 1950, pp. 380–84.

3. Steady-State Stability of Synchronous Machines as Affected by Angle Regulator Characteristics, **C. Concordia.** *Ibid.*, vol. 67, 1948, pp. 687–90.

4. Regulating System for Dynamoelectric Machines, **G. Kron.** *U. S. Patent No. 2,692,967,* Oct. 26, 1954.

Improvement of System Stability by Switched Series Capacitors

E. W. KIMBARK, FELLOW, IEEE

Abstract—**Improvement of transient stability of a 2-circuit ac transmission link by use of switched series capacitors is analyzed by equal-area diagrams for two kinds of disturbances: 1) a permanent fault on one ac circuit, cleared by opening a section of that circuit, without reclosure and 2) a permanent fault on a dc line operating in parallel with the ac line. In both cases, the aggregate megavar rating of series capacitors required for a given transient stability limit is shown to be smaller if some capacitors are switched than if all are unswitched. Moreover, the scheme with switched capacitors reduces the angular swing of the machines and the fluctuation of load voltages during the swing.**

Introduction

THE POWER CAPABILITY of long, interregional, ac tie lines is usually limited by transient stability. Such economic factors as the high cost of long lines and the additional revenue obtainable from the delivery of additional power give strong incentive to explore all economically and technically feasible means of raising the stability limit. At least two ac circuits in parallel are required for firm transmission, but additional circuits are not likely to prove economical until after all feasible improvements have been made to the first two circuits. Rapid clearing of short circuits and sectionalization of the lines by means of intermediate switching stations have been and will be used. Bundle conductors will be used on most 500-kV interties now being planned. Series compensation is one of the most effective and economical means of raising stability limits, although high degrees of compensation accentuate problems in protective relaying, normal voltage profiles, and overvoltages during line-to-ground faults.

Series compensation is more effective and economical if part of it is switched so as to increase the degree of compensation upon occurrence of a disturbance which threatens stability. A second advantage of switched series compensation over fixed series compensation is that by decreasing the angular swing caused by a disturbance it also decreases the fluctuation of load voltages. These two advantages will be illustrated for two types of disturbance; 1) a short circuit on one of two parallel ac circuits and 2) a fault on a dc line operating in parallel with ac lines. The application of switched series capacitors to improvement of stability against each of these kinds of disturbance will be illustrated by the use of equal-area diagrams. The details of analysis differ somewhat between the two types.

Paper 31 TP 65-705, recommended and approved by the Transmission and Distribution Committee of the IEEE Power Group for presentation at the IEEE Summer Power Meeting, Detroit, Mich., June 27–July 2, 1965. Manuscript submitted March 29, 1965; made available for printing May 7, 1965.

The author is with the United States Department of the Interior, Bonneville Power Administration, Portland, Oreg.

Fault on a 2-Circuit AC Tie Line

Components of Shock

The worst threat to stability is usually a 3-phase fault at the sending end of the tie line cleared by switching out the faulted circuit. The shock to the system caused by such a disturbance consists of two parts: 1) that resulting from the brief presence of the short circuit and 2) that resulting from opening the faulted circuit and leaving it open. The separate effect of the first part is equal to that of a fault on an unloaded radial line adjacent to the sending-end bus, cleared without opening the tie line. The separate effect of the second part is equal to that of opening one circuit of an unfaulted section of the tie line.

The shock caused by the fault itself decreases with faster clearing time and with increased period of swing. Modern relays and circuit breakers permit 3-cycle (0.05 second) clearing time, which is so short that this part of the shock is usually the minor one. On interregional tie lines having transmission capability of approximately only 10 percent of the generating capability of the smaller of the two interconnected systems, the period of swing (2 to 4 seconds) is so long compared to the period of intrasystem swings (approximately 1 second) that the relative effect of the fault itself is still smaller. In such tie lines with fixed series compensation, the shock caused by opening the faulted section predominates. Indeed, the mere opening of one unfaulted circuit of a long, heavily loaded section of line could cause instability.

If the shock of opening the faulted section were eliminated or considerably decreased, a significant increase of stability limit could be achieved. This is true regardless of whether the criterion for stability limit be based on successful reclosure, no reclosure, or unsuccessful reclosure, but it is especially true for no reclosure or unsuccessful reclosure. Normally, this shock increases with the length of section switched out and could, therefore, be decreased by dividing the line into more sections, separated by more switching stations. This is a costly method. Increased degree of fixed series compensation in the individual circuits decreases the effective length of section and consequently decreases the shock but does not eliminate it.

Reprinted from *IEEE Trans. Power App. Syst.*, Feb. 1966.

Switched Series Compensation

The shock caused by disconnection of a faulted circuit can be eliminated by increasing the series compensation of the remaining circuit or circuits so that the overall transfer reactance of the tie lines remains unchanged. This means essentially that the overall series capacitive reactance must be increased by the same amount as that by which the overall inductive reactance is increased by the switching out of the faulted circuit section. One switching scheme by which this can be accomplished is shown in the following example.

Example 1

Figure 1 is a simplified diagram of a 360-mile, 2-circuit, 2-section, 525-kV ac transmission link interconnecting two regional power systems, each of which is represented by its Thevenin's equivalent circuit (EMF and impedance in series). The EMFs are assumed to remain constant in absolute value, although their phase difference δ varies during a swing. For simplicity, resistance and shunt capacitance are neglected. The effect on the stability limit of switching series capacitors upon occurrence of a disturbance will be shown. The disturbance now considered is a 3-phase short circuit at the sending end of one line circuit, cleared by disconnection of the faulted line in 0.05 second without reclosure. The initial series compensation of 50 percent is provided by capacitors in the individual line circuits. (The degrees of series compensation are stated as the ratio of series capacitive reactance to the series inductive reactance of the transmission tie lines only, excluding the reactances of the terminal systems.)

Three cases will be considered. The transient stability limit for each case will be found by the equal-area method [1] (Fig. 2).

Case 1: Fixed compensation (no switched capacitor). Reduction of the circuit of Fig. 1(a) yields

inductive reactance:	$12.5 + 2 \times 106/2 + 23.5 = 142$ ohms
capacitive reactance:	$2 \times 53/2 = 53$ ohms
net reactance:	$142 - 53 = 89$ ohms.

Amplitude of power-angle curve $(525)^2/89 = 3090$ MW = 3.09 GW. This is the upper sinusoidal curve in Fig. 2(a).

While the 3-phase short circuit is on the line no power can be transmitted. The power-angle curve for this condition is the horizontal axis in Fig. 2(a).

With the faulted circuit switched out [Fig. 1(b)], the reactances become

inductive:	$36 + 1.5 \times 106 = 195$ ohms
capacitive:	$1.5 \times 53 = 79.5$ ohms
net:	$= 115.5$ ohms.

Amplitude of power-angle curve is $(525)^2/115.5 = 2390$ MW = 2.39 GW. This is the lower sinusoidal curve in Fig. 2(a).

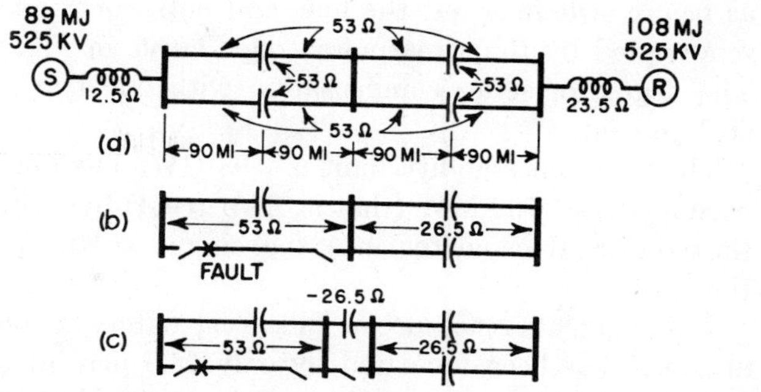

Fig. 1. Two-circuit two-section 360-mile transmission line with 50-percent series compensation of line reactance. (a) Intact: net reactance between busses is 53 ohms; between internal voltages, 89 ohms. (b) Faulted circuit section switched out: reactance increased by 26.5 ohms. (c) Like (b) but with capacitor inserted at central switching station, restoring reactance to initial value.

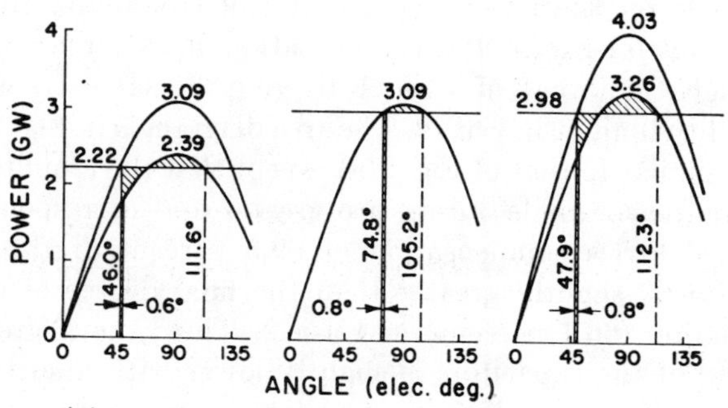

(a) (b) (c)

Fig. 2. Transient stability of 2-circuit 360-mile ac line for 3-phase fault at sending end. (a) Fixed series compensation of 50 percent. (b) Initial compensation 50 percent, switched to 67 percent. (c) Fixed compensation of 70 percent.

The transient stability limit is 2.22 GW. This is the height of the horizontal line such that the shaded areas above it and below it are equal. With the kinetic energies of the two interconnected systems as indicated in Fig. 1(a), the relative angular motion of the two equivalent machines during the 0.05 second of the fault is only 0.6°, and the area of the tall, narrow rectangle of that width is approximately only 13 percent of the total accelerating area. This confirms the statement made earlier that the opening of the faulted line constitutes a much greater part of the disturbance than does the brief presence of the fault. The reinsertion time of the capacitors, being not more than one half-cycle, is neglected.

Case 2: Switched capacitor. In Case 1, the opening of the faulted circuit increases the net reactance by $115.5 - 89.0 = 26.5$ ohms. That increase is compensated in Case 2 by switching into the circuit an additional series capacitor having reactance of -26.5 ohms and located at the bus of the intermediate switching station, as shown in Fig. 1(c). It is reasonable to suppose that the capacitor is inserted at the same time that the faulted line is switched out, because both operations are performed by opening circuit breakers of similar speed, actuated by the same protective relay. The net reactance is then the same after as before switching out the line, and both conditions are represented by the same power-angle curve in Fig. 2(b). The degree of series compensation with one line out is 66.7 percent.

The transient stability limit is 2.98 GW. The limit has been increased 760 MW (that is, 34 percent) by switching the additional capacitor into the circuit at the proper time.

The aggregate reactance of all the capacitors (as though in series) has been increased by only 12.5 percent (from $4 \times 53 = 212$ ohms to $212 + 26.5$ ohms). However, if it is assumed that the rated current per circuit is increased in proportion to the stability limit, the aggregate megavar rating of the capacitors in Case 2 is $(2.98/2.22)^2 \times 1.125 = 2.04$ times that of Case 1.

Case 3: A better comparison of the aggregate rating of the capacitors with and without a switched capacitor is made on the basis of equal transient stability limits. Case 3 is for fixed series compensation of a degree greater enough than that of Case 1 to give the same transient stability limit as in Case 2. The area diagram is in Fig. 2(c). It is similar to that of Fig. 2(a) except that the amplitudes of the power-angle curves are greater and correspond to a fixed series compensation of 69.5 percent of the line reactance, slightly greater than the final degree of compensation (66.7 percent) in Case 2. Thus, the aggregate rating of the capacitors is slightly lower with a switched capacitor than with only unswitched capacitors.

Steady-State Limits

It is apparent that the transient stability limits found in Fig. 2 are close to and perhaps even higher than the steady-state stability limits after switching. Consequently, the latter might be the real limiting factor. The approximate steady-state limits could be found by replacing the transient reactances of all machines by their equivalent synchronous reactances and then reducing the sending- and receiving-end systems to single reactances which would be greater than those shown in Fig. 2. They would be only slightly greater, however, because both terminals of the line connect to large power systems in which the aggregate generating capability is approximately ten times the capability of the tie lines. The reactances of transmission lines and transformers within such systems predominate and tend to mask the difference in machine reactances between steady-state and transient conditions. For present purposes, the steady-state stability limit is assumed to be the amplitude of the transient power-angle curve for the final condition of the circuit.

In comparing Fig. 2(a) and (b), it is clear that the insertion of a switched capacitor increases the steady-state stability limit as well as the transient limit. In comparing (b) and (c) of the same figure, we see that for equal transient limits, the switched-capacitor scheme (b) works closer to the steady-state limit than does the scheme with increased fixed compensation. Consequently, the possible saving by use of switched capacitors would be somewhat less than that calculated on the assumption that the steady-state limit was the amplitude of the final transient power-angle curve. The discrepancy would be greater for transmission from generation to a load center than for an interregional tie line.

Reclosure

No reclosure was assumed in Example 1 for two reasons: 1) rapid reclosure onto a permanently faulted line gives a lower transient stability limit than does no reclosure and 2) rapid successful reclosure does not raise the stability limit if capacitor switching is employed such that the transfer reactance is not increased by switching out a line. With such capacitor switching, there is no hurry to reclose a line except to decrease the risk of instability caused by a subsequent fault. Ample time can be allowed for arc deionization. If desired, the line can be tested before reclosure. One method of testing it would be to energize it first from one end only, preferably from that end where unsuccessful reclosure would give the least shock, and then to close the other end only if the first end closed did not trip out again.

However, as the system would be designed to be stable for the worst fault with no reclosure, and, as in a stable system, the angular separation reaches a maximum and then decreases, the best time to reclose is while the machines are swinging toward one another. An unsuccessful reclosure at this time would not cause the machines to swing apart faster but would merely make them swing together more slowly, that is, to decrease the amplitude of oscillation. For brevity, reclosure under this condition may be termed "reclosure on the backswing." In the system considered as an example, the period of swing is approximately 2.7 seconds and maximum negative relative angular velocity occurs at approximately 1.7 seconds (5/8 of the period). Simultaneous reclosure at 1.7 seconds is fast enough to give small risk of a second fault during this time, and yet unsuccessful reclosure at this time would not impair stability.

In the event of a permanent fault, generally the tie line load should be decreased as fast as feasible to a value that the receiving system could afford to lose at one time, and normal load would not be resumed until after the faulty line had been repaired.

Overcompensation or Undercompensation of Switching a Circuit Out

In the examples and in the whole paper, the series capacitors switched in when a line circuit is switched out are assumed to have such reactance that the transfer reactance of the whole transmission link is the same after switching as before switching. It is recognized, of course, that switched capacitors can improve stability even though they may make the transfer reactance after switching somewhat more or less than that before switching.

If the reactance is decreased by switching, the shock attributable to the occurrence of a short circuit can be partly or wholly cancelled in addition to the shock attributable to switching out the faulted circuit. However, unless a series capacitor is switched for faults on neighboring circuits as well as for those on the tie line itself, faults off the tie line may become the most serious disturbances.

If the reactance is increased less by switching out a line and switching capacitors in than it is by merely switching out a line, then the transient stability limit is raised less than it would be by "equal transfer-reactance switching." This procedure is recommended if the transient stability limit would otherwise exceed the steady-state limit.

When a circuit is switched out in order to clear a fault, the overall resistance and losses of the transmission link are increased, while the shunt capacitance is decreased. These effects, though neglected in this paper for the sake of simplicity, do have some effect on the power-angle curves and should be taken into account when accurate calculations are to be made. The effect of resistance on limiting the maximum feasible series compensation was analyzed in the Appendix of [2], where it was shown that the impedance angle of the transmission system should not be less than 60°. This criterion should be applied with a line section switched out. The effect of shunt capacitance is discussed in a later section of the present paper.

Fault on a DC Line in Parallel with an AC Line

Loss of a large block of load in the sending-end system accelerates the generators of that system. Loss of a large generator in the receiving-end system retards the remaining generators there. Complete and permanent loss of a dc tie line connecting the same receiving and sending systems as an ac tie line is equivalent to both loss of load in the sending system and simultaneous loss of generation in the receiving system. Any of these disturbances causes relative acceleration of the two interconnected systems and, if severe, may threaten the stability of the ac tie. In the graphical representation by an equal-area diagram, any of these disturbances shifts the horizontal "load line" upwards by an amount equal to the additional load imposed on the ac tie. The inertia constant has no effect on the stability limit if no switching is done in the ac system and if excitation control is neglected. If, however, capacitors are switched with a definite delay, increased inertia constant has a small beneficial effect. Increased inertia helps stability even more by affording the excitation system more time for building up the internal voltages.

The gain in stability limit attributable to switching series capacitors will be shown in the following example.

Example 2

Operating in parallel with the 2-circuit 2-section ac intertie already considered in Example 1 is a dc intertie transmitting 1.4 GW (see Fig. 3). What is the transient stability limit of the ac intertie for sudden loss of the dc line without and with the same switched capacitor used in Example 1?

Case 1: Without switched capacitor. The equal-area diagram is shown in Fig. 4(a). The power-angle curve is for 50 percent compensation. It is supposed that the entire

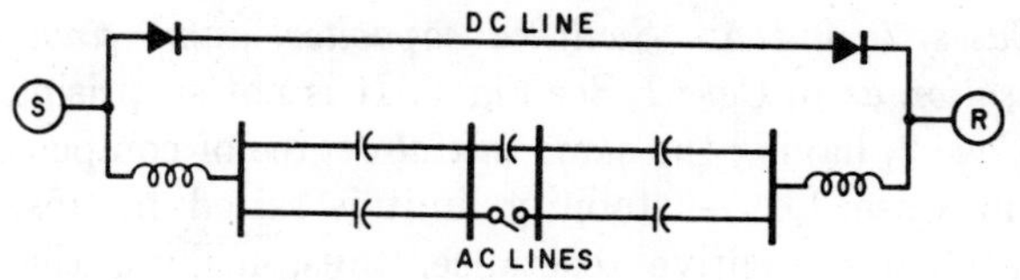

Fig. 3. Alternating-current system of Fig. 1 with addition of parallel dc line.

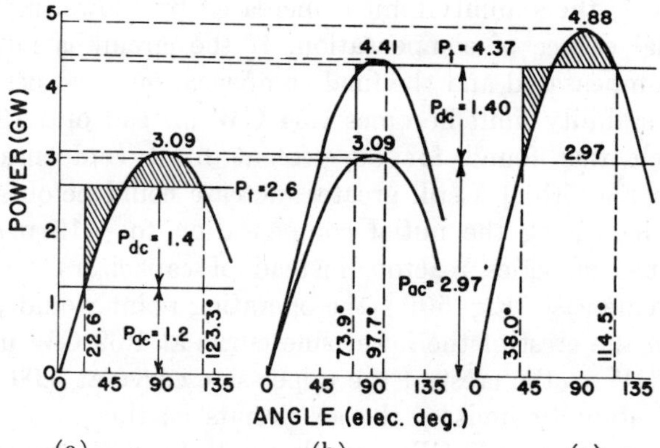

Fig. 4. Transient stability limit of 2-circuit 360-mile ac line when parallel dc line carrying 1.4 GW is lost. (a) Fixed series compensation of 50 percent. (b) Initial compensation 50 percent, switched to 75 percent. (c) Fixed compensation of 81 percent.

load of the dc line is assumed by the ac line. This is not a wholly realistic supposition. Thus, the load line is raised by 1.4 GW. The heights of the load line before and after the disturbance are found so as to make the two shaded areas equal. These heights are 1.2 and 2.6 GW, respectively. The transient stability limit is here defined as the critical *initial* load on the ac intertie. In this case, it is 1.2 GW, which is much less than the limit (2.2 GW) found in Case 1 of Example 1 for a 3-phase fault. However, the critical *final* load (2.6 GW) is greater than that found for a 3-phase fault.

Case 2: With switched capacitor. Two power-angle curves are shown in Fig. 4(b), the lower one being that used in Case 1 and the upper one representing the insertion of the additional series capacitor, which raises the degree of series compensation to 75 percent. The heights of the two load lines, again separated by 1.4 GW, are adjusted so as to make the two very small black areas equal.

For simplicity, suppose that the insertion is instantaneous. There would be a short delay, during which the operating point would move a short distance along the lower sine curve before jumping to the upper one.

Use of the switched capacitor has increased the transient stability limit to 2.97 GW, nearly equal to the limit found in Case 2 Example 1 for a 3-phase fault with switched capacitor. The transient limit is only slightly less than the steady-state limit calculated with transient reactances.

Case 3: No switched capacitor, same stability limit as in Case 2. See Fig. 4(c). The amplitude of the power-angle curve must be increased to 4.88, corresponding to fixed series compensation of 81 percent, higher than the compensation after insertion of the capacitor in Case 2.

Cases 4 and 5: Switched capacitor, same final compensation as in Case 1. See Fig. 5. It is not surprising that in Case 2, having the same *initial* degree of compensation as in Case 1, the stability limit is raised by inserting additional capacitive reactance, thus, making the final degree of compensation higher than the initial degree. It is possibly more surprising that in the present case, having the same *final* degree of compensation as in Case 1, the stability limit is increased by "lowering" the initial degree of compensation. If the circuit is initially uncompensated and the final compensation is 50 percent, the stability limit becomes 1.54 GW instead of 1.2 GW, which was found for 50 percent fixed compensation. [See Fig. 5(b).] A still greater increase could be obtained by decreasing the initial compensation to −19 percent by use of series reactors instead of capacitors. In this extreme case [Fig. 5(c)], the operating point would jump from the crest of the lower sine curve at 1.69 GW up by 1.4 GW to the crest of the upper sine curve at 3.09 GW. The stability limit of the ac circuits is, thus, 1.69 GW, an increase of 0.49 GW over that with fixed compensation [Fig. 5(a)]. This case is remarkable because there is no angular movement of the machines.

Operation in the case of Fig. 5(c) is right at the transient stability limit and would be above the steady-state stability limit if the machines were represented in the steady state by a reactance higher than transient reactance. Figure 6 shows a case which is more realistic in that the system is operating with some margin of stability. For the same power on the dc line as in Fig. 5(b) or (c), the initial compensation must be less than in either of those cases. Again, there is no angular movement.

The correct values of switched capacitive reactance for minimizing the angular swing in the event of a fault on one circuit of the ac line is nearly that value which makes the transfer reactance with a circuit switched out for clearing the fault equal to that of the initial, intact system. Provided that the time of clearing the fault is very short in comparison with the period of swing, the correct value is almost independent of the type of fault or its place of occurrence within the section. Generally, each section would have its own switched compensation of value fitted to the length of the section, although Fig. 1(c) illustrates the possibility of sharing the same switched capacitor between two equal line sections.

The choice of best value of switched series compensation for minimizing the angular swing caused by loss of a dc line in parallel with ac lines is more uncertain because it depends on the power carried by the dc line and on whether one pole or two poles are lost. Conceivably, controls could be arranged to vary the amount of compensation inserted in accordance with the need, but it is more likely that a single, compromise value would be used. In some of the foregoing examples, the compensation switched on loss of the dc line was chosen to be the same as that inserted in the event of a fault on the ac line.

Calculations similar to those represented by Fig. 4(a) and (b) for transient stability limit with and without a

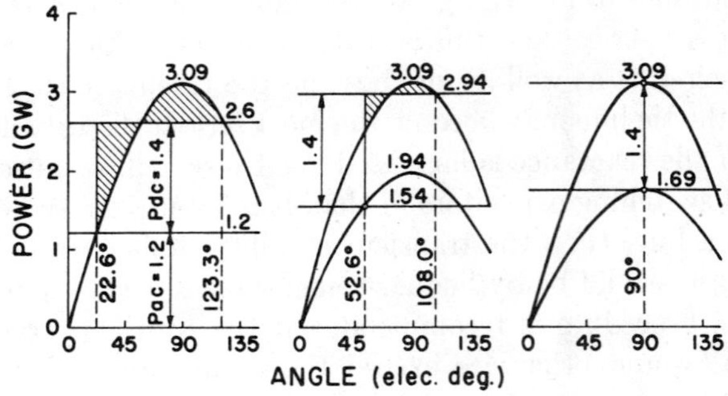

(a) (b) (c)

Fig. 5. Transient stability limit of 2-circuit 360-mile ac line when parallel dc line carrying 1.4 GW is lost. (a) Fixed series compensation of 50 percent. (b) Compensation switched from 0 to 50 percent. (c) Switched from −19 to 50 percent.

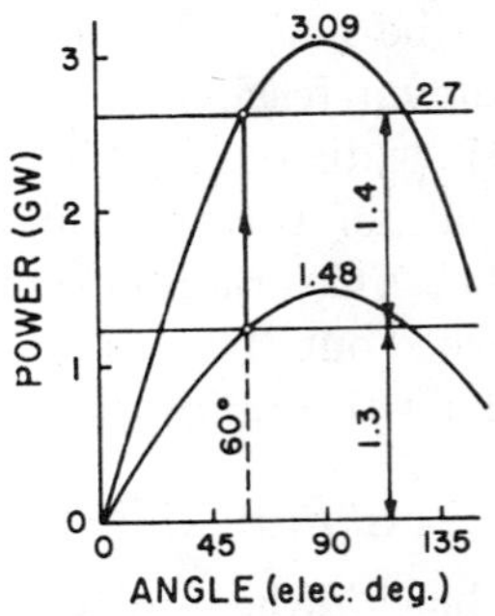

Fig. 6. Loss of parallel dc line carrying 1.4 GW; compensation switched so that there is no oscillation; system operating with margin of stability.

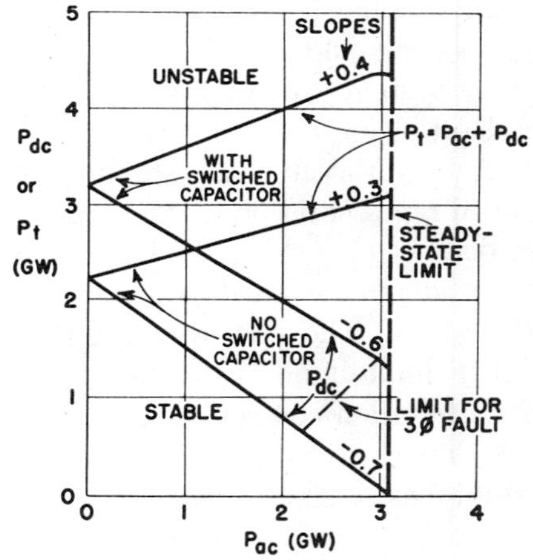

Fig. 7. Direct-current power and total power as functions of ac power at transient stability limit for loss of dc line with and without switched series compensation in ac lines.

switched capacitor of reactance −26.5 ohms were made for loss of a dc line carrying various amounts of power, and the results are shown in Fig. 7. The curves are practically straight lines except at the extreme right-hand ends. If each of the three transmission lines carries 1.4 GW and if series compensation is switched in the event of a fault, the system operates within its steady-state stability limit and within its transient stability limit for a permanent fault either on both poles of the dc line or on all three phases of either ac line.

Fluctuation of Load Voltage

Large fluctuation of voltage across loads is objectionable. During a disturbance which threatens stability, voltages decrease as the machines swing apart, and, if the system is stable, increase as the machines swing back together. If the voltage at any load bus decreases to 75 percent of normal voltage or below, induction motors probably will be disconnected by release of magnetic contactors. The greatest dip in voltage in a 2-machine system occurs at the "electrical center," the point connected by equal reactances to the internal EMFs of the machines.

As stated previously, the use of series capacitors switched into the circuit after occurrence of a disturbance decreases the fluctuation of voltage. Let us assume a load to be connected at the intermediate switching station of the ac system shown in Figs. 1 and 3. More specifically, let it be on the right-hand side of the switchable capacitor. Let the dip in voltage at this load be computed for each of the disturbances already considered in Examples 1 and 2. For simplicity, let the voltage be calculated as though the load were negligible. The calculation may be done graphically by use of a phasor diagram in which phasors representing the two internal voltages are drawn from a common origin, somewhat as in Fig. 8, with the appropriate initial or maximum angle between them. A third phasor, drawn from that origin to a point near the center of the straight line joining the tips of the first two phasors, represents the load voltage.

In Example 1, a 3-phase short circuit was assumed to occur at the point indicated in Fig. 1. In Cases 1 and 3 [Fig. 2(a) and (c)], with fixed series compensation, the load voltage at the maximum angle of the critical swing (at the stability limit) decreases to 62 and 61 percent, respectively, of its initial value. In Case 2 [Fig. 2(b)], with switched series compensation, it decreases to only 77 percent of its initial value.

In Example 2, in which the dc line carrying 1.4 GW is switched out, the effect of switched compensation is even better. In Cases 1 and 3 [Fig. 4(a) and (c)], with fixed compensation, the voltage dips to 50 and 64 percent, respectively, of its initial value; while in Case 2 [Fig. 4(b)], with switched compensation, it dips to only 99 percent, and in Case 5 [Fig. 5(c) or Fig. 6], it does not dip at all. In Case 5, the amount of switched compensation is exactly right for the assumed loadings of the ac and dc circuits.

In accordance with the assumptions made, the load voltage after a swing back to the minimum angle would rise a few percent above its initial value. In practice, however, the generators would be equipped with voltage regulators and excitation systems, which, during a slow swing, would have time to considerably increase the generator voltages, with the result that on the back swing the load voltage might rise to 125 percent or more of its initial value. Such rises of load voltage would be reduced by the use of switched compensation and would disappear in those cases where there is no angular disturbance.

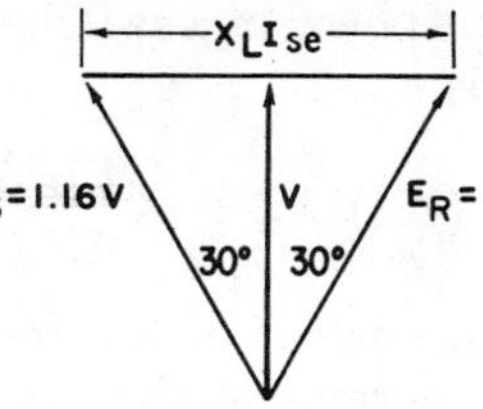

Fig. 8. Voltage phasor diagram assumed at full load.

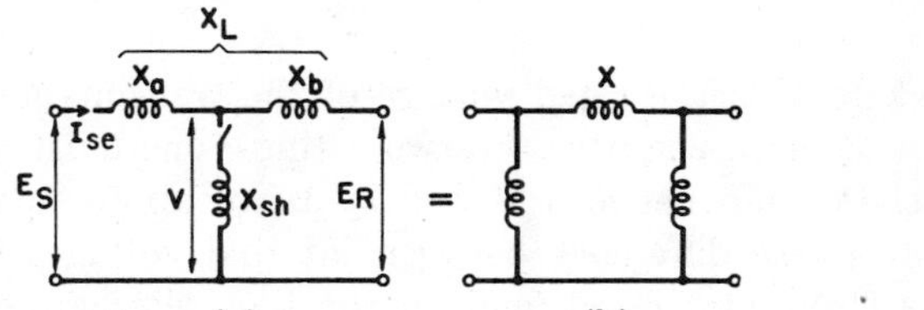

Fig. 9. Equivalent T and π circuits.

Switched Shunt Capacitors or Reactors

Switching a shunt capacitor ON or a shunt reactor OFF raises the voltage and decreases the transfer reactance, as will be shown. Thus, the effect of switching either of these shunt elements is roughly equivalent to switching a series capacitor into the circuit.

For simplicity, though with sacrifice of accuracy, let series resistance and distributed shunt capacitance be neglected, and let the entire network between internal voltages of the sending- and receiving-end machines be represented by series inductive reactance at some intermediate point of which is connected the lumped shunt reactance X_{sh} under discussion. This gives a T circuit with series arms X_a and X_b as shown in Fig. 9(a). When converted to an equivalent π [Fig. 9(b)], the reactance of the architrave, which is also the transfer reactance between internal voltages, is

$$X = X_a + X_b + \frac{X_a X_b}{X_{sh}}. \tag{1}$$

If X_{sh} is positive (inductive) it increases X, and X is, therefore, decreased by switching X_{sh} off. If X_{sh} is negative (capacitive), it decreases X, and X is, therefore, decreased by switching it on. It can be shown that the effect of a given shunt element is greatest when placed at the electrical center, where $X_a = X_b = X_L/2$. There we have

$$X = X_L\left(1 + \frac{X_L}{4X_{sh}}\right). \tag{2}$$

How does the required rating of a shunt capacitor compare with that of a series capacitor producing an equal effect in reduction of reactance? The net reactance with a series capacitor of (negative) reactance X_{se} is

$$X = X_L + X_{se} \tag{3}$$

so that for equal effect we must have

$$X_{se} = \frac{X_L^2}{4X_{sh}}. \tag{4}$$

Now the rated reactive power of a switched series capacitor having equal effect is

$$Q_{se} = (kI_{se})^2 X_{se} = \frac{(kI_{se})^2 X_L^2}{4X_{sh}} \tag{5}$$

where I_{se} is the rated current of the transmission circuit and kI_{se} is the rated current of the capacitor. The rated reactive power of a shunt capacitor is

$$Q_{sh} = \frac{V^2}{X_{sh}} \tag{6}$$

where V is the rated voltage of the transmission circuit. In order to compare the two ratings, one must assume a relationship between V and I. As an approximation, let the phase difference between internal voltages be taken as 60° under rated load on the line. (Perhaps only 30° of this appears across the line itself.) Then the voltage vector diagram is an equilateral triangle (Fig. 8), the sides of which represent the two internal voltages $1.16V$ and their difference $X_L I_{se}$, where X_L is the transfer reactance with either capacitor switched out. Then (6) becomes

$$Q_{sh} = \frac{(0.866 X_L I_{se})^2}{X_{sh}} \tag{7}$$

and from (5) and (7) we have

$$\frac{Q_{sh}}{Q_{se}} = \frac{(0.866 X_L I_{se})^2}{X_{sh}} \cdot \frac{4X_{sh}}{(kI_{se})^2 X_L^2} = \frac{3}{k^2}. \tag{8}$$

The factor k will range from 0.7 to 1. Thus the required megavar rating of shunt capacitors at the most effective location is from 3 to 6 times that of the equivalent series capacitors.

In general, then, switched shunt capacitors would prove less economical than switched series capacitors of equal effectiveness. However, if shunt capacitors or inductors are provided for control of voltage or reactive power, it may be economical to switch them during a disturbance so as to improve stability. This could be done under control of time-delay undervoltage relays or under control of fault-detector relays. A spread between dropout and pickup voltages should be provided in order to prevent hunting. Any time delay in switching shunt elements would probably be longer than that for switching series elements, which would further reduce the effectiveness of the shunt elements.

Some studies made at Bonneville Power Administration showed that switching off 500 MVAr of shunt reactors on an interregional tie line with 0.17 second delay would increase the stability limit of the line by 100 to 250 MW. Alternatively, additional *fixed* series compensation of 110 Mvar would produce an equal increase. Here the ratio of shunt reactor rating to series capacitor rating was found to be 500/110 = 4.5.

Conclusions

1) The use of switched series capacitors in addition to unswitched capacitors prevents the steady-state stability limit from being reduced by the switching out of one circuit of any section of a two-circuit link in order to clear a fault. It also raises the transient stability limit to a value almost equal to the steady-state limit.

2) If switched capacitors are thus used, rapid reclosure for raising the transient stability limit becomes unnecessary.

3) If the stability of an ac transmission link is threatened by the loss of a parallel dc line, its steady-state and transient stability limits can be increased by rapid insertion of switched series capacitors, enabling the ac line to assume and temporarily carry without instability the additional load initially carried by the dc line.

4) For a given transient stability limit, the aggregate rating of series capacitors required is somewhat less if some are switched than if all are unswitched.

5) The use of switched series capacitors, in comparison with fixed series capacitors giving the same transient stability limit, reduces the angular swing and, hence, reduces the fluctuation of voltage at loads, especially at those near the electrical center.

6) Switching shunt capacitors on or switching shunt reactors off during a disturbance also raise stability limits, but the megavar rating of shunt capacitors or reactors required is 3 to 6 times the rating of switched series capacitors for the same increase of stability limit.

References

[1] E. W. Kimbark, *Power System Stability*, vol. I. New York: Wiley, 1948, ch. IV, pp. 122-139.

[2] G. D. Breuer, H. M. Rustebakke, R. A. Gibley, and H. O. Simmons, Jr., "The use of series capacitors to obtain maximum EHV transmission capability," *IEEE Trans. on Power Apparatus and Systems*, vol. 83, pp. 1090–1102, November 1964.

[3] R. S. Seymour and E. C. Starr, "Economic aspects of series capacitors in high-voltage transmission," *Trans. AIEE*, vol. 70, pt. II, pp. 1663–1670, 1951. (Suggests switching capacitors to conserve transfer reactance and overloading capacitors for short times.)

[4] E. C. Starr and R. D. Evans, "Series capacitors for transmission circuits," *Trans. AIEE*, vol. 61, pp. 963–973, 1942.

[5] S. B. Crary and L. E. Saline, "Location of series capacitors in high-voltage transmission systems," *Trans. AIEE* (*Power Apparatus and Systems*), vol. 72, pp. 1140–1151, December 1953.

Discussion

M. L. Shelton (Bonneville Power Administration, Portland, Oreg.): Dr. Kimbark's paper shows the effect which switching series capacitors in high voltage transmission lines has on stability by illustrating the effect of switching series capacitors on the value of X in the equation $P_{max} = E_1E_2/X$. His paper demonstrates that a definite gain in stability is achieved by the use of switched series capacitors in long-distance high-voltage transmission lines by effectively decreasing the transfer impedance during transient conditions.

A further gain in stability by the method described in the paper, but not mentioned by Dr. Kimbark, is obtained from the increased voltages E_1 and E_2 behind generator reactances.

At 500 kV, the surge impedance loading point is high enough that for normal operation the line loading will be close to, if not below, surge impedance loading. If the lines are highly series compensated, the surge impedance loading goes up until the normal operating point might be well under surge impedance loading. It

Manuscript received May 26, 1965.

can readily be shown that highly compensated 500-kV lines (Case 1) will, at normal loading, furnish capacitive reactive power to the system they are tied into. This capacitive reactive power will have to be absorbed by the generators at each end, causing the generator exciters to operate at a point which would cause the generators to be close to unity power factor or even underexcited.

Now compare Case 1 and Case 2 where the 500-kV lines are not so heavily series compensated. In the latter case, the reactive loss in the lines tends to take care of the reactive power which had to be absorbed by the generators in Case 1. In Case 2, where the lesser value of series compensation is used, the generators will be operating at a more lagging power factor and furnishing reactive power to the system. The exciters will then be operating at a higher voltage which, when referred to the equation $P_{\max} = E_1E_2/X$, means that E_1 and E_2 will be higher. That is, the voltage behind transient reactances will be higher with the lower value of series compensation.

Dr. Kimbark assumed, for simplicity of illustration, that the voltage behind transient reactance was constant and equal at each generator in the two machine problems which he used for illustration; thus, his use of E^2 in the previous equation. Had he taken advantage of the higher voltage behind the generator reactances, which would have resulted from starting at the lower value of series compensation, he could have shown even greater benefits of switching to higher compensation during transient conditions.

In an actual fault case, the excitation and the voltage behind the generator reactance would not remain constant. Instead, the exciters would tend to increase the generator internal voltage in order to keep bus voltage up. If the exciters are already riding at a higher point, as mentioned in Case 2, there will be a definite gain in the stability limit.

It would seem that any means of keeping bus voltages and generator internal voltages up during a disturbance should add to the system stability. The use of shunt inductive reactances on 500-kV lines would have just this effect. In studies at Bonneville Power Administration for the effects of shunt reactors on the stability of the Pacific Northwest-Pacific Southwest intertie, adding shunt reactors and restoring previous line voltages had a definite benefit on stability. Switching the reactors off during a swing further added to stability. When checks are made to pinpoint the relative worth of switched shunt reactors vs. unswitched reactors, it was found that the effect of raising the internal voltage of the generators made up approximately 40 percent of the gain in stability while switching the reactors off during a swing added approximately 60 percent of the total gain.

The switching of series capacitors from a lower to a higher value or the addition of shunt reactors on the 500-kV lines will tend to raise the generator excitation and increase the reactive output of the generators with an accompanying raise in stability. There is, of course, a limit as to how far both of these methods can be carried. The generator reactive capabilities determine one limit. Proper bus voltages, insulation levels, economics, exciter ceilings, type of disturbance, etc., determine other limits.

S. J. Marmaroff (American Electric Power Service Corporation, New York, N. Y.): The author has suggested an interesting extension to the use of series capacitors. The use of switched series capacitor banks, however, raises a question as to what type of relaying scheme might be expected to function properly in each of the numerous possible system states. Perhaps the author would care to comment on the relaying techniques which might be employed.

The author presents a basic series capacitor connection scheme with the capacitors located in the transmission lines and then suggests switching additional blocks of capacitors to effect "equal transfer-reactance switching" when clearing a faulted circuit. The author then indicates that, for the same transient stability limit, the use of switched capacitors gives a reduction in required MVAr rating as compared with a simple in-line unswitched capacitor scheme. Although we agree with the author's contention that switched series capacitors are a feasible means of increasing stability limits, we also believe that considerable additional benefit can be realized by redesigning the basic series capacitor connection scheme to place the capacitors "in the bus" as shown in Fig. 10. With this arrangement under normal operating conditions, the capacitor banks will appear to be connected in parallel. When a faulted line section is cleared, one of the parallel capacitor banks is disconnected, and the capacitive reactance in series with the line transfer impedance is automatically increased. With fifty percent series compensation (all lines in service) "equal transfer reactance switching" is achieved without the use of additional switched capacitors, and, for any compensation level, the transient stability limit is increased beyond that realized by connecting the series capacitors in the line sections. Thus, it can be seen that switched capacitors are required to give capacitor-in-the-line schemes a transient stability limit equal to that of capacitor-in-the-bus schemes. Figure 11 shows the required

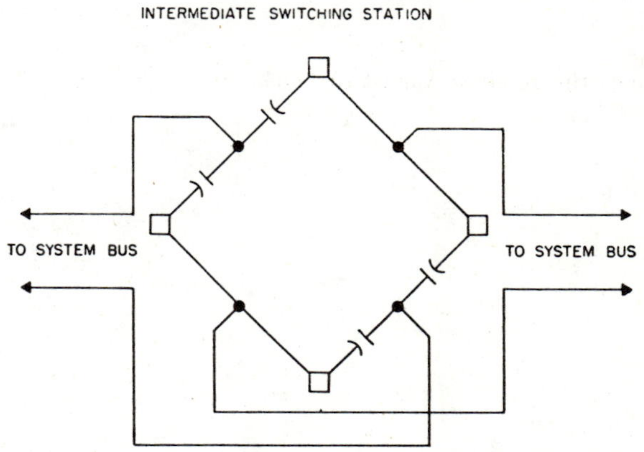

Fig. 10. Switching arrangement for capacitors connected "in the bus."

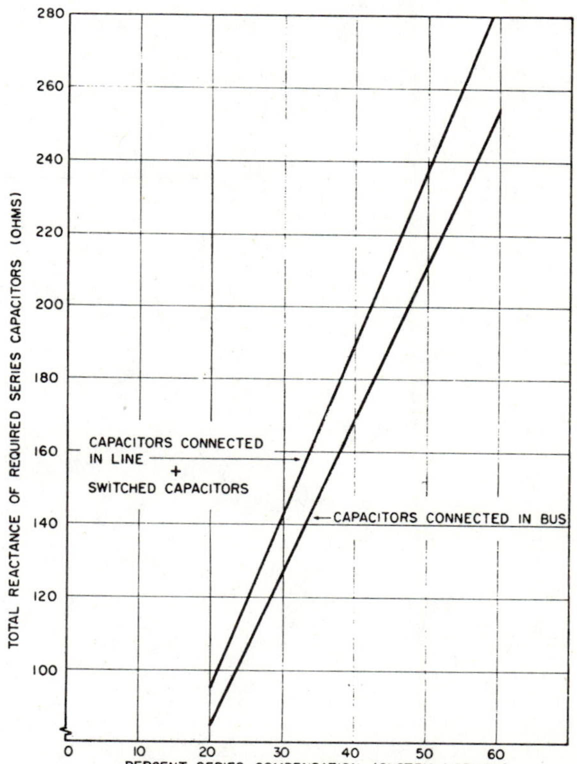

Fig. 11. Relative amounts of series capacitors for two connection schemes, each having the same transient stability limit at the same

capacitor reactance for both schemes applied to the author's Example 1, where for any percent compensation the two schemes have the same power transfer capability. If we assume current proportional to power transfer, the abscissa of Fig. 11 translates directly into MVAr and thus dollars, and we see that connecting capacitors in the bus will result in approximately eleven percent savings in capacitive MVAr. The in-bus arrangement permits latitude in selecting only one of the transfer impedances (system normal or one line out), but the addition of switched capacitors to this basic scheme could conceivably result in design flexibility as well as economy.

E. W. Kimbark: Mr. Shelton calls attention to an important point which I had overlooked, and I am grateful to him for having done so.

Mr. Marmaroff asks for comments on relaying techniques which might be employed in conjunction with switched series capacitors, and he advocates the use of four capacitors in a ring bus at the intermediate switching station.

Manuscript received August 13, 1965.

If the switched capacitors are located at one of the terminals of a circuit which becomes faulted, the same relays which determine that the circuit is faulted can be used to insert the capacitors. Otherwise, a signal must be sent from those relays to the site of the capacitors. Lines with a high degree of series compensation present difficulties to distance relaying, whether or not any series capacitors are switched, because of uncertainties regarding the flashing of protective gaps. However, phase comparison relays may be expected to operate properly under most conditions.

The switching scheme having four capacitors in a ring bus seems attractive because it switches series capacitors with the same circuit breakers as those which switch lines and it does so simultaneously. The same scheme was shown in [2] of the paper as Fig. 2(f). With initial series compensation of 50 percent, it preserves constant transfer reactance when any one of the four circuit sections is switched out. However, it has some shortcomings. It cannot be used to insert series capacitors in the event of a fault on a parallel dc line (Example 2 of the paper). Moreover, at 50 percent series compensation, the loop consisting of the two circuits of one section, closed through a solid bus at one end and through the capacitors of the ring bus at the other end, is series-resonant. Consequently, a small unbalance of phase impedances could cause large circulating currents around the loop and greatly increase the losses.

SERIES CAPACITOR SWITCHING TO QUENCH ELECTROMECHANICAL TRANSIENTS IN POWER SYSTEMS

Richard H. Webster Ajit P. Mane
Pacific Gas and Electric Company
245 Market Street
San Francisco, California 94111

Professor O.J.M. Smith
Department of Electrical Engineering
and Computer Sciences and
Electronics Research Laboratory,
University of California, Berkeley, Calif. 94720

ABSTRACT

A method of quenching transient oscillations of a generator feeding a load over long lines was implemented in the laboratory. A disturbance that caused instability was repeated with this control, and the system remained stable.

The tests consisted of disconnecting (and later reconnecting) one of the two parallel transmission lines between a 5 kVA generator and the "infinite" bus. The oscillograms showed that nearly deadbeat transients were obtained by more than one insertion and bypassing of a series capacitor in the transmission line.

The switching decision algorithm was implemented using semiconductor elements. The circuit was operated from a noise-free and reliable signal proportional to the change in the generator rotor velocity. An unorthodox method of obtaining the rotor velocity signal is described in the companion paper, "Rotor Velocity Deviation Measurement from Power Balance Analog."

INTRODUCTION

A modern, interconnected power transmission system exhibits a very complex and nonlinear dynamic behaviour. Whenever a sudden change occurs in such a system, there are transients. Due to the peculiar nonlinear nature of the system, if the changes in the energy levels in the system during the transient are large enough, the system becomes unstable. Stability can be maintained, however, if sufficient damping is provided by manipulating the power flows between generating stations and loads during the transient,[1,2]

A method of controlling system transients by switched series capacitors was implemented. Fig. 1 shows the shaft velocity and the generator power output after losing one circuit of a double circuit transmission line. The generator lost synchronism. The same experiment, under identical conditions, was repeated using the method of control described here.

Fig. 2 shows the shaft velocity and the generator power output. A nearly deadbeat transition from the pretransient state to the posttransient state was obtained and the generator did not lose synchronism.

The use of the control proved to be highly successful at all levels of load and for disturbances which shifted the target angle of the generator shaft ahead of the pretransient angle or behind it.

Whenever a transient was detected the control was activated for several seconds. A series capacitor was inserted whenever the shaft velocity was more than the synchronous velocity, and the capacitor was bypassed as the shaft velocity became less than the synchronous velocity (see Fig 3). The switching decision was based on a velocity signal obtained by integrating a signal proportional to the computed accelerating power, as explained in the companion paper, "Rotor Velocity Deviation Measurement from Power Balance Analog."

The method of control employed and the rationale behind it, the results of the tests carried out in the laboratory to study the system performance, and in conclusion, proposals for further work in the same field, are presented.

The papers by E. W. Kimbark[3] and Otto J. M. Smith[4,2] have a direct bearing on the approach taken in this paper.

Paper 70 TP 677-PWR, recommended and approved by the Power System Engineering Committee of the IEEE Power Group for presentation at the IEEE Summer Power Meeting and EHV Conference, Los Angeles, Calif., July 12-17, 1970. Manuscript submitted September 15, 1969; made available for printing May 4, 1970.

GLOSSARY OF SYMBOLS

θ	=	instantaneous angle between the vectors V_g and V_b
$\dot{\theta}$	=	$d\theta/dt$
$\ddot{\theta}$	=	$d\dot{\theta}/dt$
$\dddot{\theta}$	=	$d\ddot{\theta}/dt$
θ_i	=	pretransient angle between the vectors V_g and V_b
θ_n	=	posttransient angle between the vectors V_g and V_b
$\dot{\theta}_m$	=	maximum velocity during the first transient swing
f(t)	=	a function of time
k	=	critical reactance/resistance ratio
P_1	=	total losses at the generator node
P_s	=	shaft power
R_1	=	total instantaneous resistance of the generator, capacitor, and line
V_b	=	voltage of the infinite bus
V_g	=	voltage behind the synchronous reactance
w	=	instantaneous shaft angular velocity
w_0	=	pretransient value of shaft angular velocity
X_c	=	reactance of the series capacitor
X_i	=	total instantaneous reactance of the generator and line
X_1	=	total instantaneous reactance of the generator, series capacitor, and line $(X_i - X_c)$
X_s	=	synchronous reactance of the generator

METHOD OF CONTROL AND DECISION FUNCTIONS

A Review of the Plant Model

All of the methods of damping system transients depend on schemes to properly control the power flow between the generators and the loads.

In quasi-stationary state, the equation relating the power flow between a generator and the infinite bus into which it feeds, with the other variables, is approximately,

$$P_s - P_1 = \frac{V_g V_b}{X_1} \sin \theta.$$

Reprinted from *IEEE Trans. Power App. Syst.*, Mar./Apr. 1971.

The stability is determined by the maximum value of θ during a transient swing. Therefore, to stabilize the system θ and $\dot{\theta}$ have to be controlled. However, the plant to be controlled (generator rotor) is a double integrator, so only $\ddot{\theta}$ can be directly manipulated. There are various means for this purpose as suggested by the form of the equation (1).[1]

The Method of Control Employed

The line reactance can be manipulated by switching some series capacitors in and out of the transmission line. This corresponds to changing X_1 in Eq. (1) in steps or in "bang-bang" mode.

Basic Principle

The rationale of the control scheme can be explained on the basis of the swing curves in Fig. 4. Assume the system had both the parallel lines in service and delivered power P_s with the angle at θ_i. If one of the parallel transmission lines is suddenly switched off, the transmitted power drops suddenly to a lower level. As shown in Fig. 4, if the capacitor were not inserted the generator rotor would gain so much momentum during its advance from θ_i to θ_n that the system would become unstable. However, if the series capacitor is inserted immediately after the line is lost, the forward acceleration is reduced as the drop in transmitted power is reduced. If the capacitor value is such that the value of $(X_i - X_c)$ is held low, the equal area criterion is satisfied when the shaft angle reaches B. This is also approximately the angle when the rotor velocity deviation becomes zero. If the capacitor were left in the line, the rotor would tend to gain velocity in the reverse direction. However, at this point if the capacitor is removed so that the power transmitted is determined by the reactance of one line alone, the rotor again accelerates in the forward direction. If nothing is done it will tend to continue accelerating and overshoot the new target angle θ_n which is the equilibrium point where the power P_s can be transmitted over one line only. In order to prevent this overshoot, at the point C the capacitor is inserted once more, increasing the power transmission capability. The transmitted power suddenly becomes more than the power input to the rotor shaft and the rotor decelerates. The capacitor is lastly switched out exactly where the rotor velocity becomes zero, say at point F. If the point F can be made to coincide with the new target angle θ_n then one can steer the state of the system from initial state $(\theta_i, \dot{\theta} = 0)$ to the new target state $(\theta_n, \theta = 0)$ in a deadbeat manner.

Functional Components of the Realization

The signal flow block diagram of the experimental system is shown in Fig. 5.

A constant power D.C. motor drives the 3-phase synchronous generator, and feeds power over two parallel transmission lines to the infinite bus. The measured variables are the generator output power, voltage and current. A computational circuit (analog) computes the instantaneous rotor velocity and acceleration (see the companion paper, "Rotor Velocity Deviation Measurement by Power Balance Analog").

The acceleration signal $\ddot{\theta}$ is fed into the Inhibitor unit. It consists of a differentiator and a toggled monoflop timer. Whenever a sudden change in power level and hence in the rotor acceleration takes place, the differentiator triggers the timer. The timer output is a binary signal which remains "on" for a few seconds. It feeds into an AND gate. Effectively, the Inhibitor recognizes the onset of a transient and arms the capacitor switching control circuit.

The velocity deviation signal $\dot{\theta}$ is fed to a comparator and an exponential reference circuit, whose operation is described later. The comparator is an "infinite gain" device which responds to zero crossings of its input giving binary output.

The capacitor switching can take place only when the Inhibitor-timer is "on," i.e. during a transient when the output of the comparator gives a switching signal. This design has the advantage of a higher sensitivity in the comparator circuit yielding better accuracy and yet preventing sporadic malfunctioning of the control scheme due to the binary noise on the comparator output in the steady state.

Choice of the Value of X_c

There is a restriction on the value of the reactance of the series capacitor which can be used for the mode of control described above. If a deadbeat response is desired there should not be any overshoot in the angle excursion. It is obvious from Fig. 4 that if we desire to place the point B, i.e. the angle at which the rotor attains zero velocity deviation in the first swing, between θ_i and θ_n, then the swing curve for 1 line and capacitor must have a high enough peak. By definition the posicast design must have a value of X_c such that the point B described before must lie at θ_n. This is one extreme case. On the other hand, if X_c is increased so much that the effective value of line reactance X_1 is below the critical value, i.e.

$$X_1 = X_i - X_c \leqslant R_1 k$$

then the system has negative damping [5,6].

With any value of X_c between these two values, the control will cause the first swing to terminate at less than the target angle. Values of X_c less than that for posicast control will cause the first swing to overshoot the target.

In the laboratory experiment the value of X_c was chosen by successive trials until a satisfactory transient performance was obtained.

Problem of Chatter

A simple algorithm using the sign of the velocity deviation can be used for control.

$$\begin{aligned} &\text{i.e. insert } X_c \text{ if } w - w_0 > 0 \\ &\text{bypass } X_c \text{ if } w - w_0 < 0 \end{aligned} \tag{2}$$

This algorithm functions very well during the first swing (of Fig. 4), i.e. until the angle reaches B and the capacitor is bypassed. As the rotor has not yet reached the target, there would be a chatter as the rotor advances towards the target, due to the "infinite gain" algorithm for switching. It is not desirable in practice to have such a chatter. Mathematically this is an elementary relay control on velocity. What was desired was an angle control.

Solution Using Exponential Reference

A satisfactory solution to the problem of chatter is as follows. Let the algorithm (Eq. (2)) be changed to inserting the capacitor when the velocity exceeds a fraction of the previous maximum value as shown in Fig. 6, i.e.

$$\begin{aligned} &\text{insert the capacitor if } w - w_0 > \dot{\theta}_m f(t) \\ &\text{bypass the capacitor if } w - w_0 < 0 \end{aligned} \tag{3}$$

Consider f(t) = exp(-t/T) in the algorithm (Eq. (3)). It was implemented and found satisfactory. This function if plotted in Fig. 6 would pass through the points $\dot{\theta}_m$, C, and E. This function is responsible for the successive reductions in maximum velocity. During the second forward swing the velocity was allowed to increase till it exceeded the value given by this exponential signal, i.e. until C. Then the capacitor was inserted; the rotor slowed down to zero velocity at D and the capacitor was again bypassed. A similar cycle was repeated at E and F. At F the state of the system came so near the target state (θ_n, 0) that the velocity did not rise up to the value of the exponential reference; hence the capacitor was not inserted and the system underwent small damped oscillations about the target state (θ_n, 0).

It is obvious that in order to have near deadbeat performance, the residual velocity when the rotor angle reaches θ_n must be minimized. On the other hand the number of insertion and bypass cycles have to be minimized due to practical considerations. This calls for a compromise solution. The amplitude of the residual oscillations about the target can be adjusted by choosing a suitable value of the decay time constant.

This particular exponential function was chosen because of the several advantages it offers. A control strategy should be such that when the system state is far away from the target it will take fast or rapid steps and then become more and more cautious as it approaches the target. The exponetial function fits these requirements quite well. It also controlled the envelope of both the angle and the velocity transients with only velocity information and velocity decision function. Moreover the exponential function is easy to realize using analog circuitry.

RESULTS

Figure 7 shows a transient created by the loss of one of the two parallel circuits. This transient was so severe that the machine angle actually became larger than 90°, and it took 30 oscillations before the transient was damped to 10%. Fig. 8 shows the same transient with the addition of the series capacitor control. The initial velocity swing was reduced to 56% of the uncontrolled value. The machine angle moved unidirectionally toward its target until it arrived there in less than one second with little overshoot. Large machine angle swings and a possible instability were avoided. The points marked A, B, C, D, E, and F on Fig. 8 correspond to those on Fig. 4, 6.

The control was also used to quench a transient caused by gaining a second transmission line (see Fig. 9). The same switching algorithm including the exponential reference was in operation. Since this transient is basically a move toward greater stability, it is significant only to check that the control does manage to help it considerably even without changing switching algorithms.

CONCLUSION

It has been shown on an actual machine that switched series capacitors can be a very powerful tool to improve the transient stability of power systems. E. W. Kimbark [3] has shown that switched series capacitors can be used to improve the posttransient steady-state stability. It is logical to expect that a coordinated operation of both these schemes will be desirable (using two series capacitors in series). At the time of the transient, one capacitor would be inserted and left in to reduce the posttransient rotor target angles while keeping the value of the net reactance to resistance ratio X_1/R_1 above the critical value, to avoid negative damping. [5,6] The second series capacitor would be inserted and removed at appropriate instants to reduce the overshoot of the transient machine angles. This combined scheme would provide an effective damping of the transient oscillations as well as a good margin of posttransient stability. The most valuable feature of such a scheme is that it not only improves posttransient stability, but also prevents tripping of other heavily loaded lines by reducing the rotor angle swings in the control area.

ACKNOWLEDGEMENTS

The research work on which this paper is based, was supported at the Electronics Research Laboratory of the University of California, Berkeley through the research grants given by the Bonneville Power Administration (U.S. Department of Interior) Portland, Oregon and the Pacific Gas and Electric Co., San Francisco, California.

REFERENCES

1. W. A. Mittlestadt, "Four Methods of Power System Damping," IEEE TRANS. POWER APPARATUS AND SYSTEMS, vol. PAS-78, no. 5, pp. 1323-1329, May 1968.

2. Otto J. M. Smith, "Optimal Transient Removal in a Power System," IEEE TRANS. POWER APPARATUS AND SYSTEMS, vol. PAS-84, nos. 5 and 6, pp. 361-374 and 528-531, May and June 1965.

3. E. W. Kimbark, "Improvement of System Stability by Switched Series Capacitors," IEEE TRANS. POWER APPARATUS AND SYSTEMS, vol. PAS-85, pp. 180-188, February 1966.

4. Otto J. M. Smith, "Power System Transient Control by Capacitor Switching," IEEE TRANS. POWER APPARATUS AND SYSTEMS, vol. PAS-88, no. 1, pp. 28-35, January 1969.

5. J. W. Butler and C. Concordia, "Series Capacitor Application Problems," AIEE TRANSACTIONS, August 1937.

6. A. M. El-Sarfi, "Selbsterregte Schwingungen der Synchronmaschine unter dem Einfluss von ohmschen Reihenwiderstaenden und von Reihenkondensatoren," ETZ-Bc. 87 (1966) H.1.

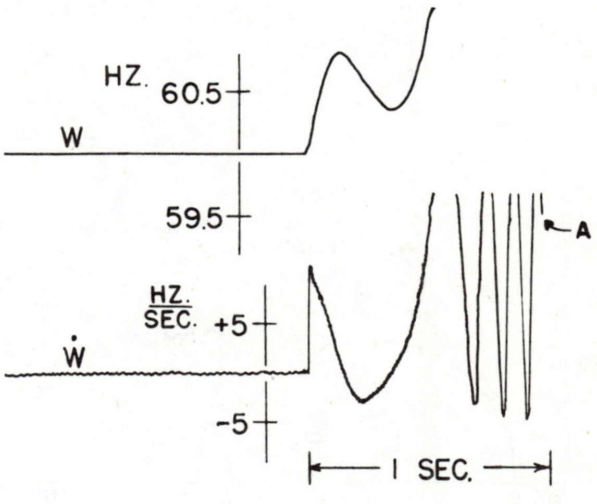

Fig. 1. Lost line without control.
w = Rotor velocity
ẇ = Rotor acceleration
Circuit breaker tripped at A after slipping four poles.

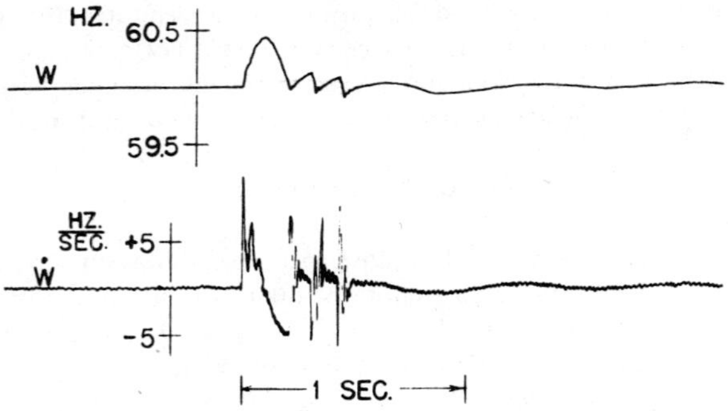

Fig. 2. Lost line with control.
w = Rotor velocity
ẇ = Rotor acceleration
The frequency deviation has been halved by the control. The initial disturbance was the same as for Fig. 1.

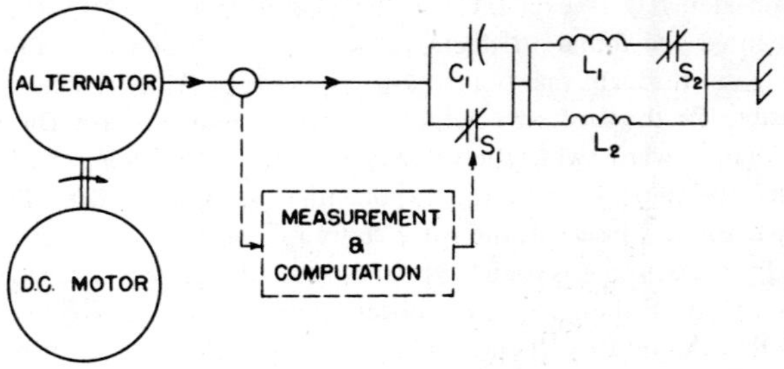

Fig. 3. Single line power flow diagram showing motor generator set, power measurement, series capacitor, transmission lines, and infinite bus. Component values were X_{c1} = 0.193 perunit; $X_{L1} = X_{L2}$ = 0.55 perunit; X_s = 0.88 perunit. S_2 creates a distubance by switching out one of the two transmission lines. S_1 is the bypass switch which is normally closed and which opens responsive to the control.

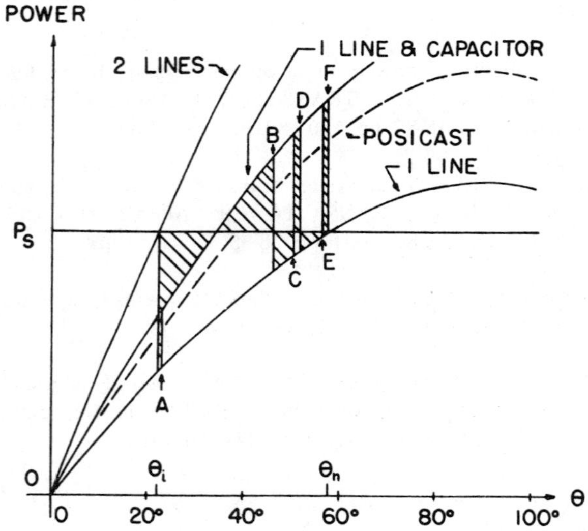

Fig. 4. Basic principle of the control is insertion of the capacitor at states A, C, and E, and bypassing of the capacitor at states B, D, and F.

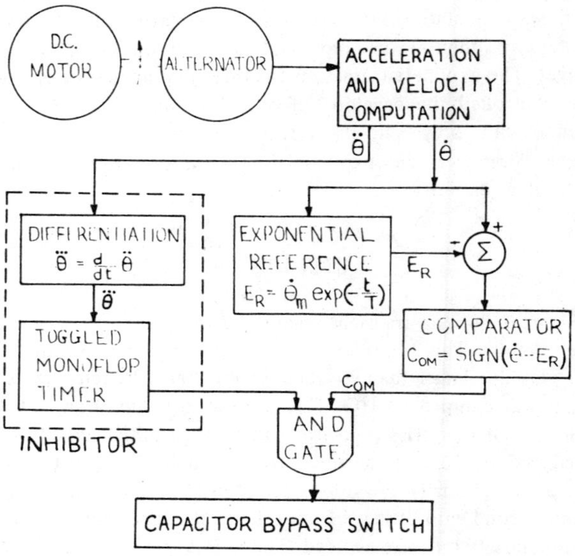

Fig. 5. Signal flow diagram for analog tachometer and switching control.

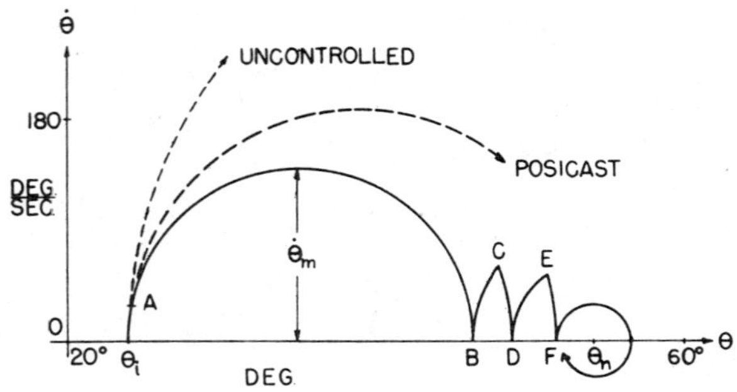

Fig. 6. Phase plane plot of state trajectories for the transient control shown in Fig. 4. The capacitor was inserted at states A, C, and E, and removed at states B, D, and F. The residual oscillation is due to the θ_n − F difference.

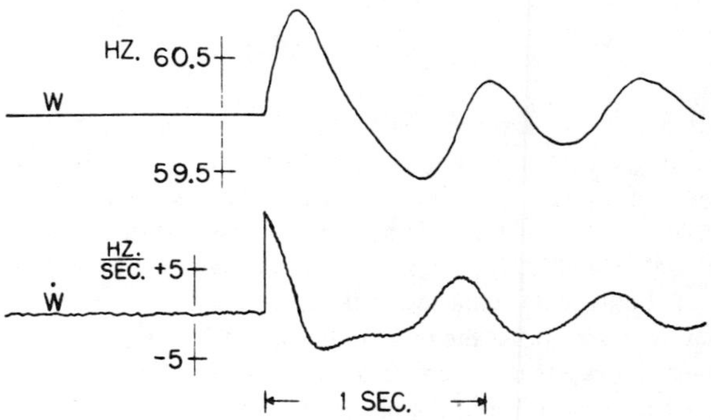

Fig. 7. Lost line without control
w = Rotor velocity
ẇ = Rotor acceleration

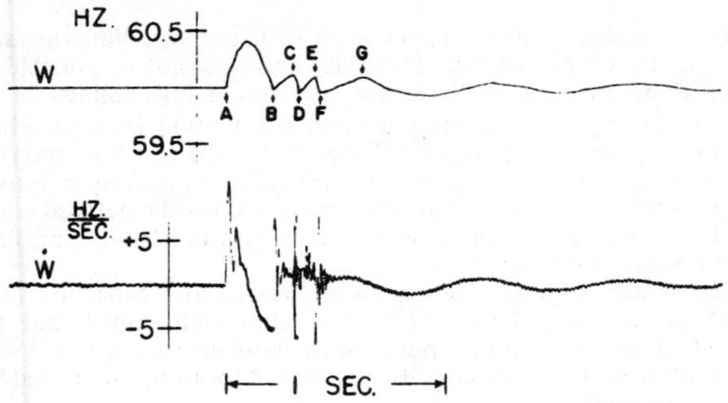

Fig. 8. Lost line with control.
w = Rotor velocity
ẇ = Rotor acceleration
Initial disturbance was the same as for Fig. 7. The velocity deviation has been halved by the control.

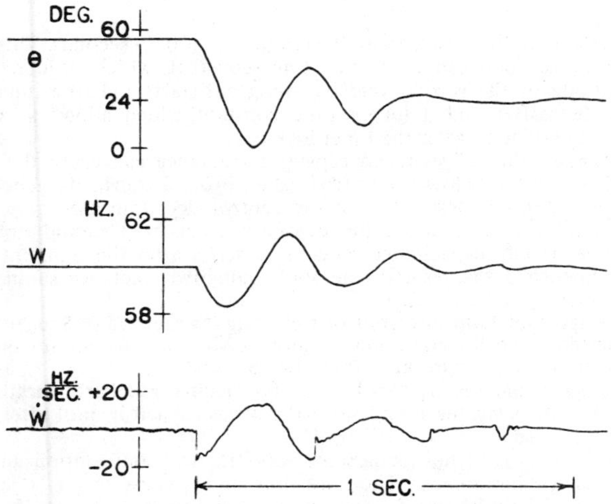

Fig. 9. Gained line with control.
w = Rotor velocity
ẇ = Rotor acceleration
The capacitor was inserted three times in the one second control interval.

Discussion

B. S. Ashok Kumar and **K. Parthasarathy** (Indian Institute of Science, Bangalore-12, India): The authors have presented a comprehensive discussion of the series capacitor switching, backed by experimental results. In recent years, much progress has been accomplished on the application of series and shunt compensation for the improvement of transient stability. The several aspects presented, such as generation of an exponential function and its control for overcoming the inherent problem of 'Chatter' in capacitor switching appears to provide a satisfactory control of electromechanical transients.

However, a few comments on the paper appear to be appropriate.

It may be in order for the authors to present some details as regards the transient control characteristics with respect to the 'Time Constant' of the exponential function. This we believe will be very helpful for comparitively assessing this mode of switching.

The authors have referred to the problem of negative damping effects when the effective line reactance is below the critical value. Normally, for short transmission lines the transient could be controlled, effectively by sequential or posicast switching, and the series capacitance required for this control will be below 20-25 percent. The problem of negative damping cannot arise in such cases.

Manuscript received July 27, 1970.

For longer lines, two capacitors, one for improving the power angle curve and the other for control of transients will be made use of. In such cases, the overall series compensation may cause the line reactance to fall below the critical value.

However, we feel that the problem of over voltage along the line is more critical than the above. From the power frequency over-voltage point, the total series compensation must be such as to maintain the line reactance well above this critical value ($R_1 K$)

Attempts have already been made to coordinate the different methods of improving the power system stability. The method as suggested by Dr. Saugen and Mr. Mittelstadt [7] to improve transient stability, is to control certain parameters to force the derivative of an error function to be as negative as possible. One among them is,

$$\phi_{YB} = -\sum_{i=1}^{N}\sum_{J=1}^{N} \frac{\epsilon_i E_i\, E_J\, d\,(Y_{iJ}\, B_{iJ})/dt}{M_i} \tag{1}$$

Where, Y_{iJ} and B_{iJ} are defined by the network equations

$$P_i = E_i^2\, G_i + \sum_{i=1}^{N} E_i E_J\, Y_{iJ}\, B_{iJ}\,.$$

$$B_{iJ} = \mathrm{Cos}\ (\alpha) = \mathrm{Cos}\ (T_{iJ} - A_i + A_J)\,.$$

Equation (1) represents the changes in the network.

These authors have so far not framed any physical constraints to achieve the control. The equation for $d(Y_{iJ} B_{iJ})/dt$ can be rewritten as.

$$d(Y_{iJ}\, B_{iJ})/dt = \dot{Y}_{iJ}\, \mathrm{Cos}\,\alpha - Y_{iJ}\,(\dot{T}_{iJ} - \dot{A}_i + \dot{A}_J)\ \mathrm{Sin}\ \alpha \tag{2}$$

Y_{iJ} and B_{iJ} in the above equation are not continuous variables with respect to time, and their derivatives represent only discrete switching operations. The above equations even though fail to give information about the quantitative switching suggest the following criteria. From equation (1) we could see that ϕ_{YB} is negative definite if,

Y_{iJ} increases for $-90° < \alpha < 90$
and decreases for $-90 > \alpha > 90$
T_{iJ} decreases for $0 < \alpha < 180°$
and increases for $0 > \alpha > 180°$

for all positive values of ϵ_i and vice versa. This control scheme may be compared experimentally with the scheme suggested by the authors.

The system has been analysed assuming a sudden switch-off of one of the parallel transmission lines on the simulated model. In practice however, the situation will be different. The transmission line switching will be preceded normally by a more severe period of fault condition. If the series capacitor is switched in during this fault condition (the analog tachometer switching control is designed to operate when there is acceleration of motor), the effect may be more severe particularly from the point of fault level and transient over voltages.

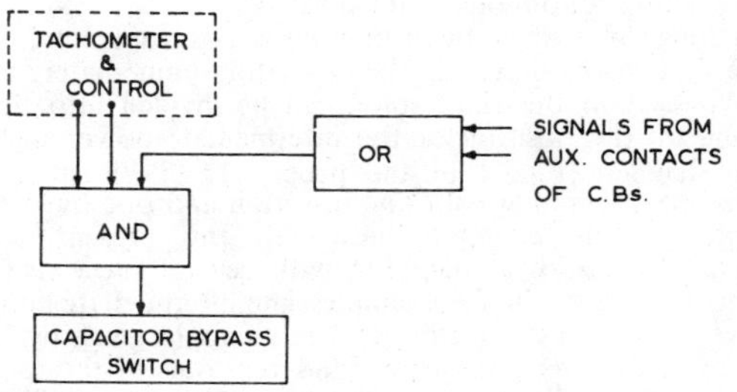

Fig. 10

This problem could be taken care of by slightly modifying the design of tachometer switching control as shown in Fig. 10. An 'OR' gated signal from the auxilliary contacts of the circuit breakers may be fed to the final 'And' gate which controls the capacitor bypass switch.

This modification inhibits the operation of the bypass switch till one of the circuit breakers opens.

REFERENCE

[7] W. A. Mittelstadt and J. L. Saugen, 'A method of improving transient stability using controllable parameters' IEEE Transactions, PAS, Vol. 89, No. 1, Jan. 70 PP 23-27.

E. R. Taylor (Westinghouse Electric Corporation, East Pittsburgh, Pa. 11512): For those who are seeking functional applications for series capacitors, the authors present some new thoughts on uses for series capacitors for improving stability. However, subharmonic transient responses of a system as series capacitors are inserted may raise questions of other concerns about rapid multiple bypassing and reinsertion, particularly should the transmission line contain other series capacitors which are not being switched. The insertion of a large series capacitor at a large angle in the system swing gives rise to subharmonic line and ground currents of considerable magnitudes. The subharmonic line currents can produce high subharmonic voltages across the series capacitors, and protective gaps across the series capacitors must be set sufficiently high that they do not sparkover and take out the capacitors during the time when they are most needed to maintain stability. The subharmonic ground currents are a consideration from the standpoint of ground fault relaying. Would the authors comment further on the values of X_c they found necessary for stability control?

Another consideration with respect to subharmonic currents is the possibility, especially with relatively small synchronous machines, of subharmonic oscillation buildup should circuit losses not provide sufficient damping. [1] This is due to the synchronous machine acting like an induction generator at the subharmonic frequencies. Should the series capacitors be inserted for a sufficient number of cycles in a system with insufficient damping, the subharmonic voltages across the series capacitors may again be a concern from the standpoint of the protective gap settings. Possibly the authors may wish to comment on the maximum insertion time they envision to achieve stability utilizing their proposition.

[1] "The Prediction and Control of Self-Excited Oscillations Due to Series Capacitors in Power Systems," L. A. Kilgore, L. C. Elliott, E. R. Taylor, Trans. Paper 626, presented at the IEEE Summer Power Meeting and EHV Conference, Los Angeles, Calif. July 12-17, 1970.

Manuscript received August 4, 1970.

E. W. Kimbark and **W. A. Mittelstadt** (Bonneville Power Administration, Portland, Ore. 97208): It appears that automatic switching of series capacitors can be a good method of damping (quenching) electromechanical swings of large amplitude, leaving to the supplementary signals of voltage regulators the duty of damping small oscillations and preventing them from becoming big ones. Capacitor switching would be used only on the occasion of a large disturbance and would be discontinued, as suggested by the authors when the amplitude of oscillation became less than some preselected value; whereas voltage regulators with their supplementary signals would be acting continuously and always.

Quenching of swings by switching series capacitors on a two-machine system, by inserting the capacitors immediately and leaving them bypassed in the final state, can be divided into three classes depending on the position of the intermediate power-angle curve of the area diagram (Fig. 4 of the paper). 1) "Posicast" control [4] gives dead-beat damping with one insertion and one bypassing of the capacitor. 2) The example shown in the present paper gives approximately dead-beat damping with several successive insertions and removals – three in the example. The intermediate curve is above the curve for Posicast control. 3) The example given by Mittelstadt [1] also involves several insertions and removals of the capacitor but gives underdamping and hence overshoot. The intermediate curve is below the Posicast curve. To sum up, one can get underdamping, critical damping, or overdamping, depending on the reactance of the switched capacitor. The Posicast method gives critical damping, the fewest switch operations, and quickest arrival to the new steady state. For reasons of economy, the underdamped case may be the most important one.

Figures 4 and 6 of the paper seem to imply switching operations separated by very short time intervals. This may not be possible with the kind of switches available for high-power high-voltage circuits. However, it appears that the same end result could be accomplished with fewer switching operations. For example, the two brief insertions indicated by CD and EF on the figures could be combined into one longer insertion. Or, the initial insertion at A could be delayed enough to balance the positive and negative areas with only one bypassing as in the Posicast method.

What is the authors' opinon about leaving the capacitor in the circuit in the final state? Also, have they determined that their method of control gives correct results with disturbances of various magnitudes, such as short circuits in different locations or cleared with different speeds?

Manuscript received August 3, 1970.

Richard H. Webster, Ajit Mane, and **Otto J. M. Smith:** The authors wish to thank B. S. Ashok Kumar and K. Parthasarathy for their many comments and questions. The exponential reference function in the experiments was

$$f(t) = \frac{\dot{\theta}_m}{3} \exp \frac{t}{T_1}$$

The results in Figures 2, 8 and 9 were for T_1 of 0.75 seconds. This is a compromise between a shorter time constant which reduces the magnitude of the second velocity swing but starts a chatter immediately thereafter and a longer time constant which allows a larger residual oscillation after the timer lock-out.

A large value of switched capactive reactance can cause the line reactance to fall below the critical value. From a transient-quenching viewpoint, this is desirable. For the control algorithm used here, the total line reactance when the capacitor is inserted should not go negative. If any capacitance remains in series after the transient has been quenched the resultant network inductive reactance should be more than the critical value.

We have not built any control following the method of Saugen and Mittelstadt, and do not know of anyone who has. We are unable to comment on a comparison of these two methods.

We agree that the logic in Figure 10 of Kumar and Parthasarathy is desirable to delay the insertion of the series capacitor until after the fault is cleared.

Mr. E. R. Taylor has made some very practical contributions in the field of subharmonics. The insertion and bypassing of a series capacitor should be timed to occur at the maximum of the 60 Hz line current. This minimizes the voltage offset on the capacitor. Vacuum switches with tungsten contacts and separate timing in the three phases are satisfactory in this application. Since all three phases will close or open within 6 milliseconds, ground current relays will not be affected. The first insertion of the series capacitor will excite a subharmonic oscillation. There is a most desirable phase on each subharmonic wave for each subsequent switching to minimize the resulting subharmonic oscillation. Kilgore, Elliott, and Taylor have suggested a method for damping subharmonic oscillations using a damping resistor and a 60 H_z parallel-resonant blocking trap.

The most desirable value of X_c is that which would produce a Posicast response (dead beat in one-half period), but this value varies with system configuration and loading. The algorithm of this paper will work for any X_c less than X_i, the total instantaneous reactance of the generator and line. The larger the value of X_c, the higher the cost; however, a larger value enables the system to recover from a more severe disturbance.

The authors expect that the maximum total interval for control would be 5 seconds on a typical large system. If the capacitors are heat rated for 10 seconds, this should be adequate.

The authors thank Dr. Kimbark and Mr. Mittelstadt for their summary of the three catagories of switched capacitor control. Vacuum switches are quite suitable for the switching times encountered in practice. Phasing can be accomplished if the actuator is directly coupled to the switch.

The design in this paper was intended to be an optimum with constraints on equipment, type of algorithm, and control time. Any deviation from this algorithm will degrade the performance in some manner. For example, combining two switch intervals into one will cause a larger velocity deviation. A delay in the first insertion is very bad because more information is required for the delayed Posicast control, and this is (1) more expensive, (2) more uncertain, and (3) has larger velocity deviations. The authors propose that the switched capacitor should always be removed before it burns up, and for cost reasons, the capacitors should be short-time intermittant-duty rated.

Manuscript received September 14, 1970.

THE PREDICTION AND CONTROL OF SELF-EXCITED OSCILLATIONS DUE TO SERIES CAPACITORS IN POWER SYSTEMS

L. A. Kilgore L. C. Elliott E. R. Taylor
Westinghouse Electric Corporation
East Pittsburgh, Pennsylvania

ABSTRACT

This paper covers the analysis of the problem of self-excited current oscillations which may occur when synchronous machines are closely coupled to transmission lines which are highly compensated with series capacitors. A mathematical analysis is made. Corrective measures are suggested and the results of confirming tests are discussed. Two typical system configurations which would be likely to produce these self-excited oscillations are used as examples for detailed calculations.

INTRODUCTION

The need for transmitting larger and larger blocks of power over long transmission lines is rapidly becoming the major engineering problem facing utility engineers today. The power requirements of densely populated areas is increasing at a fast pace while right-of-way space for transmission lines is becoming increasingly more difficult to obtain. One method of materially increasing the transmission capability of long high voltage and extra high voltage lines is by the use of series capacitors. This method is presently being used on a number of lines in the western part of the United States and from a review of the future power requirements of the country, it becomes evident that the series capacitor will become an important tool for helping the transmission engineer move power from the point of generation to the consumer.

There are some applications of series capacitors which can present problems to the utility. One of these problems which, in the past, has not been too well understood is subharmonic resonance of systems where synchronous machines are closely coupled to transmission lines highly compensated with series capacitors. It is the purpose of this paper to acquaint the prospective user of this equipment with a simplified method of calculating whether or not self-excited subharmonic currents can be expected, and to acquaint him with corrective measures which may be taken in the event they are found necessary.

In the application of series capacitors to feeders or to individual machines such as a synchronous condenser[1], it has often been necessary to use a shunt resistance across the series capacitor to damp out oscillations. This is particularly the case if the machines are to be started with the capacitor in the circuit.

In the application of series capacitors to lines connecting large systems, it has, to date, apparently not been necessary to provide any damping other than the natural resistance of the lines. However, the possibility of getting self-excited oscillations in special cases make it worthwhile to have a general method of checking whether self-excited oscillations are possible in a given case, and to know what to do if oscillations are predicted.

Series capacitors reduce the ratio of x to r in the primary circuit and, hence, cause some increase in the tendency to hunt. This is generally negligible and it will be dealt with separately from subharmonic resonance in Appendix I.

Paper 70 TP 626-PWR, recommended and approved by the Transmission and Distribution Committee of the IEEE Power Group for presentation at the IEEE Summer Power Meeting and EHV Conference, Los Angeles, Calif., July 12-17, 1970. Manuscript submitted February 16, 1970; made available for printing June 1, 1970.

ANALYSIS OF SUBHARMONIC RESONANCE

A series capacitor applied to cancel a portion of the inductive reactance of the system will always result in a natural frequency (f_o) less than line frequency and this will be referred to as a subharmonic.

An oscillation at the subharmonic frequency will tend to be damped by the line and transformer resistance. However, any rotating machinery on the system will feed energy into the positive sequence subharmonic currents. This is due to the fact that the positive sequence subharmonic currents set up a field which rotates at a speed less than the machines, which are running at line frequency or with only a small slip for induction motors.

Visualizing the subharmonic currents in any machine as separate positive and negative sequence components at a frequency f_o, it is evident that the effects of the rotation of the machine is to act like an induction generator feeding energy into the positive sequence subharmonic because the rotor is running faster than the positive sequence subharmonic excitation. The negative sequence current running backward acts like a brake on the rotor, absorbing energy and helping to damp this component which usually disappears rapidly after a transient. However, the positive sequence might be poorly damped or it might be negatively damped if the machines could feed in more energy than is absorbed by the resistance loss in the lines and transformers.

If the machines are symmetrical on both axes, they will act like simple induction generators where the primary power generated is $I_o^2(R_r/S)$ where I_o is the oscillatory positive sequence subharmonic component of current, R_r is rotor resistance transferred to the primary, and s is the per unit slip for the subharmonic:

$$s = \frac{f_o - f}{f_o} \tag{1}$$

where f is line frequency.

For most purposes R_r can be taken as $2(r_2 - r_1)$ where r_2 and r_1 are the negative and positive sequence resistances, respectively. For the purpose of calculating f_o, the effective reactance of the machine is approximately equal to the negative sequence reactance (x_2). For machines with high resistance damper windings, the effective reactance is approximately equal to the transient reactance (x_d'). A more exact method for obtaining the R_r and x of the machine would theoretically be to apply a voltage to the locked machine at a frequency equal to slip frequency and of such magnitude as to give the current level desired. An equivalent test would be to measure torque and current at the right slip frequency. These tests would be very difficult to make but, fortunately, the machine designer can calculate these constants just as he would for starting calculations.

For special cases a correction factor K_u should be made to the effective R_r/s to allow for the dissymmetry on the two axes. Most synchronous machines have some dissymmetry on the two axes. The effect of this dissymmetry is to cause other components of current, which absorb more energy, in effect multiplying $I_o^2(R_r/s)$ by a factor K_u. It will be shown that for turbine generators and salient pole machines with low resistance dampers K_u can be taken as one. Values for K_u will be given along with typical values of R_r for different types of machines. The value of K_u is also affected by the ratio of total system reactance to the machine reactance as shown in Figure 1.

For a single machine connected to a very large system through a

Reprinted from *IEEE Trans. Power App. Syst.*, May/June 1971.

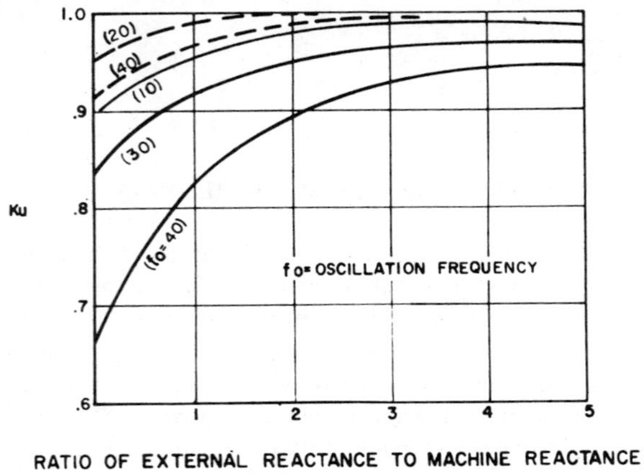

Fig. 1. Correction factor (K_u) to the effective negative rotor resistance (R_r/s).
$K_u = 1.0$ for turbine generators and waterwheels with copper damper windings.
-- Salient-pole machines with brass damper windings.
— Machines with high resistance damper windings such as high starting or low inrush synchronous motors or machines with nonconnected damper windings.

compensated line one would need only to compare the effective negative resistance $K_u(R_r/s)$ with the total positive resistance including line, transformers, and machine stator resistance. However, for a large complex system the total system may be represented on an analogue or digital computer and solutions obtained for the positive sequence subharmonic frequency currents (I_o) in each branch of the system resulting from a transient such as capacitor reinsertion. If there are several capacitors, there can be more than one natural frequency and for each the current distribution must be checked separately.

Having solved for the harmonic current distribution, one needs only to tabulate the $I_o^2 r$ losses in each branch and the $I_o^2 K_u(R_r/s)$ values for the machines to determine if at this frequency there is a net energy feed into the system, which would indicate sustained or increasing current oscillations.

To get the exact current distribution between machines with varied rotor resistances requires representation of these negative resistances in the solution of the network. Most analogue and digital systems are not set up to deal with negative resistances. A very good approximation can be obtained by representing the approximate machine R/s values as positive resistance. This will give the right distribution of current between machines and the $-I_o^2(R/s)$ values can then be determined at the observed oscillation frequency.

The effect of shunt loads on the system is generally to absorb energy at the oscillation frequency, and to neglect this effect is conservative. Small motors below 5 HP usually have sufficient stator and lead resistance to cancel the negative resistance R_r/s of the rotor. Larger motors may be grouped and represented as an equivalent machine.

Resistive loads can be grouped and represented as a shunt resistance. As an example, shunt resistive loads near a generator equal to 50 percent of its capacity, when paralleled with the generator reactance of 20 percent at a natural frequency of 30 Hz will produce a loss equivalent to a 2 percent series resistance. This value appreciably reduces the combined negative resistance of the machine and load.

TYPICAL MACHINE CONSTANTS

Typical values of rotor resistance for different types of machines are given in Table I. The values may also be approximated

TABLE I
TYPICAL VALUES OF ROTOR RESISTANCE

TYPE OF MACHINE	DAMPER MATERIAL	R_r p.u.
Waterwheels or Condensers with Connected Damper or High Pull-in Torque Synchronous Motors	Cu Brass	.01 .03
Synchronous Motors Low Inrush or High Starting Torque	High Res.	.07
Waterwheels with nonconnected Dampers or no Dampers (5% current)		.07*
Turbine Generators Conventional above 50 MVA		.030*
Large Conductor Cooled Generators, 2 Pole		.020*
Large Conductor Cooled Generators, 4 Pole		.025*

*Based on 30 Hz slip.

from the negative sequence resistance. For waterwheel generators with connected damper windings and for synchronous condensers, one can use $R_r = 2(r_2 - r_1)$ where r_2 and r_1 are the negative and positive sequence resistances, respectively, of the machine.

For salient pole machines with copper damper windings ($r_2 - r_1$) was increased by about 2 times due to eddy currents. Hence for machines with copper damper windings only, a better approximation is $R_r = (r_2 - r_1)$. For brass dampers the effect of eddy currents is much less, and for high resistance bars it is negligible. Typical values of rotor resistance R_r for machines with connected damper windings are: .01 p.u. for copper dampers, .03 p.u. for brass and .07 p.u. for higher resistance dampers. For non-connected damper windings (no connection between poles) the quadrature currents flow down the end of the pole and through the spider end plates. Since iron circuits are involved, it will have a non-linear rotor resistance. For the purposes of this discussion, it will be assumed that the subharmonic oscillating current should be limited to not more than 5 percent of rated current. At 5 percent current the end effects have been calculated for one construction to add about .06 p.u. to the value of R_r; hence, a typical value of R_r for a non-connected copper damper is .07 p.u. For water-wheel machines without dampers the calculations are less exact but an R_r of .07 p.u. would appear to be of the right order.

For turbine generators some of the induced rotor currents are in the solid iron pole center where resistance is again non-linear. There is a maximum R_r at a low current of about 20 percent. This maximum resistance is calculated for 30 Hz slip. For typical large conductor-cooled machines, R_r has been determined to be .02 p.u. For conventional turbine generators above 50 MVA, an R_r of .03 p.u. was found to be typical.

It is desirable to relate R_r to r_2 for turbine generators; however, both are affected by current magnitude. The maximum value of R_r/s has been calculated to occur at a subharmonic current of about 20 percent. This value of current will have the same depth of penetration into the rotor at 24 Hz slip as a negative sequence current at rated value. However, at this lower frequency the depth of penetration in the non-magnetic material of the rotor will be nearly twice as great as for the negative sequence currents. Since the loss in the magnetic steel usually predominates, the use of an R_r of $2(r_2 - r_1)$ will be pessimistic, and an R_r of $1.7(r_2 - r_1)$ would be more typical for values of f_o in the range of 20 to 40 Hz. For frequencies below 20 Hz, R_r will be closer to the value $2(r_2 - r_1)$. The above relations assume the r_2 values have been determined at a rated negative sequence current. Calculated values of r_2 are likely to be more accurate than any test values.

EXAMPLES

Two examples will be given to illustrate the method. These were chosen to be cases which might give trouble. Methods of correcting such possible troubles are suggested.

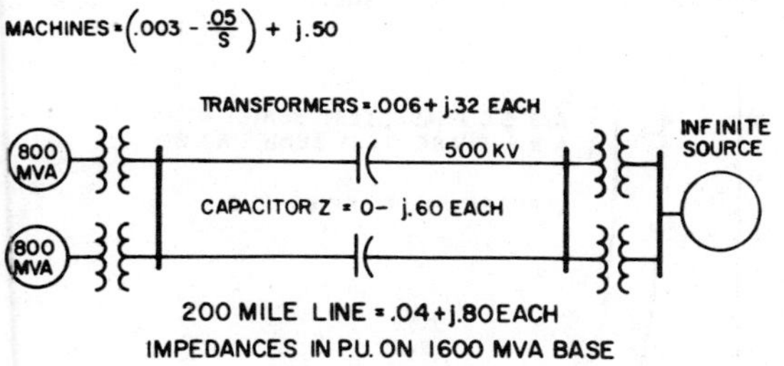

Fig. 2. Circuit for example, Case I.

CASE I: Figure 2 shows a mine mouth station connected to a large system through two 500 kV lines with 75 percent compensation. For both the machines and the lines, the total inductive reactance is .97 p.u. on a 1600 MVA base and the total series capacitive reactance is .30 p.u. The natural frequency f_o, of this circuit will then be

$$f_o = \sqrt{\frac{x_c}{x_l}}\ (f) = \sqrt{\frac{.30}{.97}}\ (60) = 33 \text{ Hz}$$

and the slip will be

$$s = \frac{33 - 60}{33} = -.82 \text{ p.u.}$$

With a typical turbine generator rotor resistance of .020 p.u., R_r/s is .0245 and is negative for the subharmonic. With a total positive resistance of the system of .0275 p.u. a net positive resistance (r) of .003 results; thus any subharmonic oscillations will be damped out.

In the event both lines are in service and one machine is off the line, the total inductive reactance is 1.38 p.u. The natural frequency will now be

$$f_o = \sqrt{\frac{.3}{1.38}}\ (60) = 28 \text{ Hz}$$

and the slip will be s = 32/28 = -1.14 p.u. The effective R_r/s for one machine on a 1600 MVA base is

$$(2)\ \frac{.020}{-1.14} = -.035 \text{ p.u.}$$

The positive resistance is .029 p.u. leaving a net negative resistance(r) of .006 p.u. Thus, subharmonic oscillations can be expected to build up.

This condition can be corrected by a shunt resistor (R) of 68 p.u. across the two capacitors or a 136 p.u. resistor across the capacitor in each of the two lines. If these values are used to stabilize the one machine case the losses at 60 Hz under the full 1600 MVA rating will be (quite accurately) $x^2/R = .30^2/68 = .0013$ p.u. If this loss is objectionable, a parallel reactor and capacitor, tuned to 60 Hz can be used in series with the shunting resistor to reduce the loss by blocking the 60 Hz while absorbing the subharmonic oscillating current. In order to keep the cost of the damping circuit components at a minimum the optimum is obtained when the resistance and the reactance of the paralleled capacitor and reactor are both equal, at frequency f_o, to one-half the single shunt resistor value. The KVA rating of the reactor and capacitor will normally be small and in the above example will be something less than one percent of the series capacitor KVA size.

Since the machine constants cannot be determined exactly and since it is usually desirable to have an appreciable net damping for transients, even more net positive damping should be provided than in the above example. The time constant of an oscillating circuit provides a measure of damping and is 2(L/r) seconds or $x/\pi r$ measured in cycles, where x is the reactance at the oscillating frequency f_o. Hence, the damping measured in per unit per cycle change is $\pi r/x$. At the oscillating frequency, x is equal to $x_c(f/f_o)$, where x_c is the capacitive reactance at 60 Hz. The damping for the positive sequence oscillations at frequency f_o may, then, be expressed as a per unit per cycle decrease equal to

$$d = \frac{\pi r f_o}{x_c f} \text{ p.u.} \qquad (2)$$

where r is the net circuit resistance.

CASE II: If a line between large systems has two or more series capacitor banks in series and a generating station, synchronous condenser, or large motor load is tapped between the two banks, a likely case for self-excited oscillations is established.

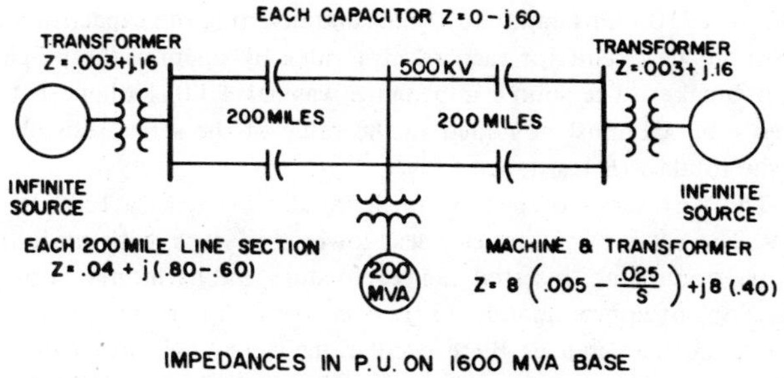

Fig. 3. Circuit for example, Case II.

Figure 3 shows a typical case with two 200 mile, 500 kV line sections, 75 percent compensated. The constants are given on the line rating of 1600 MVA. For this example the constants for a 200 MVA machine with its transformer is shown converted to 1600 MVA. As viewed from the tap, the line and system impedance is (.0115+j.28-j.15). When the machine and transformer are added the 60 Hz reactance is (j 3.48 - j .15).

By adding the stator and transformer resistances to the line resistance of Figure 3, a total resistance of 0.0515 p.u. on the 1600 MVA base is obtained. For a machine with an R_r of .03 p.u. on its own base (200 MVA), the net resistance is (.0515/8) - (.03/s). For an s of 4, the net resistance is a negative .001 p.u. Thus, subharmonic oscillations can be expected.

This condition can also be solved by shunt resistors across the series capacitor banks, but more losses will result than in the previous example. One corrective measure would be a small resistor slightly larger than the negative resistance, or about .0012 p.u. on a 200 MVA base, in series with the 200 MVA tap. If this loss is undesirable a small capacitor can be placed in parallel with a .0024 p.u. series resistor allowing the 60 Hz current to flow with negligible loss, still providing the desired damping.

One possible contribution to negative resistance in the tap would be the starting of large motors. Even a 5000 KVA synchronous motor started across the line would require additional damping, equivalent to 0.002 p.u. series resistance on the 200 MVA base. Appendix III describes the general method for determining the negative resistance effects of motors started in parallel with other running machines.

LABORATORY TESTS

Tests were conducted in the Laboratory to verify the mathematical analysis. A 750 horsepower, 2300 volt, salient pole, synchronous motor with a separate d-c field supply was available for these tests. This machine has an extremely high resistance damper winding giving maximum motor asymmetry. A one ohm capacitor was connected in series with a variable resistor and the two were connected in series with the motor. A circuit breaker was connected across the resistor and capacitor which would allow simulation of the insertion of a capacitor in a power line one end of which is connected to a generator and the other end to an effectively large system. This circuit is shown in Figure 4.

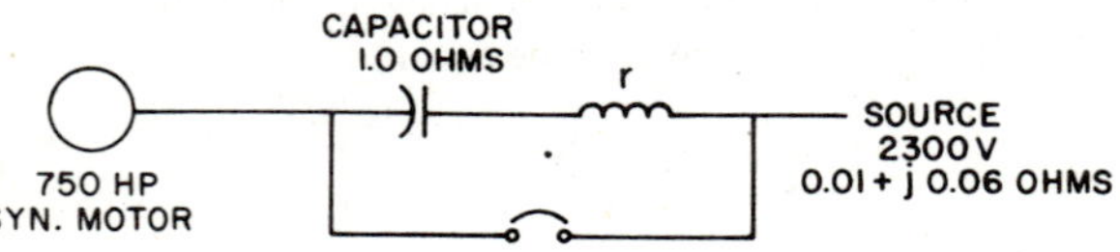

Fig. 4. Test circuit for application of a series capacitor in series with a synchronous machine.

The unloaded motor was started without the capacitor in the circuit. A series of tests were then made by varying the resistor between 0.910 ohms and 0.265 ohms and inserting the capacitor and resistor in the circuit for each resistor value by opening the by-pass circuit breaker. The source impedance was .01 + j.051 ohms so the source r of .01 must be added to the value of the series resistor to give the total series resistance.

The first series of tests were made starting with a total R of 0.919. The value of the resistor was lowered in steps, with capacitor and resistor being inserted on each step. Subharmonic current oscillations of approximately 36 Hz occurred on insertion, but at the higher circuit resistances these oscillations were small in magnitude and damped out in a very few cycles. Hunting (about 2 Hz) of the machine was quite evident at the higher resistances; however, the last evidence of hunting occurred with a series resistance of about 0.7 ohms or less. At about 0.32 ohms the 36 Hz subharmonic current failed to damp out and was continuous. Figure 5 is a typical oscillograph of the motor current showing the 36 Hz subharmonic current. Below 0.32 ohms a slight system disturbance would cause a rapid increase in the magnitude of the subharmonic current, thereby causing a major increase in voltage across the capacitor which resulted in the protective gap operation, taking the capacitors out of the circuit. The curve in Figure 6 shows the time required for the subharmonic oscillations to be damped out to one-third maximum value for various values of resistances.

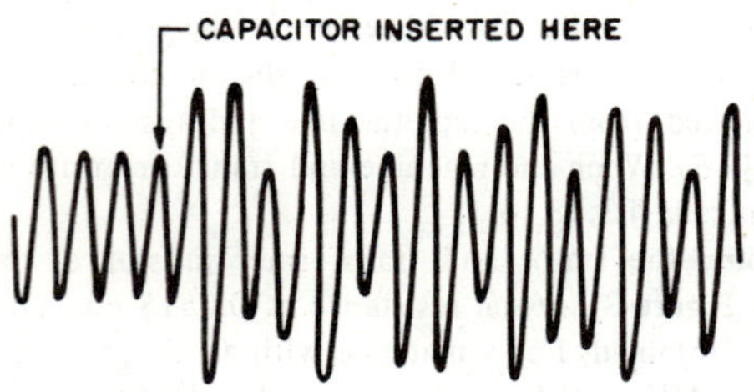

Fig. 5. Self-excited subharmonic oscillation current obtained from circuit in Figure 4. The frequency (f_o) is about 36 Hz.

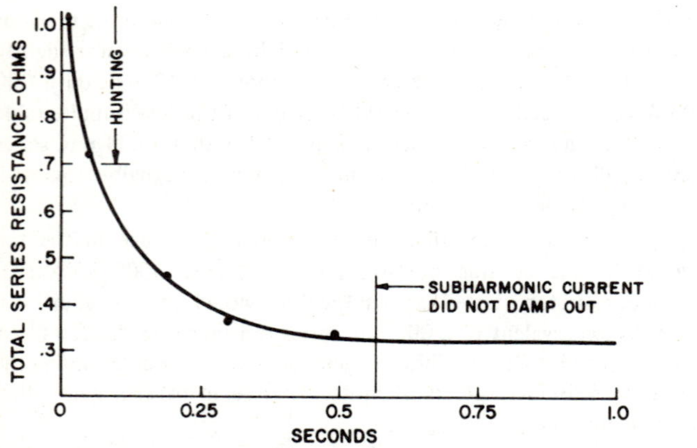

Fig. 6. Subharmonic oscillation damping time for the test case at various series resistance values.

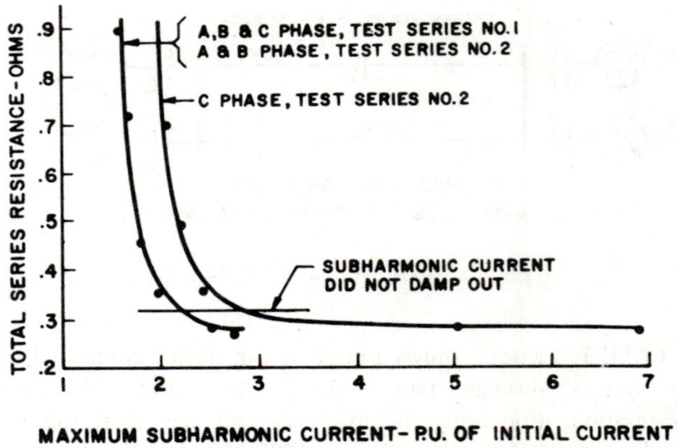

Fig. 7. The maximum subharmonic oscillation currents obtained from test case for various values of series resistance. For test No. 2 C phase of the breaker was delayed in opening, inserting the series capacitor about one-quarter of a cycle (60 Hz base) behind A and B phases.

A second series of tests were conducted to simulate the clearing of protective gaps across series capacitor banks where reinsertion of each phase was not exactly at the same time. The C phase pole of the breaker was delayed about one-quarter of a cycle on a 60 Hz base on opening for insertion of the capacitor. The magnitude of the A and B phase subharmonic current remained essentially the same as that in the first series of tests but C phase current showed a material increase in magnitude. This can be seen from the curves in Figure 7. The damping time of all three phases did not vary materially from the results shown in Figure 6 for the first series of tests.

CHECKING K_u BY TEST

The equivalent circuits for the test machine are given in Figure 8. The values of resistance and reactance are in per unit on a 60 Hz and a 560 kW base. The circuits consist of a direct and a quadrature axis equivalent. The direct-axis equivalent circuit consists of the direct-axis magnetizing reactance of 0.9 p.u. in parallel with a field impedance of (.007/s)+j.16 p.u. and a direct-axis damper impedance of (.26/s)+j.06 p.u., all in series with the stator leakage reactance of

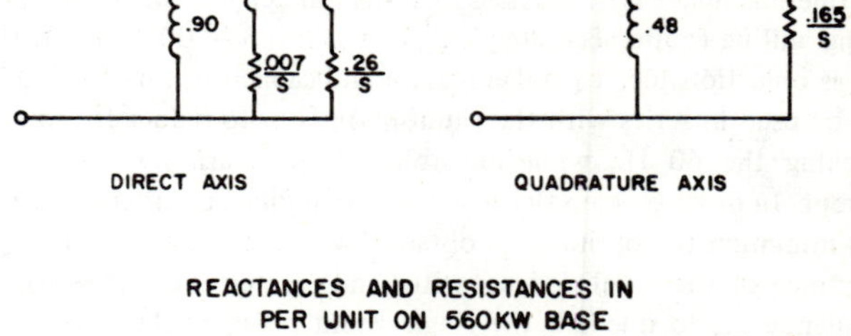

Fig. 8. Equivalent circuit for the test machine. All reactances and resistances are in per unit on a 560 kW base.

.07 p.u. The quadrature axis equivalent circuit consists of the quadrature-axis magnetizing reactance of 0.48 p.u. in parallel with the quadrature-axis damper impedance of (.165/s)+j.04, all in series with the stator leakage reactance.

Since the subharmonic oscillations were recorded at about 36 Hz, the slip is 60 - 36 = 24 Hz. At a slip of 24 Hz, with a unit volt per cycle applied to each of the equivalent circuits a direct-axis current (I_d) of 5 is obtained and a quadrature-axis current (I_q) of 2.5 is obtained. The reduction of these equivalent circuits give the direct axis an r_d of .011 p.u. and in the quadrature-axis an r_q of .088 p.u. The average of r_d and r_q is .05 p.u. K_i, the ratio of I_d to I_q, is 2.0, and K_r, the ratio of r_d to r_q, is .12. From Equation (10), Appendix II, K_u, the correction factor for rotor asymmetry, is .53. The value of the capacitive reactance was one ohm or .105 p.u.

The average machine reactance may be taken as the reciprocal of the average of the direct and quadrature axis currents. Thus

$$x = \frac{2}{I_d + I_q} = .265 \text{ p.u.}$$

The total circuit inductive reactance at 60 Hz for an external reactance of .01 p.u. is .275 p.u. Thus, the subharmonic oscillating frequency is

$$f_o = \sqrt{\frac{.105}{.275}}\,(60) = 37 \text{ Hz}$$

which, within the accuracy of the measurements, is in agreement with the test frequency.

For the frequency f_o the slip (s) is (60 - 37)/37 or .62 p.u. The effective negative rotor resistance for this frequency and slip is

$$K_u \frac{R_r}{s} = .53 \ \frac{.05}{.62} = .042 \text{ p.u.}$$

This is .40 ohms on the 560 kW base. The test value was .32 ohms plus .01 ohms in the source and .055 ohms in the machine stator. This gives a total of .385 ohms vs. a calculated .40 ohms, which is within the accuracy of the machine constants.

CONCLUSIONS

The use of series capacitors as a transmission tool has not been fully developed. If such applications are not to be limited by a concern for possible sustained harmonic oscillations a method of analysis and a method of correction are needed. Up to the present time it has, apparently, not been necessary to provide additional damping resistance to prevent self-excited subharmonic current oscillations on lines that are highly compensated with series capacitors. In special cases, however, the possibility of getting these subharmonic oscillations is very real, and this paper provides a practical method of determining the possibility of these sustained oscillations. This analysis is applicable to different system configurations as well as the individual machines.

The method involves determining a negative resistance (R_r/s) for each large rotating machine involved in that part of the system being investigated. In some cases it is shown that a solution requires the application of a correction factor (K_u) to allow for the dissymmetry in the direct and quadrature axes. This effect can be neglected for all machines except salient pole machines with non-connected damper windings or machines with high resistance damper windings.

Typical values of R_r are given in Table I, and approximate relations between R_r and the negative sequence resistance (r_2) have been developed for use where the r_2 values are accurately known.

A simple corrective measure to absorb the energy of the subharmonic current is the application of a shunt resistor across the series capacitor. In case the 60 Hz energy loss of the resistor may be uneconomical, a simple circuit may be installed to block the 60 Hz current through the resistor thereby reducing the power loss to a tolerable value.

The effect of series capacitors on hunting of synchronous machines is shown in Appendix I to be negligible in the normal range of compensation for transmission lines.

GLOSSARY OF SYMBOLS

- E_d' internal voltage back of the machine transient reactance
- f normal line frequency, 60 Hz
- f_o natural frequency of a circuit or the self-excited oscillation (subharmonic) frequency
- I_d direct axis current of a synchronous machine
- I_q quadrature axis current of a synchronous machine
- I_o positive sequence subharmonic oscillating component of current
- K_c constant defined as $x_c/(x_d' + x_e)$
- d damping factor
- K_i ratio of the direct axis current to the quadrature axis current, or I_d/I_q
- K_r ratio of direct axis resistance to the quadrature axis resistance, r_d/r_q
- K_u correction factor to the effective rotor resistance (R_r/s) for dissymmetry of the two axes of synchronous machines
- L inductance of a line, system, etc.
- r resistance of a conductor, external resistor, etc.
- r_1 positive sequence circuit resistance
- r_2 negative sequence circuit resistance
- r_d direct axis resistance of a synchronous machine
- r_q quadrature axis resistance of a synchronous machine
- R total circuit resistance
- R_r rotor resistance of a synchronous machine reflected to the stator
- s rotor slip of a machine
- T per unit torque of a synchronous machine
- S per unit speed of a synchronous machine
- x net circuit reactance
- x_c series capacitive reactance
- x_d' transient reactance of a machine
- x_e inductive reactance external to a machine
- x_2 negative sequence reactance

APPENDIX I
ANALYSIS OF HUNTING

Hunting is a low frequency oscillation involving a swing of the rotor about an average position. The damper winding usually provides sufficient damping such that these oscillations are not sustained unless excited by a regulator or by governor action. However, the stator circuit resistance does contribute some negative damping to this type of oscillation. This effect can be shown to be a function of r/x. The effect of series capacitors in reducing the reactance is to increase this negative damping effect, however small. To evaluate this effect one can again resolve the currents into components. This time the stator current will have components at frequencies equal to line frequency plus and minus the hunting frequency.

The damping coefficient for a rotating machine is given as the ratio of the change of per unit torque to the change of per unit speed ($\Delta T/\Delta S$). The negative damping coefficient due to stator circuit resistance (r_1) can be estimated on the assumption of constant flux linking the field winding[3] as

$$\frac{\Delta T}{\Delta S} = \frac{r_1 (E_d')^2}{r_1^2 + (x_d' + x_e)^2} \tag{3}$$

where r_1 is the primary resistance; x_d' the transient reactance; x_e the external inductive reactance; and E_d' the voltage back of transient reactance. This is equivalent to the slope of a curve of braking torque, due essentially to constant flux linking the field winding, vs. speed. This negative damping coefficient is subtracted from the positive rotor damping due to the stator flux cutting the rotor damping circuit. The lowest rotor positive damping occurs normally at light load where the coefficient can be estimated as

$$\frac{\Delta T}{\Delta S} \cong \frac{(E_d')^2}{2r_q} \tag{4}$$

where r_q is the quadrature damping resistance at the oscillation frequency and is approximately equal to two and one-half times the negative sequence resistance r_2 minus r_1.

Since the net frequency given by the line frequency minus the hunting frequency results in a higher capacitive reactance than that given by the sum of the line and hunting frequencies, the net reactance of the primary circuit is reduced more by the capacitive reactance value obtained from the difference of the line frequency and the hunting frequency than the sum. This has the effect of increasing the slope of the braking torque vs. speed curve between the speed points corresponding to the line frequency plus and minus the hunting frequency.

Hunting frequencies (Δf) are seldom over 2 Hz or Δ equal to .03 for 60 Hz line frequency. Let a factor K_c be defined as follows:

$$K_c = \frac{x_c}{x_d' + x_e} \tag{5}$$

where x_d' and x_e have been defined above and x_c is the series capacitive reactance at 60 Hz.

In order to simplify the development of a factor representing the increase in negative damping due to the addition of series capacitive reactance in the primary circuit, it is necessary to make several simplifying assumptions:

1. The hunting frequency (Δf) is small compared to the line frequency (f) hence, $1/(1+\Delta)$ will be approximately equal to $(1-\Delta)$.

2. The resistance squared $(r_1)^2$ is small compared with the square of the net circuit reactance.

The net circuit reactance at 60 Hz in terms of K_c is $(1 - K_c)(x_d' + x_e)$. As indicated previously, for any hunting frequency Δf, the negative damping is, in per unit, equal to the slope $\Delta T/\Delta S$ between the two points determined by the sum and difference of the line and hunting frequencies, or between the points at $(1+\Delta)f$ and $(1-\Delta)f$. At $(1+\Delta)f$ the total reactance is

$$(1+\Delta)(x_d' + x_e) - (1-\Delta)x_c.$$

In terms of K_c, as defined above, the net reactance at $(1+\Delta)f$ is

$$(1+\Delta)(x_d' + x_e)\left[1 - K_c\left(\frac{1-\Delta}{1+\Delta}\right)\right]$$

The torque due to the rotor flux only, at $(1+\Delta)f$ is

$$\frac{\left(\frac{r_1}{1+\Delta}\right)(E_d')^2}{\left(\frac{r_1}{1+\Delta}\right)^2 + (x_d' + x_e)^2\left[1 - K_c\left(\frac{1-\Delta}{1+\Delta}\right)\right]^2}$$

At $(1-\Delta)f$, the form of the above equation for the net circuit reactance and torque are the same except the plus Δ becomes a minus Δ. Expanding the factor

$$\left[1 - K_c\left(\frac{1-\Delta}{1+\Delta}\right)\right]^2$$

using the previous simplifying assumptions, one obtains

$$\begin{aligned}[1 - K_c(1-2\Delta)]^2 &= (1 - K_c + 2\Delta K_c)^2 \\ &= (1-K_c)^2\left[1 + \left(\frac{2\Delta K_c}{1-K_c}\right)\right]^2 \\ &\cong (1-K_c)^2\left[1 + \left(\frac{4\Delta K_c}{1-K_c}\right)\right]\end{aligned}$$

Since the total per unit increment of speed (ΔS) is 2Δ, the difference in torques at $(1+\Delta)f$ and $(1-\Delta)f$ divided by 2Δ yields $\Delta T/\Delta S$, or

$$\frac{\Delta T}{\Delta S} = \left(\frac{-r_1 (E_d')^2}{r_1{}^2 + (x_d' + x_e)^2}\right)\left(\frac{1+3K_c}{(1-K_c)^3}\right) \tag{6}$$

which is essentially the negative damping for no capacitive reactance, times the factor

$$(1+3K_c)/(1-K_c)^3.$$

This relationship, Equation (6) for the negative damping due to rotor flux alone must be compared with the positive damping due to the stator flux, Equation (4), to determine if the net damping is positive.

For typical machine constants Equation (4) gives a positive damping coefficient of about 10. With typical values of $(x_d' + x_e)$ of .09 p.u. and r_1 of .06 p.u.

$$\frac{\Delta T}{\Delta S} = .075\left[\frac{(1+3K_c)}{(1-K_c)^3}\right] \tag{7}$$

For this to be equal to 10 would require K_c to be about 0.7. Since the line reactance is seldom greater than twice the machine reactance, a K_c value of 0.7 represents an x_c greater than line reactance. Thus in any practical case more than 100 percent line reactance will be required to cause hunting. The exception to this would be waterwheel generators with no damper windings where, at light load, the damping might approach zero. Such a machine, however, would probably hunt at no load, even without series capacitors.

For K_c to be as much as 0.5, representative of series compensation of approximately 100 percent, and therefore, generally not typical, a negative damping coefficient of 2 is obtained from Equation (7). This value of negative damping only cancels about 20 percent of the positive damping of machines with typical damper windings at no load, and therefore, hunting of these machines should be little problem with series compensation.

APPENDIX II
DERIVATION OF K_u

The unsymmetrical rotor gives different values I_d and I_q for the currents on the two axes and these will encounter different values of rotor resistance (r/s) on the two axes; that is, r_d/s and r_q/s. These would be the resistances transferred to the primary. The primary loss is then

$$\left(\frac{I_d{}^2 + I_q{}^2}{2}\right)r_1 = \left[\frac{(I_d + I_q)}{2}\right]^2 r_1 + \left[\frac{(I_d - I_q)}{2}\right]^2 r_1 \tag{8}$$

and the rotor input to the stator is the torque times the speed. This as given by Park's formula, assuming constant primary excitation, is

$$\left(\frac{I_d{}^2 r_d + I_q{}^2 r_q}{2}\right)\left(\frac{f_o}{f - f_o}\right) \tag{9}$$

Now, if we define $K_i = \frac{I_d}{I_q}$ and K_r as $\frac{r_d}{r_q}$

Then $[(K_i{}^2 K_r) + 1]\, r_q = (K_i{}^2 + 1)\, r_1$

We will use $\frac{r_d + r_q}{2}$ as $R_r = (\frac{(K_r + 1)}{2}) r_q$ or

$$r_q = (\frac{2}{1 + K_r}) R_r$$

the point where energy fed out just equals energy absorbed, hence the balancing

$$r_1 = \frac{2\,[1 + (K_i^2 K_r)]}{(1 + K_i^2)(1 + K_r)} (\frac{R_r}{s}) \quad (10)$$

If we define K_u such that r_1 is $K_u(R_r/s)$ then

$$K_u = \frac{2\,[1 + (K_r K_i^2)]}{(1 + K_i^2)(1 + K_r)} \quad (11)$$

APPENDIX III
THE EFFECT OF MOTOR STARTING ON SUBHARMONIC OSCILLATIONS

The subharmonic oscillations associated with motor starting on a feeder with series capacitors has been discussed in Reference (2). The effect of starting a large motor which is parallel with other machines on a transmission line, or on a tap, as in the last example, is to increase the negative resistance.

This effect can be visualized by considering that the starting motor must pass through the speed corresponding to f_o, and at that point will have zero slip. Since the slip (s) becomes zero, the effective rotor resistance becomes infinity. Physically, this means that, at other speeds, the effective rotor circuit resistance can have any value from its value at full speed on up to infinity. When viewed from the machine bus, the rotor circuit reactance is in parallel with its own magnetizing reactance and the reactance (x_2) of the other parallel machines. The maximum negative resistance of this combination occurs when the effective rotor circuit (rotor resistance over the slip or R_r/s) is equal to the parallel reactance at the oscillation frequency. At this slip and frequency the combined machines will have an equivalent resistance and equivalent reactance equal to one-half the parallel reactance at f_o. However, for small machines, the minimum rotor resistance at full speed is still so high that when paralleled with the reactances of other machines at the oscillation frequency, the maximum effective negative resistance is not reached. Hence, the effective negative resistance of small machines can be determined as previously discussed for generators at full speed. Thus the consideration passes out of the motor starting problem.

REFERENCES

1. R. L. Witzke and E. L. Michelson, "Technical Problems Associated with the Application of a Capacitor in Series with a Synchronous Condenser", AIEE TRANS., vol. 70, pp. 519-525, 1951.
2. C. F. Wagner, "Self-Excitation of Induction Motors with Series Capacitors", AIEE TRANS., vol. 60, pp. 1241-47, 1941.
3. L. A. Kilgore and E. C. Whitney, "Spring and Damping Coefficients of Synchronous Machines and Their Application", AIEE TRANS., vol. 69, pp. 226-230, 1950.
4. C. F. Wagner, "Machine Characteristics", Transmission and Distribution Reference Book, Westinghouse, pp. 145-194, 1950.
5. H. M. Rustebakke and C. Concordia, "Self-Excited Oscillations In A Transmission System Using Series Capacitors", IEEE 70 TP 69-PWR.

SELF EXCITED TORSIONAL FREQUENCY OSCILLATIONS WITH SERIES CAPACITORS

Colin E. J. Bowler Member, IEEE, Donald N. Ewart Member, IEEE, and
Charles Concordia Fellow, IEEE

Abstract—The application of series capacitors to compensate long-distance high-voltage transmission lines introduces potential modes of dynamic instability caused by interactions with synchronous machines and turbine-generator shaft torsional resonance. This paper describes the modes of instability and analytical methods used to predict the combinations of parameters which may lead to instability. Insight into the phenomena of modal interaction, a factor causing one of the unstable modes, is provided.

INTRODUCTION

The application of series capacitors to compensate the inductive reactance of long transmission lines for improved stability is being pursued by several utilities in the United States and many other countries.

A recent incident of generator damage has prompted further investigation into the dynamic behavior and stability of series-capacitor-compensated power systems.

The contribution of this paper is showing that there are regions of instability occasioned by the dynamic interaction of the power system and the torsional mechanical modes of a turbine generator. It will be shown that when the line natural frequency of the generator-transmission system seen by the generator rotor is close to a torsional natural frequency the system may become unstable even in otherwise stable regions. Actual loss of stability depends on the system external resistance, the mechanical damping and the rotor resistance. Procedures for calculating the degree of stability of the system as well as the influence or corrective measures are described in this paper.

From a physical point of view the fundamental phenomenon of electrical self-excitation of a rotating machine (motor or generator) may be regarded (at least in the frequency range of interest) as operation as an induction generator. For a constant rotor speed and linear system characteristics the self-excited currents may be initiated and grow quite independently of any applied armature voltage or field excitation. If there are no applied voltages the machine will develop a torque which would in practice grow to some constant value. However if there are applied voltages, the self-excited currents will interact with the normal rated frequency currents to generate a pulsating torque, which would be primarily at slip frequency. It is in principle this pulsating torque which may excite significant torsional movement of the rotor, if the slip frequency is near torsional resonance. Then, in turn, the now appreciable voltages generated by the rotor oscillations will themselves produce relatively large currents which may then be said to reinforce the self-excited currents.

Previous literature concerned with the dynamic stability of series compensated systems [1,2,3] has not accounted for the torsional mechanical modes but only for the possible oscillations of the rotor as a whole against the system electrically. It centered principally on the instability caused by potential line current resonance for which the generator appears to have negative resistance due to induction generator action. When this negative resistance exceeds the positive resistance of the line, at the resonant frequency, instability was assured. This problem is more pronounced as compensation becomes large since the negative resistance of the generator increases with decreasing slip.

Paper T 73 218-5, recommended and approved by the IEEE Power Generation Committee of the IEEE Power Engineering Society for presentation at the IEEE PES Winter Meeting, New York, N.Y., January 28–February 2, 1973. Manuscript submitted September 21, 1972; made available for printing December 13, 1972.

It was also shown that for high armature circuit resistance the line natural frequency as seen by the generator rotor (natural freqeuncy compliment) becomes close to the natural hunting frequency of the generator with respect to the rest of the system.

The torsional interaction instability described in this paper is very similar in principle to this latter case since the first torsional oscillatory mode is similar in character to the mode in which the turbine generator oscillates as a whole with respect to the system.

Systems Studied

Three examples are used here to show the regions of stability from a parametric study of the system variables. The first example, a simplified hydro-generator system, Fig. 1, extends the results of [1] to show continuity between modes of oscillation and broaden the base for understanding these phenomena. It serves to highlight the need for a comprehensive analysis approach.

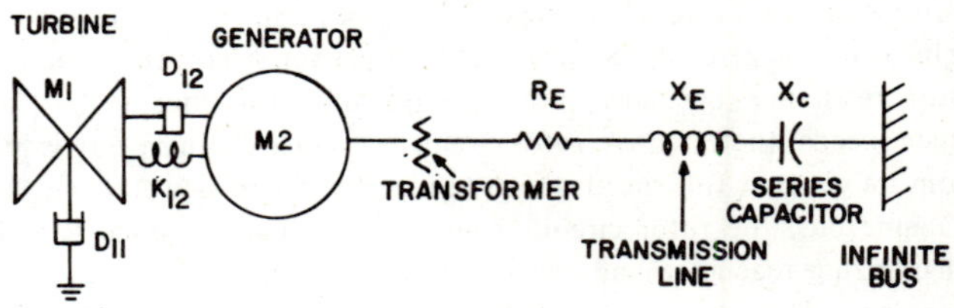

Fig. 1. Radial series compensated transmission line with generator electro-mechanical dynamics.

The second example is a duplication of the analysis above but represents a tandem-compound fossil generating unit. This case appears to be much worse from a stability point of view than the hydro system considered.

The third example is a departure from the synchronous machine model to a simple system of coupled resonant circuits. The point here is to show that this problem of mode interaction is associated with a more general physical phenomenon.

Method of Analysis

The method used to investigate the stability of these three examples is an eigenvalue analysis of the Nth order differential equations describing the system dynamics for small perturbation about the operating point.

Equation 5 of Appendix A shows that the eigenvalues (λ) or roots of the system are obtainable by solving for λ when $|A-\lambda I| = 0$, where A is the matrix of coefficients of the N first order differential equations of the system. These eigenvalues represent the character of the normal modes of the small-signal time response where the time response of a particular state variable is expressible as some linear combination of the eigenvalues. This condition is given mathematically as:

$$x_n = \sum_1^N a_n \; e^{-\lambda_n t} = \sum_1^N a_n \; e^{-(\sigma_n + j\omega_n)t}$$

σ_n and ω_n are the real and imaginary parts of the eigenvalue λ_n; σ_n is the decrement factor and ω_n is the natural frequency of the oscillatory mode. ω's are either zero or occur as complex conjugate pairs.

These roots are best portrayed in the two-dimensioned complex-frequency plane or s-plane as shown in Fig. 2. Also shown in the character of the time response associated with certain locations of the roots in the s-plane.

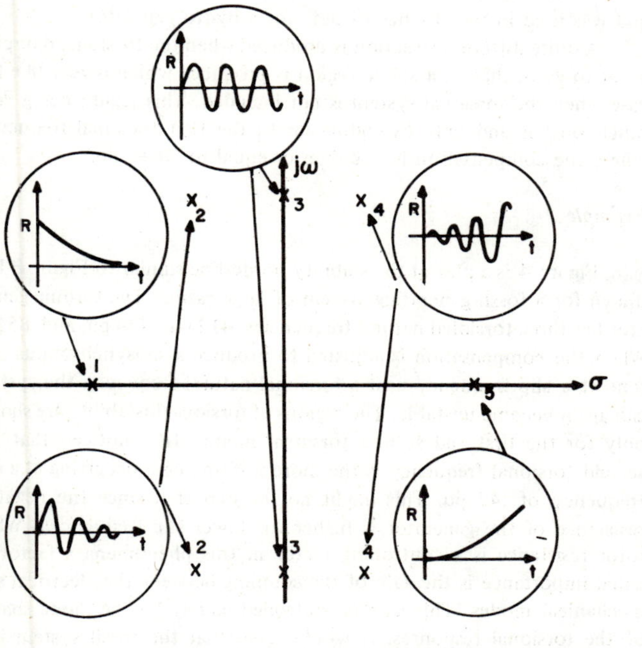

Fig. 2. S-plane showing possible root locations with time response. Roots in the left-half plane are stable and roots in the right half plane are unstable.

In summary, roots located on the left half plane ($-\sigma$) are positively damped and roots in the right half plane ($+\sigma$) are negatively damped.

1. For ω=0, σ<0 a damped unidirectional component is produced.
2. $\omega \neq 0$, σ<0 a damped sinusoid is produced.
3. ω=0, σ=0 a constant amplitude sine wave characterizing the borderline of stability.
4. $\omega \neq 0$, σ>0 undamped sinusoid increasing without bound.
5. ω=0, σ>0 undamped unidirectional time response.

The results portrayed in this paper concern the roots of Type 3 representing the stability borderline and also the migration of roots in the s-plane (root locus) due to change of a system parameter.

To study the problem at hand the schematic representation of Fig. 1 depicts the extent of the system model.

In the analysis the effects of the voltage regulator has been neglected as have the effects of line shunt capacitance and possible shunt reactance. Even though the dynamics of voltage regulators are important due to the application of speed stabilizers [4] the problem is more readily understandable by their neglect; for a practical system study they should be incorporated.

Equation 6 of Appendix A shows that the system of Fig. 1 will result in twelve eigenvalues; 4 of these will be real, and the remaining 8 occur in complex conjugate pairs. A typical choice of parameters results in four real damping constants representing the rotor circuit time constants; two line natural frequencies and two natural frequencies for the inertias.

The natural frequency of line current oscillations proportional to $\sqrt{Xc/X_E}$ appears as two frequencies when viewed from the rotor of the synchronous machine because the rotor is travelling more or less at synchronous speed. These two natural frequencies are referred to as the line natural frequency compliments given by $\omega_1 = 60 - \sqrt{Xc/X_E}$, $\omega_2 = 60 + \sqrt{Xc/X_E}$ the lower natural frequency being responsible for the torsional interaction. The first mode of oscillation of the inertias corresponds to the local oscillation of the two inertias swinging in unison against the system electrical stiffness: this is the familiar hunting frequency. The second mode of oscillation of the inertias is characterized by the inertias swinging against one another due to the torsional stiffness of the connecting shaft. This mode is the first mechanical torsional mode of vibration, its undamped natural frequency given by:

$$\omega_{T1} = \sqrt{\frac{K12}{Meq}}$$

where

$$M_{eq} = \frac{M1 * M2}{M1 + M2}$$

For the case of more than two turbine generator inertias there are correspondingly more modes of oscillation, in fact equal to the number of connected inertias.

Three forms of dynamic instability can be associated with the system of Fig. 1.

1. Steady-State Instability

Instability in which rotor angle oscillations with respect to the infinite bus builds up with time. In this form, the mass of the turbine and the generator move together at the hunting frequency which may range from .2 to 2 Hz depending on the inertia of the set and the impedance of the transmission system. Power system speed stabilizers are sometimes used with the excitation system to reduce the likelihood of this form of instability.

2. Electrical Self-Excitation

Instability involving a resonance between the net series inductance and capacitance in the circuit. Depending on the magnitude of these parameters, a potential resonance may exist somewhere between 20 and 40 Hz. The resonance can become unstable and break into self-excited oscillations if the net negative resistance of the generator, viewed as an induction generator at the resonant frequency, exceeds the net positive circuit resistance [1,2].

3.Mode Interaction

Instability caused by the bilateral (two-way) coupling of the shaft torsional modes with the mode described in 2 above. This form will occur when both the torsional mode and the electrical mode are, by themselves, stable. The frequency of oscillation will be close to that of the torsional mode.

It is this last type of instability that is of principal concern, although the analysis used detect the other two modes as well.

The two synchronous machine examples will use the following typical parameters shown in Table 1. The hydro system data is similar to Ref. 1. The fossil generating system data is given on a 500 KV 500 MVA base, the transmission system is a typical 500 KV 70% compensated 170-mile bundled conductor line.

EXAMPLES

Example 1

Figure 3 is a plot of the stability borderline of the Hydro System as a function of the transmission line resistance and capacitive reactance for four system conditions.

Without the torsional dynamics represented the curve is similar to those of reference (1) where for small values of resistance the stability

TABLE 1

Parameter		Example 1 Hydro Sys.	Example 2 Fossil Sys.
XFD	Field Leakage Reactance	.25	.10197
X1D	D Axis Rotor CCT Reactance	.16	.08471
XL	Armature Leakage Reactance	.153	.16
XAD	D Axis Magnetizing Reactance	.847	1.64
X1Q	Q Axis Rotor CCT Reactance	.26	.12532
X2Q		0	.07352
XAQ	Q Axis Magnetizing Reactance	.507	1.56
RFD	Field Resistance	.000392	.000911
R1D	D and Q Axis Rotor	.027	.04591
R1Q	Circuit Resistances	.034	.00388
R2Q		00	.05027
AR	Armature Resistance	.005	.000696
RE	Line Resistance	.03288	.012
XE	Line Reactance	1.1224	.4976
M1		2.4825	.37674
M2	Turbine Generator	4.1375	.67017
M3	Inertias	0	1.5124
M4		0	.008
K12	Turbine Generator	22.5	47.416
K23	Shaft Stiffness	0	61.790
K34		0	4.540
D11		1.0	.55
D22	Turbine Generator	0	.449
D33	Load Damping	0	0
D44		0	0
D12	Turbine Generator	.6	.6
D23	Torsional Damping	0	.6
D34		0	.05

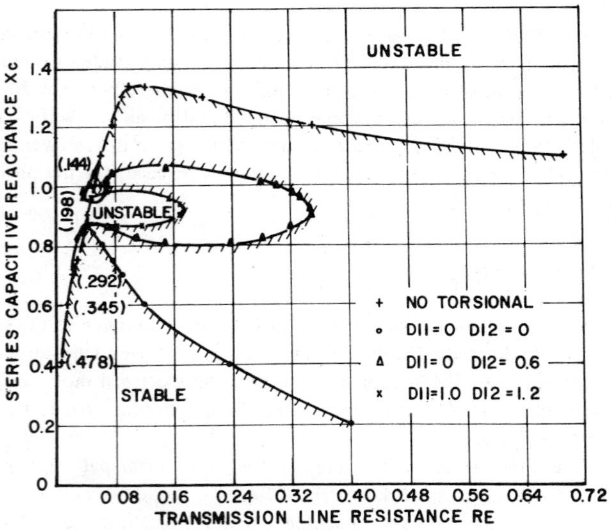

Fig. 3. Plot of stability borderline for the Hydro System as function of mechanical damping. The shaded side of curve is the stable region, Electrical slip frequency indicated in parenthesis.

limit is independent of load and is a dynamic instability of Type 2. For large values of resistance the system again becomes unstable and is an instability of Type 1. Comparison of this curve with Reference (1) for the same load condition will show some differences, this is due to our neglect of the shunt reactive elements of the line. This curve is presented to provide continuity between the paper and Reference 1 which used a different procedure to detect instablity.

The remaining three curves include the torsional representation and it can be seen that its inclusion introduces an extra region of unstable operation. This is an instability of Type 3 occurring at the torsional frequency of .198 pu which corresponds approximately to the system electrical resonant frequency compliment at Xc = .9. It is seen that this extra region of instability is dominated by the mechanical damping, generally a very small quantity. From the two curves plotted with D12 = .6 and D12 = 1.2 a large sensitivity to damping is noted in the stability boundary.

A value of D12 of .6 → 1.2 is typical of fossil turbine generators and was used in the absence of data for a hydro generator.

A quite different situation is produced when the torsional damping is set to zero; that is, a stable region is produced which is very like the case when the torsional system is not included. This region being very much smaller and only extending up to the first torsional frequency where the compensation is less than or equal to Xc = .86.

Example 2

Figure 4 is a plot of the stability borderline similar to Figure 3 but drawn for a fossil generating system of large rating. The turbine generator has three torsional natural frequencies .412 pu, .474 pu, and .85 pu. When the compensation is adjusted to produce a subsynchronous current at a slip frequency near a torsional natural frequency the system can again become unstable. The regions of torsional instability are shown only for the first and second torsional modes. It is noticed that the second torsional frequency is the more limiting case occurring at a slip frequency of .47 pu. This might not be expected since the negative resistance of the generator is higher for lower slip frequencies. While rotor resistance is a controlling factor in this phenomena a factor of equal importance is the gain of the coupling between the electrical and mechanical modes. This gain is controlled largely by the mode shapes of the torsional responses. It would seem that the fossil system is a more severe problem than the hydro system because coincidence of the slip frequency and the torsional frequency can occur at realistic values of line compensations.

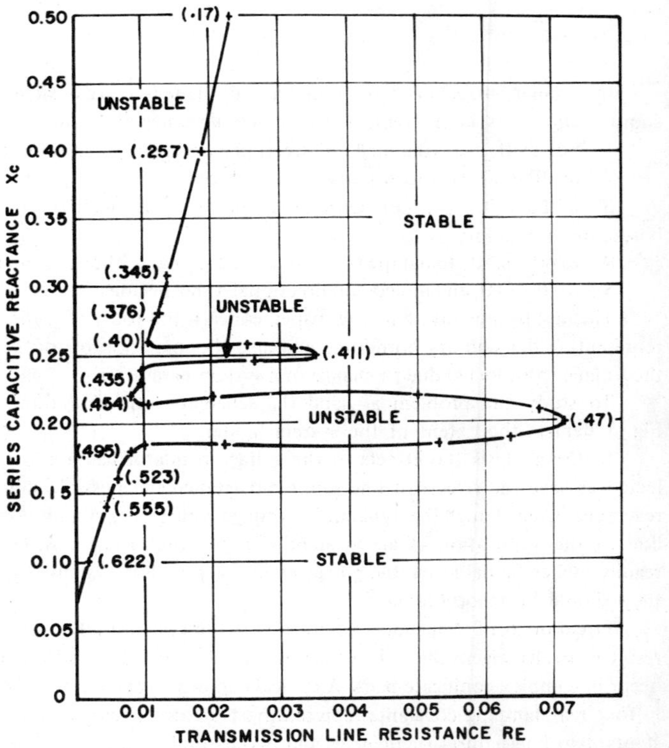

Fig. 4. Plot of stability borderline for fossil system. Electrical slip frequency pu in parenthesis.

Figure 5 is a root locus plot for the fossil system drawn for RE = .0125, 0 < Xc < .5 to indicate the relative stability of the torsional system and the electrical system as functions of series compensation. To facilitate interpretation, the real part of the torsional roots were plotted as a function of electrical frequency rather than the torsional frequency.

This figure shows that the real parts of the roots nominally associated with the torsional modes move into the right half plane and back as the oscillation frequency of the electrical mode approaches and passes those of the torsionals. Instability of Type 3 is indicated on this figure for values of Xc which produce real parts of roots in the right half plane. Thus, an instability can be created by the act of coupling two otherwise stable resonant modes.

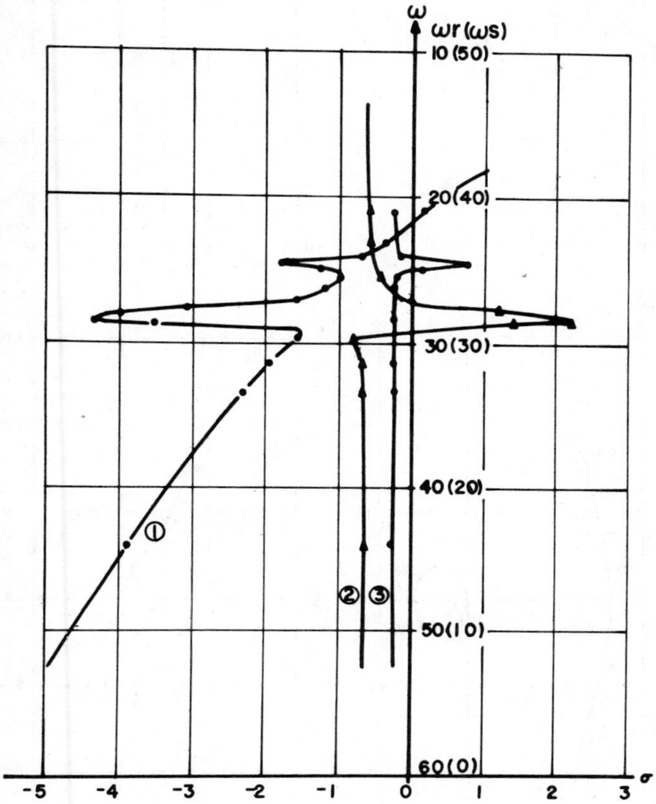

Fig. 5. Root locus plot for fossil system RE = .0125
(1) Electrical System Root Locus
(2) 2nd Torsional Damping Locus vs. Electrical Frequency
(3) 1st Torsional Damping Locus vs. Electrical Frequency

Example 3

In order to provide additional physical insight into the behavior of coupled resonant elements, a stability analysis of a system of two RLC circuits was performed. The circuit studied is shown in Fig. 6. Both of the RLC circuits illustrated contain a generator which produces a voltage proportional to the instantaneous current in the other circuit. This is the only coupling between the two circuits.

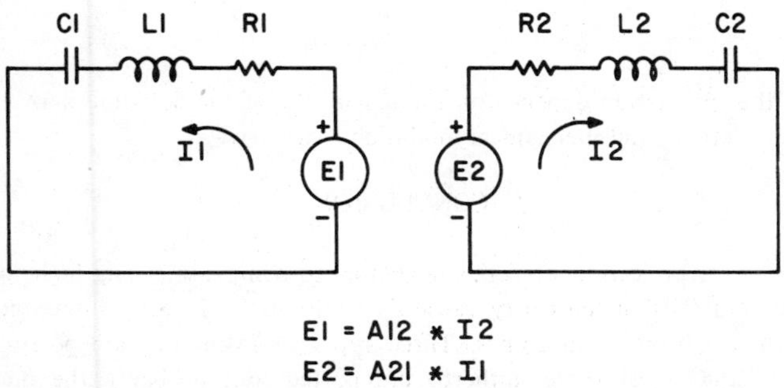

Fig. 6. Bilaterally Coupled RLC Circuits. Voltage Source in each circuit is proportional to current in other circuit.

The analysis consisted of determining the eigenvalues of this system as the resonant frequency of Circuit 2 was swept past the resonant frequency of Circuit 1. For simplicity without loss of generality, L1, C1 and L2 were kept at 1.0. In any particular run, R1, R2, A12 and A21 were held fixed while C2 was varied from .95 to 1.05. Values of R1, R2, A12 and A21 were changed from run-to-run. The questions which apply are developed in Appendix B.

In the absence of coupling between the circuits, (i.e. A12=A21=0) the eigenvalues simply represent the resonant frequency and damping of the two circuits. For example, with R1=R2=.01 ohms, the resonant frequency, ω, of Circuit 1 will be nearly exactly 1 radian/sec. (1 cycle in 6.28 seconds) and the damping constant, σ; will be .005 sec.^{-1}. Thus, one pair of eigenvalues will appear at .005±j1.000. The resonant frequency of the Circuit 2 will change from 1.026 to .976 Rad/sec. as C2 is swept from .95 to 1.05. The damping constant will remain at .005 sec^{-1}. A plot of these eigenvalues is shown in Figure 6-a. If a time response were obtained, it would show that both circuits exhibit damped sinusoidal behavior.

Next a case of unilateral coupling was investigated. Unilateral coupling means that either A12 or A21 is zero but the other is finite. Interestingly, the eigenvalues in the case are the same as if no coupling exists, that is, Figure 6-a applies to unilateral coupling as well as to zero coupling. In this case, however, a time response would show the presence of two frequencies in the circuit into which the coupling was made.

When bilateral coupling exists, however, the locus of eigenvalues as C2 is varied indicates that both eigenvalues shift. This is illustrated in Figure 6-b for a case in which A12=A21=.02. Examination of this plot indicates that the eigenvalues tend toward one another in frequency as C2 in increased toward 1.0, and that they become equal at C2 equal about .962. Further increases in C2 toward 1 cause the eigenvalues to separate in damping. When C2 is between about .968 and 1.035, one of the eigenvalues is to the right of the imaginary axis indicating undamped sinusoidal performance. This case clearly demonstrates the phenomena which is of concern in the torsional-electrical system interaction; namely the two resonant elements both of which are damped by themselves, can exhibit undamped or unstable behavior when bilateral coupling exists.

Further studies, not illustrated, showed that the magnitude of interactive effect is proportional to the square root of the product of the coupling coefficients. It has been found that changing the sign of both coefficients resulted in no change in the eigenvalues.

Changing the sign of one of the coupling coefficients however had a major effect as illustrated on Figure 6-c. In this case the eigenvalues tended to move apart in frequency. One eigenvalue comes down from the top and approaches .005+j1.0 as C2 gets large while the other starts at .005+j1.0 and moves down. A study of this behavior will convince one that it is not possible to associate an eigenvalue with one or the other of the circuits.

Figures 6-d through 6-g illustrate cases in which the resistance of one circuit is .02 while the other is zero. Both positive and negative coupling is shown. Figure 6-f shows a case which is similar to that illustrated on Figure 5.

The following generalizations may be made from the results of the analysis of coupled LRC circuits.

1. Bilateral coupling of two resonant elements tuned to about the same frequency causes the eigenvalues, or natural modes, to shift.

2. The magnitude of the shift depends on the square root of the product of the coupling coefficients. (If the coupling is unilateral, no shift results).

3. The direction of shift depends upon the sign of the product of coupling coefficients.

a. If the sign is positive, the eigenvalues tend toward one another in frequency but away from one another in damping.

b. If the sign is negative, the eigenvalues move away from one another in frequency but toward one another in damping.

It is concluded that the process of eigenvalue shift as resonant elements are coupled bilaterally is a fundamental physical phenomenon.

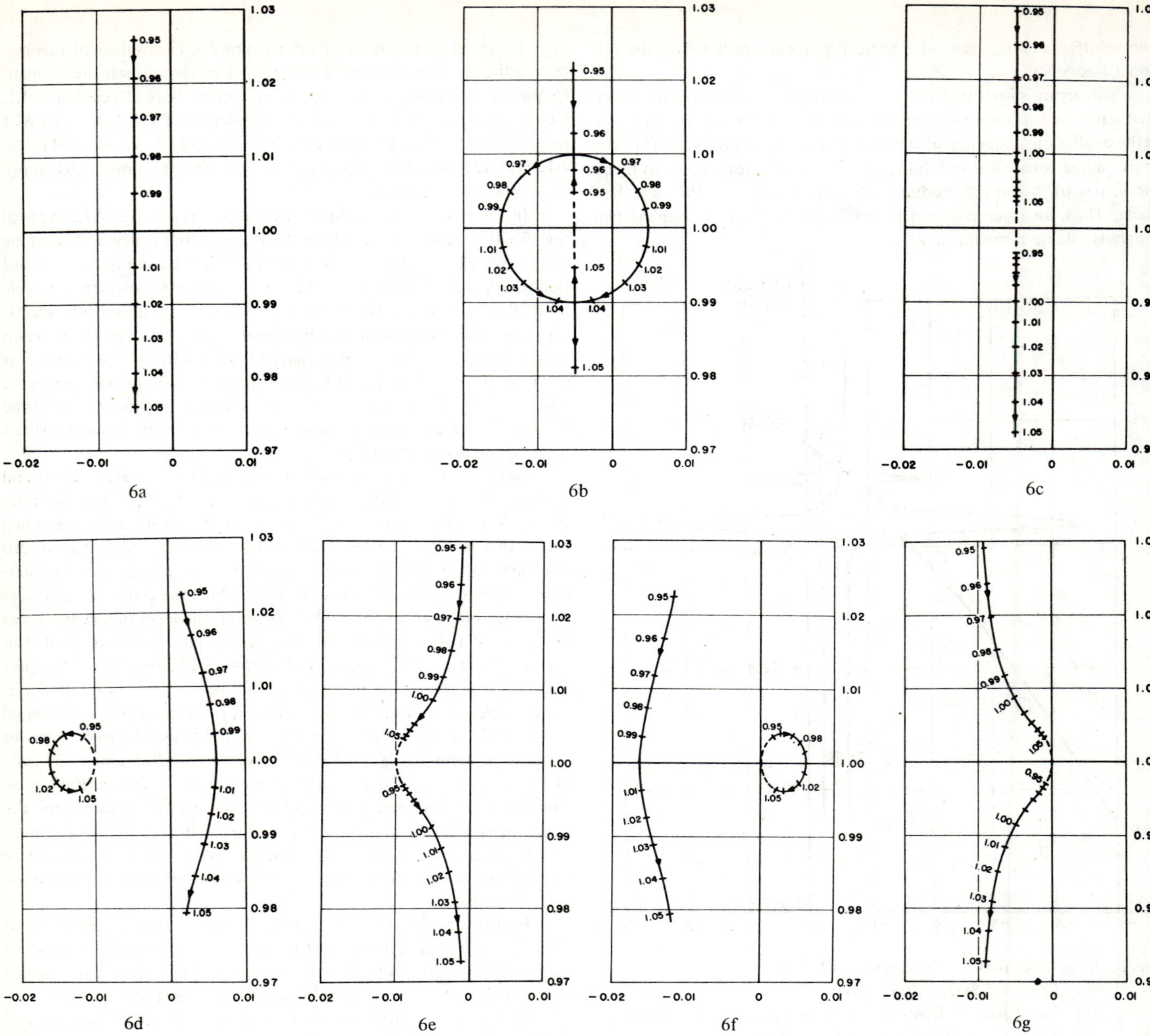

Fig. 6a-g. Locus Of Eigenvalues Of Coupled RLC Circuits As C2 Is Varied From .95 to 1.05. Axes are (damping) horizontally and (frequency) vertically. Only positive imaginary roots are shown.

6a. No coupling or unilateral coupling.
R1=R2=.01 ohms.
6b. Positive Bilateral Coupling.
R1=R2=.01 ohms A12=A21-.02
6c. Negative Bilateral Coupling
R1=R2=.01 ohms A12=.02, A21=-.02
6d. Positive Bilateral Coupling
R1=.02, R2=0 ohms A12=A21=.02
6e. Negative Bilateral Coupling
R1=.02, R2=0 ohms A12=.02, A21=-.02
6f. Positive Bilateral Coupling
R1=0,R2=.02 ohms A12=A21=.02
6g. Negative Bilateral Coupling
R1=0,R2=.02 ohms A12=.02, A21=-.02

Similar effects have been described when mass-spring oscillators are coupled and indeed when atoms are brought into close proximity to one another (Ref. 6). This latter effect is the basis of the Pauli Exclusion Principle and the theory of semi-conductor operation.

Summarizing this example, the bilaterally coupled RLC circuits behave like the torsional-electrical system when the product of the coupling coefficients is positive. In both cases, two resonant elements, both of which are damped when uncoupled, can exhibit undamped performance when their frequencies are close to one another. The nature of the interaction depends on the magnitude of the coupling between the electrical and mechanical modes and vice versa.

CONCLUSIONS

1. Application of series-capacitors to compensate long high-voltage transmission lines may cause dynamic instabilities by interaction with synchronous machines. Three types of instability are possible.

2. One of these, hitherto unreported, but probably the most

serious, involves an interaction between an electrical resonant mode and the torsional resonant modes of turbine generator shafts.

To study this mode, it is necessary to account for the electrical as well as the mechanical parameters of the system.

3. Eigenvalue analysis, as outlined in the paper, provides a valuable means to study all three types of instability.

4. The interaction of coupled resonant modes is a fundamental physical phenomenon, (but generally unrecognized by engineers).

5. Additional mechanical damping would appear to be the key to maintaining torsional stability if electrical system resistance is to be minimized.

6. An analytical technique has been demonstrated to further develop the art of successful application of series compensation to transmission systems. Torsional critical frequency dynamic interaction is just one more problem to be faced when designing systems of this type.

APPENDIX A

The use of an eigenvalue analysis to solve the dynamic stability problem requires the definition of system equation of state. The system equations are written from the point of view of an observer on the generator rotor and the coordinates used are based on the two reaction theory of the electrical machines. The system equations initially take the form of

$$[H][\dot{X}]=[F][X] \quad \text{for the homogeneous solution of the differential equation.} \qquad 1.$$

by simple manipulation the state equations can be obtained of the form

$$[\dot{X}]=[H^{-1}][F][X] = [A][X] \qquad 2.$$

where A is the state matrix.

To find eigenvalues we require the solutions for [X] of the form

$$[X] = [\bar{X}]e(\lambda)t \qquad 3.$$

By substitution of 3 into 2 it is obtained that

$$[A-\lambda I][\bar{X}]=0 \qquad 4.$$

Equation 4 has solutions for $X \neq 0$ when

$$|A-\lambda I|=0 \qquad 5.$$

The solution of 5 is the general eigenvalue problem where the λ's are the system eigenvalues.

The problem at hand can be written where

$$[X]^T=(ifd,ixd,i1q,i2q,Ved,Veq,id,iq,\delta 1,\delta 2,n1,n2] \qquad 6.$$

is the system state vector chosen for analysis.

APPENDIX B

Equations for Example 3

Letting

x_1 = voltage across capacitor $C_1 = \frac{1}{C1} \int I_1 dt$

x_2 = Flux linkage in $L1 = L1 \cdot I_1$

x_3 = Voltage across capacitor $C2 = \frac{1}{C2} \int I_2 dt$

x_4 = Flux linkage in $L2 = L2 \cdot I_2$

The differential equations which apply to this system, written in state-space form are as follows:

$$\begin{bmatrix} \dot{x}_1 \\ \dot{x}_2 \\ \dot{x}_3 \\ \dot{x}_4 \end{bmatrix} = \begin{bmatrix} 0 & \frac{-1}{L1*C1} & 0 & 0 \\ 1 & -\frac{R1}{L1} & 0 & \frac{A12}{L2} \\ 0 & 0 & 0 & \frac{-1}{L2*C2} \\ 0 & \frac{A21}{L1} & 1 & -\frac{R2}{L2} \end{bmatrix} \begin{bmatrix} x_1 \\ x_2 \\ x_3 \\ x_4 \end{bmatrix}$$

ACKNOWLEDGEMENT

The authors wish to thank Mrs. Valerie Monaghan for her efforts in the preparation of computer programs and making runs associated with this work.

REFERENCES

1. *Self-Excited Oscillations In A Transmission System Using Series Capacitor,* H.M. Rustebakke, C. Concordia, Paper 70 TP 69-PWR, IEEE Trans. of Power Systems, Vol. PAS-89, No. 7.
2. *The Prediction and Control of Self-Excited Oscillations Due To Series Capacitors In Power Systems* L.A. Kilgore, L.C. Elliott, E.R. Taylor, IEEE Trans. on Power Systems, Paper 70 TP 626-PWR,
3. *Negative Damping of Electrical Machinery,* C. Concordia, G.K. Carter, AIEE Transactions, Paper 41-12, March 1941, Vol. 60.
4. *Static Exciter Stabilizing Signals on Large Generation – Mechanical Problems,* W. Watson, M.E. Coulter, Paper No. T 72 228-0, Presented at the 1972 IEEE Winter Power Conference.
5. *Two Reaction Theory of Synchronous Machines, Part 1 and 2.,* R.H. Park, AIEE Trans., Vol. 48, No. 2, July, 1929.
6. *Integrated Circuits,* 1965 Mc-Graw-Hill, Edited by R.M. Wagner, Jr.

Discussion

Richard G. Farmer (Arizona Public Service Company, Phoenix, Arizona 85036): The authors are to be complimented on their paper which describes one aspect of the subsynchronous resonance problem facing transmission system designers when series capacitors are applied. The transient torque applied to turbine-generator shafts following system disturbances on series compensated lines is equally serious. It has been demonstrated analytically that oscillating torques can build up in turbine-generator shafts to a level exceeding the yield point.

Arizona Public Service is a participant in the Navajo Project which is faced with the subsynchronous resonance problem. The turbine-generators have 5 torsional modes with corresponding electrical system frequencies ranging from 11.8 to 44.5 HZ. These generators will be connected to a 500 KV transmission system containing 1800 miles of series compensated lines. The resonant frequencies of the electrical system range from 20 to 50 HZ. Considering all of the combinations of system conditions that might exist, it is very likely that at some time the electrical system frequency will correspond to a torsional frequency.

The eigenvalue program, described in this paper, has been employed to seek out the system conditions that would yield instability. Fortunately, General Electric had mechanical damping test results for a twin to the Navajo turbine-generators. I would like the authors to comment on how the eigenvalue program would be applied if mechanical damping data was not available.

Conclusion 5 of the paper refers to additional mechanical damping. Is there a practical method to obtain additional mechanical damping?

The authors state that the dynamics of the voltage regulator are important due to the application of speed stabilizers. This statement leads me to ask if a speed signal could be applied to eliminate instability.

Manuscript received February 7, 1973.

Walter H. Croft (Arizona Public Service Company. Phoenix, Arizona 85036): The authors have presented an interesting analysis of the relationship of externally generated oscillations to the torsional oscillations of turbine and generator shafts. Since the quantity RE, the transmission line resistance, has considerable influence on the results, as shown in figure 3 and 4, it appears that in application to a practical

Manuscript received February 20, 1973.

case, either after-the-fact analysis or to determine some threshhold in a design problem, that the value of RE should be known more accurately than is originally used in practice by utility engineers. Since RE can vary widely with temperature or more correctly with its meteorological environment at a given time, as well as the load on the line if a loaded line is considered, it would seem that it may be desirable to have as correct as possible a figure for RE. Going from a hot summer day to a cold winter night could very readily change a borderline condition from stable to unstable. In a long line a correct value of RE is a composite as the line may traverse a route where the geography and meteorological conditions vary widely for a given hour as well as a given season of the year.

Would the authors please comment on the necessity of using more sophisticated values of line resistance than is customarily used by power engineers in their day to day work?

L. A. Kilgore, E. R. Taylor, and **D. G. Ramey** (Westinghouse Electric Corp., East Pittsburgh, Pa. 15112): An earlier paper by Kilgore et al discussed the induction generator effect in which the interaction of the subsynchronous stator currents with the rotor flux feeds energy into current oscillations at any subsynchronous frequency. It also described means of correcting for this induction generator effect in the rare cases where corrections are required.[1]

Recent experience[2,3] has shown that there can be an additional contribution to the energy feedback into the electrical system at a particular subsynchronous frequency. This contribution arises when one of the subsynchronous frequencies in the electrical system coincides with the fundamental frequency compliment of one of the mechanical torsional frequencies of the turbine-generator. In such a case, an oscillating torque at the mechanical torsional frequency is produced by the interaction of the subsynchronous electrical current and the flux produced by the generator field. Also, for small amplitude oscillations the generator rotor motion induces electrical voltages at frequencies $f_o - f_m$ and $f_o + f_m$ into the electrical system. By this mechanism oscillations in both the electrical system and mechanical system receive energy from the fundamental frequency operation of the turbine generator. The authors' paper gives one method of analyzing this coupling effect of the mechanical system with the electrical system.

The same results, practically, may be calculated by obtaining steady state network solutions for both the electrical and mechanical systems separately over the range of frequencies. The effective reduction in mechanical damping due to induced oscillating currents may be seen by comparing the torque produced by a mechanical oscillation acting through the electrical circuit with that required to sustain the oscillation. This calculation is quite accurate because the main coupling between the two circuits involves interactions between the oscillations and the fundamental frequency rather than between the two oscillating frequencies.

This paper provides an indication of the severity of the torsional interaction for two simple systems. However, to avoid the impression that these problems may rule out the real economic advantages of series compensation, it is important to note that there are practical answers to these difficulties where they arise, and in many systems, the problem does not exist.

It should be further noted that there are now transmission systems with series compensated lines which have been in operation for some time that do not experience this problem. For example, large systems tied with compensated lines do not experience problems because of the divergence of loads, and because the numerous machines in these systems have different mechanical frequencies which damp and dissipate any tendency to oscillate at any one frequency. Likewise, water wheel generators, even though directly feeding into series compensated lines,[4] should not experience sustained shaft subsynchronous oscillations since their mechanical torsional frequencies generally are about 10 Hz or lower, corresponding to a 50 Hz and higher electrical frequency. Such a high electrical frequency requires higher amounts of compensation than presently used, and are not encountered even with 70 percent compensation.

In situations with remote generating plants feeding directly into series compensated lines, a careful study should be made of this problem, and in some cases special measures similar to those discussed in previous works may be necessary. In the simple systems of such situations, it may be practical to avoid the exact electrical frequencies that correspond to the mechanical torsional frequencies of concern. With multiple compensated lines it may not be possible to avoid all the frequencies of concern, and in such cases it may be necessary to provide electrical damping circuits.

An additional consideration not covered in this paper is that of subsynchronous frequency torques produced by the oscillating currents following short circuits. The study of short circuit torques can require a careful study with an accurate simulation of the machines and lines. Again, however, there can be practical solution where these are a problem.

REFERENCES

1. L. A. Kilgore, E. R. Taylor, L. C. Elliott, "The Prediction and Control of Self-Excited Oscillations Due to Series Capacitors in Power Systems," *IEEE Trans on Power App. & Sys.* Vol. PAS 90, No. 3, May/June 1971. pp. 1305-1311.
2. S. Goldberg, W. R. Schmus, "Subsynchronous Resonance and Torsional Stresses in Turbine-Generator Shafts", to be presented at the 1973 Winter Meeting of the IEEE Power Engineering Society.
3. J. W. Ballance, S. Goldberg, "Subsynchronous Resonance in Series Compensated Transmission Lines," to be presented at the 1973 Winter Meeting of the IEEE Power Engineering Society.
4. L. W. Lloyd, E. R. Taylor, F. G. Berg, S. M. Merry, "A study of Reinsertion Transient Voltages and Currents for Series Capacitors on USBR Glen Canyon – Flagstaff 345 kV Lines," presented at the 1971 Summer Meeting of the IEEE Power Engineering Society.

Manuscript received February 13, 1973.

P. L. Dandeno and **P. Kundur** (Analytical Planning Ontario Hydro, Ontario, Canada): This paper has drawn attention to the problem of subharmonic torsional oscillations in turbinegenerators. As reported in reference 4 of the authors paper, problems can occur on turbine-generators which are excited by high gain, low time constant thyristor excitation systems. This is particularly so where stabilization of the excitation system is achieved by directly utilizing change of shaft speed.

All of the natural torsional oscillation frequencies must be considered in the analysis of such generators and their excitation systems. Typical values for Tandem Compound 3600 RPM sets with twp LP stages, and of about 500 MW capacity appear to be close to 15, 25, 30 and 45 Hz. For 1800 RPM Tandem Compound units in the 700 MW class, with three LP stages, these values can be as low as 10 Hz and range up to 25 Hz. By a judicious choice of shaft speed location it has been found that amplification of these torsional natural turbinegenerator oscillations, fed through thyristor excitation systems, can be avoided. The torsional modes are such that the magnitudes decline to zero or reverse their sign depending upon the shaft location considered. Such node points have been found to occur at or near shaft coupling flanges.

An examination of Figures 4 and 5 of the paper indicates that the torsional natural frequencies of concern to the authors are about 25 and 28 Hz, with their reported value of 52 Hz not causing any problem.

Would the authors care to comment on the type of fossil unit considered in their analysis, and whether one might ever be concerned with instability created by the joint effect of electrical and torsional resonant modes much above or much below 25-28 Hz?

Manuscript received February 27, 1973.

C.E.J. Bowler, D.N. Ewart, and **C. Concordia**: The discussions have brought up many interesting points.

1) Both Mr. Farmer, and Messrs. Kilgore, Taylor, and Ramey, have mentioned the transient torques following disturbances as an additional factor to be considered. We are well aware of this and have made studies of transient cases. The situation is much like the usual power system stability problem, where it is necessary to first design for steady state stability, and then in addition assure transient stability for all reasonably severe disturbances. This present paper was limited in scope to discussion of the steady-state case.

2) Mr. Farmer has asked how the eigenvalue method would be applied if mechanical damping data were not available. In actual fact very little mechanical damping data are available, and the damping may have to be estimated from the few tests that have been made or calculated from physical considerations. Indeed, Mr. Farmer's question applies with equal force regardless of the mathematical technique employed to analyze the problem. The answers will only be as good as the data used.

3) With regard to additional mechanical damping, this is being studied, but no final conclusions have been arrived at.

4) With regard to speed signals to eliminate instability, this is of course, theoretically possible. The question of whether or not one would

Manuscript received April 18, 1973.

wish the safety of the unit to depend on continuous control action seems important. We are presently in the process of studying the most effective corrective means, which could involve a combination of more than one method.

5) Messrs. Kilgore, Taylor, and Ramey mention the possibility of a steady-state analysis of the electrical and mechanical systems separately. Actually, there appear to be at least four methods of calculating the steady-state stability conditions.

(a) The eigenvalue method used in the present paper, or simply finding the roots of the characteristic equation of the whole electromechanical system.

(b) The use of the Routh or similar criterion to determine the boundaries of stability regions, as used in references 3 of the paper.

These two (a and b) are the only truly fundamental methods, and the following two methods have to be justified by reference to one of them.

(c) The frequency-scanning method used in reference 1 of the paper.

(d) The calculation of the synchronizing and damping torques produced by the generator motion interacting with the electrical network, as employed in reference 1 of the closure. The method mentioned by Mr. Kilgore et al is an approximation to this last method, that simply tacitly assumes (which is not always the case) that the torsional natural frequencies are not affected by the electrical network.

The choice of the method to be used is in a sense a matter of individual taste, but is conditioned by the mathematical procedures and computer programs already developed and readily available, as well as by one's concept of what may be the ultimately most efficient method.

With further regard to simplified and approximate methods, Mr. Kilgore took the authors of reference 1 of the paper to task for using a rather complicated method, and including the effect of the rotor motion, and suggested that a much simpler analysis was adequate. We believe events have shown that it may in the long run be better to stick with a more fundamental and comprehensive viewpoint at least until the problem is better understood.

(6) We could not agree more with Mr. Kilgore that there are practical answers to the difficulties that may be encountered, and are working to that end.

7) We agree also that interties between large systems should not experience the problem of self-excitation. However, this may not be so much because of diversity of loads and numerous machines as simply because of the relative size of the equivalent machines. Further, we agree with the remarks concerning hydraulic-turbine-driven generators.

8) Mr. Crofts has brought up the question of more accurate values of line resistance. It is of course, good to have accurate data, and line resistance is one of the most important items. However, certain other equally important data, such as mechanical damping, may be even less well known. We must strive for improvement in all of the data.

9) The excitation systems problem mentioned by Messrs. Dandeno and Kundur is another aspect where it becomes important to understand better the mechanical characteristics of generating units. In this problem the possible tendency toward lower torsional frequencies with larger machines also has a bearing. With the present rather limited experience we believe it would not be wise to assume that the self-excitation problem with torsional resonance is confined to any particular range of frequencies. At least a preliminary look would seem justified in practically all cases.

Finally, we wish to thank all of the discussors for their contributions.

REFERENCES

1. Self-Excited Oscillations of Capacitor-Compensated Long-Distance Transmission Systems by C. Concordia, R. B. Bodine, and Gabriel Kron AIEE Transactions, Vol. 62, 1943, pp 41-44.

Suppression of Ground-Fault Arcs on Single-Pole-Switched EHV Lines by Shunt Reactors

E. W. Kimbark, *Fellow IEEE*

Summary: Arcing line-to-ground faults isolated by single-pole switching are maintained by capacitive coupling between the faulted and unfaulted phases. Such faults are difficult to extinguish on long EHV (extra-high-voltage) lines because the fault current is proportional to both length and voltage of the line. A promising remedy consists of neutralizing the capacitive coupling by shunt reactors, which are required anyhow on many lines for compensating the normal charging current.

Single-circuit a-c tie lines between power systems ordinarily cannot be regarded as having firm-power capability because 3-pole opening and reclosing cannot be accomplished quickly enough to retain synchronism. Because about 90% of the faults on high-voltage steel-tower transmission lines without overhead ground wires are of the one-line-to-ground type, nearly all of these being transitory, it would seem reasonable to rate the transmission capability of such a tie line by its stability limit for one-line-to-ground faults, *provided that such faults could be successfully cleared and reclosed by single-pole switching*.

Effective single-pole switching would increase the reliability of a line approximately as much as would the addition of overhead ground wires and at a much lower cost.

Arc Extinction with Single-Pole Switching

When one conductor of a 3-phase line is opened at both ends in order to clear a ground fault, this faulted conductor is capacitively and inductively coupled to the two unfaulted conductors, which are still energized at approximately normal circuit voltage and carrying load current. This coupling has two effects:

1. Before extinction of the fault arc, it feeds current to the fault and maintains the arc.
2. After the arc current becomes zero (as it does twice per cycle), the coupling causes a recovery voltage across the arc path. If the rate of rise of recovery voltage is too great, it will reignite the arc.

Of the two types of coupling, the capacitive coupling is the more important. Its importance increases with increase of circuit voltage, and it is the only type of coupling considered in detail in this paper.

The arc on the faulted conductor after it has been switched off is called the secondary arc. Extinction of the secondary arc depends on its current, recovery voltage, length of arc path, wind velocity, and perhaps on other factors. Recovery voltage and length of arc path both increase with circuit voltage and thus the effect of one factor may partially offset the other. This leaves the magnitude of the secondary arc current as the most significant index of whether the arc will be self-extinguishing. For given interphase capacitance, the secondary arc current is proportional to the circuit voltage and to the length of the line section that is switched out. Hence, the length of section on which single-pole switching can be employed successfully is inversely proportional to the circuit voltage. The situation is unfavorable on EHV lines because the circuit breakers are expensive, and it would be desirable to make the sections even longer than is customery at lower voltages. At 500 kv, the estimated permissible length is about 50 miles.

Proposed Method of Arc Suppression

Since coupling of the faulted conductor to the sound conductors through the shunt capacitive reactance between phases is the chief cause of the secondary arc current and recovery voltage, it is proposed to neutralize this capacitive reactance by means of lumped shunt inductive reactance, equal and opposite to the capacitive reactance. The proposed scheme is analogous to the use of a Peterson coil, and both might well be called ground-fault neutralizers. However, the scheme now proposed would neutralize the capacitance between phases, amounting to $C_1 - C_0$ per phase; that is, the difference

Paper 64-56, recommended by the IEEE Transmission and Distribution Committee and approved by the IEEE Technical Operations Committee for presentation at the IEEE Winter Power Meeting, New York, N. Y., February 2–7, 1964. Manuscript submitted November 4, 1963; made available for printing December 10, 1963.

E. W. KIMBARK is with the Bonneville Power Administration, Portland, Oreg.

Reprinted from *IEEE Trans. Power App. Syst.*, Mar. 1964.

between the positive-sequence and zero-sequence capacitances, whereas the Peterson coil neutralizes the capacitances to ground, $3C_0$. A further difference is that, whereas one Peterson coil can suppress ground faults anywhere on an entire transmission network, the reactors used with single-pole switching must be used on every transmission line too long for secondary arc extinction without them and must be switched with the line. Thus, a large number of reactors might be required.

However, many EHV lines require shunt reactors for wholly or partly compensating the normal, positive-sequence charging current. By proper connection, these reactors can be made to serve the additional purpose of ground-fault suppression at a moderate additional expense.

Connection Schemes and Susceptances of Shunt Reactors

Equivalent Circuits of Shunt Capacitances

To understand the ways in which shunt reactors may be connected, one must first know the equivalent circuits for the shunt capacitances of a 3-phase line. The most general equivalent circuit for the capacitances between four conductors (line conductors *a*, *b*, *c*, and ground) has a branch joining each pair of conductors, Fig. 1(A). The three capacitances of the delta terminating on *a*, *b*, and *c* can always be replaced by an equivalent Y as in Fig. 1(B). These two circuits are valid for both balanced and unbalanced lines, though the labels on the capacitances are valid only for balanced lines. For balanced lines, the 4-branched star circuit of Fig. 1(C) is also correct.

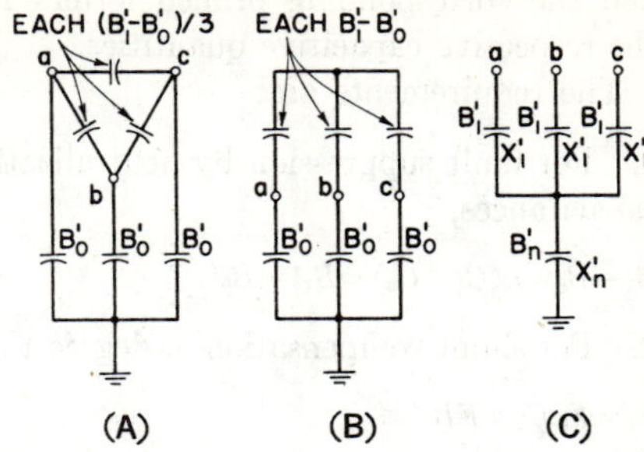

Fig. 1. Equivalent circuits of the shunt capacitances of a 3-phase transmission line

The capacitances of all these circuits are labeled in terms of the zero-sequence capacitive susceptance B_0' or reactance X_0' and the positive-sequence capacitive susceptance B_1' or reactance X_1' of the line. The correctness of the labeled values is most readily checked in Fig. 1(B), as follows: If zero-sequence voltages are applied from *a*, *b*, *c* to ground, charging current flows only in the grounded capacitances, each of which should therefore have susceptance B_0'. If positive-sequence voltages are applied, the grounded Y and the ungrounded Y are effectively in parallel. Hence, the total susceptance per phase must be B_1', and the ungrounded capacitances must make up the difference $B_1'-B_0'$. In Fig. 1(C), with positive-sequence voltage applied, currents are confined to the Y-connected capacitances, which should therefore have reactances X_1'. With zero-sequence current I_0' applied, the zero-sequence voltage is

$$V_0=jX_0'I_0'=jX_1'I_0'=jX_n'3I_0' \tag{1}$$

whence

$$X_n'=\frac{X_0'-X_1'}{3} \tag{2}$$

and

$$B_n'=\frac{3B_0'B_1'}{B_1'-B_0'} \tag{3}$$

Connections of Reactors

The shunt-reactor schemes of Figs. 2(A) through 2(C) are analogous to the equivalent-capacitance circuits of Figs. 1(A) through 1(C), respectively. Fig. 2(D) indicates a 3-phase reactor having mutual reactance X_m between phases. Its equivalent circuit in Fig. 2(E) has the same form as the star connection of Fig. 2(C), hence the analysis to be made of the latter is applicable to the former also.

Comparison of Schemes

The aggregate reactive-power rating of the group of shunt reactors is the same in the schemes of Figs. 2(A), 2(B), and 1(D), being that required for shunt compensation under normal conditions. In the 4-reactor scheme of Fig. 2(C), the aggregate rating of the three main reactors has that same value. In addition, there is the fourth or grounded reactor which is energized only during a line-to-ground fault and which, for high degrees of shunt compensation, has a momentary rating equal to only a small per cent of the aggregate rating of the three main reactors.

If the cost of reactors were directly proportional to their ratings, all schemes would cost approximately the same. But, because the cost per kilovar decreases with increasing rating, the scheme having the fewest and biggest reactors costs the least, provided that the reactors are not too big for shipping and handling. The number of high-voltage bushings also significantly affects the cost.

The numbers of reactors and bushings required are as listed, assuming single-phase construction except for Fig. 2(D).

Scheme of Figures	*2(A)*	*2(B)*	*2(C)*	*2(D)*
Number of:				
Reactors	6	6	4	1
High-voltage bushings	9	6	3	3
Medium-voltage bushings	0	3	0	0
Low-voltage bushings	0	0	4	0

On these grounds, the 3-phase reactor of Fig. 2(D) is preferable if it is not too big. If it is, the 4-reactor scheme of Fig. 2(C) is the next choice. In both these schemes, however, the degree of shunt compensation cannot be switched without losing the correct neutralization of interphase capacitance for fault suppression. The 6-reactor scheme of Fig. 2(B) has the advantage that the grounded reactors can be switched on or off or their reactance varied without affecting the neutralization.

The 6-reactor scheme of Fig. 2(A), as compared with that of Fig. 2(B), requires three additional high-voltage bushings and better insulation for the delta-connected reactors, and it has no compensatory advantages. The scheme of Fig. 2(A) will be dismissed, therefore, from further consideration.

Required Susceptances or Reactances

Because the reactors are in parallel with the line capacitances, it is more convenient to express their values in terms of susceptance *B* than reactance *X*. Unprimed letters *B* and

X will be used to denote inductive susceptance and reactance, and the corresponding primed letters B' and X' stand for the respective capacitive quantities.

The requirements are:

1. For fault suppression by neutralization of the interphase capacitances,

$$B_1 - B_0 = \omega(C_1 - C_0) = B_1' - B_0' \qquad (4)$$

2. For shunt compensation of degree F,

$$B_1 = F\omega C_1 = FB_1' \qquad (5)$$

From equations 4 and 5, it follows that

$$B_0 = B_0' - (1-F)B_1' \qquad (6)$$

The required susceptances or reactances of the reactors are found by using equations 4, 5, and 6 in conjunction with the appropriate pair of equations which follow.

The susceptances of the reactors in Fig. 2(B) in terms of zero- and positive-sequence susceptances are

$$B_g = B_0 \qquad (7)$$

$$B_u = B_1 - B_0 \qquad (8)$$

In Fig. 2(C), the susceptances are

$$B_p = B_1 \qquad (9)$$

$$B_n = \frac{3B_0B_1}{B_1 - B_0} \qquad (10)$$

and the corresponding reactances are

$$X_p = X_1 \qquad (11)$$

and

$$X_n = \frac{X_0 - X_1}{3} = \frac{B_1 - B_0}{3B_0B_1} \qquad (12)$$

For the *6-reactor scheme*, Fig. 2(B), equations 8 and 4 yield:

$$B_u = B_1' - B_0' \qquad (13)$$

while 7 and 6 yield

$$B_g = B_0' - (1-F)B_1' = FB_1' - B_u \qquad (14)$$

Equation 13 shows that the ungrounded reactors must resonate with the interphase capacitive susceptance. These reactors provide a degree of shunt compensation

$$F_{\min} = \frac{B_1' - B_0'}{B_1'} = 1 - \frac{B_0'}{B_1'} = 1 - \frac{X_1'}{X_0'} \qquad (15)$$

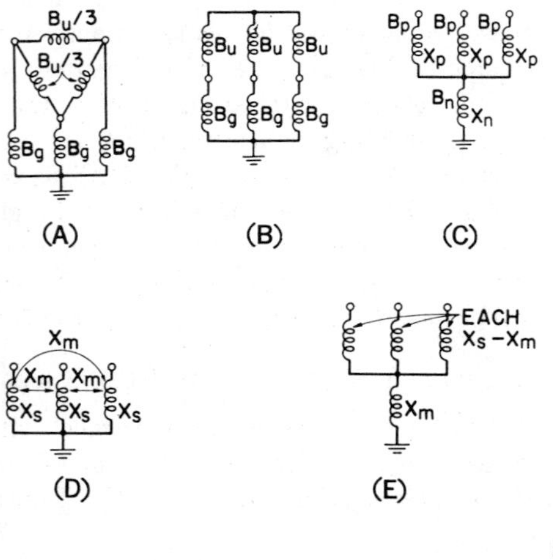

Fig. 2. A, B, C, D — Possible connections of shunt reactors for fault suppression and compensation of charging current. E—Equivalent circuit of D

which, for a typical EHV line without ground wires, is about 30%. If a higher degree F of shunt compensation is desired, it may be obtained by the addition of grounded reactors having the susceptance given by equation 14. For 100% shunt compensation $F=1$), this susceptance is simply $B_{g100} = B_0'$. The grounded reactors play no part in fault suppression.

For the *4-reactor scheme*, Fig. 2(C), equations 9 and 5, yield

$$B_p = FB_1' \qquad (16)$$

while 10, 6, 5, 4 yield:

$$B_n = \frac{3FB_1'[B_0'(1-F)B_1']}{B_1' - B_0'} = B_n'[1-(1-F)B_1'B_0'] \qquad (17)$$

Equation 16 shows that the phase reactors furnish the desired degree F of shunt compensation. The minimum degree, however, is given by equation 15. At this degree B_n, given by equation 17, vanishes, showing that the phase reactors must have ungrounded neutral. At any greater degree, B_n is positive and the neutral reactor is required for fault suppression. Its susceptance depends on F, as shown by equation 17. For $F=1$, it becomes

$$B_{n100} = \frac{3B_0'B_1'}{B_1' - B_0'} = B_n' \qquad (18)$$

The corresponding reactance is

$$X_{n100} = \frac{X_0' - X_1'}{3} = X_n' \qquad (19)$$

How Shunt Reactors Suppress Fault Arcs

The manner in which a Peterson coil suppresses arcing line-to-ground faults on an otherwise-ungrounded network is well known from both the theoretical and practical standpoints. Shunt reactors operate on the same principle when suppressing line-to-ground faults on conductors isolated by single-pole switching from the rest of a well-grounded network. In both cases, parallel resonance between distributed shunt capacitance of one or more lines and lumped shunt inductance is employed.

As previously stated, the addition of the resonant shunt inductance decreases (1) the current in the fault and (2) the voltage across the fault path after the current ceases.

Theoretically, in a lossless parallel LC (inductance-capacitance) circuit, the net current can be made zero by proper tuning. Actually, in either form of resonant ground-fault suppression, the fault current does not become zero because of (1) imperfect tuning, (2) losses, or (3) harmonics. However, the neutralized current is only 10% to 20% of the unneutralized, capacitively fed fault current. The lower value seems probable with single-pole switching.

The parallel LC circuit continues to oscillate after the fault arc is extinguished. In a perfectly tuned, lossless LC circuit, the frequency and amplitude of the free voltage oscillation would exactly match those of the forced oscillation existing before extinction of the arc. Consequently the faulted conductor would remain at ground potential after extinction of the arc, and there would be no recovery voltage across the arc path. Actually, there would be some difference in frequency due to imperfect tuning and some decrease of amplitude caused by losses. Consequently, the amplitude of the power-frequency recovery voltage would slowly increase. Moreover, because a parallel resonant circuit has unity power factor, the instantaneous recovery voltage increases very slowly in the first quarter cycle after arc extinction at a normal current zero, thereby making arc reignition unlikely.

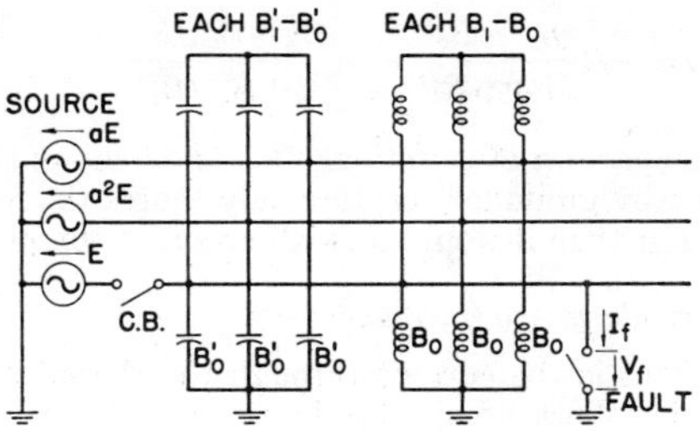

Fig. 3. Equivalent of source and faulted line section, isolated by single-pole switching, series impedance being neglected

By contrast, conditions in the capacitively fed, unneutralized secondary fault arc make extinction more difficult. The fault current and the a-c recovery voltage are in quadrature, so that when the current is broken at a natural zero, the alternating voltage is at its crest value. Consequently, there is a d-c component of voltage due to charge trapped on the isolated conductor. The resultant voltage of the a-c and d-c components is zero at the moment of interruptions, giving a completely offset voltage wave. Half a cycle later, however, the instantaneous recovery voltage reaches twice the crest value of its a-c component. Delayed restriking of the arc is probable.

Because of the low recovery voltage and high power factor of the neutralized fault arc, the current that can extinguish itself is 15 to 30 times higher than in an unneutralized arc.

Arc Current That Can Be Extinguished

In order to estimate the length of line section on which single-pole switching can successfully clear line-to-ground faults, both with and without inductive neutralization, we can utilize (1) the values of arc current that are self-extinguishing on ungrounded systems and systems grounded through Peterson coils and (2) the values of unneutralized arc current on lines having successful single-pole switching.

Peterson Coils

Extensive experience with Peterson coils[1] has shown that the maximum-fault arc currents that are self-extinguishing are:

1. Unneutralized, 5 to 10 amperes.
2. Neutralized, about 150 amperes.

Experience has shown also that faults are successfully cleared with the Peterson coil detuned by 10% or 20%. There is still a resonant circuit which decreases the rate of rise of recovery voltage, even though the frequency of that circuit differs somewhat from the power-system frequency. In a system grounded through a Peterson coil, the detuning arises from switching lines on or off. This cause of detuning will not occur with the scheme herein proposed in which the neutralizing reactors would be switched with the associated line sections. The cited fact shows, however, that, since tuning is not critical, the reactors need not be tunable, and no allowance need be made for differences between capacitances of the three conductors of the untransposed line.

Single-Pole Switching

There are two factors in single-pole switching that differ from Peterson coil operation:

1. The primary arc current is greater than the secondary arc current and may, therefore, establish greater initial ionization.
2. The time in which the secondary arc is extinguished is important so that reclosure rapid enough to give a reasonably high stability limit will be successful.

Information on the unneutralized secondary arc current that permits successful reclosure has been compiled by Maikopar[2] from installations of single-pole switching in many parts of the world. It indicates that with 0.4-second dead time, there is a high probability of successful reclosure if the secondary arc current does not exceed 18 amperes. However, the actual secondary arc current without fault neutralization may be as high as twice the value computed from the voltage and shunt capacitive susceptance. Therefore, we may take the value of 9 amperes of so-computed unneutralized secondary arc current as the maximum permissible value. This value agrees well with the 5 to 10 amperes fault current permissible on an ungrounded system.

As no information has been found on maximum permissible neutralized secondary arc current for single-pole switching, the value obtained from Peterson coil experience (150 amperes) may be used as a rough guide. Even if this be reduced to 75 amperes to give a factor of safety, and if it be assumed that neutralization reduces the secondary arc current to only 20% (instead of 10%) of its unneutralized value, the maximum permissible lengths of section of a typical 500-kv line without ground wires, computed in the Appendixes, are:

1. Unneutralized, 45 miles.
2. Neutralized, 1,800 miles.

The former length is uneconomically short, whereas the latter is much longer than would be used in practice.

Conclusions

It appears from theory and from experience with Peterson coil grounding that the use of shunt reactors for neutralizing the interphase shunt capacitance C_1–C_0 of an EHV transmission line could increase, by a factor of 40 or more, the length of line on which line-to-ground transitory faults could be successfully cleared by single-pole opening and 30-cycle reclosure. Moreover, the same shunt reactors would supply part of the shunt compensation that is usually needed on EHV lines. Thus, fault suppression would be provided at a small additional cost. The scheme merits field tests of its effectiveness.

Successful clearing of transitory line-to-ground faults would greatly improve the reliability of EHV lines without ground

Fig. 4. Simplification of Fig. 3 applicable to first set of emf's

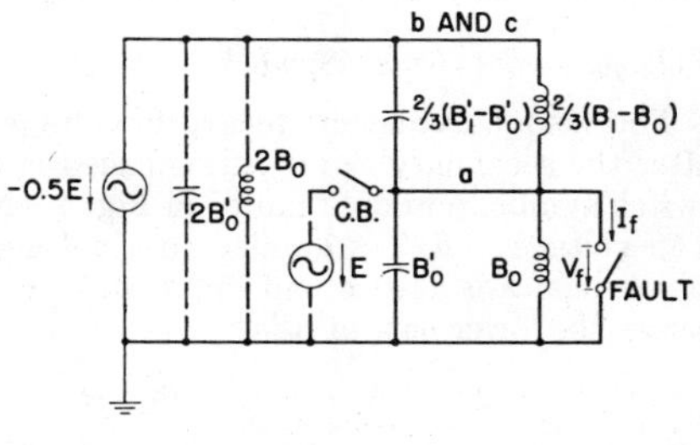

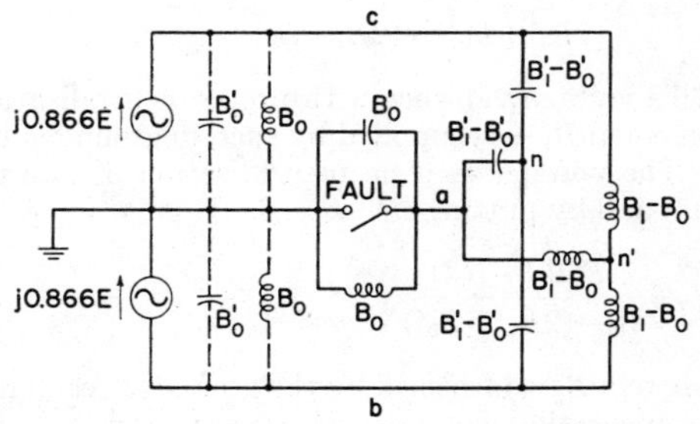

Fig. 5. Simplification of Fig. 3 applicable to 2nd set of emf's

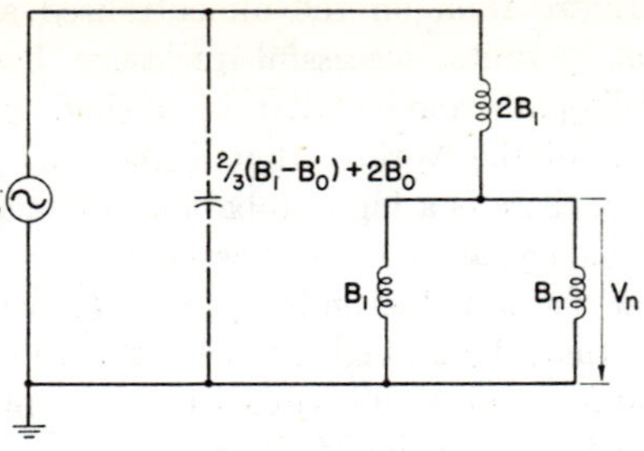

Fig. 6. Equivalent circuit for finding the voltage across the grounded reactor of the 4-reactor scheme during an isolated line-to-ground fault

wires and would make it reasonable to rate the transmission capability of single-circuit a-c ties by their transient stability limits for such faults.

A more detailed analysis of the subject appears in the Appendixes. Appendix I derives equations for the secondary arc current and the recovery voltage of lines with and without neutralization and for the voltage and reactive power ratings of the shunt reactors. Appendix II computes numerical values of these quantities for four typical 500-kv lines.

Appendix I. Circuit Analysis for Single-Pole Switching

Except as noted, longitudinal voltages and impedances are neglected. Consequently, the several shunt elements—the distributed capacitance, the reactors when used, and the line-to-ground fault when present—may be regarded as in parallel; see Fig. 3. The 6-reactor scheme (grounded Y and ungrounded Y) is assumed except when noted to the contrary, and the line capacitance is represented by a like connection. The source is assumed to have balanced, positive-sequence electromotive forces (emf's): $E_a=E$, $E_b=a^2E=-0.500E-j0.866E$, and $E_c=aE-=0.500E+j0.866E$.

The emf's may be resolved into two sets: the real components and the imaginary ones. The first set is $E_a'=E$, $E_b'=E_c'=-E/2$. The second set is $E_a''=0$, $E_c''=-E_b''=j0.866E$. Simplification of the circuit on which the first set acts yields Fig. 4, and simplification of the circuit on which the second set acts yields Fig. 5. From the symmetry of the circuit of Fig. 5 with respect to phase a, it is apparent that phase a is at ground potential. Therefore, there is no voltage across the fault when the fault path is open and no current in it when closed. Thus, the second set of emf's contributes nothing to the fault current or recovery voltage and may be ignored. In computing fault current and voltage, we need consider only the circuit of Fig. 4, disregarding the branches drawn in broken lines, which are directly across the source.

Voltage Across Switch Symbol

The source-frequency recovery voltage across the arc path after the secondary arc is extinguished is the voltage across the switch symbol marked "fault" in Fig. 4 when the switch is open. The voltage $-E/2$ is applied to a voltage divider, each part of which has capacitance and inductance in parallel. The voltage across the lower part of it is

$$V_f=-\frac{E}{2}\cdot\frac{(2/3)[(B_1'-B_0')-(B_1-B_0)]}{(2B_1'/3+B_0'/3)-(2B_1/3+B_0/3)}$$

$$=-\frac{E(B_1'-B_0')-(B_1-B_0)}{(2B_1'+B_0')-(2B_1-B_0)} \qquad (20)$$

This voltage vanishes with proper neutralization of $B_1'-B_0'$ by an equal B_1-B_0 supplied by each ungrounded reactor.

The voltage with no neutralization of interphase capacitance is found by putting $B_1-B_0=0$. It is

$$V_f=\frac{E(B_1'-B_0')}{B_0-(2B_1'+B_0')} \qquad (21)$$

wherein B_0 is furnished by the grounded reactors. With no shunt compensation whatever, $B_0=0$ also, and

$$V_f=-E\frac{B_1'-B_0'}{2B_1'+B_0'}=-\frac{E1-B_0'/B_1'}{2+B_0'/B_1'} \qquad (22)$$

Comparison of equations 21 and 22 shows that shunt compensation by grounded reactors only causes the recovery voltage to be *higher* than it would be with no shunt compensation.

Secondary Arc Current

This is the current through the closed fault switch of Fig. 4, which short-circuits the lower part of the voltage divider. It is

$$I_f=-j\frac{E}{2}\cdot\frac{2}{3}[(B_1'-B_0')-(B_1-B_0)]$$

$$=(-jE/3)[(B_1'-B_0')-(B_1-B_0)] \qquad (23)$$

The fault current becomes zero under the same condition under which the fault voltage became zero; namely, by correct neutralization of $B_1'-B_0'$ by B_1-B_0.

Without neutralization, the fault current is

$$I_f=(-jE/3)(B_1'-B_0') \qquad (24)$$

It is not affected by grounded shunt reactors.

6-Reactor Scheme

Voltage of the neutral of the ungrounded Y-connected reactors—6-reactor scheme with respect to ground during isolated line-to-ground fault—is dealt with in this section.

During the fault, voltage $-E/2$ of the real set is impressed on the ungrounded reactors, marked $(2/3)$ (B_1-B_0) in Fig. 4. The marked value resulted from a simplification of the connection of two reactors from combined phases b and c to neutral point n'; one reactor thence to phase a, which is grounded by the fault. Thus the voltage from n' to ground is

$$V_n=-\frac{E}{2}\cdot\frac{2}{3}=-\frac{E}{3} \qquad (25)$$

where E is the line-to-neutral voltage of the source. V_n is not affected by the imaginary components of source emf. Equation 25 shows that the neutral bushings of the three ungrounded reactors would be momentarily subjected to a third of normal line-to-neutral voltage.

4-Reactor Scheme Isolated Fault

Voltage across the grounded reactor of the 4-reactor scheme during an isolated line-to-ground fault is described by equation 26. The circuit for real components of voltage as in Fig. 4 but with the reactors connected as in Fig. 2(C) simplifies to the circuit of Fig. 6. The voltage across the grounded reactor is

$$V_n=-\frac{E}{2}\cdot\frac{2B_1}{(3B_1+B_n)}=-\frac{EB_1}{3B_1+B_n}$$

$$=-\frac{EX_n}{X_1+3X_n}=-\frac{E(X_0-X_1)}{3X_0} \qquad (26)$$

where $X_n=(X_0-X_1)/3$.

4-Reactor Scheme Unisolated

Voltage across the grounded reactor of the 4-reactor scheme during an unisolated line-to-ground fault before the circuit breakers open is calculated herewith.

The calculation consists of two parts. First, the voltage V_n across the grounded reactor is expressed as a fraction of the zero-

Table I. 500-Kv 60-Cycle 3-Phase Lines Used for Illustration

Line number	1	2	3
Number of Conductors per Phase	2	1	2
Name of Conductor	Falcon	Special Expanded	Chukar
Outer Diameter of Conductor, Inches	1.545	2.50	1.602
Separation Between Conductors of Same Phase, Inches	16	—	60
Spacing Between Centers of Adjacent Phases, Feet	37	33.5	40.7

Table II

	Line Number			
	1	2	3	4
Postive-Sequence Shunt Capacitive Susceptance, B_1', Micromhos/Mile	6.65	5.61	7.52	5.61
Zero-Sequence Shunt Capacitive Susceptance, B_0' Micromhos/Mile	4.59	3.92	5.19	4.42
Ratio B_0'/B_1'	0.69	0.70	0.69	0.79
Normal Charging Current, Amperes/Mile	1.92	1.62	2.17	1.62
Normal Charging Reactive Power, Mvar's/Mile	1.66	1.40	1.88	1.40
Current in Secondary Arc, Amperes/Mile Without Neutralization	0.20	0.16	0.22	0.12
Normal-Frequency Recovery Voltage, Kv, Without Neutralization	33	32	33	22
Length of Line, Miles, for which Secondary Arc Would Be Self-Extinguishing:				
Without neutralization, 9 amperes	45	55	40	78
With Neutralization, 75 amperes	1,850	2,250	1,650	3,300
Grounded Reactor of 4-Reactor Scheme				
Voltage V_n Kv for F=1:				
During Isolated Fault	30	29	30	20
During Unisolated Fault, $Z_0/Z_1=$ 1	29	28	29	20
2	—	—	—	30
3	54	52	54	—
Momentary Rating Q_n, Kvar's/Mile for F=1;				
During Isolated Fault	40	33	45	26
During Unisolated Fault, $Z_0/Z_1=$ 1	38	32	43	25
2	—	—	—	58
3	130	116	144	—

sequence voltage of the line at the point where the group of reactors is connected. It is

$$\frac{V_n}{V_0}=\frac{3B_n}{B_1+3B_n} \tag{27}$$

Second, the zero-sequence voltage V_0 is expressed as a fraction of the normal line-to-neutral voltage E. Now we must take into account the series impedances which were neglected when the fault was isolated. Use of the well-known equivalent circuit of a line-to-ground fault—series connection of the three sequence networks—gives:

$$\frac{V_0}{E}=\frac{Z_0}{Z_1+Z_2+Z_0}=\frac{Z_0/Z_1}{2+Z_0/Z_1} \tag{28}$$

at the point of fault. Elsewhere V_0 is less. The highest V_0 at the fault is obtained if the fault is distant from a grounding point, in which case Z_0/Z_1 is substantially that of the line itself. The highest V_0 on the shunt reactors occurs if the fault is near the reactors and far from the grounding points excluding the reactors, and this condition is assumed in some of the calculations of V_n in Appendix II, Z_0/Z_1 being taken as 2 or 3. However, as some reactors may be near grounding points, V_n is calculated also for $Z_0/Z_1=1$.

Reactive Power in Grounded Reactor of 4-Reactor Scheme

$$Q_n=\frac{V_n^2}{X_n}=V_n^2B_n \tag{29}$$

Appendix II. Numerical Calculation for Typical 500-Kv Lines

Description of Lines

Computations of normal charging current and reactive power, of secondary fault current and recovery voltage for a line-to-ground fault without neutralization, and of the shunt compensation required for fault neutralization are made for three different designs of 500-kv lines described in Table I.

None of the lines in Table I has ground wires, and the average height of the conductors above ground is assumed to be 50 feet, corresponding to a minimum ground clearance of 35 feet. In order to determine the effect of ground wires, similar computations were made for a variant of line 2:

Line 4 is like line 2 except that two ground wires are added. They are assumed to be 33.5 feet apart and 16.8 feet above the line conductors.

Results of the computations are given in Table II. The ratio B_0'/B_1' is very nearly the same for each of the three basic designs, being about 0.7. The addition of ground wires increases this ratio to approximately 0.8, but ground wires presumably would not be used with single-pole switching.

Nomenclature

$a=\exp(j2\pi/3)=(1+j\sqrt{3})/2$
$B=1/X=$ inductive susceptance of the shunt reactors of a section of line
- $B_0=$ zero-sequence
- $B_1=$ positive sequence
- $B_g=$ inductive susceptance of each grounded reactor of the 6-reactor scheme
- $B_{g100}=$ value of B_g required for $F=1$
- $B_u=$ inductive susceptance of each ungrounded reactor of the 6-reactor scheme
- $B_n=$ inductive susceptance of the grounded reactor of the 4-reactor scheme
- $B_{n100}=$ value of B_n required for $F=1$
- $B_p=$ inductive susceptance of each ungrounded reactor of the 4-reactor scheme

$B'=\omega C=$ distributed capacitive susceptance of a section of line
- $B_0'=$ zero-sequence
- $B_1'=$ positive-sequence
- $B_n'=$ capacitive susceptance of grounded leg of 4-branched star equivalent circuit

$C=$ distributed shunt capacitance of a section of line
- $C_0=$ zero-sequence
- $C_1=$ positive-sequence

$E=$ rms positive-sequence line-to-neutral emf of the source
- E_a, E_b, $E_c=$ line-to-neutral phase emf's
- E_a', E_b', $E_c'=$ real components of E_a, E_b, E_c
- E_a'', E_b'', $E_c''=$ imaginary components

$F=B_1/B_1'=$ degree of shunt compensation, per unit
- $F_{min}=$ at least degree of shunt compensation when reactors are used for fault neutralization.

$I_f=$ rms current of secondary arc, isolated line-to-ground fault
$I_0'=$ zero-sequence charging current of line
$j=\sqrt{-1}$
$Q_n=$ reactive power consumed by neutral (grounded) reactor of the 4-reactor scheme during a line-to-ground fault
$V_f=$ rms source-frequency recovery voltage across path of secondary arc
$V_n=$ rms source-frequency voltage from neutral of reactor group to ground
$V_0=$ zero-sequence voltage of line at shunt reactors or at the fault
$X=\omega L=$ inductive reactance of shunt reactors, with subscripts as for B. In addition,
- $X_s=$ self-reactance per phase of 3-phase reactor
- $X_m=$ mutual reactance between phases of 3-phase reactor

$X'=1/B'=$ shunt capacitive reactance of line, with subscripts as for B'
$Z=$ impedance of system seen from point of fault
- $Z_0=$ zero-sequence
- $Z_1=$ positive-sequence
- $Z_2=$ negative-sequence

$\omega=2\pi f=$ angular frequency, in radians per second

References

1. NEUTRAL GROUNDING IN HIGH-VOLTAGE TRANSMISSION (book), **R. Willheim, M. Waters.** Elsevier, Amsterdam, The Netherlands, 1956, esp. pp. 612–13 and 631.

2. MINIMUM TIME OF AUTOMATIC RECLOSING, **A. S. Maikopar.** *Electric Technology, U.S.S.R.*, Pergamon Press, New York, N.Y., 1960, pp. 302–15, esp. Fig. 3, p. 311.

3. SINGLE-PHASE SWITCHING OF TRANSMISSION LINES USING REACTORS FOR EXTINCTION OF THE SECONDARY ARC, **N. Knudsen.** *Paper No. 310*, CIGRE, Paris, France, 1962.

Note: This last reference was called to the author's attention by a reviewer after submission of the present paper to the IEEE.

A Method for Reducing Dead Time for Single-Phase Reclosing in EHV Transmission

HAROLD A. PETERSON, FELLOW, IEEE, AND NARAYAN V. DRAVID, MEMBER, IEEE

***Abstract*—Most faults on transmission lines are single line to ground (SLG). For such faults single-phase reclosing would result in an improvement in transient stability over the alternative of three-phase reclosing. The improvement is offset by the capacitive coupling from the other two energized conductors because it tends to sustain current in the original fault arc path, thereby lengthening the time required for arc deionization. An increase in dead time is therefore required which decreases the net improvement. This paper presents a novel method of nullifying the capacitive coupling effect. The method requires the addition of a capacitor, connected across the terminals of each breaker pole, proportional to the particular line being switched.**

Paper 68 TP 72–PWR, recommended and approved by the Transmission and Distribution Committee of the IEEE Power Group for presentation at the IEEE Winter Power Meeting, New York, N. Y., January 28–February 2, 1968. Manuscript submitted September 18, 1967; made available for printing December 11, 1967.

H. A. Peterson is with the Department of Electrical Engineering, University of Wisconsin, Madison, Wis. 53706.

N. V. Dravid was with the Department of Electrical Engineering, University of Wisconsin, Madison, Wis. He is now with the Bonneville Power Administration, Portland, Ore. 97208.

Introduction

THE ADVANTAGES of high-speed automatic reclosing of circuit breakers on transmission lines following nonpermanent faults are well understood and documented in the technical literature [1].[1] A significant gain in transient stability is the principal benefit to be derived. The importance of this gain in transient stability varies from one case to another. In some cases high-speed automatic reclosing may be necessary to achieve the desired margin of stability consistent with the criteria of loading and fault conditions to be met.

Most applications of high-speed automatic reclosing have been three phase. That is, for any kind of fault on the transmission line, all three poles of both circuit breakers (one at each end) are opened after two or three cycles. After a further delay of 10 or 12 cycles to permit the fault arc path to deionize (dead time), both circuit breakers reclose thus restoring the system to its original capability provided the fault is nonpermanent.

[1] Following [1] is an exhaustive bibliography pertaining to arc deionization time and high-speed reclosing in its various aspects.

Reprinted from *IEEE Trans. Power App. Syst.*, Apr. 1969.

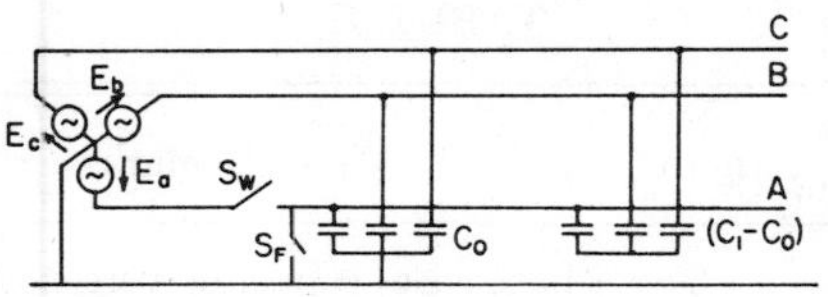

Fig. 1. Lumped capacitance three-phase system.

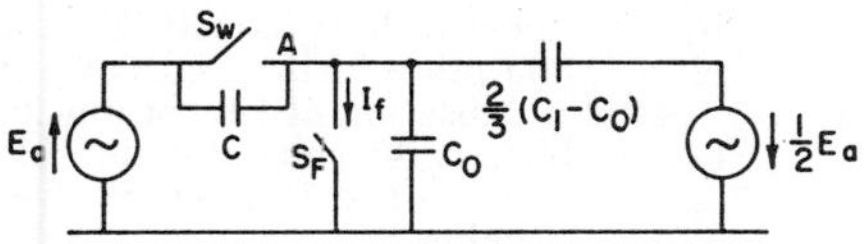

Fig. 2. Equivalent circuit for system of Fig. 1.

There are two major reasons for favoring three-phase reclosing.

1) Relaying is simpler since the faulted phase or phases need not be identified.

2) With all three phases of a transmission line open at both ends, there is no voltage available on the line to sustain a current in the original fault arc path. Thus the required dead time is kept to a minimum.

As is well known, most faults on transmission lines are single line to ground (SLG). For such faults, opening and reclosing the faulted phase only would result in an improvement in transient stability over the alternative of switching all three phases. The obvious improvement is partially offset, however, by a less obvious effect, namely, that of capacitive coupling from the still energized conductors. This effect tends to sustain current in the original fault arc path and therefore lengthens the time required for deionization. Thus an increase in dead time is incurred and the corresponding delay in reclosing decreases the net improvement.

It is possible in certain cases that the additional gain in transient stability achievable by single-phase reclosing would warrant the attendant increase in complexity of relaying. If, in addition, the required dead time for single-phase reclosing can be made comparable with that for three-phase reclosing, then there is additional incentive for thinking more favorably of it.

It is the purpose of this paper to disclose a method, believed to be new, for nullifying the capacitive coupling effect which tends to increase the required dead time when single-phase high speed automatic reclosing of transmission lines is employed [2].

Basic Concepts

Consider a three-phase transmission line connected to an ideal source at one end, the other end being open. To explain the basic idea we may approximate the transmission line by lumped capacitances, as seen from the source, as in Fig. 1 where C_1 is positive sequence capacitance and C_0 is zero sequence capacitance for the line length under consideration. An L–G fault (represented by switch S_F in closed position) occurs on phase A and the switch S_W is opened to clear the fault. The system can be further simplified in the form of an equivalent circuit shown in Fig. 2 by means of Thevenin's theorem, retaining the identity of the switches S_F and S_W. A capacitor C has been connected across the switch S_W. If C is equal to zero, then there exists a fault current I_f given by

$$I_f = -E_a\omega(C_1 - C_0)/3 \tag{1}$$

where ω is the system frequency in rad/s. Thus a current in the fault path is sustained in spite of phase A being open and clearly this current is due to the capacitive coupling that phase A has with phases B and C. This current is generally known as secondary arc current.

If the capacitor C is chosen to be equal to $(C_1 - C_0)/3$, then the voltage source E_a (see Fig. 2) will circulate a current through the fault of value $I_f \angle 180°$. In that case the net fault current will be zero and the fault will be cleared.

The fundamental frequency component of voltage on the cleared conductor (or phase) V_{FF} will be the voltage across the capacitor C_0 (see Fig. 2) when the switch S_F is open. For a successful reclosing operation V_{FF} should be zero or very small so that the arc is not sustained. For the value of C, as mentioned above, V_{FF} will be zero for the simplified system of Fig. 1. If the Thevenin impedance as seen from the fault terminals, for any fault location on the line, is nearly constant, then the secondary arc current will be proportional to V_{FF} and, therefore, any method that reduces V_{FF} will reduce the secondary arc current.

The transmission line selected for this study has the value of C_1/C_0 equal to 1.845. Using this value and the system of Fig. 1, magnitudes of V_{FF} obtained for different types of faults are shown in Table I. In arriving at these magnitudes, it is assumed that for any particular type of fault only the breakers in the phases involved in the fault are opened to clear the fault. It is apparent, therefore, that with capacitor C equal to $(C_1 - C_0)/3$ the greatest benefit is derived in the SLG-fault case. For the simple system of Fig. 1 it can also be shown that V_{FF} cannot be reduced to zero by this method for any value of C for the line-to-line-to-ground (LLG)-fault case. An attempt to minimize V_{FF} for the LLG case reduces the benefit for the SLG-fault case. For the three-phase fault case, obviously, the capacitors should be equal to zero for the greatest benefit. The foregoing summary may have implications in so far as the relaying requirements for this scheme are concerned. However, since a great majority of the faults are of SLG type it seems probable that the value of the capacitors used in any practical case would be $(C_1 - C_0)/3$ with the understanding that for faults other than the SLG type all the three phases are completely isolated from the source to clear the fault. This will make relaying somewhat less complicated.

TABLE I

Type of Fault	V_{FF} with $C = 0$ (pu)	V_{FF} with $C = (C_1 - C_0)/3$ (pu)
SLG	0.18	0.0
LLG	0.22	0.146
Three-phase	0.0	0.1325

System Selected for Analysis

Fig. 3 indicates the system that was selected for the analysis. V_S and V_R are initial L–N voltages at the ends of the transmission line and δ is the angle between them under balanced operating conditions. (In this study the source negative sequence reactance is assumed to be the same as the source positive sequence reactance. Therefore the terms "positive sequence reactance" and "α component reactance" for the source will be used interchangeably.) $X_{\alpha S}$, $X_{\alpha R}$, X_{0S}, and X_{0R} are the α component reactances and zero component reactances for the sending-end and receiving-end sources. Capacitors, each of value $C/2$, are shown connected across the breakers. It is understood that, in the analysis, only the capacitors in the phase with open breakers will appear.

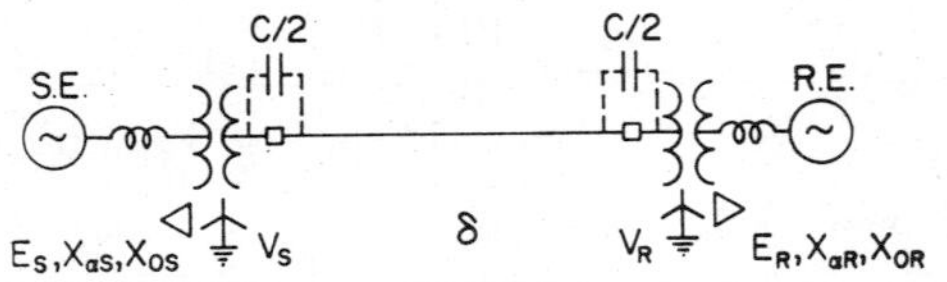

Fig. 3. Single-line diagram of system studied.

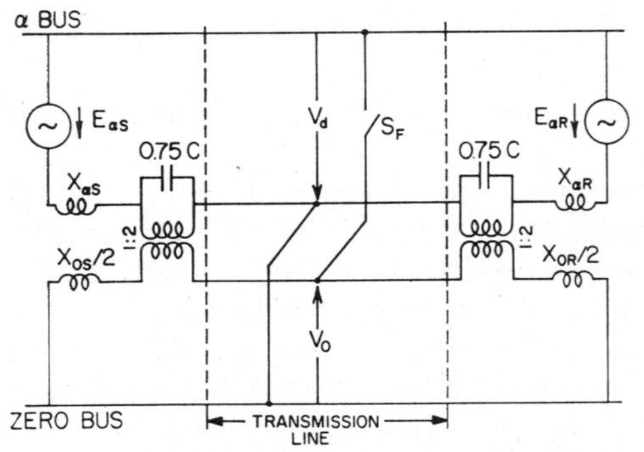

Fig. 4. α, β, 0 component representation of system.

Transmission Line Constants

The transmission line selected for this study had the following constants:

positive sequence resistance	$R_1 = 0.03923\ \Omega/\text{mi}$
positive sequence inductance	$L_1 = 1.458$ mH/mi
positive sequence capacitance	$C_1 = 0.0204\ \mu$F/mi
zero sequence resistance	$R_0 = 0.42988\ \Omega/\text{mi}$
zero sequence inductance	$L_0 = 3.884$ mH/mi
zero sequence capacitance	$C_0 = 0.01107\ \mu$F/mi
nominal L–L voltage	= 500 kV
	= 1.0 per unit voltage
surge impedance loading	= 935 MW
	= 1.0 per unit power.

System Representation

α, β, and 0 components [3] have been used to represent the system for this study. Since only phase A is involved in the unbalances (open breakers and L–G fault), this set of components is most economical because only α and 0 networks are required for the representation. Fig. 4 shows the system representation in terms of α, β, and 0 components. This is arrived at by first writing the equations of unbalances in terms of the components and then simplifying them to obtain the conditions of α, β, and 0 network interconnections. To satisfy these conditions simultaneously it will be necessary to halve all the impedances in the zero network and to use ideal 1:2 transformers for the interconnections. The switch S_F in the closed position represents the L–G fault on phase A although the current through the switch is only two thirds of the fault current. In arriving at Fig. 4 it was assumed that the transmission line is transposed and that the sources are symmetrical.

This representation was used while making the study using a digital computer program and also on an ac network calculator. Cases studied on the latter were mainly used to verify the results obtained from the former. Two-port network theory (*ABCD* constants as known to power engineers) was found to be most convenient for analysis of the circuit using a digital computer (see Appendix).

TABLE II

Case	Figure	Remarks
1	5	base case; capacitor versus shunt reactor
2	6	δ varied; 0–45°
3	7	$X_{\alpha S}$, $X_{\alpha R}$ varied; 0–0.6 pu
4	8	$X_{0S}/X_{\alpha S}$, $X_{0R}/X_{\alpha R}$ varied; 0.5–1.4
5	9	V_S/V_R varied; 0.95–1.1
6	10	line length varied; 50–300 miles
7	11	300 miles of line; 50 percent series compensation at 1.0 per unit power
8	12	base case; capacitor and shunt reactors varied; ±10 percent from optimum value

Summary of Cases Studied

A base case, with the following parameters, was studied.

transmission line length	200 miles
V_S, V_R	1.0 per unit
δ	25° (1.06 per unit power)
$X_{\alpha S}$, $X_{\alpha R}$	0.25 per unit
$X_{0S}/X_{\alpha S}$, $X_{0R}/X_{\alpha R}$	0.5.

A number of cases were studied by varying the parameters of the base case, one at a time, including the effect of using series capacitors for longer lines. Use of shunt reactors has been suggested as an alternative method of reducing capacitive coupling [4], [5]. This method was compared with the proposed method for the base case. A study was also made to determine the sensitivity of capacitor and shunt reactor values used in the two methods for the reduction of capacitive coupling effects. Table II is a summary of the cases studied. Only the parameters being varied are shown in the table, other parameters being those of the base case.

Thevenin Impedance

A separate computer program was written to evaluate the Thevenin impedance as seen from the fault terminals. It was observed that this impedance increased by less than 4.5 percent when the fault location was moved from line end to the midpoint of the line. Therefore only V_{FF} rather than the fault current was evaluated for all the cases (switch S_F in Fig. 4 being open). This, incidentally, resulted in a less complicated program.

Unless otherwise stated, it is assumed that the capacitors or the shunt reactors (Y ungrounded at both ends of the line), employed for the reduction of capacitive coupling, were chosen (optimum value) in order to minimize V_{FF} at the midpoint of the line.

Discussion of Results

Comparison of curves *a* and *b* of Fig. 5 shows that the proposed method reduces V_{FF} substantially. Curves *b* and *c* of Fig. 5 show that this method yields results comparable to the alternative of using shunt reactors. Effectiveness of the proposed method in completely nullifying the capacitive coupling may be seen by comparing curves *d* and *e* of Fig. 5. Results obtained from network calculator studies corresponding to cases of curves *a* and *b* of Fig. 5 are also shown for comparison.

With increasing δ or source X_α, the magnitudes of E_S, E_R, and the angle between them increase, thereby increasing the

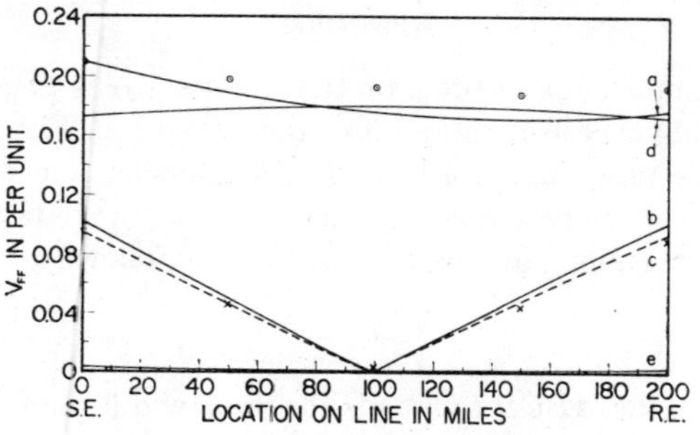

Fig. 5. Base case—V_{FF} at various locations on line. Curve a—$C = 0$; curve b—C = optimum value; curve c—shunt reactor = optimum value; curve d—$\delta = 0°$, $C = 0$; curve e—$\delta = 0°$, C = optimum value; ⊙, ×—data from network calculator studies corresponding to curves a and b, respectively.

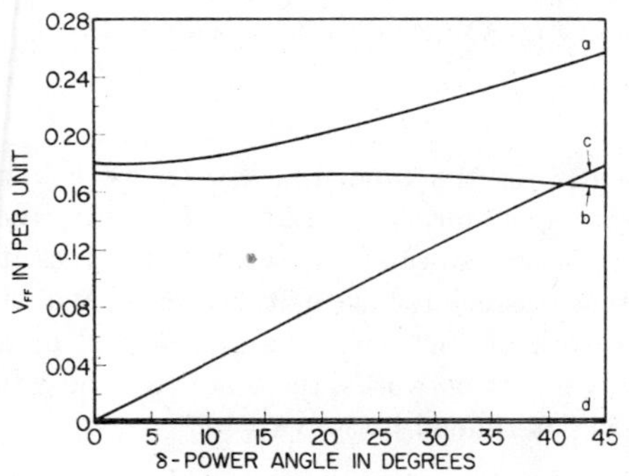

Fig. 6. Effect of power-angle variation. For $C = 0$, curve a—maximum V_{FF}, curve b—minimum V_{FF}; for C = optimum value, curve c—maximum V_{FF}, curve d—minimum V_{FF}.

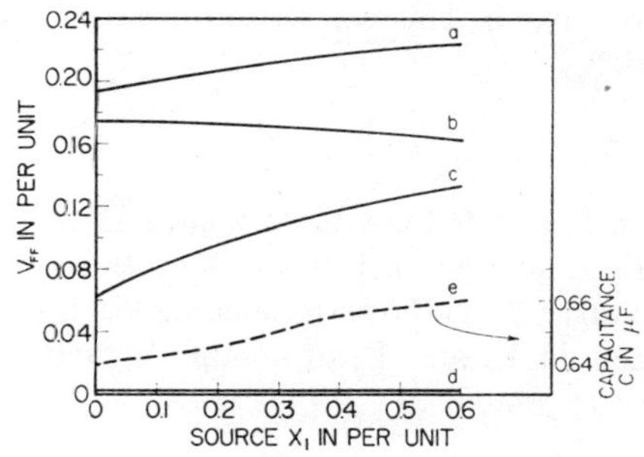

Fig. 7. Effect of source positive sequence reactance variation. Curves a–d—same as for Fig. 6; curve e—optimum values of C.

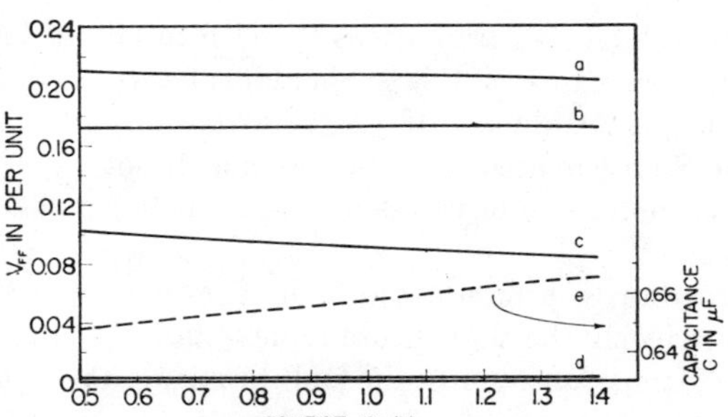

Fig. 8. Effect of source zero sequence to positive sequence reactance ratio variation. Curves a–e same as for Fig. 7.

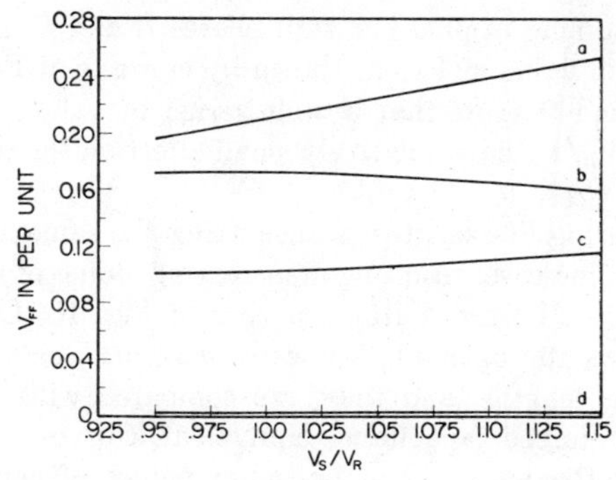

Fig. 9. Effect of transmission line sending-end to receiving-end voltage ratio variation. Curves a–d same as for Fig. 6.

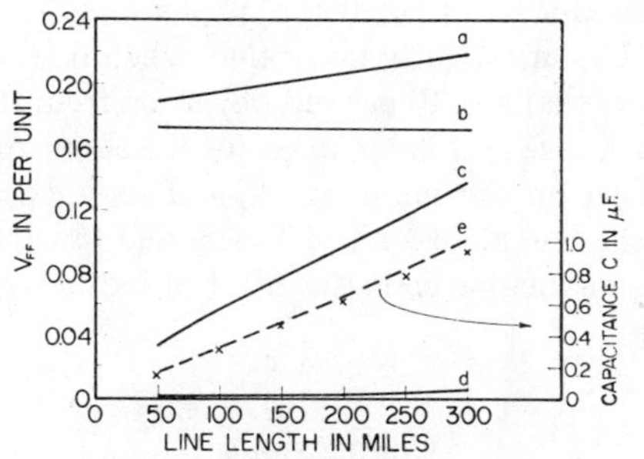

Fig. 10. Effect of line length variation, constant power. Curves a–e same as for Fig. 7. ×—capacitor values based on lumped capacitance system.

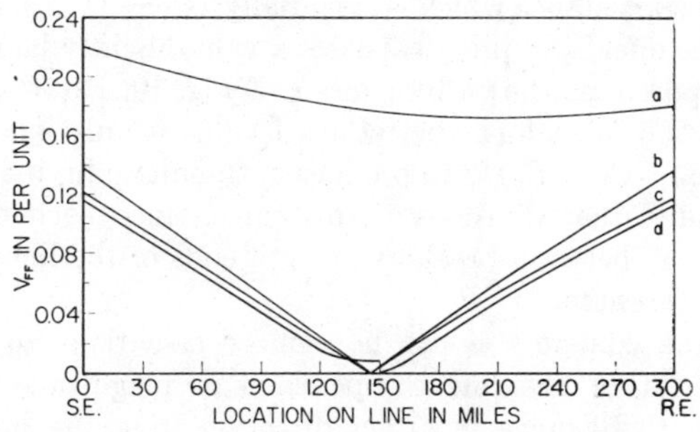

Fig. 11. Effect of series capacitors V_{FF} at various locations on line. No compensation: curve a—$C = 0$, curve b—C = optimum value; 50-percent compensation, C = optimum value, series capacitor in phase A: curve c—shorted, curve d—not shorted.

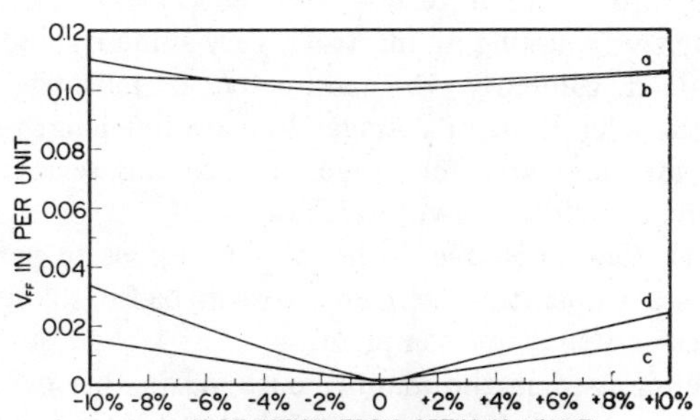

Fig. 12. Base case—effect of deviation from optimum value of capacitance: curve a—maximum V_{FF}, curve c—minimum V_{FF}; and from inductance: curve b—maximum V_{FF}, curve d—minimum V_{FF}.

inductive coupling of phase A with phases B and C, and hence the maximum value of V_{FF} as shown by curves c of Figs. 6 and 7. Figs. 8 and 9 show that a wide range of values of source X_0/X_α and V_S/V_R have relatively small effect on the magnitude of V_{FF}.

For increasing line lengths, transmitting the same amount of power, δ also increases resulting in increasing values of maximum V_{FF} as may be observed from curve c of Fig. 10. Curve e of Fig. 10 shows the optimum values of capacitances needed for different line lengths, and these are compared with those obtained from lumped capacitance equivalent. Curves e of Figs. 7 and 8 show the variation of optimum values of capacitances required as function of source X_α and source X_0/X_α.

For the same value of transmitted power, use of series capacitors will decrease δ and thereby decrease the value of V_{FF} as may be seen by comparing curve b with curves c and d in Fig. 11. When the series capacitor in phase A is not shorted, the values of V_{FF} are slightly lower than when it is shorted.

Fig. 12 shows that a ± 10 percent deviation from the optimum value of capacitance and inductance (of the shunt reactors) has very little effect on the minimum V_{FF}. This is because at the midpoint of the line almost all of V_{FF} is due to capacitive coupling whereas at the line ends the effect of inductive coupling is considerable.

Conclusions

A single capacitor connected across each pole of the circuit breakers used to switch transmission lines can be very effective in nullifying the capacitive coupling effect inherent in single-phase switching followed by high-speed reclosing. Each phase requires a capacitance which is essentially $(C_1 - C_0)/3$. Because of long-line effects, in practical cases a value slightly larger than this is required, as shown by curve e of Fig. 10. However, even for a line 200 miles long the required value is only 4.0 percent greater than $(C_1 - C_0)/3$ to produce optimum nullifying effects. It was found that divisions of this capacitance, each of value $(C_1 - C_0)/6$, between breakers at both ends of the line yielded best overall results.

While the voltage V_{FF} can be reduced essentially to zero at the midpoint of the line, it increases in magnitude, almost linearly, with distance in either direction from the midpoint. In all cases studied, however, the voltages V_{FF} were lower with the proper capacitor across breaker pole than with no capacitor. The parameters of most significance affecting the magnitude of voltages V_{FF} are line length and line loading. For a given line loading, series compensation tends to reduce the voltages V_{FF}.

For the case of shunt reactor compensation to achieve the same capacitive coupling nullification, very similar values of V_{FF} were obtained. Similarly, the parameters of most significance affecting the magnitude of voltages V_{FF} are line length and line loading. Again, similarly, for a given line loading, series compensation tends to reduce the voltage V_{FF}.

Shunt reactors, of course, have other functions to perform in normal system operation so a comparison based solely on the magnitudes of V_{FF} is not complete.

A capacitor used in the manner outlined in this paper is, of course, obviously helpful in reducing the rate of rise of recovery voltage across the circuit breaker terminals for all SLG faults on the transmission line. Quantitative evaluation of transient overvoltages accompanying the opening and subsequent reclosing of circuit breakers is left as a problem for continuing study.

Appendix

The digital computer program to calculate V_{FF} is based on the principle of superposition and uses the network in Fig. 4. Input data to the program consist of the transmission line constants and values of the base case parameters (or any variations thereof). The program is briefly described by the following steps.

Step 1

Compute the $ABCD$ constants for the α and 0 components of the line from the line constants.

Step 2

Using V_S, V_R, δ, $X_{\alpha S}$, $X_{\alpha R}$, and the $ABCD$ constants of the α line component, compute $E_{\alpha S}$, $E_{\alpha R}$, and the angle between them. This computation assumes balanced system operation and, therefore, the capacitors across the breakers (across the coupling transformers in Fig. 4) are short circuited.

Step 3

Choose values for the capacitors across the coupling transformers. Express, in terms of $ABCD$ constants, the network consisting of the transformer, the capacitor across it, and the source reactance associated with it, for each end of the line. Now the network of Fig. 4 can be represented in terms of a total of four two-port networks, the other two being the α and 0 equivalents as determined in Step 1.

Step 4

For $E_{\alpha R}$ equal to zero, compute the currents and voltages associated with this network, especially the voltages at the ends of the α and 0 line components. (It is optional, in this step, to reduce the network to two or one two-port networks with respect to the active source. This will simplify the program for solving the network.)

Step 5

Using the distributed constants approach, compute the voltages V_α at various specified locations on the α line component and the voltages V_0 at the corresponding locations on the 0 line component. Add V_α and V_0 to obtain V_a (voltage of phase A with respect to ground).

Step 6

Repeat Step 4 for $E_{\alpha S}$ equal to zero and repeat Step 5.

Step 7

Add the voltages V_a from Steps 5 and 6 to obtain the total voltage V_a. Check to see if this is minimum at a specified location on the line, e.g., midpoint. If yes, print the values of V_a and proceed to Step 1 to accept new set of data. If not, go to Step 3 and choose another set of values for the capacitors and continue the process.

The program to compute the fault Thevenin impedance is based on reducing the three-phase representation of the system into a single impedance across the fault terminals, through a succession of wye-delta transformations. Here the transmission line is represented in terms of lumped constants which yield the exact equivalent terminal impedances for fundamental frequency.

References

[1] IEEE Working Group on Arc Deionization Times, "Arc deionization times on high speed three-pole reclosing," *IEEE Trans. Power Apparatus and Systems* (*Special Supplement*), pp. 236–253, 1963.

[2] N. V. Dravid, "Analysis of a new scheme for decreasing arc extinction time required for single pole reclosing in EHV systems," Ph.D. dissertation, University of Wisconsin, Madison, 1968.

[3] E. Clarke, *Circuit Analysis of Power Systems*, vol. 1. New York: Wiley, 1943, ch. X.

[4] N. Knudsen, "Single phase switching of transmission lines using reactors for secondary arc," CIGRE, Paper 310, 1962.

[5] E. W. Kimbark, "Suppression of ground-fault arcs on single-pole-switched EHV lines by shunt reactors," *IEEE Trans. Power Apparatus and Systems*, vol. 83, pp. 285–290, March 1964.

Discussion

Eric T. B. Gross (Rensselaer Polytechnic Institute, Troy, N. Y. 12181): "Dead" time is required in single-pole reclosing in order to allow the arc path to deionize. With long line sections and very high voltages, the "residual current" through capacitive coupling between the opened phase and the two other still energized phases may maintain an arc to ground. This problem has been under study for 20 years or more [1].

The main problem is the complete elimination of the residual arc from the point of fault. An early suggestion [6] comprises a water-jet ejector which grounds the faulted phase at the substation end of the line temporarily, thereby establishing a preferred path for the residual current which results in quick arc extinction. It is believed that this method has not been widely used abroad, mainly because residual arcs of long duration do not develop if the line sections are relatively short.

A very much simpler solution is a variation of the water-jet method. The water jet is replaced by a high-speed, single-pole grounding switch using vacuum breaker elements. Such vacuum switches have been installed in two systems in this country at 230 kV operating voltage in a basically similar application involving differential protection of transformers tapped directly on the HV line. It is obvious that such simple vacuum switches provide a considerably cheaper solution than the installation of HV capacitors of considerable size.

References

[6] M. Muller, F. Gygaz, C. Hahn, and P. Baltensperger, "Protection of E.H.V. systems, taking into account single-phase automatic reclosure on very long lines," *Brown Boveri Rev.*, vol. 45, pp. 243–253, 1958.

[7] R. Willheim and M. Waters, *Neutral Grounding in High Voltage Transmission*. Amsterdam: Elsevier, 1956.

Manuscript received February 16, 1968.

E. W. Kimbark (Bonneville Power Administration, Portland, Ore. 97208): The authors are to be congratulated for having invented a simple, ingenious, new method of cancelling the effect of capacitive coupling between the conductor isolated by single-pole switching and the two live conductors of a three-phase transmission line. A method previously proposed utilizes specially connected shunt reactors.

Both methods seem to be almost equally effective when judged by calculations of steady-state secondary arc current and recovery voltage. However, they may differ in their transient behavior. On a line with no shunt reactors, the recovery voltage consists of a sine wave of power frequency and a dc voltage due to trapped charge. On a line with shunt reactors, as pointed out by Dr. T. Udo, the recovery voltage consists of a sustained sine wave of power frequency and a damped sine wave of natural frequency, the latter being generally of lower frequency than the power frequency. The two frequencies beat. As a consequence of the beats and of the initial conditions, the first few voltage peaks are lower than the steady-state peak voltage, and this circumstance may aid extinction of the arc. This inference should be verified by laboratory tests. Of course, a line employing the new method may have shunt reactors, presumably connected in grounded Y, installed for other reasons, and, if it has, the recovery voltage will have beats.

Manuscript received February 20, 1968.

Because the two methods may prove almost equally effective, it becomes important to compare them with respect to the cost of the additional devices which may be required for satisfactory arc extinction. I have calculated the reactive-power rating of the required capacitors or inductors, and have estimated their costs, using the 200-mile, 60-Hz, 500-kV line of the base case of the paper for this comparison. The results are as follows:

New method, using one bank of series capacitors, rated

58.5 Mvar, cost at $5.80/kvar	$340 000
3 circuit switchers	180 000
	$520 000.

Former method with minimum shunt compensation (45.8 percent)

3 Y-connected ungrounded shunt reactors, 175.5 Mvar at $3.95/kvar	$693 000.

Former method with 75 percent shunt compensation

additional neutral reactor, 7.6 Mvar at $3.95/kvar	$ 30 000
substitution of 69-kV bushings for 15-kV bushings and increased insulation of main reactors	60 000
total estimated cost	$ 90 000

The estimated incremental cost of fault-neutralizing equipment is considerably lower than that of the new method if shunt compensation is already needed for other purposes. A 200-mile 500-kV line would normally require shunt reactors.

Both the neutralizing capacitors and the grounded reactor of the four-reactor scheme are energized only during the "dead time" of less than one second and this circumstance permits some economy by overworking the neutralizing equipment.

The new method requires additional switches for inserting a neutralizing capacitor when clearing a line-to-ground fault but not for other types of faults, whereas the neutralizing reactors of the former method are left connected to the line. Methods of switching are not discussed in the paper. Have the authors chosen, at least tentatively, the simplest and most economical switching scheme?

Unfortunately, neither of the methods proposed for neutralizing capacitive coupling can also neutralize inductive coupling to the faulted conductor, and therefore inductive coupling, though smaller than capacitive coupling, becomes the "bottle neck." If single-pole switching is ever required on a line so long and of so high voltage that inductive coupling maintains the fault arc too long, perhaps the best remedy is to provide a set of rapid grounding switches at each end.

W. A. Lewis (Illinois Institute of Technology, Chicago, Ill. 60616): The possibility of using single-pole reclosing for protection against single line-to-ground faults is receiving considerable consideration for EHV lines, in the range of 500 kV and higher, because in several cases it may be desirable to initiate a development with a single line, at EHV, instead of two or more lines at lower voltages, when

Manuscript received February 21, 1968.

the ultimate project fully justifies the higher voltage. If, by the use of single-pole reclosing, adequate service reliability can be obtained with the initial stages of the single line, the single-pole reclosing will be quite advantageous. With these ideas in mind, the author of this discussion has also carried on an extensive investigation of this problem for the Bonneville Power Administration (BPA), except that the investigation did not include the possibility of a capacitor across the circuit breakers to be opened, as is suggested by the authors of the present paper.

The analysis for BPA has included the effect of shunt reactors for capacitance compensation, over a substantial range. As has been pointed out by earlier investigators, various arrangements, such as a bank of three shunt reactors in Y connection with a fourth reactor between the neutral point of the connection and ground, offers the possibility that the voltage induced by capacitive coupling between the sound conductors and the isolated faulted conductor can be completely neutralized, and in addition the positive sequence capacitance can be substantially reduced, thereby increasing the impedance for the secondary arc current. Theoretically, by increasing the size of the reactors, all of the line capacitance can be compensated, and the secondary arc current should be reduced to zero. The analysis has shown, however, that when an attempt is made to reach this condition, the resulting network becomes very sensitive to slight variations from the ideal values of shunt reactors, and even small departures from resonance can produce extremely high recovery voltages across the arc path in the event that the arc is extinguished. The difficulty of obtaining exact resonance, and the fact that the line characteristics will change somewhat because of temperature and weather changes, suggest that the attempt to reduce the secondary arc current to zero by this means will prove impractical. Accordingly, the practical approach appears to be to neutralize the capacitive induction, and then use shunt reactors to minimize the effect of magnetic induction, and thereby reduce the secondary arc current to a reasonably small value. By this process it appears that the secondary arc current can be made small enough, without excessive reactor costs, that rapid extinction of the secondary arc can be obtained unless the line sections involved are quite long.

With regard to the other factors treated by the authors and presented in Figs. 5–12, the investigation that I have conducted is in substantial agreement. In conducting this investigation, a somewhat more general digital computer program was developed that is capable of treating long distributed lines in detail, with the rest of the system represented by equivalents, not only for the single-pole reclosing case, but for other types of multiple unbalances, such as simultaneous faults on different phases of the two parallel lines, and other problems that have previously been difficult to handle, although generally of infrequent occurrence. Thus, the possibility exists that the methods developed for this problem may have application to other problems that will gradually become more prominent as the protection against the more common types of short circuits is further improved.

At this time it is difficult to reach any conclusion regarding the suitability of the capacitor connected across the open pole of the circuit breaker. For the purpose of extinguishing the secondary arc, it appears that this device will be quite effective. On the other hand, it appears possible that there might be some undesirable side effects resulting from the fact that the faulted conductor is not completely isolated, and it does not appear possible, without additional circuit breakers, to open the capacitor circuits without first reclosing the circuit breakers to bypass the capacitors. Although the capacitor currents will be small, the voltages involved are still quite high, and even to interrupt the capacitor currents would require a substantial capacitance switching device unless the circuit breakers can always be reclosed first. Whether this arrangement can be tolerated in practice is one of the questions that must be carefully investigated in determining whether the capacitor arrangement is to be applied.

Harold A. Peterson and **Narayan V. Dravid:** We greatly appreciate the comments made by the several discussers of our paper. A number of pertinent facts have been brought out, many of which were included in Dr. Dravid's Ph.D. dissertation, but not in our paper.

Single-pole grounding switches, of the vacuum-break type as suggested by Dr. Gross, might well be more suitable than any other means, especially for the longer transmission lines. This thought appears to be shared also by Dr. Kimbark, as stated in his last paragraph.

Comments with regard to the transient voltages by Dr. Lewis and Dr. Kimbark point up the fact that this aspect of the problem requires careful study in order to finalize conclusions. While certain generalities can be pointed out, as has been done, each application should be considered as a separate problem.

Costs were not within the scope of the thesis. As regards costs of alternatives, Dr. Kimbark's figures are very revealing. It would appear that for long lines, the incremental cost of achieving secondary fault current neutralization by means of shunt reactors is so small that the capacitor scheme would not be competitive. For lines of intermediate length, especially when shunt reactor compensation is not otherwise required, the comparison might be more favorable.

Dr. Lewis' comments reflect his depth of analysis of the general problem under discussion. We are glad to have his substantial agreement with our results. Perhaps at some future time we may expect to have available through publication more of his results, including simultaneous faults on multicircuit transmission lines. Field tests, finally, would be most revealing as to the relative adequacy and suitability of the various alternatives proposed.

Manuscript received March 11, 1968.

Digital Analysis of Single-Pole Switching on EHV Lines

R. BRUCE SHIPLEY, FELLOW, IEEE, HENRY J. HOLLEY, MEMBER, IEEE, AND DOROTHY W. COLEMAN, ASSOCIATE MEMBER, IEEE

Abstract—**Long distance EHV transmission and generator installations at remote locations have intensified stability problems. Transmission necessary to improve stability is expensive. Single-pole switching is being reviewed as an alternate. Phase-to-ground fault current in the open phase of the line is reduced by reactor compensation to make arc extinction possible. A means for increasing the line capability when one phase is open is investigated. The effectiveness is analyzed in detail using a digital computer.**

Paper 31 C 69, recommended and approved by the Power System Engineering Committee of the IEEE Power Group for presentation at the PICA Conference, Pittsburgh, Pa., May 15–17, 1967. Manuscript submitted February 6, 1967; made available for printing February 19, 1968.

The authors are with the Tennessee Valley Authority, Chattanooga, Tenn.

Introduction

SURGE impedance loading on a multiconductored 500-kV line approximately equals the capacity of modern 1 000 000-kW generators. The economic attractiveness of installing a remote generator on a radial line becomes apparent. When the

Reprinted from *IEEE Trans. Power App. Syst.*, Aug. 1968.

sudden disconnection of a 1 000 000-kW generator from the system can be tolerated, rather large installation economies can be realized. However, it may still be desirable to minimize the frequency of the disturbances, provided this can be done without excessive expenditures. One possible method is single-pole switching, a method that has gained little popularity in this country because of the uncertainty of arc extinction on long high-voltage lines. However, Kimbark and others have introduced methods of reactor compensation that seem to offer the possibility of reducing the arc current to safe values [1], [2]. Analysis of the problem is made more complex when the line is not transposed. Complex inductive and capacitive coupling in the line make it necessary to use modern techniques of analysis if reasonably accurate results are to be achieved.

It is possible that the length of the line and the system constants may be such that stability cannot be maintained when one phase of the circuit is out of service for the length of time required for arc extinction. Power transfer over the two remaining phases can be increased by installing series capacitor compensation in the phase conductors. Considerable literature is available on this subject. Other methods of increasing power transfer include transmission system additions, as well as the reduction of equipment impedances.

A capacitor in the neutral of the generator transformer bank as a means for increasing power transfer is academically studied. This is believed to be a new consideration that has not been previously investigated. Theoretically, it can increase the power transfer in the two good phases by approximately the same amount that is lost by the removal of one phase. Further, a ground fault on the line does not significantly alter the power transfer. Therefore, the shock to the system is greatly reduced and the probability of retaining synchronism is greatly increased. Such an installation would raise many uninvestigated problems of protection, increased insulation levels, ferroresonance, etc. that are beyond the scope of this study.

Power System Investigated

The power problem investigated is that of connecting a 1100-MW unit into a large system over a 90-mile, 500-kV transmission line. Engineering conditions are sufficiently typical that the results may be of interest to others. The generator step-up transformer has an impedance of 17.5 percent on a 1200-MVA base. The 500-kV line is of typical three conductor per phase construction as described in Holley *et al.* [3]. The 1100-MW generator has a subtransient reactance of 16 percent and a transient reactance of 30 percent. The sequence impedances of the receiving end system are $Z_1 = Z_2 = 12.9$ percent and $Z_0 = 19.6$ percent on a 1000-MVA base. For transient stability studies, the receiving end system is represented in considerable detail, including the 500-kV interchange lines and the power systems to the south and west of the Tennessee Valley Authority (TVA).

Philosophy of Approach

Transient Stability

The transient stability of the system under single-pole reclosing conditions is investigated by the use of a stability program produced by TVA [4]. This program uses only positive-sequence impedances. To study a single phase-to-ground fault it is necessary to connect the fault point to ground through an impedance. The impedance is the sum of the negative- and zero-sequence impedances of the entire system looking in at the fault point. Any power flow unbalance produced by connecting this impedance causes the swing condition to begin.

After the breaker poles at each end of the faulted phase open, it is necessary to represent the open-phase condition in the positive-sequence network. This is accomplished by inserting an impedance in series with the positive-sequence line impedance. The exact procedure to be followed is a function of the reactor compensation that is to be applied to the line. If the line is completely compensated, sufficient reactors are applied at each terminal of the line to cancel the effect of all shunt capacitance in all sequence networks. For an untransposed line these reactors will not all be the same size.

Considering a completely reactor-compensated transposed line, the impedance that must be inserted in series with the line positive-sequence impedance is determined as follows. Open the three phases at only one end of the line. Determine the series negative-sequence impedance and the series zero-sequence impedance, including the line and the system. Parallel the two impedances to give the desired value. In the stability study after the breaker poles open, the line impedance is simultaneously modified to represent the open phase, and the study is continued for the required time.

Note that when the zero- and negative-sequence impedances are used in the faulted and open-phase calculations, they apply to system values and not just line values. Thus they involve generator impedances. In a stability study there is always a question about the value that should be used for generator impedances since they are functions of the speed of the disturbance. Transient impedances are used in the stability study. Subtransient impedances are used to determine short-circuit impedances that are inserted in the positive-sequence network. This introduces certain approximations into the study. However, these approximations should be as good as the usual approximations associated with a stability study.

Reactor Compensation

The capacitances of the line can be compensated for secondary arc current reduction by connecting the reactors in a number of different ways. The simplest compensation requires six reactors at each end of the line. These six reactors are tuned approximately to resonance with the six line capacitances. Of course, three line capacitances exist between each of the three phase wires, while three more capacitances exist from each phase wire to ground. If only the interphase capacitances are tuned to resonance, the ground fault on the open conductor will have near zero current only when the line conductors have no induced longitudinal voltage. Actually, the line conductors always have a longitudinal voltage, and the fault current can be further reduced by also compensating the phase-to-ground capacitance. Naturally, different degrees of compensation can be achieved by placing the reactors at only one end of the line, etc. (see Kimbark [1]).

To gain insight into the compensation problem, consider the following. The number of circuit elements required to represent a three-phase untransposed line carrying unbalanced currents tends to obscure the problem. A line made up of only two conductors will serve to simplify it. Fig. 1 shows such a line with one of the conductors energized at each end, while the other conductor is open at its two ends. Assume that the conductor is energized in such a way that it carries no longitudinal current and therefore there is no longitudinal voltage induced along the open conductor. Currents do flow through the capacitors to earth. When the switch is closed, placing the fault at one end of the line, current flows only in the fault and in the line-to-line capacitances if no voltage drop along the conductor is assumed. Therefore, for this case, the fault current can be made

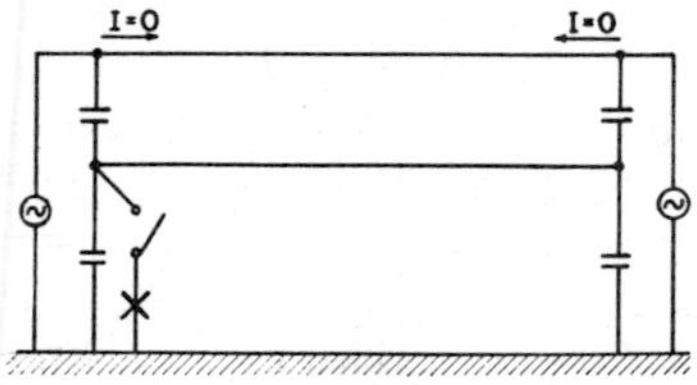

Fig. 1. Simplified circuit with no longitudinal voltages.

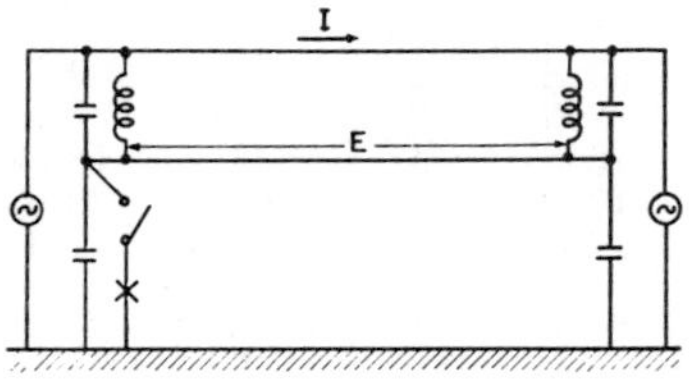

Fig. 2. Simplified circuit considering longitudinal voltages.

zero by simply neutralizing the line-to-line capacitances. So long as no longitudinal voltages exist, the line capacitances can be neutralized by installing one or more reactors between the two conductors. This condition reduces the ground fault current to zero regardless of the location of the ground fault on the open conductor.

Fig. 2 shows the phase capacitances compensated at each end, but with longitudinal voltage induced in the open conductor by current flowing in the other. For this condition, it is evident that the induced voltage is available to circulate current through the open conductor and through its capacitances to ground. Thus, when the fault is placed at one end, a voltage exists across the capacitances to earth at the other end, with a consequent flow of current. The return path for this current is through the fault. Thus, to reduce the fault current, it is necessary to compensate the line-to-ground capacitance. From Fig. 2 it is evident that for the lumped representation, reactor compensation of the capacitances to ground at the two ends will result in no fault current regardless of the location. In the actual case, the problem is more complex: exact compensation cannot be expected. It would be theoretically possible for the phase-to-ground compensation to be slightly capacitive while the phase-to-phase compensation was slightly inductive, resulting in series resonance and high voltage on the open conductor. Complete compensation using lumped reactors can result in other obscure troubles. Consider lumped reactors sized to compensate all the phase-to-ground capacitance and located at one end of the circuit. Further, consider no fault on the open conductor and no self-impedance in the open conductor. Now the longitudinal voltage appears across a series resonant circuit and will result in high voltages and high currents on the open conductor. For these reasons it is best to avoid attempts at perfect compensation. However, when the circuit is not completely compensated, the fault current on the open conductor becomes a function of the current in the other conductor. It is evident that the fault current will be a function of the fault location along the conductor. Several of the more usual cases have been calculated and are given in the Section *Results*.

Arc extinction could be more effectively accomplished by the installation of devices at each end of the line-to-ground open phase. Then, before reenergizing the open phase, it would be necessary to open these devices. Naturally, costs and additional equipment that might fail are the objectionable features of this scheme.

Increasing Stability Limit by Use of a Neutral Capacitor

For radial systems that lose stability during the time required for arc extinction in a single-pole reclosing operation, it is theoretically possible to correct the trouble by installing a capacitor in the neutral of the sending end transformer. For the case of a transposed line with all the line capacitances completely compensated, the following conditions can be established. Let the neutral capacitor be sized so that it resonates with the zero-sequence reactance. Now the zero-sequence reactance at any point along the line with all conductors normally connected will be infinite. That is, current cannot be forced to flow into a ground fault because the three-sequence networks are connected in series for phase-to-ground faults. Actually, circuit resistance prevents resonance and some ground current does flow.

To represent an open conductor in the sequence networks, the negative-sequence and the zero-sequence networks are placed in parallel and then in series with the positive-sequence network. The zero-sequence impedance to be used in this case is the series impedance of the zero-sequence circuit, and this impedance is near zero. Thus the negative-sequence impedance is shunted by a near-zero impedance. Therefore, the positive-sequence network is virtually unchanged, and the power transfer of the system is approximately the same whether the conductor is open or not. Power system stability under these conditions is virtually unaffected by single-pole operations.

Results

Transient Stability Study

The following condition proved stable for the system. A phase-to-ground fault was placed on the sending end of the line for four cycles. The faulted phase was then opened at both ends. Thirty cycles from the beginning of the fault the phase was successfully reclosed. If the successful reclosure was delayed two additional cycles, the system was unstable. Thus, almost no stability margin existed on the normal system.

Reactor Compensation

According to published literature on single-pole reclosing, two important factors are the following [5]:

1) the current flow (secondary arc current) in the arc after the breaker poles at each end of the faulted phase open;

2) the 60-Hz voltage (coupling voltage) that appears at the fault point after the secondary arc is extinguished.

These values have been calculated at points 1–5 (see Fig. 3) for several cases.

Table I shows a comparison between these quantities for an untransposed line and for a line transposed as in Fig. 3(b). There is no reactor compensation in either case. All quantities tabulated are calculated by holding the generated voltages at the maximum swing angle observed from calculated stable cases. This produces severe switching conditions.

Table II shows the effect of nominally compensating the line 100 percent, using three different reactor arrangements. In all three cases the line is transposed as in Fig. 3(b). Note that with reactors at the sending end only, or with reactors at each end, the voltages along the open phase are extremely high, indicating resonance effects. Actually, the case with reactors at the middle of the line is also in partial resonance, although not as noticeably as in the other two cases.

The high-voltage condition for the case with the reactors at the sending end is due to series resonance. The end shunts at one

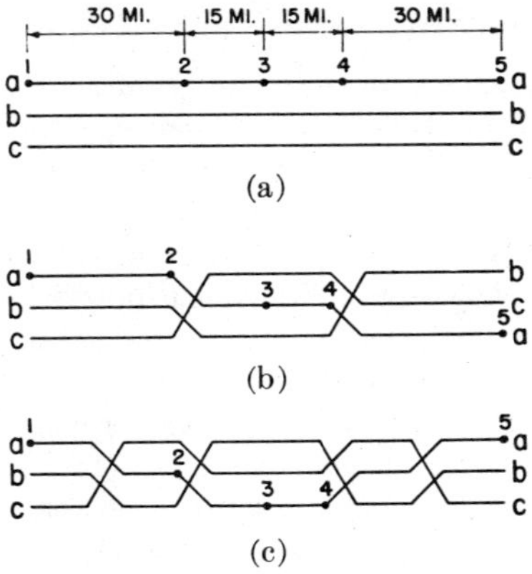

Fig. 3. Methods of transposition and points at which voltages and currents were calculated.

TABLE I

EFFECT OF TRANSPOSITIONS ON SECONDARY ARC CURRENTS AND COUPLING VOLTAGES

	Untransposed		Transposed	
Point	Coupling Voltage (kV, rms)	Secondary Arc Current (A, rms)	Coupling Voltage (kV, rms)	Secondary Arc Current (A, rms)
1	63.8	41.1	50.7	33.0
2	38.0	24.3	27.2	17.5
3	26.8	17.1	18.0	11.5
4	19.8	12.7	20.2	13.1
5	30.6	19.8	42.1	27.3

TABLE II

EFFECT OF NOMINAL 100 PERCENT REACTOR COMPENSATION ON SECONDARY ARC CURRENTS AND COUPLING VOLTAGES

	Reactors at Sending End		Reactors at Midpoint		Reactors at Each End	
Point	Coupling Voltage (kV, rms)	Secondary Arc Current (A, rms)	Coupling Voltage (kV, rms)	Secondary Arc Current (A, rms)	Coupling Voltage (kV, rms)	Secondary Arc Current (A, rms)
1	1512	20.5	82.4	0.47	616	3.9
2	1504	19.8	67.3	0.39	621	3.9
3	1496	19.5	65.6	0.38	625	3.9
4	1485	19.2	67.7	0.39	629	3.9
5	1461	18.8	86.1	0.50	640	4.0

TABLE III

EFFECT OF ADDITIONAL TRANSPOSITIONS ON SECONDARY ARC CURRENTS AND COUPLING VOLTAGES

	Reactors at Each End		Reactors at Midpoint	
Point	Coupling Voltage (kV, rms)	Secondary Arc Current (A, rms)	Coupling Voltage (kV, rms)	Secondary Arc Current (A, rms)
1	49.1	0.31	68.2	0.42
2	20.2	0.13	38.7	0.24
3	4.8	0.03	23.1	0.15
4	10.6	0.07	7.5	0.05
5	40.0	0.25	22.6	0.14

TABLE IV

EFFECT OF COMPLETE INTERPHASE COMPENSATION AND 45 PERCENT PHASE-TO-GROUND COMPENSATION ON SECONDARY ARC CURRENTS AND COUPLING VOLTAGES

Point	Coupling Voltage (kV, rms)	Secondary Arc Current (A, rms)
1	44.6	13.1
2	16.1	4.7
3	0.3	0.0
4	15.7	4.6
5	46.5	13.7

end become inductive when the reactors are connected in parallel. The end shunts at the other end remain capacitive. The voltage induced in the open phase appears across a series resonant circuit, resulting in high voltage.

The high-voltage condition for the cases of reactors at the middle of the line or at the two ends of the line is due to improper transpositions. Table III shows the effect of transposing the line as in Fig. 3(c). Note that the distribution of voltages along the open conductor is fairly symmetrical for the case with reactors at each end. This distribution is not symmetrical with reactors at the middle, because the presence of the reactors introduces a discontinuity in the line.

The voltage distribution along the line can be improved by adjusting the reactors so that the line is exactly 100 percent compensated. However, due to the fact that transmission line capacitances are difficult to calculate accurately and also because capacitances change due to weather and seasonal variations, no attempt is made to use exactly 100 percent compensation. Furthermore, a small variation in the reactance of the reactors when the compensation is near 100 percent can cause overvoltages. For example, the calculations of Table III are made with six reactors at each end. The three phase-to-ground reactors at each end have a reactance of 3840 ohms. If a phase-to-ground reactor at the sending end is changed by 2.6 percent to 3740 ohms, the voltage at the sending end will rise from 49.1 kV to 590 kV. Another case where extremely high voltages will occur with 100 percent compensation is the case of reclosing one phase of a breaker at one end of the line into a ground fault on that phase. Therefore, it is felt that compensation in the neighborhood of 100 percent should be avoided.

Preliminary cost estimates indicate that the most economical reactor installation will be a four-legged reactor bank located

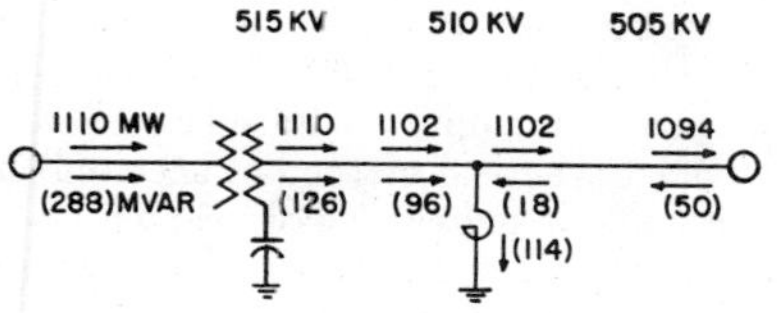

Fig. 4. Normal power flow and voltage conditions.

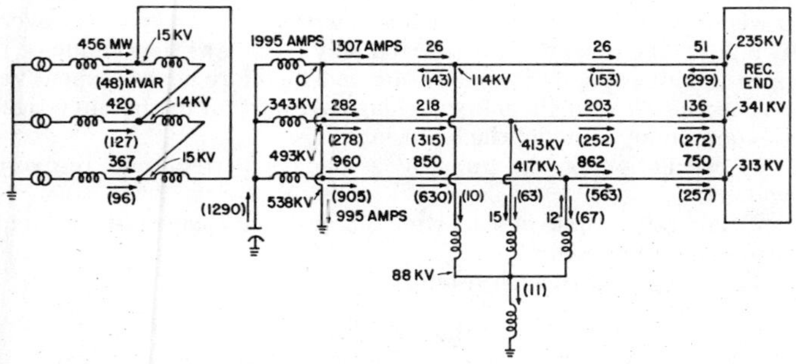

Fig. 5. Power flow and voltage conditions with phase-to-ground fault at sending end.

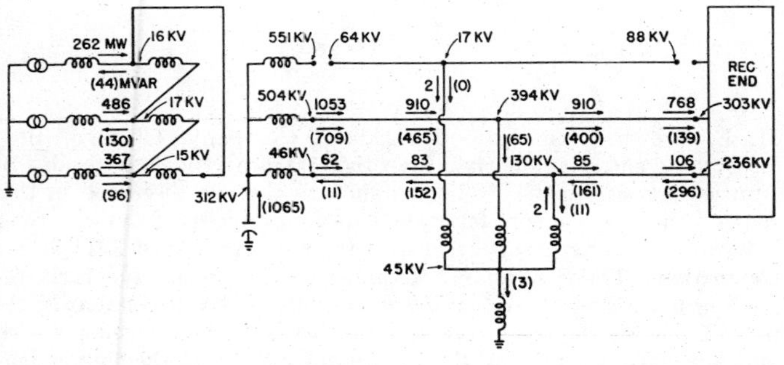

Fig. 6. Power flow and voltage conditions after faulted phase has been isolated by single-pole operations at each end.

at the middle of the line. The phase-to-neutral reactors should be 2500 ohms, so that a reactor with an adequately insulated neutral can be used as a spare with this bank or with other 100-Mvar banks on the system. The ohmic value of the neutral reactor is determined by finding the neutral ohms required to obtain approximately zero voltage at the middle of the open phase, and is found to be approximately 588 ohms. With 2500 ohms from phase to neutral and 588 ohms from neutral to ground, the interphase capacitances are compensated approximately 100 percent and the phase-to-ground capacitances are compensated about 45 percent. Due to the fact that the overall compensation is not 100 percent, resonance problems caused by reactor and capacitance variations will be avoided. Table IV shows the secondary currents and coupling voltages obtained using this reactor bank.

Capacitors in Transformer Bank Neutral

Fig. 4 shows circuit conditions existing when normally carrying 1100 MW. Fig. 5 shows the currents and voltages that exist on the system when one phase is faulted to ground. Fig. 6 shows the currents and voltages that exist on the system when one phase is open. It should be noted that the power transfer is almost unaffected by the fault or by opening the phase. The neutral capacitor can drastically affect current and voltage distribution not only for the single phase-to-ground fault but also for other faults which are not included in this study.

Conclusions

1) To use balanced reactors to suppress secondary arc currents, it is desirable to completely transpose the transmission line once. Equal reactor banks at each end of the line or one reactor bank at the middle of the line give approximately the same performance. The choice should be made on the basis of system economics.

2) If the capacitance of the line is compensated 100 percent, the line should be completely transposed twice. With 100 percent reactor compensation, small variations in reactor constants can cause resonance problems. Also certain fault conditions can cause resonance problems.

3) According to published literature, a 90-mile line does not require 100 percent compensation to extinguish the secondary arc [5], [6]. As line length increases, the coupling voltage increases, requiring a greater degree of compensation to reduce the secondary arc currents to a level that can be extinguished. If the secondary arc current is to be limited to a fixed value with a fixed percent compensation, then the maximum length of the line is fixed. If there is a maximum allowable coupling voltage, this will also determine the maximum line length that can be compensated for a given operating condition.

References

[1] E. W. Kimbark, "Suppression of ground-fault arcs on single-pole-switched EHV lines by shunt reactors," *IEEE Trans. Power Apparatus and Systems*, vol. 83, pp. 285–290, March 1964.

[2] N. Knudson, "Single-phase switching of transmission lines using reactors for extinction of the secondary arc," CIGRE, Paper 310, 1962.

[3] H. Holley, D. Coleman, and R. B. Shipley, "Untransposed EHV line computations," *IEEE Trans. Power Apparatus and Systems*, vol. 83, pp. 291–296, March 1964.

[4] R. B. Shipley, N. Sata, D. W. Coleman, and C. F. Watts, "Direct calculation of power system stability using the impedance matrix," *IEEE Trans. Power Apparatus and Systems*, vol. PAS-85, pp. 777–782, July 1966.

[5] Y. Ozaki, "Extinction of secondary arc current in a single-pole reclosing system," *J. IEE* (Japan), vol. 84, p. 909, 1964.

[6] A. S. Maikopar, "Minimum time of automatic reclosing," *Elec. Tech.* (USSR). New York: Pergamon Press, 1960, pp. 302–315.

Discussion

Edward W. Kimbark (Bonneville Power Administration, Portland, Ore.): This paper is not merely one on computational techniques but rather one whose authors show much insight and inventiveness on an engineering problem of current importance. It advances the knowledge of single-pole switching in at least three respects:

1) It demonstrates the importance of balance of interphase coupling and of symmetry of the compensating reactors with respect to the middle of the line for minimizing secondary arc current and recovery voltage.

2) It points out that the secondary arc current can be reduced by shunt-reactor compensation of the line-to-ground capacitances (C_0), but also shows that such compensation increases the recovery voltage, especially if this compensation is 100 percent.

3) It suggests the possibility of series-compensating the zero-sequence reactance for increasing the power transmitted while one conductor is open for clearing the fault.

I shall discuss each of these three points in turn and then suggest which aspects of single-pole switching require further investigation.

1) An investigation of the effects of unbalances in transposed and untransposed lines would be very laborious unless a suitable com-

Manuscript received May 8, 1967.

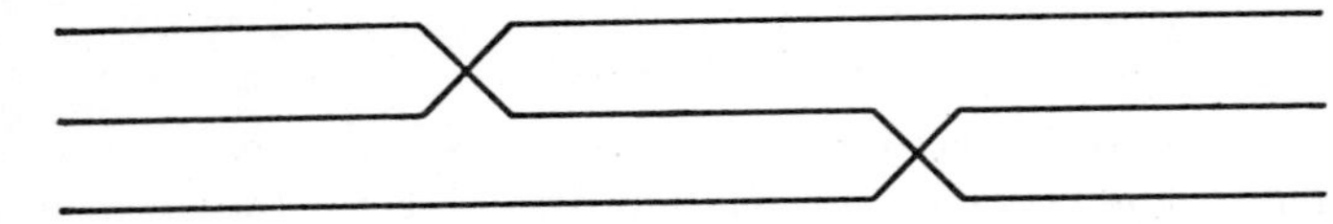

Fig. 7. Partial transposition.

puter simulation were used. A digital computer with a program in which the transmission line is represented in terms of its unbalanced self- and mutual-series impedances and shunt admittances appears to be the most feasible method because simple methods of representing such a line on a network analyzer have not been devised.

The advantage of a line with balanced interphase capacitances is that these can be well compensated with equal reactors. Equal reactors have a definite advantage in minimizing the number or complexity of spares.

Having shown the advantage of a balanced line, we turn to a consideration of methods of balancing an unbalanced line. The only one considered in the paper is full transposition.

Most HV transmission lines have parameters that are symmetrical about the middle conductor, the commonest configuration having equally spaced conductors in a horizontal plane. The parameters of such a line can be balanced by transposing each outer conductor with the inner conductor. One cycle of such transposition is shown in Fig. 7. This kind of transposition is perhaps cheaper than the kind shown in the paper because only two, instead of three, conductors are involved in each transposition.

Another way of balancing a line is to add lumped elements. Since the capacitance between the two outer conductors is less than that between the middle and each outer conductor, the line-to-line capacitances can be balanced by connecting one or more shunt capacitors between the two outer conductors.

What transposition was assumed in Table IV?

2) The paper points out that two quantities deemed important in extinction of the secondary arc are current and recovery voltage. The relative importance of these two factors is perhaps not yet fully understood. In Peterson-coil grounding, with which there has been much experience, correct tuning diminishes both current and recovery voltage. In single-pole switching or in laboratory simulations thereof, it is possible to vary the two factors independently. Such experiments were reported in a recent paper from CESI [7] and seem to show that recovery voltage is the more important factor. With recovery voltages of 30 kV or less, the arc extinction time is practically independent of the current.

Reactor compensation of the interphase capacitances, of course, diminishes both the arc current due to capacitive coupling and the recovery voltage due to the same, but does not affect the current and voltage due to magnetic coupling. It occurred to me some time ago, as it did to the authors, that, as the voltage magnetically induced in the isolated conductor circulates a current through the ground fault and line-to-ground capacitance, the current in the fault could be reduced by shunt-reactor compensation of the line-to-ground capacitances. Unfortunately, however, at the same time it increases the recovery voltage, especially but not solely because of the series resonance that occurs when this compensation is complete, as is so ably explained in the paper. Now, because arc current appears to be a less important factor than recovery voltage is in maintaining (or reigniting) the arc, it appears undesirable to completely compensate line-to-ground capacitances. In other words, the conclusions of the CESI paper [7] reinforce this paper's conclusions.

3) The possibility of compensating the zero-sequence reactance is fascinating. I had previously speculated on the possibility of using low-impedance insulated overhead ground wires with series capacitors for this purpose. The method suggested by the authors—namely, grounding the neutral point of the sending-end transformer bank through a capacitor—has the disadvantage of considerably raising the voltages to ground both of the line conductors and of the transformer neutral point during a line-to-ground fault, as shown in Fig. 5, and with an open conductor, as shown in Fig. 6. This increase of voltage is particularly objectionable in an EHV system.

An alternative would be to connect a capacitor or capacitors in series with a tertiary winding which would carry only zero-sequence current. This would allow the neutral of the Y-connected HV windings to be solidly grounded and would permit the use of graded insulation in those windings.

The principal object of series-compensating the zero-sequence impedance as proposed is to enable the transmission line to carry the same power with the same phase difference between terminal voltages whether or not one conductor is isolated for clearing a fault. I suggest, as the most practical method of accomplishing this object, the provision of a bank of switched series capacitors in the transmission line. These capacitors would be normally bypassed but would be inserted simultaneously with the disconnection of the faulted conductor. These series capacitors would reduce the loop reactance of all three sequence networks equally. There are two reasons why these capacitors should not be left permanently inserted. 1) It is better for transient stability to have no increase in line angle when the conductor is disconnected, because such an increase always results in overswing. 2) The capacitors, if inserted only as long as the conductor is open, may have a short-time rating and therefore be less expensive.

The aspects of single-pole switching that warrant further investigation are, in my opinion, the following:

1) calculation of the *transient* secondary current and recovery voltage;

2) laboratory tests of arc extinction in which compensatory shunt reactors are represented;

3) field tests with such reactors.

REFERENCES

[7] A. Balossi, M. Malaguti, and P. Ostano, "Laboratory full-scale tests for the determination of the secondary arc extinction time in high-speed reclosing," presented at the IEEE Summer Power Meeting, New Orleans, La., July 10–15, 1966.

H. E. Brown (Commonwealth Edison Company, Chicago, Ill.): It is very encouraging that investigators are reporting results of frontier investigations. Problems, such as the one discussed in this paper, that now confront large systems are problems of the very near future to smaller systems because of the rapid growth of EHV interconnections. These investigations must be carried on, and early reporting is important so that research will not be unnecessarily repeated. The authors are to be congratulated on their excellent work and especially for making the nature and results of their investigations available.

The authors recognize all the objections to this method of increasing stability and wisely avoid attempting to discuss the questions that are raised: "problems of protection, increased insulation levels, ferroresonance, etc. . . . ." Other investigators should address themselves to selected items in the list of problems.

A great deal has been said and written about the necessity of adequate documentation of a program. The authors have obviously again used their large library of matrix operations subroutines to put together a program that has produced some very valuable results. It would seem that this study would never be repeated and therefore documentation of the program would be unjustified. Is this assumption correct? It would be interesting to know approximately how many subprograms were called from the library tape in order to produce the program. Have the authors developed a SUPER-FORTRAN that permits easy assembly of a program of this type?

In the section *Philosophy of Approach* do the negative- and zero-sequences impedances refer to Thevenin impedances?

Manuscript received May 29, 1967.

W. A. Lewis, (Chicago, Ill.): This paper brings out important information regarding the possibility of using single-pole switching on EHV lines in the United States, and extends the probability that such switching will be used on the future EHV lines in this country.

There are certain sections of the paper that could profitably be somewhat clarified, and this discussion requests that certain addi-

Manuscript received June 5, 1967.

tional information be provided. The authors give a description of the method of representing an open-phase condition in the analysis of the positive-sequence network for stability. The procedure described is valid if there is only a single series unbalance in the system at the time in question. However, in the proposed application, one pole of the circuit breaker at each end of the line must be opened to clear the fault, and the possibility may also exist that the fault persists even after the two circuit breaker poles are opened. For this condition, the representation of the negative- and zero-sequence networks does not reduce to the single equivalent impedance as described, but involves a network of impedances interconnecting the various points of fault in the positive-sequence network. In some cases, the additional effects of the second and third unbalances may not be very large, because the opening of the first circuit breaker will introduce a large impedance in one phase that will limit the current in that phase, and the current will not be reduced much further by the opening of the second circuit breaker. Also, it is to be hoped that the arc will extinguish, so that the third unbalance will disappear. However, when the compensation is not complete, the opening of the first circuit breaker would still leave an appreciable current flowing through the second circuit breaker pole, and this would have to be reduced to zero by the insertion of an opening at the second location. Therefore, the more complex representation would be necessary.

Also, as has been brought out in the paper, there are likely to be conditions where the arc is not extinguished unless compensation is provided. Although the arc current may be small, the representation of the arc as a third unbalance is necessary if the voltage of the unfaulted conductor is to be properly calculated. Since both the secondary arc current and the voltage that appears across the arc when it is extinguished are important factors in the process of extinction of the arc, there will be cases when this representation is needed also.

The authors express some concern regarding the appropriate generator impedance to use in the study. Although it is recognized that further work is needed in the development of appropriate synchronous machine equivalents for stability studies, it seems appropriate in the present state of the art that transient reactance should be used to represent the generators for this type of study. Although subtransient reactance is regularly used in the determination of short-circuit currents, this is done because the short-circuit currents are usually desired at the first instant of short circuit. If their values were desired a few cycles later, it would then be appropriate to use the transient reactance. Since the primary purpose of this paper is to study particularly the conditions after the opening of the single poles of the circuit breakers, it is clear that the results are desired a few cycles after the fault has taken place. Therefore, for the calculation of the fault currents, the use of transient reactance is the most appropriate available. Also, it has been traditional to use transient reactance in stability studies as being the most appropriate compromise, taking into account that the subtransient effects die out so quickly that they have only a minimum effect on the stability swings. Therefore, it is the opinion of this writer that it would be appropriate to use transient reactance for the generators, at least for all conditions after the initial determination of the fault, before any circuit breakers have opened. Of course, the comments just made apply only to the positive-sequence network. For the negative-sequence network, it would be appropriate to use the generator negative-sequence reactances, which are regularly expressed in terms of subtransient reactances in the direct and quadrature axes, because of the continually changing effects of the negative-sequence current on the generators.

In all of the diagrams used to illustrate the paper, the circuits are represented as either nominal-pi or nominal-T circuits, with the line capacitance lumped in two segments at the ends of the line or in one segment at the middle of the line, depending upon whether a pi or T circuit is used. This leads to the inference that lumped circuits have been used in the calculations. However, it is my understanding that the authors have actually used several short segments of line connected in series, so that the results very closely approach the values that would be obtained when the lines are represented by distributed parameters. If this is the case, it would be desirable to make it clear.

The authors of the paper were involved in an earlier study of a somewhat similar system, in which the effects of omitting transpositions were investigated, particularly when the line was equipped with series capacitors. The results of that study seemed to indicate that, at least when series capacitors were used, the unbalance caused by omitting transpositions could create substantial negative-sequence currents that might be large enough to cause difficulties in operation and possibly objectionable rotor heating in the generators. Since, on long high-voltage lines, there is considerable likelihood that series capacitors will be used, there is some question whether it would be appropriate to operate such lines without at least one barrel of transposition. For the lines that are short enough not to require series capacitors, the effect of the unbalance is less pronounced, and therefore it is more likely that all transpositions might be omitted. However, when considering the longer lines, where series capacitors might properly be used, it would seem appropriate that one barrel of transposition should be considered as a starting point, rather than a line with no transpositions whatever.

In this connection, this writer has carried on similar work in investigating the use of single-pole reclosing on lines of various high voltages and various lengths. This work seems to indicate that reasonable values of series capacitance, up to approximately 70 percent of the line reactance, do not have any major effect on the secondary arc phenomenon so long as the resonance condition caused by full compensation of the line-to-ground capacitance is avoided. In reviewing my own work in conjunction with that done by the authors it seems possible to summarize a general conclusion that the phase-to-phase capacitance coupling between the conductors of the transmission line should be compensated approximately 100 percent, so that there will be no secondary arc current after the single poles of the breakers are opened, as a result of capacitance coupling alone. Therefore, the secondary arc current will be produced almost entirely by the magnetic induction caused by load current in the unfaulted phases. If the maximum load is such that the extinction of this component of secondary arc current is doubtful, further compensation of the phase-to-ground coupling may be needed, until the secondary arc current is reduced to a sufficiently low level that the arc will go out. Although the data on arc extinction under these conditions are rather meager, it is reasonable to expect that the arc will be self-extinguishing well before the compensation for line-to-ground capacitance approaches a resonant condition. Therefore, it appears practical to provide compensation that will result in arc extinction without serious overvoltages, except for very long lines.

According to the calculations of this writer, if the phase-to-phase capacitive coupling is not compensated approximately 100 percent, the uncompensated component of the secondary arc current resulting from the uncompensated capacitance will combine with the inductively induced secondary arc current, caused by load in the parallel conductors, in such a way that the secondary arc current for a fault at one end of the line will be the sum of these two components and at the other end of the line will be the difference. However, if the phase-to-phase capacitance component is completely compensated, the secondary arc current resulting from load current alone will have about the same magnitude for a fault at either end of the line. When this is the case, the problem of arc extinction will be less severe for the same degree of compensation of the line-to-ground capacitance.

For the longer lines, some shunt reactor compensation is likely to be needed for voltage regulation and to prevent excessive voltage rise when one end of the line is opened. In many cases, it appears reasonable that the values of shunt reactance required for this purpose will be nearly the same as those required to produce compensation for the phase-to-phase capacitance. Therefore, the same total value of reactive compensation may serve both purposes. However, if these reactors are used to compensate for the line-to-line capacitance, their neutral must be isolated from ground or connected to ground through the value of reactance used to compensate for the line-to-ground capacitance. Therefore, when single-pole switching is used, the principal reactors require a higher level of insulation at the neutral than would be required if single-pole switching were not attempted. Since circuit breakers for EHV lines can be readily adapted to single-pole switching, it appears that the major part of the cost for the use of single-pole switching will lie in the extra cost of insulating the neutrals of the reactors. There will be some extra cost for the reactor connecting the neutral of the phase reactors to ground, and for additional relaying, but these costs would probably be only a small percentage of the total.

The authors have suggested that the use of single-pole switching will improve stability, and substantial improvement can be obtained. They also suggest the possibility of further improvement by the use of series capacitors in the neutral connection. On the basis of the results presented, it appears that the use of the capacitor in the neutral will result in extremely high voltages, resulting in rather large cost increases to take care of insulation requirements. It therefore appears doubtful to this writer that the use of the series capacitor in the neutral would be attractive from an economic standpoint, even if it were not rejected for other reasons of overvoltage. Also, other methods of producing a similar increase in stability limits should be considered from the economic standpoint. One of these might very well be the use of three-pole reclosing, accompanied by the insertion of additional series capacitor compensation at the moment of reclosure, in line with the suggestions that have been made by Kimbark [1]. In the final analysis, the choice among various alternatives will likely be made on an economic basis.

Robert O. Bigelow (New England Electric System, Boston, Mass.): The authors have described computer applications for a type of power system operation which is quite common in Europe but is presently rather unusual in the United States, namely, single-pole switching and reclosing of transmission lines. They have presented an excellent analysis of a specific problem involving transpositions and shunt reactors to limit arc currents and thereby allow fast reclosing. This is a problem which is of primary concern on long lines with high voltages such as 500 kV.

We in New England are also actively engaged in studies for applying single-pole switching to an element of the main bulk transmission system. Our problems are somewhat different, however, because our lines are shorter and our transmission voltage is 345 kV. As a result, we do not require shunt reactors to limit secondary arc currents, and therefore have not had to concern ourselves too greatly with the specific problem described in this paper. We have found, however, that single-pole switching can introduce another problem where much work needs to be done in the application of computers. This is the problem of circulating zero-sequence currents in parallel underlying systems. These will flow when one phase of a line is open and can be of sufficient magnitude to cause false operation of ground relays at a time when we can least afford such misoperations. The problem will be further aggravated by mutual coupling if lines of the underlying system (115 kV in our case) are on the same right-of-way as the line being switched on a single-pole basis.

Referring to the stability studies described, I am surprised at the tight limits that seem to apply to the system described in the authors' paper. I note that stability following a single line-to-ground fault requires four-cycle single-pole clearing and 30-cycle reclosing and, further, that instability is expected if reclosing is delayed another 2 cycles to 32 cycles. This leaves very little margin for error.

It has been our conclusion, from numerous stability studies, that once a fault is cleared, a large amount of synchronizing power can flow over a line even if one phase is open, providing there are strong ground sources at both ends. For example, in our case, we have a 700-MVA steam generator connected to the system via a single 37-mile 345-kV line with a relatively weak underlying 115-kV system in parallel. For all cases the machine was unstable with any type of three-pole switching and reclosing, with or without a fault. With single-pole switching and reclosing on single-line faults, however, the reclosing could be delayed indefinitely and stability maintained. In other words, the timing for reclosing would be dictated by considerations other than stability, such as negative-sequence currents in the generator rotor or zero-sequence currents in ground relays on parallel circuits. We therefore plan to delay reclosing to at least one second to allow plenty of time for arc extinction.

Our results seem to be corroborated by current practice in France where we understand that the Electricité de France system is using single-pole switching on all lines at 380 kV with reclosing time uniformly set at 1.5 seconds. They report that this is satisfactory for their stability requirements.

Would the authors comment on what features of their system are causing such severe transient stability limitations, and also indicate how close to steady-state stability limits they may be operating under the conditions described in the paper?

Manuscript received June 26, 1967.

R. B. Shipley, H. Holley, and D. Coleman: The authors are very grateful for the efforts expended in preparing the excellent discussions by Dr. Kimbark, Mr. Brown, Dr. Lewis, and Mr. Bigelow. There is a minor amount of unavoidable overlap in some of the discussions. Therefore, we will avoid repetition in the closure by not adhering rigidly to answering each discusser completely independently.

Dr. Kimbark and Dr. Lewis recommend that the interphase capacitance be completely compensated and the line-to-ground capacitance compensated to the extent that the secondary arcs will be extinguished. We agree with this recommendation for most cases. An exception might be made when a section of line crosses a high-hazard area or an area of excessive contamination. In such a case, it may be desirable to adjust the compensation to minimize the secondary arc current in that particular area.

We agree with Dr. Kimbark that switching series capacitors into the line during the swing interval following a phase-to-ground fault appears to be the most practical way to transmit the same amount of power regardless of whether a breaker pole is opened or closed. Preliminary calculations indicate that the series compensation required will be more than 100 percent of the positive-sequence reactance of the line. This makes it almost mandatory to bypass the series capacitors during normal operation. Dr. Lewis remarks that series compensation of up to about 70 percent does not appreciably alter the secondary arc phenomenon. If he has made studies with larger amounts of compensation, perhaps the results should be published since they would be applicable in this case.

We agree with Dr. Kimbark that results from field tests with compensatory reactors would be very valuable at this time. Maikopar and Baliakov [8] show the results of field tests made on reactored lines. However, in their case the reactors are solidly grounded and not sized specifically to neutralize the interphase capacitance.

The type transposition used in Table IV was a complete roll. It is correct that this can be simulated as Dr. Kimbark suggests, and this is discussed in Holley *et al.* [3].

In reply to Mr. Brown, we did use our large group of matrix programs to solve this problem. We do not plan to try to document these programs because of the large number of possible sequences used to solve problems. These programs have been discussed philosophically in a conference at Stockholm, Sweden [9]. We used about 15 programs to solve this problem. On the old 704 computer we could call in any program in any sequence. We are now attempting to convert these programs for the 360 computer.

Dr. Lewis is correct in pointing out that the representation of the negative- and zero-sequence networks does not reduce to a single equivalent impedance to be inserted in the positive-sequence network in those cases where the compensation is not complete. In those cases a pi circuit must be inserted in the positive-sequence network. The procedure we use for determining this pi circuit is as follows. The open poles of the circuit breakers introduce discontinuities between the source and line terminals of the circuit breakers in the sequence networks. Perfect three-winding transformers with unity-turns ratios are connected into these sequence networks with a winding for each sequence network. After a network reduction is made to eliminate the extraneous terminals, the result will be a pi circuit with the architrave connected between the breaker source terminals in the positive-sequence network and the end shunts (generally unequal) connected from the breaker source terminals to the zero-potential bus in the positive-sequence network. If the line is long enough that a ground fault on the disconnected phase appreciably alters the power transfer, the sequence networks, in addition to being connected by the three-winding transformers, would be connected together at the fault point in the usual manner for line-to-ground faults and the extraneous terminals eliminated by a network reduction to obtain a pi circuit to insert in the positive-sequence network.

The symmetrical component approach was used only in the stability calculations. Secondary arc currents and the voltages following arc extinction were calculated with all three phases of the line represented. The three-phase representation was determined by connecting short segments of the line in series and eliminating the extraneous terminals with network reductions.

Dr. Lewis states that transposition might be omitted when a line is not long enough to require series compensation. We believe that transpositions should be made in those cases where stability condi-

Manuscript received May 9, 1968.

tions require that arcs be extinguished rapidly. Table I of the paper shows that transpositions appreciably reduce the secondary arc currents and coupling voltages for an unreactored line. Additional calculations that we have made show that the reduction in currents and voltages is larger when the lines are compensated with balanced reactors.

As mentioned in the paper, compensation of the zero-sequence network was studied academically. Our purpose was to show that the power transfer remained about the same whether the line was unfaulted, faulted to ground, or had one conductor open. Had a phase-to-ground fault been placed on the other end of the line, voltages on the line would have been on the order of 5 per unit because the zero-sequence impedance of the line and transformer canceled out a large part of the capacitive impedance in the neutral. One way of avoiding this overvoltage condition is to terminate the line in a wye winding of a wye-delta transformer bank so as to separate the zero-sequence network of the system from the zero-sequence network of the line. Then the capacitive reactance inserted in the neutral of the sending-end transformer must be made much larger than the zero-sequence impedance of the line and transformers. A reactor sized to resonate with the zero-sequence network must then be inserted in the neutral of the receiving end transformer. With this method of compensation, the neutral voltage shift will be about rated phase-to-ground voltage regardless of the location of the fault. Whether to connect the capacitor and reactor into the transformer neutrals of the two-winding wye-delta transformers or to use three-winding wye-wye-delta transformers with reduced insulation in the high-voltage winding is a matter of economics. We have not made an economic comparison of these two alternatives.

We agree with Mr. Bigelow that circulating zero-sequence currents caused by the single-pole operation can cause problems. In our case, the line on which single-pole reclosing is used is terminated on a bus with other 500-kV lines and a large wye-delta-wye transformer bank with the low-voltage wye winding grounded through a neutral reactor. This effectively screens the low-voltage circuit from the unbalance in the high-voltage one. However, adjustments have been made in the ground relays at other 500-kV buses that are electrically close to the bus on which the line in question is terminated.

Our stability problem is due to the fact that when a phase of the line is opened, the transient stability limit is only about two-thirds of the generator rating. This requires the open phase of the line to be reclosed in about 30 cycles. We feel that the assumptions made in the stability studies cause the results to be somewhat conservative. The secondary arc currents and coupling voltages are calculated for the maximum angle across the line and are therefore conservative. However, the CESI paper [7], mentioned by Dr. Kimbark, indicates that the maximum arcing time will be less than 0.3 second. Therefore, because of the conservative nature of the calculations, we feel that we have more margin than the two cycles mentioned in the paper.

The steady-state stability limit of the generator, taking saturation into account and assuming manual voltage regulation, is about 110 percent of the generator rating at unity power factor and greater than 110 percent of the generator rating for lagging power factors. However, the machine will be equipped with a continuously acting voltage regulator and if this is taken into account, the steady-state stability limit will be about 160 percent of the generator rating [10].

References

[8] A. S. Maikopar and N. N. Baliakov, "Arcing short-circuits on 400 kV transmission lines and methods of dealing with them," *Elec. Tech.* (USSR), vol. 4, pp. 27–38, December 1958.

[9] R. B. Shipley and D. Coleman, *1966 Proc. Power Systems Computational Conf.*, pt. 2, "Network analysis," pp. 4–10.

[10] C. Concordia, "Steady-state stability of synchronous machines as affected by voltage-regulator characteristics," *AIEE Trans.*, vol. 63, pp. 215–220, May 1944.

SINGLE-POLE SWITCHING ON TVA'S PARADISE-DAVIDSON 500-KV LINE DESIGN CONCEPTS AND STAGED FAULT TEST RESULTS

Leo Edwards, J. W. Chadwick, Jr., H. A. Riesch, and L. E. Smith
Tennessee Valley Authority, Chattanooga, Tennessee

ABSTRACT

Single-pole switching is selected for a radial 500-kV line which connects a 1,278-MVA generator into the transmission system. Line protective relays providing selective tripping and special compensation assisting with secondary arc current extinction are described. System planning criteria and design concepts for 500-kV single-pole switching are verified for the first time by staged fault tests conducted on an operating 500-kV system.

INTRODUCTION

The Tennessee Valley Authority's Paradise Steam Plant is located in the coal fields of western Kentucky at the fringe of the TVA service area and requires that energy be transported some distance to load centers. The plant's capability of 2,722 MVA is provided by two 722-MVA units and the recently installed 1,278-MVA cross-compound steam-turbine generator. The two 722-MVA units connect into the 161-kV transmission system through four transformers. The new generator connects into the 500-kV transmission system through a 1,260-MVA transformer bank and over a 500-kV radial transmission line to the Davidson Substation near Nashville, Tennessee.

The Davidson Substation is a major focal point for 500-kV system connections, as shown in Fig. 1, and supplies the heavily loaded 161-kV network in the Nashville area through a 1,200-MVA transformer bank.

The connecting line between Paradise and Davidson is 93 miles (150 km) in length. Synchronous stability is maintained for nonsustained single-phase-to-ground faults on this radial connection with single-pole switching of 500-kV power circuit breakers.

Because of the increasing interest in single-pole switching for EHV lines, an actual application at the 500-kV voltage level with supporting field test data is presented.

Selection of Single-Pole Switching

The 1,150-MW rating of the new generator slightly exceeds the surge impedance loading of a 500-kV line; therefore, to achieve an economy of 11 million dollars, the possibility of using only one 500-kV connection to the main transmission grid was considered. The reliability of such a connection was investigated because it was desirable that it approach the reliability of two 500-kV lines until system load requirements show the need for a second 500-kV connection.

A study of outage records of our 161-kV lines and results of special nanosecond model studies of the 500-kV line design indicated that 0.77 outages per year due to lightning could be expected for each 100 miles of 500-kV line. It was evident that 75 percent or more of these momentary outages would involve only one phase and ground. Provided synchronous stability could be maintained during single-pole switching for single-phase-to-ground faults, only one permanent outage each five years should be expected for this radial connection. Possible loss of the unit once in five years due to insufficient transmission capability provided an acceptable degree of reliability.

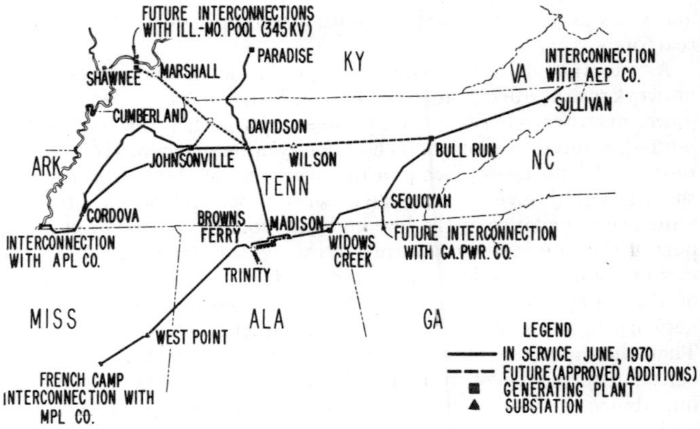

Fig. 1. TVA's 500-kV transmission system

Knudsen and Kimbark showed that shunt reactor banks generally required to absorb excessive line charging kVA could theoretically be modified to compensate for both interphase and phase-to-ground capacitive coupling.[1,2] These modified reactor installations could effectively reduce the magnitudes of secondary arc currents and enhance secondary arc current extinction.

Results of TVA's transient stability studies indicated that synchronous stability could be maintained during single-phase-to-ground faults provided: (a) the fault could be recognized, (b) the faulted phase isolated, and (c) successful reclosure accomplished within one-half second. Calculations by Shipley, Holley, and Coleman indicated that approximately 100 percent of the phase-to-phase and 45 percent of the phase-to-ground capacitive coupling should be compensated with a four-legged reactor combination and the equivalent of one transposition of phase conductors.[3] This would limit the secondary arc current to 20 amperes and the arc path recovery voltage to 50 kV. The four-legged reactor combination required that an appropriate reactor be installed in the neutral of the 100-MVA reactor bank already planned for this connection.

Single-pole switching was selected to isolate and clear single-phase-to-ground faults only. Conventional three-pole switching was selected and automatic reclosing canceled for other line faults, including single-phase-to-ground faults which evolve into multi-phase faults before being cleared.

Paradise-Davidson Connection

The Paradise-Davidson connection is shown in Fig. 2. At Paradise two 500-kV breakers are installed to provide a spare. Each breaker can operate in the single-pole mode; however, only one breaker is in service at a time. At the Davidson Substation one substitute breaker is provided for the line breakers; therefore, to prevent additional complexities, single-pole switching is not used with the substitute breaker. To provide more secure operations, bus potential transformers supply the single-pole relays at both Paradise and Davidson.

The actual compensation on the line is less than the calculated value because: (a) for economic

Paper 71 TP 147-PWR, recommended and approved by the Transmission and Distribution Committee of the IEEE Power Engineering Society for presentation at the IEEE Winter Power Meeting, New York, N.Y., January 31-February 5, 1971. Manuscript submitted September 15, 1970; made available for printing December 21, 1970.

Reprinted from *IEEE Trans. Power App. Syst.*, Nov./Dec. 1971.

reasons, the reactor is located at the receiving end rather than at the midpoint of the line, and (b) for convenience, the conductors are rolled as shown in Fig. 3, thereby providing less than ideal compensation.

Except for the four-legged reactor and the transposition of phase conductors, other 500-kV station and line equipment is identical to that already reported.[4,5]

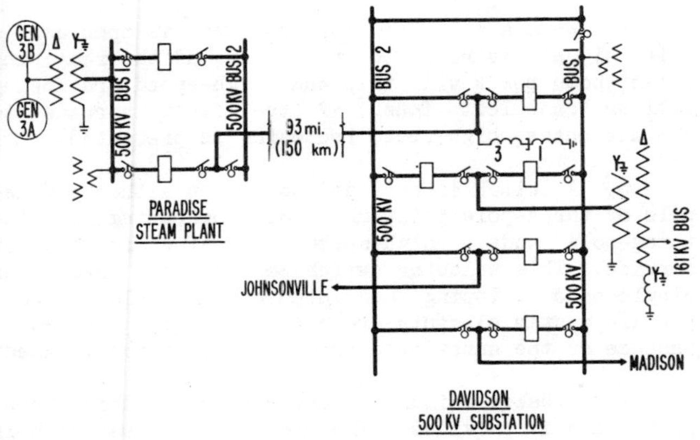

Fig. 2. Paradise and Davidson 500-kV bus switching arrangements

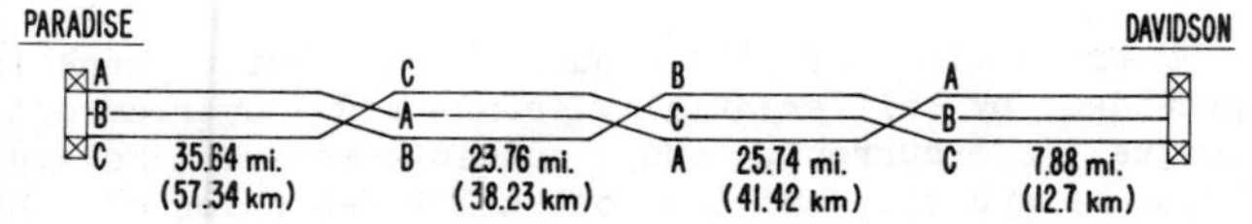

Fig. 3. Paradise-Davidson 500-kV line showing transposition of phase conductors

Compensating Reactors

The three oil-immersed phase reactors are rated 33.3 MVA each and are similar to those described in reference 4 except for the higher neutral insulation required to accommodate a neutral reactor.

To provide the desired compensation, a value of 588 ohms was specified for the neutral reactor.[3] Its 60 Hz voltage requirements were found not to exceed 69 kV; therefore, a ten-second rating of 117.5 amperes was determined. A 92-kV nominal voltage rating with a 450-kV BIL was selected because TNA studies showed a 350-kV BIL to be marginal if three-pole high-speed reclosing is ever required at Davidson to supply charging vars to the system following a generator trip at Paradise. The 25-ampere continuous current rating determined the equivalent KVA rating.

Line Protective Relays

Redundant line protection is provided by separate and independent electromechanical[6] and solid-state[7] relay schemes. This type of protection improves circuit reliability by providing continuous operation in the single-pole mode with either set of relays out of service.

Protection for interphase line faults is provided by a directional comparison, carrier blocking distance relay scheme, using two forward-looking zones and one reverse zone. Ground distance relays with two forward-looking zones are used for ground fault protection.

The electromechanical phase relays use conventional mho-type elements. Zone 1 provides direct instantaneous tripping. Zone 2 provides instantaneous tripping via the carrier auxiliary relays as well as time delay tripping. Zone 3 starts carrier and provides time delay tripping. For single-phase-to-ground faults at certain locations, it was necessary to install "blinders" on the phase relays at Davidson only. Characteristics of the electromechanical phase relays are shown in Fig. 4A.

Phase-to-ground fault protection is provided by conventional mho-type ground-distance relays. Zone 1 provides direct instantaneous tripping. Zone 2 provides instantaneous tripping via the carrier auxiliary relays as well as time delay tripping. Carrier is started by a nondirectional overcurrent relay. To prevent unintentional operations, "blinders" are installed on the ground relays at Paradise only. Characteristics of the electromechanical ground relays are shown in Fig. 4B.

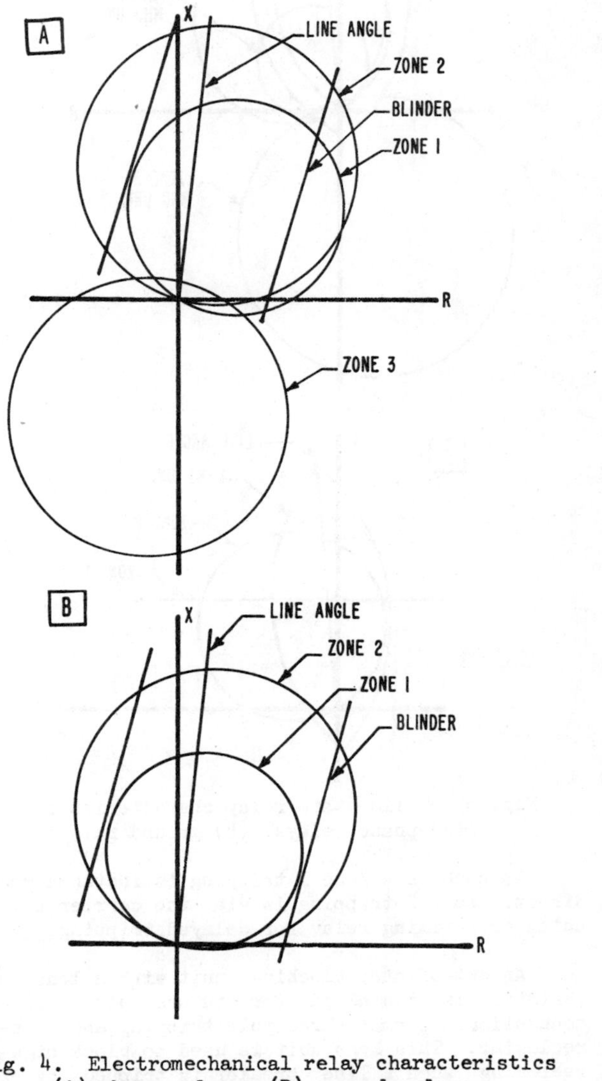

Fig. 4. Electromechanical relay characteristics (A) phase relays, (B) ground relays

The solid-state relay scheme uses mho units for phase fault protection and further restricts the tripping area with a "lens" unit. Zone 1 tripping is initiated for faults within the area bounded by Zone 1 mho unit that fall within Zone 2 mho unit or Zone 2 lens unit. Zone 2 tripping is confined to the area bounded by Zone 2 mho unit or Zone 2 lens unit. Zone 3 starts carrier and provides time delay tripping. Characteristics of the solid-state phase relays are shown in Fig. 5A.

Phase-to-ground fault protection is provided by mho, ohm, and lens units. Zone 1 tripping occurs for faults within the area bounded by Zone 1 ohm unit,

Zone 2 mho unit, and the lens unit. Zone 2 tripping occurs for faults within the boundaries of Zone 2 mho unit and the lens unit. Carrier is started by a nondirectional overcurrent unit. Characteristics of the solid-state ground relays are shown in Fig. 5B.

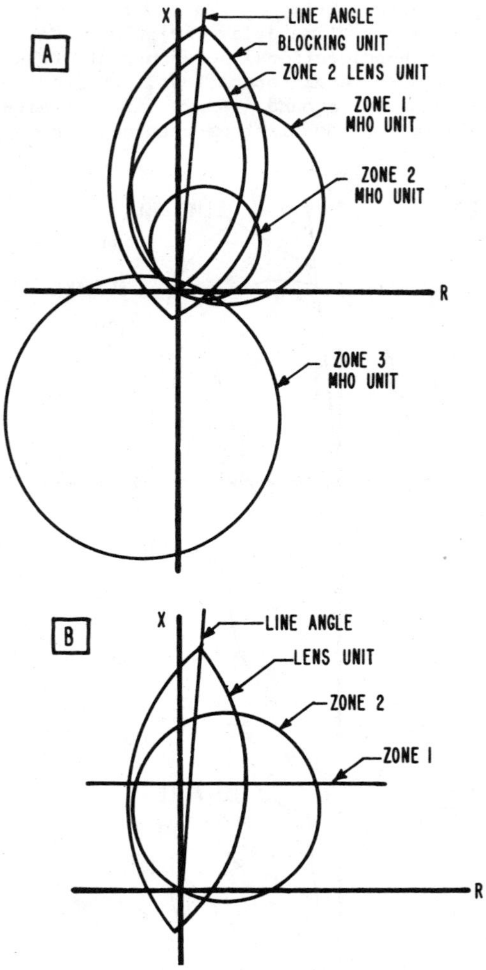

Fig. 5. Solid-state relay characteristics (A) phase relays, (B) ground relays

In each case Zone 1 tripping is instantaneous and direct. Zone 2 tripping is via the carrier auxiliary units or a timing relay for delayed tripping.

An out-of-step blocking unit with a lens characteristic is installed for future use when system connections permit three-pole tripping and high-speed reclosing. This lens unit is used to block high-speed reclosing when a line breaker is tripped by Zone 2 during an out-of-step condition; however, it is ineffectual while single-pole tripping and reclosing are used on this circuit.

The basic criterion for successful single-pole tripping is the positive indentification of the faulted phase. Therefore, several additional requirements of the relay schemes are listed.

1. Operation by any of the phase relays will trip all poles of the breaker, cancel high-speed reclosing, and initiate shutdown of the unit at Paradise.

2. Instantaneous single-pole ground trip does not cancel high-speed reclosing.

3. Time delay ground relay operation trips all poles of the breaker, cancels high-speed reclosing, and initiates shutdown of the unit at Paradise.

4. When a pole of the breaker is open, the ground relays are prevented from starting carrier to provide instantaneous tripping of the two remaining poles for subsequent Zone 2 ground relay operations.

5. When a pole of the breaker is open and a trip signal is received by either of the remaining poles, both poles will trip and high-speed reclosing will be canceled. Thus, if the fault spreads to another phase, high-speed reclosing is prevented.

6. A switch is provided to select either single-pole or three-pole tripping. When operating in the three-pole mode, high-speed reclosing is out of service. This selector switch permits the removal of single-pole tripping and high-speed reclosing when certain system elements are out of service for maintenance or the substitute breaker at Davidson is used.

7. Breaker failure (breaker pole failure) backup protection is provided. The protection at Davidson is "local" backup because of the bus switching arrangement used. The protection at Paradise is "remote" backup using dual-channel transfer trip to Davidson.

Protection of the four-legged reactor bank is provided by differentially connected overcurrent, neutral overcurrent, and sudden-pressure relays. These relays trip and lock out all three poles of the breaker at Davidson and trip and lock out the breaker at Paradise by dual-channel transfer trip.

In recognizing that proper operation of protective relays is the key to successful performance, miniature system tests were made at the manufacturer's facilities on the solid-state relays. These tests resulted in several changes in the relaying scheme and characteristics of elements, such as the combinations used in phase tripping and changes in logic circuitry. Also, the transient response of these relays was checked to the extent possible on the manufacturer's facilities.

Miniature system tests were not made on the electromechanical relay scheme for several reasons: (1) existing schemes were not acceptable; (2) time did not permit manufacturers to design a scheme, test it, and make design changes; and (3) the transient response of electromechanical relays was questionable since in most applications their behavior is immaterial after a trip signal is initiated. An electromechanical relay scheme we proposed was installed, and subsequent minor modifications made as a result of the staged fault tests are discussed later in the text.

Arrangements for Testing

Since single-pole switching had not been used at the 500-kV level, staged fault tests on the operating connection were required. Therefore, within the period that the generator was operating, but before it was placed into commercial operation, a test program was initiated to provide answers to the following unknown aspects of single-pole switching: (1) Is single-pole switching at the 500-kV voltage level applicable? (2) Is special compensation required for secondary arc current extinction on a circuit of this length? (3) Are results of digital transient stability studies using conventional machine representation without damping factors representative? (4) Are special solid-state and electromechanical line protective relays acceptable for phase selectivity?

It was not possible to place different types of faults around a string of suspension insulators or

Fig. 6. 500-kV transmission structure adjacent to Davidson showing the ground fault bus

between arcing horns at the insulator assemblies since this would require an outage on the generator to replace damaged insulators for each test. The fault structure shown in Fig. 6 was designed for the application of phase-to-ground faults. Such structures were connected to the 500-kV line at the first structures adjacent to the Davidson and Paradise switchyards, and were operated as a portion of the line until the test program was completed. Connections to the 500-kV line were made with single 2035 kcmil ACSR conductor connected to the C-phase bundle conductor of the line. Each ground fault bus was a 2/0 copper conductor connected to a ground at the transmission line structure. The two parallel conductors were spaced a distance of 11.5 feet (3.5 m), which is the equivalent length of twenty-four 5-3/4-inch by 10-inch standard suspension insulators. Single-phase-to-ground faults were applied with a 0.004-inch (0.1 mm) diameter steel music wire, approximately 12 feet (3.66 m) in length, which was fired with a crossbow to draw a vertical arc between the two parallel conductors. It is believed that faults at line terminals of this connection represent the most severe conditions since ground faults adjacent to the switchyard at Paradise are most severe for transient stability, and faults at the Davidson terminal shunt the compensating reactor to minimize its effect. It is expected that the compensating reactor will be at least partially effective for faults at all other locations on the line.

Studies had shown that fault magnitudes of 8,680 MVA or 10,000 amperes at 500 kV would be experienced for single-phase-to-ground faults on the line connection at Davidson, and that 56 percent of the fault current was confined to the 500-kV neutrals of the transformer banks at Davidson and Paradise. Similarly, fault magnitudes of 6,860 MVA or 7,920 amperes at 500 kV were expected for single-phase-to-ground faults on the line connection at Paradise and 87 percent of the ground current would be contributed by the 500-kV transformer bank at Paradise. Calculations also indicated that 52 percent of the residual current would be confined to the 500-kV neutral of the Davidson transformer bank when the faulted phase was isolated. Therefore, it was anticipated that the most severe conditions affecting other line relays would be for single-phase-to-ground faults near the Davidson Substation.

Instrumentation for Tests

The true magnitude of the secondary arc current after the operation of breaker poles was of primary importance. A "full scale reading" of about 50 amperes was desirable so that an expected magnitude of about 25 amperes could be read easily. This was accomplished by using a saturating current transformer consisting of a nickel-ferrite toroid core wound with 100 turns of enameled wire. Dimensions of the core were not critical; however, they were large enough to slip over a section of line hose ("snake") which provided primary insulation. The saturation current point of the current transformer was established by adjusting the secondary load resistance and, therefore, the volts per turn, so that a sine wave was obtained at 50 amperes with one primary turn. There was no "current" wave distortion below 50 amperes; however, above this value the ratio of the current transformer was meaningless. An amplifier was used to drive an oscillograph galvanometer which was calibrated by using primary current. Since there is a pulse generated when the flux reverses, as in a pulse transformer, a series group of silicon diodes with one polarity was connected in parallel with another series group of diodes of opposite polarity to absorb all pulses above two volts where the readings were meaningless. The recovery time of the saturating current transformer was checked by noting any change of ratio in the current transformer when a surge current of about 5000 amperes peak value was passed through the primary along with 20 amperes at power frequency. Recovery time of the saturating current transformer was less than two microseconds under any condition of surge polarity or amplitude up to 5000 amperes.

The recovery voltage across the arc path was second in importance. Special capacitance voltage dividers of 28 pF were connected to the C-phase conductor at the line terminals to accurately measure recovery voltages during single-phasing conditions and switching surges upon reclosing. The recovery voltage was recorded on an oscillograph by expanding the normal line voltage trace. The expanded recovery voltage was obtained by using an amplifier gain of four or five times that used for normal line voltage. The maximum trace deflection was limited to 1-1/2 - 2 inches peak-to-peak by using a current limiting amplifier.

Special instrumentation for the staged fault tests included eight oscillographs with 91 active oscillograph elements to record information at each end of the transmission line. One oscillograph was located near the fault structure and recorded the following special data: (1) 500-kV bus voltage, (2) secondary arc current, (3) total fault current, (4) expanded recovery voltage, (5) C-phase power circuit breaker recovery voltage, (6) 500-kV line voltage, and (7) 60 Hz timing wave. All eight oscillographs were started simultaneously about one second before each fault was applied.

Special precautions were taken against possible "rise of ground potential." All instrument leads were treated as though a difference of potential of 20 kilovolts could exist. The current transformer leads were grounded only at the measuring point. Triaxial test cable was used with multiple grounds spaced 100 feet apart. Care was taken that no shield grounds were present at current transformers, the cores of which were insulated heavily against adjacent grounded parts and the earth itself.

The trip coil of the crossbow which fired the fault initiating wire was also isolated with a 1 to 1 transformer with over 20-kV insulation. This was constructed by using a bushing-type current transformer as the secondary and winding a large diameter primary to provide air insulation. This also gave a simply constructed 120/120-volt ratio, 500 VA power supply and prevented possible high currents and voltages in the crossbow trip circuit.

Two high-speed 16 millimeter cameras were used which provided color films run at 600 to 1200 frames per second. A neon lamp with isolated power supply provided short pulses of light at the power frequency rate. This neon lamp gave marker points near the sprocket holes so that an accurate time lapse was recorded independently of the variable film speed. The camera nearest the fault, with its line of vision parallel to the fault structure, had its aperture partially opened to look directly at the arc path within the high intensity luminous cloud. The other camera, with its line of vision nearly perpendicular to the fault structure, had its aperture fully opened to record relative movement of the cloud created by the primary arc.

Staged Fault Tests

While the generator was operating, but before it was placed into commercial service, thirteen C-phase-to-ground faults were applied to the operating connection. Between June 25, 1969, and March 17, 1970, six faults were applied at Davidson and seven at Paradise with at least one test being performed each month, except for the winter months of December - February. The maximum and minimum generation on the Paradise No. 3 unit during these tests were 990 MW, 245 Mvar, and 390 MW, 185 Mvar, respectively. The thirteen staged faults resulted in shutdown of the generator four times, one of which was anticipated when the special reactor compensation at Davidson was removed purposely.

On the first test at Paradise (Test No. 4), generator shutdown was caused by malfunction of the electromechanical relays. Temporary modifications were made in the phase protective relays at Paradise as an expedient to obtain the required information on secondary arc current extinction and transient stability of the generator. The solid-state relays were not available except for the last two tests. The entire complement of electromechanical and solid-state relays, however, was in service for Tests 12 and 13.

Secondary Arc Current Extinction

The effectiveness of a four-legged reactor installation as special compensation to enhance secondary arc current extinction is apparent in the comparison of Figs. 7-10 with Fig. 13.

Sample oscillograms of staged faults at Paradise are shown in Figs. 7 and 8. Fig. 7 is from Test 5 with the generator at Paradise supplying 710 MW. Fig. 8 is from Test 10 with 905 MW being supplied from the generator. Secondary arc current of 6.5 amperes and recovery voltage of 24.5 kV are shown in Fig. 7 as compared to 7.5 amperes and 28.5 kV in Fig. 8. However, restrikes across the arc path are apparent for 5.5 cycles before the secondary arc current diminishes to zero with 710 MW of generation and 3.5 cycles with 905 MW of generation. Weather conditions probably account for the secondary arc currents diminishing more rapidly at the higher level of generation since the fault with 710-MW of generation was made on a cool September evening; whereas, the test at 905-MW level was made during cold weather with temperatures around 0°F and 11 inches of snow.

Sample oscillograms of staged faults at Davidson are shown in Figs. 9 and 10. Fig. 9 is from Test 2 with the generator at Paradise supplying 525 MW. Fig. 10 is from Test 9 with 945 MW being supplied from the generator. Secondary arc current of 7.5 amperes and recovery voltage of 37 kV are shown in Fig. 9 as compared to 12.5 amperes and 59 kV in Fig. 10. The restrikes across the arc path are apparent for 4.2 cycles before the secondary arc current diminishes to zero with 525 MW of generation and 7.5 cycles with 945 MW generation. It is believed that the additional time required for the secondary arc current extinction in Fig. 10 results from the additional generation at Paradise increasing the recovery voltages by inductive coupling. The expanded recovery voltage trace shows a modulating frequency of 15 Hz. The breaker recovery voltage in Fig. 10 is in error because test connections were changed permitting a "sneak" connection. The trace shows a recording of the difference of potential between two grounds during the flow of fault current.

The sample oscillogram shown in Fig. 11 is a duplicate of Fig. 9, reproduced as reference for the arc path characteristics shown in Fig. 12. These characteristics are shown at one-cycle intervals selected from film taken at approximately 22 frames per cycle.

The sample oscillogram of a staged fault at Davidson shown in Fig. 13 is from Test 3 when Paradise No. 3 was generating 600 MW and the compensating reactor was purposely removed from service to determine its effectiveness. Secondary arc current of 22.5 amperes is recorded initially, after breaker poles on the faulted phase open, and increases to a constant maximum magnitude on restrikes after a few cycles. The secondary arc current is extinguished for approximately one-half cycle before the Davidson breaker pole recloses. This oscillogram shows that special compensation is necessary on this 500-kV line since reclosing within one-half second is required to maintain machine stability.

Fig. 14 shows the arc path characteristics for the staged fault test oscillogram of Fig. 13. These photographs record the arc in cycles from the inception of the fault until the breaker pole at Davidson recloses. The luminous cloud created by fault current begins breaking up at approximately nine cycles; and after the gaseous path has apparently cooled, restriking continues until the arc current is extinguished one-half cycle before reclosing. The strike across the arc path upon reclosing occurs during resistor preinsertion before main contacts of the breaker close.

Single-Phase Operation

The staged fault test at Davidson with 945 MW generation on Paradise No. 3 represented the most severe condition tested. A residual line current of 1350 amperes was measured at Paradise, which agreed with the $3I_0$ measured in the 500-kV neutral of the transformer bank at Paradise. Measurements showed distribution of these currents as 54 percent in the 500-kV neutral of the transformer bank at Davidson, 5 percent in the compensating reactor neutral, and 41 percent in the other 500-kV transmission lines at Davidson.

Relay Performance

Tables I and II summarize the test conditions and breaker operating times for the 13 fault tests. All the faults were C-phase-to-ground. The first 11 tests were made with only the electromechanical relays in service. The last two tests were made with the electromechanical and solid-state relays in service. In columns 5 and 6 of Table I, "normal" indicates that the carrier relay contacts were in the trip circuit. For some tests, the carrier relay contacts were by-passed so that Zone 2 of the ground relay tripped

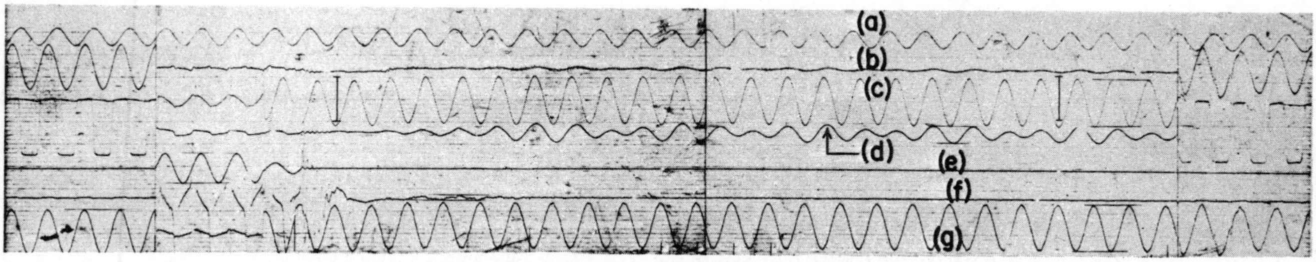

*Fig. 7. Oscillogram record at Paradise for a C-phase-to-ground fault at Paradise Generation 710 MW, Voltage 528 kV.

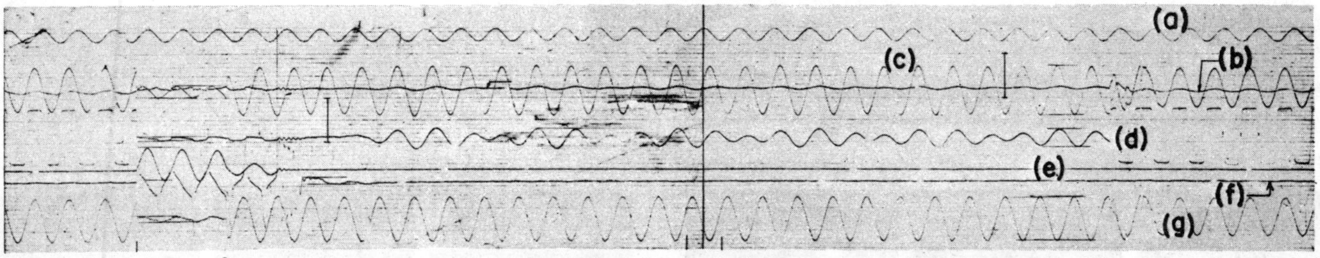

*Fig. 8. Oscillogram record at Paradise for a C-phase-to-ground fault at Paradise Generation 905 MW, Voltage 528 kV.

*Trace Identification for Figs. 7 and 8

(a) Crossbow Trip Voltage. (b) Davidson Line - C-Phase Voltage (C. T. Capacitance Tap). (c) Recovery Voltage, C-Phase Breaker. (d) Expanded Recovery Voltage - Line (5 Times Trace b). (e) Total Fault Current. (f) Secondary Arc Current. (g) 500-kV Bus - C-Phase-to-Neutral Voltage (Capacitance Divider).

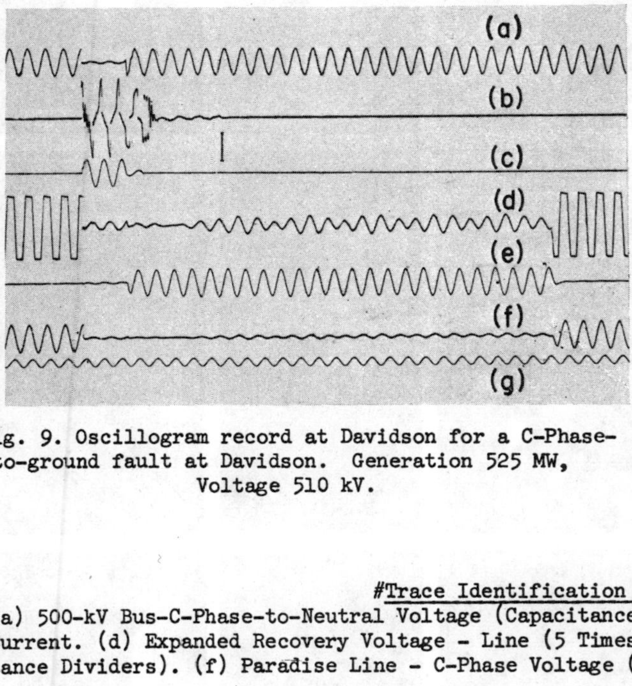

#Fig. 9. Oscillogram record at Davidson for a C-Phase-to-ground fault at Davidson. Generation 525 MW, Voltage 510 kV.

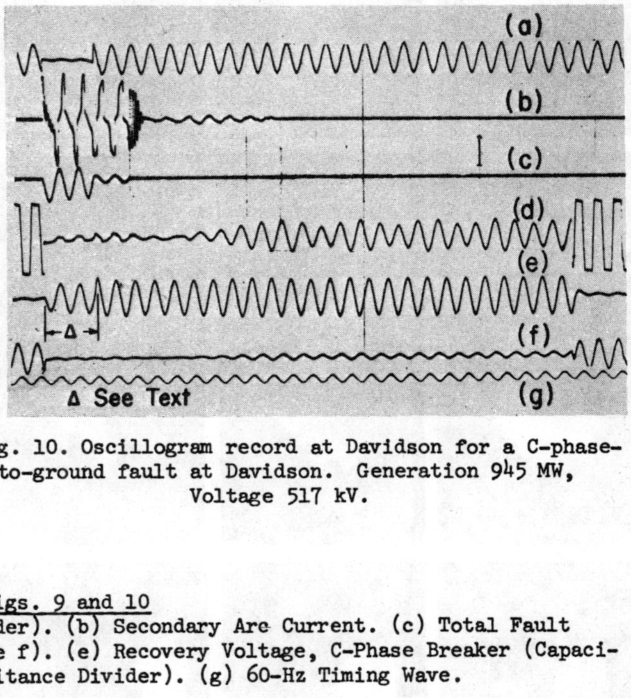

#Fig. 10. Oscillogram record at Davidson for a C-phase-to-ground fault at Davidson. Generation 945 MW, Voltage 517 kV.

#Trace Identification for Figs. 9 and 10

(a) 500-kV Bus-C-Phase-to-Neutral Voltage (Capacitance Divider). (b) Secondary Arc Current. (c) Total Fault Current. (d) Expanded Recovery Voltage - Line (5 Times Trace f). (e) Recovery Voltage, C-Phase Breaker (Capacitance Dividers). (f) Paradise Line - C-Phase Voltage (Capacitance Divider). (g) 60-Hz Timing Wave.

direct, the same as Zone 1, regardless of any carrier blocking signal.

For Test 1, the breaker reclosing timer was set for 20 cycles and normal relay protection was in service at both line terminals. In this test, the Paradise line terminal was slow in tripping. Analysis of the test results indicated that "blinder" units for the ground distance relays at Paradise were just on the verge of operating. This accounts for the delay in operation of these relays on Test 1. The settings of the "blinder" units were changed; and the contacts of these relays in the trip circuit of the faulted phase were bypassed and their operations monitored with oscillogram traces on subsequent tests. Under fault conditions, the contacts of the "blinder" units were at least one cycle slower in closing than the other relay contacts in the trip circuits. Since tripping was satisfactory for all tests with the "blinder" unit contacts bypassed, it was decided to leave them bypassed for normal operation.

Following Test 1, the breaker reclosing timer was increased from 20 cycles to 25 cycles to allow more time for arc extinction after breaker poles on the faulted phase opened.

In Test 4, the breaker at Paradise tripped 3-pole. Although there were no phase relay targets, it was suspected that one of the phase relays had operated to initiate 3-pole tripping of the breaker. For Tests 5, 7, 9, 10, and 11 the phase relays at Paradise were removed from service and operation of the phase relay contacts was monitored. In Test 10, the

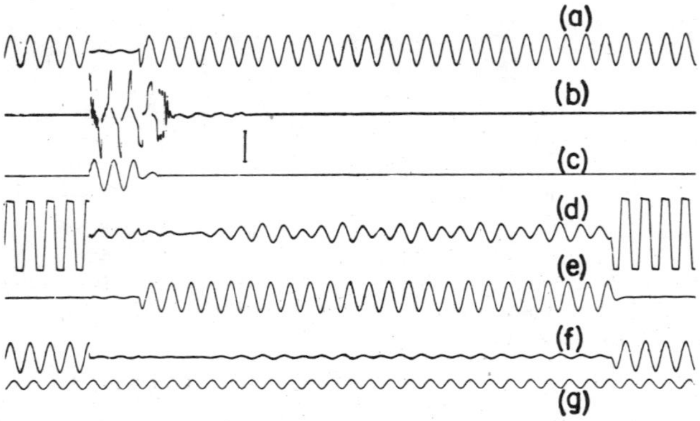

*Fig. 11. Oscillogram record at Davidson for a C-phase-to-ground fault at Davidson - reactors in

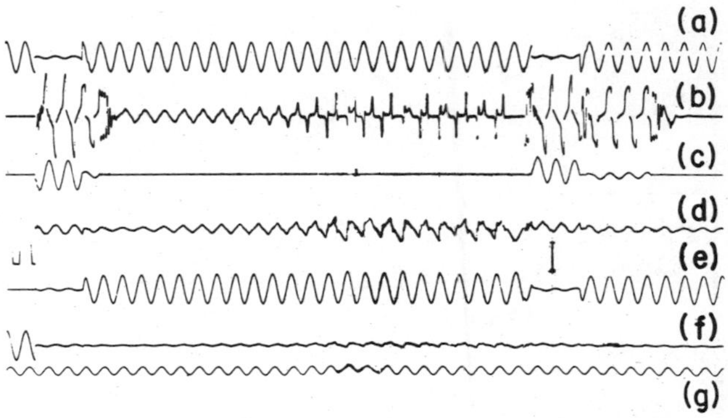

*Fig. 13. Oscillogram record at Davidson for a C-phase-to-ground fault at Davidson - reactors out

* Oscillogram traces (a)-(g) of oscillograms shown in Fig. 11 and 13 are identical to those in Fig. 9 and 10.

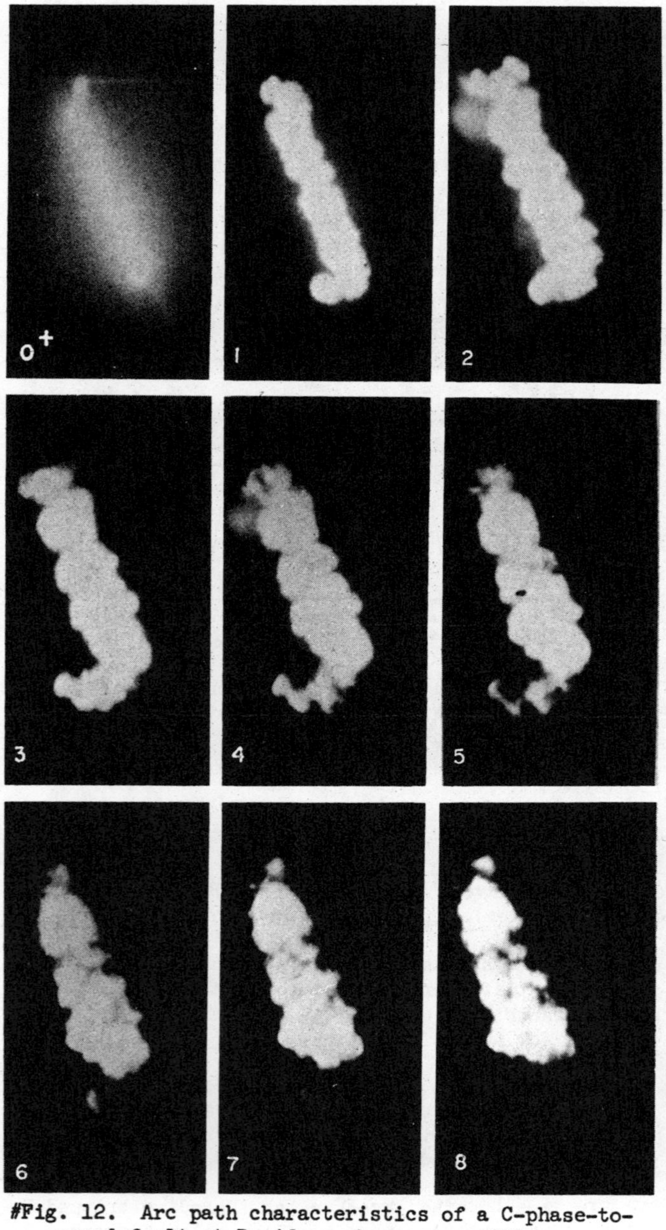

#Fig. 12. Arc path characteristics of a C-phase-to-ground fault at Davidson shown on oscillogram record in Fig. 11 - reactors in

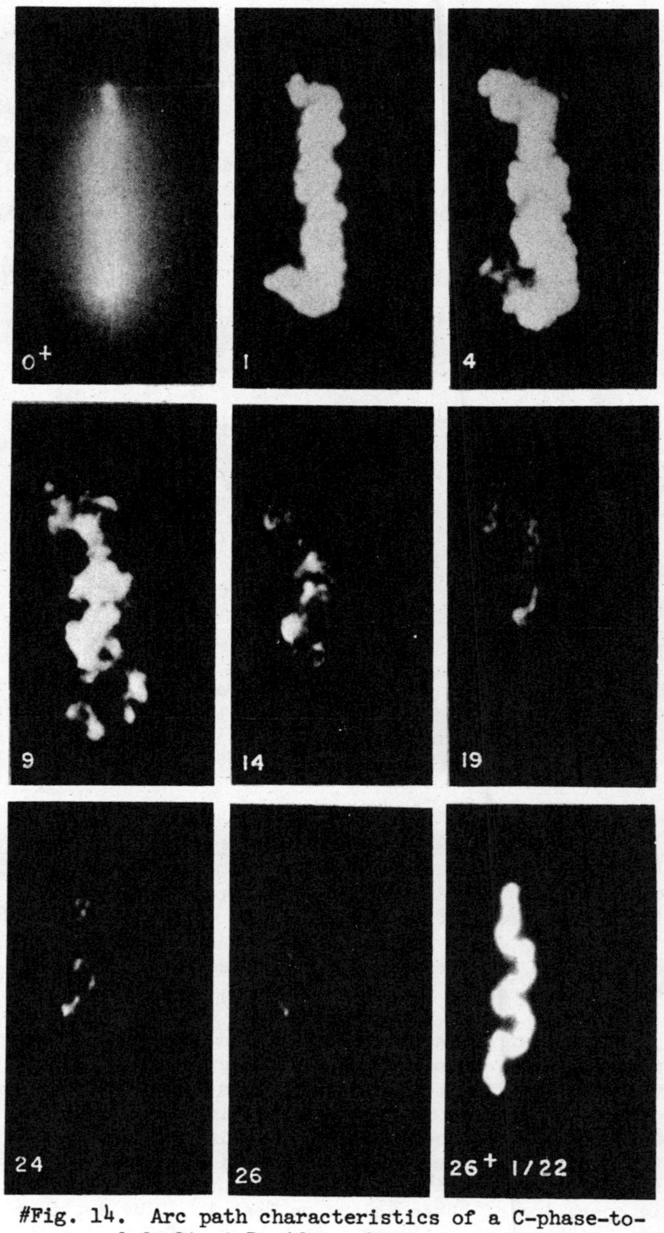

#Fig. 14. Arc path characteristics of a C-phase-to-ground fault at Davidson shown on oscillogram record in Fig. 13 - reactors out

#Numbers shown indicate time in cycles on a 60 Hz base following the inception of a C-phase-to-ground fault.

SUMMARY OF FAULT TESTS ON PARADISE-DAVIDSON 500-KV LINE

TABLE I - TEST CONDITIONS

Test No.	Fault Location	Line Load at Davidson	Line Reactors at Davidson	Grd. Relay Zone 2 at Davidson	Grd. Relay Zone 2 at Paradise	Blinder Unit Contacts at Paradise	Phase Relays at Paradise	Remarks
1.	Davidson	750 MW	Yes	Normal	Normal	Normal	In service	Fault not cleared single-pole. Tripped 3-pole on reclosing test.
2.	Davidson	500 MW	Yes	Normal	Trip direct	Bypassed	In service	Single-pole tripping and reclosing.
3.	Davidson	580 MW	No	Normal	Trip direct	Bypassed	In service	Fault not cleared single-pole. Tripped 3-pole on reclosing
4.	Paradise	940 MW	Yes	Trip direct	Normal	Bypassed	In service	Davidson tripped and reclosed single-pole. Paradise tripped 3-pole.
5.	Paradise	680 MW	Yes	Normal	Normal	Bypassed	Out of service	Single-pole tripping and reclosing.
6.	Davidson	620 MW	Yes	Normal	Normal	Bypassed	In service	Single-pole tripping and reclosing.
7.	Paradise	680 MW	Yes	Normal	Normal	Bypassed	Out of service	Single-pole tripping and reclosing.
8.	Paradise	690 MW	Yes	Normal	Normal	Bypassed	In service	Davidson tripped and reclosed single-pole. Paradise tripped 3-pole.
9.	Davidson	900 MW	Yes	Normal	Normal	Bypassed	Out of service	Single-pole tripping and reclosing.
10.	Paradise	875 MW	Yes	Normal	Normal	Bypassed	Out of service	Single-pole tripping and reclosing.
11.	Paradise	400 MW	Yes	Normal	Normal	Bypassed	Out of service	Single-pole tripping and reclosing.
12.	Davidson	880 MW	Yes	Normal	Normal	Bypassed	In service	Single-pole tripping and reclosing.
13.	Paradise	880 MW	Yes	Normal	Normal	Bypassed	In service	Single-pole tripping and reclosing.

SUMMARY OF FAULT TESTS ON PARADISE-DAVIDSON 500-KV LINE

TABLE II - BREAKER OPERATING TIME

Test No.	Fault Location	Recl. Timer Setting-Cycles*	At Davidson Trip Cycles*	At Davidson Main Contacts Open-Cycles*	At Davidson Trip Cycles*	At Paradise Trip Cycles*	At Paradise Main Contacts Open-Cycles*	At Paradise Trip Cycles*	Dead Time# on Line Cycles*
1.	Davidson	20	2.50	19.75	3.25	20.00	21.00	10	2.25
2.	Davidson	25	2.50	23.80	-	3.40	26.60	-	22.90
3.	Davidson	25	2.50	24.00	2.60	3.35	26.00	3.50	23.15
4.	Paradise	25	3.25	23.75	-	2.65	-	-	23.75
5.	Paradise	25	4.00	25.20	-	2.60	25.4	-	24.00
6.	Davidson	25	2.90	24.60	-	4.40	25.15	-	23.10
7.	Paradise	25	5.00	24.27	-	2.93	25.56	-	23.49
8.	Paradise	25	4.58	24.28	-	2.60	-	-	24.28
9.	Davidson	25	2.64	24.65	-	4.20	25.84	-	23.09
10.	Paradise	25	4.06	24.38	-	2.56	25.44	-	23.94
11.	Paradise	25	----	No record	----	2.97	25.72	-	-
12.	Davidson	25	2.75	24.50	-	2.75	26.00	-	24.50
13.	Paradise	25	3.00	24.75	-	3.00	25.50	-	24.75

* Time measurements are on a 60 Hz base.

\# Dead time is the time from last main contact opening to first main contact closing (breakers at both locations are equipped with pre-inserted resistors on closing and opening).

monitoring circuit showed that the Zone 2 unit of the C-A phase relay operated about one cycle after initiation of the fault. Correct calibration of the relay was verified and special tests were made to determine whether asymmetrical fault currents would cause operation of a phase relay at a lower value of current than for symmetrical currents. These tests were not conclusive since it was not possible to duplicate the exact fault currents and voltages. As a temporary expedient, carrier tripping by Zone 2 of the electromechanical phase relays was delayed 4 cycles and the relays were placed in service for Tests 12 and 13. The relay manufacturer has analyzed the relay torque developed during transient conditions, taking into account memory action. It is believed that the memory action caused the transient overreach of the Zone 2 units. Changes in the relay design are being considered to overcome this difficulty.

It will be noted from Table II that the solid-state relays cleared the fault in 3 cycles at both line terminals in Tests 12 and 13. Zone 1 of the electromechanical ground relay also operated at the terminal where the fault was placed. At the remote terminal, the solid-state relays tripped the breaker before the electromechanical ground relay closed its contacts. Fig. 15 shows an oscillogram which was recorded at Paradise for Test 13. It shows the clearing and reclosing time of the breaker, the changes in the phase currents and voltages, and the transformer neutral currents existing while one phase of the line is open.

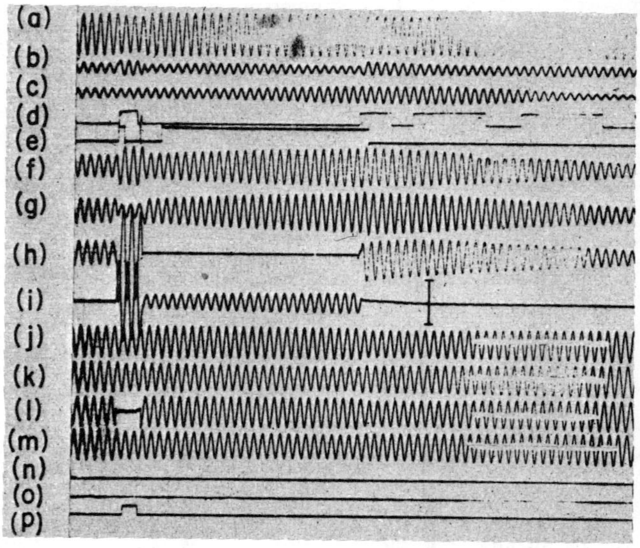

Fig. 15. Oscillogram record at Paradise for a C-phase-to-ground fault at Paradise

(a) Gen. 3, A-B Phase Voltage. (b) Gen. 3A, A-Phase Current. (c) Gen. 3B, B-Phase Current. (d) Davidson Line - Received Carrier (E.M.). (e) Davidson Line - Received Carrier (S.S.). (f) Davidson Line - A-Phase Current (E.M.). (g) Davidson Line - B-Phase Current (E.M.). (h) Davidson Line - C-Phase Current (E.M.). (i) 500-kV Transformer Neutral Current. (j) 500-kV Bus, A-Phase-Neutral Voltage. (k) 500-kV Bus, B-Phase-Neutral Voltage. (l) 500-kV bus, C-Phase-Neutral Voltage. (m) Davidson Line - B-Phase-Neutral Voltage. (n) PCB Trip Coil 1, A-Phase. (o) PCB Trip Coil 1, B-Phase. (p) PCB Trip Coil 1, C-Phase.

CONCLUSIONS

1) Single-pole switching at 500 kV is verified for the first time by staged fault tests conducted on an operating 500-kV system.

2) Special compensation consisting of a four-legged shunt reactor installation is effective in assisting with secondary arc current extinction.

3) Without special compensation, single-pole tripping and reclosing within one-half second is unsuccessful on a 500-kV radial line 93 miles (150 km) in length.

4) Synchronous stability of a large generator is maintained during single-pole switching provided reclosing is accomplished within one-half second.

5) Switching surges and breaker transient recovery voltages for single-pole switching are no more severe than for three-pole switching provided closing and opening resistors are used.

6) Solid-state line protective relay systems properly identify the faulted phase and, with two-cycle breaker operating time, provide simultaneous three-cycle clearing at both ends of the protected line. Electromechanical relay systems require up to two additional cycles of clearing time at the far end of the protected line.

7) Line protective relaying schemes external to the protected line section operate properly, and erroneous relay operations do not occur because of the open phase condition on the Paradise-Davidson connection.

8) The surge protective packages supplied with the solid-state line protective relays are very effective during both staged fault and special arcing disconnecting switch tests.

REFERENCES

(1) N. Knudsen, "Single-phase switching of transmission lines using reactors for extinction of the secondary arc," CIGRE, Paper 310, 1962.

(2) E. W. Kimbark, "Suppression of ground-fault arcs on single-pole-switched EHV lines by shunt reactors," IEEE Trans. Power Apparatus and Systems, vol. 83, pp. 285-290, March 1964.

(3) R. B. Shipley, H. Holley, and D. Coleman, "Digital analysis of single-pole switching on EHV lines," IEEE Trans. Power Apparatus and Systems, vol. 87, pp. 1679-1687, August 1968.

(4) R. M. Milton, H. H. Leech, and R. C. St. Clair, "Tennessee Valley Authority's 500-kV system-stepdown substation design," IEEE Trans. Power Apparatus and Systems, vol. 85, pp. 36-46, January 1966.

(5) A. C. Pfitzer and G. M. Wilhoite, "Tennessee Valley Authority's 500-kV system-transmission line design," IEEE Trans. Power Apparatus and Systems, vol. 85, pp. 28-35, January 1966.

(6) A. R. van C. Warrington, "Graphical method for estimating the performance of distance relays during faults and power swings," AIEE Transactions, vol. 68, pp. 608-621.

(7) C. G. Dewey, C. A. Mathews, and W. C. Morris, "Static mho distance and pilot relaying--principles and circuits," IEEE Trans. Power Apparatus and Systems, vol. 82, pp. 391-400, June 1963.

Discussion

G. D. Rockefeller (Westinghouse Electric Corp., Newark, N.J.): It isn't often that a 1278 MVA generator and a 93 mile 500 kV line are available for testing over a period of 9 months! The authors and associates seem to have made considerable capital of this opportunity. Too bad an entire paper wasn't devoted to relaying aspects.

With the possible exception of series capacitors, single-pole tripping of long EHV lines represents the greatest challenge in relaying system design. In particular, reliable faulted-phase selection for the myriad of possible situations represents a typical "iceberg" problem – looks small from the surface.

The authors suspected undesired phase distance operation for test 4 and found one in test 10 – the CA relay operated for a CG fault, initiating 3-pole tripping and reclose blocking. Load current effects most probably account for this misoperation, both increasing the current level and causing an unfavorable phase shift.

In the absence of load current and with a CG fault the CA phase relay (uses V_{CA} & $I_C - I_A$) sees an apparent impedance at about 115° (second quadrant). Load current flow towards Davidson shifts Z_{CA} at Paradise into the first quadrant where the blinders (if used) operate.

Note that trouble was encountered on tests 4 and 10, the two Paradise faults with the highest line loading! For the Davidson faults the higher fault current level and opposite load flow direction presumably combine to keep the Davidson relays out of trouble.

Fig. 15 shows the effects of load-fault current super-position. The sound phases A & B have markedly different currents during the fault. Phase C, trace (h) shows that the load current is substantial compared to the fault current.

The basic problem: how to keep relays sensitive to distant faults from unintended operation for nearby faults.

The spurious recovery voltage in Fig. 10, trace e, during the fault was explained. What about the Fig. 7, trace c potential?

The authors cite Ref. 3 as indicating how the secondary arc current can be limited to 20A & 50 kV by partial neutralization. Regarding Fig. 13 (Test 3) where the neutral reactor was removed, the authors state that "secondary arc current of 22.5 amperes is recorded initially . . . . " One might conclude perhaps erroneously, that the 20A limit is too high or marginal for prompt arc deionization. What is the authors' opinion?

Manuscript received January 27, 1971.

C. Concordia (General Electric Company, Schenectady, N.Y. 12305): We were very pleased to see this report of a very practical application of single-pole switching, as we have felt for a long time that single-pole switching has been rather neglected as a means for improving over-all system reliability.

Based on the knowledge that most faults are single phase, it has long been recognized that single-pole switching would theoretically result in considerable improvement for those cases. However, the emphasis has for the most part been on the more severe disturbances, and it has also been recognized that a really reliable power supply requires at least two parallel paths, in which case the benefit of single-pole switching sharply decreases. Nowadays, with decreasing availability of rights-of-way, and with the consequent use of the highest voltages, there appear to be many cases where only one line is used. Even though this may seem rather strange, with the present emphasis on reliability, the practical choice may be between one line or none.

We believe that single-pole switching gained a (not entirely deserved) poor reputation on the basis of experience many years ago, when it was not fully appreciated that the required (but also the permissable) dead time for successful reclosing was considerably longer than for three-phase switching. Further, there was the feeling that the system must withstand the more severe three-phase fault anyway. Later work showed the much longer dead times required at high voltages, and so still further discouraged the application of single-pole switching.

The report by Professor Knudsen showed very clearly the possibility of more or less decoupling the faulted phase, and so drastically reducing the required dead time. However, he also showed that the economics were favorable primarily if the shunt compensation were needed anyhow, which was not always the case at the time he did his work.

The present application accepts the fact that a multi-phase fault cannot be survived, and make the best of it. However, even in other less clear-cut cases, with the modern tendency to look, not merely at the most severe cases, but rather at the whole spectrum of disturbances from a probabalistic point of view, we feel that it may often be found that single-pole switching is worth while.

Manuscript received February 10, 1971.

D. E. Hedman (Power Technologies, Inc., Schenectady, N.Y. 12301): This application of single-pole reclosing on a 500 kV reactor compensated line is the first application of this type in the world and presentation of these field test results make a valuable contribution to the industry. The planning for the compensating neutral reactor requirements were presented in an earlier paper prepared by TVA authors (Reference 3 of the paper). The recommendations from that earlier paper appear to have been followed in this installation even though the specific line lengths and transpositions used here were not presented in that earlier paper. Can the authors comment if specific analytical calculations were made on the Paradise-Davidson configuration for evaluation prior to operation of the system?

Do the authors feel that the phase transpositions are required on a line of this length to have successful single phase reclosing?

Manuscript received February 16, 1971.

H. T. Seeley (1700 Spottswood Place, Lynchburg, Va. 24503): It is gratifying to see a well-planned and thoroughly-proven adoption of single-pole switching with modern relaying techniques, as a tool in effecting a major saving in transmission investment, with frank acceptance of the reliability limitations that result from single-line operation.

In some previous applications of single-pole switching, there has been concern about possible misoperation of ground relays on sound lines, due to the circulation of zero-sequence current through those lines during the open-pole interval. The authors' consideration of this problem is shown by their mention – under "Arrangements for Testing" – of single-phase-to-ground faults near the Davidson station as the most severe conditions. Would they please state whether any change in ground relay settings on other lines was considered necessary, or found necessary, by reason of the use of single-pole switching?

The authors are to be congratulated on their concise yet thorough coverage of a complex subject. Later investigators will probably benefit from the details given on the instrumentation which gave such good records of secondary arc current and expanded recovery voltage. On the basis of those records it is suggested that under most conditions, earlier reclosing with the 4-legged reactor would be successful if initiated by the reappearance of some critical value of recovery voltage persisting for a few cycles.

Manuscript received February 13, 1971.

L. E. Goff (General Electric Company, Philadelphia, Pa. 19141): We wish to compliment the authors of this paper for their clear and concise reporting on a very complex subject and a test program of such imposing magnitude.

It is naturally gratifying to us as relay engineers to have been involved in the development stages of the relaying schemes described here and then to see their merits proved by such an exhaustive series of tests. Being relay oriented, we would have liked to have seen a more detailed discussion of the operating principles of the single-pole relay schemes, and of the problems encountered in their development but we are sympathetic with the authors' need to avoid too much detail in presenting a subject of this magnitude.

The impedance characteristics of the static phase relays, shown in Figure 5A of the paper, deserve further comment. To those familiar with the application of distance relays for transmission line protection it is common knowledge that a phase distance relay may operate on nearby single-phase-to-ground faults, and further that a ground distance relay associated with an adjacent phase may also operate. When three-pole tripping is involved such operations are not usually considered objectionable. However, on the single-pole scheme described in this paper operation of any phase relay, or more than one ground relay, trips all poles. Therefore, an extremely important consideration in the choice and adjustments of the relays associated with this single-pole scheme is that no phase relay should operate on nearby single-phase-to-ground faults, and that only the ground relay on the faulted phase should operate.

Most of the investigations and tests on the model power system made by the manufacturer, which the authors mention, were aimed at determining the performance of the phase relays on single-phase-to-ground faults and devising means of preventing their operation on such faults. Traditionally a graphical method[1] has been used to estimate the response of phase relays to ground faults, but this approach cannot conveniently accommodate the effects of load current. The digital computer proved to be useful in this area and was used extensively during the investigation to determine impedances seen by the phase relays

Manuscript received February 18, 1971.

during ground faults taking into account magnitude and direction of load flow. While the results of this phase of the investigation were very helpful they were not accepted as final because effects of transient response of the relays were not included. This is where the model power system proved its worth. With this tool actual line and system conditions were simulated and the phase relay response, including the effects of transients and load flow, was checked for ground faults at several points on the protected line.

It was as a result of these investigations that the steady-state phase relay characteristics shown in Figure 5A of the paper were finally selected. The objective was of course to minimize exposure of the phase relays to operation on a single-phase-to ground fault. The directional zone-2 phase lens unit while minimizing the tripping area provides insufficient tolerance to arc resistance for phase faults near the relay location. The short reach mho unit, designated as the zone-2 mho unit in Figure 5A, was included to detect these close-in phase faults that might fall outside the lens characteristic. The extensive simulator tests by the manufacturer and the authors' staged fault tests indicate that the objectives of this static phase relay characteristic have been realized.

REFERENCE

[1] "Graphical Method of Estimating the Performance of Distance Relays During Faults and Power Swings", A. R. van C. Warrington, AIEE Transactions, Vol. 68, 1949.

K. H. Engelhardt and **D. G. McFarlane** (British Columbia Hydro and Power Authority, Vancouver, B.C., Canada): The authors are to be congratulated on a most comprehensive presentation of single-pole reclosing application. The paper demonstrates the necessity for staged fault tests on such a sophisticated system, to confirm the correctness of both system and relaying application and design, as well as to detect possible errors or mistakes and provide data for their remedy.

We are most interested in this paper since our company is considering a similar application of single-pole reclosing on our 500 kV system in the near future.

System Considerations

In the discussion of outages it is suggested that over 75% of all faults due to lightning would be single-line to ground. Using an outage rate of 0.77 outages per year per 100 miles, it can be shown that the one permanent outage assumed to occur every five years would be due to the 25% of faults which were more severe than line to ground. Our experience on 500 kV lines is that almost all the faults involve only one-phase and ground and are temporary faults. Therefore the single-pole reclosing should be almost 100% successful.

All of the fault tests reported with the neutral reactor in service show very rapid extinction of the secondary arc. The tests were performed with staged power faults and did not include the effect of lightning stroke discharge or discharge time. The system appears to have been in service since early 1969, have the authors any field experience on actual system faults or do they foresee any possible failures due to lightning stroke duration with the reclose time chosen?

One of the conclusions points out the necessity of closing resistors on the circuit breakers. This requirement is questionable for the single-pole switching, since such switching should produce no higher switching surges than the three-pole energization of a de-energized line.

Protective Relaying

In the basic choice of relays, could the authors elaborate why one of the two relay schemes utilizes electromechanical components, rather than equipping both lines of protection with solid-state relays?

With regard to the application of ground distance relays, are the ground fault resistances encountered always low enough as to be within the R-axis reach of the relays? Reference paper (5) on transmission line design indicates that the groundwire is isolated from the transmission towers. Without connection capable of lumping several towers together, the footing resistance of a single tower might easily bring the fault outside the relays' operating locus (tower footing resistance values of 100 ohms have been measured on rocky ground.)

It is noted that a directional comparison on-off carrier blocking scheme is used, i.e. the carrier is normally off, and is keyed "on" in a blocking mode for external faults. Such a scheme lacks a continuous monitoring feature of the carrier. What criteria determined this application, rather than a scheme utilizing a carrier which is continuously on (and capable of being monitored), and on internal faults either keyed off or shifted to a different frequency (shift from "guard" to "trip")?

One of the staged tests showed an incorrect operation of a phase-to-phase distance relay on a SLG fault, resulting in incorrect three-pole tripping. Is no "ground preference" logic provided in the scheme, which would inhibit operation of phase-to-phase distance relays on SLG faults?

With regard to protection of the shunt reactors, what means are used to detect faults in the neutral leg, which, except during system faults, will draw no current?

Compensating Reactors

The three-single-phase reactors presumably have a neutral insulation equivalent to the neutral reactor, namely 92 kV 60 Hz and 450 kV BIL. It appears the 450 kV BIL was based on switching surges during high-speed three-pole reclosing. For such a system operation the neutral reactor could be bypassed with a disconnect switch. Could the authors please comment further on the characteristics of the phase reactors and the neutral reactor. Are they iron core or air core and what is their saturation level? Has there been any attempt to provide surge protection for the neutral reactor or the phase reactors?

It would be interesting to know what system conditions established the 10 second rating of 117.5 amperes for the neutral reactor. It would seem that the maximum 60 Hz overvoltage occurs for a single-line to ground fault which should be cleared well within one second even with backup relaying.

In establishing the kVA rating of the neutral reactor some continuous unbalance has to be assumed. With the use of fully transposed lines and balanced phase reactors the neutral unbalance would be quite small. We have been studying a similar installation on future 500 kV line section without transpositions. To determine the kVA rating a maximum Vo of 10% was assumed. How did TVA establish the 25 ampere continuous rating?

Staged Tests

The staged fault tests certainly proved the effectiveness of using a neutral reactor to compensate the transmission line capacitance, and resulted in a very successful operation of the single-pole reclosing scheme. A comparison of the test results with the calculated results of the digital analysis would have been most illustrative. Only one test without the neutral reactor was reported. This test was unsuccessful but appears from Fig. 13 to be close to a successful reclose. With a slightly longer reclose time successful operation could have been obtained and by lowering the plant output the same stability margin would be retained.

It was pointed out that earlier stability studies proved the requirement of reclosing within one-half second to maintain stability. Yet no test appears to have been performed at the full plant output to prove this point. Was there any attempt to provide test information to compare the stability margin with digital computer studies?

Manuscript received February 18, 1971.

E. W. Kimbark (Bonneville Power Administration, Portland, Oreg. 97208): The scheme of single-pole switching on TVA's Paradise-Davidson line makes history in at least two respects: it is the first use of single-pole switching on a 500-kV line, and it is the first line of any voltage to use specially connected shunt reactors for neutralizing capacitive coupling between phases as an adjunct to single-pole switching.

The series of tests made on this line clearly shows the beneficial effect of the shunt reactors in hastening arc extinction. In the one test made with the reactors disconnected, extinction occurred too late and the fault arc restruck when the first circuit breaker reclosed. All the other tests showed successful reclosure except those in which the protective relays misoperated.

Bonneville Power Administration is planning to use single-pole switching on its 525-kV 144-mi (232-km) line now under construction from the Dworshak hydroelectric powerplant on the Clearwater River above Lewiston, Idaho, to Hot Springs, Montana. There will be a bank of four shunt reactors at each end of the line. Normally this line will not operate radially but occassionally it may do so.

In situations such as that described in the paper, in which a generating plant feeds radially into a network over a single transmission circuit, an alternative to single-pole switching that might be considered is the use of a shunt-resistance brake in conjunction with three-pole switching. Was this alternative studied by the engineers at TVA?

Manuscript received February 22, 1971.

Hung Jen Li (Westinghouse Electric Corporation, Newark, N.J.): This paper is highly interesting, it has verified the application of single pole

Manuscript received June 1, 1971.

switching techniques for a 500 kV system. The line protective relaying systems are of the utmost interest to us.

The authors indicate that in order to provide more secure operations, bus potential transformers supply the single pole relays at both terminals. For a sound phase-to-ground fault during single phasing period, zero-sequence potential and current may have a phase angle relation which produces opening torque to a directional overcurrent ground relay with potential polarization if line side potential devices are used. However, there are no directional overcurrent ground relays used in these schemes. Therefore, what are the reasons or considerations for the use of bus side potential devices?

Experience reported by the authors seems to indicate that the blinders are necessary for improving the angle discrimination of mho units to minimize improper fault-type determination. Why were the blinders only installed with the phase relays at Davidson and only the ground relays at Paradise?

Fig. 5A and related text are confusing.

(1) The labels of "zone 1 mho unit" and "zone 2 mho unit" look like they should be reversed.

(2) Descriptions of the characteristics look like they should be "zone 1 tripping is initiated for faults within the area bounded by zone 1 mho unit that fall within zone 2 mho unit *and* zone 2 lens unit. Zone 2 tripping is confined to the area bounded by zone 2 mho unit *and* zone 2 lens unit."

Item 4 under "Line Protective Relays" listing the additional requirements of the relay schemes is for sound phase protection.

(1) Are there any special means or considerations in the scheme to block for external faults during the single phasing period since ground carrier start circuitry has been blocked?

(2) In this regard, it is noted that no tests were run with phase A or B open, simulating an internal C-G fault during the period following an A-G or B-G single pole trip. Do the authors have an opinion as to the likelihood of such an occurrence in practice?

An additional special requirement should also be considered. When a pole of the breaker is open, a time delay 3-pole tripping should be initiated to prevent from over heating to the generators due to the unbalanced load current in case the opened pole does not reclose properly by some reason. Have the authors considered this point? Finally, does one reclosing relay control all breaker poles?

L. Edwards, J. W. Chadwick, Jr., H. A. Riesch, and L. E. Smith: The authors appreciate the complimentary remarks of the discussers and their interest in this application of single-pole switching. Since many of the inquiries are similar, the discussers' questions will be answered by categories.

System Design Considerations

Mr. Kimbark asked whether the use of a shunt-resistance brake in conjunction with conventional three-pole switching was investigated as an alternate to single-pole switching. This was not studied because of the obviously higher cost. Use of the compensating reactor with single-pole switching seemed to be an economical means to stability for the radial connection described, and this was explored in detail. In case other alternatives were required, however, all equipment including the line protective relays was designed for conventional three-pole switching.

With reference to Mr. Hedman's question, no specific analytical calculations were made on the actual Paradise-Davidson configuration for evaluation prior to operation of the system. The four-legged reactor was planned for connection at midpoint of the line with one complete line transposition between the reactor and each line terminal. Relocation of the reactor to the line terminal at Davidson and the selection of phase transpositions were matters of economics. It was anticipated that changes in the line compensation would result in slightly greater magnitudes of secondary arc currents and restriking voltages, but without practical limits as guidelines from an operating 500-kV system, the cost of ideal compensation could not be justified.

Mr. Hedman asked if the authors feel that phase transpositions are required on a line of this length in order to have successful single-pole reclosing. Since results of our staged-fault tests were conclusive, additional calculations were not made to compare with these or to evaluate the phase transpositions actually used.

We believe these phase transpostions contribute considerably in altering those components of secondary arc currents and recovery voltages attributable to inductive coupling. This, of course, increases the probability of successful reclosures. As noted in the text and Figures 9 and 10 of the paper, secondary arc currents and recovery voltages vary with line loading; therefore, these phase transpositions should be most effective during phase-to-ground faults with maximum output on the generator. In addition, any number of complete line transpositions serve to (1) decrease the dissymmetry in line capacitances, thereby permitting the use of more equal phase reactors, (2) make the compensation provided by the neutral reactor more equally effective for single-phase-to-ground faults on any phase, and (3) balance line impedances which simplify relaying problems.

Only one complete line transposition for this connection was selected after plans were changed to relocate the reactor to Davidson. This line transposition was achieved by rolling one outside phase beneath the others at favorable locations along the right of way. The additional rearrangement of phases near Davidson, as indicated in Figure 3 of the paper, was made to maintain the same relative position of phase conductors emanating at substations on North-South intra-system ties within the TVA grid.

Messrs. Engelhardt and McFarlane may be correct in stating that we should expect approximately 100 percent successful reclosures; however, when single-pole switching was investigated for this connection, 500-kV facilities had been in service in this country only two years. It was expected that many years of operating experience would be required before an acceptable lightning outage rate could be established. Our statistical and oscillographic records show that the 500-kV system has experienced 22 line operations resulting from lightning since 1965. Twenty of these were single-phase-to-ground faults, but two involved two phases and ground. This ratio of single-phase to other types of faults does exceed our original expectations.

Relative to Messrs. Engelhardt and McFarlane's question about lightning, we anticipate no difficulties from lightning stroke discharge time. For many years we have been using a dead time of 12 cycles at 161 kV, and for the past five years we have been using a dead time of 20 cycles for 500-kV line faults. At the 161-kV voltage level, the fault is very rarely reestablished after being properly cleared; however, occasionally we have experienced failures to reestablish a connection because of multiple lightning strokes. Our records show, for example, where a line relayed initially by lightning, reclosed successfully by high-speed reclosure, but then after five to ten cycles the line tripped again for a subsequent lightning stroke. As for the Paradise-Davidson 500-kV connection, it has been operating since early 1969 without outages resulting from faults on the 500-kV line. The lack of actual faults tends to emphasize the importance of staged-fault-test evaluations for innovations adaptable to EHV transmission.

Messrs. Engelhardt and McFarlane raised the question as to whether closing resistors are necessary on circuit breakers used for single-pole switching. At the 500-kV and higher voltage levels, closing resistors are used to limit magnitudes of switching surges and thereby permit the use of lower and more economical insulation levels. The importance of the preinserted resistors on closing at 500 kV seems apparent when minimum dead time is required with single-pole switching for stability considerations. Unnecessarily high switching surges could very well result in unsuccessful reclosures. Nevertheless, the intent of the authors was to point out that breakers used on this connection were equipped with resistors in both the closing and opening circuits, and the alignment of resistors and main contacts was not altered during the staged-fault tests.

Compensating Reactor

Messrs. Engelhardt and McFarlane had several questions concerning the four-legged reactor bank. We appreciate these questions and offer the following explanations.

The low-voltage terminals of the compensating reactors are rated 92 kV, 450-kV BIL to match the neutral reactor. Except for higher neutral insulation, these phase reactors are similar to others used on our 500-kV system. We believe the 1.6-percent additional cost for increasing the neutral insulation from 110- to 450-kV BIL would not have been appreciably different if 350-kV BIL had been selected. The higher insulation, of course, eliminated the need for a bypass switch around the neutral reactor or the use of a coupling capacitor to drain high-frequency transients accompanying high-speed, three-pole reclosing at Davidson during periods when the machine at Paradise is out of service.

The phase reactors are of the gapped-iron core design with such saturation characteristics that their impedances do not change appreciably between 100-130 percent rated voltage. Arresters are used to protect these windings from lightning, but they are not used as first-line protection against switching surges.

The neutral reactor was purchased according to AIEE 32, Standard for Neutral Grounding Devices. It is agreed that all faults should be cleared well within one second; however, the minimum standard short-time current rating of 10 seconds provides a satisfactory margin for the protection of these facilities.

A continuous-current rating of 15 amperes was determined for the neutral reactor by manufacturing tolerances of ±5 percent in shunt reactor impedances and a ±5 percent unbalance in system operating voltages. However, the equivalent kVA rating of the neutral reactor was established by conservatively using 25 amperes.

Manuscript received June 1, 1971.

The neutral reactor is a standard item consisting of an oil-immersed winding, and it is protected with a lightning arrester. Since the device is required to carry current only during fault conditions, an open circuit is not anticipated. Similar to other neutral reactors in service on the TVA system, protection is not provided for turn-to-turn or turn-to-ground faults.

Relaying

In reply to Mr. Seeley's question, no changes were required in relay settings on other 500- or 161-kV lines because of single-pole switching. All adjacent 500- and 161-kV lines have carrier-blocking directional comparison relay schemes which functioned correctly during the staged-fault tests. This is attributed to the grounding practices used on our interconnecting transformers. These three-winding transformers have their 500- and 161-kV windings connected in wye. The 500-kV windings are solidly grounded and provide low impedance paths for residual currents; however, the 161-kV windings are grounded through neutral reactors to limit the transfer of zero-sequence currents between voltage systems. The ground-current distribution during single-phase operation is, therefore, confined almost entirely to the 500-kV transmission system. This also points out the necessity, from an operating standpoint, for having the receiving end transformer bank in service when single-pole switching is used.

Mr. Seeley suggests that early reclosing with the four-legged reactor would be successful if initiated by the reappearance of some critical value of recovery voltage persisting for a few cycles. Oscillograms of our staged faults show the magnitude and the duration of secondary arc currents and corresponding magnitudes of recovery voltages after secondary arc extinction as functions of load currents. Values of these recovery voltages vary between 24 and 59 kV. Possibly the sample oscillograms presented were misleading; nevertheless, when it is realized that recovery voltages are functions of load currents, their use to determine minimum dead time for reclosure seems questionable.

Mr. Rockefeller correctly points out that single-pole switching presents a challenge in relay-system design, particularly in determining the simplest means of preventing phase relays from operating for single-phase-to-ground faults. The unwanted CA electromechanical relay operations during the staged-fault tests might be described as resulting from the effects of load current, but we now believe this to be a very indirect cause. The principal culprit is the difference in time constants associated with the polarization potential and the potential restraint circuits of the relay. The relay contact closed about one cycle after fault initiation, and it is believed that memory action caused the transient overreach. This is an example of misapplication of a relay since the device is a Zone 2 element having memory action without the improved transient overreach characteristics of a Zone 1 element. On a normal 3-pole trip this would be of no significance; however, in a 1-pole trip scheme it is critical. A four-cycle delay was inserted in the carrier-phase trip at Paradise and this produced the desired results. The additional delay in phase trips at Paradise is not significant since a 3-pole trip results in generator shutdown and reclose cancellation.

Regarding the question raised by Messrs. Engelhardt and McFarlane on our choice of two relay schemes – one electromechanical and one solid state – this was based on experience. Electromechanical relays have proven reliable over the years, whereas in 1967 our operating experience with solid-state relays had been limited to two years. During this period we experienced some erroneous operations with solid-state relays, and these difficulties had not been resolved at the time the Paradise-Davidson connection was planned. These solid-state relays were specially designed and tested for this application and also for operation with conventional three-pole switching. They performed perfectly during the tests and based on their performance, two sets of solid-state relays would have been a prudent choice.

As pointed out by Messrs. Engelhardt and McFarlane, excessive tower-footing resistances could result in failure of ground-distance relays to recognize all ground faults. On the Paradise-Davidson connection there are 435 structures, and 96.8 percent of these have footing resistances less than 22 ohms. The three highest measured values are 45, 35, and 24 ohms, respectively, and we anticipate no difficulty with these. Although the overhead ground wires are insulated so they can be used as a communications medium, a 7/8-inch gap which flashes over between 15-22 kV is provided at each structure to prevent interference with their normal function as ground wires. It was expected that these gaps would flash during staged-fault tests, and actual flashover was observed at adjacent structures and within the switchyards.

Messrs. Engelhardt and McFarlane expressed interest in our selection of carrier-blocking relay schemes. The use of carrier-blocking or a continuous-carrier relay scheme is largely a matter of preference. We have used carrier-blocking schemes for many years with satisfactory performance. Our preference for carrier blocking is that it is inherently "fail-safe," and its conservation of the frequency spectrum is increasingly important. Our use of the continuous-carrier signal has been almost entirely limited to transfer-trip schemes. We have had several erroneous operations with these, and the operations have been difficult to analyze because the difficulties have been attributed to spurious signals or noise.

Relative to the operation of the phase relay at Paradise for single-line-to-ground faults there, Messrs. Engelhardt and McFarlane questioned the omission of "ground preference logic" in our line protective schemes. We do not have "ground preference logic" with either the solid-state or electromechanical relays and elected not to use it. In fact, "ground preference" has not been used in our relaying for many years. In weighing the merits of "ground preference," we had much rather depend on correct relay operations.

Mr. Li asked why bus potential transformers rather than line-side potential devices were selected for our single-pole reclosing scheme. The choice of bus potential was determined by bus switching arrangements rather than specifically for the relay schemes. Our standard 500-kV main-and-transfer-bus switching arrangement, which requires bus-potential devices, is used at Davidson. Only two 500-kV breakers are installed at Paradise now, but ultimately the main-and-transfer-bus switching arrangement is to be used there. Bus potential transformers were required, but they would have been selected anyway because they offer great simplicity in single-pole relaying schemes.

We offer the following in answer to Mr. Li's specific questions concerning our single-pole relaying scheme.

1. For those familiar with the R-X diagram plottings of relay characteristics, the labels of "Zone 1, mho unit" and "Zone 2, mho unit" in Figure 5 of the paper may appear reversed, but these labels and the related text are correct as described. The Zone 2, mho unit is required for this scheme because "close-in" faults involving some fault resistance may not be seen by the Zone 2 lens unit to initiate carrier relaying or instantaneous tripping similar to conventional Zone 1 trips.
2. There are no special means in the relaying scheme to block for external faults during the single-phasing period or other second-contingency conditions.
3. One reclosing relay controls all breaker poles.
4. The machine at Paradise is self-protecting with its own protective devices, including negative-sequence relays. Breaker failure protection is provided to shut down the machine at Paradise, cancel reclosing, and transfer trip Davidson. If the open pole at Paradise does not reclose properly within a prescribed time, the other two poles are tripped, the unit is shut down, and transfer trip to Davidson is initiated.
5. If the open pole at Davidson does not reclose properly, local backup protection is initiated and the machine is tripped by dual-channel, transfer-trip facilities.

We wish to thank Mr. Goff for elaborating on some important details of the static-relay scheme development. The model power system tests and the digital computer analysis were invaluable in arriving at a satisfactory solution to all the problems that arose. Mr. Goff's comments also provided answers to some of Messrs. Rockefeller and Li's questions.

It is surprising that none of the discussers inquired about the performance of sudden-pressure relays used for protection of the transformers at Paradise and Davidson. The step-up transformer bank at Paradise and the receiving-end transformer bank at Davidson are dissimilar and they were purchased from different manufacturers. Each transformer, however, is protected by a General Electric, type J, sudden-pressure relay. Although we were somewhat apprehensive concerning the operation of these protective devices, we can report that they did not operate for any of the staged-fault tests.

Staged-Fault Test Results

Messrs. Engelhardt and McFarlane asked if any attempt was made to determine the margin of stability for the radially connected generator during our staged-fault tests. We elected not to instrument the cross-compound machine to determine angular displacement of rotors since this unit had not been released for commercial operation. Instead, the relative movement of unit 3 was measured indirectly. This was achieved by comparing phase relationships of the A-B phase generator terminal voltage with the A-phase-to-neutral bus voltage at Paradise. These quantities were recorded on the same oscillograph and the 161-kV bus voltage trace (timing trace) was pipped every time the A-B phase generator voltage passed through zero in the positive direction.

The reasonableness of this comparison can be realized by visualizing the generator connections at Paradise. Units 1 and 2 connect to their respective generator transformers and over seven 161-kV bundled-conductor lines to the 161-kV transmission grid. Similarly, independent of other units, unit 3 connects into the 500-kV transmission system at Davidson over a radial 500-kV line. Station service for all units, however, is supplied from the 161-kV system at Paradise. Electrically, unit 3 is far removed from units 1 and 2.

Table III shows the relative movement of generator No. 3 for staged-fault tests at Paradise and Davidson. Other pertinent data for each of these tests are given in Tables I and II of the paper. Because of the many problems encountered in getting a unit of this size operational, full output of the 1150-MW generator was not available during the test period.

TABLE III – RELATIVE MOVEMENT OF PARADISE GENERATOR NO. 3 DURING STAGED FAULTS WITH SINGLE-POLE SWITCHING

Test No.	Fault Location	-Generation- Unit No. 3 @ Paradise MW	Mvar	Relative Swing of Unit No. 3 @ Reclosure
11	Paradise	390	85	5.4°
7	Paradise	720	140	13.0°
10	Paradise	905	210	24.6°
2	Davidson	525	170	8.4°
6	Davidson	660	425	6.5°
9	Davidson	945	233	32.2°

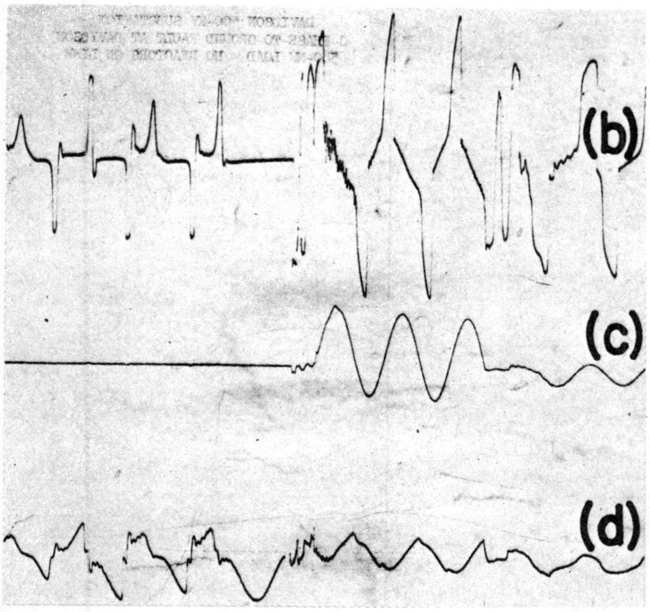

Fig. 13A. An enlargement of trace c of Figure 13 in the paper.

Mr. Rockefeller correctly observed that trace c in Figure 7 of the paper is in error. Connections to instruments used to measure breaker recovery voltages permitted the recording of differences of potential between two grounds during the flow of fault current. This error is present to some degree for all staged-fault tests. However, these traces show correctly the values of breaker recovery voltages at the instant of breaker opening and thereafter. This points out the unexpected difficulties with instrumentation regardless of the precautionary methods employed.

A number of the discussers expressed interest in the staged-fault test that was made with the compensating reactors disconnected. Comments concerning this particular test imply that our results were misinterpreted. Perhaps we were unduly brief in our presentation, but unfortunately some very significant information is not apparent on the sample oscillogram of Figure 13 in the paper. The minor oscillations present for 0.41 cycles before main contact closure at Davidson are no longer visible on trace c of Figure 13 after reducing the original oscillogram for our presentation. In Figure 13A, with trace identification the same as Figure 13 of the paper, a small area of the original oscillogram is enlarged to show these oscillations. The significance of this test justifies a more detailed interpretation of our test results.

With reference to Table II and Figures 13-14 of the paper, the original oscillogram and shots from two high-speed cameras which recorded pictures at 22 frames each cycle permits the following analysis of test 3. Each period (60 Hz base) was recorded on the high-speed film independently of film speed.

The C-phase poles of breakers at Davidson and Paradise opened their main contacts at 2.5 and 3.35 cycles, respectively, after the inception of a fault at time zero. Continuous secondary arc currents were apparent at 14½ cycles. Thereafter, as the instantaneous values of restriking voltages approached maximum during each subsequent half cycle, restrikes were apparent until 24.91 cycles had elapsed. At this time, as perhaps implied by trace b of Figure 13, some of the discussers interpreted this as secondary arc extinction caused by natural forces. This is not the case at all. The extended zero-current pause, apparent between restrikes at previous half cycles, is present immediately following current zero at 24.91 cycles. A restrike did not occur as the recovery voltage approached its maximum negative value, so indicated in trace d, since the 400-ohm resistor was preinserted just prior to this, preparatory to reclosing main contacts of the breaker. Between 24.91 and 26.05 cycles, the combining effects of the resistor and the phase relationship of system voltage and line-side restriking voltage did not permit additional restrikes. During this period, however, films showed the secondary arc path illuminating, very similar to the arc characteristic labeled 26 in Figure 14. At precisely 26.05 cycles the arc was reestablished by a system fault through the 400-ohm resistor. The magnitude of fault current was approximately 200 amperes. This current was extinguished, reestablished, and extinguished again before the main contacts shunted the preinserted resistor at precisely 26.46 cycles. This reestablished the primary arc of 7940 amperes indicated in trace c of Figure 13. Unfortunately, only the higher magnitude of system fault current is recognizable in the reduction of the original oscillogram. However, saturation of the current transformer measuring secondary arc currents is apparent at 26.05 cycles in trace b of Figure 13; and the arc characteristic labeled 26+ in Figure 14 shows reestablishment of the system fault at 26.05 cycles, exactly 0.41 cycles before main contacts close. As noted in Table II of the paper, the Paradise breaker pole did not reclose until 29.35 cycles had elapsed.

Mr. Rockefeller comments that from our test results it could be concluded, perhaps erroneously, that secondary arc currents exceeding 20 amperes are too high or marginal for prompt arc deionization. It was not the intent of the authors to purport that values of secondary arc currents and recovery voltages must be limited to 20 amperes and 50 kV, respectively, for successful single-pole switching at the 500-kV voltage level. The Paradise-Davidson connection is a special application of single-pole switching where special compensation is required to enhance secondary arc current extinction and to permit successful reclosure within ½ second for stability reasons.

It is noted in the text and Figure 10 of the paper that the secondary arc was extinguished with a restriking voltage of 59 kV. For faults on this line with full generation on unit 3 at Paradise, greater magnitudes of secondary arc currents and recovery voltages are expected. With the compensation applied, secondary arc currents below 20 amperes are anticipated, but restriking voltages should approach 70 kV.

There are no doubts in the authors' minds that secondary arc currents exceeding 20 amperes can be extinguished provided stability presents no problem and "dead times" considerably greater than one minute are employed. It is noted in Figure 14 of the paper that natural forces remove the gaseous cloud caused by the major arc, and within approximately 1/5 second there remains nothing more than a narrow ionized path permitting the flow of secondary currents. These currents are reestablished as the instantaneous values of restriking voltage approach maximum in both the positive and negative half-cycles. It is the opinion of the authors that such ionized paths will be elongated by natural forces if sufficient "dead time" permits. Our major concerns included both dispelling the secondary arc path and reclosing within a reasonable time to maintain stability.

Messrs. Engelhardt and McFarlane suggest that test 3, without the compensating reactor, would have been successful if slightly longer re-close time had been used and plant output lowered to provide a satisfactory margin for stability. This is a reasonable conclusion which could be drawn from the secondary arc currents of trace b in Figure 13 of the paper. However, a more detailed analysis of test 3 is presented in this closure. To us the compensating reactor provided both practical and economical means of meeting our established criteria for this project.

IMPROVED RELIABILITY OF BULK POWER SUPPLY BY FAST LOAD CONTROL

R. H. PARK
President
Fast Load Control, Inc.
Brewster, Massachusetts

INTRODUCTION

This paper makes the point that it is feasible to apply what are termed "advanced-type fast load control procedures" to make a major improvement in eliminating instability and improving reliability by substantially preventing at the source the occurrence of major system swings.

Although some of the procedures proposed have been discussed individually as means of reducing instability difficulties, the controlled combination of the procedures produces a new concept, and the possibility of major improvements, that appears not to have been recognized before.

As a means to improve bulk power supply reliability, strong system interconnection such as has been mainly advanced as the soundest approach, offers the advantage of inherent, built-in and, as one might say, static strength.

Fast load control introduces the supplementary element of favorably effected, rapid dynamic response to adverse events.

Advanced-type fast load control procedures set forth in the paper can be used as a way to reduce need for construction of new lines, while at the same time favorably affecting reliability.

CONCEPT OF FAST LOAD CONTROL

A key deficiency of present-day power systems resides in failure to provide or use means of rapidly controlling transmission circuit loadings other than by way of generator dropping and the opening of transmission circuits.

Due to lack of versatile fast means of control, adverse line loadings can develop in the event of system disturbances, with resultant loss of generation and the opening of system ties.

Development of increased line loadings can be minimized, and opportunity to design and operate systems on the basis of increased scheduled loadings can be provided, by employment of fast load control procedures which respond, first, to fault relaying and events leading to the opening of generator and tie-line breakers, and additionally, on a feed forward basis, to predisturbance conditions and the nature of disturbance producing and produced events, by:

1. In system areas having excess power supply, operating to:
   a. Momentarily apply resistive loads at, or electrically near to, generator terminals, with the benefit of feedback control responsive to generator speed and acceleration and the integral of speed change.
   b. Optionally provide for fast tripping of generators, feedback controlled braking, and reconnection in a fraction of a second or a few seconds, at a preferred displacement angle, as a means of dealing with chance of delayed fault clearance, with provision so that, in the event of delayed clearance, reconnection of generators would be blocked

Reprinted with permission from *American Power Conf. Proc.*, 1968.

selectively when the effect would be adverse.

c. Initiate, coincidentally with braking, programs of steam turbine control and intercepting valve closure, momentary bleed or bypass of steam, and, dependent on the programmed extent of sustained valve closure, rapid reduction of steam generation and closure of throttle valves ahead of or within superheater systems.

d. Initiate, conincidentally with braking, programs of reduction of water acceptance of waterwheels of hydro units, and/or tripping of generators of hydro units, optionally on the basis of resynchronization within a fraction of a second or a few seconds.

2. In system power deficient areas, operating to:

a. Initiate programs of rapid opening of selected feeder circuit breakers, in selected load areas, to deenergize loads for periods up to ½ second, and to disconnect within customers' premises temporarily interruptible loads, such as heating, air conditioning, air compressors and, in some cases, pumps, in response to loss of voltage; thereafter reconnect loads on a staggered time basis over a period of up to 15 minutes.

b. Simultaneously initiate programs of rapid control valve opening in steam-electric plants having spinning reserve capability, and, in the case of hydro spinning reserve, programs of fast increase of water acceptance of waterwheels.

c. Modify tie-line feedback-type load controls and rapidly initiate programs of starting and loading of nonspinning reserve generation.

3. In relation to tie lines, initiate programs of opening selected tie-line breakers, followed, within a second or a few seconds, by reclosing at an advantageously determined phase angle.

Control procedures of a type that affect system load rapidly enough to prevent initial development of system instability can be termed methods of "fast load control." Because of this capability, plus ability to avoid system weakening and customer inconvenience, coordinated application of the fast acting load control procedures enumerated under (1), (2), and (3) above can be termed means of effecting "advanced-type fast load control."

ADVANTAGES OF ADVANCED-TYPE FAST LOAD CONTROL

As noted in Volume II of the FPC's June 1967 report, "Prevention of Power Failures," the objective of providing due assurance of bulk power supply reliability implies a need to design power systems so as to avoid concentrations of generating and transmission capacity and switching facilities, and to maintain at all times substantial margins of transmission capability above scheduled transmission line flows, in order to tolerate power swings resulting from facility outages and system disturbances.[51]

Nevertheless, in consideration of economic factors, major power system and regional pools have planned employment of concentrations of transmission capability as via, for example, the overlaying of

345-kV transmission with 765-kV lines, 230-kV with 500, and 138 with 345.

However, on the basis of use of only conventional slow load control procedures, the practical effectiveness of inherently very high power transfer transmission circuits is considerably diminished due to a need to limit scheduled line loadings to contend with swings that develop in the event of contingencies.

Transmission circuit power swings incorporate oscillatory and sustained components, the latter developing in the event of generator trips, sudden load losses, and other than momentary transmission line circuit breaker openings.

Conventional slow load control is entirely unable to counter oscillatory components, and operates to correct sustained components of load change too slowly to influence system stability.

In this situation, employment of advanced-type fast load control procedures, by providing a fast acting way to restrain both oscillatory and sustained components of power swings, affords a means of achieving either a predetermined or an increased degree of bulk power supply reliability with reduced need to construct added lines, and therefore can serve as a way to lessen conflict with growing public interest in land conservation and the preservation of natural and historical environments.

Advanced-type fast load control procedures offer special advantages as a means of dealing with very severe contingencies, such as the ability to be used to cancel the effects of delayed fault clearance and to counter the effects of entire loss of a main corridor of transmission.

In general, procedures that can best be employed as a means of countering tendencies to system separation and internal system breakup, in the event of severe contingencies, will vary depending on such factors as distances between major load centers, the size and location of cities, whether systems are centrally or peripherally located, whether or not lower voltage transmission is in process of being overlaid with only a few lines of considerably higher voltage, and if it is, whether or not the higher voltage transmission is of the single backbone circuit type, or comprises either a parallel line or a network type that provides parallel paths of power flow.

It should always be borne in mind that it was the process of widespread system interconnection that brought into existence the problem of cascading instability.

The position has been advanced that the best remedy for cascading is to construct enough lines to provide very strong system internal connections and interconnections. However, whereas it is both possible and desirable to design systems and regional pools that will tend to strongly resist internal cascading, the holding together of pools despite severe contingencies is another matter. Further, the position that strong internal connections will provide a way to avoid breakup of pools or individual systems always has to be viewed in relation to the severity of the contingency that can be encountered and successfully handled.

There has been rather general acceptance of the view that designing power systems so as to entirely prevent separations leading to cascading breakup is not feasible,[51] especially in relation to the perhaps remote but yet possible event of nuclear attack.

In relation to separations, an advantage in employment of versatile fast load control procedures resides in a favorable potentiality for adaptation to use as a way to minimize the extent of cascading separations and to favorably influence types of separation and service restoration following separations.

EARLIER METHODS OF FAST LOAD CONTROL

In recent years, especially with the widespread introduction of system interconnection via EHV lines, it has become increasingly the practice to rely on programmed fast generator dropping responsive to transmission line faults as a method of preventing development of system instability in the event of facility outages and system disturbances.

Generator dropping, system separating, and the opening of interconnections, when employed rapidly as a way to prevent system instability, can be classed as types of "fast load control" procedure in that these procedures bring about fast changes in system load. However, they also act to weaken a system just at the time when it is desirable to conserve strength, this being especially true when, as is presently conventional, tripout of a steam turbine-driven generator typically introduces a delay of several hours in reloading because of the need of going through the process of hot restart.

In addition to generator dropping and tie-line opening, studies have been made of use of dynamic braking, absent simultaneous employment of any supplementary fast control measures, as a way to counter generator rotor and related transmission system power swings in the event of faults (References 11-15, 17, 18, 20, 21, 24, 26, 41, 43 and 44). Braking installations have been reported in use in the USSR, and a Canadian hydro station is presently in process of being equipped for dynamic braking.[41]

Also, several United States power producers have been giving consideration to use of fast momentary steam turbine valve closing as a swing retardant.

Notwithstanding early development of the concept of use of dynamic braking and fast control of prime mover driving power as aids to stability,[1-10] there has been delay in appreciation of the merit of use of these measures in combination.[27, 29] Further, the potentiality of momentary and short-term load shedding as a useful supplement to area power generation boost procedures seems to have been overlooked.

ADVANCED-TYPE FAST LOAD CONTROL

Implementation

Provision to allow simultaneous accomplishment of dynamic braking and fast steam turbine valving makes it possible to very rapidly initiate and sustain a net reduction in generator export power.

Also, with provision for use of favorable types of steam bypassing procedure, it becomes possible to sustain a considerable period of operation in which a generator is connected to a system, but operates at low or zero export power output, and yet effect rapid reloading in a matter of a fraction of a minute.

It is also possible to employ braking and water gate closing to reduce very rapidly export power of hydroelectric units.[29]

In the converse case of power deficiency, load shedding can perform the converse of dynamic braking.

Load shedding, as commonly practiced, in response to reduced frequency, is essentially a backup procedure, rather than a preventive measure, because the drop in frequency in a major interconnected system is an indication that parts of the system have already separated.

To be effective as a means of fast load

control adapted to prevent system separation, load shedding must be utilized in a controlled manner directly in response to the initial fault or other disturbance. Controlled load shedding of this type can be provided, without significant inconvenience to customers,[58] by momentarily opening feeder circuit breakers controlling the supply of power to power users located within load areas that have been advantageously selected in response to conditions and events, as a way to shed load for a period of to one half second, and to thereby counter deceleration of generators located in or electrically adjacent to areas of power deficiency.

To an extent that would require determination, provision can be made so that the condition of temporary loss of voltage in customers' premises, produced by the momentary opening of feeder circuit breakers in response to centrally executed control, would cause interruption of power supply to certain classes of inherently temporarily interruptible load, followed by automatic reconnection on a staggered time basis over a period of minutes, responsive to the operation of inexpensive time delay relays controlling the energization of contactors and circuit breakers.

In the event of evidence of a sufficiently severe emergency, more than one momentary load interruption could be effected, while in case of need, including but not limited to development of underfrequency, centrally selected feeder breakers could be opened for periods of minutes, rather than a fraction of a second, and thereafter reclosed, or could be opened repetitively.

Coincidentally with load feeder circuit breaker opening, steam valve and water gate opening signals can be applied to steam and hydraulic turbines and suitably selected generating stations.

While loads are in process of coming back on the line, conventional power-frequency tie-line control, supplemented by disturbance responsive automatic means of rescheduling of transmission line power interchange, can operate to suitably further control turbines and steam generators of spinning reserve generation, and to automatically initiate the starting and loading of gas turbine-driven and hydroelectric generators.

Dynamic braking can be considered as a means of compensating for the delay in response of a prime mover to driving power reduction signals.

Similarly, momentary and short-term load shedding, to whatever extent determined to be allowable and provided, can comprise a means of compensating for the time required for operation of tie-line power-frequency control, for delay in response of prime movers and steam generators to generator output power boost signals, and for the time required to start and load fast starting, nonspinning reserve generation.

With employment of generator voltage level braking resistors having a power acceptance in excess of generator rating, it is possible to momentarily disconnect a generator from a bus, brake, and reconnect in less than a second, as a way of dealing with a condition of delayed fault clearance.

In an alternate approach, a steam-electric generator can be disconnected from a bus, have braking applied at generator auxiliary transformer voltage level, while being subject to fast turbine valve closing, and be reconnected in several seconds, after the generator rotor has advanced several complete revolutions in synchronous space.

In still another approach, it can be feasible in some cases to employ repeat-cycle-type braking[28] as a satisfactory means of dealing with delayed fault clearance.

Effectiveness of dynamic braking is increased by use of feedback controlled timing of braking load application and removal.[21, 28, 29]

Initiation of braking can take place in response to an indication of a fault or any other system disturbing event, and can also be made supplementarily or independently responsive to a suddenly developed increase in generator rotor speed.

Disconnection of braking load is preferably accomplished by a control system including elements which respond to a combination of speed, acceleration, and rotor angle, and it will normally be desirable to make provision for reconnection and disconnection one or more times if and when needed.

To avoid need for very fast and accurate timing of braking load removal, magnitude of braking load should be controlled in response to predisturbance generator load. Especially when a disturbance producing event comprises a line fault, it can also be useful to control magnitude of braking in response to prefault transmitted load, and to fault conditions and postfault events.

In view of the fact that steam turbine driving power reduction takes place only subject to a delay of about one half second, and also because, in order to counter hazard of overspeed, rate of upward adjustment of driving power subsequent to initial momentary valve closure is limited, valve lifting can require a period of some seconds, and it is therefore advantageous to execute control functions mainly on a feed-forward basis.

Parameters of feed-forward-type steam turbine driving power control responsive to a disturbance inducing event can advantageously relate to the magnitude and direction of predisturbance transmitted load, the predisturbance amount of local radially connected load, the number of generators in service in a generating station and their predisturbance active and reactive loads, the change in transmitted loads due to a disturbance, the occurrence or nonoccurrence of a successful faulted line reclosure, and the event of delayed fault clearance.

The same general type of feed-forward fault responsive controls are adapted for use in determining extent, duration, and desirable area of accomplishment of momentary and short-term load shedding.

A significant element that can usefully be incorporated in feed forward controls comprises the predisturbance status of circuits that would tend to significantly affect processes of power redistribution.

In general, to an extent that would be determined, and especially in first employment, control system complexity would advantageously be appropriately restricted, in the interests of simplicity, and to provide maximum assurance of reliable operation.

In the case of essentially radial transmission systems, controls can inherently be relatively simple. In more complex situations, even provision of only simple fast acting controls can operate to effect a notable improvement in system performance.

However, the magnitude of the November 9, 1965 Northeast blackout can be regarded as primarily due to provision of too simple types of automatic control.

Thus, automatic fast load shedding in Ontario, responsive to loss of the lines from Beck to the Hamilton-Toronto area, would have limited the scope of the Ontario blackout.

Fast load control applied at Beck and in

steam plants in the Niagara Falls-Buffalo area, or perhaps merely fast isolation of Beck, would have confined the blackout to Canada.

Absent provision for fast control in western New York, availability in central and eastern upper New York could have prevented spread of the blackout to those areas, and absent suitable controls there, system separation could have prevented spread to New England and New York City.

Therefore, as has already been generally accepted, there can be such a thing as too much simplicity.

Whereas an increase in the complexity of a control system increases the chance of difficulty resulting from failure of components, as taking account of the chance of a delayed fault clearance, allowance can be made for failure in the process of system design. Presumably, simultaneous occurrence of two very unusual events, of which one would be delayed fault clearance and the other some other unusual control system failure, can be accepted as a possible but improbable situation that could cause interconnected system or regional pool breakup, but which, in any case, with the avoidance of need for steam generator shutdowns, would be of reduced consequence.

Equipment Aspects—Dynamic Braking

In the case of steam turbines, fast control and intercepting valve closing can be executed in 0.20 second, and will result in major reduction of turbine driving power in 0.5 second, from which it follows that when dynamic braking is used in combination with fast turbine valve closing, the period during which braking primarily needs to be applied can be as little as 0.5 second.

In contrast, in hydroelectric plants, except where means of fast water bypass are available, delays in accomplishment of a substantial reduction in prime mover driving power can range up to 5 seconds. This being the case, it might be advantageous in some cases to employ line-fault-initiated programs of joint employment of braking and generator dropping, with dropping employed only in the event of a permanent-type fault. However, it is also feasible to employ fast prime mover control as an option to generator dropping, since, regardless of generator size and type, the problem of providing a resistor that can momentarily accept a requisite braking load can be dealt with by employing a sufficiency of metal such that, notwithstanding a sequence of load cycles, and without time for cooling, the energy delivered to the resistor can be accepted in the form of sensible heat.

Whereas series braking [7] could be employed as a way of dealing with delayed fault clearance, either momentary generator tripping with shunt braking applied during the period of generator isolation or multiple cycle-type [28] shunt braking appear to represent more versatile and simpler procedures.

In use of momentary trip-type braking, provision can be made to always open high-voltage bus backup breakers, in the event of other than a line to ground fault, without delaying to determine whether or not delayed fault clearance has in fact occurred.[43]

In the case of steam-electric units, it is feasible to effect a process of momentary generator trip with employment of combined fast valving and dynamic braking, in which braking is applied at generator service transformer voltage level; in such case the power, although not the energy acceptance, of the braking resistor would be reduced, and reconnection of a generator initially operating at full load would take place only after the generator rotor had rotated several times in synchronous space.

In the case of braking at generator voltage level, a braking installation would preferably incorporate isolated phase design principles.

To guard against chance of resistor burnup, provision should be made in braking load application circuits for a backup breaker plus a thermally activated means of automatic shortout of the resistor in the event of development of excessive temperature.

To ensure effective feedback control, braking load control breakers will preferably both close and open in two cycles. Since braking cycles would be of short duration, circuit breakers can advantageously be adapted to single-phase operation. This would provide a convenient way of varying the magnitude of braking effects, yet in view of the magnitude of applied load, and the short duration of application, generator rotor heating would not present a problem.

Energy acceptance of resistors will preferably be such as to allow execution of several, say three or four, successive braking cycles without time for cooling.

In the case, of momentary generator isolation and momentary tie-line opening type braking, it is necessary to employ a fast acting speed and acceleration responsive control which will reclose the breakers that have reopened at a preferred value of speed and rotor displacement angle. This implies a need to employ generator output and tie-line breakers which will reclose rapidly, preferably in 2 or 3 cycles, in response to a reclose signal. At least one high-voltage circuit breaker manufacturer already produces breakers which will meet this reclosing specification.

At least one circuit breaker manufacturer is presently tooling up to produce generator voltage level circuit breakers which will both close and open within 2 to 3 cycles, and it appears that the breakers in question will have a momentary current carrying capacity adapted to apply and remove resistive loads of a size up to those suitable for braking of generators of 1250-MVA capacity.

It is feasible to use conventional generator grounding resistors as energy absorbing components of braking resistors, but it also appears to be possible to design resistors which are especially well adapted to braking service.

It is estimated that the installed cost of a conservatively designed isolated phase braking resistor installation, plus controls, for a 1000- to 1250-MVA generator, would be of the order of $250,000, when intended for employment at generator voltage level, and substantially less when planned for use at generator service transformer voltage level.

Short Duration Load Shedding

Extent and duration of momentary load shedding can be controlled by selective tripping of distribution feeder circuit breakers,[33] and by varying delay in reclosure responsive to the type and location of disturbance producing event, pre-event conditions,[27, 53] and events taking place subsequent to an initiating event. Controls could include power-surge-type relays,[53] or could depend on the response of frequency and voltage sensitive devices to conditions obtaining during system swings.

Short-term load shedding of any desired duration can be effected by the transmission of signals into load areas to cause circuit breakers or contactors to disconnect noncritical loads.

TABLE I
METHODS OF HANDLING LARGE SUSTAINED REDUCTIONS IN DRIVING POWER

Procedure	Problems Capable of Being Solved
Electrically controlled relief valves preferably located at the hot end of superheaters and reheaters	1, 3, 4
Rapid fuel reduction and other boiler or reactor adjustments [45 47 48 56]	2, 3
Steam bypassing, including in one approach bypassing both around the high-pressure turbine to the cold reheat line and from the hot reheat line to the condenser [19 31 36 56]	1, 2, 3, 4, 5
Throttling ahead of the superheater [23 40]	6
Reloading within 10 to 15 minutes [32 47]	6
Single admission variable pressure operation [38]	6

While various means of signal transmission could be employed, including radio and telephone circuits, as previously noted, a simple way to signal can consist of providing a response to the momentary opening of feeder breakers. Vacuum-type reclosers can be supplied which would open in two cycles, and which can automatically reclose in as little as three additional cycles. However, normally it would probably be desirable to reclose in one half second.

In some cases it could be desirable to control the amount of load shed on a short-term basis, or the duration of shedding, responsive to the duration of momentary interruption, or in response to a sequence or sequences of interruptions.

Cost of time delay relays adapted to delay reclosing of contactors after tripout is under $3. Therefore, where loads are already contactor or circuit-breaker controlled, such as would apply in the case of office building and school heating and air conditioning loads, large air compressors and large pumping installations, expense would be negligible.

In contrast, residential heating loads, which are presently mainly controlled only by slow acting thermal relays, would usually require installation of a contactor, the cost of which would appreciably increase the cost per kW of the load affected. However, the increase in cost might not be enough to rule out provision for short-term load disconnection.

Fast Turbine Valve Closing

Problems involved in fast momentary turbine valve closing [6,8,9,11,42] are minor. However, realization of full potentiality of fast control of prime mover driving power can be attained only by employment of rapidly executed sustained driving power changes.[25,26,30,36,39]

In the case of minor system disturbances, sustained driving power changes, if used, could be held in the range of 10 to 20 percent down or up, in which case it is easily possible to execute coordinated control of steam generators. When large sustained reductions in driving power are involved, it will, in general, become necessary to take measures to avoid difficulties which can otherwise result, due to:

1. Blowing of safety valves, with the possible consequence of sustained leakage.
2. Excessive loss of steam through relief to atmosphere.
3. Excessive excursions in temperature of superheaters and reheaters.
4. Difficulties in rapidly reducing fuel supply.
5. Difficulties in reducing the rate of steam delivery of nuclear steam generators.

6. Excessive turbine thermal fatigue.

Procedures adapted to allow adequately handling these six problems are available, as shown in *Table I*.

As will be seen from *Table I*, bypassing steam represents the most versatile approach, while also it makes possible very fast restoration of driving power.

Where sustained driving power reductions not exceeding 50 to 60 percent are involved, it is possible to employ a valve closing program in which only the intercepting valve is rapidly closed, following which it is opened, and the control valve simultaneously closed on a coordinated basis.

In the case of drum-type boilers, special provision of superheater headers would be needed to make possible throttling ahead of some portion of the superheater system.

A study has shown that a fast step reduction of load of a base-load turbine from 100 to 50 percent of rating will tend to cause no worse turbine thermal fatigue effects than those that are likely to be encountered in a conventional hot restart,[50] which would imply that throttling ahead of the superheater could be dispensed with in most situations.

Also, in the matter of need to provide throttling as a way to guard against development of turbine thermal fatigue, a point to be borne in mind is that loss of 100 percent export load, with turbine pressure, temperature, and speed maintained, must occur repeatedly before cracking develops.[38,55]

In relation to the problem of providing bypass valves that will not leak,[52] it appears that highest commercial pressure and temperature steam valves can be supplied which are effective and leak-free in fossil-fuel-fired steam generator steam bypass service.[31] On the other hand, it is also possible to use either pilot-operated or spring-loaded, air-lifted, electrically-controlled relief valves with provision of a pair of shutoff valves on either side of the relief unit to allow servicing without need for boiler shutdown.

Control valves of commercial turbines can be both closed and opened at moderate speed under servo control, and at least one United States turbine producer employs servo means of intercepting valve control which could be adapted relatively easily to control the rate at which the valve would open.

Provision for fast closing and opening of control and intercepting valves under servo control can be made with employment only of straight-forward design changes.

Added steam generator costs required to provide implementation of sustained-type fast turbine valve closing would dedend on the extent of control wanted and on decisions as to provision for bypassing steam.

Response Delay Compensated Fast Boosting of Turbine Output

As has been recognized, approaches to the problem of boosting the capacity of base-load turbines can be employed which do not adversely affect efficiency at best point.[34] Steam bypassing around the high-pressure turbine, such as has already been cited as advantageous in relation to implementation of fast valve closing, can be so employed,[22] if allowed for in turbine design.

Fast Feedback-Type Prime Mover Controls

To an extent that will require study, a role presumably will be found for feedback-type fast prime mover control.[16,29,35,46]

On-Line Computers and Relays

In a first and, as it appears for many purposes, adequate approach, feed forward aspects of prime mover fast load control can be provided merely by use of already available fault and other relays, adapted to respond to events that will cause or are caused by a system disturbance, supplemented by controls responsive to simple local analog-type memories of prefault transmitted and generated load, and the status of such lines and generators as would tend to significantly influence redistribution of load consequent on a group of possible system disturbing events. To the extent that a more sophisticated approach should prove warranted, digital computers could be employed, primarily as a way to accumulate information as to the nature of predisturbance conditions, with the information so obtained employed as a way of continually readjusting the settings of fast acting analog-type controls.[49,57]

This is not to suggest exclusion of further computer use, as perhaps preferably by modulation of analog controls in relation to computer response to conditions obtaining during a disturbance.

Analog-Type Controls

As regards analog controls and devices, electrohydraulic governors develop electrical indications of speed which can be used to operate relays of adequate sensitivity and speed of response. Integration of speed signals can provide a reliable way of evaluating changes in generator rotor angular position, and initial angle can be related to transmitted and generated load.

In one approach, an index of generator acceleration can be sought in terms of changes in generator power output. However, this procedure fails to take account of generator rotor braking torques dependent on the d-c and negative phase sequence component of armature current; therefore, direct measurement of acceleration, as with employment of a drag-cup-type generator, would be preferable.

In relation to providing response to generated and transmitted load, where only response to preinitial event load is needed, conventional types of watt transducers will meet requirements, while to the extent that any conventionally used devices would offer inadequate speed of response, Hall effect or other fast acting transducers can be substituted.

Analog-type memories of loads can comprise capacitors charged from watt transducers through resistors, and relay contacts which open in response to the occurrence of a system disturbance producing event.

CONCLUDING REMARKS

There appears to be no major problem in obtaining equipment needed for implementation of fast load control.

Whereas, in the interest of thoroughness, an attempt has been made to project overall long run possibilities in use of advanced-type fast load control procedures, and while, as always tends to be true, attainment of maximum potentiality would involve a degree of complexity, in many situations it is possible to employ simple approaches which offer major advantages.

In specific situations, cost of providing against loss of intra- and intersystem stability, by use of such approaches, can be low relative to cost of accomplishment by other means.

REFERENCES

1. Fortescue, C. F., Wagner, C. F. and Griscom, S. B., discussion of "Power System

Transients" by V. Bush and R. D. Booth, *Trans. AIEE*, **44**, 98-103 (1925) February.

2. Fortescue, C. F., "Transmission Stability. Analytical Discussion of Some Factors Entering into the Problem," *Trans. AIEE*, **44**, 984-1003 (1925) September.
3. Evans, R. D. and Wagner, C. F., "Studies of Transmission Stability," *Trans. AIEE*, **45**, 51-94 (1926) February.
4. Griscom, S. B. and Wagner, C. F. (assignors to Westinghouse Electric and Manufacturing Co.) "Regulator System," U.S. Patent 1,935,292, November 14, 1933.
5. Skeats, W. F. (assignor to General Electric Co.) "System of Electrical Distribution," U.S. Patent 1,834,807, December 1, 1931.
6. Buell, R. C., *et al.*, "Governor Performance During System Disturbances," *Trans. AIEE*, **50**, 354-69 (1931) March.
7. Bergvall, R. C., "Series Resistance Method of Increasing Transient Stability Limit," *Trans. AIEE*, **50**, 490-97 (1931) June.
8. Mayer, R. I., "Fault Initiated Control of Steam Turbines as a Means of Increasing Stability of Power Systems," *Elektrichestvo*, **No. 13**, 27-32 (1934).
9. Zdanov, R. C., "Stability of Electric Power Systems." Moscow: Gosenergoisdat, 1938.
10. Hamilton, F. A., Jr. (assignor to General Electric Co.) "Control Equipment," U.S. Patent 2,285,203, June 2, 1942.
11. Crary, S. B., "Power System Stability," Vol. 2, 186-97. New York: John Wiley and Sons, 1947.
12. Cahen, F., "Braking of Hydro-Electric Generator Groups by Rapid Switching of Liquid Resistors Across the Terminals of the Alternators," *Rev. Gen. Elec.*, **56**, 257-67 (1947) June.
13. Rathsman, B. G., "Report of the International Study Committee No. 13: System Stability and Voltage-Load-Frequency Control," *CIGRE Paper No.* **336**, presented at the International Conference on Large Electric Systems, Paris, France, June 1952.
14. Crary, S. B. and Saline, L. E., "Comparison of Methods for Improving the Stability of High Voltage Transmission Systems," *CIGRE Paper No.* **307**, presented at the International Conference on Large Electric Systems, Paris, France, June 1952.
15. deMello, F. P., "Some Technical and Economical Aspects of Power Transmission in Brazil," *Proc. World Power Conf.*, **2**, 311-37 (1954).
16. Gornshtein, V. M., "Some Possibilities of Improving the Stability of Power Systems with Weak Couplings," *Elektrichestvo*, **No. 5**, 27-31 (1955).
17. Azariev, D. I., "Increase of the Transmitting Capacity of Transmission Lines," *Elektrichestvo*, **No. 6**, 1-7 (1955).
18. Azariev, D. I., *et al.*, "Increasing the Reliability of Operation of Power Systems and Long-Distance Transmission Lines," *CIGRE Paper No.* **318**, presented at the International Conference on Large Electric Systems, Paris, France, June 1958.
19. Powell, E. M., "Engineering the Eddystone Steam Generator for 5000-Psig, 1200-F Steam," *Trans. ASME*, **79**, 1447-57 (1957) August.
20. Venikov, V. A., Discussions, *Proc. Intern. Conf. Large Elec. Systems, 17 Convention*, **I**, 678-80 (1958).
21. Glebov, I. A., Kashtelan, V. E., and Siryi, N. S., "Electric Braking of Synchronous Generators Supplying Long-Distance Power Transmission Lines," *Elektrichestvo*, **No. 6**, 7-10 (1958).
22. Downs, J. E. and Miller, E. H., "Large Steam Turbine-Generators with Spinning Reserve Capacity," *Proc. Amer. Power Conf.*, **22**, 248-55 (1960).
23. Strohmeyer, C., Jr. (assignor to Gilbert Associates, Inc.) "Devices for Improving Operating Flexibility of Steam-Electric Generating Plants," U.S. Patent 3,226,-932, January 4, 1966.
24. Kostenko, M. P., Discussions, *Proc. Intern. Conf. Large Elec. Systems, 18 Convention*, **I**, 770-72 (1960).
25. Murganov, B. P., "Regulation of Power of Turbogenerators in Power Systems," *Teploenergetika*, **No. 2**, 9-13 (1961).
26. Croft, W. H. and Hartley, R. H., "Improving Transient Stability by Use of Dynamic Braking," *Trans. AIEE*, **81, Pt. III**, 17-26 (1962) April.
27. Park, R. H., "Means for Maintaining Stability of Power Transmission Systems During a Fault," U.S. Patent 3,051,842, August 28, 1962; Park, R. H., "Means for Maintaining Stability of Power Transmission Systems Not-Withstanding a Fault," U.S. Patent 3,234,397, February 8, 1966; discussion of Reference 26.
28. Trofimenko, D. Ye., "The Stability of a Hydro-Electric Generator with Electric Braking," *Elec. Technol. USSR*, **1**, 70-78 (1962) August.
29. Gornshtein, V. M. and Luginskii, Ya. N., "The Use of Repeated Electrical Braking

and Unloading to Improve the Stability of Power Systems," *Elec. Technol. USSR*, **2,** 292-302 (1963) January.

30. Sokolov, N. I., Gurevitch, U. E., and Khvoschinskaya, Z. G., "An Analogue Computer Study of the Joint Operation of Large Turbogenerators," *Elektrichestvo,* **No. 10,** 5-13 (1963).
31. "The Eddystone Story," *Elec. World,* **155,** 87-110 (1963) March 11.
32. Ganther, C. E. and Mallett, A. A., "Operation of Generating Units During Serious System Disturbances," *Trans. IEEE, Power Apparatus Systems,* **PAS-82,** Spec. Suppl., 679-83 (1963).
33. Pope, J. R. (assignor to Westinghouse Electric Corp.) "Load Shedding Apparatus," U.S. Patent 3,271,580, September 6, 1966; Rockefeller, G. D., Jr., "Load Computing and Shedding Apparatus," U.S. Patent 3,300,648, January 24, 1967; Rockefeller, G. D., "A Selective Load Shedding System," *Westinghouse Engr.,* **23,** 187-89 (1963) November.
34. U.S. Federal Power Commission, "National Power Survey. Part 2. Advisory Reports," 4-5. Washington, D.C.: Govt. Print. Office, 1964.
35. Fawcett, D. V., "25/41-MW Cyclic Steam Power Plant Serving Hot Strip Steel Finishing Mill," *Trans. IEEE Power Apparatus Systems,* **PAS-84,** 327-33 (1965) April.
36. Shcheglyaev, A. V., *et al.,* "Application of Depressurizing Arrangement in Boiler Turbine Units," *Teploenergetika,* **12,** 2-9 (1965) January.
37. Brown, R. D. and McClymont, K. R., "A Power Sensing Relay for Predicting Generation Instability," *Trans. IEEE Power Apparatus Systems,* **PAS-84,** 219-24 (1965) March.
38. Hanzalek, F. J. and Ipsen, P. G., "Boiler-Turbine Coordination During Start-Up and Loading of Large Units," *ASME Paper* **65-WA/PWR-9;** abstracted in *Mech. Eng.* 88, 71 (1966) January.
39. Kashtelian, V. E., Urevitch, E. I., and Hertzenberg, G. R., "High Speed Regulation of Steam Turbines Improves Power System Stability," *Elektrichestvo,* **No. 4,** 1-8 (1965).
40. Strohmeyer, C., Jr., "Starting and Operating Once-Through Steam Generators," *ASME Paper* **65-WA/PWR-11;** abstracted in *Mech. Eng.,* **88,** 72 (1966) January.
41. Ellis, H. M., *et al.,* "Dynamic Stability of the Peace River Transmission System," *Trans. IEEE Power Apparatus Systems,* **PAS-85,** 586-600 (1966) June; see Ellis, H. M. and Hardy, J. E., Manitoba EHV Symposium, September 1966.
42. deMello, F. P., Ewart, D. N., Temoshok, M., and Eggenberger, M. A., "Turbine Energy Controls Aid in Power System Performance," *Prac. Amer. Power Conf.,* **28,** 438-45 (1966).
43. Lokay, H. E. and Bolger, R. L., "Transient and Dynamic Stability Analysis for the Kincaid Mine-Mouth Generating Plant," *Proc. Amer. Power Conf.,* **28,** 1086-94 (1966).
44. Stewart, H. R., "Switching Surge Magnitudes, BIL's, Stability Limits and Relaying Problems in an EHV Design," *Proc. Amer. Power Conf.,* **28,** 958-65 (1966).
45. Ward, A. A. and Stubbins, R. L., "Control Testing on Rapid Load Changes of the Supercritical Boiler at Alamitos Steam Station Unit 5," *ASME Paper* **66-WA/PWR-2;** abstracted in *Mech. Eng.,* **89,** 73 (1967) April.
46. Venikov, V. A., *et al.,* "Turbine Regulation as a Means of Improving Power System Transients," *Elektrichestvo,* **No. 2,** 13-21 (1967).
47. Hillery, R. H. and Holdup, E. D., "Load Rejection Testing of Large Thermal-Electric Generating Units," *IEEE Conference Paper* **31-PP-67-25,** presented at the 1967 Winter Power Meeting, New York, January 29-February 3, 1967.
48. Durrant, O. W. and Siegfried, E. P., "Operation of Drum and Once-Through Boilers During Electric Power System Emergencies," *ASME Paper* **67-PWR-15;** abstracted in *Mech. Eng.,* **89,** 62 (1967) December.
49. DyLiacco, T. E., "The Adaptive Reliability Control System," *Trans. IEEE Power Apparatus Systems,* **PAS-86,** 517-31 (1967) May.
50. Schabtach, C., General Electric Co., unpublished information, September 1966.
51. U.S. Federal Power Commission, "Prevention of Power Failures. Volume I—Report of the Commission," 50-51, 58-62, 68-69. "Volume II—Advisory Committee Report. Reliability of Electric Bulk Power Power Supply," 13-16. Washington, D.C.: Govt. Print. Office, 1967.
52. Shubenko-Shubin, L. A., *et al.,* "The Use of Desuperheaters in a Boiler-Turbine Unit," *Teploenergetika,* **14,** 21-14 (1967) April; Translation in *Thermal Eng.,* **23,** 28-32 (1967).
53. Iofiev, B. I., "The Effectiveness of Starting Elements for Automatic Load Dropping on a Transmission System During Faults to Retain Stability," *Elektrichestvo,* **No. 7,** 48-54 (1967).

54. Timo, D. P. and Sarney, G. W., "The Operation of Large Steam Turbines to Limit Cyclic Thermal Cracking," *ASME* Paper **67-WA/Pwr-4;** abstracted in *Mech. Eng.*, **90,** 75 (1968) February.

55. Akers, H. T., Dickinson, J. D. and Skooglund, J. W., "Operation and Protection of Large Steam Turbine Generators Under Abnormal Conditions," *IEEE Conference Paper* **31-TP-67-601.**

56. Strohmeyer, C. S., Jr., "Combined Digital Analog Control Approach for Sibley Station Unit No. 3," IEEE Paper presented at the ASME-IEEE Joint Power Generation Conference, Detroit, Mich., September 24-28, 1967. See also "Balanced Circuit Protects Once-Through Boilers Following Tripout," *Elec. Light Power,* **45,** 104-07 (1967) November.

57. Couvreur, M., "Logic-Adaptive Process for Control and Security in Interconnected Power Systems," *IEEE Conference Paper* **68 TP 39-PWR,** presented at the 1968 Winter Power Meeting, New York, January 28-February 2, 1968.

58. Money, S. H. and Harris, J., "Autoreclosing Switchgear in Distribution Practice," *Proc. IEEE,* **115,** 290, col. 2, (1968) February.

Improvement of Power System Stability by Changes in the Network

EDWARD W. KIMBARK, FELLOW, IEEE

***Abstract*—Changes in the network such as connecting shunt-resistor brakes, disconnecting generators or load, and inserting series capacitors, each rapidly accomplished under suitable control, can counteract other sudden, undesired changes in the network (short circuits or accidental loss of loads or generators) which threaten the stability of interties. The countermeasures taken should match the disturbance with respect to location, amount, duration, and dependence on angular difference. Such measures can improve stability of interties more economically than can construction of additional circuits.**

I. Introduction

A. *Scope of Paper*

THIS PAPER discusses the improvement of stability by opening and closing switches, i.e., by discontinuous changes in the electric power network as opposed to continuous or modulating action exemplified by generator voltage regulation. Familiar examples of the kind of action to be discussed are the opening of switches to clear a short circuit and the reclosure of the same switches to restore the line to service. These are too well known to need discussion. However, the following actions, though not all new, are little used:

1) braking resistors
2) generator dropping
3) load shedding
4) intentional short circuit
5) switched series capacitors
6) switched shunt reactors or capacitors
7) boosting power of a dc line.

These will be discussed, as well as the general theory of coordinating these corrective measures with the nature and location of the disturbance.

B. *Type of Power System Considered*

Consider an ac power system consisting of several areas, each having strong internal transmission systems between their respective generating plants and load centers, these areas being interconnected by relatively weak tie lines. The principal stability problem of such a system is that of the stability of the interties.

C. *Disturbances Considered*

1) short circuit
2) opening of an ac intertie
3) blocking of a dc intertie
4) loss of a generator
5) loss of a large load.

D. *Alternative Methods of Improving Stability*

1) The prevalent opinion seems to be that instability of the interties can be avoided by building stronger interties, that is, more lines or higher voltage lines. This solution is very effective but very expensive. Moreover, it is unnecessary if equally effective but cheaper methods are available.

2) The method advocated in this paper is to employ some of the devices listed in Section I-A, properly controlled, so as to avoid excessive demands on the interties. In conjunction with them there should also be employed 1) faster control of prime movers as described by Park [1] and 2) high-speed excitation systems with supplementary signals for providing strong damping of swings. These two features are not discussed in this paper.

3) Still another method is to interconnect the areas only through dc asynchronous ties. This method merits serious consideration but is not discussed in this paper.

E. *Proposed Policies*

In the power system consisting of strong areas connected by weak ties, suppose the following policies to be adopted.

1) Each area has the responsibility of seeing that its internal disturbances do not disrupt the interties or impair service in other areas.

2) Both terminal areas are responsible for counteracting disturbances on the interties between those areas.

These responsibilities are fulfilled if each area keeps its generation running at rated speed (frequency) and with the time integral of its speed deviation (which is angle deviation) kept within close limits, say one radian.

F. *Equilibrium*

During steady operation of a power system there is equilibrium between the mechanical power input of each generator and the sum of the losses and the electrical power output of that generator. Similar equilibrium exists for each generating plant, for each area, and for the entire power system.

This equilibrium is destroyed by a disturbance. A short circuit near a generating plant greatly decreases the electric power output of the plant and increases its losses by a lesser amount, but it has no immediate effect on the mechanical input. The excess of input over losses and output (called accelerating power) goes into increasing the kinetic energy of the generating units in the plant. Similarly, excess of output and losses over input (decelerating power) decreases the kinetic energy.

Accelerating power divided by the inertia constant equals angular acceleration. The time integral of angular acceleration is the angular speed deviation. The time integral of angular speed deviation is the change of angle relative to a reference axis

Paper 68 TP 701–PWR, recommended and approved by the Power Generation and Power System Engineering Committees of the IEEE Power Group for presentation at the 1968 Joint Power Generation Conference, San Francisco, Calif., September 15–19. Manuscript submitted June 7, 1968; made available for printing June 12, 1968.

The author is with the Bonneville Power Administration, Portland, Ore.

Reprinted from *IEEE Trans. Power App. Syst.*, May 1969.

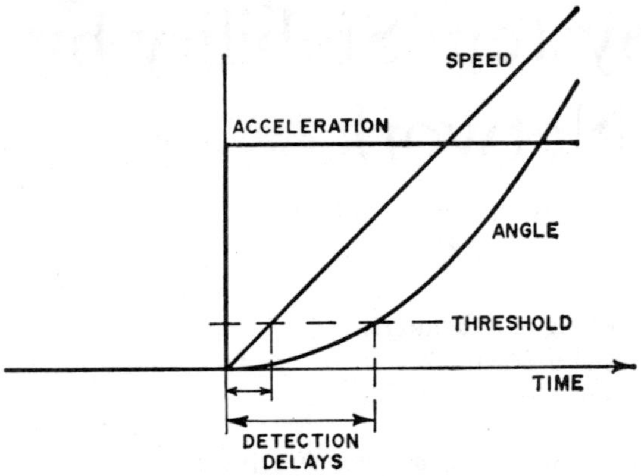

Fig. 1. Angle and angular speed as functions of time for a step change of angular acceleration.

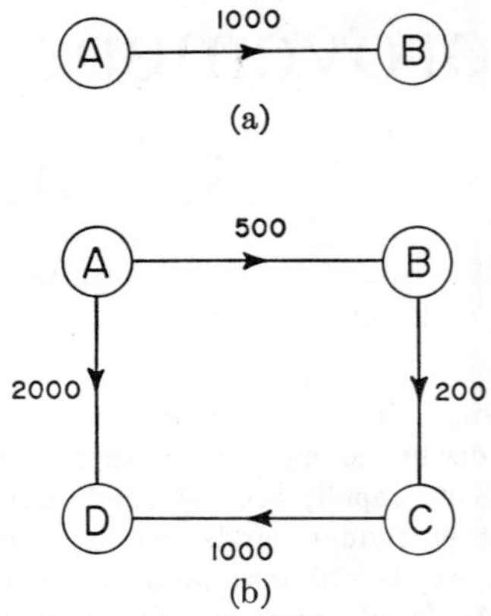

Fig. 2. Systems of interconnected areas. (a) Two areas. (b) Four areas.

rotating at rated speed. If the relative angle between two areas become too great, synchronism is lost.

Note that the first sign of impending trouble is acceleration. The speed deviation appears later and the angular change still later (Fig. 1). Therefore, detection of angular acceleration is the most promising way of quickly initiating preventive action.

Disequilibrium of power can occur as a result of either a change of mechanical input or of a change of electrical output. However, from a practical viewpoint, it is obvious that changes in electrical output can occur almost instantaneously because of changes in the network, whereas the mechanical input cannot change nearly so fast. A short circuit or an open circuit could cause the electrical output to change instantly from rated full load to zero, whereas a change of like amount in the mechanical input would require at least a few seconds. Although a sudden change in output should properly be counteracted immediately by a sudden change of input of equal amount, this cannot always be accomplished in practice rapidly enough to prevent instability. This fact suggests that, after disequilibrium has been caused by a sudden change in the network, it must be counteracted by another sudden change in the network. The second sudden change may be undone when the necessity for it has disappeared either through slower readjustment of mechanical input or through the disappearance of the first change in the network.

G. *Brief and Prolonged Disturbances*

Some disturbances produce a brief lack of equilibrium, lasting typically a fraction of a second, for example, a short circuit cleared in the normal time. Such a fault near a generating plant accelerates the generators. It can be counteracted by a short application of a shunt-resistor brake located at the generating plant. The time of application of the brake need not be equal to the duration of the fault, but the energy absorbed by the brake should be equal to or somewhat greater than the excess kinetic energy of the generators.

Some other disturbances produce a sustained disequilibrium, for example, the loss of a large load. The countermeasure should likewise be sustained and, in the event of the loss of a large load, it is logically the disconnection of an amount of generation equal to the lost load.

In the last example, prolonged application of a braking resistor would be effective, but it would be uneconomical to provide resistors of long-time rating. In the earlier example, dropping of generation would be inappropriate unless it could be restored to service rapidly and with accurate timing.

H. *Classification of Areas*

Fig. 2(a) shows a power system of two interconnected areas. The arrow head shows the direction of power flow on the intertie, and the numeral shows the amount of power flow in megawatts. Area A is the sending area and area B the receiving area.

Fig. 2(b) shows a power system of four interconnected areas. Area A is clearly a sending area and D a receiving area, but B and C both send and receive. These can be designated according to the net outflow or inflow; thus B is predominantly a receiving area and C a sending area. The terms *exporting* and *importing* can be used instead of *sending* and *receiving*.

A short circuit on an intertie reduces the power flow on that intertie. The area at the sending end of that intertie becomes an area of excess generation, and the area at the receiving end becomes an area of deficient generation. Similarly, loss of generation in an area would produce a deficiency of generation in that area, and loss of a load would result in an excess of generation. Areas of excess generation accelerate and those of deficient generation decelerate.

I. *Principles of Counteracting Disturbances*

A brief disturbance should be counteracted by a brief countermeasure. A prolonged disturbance should be counteracted by a prolonged countermeasure. A disturbance in a given area should be counteracted by a countermeasure in the same area.

A disturbance on an intertie should be counteracted preferably by a countermeasure on that intertie or alternatively by countermeasures in the areas at the two ends of that intertie. Although a countermeasure at one end only would aid stability, it would also cause a deviation in average frequency of the system and would let the disturbance affect other areas more than if countermeasures were taken at both ends.

For example, a short circuit on intertie AD of Fig. 2(b) followed by clearing and successful reclosure might be counteracted by a brief application of braking in area A, which otherwise would have excess power during the fault. The strain on tie AD is reduced by this braking. However, area D is power-deficient during the fault and drops in frequency and thus causes increased power flow from areas B and C to area D. Moreover, the system as a whole is power-deficient, and in the end its frequency becomes subnormal. It would be better not only to brake in area A but also to shed some load briefly in area D so as to maintain normal frequency there and avoid propagation of the disturbance to areas B and C.

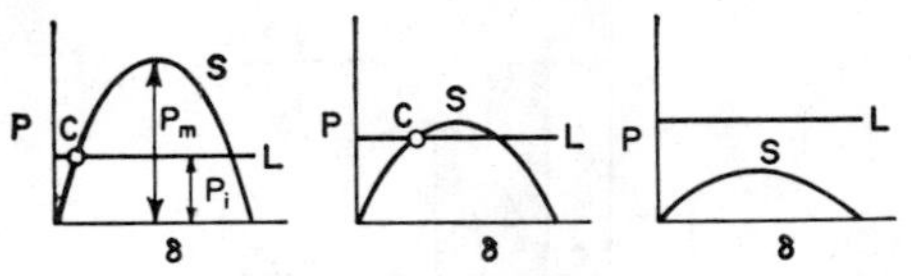

Fig. 3. Power angle diagrams of a two-machine system. (a) Very stable. (b) Precarious. (c) Unstable.

J. The Power-Angle Diagram of a Two-Area System

The power-angle diagram is applicable to a two-machine system and, by extension, to a two-area system [Fig. 2(a)]. Conceptually, it aids in the understanding of the stability of a multiarea system, such as that of Fig. 2(b), as approximated by a two-machine system.

The simplest system for which to develop such a diagram comprises one finite machine of inertia constant M, mechanical input P_i, and internal voltage E_1 feeding power P_u over a circuit of series inductive reactance X (including the transient reactance of the finite machine) to an infinite bus of voltage E_2. The power output of the finite machine, which is also the power input to the infinite bus, is given by

$$P_u = P_m \sin \delta \tag{1}$$

where

$$P_m = \frac{E_1 E_2}{X} \tag{2}$$

and δ is the phase angle by which E_1 leads E_2. This power is shown by the sinusoidal curve S in Fig. 3. The mechanical input P_i to the finite machine is independent of δ and is represented by the horizontal line L. The accelerating power is given by

$$P_a = P_i - P_u = P_i - P_m \sin \delta \tag{3}$$

which is the difference in ordinates of line L and sine curve S. It is more convenient to use L and S than to draw a new curve of P_a versus δ.

In the steady state the acceleration is zero, and operation is at point C of Fig. 3, an intersection of L and S. In Fig. 3(c) there is no such intersection, and the system represented could never reach a steady state.

The differential equation of angular motion of the finite machine is

$$M \frac{d^2\delta}{dt^2} = P_a. \tag{4}$$

It is shown in [2] that a system of two finite machines having inertia constants M_1 and M_2, internal voltages E_1 and E_2, mechanical power inputs P_{i1} and P_{i2}, electrical power outputs P_{u1} and P_{u2}, and angular positions δ_1 and δ_2, connected through any linear two-port network (including the transient reactances) the equivalent π of which network has nodal transfer admittance $Y_{12} \angle \theta_{12}$ and nodal self-admittances $Y_{11} \angle \theta_{11}$ and $Y_{22} \angle \theta_{22}$, is equivalent to a system of one finite machine and an infinite bus and having M, δ, P_I, P_M, and γ as given in the Appendix. Equation (3) is replaced by (5) which follows and (4) is still valid.

$$P_a = P_I - P_M \sin (\delta - \gamma). \tag{5}$$

Again, the accelerating power is represented by the difference of the ordinates of a horizontal line L and a sine curve S. Now, however, the height P_I of the horizontal line is no longer the mechanical power input to the machines of the sending area; it is approximately that input less the local loads, and hence is the steady-state power delivered from that area to the intertie. Similarly, the ordinate of S is approximately the total electric power output of the machines in the sending area less the local loads, and hence is approximately the power delivered from that area to the intertie as a function of angle δ.

It is shown in [2] that the geometrical area between the curves S and L starting from a point of zero relative speed is proportional to the square of the relative speed. That is,

$$\text{area} = \int P_a \, d\delta = \frac{M}{2} \left(\frac{d\delta}{dt} \right)^2. \tag{6}$$

TABLE I
CHANGES WHICH AFFECT STABILITY

Impair Stability	Improve Stability
Changes in Network which Affect Stability	
By decreasing the amplitude of the sine curve	By increasing the amplitude of the sine curve
Occurrence of short circuit on tie line	Removal of short circuit
Opening a tie line	Reclosing the tie line
Bypassing a series capacitor	Inserting a series capacitor
Connecting a shunt reactor	Disconnecting a shunt reactor
Disconnecting a shunt capacitor	Connecting a shunt capacitor
By raising the load line	By lowering the load line
Disconnecting a shunt resistor in the sending area	Connecting a shunt resistor in the sending area
Connecting a shunt resistor in the receiving area	Disconnecting a shunt resistor in the receiving area
Sudden disconnection (loss) of a large load in the sending area	Disconnecting (shedding) load in receiving area
Sudden disconnection (loss) of generators in the receiving area	Disconnection (dropping) of generators in the sending area
Sudden increase of load in the receiving area (improbable)	Sudden increase of load in the sending area (not feasible)
Blocking (loss) of a dc line in parallel with the ac ties	Sudden boosting of power on the dc line
Changes in Steam or Hydraulic System which Affect Stability	
By raising the load line	By lowering the load line
Close throttle or gate in receiving area	Close throttle or gate in sending area
Open throttle or gate in sending area	Open throttle or gate in receiving area (spinning reserve)

K. Effect of Disturbances on the Lines of the Power-Angle Diagram

As indicated in Fig. 3, stability is impaired either by raising the load line L or by decreasing the amplitude of the sine curve S. Stability is enhanced by lowering L or by increasing the amplitude of S.

Occurrences in a terminal area or on a dc intertie affect the load line only or predominantly. Occurrences on an ac intertie affect the sine curve only (see Table I).

A disturbance which raises the load line is logically counteracted by one which lowers the load line. For example, loss of generation in the receiving area is counteracted by shedding load there. A disturbance which decreases the amplitude of the sine curve is logically counteracted by a measure which increases the amplitude. For example, the opening of one of two parallel circuits of an ac intertie is counteracted by inserting additional series compensation in the remaining circuit.

II. Countermeasures

In this section each of several countermeasures will be discussed individually.

A. Resistor Braking

One of the most useful devices is a bank of resistors located near a large generating plant in a sending area and connected in shunt with the three-phase bus through a suitable switch that is normally open. A short circuit on the bus or on any of the lines near the bus raises the load line and accelerates the generators. A suitable control device detects increase of kinetic energy exceeding a suitable threshold value and initiates closure of the switch to connect the resistors. Thereupon the load line drops again, preferably to a height slightly below its original height.

If the short circuit is cleared in normal time (say in 50 ms), the switch connecting the brake may do so after the fault has been cleared. The brake application is still required for bringing the speed and angle deviations back nearly to zero. A short application (0.5 to 1.0 seconds) should be used with energy absorption equal to or slightly greater than the excess kinetic energy due to the fault. In some systems two or more successive applications are required on successive swings.

Brakes are given a short-time rating, perhaps a 2-second total, making the cost much less than that of a brake of equal power rating and continuous duty. During this short time nearly all the energy absorbed raises the temperature of the resistors; the heat lost by radiation or convection is negligible.

The dose of braking (that is, the energy absorbed by the brake) should be adjusted according to the excess kinetic energy. This, in turn, depends on the severity of the short circuit (location and type), the time of clearing, and the initial power output of the plant. The dose can be adjusted by use of two or more sections in parallel controlled by separate switches or by the time of application. Park [1] suggests that the power can be adjusted by closure of 1, 2, or 3 poles of the switch.

If clearing of the short circuit is delayed, the brake must be disconnected before it reaches its maximum safe temperature, and some generation should be dropped at the same time.

The application of a brake tends to drop the voltage, and the power taken by the brake varies as the square of the voltage. The voltage can be prevented from dropping by use of capacit rs in parallel with the resistors. The capacitors, like the resistors, are overworked for a short time, thereby costing less than if the same reactive power were to be supplied continuously.

For equal energy absorption resistors cost less when insulated for a lower voltage. Protection should be provided against excessive temperature and against a short circuit in the brake.

Resistor brakes are planned for the Portage Mountain plant of the British Columbia Hydro and Power Authority [6].

B. Generator Dropping

The best use of generator dropping is in a sending area to counteract the loss of a large load in that area. The loss of load raises the load line, whereas generator dropping lowers it. The amount of generation dropped should roughly equal the amount of load lost.

An analysis of the transient swings is based on Fig. 4. Initial operation is at $C1$, the intersection of $L1$ and S. Loss of a large load raises the load line to $L2$. Obviously the system would be unstable if no measures were taken. The angle δ increases. Generation is dropped as soon as possible, lowering the load line to $L3$. The angle at this moment is δ_1. The excess kinetic energy of the remaining generators is then proportional to

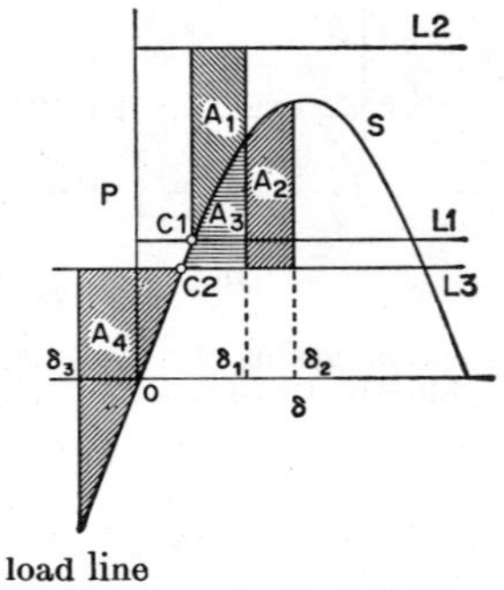

$L1$ initial load line
$L2$ load line after loss of a large load
$L3$ load line after loss of load and generator dropping

Fig. 4. Equal-area analysis of generator dropping.

positive area A_1 between S and $L2$. These generators continue to swing ahead of δ^2, where negative area A_2 equals the positive area A_1. The speed deviation is then zero but immediately becomes negative, and the generators swing to δ_3, where $A_2 + A_3 + A_4 = 0$. In the absence of damping the remaining generators would continue to oscillate between δ_2 and δ_3. However, there is damping torque which finally brings the generators to rest at $C2$, the intersection of $L3$ and S.

Generator dropping is sometimes used as a cheap substitute for braking as a countermeasure for short circuits in the sending system, but it cannot be as accurately controlled as braking.

C. Load Shedding

Load shedding is required to counteract loss of generation in a receiving area. Power companies should encourage the acquisition of loads that, by contract provision, are interruptible for short times without notice in return for a low energy charge. Among loads that can be interrupted with only minor inconvenience to the customer are aluminum reduction plants, air conditioning, and pumping where storage is provided in tanks or reservoirs.

Load shedding followed by restoration can be obtained by any of the following methods:

1) opening of one or more selected feeders followed by reclosure

2) temporary depression of voltage by an intentional short circuit followed by its clearing

3) temporary depression of voltage by connection of a shunt reactor followed by its disconnection.

In methods 1) and 2) some load will be shed for a longer time than the rest because of operation of undervoltage releases on motor starters.

The usual control of load shedding by underfrequency relays is too slow for preservation of transient stability. Transferred signals from generator circuit breakers or from fault-clearing relays on interties should be used, subject to supervision by the power output of the generators or of the power delivered by the interties.

D. Switched Series Capacitors [3]

1) *Fault on an AC Intertie*: Suppose that a short circuit occurs at point × on one circuit of the two parallel circuits of the ac intertie of Fig. 5(a) and is cleared without reclosure. The power-angle diagrams are shown in Fig. 5(b) and (c). In Fig. 5(b) initial operation is at point $C1$, the intersection of load line $L1$ and sine curve $S1$ which pertains to the intact system. Occurrence of the short circuit depresses the sine curve

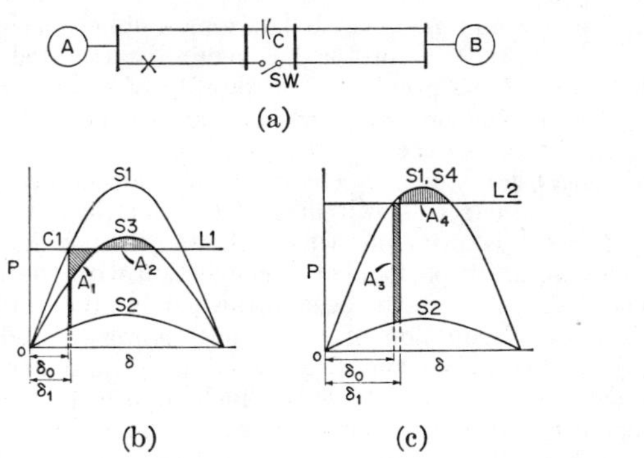

*S*1 all lines in service
*S*2 short circuit on one line
*S*3 faulted section disconnected
*S*4 faulted section disconnected and capacitor C inserted

Fig. 5. (a) Two-circuit ac intertie. (b) Area diagram without switched series capacitor. (c) Area diagram with switched series capacitor for fault cleared with no reclosure.

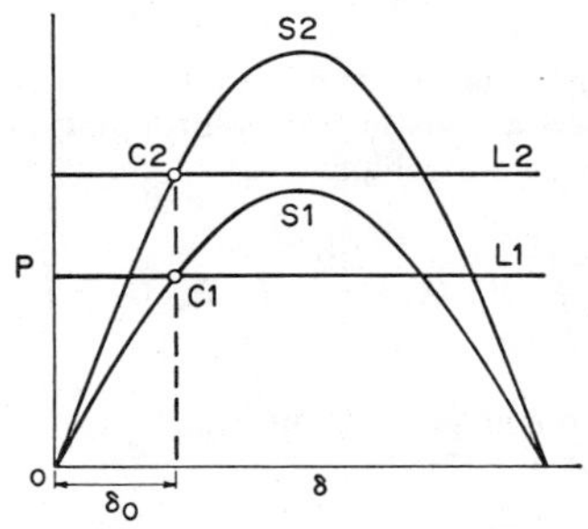

Fig. 6. Power-angle diagram for insertion of series capacitors in ac interties in the event of loss of a parallel dc intertie.

to *S*2. Opening the faulted circuit raises it to *S*3. The transient stability limit is found by adjusting the height of *L*1 for equality of areas A_1 and A_2, while simultaneously adjusting the clearing angle δ_1 to conform to a fixed clearing time and variable acceleration.

Fig. 5(c) shows the effect of inserting a series capacitor dimensioned to give the same transfer reactance between areas *A* and *B* as that given by the intact system, and hence to give sine curve *S*4 coinciding with *S*1. The capacitor is inserted at the instant of opening the circuit. Line *L*2 is adjusted for equality of areas A_3 and A_4 and is higher than *L*1, indicating a higher transient stability limit.

2) *Permanent Fault on a DC Intertie*: The effect of a permanent fault on a dc intertie is to block the converters at both ends. If the dc intertie is operating in parallel with ac interties, loss of the former tends to shift power to the latter. In other words, it raises the load line, say from *L*1 to *L*2 in Fig. 6. A signal from either converter to series-capacitor stations on the ac interties inserts additional series compensation, thereby raising the sine curve from *S*1 to *S*2. If the correct amount of capacitive reactance is inserted, the new operating point *C*2 is attained without oscillation, a desirable but unnecessary condition. If the fault on the dc line proves to be transitory, the inserted capacitors are bypassed.

Switched series capacitors will be used on the Pacific Northwest-Southwest interties to counteract the loss of a parallel dc line or the loss of one of two parallel ac lines.

3) *Damping*: Series capacitors in an intertie can be used for damping swings [4]. Capacitors used for this purpose should be inserted while the areas at the two ends of the intertie are swinging apart and should be bypassed while the two areas are swinging toward one another. The control quantity may well be the rate of change of absolute value of power on the intertie.

This method of damping should probably be confined to rapid damping of swings of large amplitude. Supplementary speed signals to generator-voltage regulators appear to be a more economical way of preventing small oscillations from growing. Smith [5] has proposed switching series capacitors after a disturbance so that the new operating angle is reached without overswing.

E. Switched Shunt Capacitors or Reactors

By means of a *T*-to-π transformation it can be shown that switching a shunt capacitor onto the middle of an ac intertie or switching a shunt reactor off from there is equivalent to insertion of a series capacitor [3]. Series capacitors are preferable unless the shunt elements are required for another purpose, usually for control of voltage.

F. Boosted Power on DC Line

This measure would relieve parallel ac lines of some of their load and make them more stable. For it to be effective, either the converter must have substantial overload capacity, or the dc line must be operated well below its full load. There is still another difficulty. When the direct current is boosted, the converter draws more reactive power and, unless the ac systems have a high short-circuit capability at the dc terminals, the additional draft of reactive power depresses the alternating and direct voltages, with the result that the per-unit increase of power may be much smaller than the per-unit increase of current. To overcome this difficulty either additional shunt capacitors must be switched on rapidly or series capacitors must be used to compensate the inductive reactance of the ac system.

III. Conclusion

The time has come when the various means of improving power system stability should be viewed as a coordinated system instead of as individual measures applied here and there. Park has worked earnestly toward that end and has evolved methods of braking, load shedding, and generator dropping on the electrical side coordinated with fast valving and bypassing of steam or water on the mechanical side [1]. Coordination between power-system areas is also required, and this will require joint studies and cooperation in design and operation of interconnected power systems under many different ownerships.

Measures such as switching of series capacitors on the tie lines between areas should be joined with those measures advocated by Park.

There is a large field for development of control systems for all these measures. Coordination of the various measures does not necessarily imply central control by a digital computer, though some day this may come to pass. Much can be accomplished by local, analog-type devices, analogous to protective relays.

Appendix

System of a Finite Machine Connected to an Infinite Bus Equivalent to a Given System of Two Finite Machines Connected Through Any Linear Network

$$M = \frac{M_1 M_2}{M_1 + M_2} \tag{7}$$

$$\delta = \delta_1 - \delta_2 \quad (8)$$

$$P_I = \frac{M_2(P_{i1} - E_1^2 G_{11}) - M_1(P_{i2} - E_2^2 G_{22})}{M_1 + M_2} \quad (9)$$

$$P_M = E_1 E_2 Y_{12} \frac{\sqrt{M_1^2 + M_2^2 - 2M_1 M_2 \cos 2\theta_{12}}}{M_1 + M_2} \quad (10)$$

$$\gamma = \tan^{-1}\left(\frac{M_1 + M_2}{M_1 - M_2} \tan\theta_{12}\right) - 90° \quad (11)$$

where $G_{11} = Y_{11} \cos\theta_{11}$ and $G_{22} = Y_{22} \cos\theta_{22}$ are the nodal self-conductances.

References

[1] R. H. Park, "Improved reliability of bulk power supply by fast load control," presented at the American Power Conf., Chicago, Ill., April 24, 1968.

[2] E. W. Kimbark, *Power System Stability*, vol. 1. New York: Wiley, 1948, ch. 4, pp. 132–135.

[3] E. W. Kimbark, "Improvement of system stability by switched series capacitors," *IEEE Trans. Power Apparatus and Systems*, vol. PAS-85, pp. 180–188, February 1966.

[4] W. A. Mittelstadt, "Four methods of power system damping," *IEEE Trans. Power Apparatus and Systems*, vol. PAS-87, pp. 1323–1329, May 1968.

[5] O. J. M. Smith, "Power system transient control by capacitor switching," *IEEE Trans. Power Apparatus and Systems*, vol. PAS-88, pp. 28–35, January 1969.

[6] H. M. Ellis, J. E. Hardy, A. L. Blythe, and J. W. Skoogland, "Dynamic stability of the Peace River transmission system," *IEEE Trans. Power Apparatus and Systems*, vol. PAS-85, pp. 586–600, June 1966.

Discussion

W. A. Mittelstadt (Bonneville Power Administration, Portland, Ore.): Dr. Kimbark presents a logical approach for introducing network changes to improve power system stability. The prolonged network change which he recommends serves two important purposes. First, the countermeasure maintains a strong steady-state stability limit and second, the required dynamic change in the state is correspondingly reduced, thereby minimizing the strain on other types of stability control.

The brief or short-term network changes discussed are generally more difficult to properly control than sustained changes since two or more switching times must be determined. Criteria have been developed for short-term dynamic braking at generator sites using locally measurable parameters [7]. Network changes which are made at points remote from generating plants generally always involve a compromise. This compromise comes between stabilizing each regional area and balancing interarea deficiencies. For example, the countermeasure given in the example in Section I-I may act in opposition to the requirements of certain generators in areas *A* or *D*, (Fig. 2). This, however, does not present a serious problem if each individual area has good transient stability. The recognition of the importance of generator angular acceleration as a fundamental input quantity cannot be overemphasized because of the speed by which control action must be summoned. Fast and economical transducers capable of measuring or estimating this parameter are necessary.

References

[7] W. A. Mittelstadt, "A method of improving power system transient stability using controllable parameters," presented at the IEEE Winter Power Meeting, New York, N. Y., January 26–31, 1969.

Manuscript received August 2, 1968.

R. G. Farmer and **A. C. Lyle** (Arizona Public Service Company, Phoenix, Ariz.): The author has again contributed significantly to the study of the problem of instability of power systems and is to be commended on the clear and concise method of presenting possible solutions to the stability problem.

As the author states, a change in the mechanical input could have the same effect as switching of the electrical circuit. A means of applying this principle which shows promise is fast valving. Fast valving has been studied in connection with the Four Corners project, and curves have been furnished by the turbine supplier which show a reduction of mechanical power of 400 MW in 0.5 second by the application of fast valving to a cross-compound unit generating 795 MW. The 0.5 second includes 5 cycles for the detection of a condition for which fast valving is required [8].

In Section II-B, under generator dropping, the author states that at the point where δ is equal to δ_2 the speed deviation is zero because area A_1 is equal to area A_2. We feel that it should be pointed out that this is true only if the inertia constant of the machines dropped is very much smaller than the combined inertia constants of the systems under consideration.

Rearranging (6) and applying limits from Fig. 4,

$$\frac{2}{M} \int_{\delta_0}^{\delta_2} P_a \, d\delta = \left(\frac{d\delta}{dt}\right)^2. \quad (12)$$

If we designate M_1 as the inertia constant before dropping generation and M_2 as the inertia constant after dropping generation, we can write the following relationships:

$$\frac{2}{M_1} \int_{\delta_0}^{\delta_1} P_a \, d\delta + \frac{2}{M_2} \int_{\delta_1}^{\delta_2} P_a \, d\delta = \left(\frac{d\delta}{dt}\right)^2 \quad (13)$$

where δ_0 is equal to δ at point C_1 in Fig. 4.

For $d\delta/dt = 0$ at $\delta = \delta_2$:

$$\frac{1}{M_1} \int_{\delta_0}^{\delta_1} P_a \, d\delta = \frac{1}{M_2} \int_{\delta_1}^{\delta_2} - P_a \, d\delta. \quad (14)$$

Again referring to (6) and Fig. 4 the following expression can be written from (14):

$$\frac{1}{M_1} A_1 = \frac{1}{M_2} A_2 \quad (15)$$

$$A_2 = \frac{M_2}{M_1} A_1. \quad (16)$$

Therefore, it can be seen that the equal-area criterion needs to be modified in accordance with (15) for generator dropping. Apparently the author assumed M_1 approximately equal to M_2 in his analysis.

References

[8] R. G. Farmer, R. H. Hartley, M. H. Kent, and L. M. Wheeler, "Four Corners project stability studies," presented at the ASME/IEEE Joint Power Generation Conf., San Francisco, Calif., September 15–19, 1968.

F. R. Schleif (Bureau of Reclamation, Denver, Colo.): Dr. Kimbark has set forth a timely and revealing analysis of needs and techniques for improvement of power system stability. The analysis is much more than a solution technique for system studies. The coordination philosophy it presents for the preservation of stability against transients would significantly aid the system interconnections under present conditions.

The analysis is well conceived and presented, and the few comments following are offered as extensions of the philosophy. In

Manuscript received September 23, 1968.

Section II-D3 the author has treated the subject of damping separately, as though it were to be distinguished from the improvement of transient stability. However, the result from both categories is to minimize the increase of the angle after a disturbance. Notwithstanding the fact that he has referred to speed signal supplementary control of excitation as an economical means of preventing small oscillations from growing, potential of this excitation control goes much further. It achieves damping by opposition of all swings including some reduction of the first swing. The amount of reduction which can be accomplished depends upon the speed of response of the excitation system and the rate of the system swing. Our own studies have indicated that for swing rates typical of interarea ties in the order of 2 rad/s or less and for modest response rates of rotating exciter systems, the first swing may be reduced by more than ten percent by influencing only 50 percent of the generating capacity. With high-speed static excitation systems and supplementary control, such as that planned for the generators of the Grand Coulee third powerplant, the improvement of transient stability is still more impressive.

The response of excitation to speed or frequency signal supplementary control naturally becomes discontinuous at the exciter ceiling. Since the author remarks that the switching of series capacitors may be confined to large swings, it would be logical to initiate this switching for the approximate magnitude of the swing at which the excitation ceiling is reached by influence of the supplementary control.

In Dr. Kimbark's analysis he has shown that for system disturbances the most advantageous corrective measures are those which are as nearly the counterpart of the disturbance as is feasible. For example, loss of generation sufficient to threaten the stability of an interarea tie can best be offset by the interruption of load in the same area. This leads directly to the philosophy that each utility or area not only has a responsibility to initiate the corrective measure for its disturbance but also has the most effective opportunity to offset it. Thus a logical application procedure for this excellent work would be for the operating committees already concerned with coordination of load shedding for slower threats to system continuity to also promote application of the means described in this paper for preserving transient stability.

Manuscript received September 26, 1968.

Glenn W. Bills (Autonetics Division of the North American Rockwell Corporation, Anaheim, Calif.): I heartily concur with the statement that "Such measures can improve stability of interties more economically than can construction of additional circuits." It has been the practice of the most progressive as well as the most conservative electric utilities to design and build transmission networks to withstand extremely severe and infrequent disturbances. This practice does not prevent system-wide breakups because of unforseen or extremely unlikely disturbances. For this reason fast control by system switching should prevent major system breakups and perhaps reduce the number of transmission circuits required for bulk power security.

Fast control cannot insure that some system loads will not be disconnected or reduced, but it should insure that complete system shutdown will not occur. However, overbuilding transmission networks will not insure that system-wide disruptions will not occur although such a policy will usually minimize interruption of individual loads.

Future power system studies will probably show that fast control will usually be more economical than overbuilding. The question of how fast control is to be achieved, i.e., by local control action or by a combination of local and central control action, has yet to be answered. At this time it does not appear to me that local control alone will always be adequate for fast control of system stability.

Sensors, data acquisition equipment, digital communication, data processing, and control techniques exist which make fast control by a central computer possible. However, the cost of the required computer and communication network must be equal to or less than the power transmission facilities needed to achieve the same reliability.

It is encouraging to have such well-known authorities as Mr. Almon, Dr. Kimbark, and Dr. Park publish papers advocating fast control for electric power systems. Dr. Kimbark has clearly presented the case for improving system stability by local control measures. The next step is to determine whether local control is adequate to prevent system instability for all circumstances, or whether it will have to be supplemented by central computer control.

Manuscript received September 30, 1968.

R. H. Park (R. H. Park Company, Brewster, Mass.): In the matter of network changes which employ the bus arrangement shown in Fig. 7, the Commonwealth Edison Company has introduced at its Kincaid station an apparently novel combined fault clearance and network change procedure. Four breakers are cleared simultaneously in response to all multiphase faults occurring within a predetermined zone on either line A or B. The tripping of four breakers (e.g., breakers 1–4 for a fault on line A) with the effect of tripping off one generator has the advantage over merely tripping breakers 2–4, since thus the loss of a second generator due to development of instability is avoided, despite the accidental occurrence of a delay in the opening of either breaker 2 or 3. Whereas the bus arrangement of Fig. 7 is a ring arrangement, the network change concept involved can be applied to other bus arrangements., e.g., a breaker and a half scheme, and might be termed "advanced backup-type fault clearance."

Dr. Kimbark has cited the point that the utility of rapidly executed network changes can be increased if coordinated with fast control of prime movers and favorably controlled excitation systems. Aside from whether or not the use of advanced backup-type fault clearance is contemplated, with a combination of very fast closing high-voltage breakers and feedback-controlled dynamic braking applied at generator voltage level, it is possible to disconnect a generator and resynchronize rapidly (on the order of 0.5 second). In such a process generator isolation can be accompanied by a programmed, simultaneous rapid closure of intercept and/or control valves, with the turbine stop valves held open.

When the generator is disconnected, it will initially accelerate at a rate proportional to its preisolation load. Subsequently a braking procedure with a load related to but exceeding preisolation load can be applied such that the instant (or instants) of initiation of the braking load control breaker closing (or closings) is (or are) responsive both to the occurrence of the fault and the extent of the generator rotor speed increase.

Speed increase can be established either from an electrical speed signal or by integrating an electrical acceleration signal. Displacement of the generator rotor in synchronous space can be determined by integrating the speed increase signal. Braking load removal can be initiated in response to speed. Generator reconnection can be initiated in response to values of predicted speed and rotor angle, determined with allowance for the time lag of closure of the high-voltage breakers involved, following a signal. For accuracy, the lag should be low, say in the range $2\frac{1}{2}$ to 3 cycles.

Provision can be made so that reconnection would be inhibited in the improbable but possible event of failure of braking load breakers to close or open, and/or if for any reason there were failure to arrive at a prediction of a favorable generator reconnect initiation condition. Provision can also be made so that failure of high-voltage breakers to open normally would influence reconnection. In particular it could be provided that if, for a fault on line A, breaker 2 failed to clear normally, breaker 3 would not reclose (Fig. 7). It could be provided that breaker 4 would reclose despite failure of breaker 2 to clear, but failure of breaker 3 to clear would cause inhibition of generator reconnection.

Furthermore, provided that braking allowed time enough for arc extinguishment, there could be the optional provision that breaker

Manuscript received September 17, 1968.

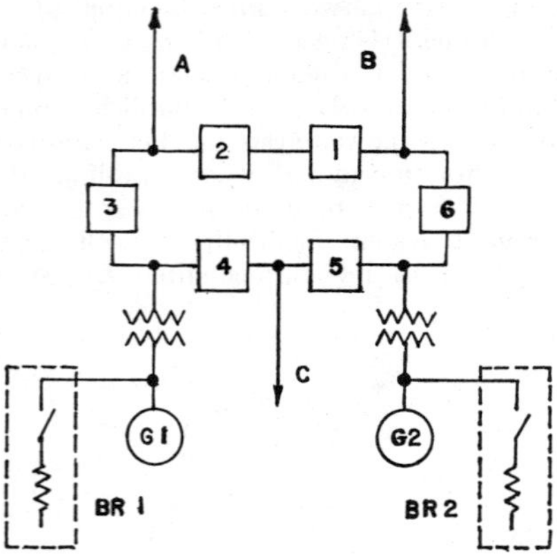

Fig. 7.

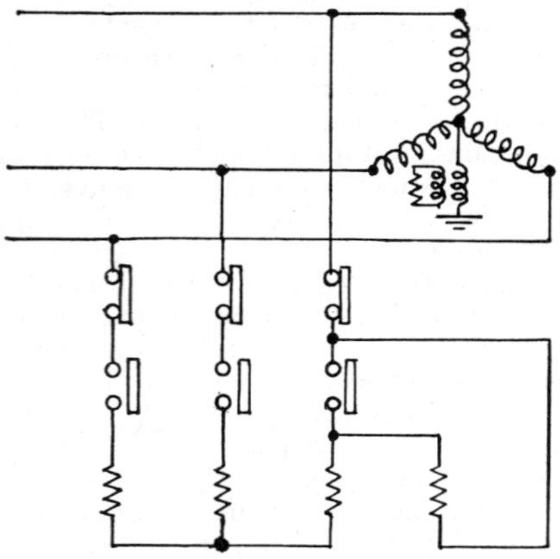
Fig. 8.

3, having cleared normally, could be reclosed with breaker 4 remaining open. In this case, if a fault reappeared, a braking load could be reapplied, and assuming that breaker 3 cleared normally, breaker 4 could then be automatically closed; provided only that the result of the braking cycle did not fail to meet requirements.

On the other hand, it would appear that in most cases it would be preferable to initially close breaker 4, leaving breaker 3 open and reclosing it only after a predetermined time lapse or at a preferred portion of a swing curve and, optionally, only subsequent to a test to see if a fault reappeared when line A was energized from the remote end.

Consider next the control of steam turbine valving. If the application of a braking load sufficient to cause deceleration were verified, it would be safe to interrupt a process of control and intercepting valve closure, and at the same time optionally initiate a prefault condition responsive sustained change in turbine driving power which, if wanted, could be made subject to automatic alteration due to post fault events.

Various modifications are possible.

As it would appear, there are advantages in using procedures which initiate a process of rapid closure of the turbine control and intercept valves when a fault first occurs or reoccurs. Thereafter, events leading to resynchronization at a controlled level of turbine driving power will take place conditioned on the fact that an inherently improbable condition of defective performance of high-voltage breakers and generator braking systems fails to occur.

While for a fault on line A high-power braking would be applied to generator 1, a lesser degree of braking and a controlled amount of momentary and/or sustained valve closing can be simultaneously applied to generator 2.

For a fault on line C it is possible to trip breakers 3–6, to apply a braking load to each generator, and to reclose breakers 3 and 6 and/or 4 and 5 (Fig. 7), in a fraction of a second, except if either breaker 4 or 5 failed to open normally, or if either braking system BR_1 or BR_2 failed to function properly. This would prevent the loss of more than one generator, notwithstanding delay in fault clearance.

Note that provision can also be made to deal with a condition of simultaneous multiphase faults on two lines and yet avoid a need to trip more than one generator, despite delay in the opening of any one breaker.

Although Fig. 7 illustrates a station incorporating only two generators, in individual situations provision for fast sustained partial turbine valve closing and steam bypassing will allow adding a generator or generators without introducing a need to construct additional lines. Yet no generator will need to be tripped, except in the unusual event of delay in fault clearance, or if braking were incorrectly effected, in which case only one unit would require tripping.

When a means of braking load application is to be provided, several considerations enter. The desirable feature of a fully enclosed resistor that would be constantly energized to ground, thereby avoiding the chance of a fault at the instant of braking load application can be arrived at by providing an isolated phase construction that can be used at generator, autotransformer tertiary, or high-voltage step-down transformer secondary voltage level.

In a form which allows for considerable variation in magnitude of braking load, the braking resistor could have the internal connections shown in Fig. 8, but also, simpler alternate arrangements would be possible.

A rugged, compact, and low-cost resistor can be constructed in the form of an assembly of coaxial resistors for operation in the range of generator voltages or below.

The concept of benefiting system stability by fast switching of transmission line series capacitors advanced by Dr. Kimbark would appear to warrant economic evaluation; it is assumed that studies on this score are in process.

The use of fast valving coordinated with dynamic braking and steam bypassing can provide a favorable way to affect both transient and post fault steady-state stability. This could lead to more general employment of series capacitors than has heretofore been made.

Most transmission lines equaling or exceeding a voltage of 345 kV are being designed to operate well below the thermal rating of the conductors.

The use of fast valving, dynamic braking, and steam bypassing, together with series capacitors, and the optional use of fast capacitor switching (dependent on economic aspects) offer a way to allow increased loading of lines.

In relation to steam turbines, what has been outlined, while involving certain complexities, notably offers the advantage of providing a way to improve system stability which avoids the complex and to some degree always hazardous problems associated with steam turbine tripoffs initiated by system difficulties. Trip-offs, when employed to deal with system difficulties, besides introducing station problems, are capable of compounding a system difficulty, since there is the chance of simultaneous (or nearly simultaneous) tripping of units at more than one generating station.

Edward W. Kimbark: I wish to thank the discussers for their approval of the general scheme set forth in the paper for the enhancement of stability of ac regional interties and also for the valuable supplementary information that they have contributed.

Mr. Mittelstadt stresses the importance of accelerometers for initiating rapid corrective action. The device described by Shier and Blythe [9] (originally devised for supplying a supplementary signal to voltage regulators to increase the damping of oscillations of machines) appears to be suitable also for controlling the application of measures such as braking for improving transient stability.

Mr. Farmer and Mr. Lyle are entirely correct in their contention that the equal-area criterion should be modified to take change of inertia into account when the criterion is applied to

Manuscript received October 17, 1968.

generator dropping. Also, for more rigor, the curve S in Fig. 4 should be replaced by two curves: one for initial conditions, and another for the condition after disconnecting some generators. Both of these, however, when taken as corrections to a method which is already an approximate one, become negligible, when applied to the type of power systems postulated in Section I-B, and when the fact is considered that only a small fraction of the total generating capacity of any area would be dropped.

I agree completely with Mr. Schleif's statement that the damping due to a properly controlled excitation system, especially to one of fast response, can effect an important improvement in transient stability. Indeed, improvement of excitation systems may turn out to be one of the most economical methods of improving stability. The important role that Mr. Schleif has already played in promoting the use of supplementary signals to voltage regulators for increasing damping of swings gives special weight to his comment on this subject [10].

Mr. Bills feels that central control computers will be required to obtain the best results from the countermeasures advocated in this paper. My view is that much can be accomplished by local control, at lower cost and with greater reliability, and that central control will supplement local control at some later time, but it will not supersede it. After central computers, extensive telemetering, and control channels are in use for other more necessary purposes, stability control can be added at a reasonable incremental cost in order to achieve some improvement in performance over what can be done by local control alone.

Mr. Park discusses several measures that either were not mentioned in the paper or were barely mentioned. These are fast backup tripping and fast valving and bypassing of steam. Fast backup tripping, when combined with independent relays and instrument transformers, can do much to avoid system instability. This would otherwise be almost sure to occur in the event of a stuck breaker or a relay failure at a large generating plant, with the customary slow backup clearing. The fact that a generator is tripped off in this scheme is not a disadvantage but is generally favorable to stability; in some plants it is necessary.

The importance of fast valving and bypassing of steam, coordinated with resistance braking, becomes apparent from consideration of facts stated in the paper on the Four Corners powerplant, where generator dropping was determined to be necessary [8]. That paper states that, after a generator has been dropped, six to nine hours are estimated as the time required to put it in service again. Fortunately, hydroelectric units can be put back on the line within a few minutes after dropping. Mr. Park has made a thorough study of braking and of fast valving and bypassing of steam for improving power system stability [1]; consequently, his comments greatly add to the value of the paper.

References

[9] R. M. Shier and A. L. Blythe, "Field tests of dynamic stability using a stabilizing signal and computer program verification," *IEEE Trans. Power Apparatus and Systems*, vol. PAS-87, pp. 315–322, February 1968.

[10] F. R. Schleif, H. D. Hunkins, G. E. Martin, and E. E. Hattan, "Excitation control to improve powerline stability," *IEEE Trans. Power Apparatus and Systems*, vol. PAS-87, pp. 1426–1434, June 1968.

FAST VALVING AS AN AID TO POWER SYSTEM TRANSIENT STABILITY
and
PROMPT RESYNCHRONIZATION AND RAPID RELOAD AFTER FULL LOAD REJECTION

E. W. Cushing, Jr.
Montaup Electric Co.
Fall River, Mass.

G. E. Drechsler
Northeast Utilities Service Co.
Berlin, Conn.

W. P. Killgoar
Boston Edison Co.
Boston, Mass.

H. G. Marshall
Central Maine Power Co.
Augusta, Maine

H. R. Stewart
New England Electric System
Westboro, Mass.

ABSTRACT

As steam turbine generator units increase in rating, their inertia constants tend to decrease and their reactance constants to increase, thus aggravating the power system transient stability problem. There is also a tendency for the complete unit to take longer to restart and reload when unscheduled disconnection from the system takes place. Counter measures are needed and may be available within the power plant. A study of these measures was made by the authors acting as a Working Group of New England's NEPLAN Planning Committee's System Design Task Force. The paper presents the Working Group's findings as widely supported by the major utilities in New England.

INTRODUCTION

a. Physical Limitations of Fault Clearing Equipment

In recent years, with the interconnection of many large power systems, the trend has been toward higher voltage transmission lines and larger generating units. These interconnected power systems have proven to be both reliable and economical.

Several problems have been encountered, however, from this method of operation, leading to increased difficulty in maintaining system stability.

The large generating plants, in most cases, are located a considerable distance from load centers for many economic and political reasons. The power must then be transmitted to the load centers via high voltage transmission lines and today, power companies are finding it increasingly difficult to obtain right-of-ways for these overhead high voltage lines. Many lines are delayed years after planned in-service date and some plans are even abandoned. These problems can have an adverse effect on the stability of the system.

Another factor affecting stability is the relatively low inertia of new generating units. During system short circuits, they tend to speed up faster than most older and smaller units and, therefore, are less stable. The accelerating power of a turbine-generator unit is equated to the mechanical power input to the turbine minus the electrical power output of the generator. In the past, in order to prevent a unit from accelerating beyond the point of no return, means were used of increasing the electrical power output of the unit during fault conditions. This was done either by decreasing the fault clearing time (relaying time plus breaker operation time) or increasing the number of transmission lines out of a station, or both. The stage has now been reached where fault clearing times have been decreased as far as modern technology will permit - four cycles normal clearing time and eight to eleven and one-half cycles delayed clearing time for 345 kV transmission lines. In some cases, however, these virtually irreducible clearing times are insufficient to prevent the instability of a unit.

In light of this problem, the System Design Task Force of the New England Planning Committee established the Stability Equipment Working Group. The objective of this group was to determine availability, feasibility, and applicability of new methods for maintaining and/or improving system stability. The methods taken under consideration were braking resistors, short-circuit current limiting devices, turbine fast valving and full load rejection systems.

b. Methods for Maintaining Stability

1. Braking Resistors

The stability of power stations remotely located from the major load centers is jeopardized by transmission line faults. When a fault occurs on the system, the voltage falls instantaneously preventing power from being transmitted beyond this point. When no power can be transmitted and there is no local load to absorb power, it all goes into accelerating the machines.

Braking resistors act as a substitute for the system load. During a fault, the resistors are applied to the terminals of the generator through circuit breakers by means of an elaborate control scheme. The control scheme determines how much resistance must be applied, the phase-to-phase proportionment of this resistance, and the duration of application. The braking resistors remain on for a matter of cycles both during fault clearing and after system voltage is restored. The control scheme must also be able to derive a quantity which distinguishes between various types of faults, and is also dependent on their location.

British Columbia Hydro recently installed braking resistors at their new Peace River plant.

2. Short-Circuit Current Limiters

Short-circuit current limiters are being used to limit the short-circuit duty for distribution lines. They are also being proposed for use in long transmission lines to modify favorably the transfer impedance during fault conditions.

It is proposed to study the application of these devices on transmission lines emanating from a generating station to determine what effect, if any, this will have on the stability of that particular station.

3. Turbine Fast Valving

The two methods previously discussed have both been an attempt at replacing the system load so as to increase the electrical output of the generator during fault conditions (which for a close-in three phase fault at a remote site would otherwise be zero). Another method of improving the stability of a unit is to decrease the mechanical input power to the turbine. This can be accomplished by means of fast valving.

Paper 71 TP 705-PWR, recommended and approved by the Power Generation Committee of the IEEE Power Engineering Society for presentation at the 1971 Joint IEEE-ASME Power Generation Conference, St. Louis, Mo., September 19-23, 1971. Manuscript submitted June 11, 1971; made available for printing July 13, 1971.

Reprinted from *IEEE Trans. Power App. Syst.*, Nov./Dec. 1971.

Briefly, during a fast valving operation, the interceptor valves are rapidly shut (in 0.1 to 0.2 sec.) and immediately reopened (within approximately 11 sec.). This procedure will increase the critical switching time long enough so that in more cases the unit will remain stable for faults with stuck breaker clearing times. (Critical switching time is the time in which a fault must be cleared in order to prevent instability.)

There are several units in operation today in the United States which employ fast valving, and there are still more under construction which will use it.

c. Method for Improving Stability: Full Load Rejection Technique

As previously mentioned, fast valving combined with high speed backup clearing times will suffice in maintaining stability in many cases. However, there are still cases where even fast valving cannot provide the necessary benefits to maintain stability. In such cases, the normal procedure is to automatically trip the unit off the line. Even though the unit does not go into a maintenance cycle, which is often the case, it usually takes several hours to get the unit back in operation. The loss of a major unit even for several hours can seriously jeopardize the remaining system.

Under extreme conditions as this, where a unit must be separated from the system, a full load rejection system could be utilized. For this purpose, the unit may have to be equipped with a large steam by-pass system. When the generator breakers are tripped, the stop valves would be rapidly closed allowing the steam to flow through the by-pass system to the condenser. Then the stop valves would be partially reopened to permit enough steam to flow through the turbines to operate the unit at auxiliary load. After the system has recovered from the fault, the unit could be resynchronized and reloaded.

The major disadvantage to the full load rejection scheme is the added cost of such a large by-pass system. One would have to carefully weigh the operational advantage against this extra cost before installing such a system.

The preceeding paragraphs attempted to summarize the various methods of improving stability that the Working Group has been exploring. The remainder of this paper is in two parts. The first part describes the method of fast valving for various type units, and the problems that can be encountered. The second part deals with the different types of full load rejection systems, and some of the problems encountered with this mode of operation.

PART I - FAST TURBINE DRIVING POWER REDUCTION

GENERAL

Fast valving is a natural extension of the inherent capabilities of the modern electro-hydraulic (EH) turbine control system. The EH control system utilizing solid state electronics, high pressure hydraulic actuator system and fast response turbine valves achieves extremely rapid control necessary to limit turbine overspeed following a full rejection. Normal speed control and valve stroking rates are not fast enough to limit overspeed to a desirable level following a major load rejection (greater than 20 to 40% of full load). The EH overspeed control preanticipates rotor acceleration by monitoring the mismatch between turbine driving power and generator load and initiates fast

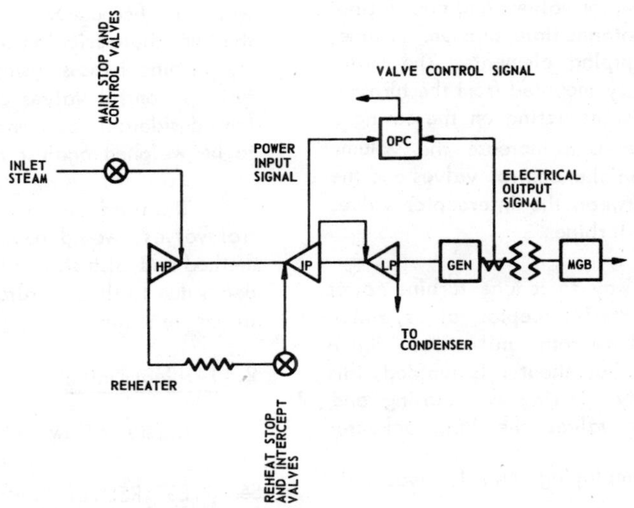

Fig. 1 Steam Flow Path - Fossil Fired Unit

closure of control and intercept valves (within .10 to .20 sec. for full stroke) when the mismatch exceeds the limit (approximately 40%). On some units, high-pressure and reheat stop valves may also be tripped closed and automatically reset if peak rotor overspeed remained within normal limits. The fast closure is achieved by means of special solenoid operated disk dump valves which when actuated release the hydraulic fluid from the bottom of the pistons causing the valves to close rapidly. The EH overspeed control systems have been modified to incorporate the fast valving function. Upon detection of a fault condition of sufficient severity (turbine power output and power-load unbalance both exceeding certain preset limits), fast closure of interceptor valves* is initiated followed by immediate reopening, thereby momentarily reducing unit accelerating power and improving unit stability. This valving procedure is generally referred to as "fast valving control". Normal speed governor action also acts to limit rotor acceleration but has little effect in the critical time interval for first swing stability (first one second). A fault condition is distinguished from an instantaneous load rejection by monitoring main generator circuit breaker status or turbine power-generator current mismatch.

Fast valving logic has also been applied on units with mechanical hydraulic control; however, it is somewhat more difficult to implement and may require modifications to valve actuators and other equipment.

VALVING PROCEDURES

1. Fossil-Fired Units:

The steam flow path through a modern fossil-fired, tandem-compound, single-reheat turbine generator unit is shown in Figure 1 High pressure steam enters the high-pressure turbine through the main inlet stop valves, control valves, and connecting piping. Reheat steam flows into the intermediate-pressure turbine through the reheat stop valves, the reheat interceptor valves, and connecting piping. Steam leaving the intermediate pressure turbine flows through the crossover piping elements. The main and reheat valves are separately mounted from the turbine casings to minimize piping forces acting on the casings. However, this arrangement tends to increase the volume of uncontrolled steam between the control valves and the high-pressure turbine and between the interceptor valves and the intermediate-pressure turbine.

The least complicated way to reduce turbine power rapidly is to "fast valve" only the interceptor valves, which control approximately 70% of the total unit power. Since the stored energy effect of the reheater is avoided, this major portion decays quickly. During the closing and reopening of the interceptor valves, the large reheater volume acts as a cushion, absorbing steam flow ahead of the interceptor valves for several seconds resulting in a negligible effect on throttle flow. Reheater safety valves may operate limiting reheat pressure to a level approximately 10% above full load pressure. Electro matic type relief valves of sufficient capacity can be provided to avoid lifting spring loaded safety valves and potential reseating problems.

In recent fast valving tests conducted on a fossil reheat unit, reheat safety valves did not lift provided that interceptor valves were reopened immediately with no additional delay*. Immediate reopening of intercept valves is most essential from the standpoint of reheater metal protection. Reopening after a fast closure is inherently delayed for approximately one second (to allow for restoration of oil to the hydraulic cylinder) and then takes approximately 10 seconds for full stroke. Indications are that this is within acceptable reheater limits.

Present equipment designs require that the turbine return to its prior power output following a fast valving action. However, a controlled load runback at maximum rate of 100% per 10 to 30 sec. may also be initiated concurrently with or following fast valving, to attain a partial power output.

The alternate scheme of closing control valves and intercept valves would cause rapid decrease of high pressure turbine power as well as intermediate and low pressure power and hence greater stability benefit. However, the sudden stoppage of main inlet steam flow would impose a severe transient on the boiler and control systems. Potential problems include the lifting of boiler safety valves, combustion and feedwater control problems, turbine thermal stresses, transient thrust unbalance and overloading of the last few HP stages. A significant problem may be that of discriminating the fast valving against a full turbine trip which is thoroughly interlocked with the fuel system for the immediate tripping of fires. A fast acting high capacity turbine bypass system, closely coordinated with the turbine control valves could be of potential benefit, but the considerable cost and increased complexity would have to be weighed against the potential stability gains.

The third possible alternate, closing only the control valves, would have the disadvantages of the above method and also slow intermediate and low pressure power decay due to the uncontrolled expansion of the steam stored in the reheater.

2. Nuclear Units:

The steam flow path of a nuclear, tandem-compound,

* Alternate valving procedures employing control valves will be discussed later.

* See "TEST RESULTS", Page 2521.

reheat turbine generator unit is quite similar to that shown in Fig. 1 except steam leaving the high-pressure turbine flows through a moisture separator-reheater, through the reheat stop and interceptor valves and then directly to the low pressure turbine.

As with a fossil unit, fast valving of the interceptor valves is employed as the most practical valving procedure. Dynamic response is somewhat improved due to the location of interceptor valves close to the low-pressure turbine inlets resulting in a smaller volume of uncontrolled steam.

The remainder of the paper will consider fast valving of intercept valves only.

FAST VALVING LOGIC

The interface between fast valving logic and overspeed control is readily apparent in the logic diagram presented in Fig. 2.

This logic is incorporated in the overspeed protection controller (OPC) depicted in Fig. 1. Mechanical power input to the generator is measured with a reheat pressure transducer and compared to electrical power output of the generator which is measured with a Hall watt transducer. Assuming electrical power is suddenly lost, a power/load unbalance exists. Referring to Fig. 2, if the unbalance (reheat pressure minus kW) exceeds A% and the reheat and kW transducers have not failed and reheat pressure exceeds B%, either of two courses of action take place:

(1) If the main generator breaker is closed indicating a system fault condition and partial load loss, fast valving of intercept valves is initiated. At the same time, the closing signal initiates intercept valve reopening through a time delay (adjustable) provided the MGB has remained closed.

(2) If the MGB is open, a signal is sent to the overspeed control logic to close turbine control and intercept valves.

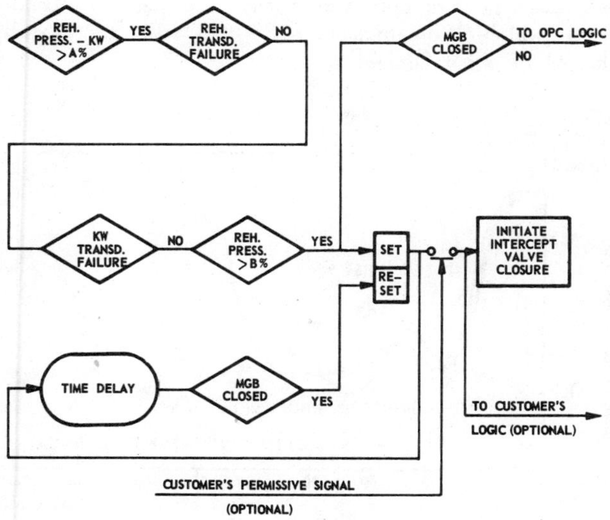

Fig. 2 Fast Valving Logic Diagram (Westinghouse)

This typical logic diagram represents the standard logic offered by one turbine generator manufacturer. As indicated, certain optional features can be provided as an output to the user's logic circuitry (for example to initiate switching of braking resistors) or a permissive signal input from the user's logic which could delay or block fast valving depending on fault characteristics. User interfaces are not permitted with any other portion of the circuitry since such could jeopardize the integrity of the overspeed control. It may prove desirable to incorporate additional logic into the fast valving such as another power load unbalance level to further discriminate against faults of certain severity which do not require fast valving or to introduce a time delay to allow for normal breaker clearance. However, studies indicate that additional time delays of even a few cycles can substantially counteract the stability benefits gained through fast valving. In many cases, perhaps the selectivity of the logic must be refined to minimize the delay. It is recommended that stability studies be run on each unit to properly evaluate the potential gains of fast valving and the influence of increased logic complexity and accompanying time delay.

TURBINE RESPONSE TO FAST VALVING

The turbine power decay following fast valving is inherently limited by several particular characteristics of large steam turbines. First of all, there is a control system time delay involved to detect the the power-load unbalance, to amplify the control signal, and to get the extremely large intercept valves closed. Furthermore, the power decay is a function of the valve flow characteristic and of the energy in the uncontrolled expansion of steam downstream of the valves.

The control system delays are minimized with the EH control system to approximately 0.10 seconds. Valve actuator closing times have also been reduced to a minimal value of 0.10 to 0.20 seconds for full stroke. In this short interval, the valve moving parts equivalent to the weight of an automobile are accelerated from standstill to a maximum velocity and then decelerated to an acceptable impact velocity.

Typical valve flow characteristics for both plug and butterfly type intercept valves are presented in Fig. 3a. Note that with the butterfly type, which is utilized on some nuclear units, effective steam throttling commences as the valves begin to close, whereas the plug type must close perhaps 25 percent of the total stroke before effective throttling

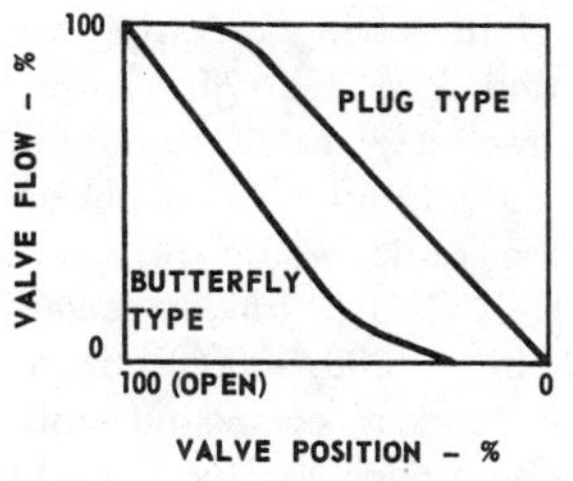

Fig. 3a Typical Intercept Valve Flow Characteristics

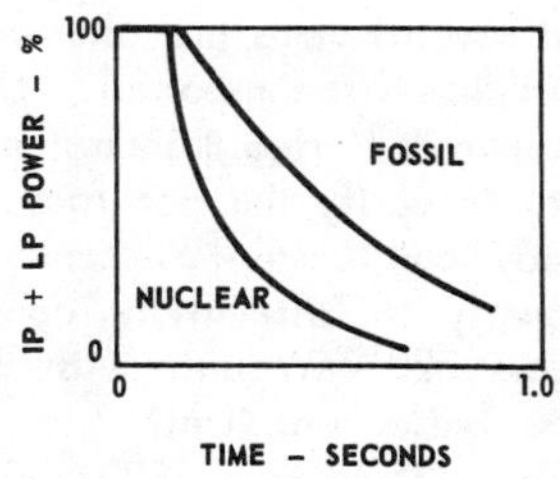

Fig. 3b Typical Turbine Power Decay Curves

begins (dead travel). The situation is just the reverse on the opening stroke and it is significant to note that for the plug type utilized in fossil units, considerable reheater flow may effectively be established with very little stroke and is thus very beneficial for reheater protection.

The sum total of all these effects is incorporated in the turbine power decay curves presented in Fig. 3b for typical fossil and nuclear units. Note that power decay is somewhat more rapid for nuclear units because the intercept valves are located right at the low pressure turbine inlets thus minimizing the amount of uncontrolled steam and also, the entrapped steam being at lower pressures has less stored energy compared to a fossil-fired unit. Additional benefit is also derived from the use of butterfly type intercept valves with essentially no initial dead travel in the closing direction.

EFFECT OF FAST VALVING ON STEAM GENERATOR, AND CONTROLS

The general consensus is that the steam generator (nuclear and fossil) and controls will ride through the load changes with very little disturbance provided that intercept valves are fully open within approximately 11 seconds from initiation of fast valving. Reheater tube protection appears to be the limiting factor on fossil reheat units. If power operated relief valves are utilized on the reheater outlet and automatically opened, periods longer than 11 seconds could be tolerated. It is recommended that specific units being considered for fast valving be analyzed to determine the potential life reduction of the reheater based on reheater surface location, flow-interruption time, relief flow, etc.

Conventional steam generator controls and controlled elements would not have time to react appreciably to such a rapid momentary reheat flow interruption. However, in the initial design stages, the entire plant control and protection scheme must be studied and perhaps some minor modifications made to assure compatibility with fast valving control. For instance, the coordinated or integrated type control systems which utilize megawatt error must have additional logic incorporated in order to respond properly. Also, when a turbine stage pressure is utilized for control purposes, the fluctuations accompanying fast valving must be taken into account.

TEST RESULTS

The application of fast valving is new and although several units with this feature have recently been installed and several more are planned, the field testing of only two units has been reported. Units 2 and 3 of Georgia Power Company's Harlee Branch Plant were recently tested to obtain data to verify the accuracy of the calculated turbine power decay and obtain assurance that the units would respond properly to fast valving control. Unit 2 is a tandem-compound 320 MW unit with a 2400 lb - 1000/1000°F drum type boiler and Unit 3 a 480 MW tandem compound unit with a 3500 lb - 1000/1000°F once-through boiler. Field test data of generator power output vs. time agreed quite well with computer analytical calculations as shown in Fig. 4. Included also is the manufacturer's predicted turbine mechanical power decay characteristic employed in the computer calculations. Note that generator output has almost fully recovered within 120 cycles (2 seconds) due to the extreme non-linear characteristics of this particular plug type intercept valve. Test results also indicate that reheat safety valves should not operate unless the unit is carrying load in excess of rated capacity.

GAINS IN TRANSIENT STABILITY

Generalized system stability studies have been run to evaluate the potential gains in system transient stability with fast valving. Typical increases in critical fault clearing times for fossil units are one to two cycles. For nuclear units with relatively higher unit inertia and with quicker turbine power decay, an increase of two to three cycles may be expected. For maximum gain, all units in the power plant should employ fast valving control.

As mentioned previously, to achieve maximum stability gains, it is very important to initiate intercept valve closure with a minimum of delay after the fault begins. The gains mentioned above include a minimal control system delay of approximately 0.10 sec. (6 cycles) required to sense the power-load unbalance. It has been estimated that additional time delay of only three or four cycles can reduce the stability gain by about one cycle.

POTENTIAL PROBLEMS

From the theoretical viewpoint emphasized thus far, fast valving looks quite attractive. However, there are several areas of concern with some degree of uncertainty which must be considered in evaluating the application of fast valving particularly for large units and more so for nuclear applications. A brief summary of potential problems revealed to date:

(1) Increased complexity of controls may decrease unit reliability. Several false unit trips have been experienced to date on a total of five units of different manufacture due primarily to electrical component failures in the fast valving circuitry.

(2) Increased interceptor valve maintenance due to more frequent operation and more severe duty. The increased duty results from the high energy dissipation in the valve when close-opened with full load reheat pressure ahead of the valve. Prime concern is for the large plug type valves on nuclear units which may reach 34 inch nominal size, weighing approximately 25 tons with an overall height of about 20 feet.

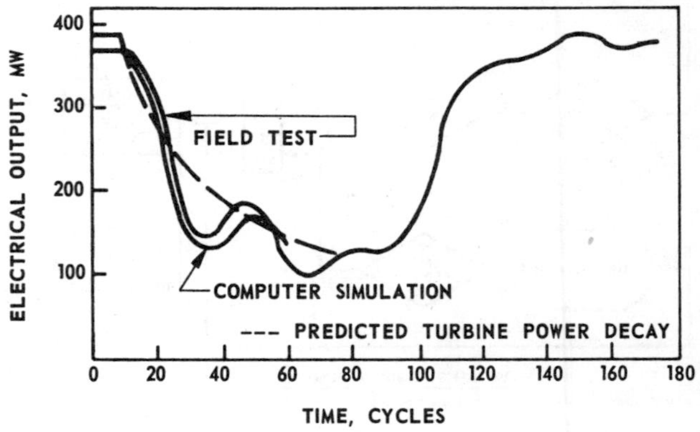

Fig. 4 Fast Valving Test Results - Generator Output Versus Time (Reference 13)

(3) Reheat safety valves may lift on some fossil units depending primarily on reheater volume. As previously discussed, power operated relief valves of sufficient capacity can be provided to avoid lifting spring loaded safeties. On nuclear units with moisture separator, reheater volumes in the normal range (MSR time constants from 2 to 5 seconds) it is predicted that safety valves will blow when fast valving occurs above 75% of full load in many cases, unless additional power operated relief valve capacity is supplied.

(4) On nuclear units, some uncertainty exists regarding the satisfactory design and operation of the complex moisture separator-reheater drainage system during transient conditions. Difficulties are still being experienced under steady state conditions and it will take some time yet before firm design recommendations can be developed to assure proper operation during transient conditions including fast valving operations.

In view of the potential problem areas together with the very limited operating experience to date, some turbine generator manufacturers have recently recommended not to apply fast valving to large units, particularly nuclear until additional experience has been gained on smaller units. Also, when fast valving is applied, the number of such operations should be held to a bare minimum. Preferably all unnecessary operations should be eliminated through the use of sufficiently restrictive logic. However, achieving this with a minimum of complication and time delay may in itself present a rather formidable problem.

PART II - UNIT LOAD REJECTION AND RAPID RELOAD CAPABILITY

GENERAL

If all efforts to maintain stability fail, and a unit is separated from its load, the capability to continue operating and furnishing its own auxiliary load can be of prime importance. The unit then is available for immediate resynchronization and reloading once the system upset is cleared. Full power output can be obtained in a matter of 15 to 20 minutes as opposed to a delay of perhaps one hour but extending up to several hours for some units.

The desirability of providing this capability must be evaluated on an individual unit basis, and will largely depend on such factors as system reserve capability, interconnection arrangements with other systems, transmission design from the unit, black start capability, added cost and complexity to the unit, and savings in outage time compared to a full trip-out.

In many cases all that is required to maintain a unit on auxiliary load is a new philosophy and some minor control changes. In some instances, however, the problem is not that simple requiring more extensive changes to implement and also to minimize the reduction in equipment life which may be quite significant for such a severe cycle. Major problems will be reviewed along with the capabilities and limitations of major plant equipment to establish the general feasibility and practicality of maintaining auxiliary load and then rapid reloading after a full load rejection.

TURBINE GENERATOR

Turbine generators have generally been designed to accept a full load rejection without experiencing a complete trip-out. This capability is basically the first line of defense against excessive overspeed. On modern reheat units with high exhaust end loadings and water cooling in the generators, resulting in low rotor inertia per kW output, acceleration on loss of full load is in the order of 15 to 20% of rated speed per second. Rotor overspeed control thus presents a major challenge to turbine designers, a challenge which has been met through the application of solid state electronics, high pressure hydraulic actuation system and fast response valves to achieve the extremely rapid control necessary to limit overspeed to a value less than the primary trip setting (typically 110% of rated speed) following full load rejection.

If generator load is suddenly lost, the electrohydraulic (EH) control system senses the power/load unbalance and initiates fast closure of control and intercept valves (within 0.10 to 0.20 sec. full stroke). The unit will accelerate to a peak speed below trip speed (typically 1/2 to 1% below). Then as the unit decelerates to a given speed (typically 102%) the turbine control system will modulate intercept valves to blow down the reheater and finally the unit will settle down at rated speed with control and intercept valves opened slightly to carry auxiliary load. The rapid load decrease and subsequent reloading causes rapid cooling followed by heating of large metal parts (rotors and casings) and it is the resulting thermal stresses that have the greatest influence on the recommended practices. The limiting thermal stresses occur in the high temperature regions of the high pressure and reheat sections of the turbine, with the rotor stresses generally leading casing thermal stresses. Differential expansion between rotating and stationary parts may also be limiting and requires special design considerations to assure adequate axial clearances and requires close monitoring.

Desirable equipment design features and operating practices are essentially the same as employed on cycling units, the common goal being to minimize thermal stresses and maximize fatigue strength of materials to achieve satisfactory equipment life.

The following means can be employed to reduce thermal stresses, all of which generally attempt to reduce the amount and the rate of steam temperature changes at the inlet to and within the high temperature regions of the turbine:

1. Full Arc Admission:

Most large modern reheat machines installed in the U.S.A. operate with constant boiler pressure and utilize partial arc admission to achieve efficient part load operation. A typical unit would experience a change in first stage temperature from full load to auxiliary load as indicated in Fig. 5. By employing full arc admission this temperature change is reduced appreciably as evident in Fig. 5. The modern EH turbine control systems have the capability to automatically transfer from partial arc to full arc control. This transfer should be incorporated into the full load rejection and reload operating procedures so that such operations are accomplished in the full arc mode.

2. Variable Pressure Operation:

Following a trip back to auxiliary load first stage temperature can be raised appreciably by lowering boiler pressure as shown in Fig. 5. The low pressure causes turbine control valves to open wider, thereby reducing throttling and temperature drop across the control valves. This "depressurization" can be readily achieved with either the once thru or drum type boiler which generally operate under variable pressure during startup.

3. Steam Temperature Control:

Superheat and reheat temperature may typically fall off 150°F within 15 minutes after tripback to auxiliary load as shown in Fig. 5. This external temperature drop directly lowers internal turbine first stage temperature by the same order of magnitude. Actually, with constant pressure boiler operation, first stage temperature drop will exceed initial temperature variation due to increased throttling across turbine control valves as can be seen in Fig. 5. Although Fig. 5 has been drawn for a supercritical unit, this temperature decay (as well as other general characteristics) can be considered typical of drum type units as well.

Steam temperature control with conventional boiler systems is quite limited at such low steam flows (5 to 10% of full load). Limiting factors include the need to protect superheater and reheater elements from overheating due to poor steam flow distribution. Gas temperature must be held to 1000°F at critical locations and firing rate is thus restricted to about 5 to 10%. Also, furnace purge requirements dictate a minimum of 30% continuous air flow creating a high air-flow to firing-rate unbalance.

Once thru units which are equipped with bypass systems are generally able to achieve better steam temperature control than drum type units at low loads. Temperature decay at auxiliary load may in some cases be less than 50°F as can be noted in the field test data in Figs. 6 and 7. For drum type units, temperatures at low loads are normally random and uncontrolled unless special design provisions are made. One possibility is to incorporate a bypass system. To be truly effective on a drum type unit, the European type "bypass thru the reheater" may be required. The conventional U.S.A. bypass system utilized primarily on once thru units does not bypass steam through the reheater and although superheat flow is increased by bypassing steam around the turbine to the flash tank and/or condenser, reheater flow is not affected imposing some firing rate restriction. Though attractive in principle, the advantage of this system must be evaluated against the increased cost and complexity of the system and also, it is not yet available from U.S.A. manufacturers.

In addition to the above, thermal stresses can be further reduced through the use of optimum rotor and casing geometry including thinner casing sections and smaller rotor diameters (where possible) and increased fillets. Smooth rotor packings and/or increased radial seal clearances may also prove advantageous but at the expense of an increase in heat rate.

The extent of application of the above features will depend heavily on individual unit characteristics, normal operating modes and associated economic factors.

A most significant parameter to be considered in the early stages of design is the allowable reduction in unit life for such operations. An estimate must be made of the total expected number of such operations and incorporated in the total prediction of the number and types of starting and load changing cycles. Presumably, the number of load rejection cycles can be held to a

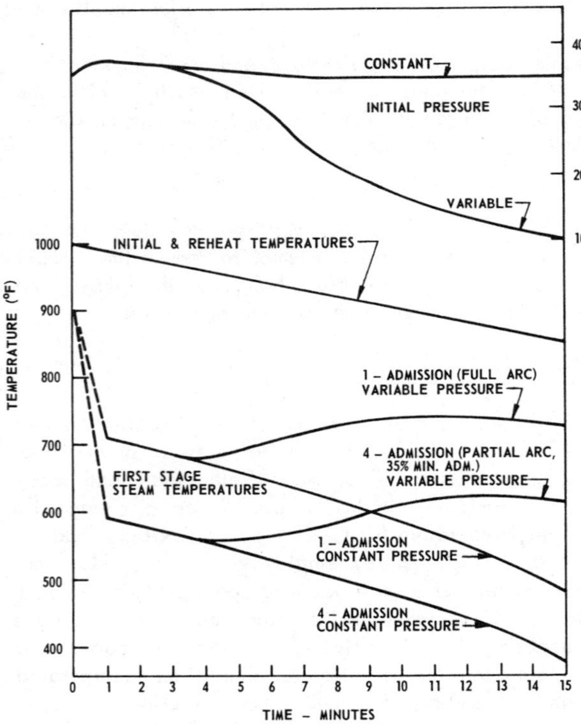

Fig. 5 *Steam Temperature Variation in First Stage Shell Following Tripback to Auxiliary Load (Reference 11)*

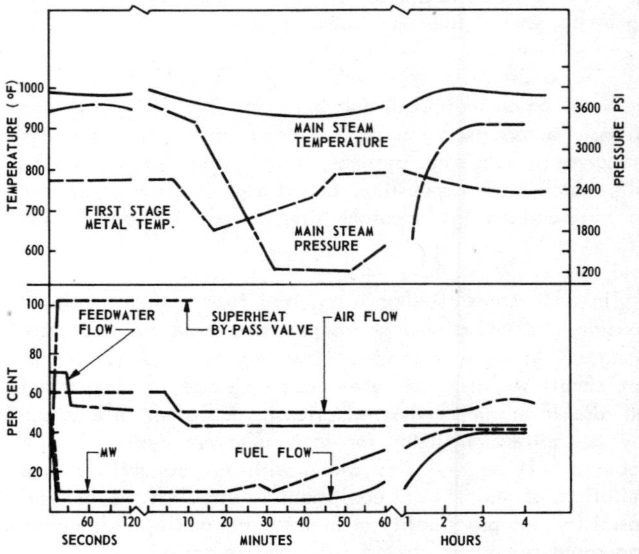

Fig. 6 *Tripback to Auxiliary Load - Supercritical Unit, Reload from Reduced Pressure (Reference 6)*

maximum of one or two per year where life reductions in the order of 1/4 to 1% of the total fatigue capacity of the rotor are reasonable. Such values are readily attainable through application of the techniques discussed above. On the other hand though, rotor life reductions of up to 4% per cycle can be incurred if proper attention is not given.

FOSSIL STEAM GENERATORS

Whereas turbine-generators are basically designed to accept a full load rejection down to auxiliary load, boilers, auxiliaries and controls are not unless initially specified and require some special design provisions and operating procedures. Major problems include:

Rapid cutback in firing rate to a minimum of 5 to 10% without loss of ignition.
Minimizing thermal overstresses and resulting equipment life reduction in heat absorption sections during the fuel reduction and rapid reloading.
Preventing spring loaded safety valves from lifting.
On drum type units, maintaing drum level within the high/low limits.
Maintaining auxiliary equipment such as fans and feedwater pumps in service.
Designing and tuning boiler control system for fast, stable response incorporating special predetermined control actions.

The above considerations will be reviewed in general with distinctions made between unit types where pertinent.

Upon full load rejection, turbine control and intercept valves close within approximately 0.20 to 0.30 seconds including 0.10 seconds for electro-hydraulic control system response time. Steam flow through the superheater and reheater ceases causing an immediate rise in main steam pressure equivalent to the normal pressure drop due to flow through the superheater for the drum type unit and through boiler and superheater for the once thru unit. Pressure then continues to rise at a slower rate depending on firing rate.

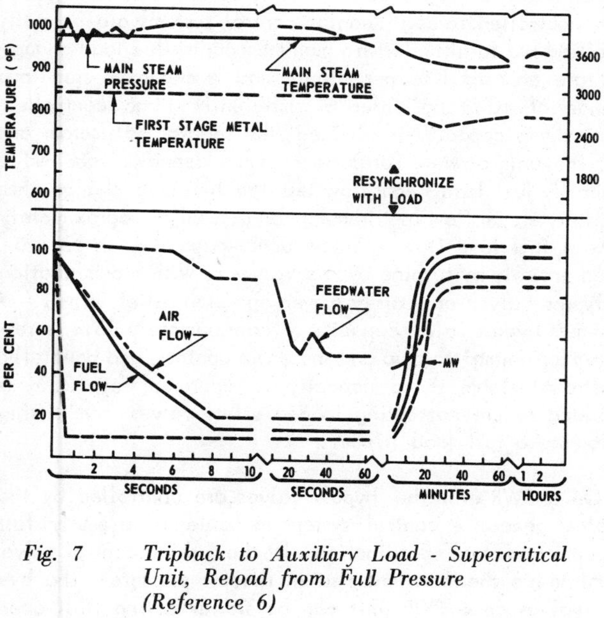

Fig. 7 Tripback to Auxiliary Load - Supercritical Unit, Reload from Full Pressure (Reference 6)

For a gas or oil fired unit equipped with a burner control and monitoring system, the preferred procedure is to trip the majority of burners in service but to maintain at least 20 per cent of the burners providing a total fuel input of 5 to 10 per cent of full load rate. Typically, a unit may be equipped with a total of 24 burners with a 4/1 turndown capability. 20 burners could be tripped and the remaining 4 operated at approximately half capacity. Tripped oil burners must be purged (within 15 to 30 seconds) and retracted to prevent coking and overheating. Air should be reduced to the normal purge flow (25 to 35%) for continuous furnace purge.

A coal fired unit would presumably have to be runback to a firing rate less than the normal minimum coal input requiring that all coal be tripped and auxiliary fuel introduced. Some auxiliary fuel may have to be burned continuously to assure adequate furnace protection since the time interval to introduce fuel is so short.

The more conventional procedure employed in most of the load rejection testing to date has been to trip all fuel on the initiation of a turbine trip-back and then to wait for at least a five minute furnace purge period before refiring. The furnace purge time can be started at the time of the fuel trip assuming air flow will be maintained above the minimum of 25% of full load flow. This method has proven quite satisfactory for drum type units which have sufficient thermal storage to allow the operator to re-establish auxiliaries and firing with some margin for delays. The thermal metal storage is sufficient to maintain relatively high steam temperature for approximately 5 to 20 minutes depending on individual unit characteristics. However steam temperature will then drop severely if firing has not been reinitiated and the turbine should be tripped to avoid a severe thermal shock. Once thru units have considerably less thermal storage and may be difficult to "boot-strap" particularly if the feed pump is steam turbine driven.

The firing cutback cannot be achieved fast enough to avoid lifting safety valves. On a once thru unit, quick opening of the bypass valves will prevent safeties from lifting. Following relief of excess pressure, the bypass is returned to normal pressure control. A drum type unit can be provided with power operated relief valves of perhaps 10 to 25% capacity which can be popped open within 0.10 to 0.20 seconds to prevent spring loaded safety valves from lifting. Such valves may be provided with an isolation valve for on-line maintenance. If such valves are exhausted to atmosphere, silencers should be considered. Bypass valves if provided on drum type units for improved start-up and low-load temperature control can also serve this pressure relief function.

For both the drum and once-thru boilers, a predetermined procedure must be established for the control of feedwater flow following the fuel input reduction. A drum type boiler experiences an almost instantaneous drop in drum level of several inches due to shrinkage (collapse of bubbles) and steam discharge through bypass or relief valves, requiring that feedwater flow be initially maintained at a high rate and then reduced gradually. Drum level should then be controlled at a safe low level to accommodate expected swell when firing is increased. For once thru units, a minimum flow of feedwater is required for continuation of firing. If all fuel is tripped, the feedwater flow may also be tripped. If a steam turbine driven feedpump is employed, normally supplied with extraction steam from the main unit, an auxiliary steam source must be provided to automatically take over on low extraction pressure which decays rapidly with closure of turbine control and intercept valves.

BOILER CONTROLS

It is essential that an overall systems approach be undertaken in the initial design stages to develop satisfactory control procedures and to incorporate essential logic. The approach of tripping all fires presents the minimum complication control-wise. However, if the preferred procedure of tripping back to a minimum firing rate is to be followed, an automatic burner system is important. A coordinated or integrated type control system which assures close coordination between boiler and turbine control is also highly desirable. This type of system now standard on all once thru type units is becoming a necessity on large drum type units as well. Advanced control systems capable of fast response have been developed and applied in the U.S.A. However, in most instances, the utilities have not been able to spare the time or load to permit the proper control system tuning essential to the utilization and development of the full capability of these systems.

OPERATION AT AUXILIARY LOAD WITH RAPID RELOAD

Assuming the unit has been run back to auxiliary load, turbine first stage steam temperature will have decayed according to the characteristic in Fig. 5. the magnitude depending on boiler temperature characteristics, pressure control and mode of turbine valve control. The critical turbine first stage region will then undergo forced cooling reaching a new equilibrium state in approximately one hour. During this period, thermal stresses will reach a peak in approximately 10 to 15 minutes. It is quite advantageous, therefore, if load can be restored within 5 to 10 minutes, thus avoiding the peak stresses and allowing reloading at maximum rates. If reloading is not expected within approximately five minutes, on a once through unit it is recommended that the superheater be depressurized (typically 1000 psig for supercritical unit and 500 psig for subcritical) to reduce throttling, increasing turbine first stage temperature as reviewed previously. Typically depressurization takes approximately 15 minutes, the rate being limited primarily by boiler thermal stress considerations. Depressurization does increase reloading time but is not too great a penalty for a once thru unit since only the superheater is affected. Typically, repressurization is accomplished over a 15 minute period while going from 10% to 25% load. On a drum type unit, however, the entire unit, drum and superheater would have to be depressurized, consequently causing a greater delay in reloading.

It is normal practice to provide gas temperature probes to measure the gas temperatures entering the superheater during startup. Such probes are utilized to control or limit fuel input during operation at auxiliary load to prevent overheating the superheater and reheater metals. Also, as noted earlier, a properly designed bypass system provides good final steam temperature control at high temperature levels at low load.

Resynchronization can be accomplished within minutes after load is available. Reloading, however, is limited by both turbine and boiler thermal stresses and boiler firing capability, the limiting factor depending on the type of unit and mode of operation. Generally, the turbine limits the rate of load increase except where reload is accomplished within a very short time (5 minutes or less) or where variable pressure operation is employed.

With the widespread application of computers to plant supervision and control, it is highly recommended that pertinent metal and steam temperature be monitored and thermal stresses calculated by the computer for turbine and boiler to assure that the limitations and capabilities of both are closely adhered to.

A once thru unit can be overfired at very high rates within the capability of fuel and boiler feed equipment to produce the most rapid load pickup. Once thru units operating with variable pressure have been overfired as much as 200% with no sign of tube overheating or thermal stress problems. Drum type units are limited in reloading by drum level swell and if pressure has been decreased, by rate of saturation temperature increase to limit drum thermal stresses. The forced circulation unit has a greater capability of up to 400°F/hr. saturation temperature increase compared to 150 to 300°F/hr. for the natural circulation unit.

On all types of units, reloading may be further limited by the time required to reinsert, light up and prove the tripped burners. Automatic burner controls will minimize this delay.

For a typical large modern fossil unit designed to trip back to auxiliary load, 15 to 20 minutes is a reasonable time to return to full load assuming resynchronization is initiated right away. If boiler fires are tripped out completely, an additional delay of approximately 10 minutes can be expected.

NUCLEAR UNITS

Full load rejection capability down to auxiliary load is an optional feature on nuclear units. This is accomplished by controlled dumping of turbine throttle steam directly to the condenser or to the condenser and atmosphere. The steam dump basically maintains a short term load on the reactor, allowing the reactor to automatically cut back power at maximum controlled rates. Reactor control systems are normally specified and furnished with a maximum controlled load changing rate of 5 to 10% per minute and a maximum step load change of 10% and unless a sufficiently large controlled steam dump capacity is utilized, the reactor will scram and shut the unit down. Total steam dump capacity required is generally full load steam flow less the 10% step change, less auxiliary steam flow requirement for a total of approximately 85% of full load flow. Steam dump capacity can be provided entirely by turbine bypass valves or with a combination of bypass valves and safety power operated relief valves. A minimum bypass valve capacity of approximately 15% is required for normal start up and shutdown control, and generally additional steam dump capacity of approximately 25% is provided to prevent spring loaded safety valves from lifting following a full load tripout.

On a BWR unit the bypass valves are controlled by the turbine generator control system to achieve extremely fast response and close coordination with the turbine control valves to minimize the transient rise in reactor pressure. The bypass valves on a PWR unit can be slower acting (full open in 3 to 5 secs.) and are controlled by the reactor control system. The dumping of steam is reduced as rapidly as the reactor control can reduce core power by inserting control rods and also, in the case of the BWR unit, reducing circulation flow. Pressure breakdown orifices must be provided in the bypass lines along with special baffling and desuperheating in the condenser to dissipate the tremendous additional input.

In addition to the steam dump system, careful consideration must be given to the feedwater system to assure that the criteria of the nuclear steam supply system are met during this transient.

FULL LOAD REJECTION TEST RESULTS

Full load rejection to auxiliary load has been successfully verified by many utilities on drum type fossil units. Extensive test results were reported in Reference 14.

Several once thru units have also successfully tripped back to house load although experience in the U.S.A. has been very limited. To our knowledge, no such tests have been conducted on once thru units by U.S. utilities, but domestic manufacturers have reported test results on two European once thru supercritical units supplied through their licensees. Typical results from a series of tests on one of these units are shown in Figs. 6 and 7.

In the first test shown in Fig. 6, the main circuit breaker was opened, the fuel tripped back to a predetermined minimum, and the unit held at full pressure to determine whether any problems would arise. The initial and only difficulty was that the turbine first stage metal temperature started to decrease at a greater than desirable rate. At this time, the superheater was depressurized and then the first stage temperature returned to its initial level where it remained until the unit was repressurized. Note that the unit operated for several minutes carrying house load on the storage in the superheater as the pressure was reduced. During the test, rapid opening of the superheater bypass valves was initiated by the circuit breaker open signal preventing the boiler safety valves from lifting. After the initial opening, the valves were transferred to pressure control.

Several additional tests were conducted on this supercritical unit, but for the remainder, the unit was tripped to full pressure operation and then reloaded without delay, as shown on Fig. 7. Where the unit can be loaded within five minutes, it is recommended not to depressurize but to reload the unit rapidly from full pressure.

Considerably more experience has been accumulated by European utilities where it is common practice to design all units for trip back to house load. Most of these units use turbine bypass systems, variable pressure operation, and other design features to facilitate cycling operation and at the same time full load rejection and rapid reload. Typically, once thru units so designed have resynchronized and reloaded to 30% load within 9 minutes following full load rejection as reported in Reference 5. A good description of the European turbine bypass system and other design practices are also covered in this Reference.

Successful load rejection tests up to full load were recently conducted on a large nuclear unit on the Northeast Utilities system. This 650 MWe BWR unit entered full power operation in March 1971 and is equipped with a 105% capacity turbine bypass system. In the test depicted in Fig. 8., the main generator circuit breaker was opened at full load, and the unit remained connected to house load (approximately 4-1/2% of full load). The unit was resynchronized within 5 minutes and then reloaded. The precise coördination of the turbine bypass valve opening with the rapid closing of control and intercept valves is evident from the cam position traces in Fig. 8. Note that turbine overspeed was limited to less than 8% of rated speed while reactor pressure rise was less than 1% of rated pressure. Main steam flow experienced only minor fluctuations and reactor water level held very steady. Reactor power was initially reduced 15% with select rod insert and then further reduced to 50% by manually decreasing recirculation flow. Turbine control and intercept valves reopened after about 5-1/2 secs. as turbine speed decreased to 105%. The operator manually returned the unit to rated speed (mechanical hydraulic control system) and resynchronized within 5 minutes. Full load was attained within approximately 50 minutes although in an emergency it could have been accomplished within 15 minutes.

CONCLUSIONS

As previously discussed, the physical characteristics of presently available fault clearing equipment, which allow fault durations of from 8 to 11.5 cycles for delayed breaker clearing, appear to impose too severe a restriction on system design.

Additional automatic equipment is desirable to mitigate the effects of these long duration faults. The application of braking resistor systems, short-circuit limiting couplers or rapid turbine driving power reduction systems or combinations of these systems hold the promise of economically feasible solutions to maintaining first swing transient stability.

The application of fast turbine driving power reduction by momentary closure of turbine intercept valves will produce substantial improvements in first swing stability. As measured by allowable increases in critical fault clearing times, typical gains are 1 to 2 cycles for fossil units and 2 to 3 cycles for nuclear units. The fast closure of intercept valves will have very little effect on the steam generator, its controls or auxiliaries if the reheater flow interruption time is limited to 10 seconds for either fossil or nuclear units.

In the event a particularly severe system disturbance causes a unit to become completely separated from the system, it can be designed, at extra cost, to withstand a full load rejection and continue to operate at auxiliary load. With this approach the unit does not go through its normal shutdown and startup cycles and full power may be available within minutes rather than hours normally required for the conventional unit trip.

The feasibility and economics of providing this feature will have to be considered on an individual unit basis. Consideration must be given to the effect of these major load changes on equipment life, control of auxiliary equipment such as fans and feedwater pumps and the effects of transients in the steam and feedwater systems on other equipment such as extraction point auxiliaries. In many cases, however, only minor changes in control circuitry are required to implement this very desirable function.

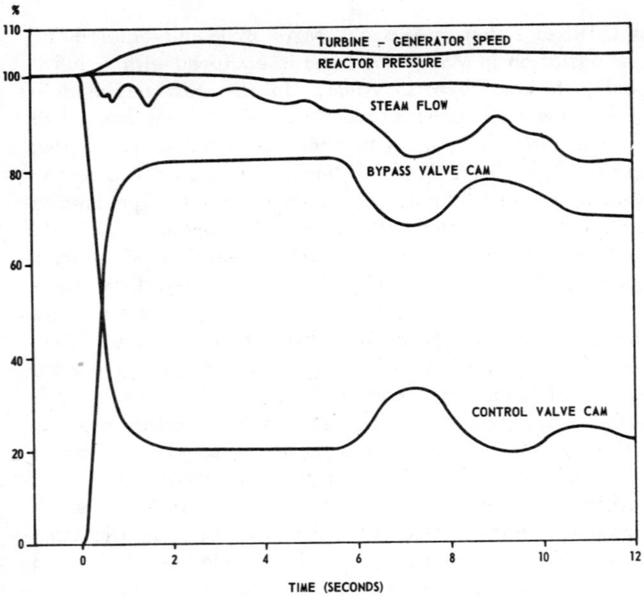

Fig. 8 Tripback to Auxiliary Load - Nuclear Unit

ACKNOWLEDGEMENTS

The authors wish to express their appreciation to the many engineers in the Allis Chalmers Power Systems, Babcock and Wilcox, Bailey Meter, Brown Boveri, Combustion Engineering, Foster Wheeler, General Electric, Hagan Controls, Riley Stoker, and Westinghouse Electric companies, and in the sponsoring utility industry in New England, whose information and advice have been so helpful in making this paper possible.

REFERENCES

1. "Turbine Energy Controls Aid in Power System Performance," F. B. Demello, D. N. Ewart, M. Temoshok and M. A. Eggenberger presented at American Power Conference April 16, 1966.

2. "Improved Reliability of Bulk Power Supply by Fast Load Control," R. H. Park, presented at American Power Conference April 24, 1968.

3. "Four Corners Project Stability Studies," R. G. Farmer, M. H. Kent, R. H. Hartley and L. M. Wheeler, IEEE Paper No. CP-708-PWR presented at ASME/IEEE Joint Power Generation Conference San Francisco, California September 15-19, 1968.

4. "The Decision and Use of Breaking Resistors," R. H. Park, IEEE Paper No. 69C-12 published in proceedings IEEE Resources Roundup April 16-18, 1969 Phoenix, Arizona.

5. "Safe Cycling of High Pressure Steam Turbines," Werner Trassl, presented at American Power Conference, April 22, 1969.

6. "Operation and Control of Once-Through Boilers During Electric Power System Emergencies," Oliver W. Durrant, R. P. Siegfried, presented at IEEE Section meeting Dallas, Texas October 21, 1969 and at ISA Power Instrumentation Symposium, Kansas City, Mo. May 20, 1970.

7. "The Two-Phase Accelerometer," J. Law, Jr., D. W. Movotney, IEEE P.A. and S. Transactions Paper 63-1009.

8. "Effects of Excitation, Turbine Energy Control and Transmission on Transient Stability," P. G. Brown, E. H. Lenfest, F. P. Demello and R. J. Mills, IEEE Transactions Paper 70TP203-PWR.

9. "Rapid Response and Maneuverability are Obtainable from Supercritical Plants," F. H. Fenton, J. V. Pigford, presented at ISA Power Instrumentation Symposium, Kansas City, Mo., May 20, 1970.

10. "Effects of Future Turbine-Generator Characteristics on Transient Stability," H. E. Lokay and P. O. Thoits, IEEE Paper 71CP75-PWR.

11. "Boiler-Turbine Coordination During Startup and Loading of Large Units," F. J. Hanzalek, P. G. Ipsen, presented at the Winter Annual Meeting ASME November 7-11, 1965.

12. "Fast Valve Control Can Improve Turbine-Generator Response to Transient Disturbances," O. J. Aanstad, H. E. Lokay - Westinghouse Engineer July 1970 issue.

13. "Fast Intercept Valving Aids Units Stability," J. Vandergrift, J. R. Woodall, Electrical World, July 13, 1970.

14. "Load Rejection Testing of Large Thermal-Electric Generating Units," R. H. Hillery and E. D. Holdup, IEEE Transactions Paper 31PP67-25, June 1968 P.A. and S.

15. "Keep Generators Running - Improve Reliability at Little Cost," C. F. Paulus, IEEE Paper 70CP219-PWR.

16. "Need for a Strategy for Boiler-Turbine-Generator Operation and Control," O. W. Durrant and H. D. Vollmer, IEEE Paper 71CP-244PWR.

17. "Design of Turbines for Frequent Starting" P. G. Ipsen and D. P. Timo, presented at American Power Conference, April 22, 1969.

MODERN STABILITY AIDS FOR CALVERT CLIFFS UNITS

W. A. Morgan* Fellow IEEE | H. B. Peck** Senior Member IEEE | D. R. Holland** Senior Member IEEE | F. A. Cullen** Member IEEE | J. B. Ruzek* Member IEEE

ABSTRACT

Stability is difficult to maintain under the three-phase-fault-breaker-failure criterion with a plant of two nominal 800 MW units connected radially by two 500 kV lines to a 230 kV load network. This paper describes the application of dynamic stability aids such as: fast-valving, 500 kV breakers with independently operated poles, fast-excitation, and alternate 500 kV line arrangements, and investigates machine internal losses during the fault, alternative fault types and locations, and the setting of the load unbalance relays. The costs of the alternative stability aids are compared and a choice of load unbalance relay settings for initiating fast-valving is made to obtain stability at a minimum cost with fast-valving and 500 kV breakers having independently operated poles.

INTRODUCTION

The planning criteria of many large power pools require that stability be maintained for the condition of delayed clearing of a three-phase fault due to failure of a circuit breaker to operate properly. This is called the breaker failure ("stuck" breaker) criterion.

Preliminary stability studies by the Baltimore Gas and Electric Company using the classical stability method showed that instability would result for the two Calvert Cliffs units with the three-phase-fault-breaker-failure criterion. Detailed representation stability studies were then conducted at Schenectady, N.Y., using the General Electric Co. E-ST6 computer program. Stability was achieved for the same criterion by utilizing some of the newly available stability aids. In addition, investigation of alternative system arrangements and assumptions for making the studies, and comparison of costs for each alternative were made. Costs are for equipment only and exclude design, field installation and overhead costs.

The Calvert Cliffs plant will have two nuclear steam turbine generator units, each having a maximum capability of about 900 MW, and scheduled to be in service by 1973 and 1974, respectively. Unit 1 is being manufactured by General Electric Company, while Unit 2 is a Westinghouse machine. The steam supply systems for both units are being built by Combustion Engineering Company.

The Calvert Cliffs plant will be connected initially only to Waugh Chapel Substation, the southern juncture of the Baltimore 230 kV load network by two 500 kV circuits each about 47 miles long. The northern terminal of the Baltimore 230 kV loop connects to the PJM 500 kV network at Conastone substation. Thus, the Calvert Cliffs plant will be connected radially, and will have the 230 kV load-serving system and transformers interposed between it and the relatively low impedance 500 kV PJM network. (Figure A).

One way of meeting the breaker failure criterion is to trip one generating unit. However, it was decided to avoid this expediency if possible. Therefore, the purpose of the studies was to determine the measures necessary to maintain stability without tripping a Calvert Cliffs unit.

EXPLANATION OF THE STABILITY PROBLEM

When high voltage system faults are cleared in a delayed time of 7.5 to 12 cycles by backup breakers due to a stuck breaker, instability can result without stability aids. Figure 1 shows an example where the two Calvert Cliffs units would lose synchronism with the system for a close-in three-phase fault cleared after 7.5 cycles, (.125 seconds). This case is unstable. The Calvert Cliffs units had reached 170° with respect to the reference machine when the computer program stopped. Unit 2 pulled away from the system before Unit 1. The machines had lost synchronism with the rest of the interconnected system.

Figure 2 shows a plot of some important generator parameters for Unit 1 for the unstable case shown in Figure 1. The plot of field voltage, E_{FD}, corresponds to a response ratio of 1.0. The mechanical power input remains nearly constant, reflecting the slow action of normal governor valving control. The generator output to the system following clearing of the fault declines from near rating to zero at 0.5 seconds as the unit approaches a 180 electrical degree difference from the rest of the system.

While the fault is on, there is a large unbalance between the mechanical power input and the electrical power output of the generator. This causes the generator to speed up with respect to the rest of the system, which is reflected in the increase in angle (Figure 1). The duration of the fault and power unbalance can be long enough to create an acceleration which cannot be overcome when the fault is removed unless special stability aids are used.

FAST CLOSING OF INTERCEPT VALVE

The effectiveness of fast valving (also called early valving) is illustrated in Figure 3 for a three-phase fault on the Calvert Cliffs 500 kV bus cleared in 7.5 cycles. Compared with Figure 1 which was unstable for the same fault assumption, Figure 3 shows that stability can be maintained if fast valving is utilized. The maximum angular swings for Units 1 and 2 are 111 degrees and 114 degrees, respectively, relative to the reference plant.

Figure 4 shows a plot of the computer output results for Unit 1 for the same case illustrated by swing curves in Figure 3. The mechanical power input starts to reduce at 0.12 seconds and reaches about one-half the generator rating at about 0.7 seconds. By reducing the accelerating power input shortly after the fault has occurred, the angular advance of the generator is decreased. With a smaller angular difference with respect to the system the unit is able to release to the power system the energy stored in it during the fault period when the low terminal voltage prevented this. The generator output reaches about 1.35 per unit of rating at about 0.95 seconds as the unit starts to pull back in angular displacement toward the system.

Figure 5 shows the turbine power response to fast valving characteristic curve provided by the manufacturer of Unit 1. In this case the low pressure turbine corresponded to 70 percent of the total turbine power, and closing the intercept valve (which supplies the L.P. turbine) thus left 30 percent of

*Bechtel Corporation, San Francisco, Calif. **Baltimore Gas and Electric Company, Baltimore, Md.

Paper 70 TP 147-PWR, recommended and approved by the Power System Engineering Committee of the IEEE Power Group for presentation at the IEEE Winter Power Meeting, New York, N. Y., January 25–30, 1970. Manuscript submitted September 8, 1969; made available for printing December 4, 1969.

Reprinted from *IEEE Trans. Power App. Syst.*, Feb. 1971.

the machine rating supplied by the high pressure turbine. The curve provided by the manufacturer of Unit 2 was similar except the corresponding minimum power output is 35 percent and the intercept valve would be opened at a slower rate.

Figure 5 indicates time delays T_1 and T_2 before valve closure starts. T_2 is fixed by the time for the solenoid to act which was agreed by the manufacturers to be 0.10 second. T_1 depends upon the time for the power unbalance relay to act, estimated by the manufacturers to range beteeen 1 and 3 cycles. The studies showed that changing the total time delay $(T_1 + T_2)$ from 0.12 to 0.15 second increased the angular excursions of Unit 1 by 10 degrees and Unit 2 by 8 degrees with all other conditions being the same. The machine's electro-hydraulic governor systems inherently provide fast valving and cost of adding early initiation thereof was less than $10,000.

500 KV BREAKERS WITH INDEPENDENTLY OPERATED POLES

A three-phase fault cleared in 9 cycles would cause instability, even with both fast valving and fast excitation (to be described later), unless a unit were tripped or a third 500 kV line from Calvert Cliffs were added to the system. Although fast valving maintained stability for three-phase faults cleared in 7.5 cycles, some additional stability aid is required in order to provide some margin beyond 7.5 cycles which is the minimum total back-up time required for the relays, timer and breakers to operate.

One practical way of reducing the severity of the stability problem is to use independent pole tripping with duplicate auxiliary switches, pressure switches, and trip coils on each pole to effectively create three independent single-phase breakers. With such an arrangement, the protective relaying will operate all poles for any type of fault. However, with a contingency philosophy of only one piece of equipment failing to operate to clear a fault, a three-phase fault will be downgraded to a SLG fault within 3 cycles assuming one pole failed to interrupt. This SLG fault will then be cleared in the normal single phase to ground breaker failure overall time of 12 cycles.

The studies showed that stability could be maintained for a three-phase fault changing to SLG at three cycles, with the SLG fault persisting for the remainder of the 12 cycles. For this case, the fast valving assumed a total time delay of 0.15 second, the conventional excitation system assumed a response ratio of 1.0 and the machine internal fault loss was 15 percent, for both units. The maximum swings for Units 1 and 2 were 110 degrees and 117 degrees, respectively. These maximum angular swings are comparable to those for Figure 3, which was for a three-phase fault cleared in 7.5 cycles with only fast valving. So, the 500 kV breakers with independently operated poles may be said to extend the capability for stability of the initial two 500 kV line system from 7.5 cycle clearing, to 12 cycle clearing for three-phase faults, both with fast valving.

The independent pole tripping scheme was conceived by the Conemaugh Relay and Communication Task Force for the Conemaugh mine-mouth plant project in western Pennsylvania, of which the Baltimore Gas and Electric Company is a participant. There will be little or no extra cost as development costs were absorbed on the Conemaugh application.

EXCITATION SYSTEMS

Figure 4 shows the field voltage and terminal voltage response for Unit 1 for a conventional excitation system for response ratio of 1.0. The terminal voltage has a low range of between 50 and 60 percent for about 0.4 second. Unit 2 was similar. These correspond to General Electric Alterrex and Westinghouse Brushless. For the size of the Calvert Cliffs units, the additional costs for higher response ratios than 0.5 as listed in both manufacturer's catalogs are as follows:

1.0	–	$ 40,400
1.5	–	85,200
2.0	–	134,000

There is a new "fast" excitation system available that goes to ceiling voltage with almost no delay. Figure 6 shows the field voltage and terminal voltage characteristic for the same fault conditions as Figure 4 except for the excitation systems. The field voltage for both Units 1 and 2 go to "ceiling" in a delay of less than 0.05 second. The response ratios are 2.5 for both. Obviously, the present definition of response ratio is not adequate for this new system.[1]

The improvement in stability of this new case with fast excitation (Figure 6) compared with conventional excitation (Figure 4) amounted to a reduction in the angular swings of Unit 1 of 21 degrees and Unit 2 of 16 degrees.

With fast excitation, both units reach 100 percent terminal voltage in 0.7 second, instead of 1.0 second as in Figure 4. And, compared with Figure 4, Unit 1 terminal voltage has a low range of between 70 and 80 percent for about 0.3 second. So, the voltage performance is much better.

General Electric (Unit 1) calls their fast excitation system "Althyrex" (formerly Alterrex SiCR), while Westinghouse calls their system "Low Time-Constant, Brushless."

A stabilizing signal responsive to some function of speed must be applied to these fast excitation systems. Otherwise, negative damping has been found to occur when stability studies have been extended beyond one second into the dynamic stability range.

Stability was not achieved with fast excitation as an alternative to fast valving. Although fast excitation gave better voltage performance and an extra margin of stability that may be required in some applications, in this instance it could not be recommended as being required in order to maintain stability as an alternative to the other stability aids. The principal factor ruling against its application was the additional cost of about $296,000 for Unit 1 and $330,000 for Unit 2 for relatively small improvement.

500 KV TRANSMISSION SYSTEM ARRANGEMENT

As indicated previously, if a third 500 kV transmission line from Calvert Cliffs Plant were connected to Chalk Point, no other stability aids would be required. However, this would be the most expensive way of achieving stability because the 16 mile line would cost several million dollars.

LOAD UNBALANCE RELAY SETTING

During the course of the studies, it was found that a SLG fault cleared in 12 cycles, required fast valving to maintain stability with excitation response ratios of 0.5 each, or, without fast valving, the required response ratios were 1.0 and 2.0, for Units 1 and 2, respectively. Also, it was found that a three-phase fault converted to SLG at 3 cycles and finally cleared in 12 cycles, with breakers having independently operated poles, and fast valving, required a response ratio of 1.0 for both units.

Therefore, the question arose: Should fast valving be applied for the more frequent SLG faults, or should it be applied only for three-phase faults that may occur, say, once in twenty years. The additional cost if fast valving were applied only to three phase faults, would be about $93,600, the difference between response ratios of 1.0 and 2.0 for Unit 2.

The choice is made by the setting of the load unbalance relay which actuates the fast valving operation. This relay uses transducers to compare the inlet steam pressure (power input) with the generator three-phase power output. The setting of this relay is adjustable from 25 to 100%. In the Calvert Cliffs studies, a three-phase fault on the 500 kV system caused a load unbalance of about 95 percent, whereas a SLG fault caused an unbalance of about 34 percent.

The setting of the load-unbalance relay may be changed, or the fast valving discontinued, as future changes in the 500 kV system arrangement occur.

The frequency of SLG faults due to lightning was estimated[2] for the two 500 kV Calvert Cliffs-Waugh Chapel lines to be one outage in one to five years, depending upon the tower footing resistance and other design parameters of the line. With this low frequency, it was decided to use a load unbalance relay setting of 25 percent which would allow fast valving for SLG, as well as three-phase, faults. This resulted in a savings of $93,600 in keeping Unit 2 response ratio at 1.0, instead of 2.0 that would be required otherwise.

CHARACTERISTICS OF THE GENERATOR UNITS

The two Calvert Cliffs units are both nuclear units of the pressurized water reactor type, and are tandem compound, 1800 rpm, with brushless excitation systems. The pertinent stability data for comparison purposes are as follows:

	Unit 1	Unit 2
MVA	1,020	1,003
P.F.	0.90	0.90
Inertia Constant, H.	4.40	3.35
Transient Reactance, p.u.	0.0321	0.0419
Short Circuit Ratio	0.64	0.64

The transient reactance values are on a 100 MVA base, and the inertia constants are stated in MW-Sec/MVA. The transient reactance corresponds to 95% of the rated current values to consider saturation effects.

The amount of machine internal loss for a three-phase fault was found to be different for the two machines. The calculated values for alternative methods of calculation and fault location are derived in the Appendix. The effect of changing this value was tested during the studies, but for final design purposes a value of 15 percent of initial machine loading was used for both units.

While Unit 2 would be, by observation, less stable than Unit 1 considering the above inertia and reactance values, it was found to be faster to respond to fast valving. Figures 7a and 7b show a comparison of without and with fast valving for the same fault condition, respectively. With fast valving the second swing of Unit 2 is less than Unit 1 at about 1.6 seconds when the intercept valve is still effectively closed, and Unit 2 has a lower range of oscillations in subsequent swings. This reflects the lower inertia of Unit 2 and its greater susceptibility to stability aids. Thus, there is hope that new stability aids will be effective in the future even though the machines themselves are forced into lower inertia and higher reactance characteristics (lower short circuit ratios) as larger sizes are designed.[3]

CONCLUSIONS

Following are the salient conclusions of the dynamic stability study:

1. Fast or early valving of the intercept valve makes it possible to maintain stability with only the two 500 kV Calvert Cliffs-Waugh Chapel lines with conventional excitation systems with response ratios of 1.0 for a SLG fault cleared in 12 cycles, or a three-phase fault cleared in 7.5 cycles. There is little additional cost.

2. 500 kV circuit breakers having independently operated poles and appropriate relaying connections, together with fast valving, make it possible to maintain stability for three phase faults cleared in a delayed time of 12 cycles. It is assumed that the three-phase fault is changed to a SLG fault at the end of 3 cycles insofar as the system is concerned on the basis that only one of the poles would have failed and the other two would have operated. There is little or no additional cost.

3. The load-unbalance relays for initiating fast valving may be set over a range of 25 to 100 percent. The setting may be changed or the load-unbalance relay disconnected at any time. It was decided to initially set the load-unbalance relays for 25 percent to allow the application of excitation response ratio of 1.0 for both units which will maintain stability for both SLG and three-phase faults with delayed clearing.

4. Conventional excitation systems (General Electric Alterrex; Westinghouse-Brushless) with a response ratio of 1.0 will be adequate for stability in conjunction with fast valving and 500 kV breakers with independently operated poles. Fast excitation (General Electric - Althyrex; Westinghouse - Low Time Constant, Brushless) improved stability and voltage performance during the transient period. However, this stability aid did not replace other alternatives and cost more and so was not recommended.

5. A third 500 kV transmission line from Calvert Cliffs would make the other stability aids unnecessary. However, the high cost of this alternative could not be justified initially for the purpose of maintaining stability.

6. This study indicates that system stability may be maintained in the future if modern aids are applied even though there may be a trend toward less stable machine characteristics as generator sizes increase.

APPENDIX

Machine Internal Fault Loss

During close-in three-phase faults the losses in the generator due to "trapped" flux can be appreciable. In the classical method of transient stability studies, the effect of machine internal loss is ignored. Because these losses consume power during the period when the fault is on, the angular swing of the generator is reduced over what it would be if machine internal losses were ignored.

These losses are due to both AC and DC current components flowing through positive and negative sequence machine resistances. The amount of DC loss depends upon the offset current when the fault is applied. A three-phase fault has a fixed amount of DC losses since the sum of the flux in the three-phases is a constant for balanced conditions. For a two-phase or one-phase-to-ground fault, the DC losses depend upon the value of the flux in the faulted phase or phases when the fault is applied. The most severe test from a stability standpoint for a one-phase-to-ground fault condition is to assume zero machine internal loss in the faulted phase.

For three phase faults directly on the generator terminals, the computer program manual suggests setting the electrical power output to 0.15 times Initial Value of Mechanical Power Input to provide for the I^2R losses in the rotor. It was suggested by General Electric that this rule be applied also for faults on the Calvert Cliffs 500 kV bus, but not for faults further removed from the generators.

Upon investigating the background for the above general assumptions it was found that there was no positive agreement as to the value that should be used to represent machine internal losses in a magnitude versus location form. Accordingly, several values were tried to test the sensitivity of stability to this assumption. An increase of 4 percent with all other conditions unchanged caused a decrease in angular swing of both units of 10 degrees.

In checking into the theory and calculation methods for estimating the machine internal losses, one reference[4] shows that the loss varies as the 1.8 power of the current while the General Electric Company recently estimated that the power should be at least 2.0. In using the prescribed calculation procedures, it was found that not only is there a large difference depending upon the individual machine characteristics, but that it would be appropriate to also consider machine internal losses for faults on the 500 kV receiving station at Waugh Chapel, 47 miles away, as well as on the Calvert Cliffs 500 kV bus.

The following tabulation shows the values that were calculated for the two generators for two fault locations using the assumptions of loss varying as the 2.0 and 1.8 power. These values are average for three cycles.

Calculated Total Losses to Point of Fault
(Percent of Full Load Power)

	Unit 1	Unit 2
Fault at 500 kV Calv. Cliffs Bus		
Loss varies as 2.0 power	21.5	26.3
Loss varies as 1.8 power	14.2	16.5
Fault at 500 kV Waugh Chapel Bus		
Loss varies as 2.0 power	13.9	17.1
Loss varies as 1.8 power	10.5	12.0

Machine internal fault losses should be applied for three-phase faults, only. For single-line-to-ground (SLG) faults, it is conceivable that the faulted phase might not be the one that would have the trapped armature flux as explained previously. However, a three-phase fault would always have this loss because the trapped flux is bound to be included in one or two of the three phases involved. Considering that less than two percent of the faults would be three phase but that their effect on stability is greatest, it appeared reasonable to use values of up to 15 percent in the studies for internal machine losses. In addition, there was a loss of about 4 percent for resistance losses to the point of the fault for the generators' isolated phase bus runs, the unit transformers, and the 500 kV connections to the switchyard bus.

REFERENCES

1. C. Concordia and M. Temoshok, "Generator Excitation Systems and Power System Performance," 67CP536, presented at the 1967 IEEE Summer Power Meeting, Portland, Oregon, July 9-14, 1967.

2. J. M. Clayton and F. S. Young, "Estimating Lightning Performance of Transmission Lines," IEEE Transactions on Power Apparatus and Systems, 1964, pages 1102 to 1110.

3. L. A. Kilgore and H. E. Lokay, "Future Generator Design Features and System Requirements," 68CP707, presented at the 1968 ASME/IEEE Joing Power Generation Conference, San Francisco, Calif., Sept. 15-19, 1968.

4. *Electrical Transmission and Distribution Reference Book* page 176-7, Westinghouse Electric Corporation, East Pittsburgh, Penn., 1964.

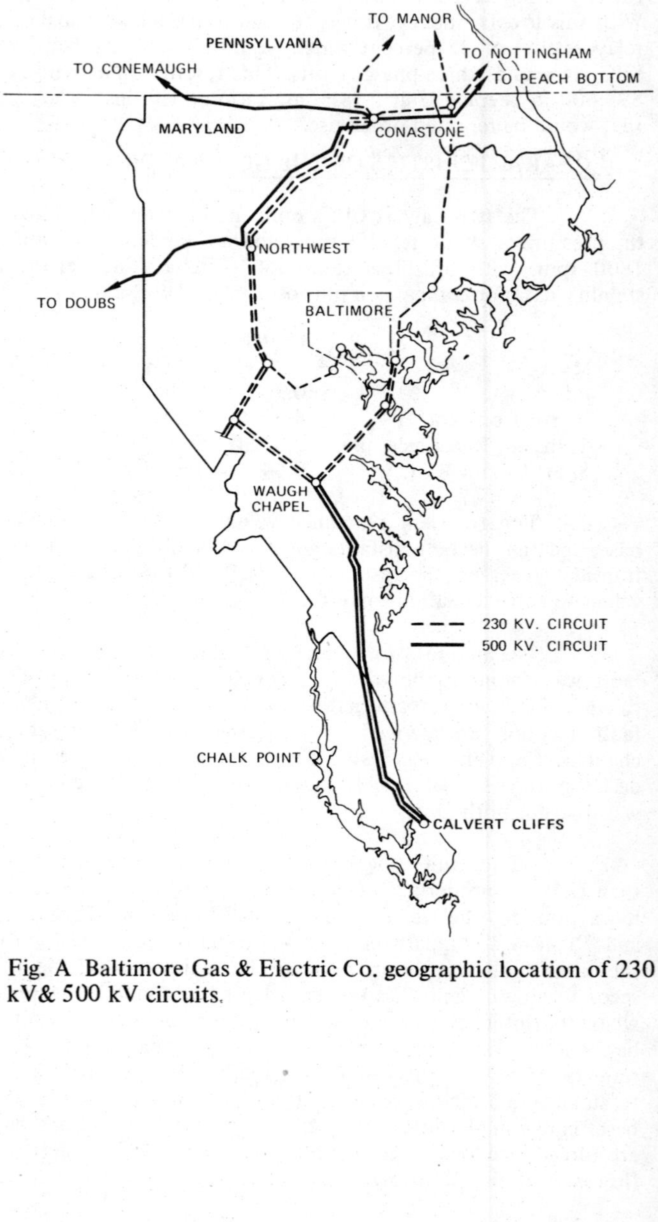

Fig. A Baltimore Gas & Electric Co. geographic location of 230 kV& 500 kV circuits.

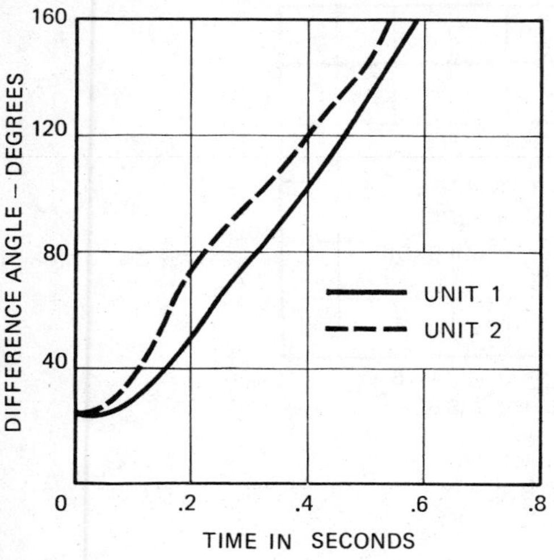

Fig. 1 Swing curves for three-phase fault at Calvert Cliffs on 500 kV line cleared in backup time of 7.5 cycles without stability aids.

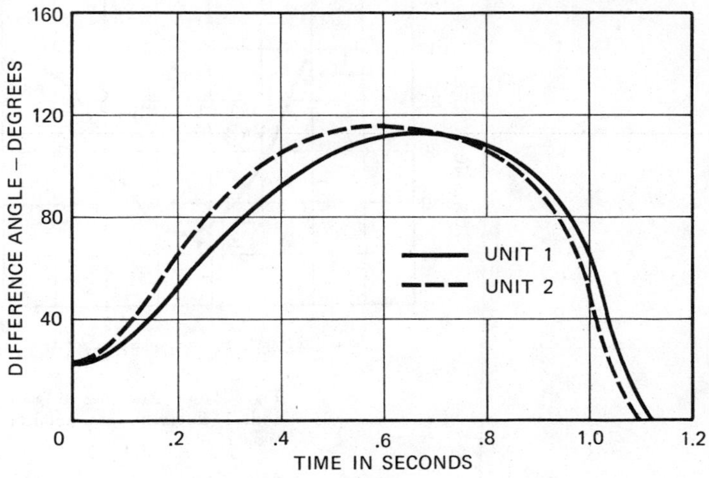

Fig. 3 Swing curves for same fault conditions as Fig. 1 except with aid of fast valving.

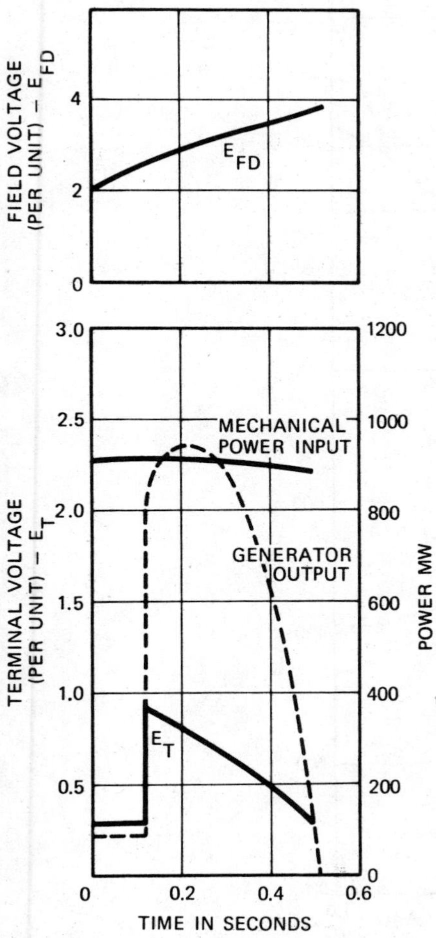

Fig. 2 Response of Unit 1 field voltage, mechanical power input, generator output and terminal voltage for same case as Fig. 1.

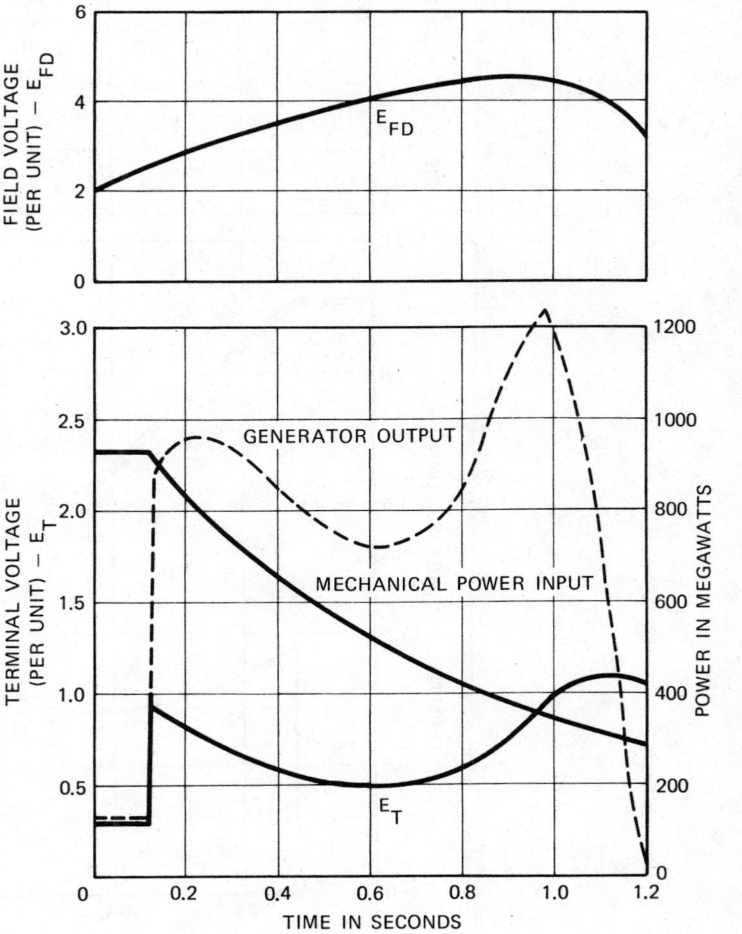

Fig. 4 Response of Unit 1 field voltage with conventional excitation response ratio of 1.0, mechanical power input, generator output and terminal voltage for same case as Fig. 3, showing effect of fast valving.

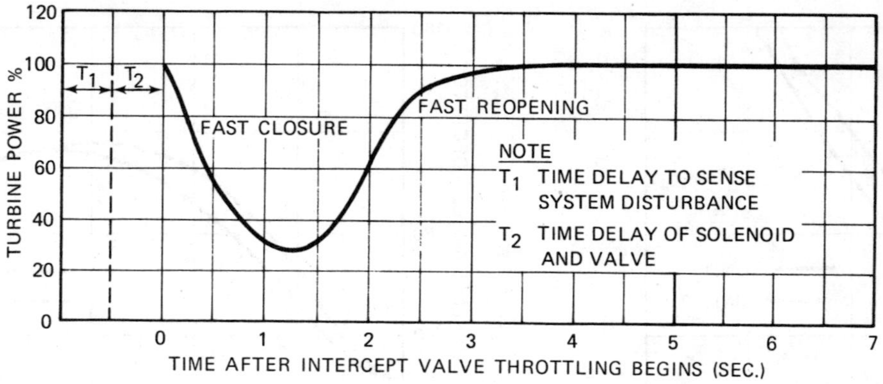

Fig. 5 Power Response to fast (early) valving for Calvert Cliffs Unit 1, intercept valve application, only. Curve for Unit 2 is similar.

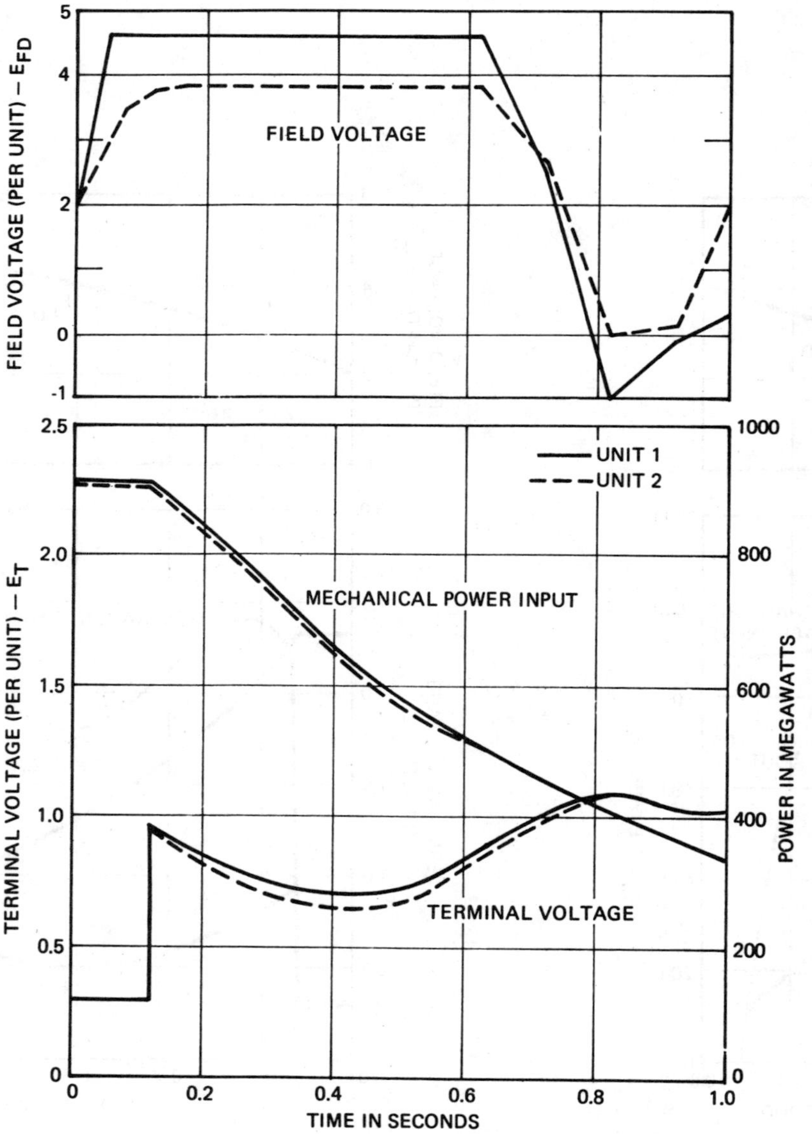

Fig. 6 Response of both Units 1 and 2 field and terminal voltage with "fast" excitation having a 2.5 response ratio, for same fault and fast valving assumptions as Fig. 4.

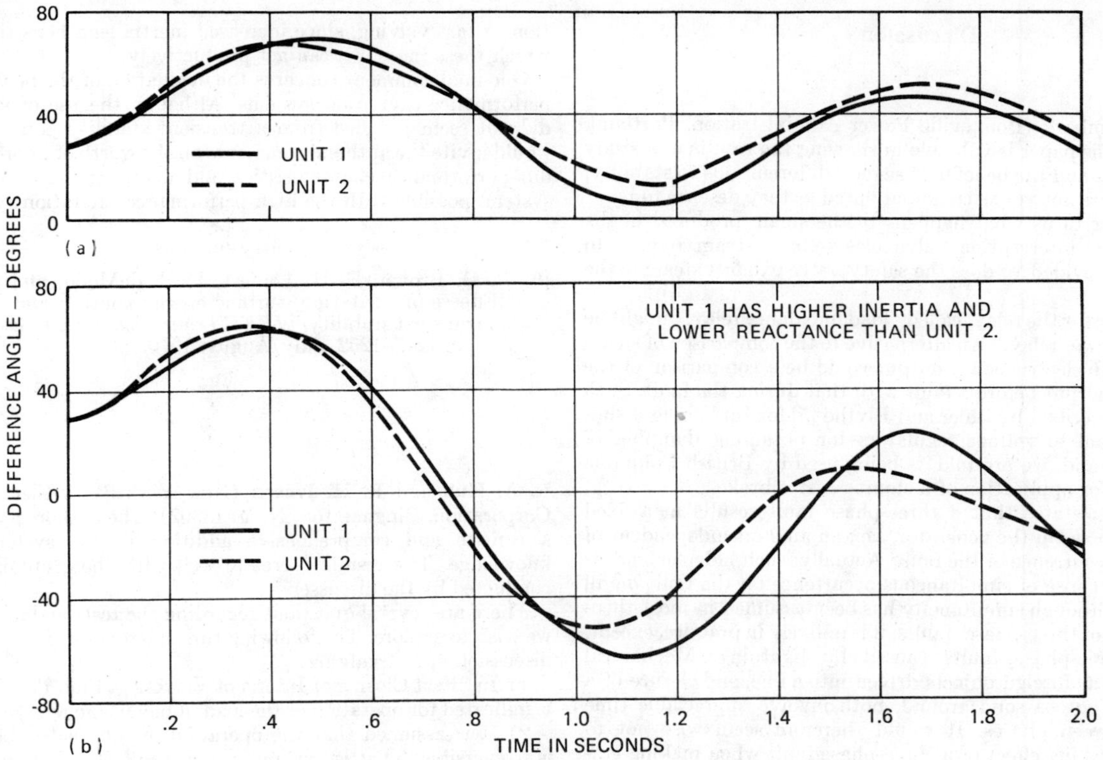

Fig. 7 (a) Swing curves for three-phase fault cleared in 4.5 cycles during light load without fast valving. (b) Same, except with fast valving for both units, showing how Unit 2 with lower inertia responds better to control.

Discussion

Edward W. Kimbark (Bonneville Power Administration, Portland, Ore. 97208): This paper is valuable in showing the results of a study of both the costs and the benefits of several different aids to stability, some of which are not yet as commonly used as they deserve to be.

It is not clear to us what happens to the steam produced in the boiler after the intercepting valve closes. Is a steam bypass to the condenser provided, or does the safety valve exhaust steam to the atmosphere?

A better name for the relay, called "load unbalance relay," might be "accelerating power relay." An alternative to the comparison of steam power input with electric power output could be a comparison of the electric power output before a fault with that during the fault. Such a device was described by Shier and Blythe [5] for furnishing a supplementary signal to voltage regulators for obtaining damping of angular swings and, we are told, is being used by British Columbia Hydro system for application of a shunt-resistor brake.

The appendix states that a three-phase fault results in a fixed amount of dc losses in the generator, i.e., an amount independent of the instant of occurrence of the fault. Actually, such independence is based on the premise of simultaneous occurrence on the fault on all three phases. Although simultaneity has been assumed in most theoretical analyses of three-phases faults, it is unlikely in practice, except, perhaps in three-phase faults caused by lightning. Mechanical accidents, such as foreign objects driven into a line, and closure of a circuit breaker onto a solid ground, both involve appreciable time differences between phases. It would therefore seem wise not to count on the braking effect of a three-phase fault when making stability studies.

References

[5] R. M. Shier and A. L. Blythe, "Field tests of dynamic stability using a stabilizing signal and computer program verification," *IEEE Trans. Power App. Syst.*, vol. PAS-87, pp. 315–322, February 1968.

Manuscript received January 21, 1970.

F. P. deMello (Power Technologies, Inc., Schenectady, N. Y. 12301): The authors are to be commended for a clear and interesting exposition of a real life study, evaluating alternate design measures necessary to assure, from a transient stability viewpoint, reliable supply from a large generating station.

The quantitative evaluations in terms of performance and costs are both informative and interesting. The significant point concerned the trade-off between a few thousand dollars in fast valving options versus deferring investment of millions of dollars in transmission. Even the more expensive high-performance excitation systems could be viewed as relative bargains in terms of the gains they provide. These results of a specific study are very much in line with general findings of a parametric study [6].

In regard to the discussion of the effects of generator characteristics, such as inertia, we offer the thought that lower inertia does not necessarily represent greater susceptibility to stability aids.

The behavior of the rotor angle excursion of the lighter unit 2 should be interpreted as being due not only to the restoring forces provided by the system and fast valving but also to the synchronizing energy provided by unit 1 in response to deviations in angle between two units. Hence, in order to make a judgment on susceptibility to stability aids, one would have to consider the two units in situations where they are independent of each other.

Our own studies have revealed that higher inertia is beneficial to stability. It produces compounding effects, in that the heavier the machine, the more effective can be other stability aids such as excitation or fast valving, since increased inertia lengthens the time over which these measures can act productively.

One final comment concerns the evaluation of the benefits of high-performance excitation systems. Although the use of such systems did not seem justified from a transient stability point of view, we would invite the authors' comments on the worth of improved voltage dip performance characteristics and damping performance of the system, possible with the high-performance excitation system.

References

[6] P. G. Brown, E. H. Lenfest, F. P. deMello, and R. J. Mills, "Effects of excitation, turbine energy control, and transmission on transient stability," *IEEE Trans. Power App. Syst.*, vol. PAS-89, pp. 1247–1252, July/August 1970.

Manuscript received February 2, 1970.

R. A. Hull and **R. H. Iveson** (New York State Electric and Gas Corporation, Binghamton, N. Y. 13902): The authors have prepared a concise and comprehensive addition to the system planner's knowledge. The results correlate well with those obtained in tests conducted by the discussers.

There are several questions regarding the tests and analysis which we wish to explore. The following three factors form the basis for our discussion of fast valving.

1) In "Fast Closing of Intercept Valves," a time T_1 of 1 to 3 cycles is indicated for operation of the load unbalance relay.

2) It is assumed that the operation of the load unbalance relay is irreversible. That is, once initiated, the fast-valving cycle will be completed.

3) Normally cleared three-phase faults on the 500-kV and either normally or delayed cleared faults on the adjacent 230-kV system north of Waugh Chapel may cause reduction in unit output greater than 25 percent. The discussers have observed transient tests of three-phase faults in which reductions of machine output of 25 percent were seen for machines remote from the fault.

With this background, the following questions are posed.

1) Would not the intercept valve operation be initiated for: a) normally cleared three-phase faults on the 500-kV lines; b) normally and/or delayed cleared three-phase faults on the adjacent 230-kV system?

2) If application of the load unbalance relay should become widespread, is it possible that the load unbalance relays of several large machines could be actuated by a single heavy three-phase fault, thus causing cycling of the several intercept valves and accentuating the severity of the system disturbance by seriously curtailing the power capability in the area? For example, if load unbalance relays had been applied to other nuclear units in the adjacent area, did the studies show whether the relays would have operated?

An alternative arrangement is to use only the breaker failure protective scheme on the generator breaker(s) to initiate fast valving on that particular generator. This would require an additional time delay of the breaker failure timers before fast valving could be initiated, but in many cases this would still be a means of retaining unit stability.

A less selective but faster scheme would initiate fast valving only when a trip signal appears on the generator breaker(s) or the breakers adjacent to the generator breakers. This scheme would not be nearly as secure as the former scheme, but at least it limits fast-valving operations to those faults on lines adjacent to the station.

It is the discussers' impression that fast valving may impose steam overpressures as well as steam bypass and suppression pool capability requirements which, if not accommodated, can result in a reactor scram. Would the authors comment on the overpressure and bypass capabilities considered in their analysis?

In the appendix, calculated values for machine internal losses are given based on an average for 3 cycles. Since in the delayed cleared faults, times from 7.5 to 12.0 cycles were examined, did the authors calculate the average losses for these more lengthy faults and, if so, what were their results?

Manuscript received February 6, 1970.

W. R. Brownlee (Southern Services, Inc., Birmingham, Ala. 35202): This paper shows a careful approach to the problems of designing adequate stability into a new power station, based on severe conditions such as an inoperative circuit breaker. In the judgment of some the probability of a three-phase fault adjacent to a power station, combined with the failure of a breaker to clear a short circuit, is so small that it is not imperative to guard against it. However, there is growing recognition among reliability groups in the United States that this particular situation, involving, as it does, merely one failure of equipment, must be safeguarded against because of the severity of the result if a significant plant becomes unstable with the rest of the system.

Studies made on portions of the Southern Company system indicate that the authors are correct in their conclusion that an excitation response ratio of about 1.0 (compared with the more conventional 0.5) is about as much of an increase as can be justified in terms of this stability problem. Also, our studies indicate that intercept-valve closing offers considerable promise for such applications.

The authors describe independent pole tripping of circuit breakers such that a portion of the fault time during the failure of a breaker to operate can be limited to a single phase-to-ground fault, thus reducing stability requirements. Independent pole tripping would be a consummation greatly to be desired, but it is very doubtful that full assurance is obtainable on its achievement. That is, other contingencies not presumed in the manufacturer's analysis can account for the failure of any pole of a breaker to clear.

The authors make several references to a total clearing time of 7.5 cycles. Relay coordination tests made on the Southern Company system indicate that with solid-state relaying and two-cycle circuit breakers, the shortest assured exposure to a three-phase fault is about 9 cycles. This means that some other expedients may be required if fast valving alone will not permit a total time of 9 cycles.

The authors are to be commended for their courage in applying fast intercept valving to nuclear units. The application is no doubt made less complex because both the Westinghouse and the General Electric units are connected to pressurized water steam systems of Combustion Engineering, Inc. In contrast, a full steam bypass for a boiling water unit has been quoted at over $1.5 million. When full consideration is given to the multitude of circumstances and possible conditions which will cause a nuclear unit to be shut down, the odds are at least ten to one that the unit will be shut down on the first needed actuation of its intercept-valve closing scheme. A series of full-scale tests for debugging purposes, when the unit is ready for commercial operation, could reduce this probability.

Manuscript received February 9, 1970.

M. L. Crenshaw and **C. C. Young** (General Electric Company, Schenectady, N. Y. 12305): The use of a three-phase fault breaker-failure criterion for the planning of a modern power system presents the system planner with an interesting and challenging problem. The results which we have observed in several different studies indicate that the critical switching time is often in the range of 8 to 10 cycles, and may be as low as 6 to 8 cycles when the machine inertia is low. Because of the difficulty of achieving these switching times under stuck-breaker conditions, special techniques may have to be used to obtain stability. The authors' discussion of their experience in evaluating several of these techniques for the Calvert Cliffs units should be very interesting to any one who is involved in solving this type of problem.

We would like to offer specific comments on some of the results and the discussion presented in this paper.

1) The inertia of a unit will affect both the critical switching time and the damping of the unit. A higher inertia will increase the critical switching time, but a lower inertia will improve the damping of the unit as well as the response to changes in mechanical torque. This can be seen in Figs. 1, 3, and 7.

2) The internal losses in a machine during a fault have an important effect upon the critical switching time, and these losses should be represented for both classical and detailed representations. The magnitude of these losses for the simultaneous application of a three-phase fault to all phases can be calculated using a formula. The important consideration is not whether a power of 1.8 or 2.0 should be used in computing the losses, but whether or not to assume a simultaneous fault because this assumption has an important effect upon the critical switching time. A simultaneous fault will give the highest losses, therefore, the highest critical switching time. This has been observed on staged tests where the system has been more stable than it has been in actual operation.

It is entirely possible for a fault to progress rapidly from a single-phase fault to a three-phase fault in such a way that little or no dc offset current is produced in the generator stator windings. Since the dc offset contributes a significant amount of the fault torque, the average fault torque is quite sensitive to this effect. In one case, we found that the average fault torque would be reduced by about two thirds if the dc offset were negligible.

We believe that it is more realistic to assume that a three-phase fault progresses from a single-phase fault. Therefore, we suggest that the fault torque due to the dc component be reduced from that computed by the usual formulas, and that this component be arbitrarily reduced by about one half.

3) The behavior of the excitation system for unit 2 as shown in Fig. 6 of the preprint is not correct. The terminal voltage is below the desired value for 0.8 s, yet the field voltage is on a minimum limit. Since the symptoms look like those of numerical instability, using the input data, we computed the effective time constants of the excitation system. One was approximately 0.003 s. Since the time step used was 0.0125 s, the solution was numerically unstable.

The exciter was a type 1 exciter with both limits and a nonlinear saturation function represented. When the data provided for this exciter is used in accordance with the standard IEEE representation, [7], a time step of less than 0.003 s is required. Depending upon the internal details of a program, it might be possible to rearrange the terms, factor them, and neglect any extremely small time constants. However, this is not in accordance with the procedure intended for using the IEEE representation, and it raises questions about how one might adequately represent the limits and nonlinear functions.

We deliberately adjust the time constants in our representation of fast initial response exciters in order to have an effective time constant of 0.03 s, which allows this representation to be used directly in programs and permits a reasonable time step. We would urge that all manufacturers supplying this form of excitation data state the effective time constants of their representation, in order to avoid problems of this type.

4) The authors mention that definitions are not adequate for the new fast excitation systems. A recent IEEE committee report [8] recognized and responded to this need. A new definition was proposed for the high initial response excitation system, a system attaining 95 percent ceiling in 0.1 s or less. This new definition was shown to be sufficient when coupled with the established voltage response ratio definition to adequately specify the performance requirements of these fast systems.

References

[7] IEEE Committee Report, "Computer representation of excitation systems," *IEEE Trans. Power App. Syst.*, vol. PAS-87, pp. 1460–1464, June 1968.

[8] IEEE Committee Report, "Proposed excitation system definitions for synchronous machines," *IEEE Trans. Power App. Syst.*, vol. PAS-88, pp. 1248–1258, August 1969.

Manuscript received February 13, 1970.

Wendal A. Morgan, Henry B. Peck, Donald R. Holland, Frank A. Cullen, and **John B. Ruzek:** The authors appreciate the interesting comments and questions by the discussers of this paper.

Manuscript received March 30, 1970.

Mr. Kimbark, Mr. Hull, and Mr. Iveson have asked what happens to the steam system during fast valving. The line fault conditions analyzed impose a less severe requirement than a reactor scram would. Full load rejection, or emergency shut-down procedure, calls for both the intercept and control valves to close. In a pressurized water reactor (PWR), the steam flow is partially bypassed to the condenser and partially released to the atmosphere through the main relief valve. In a boiling water reactor (BWR) that steam not bypassed to the condenser is released to the pressure suppression chamber.

In contrast, under transient fault conditions which our paper considers for applying fast valving, the intercept valve is momentarily closed while the main inlet control valves remain open. If the intercept valve would be effectively reopened in 2 or 3 s, perhaps no safety valve operation would occur. However, since the pressure drop across the high-power turbine should remain relatively constant, the steam pressure in the low-pressure turbine steam chest must be relieved if the intercept valve is closed for more than a few seconds. For extended closure of the intercept valve, the steam would be exhausted to the atmosphere through a relief valve in the PWR and bypassed to the condenser in a BWR. It should be noted that the valve may be effectively opened for steam flow at a quarter to a third of the fully open position.

Once operation of fast valving is initiated, it is irreversible. We found that fast valving is most effective when operated with minimum time delay. We conservatively based our design decision on a total time delay of 0.15 s, although the manufacturers indicated 0.12 s would be satisfactory. However, we now understand (about a year and a half since we made our studies) that a total delay time of 0.08 s would be reasonable to consider. It is possible that benefits equivalent to those found in our study now could be accomplished with fast valving, utilized only for a stuck breaker condition. We intend to explore this possibility prior to the actual placing into operation of the Calvert Cliffs units.

Mr. Hull and Mr. Iveson discuss the setting of the load unbalance relay. Initially our stability problem was severe in having such a large plant not directly connected to the 500-kV grid. Fast valving is required only until Calvert Cliffs is directly connected to the 500-kV network or until a third 500-kV line is added. Calvert Cliffs will actually be part of a more stable system than we studied, since a 400-MW fossil-fired unit not included in our studies, is now being constructed at Wagner. This may influence our load unbalance relay setting and could raise it to a point near 80 to 90 percent where it would be actuated only by three-phase faults.

The load unbalance relay setting will be set so that single line-to-ground faults on the 230-kV system at Waugh Chapel will not initiate fast valving at Calvert Cliffs. The 25-percent setting is the smallest that can be made. The actual setting we select will be between 25 and 34 percent, to be determined by more stability and relay application studies and by field monitor type tests to be conducted during the year when only unit 1 is in operation and fast valving is not required. The occurrence of three-phase faults on the 230-kV system would be of the same low probability as for the 500-kV system. We are also considering the possibility of a two-setting scheme consisting of a high setting for multi-phase faults and a low setting for single line-to-ground faults on the 500-kV system, with appropriate means of discrimination. It is unlikely that several plants would have fast valving actuated by the same fault, but this condition must be studied and the times and settings of various load unbalance relays coordinated, if fast valving is to be become widely used on systems.

Mr. deMello, Mr. Crenshaw, and Mr. Young discuss the effects of inertia on response to fast or early valving. Perhaps the difference of opinion is a matter of defining the time period under consideration. We will all agree that higher inertia is desirable for the first swing as it certainly increases the critical switching time, i.e., it gives more time to initiate corrective measures. However, the comments in our paper relative to Fig. 7 point out the apparent advantage of lower inertia in responding to fast valving for the second and subsequent swings, and evidently Crenshaw and Young agree with this observation. In at least one pool, instability has developed after the third swing in some studies; therefore, we believe that conclusions cannot always be based on stability studies extending only for one second.

Mr. Brownlee has commented on our assumption of independent pole tripping to reduce the stability requirements. Perhaps a review of the present status of this feature may give greater confidence in its application. Each breaker pole will have duplicate trip coils to obviate failure of a trip coil or a malfunction in the trip circuit. Each station will have duplicate batteries so that each of the two sets of trip coils will be supplied by a different battery. Also, each pole will have duplicate auxiliary switches and pressure switches. The dual pressure switches on each phase effectively isolate the high-pressure system into three independent systems. These pressure switches are connected into the two trip circuits to trip the breaker on loss of pressure. Verification was obtained that, for any realistic failure or rupture in the air pressure system, the breaker could be tripped before the pressure level dropped below that necessary to assure interruption. This feature is discussed in greater detail in [9].

Mr. Brownlee has indicated that the shortest assured exposure to three-phase fault is about 9 cycles on the Southern Company system with solid-state relaying and two-cycle circuit breakers, compared to 7.5 cycles in our study. Perhaps the difference is the amount of time that is allowed for coordinating margin after investigating the time range allowable for the protective relay, trip relay, timer, and circuit breaker.

Mr. Kimbark, Mr. Brownlee, Mr. Hull, Mr. Iveson, Mr. Crenshaw, and Mr. Young discuss the assumption of a three-phase fault criterion. It is agreed that in actual practice it is difficult to achieve a three-phase fault. However, many power pools require the assumption of a three-phase fault in studies. This paper includes an analysis of the dc losses in the generator so that the assumption of a three-phase fault would not be unduly conservative. An average value of 15 percent was used for the machine internal losses for three-phase faults at Calvert Cliffs cleared in three cycles. The generator's output was set at 19 percent for three cycles to meet the internal and external losses to the point of the fault. In those cases where the independently operated poles of the 500-kV breaker changed the three-phase fault to a single line-to-ground fault at the end of three cycles, the generator's output was set to 19 percent for only the first three cycles, the duration of the three-phase fault. In those cases in which the three-phase fault was cleared away in 7.5 cycles at Calvert Cliffs, a value of 10 percent was used for the average machine internal fault losses. The average value of the losses decreases for faults of longer duration, since the fault currents decrease exponentially with time. Perhaps this paper and the discussion by Crenshaw and Young may help to generate more effort in setting up a procedure for accurately representing losses in the generator during a three-phase fault.

Mr. Brownlee and Mr. deMello discuss the conclusions and results with respect to the excitation systems, and we concur with both of them. Mr. Crenshaw and Mr. Young explain the problem of proper representation of excitation characteristics for what we have called the fast excitation systems and point out the need to know the effective time constants as well as the other data intended for use with the standard IEEE representation. Their statement that Fig. 6 is not correct applies to the preprint. Fortunately, Mr. Young discovered that the original Fig. 6 did not look proper and suggested that this case be rerun, and this was done. The revised case shows a slight improvement in the terminal voltage and stability of the units, but does not affect our conclusions. Fig. 6 and the accompanying discussion have been corrected for publication.

Regarding Mr. deMello's comment on evaluation of the benefits of the high-performance fast excitation systems, our paper has pointed out that the performance is much better than conventional systems but not required for stability in this application. We agree that better damping would be obtained. It is well known that on some systems, especially those involving hydroelectric plants and long distance transmission, the high performance excitation systems have reduced power surging and significantly improved stability.

We thank all the discussers for their time in preparing their comments and hope that with this paper and closure we will have brought before the IEEE a better understanding and evaluation of today's aids to system stability.

References

[9] D. R. Holland, G. E. McNair, G. D. Paradis, J. Senuta, and W. H. Van Zee, "Conemaugh project—new concepts for 500-kV system protection," presented at the IEEE Summer Power Meeting and EHV Conference, Los Angeles, Calif., July 12–17, 1970; also *IEEE Trans. Power App. Syst.* (to be published).

Part IV DC Lines

There are now a number of dc lines operating in North American power systems representing several distinctive operating situations. One installation involves a dc line operating in parallel with an ac transmission system, and there are several other similar dc lines in various stages of planning and installation. Another line provides an asynchronous tie between a remote hydro development and a large ac system. A third line provides an asynchronous tie of limited capacity between two ac networks. There is no inherent problem of synchronous stability involved in dc transmission, of course. Power flow over a dc line is determined by rectifier and inverter controls within constraints imposed by ac system voltages and the inherent limitations of converter operation. However, dc lines can have important effects on ac system stability, and substantial effort has gone into the development of models suitable for studying combined ac and dc systems. There have also been numerous system studies to examine the influence of dc line controls on ac system behavior.

A variety of dc line control policies have been suggested as means of improving ac system stability characteristics. One of the ideas most often put forth is to improve the damping of ac system oscillations by appropriately modulating power flow over a dc line. It is also frequently proposed to reduce the undesirable effects of various ac system disturbances, such as loss of generation, by fast load control on a dc line. The interaction between ac and dc systems is quite complex, and any control objective requires thorough analysis to verify that it can be achieved and to develop control systems and parameters.

It is convenient to discuss separately two aspects of dc line modeling for stability studies: 1) the interaction of converter stations with the ac network, and 2) control system models. It is necessary to calculate real and reactive power flows in the ac system due to each of the dc line converter stations at frequent intervals throughout the period of simulation. Typically, a converter station is treated as a current source to the transmission network, with current magnitude momentarily fixed by control variables, and power factor dependent on ac system voltages. Real and reactive flows at converter stations and ac system voltages are interdependent, and are calculated simultaneously by an iterative method. Large system stability simulation demands that converter behavior be considerably simplified and idealized. For example, effects of unbalanced phase voltages are not represented, harmonic filters in the ac system are replaced by their net reactance at the ac system frequency, and variation in effective commutation reactance due to transients is ignored. These limitations on model detail do not significantly hinder system stability studies, however.

The second part of dc line modeling concerns the dynamics of dc line controls. Included are the essential rectifier and inverter firing controls and auxiliary controls utilized to achieve system objectives. There is considerable variation of control policy from one installation to another, and control system modeling is to a large extent unique for individual dc lines.

The first of three papers reprinted here analyzes certain basic control problems associated with a generator feeding the rectifier terminal of a dc line with no synchronous connections. The second paper presents models for dc lines for power system stability studies. The final paper discusses some aspects of the influence of dc lines on the stability of ac systems. The following is a listing of some related papers.

Supplementary Bibliography

DC Lines

[1] H. A. Peterson, P. C. Krause, Jr., J. F. Luini, and C. H. Thomas, "An analog computer study of a parallel AC and DC power system," *IEEE Trans. Power App. Syst.*, vol. PAS-85, pp. 191–209, Mar. 1966.

[2] H. A. Peterson and P. C. Krause, Jr., "A direct and quadrature-axis representation of a parallel AC and DC power system," *IEEE Trans. Power App. Syst.*, vol. PAS-85, pp. 210–225, Mar. 1966.

[3] T. Machida, "Improving transient stability of AC system by joint usage of DC system," *IEEE Trans. Power App. Syst.*, vol. PAS-85, pp. 226–232, Mar. 1966.

[4] G. D. Breuer, J. F. Luini, and C. C. Young, "Studies of large A-C/D-C systems on the digital computer," *IEEE Trans. Power App. Syst.*, vol. PAS-85, pp. 1107–1116, Nov. 1966.

[5] H. A. Peterson and P. C. Krause, Jr., "Damping of power swings on a parallel AC and DC system," *IEEE Trans. Power App. Syst.*, vol. PAS-85, pp. 1231–1239, Dec. 1966.

[6] A. G. Phadke, "Calculation of balanced and unbalanced fault currents in AC-DC power systems," IEEE paper 69 CP 117-PWR.

[7] D. P. Carroll and P. C. Krause, "Stability analysis of a DC power system," *IEEE Trans. Power App. Syst.*, vol. PAS-89, pp. 1112–1119, July/Aug. 1970.

[8] D. P. Carroll, "Hybrid computer simulation of multiterminal DC power transmission systems," *IEEE Trans. Power App. Syst.*, vol. PAS-89, pp. 1126–1133, July/Aug. 1970.

Calculation of Dynamic Behavior of Generators Connected to a DC Link

JACQUES J. CLADÉ, MEMBER, IEEE, AND HENRI PERSOZ

Abstract—**The dynamic performance of an alternator supplying power on a dc link is put into equations, and the equation system is resolved in order to pick out the characteristic transfer functions of the dynamic behavior of the whole system. These functions have a simple literal form and are very practical for studying all problems of regulation which concern such systems of power production and transmission. The physical signification of the most important terms is given.**

Introduction

TRANSMISSION of the power produced by a group of alternators by means of a dc line raises a certain number of new problems from the point of view of regulation, due to the special utilization conditions of the alternators and the operational characteristics of the dc line. Up to now, the two regulation techniques concerning, first, the alternators delivering current to alternating systems, and second, the dc links, have developed separately with each making allowances only for the reactions inherent in the systems concerned.

Operation of an alternator delivering current on a dc link is fundamentally different from the classical operational method where the alternator delivers current to a very high-power ac system. In particular, when the alternator delivers current on a dc line, the terminal voltage and the speed of the alternator are not imposed by the distribution system.

The result is that the regulation structure best adapted to such a power production and transmission scheme is not necessarily obtained by the juxtaposition of the two regulation structures, generally used for the alternators on the one hand and the dc links on the other.

Paper 68 TP 17–PWR, recommended and approved by the Power System Engineering Committee of the IEEE Power Group for presentation at the IEEE Winter Power Meeting, New York, N.Y., January 28–February 2, 1968. Manuscript submitted September 1, 1967; made available for printing November 1, 1967.

The authors are with Electricité de France, Paris, France.

With regard to the regulation of speed, for example, it is conceded at the present time that a classical speed regulator is very poorly adapted to the regulation of an alternator feeding into a dc link when this link is elsewhere assumed to have a constant direct current or a constant power [4], [5]. It appears, therefore, that the regulation of such a system must be revised with allowances made for the special conditions under which the alternator operates, conditions different from those usually considered. The fact that considerable reciprocal interaction exists between the performance of the alternator and that of the dc link makes it necessary to approach regulation problems in a general manner, that is, to consider the alternator and the line as forming a single unit which is the system to be regulated.

This study represents a theoretical approach to regulation problems from this point of view. It concerns the calculation of the dynamic behavior of a dc line connected at its sending end to a production system assumed to be a single alternator, and at its receiving end to an ac distribution system supplied, furthermore, by other power sources. The calculation is first conducted while assuming that the installation is without regulation system, that is, in *open loop*. The results of this calculation make it possible to obtain optimum regulation systems.

The work required of the regulation system is essentially to maintain the current or the power on the line at a set value and to keep the generator at its nominal speed. However, speed variations of the generator are without direct influence on the frequency behavior of the distribution system. For this reason, the limitations of speed variations are imposed only by the generator plant (mechanical stresses, ac supply of auxiliaries).

Voltage regulation of the alternator in this case has a different aspect from that of usual regulation. It does not control reactive power exchanges but directly controls the active power on the dc line. This power can be controlled by the grid control; therefore, it can be seen that the terminal voltage of the alternator must not be considered a regulated quantity, but is, on the contrary, an additional means for speed and power regulation. To

Reprinted from *IEEE Trans. Power App. Syst.*, July 1968.

meet these two main requirements (dc or power regulation and speed regulation), the regulation chains can act on three factors: field voltage, turbine throttle valve, and delay angle of the converters.

The purpose of this study is to calculate the transfer functions which link the direct current and the alternator rotation speed to the regulating quantities specified in the foregoing. Since the dynamic behavior of such a system is not linear, these transfer functions must be obtained by linearization and are applicable only near a particular steady state.

The transfer functions have a relatively simple literal form which makes it possible, by using certain quite allowable approximations, to express them in the form of first-degree term products. The value of each time constant is thus given as a function of the characteristics of the installation and the operational point under consideration.

Limits of Study

The equation to be discussed takes into account that three variable quantities constitute the input data of the system:

1) the field voltage of the alternator (e_f)
2) the opening of turbine throttle valve (S)
3) the delay angle of rectifiers (α_0) which will be defined more precisely later.

The ideal no-load dc voltage on the inverter side is assumed to be constant, which implies that the ac distribution system voltage, as well as the extinction angles of the inverters, are constant.

The frequency range within which the equation form is valid extends, approximately, between 0 and 5 Hz. This range is more than sufficient to study regulation problems, considering the relatively large time constants which govern the dynamic behavior of the alternator. For these frequencies, the converters can be considered continuously controlled. The alternator under consideration is assumed to be isotropic. Should this not be the case, it would be necessary to extend the Park equations used to the case of anisotropic machines.

Finally, the turbine is assumed to be ideal, i.e., it reacts instantaneously to variations in the opening of throttle valves. However, the influence of the inertia in the superheater of a steam turbine or of the water in the case of a hydraulic turbine can be approximately introduced in the form of series transmittance multiplying the transmittances, calculated in this study, relative to variations in the opening of the valves (ΔS). This additional transmittance, under transient operation, links the variations of the driving power of the turbine to those of the throttle valve.

Equation

Alternator Equation

The Park theory is used to study the dynamic behavior of the alternator. The general assumption is made that the machine is not saturated; this condition is necessary for the Park equations to be applicable. Furthermore, the alternator is assumed to be isotropic with only one damper on each shaft.

Park Equations: The Park theory assumes a system of axes attached to the rotors. It links the components of the voltage to the components of the flux or of the currents, the latter assumed to be circulating inside two imaginary windings fixed with respect to the system of axes under consideration.

It can be calculated that, within the frequency range considered in this study (less than 5 Hz), the transformation voltages are negligible relative to the rotation voltages. Resistive voltage drops are also neglected.

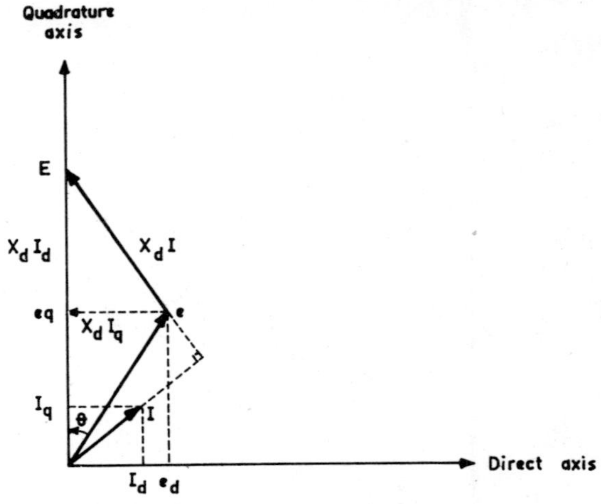

Fig. 1. Fundamental diagram of alternator.

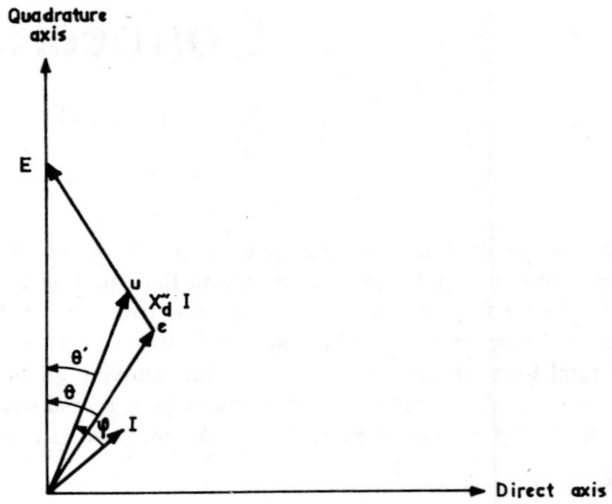

Fig. 2. Internal commutation voltage.

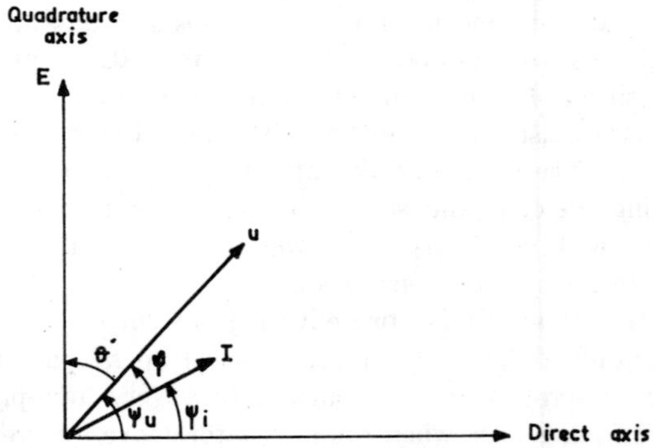

Fig. 3. Change of coordinates.

The steady state is represented by the vector diagram in Fig. 1 in which $\boldsymbol{E}$ is the electromotive force and $\boldsymbol{e}$ the terminal voltage of the machine. For a given machine and a given field voltage, the steady state is completely characterized by vector $\boldsymbol{E}/\boldsymbol{e}$ which is stated as being equal to:

$$\frac{E}{e} = \rho e^{j\theta}.$$

The Park equations are written in relative increment by bringing down each quantity being related to its steady-state value (conventionally, the symbol Δ represents a relative increment).

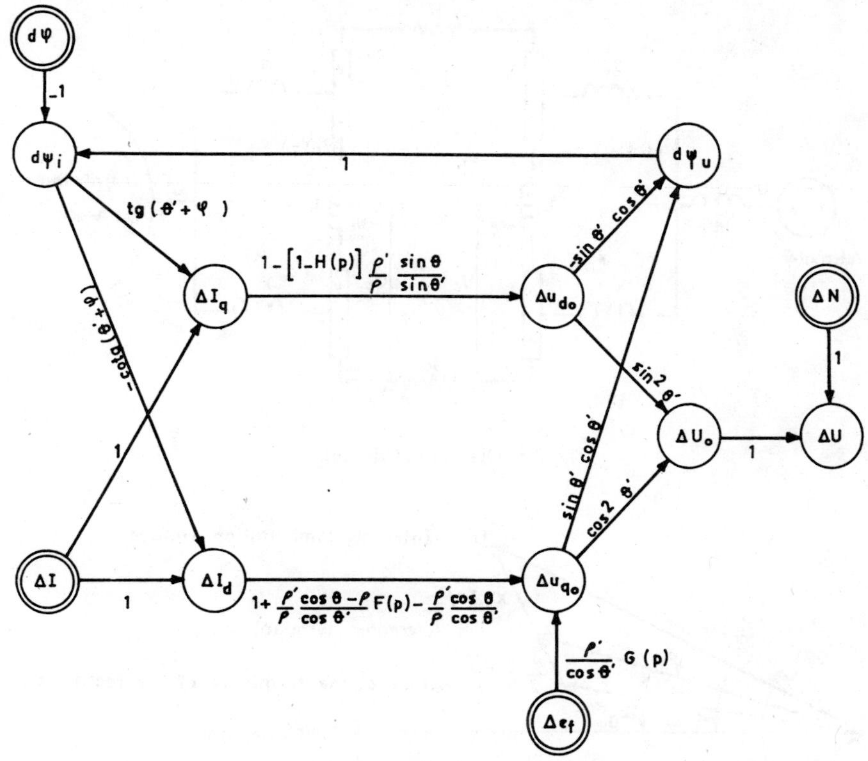

Fig. 4. Alternator flow graph. (Input values surrounded by concentric circles.)

Thus, Park equations may be written:

$$\Delta e_d = H(p)\Delta I_q + \Delta N \tag{1}$$

$$\Delta e_q = \frac{\cos\theta - \rho}{\cos\theta} F(p)\Delta I_d + \frac{\rho}{\cos\theta} G(p)\Delta e_f + \Delta N \tag{2}$$

where d applies to direct and q to quadrature components, e is the phase terminal voltage of the alternator, e_f the field voltage, N the speed of rotating unit, and p Laplace's derivation operator.

Expressions $H(p)$, $F(p)$, and $G(p)$ are given by [2]:

$$H(p) = \frac{1 + T_q''p}{1 + T_{q0}''p} \tag{3}$$

$$F(p) = \frac{(1 + T_d'p)(1 + T_d''p)}{(1 + T_{d0}'p)(1 + T_{d0}''p)} \tag{4}$$

$$G(p) = \frac{1 + T_{kd}p}{(1 + T_{d0}'p)(1 + T_{d0}''p)}. \tag{5}$$

The various time constants in these expressions are fundamental characteristics of the machine; their signification is recalled in the Appendix.

Internal Commutation Voltage: The introduction of this voltage is required as a result of the special operational conditions to which the alternator is subjected. In actual fact, where the alternator is concerned, the commutation phenomena of the rectifiers are two-phase periodic short circuits at the terminals of the machine. The result is that the voltage which remains constant during commutation is that which is located upcurrent of the subtransient reactance. Everything takes place as through the rectifiers commutated on this voltage, called the internal commutation voltage. The ideal no-load dc voltage, on the rectifier side, is proportional to it. It is therefore necessary to calculate the components of this voltage as a function of the components of currents I_d and I_q.

Taking into account that $x_d'' \simeq x_q''$ and calling u the internal (phase) commutation voltage, Fig. 2 allows the statement:

$$u_d = e_d - X_d''I_q \tag{6}$$

$$u_q = e_q + X_d''I_d. \tag{7}$$

Defining $\boldsymbol{E}/\boldsymbol{u} = \rho' e^{j\theta'}$, it follows, by differentiation of (6) and (7) and by referring to the steady state represented by Fig. 2:

$$\Delta u_d = \left[1 - (1 - H(p))\frac{\rho'}{\rho}\frac{\sin\theta}{\sin\theta'}\right]\Delta I_q + \Delta N \tag{8}$$

$$\Delta u_q = \left[1 + \frac{\rho'}{\rho}\frac{\cos\theta - \rho}{\cos\theta'}F(p) - \frac{\rho'}{\rho}\frac{\cos\theta}{\cos\theta'}\right]\Delta I_d + \frac{\rho'}{\cos\theta'}G(p)\Delta e_f + \Delta N. \tag{9}$$

Change of Coordinates: The preceding relationships are established in the form of Cartesian coordinates in a system of axes linked to the rotor. However, the operational equations of the dc line (see Fig. 7) make the alternator currents and voltages intervene in the form of polar coordinates in a system of axes linked to voltage u. It is therefore necessary to establish the relations which make it possible to carry out this change of coordinates.

The angles formed by vectors $\boldsymbol{u}$ and $\boldsymbol{i}$ with the direct axis will be respectively termed ψu and ψi and the phase delay of $\boldsymbol{I}$ relative to $\boldsymbol{u}$ will be termed ϕ (Fig. 3). This change of coordinates is developed in Fig. 4 which represents the alternator operation. On this graph the subscript o means that the voltage is considered at constant speed (influence of speed is characterized by a special branch on the left).

DC Transmission System Equations

The scheme under consideration is in conformity with Fig. 5. Each converter station includes two Graetz bridges in series.

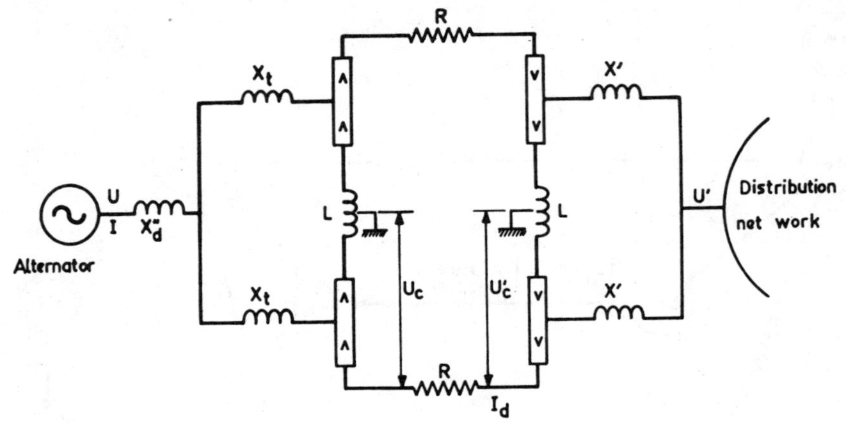

Fig. 5. Diagram of dc link.

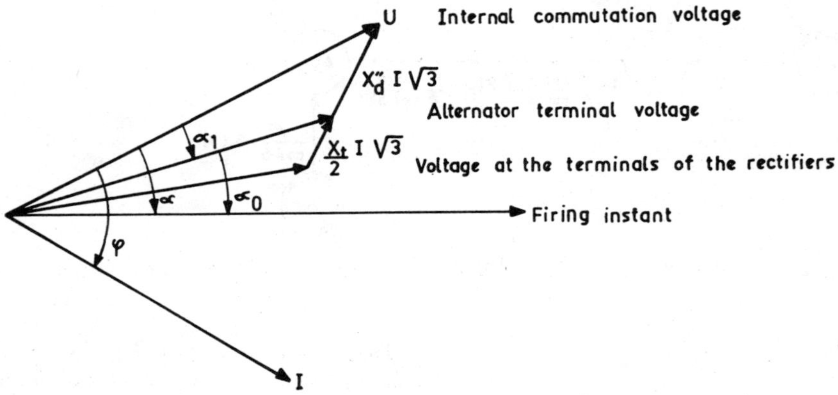

Fig. 6. Diagram of steady-state alternating voltages.

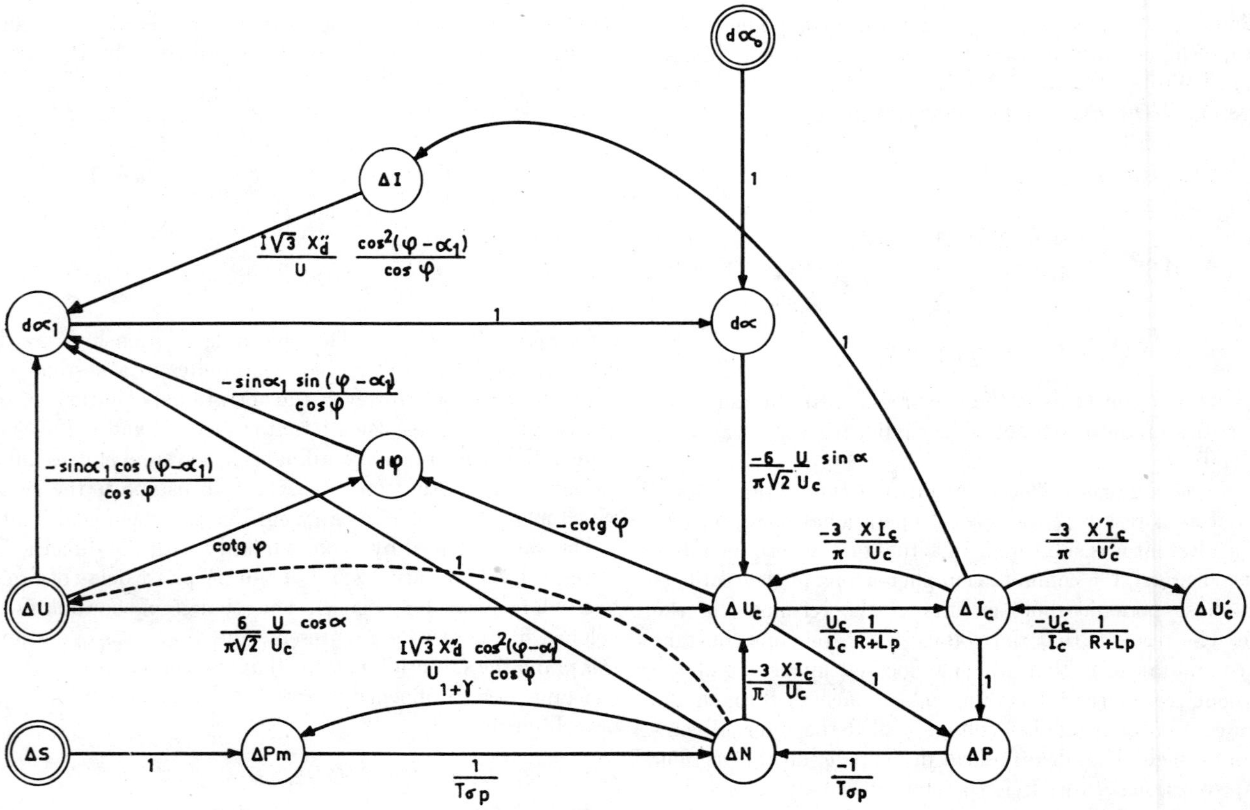

Fig. 7. Flow graph representing dc link and mechanical inertia of rotating unit.

Each bridge is supplied from a different transformer, one with star–star coupling, the other with delta–star coupling. The nomenclature used is given in the Appendix.

The fundamental diagram of the alternating voltages in steady state is given in Fig. 6 where α is the delay angle relative to the internal commutation voltage. The direct no-load voltage (rectifier side) is proportional to cos α. However, the control devices which provide a given delay angle do not dispose of the direct measurement of $\boldsymbol{u}$ which is an internal voltage of the alternator. Therefore it was assumed that the delay angle is fixed relative to the terminal voltage of the alternator, that is, the transfer functions are calculated relative to angle α_0 (see Fig. 6).

The equations of a dc link given in Cladé and Lacoste [3] are recalled here in the form of a flow graph (Fig. 7). Speed variations of the rotating unit are taken into account. They provoke voltage drop variations in the various reactances. The dc link is considered a pure resistance in series with a reactance.

Figs. 4 and 7, or the systems of the equation they represent, enable simulation of the dynamic behavior of the system on either an analog or digital computer. They also enable the study of regulation if the regulation chains are also simulated.

In the following section, Figs. 4 and 7 will be reduced, which means solving the corresponding systems of equations. This reduction, and the resulting transfer functions, enable a theoretical study, without a computer, of the dynamic behavior of the system and its regulation.

The reduction of the graphs in Figs. 4 and 7 will be at first separately conducted. However, the speed influence which acts on the ac voltage (ΔU, Fig. 4) and on the dc voltage (ΔU_c, Fig. 7) will be reported on the same graph (with dotted branch, Fig. 7). This makes it possible to dissociate the voltage variations due to flux variations (Fig. 4) and those due to speed variations of the rotating unit (Fig. 7). One dissociates, in this manner, two quite different transient behaviors.

Reduction of Graph in Fig. 4

Electromagnetic Phenomena in Alternator

In this section the influence of the subtransient behavior, which occurs only at frequencies of more than around 5 Hz, will be neglected. Equations (3)–(5) have then a simplified form:

$$(T_d'' = T_q'' = T_{d0}'' = T_{q0}'' = 0).$$

The reduction brings out transfer functions of the following form:

$$\frac{\Delta U_0}{d\phi} = G_1 \frac{1 + T_1 p}{1 + T_2 p} \qquad \frac{\Delta U_0}{\Delta I} = G_2 \frac{1 + T_3 p}{1 + T_2 p} \tag{10}$$

$$\frac{\Delta U_0}{\Delta e_f} = G_3 \frac{1}{1 + T_2 p}. \tag{11}$$

Expressions in which the following is stated:

$$G_1 = -\tan\theta' \qquad G_2 = -\tan(\theta' + \phi)\tan\theta'$$

$$G_3 = 1 - G_2 = 1 + \tan(\theta' + \phi)\tan\theta'.$$

Time constants T_1, T_2, and T_3 are expressed in the Appendix.

An interesting way to express the gains of these transfer functions is to introduce the following quantity:

$$Z = \frac{(X_d - X_d'')I\sqrt{3}}{U}$$

which represents, under nominal alternator operation, the difference in per unit value between the synchronous and subtransient reactances. It thus follows:

$$G_1 = -\frac{Z\cos\phi}{1 + Z\sin\phi}$$

$$G_2 = -\frac{Z^2 + Z\sin\phi}{1 + Z\sin\phi}$$

$$G_3 = \frac{1 + 2Z\sin\phi + Z^2}{1 + Z\sin\phi}.$$

Reduction of Graph in Fig. 7

DC Link and Alternator Speed Variations

A first reduction is made on the graph shown in Fig. 8. The literal expressions of parameters A, B, C, D, δ, λ, and μ are given in the Appendix. The characteristic equation of this graph, which represents the common denominator of the required transfer functions, is

$$E_1(p) = (1 + \gamma - T_\sigma p)(\lambda + Tp) - (B + C)(\mu + Tp). \tag{12}$$

This second-degree expression has roots sufficiently different from each other for it to be possible to consider their sum as being the greater. An approached expression of the characteristic equation is then

$$E_1(p) \simeq \left(1 + \gamma - \frac{\mu(B + C)}{\lambda} - T_\sigma p\right)(\lambda + Tp).$$

The reduction yields

$$\frac{\Delta I_c}{\Delta U_0} = \frac{AC(1 + \gamma - T_\sigma p)}{E_1(p)} \tag{13}$$

$$\frac{\Delta N}{\Delta U_0} = \frac{C(\mu + Tp)}{E_1(p)} \tag{14}$$

$$\frac{d\phi}{\Delta U_0} = \cot\!\text{an}\,\phi \cdot \frac{(1 + \gamma - T_\sigma p)[\lambda + Tp - C(\delta + Tp)] - B(\mu + Tp)}{E_1(p)} \tag{15}$$

$$\frac{\Delta I_c}{\Delta S} = -\frac{A(B + C)}{E_1(p)} \tag{16}$$

$$\frac{\Delta N}{\Delta S} = -\frac{(\lambda + Tp)}{E_1(p)} \tag{17}$$

$$\frac{d\phi}{\Delta S} = -\cot\!\text{an}\,\phi\,\frac{(\lambda + Tp) - (B + C)(\delta + Tp)}{E_1(p)}. \tag{18}$$

Total Reduction

The graph in Fig. 9 is the juxtaposition of the reduced graphs in Figs. 4 and 7. It summarizes the calculations already carried out. Each transfer function calculated in the foregoing reductions is a branch of this graph. Their literal expressions are not explained but given by (10)–(18). This graph is a first reduction of the general graph.

The final reduction of the graph in Fig. 9 would be rather cumbersome and would make necessary the use of third-degree transfer functions in which the various time constants would be difficult to explain simply. It appears preferable, at this stage of the solution, to specialize the calculations according to the frequency range under consideration, that is, to neglect the speed variations of the machine for the study of rapid phenomena and

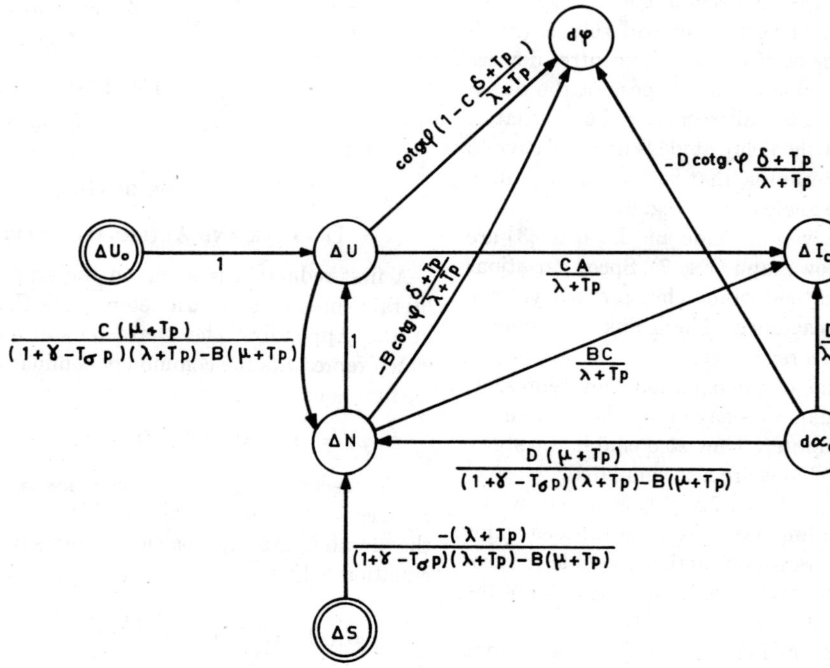

Fig. 8. Reduced flow graph of dc link.

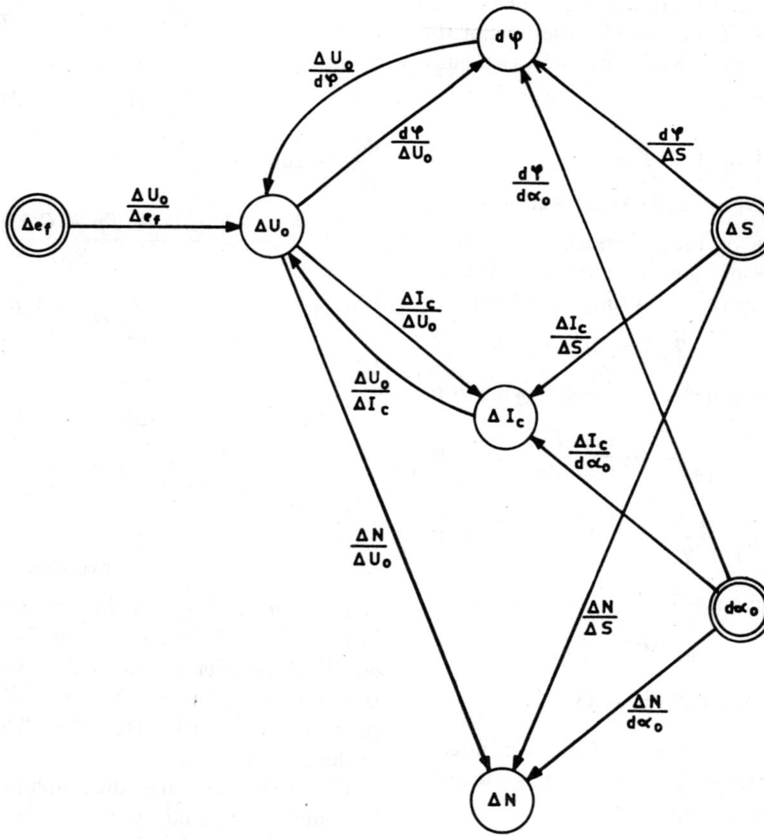

Fig. 9. Reduced flow graph of alternator feeding into dc link.

TABLE I
TRANSFER FUNCTIONS OBTAINED WHEN NEGLECTING DC LINK DYNAMIC BEHAVIOR

Regulating Quantity	Quantity To Be Regulated: Direct Current (I_c)		Speed (N)	
Field voltage (e_f)	$\frac{\Delta I_c}{\Delta e_f} = \frac{G_3E(1+\gamma - T_\sigma p)}{N_1 + R_1p + S_1p^2}$	(20)	$\frac{\Delta N}{\Delta e_f} = \frac{CG_4\mu}{N_1 + R_1p + S_1p^2}$	(21)
Throttle valve opening (S)	$\frac{\Delta I_c}{\Delta S} = \frac{G_1AB \operatorname{cotan}\phi(1+T_1p) - A(B+C)(1+T_2p)}{N_1 + R_1p + S_1p^2}$	(22)	$\frac{\Delta N}{\Delta S} = \frac{G_2E(1+T_3p) + G_1F(1+T_1p) - \lambda(1+T_2p)}{N_1 + R_1p + S_1p^2}$	(23)
Rectifier angle (α_0)	$\frac{\Delta I_c}{d\alpha_0} = \frac{AD(1+\gamma - T_\sigma p)[(1+T_2p) - G_1 \operatorname{cotan}\phi(1+T_1p)]}{N_1 + R_1p + S_1p^2}$	(24)	$\frac{\Delta N}{d\alpha_0} = \frac{D\mu[(1+T_2p) - G_1 \operatorname{cotan}\phi(1+T_1p)]}{N_1 + R_1p + S_1p^2}$	(25)

to neglect the time constant of the dc link for the study of slow phenomena. This dissociation is made possible by the great difference between the two time constants T_σ and T. This method leads to two separate types of transfer functions, the first applicable from approximately 0 to 1 Hz, obtained by considering $T = 0$ in (13)–(18), the second applicable from approximately 0.1 to 5 Hz obtained by considering $T_\sigma = \infty$ in these same formulas. At higher frequencies, as already stressed, it would be necessary to introduce the transformer voltages generated in the alternator armature and the subtransient time constants.

Total Reduction Obtained by Neglecting DC Link Time Constants

The characteristic equation of Fig. 9 can be put in the following form:

$$E_2(p) = N_1 + R_1p + S_1p^2. \tag{19}$$

Values N_1, R_1, and S_1 are given in the Appendix.

Generally, by reason of the difference between the roots, the characteristic equation can be stated

$$E_2(p) \simeq N_1\left(1 + \frac{S_1}{R_1}p\right)\left(1 + \frac{R_1}{N_1}p\right).$$

Reduction of Fig. 9 provides the required transfer functions [(20)–(25)]; these are given in Table I.

Total Reduction Obtained when Neglecting Speed Variations

The only interesting transfer functions in this case are those which represent transient behaviors at constant speed. With allowances made for the assumptions proper to this paragraph ($T_\sigma = \infty$), these transfer functions may be expressed:

$$\frac{\Delta I_c}{\Delta U_0} = \frac{AC}{\lambda + Tp} \tag{26}$$

$$\frac{d\phi}{\Delta U_0} = \operatorname{cotan}\phi \frac{(\lambda + Tp) - C(\delta + Tp)}{\lambda + Tp} \tag{27}$$

$$\frac{\Delta I_c}{d\alpha_0} = \frac{AD}{\lambda + Tp} \tag{28}$$

$$\frac{d\phi}{d\alpha_0} = D \operatorname{cotan}\phi \frac{\delta + Tp}{\lambda + Tp}. \tag{29}$$

The characteristic equation may be written:

$$E_3(p) = N_2 + R_2p + S_2p^2. \tag{30}$$

Values of N_2, R_2, and S_2 are given in the Appendix.

Generally, by reason of the great difference between roots, the following may be stated:

$$E_3(p) = N_2\left(1 + \frac{S_2}{R_2}p\right)\left(1 + \frac{R_2}{N_2}p\right)$$

The reduction of the graph yields the following transfer functions:

$$\frac{\Delta I_c}{\Delta e_f} = \frac{G_3AC}{N_2 + R_2p + S_2p^2} \tag{31}$$

$$\frac{\Delta I_c}{d\alpha_0} = \frac{DA[(1 + T_2p) - G_1 \operatorname{cotan}\phi(1 + T_1p)]}{N_2 + R_2p + S_2p^2}. \tag{32}$$

As far as the subtransient behavior of the machine is concerned, it has been stated that, in the frequency range under consideration, it did not appreciably modify the value of the transfer functions. An exception must be made for the subtransient expression, (5), of $G(p)$ which occurs in the calculation of transfer functions in the form of a multiplier term and which, for this reason, can introduce additional phase shift by approximately 15 degrees at a frequency of 5 Hz.

A more exact expression of (31) would, therefore, be:

$$\frac{\Delta I_c}{\Delta e_f} = \frac{1 + T_{kd}p}{1 + T_{d0}''p} \frac{G_3AC}{(N_2 + R_2p + S_2p^2)}. \tag{33}$$

Naturally, there are a certain number of common points between the transfer functions (20)–(25) and the functions (31) and (32) which make it possible, as will be shown in *Interpretation of Results*, to later reconstitute the transfer functions applicable throughout the frequency range under study (0 to 5 Hz).

It is interesting to discuss the properties of these transfer functions and the order of magnitude and physical significance of the various time constants. For the sake of clarity, this discussion will follow a numerical example.

NUMERICAL EXAMPLE

Numerical expressions will now be given for the transfer functions given in literal form in the section *Total Reduction*. The example selected consists of a turbogenerator delivering current on a dc line 200 km long. The main characteristics of the installation are as follows:

Synchronous reactance of the alternator, X_d	235 percent
Transient reactance, X_d'	26 percent
Subtransient reactance, X_d''	20 percent
Transient open-circuit time constant, T_{d0}'	7.7 seconds
Constant of inertia of the rotating unit, $\frac{1}{2}T_\sigma$	5 seconds

Natural characteristic slope of turbine, γ	-0.8
Direct current link time constant, T	0.087 second
Leakage reactance of transformers supplying the rectifier bridges, X_t	12 percent
Phase shift between terminal voltage of the alternator and the beginning of the commutation, α_0	0.

When the dc link time constant is neglected, the transfer functions are

$$\frac{\Delta I_c}{\Delta e_f} = 3 \frac{(1 - 40.7p)}{(1 - 125p)(1 + 0.57p)} \tag{20'}$$

$$\frac{\Delta N}{\Delta e_f} = \frac{12.5}{(1 - 125p)(1 + 0.57p)} \tag{21'}$$

$$\frac{\Delta I_c}{\Delta S} = -1.52 \frac{1 + 11.7p}{(1 - 125p)(1 + 0.57p)} \tag{22'}$$

$$\frac{\Delta N}{\Delta S} = -10.7 \frac{1 + 0.67p}{(1 - 125p)(1 + 0.57p)} \tag{23'}$$

$$\frac{\Delta I_c}{d\alpha_0} = -0.5 \frac{(1 - 40.7p)(1 + 7.13p)}{(1 - 125p)(1 + 0.57p)} \tag{24'}$$

$$\frac{\Delta N}{d\alpha_0} = -2.1 \frac{1 + 7.13p}{(1 - 125p)(1 + 0.57p)} \tag{25'}$$

It will be immediately noticed that the operation, in that case, is unstable by reason of the existence of a negative time constant. Generally speaking, this stability problem is encountered every time an alternator operates at variable speed. The very great negative time constant in the transfer function denominator is due to the fact that operation takes place near the stability limit. The instability, therefore, develops very slowly.

If the speed variations are now neglected and the time constant of the dc link is taken into account,

$$\frac{\Delta I_c}{\Delta e_f} = \frac{1.13}{(1 + 0.65p)(1 + 0.02p)} \tag{31'}$$

$$\frac{\Delta I_c}{d\alpha_0} = \frac{-0.19(1 + 7.13p)}{(1 + 0.65p)(1 + 0.02p)}. \tag{32'}$$

Interpretation of Results

In this section, the numerical value corresponding to the example selected, for each time constant mentioned, is given in parentheses.

Examination of Transfer Functions Obtained when Neglecting Time Constant in DC Line

The survey of functions (20')–(25') shows that two time constants of very different values govern the dynamic behavior of the system. The larger R_1/N_1 ($= -125$ seconds) conveys the mechanical inertia of the rotating unit, whereas the smaller S_1/R_1 ($= 0.57$ second) conveys the influence of the transient behavior of the alternator. This can be seen directly in the literal expressions forming the denominator of these transfer functions. In R_1 the term of greatest magnitude is $-T_\sigma(\lambda - G_1F - G_2E)$ (see Appendix). Therefore, the ratio R_1/N_1 is approximately proportional to T_σ and practically does not depend on T_1, T_2, and T_3 (time constants of the flux variations). The ratio S_1/R_1 is, on the contrary, a function of T_1, T_2, and T_3 and does not depend on T_σ.

It will be noted that the time constant S_1/R_1 ($= 0.57$ second) has a value less than that of the transient time constant of the alternator in short circuit ($T_d' = 0.85$ second). This is due to the great feedback engendered by the dc link: a voltage variation at the terminals of the alternator results, at constant speed, in a dc variation several times more important than one in relative value. Where the derivation time constants (time constants on the numerators) from functions (20')–(25') are concerned, it is also possible to examine, from their definition relations, the influence which they represent. It can immediately be seen in literal expressions (20)–(25) that $-T_\sigma/1 + \gamma(-40.7$ seconds) shows the influence of speed variation whereas all the other derivation terms are due to the transient behavior of the alternator.

Consequently, where transfer functions $\Delta I_c/\Delta e_f$ and $\Delta I_c/d\alpha_0$ are concerned, the influence of speed variations, shown in the example by term $(1 - 40.7p)/(1 - 125p)$, is sensitive only at low frequencies, that is, at the end of the transient rate. In the case of higher frequencies, this term can be replaced by a numerical coefficient and (20'), for example, can be stated:

$$\frac{\Delta I_c}{\Delta e_f} = \frac{0.98}{1 + 0.57p}$$

which means that, when a variation in the field voltage or the control angle of the rectifier occurs, the speed variations are too small during the first moments of the transient rate to influence the value of the direct current.

It must be pointed out, in other respects, that the frequencies beyond which speed variations are no longer negligible depend considerably on slope γ of the natural turbine characteristic, and that in the example selected, where the slope in absolute value is steep, these frequencies are exceptionally low.

As for the other transfer functions, represented by the reactions of the speed to a variation in regulating quantities ($\Delta N/\Delta e_f$, $\Delta N/\Delta S$, $\Delta N/d\alpha_0$) or the dc reactions to a variation in the opening of the admission valve, the influence of the inertia of the rotating unit is, on the contrary, fundamental, since it results in a very high time constant (-125 seconds).

Examination of Transfer Functions Obtained by Neglecting Influence of Speed Variations

These transfer functions are of interest in the study of transient behavior before the moment when the speed variations begin to have any great influence. Transfer functions (31) and (32) bring out two time constants of very different values. The larger, R_2/N_2 ($= 0.65$ second), shows the influence of the transient effect of flux variations in the alternator. The smaller value, S_2/R_2 ($= 0.02$ second), is due to the transient effect of the dc link. It is normal that S_1/R_1 ($= 0.57$ second) be very close to R_2/N_2 ($= 0.65$ second) since these two values, in actual fact, represent the same time constant due to the transient rate of the alternator. The difference is due to the fact that, in both cases, the assumptions for calculation are not the same.

It is apparent, moreover, in the literal expressions themselves, that S_1/R_1 and R_2/N_2 are very close. In actual fact, the following approximations can be made:

$$R_1 \simeq -T_\sigma(\lambda - G_1F - G_2E)$$

$$R_2 \simeq T_2\lambda - T_1G_1F - T_3G_3E.$$

Hence

$$\frac{S_1}{R_1} \simeq \frac{R_2}{N_2} \simeq \frac{T_2\lambda - T_1G_1F - T_3G_2E}{\lambda - G_1F - G_2E}.$$

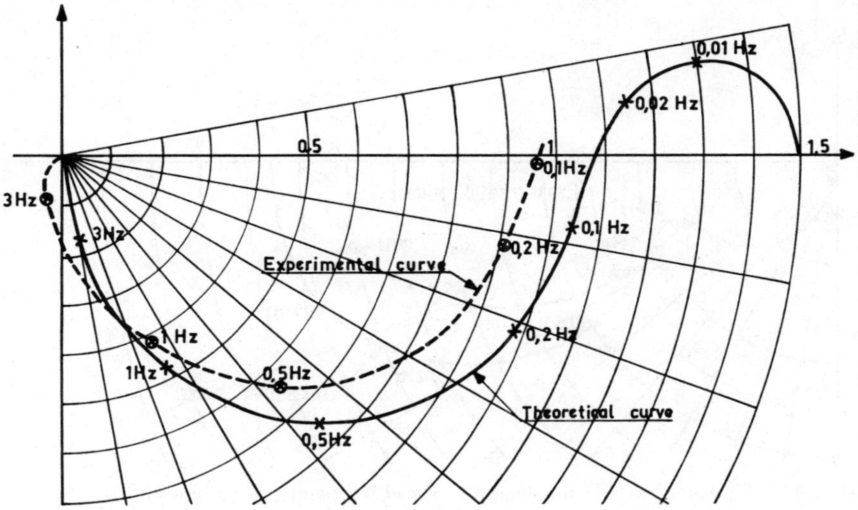

Fig. 10. Transfer loci relating direct current to field voltage of alternator.

Total Transfer Functions

To summarize, three very different time constants govern the dynamic behavior of the system under study. The two extreme time constants are obtained by different calculations applicable in different frequency ranges, whereas the central time constant is obtained simultaneously in both calculations. This is due to the fact that the two frequency ranges within which each calculation is applicable partially overlap (in the 0.1–1 Hz range). The approximately common time constant corresponds to a cut-off frequency located within this interval.

It is possible, afterwards, to reconstruct approximate transfer functions valid throughout the frequency range under study (0–5 Hz) by adding the time constant due to the dc link (itself obtained by neglecting speed variations) to the transfer functions obtained while neglecting this dc link dynamic behavior.

Thus, for the following numerical example:

$$\frac{\Delta I_c}{\Delta e_f} = \frac{3(1 - 40.7p)}{(1 - 125p)(1 + 0.57p)(1 + 0.02p)} \qquad (34')$$

$$\frac{\Delta I_c}{d\alpha_0} = \frac{-0.5(1 - 40.7p)(1 + 7.13p)}{(1 - 125p)(1 + 0.57p)(1 + 0.02p)} \qquad (35')$$

and, generally:

$$\frac{\Delta I_c}{\Delta e_f} = \frac{G_3 E(1 + \lambda - T_\sigma p)}{(N_1 + R_1 + S_1 p^2)(1 + S_2/R_2 p)} \qquad (34)$$

$$\frac{\Delta I_c}{d\alpha_0} = \frac{AD(1 + \gamma - T_\sigma p)[1 + T_2 p - G_1 \cot \phi(1 + T_1 p)]}{(N_1 + R_1 p + S_1 p^2)(1 + S_2/R_2 p)}. \qquad (35)$$

Experimental Verification

Certain transfer functions were studied experimentally to check the applicability of the assumptions and calculations made in this report. This concerns functions which link the direct current to field voltage and turbine admission valve variations. It has been possible to conduct harmonic analyses only on frequencies ranging from 0 to 1 Hz. However, an indirect check of dynamic behavior at high frequencies was achieved by the study of certain stability limits (see *Stability of System Controlled by Current Regulator*).

The experimental device used consisted of a dc model supplied by a microalternator. The possibilities of the dc model are described briefly in Cladé and Lacoste [3]. Where the alternator is concerned, it forms part of the micronetwork described in Robert [1]. Some difficulties in these tests stemmed from the fact that certain auxiliary equipments were not provided, originally, to operate at frequencies which were relatively different from 50 Hz. It follows that certain operational characteristics which do not correspond to realistic values had to be chosen. These are mentioned in the case of each test; all the others correspond to the numerical example given previously.

Harmonic Analysis—Modulation of Alternator Excitation

In this test, the slope of the natural characteristic of the turbine, γ, was -0.15. Application of (20) gives

$$\frac{\Delta I_c}{\Delta e_f} = 1.49 \frac{(1 - 11.8p)}{(1 - 16.5p)(1 + 0.328p)}.$$

The corresponding transfer locus as well as the experimental locus are shown on Fig. 10. It can be seen that the comparison is satisfactory.

Harmonic Analysis—Modulation of Valve Opening

In this test, the natural characteristic slope of the turbine was -2. The time constants inherent in the alternator were (in seconds)

$$T_{d0} = 1.32; \qquad T_d = 0.14; \qquad T = 8.$$

Application of (22) gives

$$\frac{\Delta I_c}{\Delta S} = 0.23 \frac{1 + 1.88p}{(1 + 6.6p)(1 + 0.05p)}.$$

The experimental and calculated transfer loci are superimposed in Fig. 11. Again the comparison is satisfactory.

Stability of System Controlled by Current Regulator

Rapid regulation of direct current at a given set value can be obtained either by changing the rectifier delay angles or by changing the field voltage of the alternator. A possible method of regulation consists in applying a combined effect on these two quantities, the action on these angles enables a very fast regulation and the action on the field makes it possible to reduce the consumption of reactive power and the production of harmonic currents to a minimum value. This is why it is interesting to study, separately, the characteristics and possibilities of these

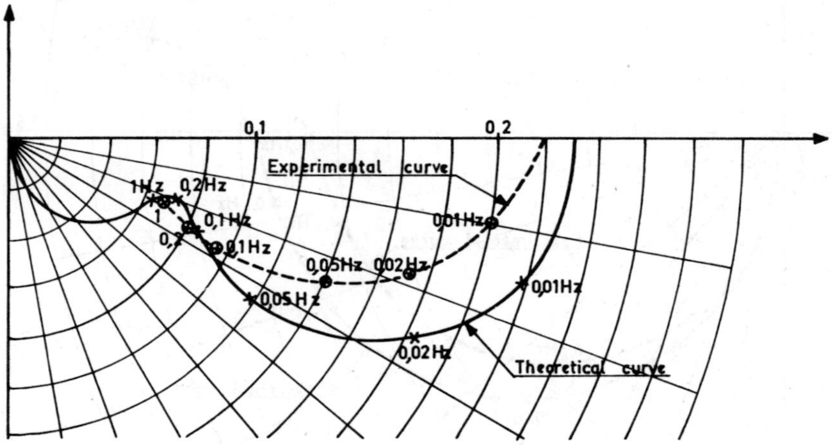

Fig. 11. Transfer loci relating generator speed to opening of turbine throttle valve.

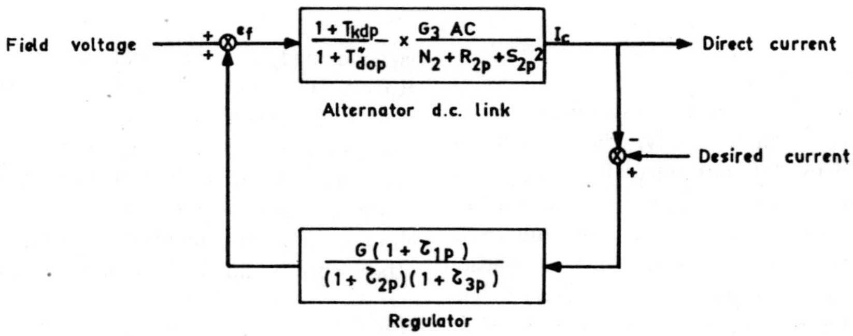

Fig. 12. Principle of current regulation.

two methods of regulation. In this section, the stability of dc regulation through action solely on the field voltage of the alternator is examined in Fig. 12.

Since the frequencies of the oscillation capable of appearing are approximately a few hertz, the speed of the rotating unit can be considered constant. The transfer function I_c/e_f is then obtained by (33). The regulation loop is assumed to have the transfer function:

$$R = \frac{G(1 + \tau_1 p)}{(1 + \tau_2 p)(1 + \tau_3 p)}$$

where τ_3 represents the time constant of the main exciter, τ_2 the time constant of the regulation chains, and τ_1 a derivation time constant. The characteristic equation of the system is

$$(1 + T_{d0}''p)(N_2 + R_2 p + S_2 p^2)(1 + \tau_1 p)(1 + \tau_2 p) + GG_3AC(1 + T_{kd}p)(1 + \tau_1 p) = 0$$

of which the signs of the roots can be studied by means of a stability criterion. The *limit of stability* curves in Figs. 13 and 14 refer to the numerical example in *Reduction of Graph in Fig. 7* in which the following has been taken (in seconds):

$$T_{kd} = 0.02 \qquad T_{d0}'' = 0.04.$$

The curve in Fig. 13 is plotted for $\tau_1 = \tau_2$. It will be noticed that time constant τ_3 is without appreciable influence on the stability, as from $\tau_3 = 0.05$ second. Below this value the stability is greatly improved, leading one to believe that an excitation with controlled rectifiers which can produce such short time constants would be very interesting, in this case, from the sole viewpoint of regulation stability.

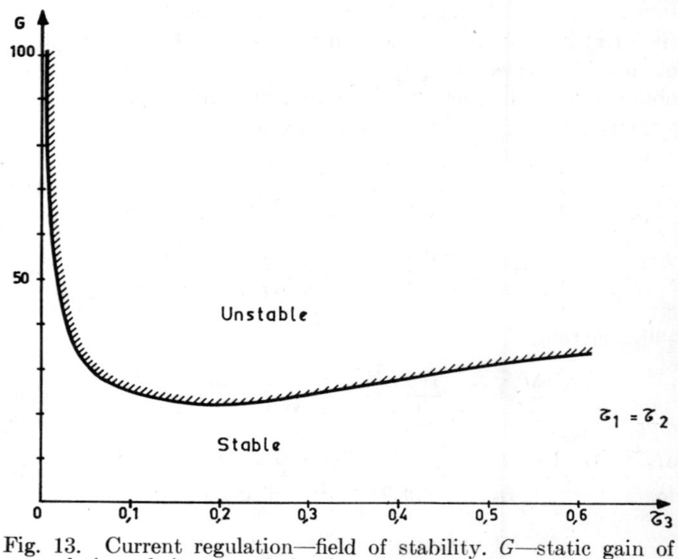

Fig. 13. Current regulation—field of stability. G—static gain of regulating chain. τ_1—derivation time constant. τ_2—time constant of regulating chain. τ_3—time constant of main exciter.

The curves in Fig. 14 show the influence of the derivation time constant τ_1 for different values of time constants τ_2 and τ_3. These curves are at a maximum for a value of τ_1 corresponding approximately to the best damping. They show, therefore, approximately the optimum value of τ_1 for each pair of values τ_2–τ_3. They also indicate that for a small given value of τ_3 (0.01), an increase of τ_2 has a good influence on stability.

Thus, it can be stated that, in the regulation case under consideration, the best damping is obtained when one of the two time constants forming the current regulation chain is very short

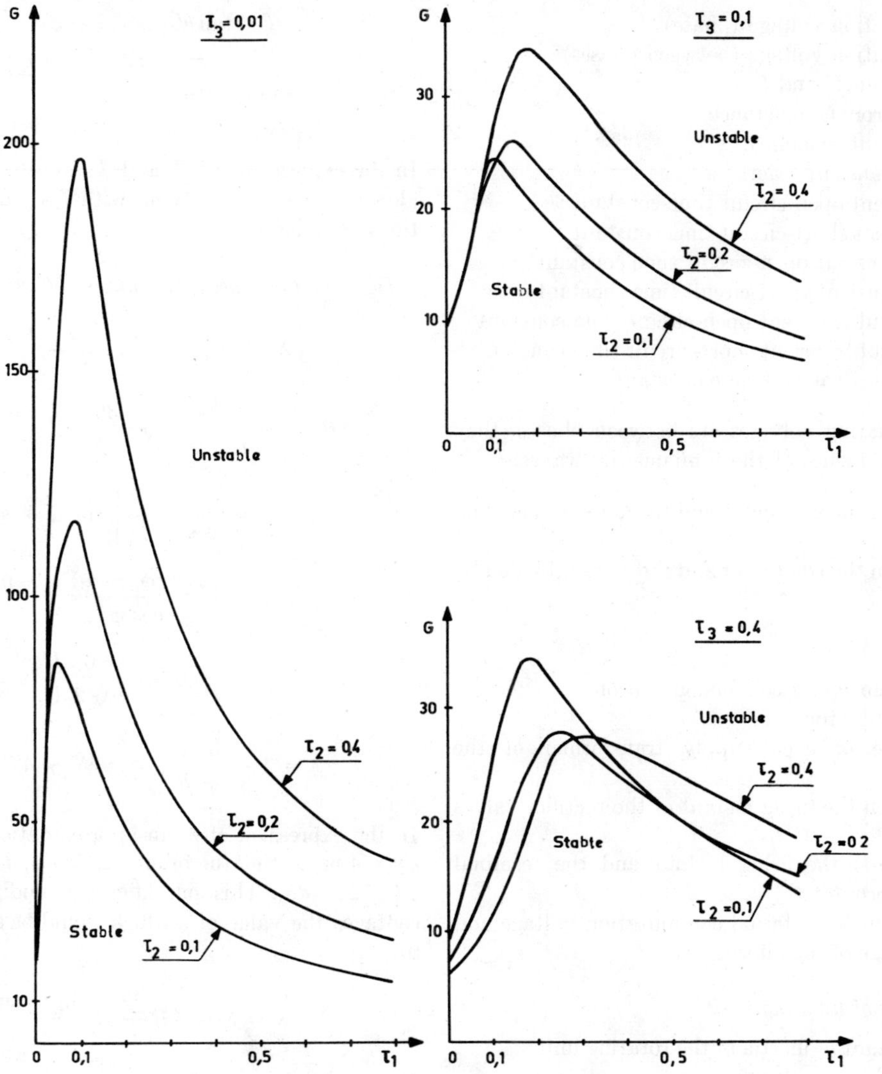

Fig. 14. Current regulator—limits of stability curves. (Symbols same as in Fig. 13.)

(τ_3 = 0.01), with the other having a considerably higher value (τ_2 = 0.4). Some experimental tests were conducted to check these stability limits. The regulation chain used was of a slightly more complicated form than that described in the foregoing by reason of the electronic features of the microalternator transient time constant. With allowances made for this complication, it is found that the theoretical stability limit for $\tau_1 = \tau_2$ and $\tau_3 = 0$ is $G = 65$ whereas the practical limit is 50. For $\tau_1 = 0.3$, $\tau_2 = 0.1$, and $\tau_3 = 0$, one finds 33 and 37 as the theoretical and practical limits, respectively.

Conclusions

1) The dynamic performance of a generator feeding into a dc line is put into equations and the resolution of the equation system thus obtained is undertaken to pick out the characteristic transfer functions of the dynamic performance of the system.

2) These transfer functions have a simple literal form; their denominator can be taken in the form of a product with three time constants. The largest of these constants is due to the mechanical inertia of the rotating unit and, for this reason, it is essentially a function of the constant of inertia of the unit and the natural characteristic slope of the turbine. The second time constant is due to the transient behavior of flux variations in the alternator. Its value is slightly less than the short-circuit transient time constant of the alternator. The shortest time constant is due to the dynamic behavior of the link.

3) The hypotheses formulated are applicable in the case of a constant air-gap alternator. In the case of a salient pole alternator, the equations relative to operation of the alternator should be in strict accuracy, modified to take into account the anisotropy of the machine.

4) The experimental verifications carried out by frequency analysis on a simulator gave satisfactory results.

5) The transfer functions obtained make possible a theoretical study of regulation problems inherent in this type of power production and transmission.

6) As an example, these transfer functions were used to study the stability of a dc regulation acting solely on the field voltage.

Appendix

The main symbols are recalled, and values of the parameters introduced in resolution of the flow graphs indicated.

Symbols

Alternator:

I rotor current
E electromotive force (simple)
e terminal voltage (simple)

u internal commutation voltage (phase)
U internal commutation voltage (between phases)
ϕ phase shift between U and I
X_d direct axis synchronous reactance
X_d' direct axis transient reactance
X_d'' direct axis, subtransient reactance
T_{d0}' direct axis transient open-circuit time constant
T_d' direct axis transient short-circuit time constant
T_{d0}'' direct axis subtransient open-circuit time constant
T_d'' direct axis subtransient short-circuit time constant
T_{q0}'' quadrature axis subtransient open-circuit time constant
T_q'' quadrature axis subtransient short-circuit time constant
T_{kd} direct axis damper (leakage) time constant.

DC Link: The alternating voltages are between phases; they are counted as effective values of the fundamental waves.

U_c dc voltage between the conductor and the mass provided by the rectifier
U_c' dc voltage between the conductor and the mass provided by the invertor
I_c direct current
P dc power
L smoothing inductance of a conversion station
R resistance of a conductor
X_t leakage reactance of each supply transformer of the rectifying bridge
α phase shift between the firing instant of the rectifier valves and voltage U
α_0 phase shift between the firing instant and the terminal voltage of the alternator
α_1 phase shift between the internal commutation voltage and the terminal voltage of the alternator.

Mechanical Equations of Rotating Unit:

T_σ double of the constant of inertia of the rotating unit
N speed of the unit
P_m driving power provided by the turbine
γ slope of the natural characteristic of the turbine
S opening of the turbine throttle valve
C_m driving torque on the turbine shaft.

Calculation Parameters

Alternator:

$\rho = E/e$
$\rho' = E/u$
θ phase shift between E and e
θ' phase shift between E and u
$G_1 = -\tan\theta'$
$G_2 = -\tan(\theta'+\phi)\tan\theta'$
$G_3 = 1 - G_2$

$$T_0 = \frac{1}{\rho' - \cos\theta}\left[\left(\frac{\rho'}{\rho}\cos\theta - \cos\theta'\right)T_{d0}' - \left(\frac{\rho'}{\rho}\cos\theta' - \rho'\right)T_d'\right]$$

$$T_1 = \frac{T_{d0}'[\tan\theta'\tan(\theta'+\phi)] + T_0}{1 + \tan\theta'\tan(\theta'+\phi)}$$

$$T_2 = \frac{T_{d0}'\left[\frac{1}{\sin\theta'\cos\theta'} + \tan(\theta'+\phi)\right] - \tan\theta' T_0}{\tan(\theta'+\phi) + \cot\theta'}$$

$$T_3 = \frac{T_{d0}'\tan\theta'\cos^2(\theta'+\phi) - T_0[\sin(\theta'+\phi)\cos(\theta'+\phi) + \tan\theta']}{\tan\theta'\cos^2(\theta'+\phi) - \sin(\theta'+\phi)\cos(\theta'+\phi) - \tan\theta'}.$$

In the expressions of T_1 and T_2, the terms with T_0 may be neglected in view of the terms with T_{d0}', thus simplifying the equations considerably.

DC Link and Speed Variations of Generator:

$$X = X_t + X_d'', \quad A = \frac{U_c}{I_c}\frac{1}{R}, \quad T = \frac{L}{R}$$

$$B = -\frac{I_c}{U_c}\left[\frac{3}{\pi}X + \frac{36}{\pi^2}X_d''\frac{\cos^2(\phi-\alpha_1)\sin\alpha}{\cos\phi}\right]$$

$$C = \frac{6}{\pi\sqrt{2}}\frac{U}{U_c}\left[\cos\alpha + \sin\alpha\sin\alpha_1\left(\frac{\cos(\phi-\alpha_1)}{\cos\phi} + \frac{\sin(\phi-\alpha_1)}{\sin\phi}\right)\right]$$

$$D = \frac{-6}{\pi\sqrt{2}}\frac{U}{U_c}\sin\alpha$$

$$\delta = 1 - \frac{3}{\pi}\frac{X'}{R}, \quad \lambda = \delta - AB, \quad \mu = \delta + A.$$

In the expression of λ, an approximation was made which consists of neglecting the branch ΔU_c, $d\phi$, $d\alpha_1$, in view of the branch ΔU_c, ΔI_c, $d\alpha_1$. This simplification modifies by a couple of percentages the value of λ which would be equal, in strict accuracy, to:

$$\lambda = \delta - AB + D\frac{\sin\alpha_1\sin(\phi-\alpha_1)}{\sin\phi}.$$

$$E = AC$$
$$F = (\lambda - c\delta)\cot\phi$$
$$H = 1 + \gamma - \frac{B\mu}{\lambda - C\delta}$$
$$K = D\delta\cot\phi$$
$$M = (\mu/\delta) - (1+\gamma)$$
$$L = (1+\gamma) - \frac{(B+C)\mu}{\lambda}$$
$$N_1 = \lambda L - G_1FH - G_2E(1+\gamma)$$
$$R_1 = T_2L\lambda - T_1G_1FH - T_3G_2E(1+\gamma) - T_\sigma(\lambda - G_1F - G_2E)$$
$$S_1 = -T_\sigma(T_2\lambda - T_1G_1F - T_3G_2E)$$
$$N_2 = \lambda - G_1F - G_2E$$
$$R_2 = T_2\lambda - T_1G_1F - T_3G_2E + T[1 - G_1(1-C)\cot\phi]$$
$$S_2 = T_2T - T_1TG_1(1-C)\cot\phi.$$

References

[1] R. Robert, "Le micro-réseau—modéle dynamique de réseaux," *S. F. E. Bull.*, vol. IV, ser. 7, no. 38, February 1954.

[2] B. Adkins, *The General Theory of Electrical Machines*. London: Chapman and Hall, 1959.

[3] J. Cladé and A. Lacoste, "Simplified study of the stability of regulation of a d.c. link," *Direct Current*, vol. 12, no. 1, 1967.

[4] F. H. Last, G. S. H. Jarrett, K. W. Huddart, G. L. Brewer, and W. G. Watson, "Isolated generator—d.c. link feasibility trials," presented at the IEE Conf. on High Voltage D.C. Transmission (Manchester, England, September 1966).

[5] J. B. C. Bower and K. J. Bradbury, "D.C. link isolated generator—real power control," presented at the IEE Conf. on High Voltage D.C. Transmission (Manchester, England, September 1966).

Digital Representation of a DC Transmission System and Its Controls

JAMES F. CLIFFORD, MEMBER, IEEE, AND ALBERT H. SCHMIDT, JR., MEMBER, IEEE

Abstract—**This paper discusses the mathematical representation of a dc transmission system and its controls. Presented for comparison are digital computer study results of the same studies performed on an intermediate power level model of a parallel ac/dc transmission system. Also presented are the mathematical representations of the several ac components of the model system.**

Paper 69 TP 2-PWR, recommended and approved by the Power System Engineering Committee of the IEEE Power Group for presentation at the IEEE PICA Conference, Denver, Colo., May 18–21, 1969. Manuscript submitted January 6, 1969; made available for printing October 1, 1969.

J. F. Clifford is with the Jackson and Moreland Division, United Engineers and Constructors, Inc., Boston, Mass. 02116, on assignment with the Edison Electric Institute AC/DC System Operation Research Project.

A. H. Schmidt, Jr., is with the Bureau of Reclamation, U. S. Department of Interior, Denver, Colo. 80202, on assignment with the Edison Electric Institute AC/DC System Operation Research Project.

Introduction

IN 1963, the Joint Task Force on DC Transmission of the Edison Electric Institute (EEI) recommended that the AC/DC System Operation Research Project be initiated. An intermediate power level (700-kW) ac/dc model system was made operational in the spring of 1967. A detailed description of the model has been given previously [1].

Among the several objectives of the research project is to demonstrate that a digital representation of dc transmission lines and their controls is adequate. By supplying the known parameters of the model system as input data for a computer program, the results of the stability studies can be compared with actual model performance.

A small system program has been written to simulate the model as closely as practical, and in spite of minor differences in model system parameters as compared with actual power sys-

Reprinted from *IEEE Trans. Power App. Syst.*, Jan. 1970.

tems, the model does provide a unique facility for checking the digital representation.

Methods for simulation of high-voltage direct current (HVDC) in ac power flow problems are not new [2], [3], and some of them have been extended to transient stability programs [4]. This paper describes the algorithms needed to incorporate into large system programs a dc transmission system simulation that has been checked out against a dynamic model system.

Program Description

Since our main object is to discuss the digital representations of a dc transmission system and its controls, the ac portion will be discussed only briefly.

The program, at this writing, is capable of calculating the dynamic response of the ac/dc system disturbances of: 1) step changes in ac loading, 2) step changes in dc line current order, 3) three-phase ac bus or line faults, and 4) ac branch switching.

Planned and currently under development are representation of faults on the dc transmission line, automatic dc line protection operation, and three-terminal line operation.

The power flow and transient stability calculations are performed in a single program. The power flow solution is basically a classical method described by Ward and Hale [6]. The dc system input data required for solution in the power flow phase consist of:

1) identification of rectifier and inverter buses
2) schedule of dc power or current
3) desired inverter dc voltage
4) dc line resistance
5) commutating resistances
6) valve arc drops
7) bridge loss resistances
8) normal firing angle of rectifier
9) constant extinction angle of inverter.

Since the model has fixed-tap commutation transformers, the actual dc line voltage and rectifier delay angle are determined in the solution without tap changer adjustment.

In addition to the usual ac power flow output, the program produces the following information for each dc terminal: 1) power flow into or out of dc terminals, 2) reactive requirements of dc terminals, 3) dc line voltage, 4) transmitted dc power, and 5) valve firing angles.

The transient stability calculations immediately follow outputting of the power flow. The ac generators are represented as indicated in the phasor diagram of Fig. 1. The variations of the field flux linkage E_q' are represented and are dependent upon the regulator–exciter control of the field voltage and the field current. Figs. 2 and 3 illustrate diagrammatically the model's generator regulator–exciter simulation and the governor–turbine simulation.

Additional input data required for solution of the dc system in the transient stability phase consist of: 1) dc line inductance and capacitance, 2) dc line smoothing reactor inductance, 3) line current control amplifier data, 4) bus numbers of buses supplying filtered source voltage for converters, and 5) data for the modulation control, when desired.

In addition to ac transient stability output, the program produces the following information for the dc system: a complete listing of initial conditions (Table I), and a listing at each ac time-step interval of: 1) real and reactive flows of each dc terminal, 2) dc line voltages and currents, 3) line current control amplifier response, 4) valve firing angles, and 5) dynamic line current order.

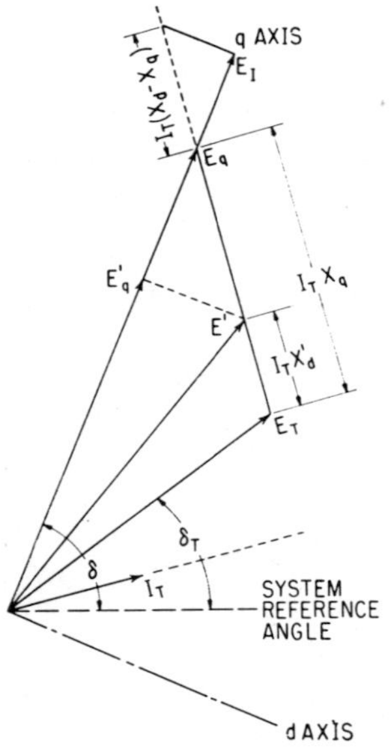

E_t terminal voltage
E' voltage behind X_d'
E_q' voltage proportional to field flux
I_t terminal current
E_q voltage behind X_q
E_i voltage proportional to field current
X_d direct axis synchronous reactance
X_d' direct axis transient reactance
X_q quadrature axis synchronous reactance
δ_t angle between generator terminal and system reference axis
angle between quadrature axis and system reference axis

Fig. 1. Generator phasor diagram.

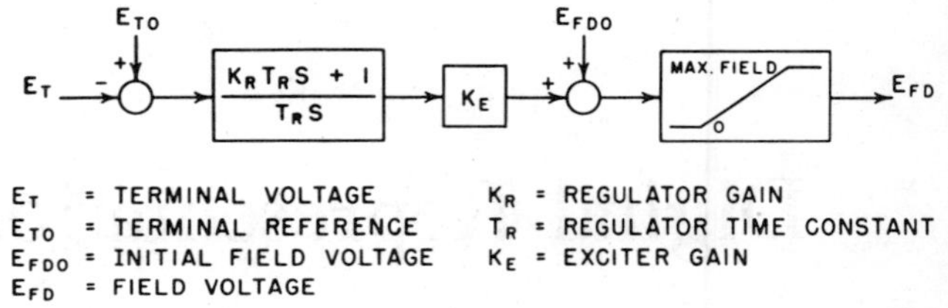

Fig. 2. EEI regulator–exciter block diagram.

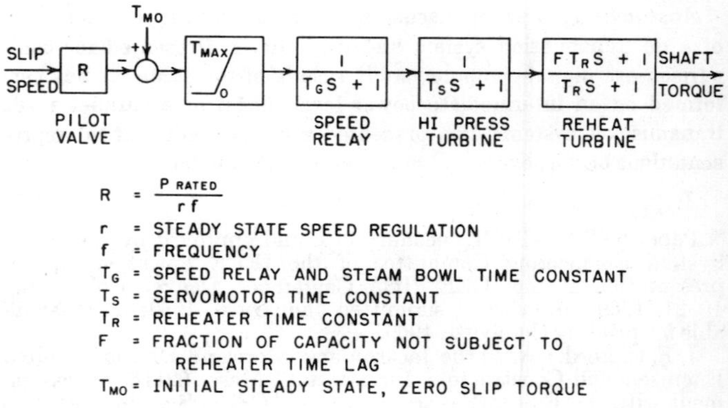

Fig. 3. EEI governor–turbine block diagram.

Optional output plots are available, including: 1) relative machine angle, 2) dc currents at rectifier and inverter terminals, 3) dc voltage at rectifier and inverter terminals, 4) ac real and reactive power at dc terminals, and 5) firing angles α and β.

TABLE I
DC System Initial Conditions—Dynamic Control of DC Line Current Turned Off

DC Line 1		
rectifier choke	= 0.110E-01 henrys	
inverter choke	= 0.110E-01 henrys	
line inductance	= 0.869E-02 henrys	
line capacitance	= 0.240E-03 farads	
line resistance	= 0.150E-00 ohms	
rectifier voltage	= 501.2	volts, dc
rectifier current	= 46.5	amperes
rectifier dc power	= 23.3	kW
rectifier arc drop	= 20.0	volts
rectifier resistive drop	= 18.8	volts
rectifier commutating reactance	= 0.7500	ohms
inverter voltage	= 494.2	volts, dc
inverter current	= 46.5	amperes
inverter dc power	= 23.0	kW
inverter arc drop	= 44.0	volts
inverter resistive drop	= 12.3	volts
inverter commutating reactance	= 0.7750	ohms
Controls—Line 1		
current order	= 46.5 amperes	
α	= 29.1 degrees	
γ_0	= 25.0 degrees	
β	= 37.7 degrees	
current margin	= 8.0 amperes	
α_{min}	= 5.0 degrees	
α_{stop}	= 100.0 degrees	
overlap	= 12.7 degrees	
Regulator Amplifiers		
$C1$	= 0.300 mF	
$C2$	= 0.150 mF	
$R5$	= 0.300 mΩ	
$R6$	= 3.300 mΩ	

DC Representation

The primary steady-state characteristics of a dc link that must be found by the power flow program are the rectified voltage as a function of the current in the link and the active and reactive power as functions of the ac voltage.

Since the main purpose of HVDC transmission is the exchange of real or active power, the concept of reactive power requirements of the conversion may seem strange. A dc bridge absorbs reactive power because the bridge current is allowed to flow only at certain intervals. Via grid control, late firing of inverter valves combined with the finite time required for current commutation results in a phase difference between the ac system voltage and current at the bridge connection. This gives rise to the mathematical concept of reactive power requirements for the dc terminals.

This reactive power is only calculated for fundamental frequency. In what follows, the term power factor of the converter is intended to mean the displacement factor, or phasor power factor, for the fundamental frequency values of ac voltage and current at the bridge connections.

Fig. 4 shows the circuit and phasor diagrams used in converter theory. V_s and X_s represent the ac system source voltage and reactance. The commutating voltage E_0 is the root-mean-square (rms) value of the filtered ac line voltage including the effect of commutating transformer tap ratio. The fundamental ac current I_b, voltage E_b and $\cos\theta_b$ at the bridge connection are calculated from the phasor diagram. All further calculations can then be performed by the existing ac program, utilizing the regular ac tap changing feature.

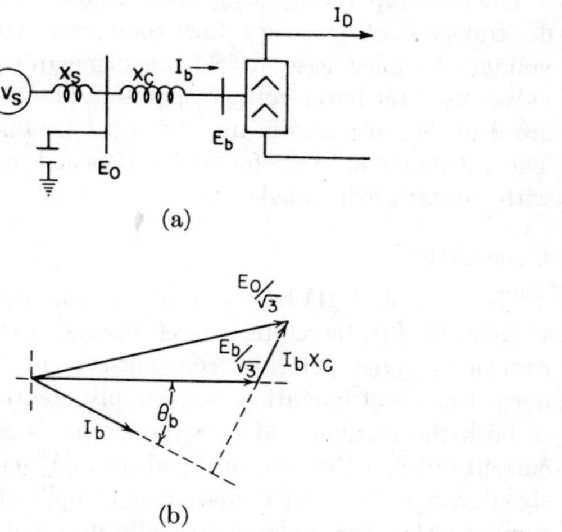

Fig. 4. Diagrams used in converter theory. (a) Circuit. (b) Phasor.

This approach allows any representation of the commutation transformer to be made as a normal ac system branch. For instance, a synchronous condenser may be connected to a special winding of the converter transformer in addition to the static capacitors on the commutating bus.

From the converter theory we obtain [2], [8]

$$D = (3\sqrt{2}/\pi)E_0 \cos\alpha - (r_{eq} + 3X_c/\pi)I_d - D_{ad} \quad (1)$$

where

- D_{ad} arc-drop voltage, sum for two valves that conduct simultaneously
- r_{eq} equivalent resistance of converter representing active losses at rated current, excluding arc drop
- X_c commutating reactance, $= \omega L_c$
- I_d direct current, average value
- α delay angle of converter
- D dc bridge voltage.

The value of the rms fundamental components of line current at the bridge is given by

$$I_b \simeq \sqrt{6}\, I_d/\pi. \quad (2)$$

The active power on the secondary bus is equal to the dc power plus the active losses, and we can write

$$P_b = \sqrt{3}\, E_b I_b \cos\theta_b = (D + D_{ad} + I_d r_{eq})I_d. \quad (3)$$

From the phasor diagram [Fig. 4(b)], we can write

$$(E_0/\sqrt{3})^2 = [(E_b/\sqrt{3})\sin\theta_b + I_bX_c]^2 + [(E_b/\sqrt{3})\cos\theta_b]^2. \quad (4)$$

Substituting (1)–(3) into (4) and rearranging, we get the expression for finding the power factor angle θ_b

$$\tan\theta_b = \sqrt{[(\pi D' + 3X_cI_d)/\pi D' \cos\alpha]^2 - 1} - 18I_dX_c/\pi^2 D' \quad (5)$$

where

$$D' = D + D_{ad} + r_{eq}I_d. \quad (6)$$

Equation (5) can be used also to find the power factor at the inverter by making the following changes in the variables:

$$D' = D - D_{ad} - r_{eq}I_d \quad (7)$$

where $\alpha = \gamma_0 =$ margin angle of commutation of inverter, and X_c, r_{eq}, I_d, and D are values for the inverter.

The arc-drop voltage D_{ad}, in high-power valves intended for dc transmission, is very low compared with the operating voltage. In most cases it can be disregarded. However, with valves used for lower voltage operation like the EEI model, the arc drop is more significant and must be taken into account. The values for r_{eq} were found from special regulation tests run with constant firing angle.

Basic Control

The control of HVDC transmission has been presented previously [8], [9], therefore, a brief discussion of constant-current control is given here. In constant-current (CC) control the normal mode of operation is to supply identical current orders to both the rectifier and inverter terminals and to modify the current order at the inverter by also supplying a current margin signal. Since the current margin is of opposite polarity to the current order, the inverter normally does not regulate current, but operates through control circuitry that maintains a minimum angle of advance (angle between the firing point and the intersection of the phase voltages undergoing commutation) and a maximum dc line voltage. The inverter then is in constant extinction angle (CEA) control.

The rectifier regulates the line current by adjusting the delay angle (angle between the intersection of the phase voltages undergoing commutation and the firing point) to produce an output dc voltage sufficient to overcome the countervoltage of the inverter and the *IR* drop in the line at the ordered current. Should the rectifier be unable to provide the ordered current at its minimum delay angle, the inverter takes over current control when its net current order (current order minus current margin) is reached, and maintains this current by adjusting its angle of advance.

In an actual system, the tap changers at each terminal ensure that the inverter will operate with CEA control and the rectifier with CC control, provided that reasonable source voltages are maintained.

DC Algorithm for Power Flow Calculations

Starting with the scheduled inverter dc power and voltage, the dc current is calculated. Next the power factor at the inverter is found, followed by the ac bridge voltage at the inverter. This voltage is inserted into the tap changer tables. Next, the rectifier-end voltage, the power factor at the rectifier end, and the ac voltage at the rectifier bridge are calculated. This ac voltage is also inserted into the tap changer tables. From the ac voltages, power factors, and dc current, the initial values of real and reactive powers P_b, Q_b, representing the dc terminals, are found.

A normal power flow solution is obtained using the P_b and Q_b loads representing the dc terminals and holding the ac voltages when the tap changing feature is available. In the event that tap changers cannot hold the desired voltages on the dc terminals, then a reduced dc voltage is held and the dc system solution is finally balanced with the ac system solution.

The balancing procedure begins by finding the new value of I_d from the new ac voltage at the inverter and the scheduled P_i. An iterative procedure is employed to make D, I_d, and E_b consistent. If the dc line is operating under constant-current control instead of constant-power control, this iterative loop is not needed. The new values of α and D are compared with the old values as a measure of the final balance. The power factor is then found and the new P_b and Q_b loads calculated. At the rectifier end, the power factor is found and the delay angle is changed to give the new dc voltage from the new I_d and E_b. The new values of P_b and Q_b are calculated and the ac solution continued.

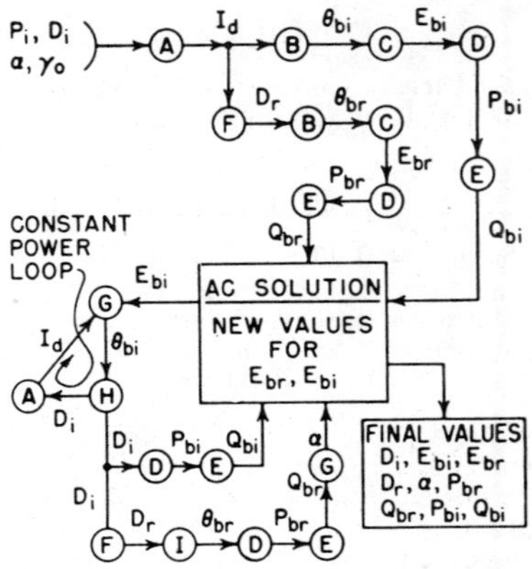

Fig. 5. Calculation scheme for balancing ac/dc system power flow.

The above procedure can be illustrated with the aid of the following set of working equations, corresponding as shown to the circled letters in Fig. 5. (Subscript r refers to rectifier quantities; subscript i refers to the inverter.)

Ⓐ $$I_d = P_i/D_i \quad (8)$$

Ⓑ $$\tan\theta_b = \sqrt{[(\pi D' + 3X_c I_d)/\pi D' \cos\alpha]^2 - 1 - 18 I_d X_c/\pi^2 D'} \quad (5)$$

Ⓒ $$E_b = \pi D'/(3\sqrt{2}\cos\theta_b) \quad (9a)$$

Ⓓ $$P_b = D' I_d, \quad \text{for inverter, } P_b \text{ made negative} \quad (3)$$

Ⓔ $$Q_b = |P_b| \tan\theta_b \quad (10)$$

Ⓕ $$D_r = D_i + I_d R_{\text{line}} \quad (11)$$

Ⓖ $$\cos\theta_{bi} = \cos\gamma_0 \sqrt{1 + (6I_d X_c/\pi\sqrt{2}\,E_b)^2 + 2(6I_d X_c/\pi\sqrt{2}\,E_b)\sin\theta_{bi}} - I_d X_c/\sqrt{2}\,E_b \quad (12a)$$

Ⓗ $$D_i = 3\sqrt{2}\,E_b\cos\theta_b/\pi + D_{ad} + I_d r_{eq} \quad (9b)$$

Ⓘ $$\cos\theta_{br} = \pi(D_r + D_{ad} + I_d r_{eq})/(3\sqrt{2}\,E_b). \quad (9c)$$

The starting point quantities are P_i, D_i, γ_0, α, and the system impedances. The order in which the calculations are carried out is indicated by the arrows in Fig. 5.

DC Control During Transients

As mentioned previously, the rectifier always tries to maintain the current order. The current regulator at the rectifier end varies the delay angle α between α min (about 5 degrees) and 180 degrees. At the same time, the inverter control operating on CEA maintains as high a dc line voltage as it can, as long as the dc current exceeds its I_{set}. When the dc current drops below this value due to a transient or the fact that the rectifier cannot maintain the I_{order}, the inverter current regulator increases the angle of advance β to lower the dc line voltage in order to maintain a reduced current flow ($I_{order} - I_{margin}$). Thus, both terminals can be on CC control during a transient or during an increase in current order resulting in a very fast response from the dc line.

The basic control scheme is shown in Fig. 6. The differential equations for the control system are solved along with the differential equations of the transmission line to reflect the dynamic

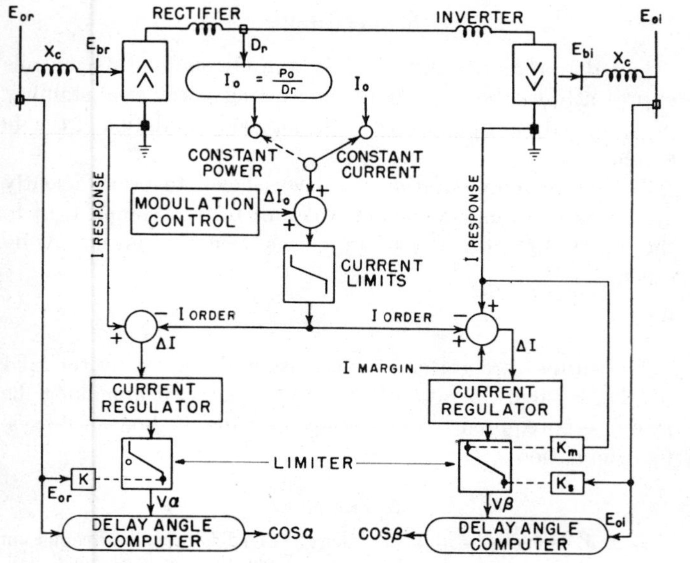

Fig. 6. Basic dc system control scheme.

behavior of the current and power factor of the dc terminals into the ac system solution. During the ac solution, the dc transmission is handled as constant-current loads whose values are adjusted each step.

The equation set used for the power flow is slightly modified for use in the stability program. The angle of advance at the inverter β is found from

$$\cos \beta = \cos \gamma_0 - \sqrt{2}\, X_c I_d / E_0. \tag{13}$$

Also,

$$\beta = \gamma_0 + \mu, \quad \mu = \text{angle of overlap}$$

$$\beta = \pi - \alpha.$$

$\beta_{\min}$ is limited to the minimum margin angle γ_0 plus the angle of overlap μ whenever the inverter is operating in CEA mode. When the inverter operates in CC mode, β is between $\beta_{\min}$ and the alpha stop (about 100 degrees). The alpha stop prevents the inverter from going into rectification by limiting $\beta_{\max}$ to about 80 degrees.

The dynamic behaviors of the firing angles α and β, respectively, are found from the transfer function of the current regulator and the delay angle computer simulation (Fig. 6). Both the rectifier and inverter current regulator are represented by the transfer function $K_g \{(1 + ST_1)/[(1 + ST_2)(1 + ST_3)]\}$. The output of the rectifier current regulator is limited between zero and KE_{or}, where K reflects $\alpha_{\min}$:

$$K = 1.0 + \cos \alpha_{\min}.$$

The output of the inverter current regulator is limited between $K_m I_d$ and $K_s E_{\text{oi}}$, where K_m reflects the current dependency of the minimum margin angle and is evaluated with initial values from the power flow solution, and K_s reflects the α_{stop}:

$$K_m = E_{\text{oi}}/I_d(\cos \gamma_0 - \cos \beta)$$

$$K_s = \cos \gamma_0 - \cos (180° - \alpha_{\text{stop}}).$$

The delay angle computer simulation solves the following for the firing angles:

$$\cos \alpha = V_\alpha / E_{\text{or}} - \cos \alpha_{\min}$$

$$\cos \beta = \cos \gamma_0 - V_\beta / E_{\text{oi}}.$$

The power factors of the bridges are a function of the delay angles, the dc current, and the ac voltages. For the rectifier:

$$\cos \theta_{br} = \cos \alpha \sqrt{1 + (6 I_d X_c / \pi\sqrt{2}\, E_b)^2 + 2(6 I_d X_c / \pi\sqrt{2}\, E_b) \sin \theta_{br}} - I_d X_c / \sqrt{2}\, E_b. \tag{12b}$$

And for the inverter:

$$\cos \theta_{bi} = \cos \beta \sqrt{1 + (6 I_d X_c / \pi\sqrt{2}\, E_b)^2 + 2(6 I_d X_c / \pi\sqrt{2}\, E_b) \sin \theta_{bi}} + I_d X_c / \sqrt{2}\, E_b \tag{12c}$$

where I_d, X_c, and E_b are values for the rectifier or inverter, respectively. The equation must be iterated to make the sine and cosine consistent.

Once the power factor is known from (12), the dc voltage is found from

$$D_r = (3\sqrt{2}/\pi) E_{br} \cos \theta_{br} - D_{\text{ad}} - I_d r_{\text{eq}} \tag{14a}$$

$$D_i = (3\sqrt{2}/\pi) E_{bi} \cos \theta_{bi} + D_{\text{ad}} + I_d r_{\text{eq}}. \tag{14b}$$

The new voltages D_r and D_i then are applied to the line and a new I_d calculated for the rectifier and inverter, which in turn changes the input to the current regulators, and so forth.

The dc system equations are solved on a smaller time step than the ac system equations, the limiting factor being the lesser of one-fourth the period of the natural frequency of the dc line when the line capacitance is represented and the smallest time constant in the current regulator amplifier. During the dc solution, the ac voltages E_{br} and E_{bi} are assumed to remain constant until the dc time equals the ac interval. Then the values of the constant-current loads representing the terminals are found and held constant while the ac system is calculated for its time interval.

Comparison Tests

The model system studied consisted of sending- and receiving-end systems each connected to a 375-kVA machine. The ac systems were connected by two ac transmission lines, one 300 miles and the other 250 miles in length. In parallel with these ac lines was a 300-mile bipolar dc transmission system. Static loads only were used in the studies.

The digital program output has been compared with actual runs made on the model. The performance of the dc portion of the program for a step change in current order is compared in Fig. 7, which shows the current response at the rectifier end of the dc transmission line. The current order was step changed from 25 to 75 amperes, and the time steps were small enough to show the transient behavior of the dc line current as calculated by the computer program. These tests were for monopolar operation of the dc line. The computer assumes an ideal, zero impedence ground return; the model uses a simulated earth return path. As would be expected, the computer output shows a faster rise time and less damping because of the ideal ground assumption.

Fig. 8 compares the ac-only behavior of the model system with the computer for a three-phase fault on the sending end of the 300-mile line. The fault was cleared in 3.6 cycles with the line reclosed after 20 cycles. The digital swing curve shows a longer period and slightly more damping than the model curve.

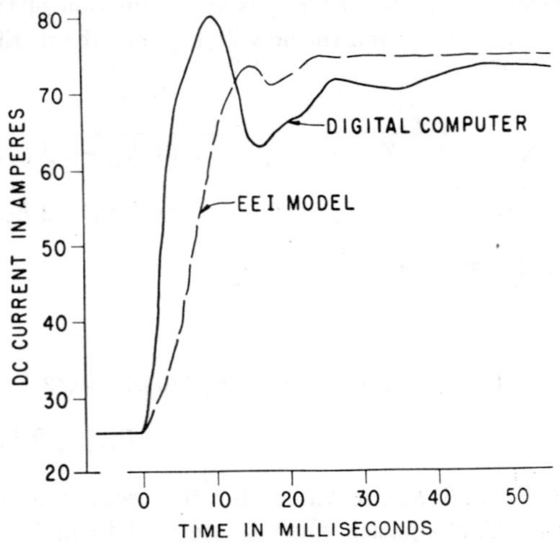

Fig. 7. Model computer comparison run for dc current order change, 25 to 75 amperes.

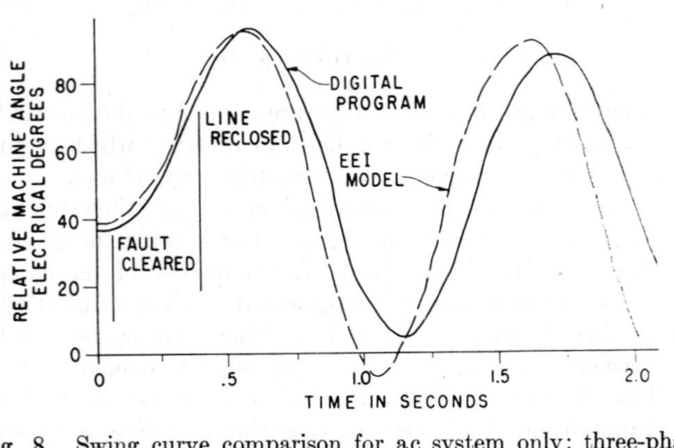

Fig. 8. Swing curve comparison for ac system only; three-phase fault on sending end.

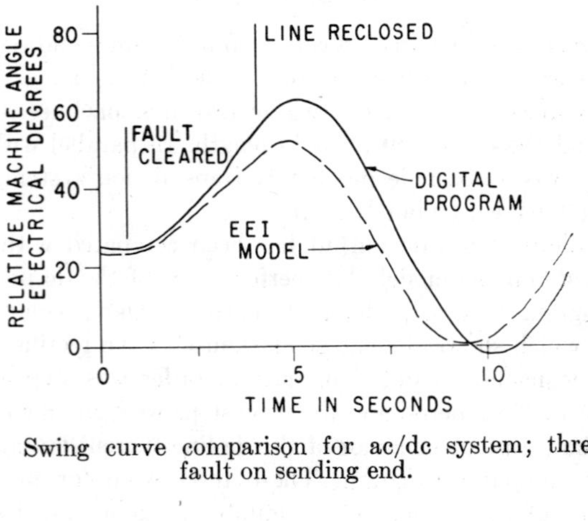

Fig. 9. Swing curve comparison for ac/dc system; three-phase fault on sending end.

A typical swing curve comparison for an ac/dc system is shown in Fig. 9. A three-phase fault was applied on the sending end of the 300-mile line. The fault was cleared in 4.1 cycles and the line reclosed in 20 cycles. In this case, the fault was cleared 0.5 cycles later than the ac-only case. The same differences in the period of the swing and in the damping as shown in Fig. 8 can be noted, indicating that the dc simulation has not contributed any appreciable discrepancy between the swing curves.

Conclusions

1) A digital representation of dc transmission and its controls as presented can be inserted into an existing ac transient stability program and can be expected to give a good simulation of the dc system.

2) The dc representation has been shown to be sufficiently accurate for several seconds of real time to be a complement to the more sophisticated ac programs recently developed by others.

Acknowledgment

The authors are grateful for the assistance given by the EEI AC/DC System Operation Research Project staff in making the model tests required to develop and test the methods of dc system simulation.

References

[1] J. J. Dougherty and V. Caleca, "The EEI ac/dc transmission model," *IEEE Trans. Power Apparatus and Systems*, vol. PAS-87, pp. 504–512, February 1968.
[2] E. Uhlmann, "The representation of an HVDC link in a network analyzer," CIGRE, Paper 404, 1960.
[3] N. G. Hingorani and J. D. Mountford, "Simulation of HVDC systems in ac load-flow analysis by digital computers," *Proc. IEE* (London), vol. 113, pp. 1541–1546, September 1966.
[4] C. C. Young and R. M. Webler, "A new stability program for predicting the dynamic performance of electric power systems," *Proc. Am. Power Conf.*, vol. 29, 1967.
[5] J. J. Dougherty and T. Hillesland, Jr., "Power system stability considerations with dynamically responsive dc transmission lines," this issue, pp. 34–45.
[6] J. B. Ward and H. W. Hale, "Digital computer solution of power-flow problems," *AIEE Trans. (Power Apparatus and Systems)*, vol. 75, pp. 398–404, June 1956.
[7] G. W. Stagg and A. H. El-Abiad, *Computer Methods in Power System Analysis.* New York: McGraw-Hill, 1968, ch. 10.
[8] Neiman, Glinternik, Emel'yanov, and Novitskii, "Dc transmission in power systems" (in Russian), transl., Israel Program for Scientific Translations, Jerusalem; available from Clearinghouse for Federal Scientific and Technical Information, Springfield, Va. 22151.
[9] C. Adamson and N. G. Hingorani, *High-Voltage Direct-Current Power Transmission.* London: Garraway, 1960.

Discussion

Gerhard Steinbrenner (Arizona Public Service Company, Phoenix, Ariz.): The authors have added to the list of available analytical tools for studying dc and/or ac/dc transmission systems. After following the description of the program capabilities, a number of questions arose which I believe deserve further discussion, and any comments by the authors will be appreciated.

1) The control system in Fig. 6 indicates two options for representing load at the terminals—constant current and constant power. Did the authors make any comparative studies, and if so, what were the results?

2) Reference [4] indicates that the limits of the current limiter are a function of the dc voltage. Fig. 6 does not indicate this. Would the authors like to comment?

3) Did the authors attempt to digitally simulate earth return for monopolar operation, and if so, did the computer results more closely approach the EEI model results (Fig. 7)? If the authors did not perform the above test, would they like to comment on the merits of such a representation?

4) In the Section *DC Control During Transients,* reference is made to a term I_{set}, which is not defined, unless I_{set} and I_{order} are used synonymously.

Manuscript received June 6, 1969.

5) I believe an indication of when per-unit values were used would increase the speed of comprehension of the reader.

The authors have indicated that they are presently studying such other aspects of digital representation of dc lines as faults on the dc line, three-terminal line operation, and line protection.

I feel sure that the authors findings on the above studies would be of great interest to the utility industry, and I hope that as their work progresses they will publish the results of their studies.

Arun G. Phadke (University of Wisconsin, Madison, Wis.): I would like to extend my congratulations to the authors for making available this additional information about the Philadelphia AC/DC Project of the EEI. I and my colleagues are familiar with some aspects of this project, and are happy to see one more paper describing the work done in Philadelphia.

I would welcome the authors' consideration and clarification of the following points:

1) It was not clear to me whether the authors' program is meant to be a general purpose program that can be used as a part of a large ac system stability program when dc transmission lines are to be included in the study, or whether it is a program primarily designed to represent the performance of the Philadelphia ac/dc model on a digital computer. Perhaps it is the authors' intention that the same program would serve in both capacities; if such is the case, is their program in small packages that may be exchanged for appropriate substitutes if (for example) the exciter, governor, or dc controls are of a type different from those shown in Figs. 2, 3, and 6?

2) It has been my impression that a detailed representation of the dc line control with its time constants T_1, T_2, and T_3 is unnecessary in a transient stability analysis of a power system. This seems to be the case, for example, if one considers the time scales of Figs. 7 and 8 (or 9). I would like to know the time-increment step size used in computing the curves of Fig. 7. Is calculation of such details a part of the stability program, or is Fig. 7 perhaps an illustration of this capability of the program not to be utilized generally in a stability study?

3) Have the authors considered any model-and-computer studies of unbalanced faults on the ac system near a converter bus? I think the model provides excellent opportunities for studies of this nature.

4) Equation (8) assumes that power is specified at the receiving end of the dc line for load-flow purposes. Occasionally, power at some intermediate location on the line may be specified. A slight modification of the computation procedure outlined by the authors should take care of such cases.

5) A comparison of Figs. 8 and 9 shows that the results from the model and from the computer program have the same order of difference between them in the two figures. However, Fig. 8 shows the model curve to have less damping, whereas Fig. 9 shows the digital program to have a smaller damping effect. Would the authors comment on this discrepancy?

6) Table I shows that the arc drop in a rectifier is 20 volts, and that in the inverter is 44 volts. I would like to know why there is this difference between the two arc drops.

7) Finally, I would like to ask the authors if they have had an opportunity to compare their approach to this problem with that of others who have published on this subject previously; and if they have made such comparisons, I would like to have their comments on the differences between their techniques and results and those of other authors.

Manuscript received June 6, 1969.

G. D. Breuer, R. V. Pohl, and **C. C. Young** (General Electric Company, Schenectady, N. Y.): We are very pleased that the authors' investigations and conclusions agree with those we have made. We agree that the method of representation of a dc system, as outlined here, can be inserted into a sophisticated ac stability program and give a good simulation of the dc system performance over a period of seconds of real time.

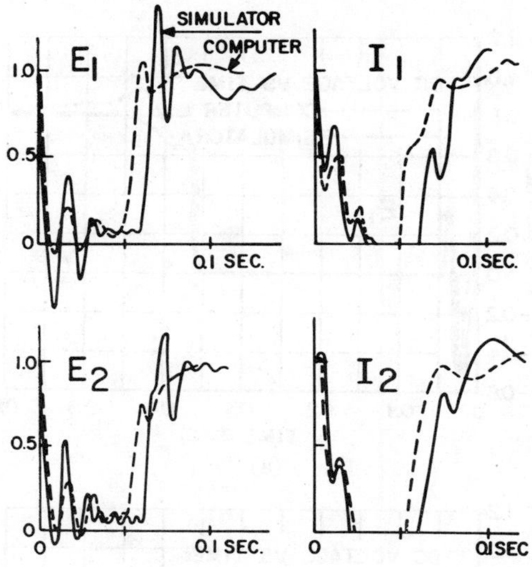

Fig. 10. Three-phase fault at rectifier.

The equations and the procedures described here are quite similar to those described by Breuer *et al.* in 1966 [10]. The basic differences in the equations are that the authors have included the effect of tube drop and converter effective resistance in their equations. This is, of course, a more general form of the equations, and since the tube drop in their model is 4 percent of the rectifier voltage and 9 percent of the inverter voltage, it is necessary to include these effects for a comparison study. Similarly, the converter losses for the model rectifier are about 3.75 percent of the rectifier power output and the model inverter losses are about 2.5 percent of the inverter power output. Since these total losses are about 4.5 times the line losses, and because of the importance of this effect upon the dc system damping, their simulation for this comparison study is also needed. In our experience, the representation of these effects is not important in the modeling of physical equipment. We would expect, for example, that the total valve drop for an 800-kV pole-to-pole rating would be about 0.25 percent of rated voltage, and that the other converter losses for a 1400-MW converter rating would be about 0.5 percent of rated power. This has led us to omit these effects from our computer program [4].

We made an equivalent series of comparisons between our ac/dc program and the General Electric ac/dc simulator [11], [12]. In making our comparisons, we adopted the philosophy that we would make the computer and simulator studies as independent as possible, and only compare our results at the conclusion of the studies. We believe that this is the most valid way of determining whether a computer program can effectively be used to predict the performance of physical systems. Three cases were studied: a three-phase fault at the rectifier terminals, a three-phase fault at the inverter terminals, and a pole-to-pole fault on the dc line. The results of the rectifier fault are shown in Fig. 10. The quantities E_1 and I_1 are the dc voltage and current at the rectifier; E_2 and I_2 are the dc voltage and current at the inverter. The fault clearing times used differed somewhat between the simulator and the computer program, which accounts for the difference in the reenergization time. The simulator exhibits less damping, but the rise and recovery characteristics are otherwise quite similar. The behavior of the dc voltages for the dc line fault are shown in Fig. 11 and the dc line currents are shown in Fig. 12. Again, there is excellent correlation between the computer and simulator results. The inverter fault also showed excellent correlation.

In assessing the role of this type of dc system representation in a computer program, we believe that it is highly suitable for evaluating the interaction of the ac and dc systems for stability studies. The representation is suitable for examining the effects of dc controls upon system behavior, and for investigating the use of supplementary stabilizing signals. When the ac representation is comparable, the combined ac/dc system behavior can be evaluated over a period of several seconds.

Manuscript received June 9, 1969.

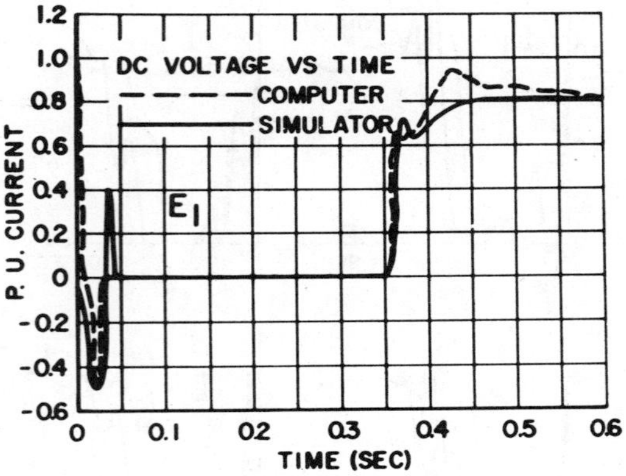

(a)

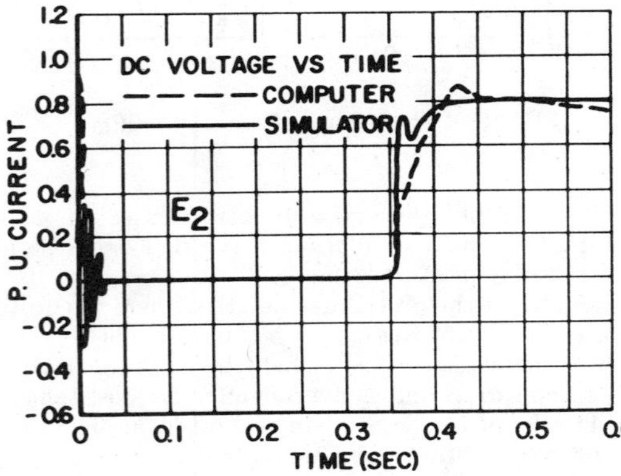

(b)

Fig. 11.

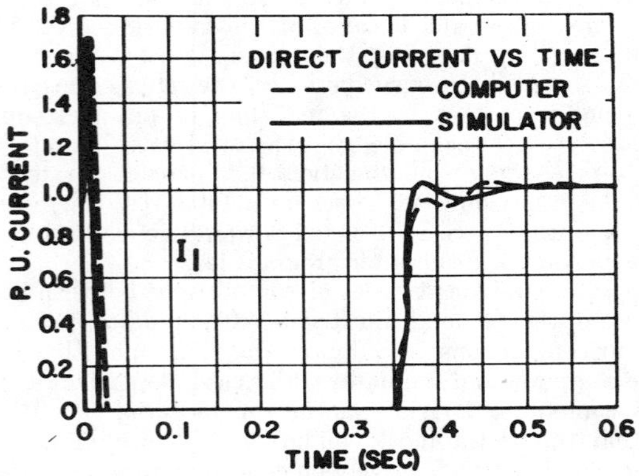

(a)

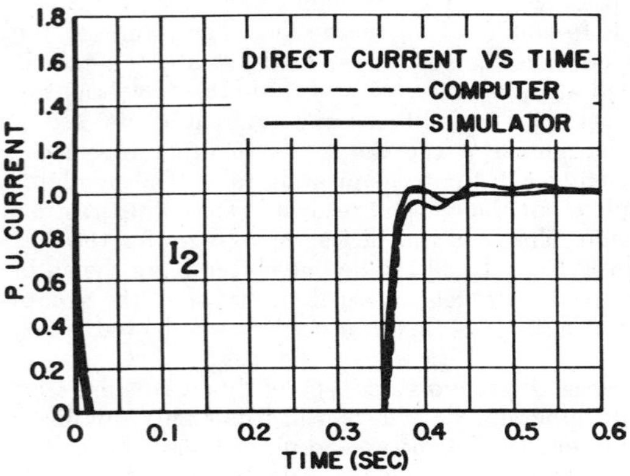

(b)

Fig. 12.

Although the representation can predict the effect of a large ac system impedance at the inverter upon the system's damping, it cannot be used to predict such problems as control system instability due to waveform distortions, or for the analysis of problems primarily associated with high-frequency or nonlinear effects. These types of studies will continue to require the use of ac/dc simulators or comparable computer programs.

References

[10] G. D. Breuer, J. F. Luini, and C. C. Young, "Studies of large ac/dc systems on the digital computer," *IEEE Trans. Power Apparatus and Systems*, vol. PAS-85, pp. 1107–1115, November 1966.

[11] G. D. Breuer, R. V. Pohl, and C. C. Young, "Comparison of HVDC simulator and digital computer programs for ac/dc system systems studies," presented at the IEEE Winter Power Meeting, New York, N. Y., January 28–February 2, 1968.

[12] G. D. Breuer, "The impact of solid state on HVDC systems," Publ. 68 C 57-PWR, presented at the IEEE EHV Transmission Conference, Montreal, P. Q., Canada, September 30–October 2, 1968.

James F. Clifford and **Albert H. Schmidt, Jr.:** The authors are gratified with the careful analysis shown by the discussers, and their valuable comments and criticisms are appreciated.

Mr. Steinbrenner asks three questions and indicates two improvements to the clarity of the paper. The control system shown in Fig. 6 indicates two options for control of the dc line—constant current or constant power. Our program will maintain constant power for the initial power flow balance regardless of which mode is desired for the transient stability study. The "load" seen by the ac system was always represented as constant current during the transient phase because the model was operating in constant-current mode.

Fig. 13 shows a very good comparison between the model performance and the digital simulation with 1) constant-current operation and 2) current modulation. In Fig. 13(b), however, the current was modulated as a function of the rate of change of the system power angle rather than as a function of the line voltage and the scheduled power. We believe that constant-current representation of the terminal "load" is all that is necessary.

The current limits in [4] are only to turn the dc line off when the ac system voltage is very low, whereas the limits shown in Fig. 6 are to prevent the constant-power control and the modulation control from driving the current order beyond the limits of +10 to 110 percent of rated dc line current.

The simulation of earth return for unbalanced or monopolar operation is not worth the effort. The difference shown in Fig. 7 will only show up for very small time steps. Normally, the time step will be much larger for usual transient stability studies. The line impedance could be slightly increased to simulate the extra damping of the earth, if desired for a particular study.

The term I_{set} should have been defined as $I_{set} = I_{order} - I_{margin}$. The paper should have indicated when per-unit values were used. The ac system and dc control simulation uses per unit; the dc line calculations are made using actual values.

To clarify the points raised by Dr. Phadke we offer the following:

1) The EEI small system program was developed primarily to simulate the ac/dc model. The dc portion of the model was simulated with small subroutines that were intended to be general purpose. These are being inserted into the large programs at the Bureau of Reclamation by modifying only the common statements to correct the data linkage. The details of the dc subroutines are described in Report S-302 from the Edison Electric Institute's AC/DC System Operation Research Project. Copies of the report may be requested from EEI headquarters in New York.

2) The time step used in computing the curves of Fig. 7 was 0.001 second for both the ac and dc systems. This is an illustration of this capability and would not normally be utilized. If the dc line capaci-

Manuscript received June 30, 1969; revised July 8, 1969.

Power System Stability Considerations with Dynamically Responsive DC Transmission Lines

JOHN J. DOUGHERTY, SENIOR MEMBER, IEEE, AND THOMAS HILLESLAND, JR., MEMBER, IEEE

Abstract—**An evaluation is made of the effectiveness of a dynamically responsive dc line in improving the steady-state and transient stability of a power system. A control system, incorporated into the Edison Electric Institute (EEI) 700-kW model, is described. Studies also reported compare the performance of a simple system consisting of two ac transmission lines with the same system when one of the ac lines is replaced by an equivalent dc line.**

Introduction

MODERN dc transmission was initiated when the 20-MW Gotland, Sweden, line was commissioned in 1954. From this modest beginning dc transmission has grown in the present decade to an installed capacity of 2000 MW in various parts of the world by the end of 1967, with an additional 4000 MW currently under construction. However, the latest milestone in the development of this new tool for the power industry was passed in February 1968 with the commissioning of a small dc transmission link between the city of Vancouver and Vancouver Island, British Columbia, Canada. This is the first dc transmission line to be integrated into an ac network. It is with this aspect of dc transmission that the Edison Electric Institute (EEI) AC/DC System Operation Research Project is primarily concerned.

Paper 69 TP 17-PWR, recommended and approved by the Power System Engineering Committee of the IEEE Power Group for presentation at the IEEE Winter Power Meeting, New York, N. Y., January 26–31, 1969. Manuscript submitted September 13, 1968; made available for printing November 7, 1968.

J. J. Dougherty is with the Philadelphia Electric Company, Philadelphia, Pa., assigned to the Edison Electric Institute AC/DC System Operation Research Project.

T. Hillesland, Jr., is with the Pacific Gas and Electric Company, Emeryville, Calif. He was with the Edison Electric Institute AC/DC System Operation Research Project, Philadelphia, Pa.

In a previous paper [1] a detailed description of the EEI's intermediate-power-level model was given, and a program of planned studies was outlined. There have been some deviations from the original plans, but most of the initial program has been carried out [2]. One of the primary objectives of the research project, and perhaps the most interesting from an overall system viewpoint, is reported in this paper.

The Basic Premise

Transient stability problems become more complex as power systems grow and interconnections increase. Although many factors are involved, a major contributor to stability problems is the relatively slow response of the ac system to major disturbances, especially when transmission lines are long. When dynamic studies were initiated on the EEI model, with typical transmission distances of 300 miles represented, it was noted that several hundred milliseconds (ms) would elapse before a disturbance on an ac bus would result in a response over an ac transmission line of such a length. However, the control system associated with a dc transmission line can vary the power flow over the line at very rapid rates and, of course, is independent of ac-system power angles. In a preliminary test, the dc terminal controls were supplied with step changes in current order to determine the ability to provide system damping during transient disturbances. The response time is so fast (approximately 10 ms for 1.0-per-unit change in rated current) that it became obvious that a dc line, under certain circumstances, could

Reprinted from *IEEE Trans. Power App. Syst.*, Jan. 1970.

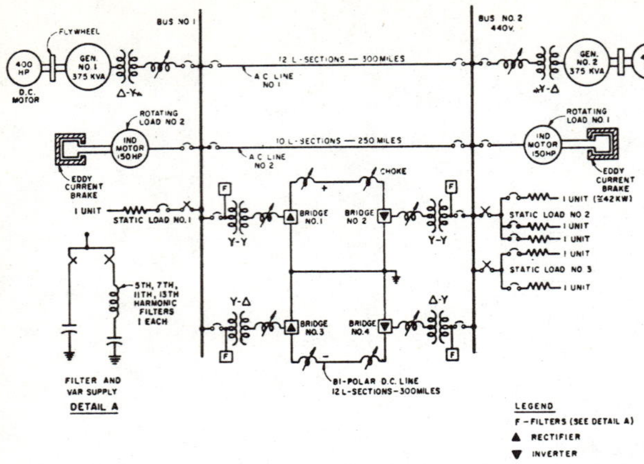

Fig. 1. Single-line diagram of EEI model system.

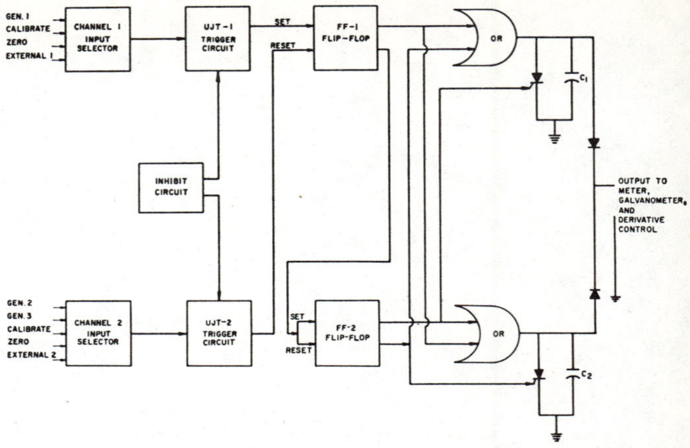

Fig. 2. Simplified diagram of machine angle computer.

completely squelch a severe system disturbance if it could be automatically instructed to make a specific change in its power flow at the proper time. From this premise several parameters of the ac system were monitored and were used to generate control signals for dc. The most successful of these efforts, using the rate of change of the power angle of machine rotors, is reported here.

System Studied

The studies were conducted on a relatively simple system of two three-phase ac lines and one bipolar dc line in parallel between two main ac buses. A single-line diagram of the components is shown in Fig. 1. Since detailed description of these components has been previously given [1], it suffices here to point out that the transient response of the model system, as indicated in earlier tests, is typical of modern large-scale turbo-alternator characteristics. The dc transmission line is equipped with conventional converter controls similar to those in service on actual transmission lines, except that they are limited essentially to the constant-current mode of operation. Inherently, the model system used in these tests has little natural system damping, despite the inclusion of fast-acting voltage regulators and modern governor action simulation. In most of the tests the dc line was loaded to one half its thermal rating and the ac lines were loaded to approximately one half their surge impedance loading. These load levels were chosen deliberately to create a system that would be only marginally stable for disturbances involving momentary interruption of one of the ac lines.

Details of Special Control Circuits

The special control circuits used in these tests consist of a machine angle computer and the derivative control unit. The first device provides a dc output voltage proportional to the phase displacement between two generator rotors or two ac voltages. The signal used for most of these tests was the angular difference between the rotors of generators 1 and 2.

Fig. 2 is a simplified diagram of the machine angle computer. Pulses of reflected light from the generator rotors discharge a light-sensitive silicon-controlled rectifier (SCR) circuit with the output fed to the machine angle computer. The pulses for the generator rotors trigger the unijunction transistor circuits, which in turn set and reset the first flip-flop FF-1. One output of FF-1 drives the circuits for the two charging capacitors, C_1 and C_2, while the other output operates the second flip-flop FF-2. This flip-flop determines the time that C_1 and C_2 can be charged.

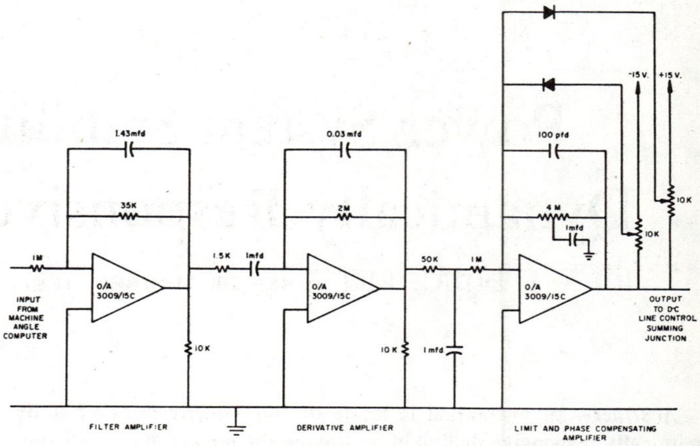

Fig. 3. Derivative control circuit ($D\theta$ control).

When one capacitor is fully charged the other is discharged through a SCR so that it can be recharged in the next cycle. The ripple in the output is reduced by filtering. Hence the output is a dc voltage proportional to the angular difference in the machine rotors.

Fig. 3 is the derivative control circuit ($D\theta$). The circuit was designed using commercially available operational amplifiers. The first stage provides additional filtering and some amplification of the machine angle signal. The second stage is the derivative unit itself, and the following stage amplifies and limits the derivative signal. For a power swing with a one-second period, the complete derivative system has a phase shift of 95 degrees, or 5 degrees greater than the normal degree displacement due to differentiation. A phase-shifting circuit was required in the last stage to compensate for the phase lag introduced by the input filtering.

In developing these controls, the overall gain of the derivative system had to be optimized, although this was not a critical adjustment. Obviously, with too low a gain, less than the desired response would be obtained. On the other hand, if the gain of the system is too high, the magnitude of variation in power flow on the dc line becomes excessive for any variation in angular difference between machine rotors, and the system becomes steady state unstable.

Limit circuits that have been incorporated in this control permit a predetermination of the minimum and maximum levels to which the responsive control can drive the dc line. The maximum limit may be necessary to prevent overloading the

TABLE I
REMOTE AC FAULTS

Figure	Series 0158 Run	Initial Angle (degrees)	Power Angle Swing First Positive Peak (degrees)	Power Angle Swing Second Positive Peak (degrees)	Final Steady-State Angle (degrees)	Damping Time* (seconds)	Remarks
4	13	38	83	89	—	Unstable	No reclosure, $D\theta$ off
5	10	38	74	72	38	>4.5†	Reclosure, $D\theta$ off
6	12	38	61	57	57	0.3	No reclosure, $D\theta$ on
7	11	38	60	39	38	1.0	Reclosure, $D\theta$ on
8	14	49	107	112	—	Unstable	Reclosure, dc line at 90 amperes, $D\theta$ off
9	15	49	98	73	49	2.7	Reclosure, dc line at 90 amperes, $D\theta$ on, limits on dc current
—	16	49	Slip-pole	—	—	Unstable	No reclosure, dc line at 90 amperes, $D\theta$ on, limits on dc current
—	21	49	Slip-pole	—	—	Unstable	No reclosure, dc line at 90 amperes, $D\theta$ off

* Time to reduce oscillation to +5 or −5 degrees from final steady-state angle.
† Damping to ±5 degrees not achieved up to indicated time, when test was terminated.

dc line, or more specifically, the converter units. Even if appreciable overload capability were provided for in the design of the converter units, it may at times be inadvisable to utilize it where stiffness of the sending-end ac system is inadequate. The low limit prevents the dynamic control from driving the dc line to zero current, which is usually an undesired condition. Similarly, too close an approach to zero current may produce a phenomenon known as discontinuous conduction in the converter units and result in undue stress on components of the converter terminal. A normal setting for the lower limit is 10 percent of the line's rated dc current.

Choice of Control Parameter

In designing an on-line control for modulating the power flow on a dc transmission line, any parameter of the ac system that varies during the system disturbance could be used as an input. The current or power level of parallel ac lines, the total power flow between buses, the magnitude of power angle between buses, or even the frequency or rate of change of frequency at a given bus could conceivably be used to generate a useful control signal. In fact, several of the above parameters were experimentally tried in model studies before the derivative control unit was developed. This control was adopted because, for the relatively simple system of the model, it proved to have the most universal application for all types of system disturbances. However, it must be emphasized that these studies should not be misinterpreted as a promotion, or even a recommendation, that such a control system be adopted. For an actual system, which will generally be far more complex, studies should be made to determine what system parameter or combination of parameters would best provide the desired control signal. The purpose of these studies is merely to demonstrate the ability of a dc line to improve the transient stability of an associated ac system, if provided with the necessary input information.

In other studies, not included in this paper, a totally different control concept was developed and tested. Those tests demonstrated the feasibility of storing various specific current orders at various points in an ac system dynamically, and only applying this information to the main dc current control for specific contingencies, such as the loss of a particular transmission line.

In the final analysis, the significant fact is the rapid controllability of a dc transmission line. The mode of utilizing this controllability will be limited only by the imagination of the engineers studying the problem.

Test Series

The tests conducted during this series might conveniently be divided into five different types, catalogued in accordance with the nature of the disturbance that initiated each group.

In each instance the tests were run with a conventional dc line instructed merely to maintain constant current, and then with a dc line being supplied with a control input that would modulate the power flow on the dc line in such a manner that the system swings caused by the initial fault were opposed.

First, tests were run in which a three-phase fault was placed at the middle of one of the ac transmission lines and the model programmed to clear this fault by tripping the breakers at the ac line terminals after a fault duration of four cycles. In some instances the fault was removed from the now deenergized ac line and reclosure of the line was attempted. In other instances, no reclosure of the faulted ac line was attempted.

Additional tests were run with the ac line faults located at either the rectifier or the inverter bus, to provide maximum insurance that normal functioning of the dc line would be disrupted for the short duration of the fault.

Similarly, relatively small step changes in load, both increases and decreases, were applied to the system to determine the damping effect of a responsive dc line for relatively small system disturbances. Finally, tests were run involving double-pole faults on the dc line itself, both with and without responsive control signals to modulate the power flow on the dc line after restoration of the line.

ANALYSES OF TEST RESULTS

General

One obvious advantage of using the first derivative of a power angle swing is that the corrective control input magnitude decreases progressively as a function of the effectiveness of the resulting modulation of dc power flow in damping the system disturbance. As a corollary of this, there is an inherent disadvantage in that the dc line must eventually return to its predisturbance value after the system swing has been damped out. On a realistic system it might at times be preferable if

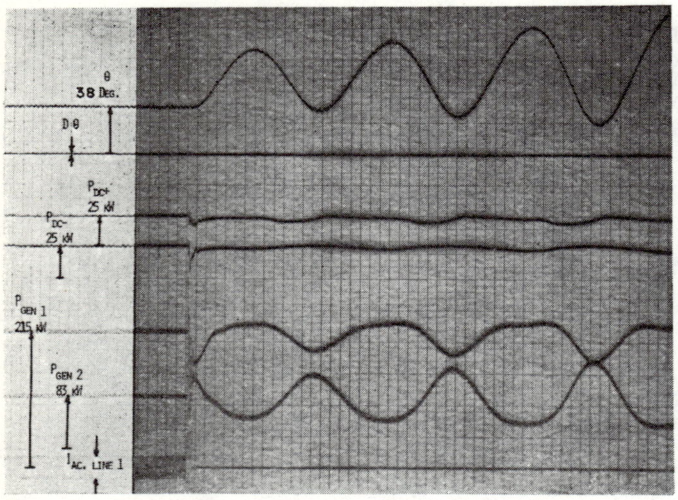

Fig. 4. Remote ac line fault, no reclosure; $D\theta$ off.

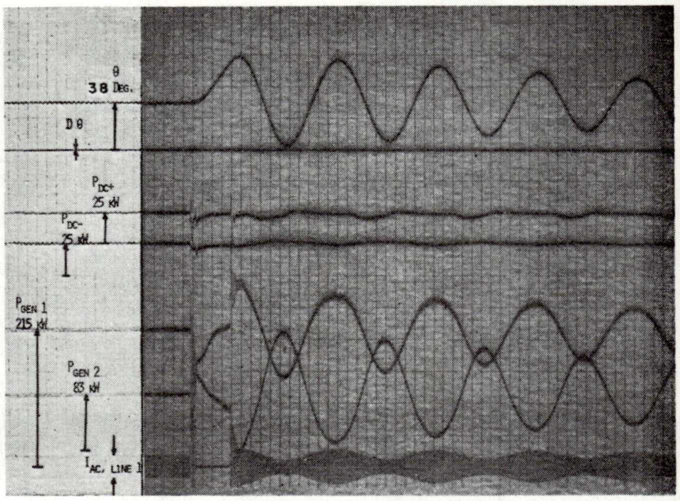

Fig. 5. Remote ac line fault and reclosure; $D\theta$ off.

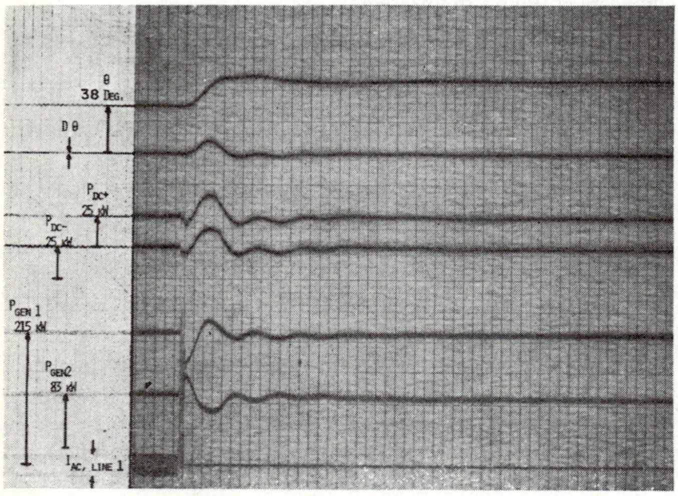

Fig. 6. Remote ac line fault, no reclosure; $D\theta$ on.

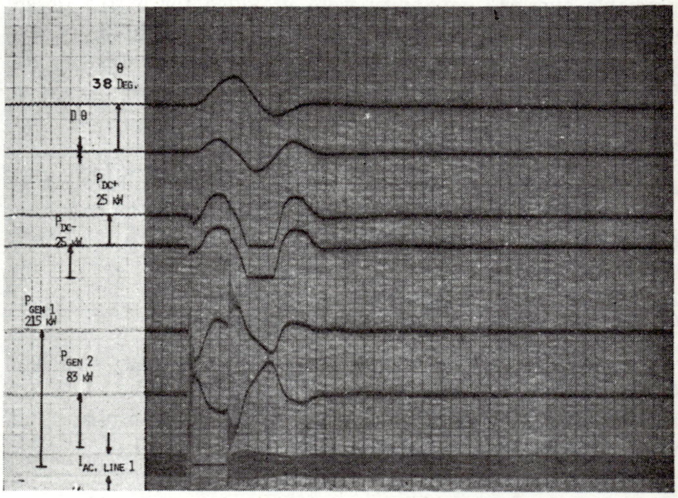

Fig. 7. Remote ac line fault and reclosure; $D\theta$ on.

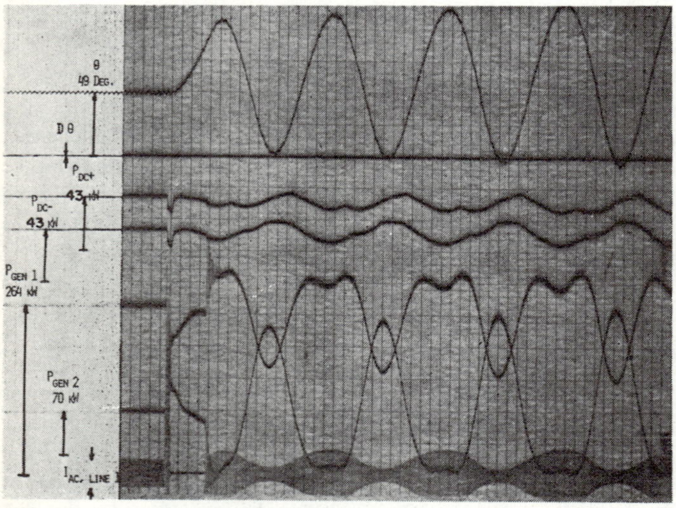

Fig. 8. Remote ac line fault and reclosure; dc line at 90 percent of rating; $D\theta$ off.

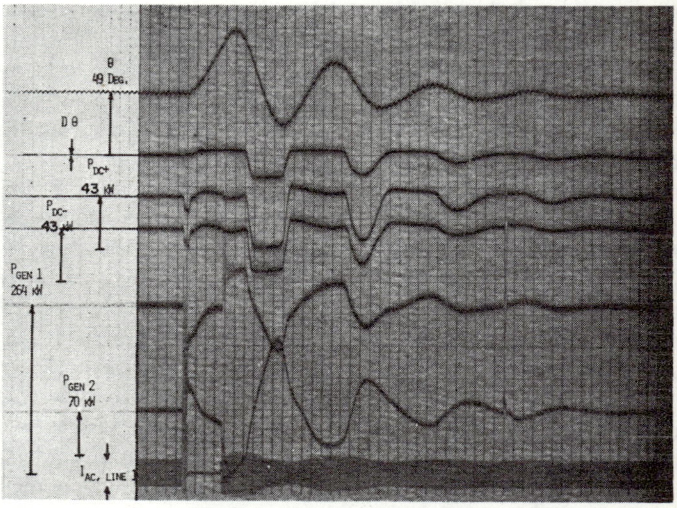

Fig. 9. Remote ac line fault and reclosure; dc line at 90 percent of rating; $D\theta$ on (10- and 110-percent limits).

some of a lost ac transmission link's capability would be picked up by a dc line to avoid possible overloading of other ac tie-lines. Such an option, of course, can be incorporated by minor modifications to the control circuitry.

Remote AC Faults

In this test series involving remote ac faults, a three-phase–ground fault was placed near the center of a parallel 300-mile ac line so that the point of fault was 125 miles from the rectifier bus. Under these conditions, no converter failures were caused by the fault itself; as can be seen in the oscillogram in Fig. 4,[1] there is a mild oscillation in dc power flow during the fault and very little other change in the dc power flow until after the system has actually gone unstable. Table I lists the various system conditions tested during this series, and lists in tabular form the degree of ac system swing between the two machine rotors on its first and second positive peaks, as well as the total time to damp out the system disturbance.

As can be seen in this table or in the oscillograms (Figs. 4–9), this type of ac line fault without successful reclosure and without a responsive dc line causes the system to go into instability as the angular swing increases on successive cycles until a pole is slipped. The oscillogram in Fig. 5 shows that for the same test conditions, a successful reclosure of the faulted ac line after 400 ms permits the system to maintain stability without assistance from the dc line. However, the natural rate of damping for this particular system configuration is low, and the system was still oscillating at the end of six seconds, when the test was terminated.

Fig. 6 gives an indication of the damping available from a suitably controlled dc line. In this oscillogram an increased current order is supplied to the dc line dynamically as a function of the first derivative of the power angle swing (shown in the oscillogram as $D\theta$). In this test, by changing the power flow on the dc line the initial power angle swing is damped to approximately the final new power angle in the first 400 ms, and a small oscillation is rapidly damped out thereafter.

Fig. 7 shows another test run on the same conditions in which the derivative control initiates system damping immediately after fault initiation. However, in this instance, successful reclosure of the previously faulted ac line is again initiated in about 400 ms, and is seen by the system as a new and disturbing influence. Consequently, continued modulation of the dc power flow is required and the incremental changes are appreciably larger than before. Nevertheless, the system is restored to stable operation in approximately 1.2 seconds, demonstrating the versatility and effectiveness of a dynamic "on-line" control.

Application of Limits

The presently used commercial converter facilities have little or no inherent overload capability. Because of this, and because of the probability that dc bulk power transmission lines may frequently be operated fully loaded, it becomes of interest to evaluate the performance of a dc line as a damper for system disturbances when the line must be limited to little or no increase beyond its predisturbance level. Figs. 8 and 9 show such an evaluation.

In Fig. 8 the same type of remote fault with 400-ms reclosure was applied on essentially the same system but with the dc line carrying 90 percent of its current rating prior to the initiation of the disturbance. The ac line loadings were maintained at about 50-percent surge impedance level. The derivative control was deactivated, and a marginally unstable system disturbance resulted. The variations in dc power flow are the consequence of variations in ac bus voltages, since the dc line is operating in a constant-current mode. In Fig. 9 the same disturbance is applied but the derivative control is operable, and limits have been added to it that prevent the dc line from exceeding 110 percent of its rating in the positive direction and prevent the current flow from being reduced to less than 10 percent on the negative swings.

As a consequence of these limitations, there is little damping on the first positive swing of the ac-system power angle. However, the decrease in dc power flow at the proper time produces appreciable damping on the first negative swing and the period of the oscillation is shortened. The system disturbance is essentially damped out in approximately 3.0 seconds. Runs 16 and 21 (Table I) show that, without successful reclosure of the ac line, the system goes unstable on the first positive swing. The $D\theta$ control is ineffective under these conditions.

This concept of limited response from the dc line may be important even if future converting equipment has large overload capability. It is quite conceivable that during a system disturbance a sending ac bus may not be sufficiently stiff to be able to supply a sudden demand for doubling the power flow on a dc line already carrying perhaps 1000 MW, and the ensuing deterioration of bus voltage could be disruptive.

AC Faults at Rectifier Bus

In this series of tests, the fault location was moved to the bus-1 end of the 300-mile ac line. With this fault location, the dc power flow is interrupted for the 4-cycle duration of the ac fault, as a consequence of the collapse of bus-1 voltage. However, no blocking of the bridges occurs and the dc power is rapidly restored to normal as soon as the fault is cleared.

In this test sequence the system was immediately unstable for such a fault, if the faulted line was not reclosed and the dc line was operated in the constant-current mode. In fact, as shown in Table II, the system was only marginally stable with a successful reclosure of the faulted line after 400 ms. However, if the $D\theta$ control was energized, the system disturbance was rapidly damped out, with or without the reclosure of the ac line. The oscillograms of these tests (not shown) were quite similar to those shown in Figs. 6 and 7 except that the magnitude of initial system swing is slightly larger due to the brief interruption of power flow on the dc line during the fault.

AC Faults at Inverter Bus

In this test series the ac fault was moved to the bus-2 end of the 300-mile ac line. In effect, the ac fault now causes a collapse of voltage at the inverter bus, which results in commutation failures in the inverter, and interruption of the dc line power flow for the duration of the ac fault. However, when the fault is cleared, commutation failures are self-clearing, and the dc power flow is restored to normal. The restoration of power flow on the dc line is not as rapid as for the rectifier bus fault, and consequently the disturbance is somewhat more severe.

As was the case with the rectifier bus fault, the system was unstable immediately when the faulted ac line was not reclosed and the dc line was operating in the constant-current mode.

[1] For identification of traces in oscillograms (Figs. 4–11), see Table VI.

TABLE II
AC Faults at Rectifier Bus

Series 0158 Run	Initial Angle (degrees)	Power Angle Swing: First Positive Peak (degrees)	Power Angle Swing: Second Positive Peak (degrees)	Final Steady-State Angle (degrees)	Damping Time* (seconds)	Remarks
25	42	96	133	—	Unstable	No reclosure, $D\theta$ off
2	42	67	67	67	0.25	No reclosure, $D\theta$ on
0	42	86	85	42	>6.0†	Reclosure, $D\theta$ off
1	42	69	45	42	1.1	Reclosure, $D\theta$ on

* Time to reduce oscillation to +5 or −5 degrees from final steady-state angle.
† Damping to ±5 degrees not achieved up to indicated time, when test was terminated.

TABLE III
AC Faults at Inverter Bus

Series 0158 Run	Initial Angle (degrees)	Power Angle Swing: First Positive Peak (degrees)	Power Angle Swing: Second Positive Peak (degrees)	Final Steady-State Angle (degrees)	Damping Time* (seconds)	Remarks
26	42	Slip-pole	Slip-pole	—	Unstable	No reclosure, $D\theta$ off
8	42	112	71	70	2.15	No reclosure, $D\theta$ on
6	42	135	140	—	Unstable	Reclosure, $D\theta$ off
7	42	107	67	42	2.12	Reclosure, $D\theta$ on

* Time to reduce oscillation to +5 or −5 degrees from final steady-state angle.

TABLE IV

Series 0158 Run	Initial Angle (degrees)	Power Angle Swing: First Positive Peak (degrees)	Power Angle Swing: Second Positive Peak (degrees)	Final Steady-State Angle (degrees)	Damping Time* (seconds)	Remarks
31	43	28‡	29‡	35	2.6	Drop load, $D\theta$ off
32	43	32‡	35‡	35	0.1	Drop load, $D\theta$ on
33	36	52	52	44	>6.0†	Pickup load, $D\theta$ off
34	36	47	44	44	0.1	Pickup load, $D\theta$ on

* Time to reduce oscillation to +5 or −5 degrees from final steady-state angle.
† Damping to ±5 degrees not achieved up to indicated time, when test was terminated.
‡ Swings in negative direction.

Even with a reclosure in 400 ms, the system was marginally unstable without dynamic response from the dc line. However, with the $D\theta$ control energized the disturbance is rapidly damped with or without the successful reclosure of the faulted ac line. Table III shows the statistical results of these tests.

Step-Load Changes

It was also desirable to evaluate the sensitivity of the derivative control for relatively minor system disturbances, and this was accomplished by making step changes in the static loads on bus 2. Table IV shows the results of such step changes both with and without the $D\theta$ control. Although the steady-state change in machine rotor angles was relatively small (8 degrees) this responsive control of the dc line was very effective in making this power angle change quickly and smoothly.

Although no stability problems are involved in disturbances of this magnitude, it does appear that an "on-line" control, which took advantage of the speed of response of the dc line, might prevent voltage flicker on ac systems due to the pickup or dropping of moderately large industrial loads.

DC Line Faults

The effectiveness of a responsive dc line was also evaluated for its ability to damp disturbances initiated by faults on the dc line itself. Since the nature of the dc line permits a line fault to involve only one pole or both poles, both types of disturbances were studied, but only the double-pole fault, which is more severe, is reported on here.

Since the model dc line is equipped with commercially available dc line fault protection equipment, this equipment was used in its normal mode. In effect, a double-pole fault was placed on the dc line and the rapidly acting voltage derivative sensing unit identified the fault and interrupted the fault by grid control action at the rectifier terminal. The control was set during these tests for a minimum dc line outage time of approximately 130 ms. Since the fault was removed during this time, successful restarts were obtained.

The data of these tests are shown in Table V and also in Figs. 10 and 11. Fig. 10 shows a relatively small but lightly damped oscillation without responsive control of the dc power flow; Fig. 11 shows a smaller and quickly damped disturbance

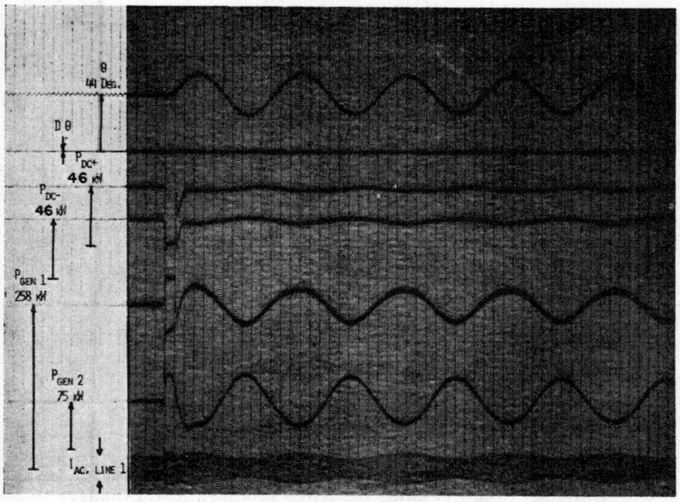

Fig. 10. Remote dc line fault, double-pole–ground with line protection operation; dc line at 100 percent of rating; $D\theta$ off.

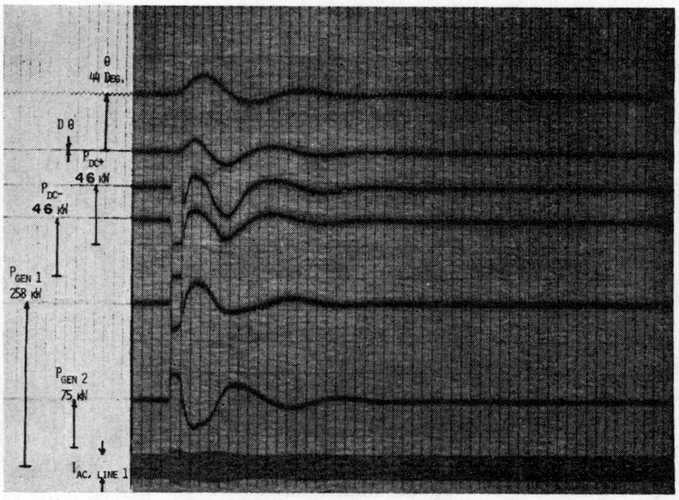

Fig. 11. Remote dc line fault, double-pole–ground, with line protection operation; dc line at 100 percent of rating; $D\theta$ on.

TABLE V

DC Line Faults at Midpoint of 300-Mile Line*

Figure	Series 0158 Run	Initial Angle (degrees)	Power Angle Swing: First Positive Peak (degrees)	Power Angle Swing: Second Positive Peak (degrees)	Steady-State Angle (degrees)	Damping Time† (seconds)	Remarks
10	19	44	61	60	44	>5.0‡	Bipolar dc line at 100 amperes, $D\theta$ off
11	20	44	60	48	44	0.54	Bipolar dc line at 100 amperes, $D\theta$ on

* Normal line protection restarting time 130 ms.
† Time to reduce oscillation to +5 or −5 degrees from final steady-state angle.
‡ Damping time to ±5 degrees not achieved up to indicated time, when test was terminated.

TABLE VI

Identification of Oscillogram Traces

Symbol	Description
θ (power angle)	Dc voltage proportional to angular difference between generator rotors
$D\theta$	Derivative of generator angle difference
$P_{DC}{}^{+}$	Power flow in positive pole of dc line
$P_{DC}{}^{-}$	Power flow in negative pole of dc line
$P_{GEN.1}$	Power output of sending-end generator
$P_{GEN.2}$	Power output of receiving-end generator
$I_{AC,\ LINE}$	Current in faulted ac line
100 ms	Time–line interval

when $D\theta$ is energized. Of particular interest here is the speed with which the dc line is restored to normal even without the $D\theta$ control. The speed with which the line is effective as a damper following its controlled outage time is also significant.

Steady-State and Transient Stability Considerations

In order to show the effectiveness of the damping characteristics of a responsive dc line, additional tests were run to compare the steady-state and transient stability limits of a simple transmission interconnection containing two ac lines only, or a combination of one ac and one dc line. For these tests three conditions were compared:

1) ac lines 1 and 2 in parallel; no dc line

2) ac line 1 replaced by the equivalent-length dc line, without responsive controls, in parallel with ac line 2

3) same as 2), but with $D\theta$ control energized.

Table VII shows the variation in loading possible on ac line 2 for the three cases. It will be noted that a constant-current dc line, substituted for ac line 1 and set for 75 kW [the power carried in 1)], reduces the allowable maximum loading on ac line 2 from 85 to 65 kW. However, when the derivative control was energized, the loading of ac line 2 could be increased to the thermal limit of the model (140 kW) without creating any steady-state problems. At 140 kW the bus angle (the electrical angle between the two ends of the ac line) was 66 degrees, compared with 32 degrees when the ac line was loaded to 85 kW.

Fig. 12 shows the variation in the transient stability for the same three cases. Here a three-phase fault of 6-cycle duration was placed on bus 1, and the line loadings were varied to determine the transient stability limit in each instance. As would be expected from the steady-state data, the lowest permissible loading for transient stability was the case of parallel ac and dc

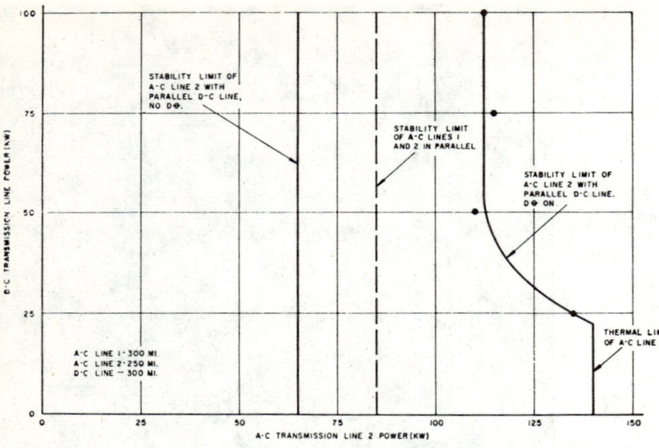

Fig. 12. Transient stability of parallel ac and dc lines for three-phase fault on bus 1.

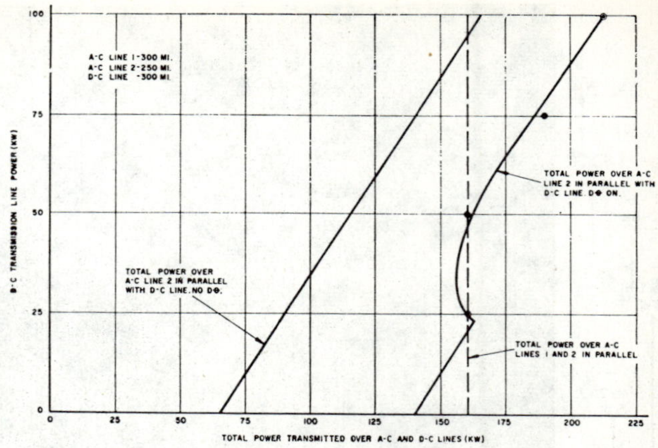

Fig. 13. Total power transmitted over ac and dc lines loaded to transient stability limit for three-phase fault on bus 1.

TABLE VII
STEADY-STATE STABILITY LIMITS OF PARALLEL AC AND DC TRANSMISSION LINES

Line Configuration	AC Line 1 300 Miles Maximum Load (kW)	AC Line 2 225 Miles Maximum Load (kW)	DC Line 300 Miles Load Set Same as AC Line 1 (kW)
Ac lines 1 and 2 in parallel, no dc line	75	85	Off
Ac line 2 in parallel with dc line, $D\theta$ off	Off	60	75
Ac line 2 in parallel with dc line, $D\theta$ on	Off	140*	75

* Thermal limit.

lines without responsive control. The highest permissible loading was, of course, an ac line in parallel with a dc line utilizing derivative control. However, the shape of this curve was somewhat unexpected, indicating that the allowable load on the ac line is not improved by reducing the flow on the parallel dc line from full rating down to the 50-percent level. Only below this point (between 50 and 25 kW) could the greater margin for damping in the dc line be converted into higher permissible loadings on the parallel ac line.

When these data are replotted, as in Fig. 13, to show the total power transfer capability of both lines, a clearer picture emerges. Under these circumstances the total power that could be transferred between the buses is plotted against the steady-state load level on the dc line. For example, ac lines 1 and 2 can carry 75 and 85 kW, respectively, and remain stable for the disturbance described. Replacing ac line 1 with a responsively controlled dc line, set for 75 kW, allows ac line 2 to be loaded to 112 kW and retain transient stability. The total power transfer has been increased by 17 percent over the 160 kW carried by two ac lines, and by 34 percent over the constant-current dc line in parallel with one ac line. With the dc line loaded to its rating, the power transfer with derivative control is 32 percent higher than the two ac line configuration and 28 percent more than one ac line in parallel with a fully loaded, constant-current dc line. Even for this two-line example, it will be noted that the total power transfer could be appreciably increased by substituting a dynamically controlled dc line for one of the two ac transmission lines.

CONCLUSIONS

1) The very rapid control of power possible on dc transmission can provide strong damping to disturbances within the power system. Tests that were conducted clearly demonstrate this ability for a variety of system disturbances.

2) Even if the source of the disturbance is a temporary outage of the dc line itself, its dynamic response provides effective damping immediately after restarting the line.

3) The $D\theta$ control used in these tests was the most effective for the EEI model system. For any actual installation, specific studies should be made to determine what parameter or combination of parameters should be used to provide system data to the dc controls.

4) For the conditions present in the EEI model, a constant-current dc line reduces the transfer capability somewhat from an all-ac system.

5) Substitution of a dynamically responsive dc line for one of two parallel ac lines will provide a marked improvement in the transfer capability, using either steady-state or transient stability as the criterion.

ACKNOWLEDGMENT

The authors gratefully acknowledge the assistance of James F. Clifford, Edward M. Gulachenski, James H. Malinowski, David W. Meikle, William C. Pokorny, and Jeffrey E. Sekerke, members of the EEI Project Staff, in conducting these tests and in the preparation of material for this paper.

References

[1] J. J. Dougherty and V. Caleca, "The EEI ac/dc transmission model," *IEEE Trans. Power Apparatus and Systems*, vol. PAS-87, pp. 504–512, February 1968.

[2] ——, "Studies on the integration of a dc transmission line into an ac network," CIGRE, Rept. 43-03, 1968.

Discussion

Olov G. Berglund (ASEA Electric, Inc., New York, N. Y.): The authors have presented a very clear and lucid demonstration of the faculty of high-voltage direct-current (HVDC) transmission to effectively damp out stability swings on ac systems. Of the many good points brought forth in the paper there is in particular one that deserves special mentioning. I refer to Figs. 8 and 9, demonstrating the ability of a dc line to effectively damp the swings on the ac system, even with very small variations in transmitted dc power.

This result confirms theoretical investigations by Dr. Uhlmann [3] of our organization. It may first appear rather surprising, but it is in fact as natural as it is significant. It is easy to understand, if it is remembered that stability swings on an ac power system are an exchange of rotating energy between various machines on the system.

The magnitude of the swing-damping measures is, therefore, not very large or important. Decisive for the damping of the power swings is rather that the damping action is rapid enough and inserted in the right moment to counteract the network's own swings. In other words, flexibility in power control and rapidity in action are the key. Here is where HVDC offers outstanding qualities, which the authors have so convincingly demonstrated.

Another interesting factor brought out by Fig. 9 is that the variation in transmitted dc power is of very short duration. The individual swings in the dc power are of the order of fractions of a second, and the total corrective action has a length of only a few seconds. This means that no thermal limitations in the dc station equipment enter into the picture.

The consequence of the above findings is that a dc link can be loaded to its rated value or very close to its rated value and still be useful for damping of stability swings on the associated ac systems. For engineers active in the HVDC technique this faculty of an HVDC transmission stands out as one of the most interesting and potentially most useful for future application of the technique.

References

[3] E. Uhlmann, "Stabilization of an ac link by a parallel dc link," *Direct Current*, vol. 9, pp. 89–94, August 1964.

Manuscript received February 13, 1969.

R. Foerst (400 kV-Forschungsgemeinschaft, Heidelberg, Germany): We welcome these investigations, since they complement extremely well the tests made in the HVDC transmission model of the 400 kV-Forschungsgemeinschaft. This model resembles very much the EEI model, and almost the same investigations have been carried out on oscillation damping. The results agree very well. Contrary to the present report, however, in our tests only a single ac line was in parallel to the dc transmission line (Fig. 14). In this arrangement transient instability can be investigated because for the first half-wave of the oscillation, after a short-circuit on the ac line, the line angle reaches extremely high values. Fig. 15 shows such an event, which is just at the limit of stability with dc current being kept constant.

During the short-circuit time, power, dc current, and the angle oscillate with operating frequency because of the fault asymmetry.

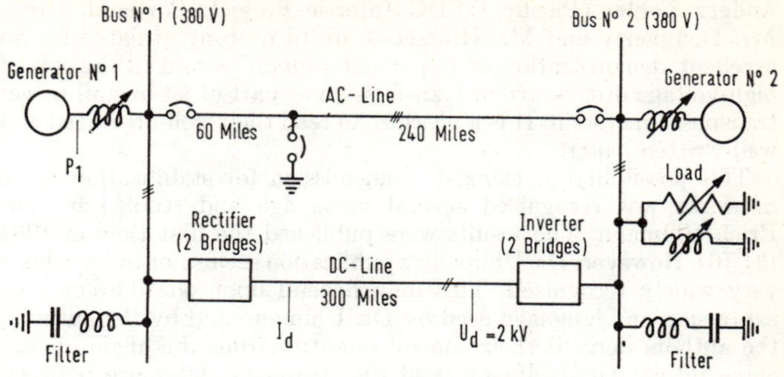

Fig. 14. Single-line diagram of 400 kV-Forschungsgemeinschaft model system.

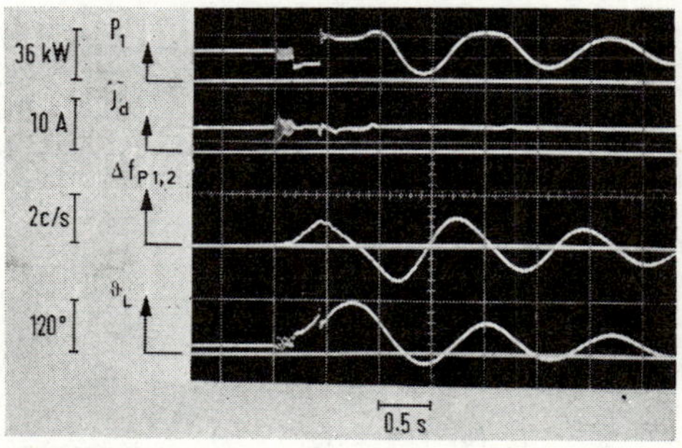

Fig. 15. Transients after one-pole short circuit on ac line followed by three-pole automatic interruption and reclosing; constant-current control of dc transmission. P_1—real power of generator 1; I_d—direct current; $\Delta f_{P1,2}$ frequency difference between rotors; ϑ_L—line angle.

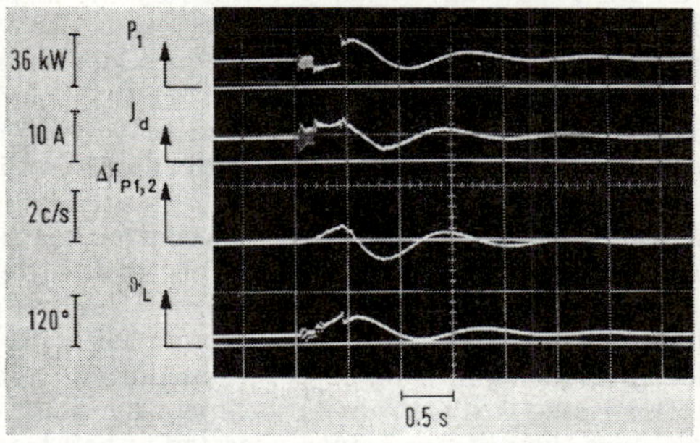

Fig. 16. Transients as in Fig. 15; dynamically responsive dc transmission.

During the interruption time, power is only transmitted by the dc line. At the moment of reclosing, the power of the ac line shows an increase because of the increase of the angle, and forces the two machine rotors back into synchronism.

Fig. 16 shows the direct current influenced by the signals of the angle ϑ_L, the first and the second derivative of the angle. The angle improves the transient stability and the steady-state stability in the case of an overloaded ac line. The first derivative improves the transient stability and the damping factor of the oscillation, and the second derivative (during fault period only) improves only the transient stability. Comparing Fig. 16 to Fig. 15 one recognizes that the angle increase is smaller and that the damping of the oscillation is better.

References

[4] R. Jötten and R. Foerst, "Model plant tests on the behavior of an HVDC transmission during ac and dc system disturbances," CIGRE, Rept. 43-08, 1968.

[5] R. Foerst and Y. Rogowsky, "Das Verhalten der HGÜ bei Ausgleichsvorgängen in den angeschlossenen Netzen," *VDE-Fachber.*, 1969.

Manuscript received February 13, 1969.

Anders Kohler (Pacific HVDC Intertie Project, Portland, Ore.): Mr. Dougherty and Mr. Hillesland are to be congratulated for an excellent demonstration of important properties and utilization of high-voltage direct-current transmission as part of an overall power transmission system. It is a pleasure to read their well-organized and well-written paper.

This possibility of using dc transmission for stabilization of ac networks was recognized several years ago and studied by Dr. Erich Uhlmann. The results were published the first time in 1964 [3], [6]. However, Dr. Uhlmann's publication seems not to have been very widely recognized. This may depend upon the difference in expression and language used by Dr. Uhlmann and by the authors; the authors derived their control quantity from the angle of machine rotors; Dr. Uhlmann used the frequency difference between the two ac networks. However, Fig. 17 and the following equations demonstrate that these are merely different expressions for one and the same quantity. The difference in machine angle θ can be expressed in the instantaneous position of machine rotors, using the electrical frequency of the two ac systems by

$$\theta = \omega_1 t - \omega_2 t = 2\pi t(f_1 - f_2) \qquad (1)$$

where

ω_1, ω_2 angular frequencies of ac voltages in networks 1 and 2

f_1, f_2 frequencies of same voltages

t time.

Derivation of (1) with respect to time gives

$$\delta\theta/\delta t = 2\pi(f_1 - f_2). \qquad (2)$$

It is a question of convenience whether the observation of the machine shaft by means of light pulses is used or the observation of the frequency of the electrical voltage is used for derivation of the control quantity.

Dr. Uhlmann's studies went beyond using the frequency difference and included time derivatives of this quantity for improved control performance. His results have been confirmed by the present simulator studies.

Although the paper is well written, a few questions are necessary for clarification. In the paper it is stated that a "conventional converter control" was used. Does this mean that in the converter control system no special efforts were made to adapt it or modify its properties to suit this specific application?

It is stated in the paper that only current control was used. Why was not the power control system utilized? This would have included an automatic correction of transmitted current for variations in direct voltage.

In Tables II–IV, summarizing observations during ac faults and load changes, nothing is stated about the operating point of the dc transmission. Was the dc transmission operated at half-load, so that the limitations in its control system were insignificant?

In the section, *Step-Load Changes* it is stated that the contribution from the dc transmission may prevent voltage flickers in the ac system due to sudden changes in loads on the ac system. This statement requires some qualification. A change in transmitted power on the dc line is always related to a change of reactive power requirements from the ac network, and this may reduce or exaggerate the voltage variations on the ac system, depending on circumstances.

During the studies of faults on the dc line, outage time for the dc transmission was set to a minimum. This worked well under the specific conditions studied in the simulator, but it should be recognized that this is not necessarily the case for all conditions. The outage time should be matched to the frequency of machine angle oscillation in the ac systems; studies elsewhere indicate that the minimum outage possible, with consideration to deionization of faults, is not necessarily the best considering stability.

In several places the authors discuss overload capabilities of converters. The authors rightly point out that even if the converters themselves have a large overload capability, this cannot always be utilized due to limitations imposed by the connected ac systems. This is illustrated by Fig. 18, displaying transmitted power- and ac-system voltage at the bus bar in the converter terminal,

Manuscript received February 14, 1969.

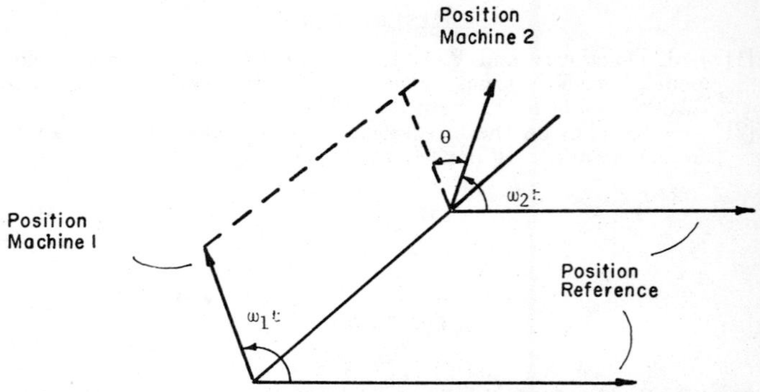

Fig. 17. Machine angles and electrical frequencies [see (1) and (2)].

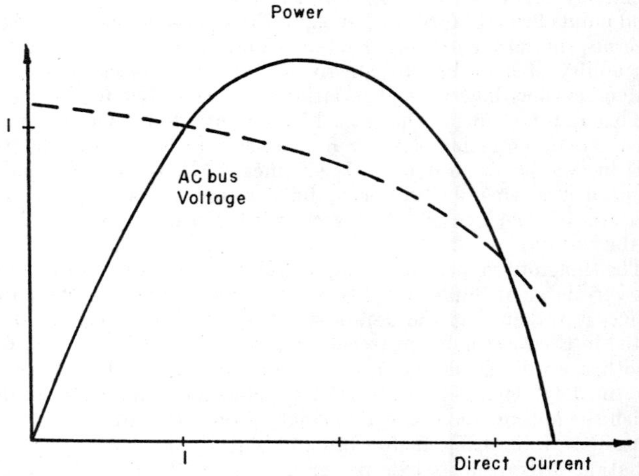

Fig. 18. Power and ac bus voltage in per-unit values at overload.

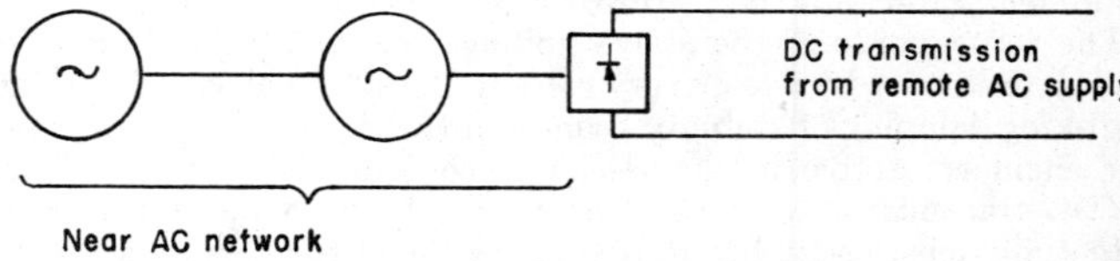

Fig. 19. Stabilization of ac system by power through dc transmission.

assuming unlimited current overload capability of the converters. Due to the voltage regulation in the ac system the power transmission has a definite maximum and transmitted power goes to zero for large currents. The figure is approximately to scale for a system where the short-circuit capacity of the ac system is four times the rated power of the dc transmission.

For future work on actual power transmission control systems, optimization will be necessary. For this purpose information is needed about the dynamic properties of the dc transmission. Such information is available. The dynamic properties of a dc transmission are not single valued, but dependent on transmission status (number of converter groups in operation) and on the operational status of the two ac systems connected to the converter. For small disturbances, the dc transmission can be considered as linear, but for large deviations the nonlinearities have to be considered.

Once the dynamic properties of the dc transmission are known and expressed in measurable quantities, the whole apparatus of the feedback control system theory can and should be applied for the optimization. In view of the interesting results reported here, it would be most interesting to learn about the result with other control concepts, briefly indicated in this report.

There is another application of dc transmission for improving stability in ac systems. Consider an ac network, into which power is fed through a dc transmission from a remote network. The dc transmission can support this ac network in case of internal stability problems. The input power can be modulated to maintain internal stability during disturbances, utilizing energies stored in the inertia of the remote ac network. The configuration is indicated in Fig. 19. This application is especially interesting when an ac network is supplied with an appreciable part of its total power through a dc transmission from a remote ac network. In this case the inertia of the remote ac network is in fact reflected through the dc transmission into the near ac network and supports the stability of this near ac network.

It is becoming obvious that dc transmission lends itself to utilization in stability problems of ac systems. It remains for the operators of ac-system electrical utilities to state the needs toward which dc transmission control should be optimized. This should be done in close cooperation with dc system designers, in order to get the best possible overall optimization.

References

[6] E. Uhlmann, "Energieübertragung mit Gleichstrom hoher Spannung," *E. u. M.*, vol. 83, no. 4, pp. 250–253, 1966.

C. H. Rush (Philadelphia Electric Company, Philadelphia, Pa.): The authors have presented an excellent paper, complementing recent discussions concerning damping signals applied to generator excitation systems [7], [8]. The approach used serves to point out that system damping sources in future planning are possible in many forms, at least one of which is external to the generator and thus compatible with generation sites that might preclude the application of more "conventional" means of system damping signals. In particular, this method might be applicable to those sites where several generators of nonsimilar response characteristics are installed.

Although the tests were necessarily restricted by the facilities available at the EEI AC/DC System Operation Research Project, answers to the following questions would be helpful in assessing the potential usefulness of this damping method:

1) Was a constant overall gain used for the derivative control system for both the large and small disturbance tests?

2) Were any comparable tests made using the derivative signal to control generator excitation?

3) The tests were apparently conducted using identical generators at each of the terminals. Would the authors comment on the effect of dissimilar system inertias at each of the two terminals?

4) The method requires a communications link between the two terminals. Will the control system time lag imposed by a 300-mile data link be pertinent system restraint?

References

[7] R. M. Shier and A. L. Blythe, "Field tests of dynamic stability using a stabilizing signal and computer program verification," *IEEE Trans. Power Apparatus and Systems*, vol. PAS-87, pp. 315–322, February 1968.

[8] W. Watson, "Stabilizing signal for improved control of generator static excitation," *Ontario Hydro Res. Quart.*, vol. 19, pp. 1–7, third quarter 1961.

Manuscript received February 14, 1969.

W. R. Schmus (Southern California Edison Company, Los Angeles, Calif.): The Southern California Edison Company is a participant in the Pacific Intertie Project including an 851-mile 750-kV dc line in parallel with a 500-kV ac system. From studies, it is apparent that system stability is the limitation on loading of the Pacific Intertie, and so we have been very interested in this work done by EEI on the possible application of the dc line to system damping.

One peculiarity of this application of the dc line was included in the paper indirectly but needs more emphasis: that damping of the ac system is only provided when the dc line is in service, which is obvious. For the Pacific Intertie Project, this is important since hypothetical permanent loss of the dc line has been one of the assumed contingencies limiting the total loading as determined by stability studies.

If a critical ac-system disturbance is found which imposes a smaller limit on the total loading but does not impair the damping ability of the dc line, the damping control feature could effectively increase the system loading limit until loss of the dc line is again limiting.

Manuscript received February 18, 1969.

John J. Dougherty and **Thomas Hillesland, Jr.**: Mr. Berlund's discussion is most welcome since he has managed to summarize in clear, concise language the most important points developed in the paper. The authors were, of course, familiar with the theoretical investigations by Dr. Uhlmann referred to in Mr. Berglund's discussion. Our results, obtained from a dynamically responsive model of a power system, are, in effect, experimental corroboration of Dr. Uhlmann's earlier work.

The discussion by Dr. Foerst was also most gratifying. We have had occasional contacts with the 400 kV-Forschungsgemeinschaft project for some years and were aware that our efforts were directed toward similar studies. It is very reassuring to find that results obtained independently and by slightly different techniques agree with each other so completely.

The authors quite agree with Mr. Kohler that there is no difference in our signal derived from machine rotor angles and that developed by Dr. Uhlmann from frequency differences between two points in an ac network. In fact, both methods have been used on the EEI model. The most significant point to be conveyed by the paper is the degree of damping available from the dc line because of its inherent speed of response. The parameter of the system used to give such instruction to the dc line may be quite different for different applications.

The "conventional converter control" referred to in the paper was a direct, commercially available control. Feedback circuits in the control were optimized after installation as is the practice in full-scale line installations. The only modification that need be mentioned is that the modulating signal from the machine angle derivative was inserted into the summing junction of the current-regulating amplifier via one of several extra inputs specified when the controls were ordered. The tests were conducted entirely with current control because a constant-power control system is not included in the model. This feature could have been added at additional cost but was not deemed necessary.

The test runs summarized in Tables II–IV (and most other tests performed on the model) were indeed run with the dc transmission operating at approximately 50 percent of rating. We are not sure what is meant by Mr. Kohler's comments that the limitations in the control system are insignificant under these conditions, but the tests shown in Figs. 8 and 9 were run with the dc line at 90 percent of its rating.

Mr. Kohler's comments on the optimum time for restarting a dc line are well taken. However, they are really more applicable to a dc line operating under the constant-current or constant-power mode. When the dc line's power flow is being controlled as a function of the system swing, a rapid reenergization is less likely to be detrimental since the control signal will reduce the dc power flow to a minimum if this is in the best interest of system stability. Although not shown in the paper, this effect has been noted in other studies at the EEI laboratory.

In reply to Mr. Kohler's question concerning other control concepts evaluated in our studies, several other system parameters were tried with varying degrees of success. The one referred to briefly under *Choice of Control Parameter* involved the use of separate monitoring circuits, which measure such quantities as ac line power flow or generator power outputs; such circuits stored continuously a dc voltage suitable for use as a step change in dc line power flow. The circuit was designed so that the monitoring function would be interrupted by an auxiliary contact of the breaker on the monitored circuit, and this started a sequence that released the stored step

Manuscript received April 28, 1969.

change to the dc line control. This concept would permit a dc line to function normally as a constant power link, ignoring minor disturbances within the ac system. However, on the occurrence of any of the specific disturbances being monitored, it would take a discrete step of corrective action, which had been previously optimized. The more general $D\theta$ control was described in detail in the paper since it was considered to be of wider interest.

Mr. Rush's discussion, particularly concerning the application of damping signals to generator excitation systems, is also most welcome. In reply to his question, the derivative signal has not been applied to control generator excitation on the EEI model. However, modifications have been made to the exciter representation and tests of this nature are planned in the near future. For the results reported in the paper, the overall gain for the derivative control system was held constant. Other gains were used and, as indicated in the paper, the gain used for the results shown was considered approximately optimized.

As Mr. Rush indicates, the tests reported on were conducted with identical generators at both terminals. However, the reasons for this were not directly connected with the function of the dc line control. Other tests, in which a smaller machine was placed at the receiving terminal, presented no difficulty to this method of system damping. In general, a smaller system inertia will cause a more rapid change in system power angle and will be matched by a higher $D\theta$ signal, which adequately compensates for it. The speed of response of the dc line is appreciably faster than any reasonable rate of change of system power angle.

In reply to Mr. Rush's question concerning the delay time for data transmission over relatively long distances, this should have no detrimental effect on the damping as shown in the paper. For 300 miles, the specific distance cited, the data transmission time is less than 2 ms. Even allowing for a verification check for such transmitted data, the total time delay should be less than 10 ms. In the results reported, the uncompensated phase shift introduced by the electronic components of the $D\theta$ control circuitry produces about a 35-ms delay in the application of the damping signal. In an actual installation this delay would be eliminated by more careful phase-shift compensation to optimize the response. Obviously, this improvement in response time by 35 ms would more than compensate for any lag introduced by data transmission.

Mr. Schmus's discussion touches on a critical point not emphasized in the paper. Obviously, permanent loss of both poles of the dc line on the Pacific Intertie could receive no assistance from dynamic controls to modulate dc power flow. The section *DC Line Faults* shows the damping available from a dc line following a momentary bi-polar outage by lightning flashover. Except for the infrequent outages involving tower failures, it is probable that most dc faults (line or terminal) will involve only one pole. Under such circumstances modulation of the power flow on the remaining pole should be capable of providing appreciable damping.

Part V

Load Shedding

The primary goal of stability analysis in system design and operation is to avoid separations in bulk power transmission systems. Any system, however, will break up when subjected to severe enough stress, and it is essential to provide for this eventuality.

The fundamental problem incurred when power islands are formed abruptly is unbalance in load and generation. Excess generation will cause a rising frequency in the island, while excess load causes a declining frequency. There may be other problems as well, such as overloaded lines and voltages outside their normal ranges.

When a small excess in generation exists, the final rise in frequency will be small. Prime mover speed-governing systems react to the overspeed and reduce turbine power. Typically, an initial generation excess of twenty percent in the island will cause a sustained overspeed of about one percent. There will also be some variation in load with frequency, so that some of the initial unbalance will disappear because of the frequency deviation. If the generation excess is a high enough fraction of the total generation in the island, overspeed protection for individual turbine-generators may come into play. This may involve automatic valve or gate closing and disconnection of generators from the transmission network. Thus it is conceivable that an excess of generation may quickly lead to an excess of load and perhaps collapse of the power island. Intuitively, large power islands are not likely to have extremely large initial percentage excesses of either load or generation.

When a small excess in load exists initially in a power island, speed-governing systems react to the declining frequency by increasing prime mover power. However, there are limitations on the sudden increase of load on steam turbines. Typically, the maximum increment of suddenly applied load which can be sustained is about ten percent of the unit rating, but this may vary considerably between plants and with operating circumstances. Rapid load pickup may be automatically limited by boiler control systems, or may be limited by a gradual deterioration of steam quality. In either case, speed-governing systems do not bring about a balance between load and generation. Load inherently diminishes with frequency, so that a stable frequency may ultimately be achieved. This is not an acceptable operating situation, however.

Overloaded power islands are characterized by low frequency, by higher than normal currents in at least some generators and transmission elements, and by lower than normal voltages for at least some buses. This combination of conditions creates a variety of stresses which have the potential for making the unbalance in load and generation worse. For example, boiler steam rates will be at their maximum at the same time that plant electrical auxiliaries are being supplied at low voltage and low frequency. Under such circumstances, it may not be possible to keep boilers operating within design limitations. Generators may be subjected to thermal stress in stator windings, in field windings because of low terminal voltage and consequent high field voltage, or both. Transformers and lines may be thermally stressed, while low voltages, high currents and low frequency may bring protective relays to the verge of operation. In this situation, any event which forces any facility out of service tends to increase the stress elsewhere and hastens the occurrence of the next forced outage. This kind of system behavior brought about the virtual complete collapse of power over most of Eastern New York and New England during the Northeast Power Failure.

The use of underfrequency relays to reduce overload by disconnecting selected load circuits provides a means of achieving power balance in isolated power systems. This section contains three papers on the application of underfrequency relays and on load shedding and load restoration policies. Additional related papers are contained in the following bibliography.

Supplementary Bibliography

Load Shedding

[1] L. P. Hadju, J. Peschon, W. F. Tinney, and D. S. Piercy, "Optimum load shedding for power systems," *IEEE Trans. Power App. Syst.*, vol. PAS-87, pp. 784–795, Mar. 1968.

[2] H. T. Akers, J. D. Dickinson, and J. W. Skooglund, "Operation and protection of large steam turbine generators under abnormal conditions," *IEEE Trans. Power App. Syst.*, vol. PAS-87, pp. 1180–1188, Apr. 1968.

[3] "Survey of underfrequency relay tripping of load under emergency conditions," IEEE Committee Rep., *IEEE Trans. Power App. Syst.*, vol. PAS-87, pp. 1362–1370, May 1968.

[4] E. L. Busby and J. W. Skooglund, "System response to severe contingencies including control systems, transmission line protection, and load management," IEEE paper CP 69-650-PWR.

[5] C. J. Durkin, Jr., E. R. Eberle, and P. Zarakas, "An underfrequency relay with frequency decay rate compensation," *IEEE Trans. Power App. Syst.*, vol. PAS-88, pp. 812–820, June 1969.

[6] R. C. Durbeck, "Simulation of five load shedding schedules," *IEEE Trans. Power App. Syst.*, vol. PAS-89, pp. 959–966, May/June 1970.

[7] M. L. Chan, R. Dunlop, and F. Schweppe, "Dynamic equivalents for average system frequency behavior following major disturbances," *IEEE Trans. Power App. Syst.*, vol. PAS-91, pp. 1637–1642, July/Aug. 1972.

[8] T. Hirayama, Y. Kato, H. Mukae, and I. Oya, "Development of AC type 'system interconnectors,'" *IEEE Trans. Power App. Syst.*, vol. PAS-91, pp. 1085–1092, May/June 1972.

Application of Underfrequency Relays for Automatic Load Shedding

H. E. LOKAY, SENIOR MEMBER, IEEE, AND V. BURTNYK, MEMBER, IEEE

Abstract—**During severe emergencies which result in insufficient generation to meet load, an automatic load shedding program throughout the affected area can prevent total area collapse. It also helps to achieve fast restoration of all affected loads. This paper describes the factors involved in applying underfrequency relays to achieve a desired "deficient generation" protection level and a calculating method to achieve optimum relay settings.**

THE APPLICATION of underfrequency relays in substations throughout the load area, preset to drop specific percent magnitudes of loads at predetermined low system frequency values, provides the simplest automatic load shedding program. Relay settings can be developed to drop approximately the minimum load to arrest system frequency decay at a safe operating level. Additional underfrequency relays can also be applied to initiate a safe and orderly system separation or shutdown if the emergency is beyond the capabilities of the load shedding program.

Rate of Frequency Decay

To apply underfrequency relays, it is necessary to know the rate of frequency decay for different magnitudes of generation deficiencies. The rate of frequency decay can then be coordinated with the system's ability to increase power generation to determine the severity of the many possible emergencies that can occur and, in turn, the load shedding program required. In today's interconnected system operation, frequency will decrease an appreciable amount only if the affected area separates from the interconnected systems. Such a separation can occur following a major disturbance which creates large power swings or tie-line flows that cause line relay operations.

Fig. 1 shows curves of change in system frequency with time for different magnitudes of overload and system inertia (H) constants. Overloads of 10 to 100 percent are shown where percent overload is as follows:

$$\text{Percent overload} = \frac{\text{Load} - \text{Sum of unit loading}}{\text{Sum of unit loading}} \times 100$$

or

$$\text{Percent overload} = \frac{\text{Deficient generation}}{\text{Remaining generation}} \times 100.$$

The curves assume no change in unit input power. It is recognized that in the times shown in Fig. 1, unit speed-governor action would be initiated by the frequency reduction and possibly some power increase available. Since the ability to increase input power depends upon unit spinning reserve allocation, control sensitivity, and energy system time constants that can differ between systems, and possibly within a single system during load variations, the governor action was omitted.

Fig. 1 is based on a reduction in load of two percent for each one-percent reduction in frequency. This value of load reduction due to frequency reduction has been listed as representative in the literature for many years. However, new work[1] indicates it may be greater. A range of 1.2–7.8 percent load reduction per one percent frequency reduction was reported. Fig. 2 illustrates the effect of different rates of load reduction per one percent frequency reduction for a 50-percent system overload condition. Final frequency values for different overloads and percent load reductions per percent frequency reduction are shown in Fig. 3.

Curves of Figs. 1 through 3 were calculated using the equations as derived in Fountain and Blackburn.[2] Other calculating methods are shown in the literature.[3] The procedure that follows for applying underfrequency relays is based on the use of curves similar to those of Figs. 1 through 3 and does not depend on the calculating method used to develop the curves.

Rates of frequency decay in Hz per second following the initial overload condition for system H constants of 4 and 7 are shown in Fig. 4. In general the rate of frequency decay rapidly becomes less than two Hz per second except for the very high overload conditions. For practical magnitudes of area overloads, the rate of frequency decay will likely be less than three immediately following the disturbance and system separation.

Permissible Frequency Reduction

The load shedding program must be coordinated with equipment operating limitations under low-frequency operation. These limitations or restrictions are primarily associated with operation of steam turbines and power house auxiliaries. In general, continuous steam turbine operation should be restricted to frequencies above 58.5 Hz (60-Hz system base), and operation below 58.5 Hz should be for a limited time only, such as ten minutes or less. Power plant auxiliaries performance from actual tests[4]–[6] indicates a limit of approximately 53–55 Hz, below which plant output begins to reduce.

Load Shedding Requirements

Load shedding programs should take advantage of the permissible safe operation below normal frequency so that only minimum load is dropped during emergencies. As shown in Figs. 1 through 3 the system frequency will settle at a level depending upon the initial system overload and percent load reduction per percent frequency reduction. The load shedding program can be designed to have the frequency settle at a value lower than permissible continuous (58.5 Hz) if the system dispatcher can survey the emergency situation and manually initiate additional

Paper 31 TP 67-447, recommended and approved by the Power System Relaying Committee of the IEEE Power Group for presentation at the IEEE Summer Power Meeting, Portland, Ore., July 9–14, 1967. Manuscript submitted April 12, 1967; made available for printing May 3, 1967.

The authors are with the Westinghouse Electric Corporation, East Pittsburgh, Pa.

Reprinted from *IEEE Trans. Power App. Syst.*, Mar. 1968.

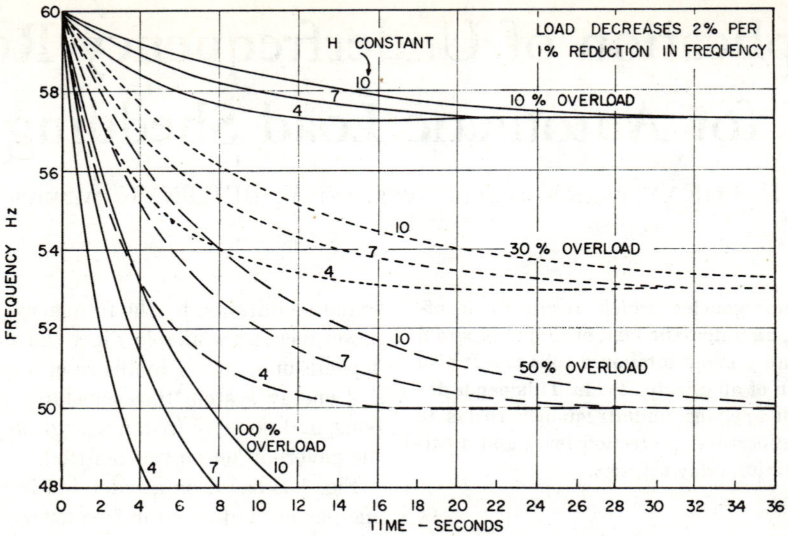

Fig. 1. Frequency decay due to system overload.

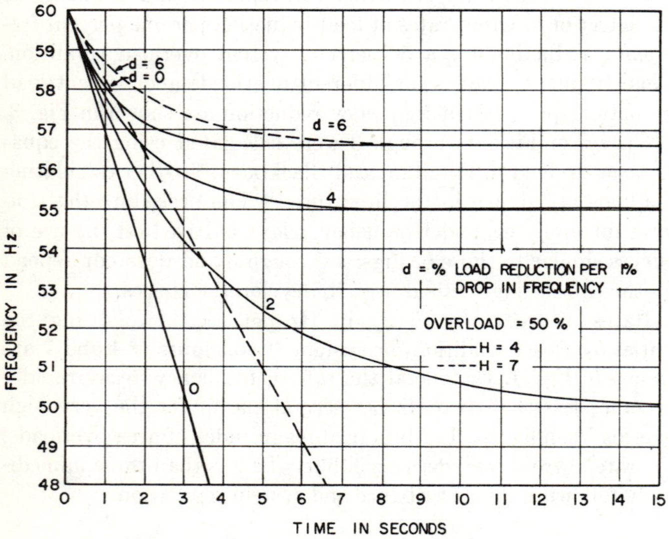

Fig. 2. Effect of load characteristic on frequency decay.

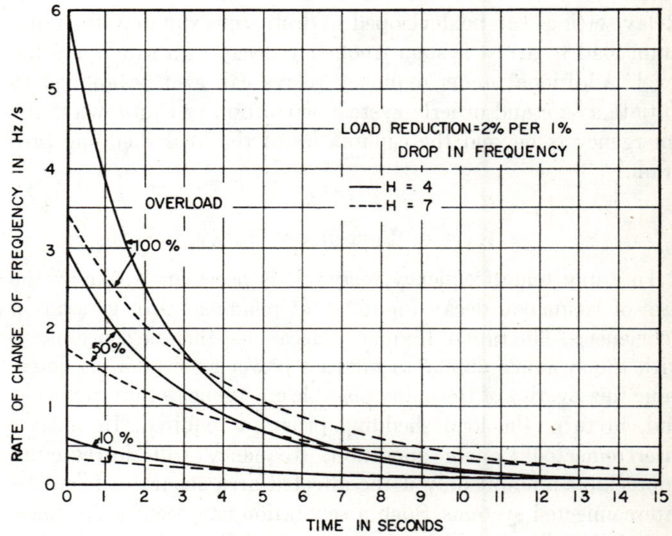

Fig. 4. Rate of frequency decay due to system overload.

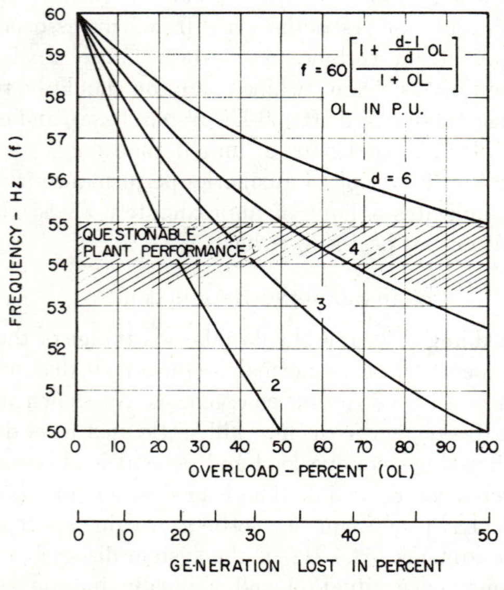

Fig. 3. Final frequency versus system overload for different percent load reductions per one percent drop in frequency (d).

load shedding and/or increase generation (such as peaking unit start up) within about ten minutes to raise frequency to permissible continuous. Governor action will also increase input power, if available, within that time. A load shedding program based on the frequency settling at 57 Hz when protecting for some maximum overload condition is reasonable.

Fig. 5 shows the amount of load required to shed for the frequency to settle at 57 Hz with different affected area overloads. The amount of load to shed is in percent of initial area load, and the overload is in percent based on the remaining area generation. The formula used for developing the curves of Fig. 5 is shown on the figure, with load and overload in per unit. Similar curves can be prepared for any settle-out frequency. Fig. 5 indicates the magnitude of overload possible without having to drop any load for different load reduction rates d. When $d = 2$ the frequency will not fall below 57 Hz for overloads up to eleven percent. However, to obtain relay coordination for load shedding steps and protect for some maximum overload condition, it may be found necessary to drop some load even for these small overload conditions.

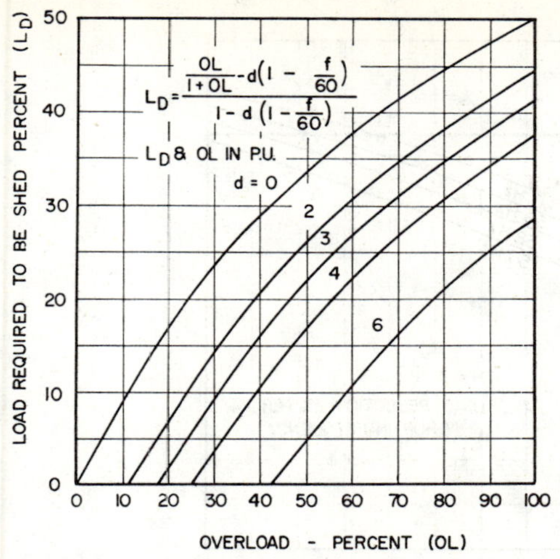

Fig. 5. Load shedding requirements to have frequency settle at 57 Hz.

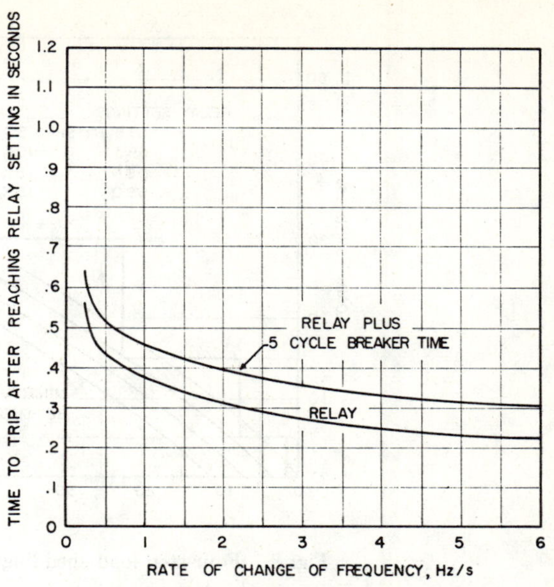

Fig. 6. Underfrequency relay trip-time characteristic.

Underfrequency Relay Characteristics

Underfrequency relays operate to close their contacts when the applied source frequency is below the relay preset value. The adjustable range for frequency settings is generally 55 to 59.5 Hz. Relay tripping times, after the preset frequency settings are reached, are a function of rate of change of frequency. The dynamic trip-time characteristic for a modern underfrequency relay is shown in Fig. 6.

With decaying frequency during the deficient generation period, the actual system frequency will be below the relay setting when the relay contacts finally close. The additional drop in frequency below its setting must be known to coordinate the settings for successive load shedding steps. The additional frequency drop must be calculated since the rate of change of frequency decay is not constant (Fig. 4) due to the load reduction with frequency reduction. Circuit breaker opening time and any additional time delay that may be used should also be included in the calculations. An example of the calculations is shown in Appendix I, and general application curves for different relay settings are shown in Fig. 7.

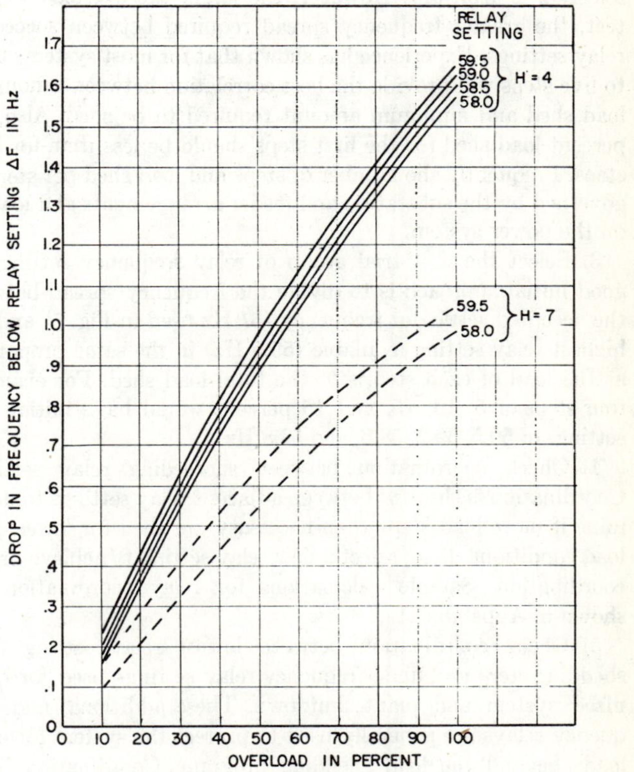

Fig. 7. Frequency drop below relay setting for various overloads.

Relay Settings, Number and Size of Load Shedding Steps

Using application curves of Figs. 5 through 7, a load shedding program can be developed. It is first necessary to select what percent overload level the load shedding program is to protect and at what value the system frequency is to settle for that overload level. It is recognized that selecting the maximum overload level is very difficult because it is not known where an area is going to separate from the interconnected systems and what generation will exist in the area. System studies can be made using digital stability programs, and line flows versus time following major disturbances can be monitored to determine likely areas of system separation. Probable overload levels can then be estimated for the areas.

A trial and error procedure is required to develop the best combination of number and size of load shedding steps and corresponding relay settings. One combination is selected, and it is then checked for relay coordination and whether it is possible to reduce the amount of excess load shed for smaller overload conditions. Additional combinations are then studied. In general, only a few trials are required to obtain the best relay settings for one set of steps and the load magnitude shed per step. Sufficient understanding is generally obtained to reduce to a few the variations in number and load shed per step. Following is a list describing the calculating procedures. An example is shown in detail in Appendix II.

1) Determine the total load required to be shed. The amount is based on maximum overload the load shedding program is to protect and the desired leveling frequency for that overload. For example, from Fig. 5, to protect for a 100-percent overload (50-percent loss of generation) and settle at 57 Hz, 44.5 percent load must be shed.

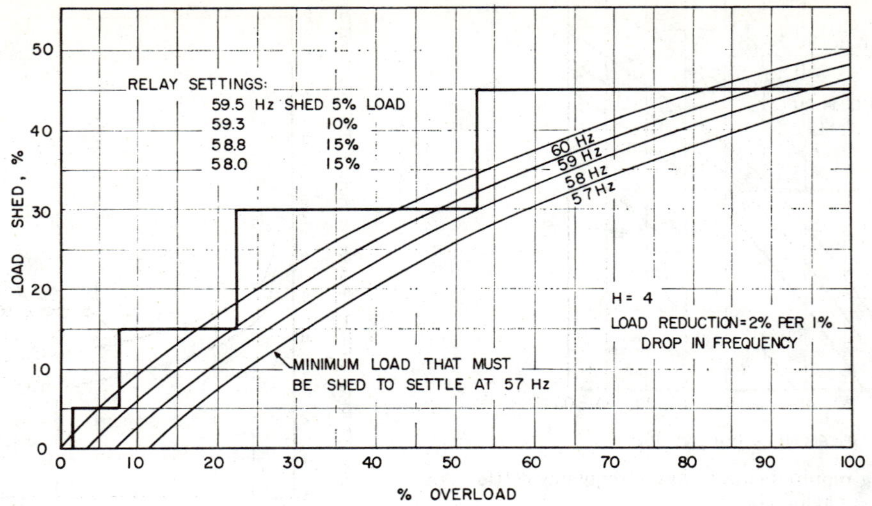

Fig. 8. Four-step load shedding program to protect up to 100 percent overload.

2) Select number of steps and load to be shed per step. The selection is arbitrary. However, the larger the overload to protect, the greater frequency spread required between successive relay settings. Experience has shown that for most systems three to five steps will provide the best correlation between amount of load shed and minimum amount required to be shed. Also, the percent load shed for the first steps should be less than for later steps. Frequently the number of steps and load shed per step are governed by the substation and feeder arrangements and loading on the power system.

3) Select the first trial group of relay frequency settings. A good initial approach is to divide the frequency spread between the accepted settle-out frequency (57 Hz used in Fig. 5) and the highest relay setting available (59.5 Hz) in the same proportion as the load of each step is to the total load shed. For example, four steps of 5, 10, 15, and 15 percent would have initial trail settings of 59.5, 59.2, 58.6, and 57.8 Hz.

4) Check coordination between succeeding relay settings. Coordination is checked between adjacent relay settings to determine if more load steps than necessary are shed for lower overload conditions. If so, select a new relay setting to achieve proper coordination. Sample calculations for relay coordination are shown in Appendix II.

5) Check coordination between lowest relay setting (load shedding step) and underfrequency relay settings used for organized system and plant shutdown. These additional underfrequency relays are generally used to protect the system for overloads beyond the load shedding program. Coordination is required to insure that the additional relays will not operate for overloads within the load shedding program.

6) Repeat steps 3 through 5 with a new set of relay settings. Calculations from the first trial will generally indicate if and how improved settings can be obtained. An indicator that can be used for improving settings is the area between the load shedding step curve and the curve of minimum load that must be shed to settle at 57 Hz as shown in Fig. 8. The minimum area when still achieving relay coordination will give the best combination of relay settings to drop the minimum excess load for smaller values of overload. The area under the load shedding step curve is easily calculated. The area under the minimum load shed curve can be calculated using the formula shown in Appendix III.

7) Repeat steps 2 through 6 with a different combination of loads per step and number of steps. The area for each combination can be compared to select the optimum combination for the load shedding program.

Fig. 8 illustrates a four-step load shedding program to protect for a 100-percent overload and to settle at 57 Hz. The best relay settings were developed as shown in Appendix II. The figure also includes minimum load shed curves for settling at 58, 59, and 60 Hz to show the relay settings in the figure and to indicate the frequency that will be obtained when the load steps are shed for a specific overload. For example, for a 20-percent initial overload, two load steps would be shed and the frequency would settle at approximately 59.5 Hz. For some overloads the frequency would rise above 60 Hz after the load is shed because more load than necessary is dropped. This is because the program is developed to protect for a maximum of 100-percent overload. Fig. 9 illustrates for the same conditions as in Fig. 8 and Appendix II, a two- and three-step load shedding program. The relay settings shown are about the best for good coordination. It illustrates that lower relay frequency settings are possible with fewer steps, but greater amount of excess load is shed for smaller initial overloads. The ratio of areas under the three-step and two-step plans compared to the four-step plan are 1.12 and 1.32, respectively.

Fig. 10 illustrates the four-step program (Fig. 8) as designed for a system with an H constant $= 4$ and percent load reduction per one percent frequency reduction $d = 2$, and how it would operate if d were equal to four instead of two. It shows that unnecessary load would be shed and indicates the importance of knowing d when developing a load shedding program. For comparison, also shown is the best combination of relay settings and load per step when designing for a d of four.

Spinning Reserve and Governor Action

In the preceding, curves and calculating techniques do not include governor action for increasing unit input power during the emergency. Because of the unknowns as to where system separation will occur, available spinning reserve in affected area, and rate of achieving the spinning reserve, the simplest procedure in developing a load shedding program is to omit governor action. This is reasonable because it will always tend to reduce the overload and amount of load shed. To include governor action Figs. 1 and 2 would have to be modified based on the rate of increase in

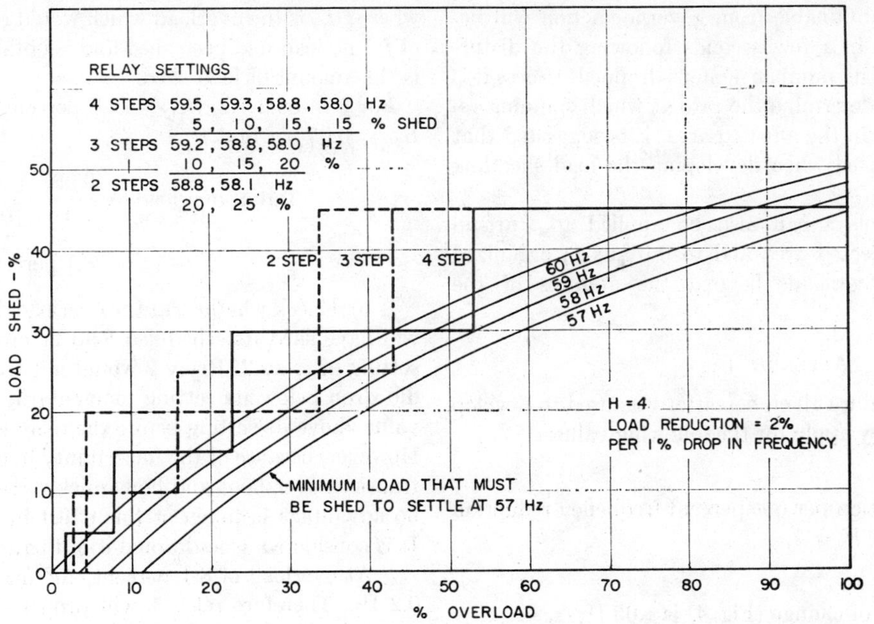

Fig. 9. Comparison of two-, three- and four-step load shedding programs to protect up to 100 percent overload.

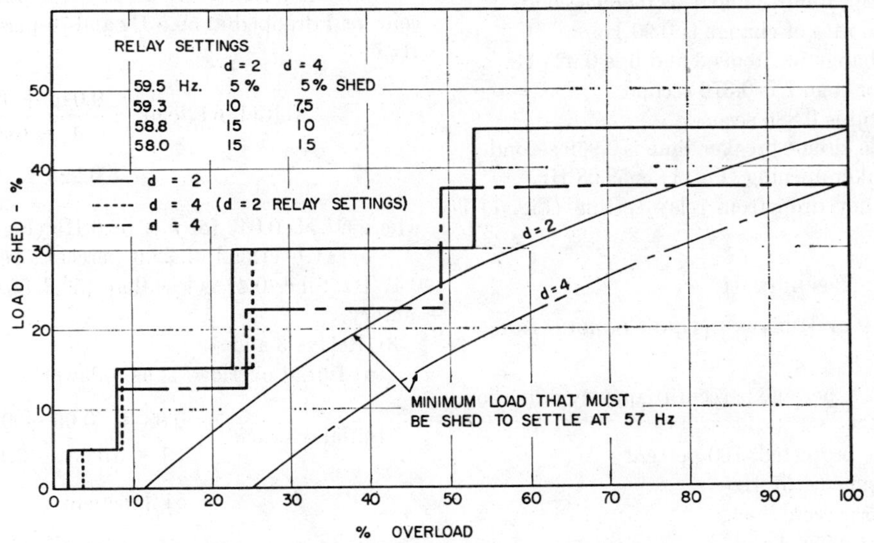

Fig. 10. Four-step load shedding programs to protect up to 100 percent overload for systems with $d = 2$ and $d = 4$, and performance of a $d = 2$ load shedding program on a system with $d = 4$.

input power. Depending upon the type of generation in the affected area, only about 20 percent of the total spinning reserve could be obtained in the first five to ten seconds following the disturbance.

Conclusion

This paper has described a procedure for the application of underfrequency relays to develop an optimum load shedding program. No consideration has been given to physical system restrictions, and no probability values have been applied to having overloads occur less than the maximum overload the load shedding program is to protect. Characteristics of only one underfrequency relay were used. The procedures described apply for other relay characteristics or for the same relay with added time delay.

The following comments can be made.

1) The rate of frequency decay rapidly reduces, and for high overloads, it becomes less than two Hz/s in just a few seconds.

2) The rate of load reduction per one percent frequency reduction greatly affects the development of the load shedding program.

3) With a decaying frequency, the frequency is below the relay setting when the relay and corresponding circuit breaker operate. This difference in frequency affects successive step relay coordination.

4) When developing the load shedding program to protect for a large overload, it is generally necessary to drop more load than necessary when smaller overloads occur.

5) The underfrequency relays for each step should be distributed throughout a complete system because points of area separation from the interconnected systems are generally unknown.

6) Spinning reserve obtainable from governor action will begin to become available in a few seconds following the disturbance and may reduce the number of steps dropped. Unless it is consistently possible to determine the rate at which spinning reserve becomes available in the affected area, it is suggested that governor action be omitted when developing the load shedding program.

7) Underfrequency relays can also be applied to start-up ready reserve in the affected area and to initiate an organized system shutdown for overloads beyond the abilities of the load shedding program.

Appendix I

The calculating procedure given hereafter was used to develop the curves in Fig. 7. They apply for the following values:

Relay setting, 59.5 Hz
Two percent load reduction per one percent frequency reduction
System inertia constant, $H = 7$
Overload, 30 percent.

1) At $t = 0$ the rate of change (Fig. 4) is 1.03 Hz/s.
2) t at 59.5 Hz (Fig. 1) is 0.51 second.
3) When $f = 59.5$ Hz the rate of change is 0.95 Hz/s.
4) Relay trip-time for item 3, Fig. 6 is 0.375 second.
5) Total t to relay trip, items 2 and 4, is 0.885 second.
6) At relay trip-time, rate of change is 0.90 Hz/s.
7) Average rate of change for items 3 and 6 is 0.925 Hz/s.
8) Relay trip time for item 7 is 0.376 second.
9) Total t to relay trip is 0.886 second.
10) Item 9 plus 5-cycle circuit breaker time is 0.969 second.
11) Frequency at breaker opening (Fig. 1) is 59.08 Hz.
12) Additional frequency drop from relay setting (Fig. 7) is 0.42 Hz.

Appendix II

Sample Calculation of Relay Setting Coordination

System inertia constant, $H = 4$
Two percent load reduction per one percent frequency reduction, $d = 2$
Maximum overload to be protected, 100 percent
Minimum settle-out frequency, 57 Hz
Total load to be shed, 45 percent
Number of load shedding steps, 4
Frequency at which last step is shed (for coordination with system shutdown relay), 57 Hz
Relay settings selected for first trial:
relay 1—59.5 Hz, 5-percent load shed
relay 2—59.2 Hz, 10-percent load shed
relay 3—58.6 Hz, 15-percent load shed
relay 4—57.8 Hz, 15-percent load shed.

A) Check Coordination Between Adjacent Relay Settings

1) Relays 1 and 2

a) Determine the overload for which the frequency would settle out just above the setting of relay 2, allowing for the load shed by relay 1.

Initial overload for which frequency would settle out at 59.2 Hz with 5-percent load dropped at 59.5 Hz can be calculated from the equation shown in Fig. 5. For convenience, the equation can be rearranged to give the following:

$$\text{Initial overload} = \frac{OL_f + L_D}{1 - L_D}$$

where OL_f is the overload which would result in a final frequency of f if no load had been shed and is obtained from Fig. 3, and L_D is the amount of load shed.

For $f = 59.2$ Hz, $OL_f = 2.7$ percent or 0.027 per unit. With $L_D = 0.05$ per unit,

$$\text{Initial overload} = \frac{0.027 + 0.05}{1 - 0.05} = 0.081$$
$$= 8.1 \text{ percent.}$$

b) Check whether, at the overload determined in a), relay 1 will have shed its scheduled load before frequency drops to the setting of relay 2. Relay 2 would not trip even if the frequency did drop below its setting momentarily and then recovered to a value above its setting before the relay contacts had fully closed. However, because of the uncertainty in estimating the amount of transient overshoot and how quickly recovery would take place, no advantage is taken of this effect in calculating the settings. It is considered as additional coordinating margin.

At overload of 8.1 percent, ΔF for relay 1 (from Fig. 7) is 0.2 Hz. Therefore relay 1 will drop its load when frequency is $59.5 - 0.2 = 59.3$ Hz. Since this is above the setting of relay 2, coordination is satisfactory.

2) Relays 2 and 3

a) The initial overload to settle out at 58.6 Hz with 5-percent load dropped at 59.5 Hz and 10-percent load dropped at 59.2 Hz is

$$\text{Initial overload} = \frac{0.045 + (0.05 + 0.10)}{1 - (0.05 + 0.10)}$$
$$= 0.229 = 22.9 \text{ percent}$$

where OL_f is 0.045 for $f = 58.6$ Hz (Fig. 3).

b) At overload of 22.9 percent, ΔF for relay 2 (Fig. 7) is 0.47 Hz. Since 0.47 is less than (59.2–58.6), coordination is satisfactory.

3) Relays 3 and 4

a) Initial overload is as follows:

$$\text{Initial overload} = \frac{0.08 + (0.05 + 0.10 + 0.15)}{1 - (0.05 + 0.10 + 0.15)} = 0.543$$
$$= 54.3 \text{ percent.}$$

b) At overload of 54.3 percent, ΔF for relay 3 is 0.96 Hz Since this is larger than (58.6–57.8), coordination appears to be unsatisfactory. However, ΔF as determined above applies for an overload of 54.3 percent during the entire operating time of relay 3. As relays 1 and 2 shed their load, the overload is reduced and ΔF will be reduced. A corrected ΔF can be estimated by calculating ΔF intervals for each of the overload periods as relays 1 and 2 operate and then summing these ΔF intervals to obtain a total corrected ΔF. This method is not completely accurate due to the fixed tripping time of the breaker but has been found to be sufficiently accurate for coordination purposes.

At the 54.3 percent overload, ΔF for relay 1 is 1.04 Hz. The relay will shed 5-percent load when frequency is 59.5–1.04 = 58.46 Hz. Overload will be reduced to 0.543–0.05 (1 + 0.543) = 0.466 or 46.6 percent.

Relay 2 sees an overload of 54.3 percent for part of its operating time and an overload of 46.6 percent for the remainder. ΔF at 54.3 percent overload is 1.02 and applies for $(59.2 - 58.46)/1.02 = 0.725$ of its operating time. ΔF of 0.89 (at 46.6 percent overload) applies for $(1 - 0.725)$ of its operating time. The corrected ΔF is therefore $0.725\,(1.02) + (1 - 0.725)\,(0.89)$

= 0.98. Relay 2 sheds 10-percent load when frequency is 59.2–0.98 = 58.22 Hz. Overload is reduced to 0.543 − 0.15(1 + 0.543) = 0.312 or 31.2 percent.

Relay 3 sees an overload of 54.3 percent between 58.6 Hz and 58.46 Hz, an overload of 46.6 percent between 58.46 Hz and 58.22 Hz, and an overload of 31.2 percent for the remainder of its operating time. ΔF at these overloads is 0.98, 0.86, and 0.59 respectively. Following a procedure similar to the above, the corrected ΔF is 0.143(0.98) + 0.279(0.86) + (1 − 0.143 − 0.279) (0.59) = 0.72. Since the corrected ΔF is smaller than (58.6 − 57.8), coordination is satisfactory.

B) Check Coordination at Maximum Overload (100 Percent)

Using the procedure described previously, the corrected ΔF for relay 4 is found to be 1.0. Relay 4 will drop its load when frequency is 57.8 − 1.0 = 56.8 Hz. This is 0.2 Hz below the desired value of 57.0 Hz and for this example is unsatisfactory.

To obtain proper coordination, the setting of relay 4 should be changed to 58.0 Hz, and the settings of relays 1 to 3 should be adjusted as necessary. Calculations in Section A indicate that settings of 59.5, 59.3, 58.8 and 58.0 should be selected for the next trial. When this group is checked for coordination, it is found to be satisfactory.

If the frequency of 56.8 at which relay 4 drops its load is considered to be satisfactory, the settings of relays 1 and 2 can be lowered. Settings of 59.3, 59.1, 58.6, and 57.8 would be found to be satisfactory.

Appendix III

The equation for the curve of the minimum load that must be shed to settle at a specified frequency shown in Fig. 5 can be rearranged to

$$L_D = \frac{OL - OL_f}{1 + OL}$$

where OL is the initial overload and OL_f is the overload that will cause frequency to settle at frequency f if no load is shed.

Area under curve

$$\text{Area} = \int_{OL_f}^{OL_m} L_D d\,(OL) = \int_{OL_f}^{OL_m} \frac{OL - OL_f}{1 + OL}\, d\,(OL)$$

where OL_m is the overload load shedding program to protect and settle at frequency f.

$$\text{Area} = \int_{OL_f}^{OL_m} \frac{OL}{1 + OL}\, d\,(OL) - \int_{OL_f}^{OL_m} \frac{OL_f}{1 + OL}\, d\,(OL)$$
$$= [(1 + OL) - \log(1 + OL) - OL_f \log(1 + OL)]_{OL_f}^{OL_m}$$

For f = 57 Hz, OL_f = 0.111 per unit, OL_m = 1.0 per unit area = 0.236.

References

[1] IEEE Committee Report, "The effect of frequency and voltage on power system load," presented at the IEEE Winter Power Meeting, New York, N. Y., January 30–February 4, 1966.

[2] L. L. Fountain and J. L. Blackburn, "Application and test of frequency relays for load shedding," *AIEE Trans. (Power Apparatus and Systems)*, vol. 73, pp. 1660–1664, February 1955.

[3] C. F. Dalziel and E. W. Steinbach, "Underfrequency protection of power systems for system relief, load shedding-system splitting," *AIEE Trans. (Power Apparatus and Systems)*, vol. 78, pp. 1227–1238, December 1959.

[4] O. D. Butler and C. J. Swenson, "Effect of reduced voltage and/or frequency upon steam plant auxiliaries," *AIEE Trans. (Power Apparatus and Systems)*, vol. 73, pp. 1628–1632, February 1955.

[5] H. A. Bauman, G. R. Hahn, and C. N. Metcalf, "The effect of frequency reduction on plant capacity and on system operation," *AIEE Trans. (Power Apparatus and Systems)*, vol. 73, pp. 1632–1637, February 1955.

[6] R. Holgate, "The effect of frequency and voltage," *AIEE Trans. (Power Apparatus and Systems)*, vol. 73, pp. 1637–1646, February 1955.

Discussion

R. H. Meyer (Northeast Utilities Service Company, Hartford, Conn.): The authors have done an excellent job of taking some of the art out of applying underfrequency load shedding schemes and replacing it with a good definitive technique based upon the engineering requirements. This paper will be a valuable guide for the utility engineer in setting up a new load shedding scheme and in expanding or checking the performance of one already working.

The value of d, defined as the load reduction due to frequency reduction, appears to be a very important factor. The final level-off frequency is very much dependent upon its value, as is the speed at which a given frequency is reached, as indicated in Fig. 2. It would seem that we need to know more about it. Although the IEEE Committee Report[1] provides some answers, the results appear to be too broad. It appears that the present analysis of this factor is incomplete and that more testing or computer simulation studies are required in order to narrow it down.

A significant portion of the total load reduction due to an impending blackout would be due to the decay in voltage which would accompany an underfrequency condition. Deliberate reduction of voltage by manual operation of load-tap-changing equipment is presently used in some places to effect an emergency reduction in load. Values that are being used range from 0.7-percent load to one percent load reduction per percent voltage drop for loads which are predominantly motor type and resistive, respectively. Thus a sizable portion of the factor d may be attributable directly to voltage and be independent of frequency change.

In many modern-day power systems a sizable portion of the var supply is obtained from static capacitors connected to substation buses or to distribution lines. The vars supplied from such capacitors would be directly proportional to frequency and to voltage squared. This is not necessarily the case with vars supplied from rotating machines, particularly those equipped with modern voltage regulating equipment. Thus it can be expected that the voltage reduction during an underfrequency operating condition may be quite uneven throughout a system due to the redistribution of var flow. Other factors, such as the saturation characteristics of power transformers and of the generators, will also have their effect on voltage and therefore load as a function of voltage.

It would seem that analyses of the separate effects of underfrequency and voltage on load, as well as that of frequency on voltage, may be required in order to yield satisfactory data in order to adequately determine the new loads and the distribution of these loads on a system that is suffering from an underfrequency condition. The use of a single factor d may be insufficient to yield reasonable results.

Manuscript received July 24, 1967.

C. Concordia (General Electric Company, Schenectady, N.Y.): In connection with this paper we should like to make three points.

1) The authors have used various values of d, the per unit variation of electrical load with frequency, and have considered values of four and higher. It is our opinion that d is in general less than two. That is, it is very difficult to find a composite load which will vary as more than the second power of frequency. In discussing this factor the

Manuscript received July 28, 1967.

authors have referred to the IEEE Committee Report.[1] We believe that it is unfortunate that this Working Group Report has included d factors which appear to us to be unreasonably large. On the other hand, the authors have not given any credit to the turbine speed governors, and we believe that in many cases governor action may materially assist in limiting the frequency drop.

2) The authors have shown very detailed calculations to determine exactly the amount of load that must be shed to recover to 57 Hz. We should point out that shedding a very little more load will permit the frequency to recover to 60 Hz. This is illustrated in Fig. 8. It would seem very much to be preferred to shed enough load so that the system will recover to 60 Hz, and we really do not understand how anyone could afford to take a chance on not shedding enough load. With proper system design the necessity for load shedding should be so infrequent, and the situation that calls for load shedding should be so severe that it will always be much more important to be sure that we have shed enough load than to figure out exactly how little need be shed. Moreover, 57 Hz is below the frequency at which operation for any extended period of time is desirable. We can only suppose that the authors have shown this low frequency in order to make load shedding more palatable by implying that only the absolute minimum amount of load need be shed.

3) It is well known that because of the inherent time lags in the operation of underfrequency relays the frequency will in general drop below the set point. If more than one block of load has to be shed this will result in a bunching together of the frequencies at which shedding actually occurs. This can be shown by the authors' calculations. In some cases this effect can be very marked. In addition, it appears unavoidable that in certain cases one may shed a block of load that need not have been shed. We believe that this possibility has to be recognized and accepted.

These three points are not intended to detract from the value of the paper, as we believe the authors have made a substantial contribution. We should like to make two additional points which we believe are rather important, even though they are not directly mentioned.

4) Intentional time delay for underfrequency relays has been discussed from time to time. We do not believe that time delay is desirable. Indeed, it is the inherent time delay which causes some of the peculiarities of application mentioned by the authors. As an exception to this, it is conceivable that time delay could be used as illustrated in the following sequence. a) The first block of load is controlled by a relay set at the first (highest) frequency with no intentional time delay. b) The second block of load is controlled by two relays—one set to trip at the second (lower) frequency, the other operating in parallel and set to trip at the first frequency with a time delay of a few seconds. c) Similarly the third block of load is controlled by two relays—one set to trip at the third frequency and the other set to trip at the second frequency with a time delay.

This arrangement would take care of those border-line cases in which shedding the first block of load will be sufficient to prevent the system from passing through the second frequency but will not be sufficient to cause complete recovery to 60 Hz. That is, if the second block of load is not shed, the frequency would tend to remain below the first set point but would not drop to the second set point. However, operation of the relay set at the first frequency with time delay would eventually shed the second block of load and allow the system frequency to recover to 60 Hz.

5) Reponse to rate of change of frequency has also been discussed. However, at least with the present inherent time delays, we have not yet been able to demonstrate any appreciable advantage to using rate of change.

H. E. Lokay and V. Burtnyk: The intent of the paper was to present a simple, straightforward calculating technique for selecting the underfrequency relay settings and the magnitude of load to be shed at each step when developing an automatic load shedding program. It also showed the effect of some of the variables in the problem. An example of developing a load shedding program is shown. However, this example is intended to illustrate procedures rather than present recommendations of design constants which depend on specific system conditions.

Manuscript received September 5, 1967.

Both Mr. Meyer and Mr. Concordia have commented on the factor d, the percent load decrease per one-percent reduction in frequency. A d value of two was used in the example, but it is not intended to be applicable to all systems. The effect of different d values was included primarily to show the importance of this factor. We fully agree with the discussers that this is an area which requires industry research. It is a critical value when developing an optimum load shedding program and also when doing system stability studies where excursions from 60 Hz can occur. It seems reasonable that, until industry studies are more complete, a conservative value of d be used, and a value of two is suggested. It is hoped that this paper will help stimulate industry study in this important area.

Mr. Meyer also mentioned that load will reduce as system voltage reduces during the emergency condition. This is very true, but it is difficult to know how voltage will change on a generalized basis and therefore it was decided not to include the effect. It would always tend to reduce the frequency drop, thereby helping to minimize the magnitude of load automatically shed.

Mr. Concordia mentioned that the load shedding program should also include the effect of spinning reserve since governor action will increase the generation on the remaining machines and help arrest the frequency decay. However, on a generalized basis, we do not know where the system will separate or the amount of spinning reserve that would be available in the isolated area. Also, it is questionable how much of the available spinning reserve can be utilized. Depending upon the type of generation, our studies indicate that only about 20 percent of the available spinning reserve is utilized in the first five to ten seconds. During severe overloads the frequency can be well below 56 Hz in this time as shown in Fig. 1. Since the contribution of spinning reserve may not be appreciable and any contribution that is obtained will always reduce the overload and hence the amount of load shed, it seems reasonable to neglect this action when selecting the relay settings. If the rate of MW pickup is known, the frequency decay curves of Fig. 1 can be modified, and the procedure for determining the load shedding steps and relay settings would be the same as described.

Mr. Concordia mentioned that no attempt should be made to minimize the load shed, but rather shed enough to restore the frequency to 60 Hz. The intent of the described load shedding program is to shed the minimum load when the maximum system overload being protected occurs and have the frequency settle at a temporary acceptable level for this condition only. A reasonable temporary acceptable level is 57 Hz. It is recognized that once the frequency has settled, additional load must be dropped or generation increased to achieve 60 Hz and permit resynchronizing with the interconnected systems. This additional load can be manually shed or, if available, can be achieved from spinning reserve or fast-start peaking generation. However, it must be achieved in the matter of a few minutes, because 57 Hz is below the minimum permissible continuous frequency operation for steam turbine generating units. For overloads less than the maximum value, more load than necessary would be dropped, and a final frequency between 57 and 60 Hz would be achieved. In fact, in the sample (Fig. 8), frequencies below 58.5 Hz, the permissible continuous value, would result only for overloads between 50 and 53 percent and for those above 90 percent. At other overloads, the frequency, even without governor action, would be above 58.5 Hz.

Mr. Concordia mentions the possibility of using intentional time delays with underfrequency relays to drop the minimum load required for any emergency. It may be possible that for some conditions good coordination can be achieved and the minimum load dropped with such a scheme. However, for large system overloads, the frequency decays so fast that additional time delays cannot be used. The method for determining the relay settings and the load to be shed per step can be used to determine if proper coordination is possible. The method described is very flexible, as it can be used with any relay characteristics, with or without additional time delay, and with different system frequency decay response and system inertia characteristics.

We wish to thank the discussers for their comments and interest in the subject, as their comments have helped bring out pertinent important points.

FREQUENCY ACTUATED LOAD SHEDDING AND RESTORATION
PART I – PHILOSOPHY

R. M. Maliszewski R. D. Dunlop
American Electric Power Service Corp.
New York, N. Y.

G. L. Wilson
Massachusetts Institute of Technology
Cambridge, Mass.

Abstract—During the last few years, there has been expanded effort throughout the country to install frequency actuated load shedding programs for the purpose of providing means to safeguard a power system should a condition of declining frequency occur. Hand in hand with a load shedding program a convenient means must be provided for restoring load after the system returns to a normal state. In this series of papers, the authors will present some of the basic principles which should be considered in the application of a load shedding and load restoration program and will describe the philosophy which led to the frequency actuated load shedding and load restoration program being implemented on the American Electric Power System. As part of this presentation, the various analytical studies which were carried out to determine the load management programs for the AEP system will be described in detail. In addition, the method of program implementation and management is presented in a companion paper.[1]

INTRODUCTION

It is technically feasible to plan, construct and operate a power system which is immune from uncontrolled, widespread cascading power interruptions, even under the most adverse set of credible conditions. This goal can be achieved by strict adherence to the basic principles of bulk power supply planning.[2] These principles are well known and have formed the basis for the simulated testing criteria used by many of the planning and coordination groups throughout the country. The goal of such a testing criteria is to provide the means by which a system can be designed and operated to meet a specific security level within which wide-spread system cascading can be avoided.

Regardless of this goal, precautionary procedures are required to safeguard the system even for those improbable disturbances which may result in the operation of the system at subnormal frequencies. Should conditions arise which would result in an excess of load on the system, a situation of declining frequency will develop. To arrest this frequency decline—for situations when the rate of change is rapid and the magnitude is significant—and thus maintain the integrity of the system, a means must be provided to rapidly balance load and available generation. This can be accomplished by shedding specific amounts of load, utilizing frequency sensitive relays.

In the usual application of a load shedding program, a predetermined amount of load is shed over a specific frequency range. Should system conditions be such that the frequency decline cannot be arrested by the predetermined load shedding program, a means should be provided to isolate the system generation in a state of operation. By safeguarding generation, the restoration of the system can be accomplished in the shortest possible time.

It must be stressed that frequency actuated load shedding is not to be employed as a planning tool. It does not replace the need for a strong bulk power transmission system nor does it avoid the consideration of severe contingencies. It is a "last ditch" precautionary operating tool, to be used to maintain the integrity of a system which may be forced, due to some unpredictable circumstance, into an imbalanced condition resulting in a declining system frequency.

Along with an automatic load shedding program, an efficient means must be provided to restore load quickly to the system when the system state is such that load can be accepted. A number of methods can be used in this regard. One method involves the use of manual load restoration. This requires substantial manpower and has the inherent disadvantage of long time delay. Another approach is to use supervisory control from a centrally directed control center. This approach is expensive, unless the supervisory control system has been justified for other reasons. In addition, it requires that the design of the load shedding program be constrained within the limits established by the availability of the supervisory control system.

A third method of load restoration that can be considered utilizes frequency sensitive relays which can detect the state of the system and automatically restore load as soon as a normal system state has been established. This approach has the inherent advantage of being self-protecting since its operation depends on the system being capable of accepting load. When the capability of the system is exceeded, load restoration automatically ceases.

The restoration approach employed in specific situations will depend to a large degree on the configuration of the system and on the controllability that is available. Regardless of the approach utilized, good coordination and protection is required in order to avoid undue burdens on the overall system.

The proper application of an automatic load shedding program requires good knowledge of the characteristics of the system. Factors, such as, the dynamic characteristics of the system generation and load, along with the system configuration need to be carefully evaluated. The uncoordinated application of a load shedding program may result in unusual stresses being imposed on the transmission system which may lead to further sub-islanding due to internal power flow distortions. In addition, since in the majority of cases it may not be possible to predict with any degree of accuracy the composition of the system which may be forced into an imbalanced condition, proper coordination of load shedding programs with neighboring systems is a requirement.

The proper application of an automatic load shedding program should be preceded by an extensive study of the system in order to observe its characteristics. Such information is invaluable in determining the form of the load shedding program and in assuring its proper application on the system. Similarly, the restoration process must be carefully investigated to assure that it does not impose burdens on the system and that it is properly coordinated on a region-wide basis.

Load Shedding and Restoration on the AEP System

The American Electric Power System serves a seven state area, including parts of Michigan, Indiana, Ohio, West Virginia, Kentucky and Tennessee which is shown on the map of Figure 1. The system has well-defined concentrations of load and generation spread out over its entire region, which are connected by a strong transmission system at 345-kv and 138-kv. The existing 3300 circuit-mile 345-kv and 7600 circuit-mile 138-kv transmission systems are being overlayed by a 1200 mile 765-kv network to be completed by 1972.

Because of the configuration and the strength of the transmission system islanding situations are not readily predictable on the AEP system. In addition, due to the nature of the load-generation pattern existing on this system, a forced separation of the transmission system during a condition of declining system frequency in order to balance load and generation, is not considered to be a practical approach. Based on extensive analytical investigations, which will be described

Paper 70 TP 672-PWR, recommended and approved by the Power System Engineering Committee of the IEEE Power Group for presentation at the IEEE Summer Power Meeting and EHV Conference, Los Angeles, Calif., July 12-17, 1970. Manuscript submitted February 16, 1970; made available for printing May 22, 1970.

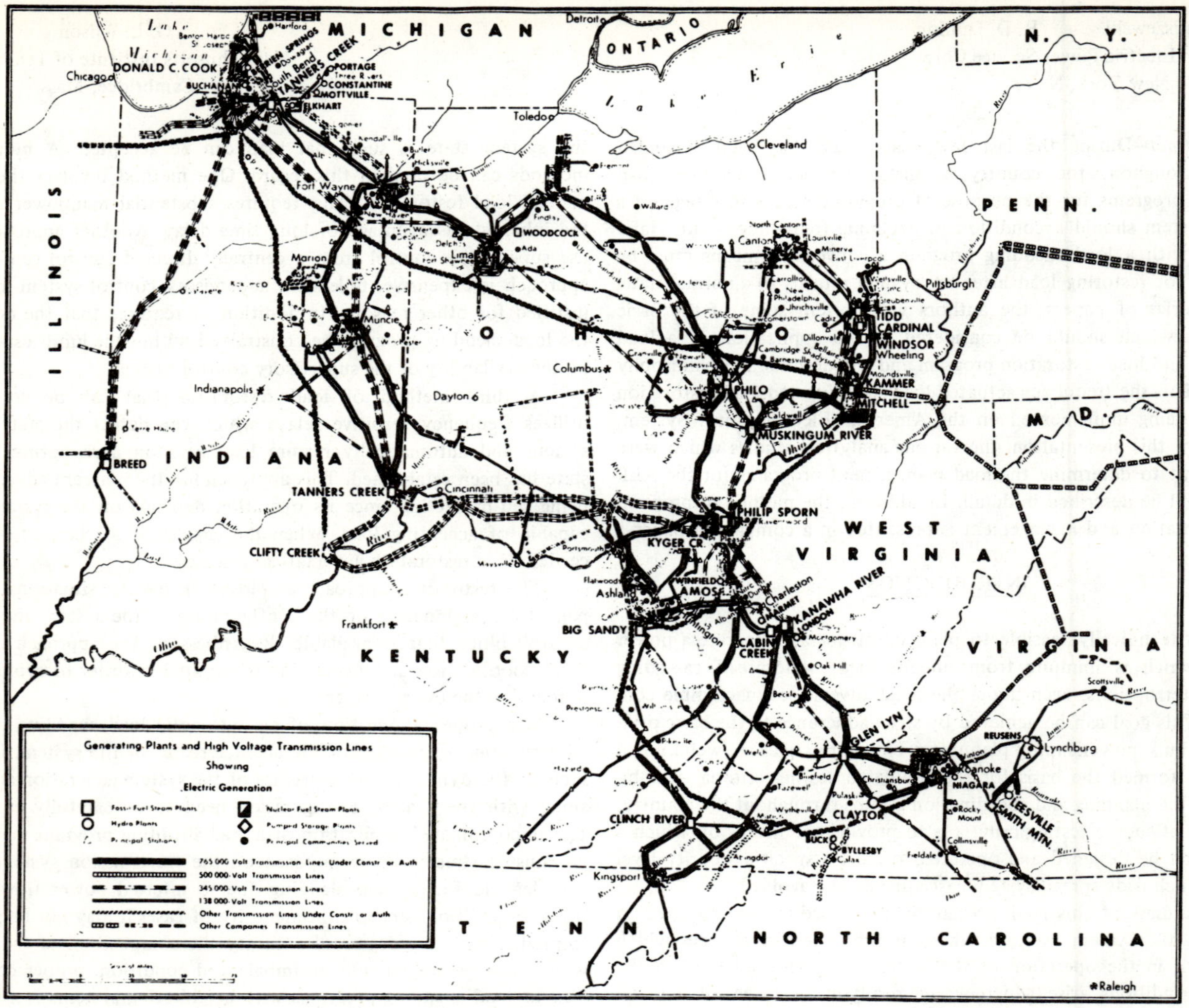

Fig. 1. Map of American Electric Power System.

in this paper, it was concluded that a program of automatic load shedding that is uniformly applied was best suited to the requirements of the AEP system. The type of load to be shed is described in a companion paper.

The AEP system is a party to the East Central Area Reliability Agreement (ECAR), involving 19 systems serving the East Central portion of the United States. Within ECAR, guidelines have been developed for the application of automatic load shedding programs. The program being implemented on the AEP system, which conforms to the ECAR guidelines, is as follows:

Frequency Trip Setting	% Load Shed
59.5 Hz	3 1/3
59.4 Hz	3 1/3
59.3 Hz	3 1/3
59.1 Hz	5
59.0 Hz	5
58.9 Hz	5

Thus, by a frequency of 59.3 Hz, 10% of the total system load will be shed. By a frequency of 58.9 Hz, an additional 15% of the total system load will be shed. To implement this program, solid-state relays will be employed.[1] These relays have a high degree of accuracy and as a result can be calibrated at frequency settings within a narrow band.

The stations on the AEP system are automatic, unattended, and do not have supervisory control, and therefore, restoration of load by manual means would necessitate a time and manpower consuming process. The restoration procedure to be employed will utilize frequency sensitive relays which can detect when the system returns to a normal state (60 Hz operation), and will restore load in controlled time delay steps. The use of solid-state load shedding relays allows for the consideration of frequency actuated load restoration in controllable steps. The technical aspects of this restoration program are described in this paper and its implementation is discussed in the companion paper.

It should be pointed out that the restoration program is a semi-automatic activity which will operate in an automatic manner to restore load when the system frequency is at or above 60 Hz. During restoration, should the frequency fall below 60 Hz, the restoration process ceases until operator intervention directs the frequency back to 60 Hz.

LOAD SHEDDING INVESTIGATIONS

The response of a system following a large load-generation imbalance may be assessed on the basis of its frequency response. A sustained excess load situation may be created through loss of generation and/or island formation with excess load. Under these circumstances the remaining generators electrically supply the system load in excess of the total mechanical input power; the result is a declining system frequency. If the spinning reserve response (increase in mechanical power input) and normal load-frequency regulation is insufficient, load shedding may be required to prevent loss of generation and total collapse of service within the island.

In a multimachine system or an island containing two or more machines the frequency response may be characterized by an average value within a band of actual values determined by the individual machines. The average value is determined by the sum total of the dynamics of all machines taken together. The band-width of actual values throughout the network is determined by the electromechanical oscillations between machines which is in turn a function of the degree of electrical coupling between machines the influence of their individual voltage and speed control response characteristics, and the size of the initial disturbance as well as subsequent disturbances due to load shedding.

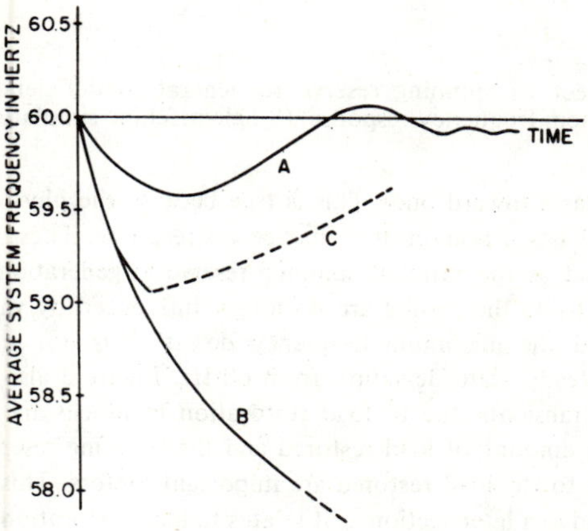

Fig. 2. Example of system frequency performance following sudden loss of generation (or addition of load).

The typical frequency response of a system with a deficiency in generation is depicted in Figure 2. If the spinning reserve is greater than the loss of generation, the response may be similar to curve A. When spinning reserve is insufficient to meet the emergency, then curve B shows a continuing decline in frequency unless some means of balancing load and generation such as load shedding is instituted in sufficient quantity to result in curve C. Curve C may also depict a situation in which spinning reserve was adequate in quantity but of such poor response quality that load shedding was required to prevent excessive frequency deviation.

It should be noted that following successful load shedding (curve C), the initial rate of increase in frequency depends on the amount by which the total load shed exceeds the deficiency in generation, taking spinning reserve response to that point into account. The character of the response from there on to steady state depends essentially on prime mover response characteristics. The steady state frequency depends on the speed droop characteristics of the system and the net change in prime mover power output required from the machines. If the prime movers must have a net increase in power output, such as may result when load shedding either does not take place or is insufficient without the aid of spinning reserve, the steady state frequency will be less than 60 Hz. However, the steady state frequency will exceed the nominal value (60 Hz.) whenever load shedding exceeds the initial deficiency.

Nature of Load Shedding Studies

To examine the dynamic performance of the system, studies were carried out in two modes. A number of tests were conducted on a simplified single machine simulation of the system, which assumes that the synchronous generators are closely coupled electrically so that intermachine oscillations are negligible. The results of these studies, covering a wide variety of load shedding applications, aided in making an informed selection of a few alternative schemes with practical merit. Based on the knowledge gained from simplified studies and on practical system considerations, those load shedding schemes which were potentially applicable were simulated on a portion of the AEP system with the system and machines represented in detail.

The authors feel that single machine simulation studies are important, for they provide not only a conceptual feel for the dynamics involved, but also indicate how the variation of load shedding and system parameters influence the average dynamic performance of the system. However, some questions relating to the influence of intermachine oscillations throughout the interconnecting network, cannot be answered on the basis of a single machine study alone. Some of the more pertinent information which detailed multimachine simulation yields is as follows:

(1) The dispersion of load shedding due to the dispersion of frequency in the network.
(2) The load shedding performance by sub-areas (representing concentrations of loads) and in particular, the performance of each area in relation to the others.
(3) The inter-area transient line flows within the island where load shedding is taking place in the presence of intermachine oscillations.
(4) The effect of realistic spinning reserve dispatch.
(5) The effect of a particular location of generation loss on the load shedding performance and intermachine oscillations.

Single Machine Simulation. When one machine is used to represent the dynamics of a multimachine system, some compromise is required. However, it is possible to design a model including the salient factors which determine overall system response following a substantial load-generation imbalance. The well-known equation of motion which relates machine speed or frequency to changes in input and output power of a synchronous machine is as follows:

$$2H\omega \frac{d\omega}{dt} = P_m - P_{elec} \tag{1}$$

where, ω is speed or frequency expressed in per unit of the rated value (2π 60 rad/sec) and P_m and P_{elec} are mechanical and electrical powers, respectively, expressed in units consistent with that of the inertia constant, H. In the single machine model P_m becomes the total mechanical power from all prime movers, P_{elec} becomes the total electric load on all machines which are electrically connected and the frequency, ω, represents the average system frequency. Since the load shedding effort is applied to limit frequency deviation to something less than five percent, the factor, $2H\omega$, is assumed constant at the rated speed value.

One key element in the analog simulation to solve equation (1) is the derivation of a "composite" model to represent the combined governor and prime mover response of several machines of differing response characteristics. This model may be adjusted to reflect the mix of non-reheat, single reheat and double reheat units and thereby aids in achieving a realistic estimate of overall system spinning reserve response.

The "composite" prime mover response model was established by assuming average response characteristics for non-reheat, single reheat and double reheat machines, respectively. One "equivalent" reheat machine was created in the following manner: 1) The non-reheat turbines were combined with that fraction of the reheat turbines which were not subject to reheat time lags. 2) The remainder of single reheat turbines and that portion of double reheat turbines which is subject to one reheat time lag were combined to form an equivalent reheat turbine. 3) The portion of power developed by double reheat turbines which is subject to a second reheat time lag was not included because its contribution to a change in prime mover output (spinning reserve) in the period of interest is small.

Multimachine Simulation. The multimachine studies were conducted using a modern multimachine transient stability program.[3] This program has the facility for calculating the apparent bus frequency at all buses in the system and a provision to simulate load shedding relays by disconnecting load at each bus in steps at specified frequencies and time delays.

The area of the AEP system that was set up in detail included approximately 3500 MW of load. The load in this portion of the system is concentrated in five major load areas ranging in size from 450 to 1100 MW. Transmission included the 765-kV, 345-kV and 138-kV networks with the load being represented at the 138-kV level. The generation within the area was approximately equal to the total load and dispatched in a manner that is typical for a peak load condition. With the exception of a few very small units, all generators were represented in detail including excitation and speed control (governor and prime mover) systems.

To initiate tests the study area was separated from the rest of the interconnected system. The choice and method of line tripping was specifically directed to minimize the initial reactive power disturbance to the area so that load bus voltages did not change appreciably as a result of island formation. Since the area was essentially balanced, additional generation was tripped to produce the desired initial level of generation deficiency. The initial disturbance in these multimachine studies took place under somewhat ideal circumstances. However, the primary purpose of these studies was to evaluate the performance of load shedding programs and their influence on system dynamic behavior. In addition, the experience gained in these studies will provide guidance in developing an efficient, consistent approach to analytical evaluations associated with the continuing review of load shedding applications in the future.

Parameters Investigated and Results

The nature of the parameters considered here may be conveniently divided into two categories: 1) those associated with the system representation, and 2) those associated with the load shedding program. Since these studies were carried out with one particular system in mind, it was possible to greatly reduce the number of system parameters. All parameters associated with the machine dynamics such as, inertia, prime mover and excitation system models, and generator electrical characteristics were fixed. The only system variable retained was the level of spinning reserve available. In the single machine studies this corresponded to a limit on the variation of input to the composite prime mover model. In the detail multimachine studies, the limits on valve travel (prime mover input) were individually set on each machine in accordance with their respective rated capabilities, and the distribution of reserve among the various units was determined by a realistic typical dispatch of generation.

The load shedding parameters considered as prime variables were as follows:

(1) relay set point
(2) step size
(3) number of steps
(4) intentional time delay

The above parameters together with system spinning reserve comprised the primary variables in the studies carried out with both the simplified single machine and the detailed multimachine simulations.

Small Disturbance Studies. A number of cases were run using the simplified model of the system to determine the influence of spinning reserve on system frequency following a small loss of generation. The results of these studies was a series of composite frequency response characteristics similar to curve A in Figure 2. The influence of spinning reserve on average frequency can most easily be shown by comparing the minimum frequency (maximum deviation) versus the ratio of spinning reserve to generation deficiency as indicated in Figure 3. For a given size deficiency, the maximum deviation from 60 Hz increases rapidly as the spinning reserve available approaches the magnitude of the disturbance (spinning reserve to generation deficiency

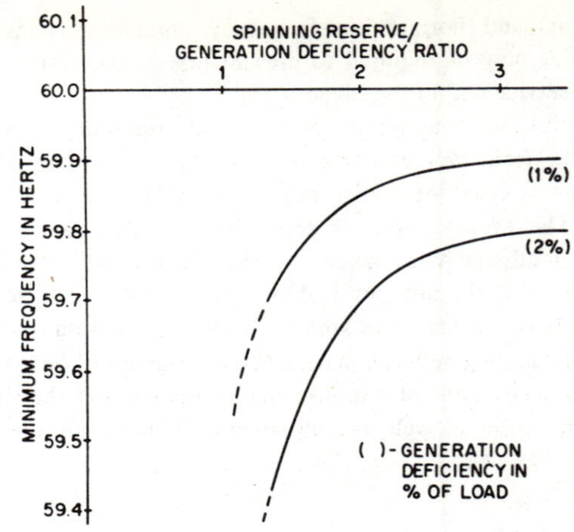

Fig. 3. Effect of spinning reserve to generation deficiency ratio on system frequency response. (Single machine simulation).

ratio decreases toward one). This is true because the physical limit on valve travel has a non-linear influence on response. These results also indicate that as the ratio of spinning reserve to generation deficiency exceeds 3 to 1, the results are no longer influenced by valve limiting effects, and the maximum frequency deviation is not much greater than the steady state deviation from 60 Hz. Figure 3 also shows that if system transients due to load restoration in blocks are to be minimized, the amount of load restored and the spinning reserve available in relation to the load restored are important factors. This aspect will be discussed in a later section as it relates to load restoration.

Single Machine Simulation Results. With the simplified simulation, many tests were performed to determine system response to single and multiple stage load shedding programs. In these studies both the step size and the frequency set point at which the load was dropped were varied. The response of the system to various size disturbances may be classified into three categories.

First, if the disturbance is sufficiently small, no load shedding will take place, i.e., both the magnitude and response capability of the available spinning reserve is sufficient to restore the generation-load balance. As the frequency set point for initial load shedding is lowered, the size of the disturbance increases for which load shedding does not take place. The nature of the frequency response for this class of disturbances was discussed in the previous section.

Both the second and third categories of disturbances involve one or more stages of load shedding. In the former, load shedding will result in an immediate arrest in frequency decline; i.e., the minimum frequency corresponds to a load shedding set point. In the latter case the disturbance is so large that load shedding at a given set point does not immediately arrest frequency decline, but the available spinning reserve is sufficiently large so that frequency decline will be arrested eventually. Here the minimum frequency is determined by spinning reserve as in the first category; however, the response characteristics of the spinning reserve may be such that additional load shedding would be required to prevent excessive frequency decline.

The most significant consideration in determining the size of the load shedding step is the potential for excess load shedding. For disturbances which just exceed the reach of the spinning reserve response, the amount of overshed approaches the step size and may, for large steps, cause a significant system frequency transient, which is characterized by a large overshoot and steady state frequency above 60 Hz.

One of the primary considerations in the choice of a multistage program is the desire to obtain extensive coverage and at the same time to limit the potential overshed to reasonable values. Figure 4 shows the potential overshed as a function of disturbance size for a two step and a six step scheme compared with the ideal case. Note that the average excess load shed is significantly reduced by increasing the

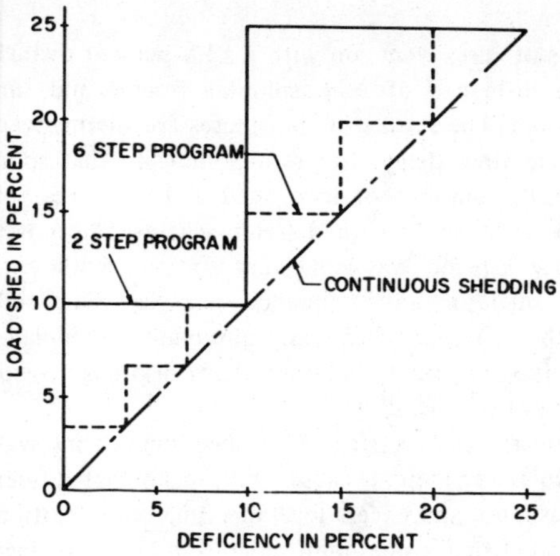

Fig. 4. Effect of size and number of load shedding steps on ideal load shedding characteristics. (No spinning reserve).

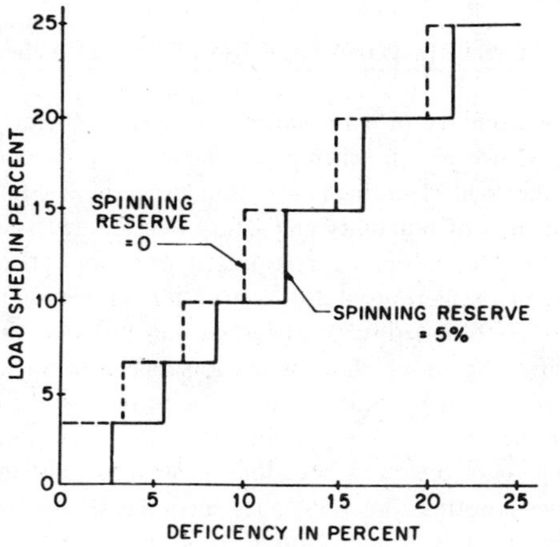

Fig. 5. Effect of spinning reserve on ideal 6-step load shedding characteristics. (Single machine simulation).

number of steps from two to six. Reducing the excess load shed is not only beneficial from the customer's point of view, but also reduces the transient overshoot and steady state frequency response which follows load shedding.

Spinning reserve is beneficial also in reducing the expected excess shedding as a function of the disturbance size. In Figure 5, the six step characteristic of Figure 4 is compared with the same program when five percent spinning reserve is available. The effect of spinning reserve is to displace the characteristic to the right; however, note that the amount of displacement decreases as the disturbance gets larger. This means that the influence of spinning reserve on load shedding performance and system response diminishes as the size of the disturbance increases. As pointed out earlier, larger disturbances result in a more rapid decline in frequency thus giving less time for spinning reserve to respond.

Multimachine Simulation Results. In the detailed multimachine simulation studies two basic load shedding programs were investigated, a two-step program (10% load at 59.3 plus 15% load at 58.9) and a six-step program as specified in the introduction. The purpose of these tests was to investigate the influence of time delay, deficiency magnitude, and spinning reserve on both system frequency response and load shedding performance in a distributed multimachine area.

Most of the tests were conducted with maximum disturbances; i.e., the loss of generation approached the maximum load shedding capability of 25 percent of system load. Some of the more significant results are compared in Figures 6-8. The frequency response shown in

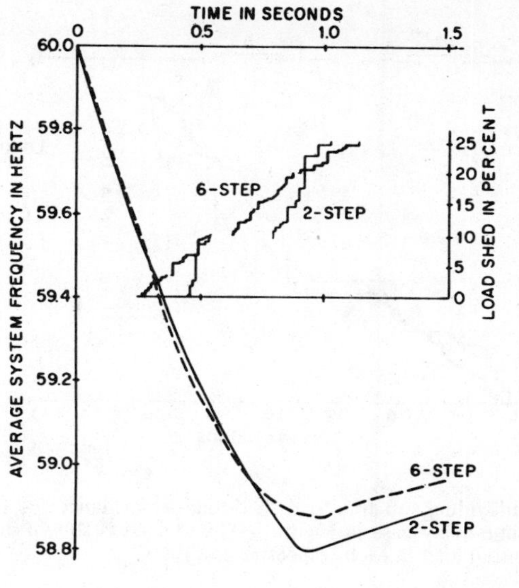

Fig. 6. Influence of 2-step and 6-step load shedding programs on system performance. (Multimachine area simulation for 23% loss of generation and 4 cycle load shedding time delay).

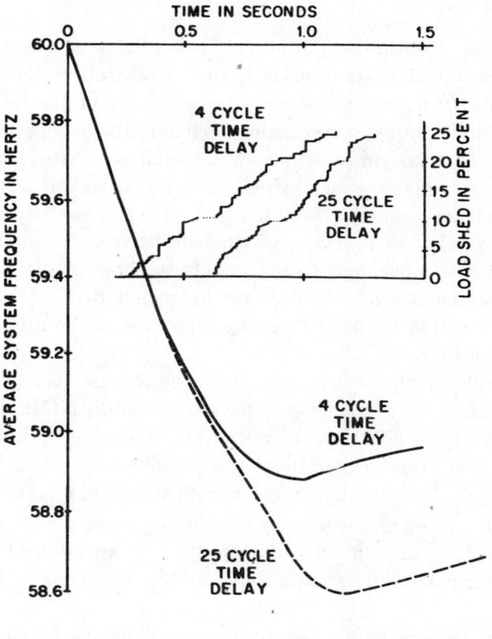

Fig. 7. Influence of time delay on system performance with 6-step program. (Multimachine are simulation for 23% loss of generation).

Figures 6 and 7 is the average frequency of the system found by integrating equation (1). The total area load shedding performance is also indicated in Figures 6 and 7.

Figure 6 illustrates the difference in performance between a 2-step and a 6-step program. In these tests, a minimum time delay (3-4 cycles) was assumed and no spinning reserve was available. Note first that the 6-step program yields a better frequency response, which is primarily caused by the load shedding effort starting at higher frequencies. Each load shedding stage is spread out rather than in discrete steps, with the width in time of a given step being a reflection of the bandwidth of frequencies which exist in the multimachine system. The time to complete a given load shedding stage increases with each successive stage because the rate of frequency decline (system deceleration) is diminished by load shedding. Increasing the number of stages from two to six results in a more uniform progression of the effort but not to the point where overlapping occurs. A careful

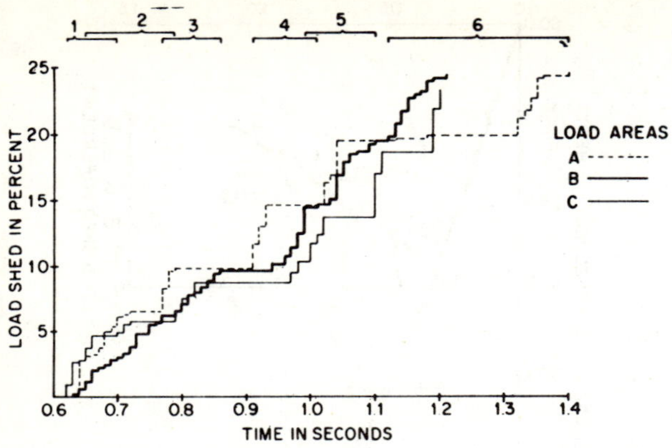

Fig. 8. Individual sub-area load shedding performance for the 25 cycle time delay case in Figure 7. (Load shed is shown in percent of initial load in each respective area).

examination of the results of the six stage program indicates that each stage is distinct; consequently, the 0.1 Hz separation between settings is adequate. For cases involving smaller disturbances, the bandwidth of frequencies was smaller and the load shedding steps became more distinct.

Figure 7 illustrates the influence of total time delay. The most significant effect of time delay is that it determines the amount by which the frequency overshoots the set point of the final stage required. Since, during a continuing decline in frequency, it is desirable to leave some margin in frequency between automatic load shedding and the next phase of action required at 58.0 Hz[1], it was concluded that time delay should be minimized to that value required by the frequency relay to properly discriminate between a switching operation and a true declining frequency. On the basis of these results, the minimum frequency should always be greater than 58.5 Hz for disturbances within the load shedding capability, if the total time delay is less than 15 cycles.

Another factor which indicates the need to minimize time delay is the tendency for overlapping in the load shedding effort. This is illustrated in Figure 8, which shows the load shedding by area for three distinct load areas in percent of the respective area load. This is the same 6-step, 25 cycle time delay case illustrated in Figure 7. Note that before one stage of shedding is completed in one area the next stage is initiated in another area. If this effect is not controlled by minimizing time delay, inter-area transient line flows could become excessive.

For the maximum disturbance cases discussed above, the frequency decay was so rapid that the effect of spinning reserve was minimal. However, spinning reserve takes on increasing importance at lower levels of disturbance.

Table 1. Influence of load shedding and spinning reserve on system performance following 13.8 percent loss of generation.

VARIABLES			RESPONSE	
S 1)	T.D. 2)	S.R. 3)	f_{min}	L.S. 4)
2	25	4%	58.82	21.7
6	25	4	58.94	19.0
2	25	0	58.80	24.7
6	25	0	58.91	19.9
6	4	0	59.09	14.4

Note: 1) Number of load shedding steps.
2) Time delay in cycles at 60 Hz.
3) Spinning Reserve in percent of initial load.
4) Total load shed in percent of initial load

A series of cases were run with a 13.8 percent disturbance to determine the influence of both spinning reserve and time delay on system response. The results of these cases are summarized in Table 1. With 25 cycle time delay, the results indicate that spinning reserve does reduce the amount of load shed and improves the frequency response for both the 2-step and 6-step schemes. Moreover, the 6-step load shedding scheme was somewhat superior with regard to both excess load shedding and frequency response. One additional case was run with a six-step scheme and minimum time delay as indicated in Table 1. It is apparent that time delay also has a significant influence on excess load shedding.

In summary, the six stage load shedding scheme was adopted in preference to fewer steps because of its beneficial influence on both frequency response and excess load shedding considerations. Since the studies indicated that a minimum separation between stages of 0.1 Hz was desirable, a greater number of stages would require load shedding to start at higher frequencies and thereby limit the influence of spinning reserve in preventing load shedding. The results also indicated that intentional time delay should be minimized.

LOAD RESTORATION INVESTIGATIONS

A semi-automatic or automatic restoration scheme can assume many forms. However, all schemes must satisfy certain criteria which ensure that the load shedding-restoration program is compatible with the high standards of reliability and safety that are necessary in today's power systems. An automatic restoration program can cause uncontrollable system oscillations if the restoration process is not carefully designed and coordinated with the shedding program. For example, consider a situation in which a restoration relay's frequency set point is such that the ensuing frequency drop due to the restoration transient encompasses the frequency set point of a load shedding relay. The load shedding relay operates, shedding some load, and the frequency rises again overshooting to a value that encompasses the restoration set point and the load is again restored. This process then continues to repeat itself, causing a relaxation oscillation which could proceed until some manual means were taken to override the automatic scheme.

In order to prevent these oscillations and to arrive at a restoration program which is compatible with interconnected system operation, two major criteria must be established:

1) Restoration Frequency Set Points – Since the restoration of load is controlled by frequency sensitive relays, the choice of frequency set points at which load is restored must be determined.
2) Size of Restoration Steps – The amount of load restoration step must be such that the resulting transient does not encompass the operation point of another shedding or restoration step.

Restoration Frequency Set Points

In general, load shedding-restoration schemes can assume one of two modes. One is a continuum mode in which the load is shed and restored in a very fine distribution of points over the frequency range below 60 Hz where shedding is most beneficial. In the limit, load would be shed in a continuous fashion if frequency were decreasing and load would be restored in a compatible and continuous fashion if frequency were increasing. In this case the frequency set points would cover the same range as the load shed points. Such a system has the effect of acting as a continuous, instantaneous governor. Criteria for designing such a system in order to prevent relaxation oscillations have been developed for the situation where the power system remains as an island during restoration. However, in order to accomplish restoration in an interconnected environment a restoration scheme in which load is continuously restored within the frequency range that shedding occurs, is not desirable. As the frequency increases, load would restore to the extent that resynchronizing of the island may be prevented. In the case where synchronizing does occur, the frequency of the previously islanded area will be forced to 60 Hz resulting in a virtually uncontrolled restoration of load, which may severely stress system generation and transmission capability.

The design of a discrete load restoration program involves a selection of relay settings and amount of load to be restored per restoration stage. Although it is presumed that a system would have islanded prior to the shedding of load, it is not necessarily true that the islanded condition is maintained after the load has been shed. It is possible, for example, that interconnections would be restored between the islanded system and the external systems. When these interconnections are made, the islanded system frequency which was either less than or greater than 60 Hz prior to interconnection will be restored to 60 Hz. A restoration program may be designed with relay settings greater than 60 Hz; however, if an interconnection is made prior to restoring all of the shed load, it would not be possible to automatically restore the remaining load since system frequency would be constrained to a very narrow band about 60 Hz.

A load restoration program in which relay settings are less than 60 Hz can also be designed. However, this case has the disadvantages noted in the discussion of a continuum restoration program.

It becomes clear then, that the restoration frequency setting should be 60 Hz. A setting less than 60 Hz is potentially dangerous in an interconnected environment, and a setting greater than 60 Hz does not allow for automatic restoration under all circumstances. It is due to the latter situation that the ±0.2 Hz accuracy mechanical relays are not recommended for use with an automatic restoration scheme.

Size of Restoration Step

With restoration relays set at 60 Hz, it is still possible for the relaxation type oscillations to occur. Coordination between the restoration program and the shedding program must be designed to prohibit overlapping of the restoration transient with the shedding relay settings. The key to preventing this is to insure that the amount of restoration per stage is more than met by spinning reserve to the extent that the resulting transient involves a small frequency swing. This transient was studied on the analog computer model, discussed in the section entitled "Load Shedding Investigations".

In Figure 3 it is seen that for a 1% restoration with at least 3% spinning reserve, the frequency drops to a minimum of 59.9 Hz, while for a 2% restoration with 6% spinning reserve, the minimum is 59.8 Hz. Since the maximum relay settings for shedding are 59.5 Hz or below, these restoration frequency minimums are well within the safe operating margin. If the spinning reserve is reduced, this margin is rapidly reduced as seen from Figure 3. In general, it is strongly advised that no more than 2% restoration be allowed per restoration stage. It is further recommended that during the automatic restoration period all units be operated with a spinning reserve not less than 3 times the restoration per stage. Thus, for a 1% restoration, 3% spinning reserve should be available, etc. To insure that oscillations do not take place, the amount of restoration per stage must be less than the amount of load shed per stage. Studies indicate that the restoration stages should be less than one-third of the load shed per shedding stage.

With all restoration relays set to restore at or near 60 Hz, the various stages must be distinguished from one another by time delay settings. These must be long enough to ensure that short duration excursions above 60 Hz, caused by minor overshedding, do not cause load to be prematurely restored. In addition, sufficient time must be provided, to allow interconnections to be closed either automatically or manually prior to the restoration of any load and any subsequent loading of those interconnections. Subsequent load restoration stages should have additional time delay to allow the restoration process to occur at the same frequency, but in a gradual manner.

System Response to Restoration

A simulation of a system with discrete load shedding and restoration was made on the Analog computer using time delays of 30 seconds for the first restoration stages and 5 second intervals for subsequent stages. A typical histogram of a shedding-restoration transient is shown in Figure 9. The initial disturbance was such that an area was islanded with a 3% deficiency in generation and remained islanded during restoration. Since the first shedding stage consisted of 5% load shed at 59.3 Hz, the frequency drops at point 1 on the diagram to 59.3 Hz where 5% load is shed. Since the load shed is 2% in excess of the generation deficiency, the system settles out to a frequency greater than 60 Hz. At point 2, the first load restoration stage times out, 30 seconds after the 60 Hz point is crossed, and 1% of the load is restored. The resulting transient causes the frequency to drop below 60 Hz, and then recover to a new steady state above 60 Hz. Thirty-five seconds after the 60 Hz point is crossed again, another 1% load is restored at point 3. This time the steady state frequency is below 60

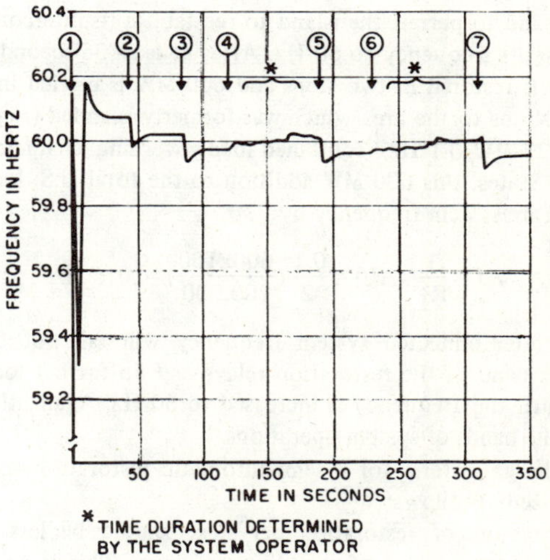

Fig. 9. Typical histogram showing load shedding and restoration transients on an islanded system following 3% loss of generation.
Load Shedding: 5% at 59.3 Hz.
Load Restoration: 1% increments at 60.0 Hz.

Hz and no further restoration will occur until generation is added to the islanded system. Thus, after all initial transients are over, the restoration of load is completely in the control of system operators through adjustment of system frequency. This fact is very important and it must be emphasized that this scheme is in fact semi-automatic. It also points out the need for coordination between systems so that the decision to pick frequency and therefore, load is by necessity a result of a cooperative effort on the part of all systems.

The time span between points 4 and 5 is dependent entirely on the rate at which generation is added to the islanded system. This time is completely in the control of system operators. If it is not desirable to restore additional load immediately, the system frequency should be maintained at a level slightly below 60 Hz. Additional load is restored, in discrete amounts, by allowing the frequency to increase to the 60 Hz point, after which load will restore at point 5 and later at point 7 in Figure 9.

The discussion above assumes that the systems which have shed load and islanded, remained so during the entire restoration process. This is not the case under interconnected system operation. There is some concern that if an interconnection is made with this scheme, the formerly islanded area would restore all the load, overload the interconnections, and cause them to trip out.

The proposed system does in fact prohibit this from occurring. Consider the situation in Figure 10a in which a section amounting to

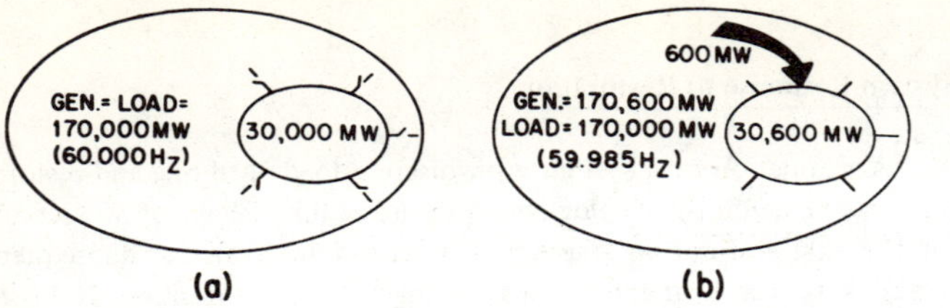

Fig. 10. Possible steady state configurations for a 200,000 MW system.
(a) Islanded situation with all ties open.
(b) Following closure of ties and restoration of 600 MW load (2 steps).

30,000 MW has been islanded from the remainder of the United States whose total generation is on the order of 200,000 MW. The islanded area is at some steady frequency below 60 Hz while the remaining 170,000 MW is operating at 60,000 Hz. Assume further that the islanded area is on an automatic restoration program in which 1% segments of load are controlled by restoration relays which will have the same 30 second initial time delay setting assumed in the previous case. These relays will restore if the frequency is greater than or equal to 60,000 ± 0.010 Hz, the accuracy of the solid-state relays. To initiate restoration and to permit the island to reestablish its interconnections, it must raise its frequency to 60 Hz. After at least 35 seconds the first 2% of load is restored in two steps and 600 MW is carried in over the interconnections to the area which was formerly islanded (Figure 10b). Due to the 2%MW/0.1 Hz[4] combined load-governing characteristic of the United States, this 600 MW addition to the total U.S. load will reduce the entire system frequency by:

$$\Delta f = \frac{1}{R_L}(\Delta L) = \frac{0.1}{2}\left(\frac{600 \times 100}{200,000}\right) = .015 \text{ Hz.} \qquad (2)$$

Thus, the interconnected system frequency will fall outside of the ± 0.010 Hz band of the restoration relays and no further load will be restored until the frequency is increased to 60 Hz. Again, the control is back in the hands of system operations.

The design criteria for a semi-automatic restoration scheme can be summarized as follows:

1. The amount of restoration per stage should be less than one third of the amount of shedding per stage. This is necessary to insure that relaxation oscillations do not occur due to the reaction between load shedding and restoration transients. Restoration steps should be no larger than 2%.
2. Each restoration stage should occur at 60 Hz with isolation between stages through time delays. The first of these should be on the order of 30 seconds or greater in order to allow tie lines to close in. It is suggested that the time delay settings between load restoration stages be no less than 5 seconds.
3. During the restoration period it is recommended that all units be operated with a spinning reserve not less than 3 times the amount of restoration per stage.
4. In the event that interconnections between the formerly islanded area and a larger system are weak, the restored load should be reconnected in the sequence that it was shed. Thus, the stages that are designed to be shed first should be restored first in order to minimize the transient in the event of any subsequent disturbance.

CONCLUSIONS

Extensive analytical investigations have been carried out to develop a frequency actuated load shedding and load restoration program for the American Electric Power System. Based on these investigations a number of general conclusions can be drawn:

1. In an interconnected environment, the application of load shedding and load restoration programs necessitates coordination among all neighboring systems in order to avoid undue stresses and possible misoperation.
2. To minimize the potential for excess load shedding, a multistage shedding program is desirable.
3. The effect of spinning reserve in arresting a frequency decline will be a function of the disturbance size. For small disturbances, spinning reserve is effective in improving system performance, whereas for large disturbances, the rate of frequency decline may be much greater than the rate of spinning reserve response.
4. The time delay associated with the operation of load shedding relays should be minimized to avoid overlapping of the shedding effort in a multi-area system and to minimize the frequency transient.
5. The frequency set point for the first stage of load shedding should be sufficiently low such that spinning reserve may be effective for small disturbances.
6. The automatic restoration of load in an interconnected environment should only be accomplished at a system frequency of 60 Hz.
7. The use of a frequency actuated load restoration scheme can be accomplished without placing a burden on the system. In reality it is a restrictive approach which is under the control of the frequency bias inherent in an interconnected system.

REFERENCES

[1] A. F. Gabrielle, S. H. Horowitz, A. Politis, "Frequency actuated load shedding and restoration. Part II—Implementation," IEEE Summer Power Meeting, July 1970.
[2] "Prevention of power failures. Vol. II—Advisory committee report on reliability of electric bulk power supply," U.S. Government Printing Office, Washington, D.C., June 1967.
[3] D. G. Ramey and R. T. Byerly, "Digital simulation of the interconnected system," IEEE PICA Conference, May 1967.
[4] N. Cohn, *Control of Generation and Power Flow on Interconnected Power Systems,* New York; John Wiley & Sons, 1966.
[5] P. Zarakas, C. J. Durkin, E. Eberle, "An Underfrequency Relay With Frequency Decay Rate Compensation," IEEE Transactions, 69 TP-57-PWR.

For combined discussion, see pp. 545-549.

FREQUENCY ACTUATED LOAD SHEDDING AND RESTORATION
PART II – IMPLEMENTATION

S. H. Horowitz A. Politis A. F. Gabrielle

American Electric Power Service Corp.
New York, N.Y.

INTRODUCTION

Power systems must be operated within limits that will ensure adequate generation and transmission capacity to avoid cascading. In developing a set of operating limits, it is important to do so within a general framework in order to ensure that the operating objectives are met. In this regard, a set of general guidelines are presented below:

1. The integrity of the transmission system will be maintained at all times without planned internal separation.
2. Maximum reasonable assistance will be given to adjacent systems experiencing difficulty. However, such assistance will be terminated – without opening interconnection circuits – when the reliable operation of the assisting system is impaired.
3. The system should be operated so that the occurrence of the next single contingency (outage of tower, circuit, unit, breaker, bus, etc.) will not result in a cascading loss of the bulk transmission system.

   The single contingency is continually geared to the current state of the system and reflects maintenance and forced outage events as they occur. In addition, when evaluating overload contingencies, the time to relieve the stress must be compared to the permissible degree of overload.
4. When the system experiences a generation-load imbalance, the principles of sound interconnected operation will be maintained by bringing under control an unscheduled tie-line power flow condition as quickly as possible.

While the above guidelines are designed to assure reliable electric service, they do not preclude the necessity of having well-defined procedures to meet emergency conditions such as a system separation and operation at subnormal frequency.

In addition, due to the nature of power system operation, it is essential that emergency procedures be coordinated with neighboring companies. In this regard, a coordinated program has been developed for the nine-state, 40,000 MW load area served by 19 power systems cooperatively seeking reliability via the mechanism of an East Central Area Reliability (ECAR) Agreement. The coordinated program adopted by American Electric Power (AEP), which conforms to the principles established by ECAR for handling a declining system frequency includes:

1. From 60.0 - 59.8 Hz

   Utilize to the extent practicable all operating and emergency reserve. The manner of utilization of these reserves will depend greatly on the behavior of the system during the emergency. For rapid frequency decline, only that capacity, on-line and automatically responsive to frequency (spinning reserve), and such items as interconnection assistance, and load reductions by automatic means are of assistance in arresting the decline in frequency.

   If the frequency decline is gradual, the system operators, particularly in the deficient area, should invoke non-automatic procedures involving operating and emergency reserves. These efforts should continue until the frequency decline is arrested or until automatic load-shedding devices operate at subnormal frequencies.
2. 59.8 Hz

   Trip automatic load-frequency control system at Operations Control Center (also trip at 60.2 Hz) and at the power plants.
3. 59.3 Hz

   Shed not less than 10% of system load with automatic load-shedding relays. No intentional time delay should be used beyond that absolutely required to avoid improper relay operation.
4. 58.9 Hz

   Shed additional load with automatic load-shedding relays in an amount not less than 15% of system load existing prior to step 3. No intentional time delays should be used beyond that absolutely required to avoid improper relay operation.
5. 58.0 Hz

   Isolate generating units without time delay.

If, at any time in the above procedure, the decline in area frequency is arrested below 59.0 Hz, that part of the system in the low frequency area should shed an additional 10% of its initial load. If, after five minutes this action has not returned the area frequency to 59.0 Hz or above, that part of the system shall shed an additional 10% of its remaining load and continue to repeat in five-minute intervals until 59.0 Hz is reached. These steps must be completed within the time constraints imposed upon the operation of generating units.

LOAD SELECTION FOR SHEDDING

To implement that portion of the above program with automatic underfrequency load shedding, the first major decision involves the type of load to select for frequency actuated control in the six-step load shedding scheme adopted by AEP – discussed in the companion paper[1] – and shown below.

Relay Trip Settings

Trip Step	Frequency	Percent Load Tripped
1	59.5 Hz	3⅓
2	59.4	3⅓
3	59.3	3⅓
4	59.1	5
5	59.0	5
6	58.9	5

Of necessity, the load selected for shedding will be different for various systems because of geographical, historical, political and reliability factors.

Since a frequency actuated, load-shedding program should not be embodied in system planning and power system design criteria, the ability to delineate specific "islands" due to cascading within the interconnected network is difficult. Moreover, as the trend toward power system integration and interconnection – together with EHV – continues to result in more closely coupled and interdependent systems, potential islands will vary in location and scope. In addition, the

Paper 70 TP 673-PWR, recommended and approved by the Power System Engineering Committee of the IEEE Power Group for presentation at the IEEE Summer Power Meeting and EHV Conference, Los Angeles, Calif., July 12-17, 1970. Manuscript submitted February 16, 1970; made available for printing May 21, 1970.

changing pattern of generation site development is also influenced by the dynamics of economic, technological and political achievements. Therefore, in order to design an underfrequency load-shedding program that must restore a balance between generation and load, under conditions that are difficult to define, it appears that a uniform distribution of load shedding should be applied. For the AEP seven-state system, this criterion eliminates controlling large blocks of industrial load such as aluminum, steel and chemicals in the 50-400 MW range. In addition, the control of large blocks of load presents technical difficulties during restoration. Consequently, it was decided to select distribution circuits for load shedding.

The installation of load-tripping devices must be compatible with present standards of service reliability. For example, where present stations or customers are being served by alternate sources which reflect the careful planning and design to enhance the integrity of the power supply, the addition of hardware that increases exposure to interruption must be minimized. However, the addition of one relay to control a distribution circuit does not materially affect customer reliability that is already primarily determined by the multitude of other elements that can cause an outage to the circuit. In addition, the choice of distribution type circuits for load shedding mitigates the discriminatory aspects of customer selection in that these circuits serve residential, commercial and industrial type customers.

A review of the distribution circuit loadings on the AEP System indicated that the average circuit peak load is about 3 mw. To accomplish effective and distributed shedding, and to meet the criterion for controlling 25% of System load (presently 9,000 MW) with a reasonable number of relays, it was decided to connect a minimum of two circuits to a relay with a load totaling at least 6 MW at peak load.

In stations with many circuits and adequate load magnitudes, two relays will be used with widely separated trip settings in order to minimize the chance of blacking out concentrated areas.

In terms of selecting circuits, a "service priority rating" was established as indicated below. As noted, six groups were established to be compatible with the six-step load-shedding program discussed in the companion paper[1].

Service Priority Rating

Outage Time	Priority	Trip Step
0	12	6
1 min. – 5 min.	11	6
6 min. – 15 min.	10	6
16 min. – 45 min.	9	6
46 min. – 90 min.	8	6
91 min. – 3 hr.	7	5
3 hr. – 6 hr.	6	4
6 hr. – 12 hr.	5	3
12 hr. – 24 hr.	4	3
1 day – 3 days	3	2
3 days –	2	1
Beyond	1	1

Loads in the Priority 8-12 category include individual medical facilities (iron lung, etc.), medical institutions, radio and TV stations, aviation agencies, communication facilities, national defense installations, power company dispatch centers, etc. Thus, circuits serving "critical" customers would not necessarily be eliminated since these customers are exposed to interruptions every day and the procedures that have been worked out locally for handling outages would still be applicable. However, the "critical" customer category is recognized in selecting the trip steps.

Thus, the choice of a seven-state, distributed load-shedding scheme controlling distribution feeders serving the load magnitude previously indicated results in an initial installation of underfrequency relays at about 400 stations. This type of program, coupled with the fact that the AEP stations are highly automated with no station operators or remote supervisory control of stations, leads to the necessity for automated load restoration procedures at 60.0 Hz. This can be accomplished by including a restoring contact with time delay – as discussed later – such that the overall scheme may be labeled as a Frequency Actuated Load-Shedding and Load-Restoration Program.

CONTROL CENTER COORDINATION OF SYSTEM RESTORATION

If a set of contingencies arise that stress the power system to a point whereby the underfrequency load-shedding relays operate to maintain a balance between generation and load, it is essential that a coordinated effort – both within the AEP System and among the AEP interconnected neighbors – be carried out to achieve a successful restoration within a minimum time period. Coordinated effort is, of course, of necessity where the "island" constitutes two or more systems. This situation reflects the increasing coupling of systems by expanded interconnections and is the more likely case – particularly in the future.

Prior to initial restoration procedures, it is important that a transmission/generation status assessment be carried out, communications established between internal and external dispatch/control centers and that power plants remain in a state of readiness awaiting coordinated direction. The assessment can be made within the AEP System by use of its transmission network data acquisition system that displays breaker status and circuit readings. This, coupled with the telemetering associated with load frequency control and economic dispatch systems together with a highly reliable and extensive communication system, should provide adequate information for internal and external system assessment.

Once the extent of the island is established, the load that was automatically shed and the restoration blocks that must be handled are available from the computer record management system for System Operation described later in this paper.

During the period of frequency decline and automatic underfrequency relay load-shedding action, the Operation Control Center and individual generator unit Load Frequency Control systems will trip out of service, thereby leaving the units under full governor control. As the frequency decline takes place, each generator unit operator will be responsible for maintaining the integrity of the unit and be in a position to load-reject the unit to maintain steam flow rather than having to resort to the "trip-button," which would require restarting. Load rejection will take place at 58.0 Hz and will maintain the unit in a state of operation supplying its own auxiliary power.

The power plant operator – while operating at a frequency other than 60.0 Hz – must rely on governor control and should not attempt to manually adjust the unit generation to correct frequency. This is vitally important since the power plant will be part of an island without knowledge of the system conditions and also without knowledge of the transmission line loadings. Thus, any attempt by a power plant operator to regulate frequency without direction from a Control Center may cause further cascading. For example, if the operator attempts to raise frequency to 60.0 Hz, the result will be to:

1. Increase power output on the unit attempting regulation.
2. Decrease power output on other units under governor control.
3. Consume the reserve of a unit being regulated.
4. Transfer power on the transmission system.
5. Possibly reclose transmission circuits automatically as the frequency difference between systems approaches permissible automatic synchronizing limits.
6. Possibly activate automatic load restoration relays.

In directing any restoration procedure, it is important to recognize the automatic features of the transmission system which include:

1. The sectionalizing that will take place when loss of potential occurs at motor operated, air-break switches.
2. The circuit reclosing that could take place when two separate systems are brought into near synchronization.

Prior to connecting load to generation – either by supervisory control, manual switching techniques, or by frequency actuated, automatic means – it is important to assess the load magnitude in terms of available generation.

As indicated in the companion paper, the resulting transient or overshoot when restoring load is minimized if the restoration block of load is less than one-third the shed block of load. Since, as indicated earlier, the shedding is carried out in $3^1/_3\%$ load steps, this would result in approximately 1% restoration steps.

A 1% restoration step is equally desirable for several other reasons. Consider the situation of a 10% automatic drop in load at 59.5 Hz. If this entire block of load were restored simultaneously at 60 Hz, the normal load-governing response of the system – using a representative[2] load frequency bias of 2% MW/0.1 Hz – would result in a drop of frequency of 0.5 Hz, thereby causing a retrip and subsequent "pumping" action. In addition, the instantaneous MW load could be as much as three times the steady-state load level due to the effects of cold-load pickup. Therefore, to avoid cold-load pickup problems, the maximum allowable load restoration block should be one-third of 10%, or $3^1/_3\%$. Thus, to avoid the limiting case (pumping action due to load-governing bias effect) with a comfortable margin, and to minimize transients, a 1% restoration block was selected.

When the "island" becomes reconnected to the interconnection, the generating capability of the entire interconnection is now available to handle load increases and, on this basis, the aforementioned 1% for the "islanded" system becomes extremely small for the interconnected system.

When all generation/transmission assessment procedures have been completed and a plan for restoration has been established – including the method for restoring frequency to 60.0 Hz (Manual load shedding, quick-start generation, other generating reserves) – the plan should be initiated with the first objective being to reconnect the "island" to the interconnected systems network by raising the frequency of the island.

As the "island" frequency approaches synchronism, transmission circuits (not necessarily interconnections because of the nature of the "island") will reclose automatically as the frequency difference between systems approaches permissible automatic synchronizing limits. However, depending upon the synchronizing devices used, this may require close and fine frequency control within the "island" which, under some conditions, may not be attainable. Therefore, to assure automatic reclosing of transmission circuits, certain key circuits should be selected for manual reclosing (remote supervisory or otherwise) to establish synchronism.

Once a stable interconnected system frequency of 60.0 Hz is established, the automatic feature of the AEP frequency actuated load-restoration program will be initiated. As noted in the companion paper, the system frequency will decay as load is picked up and the restoration sequence will be arrested until the frequency is brought back to 60 Hz. Hence, to continue restoration, the former islanded system would need to either schedule interchange power or provide adequate generation on load frequency control.

Thus, while the decisions to be made in directing a restoration plan are the same whether supervisory control or automatic techniques are employed to restore and pick up load, the automatic load restoration program waits for coordination to take place before proceeding. This is in contrast to an operator utilizing remote supervisory control who can continue to increase load pickup without the natural characteristic of the power system limiting the sequence.

RELAY SELECTION FOR SHEDDING AND RESTORATION

The implementation of the load shedding and restoration schemes will be accomplished through the use of solid-state frequency relays which are inherently more accurate than the equivalent electromechanical relay.

The static relay comes equipped with a crystal clock oscillator and a counter that determines the number of high frequency pulses that are contained in a cycle of power system voltage. If the count is greater than a predetermined value, this indicates an underfrequency condition exists. If this state lasts for a preset time, sufficient to override what may be a system transient phenomenon, the relay will give an output. This output can be delayed to override any problems associated with

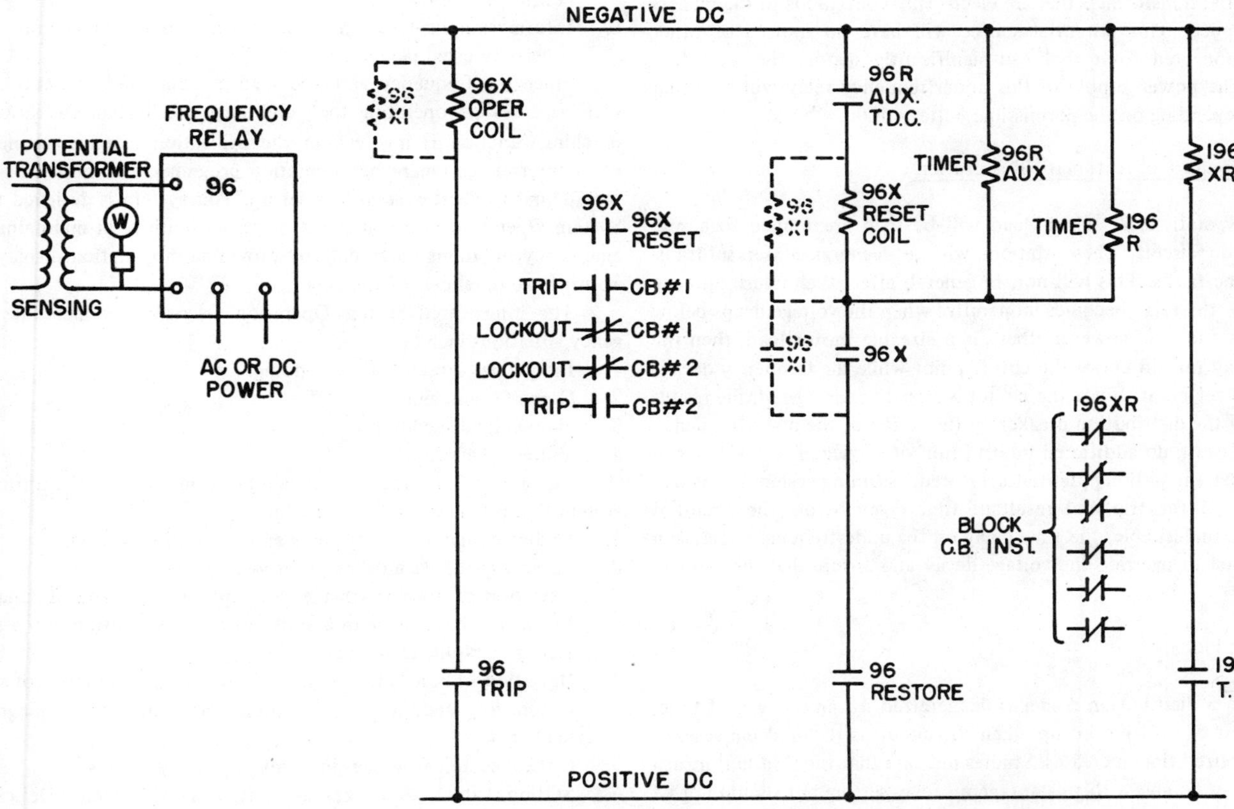

Figure 1

system transients. It is believed that 10 cycles should be sufficient. This, together with circuit breaker tripping time, will result in an overall time delay of 15 cycles as discussed in the companion paper.

The accuracy of the electromechanical relay is of the same order of magnitude as the difference between the nominal frequencies of the ECAR tripping stages which are 59.3 and 58.9 Hz. Therefore, a random distribution of shedding would occur if the electromechanical relays were used and set at the nominal tripping frequencies. The equivalence of this action can be obtained using solid-state relays and multi-stage tripping.

As discussed in the companion paper, an automated restoration program set below 60.0 Hz may result in restoring load before synchronism is established between the "island" and the interconnected system. This could, in fact, prevent resynchronization or could result in an uncontrolled restoration program. On the other hand, if automatic restoration is set above 60.0 Hz, this could result in a failure to restore load and, in addition, the fail-safe feature discussed below would be negated. Therefore, automatic restoration requires that highly accurate solid-state relays be used.

Tripping and restoration is accomplished as shown in the elementary diagram of Fig. 1. If the frequency drops below the set point, and stays there for longer than 10 cycles, the frequency relay output contact will close and energize a latching, electrically reset auxiliary relay. This auxiliary relay will then trip and lock out as many breakers in the station as was determined necessary to reduce the load by a given amount. When the frequency is restored to 60 Hz, a second, independent output contact of the frequency relay will start a timer and reset the auxiliary relay to permit the breakers to reclose, provided the frequency stays above the restoration set point for 30 seconds for the first restoration stage and an additional five seconds for each succeeding stage. The tripping and restoration timer settings are independently adjustable and may be changed if experience dictates. Another auxiliary relay may remove the instantaneous trip on the breaker to prevent tripping on cold load pickup. This will be discussed further below.

The measuring circuit of the underfrequency relay is connected to bus potential transformers that are electrically contiguous to the feeders to be tripped. This potential source will have no motors or other devices connected to it that can significantly distort the secondary voltage. The power supply of this underfrequency relay will be either ac or dc depending on the permissible battery drain.

Fail-Safe Feature of Automatic Restoration

Many stations at which load will be shed are tapped to a subtransmission circuit. These stations will be deenergized for subtransmission line faults. This will not, in general, affect the underfrequency relay since the relay becomes inoperative when the voltage drops below a preset value. If, however, there is a sizeable motor load, then the voltage may remain above the cutoff point while the frequency decays below the set point. Since the circuit is already deenergized, the resulting trip of the distribution breaker by the action of the underfrequency relay will cause no additional interruption of service. The breaker will reclose when the station potential is restored, assuming system frequency is normal. If the trip and resultant time delay before the circuit is restored is undesirable, the time delay on the underfrequency relay can be extended to override the voltage decay and ensure that the trip will be cut out.

Cold-Load Pickup

When a distribution feeder is deenergized for any length of time, the amount of load picked up when the circuit is restored increases as the deenergized time increases. Studies indicate that the resultant inrush current may be above the instantaneous relay setting but should not be above the time current curve of the time delay relays. Therefore, the underfrequency control circuit will have a provision such that when the frequency is restored to 60 Hz subsequent to an underfrequency trip, the instantaneous relays on the distribution feeders can be removed from service. Time delay protection is of course maintained. When the circuit breaker recloses, the instantaneous relays will be cut back into service.

Future Station Standard Design

In order to maintain the proper balance between load to be shed and system load as new stations are added, it has been decided to provide load shedding and restoration facilities at all new distribution stations having transformer capacities of 7500 kVA or more. This will include the mounting and wiring of the frequency relay, along with any auxiliary relays required, normally in an outdoor cabinet. The load management program (discussed later) will then determine which relays are to be connected to operate and their trip and restoration settings. At new stations of a lesser kVA only the cable will be installed and run to a convenient location so that the relays and cabinet can be easily added should the need arise.

RECORD MANAGEMENT SYSTEM FOR SYSTEM OPERATION DISPATCHING FUNCTIONS

A frequency actuated load-shedding and restoration program requires a sound management system to ensure that it remains abreast of power system dynamics. Factors that change and which must be handled include:

1. Load growth – Requires new relay installations.
2. Addition and deletion of critical customer – Changes priority listings and trip step settings.
3. Distribution system changes – Automatic two-way throwovers must be recognized to avoid negating the load-shedding feature.
4. Underfrequency relays installed as part of a standard new station package – Must be connected at some proper future date.
5. Trip Step Settings and Restoration Time Delays – Must be monitored to assure that the program stays within balance as load continues to grow.
6. Information displays – Must be provided to System Operation and other interested personnel.

Since the frequency actuated load-shedding and restoration program is a system operating tool, and since the System Operation dispatching function is involved in the utilization and management of other interrelated emergency operation programs, a computer-oriented record management system was set up. This system is designed to aid System Operation in carrying out its responsibilities in maintaining all emergency programs up to date by providing information displays and assuring the timeliness of the data.

The interrelated System Operation programs for handling emergency situations include:

1. Frequency actuated load shedding.
2. Capacity deficiency.
3. Manual load shedding
4. Voltage reduction.

The above four programs are combined in a computer information system that provides displays depicting:

1. Underfrequency relay tripping and restoration settings.
2. Circuits that have automatic throwover.
3. Deviation of tripping scheme from optimum in terms of adherence to size of shedding steps and its uniform distribution over a seven-state geographic load area.
4. Deviation of restoration scheme from optimum in terms of size of restoration steps and its uniform distribution over geographic load area.
5. Critical customer listings in terms of priority ratings.
6. Station listings and data associated with Capacity Deficiency, Manual Load Shedding, Voltage Reduction and Frequency Actuated Load-Shedding and Restoration Program.

CONCLUSIONS

A frequency actuated, load-shedding and restoration program has been developed which provides a well ordered, coordinated sequence of action to:

1. Restore a balance between generation and load following a disturbance resulting in a decline in frequency.
2. Permit reestablishing the integrity of the interconnected systems and the resumption of service under conditions which safeguard the security of the system.

These objectives are accomplished in the following manner:

1. A six-step shedding program has been instituted which operates uniformly across the AEP System on distribution type feeders. The steps are frequency spaced so that 10% of system load will be shed in three equal steps from 59.5 to 59.3 Hz, and an additional 15% will be shed in three equal steps from 59.1 to 58.9 Hz. A system of load priorities has been established, based upon permissible outage times, to permit assigning each distribution feeder to specific trip steps.
2. A 25-step automatic restoration program has been instituted so that load is restored in 1% blocks at 60 Hz in a time spaced sequence. To carry out restoration in an interconnected system environment requires coordinated procedures and operating guidelines that are the same whether the restoration is manual or automatic. An essential ingredient that is present in the automatic plan and is not present in the manual approach is the self-arresting, or self-limiting, action in handling load pickup levels. Thus, automatic restoration carried out with carefully sized load blocks offers speed and safety in an effort to reach successful restoration.
3. A computer-oriented management record system has been instituted which updates the load shedding and restoration program to account for system growth and changes in configuration.

REFERENCES

[1] R. M. Maliszewski, R. Dunlop, and G. L. Wilson, "Frequency actuated load shedding and restoration Part I – Philosophy," IEEE TRANS., 1970.

[2] Nathan Cohn, *Control of Generation and Power Flow on Interconnected Power Systems.* New York: J. Wiley & Sons, 1966.

Combined Discussion[1,2]

W. A. Flowers (Pacific Gas and Electric Company, San Francisco, Calif. 94106): The authors of the two papers are fortunate that they have the advantages of analogue simulation studies as a tool to assist in preparation and development of a frequency controlled load shedding and load restoration program. Their studies have served to illustrate the dispersion of load shedding in the various system areas due to the simultaneous differences of frequency in those areas during frequency transients. Their illustrations, Figures 6 to 8 inclusive, show that even though a load shedding program with several step frequency set points is used, in reality the load shedding occurs in a more nearly continuous manner as effected by the dispersion of system frequency. In actual load shedding this randomness of the load disconnection is further increased by the variation of the breaker tripping, caused by small inaccuracies of the relay set points.

An adequate method of system simulation was not available in 1958 when the Pacific Gas and Electric Company's load shedding program was first initiated systemwide. Therefore, a greater reliance had to be placed upon intuition and the direct observation of actual system dynamics during periods of generation overloads. Nevertheless, the system and principles developed through the simulation studies in almost all respects parallels the practices established in the P. G. and E. system. This similarity includes such features as:

1. The use of high accuracy solid state relays for both frequency controlled load shedding and frequency controlled load restoration.
2. A multi-step load shedding program using the very minimum time delay after the frequency drops below each discrete relay set point.
3. Recognition that during severe system overloads, with a major generation deficiency, frequency decline can be arrested only by very rapid disconnection of load. The response rate of spinning reserve under these conditions is so low as to be of inconsequential value.
4. Automatically restored blocks of load must be small as compared to the separated system.
5. Interconnected operation requires coordination of both load shedding and load restoration between adjacent systems. Coordination is particularly difficult in load restoration.

The similarity of the AEP and P. G. and E. programs may have developed partially because the systems are of approximately the same size and most stations are unattended and automatically operated. Under these conditions, automatic restoration of load after load shedding is attractive and possibly affords the most rapid and sound method of restoring service. The method is self regulating and experience has shown that, when properly applied, major relapses in the state of the system do not occur.

When the P. G. and E. program was first initiated it provided for 15% load shedding. The program has been revised and updated three times and now consists of an eight-step, 50% load shedding, frequency restoration program. Presently there are 180 frequency sensitive relays used. Each relay disconnects an average of 25 mw of load. In conjunction with the frequency relay, time delay reclosing relays are used to restore each block of load. There are more than 635 separately timed restoration blocks averaging 7 mw per block.

It would be of interest if the authors could furnish similar data on the number of relays used and the size of the blocks of load shed and restored per relay.

The main difference between the AEP program and the P. G. and E. program is the frequency at which load restoration occurs. On the AEP system load restoration only occurs at or above 60 Hz. On the P. G. and E. system the load is restored progressively in small steps below 60 Hz. This method has the advantage of more rapid restoration by utilizing the generating capacity which has been made available by the action of unit governors which are sensitive to the depressed frequency. By programming the restoration in gradually increasing frequency steps, the system is self regulating, yet restores load as fast as generating capacity is available and only when the frequency increases. By utilizing many small load restoration steps and by keeping a wide margin between the load shedding and load restoration set points, a condition of system pumping or oscillation is entirely avoided. A wider margin between the load shedding set points and load restoration set points is maintained by restoring the last load shed first, and progressively restoring so that the first load shed is the last load restored. If due to some unforeseen condition a pumping condition should occur, the inherent features of the automatic system will lock out the automatic restoration system.

In conjunction with the above automatic restoration plan, a ready means of manually interrupting and restoring up to 15% of the system load has been provided. The system control center can utilize this control if it is apparent that insufficient reserves are available within a separated area to restore the area to 60 Hz. After frequency recovery and reparalleling of the system, the manually interrupted load can be manually restored as system conditions permit.

There should be some concern to the operators of systems which depend on automatic load restoration at 60 Hz. Visualize a situation where a large deficiency in generation develops in an adjacent area and major load shedding results. It could be expected that at least a slightly depressed frequency would remain on the interconnected systems for some extended period. Under these conditions no load would be automatically restored at 60 Hz. Therefore, it is possible that, even though a system was exporting power to the maximum capability of its ties and had ample unloaded resources available, it still would not be capable of restoring any load shed within its control area.

Manuscript received July 20, 1970.

[1]R. M. Maliszewski, R. D. Dunlop, and G. L. Wilson, "Frequency actuated load shedding and restoration, pt I: philosophy," this issue, pp. 1452 - 1459.

[2]S. H. Horowitz, A. Politis, and A. F. Gabrielle, "Frequency actuated load shedding and restoration, pt. II: implementation;" this issue, pp. 1460 - 1464.

R. C. Durbeck and W. G. Tuel (IBM Research Laboratory, San Jose, Calif.): We commend the authors for their clear presentation of the philosophy and implementation of automatic underfrequency load shedding and restoration for the AEP System. Although occasions where load must be shed are rare, still it is important to formulate sound plans for system operation during emergencies.

Manuscript received July 29, 1970.

Shedding: It appears that the analysis of the power system response did not take "steam depletion" into account. This is an important factor to consider when steam units must respond to a large energy imbalance[1]. Figures 6 and 7, for example indicate a rather extended period of operation significantly below 60 Hz. During this time pressure and enthalpy of steam delivered by the boilers can drop significantly, thus reversing the trend of rising frequency.

Shedding loads in very small blocks at many different frequencies tends to minimize the amount of load shed, but the frequency response is poor, as pointed out by the authors. The use of intentional delays at the lower shedding frequencies may help because the additional load shed, after the frequency decay has been arrested, enhances the rate of frequency restoration.

Did the authors consider the influence of reserve distribution and the effect of frequency dependence of loads on the amount of load shed? For the single machine approximation was constant electrical power output assumed? How did this compare with the results from the detailed simulation?

The use of rate-of-change of frequency load shedding relays to drop load in proportion to the percentage of generation deficit might deserve consideration. Would the authors comment on the feasibility of such devices?

Restoration: The automatic restoration of load by sustained operation at 60 Hz appears to be a sound and sensible plan. An initial delay of 30 seconds has been suggested. Is this long enough to permit checking and closing of interconnections after the system is returned to synchronizing frequency?

Paper II describes a record management system for system dispatching functions. It is not clear whether this is an on-line function receiving direct signals from the relays themselves, or merely a catalog of the various settings. It seems to be necessary to have real time information on which relays have tripped to determine which loads have been shed during an emergency.

REFERENCE

[1] Durbeck, R. C., "Simulation of Five Load-Shedding Schedules," *IEEE Trans. Power Apparatus & Systems,* vol. PAS-89, #5, May/June, 1970, pp. 959-966.

Manuscript received August 6, 1970.

B. M. Fox (Northeast Utilities Service Company, Hartford, Conn. 06101): The authors have presented an interesting new approach to the problem of automatic load shedding. The basic ground rules as developed by the authors are that load shedding should be automatic, relatively high speed, a last ditch effort whose effect is not included in planning, and done on a local basis with no predetermined system split-ups. These concepts have certainly been accepted by the Electric Utility Industry in the United States and formed the basis for most extensive load shedding installations. The new approach in this paper, an approach that certainly deserves very close inspection by the industry, is the concept of relatively high speed automatic restoration of load following an underfrequency load shedding operation. If proper precautions can be taken to insure that the automatic restoration of load will not distress a system to the point of forcing another cascading outage, it is certainly a concept which has to be looked at and closely investigated to improve customer service.

Before making any comments as far as the philosophy of a six-step customer load shedding scheme and automatic restoration, I would like to have a few questions answered with regard to the actual simulation techniques outlined in this paper. In Equation #1 on Page 4, two important terms which can have a substantial effect on underfrequency operation appear. These two terms are inertia and electrical power. In studies of this type, I have commonly seen values of inertia of about four per unit on system base and values of system frequency damping constant of about two per unit (2% load change for 1% frequency change). Do these values agree with what the authors used in their study, and if there are any radical departures from these approximations, could we have the details which led to these variations. The authors talk about a "composite model" which approximates the overall response of their generation to simulate the effects of non-reheat and double-reheat units. It would be interesting to know the percentage breakdown as used in this composite model and also the time constants used in this simulation for each segment of the "composite model." When doing a detailed stability study with many machines and lines represented in order to study the operations during underfrequency conditions, dramatic changes can be introduced by varying the frequency and voltage characteristics of the system loads. Did the authors study the effects of varying these parameters on their proposed load shedding scheme?

It appears that there are two basic philosophies which can be used in the conceptual design of an automatic load shedding scheme. The first system, as outlined in this paper, has as its aim the tripping of the minimum amount of customer services. The other philosophy, which is widely used in the United States, is to use fewer blocks of load shedding and to admit of the possibility of overtripping by having these fewer blocks be larger in magnitude. This techniques makes note of the fact that any overtripping will allow a more rapid return to normal system frequency. With the more rapid return to normal frequency, interarea ties can more rapidly be reclosed and the system can, hopefully, regain its full integrity in a shorter time period. The two horns of the dilemma then become on the one hand tripping as few a number of customers as possible but accepting the risk of a more protracted operation at underfrequency conditions; or on the other hand, potentially overtripping and causing outages to more customers than is absolutely necessary to prevent a continuing frequency decay but to rely on the fact that the overtripping will allow a more rapid return to normal frequency conditions. It would appear that as more and more areas in the country accept the concept of automatic load restoration following an underfrequency load shedding condition, the overtripping scheme becomes less objectionable since tie lines can be closed more rapidly, resulting in rapid improvement in system integrity and reliability at the same time that customer outages are reduced to a bare minimum in duration.

Figure 6 shows the difference between the frequency response of the six-step versus the two-step system for a 23% overload condition and the total four cycles of time delay for load shedding. Would not the frequency response characteristics of these two load shedding schemes move closer and closer to being identical if a delay time of more than four cycles was used? A curve of this type with a five-cycle load shedding delay time would be interesting to see. Also, many companies use a delay time of more than 15-cycles resulting, I would think, in an almost identical frequency response for the two load shedding schemes. To the utilities that installed load shedding schemes prior to the availability of static relays, the small frequency differences between the two-step and six-step schemes become less significant due to the dispersion of load tripping caused by inherent inaccuracies in the electro-mechanical relay and its variation in operating time with rate of change of frequency.

The basic problem that can make or break the automatic load restoration scheme resolves itself into a question of the available interconnection capacity to outside the "islanded" area following a return to 60 Hz and automatic tie line closing. In the example of Figure 10, a net tie line flow of about 600 MW would flow into the "islanded" area following a 2% restoration. This flow could continue to increase if generation outside the formerly "islanded" area raised frequency back to 60 Hz while generation in the trouble area remained fixed due to having reached upper limits. The flow increase would be due to additional load being automatically restored due to raising of system frequency to 60 Hz from 59.985 Hz. At some point stability limits or line burndown limits could be exceeded resulting in an "island" again being formed and underfrequency load shedding again taking place. Certainly, if tie lines are restored with any degree of success it is difficult to visualize a 600 MW flow overloading ties to a 30,000 MW system. However, this 600 MW flow could represent not 2% of a 30,000 MW system but rather 25% of a 2,400 MW system at which point tie line limits being exceeded can become a real hazard. By way of example, a series of disturbances simultaneously occurring while some key lines were out for maintenance took place in April 1969 resulted in the separation of the State of Connecticut from the interconnected grid. At that time Connecticut normally had two-345 kV and three-115 kV interconnections to either New York or Massachusetts (The interchange capacity of these interconnections was far in excess of any reasonably projected deficiency in the state). However, because of maintenance work and line burndowns following the loss of an entire generating station, a situation resulted where only one-115 kV interconnection was available after frequency returned to 60 Hz. If all the load that had been shed was returned automatically by relay after this single tie had been closed, there is a very good chance this one tie would have tripped resulting in an overloaded island and another underfrequency operation. Is there any protection in the automatic restoration scheme to guard against such a contingency?

W. H. Underwood (Long Island Lighting Company, Hicksville, N.Y. 11801): This paper[1] makes a valuable contribution to the subject of load shedding and restoration. The attention given to load restoration is

Manuscript received August 3, 1970

particularly valuable since, to date, most attention has been focused on methods of load rejection.

One overall concern exists with respect to the frequency sensitive load restoration plan. This relates to the fixed scheme used to restore load as compared to the highly variable conditions that may exist. In the words of the paper, the load shedding (and restoration) scheme is intended to respond to "those improbable disturbances" and is a "last ditch" tool. Under such circumstances, with the possibility of internal transmission lines out, or synchronizing problems, to name two possibilities, it appears to be possible to postulate a set of circumstances under which problems would develop. Perhaps, for the AEP system all credible contingencies have been considered. It does seem, however, that the conditions for automatic load restoration are infinitely more difficult to predict than the conditions for load shedding.

Certainly the area of load restoration after intentional or unintentional load shedding, deserves more attention. A review of area outages will show that in a large proportion, some setback was experienced during attempted load restoration.

Load restoration by supervisory was ruled out as too costly. I wonder if thought was given to a permissive system employing space radio tones to achieve more control over load restoration.

In figure 4, the authors refer to continuous shedding as the ideal, yet they propose six discrete steps of load shedding. What is the disadvantage of having a continuous range of settings? Furthermore wouldn't the ideal system be one that instantaneously drops a block of load exactly equal to the amount of load that cannot be covered by remaining available generation after the contingency?

Several interesting points are made in Part I regarding spinning reserve. One point that seems to receive too little attention is what the authors term the quality of spinning reserve. The fact that there may be an adequate quantity of reserve but that its quality may be inadequate is not new but sometimes appears to be overlooked.

In the portion of the paper titled "Small Disturbance Studies," several interesting observations are made that could use some clarification. Reference is made to response limitations due to valve travel. Does this refer to valve points or the physical speed at which valves open or something else? At this point and elsewhere in the paper, reference is made to the desirability of maintaining a 3 to 1 ratio of spinning reserve to generation deficiency. Do the authors anticipate changes in spinning reserve policy in light of this ratio?

The first paragraph under "Single Machine Simulation Results" specifies three categories, but the third appears to have gotten lost. Presumably 2 and 3 are matters of degree.

One of the basic pit falls of the load restoration via frequency approach seems to be its dependence on the load governing characteristics of the United States to prevent interconnection overloads. An example is given where an island of an assumed size is synchronized to the remainder of the United States. It would seem that if the size of the island or the capacity of the tie were varied, unsatisfactory results could occur.

Part II[2] of this paper lists a general set of guidelines. Item 2 of these guidelines indicates that maximum reasonable assistance will be given to neighbors - - - however this assistance will be terminated without opening interconnection circuits – how is this accomplished?

The statement is made that the first objective of restoration is to reconnect the "island" to the interconnected system. This is followed by a statement to "assure automatic reclosing of transmission circuits, certain key circuits should be selected for manual reclosing- - -" this procedure might result in difficulty if load is picked up after a 30 second time delay at 60 hz.

In all, this is an interesting approach to load restoration. The publication of the results of the studies made are a valuable contribution to the available data regarding system performance under emergency conditions. Since this system is expected to perform under improbable conditions and calls for a fixed response subject to system frequency, it would appear to be a prime victim to Murphy's Law.

Manuscript received July 29, 1970.

S. S. Cogswell and **L. A. Soderquist** (Westinghouse Electric Corporation, East Pittsburgh, Pa. 15112): The authors have made a very thorough analysis of frequency actuated load shedding and restoration on their system, and they have presented excellent reports of their efforts in Part I, Philosophy, and Part II, Implementation. Their study procedure of this complex subject seems to have combined the right mix of simulation and practical knowledge of power system operation resulting in some concrete answers for their load shedding and restoration program.

In Part I, the authors discuss the frequency bandwidth of a system as determined by electromechanical oscillations between machines. An analysis of this bandwidth seems to be an interesting approach to the study of the degree of electrical coupling between machines and how a system will respond to an overload. If detailed simulation shows that the frequency bandwidth, characterized by machine response is a narrow one, there is a tendency for all phase angles around the system to change at the same rate resulting in little transient power flow within the disturbed area and therefore less of a tendency for further system breakup. The more heavily the lines are loaded, however, the narrower this bandwidth should be to provide margin against further isolation. We would like to ask the authors if the decision for uniformly applied load shedding was based in part, on a study of the frequency bandwidth of the system.

The authors report how they used a single machine simulation in their analysis. This appeared to be a very helpful and economical way to optimize the load shedding parameters. It seems as though the portion of the system the authors studied could be adapted very well to a single machine equivalent. This procedure, however, would not seem to be too desirable when studying an area that is not strongly tied together.

The authors' investigations have quantitatively shown how the tendency for load shedding to overlap can be reduced if the intentional time delay is reduced. It was reported that short intentional time delays minimized the overlapping. This was an interesting result that we believe can only be tested through additional simulation. It should be pointed out, however, that if the area under consideration is not a strongly tied grid, there will be a much greater tendency for overlapping of load shedding stages regardless of the intentional time delay. As the authors point out, the overlapping should be avoided because it tends to result in transient power flow within the system.

In Part II, the authors describe a computer oriented record management system for the frequency actuated load shedding and restoration program. The authors, in our opinion, have left out one important element in their discussion of the information system. That is, a provision to periodically test the updated load shedding and restoration program to determine how the power system will respond when load shedding is required.

R. M. Maliszewski, R. D. Dunlop, G. L. Wilson, S. H. Horowitz, A. Politis, and A. F. Gabrielle: The authors wish to take this opportunity to thank all of the discussers for their interest, comments and questions concerning these two papers. Mr. Flowers indicates that the similarity between the AEP and PG&E load management programs may have developed because of the similarity of the two systems. It is true that both systems are comparable in size and utilize unattended automatic stations. On the other hand, the AEP System is very highly interconnected with all of its neighbors at 138 kV, 345 kV and 500 kV. At present, we have over 86 high voltage interconnections in service with 23 neighboring companies. This difference in system configuration is reflected in the basic difference in the frequency-actuated restoration approach employed by the two systems. On the AEP System, load can not be restored at frequencies lower than 60 Hz, since this could prevent resynchronization of the islanded area with the rest of the interconnected system, or could result in uncontrolled restoration after synchronization took place.

Mr. Flowers also indicates that in the PG&E restoration program the last block of load shed is the first block of load to be restored. In the AEP approach, since restoration will take place only at 60 Hz, the first block of load shed is the first block to be restored. The concept here is to restore the system as quickly as possible to a position whereby it will be able to react to a subsequent condition of declining frequency. If the PG&E approach were utilized, should a disturbance occur while restoring load, the frequency would have to decay to the set point of the last load shed before arresting action would take place.

Mr. Flowers also asked for pertinent statistics with regard to the AEP load management program. As indicated in the paper, we are initially installing 400 relays which will disconnect an average of 6 MW of load per relay. The restoration will take place with 25 separately timed blocks, with each block restoring 1% of the load shed.

The last point Mr. Flowers makes is in regard to the situation when a large deficiency occurs in an adjacent area, which results in a slightly depressed frequency in the interconnected system, thus preventing any automatic load restoration at 60 Hz. We recognize this situation could very well exist, thus blocking the restoration of load on the AEP System. This condition is readily acceptable and would require

Manuscript received September 14, 1970.

coordination among adjacent system operators to restore the interconnected system frequency to 60 Hz when the system is capable of accepting load.

Messrs. Durbeck and Tuel inquire as to the 30-second time delay for load restoration. Synchronizing relays will operate within about 3 seconds after the phase-angle difference across the circuit is within the relay setting, thus providing sufficient margin between circuit reclosing and load restoration.

With regard to the record management system this is indicated to be an off-line function, which will catalog all existing relay settings and locations. Since the system is continually growing, a digital computer cataloging scheme is an efficient means of keeping track of current system conditions in terms of relay settings and restoration times. Real time information to assess the state of the system is available through a separate data acquisition system as mentioned in the paper.

Messrs. Cogswell and Soderquist inquire concerning the uniformly applied load shedding program. The decision to utilize this approach was based primarily on an evaluation of the configuration of the AEP System with regard to its load-generation characteristics. The system has well defined load and generation areas spread out over the entire region, however, because of its strong transmission system and interconnections, islanding situations are not readily predictable. Since underfrequency situations are difficult to define, it was judged advisable to apply load shedding in a uniform manner in order to cover completely the range of conditions that could occur. Based on the studies carried out the uniform approach does not cause any difficulties.

With regard to the use of single machine equivalents, the authors believe that this is the most efficient way of investigating a wide range of system parameters. Naturally, the effectiveness of this screening technique will be very much a function of the nature of the system and the interrelationships which exist within the system. The validity of the single machine model should be established by carrying out a number of multi-machine check cases.

The authors agree with the comment that there will be a tendency for load shedding to overlap for those systems which are not strongly tied together. Each system will have its own unique characteristic dependent on the configuration of the transmission system. Therefore, each area will need to be carefully studied before generalizations can be made.

Messrs. Cogswell and Soderquist asked if there is a provision to periodically test the updated load shedding and restoration program. There is no built-in provision for such tests but certainly recognition that such a program is an inherent feature of the AEP System will be factored into all future planning studies.

Mr. Underwood raises a concern with regard to the fixed scheme used to restore load as compared to the variable conditions that could exist. As indicated in the paper, load shedding is a "last ditch tool" which will respond to "improbable disturbances." Load restoration, on the other hand, does not fall into the same category. It is an operator-directed function, which must be carried out in a coordinated manner. It would be expected that if an islanding situation were to occur, after frequency has been arrested, no attempt would be made to restore load within the island until the cause of the disturbance is determined. Consequently, restoration should be a slow and orderly process, requiring close coordination and communication among all the system operators involved. If this is carried out in this fashion problems of restoration should be minimized.

Mr. Underwood also inquires as to whether a permissive system was considered in conjunction with the automatic load restoration approach. A space radio system was considered to provide permissive control, however, this would have added little to the safety features in view of the self-regulating aspects of the proposed restoration system and, in addition, would have doubled the overall cost of the system.

He also asks if there are any disadvantages to having a continuous range of load shedding settings. Ideally one would want to shed exactly that amount of load to match the generation deficiency existing on the system. In practice, however, in order to avoid unnecessary shedding for conditions which can be compensated for by spinning reserve, the first stage of shedding is set sufficiently low enough to be able to handle normal system transients. Beyond that point the load shedding set points are limited by the accuracies associated with the relays utilized. A continuous range of set points is not practical with present relay technology. With the six-step program described in the paper, and taking into account the frequency dispersion which will exist on the system, a continuous shedding effect is approached for a maximum disturbance, as can be seen in Figures 6 and 7.

Mr. Underwood also inquires as to the reference made to limitations imposed by valve travel. This refers to the physical limit of the valve to supply steam to the turbine. For conditions where the available spinning reserve approaches the magnitude of the deficiency, the valve will reach its upper limit and as a result the system frequency response will take on a non-linear characteristic. It is observed that when dealing with a disturbance of a magnitude approximately one-third of the amount of spinning reserve available, valve limiting influences are avoided and the frequency deviation on the system is minimized. Translating this to the restoration mode, to avoid serious frequency deviations when load is being restored, sufficient spinning reserve must be available. Based on an interpretation of Figure 3, it is suggested that a 3 to 1 ratio be maintained between the amount of spinning reserve on the overall system and the amount of load restored per stage.

Based on these observations, it is not anticipated that a change in spinning reserve policy is required in normal day-to-day operation. The need to maintain the suggested ratio would apply only for the condition when a system is being restored from an islanded state.

In the discussion of single machine simulation results (Part I), three categories of disturbances are described. The second and third categories, which are a matter of degree, both involve one or more stages of load shedding. In the second, load shedding results in an immediate arrest in system frequency. In the third category, however, the action of spinning reserve will be required to arrest system frequency.

Mr. Fox raises interesting philosophical questions the answers to which may lie as much in each company's practices and experience as in factual analysis of curves and numbers. To answer his specific questions, however, we used an inertia constant of 4.0 sec.; we did not use a load-frequency factor inasmuch as this effect is small, does not contribute to an evaluation of alternatives and its neglect leads to conservative results. We did use this factor, however, in evaluating the response of the relays to small frequency changes while restoring in an interconnected mode, and in fact, depend upon it as the means of obtaining the self-limiting feature of the frequency actuated restoration scheme. The factor we used, however, is not 2% load change for 1% frequency change but rather 2% load change for 0.1 Hz (0.17%) change in frequency. This number is in keeping with studies described in reference No. 4 and with our own measurements at our System Power Production Center in Canton, Ohio, for small changes around 60 Hz.

Messrs. Fox, Durbeck and Tuel ask about machine modeling. The composite model for the single machine studies consisted of dividing the AEP System generation into three categories; non-reheat, single-reheat and double-reheat. Non-reheat units were modeled as responding with an average 0.2 second time constant; single-reheat units had two time constants, 25% of the machine output had a 0.2 second time constant and 75% had an additional 10 second average value; double-reheat units had three time constants, 25% with 0.2 seconds, 15% with an additional 10 seconds and the remaining 60% with having a long time constant representing the second reheat cycle. With these values in mind, the composite model was established as described in the paper.

Regarding the comments of Messrs. Durbeck and Tuel, steam depletion was not included in our prime mover simulation since the studies did not place great reliance on spinning reserve to arrest frequency decline. A small amount of spinning reserve was assumed to be uniformly allocated in the composite single machine model. In the multi-machine studies, realistic dispatches were assumed; variation of reserve distribution was considered and for the system studied, it was judged that this factor has little influence.

Mr. Fox suggests that additional time delays on shedding relays would cause the response curves for the two-stage and six-stage shedding curves to coincide. This would not be the case for large disturbances in which spinning reserve supplies a small portion of system deficiency. The load characteristics would cause some coincidence of the two response curves for additional delay, but this effect would become appreciable only at frequencies well below 58 Hz. In general, the effects of additional time delay on the curves of Figure 6 would only be to delay the point of minimum frequency in time and reduce the value of minimum frequency without causing the curves to coincide.

In our opinion, the choice between a two-stage versus a six-stage scheme is not only, as Mr. Fox suggests, a choice between overshedding load. It is primarily predicated upon a desire to reduce system transients. The optimum shedding schedule would be continuous-acting, closely matching the actual system frequency decay. Conceptually, a scheme sensing a rate of change of frequency seems ideal. Messrs. Durbeck and Tuel suggest considering such a relay and we concur. For such a relay, the problem of determining the proper setting for varying island situations, the lack of widespread experience with existing devices and the desire to achieve a coordinated load shedding effort among "ECAR" companies, led us to choose instead a multi-stage schedule, actuated by frequency magnitude alone. The number of stages was determined by the accuracy of the relays. As Mr. Fox points out, even a two-stage schedule, using electromechanical relays, is multi-stage by virtue of the random spread of frequencies and setting accuracy.

Mr. Fox's point that the success of the frequency-actuated restoration program is contingent upon the available interconnection capacity to the islanded region is correct. However, the same statement is true whether the restoration program is frequency actuated or not. If, as in his example, there is an islanded region of 2400 MW, with 600-MW load shed, and a single interconnection capable of carrying only 200 MW, then in a *coordinated* effort, the system frequency within the island should not be brought to 60 Hz until sufficient interconnection capability is realized. A similar statement would hold true for a manual restoration scheme. One would not manually restore load in the formerly islanded area after the one tie line is closed if that tie line could not carry the flows to supply the load. In either case, the key is to coordinate restoration between the islanded area and the external system; in the semi-automatic case by controlling frequency, in the manual case by controlling all individuals who have the power to restore load.

Mr. Underwood asks how we can terminate any assistance given to a system experiencing difficulties, without opening the interconnection. Under the terms of our interconnection agreements and emergency operating procedures, there are a series of measures that can be taken to halt a slowly declining frequency. If these measures are not successful, the offending utility must take the necessary additional steps. The interconnections are not opened until frequency has dropped below 58.2 Hz.

With regard to reconnecting the island to the remaining systems, there should be no question of difficulty caused by the automatic pick up of load. The load restoration relays are several orders of magnitude more restrictive than the automatic synchronizing relays. If the frequencies are such that the synchronizing relays cannot operate, then certainly the restoring relays cannot. The "certain key circuits . . . . selected for manual reclosing" may be necessary to close the first line. The remaining lines will close automatically, much faster than the restoring relays can pick up load.

Naturally we hope our proposed scheme will not succumb to Murphy's Law. We appreciate the variety of system difficulties that can arise. We feel, however, that the conditions under which restoration is desirable are fairly well fixed and are amenable to study and analysis.

Part VI

Effects of Trends of Future Generator Designs

EFFECTS OF TRENDS IN LARGE STEAM TURBINE DRIVEN GENERATOR PARAMETERS ON POWER SYSTEM STABILITY

C. Concordia
Fellow
IEEE

P.G. Brown
Senior Member
IEEE

Abstract – The trend in turbine-generator characteristics as unit sizes become larger is in an adverse direction from the standpoint of power system stability. Results of specific studies as well as a general rule of thumb for evaluating this effect are given. The effectiveness of independent pole switching and other control measures available to the system planner are shown.

INTRODUCTION

With the continual growth in power systems and in interconnections, there is a corresponding continual economic pressure to install larger and larger generating units. This increasing interconnection has allowed the formation of larger and larger planning pools. In the Eastern part of the U.S.A., the interconnection is already so large (over 200,000 MVA) that the size of the largest unit is only a fraction of one percent of the total connected capacity. It has been pointed out[1] that, from a performance standpoint, the appropriate size of the largest unit in such a large and widespread interconnection should be related not to the total capacity of the interconnection but rather to the load pick-up capacity of the system tie lines. This capacity is in turn related to the transmission network voltage. Taking the KW capacity of a tie line as equal to 4 kv^2, we arrive at the very rough relation shown below in Table I. Deviations from these figures of at least two to one can be expected.

TABLE I

Network Voltage KV	Gen. Unit Size MW
230	200
345	500
500	1000
800	2500
1300	7000

Since 765 kv (800 maximum) lines are already here, we can recognize the pressure on unit size. Primarily because of mechanical limitations, this economic pressure has resulted, and will continue to result, in the need to pack more and more capacity into a given size machine. A review of these trends, and of the innovations which have made the larger ratings possible, has just been given in Reference 2.

There is no avoiding the fact that this uprating process tends in turn to result in higher reactances and lower inertia and that this tendency sometimes makes the job of designing a reliable and stable system more difficult, and more challenging. This paper deals with the effect on power system stability of probable future trends in machine characteristics and with the effective means available to the system planner for accomodating these characteristics.

Paper 71 TP 74-PWR, recommended and approved by the Power System Engineering Committee and the IEEE Power Generation Committee of the IEEE Power Engineering Society for presentation at the IEEE Winter Power Meeting, New York, N.Y., January 31-February 5, 1971. Manuscript submitted September 15, 1970; made available for printing November 25, 1970.

SYSTEM ASPECTS OF STABILITY

The first thing to point out in connection with this is that stability is a system problem, and an economic problem. In a very rough way, we can consider the over-all system reactance as consisting of the generators, transformers, transmission lines, and possible compensation. In the limit it is necessary to keep this over-all reactance within bounds, if a reasonable degree of stability is to be preserved. Of course, this limiting permissible reactance can be increased somewhat by control action, e.g., of excitation systems and turbine valves, and by further reductions in circuit-breaker opening and reclosing times.

Full advantage of control action has not yet been taken in most cases, and therefore some increases in over-all system reactance can be tolerated, if necessary, and still achieve adequate stability performance. However, the limits of such action already appear to be in sight, so that one purpose of this paper should be to look beyond the point where control has been made to give its maximum benefit. For example, thyristor excitation systems that respond almost instantly are already available, and special stabilizers have already been designed and applied to take maximum advantage of excitation control to increase system damping. Thus, we can not expect any obvious further improvements in the performance of excitation systems. On the other hand, these high-response and optimally-stabilized excitation systems have so far only been applied to a relatively small number of units, usually where the transmission system is rather weak, for example, in the cases of long-distance transmission from a generating station or of very heavily loaded interties. Thus there remains considerable opportunity to improve, or to achieve, stability by a wider application of these high-performance excitation systems, as reactance may become larger. Here we only want to make the point that we can not expect to continue indefinitely to compensate for increases in reactance by more and more powerful excitation systems. In spite of rather loose statements that have sometimes been made in the literature, both theory and experience have shown that it is not possible to completely eliminate the effects of the generator reactances by excitation control alone.

Turbine valve control to increase the transient stability limit, in the form of rapid closing and slower reopening of at least the intercept valve of a reheat turbine, is an old idea that has been much discussed recently, but has had so far very limited application. It is proper to approach this valve control with caution, even though the improvement in transient stability performance is large. The intercept valves supplied on recent new units are extremely large, and experience must be gained in a few limited applications with the required rapid closure and subsequent throttling action on reopening required of these valves for this type of control. On nuclear units there are additional considerations of the reheater and/or moisture separator drain tank systems, which require field test and experience in order to assure the practicality of early valving on these units.

From a theoretical point of view, fast valve action on the control valves as well as the intercept valves results in even larger gains. However, because of the additional duty required on the turbine control valves, and the possible repercussion on the boiler, safety valves, and boiler control, this feature is not available on turbines of the authors' company.

In principle, it could be a question of using a larger boiler, or one that could stand a higher over-pressure, and thus become a problem of

Reprinted from *IEEE Trans. Power App. Syst.*, Sept./Oct. 1971.

relative economics. From a technical standpoint, a competing method is the use of shunt and/or series braking resistors, which can be applied and removed even more quickly. If valve control is to be widely used, it would seem desirable to study the whole question thoroughly to find the optimum method of control from the economic standpoint, rather than merely to seize upon the improvement that can be realized by the simplest and least expensive intercept-valve control.

With respect to circuit breaker operating times, which are already down to 2 cycles or less, there is not very much farther to go. However, considerable improvements in performance for the milder faults (single-line-to-ground) can be made by single-pole switching, and for the most severe cases (breaker failure) by individual pole actuation.

Thus it would appear that system reactance can not be allowed to increase without limit. This still may not mean that we have to impose an arbitrary upper limit on generator reactance. It may be more economical to reduce the reactance of some other element, if the alternative is to have to accept a smaller-size generating unit in order to obtain a reduced reactance. It will usually be found most economical, first, to do as much as reasonable possible by control; next, to use the minimum available transformer reactance; next, to compensate the transmission system with series capacitors; and only then to obtain a reduced generator reactance (which might require a reduced generator unit rating) or to add more transmission lines if a further decrease is still necessary.

It is easy to see the direct effect of generator reactances and inertias on stability. In addition, there may be an indirect effect on stability arising from reduced short-time overload capability. Some major system shutdowns have occurred, not directly from instability in the classical sense of a machine losing synchronism following a severe fault, but rather from the consequences of the loss of a major transmission line or generator. This loss may change the pattern of power flow, and particularly of reactive power flow so that voltage support may place high reactive-power demands on certain generators. This situation may be perfectly stable, but may result in overloads of both armature and field currents that can be tolerated only for a limited time, and thus require automatic, rather than manual reduction.

Another form of overload limit is the permissible negative-phase-sequence current i_2, in the form of $i_2{}^2t$. This permissible value of $i_2{}^2t$ has so far been so large that it has not been a limitation on allowable fault clearing time. However, the time has come for us to review the real system requirements and not to penalize the machine design arbitrarily.

FUTURE TURBINE GENERATOR PARAMETERS

We have discussed qualitatively some of the possible adverse effects of trends in generator parameters on system stability. Now we shall look at some specific examples to see where we are, where we are going, and when, if ever, we shall run up against these limits.

The turbine-generator parameters of primary interest are the rotary inertia H, the transient reactance X'_d, and the synchronous reactance X_d (or the short-circuit ratio). We have given considerable thought to the problems of turbine-generators up to capacities of at least twice that of the present largest units on order. In that range it would appear that H will continue to decrease somewhat. For 2-pole units H may decrease from a typical value of 2.5 to about 2.0. For 4-pole units H may decrease from 4 to about 3. These values are only approximate, because present values of inertia for a given speed and rating cover a rather broad range – nearly two to one. This comes about because H is determined largely by the turbine last-stage bucket length, number of flow paths, and back-end loadings which are in turn determined by economic factors – cooling water temperatures for the particular site, condenser and turbine costs, etc. Examination of a large number of present and future designs indicates no perceptible change in H with unit size for the same back-end loading. It is possible, however, to identify two trends tending to lower the H of future units. One is the historical trend to higher turbine back-end loadings to achieve an economic balance for particular site conditions. A second, and more significant, factor is the trend toward the greater use of cooling towers, which also has the effect of increasing turbine back-end loading to achieve an economic balance. This leads to probable inertias as shown in Table II, which compares present and future units.

TABLE II

	Present	Future		
		Natural Water Bodies	Cooling Towers Evaporative	Dry
4-pole (Nuclear)	3.5 - 4.5	4	3	2.5
2-pole (Fossil)	2 - 2.75	2.5	2	1.75

With the present stress on environmental problems, there is a strong possibility that cooling towers or ponds may be predominent in the future.

The transient reactance X'_d will continue to increase. For 4-pole units possibly up to 0.5, and for 2-pole units to about 0.35, from their present values of typically 0.35 and 0.22, respectively.

The short circuit ratio has changed more slowly than the transient reactance, in part because of the natural requirements of optimum design and in part because of the desires of the user. For present-day large units there is very little incentive to reduce the short circuit ratio from the usual present values of 0.50 to 0.58 p.u., but units of twice this present largest size may have still lower short circuit ratios. We have considered the possibility of short circuit ratios in the range of 0.4 to 0.5 p.u. in our system studies.

Although we are primarily discussing the trends in generator parameters, with the usual station arrangement the step-up transformer can be regarded as practically integral with the generator. Thus, it is of interest to say a word about the trends in transformer reactance. In the case of transformers, there is no significant trend with size. Instead, the variation of reactance is with high-side voltage, as indicated in Figure 1.

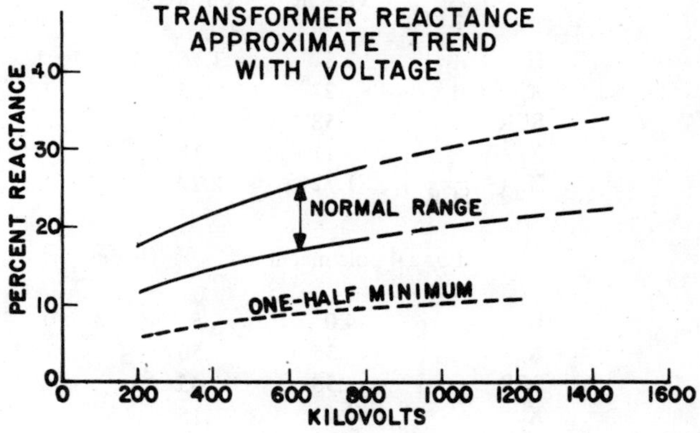

Fig. 1.

Figure 1 shows the approximate variation of reactance with voltage as presently projected. Three values are shown, the maximum and minimum of the normal range, and the smallest reactance (about one-half the normal minimum) available at a premium price. From the standpoint of stability, we should suppose that one would want a reactance at least as low as the minimum normal value, and very likely the even smaller premium value. However, stability is not always the primary limitation. Instead, the limitation may be size, shipment, or fault duty. Thus, some utilities are presently using transformers having reactances well above the normal minimum. Therefore, in some cases there may be the possibility of having reactances smaller than those of present practice, even at the highest voltages contemplated, if it becomes necessary.

The reactances of Figure 1 can be related to transformer or generator size by referring to Table I at the beginning of this paper.

What will be the effect of these changed parameters? Some of the effects are shown in Figures 2-9. However, before discussing these figures in detail, it may be worth while to make three general observations.

INDEX OF STABILITY

First, we have considered, as one measure of relative stability, the critical clearing time to a three-phase fault at the generator high-voltage bus, and found that, as a very rough rule-of-thumb, this time varies at about half the rate of each of these three principal parameters. We have expressed this mathematically as:

$$T_c \propto K_s$$

where:

$$K_s = \sqrt{\frac{H}{X'_d X_d}} \cong \sqrt{\frac{H \times SCR}{X'_d}}$$

and:

T_c	=	critical clearing time
H	=	inertia constant
SCR	=	short circuit ratio
X'_d	=	transient reactance
X_d	=	synchronous reactance $\cong \frac{1}{SCR}$

Second, as an example, we have applied this index of stability to the following two cases:

For a 2-pole machine

	Present	Future
H	2.5	1.75
X'_d	.22	.35
SCR	.58	.45
X_t	.11	.15
X_d (= $\frac{1}{SCR}$)	1.72	2.22

For a 4-pole machine

	Present	Future
H	4.0	2.5
X'_d	.35	.50
SCR	.58	.45
X_t	.11	.15
X_d (= $\frac{1}{SCR}$)	1.72	2.22

Note that we have listed the transformer reactances also, and intend to include all reactance up to the point of fault, so our index of stability becomes

$$K_s = \sqrt{\frac{H}{(X'_d + X_t)(X_d + X_t)}}$$

Then for the four cases considered, the indices are:

	Present	Future	Future/Present
2-pole	2.03	1.21	0.60
4-pole	2.18	1.27	0.58

If, for example, the critical clearing time for the present generator were 7 cycles, then it would still be somewhat more than 4 cycles for the assumed future generator, and thus still greater than the 3-cycle clearing time presently available.

Third, although all of the curves presented are for transient stability, they also imply that steady-state stability is maintained over the range of parameters shown.

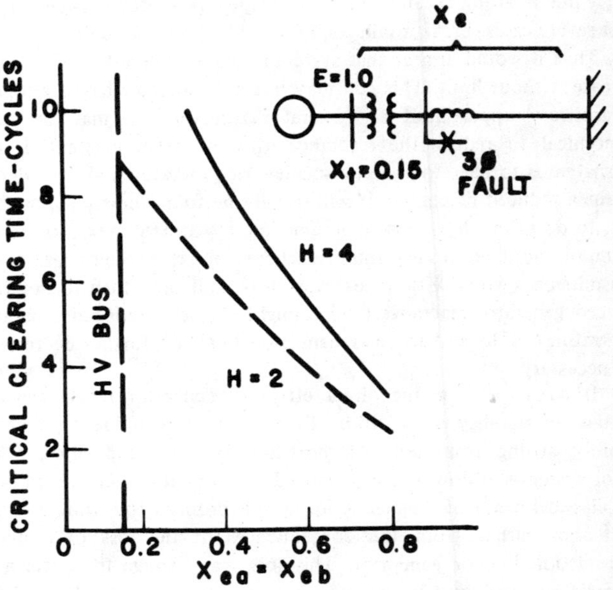

Fig. 2. Effect of inertia. Gen. SCR = .58, X'_d = .45, excitation system response = 3; rated KVA at .9 power factor. Subscripts b and a refer to system reactances before and after fault respectively.

CRITICAL CLEARING TIMES – EFFECT OF PARAMETERS

Figure 2 shows the effect on critical clearing time for a 3-phase fault of changing the inertia H from 4 to 2, or by a factor of 2. Critical clearing time (CCT) for a stub feeder fault at the generator end high-voltage bus is plotted against system reactance viewed from the terminals of the generator (and thus including the transformer). These

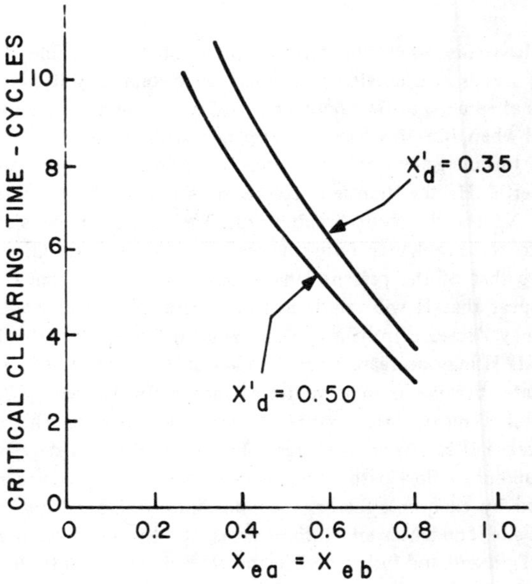

Fig. 3. Effect of transient reactance. Gen. SCR = .58, H = 4, excitation system response = 3; rated KVA at .9 p.f.

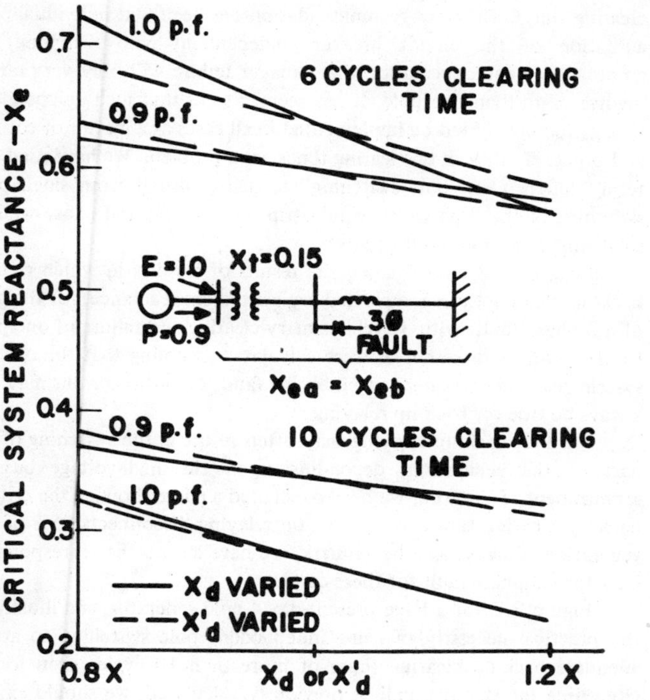

Fig. 4. Effect of generator short circuit ratio ($1/X_d$) and transient reactance (X'_d) on transient stability. Gen. nominal SCR = .58, X'_d = .42, H = 4.

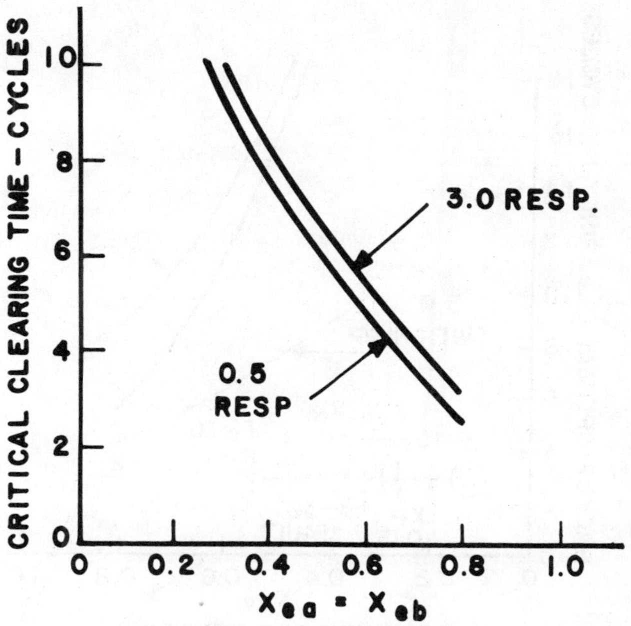

Fig. 5. Effect of excitation system response. Gen. SCR = .58, X'_d = .45, H = 4; rated KVA at .9 p.f.

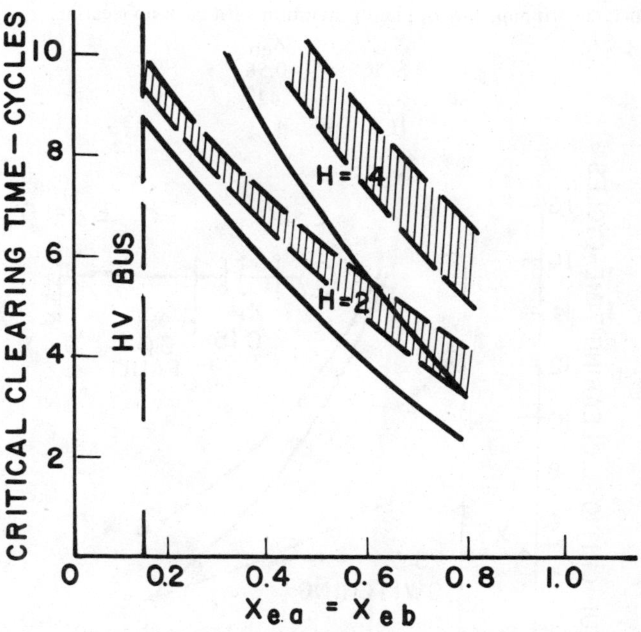

Fig. 6. Effect of early valving shown by dashed curves. Gen. SCR = .58, X'_d = .45, excitation system response = 3; rated KVA at .9 power factor.

curves were calculated with a more or less complete representation of the generator, including a high-response excitation system, and are taken from Figure 8 of Reference 3. We can see that CCT varies about as the square root of inertia, and, further, that even with H=2 the CCT stays above three cycles until X_e reaches about 0.7.

Figure 3 shows the effect on CCT of transient reactance. Increasing the transient reactance from 0.35 to 0.50, decreases the CCT by from 1 to 2 cycles.

Figure 4 shows the effect of changes in short circuit ratio as compared to the effect of changes in transient reactance X'_d. Here the stability is indicated in a different way, by plotting the critical system reactance for a 3-phase fault against the change from normal in either x'_d or X_d (=1/SCR). Curves are given for two different clearing times, 6 and 10 cycles, and two different generator power factors, 1.0 and 0.9. It is seen that the effect of a change in SCR is about the same as a change in X'_d, being slightly greater for unity power factor or slightly smaller for 0.9 power factor.

All of the curves shown so far have been for a high-response excitation system. Figure 5, again taken from Reference 3, shows the effect of changing the response ratio from 0.5 to 3.0. Here we have returned to plotting critical clearing time against system reactance. The change in critical clearing time is less than one cycle.

Figure 6 (from Reference 3) shows a more effective way of increasing the critical clearing time, namely, the rapid closing of the turbine intercept valves, followed by their reopening at a slower rate. Both of the inertias, H = 4 and H = 2, of Figure 2 are shown. Since the transient reactance of X'_d = 0.45 has been kept constant, and recalling the typical values for present and future generators that we have discussed previously, we might regard the H = 4 case as roughly a present day 4-pole generator (very slightly more severe), and the H = 2 case as roughly a future 2-pole generator. Then, by comparing the two curves, H = 4 without early valving and H = 2 with early valving, we could say that early valving tends to compensate for the changed parameters of the future machine as compared to the present machine.

Of course, all of these curves are for simplified general cases, so we can not apply the actual critical clearing times shown directly to all cases. They are shown merely as examples to justify our rough rule of thumb. Application of this rule of thumb to two typical cases showed a reduction of future CCT to about 60% of present-day values, but this will not necessarily apply to any particular system. Instead, it would seem better to use the formula itself, rather than the factor of 60%, calculating the present-day index of stability for a few cases for which the CCT is known. Even this will, of course, be only very approximate. The real answer can only be found by a stability study for each particular system configuration.

We have also called attention to the slight increase in CCT attainable by high-response excitation, and to the considerable increase attainable by early valving (of the intercept valve). Further, the curves seem to indicate that the CCT would in most cases be greater than the presently available 3-cycle clearing time.

Now we should like to discuss two related subjects, back-up relaying and overload capacity.

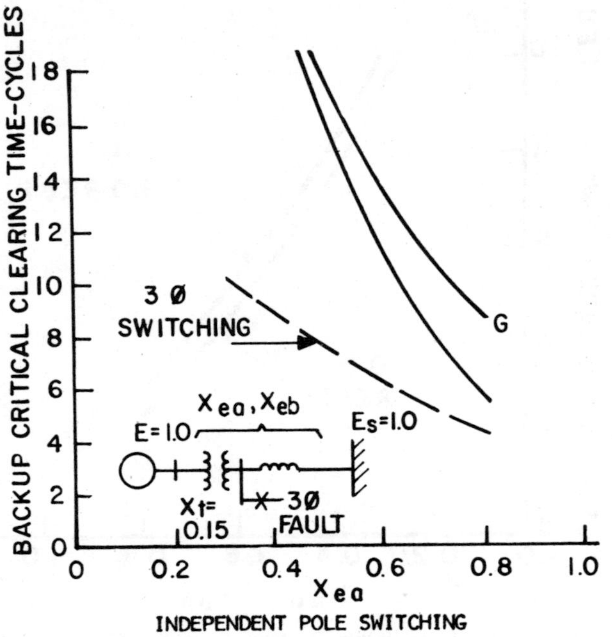

Fig. 7. Back-up critical clearing time with independent pole switching. Two phases cleared in 3 cycle primary clearing time. Curve "G" includes tripping 25% of plant generation with back-up clearing.

X_{ea} = X_{eb}
SCR = 0.58
X'_d = .42
H = 4

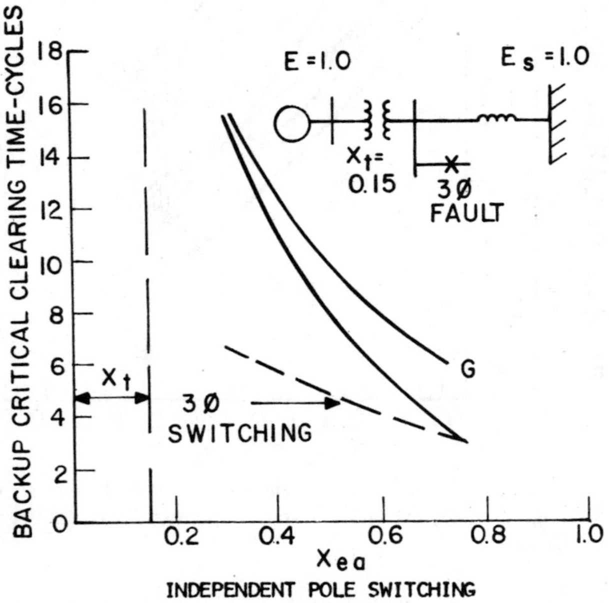

Fig. 8. Back-up critical clearing time with independent pole switching. Two phases cleared in 3 cycle primary clearing time. Curve "G" includes tripping of 25% of plant generation with back-up clearing.

X_{ea} = X_{eb}
SCR = 0.45
X'_d = 0.50
H = 2.5

BACK-UP CLEARING

There has been considerable discussion and study of stuck breakers and consequent back-up relaying within recent years. It has been made very clear that even today the critical clearing time for a severe 3-phase fault is often smaller than reasonably available back-up clearing times. Thus our recommendation has been that individual-pole actuation of the circuit breaker (independent pole switching) be employed, in the expectation that breaker failure will only very rarely involve more than one pole. It has seemed to us that such independent pole switching should be implemented in all cases as a matter of course, and especially if back-up clearing times are a problem. We have used the term "independent pole switching" to distinguish it from single-pole switching of the type which would trip only the faulted phase when a single line-to-ground fault occurs.

Figures 7, 8, and 9 show the results of a study in which critical back-up clearing times are plotted against system reactance, for the case of a 3-phase fault, with 3-cycle primary clearing and failure of one pole to clear. All of these curves were calculated assuming that the over-all system reactance was unchanged by the fault, even though this may not always be true for back-up relaying.

Further, back-up relaying may often result in the switching off of part of the generation, depending upon the high-voltage bussing arrangement. Therefore, we have concluded a curve showing the critical back-up clearing times when back-up relaying disconnects 25% of the generation. Finally, as a benchmark, we have shown the corresponding CCT for a 3-phase fault for these cases.

Figure 7 is for a large present-day 4-pole generator, and illustrates the practical necessity of using independent pole switching to attain adequate back-up clearing times of more than 10 cycles even today. (Recalling the stability indices previously calculated, we should expect these and the following curves to apply also to 2-pole units.)

Figure 8 is for the assumed future 4-pole generator, so it shows directly the probable future CCT. For a system reactance of 0.4, both critical clearing time and critical back-up clearing times are well above presently attainable values. For a high system reactance of 0.6 the critical back-up clearing time is under 6 cycles, which is not only smaller than has been attained to date but also smaller than has been projected for the future.

Figure 9 shows how the critical back-up clearing time can be increased by decreasing the primary clearing time to 2 cycles and by using early valving. Two-cycle primary clearing should be possible at this future time, through reduction of breaker operating time to 1 cycle. It is of interest to note from these results that this faster primary clearing may become necessary in order to achieve satisfactory back-up times (since we may not wish to use early valving) even though it is not necessary in itself. With early valving, the future critical back-up clearing times seem adequate even with the presently available 3-cycle primary clearing.

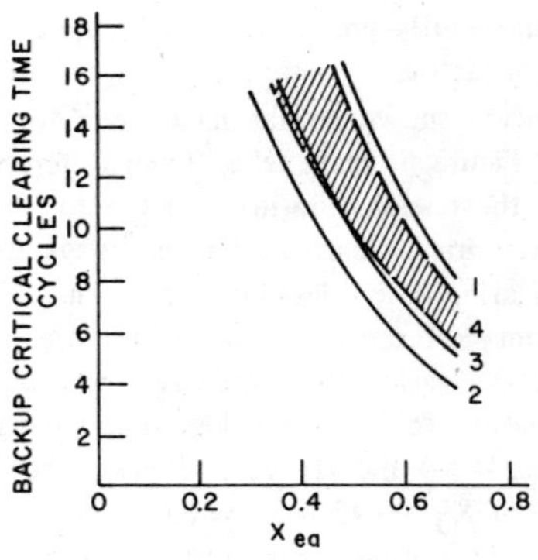

Fig. 9. Effect of primary clearing time and early valving on back-up critical clearing times achieved with independent pole switching.

Curve	Machine	Primary Clearing	Early Valving
1	SCR =. 58, X'_d =. 42, H = 4	3 cycles	No
2	SCR =. 45, X'_d =. 50, H = 2. 5	3 cycles	No
3	Same as 2	2 cycles	No
4	Same as 2	3 cycles	Yes

All of the figures that we have shown have indicated the great importance of having a sufficiently small system reactance, which can be interpreted as having a strong transmission network, but the effect is seen to be especially pronounced in the case of back-up clearing time. In fact, studying these figures we can get an approximate idea of the amount of transmission line series capacitor compensation that we would have to introduce in order to counteract the effects of the changes in generator parameters. Further, there are other measures that can be taken, in addition to the use of early valving and high-response excitation, to improve stability. Reference 4 discusses many of these in more detail, including such measures as the use of shunt and series braking resistors, and the switching of series capacitors, shunt capacitors, or shunt reactors, to improve transient stability; the use of special excitation-system stabilizers in conjunction with high-speed excitation to improve steady-state stability; and the use of synchronous condensers to improve both steady-state and transient stability.

Reference 4 also points out that, for given stability limits, over-all system reliability can be improved by systemwide monitoring and control to reduce the probability of approaching too closely to potentially unstable system conditions, and in some cases by single-pole switching to decrease the impact of the most common type of fault, the single-line-to-ground fault.

CAPABILITY DURING SYSTEM EMERGENCIES

With respect to the capacity of future generators to respond to abnormal conditions, we have mentioned two aspects: the possibility of high reactive-power demand, and the negative-phase-sequence current. The higher reactances of the future generators, together with the relatively fast switching times needed, will actually decrease the i_2^2t duty, and our studies have indicated that i_2^2t capabilities as required by the present standards can be reduced considerably, while still meeting the probable system requirements, as indeed it could be reduced even for today's generators. On the other hand, the capability of supplying large reactive power demands is a limitation even today, and is, moreover, one which has not always been fully recognized. That is, even today, a large generator can only operate at ceiling excitation for a few, perhaps ten, seconds, and then the excitation voltage must be reduced to an acceptable value to prevent damage. Even though longer times are permissible for lower excitations, this is not very significant from the point of view of system survival in an emergency. It seems to us that the significant reactive requirements of the system during such an event is that for a time of from a few minutes to a quarter or half hour, or the time necessary for the operator or automatic control to do something about the system condition. These requirements can be met, in the future as in the past, by taking them into proper account as one factor in the selection of generator rating. If the permissible time at ceiling is reduced from 10 to 5 seconds, or the permissible time at 25% above rated is reduced from a minute to half a minute, this really does not affect the problem, except to emphasize the importance of taking action immediately and automatically to bring the field current within acceptable limits, as well as the importance of predetermining the system reactive requirements under various assumed emergency conditions as part of the system planning process.

CONCLUSION

In conclusion, we have recognized the adverse effects on stability of the trends in generator parameters, we have given you a rule of thumb for evaluating the effect of these trends on critical clearing time, we have pointed out the necessity of a strong transmission network and the possibility of improving stability, if necessary, by various specific control measures, and, finally, we have indicated that the achievement of adequate stability margins seems entirely feasible within the range of unit size considered.

REFERENCES

1. C. Concordia, "Dynamic performance of large interconnected power systems following disturbances," Missouri Valley Electric Association Engineering Conference, Kansas City, Missouri, April 13-15, 1966.
2. D.B. Harrington and S.C. Jenkins, "Trends and advancements in the design of large generators," American Power Conference, Chicago, Illinois, April 21-23, 1970.
3. P.G. Brown, F.P. deMello, E.H. Lenfest, R.J. Mills, "Effects of excitation, turbine energy control, and transmission on transient stability," IEEE, Paper 70TP203-PWR, January 25-30, 1970, IEEE Winter Power Meeting, New York, N.Y.
4. C. Concordia, "Maintaining stability following system disturbance," Minnesota Power Systems Conference, Minneapolis, Minnesota, October 21-23, 1969.

Discussion

Robert H. Park (Main Street, Brewster, Mass. 02631): In the last paragraph of the paper by Concordia and Brown entitled "Effects of Trends in Large Steam Turbine-Driven Generators on Power System Stability" the statement is made that "the achievement of adequate stability margin seems entirely feasible within the range of unit sizes considered."

However, the authors are properly careful to refer in the same paragraph to "the adverse effects on stability of the trend in generation parameters," "the necessity of a strong transmission network," and "the possibility of improving stability, if necessary, by various specific control measures."

A related point that can be cited is the growing magnitude of various problems which the hard-pressed Power Industry is encountering, including rapidly escalating demands for power, difficulties in securing rights-of-way for transmission lines and sites for power plants, difficulties with new very large steam-electric generating units, especially when of the nuclear type, and perhaps above all continual increase in the dependence of the public on continuity of power supply.

Under the circumstances there has been a tendency to accept that the Power Industry needs to establish stringent criteria as to the contingencies that will be provided against, and, as a case in point, it has begun to be accepted that system plans need to be made with a view of preserving stability notwithstanding delay in the clearance of a three-phase fault occurring near to a generating station bus, and also the simultaneous or close-on sequential outage of a pair of lines taking place as a result of fault clearance.

Still another point that warrants consideration is the fact that most stability studies are carried out on the basis of load representations which are of such limited scope as to sometimes lead to important errors, and, also, often fail to include consideration of system swings following the first.

Moreover, as of today, steam-electric units tend to be purchased even before the plant location is settled, and, at any rate, decisions as to equipment design often have to be made before stability studies have been made, while, also, based on experience, it must be conceded that what the future may hold in store can depart in an adverse way from what has been anticipated, which implies a need to provide a margin of safety to allow for what cannot be foreseen.

It was because of a sense of probable developing difficulties in providing for maintenance of system stability that, beginning in 1963, the writer undertook to interest the Power Industry in evaluating the economic merit of fast valving of steam turbine type generator prime movers, and of dynamic braking applied at generator or auxiliary transformer voltage level as measures for improving system stability.

As the authors are aware, fast turbine valving was shown to be effective as means of stability improvement in 1929 and 1930 [1], but was viewed as representing a means of improving stability that was not needed and the feature of providing for fast intercepting valve closure that was found to represent a key to its success, was put to use solely as a way to limit development of excess speed in the event of either full or partial load rejection.

Manuscript received February 19, 1971; revised February 26, 1971.

This position presumably was warranted when made, as making for simplicity, but now, despite the facts that complexity will increase, and that it will prove desirable to, in some respects, alter existing turbine and steam generator control system concepts and designs, it seems to be desirable to continue to progress in the direction of increasing the versatility of fast valving procedures, in order that they will not only be available as a way to favorably affect retention of system stability, but also will operate, in greater degree than present systems, to minimize adverse effects of system break-ups that occur when stability is lost.

While it is always desirable to approach anything new with caution, it is an aspect of fast valving procedures that if found wanting they can be readily disabled and the turbine control system operated without their participation.

Hence, if properly executed, they should pose a hazard only in respect to the possibility that expected advantages might be nullified by virtue of side effects the importance of which might have been too little realized.

Thus, in this connection, providing to close intercepting valves only, can, at any rate in some types of installations, and situations, cause lifting of, and might also result in damage to, reheat pressure safety valves.

Now, on the one hand it could turn out that infrequent lifting would not be seriously objectionable, and if this were the case, there would be no need to take the trouble to prevent it. However, the reverse might turn out to be true, and in this case it would be advantageous to introduce preventative measures which could take the forms of providing for fast control valve closing, speeded up intercepting valve opening procedures, and provision of steam by-passing capability, as to which the point also applies that in some cases sufficient steam by-passing capability may represent an already built in feature of the steam generator.

In the matter of means of effecting steam bypassing, where fossil fuel steam-electric plants are concerned an European practice of bypassing around the high pressure turbine to the cold reheat line, and from the hot reheat line to the condenser, provides a maximum of flexibility and has been reported to allow the resynchronizing and reloading of good-sized units in 6½ minutes [2]. While this type of steam bypassing is expensive, its cost can be viewed as not necessarily prohibitive when viewed in the light of advantages afforded, which include benefits in respect to avoiding development of undesirable variations of turbine inlet steam temperature during startups.

Again it is possible to use power-operated relief valves and combinations of steam bypassing and fast fuel and damper control of once-through boilers, as alternate or supplemental means of adjusting to fast reductions in turbine acceptance of high pressure steam [3].

Not long past generator excitation systems were being produced in forms that are now accepted as being inadequate to meet present requirements, which include ability to damp out oscillations, and to produce both a major improvement in steady state and a limited improvement in first swing stability. As it would appear, developing acceptance of the need to successfully deal with problems of system stability has resulted in the availability of much-improved systems, and the writer would anticipate that growing appreciation of the potential advantages of providing to allow fast turbine valving will lead to a parallel improvement in turbine governing systems.

Though the writer has been an advocate of studies being made bearing on the utility of dynamic braking, and though studies have shown utility, detailed estimates have indicated that the installed cost of providing reliable systems is considerably higher than the writer had at one time had in mind [4].

Fast valve closing utilized as a means of preserving stability works in the direction of minimizing hazard of development of undesirable overspeed conditions and can be carried out in such a way that, at worst, the hazard of development of overspeed is no worse than if it were not employed. In contrast, employment of dynamic braking at generator voltage level introduces a problem in respect to how to deal with a situation in which, despite all precautions, a line-to-line fault might develop, while application of braking at auxiliary transformer voltage level produces only small though not entirely inconsequential benefits.

This is not to say that a workable solution will not be found to the problem of how to make available the considerable benefits that braking can afford, either when used alone, or in conjunction with fast valving, but rather is intended to lay stress on the need for caution in implementation of programs of braking when braking load is to be applied to generator leads.

References

[1] Buell, R.C., et al., "Governor Performance During System Disturbances," Transactions AIEE, March 1931, Vol. 50, pp. 354-369.

[2] Trassl, W., "Safe Cycling of High-Pressure Steam Turbines," American Power Conference, Vol. 31, pp. 303-331.

[3] Durrant, O.W., and Vollmer, H.D., "Need for a Strategy for Boiler-Turbine-Generator Operation and Control," IEEE paper 71 CP 244-PWR.

[4] Park, R.H., "Improved Reliability of Bulk Power Supply by Fast Load Control," American Power Conference, Vol. 30, pp. 1128-1141 (1968)

G.I. Stillman (Power Authority of the State of New York, New York, N.Y. 10019): In the first paragraph under *System Aspects of Stability,* the authors state that over-all system reactance can be considered "as consisting of the generators, transformers, transmission lines and possible compensation," and that "it is necessary to keep this over-all reactance within bounds, if a reasonable degree of stability is to be preserved." Later on, the same theme is repeated: "Thus it would appear that system reactance cannot be allowed to increase without limit. This still may not mean that we have to impose an arbitrary upper limit on generator reactance. It may be more economical to reduce the reactance of some other element . . . " The impression is created here and elsewhere in the paper that, due to generator and other system developments, overall system reactance is increasing and even encroaching on acceptable upper bounds from a stability viewpoint.

There is no doubt that generator synchronous and transient reactances have been increasing in per unit on machine-base kva as unit capacities have been increased. There is also no doubt that generator step-up transformer (GSU) reactances are increasing, not because of increased power rating but because the higher EHV values to which they step up require increased spacing between primary and secondary windings. The remainder of the transmission system reactance, however, is markedly reduced as higher EHV values are introduced and as EHV networks expand.

In attempting to assess the effect of increased generator reactance on system stability, I think there is no doubt that high-speed excitation systems work to overcome the effect of generator reactance to some degree. Figure 5 of the paper shows that the improvement in critical clearing time (CCT) to be expected by increasing speed of response from 0.5 to 3.0 is small indeed. Ellis et al [1] reported similar results in their description of the Peace River hydroelectric development, but they also reported marked improvement when using a properly stabilized, high-ceiling voltage electronic excitation system. Could the authors, in their closure, kindly provide a revised version of their figure 5 including additional curves showing the effect of static excitation systems, (a) as they are being produced today, and (b) as they may be produced in the foreseeable future?

In attempting to assess the influence of increased GSU reactance on system stability, fig. 9 of Ref. [3] of this discussion shows the ratio of 765-kv step-up and step-down transformer reactance to total system reactance as a function of distance. For a distance of 150 miles, (approximately the average length of 765-kv circuits being built today), this figure shows that 75% of total system reactance is in the EHV step-up and step-down transformers. Since the GSU transformer has 2 windings while the EHV step-down is an autotransformer, I shall assume that 45% of the 75% belongs to the GSU and 30% to the step-down autotransformer. In otherwords, for a distance of 150 miles about 45% of the total system reactance seen from generator buses is in the GSU.

Compare this value with the values of X_{ea} on the horizontal axes of figs. 2, 3 and 5 through 9 of the authors' paper. Since the authors allot 0.15 per unit from the generator bus to the HV bus, i.e. for the GSU reactance, and since the GSU reactance is about 45% of total system reactance, then for average 765-kv lengths being built today, X_{ea} is approximately 0.33 per unit on each of the above-mentioned curves and the CCT is 6 cycles or longer. Furthermore, as time goes on and the EHV lines are tapped for additional step-down stations or interconnections, the average length of 765-kv circuits will go down, thereby further decreasing total X_{ea}. So the question is, how seriously are overall system reactances encroaching on upper bounds from a stability viewpoint, leaving out the isolated instances of mine-mouth or other generation that is remote from load?

I should like to thank the authors for a stimulating paper.

References

[1] H.M. Ellis, J.E. Hardy, A.J. Blythe, J.W. Skooglund, "Dynamic Stability of the Peace River Transmission System," *IEEE Transactions on Power Apparatus and Systems,* PAS-85, No. 6, June 1966, p. 586-600.

Manuscript received February 12, 1971.

[2] V.A. Venikov and V.A. Stroev, "Power System Stability as Affected by Automatic Control of Generators," IEEE Paper 71TP21-PWR, 1971 Winter Power Meeting, New York, N.Y.
[3] G.S. Vassell & R.M. Maliszewski, "AEP 765-kv System – System Planning Considerations," ibid [1], PAS-88, No. 9, Sept. 1969, p. 1320-1328.

C. Concordia and **P.G. Brown**: Before replying to the discussions received, we want to point out that Figures 6 and 9 in the final presented paper are slightly different from the corresponding figures appearing in the preprint. The effects of early valving as originally shown seemed somewhat optimistic and we have, therefore, made further studies to include the effects of variations in the input power reductions achieved by early valving. These are indicated by cross-hatched areas of the new Figures 6 and 9.

Mr. Park has made several pertinent observations, and we are in agreement with most of them. He has noted a tendency to establish more stringent criteria, and in particular the tendency to try to maintain stability notwithstanding a stuck breaker. In the first place, we have pointed out the appropriateness of considering the beneficial effects of individual pole actuation in such cases; and, in the second place, we feel that this tendency, toward maintaining completely stable operation with a stuck breaker, cannot be justified on a realistic basis. We believe that a thorough study of the record and of the probability of incidence of various types of disturbances would show that other ways of losing the generator may well be more probable than the occurrence of a severe three-phase fault in combination with a stuck breaker, and we recommend that such a study be undertaken by the Power System Engineering Committee.

With respect to Mr. Park's discussion of fast valving, we agree in general that every effort should be made to take advantage of improvements in stability by control action. However, we do not want to use such control action as a substitute for a strong transmission network. That is, stability is only one of many aspects of system performance. We may hope that severe blackouts are a thing of the past, but they seem to have been replaced by brownouts, where there is a big advantage in strong interconnections through which energy can be supplied from relatively remote systems.

Mr. Stillman very rightly remarks that higher voltage transmission reduces reactance. Our concern is to assure that system design philosophy, in the future as in the past, continues to stress the importance of a strong network, and in particular to point out that it is not only transmission from a generating source but also the strength of the interconnections among systems that is of interest.

With respect to Figure 5, the 3.0 response curve shown is already for a thyristor excitation system.

Mr. Stillman has pointed out that transformers constitute a very significant part of total system reactance. This brings to mind the importance of having a uniform EHV voltage, as distinguished form the 230, 345, 500, 765 kV voltages presently existing in the U.S.A. Every time systems at two different voltage levels are interconnected a transformer having a reactance equal to more than 60 miles of line has to be inserted between the systems. Thus, in view of the importance of a strong over-all interconnection, it is very important to consider the plans of ones' neighbors when deciding on an appropriate next EHV level.

We wish to thank both discussors for their contributions, which have added materially to the value of the paper.

Manuscript received March 23, 1971.

EFFECTS OF FUTURE TURBINE-GENERATOR CHARACTERISTICS ON TRANSIENT STABILITY

H. E. Lokay
Westinghouse Electric Corporation
East Pittsburgh, Pennsylvania

P. O. Thoits
Westinghouse Electric Corporation
Detroit, Michigan

Abstract—Future turbine-generator unit designs, particularly for the larger ratings, are expected to have greater generator reactances and lower unit inertias. Transient stability limits are determined for the future designs in terms of critical fault clearing times for faults on the high voltage transmission lines at the power plant substation. Transient stability limits of the future designs are compared with representative existing generating unit designs. Stability performance of future turbine-generator designs will be very similar to existing designs when considering their respective breaker failure (backup) relay system capabilities. Approximate gains in transient stability critical fault clearing times are also listed when applying various feasible stability control techniques.

INTRODUCTION

Turbine-generator designs have continually shown an increase in power output per pound of material as well as per physical dimensions. Advances in metallurgy and stress calculation techniques have permitted longer low pressure turbine last row blades and increases in turbine maximum end loadings. Advances in generator cooling methods and insulation systems have permitted greater KVA ratings for the same unit physical size. Concurrently with these design advances, power system designs have permitted a trend to higher generator power factors and lower generator short circuit ratios. These design innovations have caused certain unit parameters to trend in a direction less favorable to maintaining unit (system) stability during system disturbances.

The intent of this paper is to review the two significant parameters transient reactance (X'_d) and unit inertia or 'H' constant, to determine if future machine designs will be acceptable for reliable and economical power system design. Of primary concern are unit transient stability limitations, for system faults requiring breaker-failure or backup relaying protection.

FUTURE MACHINE PARAMETERS

Figure 1 shows expected future generator saturated transient reactance values for two- and four-pole steam turbine generators. The curves are representative for the expected generator design advances in the future when the largest units shown are expected to be applied. It should be recognized that values for actual machines would not necessarily fall on the curve, but above and below, depending upon the manufacturers' standard frame sizes, type of cooling, short circuit ratio, power factor, generator voltage, etc. For the very large nuclear, 1800 rpm units, the transient reactance values are expected to approach 50 percent.

Figure 2 shows expected future turbine generator inertia (H) constant values. Curves are shown for 1800 and 3600 rpm units and they represent expected future turbine low pressure exhaust end loadings. Again, values for actual machines would not necessarily fall on the curve, depending upon turbine configuration, low pressure exhaust end loading, exhaust pressure, etc., for the specific application.

The future values shown in Figs. 1 and 2 will be used to illustrate expected unit transient stability performance.

Paper 71TP75-PWR, recommended and approved by the Power System Engineering Committee and the IEEE Power Generation Committee of the IEEE Power Engineering Society for presentation at the IEEE Winter Power Meeting, New York, N.Y., January 31-February 5, 1971. Manuscript submitted February 16, 1970; made available for printing December 10, 1970.

SYSTEM DESCRIPTION

The system used for the study was a single machine connected to an infinite bus through a transmission system and unit transformer, Fig. 3. The unit transformer was assumed to have 15 percent reactance on the machine(s) MVA base, and a turns ratio of 0.95:1.0. Unit power output was assumed 90 percent of machine MVA, and unit VAR output was varied to maintain the machine terminal voltage and infinite bus voltage at 1.0 per unit.

Faults were assumed on the line side of a circuit breaker located at the plant high-voltage bus. No fault impedance was assumed. The system reactance X_s represented multiple transmission lines, and upon clearing a permanent line fault, 75 percent of the high-voltage transmission remained or post-fault X_s equals 1.33 times prefault X_s.

Method of Analysis

Stability calculations were made using the classical representation of holding a constant voltage behind transient reactance. For later cases to determine the effect of high-response excitation systems or turbine valve control, detailed machine, excitation system, and governor control system representations were used.(1) It was recognized that the classical stability calculation method generally gives optimistic results for long duration faults; however, since the concern was a generalized comparison of differences in transient stability limits, the classical method was considered adequate.

System Reactance

In any hypothetical system study selecting the most appropriate system reactance value is a problem. To determine a representative value for evaluating future turbine-generator characteristics, a survey was made of system reactance values from the high-voltage bus of 153 generating plants. The data was obtained from short circuit studies taken at random for many power systems. A distribution curve of the X_s values in percent on each individual plant's MVA base is shown in Fig. 4. The curve indicates most plants have X_s values between 5 and 15 percent, with 76 percent of the plants having X_s less than 25 percent. Since the survey data represented all generating plants, i.e., large, small, hydro, peaking, etc., a distribution curve was also developed with all very small plants removed. The sample was reduced to 80 plants, and the distribution curve is also shown in Fig. 4. It is considered more typical for transmission systems being applied for large major power plants. The curve shape is very similar and 82.5 percent of the plants had X_s less than 25 percent.

It should not be expected that turbine generator characteristics will permit unit stability for major faults at the plant with high values of system reactance, i.e., minimum transmission systems. The many factors in system design and plant locations have not permitted the development of a system reactance value that represents an

Reprinted from *IEEE Trans. Power App. Syst.*, Nov./Dec. 1971.

Table 1

STABILITY CONTROL METHODS AND APPROXIMATE GAINS IN UNIT TRANSIENT STABILITY CRITICAL THREE-PHASE CLEARING TIMES

Method	Approximate gain
1. High response (2.0) excitation systems	½-1½ Cycles
2. Turbine valve control, Fig. 9, (all units in Plant)	1-2 Cycles
3. Assume fault remote from plant	See Fig. 10
4. Reducing unit transformer reactance	¾ Cycle for 5 percentage point reduction.
5. Reducing ratio of system post-fault reactance to system pre-fault reactance (depends highly on system conditions.)	½-1½ Cycles
6. Reduce T-G normal power output	CFC time* x 1.05 for 10% reduction in power output.
7. Reduce Generator X'_d	1 Cycle for 10 percentage point reduction.
8. Increase T-G unit 'H' constant	$\sqrt{\frac{H\ new}{H\ old}}$ multiplier i.e., multiplier = 1.15 when raising H from 3 to 4.

*As shown in paper Figs. 5-8.

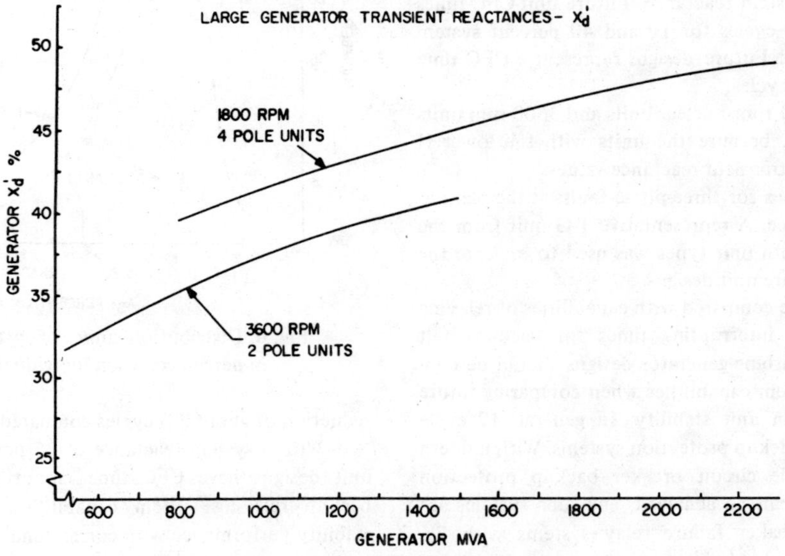

Fig. 1. Expected future large turbine generator transient reactance.

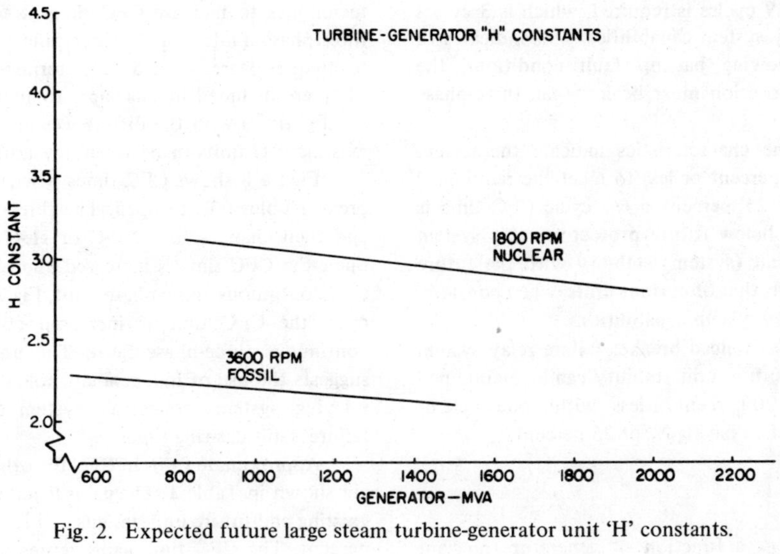

Fig. 2. Expected future large steam turbine-generator unit 'H' constants.

economic balance between transmission system and turbine-generator unit design. Based on the curve shapes in Fig. 4, a reasonable criterion suggests machine characteristics should be compatible with X_s values of 25 percent or less.

TRANSIENT STABILITY RESULTS

Three-Phase Faults

Critical fault clearing (CFC) times are plotted as a function of generator transient reactance and total unit H constant in Fig. 5, for three-phase faults and system reactance values of 10, 25 and 40 percents. The shaded areas for future 1800 rpm nuclear units and 3600 rpm units shown on Fig. 5 represent the range of future X'_d and H values indicated in Figs. 1 and 2. For comparison shaded areas representing the range of values for existing installed units are also shown. The size of the shaded areas is intended to represent the complete range of unit MVA ratings.

Currently installed designs show the CFC times range from approximately 11 cycles with a 10 percent system reactance down to 5½ cycles with 40 percent system reactance. Future unit CFC times range from about 9½ and 4 cycles for 10 and 40 percent system reactance values, respectively. Future designs represent a CFC time reduction of about 1½ to 2½ cycles.

The CFC times for 1800 rpm nuclear units and 3600 rpm units are approximately the same, because the units with the lower H constants also have the lower transient reactance values.

Figure 6 shows CFC times for three-phase faults at the plant as a function of system reactance. A representative T-G unit from the shaded areas in Fig. 5 for both unit types was used to indicate the general effect on CFC for future unit designs.

The CFC times should be compared with capabilities of relaying systems and circuit breaker interrupting times for backup fault clearing conditions. Future turbine-generator designs should be considered with future relay system capabilities when comparing future and current design effects on unit stability. In general, 12 cycle clearing is used for existing backup protection systems. With modern static relaying and two-cycle circuit breaker backup protection systems,[2] 9 cycle clearing can be achieved, and such systems are presently being applied. Breaker failure relay systems with 7½ clearing have also been proposed.[2]

Existing values of machine characteristics indicate X_s must be less than 10 percent to have a 12 cycle CFC time for use with currently installed breaker failure protection systems. For X_s of 25 percent a CFC time of about 9 cycles is required, which is 3 cycles below current protective relay system capabilities. This indicates to maintain plant stability following backup fault conditions, the current system design fault criterion must be less than three-phase fault conditions.

Future values of machine characteristics indicate the system reactance must be about 10 percent or less to meet the required 9 cycle CFC times. For X_s of 25 percent a 6½ cycle CFC time is required, which is 2½ cycles below future protective relay system capabilities. The effects on unit (system) stability with the future T-G units is almost identical to that of current units, when considering their respective backup relay system capabilities.

When applying the most advanced breaker failure relay system with 7½ fault clearing times, future unit stability can be maintained with systems having an X_s of 20 percent or less—within one cycle of meeting requirements for a system having X_s of 25 percent.

Double Line-to-Ground Faults

CFC times are plotted as a function of generator transient reactance and H constant in Fig. 7 for 2 L-G faults and a system reactance of 25 percent. Future unit designs show a CFC time reduction of about 2-3 cycles compared to existing installed units.

With a system reactance of 25 percent both existing and future unit designs have CFC times about the same as their individual backup relaying system capabilities, again indicating equivalent stability performance with current and future generating units.

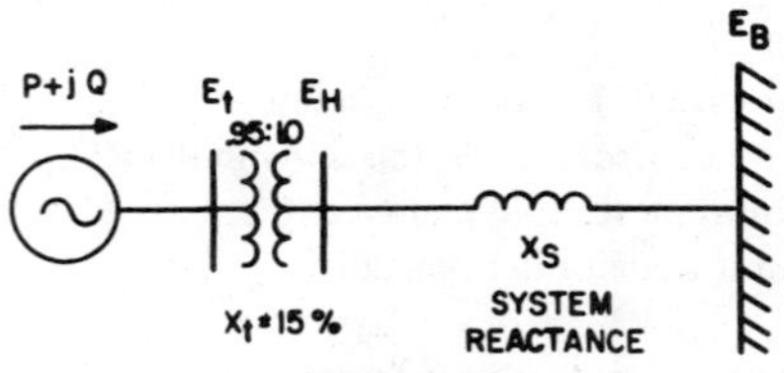

Fig. 3. Simplified system used for study.

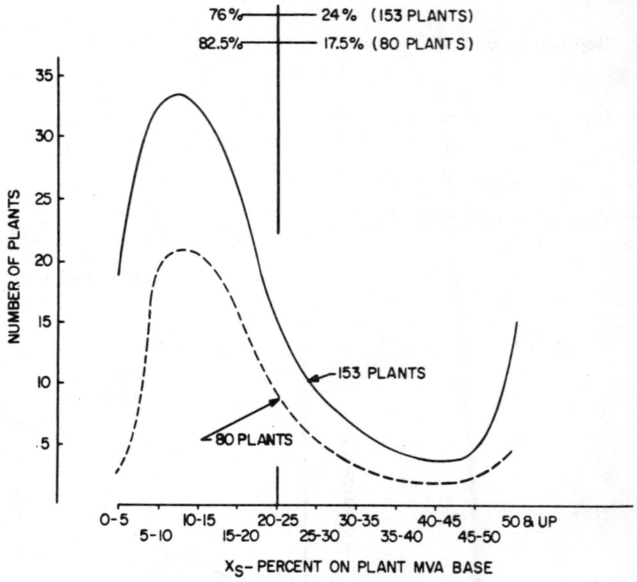

Fig. 4. Distribution curves of system reactance X_S values in percent on each individual plant's MVA base.

OTHER STABILITY CONTROL TECHNIQUES

There are increased trends in the application of stability control techniques to increase CFC times, and to achieve unit stability for three-phase faults for 'breaker failure' conditions. Several of these techniques were studied to determine the increase in CFC times. They are included in this paper to help put in perspective the change in CFC times with the differences in CFC times between future and existing T-G units or between any unit transient stability parameter.

Figure 8 shows CFC times when only one or two of the circuit breaker poles fails to open. The three-phase fault occurs for 3 cycles and then changes to a 2 L-G or 1L-G fault. When two poles fail to open the CFC time is increased approximately 1½ cycles compared to a continuous three-phase fault, Fig. 5. When only one pole fails to open the CFC time is increased about 6 cycles compared to a continuous three-phase fault. The improvement is significant and suggests the use of independent pole circuit breaker application and relaying systems to satisfy system design criteria with 'breaker failure' fault clearing times.

Approximate gains in CFC for other stability control techniques are shown in Table 1. The gains listed apply for X'_d and H values for existing and future unit designs and for system reactance values to 40 percent. The CFC time gains represent guides for system planning and are not necessarily additive if more than one technique is applied.

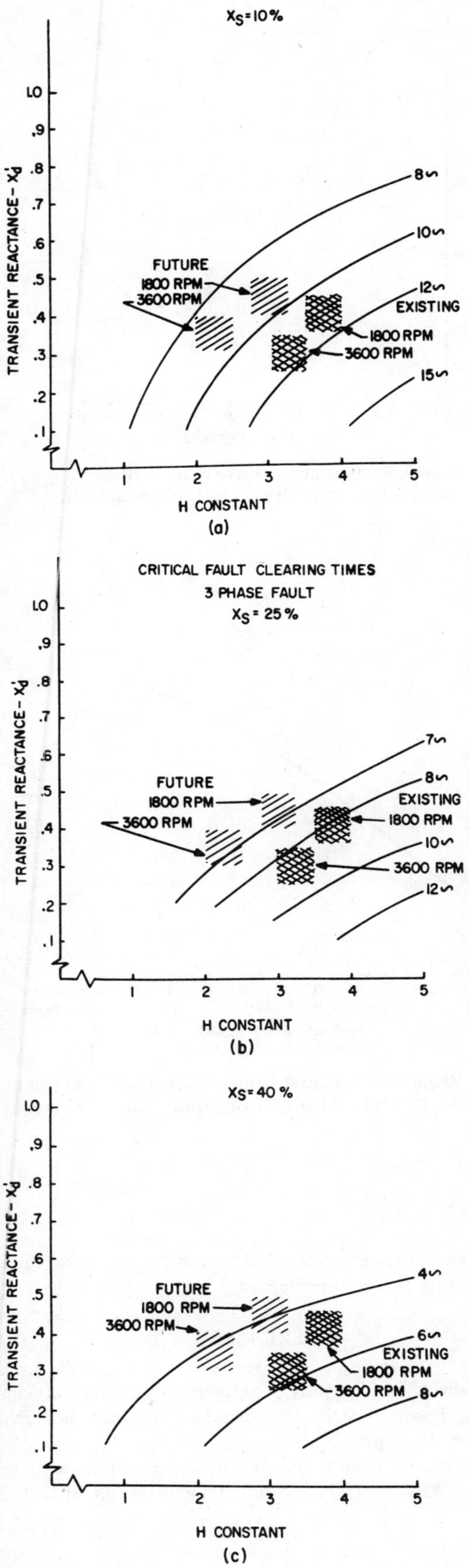

Fig. 5. Critical Fault Clearing time as a function of X'_d and H. Three-phase faults on the line side of circuit breaker at plant high voltage bus.
(A) X_s = 10% (B) X_s = 25% (C) X_s = 40%

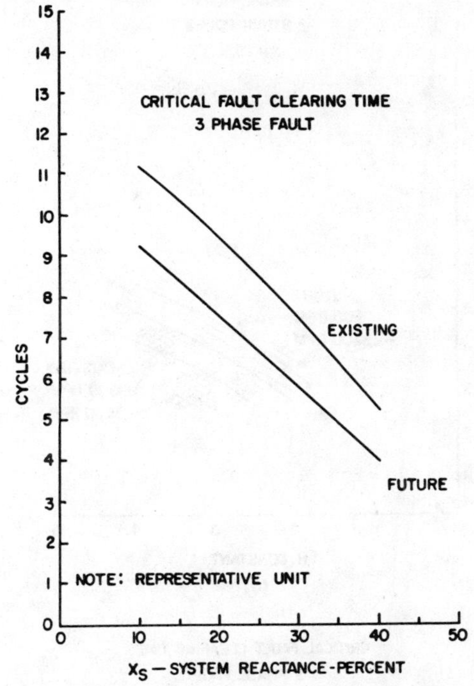

Fig. 6. Critical Fault Clearing times as a function of system reactance X_S for representative existing and future turbine generator units.

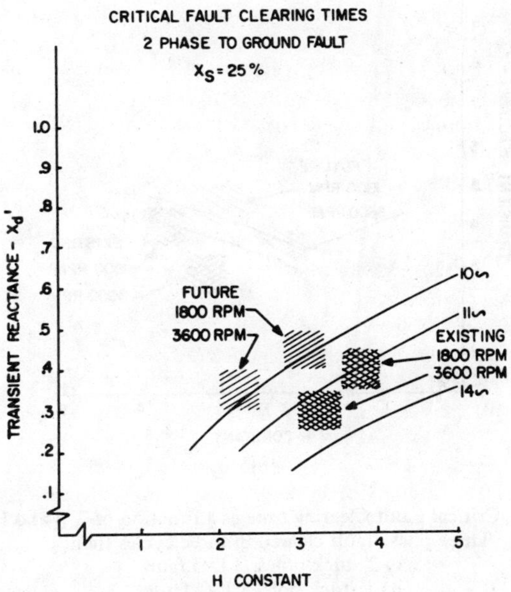

Fig. 7. Critical Fault Clearing time as a function of X'_d and H. Double line to ground faults on line side of circuit breaker at plant high voltage bus.

CONCLUDING REMARKS

1. Stability performance of future T-G designs will be very similar to existing designs when considering their respective breaker failure (backup) relay system capabilities.

2. It is generally not possible to maintain unit stability for three-phase faults for breaker failure fault clearing times and system reactances above 10 percent without using special stability control techniques. This applies to turbine-generator units with existing or future design characteristics, and their respective breaker failure relay system capabilities.

3. Unit stability can be achieved for three-phase breaker failure

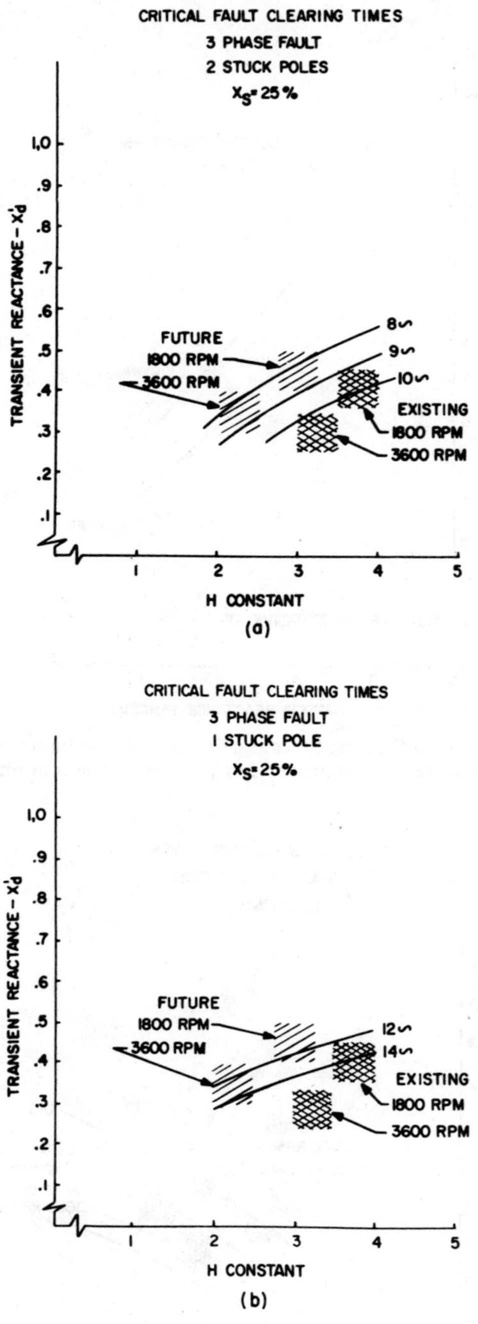

Fig. 8. Critical Fault Clearing time as a function of X'_d and H.
Three-phase fault cleared in three cycles then:
(A) 2 stuck poles, 2 L-G fault.
(B) 1 stuck pole, 1 L-G fault.

fault clear times and system reactances of 25 percent if stability control techniques are applied. Reducing the severity of the fault to 1 L-G faults through independent pole circuit breaker operation may be an effective means of meeting three-phase breaker failure fault clearing system design criterion.

4. Differences in machine characteristics, X'_d and H, that occur for specific ratings or manufacturer's T-G designs will have only minor effects on unit transient stability.

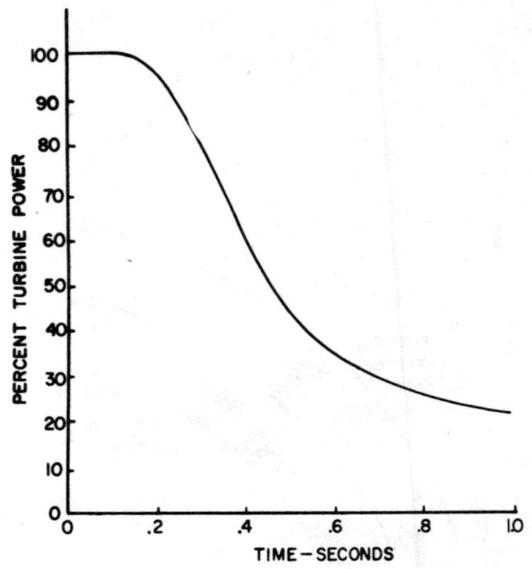

Fig. 9. Turbine input power characteristics
for turbine intercept valve control following fault.

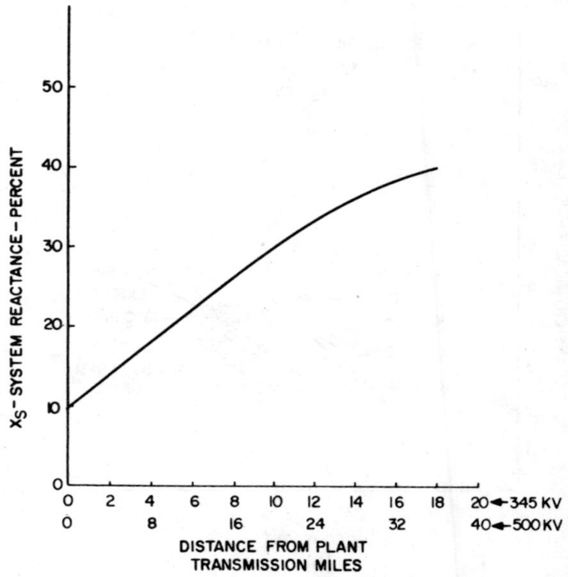

Fig. 10. Required three-phase fault location away from plant
to have 9.0 cycle critical fault clearing time–future T-G units.

5. Future turbine-generator designs will be acceptable to economic and reliable power system design.

REFERENCES

1. C. J. Baldwin, R. T. Byerly, 'Improved Digital Simulation for Analyzing Power System Disturbances,' Westinghouse ENGINEER, July 1969, pp. 109-116.
2. W. L. Hinman, 'Breaker Failure Relaying,' American Power Conference PROCEEDINGS, 1969, Volume 31, pp. 1012-1020.

Representative Bibliographies

The Application of Lyapunov Functions to Power System Stability and Related Topics

[1] G. E. Gless, "Direct method of Liapunov applied to transient power system stability," *IEEE Trans. Power App. Syst.*, vol. PAS-85, pp. 159–168, Feb. 1966.

[2] A. H. El-Abiad and K. Nagappan, "Transient stability regions of multimachine power systems," *IEEE Trans. Power App. Syst.*, vol. PAS-85, pp. 169–179, Feb. 1966.

[3] J. M. Undrill, "Power system stability studies by the method of Liapunov, Part I. State space approach to synchronous machine modelling," *IEEE Trans. Power App. Syst.*, vol. PAS-86, pp. 443–453, July 1967.

[4] ——, "Power system stability studies by the method of Liapunov, Part II. The interconnection of hydro generating sets," *IEEE Trans. Power App. Syst.*, vol. PAS-86, pp. 791–801, July 1967.

[5] Y.-N. Yu and K. Vongsuriya, "Nonlinear power system stability study by Liapunov function and Zubov's method," *IEEE Trans. Power App. Syst.*, vol. PAS-86, pp. 1480–1485, Dec. 1967.

[6] R. D. Teichgraeber, F. W. Harris, and G. L. Johnson, "New stability measure for multimachine power systems," *IEEE Trans. Power App. Syst.*, vol. PAS-89, pp. 233–239, Feb. 1970.

[7] M. A. Pai, M. A. Mohan, and J. G. Rao, "Power system transient stability regions using Popov's method," *IEEE Trans. Power App. Syst.*, vol. PAS-89, pp. 788–801, May/June 1970.

[8] J. L. Willems and J. C. Willems, "The application of Lyapunov methods to the computation of transient stability regions for multimachine power systems," *IEEE Trans. Power App. Syst.*, vol. PAS-89, pp. 795–801, May/June 1970.

[9] G. Lüders, "Transient stability of multimachine power systems via the direct method of Lyapunov," *IEEE Trans. Power App. Syst.*, vol. PAS-90, pp. 23–35, Jan./Feb. 1971.

[10] F. Williams, S. A. Louie, and G. W. Bills, "Feasibility of Liapunov functions for the stability analysis of electric power systems having up to 60 generators," *IEEE Trans. Power App. Syst.*, vol. PAS-91, pp. 1145–1157, May/June 1972.

[11] M. M. Elmetwally and N. D. Rao, "Sensitivity analysis on power system dynamic stability studies," *IEEE Trans. Power App. Syst.*, vol. PAS-91, pp. 1692–1699, July/Aug. 1972.

Optimal Generator Control and Related Topics

[1] J. L. Dinely, A. J. Morris, and C. Preece, "Optimized transient stability from excitation control of synchronous generators," *IEEE Trans. Power App. Syst.*, vol. PAS-87, pp. 1696–1705, Aug. 1968.

[2] N. Ramarao and D. K. Reitan, "Improvement of power system transient stability using optimal control: Bang-bang control of reactance," *IEEE Trans. Power App. Syst.*, vol. PAS-89, pp. 975–984, May/June 1970.

[3] K. Ganesan, J. J. Regehr, and R. J. Fleming, "Optimization of speed governor parameters in the presence of pseudo-random load disturbances," *IEEE Trans. Power App. Syst.*, vol. PAS-89, pp. 1242–1246, July/Aug. 1970.

[4] Y.-N. Yu and C. Siggers, "Stabilization and optimal control signals for a power system," *IEEE Trans. Power App. Syst.*, vol. PAS-90, pp. 1469–1481, July/Aug. 1971.

[5] S. Takata and R. Ueda, "A dual mode excitation control of synchronous machine for suppressing its hunting," *IEEE Trans. Power App. Syst.*, vol. PAS-90, pp. 1541–1545, July/Aug. 1971.

[6] S. M. Miniesy and E. V. Bohn, "Optimum network switching in power systems," *IEEE Trans. Power App. Syst.*, vol. PAS-90, pp. 2118–2123, Sept./Oct. 1971.

[7] E. J. Davison and N. S. Rau, "The optimal output feedback control of a synchronous machine," *IEEE Trans. Power App. Syst.*, vol. PAS-90, pp. 2123–2134, Sept./Oct. 1971.

[8] D. H. Kelly and A. A. Rahim, "Closed-loop optimal excitation control for power system stability," *IEEE Trans. Power App. Syst.*, vol. PAS-90, pp. 2135–2141, Sept./Oct. 1971.

[9] S. N. Iyer and B. J. Cory, "Optimum control of a turbo-generator including an exciter and governor," *IEEE Trans. Power App. Syst.*, vol. PAS-90, pp. 2142–2149, Sept./Oct. 1971.

[10] —, "Optimization of turbo-generator transient performance by differential dynamic programming," *IEEE Trans. Power App. Syst.*, vol. PAS-90, pp. 2149–2157, Sept./Oct. 1971.

[11] S. Elangoran and A. Kyppurajulu, "Suboptimal control of power systems using simplified models," *IEEE Trans. Power App. Syst.*, vol. PAS-91, pp. 911-918, May/June 1972.

[12] J. Meisel, R. D. Barnard, and R. S. Elliott, "Dynamic control of multimachine power systems based on two-step optimization over admissible trajectories," *IEEE Trans. Power App. Syst.*, vol. PAS-91, pp. 920–927, May/June 1972.

[13] H. A. M. Moussa and Y. -N. Yu, "Optimal power system stabilization through excitation and/or governor control," *IEEE Trans. Power App. Syst.*, vol. PAS-91, pp. 1166–1173, May/June 1972.

[14] Y.-N. Yu and H. A. M. Moussa, "Optimal stabilization of a multi-machine system," *IEEE Trans. Power App. Syst.*, vol. PAS-91, pp. 1174–1182, May/June 1972.

[15] A. K. DeSarkar and N. D. Rao, "Stabilization of a synchronous machine through output feedback control," *IEEE Trans. Power App. Syst.*, vol. PAS-92, pp. 159–166, Jan./Feb. 1973.

Author Index

Adler, R. B., 147

Blythe, A. L., 261, 308
Bowler, C. E. J., 400
Breuer, G. D., 365
Brown, P. G., 552
Burtnyk, V., 525
Byerly, R. T., 232, 295

Caszza, J. A., 174
Ceyhan, A., 154
Chadwick, J. W., 430
Clade, J. J., 491
Clifford, J. F. 503
Coleman, D. W., 421
Concordia, C., 189, 400, 550
Coulter, J. C., 337
Coultes, M. E., 37, 354
Crenshaw, M. L., 82
Cullen, F. A., 479
Cushing, E. W., Jr., 468

Dandeno, P. L., 28, 276, 325
Demello, F. P., 189
Dillman, T. L., 97
Dommel, H. W., 253
Dougherty, J. J., 511
Dravid, N. V., 414
Drechsler, G. E., 468
Dunlop, R. D., 533

Edwards, L., 430
Eggenberger, M. A., 103
Elliott, L. C., 393
Ellis, H. M., 261
El-Sherbiny, M. K., 212
Ewart, D. N., 20, 400

Farmer, R. G., 284

Gerhart, A. D., 343
Gibley, R. A., 365
Goodwin, C. J., 276
Gabrielle, A. F., 541
Griffin, J. D. A., 37
Gulachenski, E. M., 174

Hanson, O. W., 276
Hardy, J. E., 261
Hartley, R. H., 284
Hartung, E. C., 74
Hattan, E. E., 316
Hauth, R. L., 28
Hillesland, T., Jr., 343, 511
Holland, D. R., 479
Holley, H. J., 421
Horowitz, S. H., 541
Hunkins, H. D., 316

Iliceto, F., 154
IEEE Committee, 50, 55, 128, 164

Jones, W. D., 20

Karas, A. N., 325
Keay, F. W., 90, 97, 295
Kent, M. H., 139, 284
Kilgore, L. A., 393
Killgoar, W. P., 468
Kimbark, E. W., 378, 408, 459
Kirchmayer, L. K., 174

Lee, D. C., 37
Lenfest, E. H., 74
Lokay, H. E., 525, 558
Luini, J. F., 337, 343

Maliszewski, R. M., 533
Manchur, G. 37, 67, 349
Mane, A., 387
Marshall, H. G., 468
Martin, G. E., 316
McClymont, K. R., 67, 325
McCrackin, F. A., 139
Mehta, D. M., 212
Meloy, G. R., 74
Miller, W. J., 82
Morgan, W. A., 479
Mosher, C. C., 147
Moussa, H. A. M., 45

Nagy, I., 203

Olive, D. W., 13, 221

Park, R. H., 445
Peck, H. B., 479
Perry, H. R., 337
Persoz, H., 491
Peterson, H. A., 414
Politis, A., 541

Raczkowski, C., 97
Ramey, D. G., 121, 232

Riesch, H. A., 430
Rockfield, M. L., Jr., 343
Ross, R. J., 67
Ruckstuhl, G., 154
Rustebakke, H. M., 365
Ruzek, J. B., 479

Sato, N., 253
Schleif, F. R., 316
Schmidt, A. H., Jr., 503
Schmus, W. R., 139
Schulz, R. P., 20, 28, 82
Shier, R. M., 308
Shipley, R. B., 421
Simmons, H. O., Jr., 365
Skooglund, J. W., 97, 121, 261, 295
Smith, L. E., 430
Smith, O. J. M., 387
South, W. H., 90, 97
Stewart, H. R., 468

Taylor, E. R., 393
Temoshok, M., 82
Thoits, P. O., 560

Undrill, J. M., 174

Watson, W., 37, 325, 349, 354
Webster, R. H., 387
Wheeler, L. M., 139, 284
Wilson, G. L., 533
Wilson, R. J., 67

Young, C. C., 242
Yu, Y., 45

Subject Index

A

Althyrex excitation system, 82
Arc suppression
  single-pole switched EHV lines, 408, 414, 421, 430

C

Calvert Cliffs plant
  stability aids, 479
Capacitor compensation
  see Series capacitors
Computer program descriptions
  dynamic stability programs, 221, 232
Computer simulation
  see Digital simulation
Control systems
  see Excitation control, Speed control

D

Damping
  excitation control effect, 316
  excitation plus supplementary damping signals, 349
  salient-pole machines with static exciters, 295
DC transmission, 489
  dynamic behavior of generator feeding dc line, 491
  modeling, 503
  stability considerations, 511
Digital simulation
  dynamic stability, 221, 232, 253, 308
  numerical methods, 253
  synchronous machine transients, 13
  transient stability, 253
  see also Modeling
Dynamic stability, 187
  block diagrams of synchronous machines, 203
  computer program, 221
  damping in machines with static exciters, 295
  dc transmission, 511
  excitation plus supplementary damping signals, 349
  excitation system effects, 189, 212, 316, 325, 337, 354
  excitation with stabilizing feedback, 308
  Four Corners project, 284
  interconnected systems, 232, 276
  large-scale systems, 242
  large turbogenerator effects, 550, 558
  loading and excitation effects, 212
  Lyapunov methods, 563
  numerical methods, 253
  Peace River transmission systems, 261
  series-compensated systems, 393, 400
  stability aids, 479
  stabilizer field testing, 343

E

EHV transmission
  dynamic stability, 361, 284
  ground-fault arc suppression, 408, 414, 421, 430
  series capacitors, 365
  single-pole switching, 408, 414, 421, 430
  stability aids, 479
  stability studies, 337
Electromechanical equivalents, 174
Equivalent circuits
  see Modeling
Excitation control
  dynamic system stability, 212
  interconnected system oscillations, 276
  stability and damping improvement, 316
  stabilizer field testing, 343
  stabilizer sensing frequency deviation, 90
  synchronous machine stability, 189, 203
Excitation systems
  Althyrex system, 82
  brushless exciter with non-controlled rectifiers, 97
  damping in machines with static exciters, 295
  hydroelectric generators, 67, 74
  modeling, 50, 242
  rectifier systems, 67
  stability improvement, 337, 354
  stability of hydrogenerators, 325
  stabilizing feedback, 308
  supplementary damping signals, 349
  thyristor systems, 74, 82
  time response computation, 55
  turbine generators, 82, 97

F

Fast load control, 445, 459
Fast valving, 468, 479
Four Corners Project
  stability studies, 284

G

Generation control
  see Excitation control, Speed control
Generators
  see Hydroelectric generators, Turbine generators, Synchronous machines
Governors
  see Hydrogovernors, Speed control

H

HVDC transmission
  see DC transmission
Hydraulic turbines
  modeling, 128, 242
Hydroelectric generators
  damping in machines with static exciters, 295
  excitation effect on stability, 325
  excitation plus supplementary damping signals, 349
  excitation systems, 67, 74
Hydroelectric systems
  dynamic stability, 308

Peace River transmission system, 261
Hydrogovernors
modeling, 121, 128
steam turbine control, 103

Interconnected systems
fast load control, 445
dynamic stability simulation, 232
modeling, 242
oscillation stabilization, 276
stability aids, 479
stability improvement by discontinuous changes, 459
see also Large-scale systems

L

Large-scale systems
dynamic stability program, 221
modeling, 242
numerical methods, 253
see also Interconnected systems
Load control
see Fast load control
Load modeling, 164, 242
dynamic modeling, 139
steady-state modeling, 147
voltage-dip behavior, 154
Load restoration
frequency actuated, 533, 541
Load shedding, 523
frequency actuated, 533, 541
underfrequency relays, 525
Lyapunov methods, 563

M

Mercury-arc rectifier excitation systems
hydroelectric generators, 67
Modeling, 9
dc transmission, 503
electromechanical equivalents, 174
excitation systems, 50
hydrogovernors, 121
large-scale systems, 242
speed-control systems, 128
turbines, 128
see also Load modeling, Digital Simulation, Synchronous machine modeling

N

Numerical methods
power system stability, 253

P

Paradise–Davidson line
single-pole switching, 430
Peace River transmission system
dynamic stability, 261

R

Rectifier excitation systems
mercury-arc rectifier systems, 67
non-controlled rectifier systems, 97
see also Thyristor excitation systems
Relays
modeling, 242
see also Underfrequency relays

S

Saskatchewan–Manitoba–Ontario interconnection
oscillation stabilization, 276
SCR's
see Thyristors
Self-excited oscillations
series-compensated system, 393, 400
Series capacitors
EHV transmission, 365
self-excited oscillations, 393, 400
switching to quench electromechanical transients, 387
transient stability improvement, 378
Silicon-controlled rectifiers
see Thyristors
Simulation
see Digital simulation, Modeling
Single-pole switching, 408, 414, 421, 430
Speed-control
hydrogovernor modeling, 121
hydrogovernor systems, 103
interconnected system oscillations, 276
modeling systems and turbines, 128
steam turbines, 103
see also Fast valving
Stability
basic principles, 1
see also Dynamic stability, Synchronous machine stability, Transient stability
Stabilizers
field testing, 343
frequency-deviation sensing, 90
Steam turbines
fast valving, 468
modeling, 128, 242
speed control, 103
Subharmonic resonance
series-compensated systems, 393
Synchronous machine modeling, 242
effect on accuracy of system simulation, 28
equivalent circuit parameters determination, 45
transients simulation, 13
turbine generators, 20, 37
Synchronous machines
see Hydroelectric generators, Turbine generators
Synchronous machine stability
block diagrams, 203
damping in machines with static exciters, 295
dc link supply, 491
excitation plus supplementary damping signals, 349
excitation system effects, 189, 212, 325, 337, 343, 354
loading and excitation effects, 212

T

Thyristor excitation systems
hydroelectric generators, 67, 74
turbine generators, 82
Transient analysis
excitation systems, 55
load behavior during voltage dips, 154
synchronous machines, 13
Transient stability, 363
dc transmission, 491, 503, 511
fast load control, 445
fast valving, 468
improvement by discontinuous changes, 459
large turbogenerator effects, 550
Lyapunov methods, 563
numerical methods, 253
Peace River Transmission system, 261
series capacitors in EHV lines, 365
series capacitor switching, 378, 387
single-pole switched EHV lines, 408, 414, 421, 430
stability aids, 479

Transmission-line compensation
  see Series capacitors
Transmission systems
  see DC transmission, EHV transmission
Turbine generators
  excitation system effect on stability, 354
  excitation systems, 82, 97
  modeling, 20, 37
  system stability effects, 550
  transient stability effects, 558
Turbines
  see Hydraulic turbines, Steam Turbines

U

Underfrequency relays
  load shedding, 525
  load shedding and restoration, 533, 541
  modeling, 242

V

Valves
  see Rectifiers
Valving
  see Fast valving

Editors' Biographies

Richard T. Byerly (M'49–SM'62) received the B. S. degree in electrical engineering from Texas A & M University, College Station, and the M. S. degree in electrical engineering from the University of Pittsburgh, Pittsburgh, Pa.

He was on the Electrical Engineering Staff at the University of Pittsburgh for six years, and has been with the Westinghouse Electric Corporation, East Pittsburgh, since 1954. He has been engaged in power system studies throughout most of the past 20 years, and is presently an Advisory Engineer in the Advanced Systems Technology activity of Power Systems Planning.

Mr. Byerly is a member of the IEEE Power System Engineering Committee.

Edward W. Kimbark (M'27–SM'35–F'48–LF'63) was born in Chicago, Ill., on September 21, 1902. He studied electrical engineering at Northwestern University, Evanston, Ill., from 1920 to 1925, receiving the B.S. degree in 1924 and the E.E. degree in 1925. He then attended the Massachusetts Institute of Technology, Cambridge, from 1932 to 1937, receiving the S.M. degree in 1933 and the Sc.D. degree in 1937.

For most of his life he has been a teacher of electrical engineering. He taught at the University of California, Berkeley, from 1927 to 1929; Massachusetts Institute of Technology from 1933 to 1937 and 1942 to 1944; Polytechnic Institute of Brooklyn, Brooklyn, N.Y., from 1937 to 1939; Northwestern University, Evanston, Ill., from 1939 to 1942 and 1944 to 1950; Instituto Tecnológico de Aeronáutica, São José dos Campos, Brazil, from 1950 to 1955; and Seattle University, Seattle, Wash., where he was Dean of Engineering from 1955 to 1962. Other jobs were at the Public Service Company of Northern Illinois from 1925 to 1927, and at the Museum of Science and Industry, Chicago, from 1929 to 1932. He began to work as a part-time consultant in the Branch of System Engineering, Bonneville Power Administration, Portland, Ore., in 1956, and since 1962 has been a full-time employee. His present position is Head of the System Analysis Unit, Branch of System Engineering. He has written a 3-volume work on power system stability, a textbook on electrical transmission of power and signals, and the first volume of a 2-volume work on direct current transmission. His technical interests are in transmission theory, electric power transmission, symmetrical and similar components, power system stability, analog computation, and hvdc transmission.

Dr. Kimbark is a member of Eta Kappa Nu, CIGRE, and the Society for Computer Simulation.